Orbital Mechanics for Engineering Students

Orbital Mechanics for Engineering Students

Revised Fourth Edition

Howard D. Curtis

Professor Emeritus, Aerospace Engineering
Embry-Riddle Aeronautical University
Daytona Beach, Florida

Butterworth-Heinemann is an imprint of Elsevier
The Boulevard, Langford Lane, Kidlington, Oxford OX5 1GB, United Kingdom
50 Hampshire Street, 5th Floor, Cambridge, MA 02139, United States

Library of Congress Cataloging-in-Publication Data
A catalog record for this book is available from the Library of Congress

British Library Cataloguing-in-Publication Data
A catalogue record for this book is available from the British Library

ISBN: 978-0-12-824025-0

Publisher: Katey Birtcher
Acquisitions Editor: Steve Merken
Editorial Project Manager: Nathaniel McFadden
Production Project Manager: Maria Bernard
Cover Designer: Alan Studholme

Typeset by SPi Global, India

To my beloved wife, Mary
For her patience, encouragement, and love

Contents

Preface

The purpose of this book is to provide an introduction to space mechanics for undergraduate engineering students. It is not directed toward graduate students, researchers, and experienced practitioners, who may nevertheless find useful review material within the book's contents. The intended readers are those who are studying the subject for the first time and have completed courses in physics, dynamics, and mathematics through differential equations and applied linear algebra. I have tried my best to make the text readable and understandable to that audience. In pursuit of that objective I have included a large number of example problems that are explained and solved in detail. Their purpose is not to overwhelm but to elucidate. I find that students like the "teach by example" method. I always assume that the material is being seen for the first time and, wherever possible, I provide solution details so as to leave little to the reader's imagination. The numerous figures throughout the book are also intended to aid comprehension. All of the more labor-intensive computational procedures are accompanied by MATLAB® code.

For this, the fourth edition, I have retained the content and style of the previous editions and corrected all the errors discovered by me or reported to me by readers. Except for the new Chapter 9 on basic lunar trajectories and an expanded discussion of quaternions in Chapter 11 the book remains essentially the same. Adding the new chapter required the following reshuffling:

Topic	*This edition*	*Previous edition*
Lunar trajectories	Chapter 9	Absent
Introduction to orbital perturbations	Chapter 10	Chapter 12
Rigid body dynamics	Chapter 11	Chapter 9
Satellite attitude dynamics	Chapter 12	Chapter 10
Rocket vehicle dynamics	Chapter 13	Chapter 11

The organization of the book remains the same as that of the third edition. Chapter 1 is a review of vector kinematics in three dimensions and of Newton's laws of motion and gravitation. It also focuses on the issue of relative motion, crucial to the topics of rendezvous and satellite attitude dynamics. The material on ordinary differential equation solvers will be useful for students who are expected to code numerical simulations in MATLAB or other programming languages. Chapter 2 presents the vector-based solution of the classical two-body problem, resulting in a host of practical formulas for the analysis of orbits and trajectories of elliptical, parabolic, and hyperbolic shape. The restricted three-body problem is covered to introduce the notion of Lagrange points and to present the numerical solution of a lunar trajectory problem. Chapter 3 derives Kepler's equations, which relate position to time for the different kinds of orbits. The universal variable formulation is also presented. Chapter 4 is devoted to describing orbits in three dimensions. Coordinate transformations and the Euler elementary rotation sequences are defined. Procedures for transforming back and forth between the state vector and the classical orbital elements are addressed. The effect of the earth's oblateness on the motion of an orbit's ascending node and eccentricity vector is described, pending a more detailed explanation in Chapter 10. Chapter 5 is an introduction to preliminary orbit determination, including Gibbs' and Gauss' methods and the solution of Lambert's problem. Auxiliary topics include topocentric coordinate systems, Julian day numbering, and sidereal time. Chapter 6 presents the common means of transferring from one orbit to another by impulsive delta-v maneuvers, including Hohmann transfers, phasing orbits, and plane changes. Chapter 7 is a brief introduction to relative motion in general and to the two-impulse rendezvous problem in particular. The latter is analyzed using the Clohessy-Wiltshire equations, which are derived in this chapter. Chapter 8 is an introduction to interplanetary mission design using patched conics. Chapter 9 extends the patched conic method and the restricted three-body approach to lunar trajectory analysis. Chapter 10 is an introduction to common orbital perturbations: drag, nonspherical gravitational field, solar radiation pressure, and lunar and solar gravity. Chapter 11 presents those elements of rigid body dynamics required to characterize the attitude of a space vehicle. Euler's equations of rotational motion are derived and applied in a number of example problems. Euler angles, yaw-pitch-roll angles, and quaternions are presented as ways to describe the attitude of rigid body. Chapter 12 describes the methods of controlling, changing, and stabilizing the attitude of spacecraft by means of thrusters, gyros, and other devices. Chapter 13 is a brief introduction to the characteristics and design of multistage launch vehicles.

Chapters 1 through 4 form the core of a first orbital mechanics course. The time devoted to Chapter 1 depends on the background of the student. It might be surveyed briefly and used thereafter simply as a reference. What follows Chapter 4 depends on the objectives of the course.

Chapters 5 through 10 carry on with the subject of orbital mechanics. Chapter 6 on orbital maneuvers should be included in any case. Coverage of Chapters 5, 7, 8, and 9 is optional. However, if Chapters 8 and 9 on interplanetary and lunar missions is to form a part of the course, then the solution of Lambert's problem (Section 5.3) must be studied beforehand.

Chapter 10 is appropriate for a course devoted exclusively to orbital mechanics with an introduction to perturbations, which is a whole topic unto itself.

Chapters 11 and 12 must be covered if the course objectives include an introduction to spacecraft dynamics. In that case Chapters 5, 7, 8, and 9 would probably not be studied in depth.

Chapter 13 is optional if the engineering curriculum requires a separate course in propulsion including rocket dynamics.

The important topic of spacecraft control systems is omitted. However, the material in this book and a course in control theory provide the basis for the study of spacecraft attitude control.

To understand the material and to solve problems requires using a lot of undergraduate mathematics. Mathematics, of course, is the language of engineering. Students must not forget that the English mathematician and physicist Sir Isaac Newton (1642–1727) had to invent calculus so he could solve orbital mechanics problems in more than just a heuristic way. Newton's 1687 publication *Mathematical Principles of Natural Philosophy* ("the *Principia*") is one of the most influential scientific works of all time. It must be noted that his contemporary, the German mathematician Gottfried Wilhelm von Leibnitz (1646–1716) is credited with inventing infinitesimal calculus independently of Newton in the 1670s.

In addition to honing their math skills, students are urged to take advantage of computers (which, incidentally, use the binary numeral system developed by Leibnitz). There are many commercially available mathematics software packages for personal computers. Wherever possible they should be used to relieve the burden of repetitive and tedious calculations. Computer-programming skills can and should be put to good use in the study of orbital mechanics. The elementary MATLAB programs referred to in Appendix D of this book illustrate how many of the procedures developed in the text can be implemented in software. All the scripts were developed and tested using MATLAB version 9.2 (release 2017a). Information about MATLAB, which is a registered trademark of The MathWorks, Inc., may be obtained from

The MathWorks, Inc.
3 Apple Hill Drive
Natick, MA 01760-2089, USA
www.mathworks.com

Appendix A presents some tables of physical data and conversion factors. Appendix B is a road map through the first three chapters, showing how the most fundamental equations of orbital mechanics are related. Appendix C shows how to set up the n-body equations of motion and program them in MATLAB. Appendix D contains listings of all the MATLAB algorithms and example problems presented in the text. Appendix E shows that the gravitational field of a spherically symmetric body is the same as if the mass were concentrated at its center. Appendix F explains how to deal with a computational issue that arises in some perturbation analyses.

SUPPLEMENTS TO THE TEXT

For purchasers of the book, copies of the MATLAB M-files listed in Appendix D can be freely downloaded from this book's companion website. Also available on the companion website are a set of animations that accompany the text. To access these files, please visit https://www.elsevier.com/books-and-journals/book-companion/9780128240250.

For instructors using this book for a course, please visit www.textbooks.elsevier.com to register for access to the solutions manual, PowerPoint lecture slides, and other resources.

ACKNOWLEDGEMENTS

Since the publication of the first three editions and during the preparation of this one, I have received helpful criticism, suggestions, and advice from many sources locally and worldwide. I thank them all and regret that time and space limitations prohibited the inclusion of some recommended additional topics that would have enhanced the book.

It has been a pleasure to work with the people at Elsevier, in particular Joseph P. Hayton, Publisher; Steve Merken, Senior Acquisitions Editor; and Nate McFadden, Senior Developmental Editor. I appreciate their enthusiasm for the book, their confidence in me, and all the work they did to move this project to completion.

Finally and most importantly, I must acknowledge the patience and support of my wife, Mary, who was a continuous source of optimism and encouragement throughout the revision effort.

Howard D. Curtis
Daytona Beach, FL, United States

CHAPTER

DYNAMICS OF POINT MASSES

1

1.1 INTRODUCTION

This chapter serves as a self-contained reference on the kinematics and dynamics of point masses as well as some basic vector operations and numerical integration methods. The notation and concepts summarized here will be used in the following chapters. Those familiar with the vector-based dynamics of particles can simply page through the chapter and then refer back to it later as necessary. Those who need a bit more in the way of review will find that the chapter contains all the material they need to follow the development of orbital mechanics topics in the upcoming chapters.

We begin with a review of vectors and some vector operations, after which we proceed to the problem of describing the curvilinear motion of particles in three dimensions. The concepts of force and mass are considered next, along with Newton's inverse-square law of gravitation. This is followed by a presentation of Newton's second law of motion ("force equals mass times acceleration") and the important concept of angular momentum.

As a prelude to describing motion relative to moving frames of reference, we develop formulas for calculating the time derivatives of moving vectors. These are applied to the computation of relative velocity and acceleration. Example problems illustrate the use of these results, as does a detailed consideration of how the earth's rotation and curvature influence our measurements of velocity and acceleration. This brings in the curious concept of Coriolis force. Embedded in exercises at the end of the chapter is practice in verifying several fundamental vector identities that will be employed frequently throughout the book.

The chapter concludes with an introduction to numerical methods, which can be called upon to solve the equations of motion when an analytical solution is not possible.

1.2 VECTORS

A vector is an object that is specified by both a magnitude and a direction. We represent a vector graphically by a directed line segment (i.e., an arrow pointing in the direction of the vector). The end opposite the arrow is called the tail. The length of the arrow is proportional to the magnitude of the vector. Velocity is a good example of a vector. We say that a car is traveling eastward at 80 km/h. The direction is east and the magnitude, or speed, is 80 km/h. We will use boldface type to represent vector quantities and plain type to denote scalars. Thus, whereas B is a scalar, $\mathbf{B}$ is a vector.

Orbital Mechanics for Engineering Students. https://doi.org/10.1016/B978-0-12-824025-0.00001-5

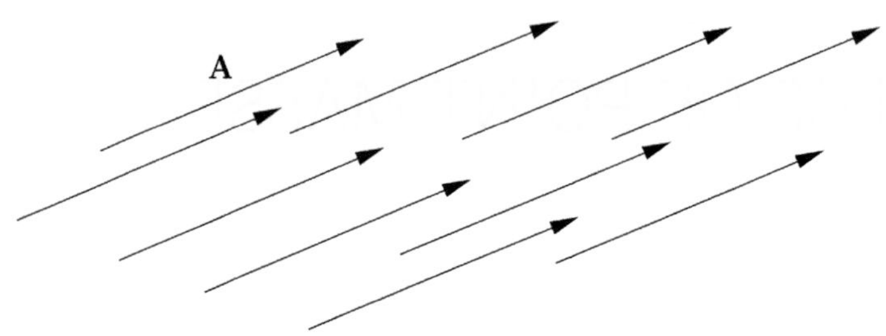

FIG. 1.1

All of these vectors may be denoted **A**, since their magnitudes and directions are the same.

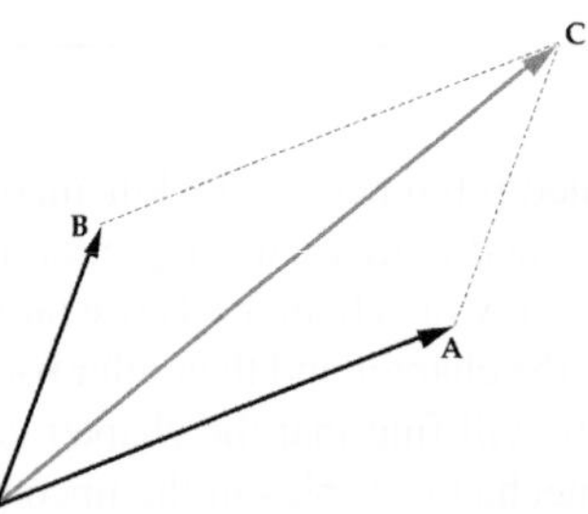

FIG. 1.2

Parallelogram rule of vector addition. $\mathbf{A} + \mathbf{B} = \mathbf{C}$.

Observe that a vector is specified solely by its magnitude and direction. If **A** is a vector, then all vectors having the same physical dimensions, the same length, and pointing in the same direction as **A** are denoted **A**, regardless of their line of action, as illustrated in Fig. 1.1. Shifting a vector parallel to itself does not mathematically change the vector. However, the parallel shift of a vector might produce a different physical effect. For example, an upward 5-kN load (force vector) applied to the tip of an airplane wing gives rise to quite a different stress and deflection pattern in the wing than the same load acting at the wing's midspan.

The magnitude of a vector **A** is denoted $\|\mathbf{A}\|$, or, simply A.

Multiplying a vector **B** by the reciprocal of its magnitude produces a vector that points in the direction of **B**, but it is dimensionless and has a magnitude of one. Vectors having dimensionless magnitude are called unit vectors. We put a hat (^) over the letter representing a unit vector. Then we can tell simply by inspection that, for example, $\hat{\mathbf{u}}$ is a unit vector, as are $\hat{\mathbf{B}}$ and $\hat{\mathbf{e}}$.

It is convenient to denote the unit vector in the direction of the vector **A** as $\hat{\mathbf{u}}_A$. As pointed out above, we obtain this vector from **A** as follows:

$$\hat{\mathbf{u}}_A = \frac{\mathbf{A}}{A} \tag{1.1}$$

Likewise, $\hat{\mathbf{u}}_C = \mathbf{C}/C$, $\hat{\mathbf{u}}_F = \mathbf{F}/F$, etc.

The sum or *resultant* of two vectors is defined by the parallelogram rule (Fig. 1.2). Let **C** be the sum of the two vectors **A** and **B**. To form that sum using the parallelogram rule, the vectors **A** and **B** are

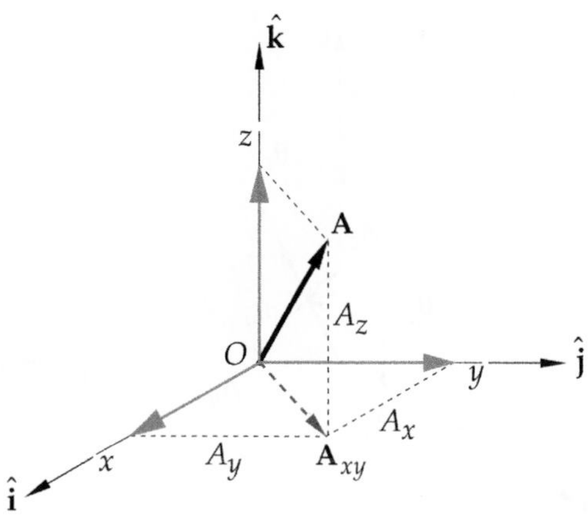

FIG. 1.3

Three-dimensional, right-handed Cartesian coordinate system.

shifted parallel to themselves (leaving them unaltered) until the tail of **A** touches the tail of **B**. Drawing dotted lines through the head of each vector parallel to the other completes a parallelogram. The diagonal from the tails of **A** and **B** to the opposite corner is the resultant **C**. By construction, vector addition is commutative; that is,

$$\mathbf{A}+\mathbf{B}=\mathbf{B}+\mathbf{A} \tag{1.2}$$

A Cartesian coordinate system in three dimensions consists of three mutually perpendicular axes, labeled x, y, and z, which intersect at the origin O. We will always use a right-handed Cartesian coordinate system, which means if you wrap the fingers of your right hand around the z axis, with the thumb pointing in the positive z direction, your fingers will be directed from the x axis toward the y axis. Fig. 1.3 illustrates such a system. Note that the unit vectors along the x, y, and z axes are, respectively, $\hat{\mathbf{i}}$, $\hat{\mathbf{j}}$, and $\hat{\mathbf{k}}$.

In terms of its Cartesian components, and in accordance with the above summation rule, a vector **A** is written in terms of its components A_x, A_y, and A_z as

$$\mathbf{A}=A_x\hat{\mathbf{i}}+A_y\hat{\mathbf{j}}+A_z\hat{\mathbf{k}} \tag{1.3}$$

The projection of **A** on the xy plane is a vector denoted $\mathbf{A}_{xy}$. It follows that

$$\mathbf{A}_{xy}=A_x\hat{\mathbf{i}}+A_y\hat{\mathbf{j}}$$

According to the Pythagorean theorem, the magnitude of **A** in terms of its Cartesian components is

$$\mathbf{A}=\sqrt{A_x^2+A_y^2+A_z^2} \tag{1.4}$$

From Eqs. (1.1) and (1.3), the unit vector in the direction of **A** is

$$\hat{\mathbf{u}}_A=\cos\theta_x\hat{\mathbf{i}}+\cos\theta_y\hat{\mathbf{j}}+\cos\theta_z\hat{\mathbf{k}} \tag{1.5}$$

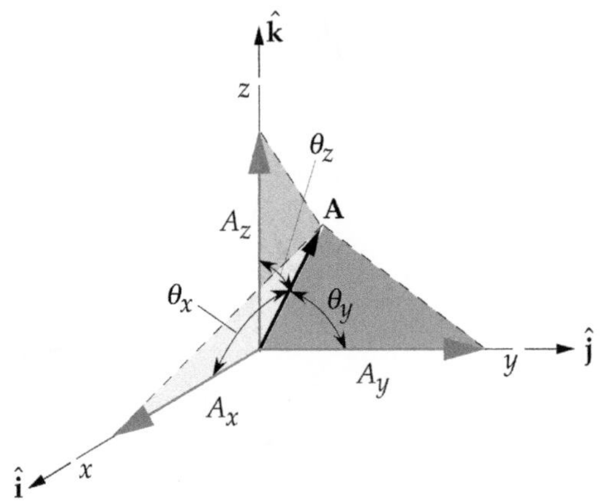

FIG. 1.4

Direction angles in three dimensions.

where

$$\cos\theta_x = \frac{A_x}{A} \quad \cos\theta_y = \frac{A_y}{A} \quad \cos\theta_z = \frac{A_z}{A} \tag{1.6}$$

The direction angles θ_x, θ_y, and θ_z are illustrated in Fig. 1.4, and they are measured between the vector and the positive coordinate axes. Note carefully that the sum of θ_x, θ_y, and θ_z is not in general known a priori and cannot be assumed to be, say, 180 degrees.

EXAMPLE 1.1

Calculate the direction angles of the vector $\mathbf{A} = \hat{\mathbf{i}} - 4\hat{\mathbf{j}} + 8\hat{\mathbf{k}}$.

Solution

First, compute the magnitude of $\mathbf{A}$ by means of Eq. (1.4),

$$A = \sqrt{1^2 + (-4)^2 + 8^2} = 9$$

Then Eq. (1.6) yields

$$\theta_x = \cos^{-1}\left(\frac{A_x}{A}\right) = \cos^{-1}\left(\frac{1}{9}\right) \Rightarrow \boxed{\theta_x = 83.62 \text{ degrees}}$$

$$\theta_y = \cos^{-1}\left(\frac{A_y}{A}\right) = \cos^{-1}\left(\frac{-4}{9}\right) \Rightarrow \boxed{\theta_y = 116.4 \text{ degrees}}$$

$$\theta_z = \cos^{-1}\left(\frac{A_z}{A}\right) = \cos^{-1}\left(\frac{8}{9}\right) \Rightarrow \boxed{\theta_z = 27.27 \text{ degrees}}$$

Observe that $\theta_x + \theta_y + \theta_z = 227.3$ degrees.

Multiplication and division of two vectors are undefined operations. There are no rules for computing the product **AB** and the ratio **A/B**. However, there are two well-known binary operations on vectors: the *dot product* and the *cross product*. The dot product of two vectors is a scalar defined as follows:

$$\mathbf{A}\cdot\mathbf{B}=AB\cos\theta \tag{1.7}$$

where θ is the angle between the heads of the two vectors, as shown in Fig. 1.5. Clearly,

$$\mathbf{A}\cdot\mathbf{B}=\mathbf{B}\cdot\mathbf{A} \tag{1.8}$$

If two vectors are perpendicular to each other, then the angle between them is 90 degrees. It follows from Eq. (1.7) that their dot product is zero. Since the unit vectors $\hat{\mathbf{i}}$, $\hat{\mathbf{j}}$, and $\hat{\mathbf{k}}$ of a Cartesian coordinate system are mutually orthogonal and of magnitude 1, Eq. (1.7) implies that

$$\begin{aligned}\hat{\mathbf{i}}\cdot\hat{\mathbf{i}}&=\hat{\mathbf{j}}\cdot\hat{\mathbf{j}}=\hat{\mathbf{k}}\cdot\hat{\mathbf{k}}=1\\ \hat{\mathbf{i}}\cdot\hat{\mathbf{j}}&=\hat{\mathbf{i}}\cdot\hat{\mathbf{k}}=\hat{\mathbf{j}}\cdot\hat{\mathbf{k}}=0\end{aligned} \tag{1.9}$$

Using these properties, it is easy to show that the dot product of the vectors **A** and **B** may be found in terms of their Cartesian components as

$$\mathbf{A}\cdot\mathbf{B}=A_xB_x+A_yB_y+A_zB_z \tag{1.10}$$

If we set **B** = **A**, then it follows from Eqs. (1.4) and (1.10) that

$$A=\sqrt{\mathbf{A}\cdot\mathbf{A}} \tag{1.11}$$

The dot product operation is used to project one vector onto the line of action of another. We can imagine bringing the vectors tail to tail for this operation, as illustrated in Fig. 1.6. If we drop a perpendicular line from the tip of **B** onto the direction of **A**, then the line segment B_A is the orthogonal projection of **B** onto the line of action of **A**. B_A stands for the scalar projection of **B** onto **A**. From trigonometry, it is obvious from the figure that

$$B_A=B\cos\theta$$

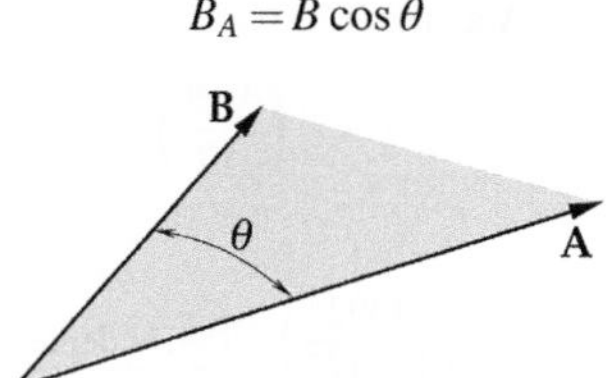

FIG. 1.5

The angle between two vectors brought tail to tail by parallel shift.

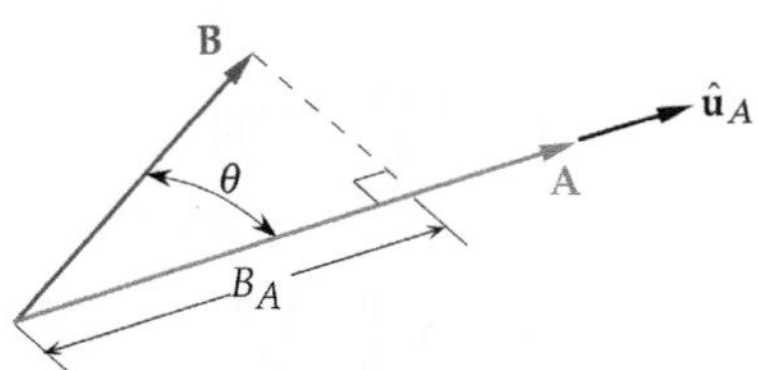

FIG. 1.6

Projecting the vector **B** onto the direction of **A**.

Let $\hat{\mathbf{u}}_A$ be the unit vector in the direction of **A**. Then,

$$\mathbf{B}\cdot\hat{\mathbf{u}}_A = \|\mathbf{B}\|\overbrace{\|\hat{\mathbf{u}}_A\|}^{=1}\cos\theta = B\cos\theta$$

Comparing this expression with the preceding one leads to the conclusion that

$$B_A = \mathbf{B}\cdot\hat{\mathbf{u}}_A = \mathbf{B}\cdot\frac{\mathbf{A}}{A} \tag{1.12}$$

where $\hat{\mathbf{u}}_A$ is given by Eq. (1.1). Likewise, the projection of **A** onto **B** is given by

$$A_B = \mathbf{A}\cdot\frac{\mathbf{B}}{B}$$

Observe that $A_B = B_A$ only if **A** and **B** have the same magnitude.

EXAMPLE 1.2

Let $\mathbf{A} = \hat{\mathbf{i}} + 6\hat{\mathbf{j}} + 18\hat{\mathbf{k}}$ and $\mathbf{B} = 42\hat{\mathbf{i}} - 69\hat{\mathbf{j}} + 98\hat{\mathbf{k}}$. Calculate
(a) the angle between **A** and **B**;
(b) the projection of **B** in the direction of **A**;
(c) the projection of **A** in the direction of **B**.

Solution

First, we make the following individual calculations.

$$\mathbf{A}\cdot\mathbf{B} = (1)(42) + (6)(-69) + (18)(98) = 1392 \tag{a}$$

$$A = \sqrt{(1)^2 + (6)^2 + (18)^2} = 19 \tag{b}$$

$$B = \sqrt{(42)^2 + (-69)^2 + (98)^2} = 127 \tag{c}$$

(a) According to Eq. (1.7), the angle between **A** and **B** is

$$\theta = \cos^{-1}\left(\frac{\mathbf{A}\cdot\mathbf{B}}{AB}\right)$$

Substituting Eqs. (a), (b), and (c) yields

$$\theta = \cos^{-1}\left(\frac{1392}{19\cdot 127}\right) = \boxed{54.77\text{ degrees}}$$

(b) From Eq. (1.12), we find the projection of **B** onto **A**.

$$B_A = \mathbf{B}\cdot\frac{\mathbf{A}}{A} = \frac{\mathbf{A}\cdot\mathbf{B}}{A}$$

Substituting Eqs. (a) and (b) we get

$$B_A = \frac{1392}{19} = \boxed{73.26}$$

(c) The projection of **A** onto **B** is

$$A_B = \mathbf{A}\cdot\frac{\mathbf{B}}{B} = \frac{\mathbf{A}\cdot\mathbf{B}}{B}$$

Substituting Eqs. (a) and (c) we obtain

$$A_B = \frac{1392}{127} = \boxed{10.96}$$

The cross product of two vectors yields another vector, which is computed as follows:

$$\mathbf{A} \times \mathbf{B} = (AB \sin\theta)\hat{\mathbf{n}}_{AB} \tag{1.13}$$

where θ is the angle between the heads of **A** and **B**, and $\hat{\mathbf{n}}_{AB}$ is the unit vector normal to the plane defined by the two vectors. The direction of $\hat{\mathbf{n}}_{AB}$ is determined by the right-hand rule. That is, curl the fingers of the right hand from the first vector (**A**) toward the second vector (**B**), and the thumb shows the direction of $\hat{\mathbf{n}}_{AB}$ (Fig. 1.7). If we use Eq. (1.13) to compute **B** × **A**, then $\hat{\mathbf{n}}_{AB}$ points in the opposite direction, which means

$$\mathbf{B} \times \mathbf{A} = -(\mathbf{A} \times \mathbf{B}) \tag{1.14}$$

Therefore, unlike the dot product, the cross product is not commutative.

The cross product is obtained analytically by resolving the vectors into Cartesian components.

$$\mathbf{A} \times \mathbf{B} = \left(A_x\hat{\mathbf{i}} + A_y\hat{\mathbf{j}} + A_z\hat{\mathbf{k}}\right) \times \left(B_x\hat{\mathbf{i}} + B_y\hat{\mathbf{j}} + B_z\hat{\mathbf{k}}\right) \tag{1.15}$$

Since the set $\hat{\mathbf{i}}\hat{\mathbf{j}}\hat{\mathbf{k}}$ is a mutually perpendicular triad of unit vectors, Eq. (1.13) implies that

$$\begin{array}{lll} \hat{\mathbf{i}} \times \hat{\mathbf{i}} = 0 & \hat{\mathbf{j}} \times \hat{\mathbf{j}} = 0 & \hat{\mathbf{k}} \times \hat{\mathbf{k}} = 0 \\ \hat{\mathbf{i}} \times \hat{\mathbf{j}} = \hat{\mathbf{k}} & \hat{\mathbf{j}} \times \hat{\mathbf{k}} = \hat{\mathbf{i}} & \hat{\mathbf{k}} \times \hat{\mathbf{i}} = \hat{\mathbf{j}} \end{array} \tag{1.16}$$

Expanding the right-hand side of Eq. (1.15), substituting Eq. (1.16), and making use of Eq. (1.14) leads to

$$\mathbf{A} \times \mathbf{B} = (A_yB_z - A_zB_y)\hat{\mathbf{i}} - (A_xB_z - A_zB_x)\hat{\mathbf{j}} + (A_xB_y - A_yB_x)\hat{\mathbf{k}} \tag{1.17}$$

It may be seen that the right-hand side is the determinant of the matrix

$$\begin{bmatrix} \hat{\mathbf{i}} & \hat{\mathbf{j}} & \hat{\mathbf{k}} \\ A_x & A_y & A_z \\ B_x & B_y & B_z \end{bmatrix}$$

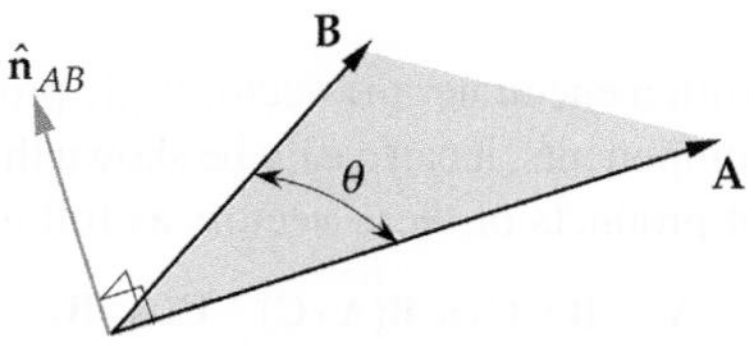

FIG. 1.7

$\hat{\mathbf{n}}_{AB}$ is normal to both **A** and **B** and defines the direction of the cross product **A** × **B**.

Thus, Eq. (1.17), can be written as

$$\mathbf{A}\times\mathbf{B}=\begin{vmatrix}\hat{\mathbf{i}} & \hat{\mathbf{j}} & \hat{\mathbf{k}}\\ A_x & A_y & A_z\\ B_x & B_y & B_z\end{vmatrix} \tag{1.18}$$

where the two vertical bars stand for the determinant. Obviously, the rule for computing the cross product, though straightforward, is a bit lengthier than that for the dot product. Remember that the dot product yields a scalar whereas the cross product yields a vector.

The cross product provides an easy way to compute the normal to a plane. Let **A** and **B** be any two vectors lying in the plane, or, let any two vectors be brought tail to tail to define a plane, as shown in Fig. 1.7. The vector $\mathbf{C}=\mathbf{A}\times\mathbf{B}$ is normal to the plane of **A** and **B**. Therefore, $\hat{\mathbf{n}}_{AB}=\mathbf{C}/C$, or

$$\hat{\mathbf{n}}_{AB}=\frac{\mathbf{A}\times\mathbf{B}}{\|\mathbf{A}\times\mathbf{B}\|} \tag{1.19}$$

EXAMPLE 1.3

Let $\mathbf{A}=-3\hat{\mathbf{i}}+7\hat{\mathbf{j}}+9\hat{\mathbf{k}}$ and $\mathbf{B}=6\hat{\mathbf{i}}-5\hat{\mathbf{j}}+8\hat{\mathbf{k}}$. Find a unit vector that lies in the plane of **A** and **B** and is perpendicular to **A**.

Solution

The plane of vectors **A** and **B** is determined by parallel-shifting the vectors so that they meet tail to tail. Calculate the vector $\mathbf{D}=\mathbf{A}\times\mathbf{B}$.

$$\mathbf{D}=\begin{vmatrix}\hat{\mathbf{i}} & \hat{\mathbf{j}} & \hat{\mathbf{k}}\\ -3 & 7 & 9\\ 6 & -5 & 8\end{vmatrix}=101\hat{\mathbf{i}}+78\hat{\mathbf{j}}-27\hat{\mathbf{k}}$$

Note that **A** and **B** are both normal to **D**. We next calculate the vector $\mathbf{C}=\mathbf{D}\times\mathbf{A}$.

$$\mathbf{C}=\begin{vmatrix}\hat{\mathbf{i}} & \hat{\mathbf{j}} & \hat{\mathbf{k}}\\ 101 & 78 & -27\\ -3 & 7 & 9\end{vmatrix}=891\hat{\mathbf{i}}-828\hat{\mathbf{j}}+941\hat{\mathbf{k}}$$

C is normal to **D** as well as to **A**. **A**, **B**, and **C** are all perpendicular to **D**. Therefore, they are coplanar. Thus, **C** is not only perpendicular to **A**, but it also lies in the plane of **A** and **B**. Therefore, the unit vector we are seeking is the unit vector in the direction of **C**. That is

$$\hat{\mathbf{u}}_C=\frac{\mathbf{C}}{C}=\frac{891\hat{\mathbf{i}}-828\hat{\mathbf{j}}+941\hat{\mathbf{k}}}{\sqrt{891^2+(-828)^2+941^2}}$$

$$\boxed{\hat{\mathbf{u}}_C=0.5794\hat{\mathbf{i}}-0.5384\hat{\mathbf{j}}+0.6119\hat{\mathbf{k}}}$$

In the chapters to follow, we will often encounter the vector triple product, $\mathbf{A}\times(\mathbf{B}\times\mathbf{C})$. By resolving **A**, **B**, and **C** into their Cartesian components, it can easily be shown that the vector triple product can be expressed in terms of just the dot products of these vectors as follows:

$$\mathbf{A}\times(\mathbf{B}\times\mathbf{C})=\mathbf{B}(\mathbf{A}\cdot\mathbf{C})-\mathbf{C}(\mathbf{A}\cdot\mathbf{B}) \tag{1.20}$$

Because of the appearance of the letters on the right-hand side, this is often referred to as the "bac–cab rule."

EXAMPLE 1.4

If $\mathbf{F}=\mathbf{E}\times\{\mathbf{D}\times[\mathbf{A}\times(\mathbf{B}\times\mathbf{C})]\}$, use the bac–cab rule to reduce this expression to one involving only dot products.

Solution

First, we invoke the bac–cab rule to obtain

$$\mathbf{F}=\mathbf{E}\times\left\{\mathbf{D}\times\overbrace{[\mathbf{B}(\mathbf{A}\cdot\mathbf{C})-\mathbf{C}(\mathbf{A}\cdot\mathbf{B})]}^{\text{bac–cab rule}}\right\}$$

Expanding and collecting terms leads to

$$\mathbf{F}=(\mathbf{A}\cdot\mathbf{C})[\mathbf{E}\times(\mathbf{D}\times\mathbf{B})]-(\mathbf{A}\cdot\mathbf{B})[\mathbf{E}\times(\mathbf{D}\times\mathbf{C})]$$

We next apply the bac–cab rule twice on the right-hand side.

$$\mathbf{F}=(\mathbf{A}\cdot\mathbf{C})\overbrace{[\mathbf{D}(\mathbf{E}\cdot\mathbf{B})-\mathbf{B}(\mathbf{E}\cdot\mathbf{D})]}^{\text{bac–cab rule}}-(\mathbf{A}\cdot\mathbf{B})\overbrace{[\mathbf{D}(\mathbf{E}\cdot\mathbf{C})-\mathbf{C}(\mathbf{E}\cdot\mathbf{D})]}^{\text{bac–cab rule}}$$

Expanding and collecting terms yields the sought-for result.

$$\boxed{\mathbf{F}=[(\mathbf{A}\cdot\mathbf{C})(\mathbf{E}\cdot\mathbf{B})-(\mathbf{A}\cdot\mathbf{B})(\mathbf{E}\cdot\mathbf{C})]\mathbf{D}-(\mathbf{A}\cdot\mathbf{C})(\mathbf{E}\cdot\mathbf{D})\mathbf{B}+(\mathbf{A}\cdot\mathbf{B})(\mathbf{E}\cdot\mathbf{D})\mathbf{C}}$$

Another useful vector identity is the "interchange of the dot and the cross":

$$\mathbf{A}\cdot(\mathbf{B}\times\mathbf{C})=(\mathbf{A}\times\mathbf{B})\cdot\mathbf{C} \tag{1.21}$$

It is so-named because interchanging the operations in the expression $\mathbf{A}\cdot\mathbf{B}\times\mathbf{C}$ yields $\mathbf{A}\times\mathbf{B}\cdot\mathbf{C}$. The parentheses in Eq. (1.21) are required to show which operation must be carried out first, according to the rules of vector algebra. (For example, $(\mathbf{A}\cdot\mathbf{B})\times\mathbf{C}$, the cross product of a scalar and a vector, is undefined.) It is easy to verify Eq. (1.21) by substituting $\mathbf{A}=A_x\hat{\mathbf{i}}+A_y\hat{\mathbf{j}}+A_z\hat{\mathbf{k}}$, $\mathbf{B}=B_x\hat{\mathbf{i}}+B_y\hat{\mathbf{j}}+B_z\hat{\mathbf{k}}$, and $\mathbf{C}=C_x\hat{\mathbf{i}}+C_y\hat{\mathbf{j}}+C_z\hat{\mathbf{k}}$ and observing that both sides of the equal sign reduce to the same expression.

1.3 KINEMATICS

To track the motion of a particle P through Euclidean space, we need a frame of reference, consisting of a clock and a nonrotating Cartesian coordinate system. The clock keeps track of time t, and the xyz axes of the Cartesian coordinate system are used to locate the spatial position of the particle. In nonrelativistic mechanics, a single "universal" clock serves for all possible Cartesian coordinate systems. So when we refer to a frame of reference, we need to think only of the mutually orthogonal axes themselves.

The unit of time used throughout this book is the second (s). The unit of length is the meter (m), but the kilometer (km) will be the length unit of choice when large distances and velocities are involved. Conversion factors between kilometers, miles, and nautical miles are listed in Table A.3.

Given a frame of reference, the position of the particle P at a time t is defined by the position vector $\mathbf{r}(t)$ extending from the origin O of the frame out to P itself, as illustrated in Fig. 1.8. The components of $\mathbf{r}(t)$ are just the x, y, and z coordinates,

$$\mathbf{r}(t)=x(t)\hat{\mathbf{i}}+y(t)\hat{\mathbf{j}}+z(t)\hat{\mathbf{k}}$$

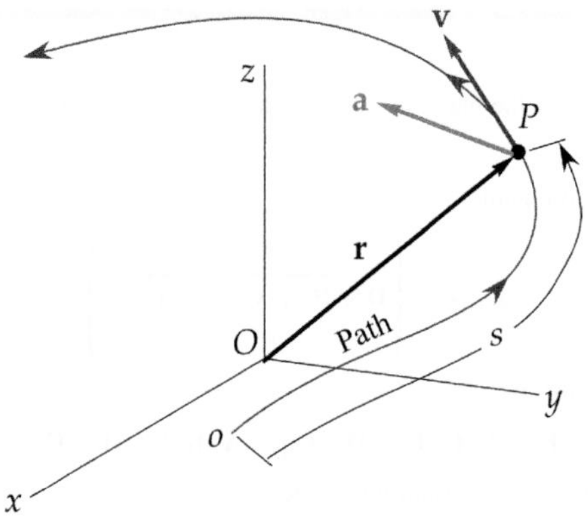

FIG. 1.8

Position, velocity, and acceleration vectors.

The distance of P from the origin is the magnitude or length of $\mathbf{r}$, denoted $\|\mathbf{r}\|$ or just r,

$$\|\mathbf{r}\| = r = \sqrt{x^2 + y^2 + z^2}$$

As in Eq. (1.11), the magnitude of $\mathbf{r}$ can also be computed by means of the dot product operation,

$$r = \sqrt{\mathbf{r} \cdot \mathbf{r}}$$

The velocity $\mathbf{v}$ and acceleration $\mathbf{a}$ of the particle are the first and second time derivatives of the position vector,

$$\mathbf{v}(t) = \frac{\mathrm{d}x(t)}{\mathrm{d}t}\hat{\mathbf{i}} + \frac{\mathrm{d}y(t)}{\mathrm{d}t}\hat{\mathbf{j}} + \frac{\mathrm{d}y(t)}{\mathrm{d}t}\hat{\mathbf{k}} = v_x(t)\hat{\mathbf{i}} + v_y(t)\hat{\mathbf{j}} + v_z(t)\hat{\mathbf{k}}$$
$$\mathbf{a}(t) = \frac{\mathrm{d}v_x(t)}{\mathrm{d}t}\hat{\mathbf{i}} + \frac{\mathrm{d}v_y(t)}{\mathrm{d}t}\hat{\mathbf{j}} + \frac{\mathrm{d}v_z(t)}{\mathrm{d}t}\hat{\mathbf{k}} = a_x(t)\hat{\mathbf{i}} + a_y(t)\hat{\mathbf{j}} + a_z(t)\hat{\mathbf{k}}$$

The derivatives of $\hat{\mathbf{i}}$, $\hat{\mathbf{j}}$, and $\hat{\mathbf{k}}$ are zero since axes of the Cartesian frame have fixed directions. It is convenient to represent the time derivative by means of an overhead dot. In this shorthand notation, if ( ) is any quantity, then

$$\dot{(\ )} = \frac{\mathrm{d}(\ )}{\mathrm{d}t} \quad \ddot{(\ )} = \frac{\mathrm{d}^2(\ )}{\mathrm{d}t^2} \quad \dddot{(\ )} = \frac{\mathrm{d}^3(\ )}{\mathrm{d}t^3} \quad \text{etc.}$$

Thus, for example,

$$\mathbf{v} = \dot{\mathbf{r}}$$
$$\mathbf{a} = \dot{\mathbf{v}} = \ddot{\mathbf{r}}$$
$$v_x = \dot{x} \quad v_y = \dot{y} \quad v_z = \dot{z}$$
$$a_x = \dot{v}_x = \ddot{x} \quad a_y = \dot{v}_y = \ddot{y} \quad a_z = \dot{v}_z = \ddot{z}$$

The locus of points that a particle occupies as it moves through space is called its path or trajectory. If the path is a straight line, then the motion is rectilinear. Otherwise, the path is curved, and the motion is called curvilinear. The velocity vector $\mathbf{v}$ is tangent to the path. If $\hat{\mathbf{u}}_t$ is the unit vector tangent to the trajectory, then

$$\mathbf{v} = v\hat{\mathbf{u}}_t \tag{1.22}$$

where the speed v is the magnitude of the velocity $\mathbf{v}$. The distance ds that P travels along its path in the time interval dt is obtained from the speed by

$$ds = v\,dt$$

In other words,

$$v = \dot{s}$$

The distance s, measured along the path from some starting point, is what the odometers in our automobiles record. Of course, $\dot{s}$, our speed along the road, is indicated by the dial of the speedometer.

Note carefully that $v \neq \dot{r}$ (i.e., the magnitude of the derivative of $\mathbf{r}$ does not equal the derivative of the magnitude of $\mathbf{r}$).

EXAMPLE 1.5

The position vector in meters is given as a function of time in seconds as

$$\mathbf{r} = (8t^2 + 7t + 6)\hat{\mathbf{i}} + (5t^3 + 4)\hat{\mathbf{j}} + (0.3t^4 + 2t^2 + 1)\hat{\mathbf{k}}\ (\text{m}) \tag{a}$$

At $t = 10$ s, calculate (a) v (the magnitude of the derivative of $\mathbf{r}$) and (b) $\dot{r}$ (the derivative of the magnitude of $\mathbf{r}$).

Solution

(a) The velocity $\mathbf{v}$ is found by differentiating the given position vector with respect to time,

$$\mathbf{v} = \frac{d\mathbf{r}}{dt} = (16t + 7)\hat{\mathbf{i}} + 15t^2\hat{\mathbf{j}} + (1.2t^3 + 4t)\hat{\mathbf{k}}$$

The magnitude of this vector is the square root of the sum of the squares of its components,

$$v = \sqrt{1.44t^6 + 234.6t^4 + 272t^2 + 224t + 49}$$

Evaluating this at $t = 10$ s, we get

$$\boxed{v = 1953.3\text{m/s}}$$

(b) Calculating the magnitude of $\mathbf{r}$ in Eq. (a) leads to

$$r = \sqrt{0.09t^8 + 26.2t^6 + 68.6t^4 + 152t^3 + 149t^2 + 84t + 53}$$

The time derivative of this expression is

$$\dot{r} = \frac{dr}{dt} = \frac{0.36t^7 + 78.6t^5 + 137.2t^3 + 228t^2 + 149t + 42}{\sqrt{0.09t^8 + 26.2t^6 + 68.6t^4 + 152t^3 + 149t^2 + 84t + 53}}$$

Substituting $t = 10$ s yields

$$\boxed{\dot{r} = 1935.5\,\text{m/s}}$$

If $\mathbf{v}$ is given, then we can find the components of the unit tangent $\hat{\mathbf{u}}_t$ in the Cartesian coordinate frame of reference by means of Eq. (1.22):

$$\hat{\mathbf{u}}_t = \frac{\mathbf{v}}{v} = \frac{v_x}{v}\hat{\mathbf{i}} + \frac{v_y}{v}\hat{\mathbf{j}} + \frac{v_z}{v}\hat{\mathbf{k}} \quad \left(v = \sqrt{v_x^2 + v_y^2 + v_z^2}\right) \tag{1.23}$$

The acceleration may be written as

$$\mathbf{a} = a_t \hat{\mathbf{u}}_t + a_n \hat{\mathbf{u}}_n \tag{1.24}$$

where a_t and a_n are the tangential and normal components of acceleration, given by

$$a_t = \dot{v}\,(= \ddot{s}) \quad a_n = \frac{v^2}{\rho} \tag{1.25}$$

where ρ is the radius of curvature, which is the distance from the particle P to the center of curvature of the path at that point. The unit principal normal $\hat{\mathbf{u}}_n$ is perpendicular to $\hat{\mathbf{u}}_t$ and points toward the center of curvature C, as shown in Fig. 1.9. Therefore, the position of C relative to P, denoted $\mathbf{r}_{C/P}$, is

$$\mathbf{r}_{C/P} = \rho \hat{\mathbf{u}}_n \tag{1.26}$$

The orthogonal unit vectors $\hat{\mathbf{u}}_t$ and $\hat{\mathbf{u}}_n$ form a plane called the osculating plane. The unit normal to the osculating plane is $\hat{\mathbf{u}}_b$, the binormal, and it is obtained from $\hat{\mathbf{u}}_t$ and $\hat{\mathbf{u}}_n$ by taking their cross product:

$$\hat{\mathbf{u}}_b = \hat{\mathbf{u}}_t \times \hat{\mathbf{u}}_n \tag{1.27}$$

From Eqs. (1.22), (1.24), and (1.27), we have

$$\mathbf{v} \times \mathbf{a} = v\hat{\mathbf{u}}_t \times (a_t \hat{\mathbf{u}}_t + a_n \hat{\mathbf{u}}_n) = va_n(\hat{\mathbf{u}}_t \times \hat{\mathbf{u}}_n) = va_n \hat{\mathbf{u}}_b = \|\mathbf{v} \times \mathbf{a}\| \hat{\mathbf{u}}_b$$

That is, an alternative to Eq. (1.27) for calculating the binormal vector is

$$\hat{\mathbf{u}}_b = \frac{\mathbf{v} \times \mathbf{a}}{\|\mathbf{v} \times \mathbf{a}\|} \tag{1.28}$$

Note that $\hat{\mathbf{u}}_t$, $\hat{\mathbf{u}}_n$, and $\hat{\mathbf{u}}_b$ form a right-handed triad of orthogonal unit vectors. That is

$$\hat{\mathbf{u}}_b \times \hat{\mathbf{u}}_t = \hat{\mathbf{u}}_n \quad \hat{\mathbf{u}}_t \times \hat{\mathbf{u}}_n = \hat{\mathbf{u}}_b \quad \hat{\mathbf{u}}_n \times \hat{\mathbf{u}}_b = \hat{\mathbf{u}}_t \tag{1.29}$$

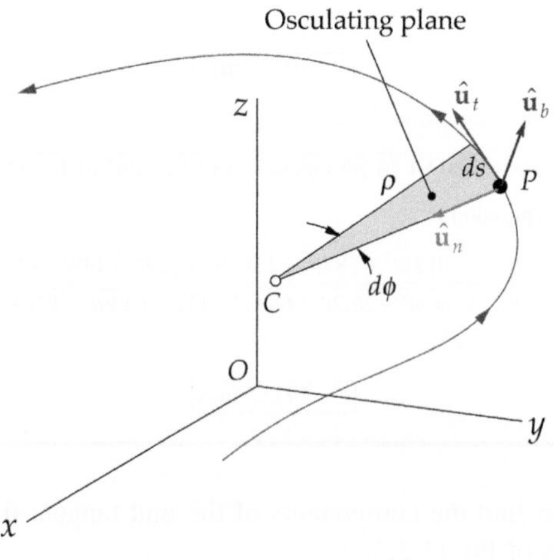

FIG. 1.9

Orthogonal triad of unit vectors associated with the moving point P.

The center of curvature lies in the osculating plane. When the particle P moves an incremental distance $\mathrm{d}s$, the radial from the center of curvature to the path sweeps out a small angle, $\mathrm{d}\phi$, measured in the osculating plane. The relationship between this angle and $\mathrm{d}s$ is

$$\mathrm{d}s = \rho \mathrm{d}\phi$$

so that $\dot{s} = \rho\dot{\phi}$, or

$$\dot{\phi} = \frac{v}{\rho} \tag{1.30}$$

EXAMPLE 1.6

Relative to a Cartesian coordinate system, the position, velocity, and acceleration of a particle P at a given instant are

$$\mathbf{r} = 250\hat{\mathbf{i}} + 630\hat{\mathbf{j}} + 430\hat{\mathbf{k}}\,(\mathrm{m}) \tag{a}$$

$$\mathbf{v} = 90\hat{\mathbf{i}} + 125\hat{\mathbf{j}} + 170\hat{\mathbf{k}}\,(\mathrm{m/s}) \tag{b}$$

$$\mathbf{a} = 16\hat{\mathbf{i}} + 125\hat{\mathbf{j}} + 30\hat{\mathbf{k}}\,\left(\mathrm{m/s^2}\right) \tag{c}$$

Find the coordinates of the center of curvature at that instant.

Solution

The coordinates of the center of curvature C are the components of its position vector $\mathbf{r}_C$. Consulting Fig. 1.9, we observe that

$$\mathbf{r}_C = \mathbf{r} + \rho\hat{\mathbf{u}}_n \tag{d}$$

where $\mathbf{r}$ is the position vector of the point P, ρ is the radius of curvature, and $\hat{\mathbf{u}}_n$ is the unit principal normal vector. The position vector $\mathbf{r}$ is given in Eq. (a), but ρ and $\hat{\mathbf{u}}_n$ are unknowns at this point. We must use the geometry of Fig. 1.9 to find them.

We begin by seeking the value of $\hat{\mathbf{u}}_n$, using the first of Eqs. (1.29),

$$\hat{\mathbf{u}}_n = \hat{\mathbf{u}}_b \times \hat{\mathbf{u}}_t \tag{e}$$

The unit tangent vector $\hat{\mathbf{u}}_t$ is found at once from the velocity vector in Eq. (b) by means of Eq. 1.23,

$$\hat{\mathbf{u}}_t = \frac{\mathbf{v}}{v}$$

where

$$v = \sqrt{90^2 + 125^2 + 170^2} = 229.4\ \mathrm{m/s} \tag{f}$$

Thus,

$$\hat{\mathbf{u}}_t = \frac{90\hat{\mathbf{i}} + 125\hat{\mathbf{j}} + 170\hat{\mathbf{k}}}{229.4} = 0.39233\hat{\mathbf{i}} + 0.54490\hat{\mathbf{j}} + 0.74106\hat{\mathbf{k}} \tag{g}$$

To find the binormal $\hat{\mathbf{u}}_b$ we insert the given velocity and acceleration vectors into Eq. (1.28),

$$\hat{\mathbf{u}}_b = \frac{\mathbf{v}\times\mathbf{a}}{\|\mathbf{v}\times\mathbf{a}\|} = \frac{\begin{vmatrix} \hat{\mathbf{i}} & \hat{\mathbf{j}} & \hat{\mathbf{k}} \\ 90 & 125 & 170 \\ 16 & 125 & 30 \end{vmatrix}}{\|\mathbf{v}\times\mathbf{a}\|} = \frac{-17{,}500\hat{\mathbf{i}} + 20\hat{\mathbf{j}} + 9250\hat{\mathbf{k}}}{\sqrt{(-17{,}500)^2 + 20^2 + 9250^2}} \tag{h}$$

$$= -0.88409\hat{\mathbf{i}} + 0.0010104\hat{\mathbf{i}} + 0.46731\hat{\mathbf{k}}$$

Substituting Eqs. (g) and (h) back into Eq. (e) finally yields the unit principal normal

$$\hat{\mathbf{u}}_n = \begin{vmatrix} \hat{\mathbf{i}} & \hat{\mathbf{j}} & \hat{\mathbf{k}} \\ -0.88409 & 0.0010104 & 0.46731 \\ 0.39233 & 0.5449 & 0.74106 \end{vmatrix} = -0.25389\hat{\mathbf{i}} + 0.8385\hat{\mathbf{j}} - 0.48214\hat{\mathbf{k}} \tag{i}$$

The only unknown remaining in Eq. (d) is ρ, for which we appeal to Eq. (1.25),

$$\rho = \frac{v^2}{a_n} \tag{j}$$

The normal acceleration a_n is calculated by projecting the acceleration vector $\mathbf{a}$ onto the direction of the unit normal $\hat{\mathbf{u}}_n$,

$$a_n = \mathbf{a} \cdot \hat{\mathbf{u}}_n = \left(16\hat{\mathbf{i}} + 125\hat{\mathbf{j}} + 30\hat{\mathbf{k}}\right) \cdot \left(-0.25389\hat{\mathbf{i}} + 0.8385\hat{\mathbf{j}} - 0.48214\hat{\mathbf{k}}\right) = 86.287\,\text{m/s}^2 \tag{k}$$

Putting the values of v and a_n from Eqs. (f) and (k) into Eq. (j) yields the radius of curvature,

$$\rho = \frac{229.4^2}{86.287} = 609.89\,\text{m} \tag{l}$$

Upon substituting Eqs. (a), (i), and (l) into Eq. (d), we obtain the position vector of the center of curvature C,

$$\begin{aligned} \mathbf{r}_C &= \left(250\hat{\mathbf{i}} + 630\hat{\mathbf{j}} + 430\hat{\mathbf{k}}\right) + 609.89\left(-0.25389\hat{\mathbf{i}} + 0.8385\hat{\mathbf{j}} - 0.48214\hat{\mathbf{k}}\right) \\ &= 95.159\hat{\mathbf{i}} + 1141.4\hat{\mathbf{j}} + 135.95\hat{\mathbf{k}}\ (\text{m}) \end{aligned}$$

Therefore, the coordinates of C are

$$\boxed{x = 95.16\,\text{m} \quad y = 1141\,\text{m} \quad z = 136.0\,\text{m}}$$

1.4 MASS, FORCE, AND NEWTON'S LAW OF GRAVITATION

Mass, like length and time, is a primitive physical concept: it cannot be defined in terms of any other physical concept. Mass is simply the quantity of matter. More practically, mass is a measure of the inertia of a body. Inertia is an object's resistance to changing its state of motion. The larger its inertia (the greater its mass), the more difficult it is to set a body into motion or bring it to rest. The unit of mass is the kilogram (kg).

Force is the action of one physical body on another, either through direct contact or through a distance. Gravity is an example of force acting through a distance, as are magnetism and the force between charged particles. The gravitational force F_g between two masses m_1 and m_2 having a distance r between their centers is

$$F_g = G\frac{m_1 m_2}{r^2} \tag{1.31}$$

This is Newton's law of gravity, in which G, the universal gravitational constant, has the value $G = 6.6742(10^{-11})\text{m}^3/(\text{kg}\cdot\text{s}^2)$. Due to the inverse-square dependence on distance, the force of gravity rapidly diminishes with the amount of separation between the two masses. In any case, the force of gravity is minuscule unless at least one of the masses is extremely big.

The force of a large mass (such as the earth) on a mass many orders of magnitude smaller (such as a person) is called weight, W. If the mass of the large object is M and that of the relatively tiny one is m, then the weight of the small body is

$$W = G\frac{Mm}{r^2} = m\left(\frac{GM}{r^2}\right)$$

or

$$W = mg \tag{1.32}$$

where

$$g = \frac{GM}{r^2} \tag{1.33}$$

g has units of acceleration (m/s^2) and is called the acceleration of gravity. If planetary gravity is the only force acting on a body, then the body is said to be in free fall. The force of gravity draws a freely falling object toward the center of attraction (e.g., center of the earth) with an acceleration g. Under ordinary conditions, we sense our own weight by feeling contact forces acting on us in opposition to the force of gravity. In free fall, there are, by definition, no contact forces, so there can be no sense of weight. Even though the weight is not zero, a person in free fall experiences weightlessness, or the absence of gravity.

Let us evaluate Eq. (1.33) at the surface of the earth, whose radius according to Table A.1 is 6378 km. Letting g_0 represent the standard sea level value of g, we get

$$g_0 = \frac{GM}{R_E^2} \tag{1.34}$$

In SI units,

$$g_0 = 9.807 \text{m/s}^2 \tag{1.35}$$

Substituting Eq. (1.34) into Eq. (1.33) and letting z represent the distance above the earth's surface, so that $r = R_E + z$, we obtain

$$g = g_0 \frac{R_E^2}{(R_E + z)^2} = \frac{g_0}{(1 + z/R_E)^2} \tag{1.36}$$

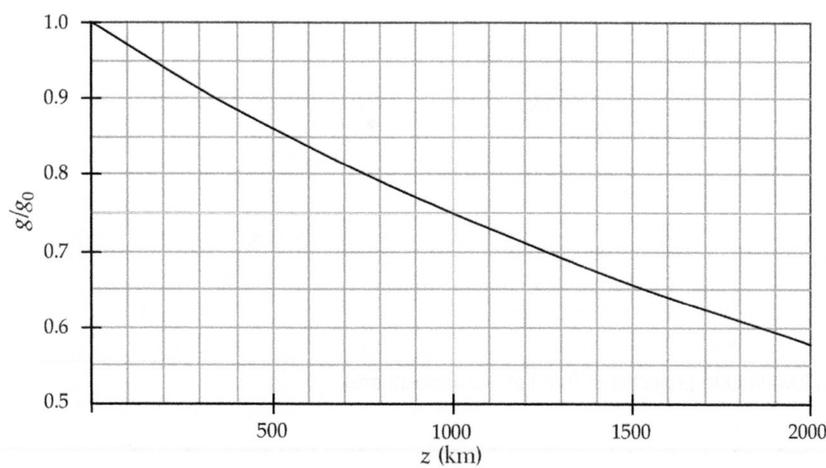

FIG. 1.10

Variation of the acceleration of gravity with altitude.

Commercial airliners cruise at altitudes on the order of 10 km (6 miles). At that height, Eq. (1.36) reveals that g (and hence weight) is only three-tenths of a percent less than its sea level value. Thus, under ordinary conditions, we ignore the variation of g with altitude. A plot of Eq. (1.36) out to a height of 2000 km (the upper limit of low earth orbit operations) is shown in Fig. 1.10. The variation of g over that range is significant. Even so, at space station altitude (400 km), weight is only about 10% less than it is on the earth's surface. The astronauts experience weightlessness, but they clearly are not weightless.

EXAMPLE 1.7

Show that in the absence of an atmosphere, the shape of a low-altitude ballistic trajectory is a parabola. Assume the acceleration of gravity g is constant and neglect the earth's curvature.

Solution

Fig. 1.11 shows a projectile launched at $t = 0$ s with a speed v_0 at a flight path angle γ_0 from the point with coordinates (x_0, y_0). Since the projectile is in free fall after launch, its only acceleration is that of gravity in the negative y direction:

$$\ddot{x} = 0$$
$$\ddot{y} = -g$$

Integrating with respect to time and applying the initial conditions leads to

$$x = x_0 + (v_0 \cos \gamma_0)t \tag{a}$$

$$y = y_0 + (v_0 \sin \gamma_0)t - \frac{1}{2}gt^2 \tag{b}$$

Solving Eq. (a) for t and substituting the result into Eq. (b) yields

$$\boxed{y = y_0 + (x - x_0)\tan \gamma_0 - \frac{1}{2}\frac{g}{v_0^2 \cos^2 \gamma_0}(x - x_0)^2} \tag{c}$$

This is the equation of a second-degree curve, a parabola, as sketched in Fig. 1.11.

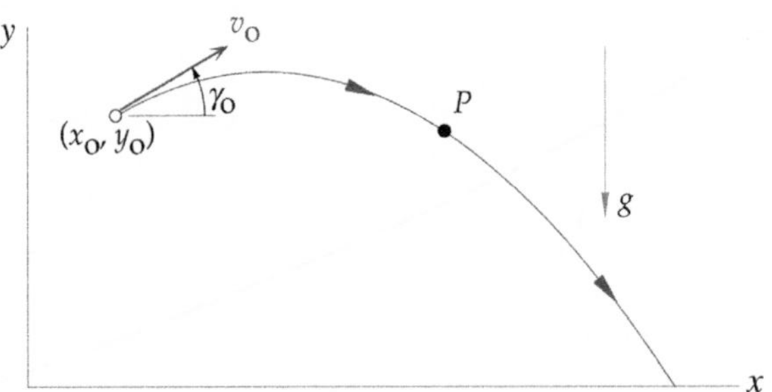

FIG. 1.11

Flight of a low-altitude projectile in free fall (no atmosphere).

EXAMPLE 1.8

An airplane flies a parabolic trajectory like that in Fig. 1.11 so that the passengers will experience free fall (weightlessness). What is the required variation of the flight path angle γ with speed v? Ignore the curvature of the earth.

Solution

Fig. 1.12 reveals that for a "flat" earth, $d\gamma = -d\phi$. That is,

$$\dot{\gamma} = -\dot{\phi}$$

It follows from Eq. (1.30) that

$$\rho\dot{\gamma} = -v \tag{1.37}$$

The normal acceleration a_n is just the component of the gravitational acceleration g in the direction of the unit principal normal to the curve (from P toward C). From Fig. 1.12, then,

$$a_n = g\cos\gamma \tag{a}$$

Substituting the second of Eqs. (1.25) into Eq. (a) and solving for the radius of curvature yields

$$\rho = \frac{v^2}{g\cos\gamma} \tag{b}$$

Combining Eqs. (1.37) and (b), we find the time rate of change of the flight path angle,

$$\boxed{\dot{\gamma} = -\frac{g\cos\gamma}{v}}$$

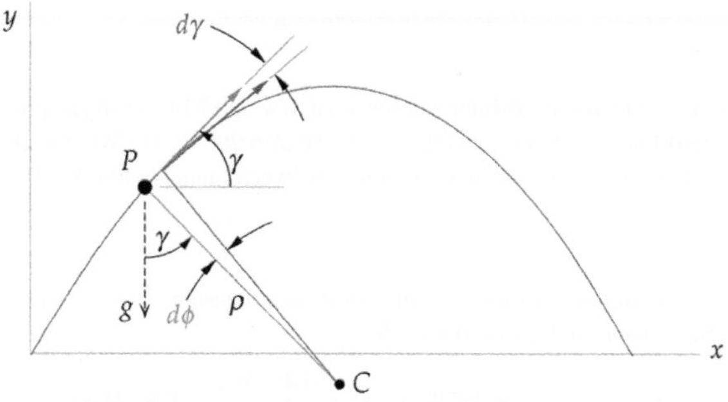

FIG. 1.12

Relationship between $d\gamma$ and $d\phi$ for a "flat" earth.

1.5 NEWTON'S LAW OF MOTION

Force is not a primitive concept like mass because it is intimately connected with the concepts of motion and inertia. In fact, the only way to alter the motion of a body is to exert a force on it. The degree to which the motion is altered is a measure of the force. Newton's second law of motion quantifies this. If the resultant or net force on a body of mass m is $\mathbf{F}_{net}$, then

$$\mathbf{F}_{net} = m\mathbf{a} \tag{1.38}$$

In this equation, **a** is the absolute acceleration of the center of mass. The absolute acceleration is measured in a frame of reference that itself has neither translational nor rotational acceleration relative to the fixed stars. Such a reference is called an absolute or inertial frame of reference.

Force is related to the primitive concepts of mass, length, and time by Newton's second law. The unit of force, appropriately, is the Newton, which is the force required to impart an acceleration of 1 m/s^2 to a mass of 1 kg. A mass of 1 kg therefore weighs 9.807 N at the earth's surface. The kilogram is not a unit of force.

Confusion can arise when mass is expressed in units of force, as frequently occurs in US engineering practice. In common parlance either the pound or the ton (2000 lb) is more likely to be used to express the mass. The pound of mass is officially defined precisely in terms of the kilogram, as shown in Table A.3. Since 1 lb of mass weighs 1 lb of force where the standard sea level acceleration of gravity (Eq. 1.35) exists, we can use Newton's second law to relate the pound of force to the Newton:

$$1\,\text{lb (force)} = 0.4536\,\text{kg} \times 9.807\,\text{m/s}^2 = 4.448\,\text{N}$$

The slug is the quantity of matter accelerated at 1 ft/s^2 by a force of 1 lb. We can again use Newton's second law to relate the slug to the kilogram. Noting the relationship between feet and meters in Table A.3, we find

$$1\text{ slug} = \frac{1\,\text{lb}}{1\,\text{ft/s}^2} = \frac{4.448\,\text{N}}{0.3048\,\text{m/s}^2} = 14.59\,\frac{\text{kg}\cdot\text{m/s}^2}{\text{m/s}^2} = 14.59\,\text{kg}$$

EXAMPLE 1.9

On a NASA mission, the space shuttle *Atlantis* orbiter was reported to weigh 239,255 lb just prior to liftoff. On orbit 18 at an altitude of about 350 km, the orbiter's weight was reported to be 236,900 lb. (a) What was the mass, in kilograms, of *Atlantis* on the launchpad and in orbit? (b) If no mass was lost between launch and orbit 18, what would have been the weight of *Atlantis*, in pounds?

Solution

(a) The given data illustrate the common use of weight in pounds as a measure of mass. The "weights" given are actually the mass in pounds of mass. Therefore, prior to launch

$$m_{\text{launchpad}} = 239{,}255\,\text{lb (mass)} \times \frac{0.4536\,\text{kg}}{1\,\text{lb(mass)}} = 108{,}500\,\text{kg}$$

In orbit,

$$m_{\text{orbit 18}} = 236{,}900\,\text{lb (mass)} \times \frac{0.4536\,\text{kg}}{1\,\text{lb(mass)}} = 107{,}500\,\text{kg}$$

The decrease in mass is the propellant expended by the orbital maneuvering and reaction control rockets on the orbiter.

(b) Since the space shuttle launchpad at the Kennedy Space Center is essentially at sea level, the launchpad weight of *Atlantis* in pounds (force) was numerically equal to its mass in pounds (mass). With no change in mass, the force of gravity at 350 km would be, according to Eq. (1.36),

$$W = 239{,}255\,\text{lb(force)} \times \left(\frac{1}{1 + \dfrac{350}{6378}}\right)^2 = 215{,}000\,\text{lb(force)}$$

The integral of a force **F** over a time interval is called the impulse of the force,

$$\mathcal{I} = \int_{t_1}^{t_2} \mathbf{F}\,dt \tag{1.39}$$

Impulse is a vector quantity. From Eq. (1.38) it is apparent that if the mass is constant, then

$$\mathcal{I}_{\text{net}} = \int_{t_1}^{t_2} m\frac{d\mathbf{v}}{dt}dt = m\mathbf{v}_2 - m\mathbf{v}_1 \tag{1.40}$$

That is, the net impulse on a body yields a change $m\Delta\mathbf{v}$ in its linear momentum, so that

$$\Delta\mathbf{v} = \frac{\mathcal{I}_{\text{net}}}{m} \tag{1.41}$$

If $\mathbf{F}_{\text{net}}$ is constant, then $\mathcal{I}_{\text{net}} = \mathbf{F}_{\text{net}}\Delta t$, in which case Eq. (1.41) becomes

$$\Delta\mathbf{v} = \frac{\mathbf{F}_{\text{net}}}{m}\Delta t \quad (\text{if } \mathbf{F}_{\text{net}} \text{ is constant}) \tag{1.42}$$

Let us conclude this section by introducing the concept of angular momentum. The moment of the net force about O in Fig. 1.13 is

$$\mathbf{M}_O)_{\text{net}} = \mathbf{r} \times \mathbf{F}_{\text{net}}$$

Substituting Eq. (1.38) yields

$$\mathbf{M}_O)_{\text{net}} = \mathbf{r} \times m\mathbf{a} = \mathbf{r} \times m\frac{d\mathbf{v}}{dt} \tag{1.43}$$

But, keeping in mind that the mass is constant,

$$\mathbf{r} \times m\frac{d\mathbf{v}}{dt} = \frac{d}{dt}(\mathbf{r} \times m\mathbf{v}) - \left(\frac{d\mathbf{r}}{dt} \times m\mathbf{v}\right) = \frac{d}{dt}(\mathbf{r} \times m\mathbf{v}) - (\mathbf{v} \times m\mathbf{v})$$

Since $\mathbf{v} \times m\mathbf{v} = m(\mathbf{v} \times \mathbf{v}) = \mathbf{0}$, it follows that Eq. (1.43) can be written

$$\mathbf{M}_O)_{\text{net}} = \frac{d\mathbf{H}_O}{dt} \tag{1.44}$$

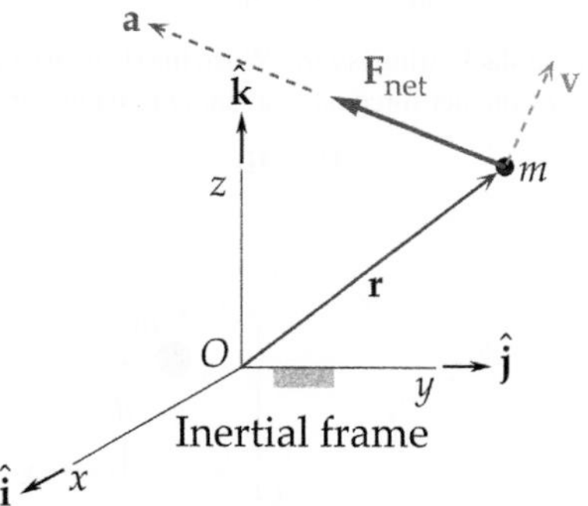

FIG. 1.13

The absolute acceleration of a particle is in the direction of the net force.

where $\mathbf{H}_O$ is the angular momentum about O,

$$\mathbf{H}_O = \mathbf{r} \times m\mathbf{v} \tag{1.45}$$

Thus, just as the net force on a particle changes its linear momentum $m\mathbf{v}$, the moment of that force about a fixed point changes the moment of its linear momentum about that point. Integrating Eq. (1.44) with respect to time yields

$$\int_{t_1}^{t_2} \mathbf{M}_O)_{\text{net}} = \mathbf{H}_O)_2 - \mathbf{H}_O)_1 \tag{1.46}$$

The integral on the left is the net angular impulse. This angular impulse-momentum equation is the rotational analog of the linear impulse-momentum relation given above in Eq. (1.40).

EXAMPLE 1.10

A particle of mass m is attached to point O by an inextensible string of length l, as illustrated in Fig. 1.14. Initially, the string is slack when m is moving to the left with a speed v_0 in the position shown. Calculate (a) the speed of m just after the string becomes taut and (b) the average force in the string over the small time interval Δt required to change the direction of the particle's motion.

Solution

(a) Initially, the position and velocity of the particle are

$$\mathbf{r}_1 = c\hat{\mathbf{i}} + d\hat{\mathbf{j}} \quad \mathbf{v}_1 = -v_0\hat{\mathbf{i}}$$

The angular momentum about O is

$$\mathbf{H}_1 = \mathbf{r}_1 \times m\mathbf{v}_1 = \begin{vmatrix} \hat{\mathbf{i}} & \hat{\mathbf{j}} & \hat{\mathbf{k}} \\ c & d & 0 \\ -mv_o & 0 & 0 \end{vmatrix} = mv_o d\hat{\mathbf{k}} \tag{a}$$

Just after the string becomes taut,

$$\mathbf{r}_2 = -\sqrt{l^2 - d^2}\,\hat{\mathbf{i}} + d\hat{\mathbf{j}} \quad \mathbf{v}_2 = v_x\hat{\mathbf{i}} + v_y\hat{\mathbf{j}} \tag{b}$$

and the angular momentum is

$$\mathbf{H}_2 = \mathbf{r}_2 \times m\mathbf{v}_2 = \begin{vmatrix} \hat{\mathbf{i}} & \hat{\mathbf{j}} & \hat{\mathbf{k}} \\ -\sqrt{l^2 - d^2} & d & 0 \\ mv_x & mv_y & 0 \end{vmatrix} = \left(-mv_x d - mv_y\sqrt{l^2 - d^2}\right)\hat{\mathbf{k}} \tag{c}$$

Initially, the force exerted on m by the slack string is zero. When the string becomes taut, the force exerted on m passes through O. Therefore, the moment of the net force on m about O remains zero. According to Eq. (1.46),

$$\mathbf{H}_2 = \mathbf{H}_1$$

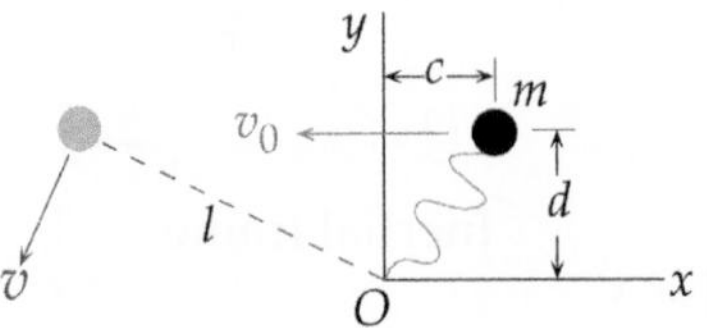

FIG. 1.14

Particle attached to O by an inextensible string.

Substituting Eqs. (a) and (c) yields

$$v_x d + \sqrt{l^2 - d^2} v_y = -v_o d \tag{d}$$

The string is inextensible, so the component of the velocity of m along the string must be zero:

$$\mathbf{v}_2 \cdot \mathbf{r}_2 = 0$$

Substituting $\mathbf{v}_2$ and $\mathbf{r}_2$ from Eq. (b) and solving for v_y, we get

$$v_y = v_x \sqrt{\frac{l^2}{d^2} - 1} \tag{e}$$

Solving Eqs. (d) and (e) for v_x and v_y leads to

$$v_x = -\frac{d^2}{l^2} v_o \quad v_y = -\sqrt{1 - \frac{d^2}{l^2}} \frac{d}{l} v_o \tag{f}$$

Thus, the speed, $v = \sqrt{v_x^2 + v_y^2}$, after the string becomes taut is

$$\boxed{v = \frac{d}{l} v_o}$$

(b) From Eq. (1.40), the impulse on m during the time it takes the string to become taut is

$$\mathcal{I} = m(\mathbf{v}_2 - \mathbf{v}_1) = m\left[\left(-\frac{d^2}{l^2} v_o \hat{\mathbf{i}} - \sqrt{1 - \frac{d^2}{l^2}} \frac{d}{l} v_o \hat{\mathbf{j}}\right) - \left(-v_o \hat{\mathbf{i}}\right)\right] = \left(1 - \frac{d^2}{l^2}\right) m v_o \hat{\mathbf{i}} - \sqrt{1 - \frac{d^2}{l^2}} \frac{d}{l} m v_o \hat{\mathbf{j}}$$

The magnitude of this impulse, which is directed along the string, is

$$\mathcal{I} = \|\mathcal{I}\| = \sqrt{1 - \frac{d^2}{l^2}} m v_o$$

Hence, the average force in the string during the small time interval Δt required to change the direction of the velocity vector turns out to be

$$\boxed{F_{avg} = \frac{\mathcal{I}}{\Delta t} = \sqrt{1 - \frac{d^2}{l^2}} \frac{m v_o}{\Delta t}}$$

1.6 TIME DERIVATIVES OF MOVING VECTORS

Fig. 1.15(a) shows a vector **A** inscribed in a rigid body *B* that is in motion relative to an inertial frame of reference (a rigid, Cartesian coordinate system, which is fixed relative to the fixed stars). The magnitude of **A** is fixed. The body *B* is shown at two times, separated by the differential time interval dt. At time t + dt, the orientation of vector **A** differs slightly from that at time t, but its magnitude is the same. According to one of the many theorems of the prolific 18th-century Swiss mathematician Leonhard Euler (1707–1783), there is a unique axis of rotation about which *B*, and therefore **A**, rotates during the differential time interval. If we shift the two vectors **A**(t) and **A**(t + dt) to the same point on the axis of rotation, so that they are tail to tail, as shown in Fig. 1.15(b), we can assess the difference d**A** between them caused by the infinitesimal rotation. Remember that shifting a vector to a parallel line does not change the vector. The rotation of the body *B* is measured in the plane perpendicular to the instantaneous axis of rotation. The amount of rotation is the angle dθ through which a line element normal to the rotation axis turns in the time interval dt. In Fig. 1.15(b) that line element is

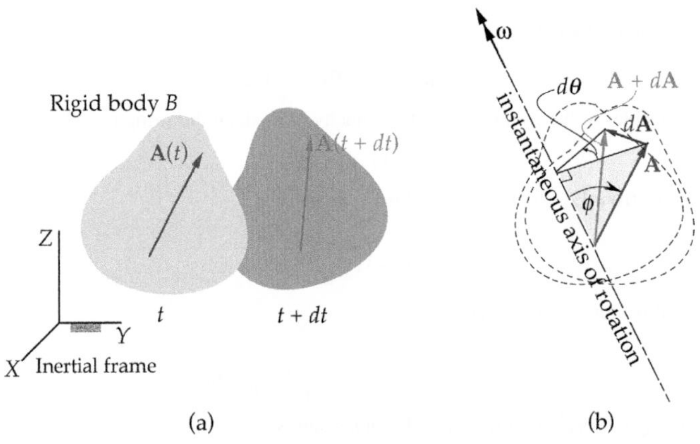

FIG. 1.15

Displacement of a rigid body. (a) Change in orientation of an embedded vector **A**. (b) Differential rotation of **A** about the instantaneous rotation axis.

the component of **A** normal to the axis of rotation. We can express the difference d**A** between **A**(t) and **A**(t + dt) as

$$\mathrm{d}\mathbf{A} = \overbrace{[(\|\mathbf{A}\| \cdot \sin\phi)\mathrm{d}\theta]}^{\text{magnitude of dA}}\hat{\mathbf{n}} \tag{1.47}$$

where $\hat{\mathbf{n}}$ is the unit normal to the plane defined by **A** and the axis of rotation, and it points in the direction of the rotation. The angle ϕ is the inclination of **A** to the rotation axis. By definition,

$$\mathrm{d}\theta = \|\boldsymbol{\omega}\|\mathrm{d}t \tag{1.48}$$

where $\boldsymbol{\omega}$ is the angular velocity vector, which points along the instantaneous axis of rotation, and its direction is given by the right-hand rule. That is, wrapping the right hand around the axis of rotation, with the fingers pointing in the direction of dθ, results in the thumb defining the direction of $\boldsymbol{\omega}$. This is evident in Fig. 1.15(b). It should be pointed out that the time derivative of $\boldsymbol{\omega}$ is the angular acceleration, usually given the symbol $\boldsymbol{\alpha}$. Thus,

$$\boldsymbol{\alpha} = \frac{\mathrm{d}\boldsymbol{\omega}}{\mathrm{d}t} \tag{1.49}$$

Substituting Eq. (1.48) into Eq. (1.47), we get

$$\mathrm{d}\mathbf{A} = \|\mathbf{A}\| \cdot \sin\phi \cdot \|\boldsymbol{\omega}\|\mathrm{d}t \cdot \hat{\mathbf{n}} = (\|\boldsymbol{\omega}\| \cdot \|\mathbf{A}\| \cdot \sin\phi)\hat{\mathbf{n}}\,\mathrm{d}t \tag{1.50}$$

By definition of the cross product, $\boldsymbol{\omega} \times \mathbf{A}$ is the product of the magnitude of $\boldsymbol{\omega}$, the magnitude of **A**, the sine of the angle between $\boldsymbol{\omega}$ and **A**, and the unit vector normal to the plane of $\boldsymbol{\omega}$ and **A**, in the rotation direction. That is,

$$\boldsymbol{\omega} \times \mathbf{A} = \|\boldsymbol{\omega}\| \cdot \|\mathbf{A}\| \cdot \sin\phi \cdot \hat{\mathbf{n}} \tag{1.51}$$

Substituting Eq. (1.51) into Eq. (1.50) yields

$$\mathrm{d}\mathbf{A} = \boldsymbol{\omega} \times \mathbf{A}\mathrm{d}t$$

Dividing through by dt, we finally obtain

$$\boxed{\frac{\mathrm{d}\mathbf{A}}{\mathrm{d}t}=\boldsymbol{\omega}\times\mathbf{A}}\quad\left(\text{if }\frac{\mathrm{d}}{\mathrm{d}t}\|\mathbf{A}\|=0\right)\tag{1.52}$$

Eq. (1.52) is a formula we can use to compute the time derivative of any vector of constant magnitude.

EXAMPLE 1.11

Calculate the second time derivative of a vector **A** of constant magnitude, expressing the result in terms of **ω** and its derivatives and **A**.

Solution

Differentiating Eq. (1.52) with respect to time, we get

$$\frac{\mathrm{d}^2\mathbf{A}}{\mathrm{d}t^2}=\frac{\mathrm{d}}{\mathrm{d}t}\frac{\mathrm{d}\mathbf{A}}{\mathrm{d}t}=\frac{\mathrm{d}}{\mathrm{d}t}(\boldsymbol{\omega}\times\mathbf{A})=\frac{\mathrm{d}\boldsymbol{\omega}}{\mathrm{d}t}\times\mathbf{A}+\boldsymbol{\omega}\times\frac{\mathrm{d}\mathbf{A}}{\mathrm{d}t}$$

Using Eqs. (1.49) and (1.52), this can be written

$$\boxed{\frac{\mathrm{d}^2\mathbf{A}}{\mathrm{d}t^2}=\boldsymbol{\alpha}\times\mathbf{A}+\boldsymbol{\omega}\times(\boldsymbol{\omega}\times\mathbf{A})}\tag{1.53}$$

EXAMPLE 1.12

Calculate the third derivative of a vector **A** of constant magnitude, expressing the result in terms of **ω** and its derivatives and **A**.

Solution

$$\begin{aligned}
\frac{\mathrm{d}^3\mathbf{A}}{\mathrm{d}t^3}&=\frac{\mathrm{d}}{\mathrm{d}t}\frac{\mathrm{d}^2\mathbf{A}}{\mathrm{d}t^2}=\frac{\mathrm{d}}{\mathrm{d}t}[\boldsymbol{\alpha}\times\mathbf{A}+\boldsymbol{\omega}\times(\boldsymbol{\omega}\times\mathbf{A})]\\
&=\frac{\mathrm{d}}{\mathrm{d}t}(\boldsymbol{\alpha}\times\mathbf{A})+\frac{\mathrm{d}}{\mathrm{d}t}[\boldsymbol{\omega}\times(\boldsymbol{\omega}\times\mathbf{A})]\\
&=\left(\frac{\mathrm{d}\boldsymbol{\alpha}}{\mathrm{d}t}\times\mathbf{A}+\boldsymbol{\alpha}\times\frac{\mathrm{d}\mathbf{A}}{\mathrm{d}t}\right)+\left[\frac{\mathrm{d}\boldsymbol{\omega}}{\mathrm{d}t}\times(\boldsymbol{\omega}\times\mathbf{A})+\boldsymbol{\omega}\times\frac{\mathrm{d}}{\mathrm{d}t}(\boldsymbol{\omega}\times\mathbf{A})\right]\\
&=\left[\frac{\mathrm{d}\boldsymbol{\alpha}}{\mathrm{d}t}\times\mathbf{A}+\boldsymbol{\alpha}\times(\boldsymbol{\omega}\times\mathbf{A})\right]+\left[\boldsymbol{\alpha}\times(\boldsymbol{\omega}\times\mathbf{A})+\boldsymbol{\omega}\times\left(\frac{\mathrm{d}\boldsymbol{\omega}}{\mathrm{d}t}\times\mathbf{A}+\boldsymbol{\omega}\times\frac{\mathrm{d}\mathbf{A}}{\mathrm{d}t}\right)\right]\\
&=\left[\frac{\mathrm{d}\boldsymbol{\alpha}}{\mathrm{d}t}\times\mathbf{A}+\boldsymbol{\alpha}\times(\boldsymbol{\omega}\times\mathbf{A})\right]+\{\boldsymbol{\alpha}\times(\boldsymbol{\omega}\times\mathbf{A})+\boldsymbol{\omega}\times[\boldsymbol{\alpha}\times\mathbf{A}+\boldsymbol{\omega}\times(\boldsymbol{\omega}\times\mathbf{A})]\}\\
&=\frac{\mathrm{d}\boldsymbol{\alpha}}{\mathrm{d}t}\times\mathbf{A}+\boldsymbol{\alpha}\times(\boldsymbol{\omega}\times\mathbf{A})+\boldsymbol{\alpha}\times(\boldsymbol{\omega}\times\mathbf{A})+\boldsymbol{\omega}\times(\boldsymbol{\alpha}\times\mathbf{A})+\boldsymbol{\omega}\times[\boldsymbol{\omega}\times(\boldsymbol{\omega}\times\mathbf{A})]\\
&=\frac{\mathrm{d}\boldsymbol{\alpha}}{\mathrm{d}t}\times\mathbf{A}+2\boldsymbol{\alpha}\times(\boldsymbol{\omega}\times\mathbf{A})+\boldsymbol{\omega}\times(\boldsymbol{\alpha}\times\mathbf{A})+\boldsymbol{\omega}\times[\boldsymbol{\omega}\times(\boldsymbol{\omega}\times\mathbf{A})]
\end{aligned}$$

$$\boxed{\frac{\mathrm{d}^3\mathbf{A}}{\mathrm{d}t^3}=\frac{\mathrm{d}\boldsymbol{\alpha}}{\mathrm{d}t}\times\mathbf{A}+2\boldsymbol{\alpha}\times(\boldsymbol{\omega}\times\mathbf{A})+\boldsymbol{\omega}\times[\boldsymbol{\alpha}\times\mathbf{A}+\boldsymbol{\omega}\times(\boldsymbol{\omega}\times\mathbf{A})]}$$

Let XYZ be a rigid inertial frame of reference and xyz a rigid moving frame of reference, as shown in Fig. 1.16. The moving frame can be moving (translating and rotating) freely on its own accord, or it can

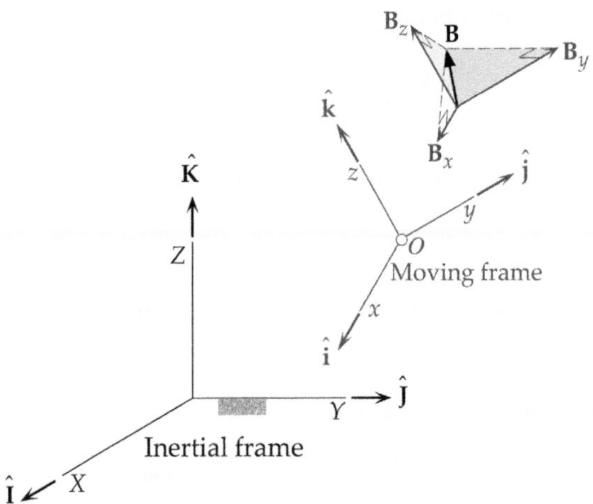

FIG. 1.16

Fixed (inertial) and moving rigid frames of reference.

be attached to a physical object, such as a car, an airplane, or a spacecraft. Kinematic quantities measured relative to the fixed inertial frame will be called absolute (e.g., absolute acceleration), and those measured relative to the moving system will be called relative (e.g., relative acceleration). The unit vectors along the inertial *XYZ* system are $\hat{\mathbf{I}}$, $\hat{\mathbf{J}}$, and $\hat{\mathbf{K}}$, whereas those of the moving *xyz* system are $\hat{\mathbf{i}}$, $\hat{\mathbf{j}}$, and $\hat{\mathbf{k}}$. The motion of the moving frame is arbitrary, and its absolute angular velocity is $\boldsymbol{\Omega}$. If, however, the moving frame is rigidly attached to an object, so that it not only translates but also rotates with it, then the frame is called a body frame and the axes are referred to as body axes. A body frame clearly has the same angular velocity as the body to which it is bound.

Let $\mathbf{B}$ be any time-dependent vector. Resolved into components along the inertial frame of reference, it is expressed analytically as

$$\mathbf{B} = B_X\hat{\mathbf{I}} + B_Y\hat{\mathbf{J}} + B_Z\hat{\mathbf{K}}$$

where B_X, B_Y, and B_Z are functions of time. Since $\hat{\mathbf{I}}$, $\hat{\mathbf{J}}$, and $\hat{\mathbf{K}}$ are fixed, the time derivative of $\mathbf{B}$ is simply

$$\frac{d\mathbf{B}}{dt} = \frac{dB_X}{dt}\hat{\mathbf{I}} + \frac{dB_Y}{dt}\hat{\mathbf{J}} + \frac{dB_Z}{dt}\hat{\mathbf{K}}$$

dB_X/dt, dB_Y/dt, and dB_Z/dt are the components of the absolute time derivative of $\mathbf{B}$.

$\mathbf{B}$ may also be resolved into components along the moving *xyz* frame, so that, at any instant,

$$\mathbf{B} = B_x\hat{\mathbf{i}} + B_y\hat{\mathbf{j}} + B_z\hat{\mathbf{k}} \tag{1.54}$$

Using this expression to calculate the time derivative of $\mathbf{B}$ yields

$$\frac{d\mathbf{B}}{dt} = \frac{dB_x}{dt}\hat{\mathbf{i}} + \frac{dB_y}{dt}\hat{\mathbf{j}} + \frac{dB_z}{dt}\hat{\mathbf{k}} + B_x\frac{d\hat{\mathbf{i}}}{dt} + B_y\frac{d\hat{\mathbf{j}}}{dt} + B_z\frac{d\hat{\mathbf{k}}}{dt} \tag{1.55}$$

The orthogonal unit vectors $\hat{\mathbf{i}}$, $\hat{\mathbf{j}}$, and $\hat{\mathbf{k}}$ are not fixed in space but are continuously changing direction; therefore, their time derivatives are not zero. They obviously have a constant magnitude (unity) and, being attached to the *xyz* frame, they all have the angular velocity $\mathbf{\Omega}$. It follows from Eq. (1.52) that

$$\frac{d\hat{\mathbf{i}}}{dt}=\mathbf{\Omega}\times\hat{\mathbf{i}} \quad \frac{d\hat{\mathbf{j}}}{dt}=\mathbf{\Omega}\times\hat{\mathbf{j}} \quad \frac{d\hat{\mathbf{k}}}{dt}=\mathbf{\Omega}\times\hat{\mathbf{k}}$$

Substituting these on the right-hand side of Eq. (1.55) yields

$$\begin{aligned}\frac{d\mathbf{B}}{dt}&=\frac{dB_x}{dt}\hat{\mathbf{i}}+\frac{dB_y}{dt}\hat{\mathbf{j}}+\frac{dB_z}{dt}\hat{\mathbf{k}}+B_x\left(\mathbf{\Omega}\times\hat{\mathbf{i}}\right)+B_y\left(\mathbf{\Omega}\times\hat{\mathbf{j}}\right)+B_z\left(\mathbf{\Omega}\times\hat{\mathbf{k}}\right)\\&=\frac{dB_x}{dt}\hat{\mathbf{i}}+\frac{dB_y}{dt}\hat{\mathbf{j}}+\frac{dB_z}{dt}\hat{\mathbf{k}}+\left(\mathbf{\Omega}\times B_x\hat{\mathbf{i}}\right)+\left(\mathbf{\Omega}\times B_y\hat{\mathbf{j}}\right)+\left(\mathbf{\Omega}\times B_z\hat{\mathbf{k}}\right)\\&=\frac{dB_x}{dt}\hat{\mathbf{i}}+\frac{dB_y}{dt}\hat{\mathbf{j}}+\frac{dB_z}{dt}\hat{\mathbf{k}}+\mathbf{\Omega}\times\left(B_x\hat{\mathbf{i}}+B_y\hat{\mathbf{j}}+B_z\hat{\mathbf{k}}\right)\end{aligned}$$

In view of Eq. (1.54), this can be written as

$$\frac{d\mathbf{B}}{dt}=\left.\frac{d\mathbf{B}}{dt}\right)_{\text{rel}}+\mathbf{\Omega}\times\mathbf{B} \tag{1.56}$$

where

$$\left.\frac{d\mathbf{B}}{dt}\right)_{\text{rel}}=\frac{dB_x}{dt}\hat{\mathbf{i}}+\frac{dB_y}{dt}\hat{\mathbf{j}}+\frac{dB_z}{dt}\hat{\mathbf{k}} \tag{1.57}$$

$d\mathbf{B}/dt)_{\text{rel}}$ is the time derivative of $\mathbf{B}$ relative to the moving frame. Eq. (1.56) shows how the absolute time derivative is obtained from the relative time derivative. Clearly, $d\mathbf{B}/dt = d\mathbf{B}/dt)_{\text{rel}}$ only when the moving frame is in pure translation ($\mathbf{\Omega}=\mathbf{0}$).

Eq. (1.56) can be used recursively to compute higher order time derivatives. Thus, differentiating Eq. (1.56) with respect to t, we get

$$\frac{d^2\mathbf{B}}{dt^2}=\left.\frac{d}{dt}\frac{d\mathbf{B}}{dt}\right)_{\text{rel}}+\frac{d\mathbf{\Omega}}{dt}\times\mathbf{B}+\mathbf{\Omega}\times\frac{d\mathbf{B}}{dt}$$

Using Eq. (1.56) in the last term yields

$$\frac{d^2\mathbf{B}}{dt^2}=\left.\frac{d}{dt}\frac{d\mathbf{B}}{dt}\right)_{\text{rel}}+\frac{d\mathbf{\Omega}}{dt}\times\mathbf{B}+\mathbf{\Omega}\times\left[\left.\frac{d\mathbf{B}}{dt}\right)_{\text{rel}}+\mathbf{\Omega}\times\mathbf{B}\right] \tag{1.58}$$

Eq. (1.56) also implies that

$$\left.\frac{d}{dt}\frac{d\mathbf{B}}{dt}\right)_{\text{rel}}=\left.\frac{d^2\mathbf{B}}{dt^2}\right)_{\text{rel}}+\mathbf{\Omega}\times\left.\frac{d\mathbf{B}}{dt}\right)_{\text{rel}} \tag{1.59}$$

where

$$\left.\frac{d^2\mathbf{B}}{dt^2}\right)_{\text{rel}}=\frac{d^2B_x}{dt^2}\hat{\mathbf{i}}+\frac{d^2B_y}{dt^2}\hat{\mathbf{j}}+\frac{d^2B_z}{dt^2}\hat{\mathbf{k}}$$

Substituting Eq. (1.59) into Eq. (1.58) yields

$$\frac{d^2\mathbf{B}}{dt^2}=\left[\left.\frac{d^2\mathbf{B}}{dt^2}\right)_{\text{rel}}+\mathbf{\Omega}\times\left.\frac{d\mathbf{B}}{dt}\right)_{\text{rel}}\right]+\frac{d\mathbf{\Omega}}{dt}\times\mathbf{B}+\mathbf{\Omega}\times\left[\left.\frac{d\mathbf{B}}{dt}\right)_{\text{rel}}+\mathbf{\Omega}\times\mathbf{B}\right]$$

Collecting terms, this becomes

$$\frac{d^2\mathbf{B}}{dt^2}=\left.\frac{d^2\mathbf{B}}{dt^2}\right)_{\text{rel}}+\dot{\boldsymbol{\Omega}}\times\mathbf{B}+\boldsymbol{\Omega}\times(\boldsymbol{\Omega}\times\mathbf{B})+2\boldsymbol{\Omega}\times\left.\frac{d\mathbf{B}}{dt}\right)_{\text{rel}} \tag{1.60}$$

where $\dot{\boldsymbol{\Omega}}\equiv d\boldsymbol{\Omega}/dt$ is the absolute angular acceleration of the xyz frame.

Formulas for higher order time derivatives are found in a similar fashion.

1.7 RELATIVE MOTION

Let P be a particle in arbitrary motion. The absolute position vector of P is $\mathbf{r}$ and the position of P relative to the moving frame is $\mathbf{r}_{\text{rel}}$. If $\mathbf{r}_O$ is the absolute position of the origin of the moving frame, then it is clear from Fig. 1.17 that

$$\mathbf{r}=\mathbf{r}_O+\mathbf{r}_{\text{rel}} \tag{1.61}$$

Since $\mathbf{r}_{\text{rel}}$ is measured in the moving frame,

$$\mathbf{r}_{\text{rel}}=x\hat{\mathbf{i}}+y\hat{\mathbf{j}}+z\hat{\mathbf{k}} \tag{1.62}$$

where x, y, and z are the coordinates of P relative to the moving reference.

The absolute velocity $\mathbf{v}$ of P is $d\mathbf{r}/dt$, so that from Eq. (1.61) we have

$$\mathbf{v}=\mathbf{v}_O+\frac{d\mathbf{r}_{\text{rel}}}{dt} \tag{1.63}$$

where $\mathbf{v}_O=d\mathbf{r}_O/dt$ is the (absolute) velocity of the origin of the xyz frame. From Eq. (1.56), we can write

$$\frac{d\mathbf{r}_{\text{rel}}}{dt}=\mathbf{v}_{\text{rel}}+\boldsymbol{\Omega}\times\mathbf{r}_{\text{rel}} \tag{1.64}$$

where $\mathbf{v}_{\text{rel}}$ is the velocity of P relative to the xyz frame (so that $\hat{\mathbf{i}}$, $\hat{\mathbf{j}}$, and $\hat{\mathbf{k}}$ are held fixed):

$$\mathbf{v}_{\text{rel}}=\left.\frac{d\mathbf{r}_{\text{rel}}}{dt}\right)_{\text{rel}}=\frac{dx}{dt}\hat{\mathbf{i}}+\frac{dy}{dt}\hat{\mathbf{j}}+\frac{dz}{dt}\hat{\mathbf{k}} \tag{1.65}$$

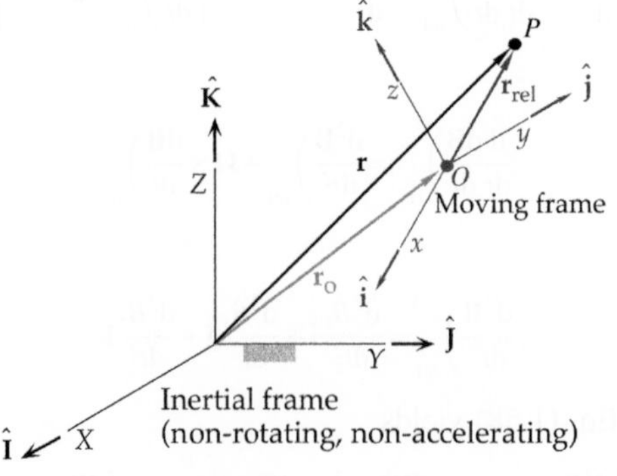

FIG. 1.17

Absolute and relative position vectors.

Substituting Eq. (1.64) into Eq. (1.63) yields

$$\boxed{\mathbf{v} = \mathbf{v}_O + \boldsymbol{\Omega} \times \mathbf{r}_{\mathrm{rel}} + \mathbf{v}_{\mathrm{rel}}} \tag{1.66}$$

The absolute acceleration $\mathbf{a}$ of P is $\mathrm{d}\mathbf{v}/\mathrm{d}t$, so that from Eq. (1.63) we have

$$\mathbf{a} = \mathbf{a}_O + \frac{\mathrm{d}^2\mathbf{r}_{\mathrm{rel}}}{\mathrm{d}t^2} \tag{1.67}$$

where $\mathbf{a}_O = \mathrm{d}\mathbf{v}_O/\mathrm{d}t$ is the absolute acceleration of the origin of the xyz frame. We evaluate the second term on the right using Eq. (1.60).

$$\frac{\mathrm{d}^2\mathbf{r}_{\mathrm{rel}}}{\mathrm{d}t^2} = \left.\frac{\mathrm{d}^2\mathbf{r}_{\mathrm{rel}}}{\mathrm{d}t^2}\right)_{\mathrm{rel}} + \dot{\boldsymbol{\Omega}} \times \mathbf{r}_{\mathrm{rel}} + \boldsymbol{\Omega} \times (\boldsymbol{\Omega} \times \mathbf{r}_{\mathrm{rel}}) + 2\boldsymbol{\Omega} \times \left.\frac{\mathrm{d}\mathbf{r}_{\mathrm{rel}}}{\mathrm{d}t}\right)_{\mathrm{rel}} \tag{1.68}$$

Since $\mathbf{v}_{\mathrm{rel}} = \mathrm{d}\mathbf{r}_{\mathrm{rel}}/\mathrm{d}t)_{\mathrm{rel}}$ and $\mathbf{a}_{\mathrm{rel}} = \mathrm{d}^2\mathbf{r}_{\mathrm{rel}}/\mathrm{d}t^2)_{\mathrm{rel}}$, this can be written

$$\frac{\mathrm{d}^2\mathbf{r}_{\mathrm{rel}}}{\mathrm{d}t^2} = \mathbf{a}_{\mathrm{rel}} + \dot{\boldsymbol{\Omega}} \times \mathbf{r}_{\mathrm{rel}} + \boldsymbol{\Omega} \times (\boldsymbol{\Omega} \times \mathbf{r}_{\mathrm{rel}}) + 2\boldsymbol{\Omega} \times \mathbf{v}_{\mathrm{rel}} \tag{1.69}$$

Upon substituting this result into Eq. (1.67), we find

$$\boxed{\mathbf{a} = \mathbf{a}_O + \dot{\boldsymbol{\Omega}} \times \mathbf{r}_{\mathrm{rel}} + \boldsymbol{\Omega} \times (\boldsymbol{\Omega} \times \mathbf{r}_{\mathrm{rel}}) + 2\boldsymbol{\Omega} \times \mathbf{v}_{\mathrm{rel}} + \mathbf{a}_{\mathrm{rel}}} \tag{1.70}$$

The cross product $2\boldsymbol{\Omega} \times \mathbf{v}_{\mathrm{rel}}$ is called the Coriolis acceleration after Gustave Gaspard de Coriolis (1792–1843), the French mathematician who introduced this term (Coriolis, 1835). Because of the number of terms on the right, Eq. (1.70) is sometimes referred to as the five-term acceleration formula.

EXAMPLE 1.13

At a given instant, the absolute position, velocity, and acceleration of the origin O of a moving frame are

$$\left.\begin{aligned} \mathbf{r}_O &= 100\hat{\mathbf{I}} + 200\hat{\mathbf{J}} + 300\hat{\mathbf{K}}\ (\mathrm{m}) \\ \mathbf{v}_O &= -50\hat{\mathbf{I}} + 30\hat{\mathbf{J}} - 10\hat{\mathbf{K}}\ (\mathrm{m/s}) \\ \mathbf{a}_O &= -15\hat{\mathbf{I}} + 40\hat{\mathbf{J}} + 25\hat{\mathbf{K}}\,(\mathrm{m/s^2}) \end{aligned}\right\}(\text{given}) \tag{a}$$

The angular velocity and acceleration of the moving frame are

$$\left.\begin{aligned} \boldsymbol{\Omega} &= 1.0\hat{\mathbf{I}} - 0.4\hat{\mathbf{J}} + 0.6\hat{\mathbf{K}}\,(\mathrm{rad/s}) \\ \dot{\boldsymbol{\Omega}} &= -1.0\hat{\mathbf{I}} + 0.3\hat{\mathbf{J}} - 0.4\hat{\mathbf{K}}\,(\mathrm{rad/s^2}) \end{aligned}\right\}(\text{given}) \tag{b}$$

The unit vectors of the moving frame are

$$\left.\begin{aligned} \hat{\mathbf{i}} &= 0.5571\hat{\mathbf{I}} + 0.7428\hat{\mathbf{J}} + 0.3714\hat{\mathbf{K}} \\ \hat{\mathbf{j}} &= -0.06331\hat{\mathbf{I}} + 0.4839\hat{\mathbf{J}} - 0.8728\hat{\mathbf{K}} \\ \hat{\mathbf{k}} &= -0.8280\hat{\mathbf{I}} + 0.4627\hat{\mathbf{J}} + 0.3166\hat{\mathbf{K}} \end{aligned}\right\}(\text{given}) \tag{c}$$

The absolute position, velocity, and acceleration of P are

$$\left.\begin{aligned} \mathbf{r} &= 300\hat{\mathbf{I}} - 100\hat{\mathbf{J}} + 150\hat{\mathbf{K}}\,(\mathrm{m}) \\ \mathbf{v} &= 70\hat{\mathbf{I}} + 25\hat{\mathbf{J}} - 20\hat{\mathbf{K}}\,(\mathrm{m/s}) \\ \mathbf{a} &= 7.5\hat{\mathbf{I}} - 8.5\hat{\mathbf{J}} + 6.0\hat{\mathbf{K}}\,(\mathrm{m/s^2}) \end{aligned}\right\}(\text{given}) \tag{d}$$

Find (a) the velocity $\mathbf{v}_{\mathrm{rel}}$ and (b) the acceleration $\mathbf{a}_{\mathrm{rel}}$ of P relative to the moving frame.

Solution

Let us first use Eq. (c) to solve for $\hat{\mathbf{I}}$, $\hat{\mathbf{J}}$, and $\hat{\mathbf{K}}$ in terms of $\hat{\mathbf{i}}$, $\hat{\mathbf{j}}$, and $\hat{\mathbf{k}}$ (three equations in three unknowns):

$$\begin{aligned}\hat{\mathbf{I}} &= 0.5571\hat{\mathbf{i}} - 0.06331\hat{\mathbf{j}} - 0.8280\hat{\mathbf{k}}\\ \hat{\mathbf{J}} &= 0.7428\hat{\mathbf{i}} + 0.4839\hat{\mathbf{j}} + 0.4627\hat{\mathbf{k}}\\ \hat{\mathbf{K}} &= 0.3714\hat{\mathbf{i}} - 0.8728\hat{\mathbf{j}} + 0.3166\hat{\mathbf{k}}\end{aligned} \tag{e}$$

(a) The relative position vector is

$$\mathbf{r}_{\text{rel}} = \mathbf{r} - \mathbf{r}_O = \left(300\hat{\mathbf{I}} - 100\hat{\mathbf{J}} + 150\hat{\mathbf{K}}\right) - \left(100\hat{\mathbf{I}} + 200\hat{\mathbf{J}} + 300\hat{\mathbf{K}}\right) = 200\hat{\mathbf{I}} - 300\hat{\mathbf{J}} - 150\hat{\mathbf{K}}\ \text{(m)} \tag{f}$$

From Eq. (1.66), the relative velocity vector is

$$\begin{aligned}\mathbf{v}_{\text{rel}} &= \mathbf{v} - \mathbf{v}_O - \boldsymbol{\Omega} \times \mathbf{r}_{\text{rel}}\\ &= \left(70\hat{\mathbf{I}} + 25\hat{\mathbf{J}} - 20\hat{\mathbf{K}}\right) - \left(-50\hat{\mathbf{I}} + 30\hat{\mathbf{J}} - 10\hat{\mathbf{K}}\right) - \begin{vmatrix}\hat{\mathbf{I}} & \hat{\mathbf{J}} & \hat{\mathbf{K}}\\ 1.0 & -0.4 & 0.6\\ 200 & -300 & -150\end{vmatrix}\\ &= \left(70\hat{\mathbf{I}} + 25\hat{\mathbf{J}} - 20\hat{\mathbf{K}}\right) - \left(-50\hat{\mathbf{I}} + 30\hat{\mathbf{J}} - 10\hat{\mathbf{K}}\right) - \left(240\hat{\mathbf{I}} + 270\hat{\mathbf{J}} - 220\hat{\mathbf{K}}\right)\end{aligned}$$

or

$$\mathbf{v}_{\text{rel}} = -120\hat{\mathbf{I}} - 275\hat{\mathbf{J}} + 210\hat{\mathbf{K}}\ \text{(m/s)} \tag{g}$$

To obtain the components of the relative velocity along the axes of the moving frame, substitute Eq. (e) into Eq. (g),

$$\begin{aligned}\mathbf{v}_{\text{rel}} = &-120\left(0.5571\hat{\mathbf{i}} - 0.06331\hat{\mathbf{j}} - 0.8280\hat{\mathbf{k}}\right)\\ &-275\left(0.7428\hat{\mathbf{i}} + 0.4839\hat{\mathbf{j}} + 0.4627\hat{\mathbf{k}}\right) + 210\left(0.3714\hat{\mathbf{i}} - 0.8728\hat{\mathbf{j}} + 0.3166\hat{\mathbf{k}}\right)\end{aligned}$$

so that

$$\boxed{\mathbf{v}_{\text{rel}} = -193.1\hat{\mathbf{i}} - 308.8\hat{\mathbf{j}} + 38.60\hat{\mathbf{k}}\ \text{(m/s)}} \tag{h}$$

Alternatively, in terms of the unit vector $\hat{\mathbf{u}}_v$ in the direction of $\mathbf{v}_{\text{rel}}$,

$$\boxed{\mathbf{v}_{\text{rel}} = 366.2\hat{\mathbf{u}}_v\ \text{(m/s)}\quad \left(\hat{\mathbf{u}}_v = -0.5272\hat{\mathbf{i}} - 0.8432\hat{\mathbf{j}} + 0.1005\hat{\mathbf{k}}\right)} \tag{i}$$

(b) To find the relative acceleration, we use the five-term acceleration formula, Eq. (1.70):

$$\begin{aligned}\mathbf{a}_{\text{rel}} &= \mathbf{a} - \mathbf{a}_O - \dot{\boldsymbol{\Omega}} \times \mathbf{r}_{\text{rel}} - \boldsymbol{\Omega} \times (\boldsymbol{\Omega} \times \mathbf{r}_{\text{rel}}) - 2(\boldsymbol{\Omega} \times \mathbf{v}_{\text{rel}})\\ &= \mathbf{a} - \mathbf{a}_O - \begin{vmatrix}\hat{\mathbf{I}} & \hat{\mathbf{J}} & \hat{\mathbf{K}}\\ -1.0 & 0.3 & -0.4\\ 200 & -300 & -150\end{vmatrix} - \boldsymbol{\Omega} \times \begin{vmatrix}\hat{\mathbf{I}} & \hat{\mathbf{J}} & \hat{\mathbf{K}}\\ 1.0 & -0.4 & 0.6\\ 200 & -300 & -150\end{vmatrix} - 2\begin{vmatrix}\hat{\mathbf{I}} & \hat{\mathbf{J}} & \hat{\mathbf{K}}\\ 1.0 & -0.4 & 0.6\\ -120 & -275 & 210\end{vmatrix}\\ &= \mathbf{a} - \mathbf{a}_O - \left(-165\hat{\mathbf{I}} - 230\hat{\mathbf{J}} + 240\hat{\mathbf{K}}\right) - \begin{vmatrix}\hat{\mathbf{I}} & \hat{\mathbf{J}} & \hat{\mathbf{K}}\\ 1.0 & -0.4 & 0.6\\ 240 & 270 & -220\end{vmatrix} - \left(162\hat{\mathbf{I}} - 564\hat{\mathbf{J}} - 646\hat{\mathbf{K}}\right)\\ &= \left(7.5\hat{\mathbf{I}} - 8.5\hat{\mathbf{J}} + 6\hat{\mathbf{K}}\right) - \left(-15\hat{\mathbf{I}} + 40\hat{\mathbf{J}} + 25\hat{\mathbf{K}}\right) - \left(-165\hat{\mathbf{I}} - 230\hat{\mathbf{J}} + 240\hat{\mathbf{K}}\right)\\ &\quad - \left(-74\hat{\mathbf{I}} + 364\hat{\mathbf{J}} + 366\hat{\mathbf{K}}\right) - \left(162\hat{\mathbf{I}} - 564\hat{\mathbf{J}} - 646\hat{\mathbf{K}}\right)\end{aligned}$$

$$\mathbf{a}_{\text{rel}} = 99.5\hat{\mathbf{I}} + 381.5\hat{\mathbf{J}} + 21.0\hat{\mathbf{K}}\ \text{(m/s)}^2 \tag{j}$$

The components of the relative acceleration along the axes of the moving frame are found by substituting Eq. (e) into Eq. (j):

$$\begin{aligned}\mathbf{a}_{\text{rel}} = &\ 99.5\left(0.5571\hat{\mathbf{i}} - 0.06331\hat{\mathbf{j}} - 0.8282\hat{\mathbf{k}}\right)\\ &+ 381.5\left(0.7428\hat{\mathbf{i}} + 0.4839\hat{\mathbf{j}} + 0.4627\hat{\mathbf{k}}\right) + 21.0\left(0.3714\hat{\mathbf{i}} - 0.8728\hat{\mathbf{j}} + 0.3166\hat{\mathbf{k}}\right)\end{aligned}$$

$$\boxed{\mathbf{a}_{\text{rel}} = 346.6\hat{\mathbf{i}} + 160.0\hat{\mathbf{j}} + 100.8\hat{\mathbf{k}}\ (\text{m/s}^2)} \tag{k}$$

Or, in terms of the unit vector $\hat{\mathbf{u}}_a$ in the direction of $\mathbf{a}_{rel}$,

$$\boxed{\mathbf{a}_{rel} = 394.8\hat{\mathbf{u}}_a (\text{m/s}^2) \quad \left(\hat{\mathbf{u}}_a = 0.8778\hat{\mathbf{i}} + 0.4052\hat{\mathbf{j}} + 0.2553\hat{\mathbf{k}}\right)} \tag{l}$$

Fig. 1.18 shows the nonrotating inertial frame of reference XYZ with its origin at the center C of the earth, which we shall assume to be a sphere. That assumption will be relaxed in Chapter 5. Embedded in the earth and rotating with it is the orthogonal $x'y'z'$ frame, also centered at C, with the z' axis parallel to Z, the earth's axis of rotation. The x' axis intersects the equator at the prime meridian (0 degree longitude), which passes through Greenwich in London, England. The angle between X and x' is θ_G, and the rate of increase of θ_G is just the angular velocity Ω of the earth. P is a particle (e.g., an airplane or spacecraft), which is moving in an arbitrary fashion above the surface of the earth. $\mathbf{r}_{rel}$ is the position vector of P relative to C in the rotating $x'y'z'$ system. At a given instant, P is directly over point O, which lies on the earth's surface at longitude Λ and latitude ϕ. Point O coincides instantaneously with the origin of what is known as a topocentric-horizon coordinate system xyz. For our purposes, x and y are measured positive eastward and northward along the local latitude and meridian,

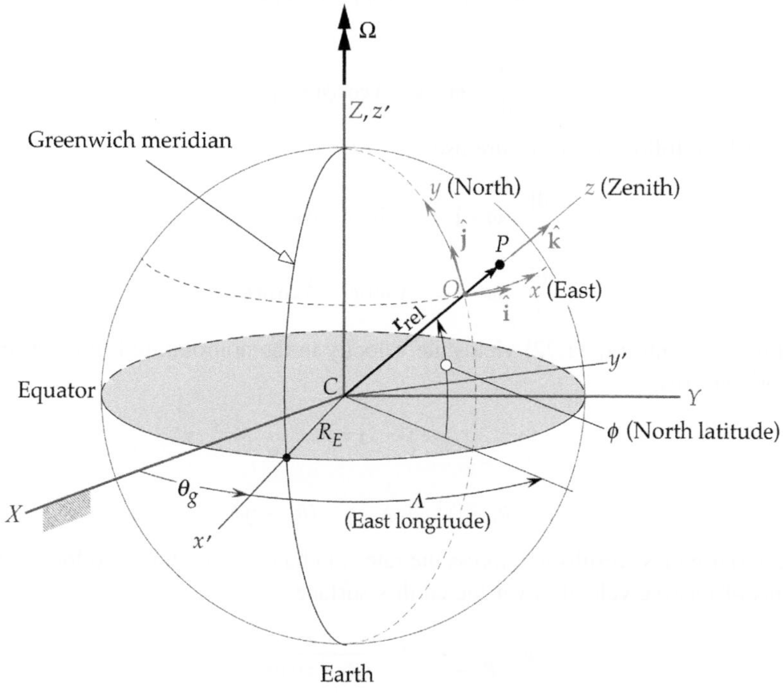

FIG. 1.18

Earth-centered inertial frame (XYZ); earth-centered noninertial $x'y'z'$ frame embedded in and rotating with the earth; and a noninertial, topocentric-horizon frame xyz attached to a point O on the earth's surface.

respectively, through O. The tangent plane to the earth's surface at O is the local horizon. The z axis is the local vertical (straight up), and it is directed radially outward from the center of the earth. The unit vectors of the xyz frame are $\hat{\mathbf{i}}\hat{\mathbf{j}}\hat{\mathbf{k}}$, as indicated in Fig. 1.18. Keep in mind that O remains directly below P, so that as P moves, so do the xyz axes. Thus, the $\hat{\mathbf{i}}\hat{\mathbf{j}}\hat{\mathbf{k}}$ triad, which comprises the unit vectors of a spherical coordinate system, varies in direction as P changes location, thereby accounting for the curvature of the earth.

Let us find the absolute velocity and acceleration of P. It is convenient to first obtain the velocity and acceleration of P relative to the nonrotating earth, and then use Eqs. (1.66) and (1.70) to calculate their inertial values.

The relative position vector can be written

$$\mathbf{r}_{\text{rel}} = (R_E + z)\hat{\mathbf{k}} \tag{1.71}$$

where R_E is the radius of the earth, and z is the height of P above the earth (i.e., its altitude). The time derivative of $\mathbf{r}_{\text{rel}}$ is the velocity $\mathbf{v}_{\text{rel}}$ relative to the nonrotating earth,

$$\mathbf{v}_{\text{rel}} = \frac{d\mathbf{r}_{\text{rel}}}{dt} = \dot{z}\hat{\mathbf{k}} + (R_E + z)\frac{d\hat{\mathbf{k}}}{dt} \tag{1.72}$$

To calculate $d\hat{\mathbf{k}}/dt$, we must use Eq. (1.52). The angular velocity $\boldsymbol{\omega}$ of the xyz frame relative to the nonrotating earth is found in terms of the rates of change of latitude ϕ and longitude Λ,

$$\boldsymbol{\omega} = -\dot{\phi}\hat{\mathbf{i}} + \dot{\Lambda}\cos\phi\hat{\mathbf{j}} + \dot{\Lambda}\sin\phi\hat{\mathbf{k}} \tag{1.73}$$

Thus,

$$\frac{d\hat{\mathbf{k}}}{dt} = \boldsymbol{\omega} \times \hat{\mathbf{k}} = \dot{\Lambda}\cos\phi\hat{\mathbf{i}} + \dot{\phi}\hat{\mathbf{j}} \tag{1.74}$$

Let us also record the following for future use:

$$\frac{d\hat{\mathbf{j}}}{dt} = \boldsymbol{\omega} \times \hat{\mathbf{j}} = -\dot{\Lambda}\sin\phi\hat{\mathbf{i}} - \dot{\phi}\hat{\mathbf{k}} \tag{1.75}$$

$$\frac{d\hat{\mathbf{i}}}{dt} = \boldsymbol{\omega} \times \hat{\mathbf{i}} = -\dot{\Lambda}\sin\phi\hat{\mathbf{j}} - \dot{\Lambda}\cos\phi\hat{\mathbf{k}} \tag{1.76}$$

Substituting Eq. (1.74) into Eq. (1.72) yields the velocity in the nonrotating frame resolved along the topocentric-horizon axes,

$$\mathbf{v}_{\text{rel}} = \dot{x}\hat{\mathbf{i}} + \dot{y}\hat{\mathbf{j}} + \dot{z}\hat{\mathbf{k}} \tag{1.77a}$$

where

$$\dot{x} = (R_E + z)\dot{\Lambda}\cos\phi \quad \dot{y} = (R_E + z)\dot{\phi} \tag{1.77b}$$

It is convenient to use these results to express the rates of change of latitude and longitude in terms of the components of relative velocity over the earth's surface,

$$\dot{\phi} = \frac{\dot{y}}{R_E + z} \quad \dot{\Lambda} = \frac{\dot{x}}{(R_E + z)\cos\phi} \tag{1.78}$$

The time derivatives of these two expressions are

$$\ddot{\phi}=\frac{(R_E+z)\ddot{y}-\dot{y}\dot{z}}{(R_E+z)^2}\qquad \ddot{\Lambda}=\frac{(R_E+z)\ddot{x}\cos\phi-(\dot{z}\cos\phi-\dot{y}\sin\phi)\dot{x}}{(R_E+z)^2\cos^2\phi} \tag{1.79}$$

The acceleration of P relative to the nonrotating earth is found by taking the time derivative of $\mathbf{v}_{\text{rel}}$. From Eqs. (1.77a) and (1.77b) we thereby obtain

$$\begin{aligned}\mathbf{a}_{\text{rel}}&=\ddot{x}\hat{\mathbf{i}}+\ddot{y}\hat{\mathbf{j}}+\ddot{z}\hat{\mathbf{k}}+\dot{x}\frac{d\hat{\mathbf{i}}}{dt}+\dot{y}\frac{d\hat{\mathbf{j}}}{dt}+\dot{z}\frac{d\hat{\mathbf{k}}}{dt}\\&=\left[\dot{z}\dot{\Lambda}\cos\phi+(R_E+z)\ddot{\Lambda}\cos\phi-(R_E+z)\dot{\phi}\dot{\Lambda}\sin\phi\right]\hat{\mathbf{i}}+\left[\dot{z}\dot{\phi}+(R_E+z)\ddot{\phi}\right]\hat{\mathbf{j}}+\ddot{z}\hat{\mathbf{k}}\\&\quad+(R_E+z)\dot{\Lambda}\cos\phi\left(\boldsymbol{\omega}\times\hat{\mathbf{i}}\right)+(R_E+z)\dot{\phi}\left(\boldsymbol{\omega}\times\hat{\mathbf{j}}\right)+\dot{z}\left(\boldsymbol{\omega}\times\hat{\mathbf{k}}\right)\end{aligned}$$

Substituting Eq. (1.74) through Eq. (1.76) together with Eqs. (1.78) and (1.79) into this expression yields, upon simplification,

$$\mathbf{a}_{\text{rel}}=\left[\ddot{x}+\frac{\dot{x}(\dot{z}-\dot{y}\tan\phi)}{R_E+z}\right]\hat{\mathbf{i}}+\left(\ddot{y}+\frac{\dot{y}\dot{z}+\dot{x}^2\tan\phi}{R_E+z}\right)\hat{\mathbf{j}}+\left(\ddot{z}-\frac{\dot{x}^2+\dot{y}^2}{R_E+z}\right)\hat{\mathbf{k}} \tag{1.80}$$

Observe that the curvature of the earth's surface is neglected by letting $R_E + z$ become infinitely large, in which case

$$\left.\mathbf{a}_{\text{rel}}\right)_{\text{neglecting earth's curvature}}=\ddot{x}\hat{\mathbf{i}}+\ddot{y}\hat{\mathbf{j}}+\ddot{z}\hat{\mathbf{k}}$$

That is, for a "flat earth," the components of the relative acceleration vector are just the derivatives of the components of the relative velocity vector.

For the absolute velocity we have, according to Eq. (1.66),

$$\mathbf{v}=\mathbf{v}_C+\boldsymbol{\Omega}\times\mathbf{r}_{\text{rel}}+\mathbf{v}_{\text{rel}} \tag{1.81}$$

From Fig. 1.18, it can be seen that $\hat{\mathbf{K}}=\cos\phi\hat{\mathbf{j}}+\sin\phi\hat{\mathbf{k}}$, which means the angular velocity of the earth is

$$\boldsymbol{\Omega}=\Omega\hat{\mathbf{K}}=\Omega\cos\phi\hat{\mathbf{j}}+\Omega\sin\phi\hat{\mathbf{k}} \tag{1.82}$$

Substituting this, together with Eqs. (1.71) and (1.77a) and the fact that $\mathbf{v}_C=\mathbf{0}$, into Eq. (1.81) yields

$$\mathbf{v}=[\dot{x}+\Omega(R_E+z)\cos\phi]\hat{\mathbf{i}}+\dot{y}\hat{\mathbf{j}}+\dot{z}\hat{\mathbf{k}} \tag{1.83}$$

From Eq. (1.70) the absolute acceleration of P is

$$\mathbf{a}=\mathbf{a}_C+\dot{\boldsymbol{\Omega}}\times\mathbf{r}_{\text{rel}}+\boldsymbol{\Omega}\times(\boldsymbol{\Omega}\times\mathbf{r}_{\text{rel}})+2\boldsymbol{\Omega}\times\mathbf{v}_{\text{rel}}+\mathbf{a}_{\text{rel}}$$

Since $\mathbf{a}_C=\dot{\boldsymbol{\Omega}}=\mathbf{0}$, we find, upon substituting Eqs. (1.71), (1.77a), (1.80), and (1.82), that

$$\begin{aligned}\mathbf{a}=&\left[\ddot{x}+\frac{\dot{x}(\dot{z}-\dot{y}\tan\phi)}{R_E+z}+2\Omega(\dot{z}\cos\phi-\dot{y}\sin\phi)\right]\hat{\mathbf{i}}\\&+\left\{\ddot{y}+\frac{\dot{y}\dot{z}+\dot{x}^2\tan\phi}{R_E+z}+\Omega\sin\phi[\Omega(R_E+z)\cos\phi+2\dot{x}]\right\}\hat{\mathbf{j}}\\&+\left\{\ddot{z}-\frac{\dot{x}^2+\dot{y}^2}{R_E+z}-\Omega\cos\phi[\Omega(R_E+z)\cos\phi+2\dot{x}]\right\}\hat{\mathbf{k}}\end{aligned} \tag{1.84}$$

Some special cases of Eqs. (1.83) and (1.84) follow.

Straight and level, unaccelerated flight: $\dot{z}=\ddot{z}=\ddot{x}=\ddot{y}=0$

$$\mathbf{v}=[\dot{x}+\Omega(R_E+z)\cos\phi]\hat{\mathbf{i}}+\dot{y}\hat{\mathbf{j}} \tag{1.85a}$$

$$\begin{aligned}\mathbf{a}=&-\left[\frac{\dot{x}\dot{y}\tan\phi}{R_E+z}+2\Omega\dot{y}\sin\phi\right]\hat{\mathbf{i}}+\left\{\frac{\dot{x}^2\tan\phi}{R_E+z}+\Omega\sin\phi[\Omega(R_E+z)\cos\phi+2\dot{x}]\right\}\hat{\mathbf{j}}\\&-\left\{\frac{\dot{x}^2+\dot{y}^2}{R_E+z}+\Omega\cos\phi[\Omega(R_E+z)\cos\phi+2\dot{x}]\right\}\hat{\mathbf{k}}\end{aligned} \tag{1.85b}$$

Flight due north (y) at a constant speed and altitude: $\dot{z}=\ddot{z}=\dot{x}=\ddot{x}=\ddot{y}=0$

$$\mathbf{v}=\Omega(R_E+z)\cos\phi\hat{\mathbf{i}}+\dot{y}\hat{\mathbf{j}} \tag{1.86a}$$

$$\mathbf{a}=-2\Omega\dot{y}\sin\phi\hat{\mathbf{i}}+\Omega^2(R_E+z)\sin\phi\cos\phi\hat{\mathbf{j}}-\left[\frac{\dot{y}^2}{R_E+z}+\Omega^2(R_E+z)\cos^2\phi\right]\hat{\mathbf{k}} \tag{1.86b}$$

Flight due east (x) at a constant speed and altitude: $\dot{z}=\ddot{z}=\ddot{x}=\dot{y}=\ddot{y}=0$

$$\mathbf{v}=[\dot{x}+\Omega(R_E+z)\cos\phi]\hat{\mathbf{i}} \tag{1.87a}$$

$$\begin{aligned}\mathbf{a}=&\left\{\frac{\dot{x}^2\tan\phi}{R_E+z}+\Omega\sin\phi[\Omega(R_E+z)\cos\phi+2\dot{x}]\right\}\hat{\mathbf{j}}\\&-\left\{\frac{\dot{x}^2}{R_E+z}+\Omega\cos\phi[\Omega(R_E+z)\cos\phi+2\dot{x}]\right\}\hat{\mathbf{k}}\end{aligned} \tag{1.87b}$$

Flight straight up (z): $\dot{x}=\ddot{x}=\dot{y}=\ddot{y}=0$

$$\mathbf{v}=\Omega(R_E+z)\cos\phi\hat{\mathbf{i}}+\dot{z}\hat{\mathbf{k}} \tag{1.88a}$$

$$\mathbf{a}=2\Omega(\dot{z}\cos\phi)\hat{\mathbf{i}}+\Omega^2(R_E+z)\sin\phi\cos\phi\hat{\mathbf{j}}+\left[\ddot{z}-\Omega^2(R_E+z)\cos^2\phi\right]\hat{\mathbf{k}} \tag{1.88b}$$

Stationary: $\dot{x}=\ddot{x}=\dot{y}=\ddot{y}=\dot{z}=\ddot{z}=0$

$$\mathbf{v}=\Omega(R_E+z)\cos\phi\hat{\mathbf{i}} \tag{1.89a}$$

$$\mathbf{a}=\Omega^2(R_E+z)\sin\phi\cos\phi\hat{\mathbf{j}}-\Omega^2(R_E+z)\cos^2\phi\hat{\mathbf{k}} \tag{1.89b}$$

EXAMPLE 1.14

An airplane of mass 70,000 kg is traveling due north at a latitude 30°N, at an altitude of 10 km (32,800 ft), with a speed of 300 m/s (671 mph). Calculate (a) the components of the absolute velocity and acceleration along the axes of the topocentric-horizon reference frame and (b) the net force on the airplane. Assume the winds aloft are zero.

Solution

(a) First, using the sidereal rotation period of the earth in Table A.1, we note that the earth's angular velocity is

$$\Omega=\frac{2\pi\,\text{radians}}{\text{sidereal day}}=\frac{2\pi\,\text{radians}}{23.93\text{h}}=\frac{2\pi\,\text{radians}}{86,160\,\text{s}}=7.292\times10^{-5}\,\text{radians/s}$$

From Eq. (1.86a), the absolute velocity is

$$\mathbf{v}=\Omega(R_E+z)\cos\phi\hat{\mathbf{i}}+\dot{y}\hat{\mathbf{j}}=\left[(7.292\times10^{-5})\cdot(6378+10)\cdot10^3\cos30^\circ\right]\hat{\mathbf{i}}+300\hat{\mathbf{j}}$$

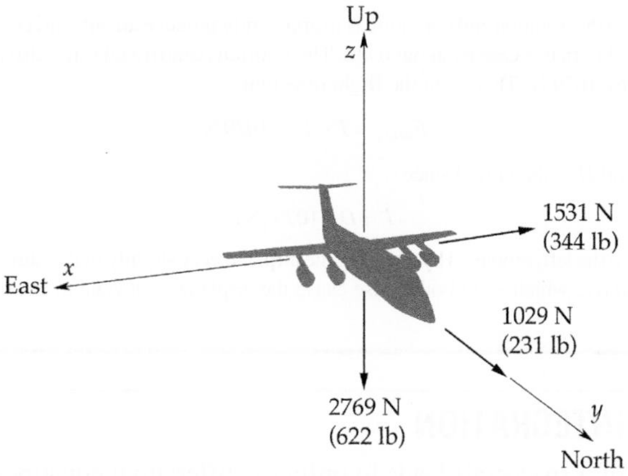

FIG. 1.19

Components of the net force on the airplane.

or

$$\mathbf{v} = 403.4\hat{\mathbf{i}} + 300\hat{\mathbf{j}}\ (\mathrm{m/s})$$

The 403.4 m/s (901 mph) component of velocity to the east (x direction) is due entirely to the earth's rotation.

From Eq. (1.86b), the absolute acceleration is

$$\begin{aligned}\mathbf{a} &= -2\Omega\dot{y}\sin\phi\hat{\mathbf{i}} + \Omega^2(R_E+z)\sin\phi\cos\phi\hat{\mathbf{j}} - \left[\frac{\dot{y}^2}{R_E+z} + \Omega^2(R_E+z)\cos^2\phi\right]\hat{\mathbf{k}} \\ &= -2\left(7.292\times10^{-5}\right)\cdot 300\cdot \sin 30°\hat{\mathbf{i}} \\ &\quad + \left(7.292\times10^{-5}\right)^2\cdot(6378+10)\cdot 10^3\cdot\sin 30°\cdot\cos 30°\hat{\mathbf{j}} \\ &\quad - \left[\frac{300^2}{(6378+10)\cdot 10^3} + \left(7.292\times10^{-5}\right)^2\cdot(6378+10)\cdot10^3\cdot\cos^2 30°\right]\hat{\mathbf{k}}\end{aligned}$$

or

$$\boxed{\mathbf{a} = -0.02187\hat{\mathbf{i}} + 0.01471\hat{\mathbf{j}} - 0.03956\hat{\mathbf{k}}\ (\mathrm{m/s^2})}$$

The westward (negative x) acceleration of $0.02187\,\mathrm{m/s^2}$ is the Coriolis acceleration.

(b) Since the acceleration in part (a) is the absolute acceleration, we can use it in Newton's law to calculate the net force on the airplane,

$$\begin{aligned}\mathbf{F}_{\text{net}} &= m\mathbf{a} = 70{,}000\left(-0.02187\hat{\mathbf{i}} + 0.01471\hat{\mathbf{j}} - 0.03956\hat{\mathbf{k}}\right) \\ &= \boxed{-1531\hat{\mathbf{i}} + 1029\hat{\mathbf{j}} - 2769\hat{\mathbf{k}}\ (\mathrm{N})}\end{aligned}$$

Fig. 1.19 shows the components of this relatively small force. The forward (y) and downward (negative z) forces are in the directions of the airplane's centripetal acceleration, caused by the earth's rotation and, in the case of the downward force, by the earth's curvature as well. The westward force is in the direction of the Coriolis acceleration, which is due to the combined effects of the earth's rotation and the motion of the airplane. These net external forces must exist if the airplane is to fly in the prescribed path.

In the vertical direction, the net force is that of the upward lift L of the wings plus the downward weight W of the aircraft, so that

$$\left.F_{\text{net}}\right)_z = L - W = -2769 \Rightarrow L = W - 2769\mathrm{N}$$

Thus, the effect of the earth's rotation and curvature is to apparently produce an outward *centrifugal force*, reducing the weight of the airplane a bit, in this case by about 0.4%. The fictitious centrifugal force also increases the apparent drag in the flight direction by 1029 N. That is, in the flight direction

$$F_{\text{net}})_y = T - D = 1029\,\text{N}$$

where T is the thrust and D is the drag. Hence

$$T = D + 1029\ (\text{N})$$

The 1531-N force to the left, produced by crabbing the airplane very slightly in that direction, is required to balance the fictitious Coriolis force, which would otherwise cause the airplane to deviate to the right of its flight path.

1.8 NUMERICAL INTEGRATION

Analysis of the motion of a spacecraft leads to ordinary differential equations with time as the independent variable. It is often impractical if not impossible to solve them exactly. Therefore, the ability to solve differential equations numerically is important. In this section, we will take a look at a few common numerical integration schemes and investigate their accuracy and stability by applying them to some problems that do have an analytical solution.

Particle mechanics is based on Newton's second law (Eq. 1.38), which may be written as

$$\ddot{\mathbf{r}} = \frac{\mathbf{F}}{m} \tag{1.90}$$

This is a second-order, ordinary differential equation for the position vector $\mathbf{r}$ as a function of time. Depending on the complexity of the force function $\mathbf{F}$, there may or may not be a closed-form, analytical solution of Eq. (1.90). In the most trivial case, the force vector $\mathbf{F}$ and the mass m are constant, which means we can use elementary calculus to integrate Eq. (1.90) twice to get

$$\mathbf{r} = \frac{\mathbf{F}}{2m}t^2 + \mathbf{C}_1 t + \mathbf{C}_2 \quad (\mathbf{F}\text{ and }m\text{ constant}) \tag{1.91}$$

$\mathbf{C}_1$ and $\mathbf{C}_2$ are the two vector constants of integration. Since each vector has three components, there are a total of six scalar constants of integration. If the position and velocity are both specified at time $t = 0$ to be $\mathbf{r}_0$ and $\dot{\mathbf{r}}_0$, respectively, then we have an initial value problem. Applying the initial conditions to Eq. (1.91), we find $\mathbf{C}_1 = \dot{\mathbf{r}}_0$ and $\mathbf{C}_2 = \mathbf{r}_0$, which means

$$\mathbf{r} = \frac{\mathbf{F}}{2m}t^2 + \dot{\mathbf{r}}_0 t + \mathbf{r}_0 \quad (\mathbf{F}\text{ and }m\text{ constant})$$

On the other hand, we may know the position $\mathbf{r}_0$ at $t = 0$ and the velocity $\dot{\mathbf{r}}_f$ at a later time $t = t_f$. These are boundary conditions and this is an example of a boundary value problem. Applying the boundary conditions to Eq. (1.91) yields $\mathbf{C}_1 = \dot{\mathbf{r}}_f - (\mathbf{F}/m)t_f$ and $\mathbf{C}_2 = \mathbf{r}_0$, which means

$$\mathbf{r} = \frac{\mathbf{F}}{2m}t^2 + \left(\dot{\mathbf{r}}_f - \frac{\mathbf{F}}{m}t_f\right)t + \mathbf{r}_0 \quad (\mathbf{F}\text{ and }m\text{ constant})$$

For the remainder of this section we will focus on the numerical solution of initial value problems only.

In general, the function **F** in Eq. (1.90) is not constant but is instead a function of time t, position **r**, and velocity $\dot{\mathbf{r}}$. That is, $\mathbf{F}=\mathbf{F}(t,\mathbf{r},\dot{\mathbf{r}})$. Let us resolve the vector **r** and its derivatives as well as the force **F** into their Cartesian components in three-dimensional space:

$$\mathbf{r}=x\hat{\mathbf{i}}+y\hat{\mathbf{j}}+z\hat{\mathbf{k}} \quad \dot{\mathbf{r}}=\dot{x}\hat{\mathbf{i}}+\dot{y}\hat{\mathbf{j}}+\dot{z}\hat{\mathbf{k}} \quad \ddot{\mathbf{r}}=\ddot{x}\hat{\mathbf{i}}+\ddot{y}\hat{\mathbf{j}}+\ddot{z}\hat{\mathbf{k}} \quad \mathbf{F}=F_x\hat{\mathbf{i}}+F_y\hat{\mathbf{j}}+F_z\hat{\mathbf{k}}$$

The three components of Eq. (1.90) are

$$\ddot{x}=\frac{F_x(t,\mathbf{r},\dot{\mathbf{r}})}{m} \quad \ddot{y}=\frac{F_y(t,\mathbf{r},\dot{\mathbf{r}})}{m} \quad \ddot{z}=\frac{F_z(t,\mathbf{r},\dot{\mathbf{r}})}{m} \tag{1.92}$$

These are three second-order differential equations. For the purpose of numerical solution, they must be reduced to six first-order differential equations. This is accomplished by introducing six auxiliary variables y_1 through y_6, defined as follows:

$$\begin{aligned} &y_1=x \quad y_2=y \quad y_3=z \\ &y_4=\dot{x} \quad y_5=\dot{y} \quad y_6=\dot{z} \end{aligned} \tag{1.93}$$

In terms of these auxiliary variables, the position and velocity vectors are

$$\mathbf{r}=y_1\hat{\mathbf{i}}+y_2\hat{\mathbf{j}}+y_3\hat{\mathbf{k}} \quad \dot{\mathbf{r}}=y_4\hat{\mathbf{i}}+y_5\hat{\mathbf{j}}+y_6\hat{\mathbf{k}}$$

Taking the derivative d/dt of each of the six expressions in Eq. (1.93) yields

$$\begin{aligned} &\mathrm{d}y_1/\mathrm{d}t=\dot{x} \quad \mathrm{d}y_2/\mathrm{d}t=\dot{y} \quad \mathrm{d}y_3/\mathrm{d}t=\dot{z} \\ &\mathrm{d}y_4/\mathrm{d}t=\ddot{x} \quad \mathrm{d}y_5/\mathrm{d}t=\ddot{y} \quad \mathrm{d}y_6/\mathrm{d}t=\ddot{z} \end{aligned}$$

Upon substituting Eqs. (1.92) and (1.93), we arrive at the six first-order differential equations

$$\begin{aligned} \dot{y}_1&=y_4 \\ \dot{y}_2&=y_5 \\ \dot{y}_3&=y_6 \\ \dot{y}_4&=\frac{F_x(t,y_1,y_2,y_3,y_4,y_5,y_6)}{m} \\ \dot{y}_5&=\frac{F_y(t,y_1,y_2,y_3,y_4,y_5,y_6)}{m} \\ \dot{y}_6&=\frac{F_z(t,y_1,y_2,y_3,y_4,y_5,y_6)}{m} \end{aligned} \tag{1.94}$$

These equations are coupled because the right-hand side of each one contains variables that belong to other equations as well. Eq. (1.94) can be written more compactly in vector notation as

$$\dot{\mathbf{y}}=\mathbf{f}(t,\mathbf{y}) \tag{1.95}$$

where the column vectors **y**, $\dot{\mathbf{y}}$, and **f** are

$$\mathbf{y}=\begin{Bmatrix} y_1\\ y_2\\ y_3\\ y_4\\ y_5\\ y_6 \end{Bmatrix} \quad \dot{\mathbf{y}}=\begin{Bmatrix} \dot{y}_1\\ \dot{y}_2\\ \dot{y}_3\\ \dot{y}_4\\ \dot{y}_5\\ \dot{y}_6 \end{Bmatrix} \quad \mathbf{f}=\begin{Bmatrix} y_4\\ y_5\\ y_6\\ F_x(t,\mathbf{y})/m\\ F_y(t,\mathbf{y})/m\\ F_z(t,\mathbf{y})/m \end{Bmatrix} \tag{1.96}$$

Note that in this case $\mathbf{f}(t,\mathbf{y})$ is shorthand for $\mathbf{f}(t,y_1,y_2,y_3,y_4,y_5,y_6)$. Any set of one or more ordinary differential equations of any order can be cast in the form of Eq. (1.95).

EXAMPLE 1.15

Write the third-order nonlinear differential equation

$$\dddot{x} - x\ddot{x} + \dot{x}^2 = 0 \tag{a}$$

as three first-order differential equations.

Solution

Introducing the three auxiliary variables

$$y_1 = x \quad y_2 = \dot{x} \quad y_3 = \ddot{x} \tag{b}$$

we take the derivative of each one to get

$$\begin{aligned} dy_1/dt &= dx/dt = \dot{x} \\ dy_2/dt &= d\dot{x}/dt = \ddot{x} \\ dy_3/dt &= d\ddot{x}/dt = \dddot{x} \overset{\text{From (a)}}{\widehat{=}} x\ddot{x} - \dot{x}^2 \end{aligned}$$

Substituting Eq. (b) on the right of these expressions yields

$$\begin{aligned} \dot{y}_1 &= y_2 \\ \dot{y}_2 &= y_3 \\ \dot{y}_3 &= y_1 y_3 - y_2^2 \end{aligned} \tag{c}$$

This is a system of three first-order, coupled ordinary differential equations. It is an autonomous system, since time t does not appear explicitly on the right-hand side. The three equations can therefore be written compactly as $\dot{\mathbf{y}} = \mathbf{f}(\mathbf{y})$.

Before discussing some numerical integration schemes, it will be helpful to review the concept of the Taylor series, named after the English mathematician Brook Taylor (1685–1731). Recall from calculus that if we know the value of a function $g(t)$ at time t and wish to approximate its value at a neighboring time $t + h$, we can use the Taylor series to express $g(t + h)$ as an infinite power series in h,

$$g(t+h) = g(t) + c_1 h + c_2 h^2 + c_3 h^3 + \cdots + c_n h^n + O\left(h^{n+1}\right) \tag{1.97}$$

The coefficients c_m are found by taking successively higher order derivatives of $g(t)$ according to the formula

$$c_m = \frac{1}{m!}\frac{d^m g(t)}{dt^m} \tag{1.98}$$

$O(h^{n+1})$ ("order of h to the $n+1$") means that the remaining terms of this infinite series all have h^{n+1} as a factor. In other words,

$$\lim_{h \to 0} \frac{O(h^{n+1})}{h^{n+1}} = c_{n+1}$$

$O(h^{n+1})$ is the truncation error due to retaining only terms up to h^n. The order of a Taylor series expansion is the highest power of h retained. The more terms of the Taylor series that we keep, the more accurate will be the representation of the function $g(t + h)$ in the neighborhood of t. Reducing h lowers the truncation error. For example, if we reduce h to $h/2$, then $O(h^n)$ goes down by a factor of $(1/2)^n$.

EXAMPLE 1.16

Expand the function $\sin(t + h)$ in a Taylor series about $t = 1$. Plot the Taylor series of order 1, 2, 3, and 4 and compare them with $\sin(1 + h)$ for $-2 < h < 2$.

Solution

The nth-order Taylor power series expansion of $\sin(t+h)$ is written

$$\sin(t+h) = p_n(h)$$

where, according to Eqs. (1.97) and (1.98), the polynomial p_n is given by

$$p_n(h) = \sum_{m=0}^{n} \frac{h^m}{m!} \frac{d^m \sin t}{dt^m}$$

Thus, the zeroth- through fourth-order Taylor series polynomials in h are

$$p_0 = \frac{h^0}{0!}\frac{d^0 \sin t}{dt^0} = \sin t$$

$$p_1 = p_0 + \frac{h}{1!}\frac{d \sin t}{dt} = \sin t + h\cos t$$

$$p_2 = p_1 + \frac{h^2}{2!}\frac{d^2 \sin t}{dt^2} = \sin t + h\cos t - \frac{h^2}{2}\sin t$$

$$p_3 = p_2 + \frac{h^3}{3!}\frac{d^3 \sin t}{dt^3} = \sin t + h\cos t - \frac{h^2}{2}\sin t - \frac{h^3}{6}\cos t$$

$$p_4 = p_3 + \frac{h^4}{4!}\frac{d^4 \sin t}{dt^4} = \sin t + h\cos t - \frac{h^2}{2}\sin t - \frac{h^3}{6}\cos t + \frac{h^4}{24}\sin t$$

For $t = 1$, p_1 through p_4 as well as $\sin(t + h)$ are plotted in Fig. 1.20. As expected, we see that the higher degree Taylor polynomials for $\sin(1+h)$ lie closer to $\sin(1+h)$ over a wider range of h.

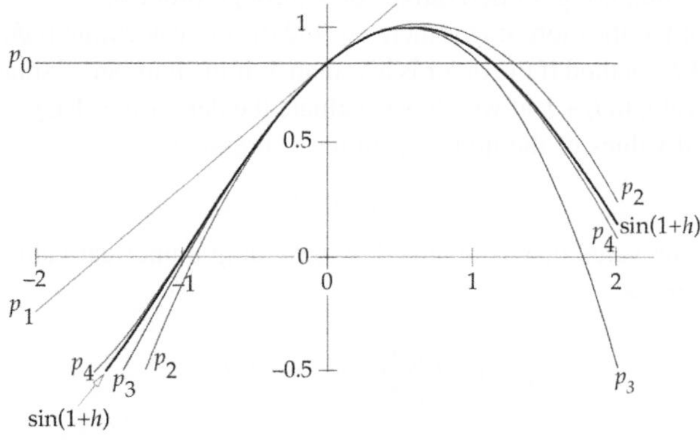

FIG. 1.20

Plots of zeroth- to fourth-order Taylor series expansions of $\sin(1 + h)$.

The numerical integration schemes that we shall examine are designed to solve first-order ordinary differential equations of the form shown in Eq. (1.95). To obtain a numerical solution of $\dot{\mathbf{y}} = \mathbf{f}(t, \mathbf{y})$ over the time interval t_0 to t_f, we divide or "mesh" the interval into N discrete times $t_1, t_2, t_3, \ldots,$ t_N, where $t_1 = t_0$ and $t_N = t_f$. The step size h is the difference between two adjacent times on the mesh (i.e., $h = t_{i+1} - t_i$). h may be constant for all steps across the entire time span t_0 to t_f. Modern methods have adaptive step size control in which h varies from step to step to provide better accuracy and efficiency.

Let us denote the values of $\mathbf{y}$ and $\dot{\mathbf{y}}$ at time t_i as $\mathbf{y}_i$ and $\mathbf{f}_i$, respectively, where $\mathbf{f}_i = \mathbf{f}(t_i, \mathbf{y}_i)$. In an initial value problem, the values of all components of $\mathbf{y}$ at the initial time t_0 together with Eq. (1.95) provide the information needed to determine $\mathbf{y}$ at the subsequent discrete times.

1.8.1 RUNGE-KUTTA METHODS

The Runge-Kutta (*RK*) methods were originally developed by the German mathematicians Carl Runge (1856–1927) and Martin Kutta (1867–1944). In the explicit, single-step *RK* methods, $\mathbf{y}_{i+1}$ at $t_i + h$ is obtained from $\mathbf{y}_i$ at t_i by the formula

$$\mathbf{y}_{i+1} = \mathbf{y}_i + h\boldsymbol{\phi}(t_i, \mathbf{y}_i, h) \tag{1.99}$$

The increment function $\boldsymbol{\phi}$ is an average of the derivative $d\mathbf{y}/dt$ over the time interval t_i to $t_i + h$. This average is obtained by evaluating the derivative $\mathbf{f}(t, \mathbf{y})$ at several points or "stages" within the time interval. The order of an *RK* method reflects the accuracy to which $\boldsymbol{\phi}$ is computed, compared with a Taylor series expansion. An *RK* method of order p is called an *RKp* method. An *RKp* method is as accurate in computing $\mathbf{y}_i$ from Eq. (1.99) as is the pth-order Taylor series

$$\mathbf{y}(t_i + h) = \mathbf{y}_i + \mathbf{c}_1 h + \mathbf{c}_2 h^2 + \cdots \mathbf{c}_p h^p \tag{1.100}$$

An attractive feature of the *RK* schemes is that only the first derivative $\mathbf{f}(t, \mathbf{y})$ is required, and it is available from the differential equation itself (Eq. (1.95)). By contrast, the pth-order Taylor series expansion in Eq. (1.100) requires computing all derivatives of $\mathbf{y}$ through order p.

The higher the *RK* order, the more stages there are and the more accurate is $\boldsymbol{\phi}$. The number of stages equals the order of the *RK* method if the order is less than 5. If the number of stages is s, then there are s times $\tilde{t}$ within the interval t_i to $t_i + h$ at which we evaluate the derivatives $\mathbf{f}(t, \mathbf{y})$. These times are given by specifying numerical values of the nodes a_m in the expression

$$\tilde{t}_m = t_i + a_m h \quad m = 1, 2, \ldots, s$$

At each of these times the value of $\tilde{\mathbf{y}}$ is obtained by providing numerical values for the coupling coefficients b_{mn} in the formula

$$\tilde{\mathbf{y}}_m = \mathbf{y}_i + h\sum_{n=1}^{m-1} b_{mn}\tilde{\mathbf{f}}_n \quad m = 1, 2, \ldots, s \tag{1.101}$$

The vector of derivatives $\tilde{\mathbf{f}}_m$ is evaluated at stage m by substituting $\tilde{t}_m$ and $\tilde{\mathbf{y}}_m$ into Eq. (1.95),

$$\tilde{\mathbf{f}}_m = \mathbf{f}\left(\tilde{t}_m, \tilde{\mathbf{y}}_m\right) \quad m = 1, 2, \ldots, s \tag{1.102}$$

The increment function $\boldsymbol{\phi}$ is a weighted sum of the derivatives $\tilde{\mathbf{f}}_m$ over the s stages within the time interval t_i to $t_i + h$,

$$\boldsymbol{\phi} = \sum_{m=1}^{s} c_m \tilde{\mathbf{f}}_m \tag{1.103}$$

The coefficients c_m are known as the weights. Substituting Eq. (1.103) into Eq. (1.99) yields

$$\mathbf{y}_{i+1} = \mathbf{y}_i + h\sum_{m=1}^{s} c_m \tilde{\mathbf{f}}_m \tag{1.104}$$

The numerical values of the coefficients a_m, b_{mn}, and c_m depend on which *RK* method is being used. It is convenient to write these coefficients as arrays, so that

$$\{\mathbf{a}\} = \begin{Bmatrix} a_1 \\ a_2 \\ \vdots \\ a_s \end{Bmatrix} \quad [\mathbf{b}] = \begin{bmatrix} b_{11} & & & \\ b_{21} & b_{22} & & \\ \vdots & \vdots & \cdots & \\ b_{s1} & b_{s2} & \cdots & b_{s,s-1} \end{bmatrix} \quad \{\mathbf{c}\} = \begin{Bmatrix} c_1 \\ c_2 \\ \vdots \\ c_s \end{Bmatrix} \tag{1.105}$$

where s is the number of stages, and $[\mathbf{b}]$ is undefined if $s = 1$. The nodes $\{\mathbf{a}\}$, coupling coefficients $[\mathbf{b}]$, and weights $\{\mathbf{c}\}$ for a given *RK* method are not necessarily unique, although research favors the choice of some sets over others. Details surrounding the derivation of these coefficients as well as in-depth discussions of not only *RK* but also the numerous other common numerical integration techniques may be found in numerical analysis textbooks, such as the one by Butcher (2008).

For *RK* orders 1–4, we list below the commonly used values of the coefficients (Eq. 1.105), the resulting formula for the derivatives $\tilde{\mathbf{f}}$ at each stage (Eq. 1.102), and the formula for the difference $\mathbf{y}_{i+1} - \mathbf{y}_i$ (Eq. 1.104). These *RK* schemes all use a uniform step size h.

RK1 (Euler's method)

$$\begin{gathered} \{\mathbf{a}\} = \{0\} \quad \{\mathbf{c}\} = \{1\} \\ \tilde{\mathbf{f}}_1 = \mathbf{f}(t_i, y_i) \\ \mathbf{y}_{i+1} = \mathbf{y}_i + h\tilde{\mathbf{f}}_1 \end{gathered} \tag{1.106}$$

RK2 (Heun's method)

$$\begin{gathered} \{\mathbf{a}\} = \begin{Bmatrix} 0 \\ 1 \end{Bmatrix} \quad [\mathbf{b}] = \begin{bmatrix} 0 \\ 1 \end{bmatrix} \quad \{\mathbf{c}\} = \begin{Bmatrix} 1/2 \\ 1/2 \end{Bmatrix} \\ \tilde{\mathbf{f}}_1 = \mathbf{f}(t_i, \mathbf{y}_i) \quad \tilde{\mathbf{f}}_2 = \mathbf{f}\left(t_i + h, \mathbf{y}_i + h\tilde{\mathbf{f}}_1\right) \\ \mathbf{y}_{i+1} = \mathbf{y}_i + h\left(\frac{1}{2}\tilde{\mathbf{f}}_1 + \frac{1}{2}\tilde{\mathbf{f}}_2\right) \end{gathered} \tag{1.107}$$

RK3

$$\begin{gathered} \{\mathbf{a}\} = \begin{Bmatrix} 0 \\ 1/2 \\ 1 \end{Bmatrix} \quad [\mathbf{b}] = \begin{bmatrix} 0 & 0 \\ 1/2 & 0 \\ -1 & 2 \end{bmatrix} \quad \{\mathbf{c}\} = \begin{Bmatrix} 1/6 \\ 2/3 \\ 1/6 \end{Bmatrix} \\ \tilde{\mathbf{f}}_1 = \mathbf{f}(t_i, \mathbf{y}_i) \quad \tilde{\mathbf{f}}_2 = \mathbf{f}\left(t_i + \frac{1}{2}h, \mathbf{y}_i + \frac{1}{2}h\tilde{\mathbf{f}}_1\right) \quad \tilde{\mathbf{f}}_3 = \mathbf{f}\left(t_i + h, \mathbf{y}_i + h\left[-\tilde{\mathbf{f}}_1 + 2\tilde{\mathbf{f}}_2\right]\right) \\ \mathbf{y}_{i+1} = \mathbf{y}_i + h\left(\frac{1}{6}\tilde{\mathbf{f}}_1 + \frac{2}{3}\tilde{\mathbf{f}}_2 + \frac{1}{6}\tilde{\mathbf{f}}_3\right) \end{gathered} \tag{1.108}$$

RK4

$$\{\mathbf{a}\} = \begin{Bmatrix} 0 \\ 1/2 \\ 1/2 \\ 1 \end{Bmatrix} \quad [\mathbf{b}] = \begin{bmatrix} 0 & 0 & 0 \\ 1/2 & 0 & 0 \\ 0 & 1/2 & 0 \\ 0 & 0 & 1 \end{bmatrix} \quad \{\mathbf{c}\} = \begin{Bmatrix} 1/6 \\ 1/3 \\ 1/3 \\ 1/6 \end{Bmatrix}$$

$$\tilde{\mathbf{f}}_1 = \mathbf{f}(t_i, \mathbf{y}_i) \quad \tilde{\mathbf{f}}_2 = \mathbf{f}\left(t_i + \frac{1}{2}h, \mathbf{y}_i + \frac{1}{2}h\tilde{\mathbf{f}}_1\right) \quad \tilde{\mathbf{f}}_3 = \mathbf{f}\left(t_i + \frac{1}{2}h, \mathbf{y}_i + \frac{1}{2}h\tilde{\mathbf{f}}_2\right) \tag{1.109}$$

$$\tilde{\mathbf{f}}_4 = \mathbf{f}\left(t_i + h, \mathbf{y}_i + h\tilde{\mathbf{f}}_3\right)$$

$$\mathbf{y}_{i+1} = \mathbf{y}_i + h\left(\frac{1}{6}\tilde{\mathbf{f}}_1 + \frac{1}{3}\tilde{\mathbf{f}}_2 + \frac{1}{3}\tilde{\mathbf{f}}_3 + \frac{1}{6}\tilde{\mathbf{f}}_4\right)$$

Observe that in each of the four cases the sum of the components of $\{\mathbf{c}\}$ is 1 and the sum of each row of $[\mathbf{b}]$ equals the value in that row of $\{\mathbf{a}\}$. This is a characteristic of the *RK* methods.

ALGORITHM 1.1

Given the vector $\mathbf{y}$ at time t, the derivatives $\mathbf{f}(t, \mathbf{y})$, and the step size h, use one of the methods *RK1* through *RK4* to find $\mathbf{y}$ at time $t + h$. See Appendix D.2 for a MATLAB implementation of this algorithm in the form of the function *rk1_4.m*. *rk1_4.m* executes any of the four *RK* methods according to whether the variable *rk* passed to the function has the value 1, 2, 3, or 4.

1. Evaluate the derivatives $\tilde{\mathbf{f}}_1, \tilde{\mathbf{f}}_2, \ldots, \tilde{\mathbf{f}}_s$ at stages 1 through s by means of Eq. (1.102).
2. Use Eq. (1.104) to compute $\mathbf{y}(t+h) = \mathbf{y}(t) + h\sum_{m=1}^{s}\tilde{\mathbf{f}}_m$.

Repeat these steps to obtain $\mathbf{y}$ at subsequent times $t + 2h$, $t + 3h$, etc.

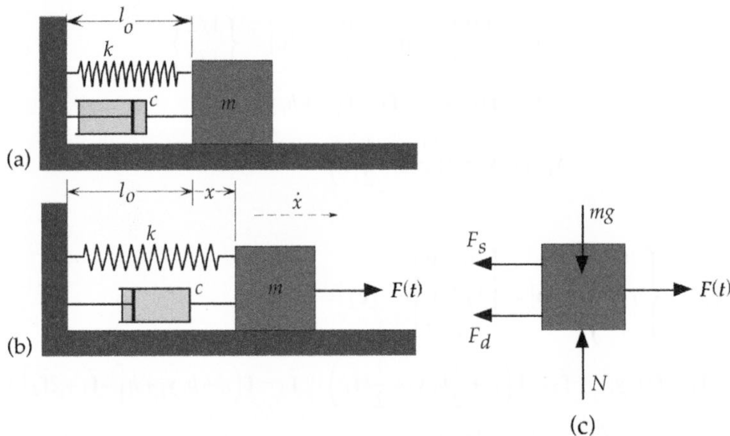

FIG. 1.21

A damped spring-mass system with a forcing function applied to the mass. (a) At rest. (b) In motion under the action of the applied force $F(t)$. (c) Free body diagram at any instant.

Let us employ the *RK* methods and Algorithm 1.1 to solve for the motion of the well-known viscously damped spring-mass system pictured in Fig. 1.21. The spring has an unstretched length l_0 and a spring constant k. The viscous damping coefficient is c and the mass of the block, which slides on a frictionless surface, is m. A forcing function $F(t)$ is applied to the mass. From the free body diagram in part (c) of the figure, we obtain the equation of motion of this one-dimensional system in the x direction.

$$-F_s - F_d + F(t) = m\ddot{x} \tag{1.110}$$

where F_s and F_d are the forces of the spring and dashpot, respectively. Since $F_s = kx$ and $F_d = c\dot{x}$, Eq. (1.110), after dividing through by the mass, can be rewritten as

$$\ddot{x} + \frac{c}{m}\dot{x} + \frac{k}{m}x = \frac{F(t)}{m} \tag{1.111}$$

The spring rate k and the mass m determine the natural circular frequency of vibration of the system, $\omega_n = \sqrt{k/m}$ (radians per second). Furthermore, the damping coefficient c may be expressed as $c = 2\zeta m\omega_n$, where ζ is the dimensionless damping factor ($\zeta \geq 0$). Making these substitutions in Eq. (1.111), we get the standard form

$$\ddot{x} + 2\zeta\omega_n\dot{x} + {\omega_n}^2 x = \frac{F(t)}{m} \tag{1.112}$$

If the forcing function is sinusoidal with amplitude F_0 and circular frequency ω, then Eq. (1.112) becomes

$$\ddot{x} + 2\zeta\omega_n\dot{x} + {\omega_n}^2 x = \frac{F_0}{m}\sin\omega t \tag{1.113}$$

This second-order ordinary differential equation has a closed-form solution, which is found using procedures taught in a differential equations course. If the system is underdamped, which means $\zeta < 1$, then it can be verified by substitution that the solution of Eq. (1.113) is

$$\begin{aligned} x = {} & e^{-\zeta\omega_n t}(A\sin\omega_d t + B\cos\omega_d t) \\ & + \frac{F_0/m}{({\omega_n}^2 - \omega^2)^2 + (2\omega\omega_n\zeta)}\left[({\omega_n}^2 - \omega^2)\sin\omega t - 2\omega\omega_n\zeta\cos\omega t\right] \end{aligned} \tag{1.114a}$$

where $\omega_d = \omega_n\sqrt{1-\zeta^2}$ is the damped natural frequency. The initial conditions determine the values of the coefficients A and B. If at $t = 0$, $x = x_0$, and $\dot{x} = \dot{x}_0$, it turns out that

$$\begin{aligned} A &= \zeta\frac{\omega_n}{\omega_d}x_0 + \frac{\dot{x}_0}{\omega_d} + \frac{\omega^2 + (2\zeta^2 - 1)\omega_n^2}{({\omega_n}^2 - \omega^2)^2 + (2\omega\omega_n\zeta)^2}\frac{\omega}{\omega_d}\frac{F_0}{m} \\ B &= x_0 + \frac{2\omega\omega_n\zeta}{({\omega_n}^2 - \omega^2)^2 + (2\omega\omega_n\zeta)^2}\frac{F_0}{m} \end{aligned} \tag{1.114b}$$

The transient term with the exponential factor in Eq. (1.114a) dies out eventually, leaving only the steady-state solution, which persists as long as the forcing function acts.

EXAMPLE 1.17

Plot Eq. (1.114a). from $t=0$ to $t=110$ s if $m=1$ kg, $\omega_n=1$ rad/s, $\zeta=0.03$, $F_0=1$ N, $\omega=0.4$rad/s and the initial conditions are $x=\dot{x}=0$.

Solution

Substituting the given values into Eq. (1.114) yields

$$x=e^{-0.03t}[0.03399\cos(0.9995t)-0.4750\sin(0.9995t)]+[1.190\sin(0.4t)-0.03399\cos(0.4t)] \tag{1.115}$$

This function is plotted over the time span 0–110 s in Fig. 1.22. Observe that after about 80 s, the transient has damped out and the system vibrates at the same frequency as the forcing function (although slightly out of phase due to the small viscosity).

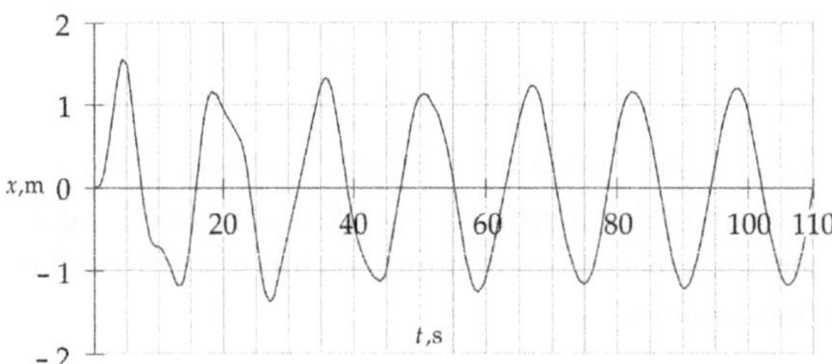

FIG. 1.22

Over time only the steady-state solution of Eq. (1.123) remains.

EXAMPLE 1.18

Solve Eq. (1.113) numerically, using the *RK* method and the data of Example 1.17. Compare the *RK* solution with the exact one, given by Eq. (1.115).

Solution

We must first reduce Eq. (1.113) to two first-order differential equations by introducing the two auxiliary variables

$$y_1=x(t) \tag{a}$$

$$y_2=\dot{x}(t) \tag{b}$$

Differentiating Eq. (a) we find

$$\dot{y}_1=\dot{x}(t)=y_2(t) \tag{c}$$

Differentiating Eq. (b) and using Eq. (1.113) yields

$$\dot{y}_2=\ddot{x}(t)=\frac{F_0}{m}\sin\omega t-\omega_n^2 y_1(t)-2\zeta\omega_n y_2(t) \tag{d}$$

Systems (c) and (d) can be written compactly in standard vector notation as

$$\dot{\mathbf{y}}=\mathbf{f}(t,\mathbf{y}) \tag{e}$$

where

$$\mathbf{y}=\begin{Bmatrix} y_1 \\ y_2 \end{Bmatrix} \quad \dot{\mathbf{y}}=\begin{Bmatrix} \dot{y}_1 \\ \dot{y}_2 \end{Bmatrix} \quad \mathbf{f}(t,\mathbf{y})=\begin{Bmatrix} y_2(t) \\ \dfrac{F_0}{m}\sin\omega t-\omega_n{}^2 y_1(t)-2\zeta\omega_n y_2(t) \end{Bmatrix} \tag{1.116}$$

Eq. (1.116) is what we need to implement Algorithm 1.1 for this problem.

We will use the two MATLAB functions listed in Appendix D.2 (namely, *Example_1_18.m* and *rk1_4.m*). *Example_1_18.m* passes the data of Example 1.17 to the function *rk1_4.m*, which executes Algorithm 1.1 for *RK1*, *RK2*, *RK3*, and *RK4* over the time interval from 0 to 110 s. In each case, the problem is solved for two different values of the time step *h*. The subfunction *rates* within *Example_1_18.m* calculates the derivatives $\mathbf{f}(t, \mathbf{y})$ given in Eq. (1.116_3). The exact solution (Eq. 1.115) along with the four *RK* solutions is nondimensionalized and plotted at each time step in Fig. 1.23.

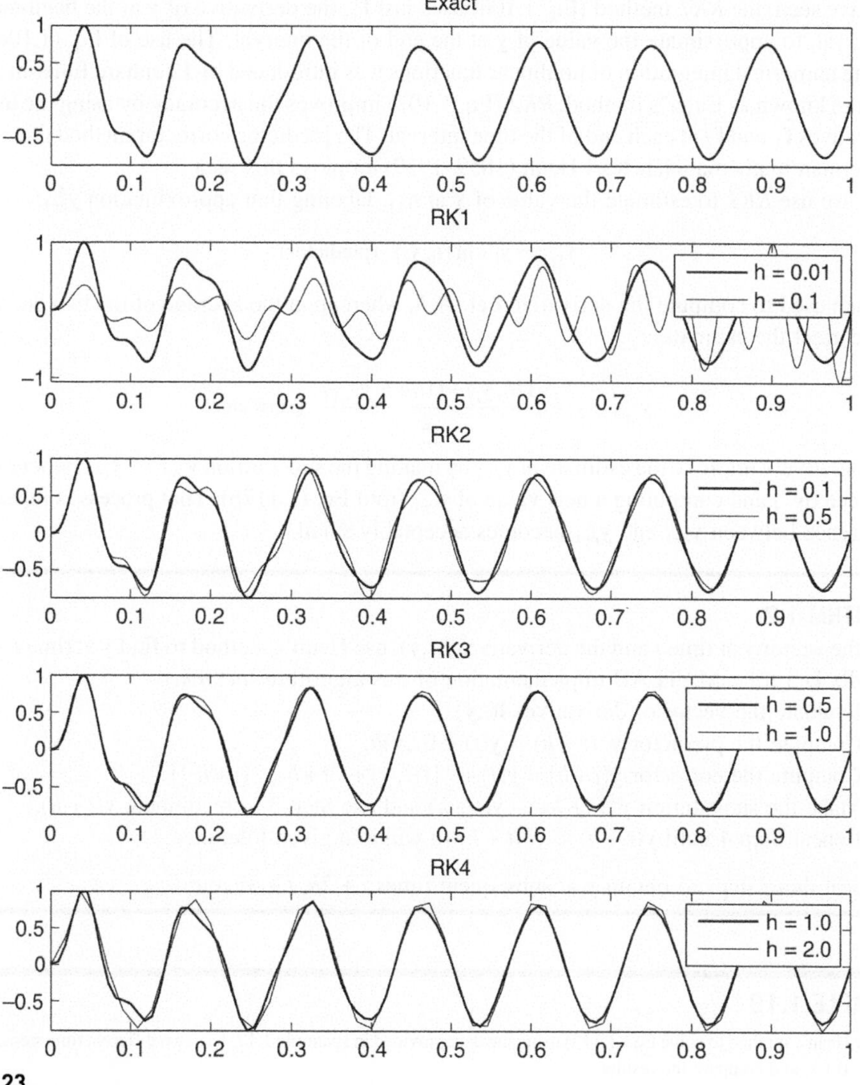

FIG. 1.23

x/x_{max} versus t/t_{max} for the *RK1* through *RK4* solutions of Eq. (1.123) using the data of Example 1.17. The exact solution is at the top.

We see that all the *RK* solutions agree closely with the analytical one for a sufficiently small step size. The figure shows, as expected, that to obtain accuracy, the uniform step size h must be reduced as the order of the *RK* method is reduced. Likewise, the figure suggests that a step size that yields inaccurate results for a given *RK* order may work just fine for the next higher order procedure.

1.8.2 HEUN'S PREDICTOR-CORRECTOR METHOD

As we have seen, the *RK1* method (Eq. 1.106) uses just $\widetilde{\mathbf{f}}_1$, the derivative of $\mathbf{y}$ at the beginning of the time interval, to approximate the value of $\mathbf{y}$ at the end of the interval. The use of Eq. (1.106) for approximate numerical integration of nonlinear functions was introduced by Leonhard Euler in 1768 and is therefore known as Euler's method. *RK2* (Eq. 1.107) improves the accuracy by using the average of the derivatives $\widetilde{\mathbf{f}}_1$ and $\widetilde{\mathbf{f}}_2$ at each end of the time interval. The predictor-corrector method due originally to the German mathematician Karl Heun (1859–1929) employs this idea.

First, we use *RK1* to estimate the value of $\mathbf{y}$ at t_{i+1}, labeling that approximation $\mathbf{y}^*_{i+1}$:

$$\mathbf{y}^*_{i+1} = \mathbf{y}_i + h\mathbf{f}\left(t_i, \mathbf{y}_i\right) \quad \text{(predictor)} \tag{1.117a}$$

$\mathbf{y}^*_{i+1}$ is then used to compute the derivative $\mathbf{f}$ at $t + h$, whereupon the average of the two derivatives is used to correct the estimate

$$\mathbf{y}_{i+1} = \mathbf{y}_i + h\frac{\mathbf{f}\left(t_i, \mathbf{y}_i\right) + \mathbf{f}\left(t_i + h, \mathbf{y}^*_{i+1}\right)}{2} \quad \text{(corrector)} \tag{1.117b}$$

We can iteratively improve the estimate of $\mathbf{y}_{i+1}$ by making the substitution $\mathbf{y}^*_{i+1} \leftarrow \mathbf{y}_{i+1}$ (where $\leftarrow$ means "is replaced by") and computing a new value of $\mathbf{y}_{i+1}$ from Eq. (1.117b). That process is repeated until the difference between $\mathbf{y}_{i+1}$ and $\mathbf{y}^*_{i+1}$ becomes acceptably small.

ALGORITHM 1.2

Given the vector $\mathbf{y}$ at time t and the derivatives $\mathbf{f}(t,\mathbf{y})$, use Heun's method to find $\mathbf{y}$ at time $t + h$. See Appendix D.3 for a MATLAB implementation of this algorithm (*heun.m*):

1. Evaluate the vector of derivatives $\mathbf{f}(t,\mathbf{y})$.
2. Compute the predictor $\mathbf{y}^*(t + h) = \mathbf{y}(t) + \mathbf{f}(t,\mathbf{y})h$.
3. Compute the corrector $\mathbf{y}(t+h) = \mathbf{y}(t) + \frac{h}{2}\{\mathbf{f}(t, \mathbf{y}) + \mathbf{f}[t+h, \mathbf{y}^*(t+h)]\}$.
4. Make the substitution $\mathbf{y}^*(t + h) \leftarrow \mathbf{y}(t + h)$ and use Step 3 to recompute $\mathbf{y}(t + h)$.
5. Repeat Step 4 until $\mathbf{y}(t + h) \approx \mathbf{y}^*(t + h)$ to within a given tolerance.

Repeat these steps to obtain $\mathbf{y}$ at subsequent times $t + 2h$, $t + 3h$, etc.

EXAMPLE 1.19

Employ Heun's method to solve Eq. (1.113) using the data provided in Example 1.17. Use two different time steps, $h = 1$ s and $h = 0.1$ s, and compare the results.

Solution

We use the MATLAB functions *Example_1_19.m* and *heun.m* listed in Appendix D.3. The function *Example_1_19.m* passes the given data to the function *heun.m,* which uses the subfunction *rates* within *Example_1_19.m* to compute the

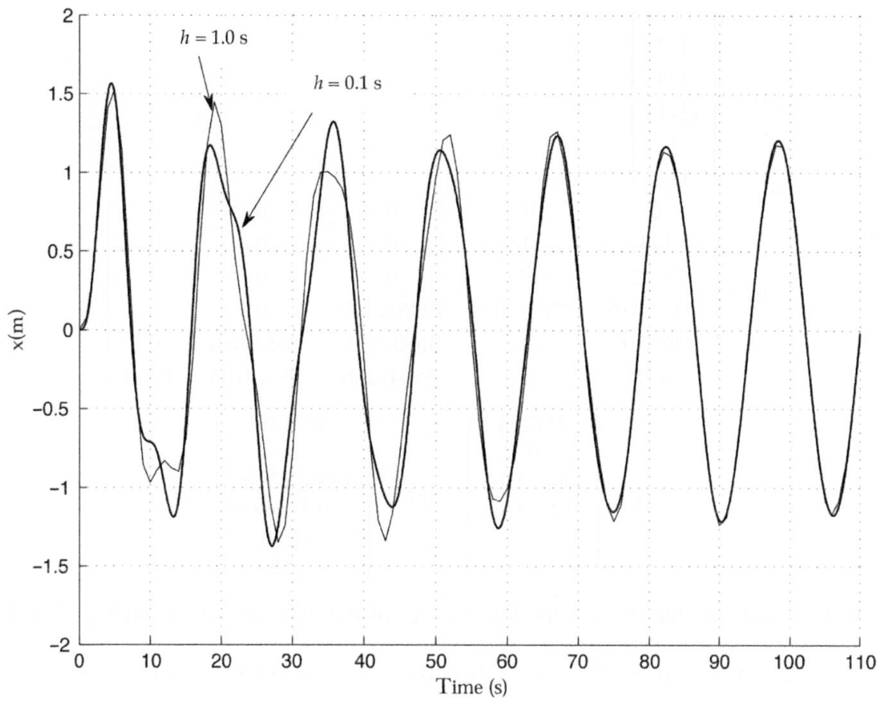

FIG. 1.24

Numerical solution of Eq. (1.123) using Heun's method with two different step sizes.

derivatives $\mathbf{f}(t, \mathbf{y})$ in Eq. (1.116_3). *heun.m* executes Algorithm 1.2 over the time interval from 0 to 110 s, once for $h = 1$ s and again for $h = 0.1$ s, and plots the output in each case, as illustrated in Fig. 1.24.

The graph shows that for $h = 0.1$ s, Heun's method yields a curve identical to the exact solution (whereas the *RK1* method diverged for this time step in Fig. 1.23). Even for the rather large time step $h = 1$ s, the Heun solution, although it starts out a bit ragged, proceeds after 60 s (about the time the transient dies out) to settle down and coincide thereafter very well with the exact solution. For this problem, Heun's method is a decidedly better choice than *RK1* and competes with *RK2* and *RK3*.

1.8.3 RUNGE-KUTTA WITH VARIABLE STEP SIZE

Using a constant step size to integrate a differential equation can be inefficient. The value of h in those regions where the solution varies slowly should be larger than in regions where the variation is more rapid, which requires h to be small to maintain accuracy. Methods for automatically adjusting the step size have been developed. They involve combining two adjacent-order *RK* methods into one and using the difference between the higher and lower order solution to estimate the truncation error in the lower order solution. The step size h is adjusted to keep the truncation error in bounds.

A common example is the embedding of *RK4* into *RK5* to produce the *RKF4(5)* method. The *F* is added in recognition of E. Fehlberg's contribution to this extension of the *RK* method. The procedure has six stages, and the Fehlberg coefficients are (Fehlberg, 1969)

$$
\{\mathbf{a}\} = \begin{Bmatrix} 0 \\ 1/4 \\ 3/8 \\ 12/13 \\ 1 \\ 1/2 \end{Bmatrix}
\qquad
[\mathbf{b}] = \begin{bmatrix} 0 & 0 & 0 & 0 & 0 \\ 1/4 & 0 & 0 & 0 & 0 \\ 3/32 & 9/32 & 0 & 0 & 0 \\ 1932/2197 & -7200/2197 & 7296/2197 & 0 & 0 \\ 439/216 & -8 & 3680/513 & -845/4104 & 0 \\ -8/27 & 2 & -3544/2565 & 1859/4104 & -11/40 \end{bmatrix} \tag{1.118}
$$

$$
\{\mathbf{c}^*\} = \begin{Bmatrix} 25/216 \\ 0 \\ 1408/2565 \\ 2197/4104 \\ -1/5 \\ 0 \end{Bmatrix}
\qquad
\{\mathbf{c}\} = \begin{Bmatrix} 16/135 \\ 0 \\ 6656/12825 \\ 28561/56430 \\ -9/50 \\ 2/55 \end{Bmatrix} \tag{1.119}
$$

Using asterisks to indicate that *RK4* is the lower order of the two, we have from Eq. (1.104)

$$
\mathbf{y}_{i+1}^* = \mathbf{y}_i + h\left(c_1^*\tilde{\mathbf{f}}_1 + c_2^*\tilde{\mathbf{f}}_2 + c_3^*\tilde{\mathbf{f}}_3 + c_4^*\tilde{\mathbf{f}}_4 + c_5^*\tilde{\mathbf{f}}_5 + c_6^*\tilde{\mathbf{f}}_6\right) \quad \text{Low-order solution}\,(RK4) \tag{1.120}
$$

$$
\mathbf{y}_{i+1} = \mathbf{y}_i + h\left(c_1\tilde{\mathbf{f}}_1 + c_2\tilde{\mathbf{f}}_2 + c_3\tilde{\mathbf{f}}_3 + c_3\tilde{\mathbf{f}}_4 + c_5\tilde{\mathbf{f}}_5 + c_6\tilde{\mathbf{f}}_6\right) \quad \text{High-order solution}\,(RK5) \tag{1.121}
$$

where, from Eqs. (1.100), (1.101), and (1.102), the derivatives at the six stages are

$$
\begin{aligned}
\tilde{\mathbf{f}}_1 &= \mathbf{f}(t_i, \mathbf{y}_i) \\
\tilde{\mathbf{f}}_2 &= \mathbf{f}\left(t_i + a_2 h, \mathbf{y}_i + h b_{21}\tilde{\mathbf{f}}_1\right) \\
\tilde{\mathbf{f}}_3 &= \mathbf{f}\left(t_i + a_3 h, \mathbf{y}_i + h\left[b_{31}\tilde{\mathbf{f}}_1 + b_{32}\tilde{\mathbf{f}}_2\right]\right) \\
\tilde{\mathbf{f}}_4 &= \mathbf{f}\left(t_i + a_4 h, \mathbf{y}_i + h\left[b_{41}\tilde{\mathbf{f}}_1 + b_{42}\tilde{\mathbf{f}}_2 + b_{43}\tilde{\mathbf{f}}_3\right]\right) \\
\tilde{\mathbf{f}}_5 &= \mathbf{f}\left(t_i + a_5 h, \mathbf{y}_i + h\left[b_{51}\tilde{\mathbf{f}}_1 + b_{52}\tilde{\mathbf{f}}_2 + b_{53}\tilde{\mathbf{f}}_3 + b_{54}\tilde{\mathbf{f}}_4\right]\right) \\
\tilde{\mathbf{f}}_6 &= \mathbf{f}\left(t_i + a_6 h, \mathbf{y}_i + h\left[b_{61}\tilde{\mathbf{f}}_1 + b_{62}\tilde{\mathbf{f}}_2 + b_{63}\tilde{\mathbf{f}}_3 + b_{64}\tilde{\mathbf{f}}_4 + b_{65}\tilde{\mathbf{f}}_5\right]\right)
\end{aligned} \tag{1.122}
$$

Observe that, although the low- and high-order solutions have different weights ($\{\mathbf{c}^*\}$ and $\{\mathbf{c}\}$, respectively), they share the same nodes $\{\mathbf{a}\}$ and coupling coefficients $[\mathbf{b}]$, and, hence, the same values of the derivatives $\tilde{\mathbf{f}}$. This is another convenient feature of the Runge-Kutta-Fehlberg (*RKF*) method.

The truncation vector $\mathbf{e}$ is the difference between the higher order solution $\mathbf{y}_{i+1}$ and the lower order solution $\mathbf{y}_{i+1}^*$,

$$
\begin{aligned}
\mathbf{e} &= \mathbf{y}_{i+1} - \mathbf{y}_{i+1}^* \\
&= h\left[(c_1 - c_1^*)\tilde{\mathbf{f}}_1 + (c_2 - c_2^*)\tilde{\mathbf{f}}_2 + (c_3 - c_3^*)\tilde{\mathbf{f}}_3 + (c_4 - c_4^*)\tilde{\mathbf{f}}_4 + (c_5 - c_5^*)\tilde{\mathbf{f}}_5 + (c_6 - c_6^*)\tilde{\mathbf{f}}_6\right]
\end{aligned} \tag{1.123}
$$

The number of components of **e** equals N, the number of first-order differential equations in the system (e.g., three in Example 1.15 and two in Example 1.18). The scalar truncation error e is the largest of the absolute values of the components of **e**,

$$e = \text{maximum of the set } (|e_1|, |e_2|, |e_3|, \cdots, |e_N|) \tag{1.124}$$

We set up a tolerance *tol*, which the truncation error cannot exceed. Instead of using the same h for every step of the numerical integration process, we can adjust the step size so as to keep the error e from exceeding *tol*. A simple strategy for adaptive step size control is to update h after each time step using a formula derived, for example, in Bond and Allman (1996),

$$h_{\text{new}} = h_{\text{old}} \left(\frac{tol}{e}\right)^{\frac{1}{p+1}} \tag{1.125}$$

where p is the lower of the two orders in an *RKFp(p+1)* method. For *RKF4(5)*, $p = 4$. According to Eq. (1.125), if $e > tol$, then $h_{\text{new}} < h_{\text{old}}$, whereas if $e < tol$, then $h_{\text{new}} > h_{\text{old}}$. A factor β is commonly added so that

$$h_{\text{new}} = h_{\text{old}} \beta \left(\frac{tol}{e}\right)^{\frac{1}{p+1}} \tag{1.126}$$

where β may be 0.8 or 0.9, depending on the computer program.

ALGORITHM 1.3

Given the vector $\mathbf{y}_i$ at time t_i, the derivative functions $\mathbf{f}(t,\mathbf{y})$, the time step h, and the tolerance *tol*, use the *RKF4(5)* method with adaptive step size control to find $\mathbf{y}_{i+1}$ at time t_{i+1}. See Appendix D.4 for *rkf45.m*, a MATLAB implementation of this algorithm.

1. Evaluate the derivatives $\tilde{\mathbf{f}}_1$ through $\tilde{\mathbf{f}}_6$ using Eq. (1.122).
2. Calculate the truncation vector using Eq. (1.123).
3. Compute the scalar truncation error e using Eq. (1.124).
4. If $e > tol$ then replace h by $h\beta(tol/e)^{1/5}$ and return to Step 1.
5. Replace t by $t+h$ and calculate $\mathbf{y}_{i+1}$ using Eq. (1.121).
6. Replace h by $h\beta(tol/e)^{1/5}$.

Repeat these steps to obtain $\mathbf{y}_{i+2}$, $\mathbf{y}_{i+3}$, etc.

EXAMPLE 1.20

A spacecraft S of mass m travels in a straight line away from the center C of the earth, as illustrated in Fig. 1.25. If at a distance of 6500 km from C its outbound velocity is 7.8 km/s, what will be its position and velocity 70 min later?

Solution

Solving this problem requires writing down and then integrating the equations of motion. Starting with the free body diagram of S, shown in Fig. 1.25, we find that Newton's second law (Eq. 1.38) for the spacecraft is

$$-F_g = m\ddot{x} \tag{a}$$

The variable force of gravity F_g on the spacecraft is its mass m times the local acceleration of gravity, given by Eq. (1.36). That is,

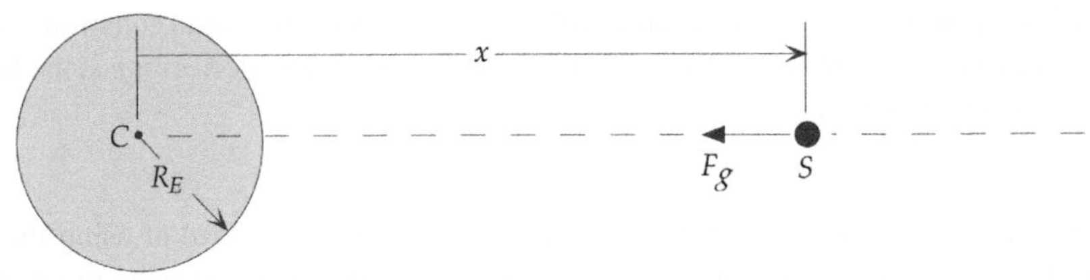

FIG. 1.25

Spacecraft S in rectilinear motion relative to the earth.

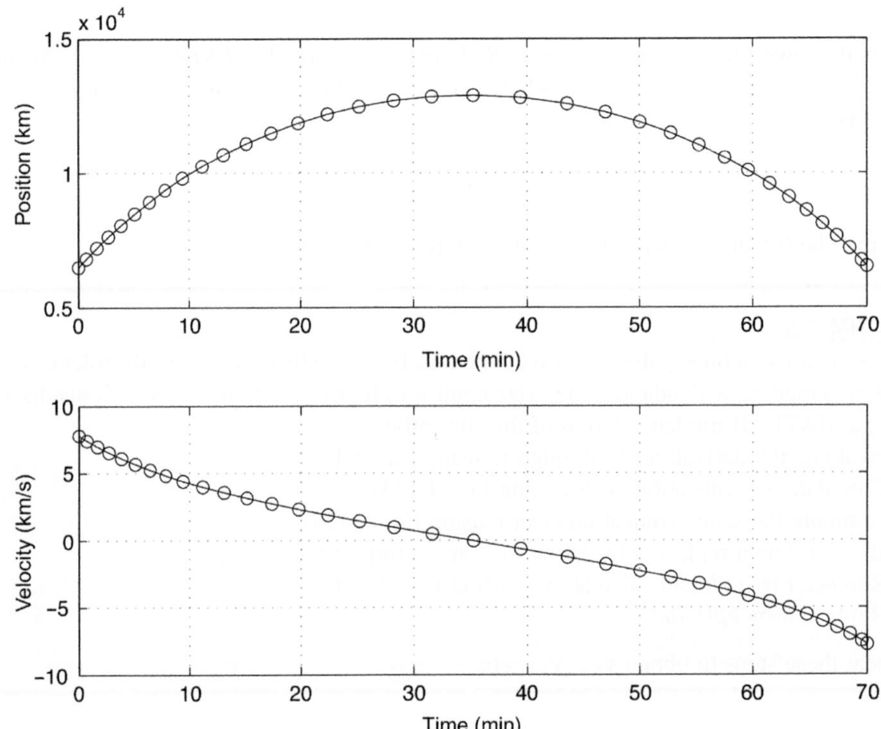

FIG. 1.26

Position and velocity versus time. The solution points are circled.

$$F_g = mg = m\frac{g_0 R_E^2}{x^2} \tag{b}$$

R_E is the earth's radius (6378 km), and g_0 is the sea level acceleration of gravity (9.807 m/s^2). Combining Eqs. (a) and (b) yields

$$\ddot{x} + \frac{g_0 R_E^2}{x^2} = 0 \tag{1.127}$$

This differential equation for the rectilinear motion of the spacecraft has an analytical solution, which we shall not go into here. Instead, we will solve it numerically using Algorithm 1.3 and the given initial conditions. For that, we must as usual introduce the auxiliary variables $y_1 = x$ and $y_2 = \dot{x}$ to obtain the two differential equations

$$\begin{aligned}\dot{y}_1 &= y_2\\ \dot{y}_2 &= -\frac{g_0 R_E^2}{y_1^2}\end{aligned} \tag{c}$$

The initial conditions in this case are

$$y_1(0) = 6500\text{km} \quad y_2(0) = 7.8\text{km/s}^2 \tag{d}$$

The MATLAB programs *Example_1_20.m* and *rkf45.m*, both in Appendix D.4, were used to produce Fig. 1.26, which shows the position and velocity of the spacecraft over the requested time span. *Example_1_20.m* passes the initial conditions and time span to *rkf4.m*, which uses the subroutine *rates* within *Example_1_20.m* to compute the derivatives $\dot{x}$ and $\ddot{x}$.

Fig. 1.26 reveals that the spacecraft takes 35 min to coast out to twice its original 6500 km distance from C before reversing the direction and returning 35 min later to where it started with a speed of 7.8 km/s. The nonuniform spacing between the solution points shows how *rkf4.m* controlled the step size such that h was smaller during rapid variations of the solution but larger elsewhere.

PROBLEMS

Section 1.2

1.1 Given the three vectors $\mathbf{A} = A_x\hat{\mathbf{i}} + A_y\hat{\mathbf{j}} + A_z\hat{\mathbf{k}}$, $\mathbf{B} = B_x\hat{\mathbf{i}} + B_y\hat{\mathbf{j}} + B_z\hat{\mathbf{k}}$, and $\mathbf{C} = C_x\hat{\mathbf{i}} + C_y\hat{\mathbf{j}} + C_z\hat{\mathbf{k}}$, show analytically that

(a) $\mathbf{A}\cdot\mathbf{A} = A^2$

(b) $\mathbf{A}\cdot(\mathbf{B}\times\mathbf{C}) = (\mathbf{A}\times\mathbf{B})\cdot\mathbf{C}$ (interchangeability of the dot and cross)

(c) $\mathbf{A}\times(\mathbf{B}\times\mathbf{C}) = \mathbf{B}(\mathbf{A}\cdot\mathbf{C}) - \mathbf{C}(\mathbf{A}\cdot\mathbf{B})$ (the bac-cab rule)

(Hint: Simply compute the expressions on each side of the = signs and demonstrate conclusively that they are the same.) Do *not* substitute numbers to "prove" your point. Use Eqs. (1.9) and (1.16).

1.2 Use just the vector identities in Problem 1.1 to show that

$$(\mathbf{A}\times\mathbf{B})\cdot(\mathbf{C}\times\mathbf{D}) = (\mathbf{A}\cdot\mathbf{C})(\mathbf{B}\cdot\mathbf{D}) - (\mathbf{A}\cdot\mathbf{D})(\mathbf{B}\cdot\mathbf{C})$$

1.3 Let $\mathbf{A} = 8\hat{\mathbf{i}} + 9\hat{\mathbf{j}} + 12\hat{\mathbf{k}}$, $\mathbf{B} = 9\hat{\mathbf{i}} + 6\hat{\mathbf{j}} + \hat{\mathbf{k}}$, and $\mathbf{C} = 3\hat{\mathbf{i}} + 5\hat{\mathbf{j}} + 10\hat{\mathbf{k}}$. Calculate the (scalar) projection C_{AB} of $\mathbf{C}$ onto the plane of $\mathbf{A}$ and $\mathbf{B}$ (see illustration below).

(Hint: $C^2 = C_n^2 + C_{AB}^2$)

{Ans.: $C_{AB} = 11.58$}

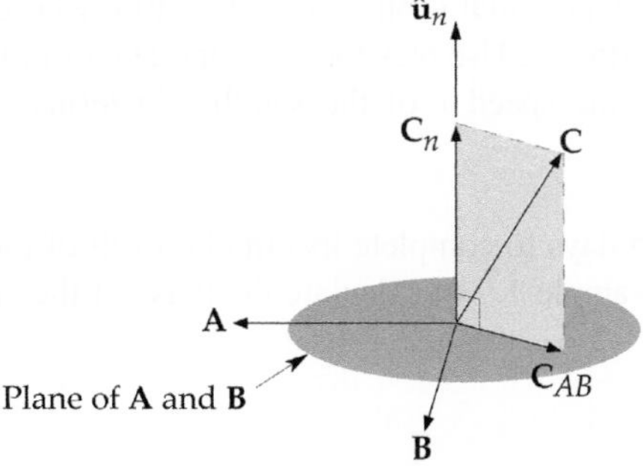

Section 1.3

1.4 Since $\hat{\mathbf{u}}_t$ and $\hat{\mathbf{u}}_n$ are perpendicular and $\hat{\mathbf{u}}_t \times \hat{\mathbf{u}}_n = \hat{\mathbf{u}}_b$, use the bac-cab rule to show that $\hat{\mathbf{u}}_b \times \hat{\mathbf{u}}_t = \hat{\mathbf{u}}_n$ and $\hat{\mathbf{u}}_n \times \hat{\mathbf{u}}_b = \hat{\mathbf{u}}_t$, thereby verifying Eq. (1.29).

1.5 The x, y, and z coordinates (in meters) of a particle P as a function of time (in seconds) are $x = \sin 3t$, $y = \cos t$, and $z = \sin 2t$. At $t = 3\,\text{s}$, determine:

(a) The velocity $\mathbf{v}$ in Cartesian coordinates.
(b) The speed v.
(c) The unit tangent vector $\hat{\mathbf{u}}_t$.
(d) The angles θ_x, θ_y, and θ_z that $\mathbf{v}$ makes with the x, y, and z axes.
(e) The acceleration $\mathbf{a}$ in Cartesian coordinates.
(f) The unit binormal vector $\hat{\mathbf{u}}_b$.
(g) The unit normal vector $\hat{\mathbf{u}}_n$.
(h) The angles ϕ_x, ϕ_y, and ϕ_z that $\mathbf{a}$ makes with the x, y, and z axes.
(i) The tangential component a_t of the acceleration.
(j) The normal component a_n of the acceleration.
(k) The radius of curvature of the path of P.
(l) The Cartesian coordinates of the center of curvature of the path.

{Partial Ans.: (b) 2.988 m/s; (d) $\theta_x = 139.7$deg; (j) $a_n = 5.398\ \text{m/s}^2$; (l) $x_C = -0.4068\,\text{m}$}

Section 1.4

1.6 An 80-kg man and 50-kg woman stand 0.5 m from each other. What is the force of gravitational attraction between the couple?
{Ans.: 36.04 μN}

1.7 If a person's weight is W on the surface of the earth, calculate the earth's gravitational pull on that person at a distance equal to the moon's orbit.
{Ans.: $275(10^{-6})W$}

1.8 If a person's weight is W on the surface of the earth, calculate what it would be, in terms of W, at the surface of

(a) the moon;
(b) Mars;
(c) Jupiter.

{Partial Ans.: (c) $2.53W$}

Section 1.5

1.9 A satellite of mass m is in a circular orbit around the earth, whose mass is M. The orbital radius from the center of the earth is r. Use Newton's second law of motion, together with Eqs. (1.25) and (1.31), to calculate the speed v of the satellite in terms of M, r, and the gravitational constant G.
{Ans.: $v = \sqrt{GM/r}$}

1.10 If the earth takes 365.25 days to complete its circular orbit of radius $149.6(10^6)$km around the sun, use the result of Example 1.9 to calculate the mass of the sun.
{Ans.: $1.988(10^{30})$kg}

Section 1.6

1.11 **F** is a force vector of fixed magnitude embedded on a rigid body in plane motion (in the *xy* plane). At a given instant, $\boldsymbol{\omega} = 2\hat{\mathbf{k}}\,\text{rad/s}$, $\dot{\boldsymbol{\omega}} = -5\hat{\mathbf{k}}\,\text{rad/s}^2$, $\ddot{\boldsymbol{\omega}} = 3\hat{\mathbf{k}}\,\text{rad/s}^3$, and $\mathbf{F} = \left(15\hat{\mathbf{i}} + 10\hat{\mathbf{j}}\right)(\text{N})$. At that instant, calculate $\dddot{\mathbf{F}}$.
{Ans.: $\dddot{\mathbf{F}} = 500\hat{\mathbf{i}} + 225\hat{\mathbf{j}}\,(\text{N/s}^3)$}

Section 1.7

1.12 The absolute position, velocity, and acceleration of *O* are

$$\mathbf{r}_O = -16\hat{\mathbf{I}} + 84\hat{\mathbf{J}} + 59\hat{\mathbf{K}}(\text{m})$$
$$\mathbf{v}_O = 7\hat{\mathbf{I}} + 9\hat{\mathbf{J}} + 4\hat{\mathbf{K}}(\text{m/s})$$
$$\mathbf{a}_O = 3\hat{\mathbf{I}} - 7\hat{\mathbf{J}} + 4\hat{\mathbf{K}}(\text{m/s}^2)$$

The angular velocity and acceleration of the moving frame are

$$\boldsymbol{\Omega} = -0.8\hat{\mathbf{I}} + 0.7\hat{\mathbf{J}} + 0.4\hat{\mathbf{K}}(\text{rad/s}) \quad \dot{\boldsymbol{\Omega}} = -0.4\hat{\mathbf{I}} + 0.9\hat{\mathbf{J}} - 1.0\hat{\mathbf{K}}(\text{rad/s}^2)$$

The unit vectors of the moving frame are

$$\hat{\mathbf{i}} = -0.15670\hat{\mathbf{I}} - 0.31235\hat{\mathbf{J}} + 0.93704\hat{\mathbf{K}}$$
$$\hat{\mathbf{j}} = -0.12940\hat{\mathbf{I}} + 0.94698\hat{\mathbf{J}} + 0.29409\hat{\mathbf{K}}$$
$$\hat{\mathbf{k}} = -0.97922\hat{\mathbf{I}} - 0.075324\hat{\mathbf{J}} - 0.18831\hat{\mathbf{K}}$$

The absolute position of *P* is

$$\mathbf{r} = 51\hat{\mathbf{I}} - 45\hat{\mathbf{J}} + 36\hat{\mathbf{K}}(\text{m})$$

The velocity and acceleration of *P* relative to the moving frame are

$$\mathbf{v}_{\text{rel}} = 31\hat{\mathbf{i}} - 68\hat{\mathbf{j}} - 77\hat{\mathbf{k}}(\text{m/s}) \quad \mathbf{a}_{\text{rel}} = 2\hat{\mathbf{i}} - 6\hat{\mathbf{j}} + 5\hat{\mathbf{k}}(\text{m/s}^2)$$

Calculate the absolute velocity $\mathbf{v}_P$ and acceleration $\mathbf{a}_P$ of *P*.
{Ans.: $\mathbf{v}_P = 156.4\hat{\mathbf{u}}_v(\text{m/s}) \quad \hat{\mathbf{u}}_v = 0.7790\hat{\mathbf{I}} - 0.3252\hat{\mathbf{J}} + 0.5360\hat{\mathbf{K}}$
$\mathbf{a}_P = 85.13\hat{\mathbf{u}}_a(\text{m/s}^2) \quad \hat{\mathbf{u}}_a = -0.3229\hat{\mathbf{I}} + 0.8284\hat{\mathbf{J}} - 0.4576\hat{\mathbf{K}}$}

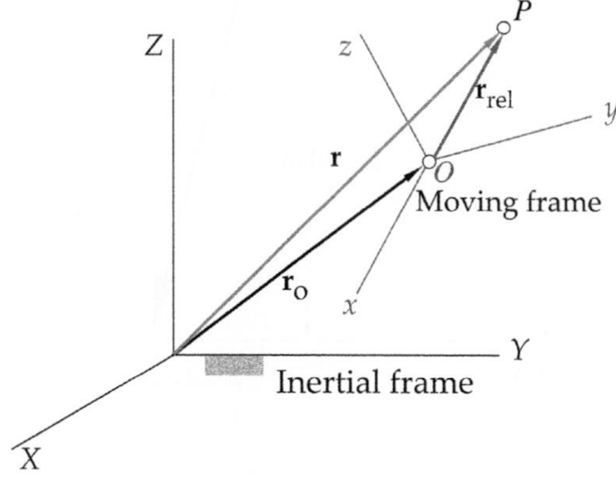

1.13 An airplane in level flight at an altitude h and a uniform speed v passes directly over a radar tracking station A. Calculate the angular velocity $\dot{\theta}$ and angular acceleration $\ddot{\theta}$ of the radar antenna as well as the rate $\dot{r}$ at which the airplane is moving away from the antenna. Use the equations of this chapter (rather than polar coordinates, which you can use to check your work). Attach the inertial frame of reference to the ground and assume a nonrotating earth. Attach the moving frame to the antenna, with the x axis pointing always from the antenna toward the airplane.
{Ans.: (a) $\dot{\theta} = v\cos^2\theta/h$, (b) $\ddot{\theta} = -2v^2\cos^3\theta\sin\theta/h^2$, (c) $v_{rel} = v\sin\theta$}

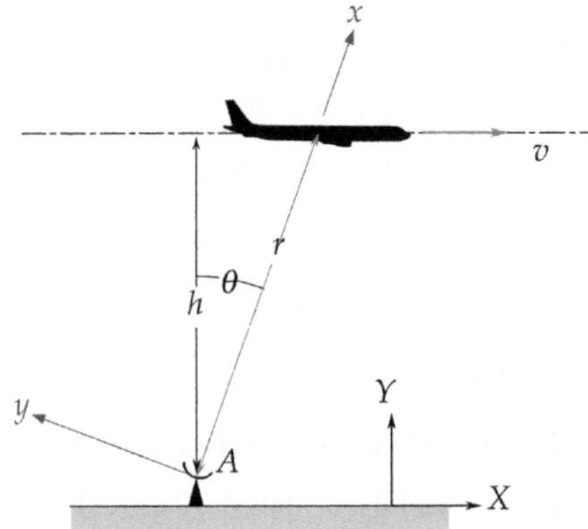

1.14 At 30°N latitude, a 1000-kg (2205-lb) car travels due north at a constant speed of 100 km/h (62 mph) on a level road at sea level. Taking into account the earth's rotation, calculate:
(a) the lateral (sideways) force of the road on the car;
(b) the normal force of the road on the car.
{Ans.: (a) $F_{lateral} = 2.026\,\text{N}$, to the left (west); (b) $F_{normal} = 9784\,\text{N}$}

1.15 At 29°N latitude, what is the deviation d from the vertical of a plumb bob at the end of a 30-m string, due to the earth's rotation?
{Ans.: 44.1 mm to the south}

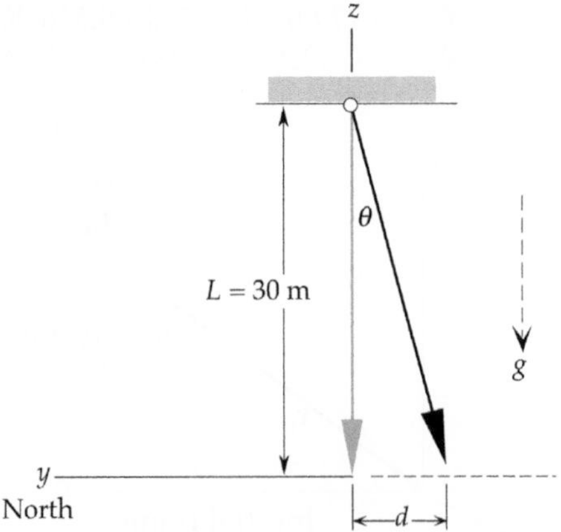

Section 1.8

1.16 Verify by substitution that Eq. (1.114a) is the solution of Eq. (1.113).

1.17 Verify that Eqs. (1.114b) are valid.

1.18 Numerically solve the fourth-order differential equation

$$\mathrm{d}^4y/\mathrm{d}t^4 + 2\mathrm{d}^2y/\mathrm{d}t^2 + y = 0$$

for y at $t = 20$, if the initial conditions are $y = 1$ and $\mathrm{d}y/\mathrm{d}t = \mathrm{d}^2y/\mathrm{d}t^2 = \mathrm{d}^3y/\mathrm{d}t^3 = 0$ at $t = 0$.
{Ans.: $y(20) = 9.545$}

1.19 Numerically solve the differential equation

$$\mathrm{d}^4y/\mathrm{d}t^4 + 3\mathrm{d}^3y/\mathrm{d}t^3 - 4\mathrm{d}y/\mathrm{d}t - 12y = te^{2t}$$

for y at $t = 3$, if, at $t = 0$, $y = \mathrm{d}y/\mathrm{d}t = \mathrm{d}^2y/\mathrm{d}t^2 = 0$.
{Ans.: $y(3) = 66.62$}

1.20 Numerically solve the differential equation

$$t\ddot{y} + t^2\dot{y} - 2y = 0$$

to obtain y at $t = 4$ if the initial conditions are $y = 0$ and $\dot{y} = 1$ at $t = 1$.
{Ans.: $y(4) = 1.29$}

1.21 Numerically solve the system

$$\dot{x} + \frac{1}{2}y - z = 0$$
$$-\frac{1}{2}x + \dot{y} + \frac{1}{\sqrt{2}}z = 0$$
$$\frac{1}{2}x - \frac{1}{\sqrt{2}}y + \dot{z} = 0$$

to obtain x, y, and z at $t = 20$. The initial conditions are $x = 1$ and $y = z = 0$ at $t = 0$.
{Ans.: $x(20) = 0.704$, $y(20) = 0.665$, $z(20) = -0.246$}

1.22 Use one of the numerical methods discussed in this section to solve Eq. (1.127) for the time required for the moon to fall to the earth after it is somehow stopped in its orbit while the earth remains fixed in space. (This will require a trial-and-error procedure known formally as a shooting method. It is not necessary for this problem to code the procedure. Simply guess a time and let the solver compute the final radius. On the basis of the deviation of that result from the earth's radius (6378 km), revise your time estimate and rerun the problem to compute a new final radius. Repeat this process in a logical fashion until your time estimate yields a final radius that is accurate to at least three significant figures.) Compare your answer with the analytical solution,

$$t = \sqrt{\frac{r_0}{2g_0R_E^2}}\left[\frac{\pi}{4}r_0 + \sqrt{r(r_0 - r)} + \frac{r_0}{2}\sin^{-1}\left(\frac{r_0 - 2r}{r_0}\right)\right]$$

where t is the time, r_0 is the initial radius, r is the final radius ($r < r_0$), g_0 is the sea level acceleration of earth's gravity, and R_E is the radius of the earth.

1.23 Use an *RK* solver such as *rkf45* in Appendix D.4 or MATLAB's *ode45* to solve the nonlinear Lorenz equations, due to the American meteorologist and mathematician E.N. Lorenz (1917–2008):

$$\dot{x} = \sigma(y - x)$$
$$\dot{y} = x(\rho - z) - y$$
$$\dot{z} = xy - \beta z$$

Start off by using the values Lorenz (1963) used in his paper (namely, $\sigma = 10$, $\beta = 8/3$, and $\rho = 28$). For initial conditions use $x = 0$, $y = 1$, and $z = 0$ at $t = 0$. Let t range to a value of at least 20. Plot the phase trajectory $x = x(t)$, $y = y(t)$, $z = z(t)$ in three dimensions to see the now-famous "Lorenz attractor." The Lorenz equations are a simplified model of the two-dimensional convective motion within a fluid layer due to a temperature difference ΔT between the upper and lower surfaces. The equations are chaotic in nature. For one thing, this means that the solutions are extremely sensitive to the initial conditions. A minute change yields a completely different solution in the long run. Check this out yourself. (x represents the intensity of the convective motion of the fluid, y is proportional to the temperature difference between rising and falling fluid, and z represents the nonlinearity of the temperature profile across the depth. σ is a fluid property (the Prandtl number), ρ is proportional to ΔT, β is a geometrical parameter, and t is a nondimensional time.)

REFERENCES

Bond, V.R., Allman, M.C., 1996. Modern Astrodynamics: Fundamentals and Perturbation Methods. Princeton University Press.

Butcher, J.C., 2008. Numerical Methods for Ordinary Differential Equations, third ed. John Wiley & Sons, West Sussex.

Coriolis, G., 1835. On the equations of relative motion of a system of bodies. J. École Polytechnique 15, 142–154.

Fehlberg, E., 1969. Low-Order Classical Runge-Kutta Formulas with Stepsize Control and Their Application to Some Heat Transfer Problems. NASA TR R-315.

Lorenz, E.N., 1963. Deterministic nonperiodic flow. J. Atmospheric Sci. 20, 130–141.

CHAPTER

THE TWO-BODY PROBLEM

2

2.1 INTRODUCTION

This chapter presents the vector-based approach to the classical problem of determining the motion of two bodies due solely to their own mutual gravitational attraction. We show that the path of one of the masses relative to the other is a conic section (circle, ellipse, parabola, or hyperbola) whose shape is determined by the eccentricity. Several fundamental properties of the different types of orbits are developed with the aid of the laws of conservation of angular momentum and energy. These properties include the period of elliptical orbits, the escape velocity associated with parabolic paths, and the characteristic energy of hyperbolic trajectories. Following the presentation of the four types of orbits, the perifocal frame is introduced. This frame of reference is used to describe orbits in three dimensions, which is the subject of Chapter 4.

In this chapter, the perifocal frame provides the backdrop for developing the Lagrange f and g coefficients. By means of the Lagrange f and g coefficients, the position and velocity on a trajectory can be found in terms of the position and velocity at an initial time. These functions are needed in the orbit determination algorithms of Lambert and Gauss presented in Chapter 5.

The chapter concludes with a discussion of the restricted three-body problem to provide a basis for understanding the concepts of Lagrange points and the Jacobi constant. This material is optional.

In studying this chapter, it would be well from time to time to review the road map provided in Appendix B.

2.2 EQUATIONS OF MOTION IN AN INERTIAL FRAME

Fig. 2.1 shows two-point masses acted upon only by the mutual force of gravity between them. The positions $\mathbf{R}_1$ and $\mathbf{R}_2$ of their centers of mass are shown relative to an inertial frame of reference XYZ. In terms of the coordinates of the two points

$$\begin{aligned}\mathbf{R}_1 &= X_1\hat{\mathbf{I}} + Y_1\hat{\mathbf{J}} + Z_1\hat{\mathbf{K}} \\ \mathbf{R}_2 &= X_2\hat{\mathbf{I}} + Y_2\hat{\mathbf{J}} + Z_2\hat{\mathbf{K}}\end{aligned} \tag{2.1}$$

The origin O of the inertial frame may move with a constant velocity (relative to the fixed stars), but the axes do not rotate. Each of the two bodies is acted upon by the gravitational attraction of the other. $\mathbf{F}_{12}$ is the force exerted on m_1 by m_2, and $\mathbf{F}_{21}$ is the force exerted on m_2 by m_1.

Orbital Mechanics for Engineering Students. https://doi.org/10.1016/B978-0-12-824025-0.00002-7

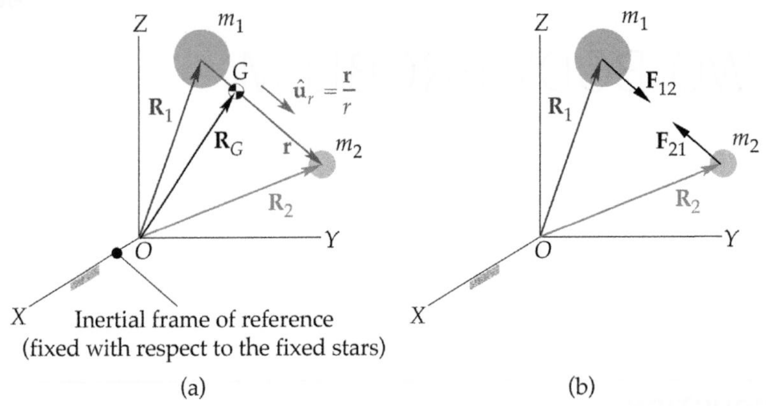

FIG. 2.1

(a) Two masses located in an inertial frame. (b) Free-body diagrams.

The position vector $\mathbf{R}_G$ of the center of mass (or *barycenter*) G of the system in Fig. 2.1(a) is defined by the formula

$$\mathbf{R}_G = \frac{m_1\mathbf{R}_1 + m_2\mathbf{R}_2}{m_1 + m_2} \tag{2.2}$$

Therefore, the absolute velocity and the absolute acceleration of G are

$$\mathbf{v}_G = \dot{\mathbf{R}}_G = \frac{m_1\dot{\mathbf{R}}_1 + m_2\dot{\mathbf{R}}_2}{m_1 + m_2} \tag{2.3}$$

$$\mathbf{a}_G = \ddot{\mathbf{R}}_G = \frac{m_1\ddot{\mathbf{R}}_1 + m_2\ddot{\mathbf{R}}_2}{m_1 + m_2} \tag{2.4}$$

The adjective "absolute" means that the quantities are measured relative to an inertial frame of reference.

Let $\mathbf{r}$ be the position vector of m_2 relative to m_1. Then,

$$\mathbf{r} = \mathbf{R}_2 - \mathbf{R}_1 \tag{2.5}$$

Or, using Eq. (2.1),

$$\mathbf{r} = (X_2 - X_1)\hat{\mathbf{I}} + (Y_2 - Y_1)\hat{\mathbf{J}} + (Z_2 - Z_1)\hat{\mathbf{K}} \tag{2.6}$$

Furthermore, let $\hat{\mathbf{u}}_r$ be the unit vector pointing from m_1 toward m_2, so that

$$\hat{\mathbf{u}}_r = \frac{\mathbf{r}}{r} \tag{2.7}$$

where r is the magnitude of $\mathbf{r}$,

$$r = \sqrt{(X_2 - X_1)^2 + (Y_2 - Y_1)^2 + (Z_2 - Z_1)^2} \tag{2.8}$$

The body m_1 is acted upon only by the force of gravitational attraction toward m_2. The force of gravitational attraction, F_g, which acts along the line joining the centers of mass of m_1 and m_2, is given by Eq. (1.31). Therefore, the force exerted on m_1 by m_2 is

$$\mathbf{F}_{12} = \frac{Gm_1m_2}{r^2}\hat{\mathbf{u}}_r \tag{2.9}$$

where $\hat{\mathbf{u}}_r$ accounts for the fact that the force vector $\mathbf{F}_{12}$ is directed from m_1 toward m_2. (Do not confuse the symbol G, used in this context to represent the universal gravitational constant, with its use elsewhere in the book to denote the center of mass.) By Newton's third law (the action–reaction principle), the force $\mathbf{F}_{21}$ exerted on m_2 by m_1 is $-\mathbf{F}_{12}$, so that

$$\mathbf{F}_{21} = -\frac{Gm_1m_2}{r^2}\hat{\mathbf{u}}_r \tag{2.10}$$

Newton's second law of motion as applied to a body m_1 is $\mathbf{F}_{12} = m_1\ddot{\mathbf{R}}_1$, where $\ddot{\mathbf{R}}_1$ is the absolute acceleration of m_1. Combining this with Newton's law of gravitation Eq. (2.9) yields

$$m_1\ddot{\mathbf{R}}_1 = \frac{Gm_1m_2}{r^2}\hat{\mathbf{u}}_r \tag{2.11}$$

Likewise, by substituting $\mathbf{F}_{21} = m_2\ddot{\mathbf{R}}_2$ into Eq. (2.10) we get

$$m_2\ddot{\mathbf{R}}_2 = -\frac{Gm_1m_2}{r^2}\hat{\mathbf{u}}_r \tag{2.12}$$

It is apparent upon forming the sum of Eqs. (2.11) and (2.12) that $m_1\ddot{\mathbf{R}}_1 + m_2\ddot{\mathbf{R}}_2 = \mathbf{0}$. According to Eq. (2.4), this means that the acceleration of the center of mass G of the system of two bodies m_1 and m_2 is zero. Therefore, as is true for any system that is free of external forces, G moves in a straight line through space with a constant velocity $\mathbf{v}_G$. Its position vector relative to XYZ is given by

$$\mathbf{R}_G = \mathbf{R}_G)_0 + \mathbf{v}_G t \tag{2.13}$$

where $\mathbf{R}_G)_0$ is the position of G at time $t = 0$. The nonaccelerating center of mass of a two-body system may serve as the origin of an inertial frame.

EXAMPLE 2.1

Use the two-body equations of motion to show why orbiting astronauts experience weightlessness.

Solution

We sense weight by feeling the contact forces that develop wherever our body is supported. Consider an astronaut of mass m_A strapped into a spacecraft of mass m_S, in orbit about the earth. The distance between the center of the earth and the spacecraft is r, and the mass of the earth is M_E. Since the only external force is that of gravity, $\mathbf{F}_S)_g$, the equation of motion of the spacecraft is

$$\mathbf{F}_S)_g = m_S\mathbf{a}_S \tag{a}$$

where $\mathbf{a}_S$ is measured in an inertial frame. According to Eq. (2.10),

$$\mathbf{F}_S)_g = -\frac{GM_Em_S}{r^2}\hat{\mathbf{u}}_r \tag{b}$$

where $\hat{\mathbf{u}}_r$ is the unit vector pointing outward from the earth toward the orbiting spacecraft. Thus, Eqs. (a) and (b) imply that the absolute acceleration of the spacecraft is

$$\mathbf{a}_S = -\frac{GM_E}{r^2}\hat{\mathbf{u}}_r \tag{c}$$

The equation of motion of the astronaut is

$$\mathbf{F}_A)_g + \mathbf{C}_A = m_A \mathbf{a}_A \tag{d}$$

In this expression $\mathbf{F}_A)_g$ is the force of gravity on (i.e., the weight of) the astronaut, $\mathbf{C}_A$ is the net contact force on the astronaut from restraints (e.g., seat, seat belt), and $\mathbf{a}_A$ is the astronaut's absolute acceleration. According to Eq. (2.10),

$$\mathbf{F}_A)_g = -\frac{GM_E m_A}{r^2}\hat{\mathbf{u}}_r \tag{e}$$

Since the astronaut is moving with the spacecraft, we have, noting Eq. (c),

$$\mathbf{a}_A = \mathbf{a}_S = -\frac{GM_E}{r^2}\hat{\mathbf{u}}_r \tag{f}$$

Substituting Eqs. (e) and (f) into Eq. (d) yields

$$-\frac{GM_E m_A}{r^2}\hat{\mathbf{u}}_r + \mathbf{C}_A = m_A\left(-\frac{GM_E}{r^2}\hat{\mathbf{u}}_r\right)$$

from which it is clear that

$$\boxed{\mathbf{C}_A = 0}$$

The net contact force on the astronaut is zero. With no reaction to the force of gravity exerted on the body, there is no sensation of weight.

The potential energy V of the gravitational force $\mathbf{F}$ between two point masses m_1 and m_2 separated by a distance r is given by

$$V = -\frac{Gm_1 m_2}{r} \tag{2.14}$$

A conservative force, like gravity, can be obtained from its potential energy function V by means of the gradient operator,

$$\mathbf{F} = -\nabla V \tag{2.15}$$

where, in Cartesian coordinates,

$$\nabla = \frac{\partial}{\partial x}\hat{\mathbf{i}} + \frac{\partial}{\partial y}\hat{\mathbf{j}} + \frac{\partial}{\partial z}\hat{\mathbf{k}} \tag{2.16}$$

For the two-body system in Fig. 2.1 we have, by combining Eqs. (2.8) and (2.14),

$$V = -\frac{Gm_1 m_2}{\sqrt{(X_2 - X_1)^2 + (Y_2 - Y_1)^2 + (Z_2 - Z_1)^2}} \tag{2.17}$$

The attractive forces $\mathbf{F}_{12}$ and $\mathbf{F}_{21}$ in Eq. (2.6) are derived from Eq. (2.17) as follows:

$$\mathbf{F}_{12} = -\left(\frac{\partial V}{\partial X_2}\hat{\mathbf{i}} + \frac{\partial V}{\partial Y_2}\hat{\mathbf{j}} + \frac{\partial V}{\partial Z_2}\hat{\mathbf{k}}\right)$$
$$\mathbf{F}_{21} = -\left(\frac{\partial V}{\partial X_1}\hat{\mathbf{i}} + \frac{\partial V}{\partial Y_1}\hat{\mathbf{j}} + \frac{\partial V}{\partial Z_1}\hat{\mathbf{k}}\right)$$

In Appendix E, we show that the gravitational potential V, and hence the gravitational force outside a sphere with a spherically symmetric mass distribution M, is the same as that of a point mass M located at the center of the sphere. Therefore, the two-body problem applies not only to point masses but also to spherical bodies (as long as, of course, they do not come into contact!).

Let us return to Eqs. (2.11) and (2.12), the equations of motion of the two-body system relative to the *XYZ* inertial frame. We can divide m_1 out of Eq. (2.11) and m_2 out of Eq. (2.12) and then substitute Eq. (2.7) into both results to obtain

$$\ddot{\mathbf{R}}_1 = Gm_2\frac{\mathbf{r}}{r^3} \tag{2.18a}$$

$$\ddot{\mathbf{R}}_2 = Gm_1\frac{\mathbf{r}}{r^3} \tag{2.18b}$$

These are the final forms of the equations of motion of the two bodies in inertial space. With the aid of Eqs. (2.1), (2.6), and (2.8) we can express these equations in terms of the components of the position and acceleration vectors in the inertial *XYZ* frame:

$$\ddot{X}_1 = Gm_2\frac{X_2 - X_1}{r^3} \quad \ddot{Y}_1 = Gm_2\frac{Y_2 - Y_1}{r^3} \quad \ddot{Z}_1 = Gm_2\frac{Z_2 - Z_1}{r^3} \tag{2.19a}$$

$$\ddot{X}_2 = Gm_1\frac{X_1 - X_2}{r^3} \quad \ddot{Y}_2 = Gm_1\frac{Y_1 - Y_2}{r^3} \quad \ddot{Z}_2 = Gm_1\frac{Z_1 - Z_2}{r^3} \tag{2.19b}$$

where $r = \sqrt{(X_2 - X_1)^2 + (Y_2 - Y_1)^2 + (Z_2 - Z_1)^2}$.

The position vector **R** and velocity vector **V** of a particle are referred to collectively as its state vector. The fundamental problem before us is to find the state vectors of both particles of the two-body system at a given time, given the state vectors at an initial time. The numerical solution procedure is outlined in Algorithm 2.1.

ALGORITHM 2.1

Numerically compute the state vectors $\mathbf{R}_1$, $\mathbf{V}_1$ and $\mathbf{R}_2$, $\mathbf{V}_2$ of the two-body system as a function of time, given their initial values $\mathbf{R}_1^0$, $\mathbf{V}_1^0$ and $\mathbf{R}_2^0$, $\mathbf{V}_2^0$. This algorithm is implemented in MATLAB as the function *twobody3d.m*, which is listed in Appendix D.5.

1. Form the vector consisting of the components of the state vectors at time t_0,

$$\mathbf{y}_0 = \lfloor X_1^0 \; Y_1^0 \; Z_1^0 \; X_2^0 \; Y_2^0 \; Z_2^0 \; \dot{X}_1^0 \; \dot{Y}_1^0 \; \dot{Z}_1^0 \; \dot{X}_2^0 \; \dot{Y}_2^0 \; \dot{Z}_2^0 \rfloor$$

2. Provide $\mathbf{y}_0$ and the final time t_f to Algorithms 1.1, 1.2, or 1.3, along with the vector that comprises the components of the state vector derivatives

$$\mathbf{f}(t, \mathbf{y}) = \lfloor \dot{X}_1 \; \dot{Y}_1 \; \dot{Z}_1 \; \dot{X}_2 \; \dot{Y}_2 \; \dot{Z}_2 \; \ddot{X}_1 \; \ddot{Y}_1 \; \ddot{Z}_1 \; \ddot{X}_2 \; \ddot{Y}_2 \; \ddot{Z}_2 \rfloor$$

where the last six components, the accelerations, are given by Eqs. (2.19a) and (2.19b).

3. The selected algorithm solves the system $\dot{\mathbf{y}} = \mathbf{f}(t, \mathbf{y})$ for the system state vector

$$\mathbf{y} = \lfloor X_1 \; Y_1 \; Z_1 \; X_2 \; Y_2 \; Z_2 \; \dot{X}_1 \; \dot{Y}_1 \; \dot{Z}_1 \; \dot{X}_2 \; \dot{Y}_2 \; \dot{Z}_2 \rfloor$$

at n discrete times t_n from t_0 through t_f.

4. The state vectors of m_1 and m_2 at the discrete times are

$$\mathbf{R}_1 = X_1\hat{\mathbf{I}} + Y_1\hat{\mathbf{J}} + Z_1\hat{\mathbf{K}} \quad \mathbf{V}_1 = \dot{X}_1\hat{\mathbf{I}} + \dot{Y}_1\hat{\mathbf{J}} + \dot{Z}_1\hat{\mathbf{K}}$$
$$\mathbf{R}_2 = X_2\hat{\mathbf{I}} + Y_2\hat{\mathbf{J}} + Z_2\hat{\mathbf{K}} \quad \mathbf{V}_2 = \dot{X}_2\hat{\mathbf{I}} + \dot{Y}_2\hat{\mathbf{J}} + \dot{Z}_2\hat{\mathbf{K}}$$

EXAMPLE 2.2

A system consists of two massive bodies m_1 and m_2 each having a mass of 10^{26} kg. At time $t = 0$ the state vectors of the two particles in an inertial frame are

$$\mathbf{R}_1^{(0)} = \mathbf{0} \qquad \mathbf{V}_1^{(0)} = 10\hat{\mathbf{I}} + 20\hat{\mathbf{J}} + 30\hat{\mathbf{K}}\,(\text{km/s})$$
$$\mathbf{R}_2^{(0)} = 3000\hat{\mathbf{I}}\,(\text{km}) \quad \mathbf{V}_2^{(0)} = 40\hat{\mathbf{J}}\,(\text{km/s})$$

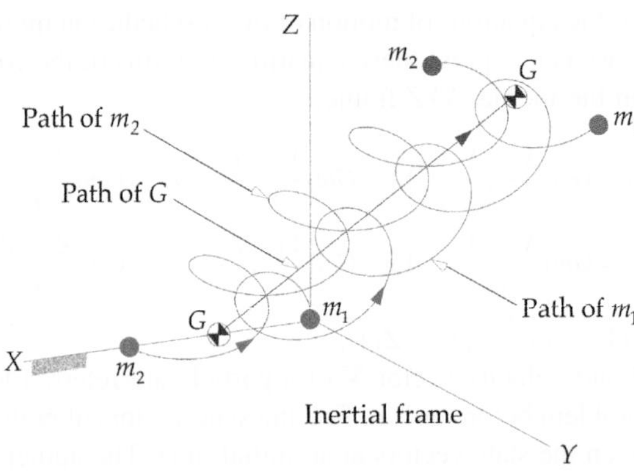

FIG. 2.2

The motion of two identical bodies acted on only by their mutual gravitational attraction, as viewed from the inertial frame of reference.

Use Algorithm 2.1 and the *RKF4(5)* method (Algorithm 1.3) to numerically determine the motion of the two masses due solely to their mutual gravitational attraction from $t = 0$ to $t = 480$ s.

(a) Plot the motion of m_1 and m_2 relative to the inertial frame.
(b) Plot the motion of m_2 and G relative to m_1.
(c) Plot the motion of m_1 and m_2 relative to the center of mass G of the system.

Solution

The MATLAB function *twobody3d.m* in Appendix D.5 contains within it the data for this problem. Embedded in the program is the subfunction *rates*, which computes the accelerations given by Eqs. (2.19a) and (2.19b). *twobody3d.m* uses the solution vector from *rkf45.m* to plot Figs. 2.2 and 2.3, which summarize the results requested in the problem statement.

In answer to part (a), Fig. 2.2 shows the motion of the two-body system relative to the inertial frame. m_1 and m_2 are soon established in a periodic helical motion around the straight-line trajectory of the center of mass G through space. This pattern continues indefinitely.

Fig. 2.3(a) relates to part (b) of the problem. The very same motion appears rather less complex when viewed from m_1. In fact, we see that $\mathbf{R}_2(t) - \mathbf{R}_1(t)$, the trajectory of m_2 relative to m_1, appears to be an elliptical path. So does $\mathbf{R}_G(t) - \mathbf{R}_1(t)$, the path of the center of mass around m_1.

Finally, for part (c) of the problem, Fig. 2.3(b) reveals that both m_1 and m_2 follow apparently elliptical paths around the center of mass.

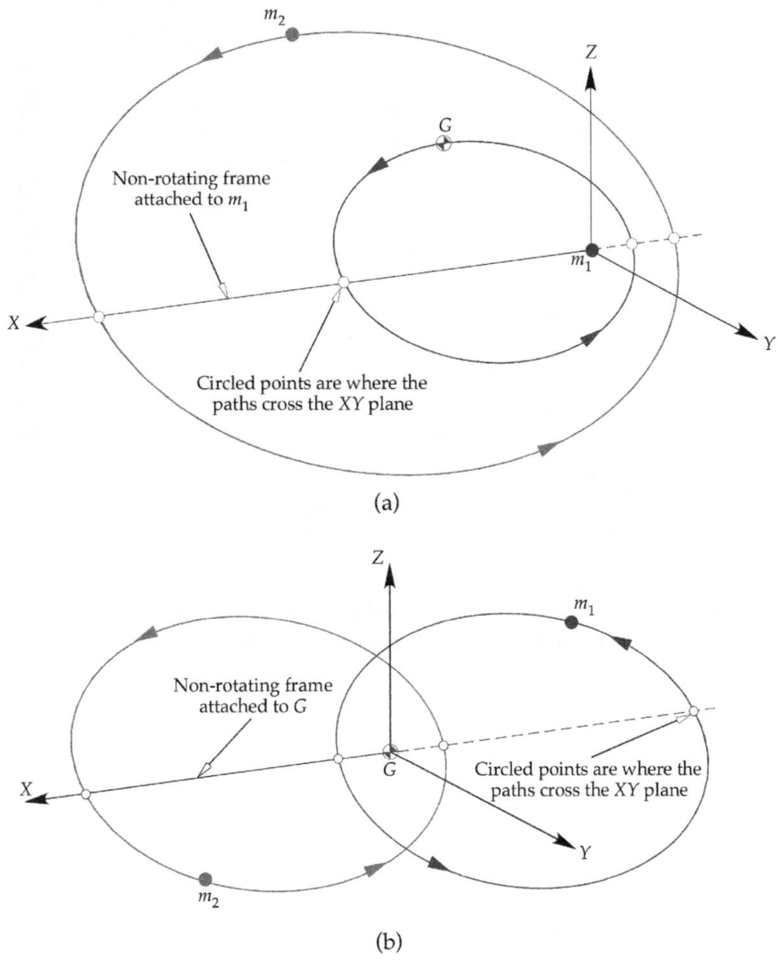

FIG. 2.3

The motion in Fig 2.2: (a) as viewed relative to m_1 (or m_2); (b) as viewed from the center of mass.

One may wonder what the motion looks like if there are more than two bodies moving only under the influence of their mutual gravitational attraction. The n-body problem with $n > 2$ has no closed-form solution, which is complex and chaotic in nature. The three-body problem is briefly addressed in Appendix C, where the equations of motion of the system are presented. Appendix C lists the MATLAB program *threebody.m* that is used to solve the equations of motion for given initial conditions. Fig. 2.4 shows the results for three particles of equal mass, equally spaced initially along the X axis of an inertial frame. The central mass has an initial velocity in the XY plane, while the other

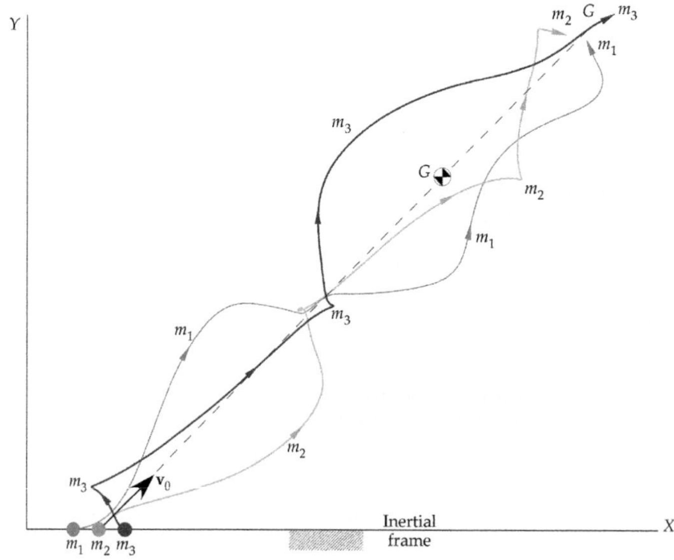

FIG. 2.4

The motion of three identical masses as seen from the inertial frame in which m_1 and m_3 are initially at rest, while m_2 has an initial velocity $\mathbf{v}_0$ directed upward and to the right, as shown.

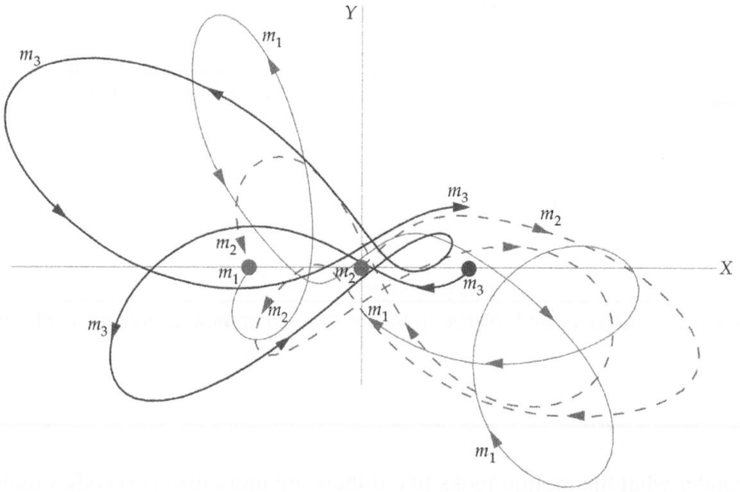

FIG. 2.5

The same motion as Fig. 2.4, as viewed from the inertial frame attached to the center of mass G.

two are at rest. As time progresses, we see no periodic behavior as was evident in the two-body motion in Fig. 2.2. The chaos is more obvious if the motion is viewed from the center of mass of the three-body system, as shown in Fig. 2.5. The computer simulation reveals that the masses all eventually collide.

2.3 EQUATIONS OF RELATIVE MOTION

Let us differentiate Eq. (2.5) twice with respect to time to obtain the relative acceleration vector,

$$\ddot{\mathbf{r}} = \ddot{\mathbf{R}}_2 - \ddot{\mathbf{R}}_1$$

Substituting Eqs. (2.18a) and (2.18b) into the right-hand side of this expression yields

$$\ddot{\mathbf{r}} = -\frac{G(m_1 + m_2)}{r^2}\hat{\mathbf{u}}_r \tag{2.20}$$

The gravitational parameter μ is defined as

$$\mu = G(m_1 + m_2) \tag{2.21}$$

The units of μ are cubic kilometers per square second. Using Eq. (2.21) we can write Eq. (2.20) as

$$\boxed{\ddot{\mathbf{r}} = -\frac{\mu}{r^3}\mathbf{r}} \tag{2.22}$$

This nonlinear second-order differential equation governs the motion of m_2 relative to m_1. It has two vector constants of integration, each having three scalar components. Therefore, Eq. (2.22) has six constants of integration. Note that interchanging the roles of m_1 and m_2 amounts to simply multiplying Eq. (2.22) through by -1, which, of course, changes nothing. Thus, the motion of m_2 as seen from m_1 is precisely the same as the motion of m_1 as seen from m_2. The motion of the moon as observed from the earth appears the same as that of the earth as viewed from the moon.

The relative position vector **r** in Eq. (2.22) was originally defined in the inertial frame (Eq. 2.6). It is convenient, however, to measure the components of **r** in a frame of reference attached to and moving with m_1. In a comoving reference frame, such as the *xyz* system illustrated in Fig. 2.6, **r** has the expression

$$\mathbf{r} = x\hat{\mathbf{i}} + y\hat{\mathbf{j}} + z\hat{\mathbf{k}}$$

The relative velocity $\dot{\mathbf{r}}_{\text{rel}}$ and acceleration $\ddot{\mathbf{r}}_{\text{rel}}$ in the comoving frame are found by simply taking the derivatives of the coefficients of the unit vectors, which themselves are fixed in the moving *xyz* system. Thus,

$$\dot{\mathbf{r}}_{\text{rel}} = \dot{x}\hat{\mathbf{i}} + \dot{y}\hat{\mathbf{j}} + \dot{z}\hat{\mathbf{k}} \quad \ddot{\mathbf{r}}_{\text{rel}} = \ddot{x}\hat{\mathbf{i}} + \ddot{y}\hat{\mathbf{j}} + \ddot{z}\hat{\mathbf{k}}$$

From Eq. (1.69), we know that the relationship between absolute acceleration $\ddot{\mathbf{r}}$ and relative acceleration $\ddot{\mathbf{r}}_{\text{rel}}$ is

$$\ddot{\mathbf{r}} = \ddot{\mathbf{r}}_{\text{rel}} + \dot{\boldsymbol{\Omega}} \times \mathbf{r} + \boldsymbol{\Omega} \times (\boldsymbol{\Omega} \times \mathbf{r}) + 2\boldsymbol{\Omega} \times \dot{\mathbf{r}}_{\text{rel}}$$

where $\boldsymbol{\Omega}$ and $\dot{\boldsymbol{\Omega}}$ are the absolute angular velocity and angular acceleration of the moving frame of reference. Thus, $\ddot{\mathbf{r}} = \ddot{\mathbf{r}}_{\text{rel}}$ only if $\boldsymbol{\Omega} = \dot{\boldsymbol{\Omega}} = \mathbf{0}$. That is to say, the relative acceleration

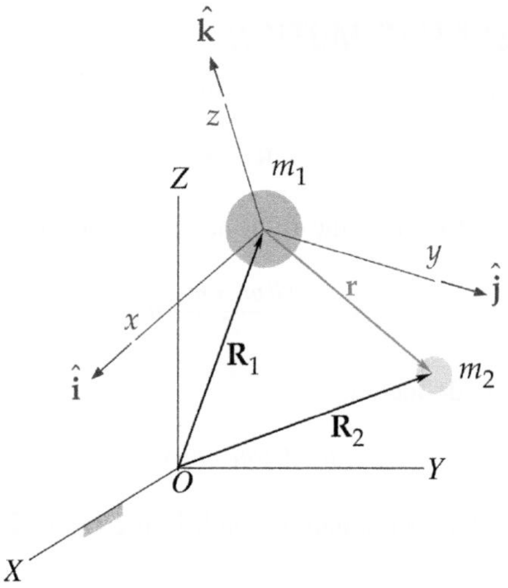

FIG. 2.6

Moving reference frame *xyz* attached to the center of mass of m_1.

may be used on the left of Eq. (2.22) as long as the comoving frame in which it is measured is not rotating.

In the remainder of this chapter and those that follow, the analytical solution of the two-body equation of relative motion (Eq. 2.22) will be presented and applied to a variety of practical problems in orbital mechanics. Pending an analytical solution, we can solve Eq. (2.22) numerically in a manner similar to Algorithm 2.1.

To begin, we imagine a nonrotating Cartesian coordinate system attached to m_1, as illustrated in Fig. 2.6. Resolve $\ddot{\mathbf{r}} = -(\mu/r^3)\mathbf{r}$ into components in this moving frame of reference to obtain the relative acceleration components

$$\ddot{x} = -\frac{\mu}{r^3}x \quad \ddot{y} = -\frac{\mu}{r^3}y \quad \ddot{z} = -\frac{\mu}{r^3}z \tag{2.23}$$

where $r = \sqrt{x^2 + y^2 + z^2}$. The components of the state vector ($\mathbf{r} = x\hat{\mathbf{i}} + y\hat{\mathbf{j}} + z\hat{\mathbf{k}}$, $\mathbf{v} = \dot{x}\hat{\mathbf{i}} + \dot{y}\hat{\mathbf{j}} + \dot{z}\hat{\mathbf{k}}$) are listed in the vector $\mathbf{y}$,

$$\mathbf{y} = \lfloor x \ \ y \ \ z \ \ \dot{x} \ \ \dot{y} \ \ \dot{z} \rfloor$$

The time derivative of this vector comprises the state vector rates,

$$\dot{\mathbf{y}} = \lfloor \dot{x} \ \ \dot{y} \ \ \dot{z} \ \ \ddot{x} \ \ \ddot{y} \ \ \ddot{z} \rfloor$$

where the last three components, the accelerations, are given by Eq. (2.23).

ALGORITHM 2.2

Numerically compute the state vector **r**, **v** of m_1 relative to m_2 as a function of time, given the initial values $\mathbf{r}_0$, $\mathbf{v}_0$. This algorithm is implemented in MATLAB as the function *orbit.m*, which is listed in Appendix D.6.

1. Form the vector comprising the components of the state vector at time t_0,

$$\mathbf{y}_0 = \lfloor x_0 \;\; y_0 \;\; z_0 \;\; \dot{x}_0 \;\; \dot{y}_0 \;\; \dot{z}_0 \rfloor$$

2. Provide the state vector derivatives

$$\mathbf{f}(t, \mathbf{y}) = \left\lfloor \dot{x} \;\; \dot{y} \;\; \dot{z} \;\; -\frac{\mu}{r^3}x \;\; -\frac{\mu}{r^3}y \;\; -\frac{\mu}{r^3}z \right\rfloor$$

   together with $\mathbf{y}_0$ and the final time t_f to Algorithms 1.1, 1.2, or 1.3.

3. The selected algorithm solves the system $\dot{\mathbf{y}} = \mathbf{f}(t, \mathbf{y})$ for the state vector

$$\mathbf{y} = \lfloor x \;\; y \;\; z \;\; \dot{x} \;\; \dot{y} \;\; \dot{z} \rfloor$$

   at n discrete times from t_0 through t_f.

4. The position and velocity at the discrete times are

$$\mathbf{r} = x\hat{\mathbf{i}} + y\hat{\mathbf{j}} + z\hat{\mathbf{k}} \qquad \mathbf{v} = \dot{x}\hat{\mathbf{i}} + \dot{y}\hat{\mathbf{j}} + \dot{z}\hat{\mathbf{k}}$$

EXAMPLE 2.3

Relative to a nonrotating frame of reference with origin at the center of the earth, a 1000-kg satellite's initial position vector is $\mathbf{r} = 8000\hat{\mathbf{i}} + 6000\hat{\mathbf{i}}$ (km), and its initial velocity vector is $\mathbf{v} = 7\hat{\mathbf{j}}$ (km/s). Use Algorithm 2.2 and the *RKF4*(5) method to solve for the path of the spacecraft over the next 4 h. Determine its minimum and maximum distance from the earth's surface during that time.

Solution

The MATLAB function *orbit.m* in Appendix D.6 solves this problem. The initial value of the vector **y** is

$$\mathbf{y}_0 = \lfloor 8000\,\text{km} \;\; 0 \;\; 6000\,\text{km} \;\; 0 \;\; 5\,\text{km/s} \;\; 5\,\text{km/s} \rfloor$$

The program provides these initial conditions to the function *rkf45* (Appendix D.4), which integrates the system $\dot{\mathbf{y}} = \mathbf{f}(t, \mathbf{y})$. *rkf45* uses the function *rates* embedded in *orbit.m* to calculate $\mathbf{f}(t, \mathbf{y})$ at each time step. The command window output of *orbit.m* in Appendix D.6 shows that

The minimum altitude is 3622 km, and the speed at that point is 7 km/s
The maximum altitude is 9560 km, and the speed at that point is 4.39 km/s

The minimum altitude in this case is at the starting point of the orbit. The maximum altitude occurs 2 h later on the opposite side of the earth.

The script *orbit.m* uses some MATLAB plotting features to generate Fig. 2.7. Observe that the orbit is inclined to the equatorial plane and has an apparently elliptical shape. The satellite moves eastwardly in the same direction as the earth's rotation.

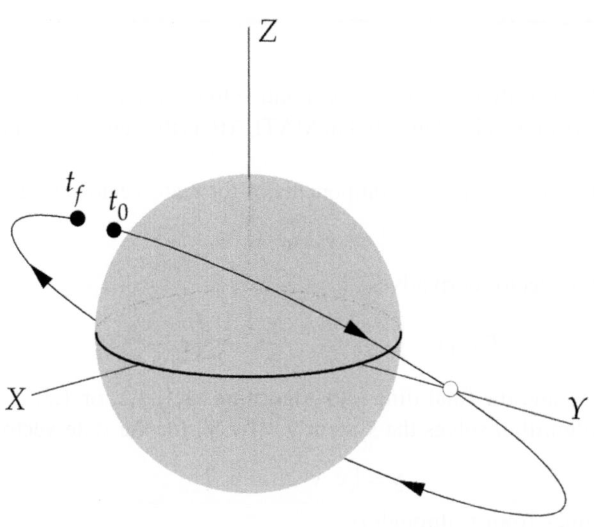

FIG. 2.7

The computed earth orbit. The beginning of the path is labeled t_0 and t_f marks the end of the path 4 h later.

As pointed out earlier, since the center of mass G has zero acceleration, we can use it as the origin of an inertial reference frame. Let $\mathbf{r}_1$ and $\mathbf{r}_2$ be the position vectors of m_1 and m_1, respectively, relative to the center of mass G in Fig. 2.1(a). The equation of motion of m_2 relative to the center of mass is

$$-G\frac{m_1 m_2}{r^2}\hat{\mathbf{u}}_r = m_2\ddot{\mathbf{r}}_2 \tag{2.24}$$

where, as before, r is the magnitude of $\mathbf{r}$, the position vector of m_2 relative to m_1. In terms of $\mathbf{r}_1$ and $\mathbf{r}_2$,

$$\mathbf{r} = \mathbf{r}_2 - \mathbf{r}_1 \tag{2.25}$$

Since the position vector of the center of mass relative to itself is zero, it follows from Eq. (2.1) that

$$m_1\mathbf{r}_1 + m_2\mathbf{r}_2 = \mathbf{0}$$

Therefore,

$$\mathbf{r}_1 = -\frac{m_2}{m_1}\mathbf{r}_2 \tag{2.26}$$

Substituting Eq. (2.26) into Eq. (2.25) yields

$$\mathbf{r} = \frac{m_1 + m_2}{m_1}\mathbf{r}_2$$

Substituting this back into Eq. (2.24) and using the fact that in this case $\hat{\mathbf{u}}_r = \mathbf{r}_2/r_2$, we get

$$-G\frac{{m_1}^3 m_2}{(m_1 + m_2)^2 {r_2}^3}\mathbf{r}_2 = m_2\ddot{\mathbf{r}}_2$$

Upon simplification, this becomes

$$-\left(\frac{m_1}{m_1+m_2}\right)^3 \frac{\mu}{r_2{}^3}\mathbf{r}_2 = \ddot{\mathbf{r}}_2 \tag{2.27}$$

where μ is the gravitational parameter given by Eq. (2.21). If we let

$$\mu' = \left(\frac{m_1}{m_1+m_2}\right)^3 \mu$$

then Eq. (2.27) reduces to

$$\ddot{\mathbf{r}}_2 = -\frac{\mu'}{r_2{}^3}\mathbf{r}_2$$

which is identical in form to Eq. (2.22).

In a similar fashion, the equation of motion of m_1 relative to the center of mass is found to be

$$\ddot{\mathbf{r}}_1 = -\frac{\mu''}{r_1{}^3}\mathbf{r}_1$$

in which

$$\mu'' = \left(\frac{m_2}{m_1+m_2}\right)^3 \mu$$

Since the equations of motion of either particle relative to the center of mass have the same form as the equations of motion relative to either one of the bodies, m_1 or m_2, it follows that the relative motion as viewed from these different perspectives must be similar, as illustrated in Fig. 2.3.

2.4 ANGULAR MOMENTUM AND THE ORBIT FORMULAS

The angular momentum of body m_2 relative to m_1 is the moment of m_2's relative linear momentum $m_2\dot{\mathbf{r}}$ (cf. Eq. 1.45),

$$\mathbf{H}_{2/1} = \mathbf{r} \times m_2\dot{\mathbf{r}}$$

where $\dot{\mathbf{r}} = \mathbf{v}$ is the velocity of m_2 relative to m_1. Let us divide this equation through by m_2 and let $\mathbf{h} = \mathbf{H}_{2/1}/m_2$, so that

$$\boxed{\mathbf{h} = \mathbf{r} \times \dot{\mathbf{r}}} \tag{2.28}$$

where $\mathbf{h}$ is the relative angular momentum of m_2 per unit mass (i.e., the specific relative angular momentum). The units of $\mathbf{h}$ are square kilometers per second.

Taking the time derivative of $\mathbf{h}$ yields

$$\frac{d\mathbf{h}}{dt} = \dot{\mathbf{r}} \times \dot{\mathbf{r}} + \mathbf{r} \times \ddot{\mathbf{r}}$$

But $\dot{\mathbf{r}} \times \dot{\mathbf{r}} = \mathbf{0}$. Furthermore, $\ddot{\mathbf{r}} = -(\mu/r^3)\mathbf{r}$, according to Eq. (2.22), so that

$$\mathbf{r} \times \ddot{\mathbf{r}} = \mathbf{r} \times \left(-\frac{\mu}{r^3}\mathbf{r}\right) = -\frac{\mu}{r^3}(\mathbf{r} \times \mathbf{r}) = \mathbf{0}$$

Therefore, angular momentum is conserved,

$$\frac{d\mathbf{h}}{dt} = \mathbf{0} \quad (\text{or } \mathbf{r} \times \dot{\mathbf{r}} = \text{constant}) \tag{2.29}$$

If the position vector $\mathbf{r}$ and the velocity vector $\dot{\mathbf{r}}$ are parallel, then it follows from Eq. (2.28) that the angular momentum is zero and, according to Eq. (2.29), it remains zero at all points of the trajectory. Zero angular momentum characterizes rectilinear trajectories whereon m_2 moves toward or away from m_1 in a straight line (see Example 1.20).

At any point of a curvilinear trajectory, the position vector $\mathbf{r}$ and the velocity vector $\dot{\mathbf{r}}$ lie in the same plane, as illustrated in Fig. 2.8. Their cross product $\mathbf{r} \times \dot{\mathbf{r}}$ is perpendicular to that plane. Since $\mathbf{r} \times \dot{\mathbf{r}} = \mathbf{h}$, the unit vector normal to the plane is

$$\hat{\mathbf{h}} = \frac{\mathbf{h}}{h} \tag{2.30}$$

By the conservation of angular momentum (Eq. 2.29), this unit vector is constant. Thus, the path of m_2 around m_1 lies in a single plane.

Since the orbit of m_2 around m_1 forms a plane, it is convenient to orient oneself above that plane and look down upon the path, as shown in Fig. 2.9. Let us resolve the relative velocity vector $\dot{\mathbf{r}}$ into components $\mathbf{v}_r = v_r\hat{\mathbf{u}}_r$ and $\mathbf{v}_\perp = v_\perp\hat{\mathbf{u}}_\perp$ along the outward radial from m_1 and perpendicular to it,

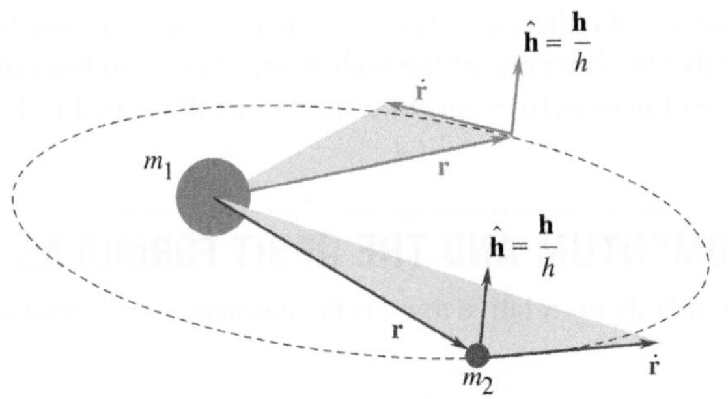

FIG. 2.8

The path of m_2 around m_1 lies in a plane whose normal is defined by $\mathbf{h}$.

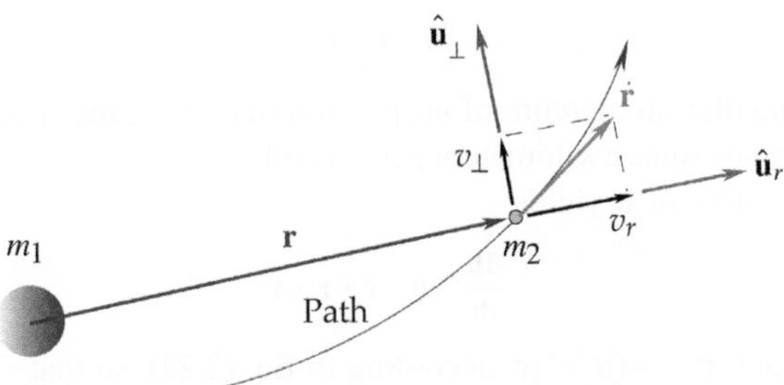

FIG. 2.9

Components of the velocity of m_2, viewed above the plane of the orbit.

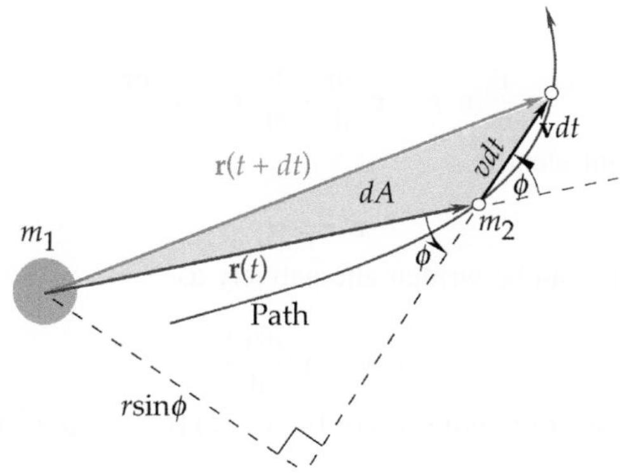

FIG. 2.10

Differential area dA swept out by the relative position vector **r** during time interval dt.

respectively, where $\hat{\mathbf{u}}_r$ and $\hat{\mathbf{u}}_\perp$ are the radial and perpendicular (azimuthal) unit vectors. Then, we can write Eq. (2.28) as

$$\mathbf{h} = \mathbf{r} \times \mathbf{v} = r\hat{\mathbf{u}}_r \times (v_r\hat{\mathbf{u}}_r + v_\perp\hat{\mathbf{u}}_\perp) = rv_\perp\hat{\mathbf{h}}$$

That is,

$$\boxed{h = rv_\perp} \tag{2.31}$$

Clearly, the angular momentum depends only on the azimuthal component of the relative velocity.

During the differential time interval dt the position vector **r** sweeps out an area dA, as shown in Fig. 2.10. From the figure it is clear that the triangular area dA is given by

$$\mathrm{d}A = \frac{1}{2} \times \text{base} \times \text{altitude} = \frac{1}{2} \times v\mathrm{d}t \times r\sin\phi = \frac{1}{2}r(v\sin\phi)\mathrm{d}t = \frac{1}{2}rv_\perp\mathrm{d}t$$

Therefore, using Eq. (2.31) we have

$$\frac{\mathrm{d}A}{\mathrm{d}t} = \frac{h}{2} \tag{2.32}$$

dA/dt is called the areal velocity, and according to Eq. (2.32) it is constant. Named after the German astronomer Johannes Kepler (1571–1630), this result is known as Kepler's second law: equal areas are swept out in equal times.

Before proceeding with an effort to integrate Eq. (2.22), recall the bac–cab rule (Eq. 1.20):

$$\mathbf{A} \times (\mathbf{B} \times \mathbf{C}) = \mathbf{B}(\mathbf{A} \times \mathbf{C}) - \mathbf{C}(\mathbf{A} \times \mathbf{B}) \tag{2.33}$$

Recall as well from Eq. (1.11) that

$$\mathbf{r} \cdot \mathbf{r} = r^2 \tag{2.34}$$

so that

$$\frac{\mathrm{d}}{\mathrm{d}t}(\mathbf{r} \cdot \mathbf{r}) = 2r\frac{\mathrm{d}r}{\mathrm{d}t}$$

But

$$\frac{\mathrm{d}}{\mathrm{d}t}(\mathbf{r}\cdot\mathbf{r})=\mathbf{r}\cdot\frac{\mathrm{d}\mathbf{r}}{\mathrm{d}t}+\frac{\mathrm{d}\mathbf{r}}{\mathrm{d}t}\cdot\mathbf{r}=2\mathbf{r}\cdot\frac{\mathrm{d}\mathbf{r}}{\mathrm{d}t}$$

Thus, we obtain the important identity

$$\mathbf{r}\cdot\dot{\mathbf{r}}=r\dot{r} \tag{2.35a}$$

Since $\dot{\mathbf{r}}=\mathbf{v}$ and $r=\|\mathbf{r}\|$, this can be written alternatively as

$$\mathbf{r}\cdot\mathbf{v}=\|\mathbf{r}\|\frac{\mathrm{d}\|\mathbf{r}\|}{\mathrm{d}t} \tag{2.35b}$$

Now let us take the cross product of both sides of Eq. (2.22) [$\ddot{\mathbf{r}}=-(\mu/r^3)\mathbf{r}$] with the specific angular momentum $\mathbf{h}$:

$$\ddot{\mathbf{r}}\times\mathbf{h}=-\frac{\mu}{r^3}\mathbf{r}\times\mathbf{h} \tag{2.36}$$

Since $\frac{\mathrm{d}}{\mathrm{d}t}(\dot{\mathbf{r}}\times\mathbf{h})=\ddot{\mathbf{r}}\times\mathbf{h}+\dot{\mathbf{r}}\times\dot{\mathbf{h}}$, the left-hand side can be written as

$$\ddot{\mathbf{r}}\times\mathbf{h}=\frac{\mathrm{d}}{\mathrm{d}t}(\dot{\mathbf{r}}\times\mathbf{h})-\dot{\mathbf{r}}\times\dot{\mathbf{h}}$$

But, according to Eq. (2.29), the angular momentum is constant ($\dot{\mathbf{h}}=\mathbf{0}$), so this reduces to

$$\ddot{\mathbf{r}}\times\mathbf{h}=\frac{\mathrm{d}}{\mathrm{d}t}(\dot{\mathbf{r}}\times\mathbf{h}) \tag{2.37}$$

The right-hand side of Eq. (2.36) can be transformed by the following sequence of substitutions:

$$\begin{aligned}
\frac{1}{r^3}\mathbf{r}\times\mathbf{h}&=\frac{1}{r^3}[\mathbf{r}\times(\mathbf{r}\times\dot{\mathbf{r}})] \quad (\text{Eq. 2.18 } [\mathbf{h}=\mathbf{r}\times\dot{\mathbf{r}}])\\
&=\frac{1}{r^3}[\mathbf{r}(\mathbf{r}\cdot\dot{\mathbf{r}})-\dot{\mathbf{r}}(\mathbf{r}\cdot\mathbf{r})] \quad (\text{Eq. 2.23 [bac} - \text{cab rule]})\\
&=\frac{1}{r^3}\left[\mathbf{r}(r\dot{r})-\dot{\mathbf{r}}r^2\right] \quad (\text{Eqs. 2.34 and 2.35a})\\
&=\frac{\mathbf{r}\dot{r}-\dot{\mathbf{r}}r}{r^2}
\end{aligned}$$

But

$$\frac{\mathrm{d}}{\mathrm{d}t}\left(\frac{\mathbf{r}}{r}\right)=\frac{r\dot{\mathbf{r}}-\mathbf{r}\dot{r}}{r^2}=-\frac{\mathbf{r}\dot{r}-r\dot{\mathbf{r}}}{r^2}$$

Therefore,

$$\frac{1}{r^3}\mathbf{r}\times\mathbf{h}=-\frac{\mathrm{d}}{\mathrm{d}t}\left(\frac{\mathbf{r}}{r}\right) \tag{2.38}$$

Substituting Eqs. (2.37) and (2.38) into Eq. (2.36), we get

$$\frac{\mathrm{d}}{\mathrm{d}t}(\dot{\mathbf{r}}\times\mathbf{h})=\frac{\mathrm{d}}{\mathrm{d}t}\left(\mu\frac{\mathbf{r}}{r}\right)$$

or

$$\frac{\mathrm{d}}{\mathrm{d}t}\left(\dot{\mathbf{r}}\times\mathbf{h}-\mu\frac{\mathbf{r}}{r}\right)=\mathbf{0}$$

That is,

$$\dot{\mathbf{r}} \times \mathbf{h} - \mu \frac{\mathbf{r}}{r} = \mathbf{C} \tag{2.39}$$

where the vector **C**, called the Laplace vector after the French mathematician Pierre-Simon Laplace (1749–1827), is a constant having the dimensions of μ. Eq. (2.39) is the first integral of the equation of motion, $\ddot{\mathbf{r}} = -(\mu/r^3)\mathbf{r}$. Taking the dot product of both sides of Eq. (2.39) with the vector **h** yields a scalar equation

$$(\dot{\mathbf{r}} \times \mathbf{h}) \cdot \mathbf{h} - \mu \frac{\mathbf{r} \cdot \mathbf{h}}{r} = \mathbf{C} \cdot \mathbf{h}$$

Since $\dot{\mathbf{r}} \times \mathbf{h}$ is perpendicular to both $\dot{\mathbf{r}}$ and $\mathbf{h}$, it follows that $(\dot{\mathbf{r}} \times \mathbf{h}) \cdot \mathbf{h} = 0$. Likewise, since $\mathbf{h} = \mathbf{r} \times \dot{\mathbf{r}}$ is perpendicular to both $\mathbf{r}$ and $\dot{\mathbf{r}}$, it is true that $\mathbf{r} \cdot \mathbf{h} = 0$. Therefore, we have $\mathbf{C} \cdot \mathbf{h} = 0$ (i.e., **C** is perpendicular to **h**, which is normal to the orbital plane). That of course means that the Laplace vector must lie in the orbital plane.

Let us rearrange Eq. (2.39) and write it as

$$\frac{\mathbf{r}}{r} + \mathbf{e} = \frac{\dot{\mathbf{r}} \times \mathbf{h}}{\mu} \tag{2.40}$$

where $\mathbf{e} = \mathbf{C}/\mu$. The dimensionless vector **e** is called the eccentricity vector. The line defined by the vector **e** is commonly called the apse line. To obtain a scalar equation, let us take the dot product of both sides of Eq. (2.40) with **r**:

$$\frac{\mathbf{r} \cdot \mathbf{r}}{r} + \mathbf{r} \cdot \mathbf{e} = \frac{\mathbf{r} \cdot (\dot{\mathbf{r}} \times \mathbf{h})}{\mu} \tag{2.41}$$

We can simplify the right-hand side by employing the vector identity presented in Eq. (1.21),

$$\mathbf{A} \cdot (\mathbf{B} \times \mathbf{C}) = (\mathbf{A} \times \mathbf{B}) \cdot \mathbf{C} \tag{2.42}$$

from which we obtain

$$\mathbf{r} \cdot (\dot{\mathbf{r}} \times \mathbf{h}) = (\mathbf{r} \times \dot{\mathbf{r}}) \cdot \mathbf{h} = \mathbf{h} \cdot \mathbf{h} = h^2 \tag{2.43}$$

Substituting this expression into the right-hand side of Eq. (2.41), and substituting $\mathbf{r} \cdot \mathbf{r} = r^2$ on the left yields

$$r + \mathbf{r} \cdot \mathbf{e} = \frac{h^2}{\mu} \tag{2.44}$$

Observe that by following the steps leading from Eqs. (2.40) to (2.44) we have lost track of the variable time. This occurred at Eq. (2.43), because h is constant. Finally, from the definition of the dot product we have

$$\mathbf{r} \cdot \mathbf{e} = re\cos\theta$$

where e is the eccentricity (the magnitude of the eccentricity vector **e**), and θ is the true anomaly. θ is the angle between the fixed vector **e** and the variable position vector **r**, as illustrated in Fig. 2.11. (Other symbols used to represent true anomaly include the Greek letters ν and ϕ and the Latin letters f and v.) In terms of the eccentricity and the true anomaly, we may therefore write Eq. (2.44) as

$$r + re\cos\theta = \frac{h^2}{\mu}$$

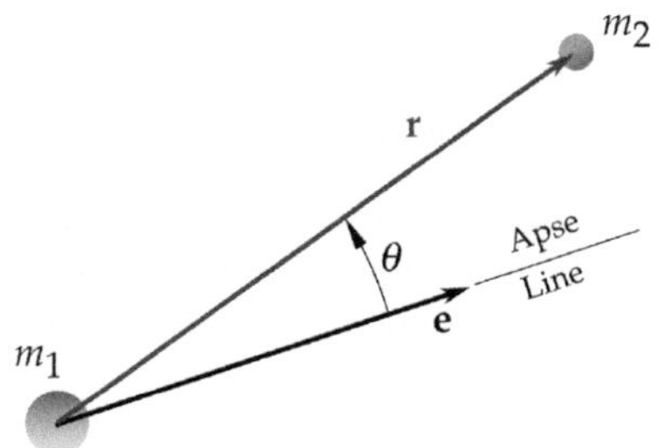

FIG. 2.11

The true anomaly θ is the angle between the eccentricity vector **e** and the position vector **r**.

or

$$\boxed{r = \frac{h^2}{\mu} \frac{1}{1 + e\cos\theta}} \tag{2.45}$$

This is the orbit equation, and it defines the path of the body m_2 around m_1, relative to m_1. Remember that μ, h, and e are constants. Observe as well that there is no significance to negative values of eccentricity (i.e., $e \geq 0$). Since the orbit equation describes conic sections, including ellipses, it is a mathematical statement of Kepler's first law (namely, that the planets follow elliptical paths around the sun). Two-body orbits are often referred to as Keplerian orbits.

In Section 2.3, it was pointed out that integration of the equation of relative motion (Eq. 2.22) leads to six constants of integration. In this section, it would seem that we have arrived at those constants (namely, the three components of the angular momentum **h** and the three components of the eccentricity vector **e**). However, we showed that **h** is perpendicular to **e**. This places a condition (namely, $\mathbf{h} \cdot \mathbf{e} = 0$) on the components of **h** and **e**, so that we really have just five independent constants of integration. The sixth constant of motion will arise when we work time back into the picture in the next chapter.

The angular velocity of the position vector **r** is $\dot{\theta}$, the rate of change of the true anomaly. The component of velocity normal to the position vector is found in terms of the angular velocity by the formula

$$v_\perp = r\dot{\theta} \tag{2.46}$$

Substituting this into Eq. (2.31) ($h = rv_\perp$) yields the specific angular momentum in terms of the angular velocity,

$$h = r^2\dot{\theta} \tag{2.47}$$

It is convenient to have formulas for computing the radial and azimuthal components of velocity, shown in Fig. 2.12. From $h = rv_\perp$ we of course obtain

$$v_\perp = \frac{h}{r}$$

Substituting r from Eq. (2.45) readily yields

$$v_\perp = \frac{\mu}{h}(1 + e\cos\theta) \tag{2.48}$$

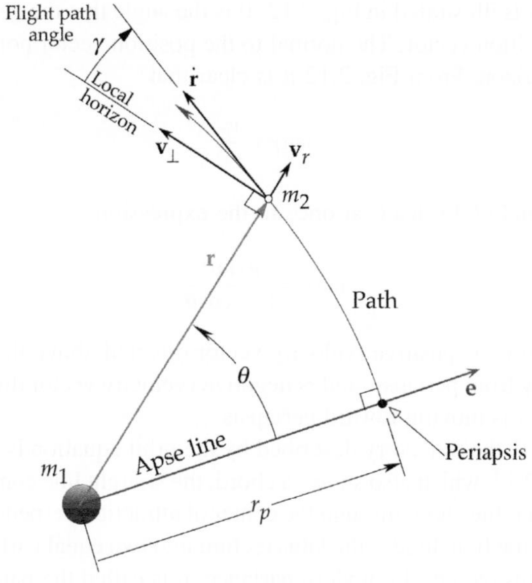

FIG. 2.12

Position and velocity of m_2 in polar coordinates centered at m_1, with the eccentricity vector being the reference for true anomaly (polar angle) θ. γ is the flight path angle.

Since $v_r = \dot{r}$, we take the derivative of Eq. (2.45) to get

$$\dot{r} = \frac{dr}{dt} = \frac{d}{dt}\left[\frac{h^2}{\mu}\frac{1}{1+e\cos\theta}\right] = \frac{h^2}{\mu}\left[-\frac{e(-\dot{\theta}\sin\theta)}{(1+e\cos\theta)^2}\right] = \frac{h^2}{\mu}\frac{e\sin\theta}{(1+e\cos\theta)^2}\frac{h}{r^2}$$

where we made use of the fact that $\dot{\theta} = h/r^2$, from Eq. (2.47). Substituting Eq. (2.45) once again and simplifying finally yields

$$\boxed{v_r = \frac{\mu}{h}e\sin\theta} \tag{2.49}$$

We see from Eq. (2.45) that m_2 comes closest to m_1 (r is smallest) when $\theta = 0$ (unless $e = 0$, in which case the distance between m_1 and m_2 is constant). The point of closest approach lies on the apse line and is called periapsis. The distance r_p to periapsis, as shown in Fig. 2.12, is obtained by setting the true anomaly equal to zero,

$$r_p = \frac{h^2}{\mu}\frac{1}{1+e} \tag{2.50}$$

From Eq. (2.49) it is clear that the radial component of velocity is zero at periapsis. For $0 < \theta < 180°$, v_r is positive, which means m_2 is moving *away* from periapsis. On the other hand, Eq. (2.49) shows that if $180° < \theta < 360°$, then v_r is negative, which means m_2 is moving *toward* periapsis.

The flight path angle γ is illustrated in Fig. 2.12. It is the angle that the velocity vector $\mathbf{v} = \dot{\mathbf{r}}$ makes with the normal to the position vector. The normal to the position vector points in the direction of $\mathbf{v}_\perp$, and is called the local horizon. From Fig. 2.12 it is clear that

$$\tan\gamma = \frac{v_r}{v_\perp} \tag{2.51}$$

Substituting Eqs. (2.48) and (2.49) leads at once to the expression

$$\tan\gamma = \frac{e\sin\theta}{1 + e\cos\theta} \tag{2.52}$$

The flight path angle, like v_r, is positive (velocity vector directed above the local horizon) when the spacecraft is moving away from periapsis and is negative (velocity vector directed below the local horizon) when the spacecraft is moving toward periapsis.

Since $\cos(-\theta) = \cos\theta$, the trajectory described by the orbit equation is symmetric about the apse line, as illustrated in Fig. 2.13, which also shows a chord, the straight line connecting any two points on the orbit. The latus rectum is the chord through the center of attraction perpendicular to the apse line. By symmetry, the center of attraction divides the latus rectum into two equal parts, each of length p, known historically as the semilatus rectum. In modern parlance, p is called the parameter of the orbit. From Eq. (2.45) it is apparent that

$$p = \frac{h^2}{\mu} \tag{2.53}$$

Since the curvilinear path of m_2 around m_1 lies in a plane, for the time being we will for simplicity continue to view the trajectory from above the plane. Unless there is a reason to do otherwise, we will assume that the eccentricity vector points to the right and that m_2 moves counterclockwise around m_1, which means that the true anomaly is measured positive counterclockwise, consistent with the usual polar coordinate sign convention.

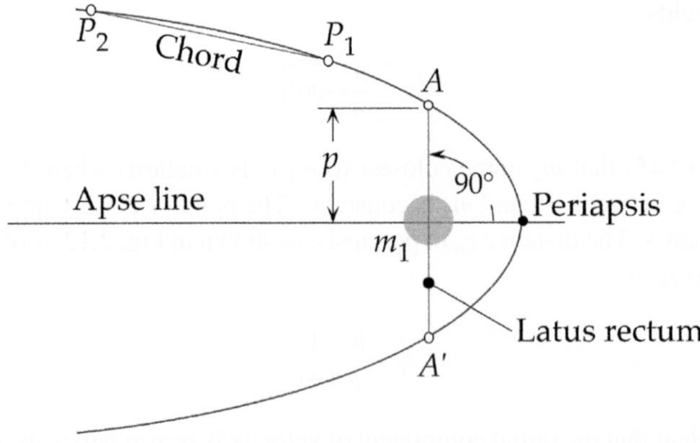

FIG. 2.13

Illustration of latus rectum, semilatus rectum p, and the chord between any two points on an orbit.

2.5 THE ENERGY LAW

By taking the cross product of Eq. (2.22), $\ddot{\mathbf{r}} = -(\mu/r^3)\mathbf{r}$ (Newton's second law of motion), with the relative angular momentum per unit mass $\mathbf{h}$, we were led to Eq. (2.39), and from that we obtained the orbit formula (i.e., Eq. 2.45). Now let us see what results from taking the *dot* product of Eq. (2.22) with the relative *linear* momentum per unit mass. The relative linear momentum per unit mass is just the relative velocity,

$$\frac{m_2\dot{\mathbf{r}}}{m_2} = \dot{\mathbf{r}}$$

Thus, carrying out the dot product in Eq. (2.22) yields

$$\ddot{\mathbf{r}} \cdot \dot{\mathbf{r}} = -\mu\frac{\mathbf{r} \cdot \dot{\mathbf{r}}}{r^3} \tag{2.54}$$

For the left-hand side we observe that

$$\ddot{\mathbf{r}} \cdot \dot{\mathbf{r}} = \frac{1}{2}\frac{\mathrm{d}}{\mathrm{d}t}(\dot{\mathbf{r}} \cdot \dot{\mathbf{r}}) = \frac{1}{2}\frac{\mathrm{d}}{\mathrm{d}t}(\mathbf{v} \cdot \mathbf{v}) = \frac{1}{2}\frac{\mathrm{d}}{\mathrm{d}t}(v^2) = \frac{\mathrm{d}}{\mathrm{d}t}\left(\frac{v^2}{2}\right) \tag{2.55}$$

For the right-hand side of Eq. (2.54) we have, recalling that $\mathbf{r} \cdot \dot{\mathbf{r}} = r\dot{r}$ and that $\mathrm{d}(1/r)/\mathrm{d}t = (-1/r^2)\mathrm{d}r/\mathrm{d}t$,

$$\mu\frac{\mathbf{r} \cdot \dot{\mathbf{r}}}{r^3} = \mu\frac{r\dot{r}}{r^3} = \mu\frac{\dot{r}}{r^2} = -\frac{\mathrm{d}}{\mathrm{d}t}\left(\frac{\mu}{r}\right) \tag{2.56}$$

Substituting Eqs. (2.55) and (2.56) into Eq. (2.54) yields

$$\frac{\mathrm{d}}{\mathrm{d}t}\left(\frac{v^2}{2} - \frac{\mu}{r}\right) = 0$$

or

$$\frac{v^2}{2} - \frac{\mu}{r} = \varepsilon \text{ (constant)} \tag{2.57}$$

where ε is a constant, $v^2/2$ is the relative kinetic energy per unit mass, and $(-\mu/r)$ is the potential energy per unit mass of the body m_2 in the gravitational field of m_1. The total mechanical energy per unit mass ε is the sum of the kinetic and potential energies per unit mass. Eq. (2.57) is a statement of the conservation of energy (namely, that the specific mechanical energy is the same at all points of the trajectory). Eq. (2.57) is also known as the vis viva ("living force") equation. It is valid for any trajectory, including rectilinear ones.

For curvilinear trajectories, we can evaluate the constant ε at periapsis ($\theta = 0$),

$$\varepsilon = \varepsilon_p = \frac{v_p^2}{2} - \frac{\mu}{r_p} \tag{2.58}$$

where r_p and v_p are the position and speed at periapsis. Since $v_r = 0$ at periapsis, the only component of velocity is $v_\perp$, which means $v_p = v_\perp = h/r_p$. Thus,

$$\varepsilon = \frac{1}{2}\frac{h^2}{r_p^2} - \frac{\mu}{r_p} \tag{2.59}$$

Substituting the formula for periapse radius (Eq. 2.50) into Eq. (2.59) yields an expression for the orbital specific energy in terms of the orbital constants h and e,

$$\varepsilon = -\frac{1}{2}\frac{\mu^2}{h^2}\left(1 - e^2\right) \tag{2.60}$$

Clearly, the orbital energy is not an independent orbital parameter.

Note that the energy $\mathcal{E}$ of a spacecraft of mass m is obtained from the specific energy ε by the formula

$$\mathcal{E} = m\varepsilon \tag{2.61}$$

2.6 CIRCULAR ORBITS ($e = 0$)

Setting $e = 0$ in the orbital equation $r = (h^2/\mu)/(1 + e\cos\theta)$ yields

$$r = \frac{h^2}{\mu} \tag{2.62}$$

That is, $r =$ constant, which means the orbit of m_2 around m_1 is a circle. Since the radial velocity $\dot{r}$ is zero, it follows that $v = v_\perp$ so that the angular momentum formula $h = rv_\perp$ becomes simply $h = rv$ for a circular orbit. Substituting this expression for h into Eq. (2.62) and solving for v yields the velocity of a circular orbit,

$$v_{\text{circular}} = \sqrt{\frac{\mu}{r}} \tag{2.63}$$

The time T required for one orbit is known as the period. Because the speed is constant, the period of a circular orbit is easy to compute.

$$T = \frac{\text{Circumference}}{\text{Speed}} = \frac{2\pi r}{\sqrt{\mu/r}}$$

so that

$$T_{\text{circular}} = \frac{2\pi}{\sqrt{\mu}} r^{3/2} \tag{2.64}$$

The specific energy of a circular orbit is found by setting $e = 0$ in Eq. (2.60),

$$\varepsilon = -\frac{1}{2}\frac{\mu^2}{h^2}$$

Employing Eq. (2.62) yields

$$\varepsilon_{\text{circular}} = -\frac{\mu}{2r} \tag{2.65}$$

Obviously, the energy of a circular orbit is negative. As the radius goes up, the energy becomes less negative (i.e., it increases). In other words, the larger the orbit is, the greater is its energy.

To launch a satellite from the surface of the earth into a circular orbit requires increasing its specific energy ε. This energy comes from the rocket motors of the launch vehicle. Since the energy of a satellite of mass m is $\mathcal{E} = m\varepsilon$, a propulsion system that can place a large mass in a low earth orbit (LEO) can place a smaller mass in a higher earth orbit.

The space shuttle orbiters were the largest man-made satellites so far placed in orbit with a single launch vehicle. For example, on NASA mission STS-82 in February 1997, the orbiter *Discovery* rendezvoused with the Hubble Space Telescope to repair and refurbish it. The altitude of the nearly circular orbit was 580 km (360 miles). *Discovery*'s orbital mass early in the mission was 106,000 kg (117 ton). That was only 6% of the total mass of the shuttle prior to launch (comprising the orbiter's dry mass, plus that of its payload and fuel, plus the two solid rocket boosters (SRBs), plus the external fuel tank filled with liquid hydrogen and oxygen). This mass of about 2 million kilograms (2200 ton) was lifted off the launchpad by a total thrust in the vicinity of 35,000 kN (7.8 million pounds). Eighty-five percent of the thrust was furnished by the SRB's, which were depleted and jettisoned about two minutes into the flight. The remaining thrust came from the three liquid rockets (space shuttle main engines (SSMEs)) on the orbiter. These were fueled by the external tank, which was jettisoned just after the SSMEs were shut down at MECO (main engine cut off), about 8.5 min after liftoff.

Manned orbital spacecraft and a host of unmanned remote-sensing, imaging and navigation satellites occupy nominally circular LEOs. An LEO is one whose altitude lies between about 150 km (100 miles) and about 2000 km (1200 miles). An LEO is well above the nominal outer limits of the drag-producing atmosphere (about 80 km or 50 miles), and well below the hazardous Van Allen radiation belts, the innermost of which begins at about 2400 km (1500 miles).

Nearly all our applications of the orbital equations will be for the analysis of man-made spacecraft, all of which have a mass that is insignificant compared with the sun and planets. For example, since the earth is nearly 20 orders of magnitude more massive than the largest conceivable artificial satellite, the center of mass of the two-body system lies at the center of the earth, and the constant μ in Eq. (2.21) becomes

$$\mu=G(m_{\text{earth}}+m_{\text{satellite}})=Gm_{\text{earth}}$$

The value of the earth's gravitational parameter to be used throughout this book is found in Table A.2,

$$\mu_{\text{earth}}=398,600\,\text{km}^3/\text{s}^2 \tag{2.66}$$

EXAMPLE 2.4

Plot the speed v and period T of a satellite in a circular LEO as a function of altitude z.

Solution

Eqs. (2.63) and (2.64) give the speed and period, respectively, of the satellite:

$$v=\sqrt{\frac{\mu}{r}}=\sqrt{\frac{\mu}{R_E+z}}=\sqrt{\frac{398,600}{6378+z}} \qquad T=\frac{2\pi}{\sqrt{\mu}}r^{\frac{3}{2}}=\frac{2\pi}{\sqrt{398,600}}(6378+z)^{\frac{3}{2}}$$

These relations are graphed in Fig. 2.14.

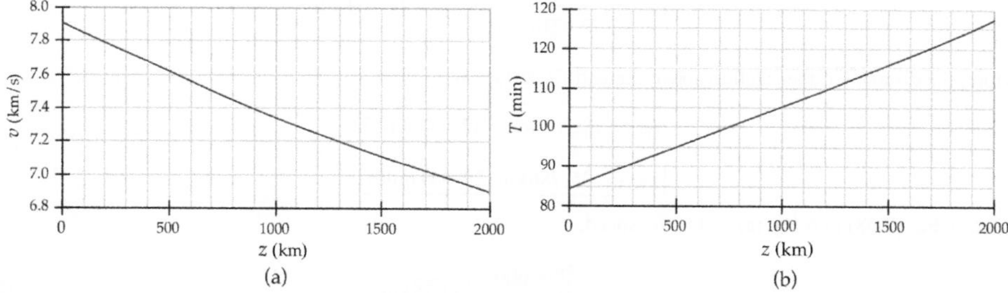

FIG. 2.14

Circular orbital speed (a) and period (b) as a function of altitude.

If a satellite remains always above the same point on the earth's equator, then it is in a circular, geostationary equatorial orbit (GEO). For GEO, the radial from the center of the earth to the satellite must have the same angular velocity as the earth itself (namely, 2π radians per sidereal day). A sidereal day is the time it takes the earth to complete one rotation relative to inertial space (the fixed stars). The ordinary 24-h day, or synodic day, is the time it takes the sun to apparently rotate once around the earth, from high noon one day to high noon the next. Synodic and sidereal days would be identical if the earth stood still in space. However, while the earth makes one absolute rotation around its axis, it advances $2\pi/365.26$ radians along its solar orbit. Therefore, its inertial angular velocity ω_E is [($2\pi + 2\pi/365.26$) radians]/(24 h). That is,

$$\boxed{\omega_E = 72.9218\left(10^{-6}\right)\,\text{rad/s}} \tag{2.67}$$

Communication satellites and global weather satellites are placed in geostationary orbit because of the large portion of the earth's surface visible from that altitude and the fact that ground stations do not have to track the satellite, which appears motionless in the sky.

EXAMPLE 2.5

Calculate the altitude z_{GEO} and speed v_{GEO} of a geostationary earth satellite.

Solution

From Eq. (2.63), the speed of the satellite in its circular GEO of radius r_{GEO} is

$$v_{\text{GEO}} = \sqrt{\frac{\mu}{r_{\text{GEO}}}} \tag{a}$$

On the other hand, the speed v_{GEO} along its circular path is related to the absolute angular velocity ω_E of the earth by the kinematics formula

$$v_{\text{GEO}} = \omega_E r_{\text{GEO}}$$

Equating these two expressions and solving for r_{GEO} yields

$$r_{\text{GEO}} = \sqrt[3]{\frac{\mu}{\omega_E{}^2}}$$

Substituting Eqs. (2.66) and (2.67), we get

$$r_{\text{GEO}} = \sqrt[3]{\frac{398,600}{\left(72.9218\times 10^{-6}\right)^2}} = 42,164\,\text{km} \tag{2.68}$$

Therefore, the distance of the satellite above the earth's surface is

$$z_{\text{GEO}} = r_{\text{GEO}} - R_E = 42,164 - 6378$$

$$\boxed{z_{\text{GEO}} = 35,786\,\text{km}\,(22,241\,\text{miles})}$$

Substituting Eq. (2.68) into Eq. (a) yields the speed,

$$v_{\text{GEO}} = \sqrt{\frac{398,600}{42,164}} = \boxed{3.075\,\text{km/s}} \tag{2.69}$$

EXAMPLE 2.6

Calculate the maximum latitude and the percentage of the earth's surface visible from GEO.

Solution

To find the maximum viewable latitude ϕ, use Fig. 2.15, from which it is apparent that

$$\phi = \cos^{-1}\frac{R_E}{r} \tag{a}$$

where $R_E = 6378\,\text{km}$ and, according to Eq. (2.68), $r = 42,164\,\text{km}$. Therefore,

$$\phi = \cos^{-1}\frac{6378}{42,164}$$

$$\boxed{\phi = 81.30^\circ} \text{ Maximum visible north or south latitude} \tag{b}$$

The surface area A visible from the GEO is the shaded region illustrated in Fig. 2.16. It can be shown that the area A is given by

$$A = 2\pi R_E^2(1-\cos\phi)$$

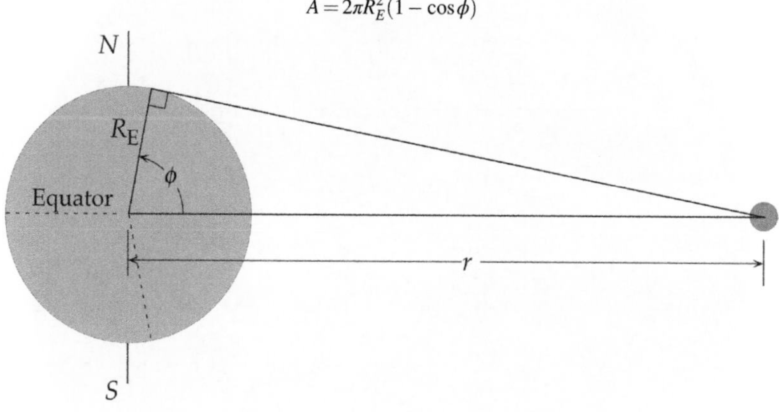

FIG. 2.15

Satellite in GEO.

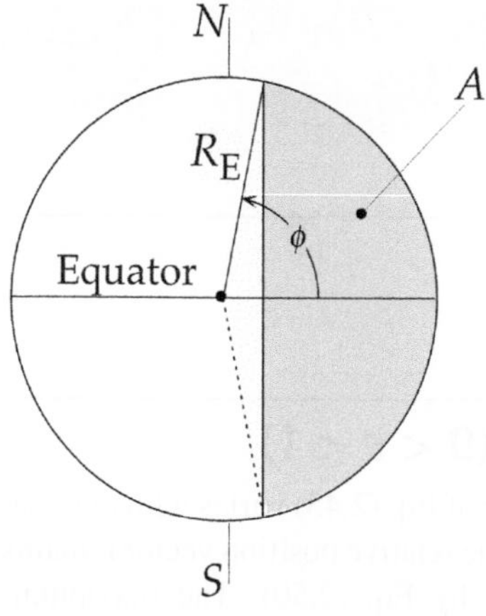

FIG. 2.16

Surface area A visible from GEO.

where $2\pi R_E^2$ is the area of the hemisphere. Therefore, the percentage of the hemisphere visible from the GEO is

$$\frac{A}{2\pi R_E^2} \times 100 = (1 - \cos 81.30°) \times 100 = 84.9\%$$

which of course means that $\boxed{42.4\%}$ of the total surface of the earth can be seen from GEO.

Fig. 2.17 is a photograph taken from geosynchronous equatorial orbit by one of the National Oceanic and Atmospheric Administration's Geostationary Operational Environmental Satellites (GOES).

FIG. 2.17

The view from GEO.

2.7 ELLIPTICAL ORBITS ($0 < e < 1$)

If $0 < e < 1$, then the denominator of Eq. (2.45) varies with the true anomaly θ, but it remains positive, never becoming zero. Therefore, the relative position vector remains bounded, having its smallest magnitude at the periapsis r_p, given by Eq. (2.50). The maximum value of r is reached when the

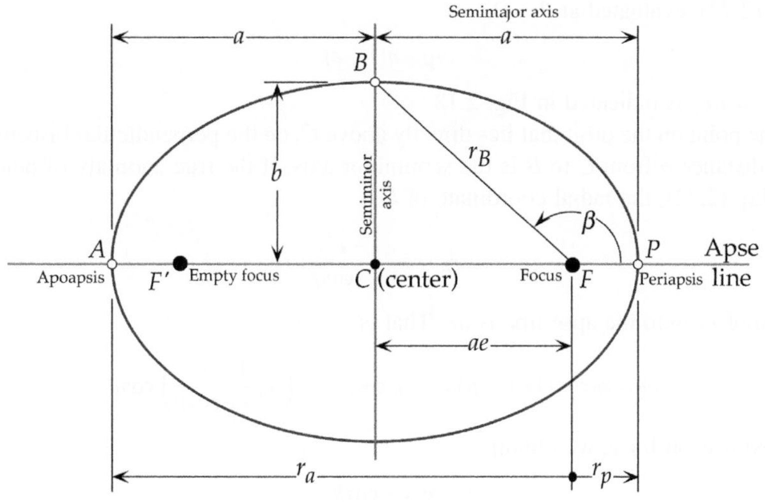

FIG. 2.18

Elliptical orbit. m_1 is at the focus F. F' is the unoccupied empty focus.

denominator of $r = (h^2/\mu)/(1 + e \cos\theta)$ obtains its minimum value, which occurs at $\theta = 180°$. That point is called the apoapsis, and its radial coordinate, denoted by r_a, is

$$r_a = \frac{h^2}{\mu}\frac{1}{1-e} \tag{2.70}$$

The curve defined by Eq. (2.45) in this case is an ellipse.

Let $2a$ be the distance measured along the apse line from periapsis P to apoapsis A, as illustrated in Fig. 2.18. Then,

$$2a = r_p + r_a$$

Substituting Eqs. (2.50) and (2.70) into this expression we get

$$a = \frac{h^2}{\mu}\frac{1}{1-e^2} \tag{2.71}$$

where a is the semimajor axis of the ellipse. Solving Eq. (2.71) for h^2/μ and putting the result into Eq. (2.45) yields an alternative form of the orbit equation,

$$r = a\frac{1-e^2}{1+e\cos\theta} \tag{2.72}$$

In Fig. 2.18, let F denote the location of the body m_1, which is the origin of the r, θ polar coordinate system. The center C of the ellipse is the point lying midway between the apoapsis and the periapsis. The distance CF from the center C to the focus F is

$$CF = a - FP = a - r_p$$

But from Eq. (2.72), evaluated at $\theta = 0$,

$$r_p = a(1 - e) \tag{2.73}$$

Therefore, $CF = ae$, as indicated in Fig. 2.18.

Let B be the point on the orbit that lies directly above C, on the perpendicular bisector of the major axis AP. The distance b from C to B is the semiminor axis. If the true anomaly of point B is β, then according to Eq. (2.72), the radial coordinate of B is

$$r_B = a\frac{1 - e^2}{1 + e\cos\beta} \tag{2.74}$$

The projection of r_B onto the apse line is ae. That is,

$$ae = r_B\cos\left(180^\circ - \beta\right) = -r_B\cos\beta = -\left(a\frac{1 - e^2}{1 + e\cos\beta}\right)\cos\beta$$

Solving this expression for e, we obtain

$$e = -\cos\beta \tag{2.75}$$

Substituting this result into Eq. (2.74) reveals the interesting fact that

$$r_B = a$$

According to the Pythagorean theorem,

$$b^2 = r_B^2 - (ae)^2 = a^2 - a^2e^2$$

which means that the semiminor axis is found in terms of the semimajor axis and the eccentricity of the ellipse as

$$\boxed{b = a\sqrt{1 - e^2}} \tag{2.76}$$

Let an xy Cartesian coordinate system be centered at C, as shown in Fig. 2.19. In terms of r and θ, we see from the figure that the x coordinate of a point on the orbit is

$$x = ae + r\cos\theta = ae + \left(a\frac{1 - e^2}{1 + e\cos\theta}\right)\cos\theta = a\frac{e + \cos\theta}{1 + e\cos\theta}$$

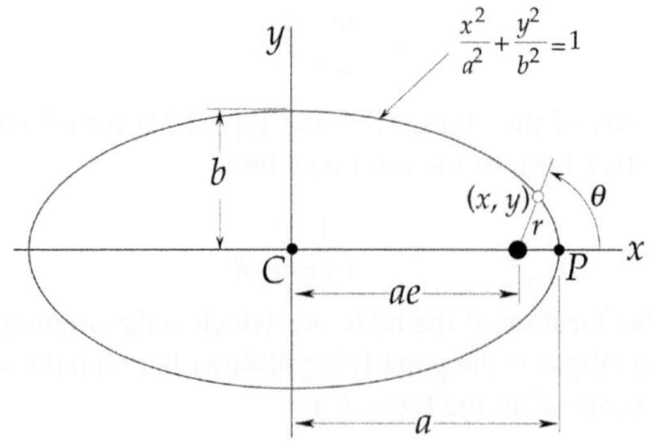

FIG. 2.19

Cartesian coordinate description of the orbit.

From this, we have

$$\frac{x}{a} = \frac{e + \cos\theta}{1 + e\cos\theta} \tag{2.77}$$

For the y coordinate, we make use of Eq. (2.76) to obtain

$$y = r\sin\theta = \left(a\frac{1-e^2}{1+e\cos\theta}\right)\sin\theta = b\frac{\sqrt{1-e^2}}{1+e\cos\theta}\sin\theta$$

Therefore,

$$\frac{y}{b} = \frac{\sqrt{1-e^2}}{1+e\cos\theta}\sin\theta \tag{2.78}$$

Using Eqs. (2.77) and (2.78), we find that

$$\begin{aligned}
\frac{x^2}{a^2} + \frac{y^2}{b^2} &= \frac{1}{(1+e\cos\theta)^2}\left[(e+\cos\theta)^2 + \left(1-e^2\right)\sin^2\theta\right] \\
&= \frac{1}{(1+e\cos\theta)^2}\left[e^2 + 2e\cos\theta + \cos^2\theta + \sin^2\theta - e^2\sin^2\theta\right] \\
&= \frac{1}{(1+e\cos\theta)^2}\left[e^2 + 2e\cos\theta + 1 - e^2\sin^2\theta\right] \\
&= \frac{1}{(1+e\cos\theta)^2}\left[e^2\left(1-\sin^2\theta\right) + 2e\cos\theta + 1\right] \\
&= \frac{1}{(1+e\cos\theta)^2}\left[e^2\cos^2\theta + 2e\cos\theta + 1\right] \\
&= \frac{1}{(1+e\cos\theta)^2}(1+e\cos\theta)^2
\end{aligned}$$

That is,

$$\frac{x^2}{a^2} + \frac{y^2}{b^2} = 1 \tag{2.79}$$

This is the familiar Cartesian coordinate formula for an ellipse centered at the origin, with x intercepts at $\pm a$ and y intercepts at $\pm b$. If $a = b$, Eq. (2.79) describes a circle, which is really an ellipse whose eccentricity is zero.

The specific energy of an elliptical orbit is negative, and it is found by substituting the angular momentum and eccentricity into Eq. (2.60),

$$\varepsilon = -\frac{1}{2}\frac{\mu^2}{h^2}\left(1-e^2\right)$$

According to Eq. (2.71), $h^2 = \mu a(1 - e^2)$, so that

$$\boxed{\varepsilon = -\frac{\mu}{2a}} \tag{2.80}$$

This shows that the specific energy is independent of the eccentricity and depends only on the semimajor axis of the ellipse. For an elliptical orbit, the conservation of energy (Eq. 2.57) may therefore be written

$$\frac{v^2}{2}-\frac{\mu}{r}=-\frac{\mu}{2a} \tag{2.81}$$

The area of an ellipse is found in terms of its semimajor and semiminor axes by the formula $A=\pi ab$ (which reduces to the formula for the area of a circle if $a=b$). To find the period T of the elliptical orbit, we employ Kepler's second law, $dA/dt=h/2$, to obtain

$$\Delta A=\frac{h}{2}\Delta t$$

For one complete revolution, $\Delta A=\pi ab$ and $\Delta t=T$. Thus, $\pi ab=(h/2)T$, or

$$T=\frac{2\pi ab}{h}$$

Substituting Eqs. (2.71) and (2.76), we get

$$T=\frac{2\pi}{h}a^2\sqrt{1-e^2}=\frac{2\pi}{h}\left(\frac{h^2}{\mu}\frac{1}{1-e^2}\right)^2\sqrt{1-e^2}$$

so that the formula for the period of an elliptical orbit, in terms of the orbital parameters h and e, becomes

$$T=\frac{2\pi}{\mu^2}\left(\frac{h}{\sqrt{1-e^2}}\right)^3 \tag{2.82}$$

We can once again appeal to Eq. (2.71) to substitute $h=\sqrt{\mu a(1-e^2)}$ into this equation, thereby obtaining an alternative expression for the period,

$$\boxed{T=\frac{2\pi}{\sqrt{\mu}}a^{3/2}} \tag{2.83}$$

This expression, which is identical to that of a circular orbit of radius a (Eq. 2.64), reveals that, like energy, the period of an elliptical orbit is independent of the eccentricity (Fig. 2.20). Eq. (2.83) embodies Kepler's third law: the period of a planet is proportional to the three-half power of its semimajor axis.

Finally, observe that dividing Eq. (2.50) by Eq. (2.70) yields

$$\frac{r_p}{r_a}=\frac{1-e}{1+e}$$

Solving this for e results in a useful formula for calculating the eccentricity of an elliptical orbit. Namely

$$\boxed{e=\frac{r_a-r_p}{r_a+r_p}} \tag{2.84}$$

From Fig. 2.18, it is apparent that $r_a-r_p=\overline{F'F}$ is the distance between the foci. As previously noted, $r_a+r_p=2a$. Thus, Eq. (2.84) has the geometrical interpretation,

$$\text{Eccentricity}=\frac{\text{Distance between the foci}}{\text{Length of the major axis}}$$

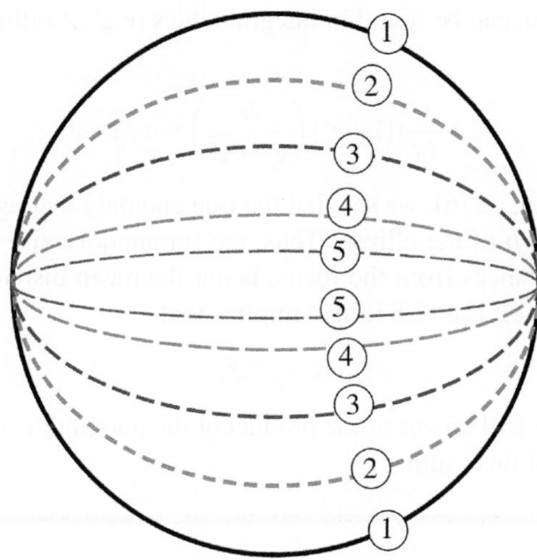

FIG. 2.20

Since all five ellipses have the same major axis, their periods and energies are identical.

A rectilinear ellipse is characterized as having a zero angular momentum and an eccentricity of 1. That is, the distance between the foci equals the finite length of the major axis, along which the relative motion occurs. Since only the length of the semimajor axis determines orbital-specific energy, Eq. (2.80) applies to rectilinear ellipses as well.

What is the average distance of m_2 from m_1 in the course of one complete orbit? To answer this question, we divide the range of the true anomaly (2π) into n equal segments $\Delta\theta$, so that

$$n = \frac{2\pi}{\Delta\theta}$$

We then use $r = (h^2/\mu)/(1 + e\cos\theta)$ to evaluate $r(\theta)$ at the n equally spaced values of the true anomaly,

$$\theta_1 = 0, \ \ \theta_2 = \Delta\theta, \ \ \theta_3 = 2\Delta\theta, \ldots, \ \ \theta_n = (n-1)\Delta\theta$$

starting at the periapsis. The average of this set of n values of r is given by

$$\bar{r}_\theta = \frac{1}{n}\sum_{i=1}^{n} r(\theta_i) = \frac{\Delta\theta}{2\pi}\sum_{i=1}^{n} r(\theta_i) = \frac{1}{2\pi}\sum_{i=1}^{n} r(\theta_i)\,\Delta\theta \tag{2.85}$$

Now, let n become very large, such that $\Delta\theta$ becomes very small. In the limit as $n \to \infty$, Eq. (2.85) becomes

$$\bar{r}_\theta = \frac{1}{2\pi}\int_0^{2\pi} r(\theta)\mathrm{d}\theta \tag{2.86}$$

Substituting Eq. (2.72) into the integrand yields

$$\bar{r}_\theta = \frac{1}{2\pi}a\left(1 - e^2\right)\int_0^{2\pi} \frac{\mathrm{d}\theta}{1 + e\cos\theta}$$

The integral in this expression can be found in integral tables (e.g., Zwillinger, 2018), from which we obtain

$$\bar{r}_\theta = \frac{1}{2\pi} a\left(1-e^2\right)\left(\frac{2\pi}{\sqrt{1-e^2}}\right) = a\sqrt{1-e^2} \tag{2.87}$$

Comparing this result with Eq. (2.76), we see that the true anomaly-averaged orbital radius equals the length of the semiminor axis b of the ellipse. Thus, the semimajor axis, which is the average of the maximum and minimum distances from the focus, is not the mean distance. Since, from Eq. (2.72), $r_p = a(1 - e)$ and $r_a = a(1 + e)$, Eq. (2.87) also implies that

$$\bar{r}_\theta = \sqrt{r_p r_a} \tag{2.88}$$

The mean distance is the one-half power of the product of the maximum and minimum distances from the focus and not one-half of their sum.

EXAMPLE 2.7

An earth satellite is in an orbit with a perigee altitude $z_p = 400\,\text{km}$ and an apogee altitude $z_a = 4000\,\text{km}$, as shown in Fig. 2.21. Find each of the following quantities:

(a) eccentricity, e

(b) angular momentum, h

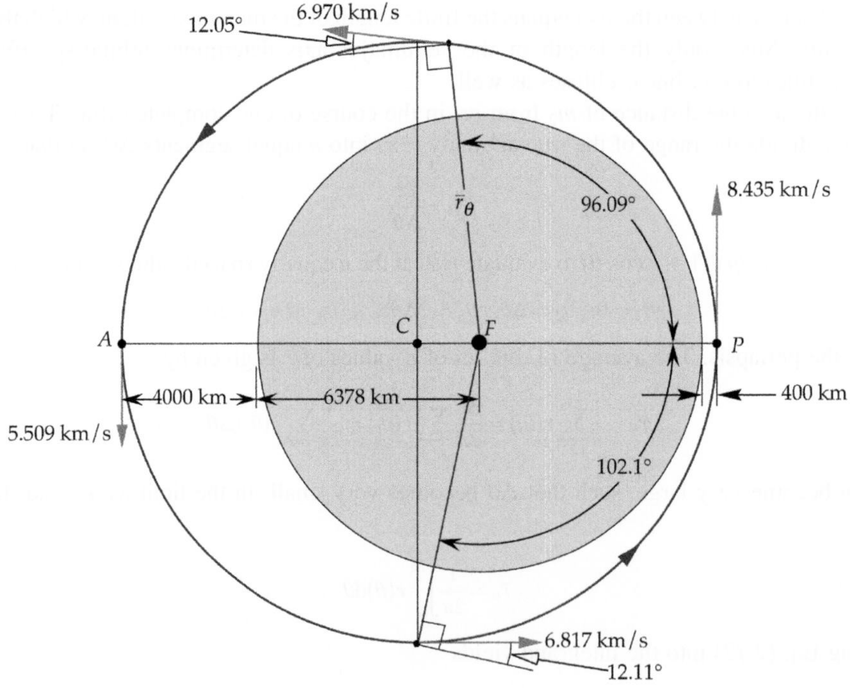

FIG. 2.21

The orbit of Example 2.7.

(c) perigee velocity, v_p
(d) apogee velocity, v_a
(e) semimajor axis, a
(f) period of the orbit, T
(g) true anomaly-averaged radius $\bar{r}_\theta$
(h) true anomaly when $r = \bar{r}_\theta$
(i) satellite speed when $r = \bar{r}_\theta$
(j) flight path angle γ when $r = \bar{r}_\theta$
(k) maximum flight path angle γ_{max} and the true anomaly at which it occurs.

Recall from Eq. (2.66) that $\mu = 398,600\,\text{km}^3/\text{s}^2$ and also that R_E, the radius of the earth, is 6378 km.

Solution

The strategy is always to seek the primary orbital parameters (eccentricity e and angular momentum h) first. All the other orbital parameters are obtained from these two.

(a) The formula that involves the unknown eccentricity e as well as the given perigee and apogee data is Eq. (2.84). We must not forget to convert the given altitudes to radii:

$$r_p = R_E + z_p = 6378 + 400 = 6778\,\text{km}$$

$$r_a = R_E + z_a = 6378 + 4000 = 10,378\,\text{km}$$

Then

$$e = \frac{r_a - r_p}{r_a + r_p} = \frac{10,378 - 6778}{10,378 + 6778}$$

$$\boxed{e = 0.2098}$$

(b) Now that we have the eccentricity, we need an expression containing it and the unknown angular momentum h and any other given data. That would be Eq. (2.50), the orbit formula evaluated at perigee ($\theta = 0$),

$$r_p = \frac{h^2}{\mu}\frac{1}{1+e}$$

We use this to compute the angular momentum

$$6778 = \frac{h^2}{398,600}\frac{1}{1 + 0.2098}$$

$$\boxed{h = 57,172\,\text{km}^2/\text{s}}$$

(c) The angular momentum h and the perigee radius r_p can be substituted into the angular momentum formula (Eq. 2.31) to find the perigee velocity v_p,

$$v_p = v_\perp)_{\text{perigee}} = \frac{h}{r_p} = \frac{57,172}{6778}$$

$$\boxed{v_p = 8.435\,\text{km/s}}$$

(d) Since h is a constant, the angular momentum formula can also be employed to obtain the apogee speed v_a,

$$v_a = \frac{h}{r_a} = \frac{57,172}{10,378}$$

$$\boxed{v_a = 5.509\,\text{km/s}}$$

(e) The semimajor axis is the average of the perigee and apogee radii (Fig. 2.18),

$$a = \frac{r_p + r_a}{2} = \frac{6778 + 10,378}{2}$$

$$\boxed{a = 8578\,\text{km}}$$

(f) Since the semimajor axis a has been found, we can use Eq. (2.83) to calculate the period T of the orbit:

$$T = \frac{2\pi}{\sqrt{\pi}} a^{3/2} = \frac{2\pi}{\sqrt{398,600}} 8578^{3/2} = 7907\,\text{s}$$

$$\boxed{T = 2.196\text{h}}$$

Alternatively, we could have used Eq. (2.82) for T, since both h and e were calculated above.

(g) Either Eq. (2.87) or Eq. (2.88) may be used at this point to find the true anomaly-averaged radius. Choosing the latter, we get

$$\bar{r}_\theta = \sqrt{r_p r_a} = \sqrt{6778 \cdot 10,378}$$

$$\boxed{\bar{r}_\theta = 8387\,\text{km}}$$

(h) To find the true anomaly when $r = \bar{r}_\theta$, we have only one choice, namely, the orbit formula (Eq. 2.45):

$$\bar{r}_\theta = \frac{h^2}{\mu} \frac{1}{1 + e\cos\theta}$$

Substituting h and e, the primary orbital parameters found above, together with $\bar{r}_\theta$, we get

$$8387 = \frac{57,172^2}{398,600} \frac{1}{1 + 0.2098\cos\theta}$$

from which

$$\cos\theta = -0.1061$$

This means that the true anomaly-averaged radius occurs at $\boxed{\theta = 96.09°}$, where the satellite passes through $\bar{r}_\theta$ on its way *from* the perigee, and at $\boxed{\theta = 263.9°}$, where the satellite passes through $\bar{r}_\theta$ on its way *toward* the perigee.

(i) To find the speed of the satellite when $r = \bar{r}_\theta$, it is simplest to use the energy equation for the ellipse (Eq. 2.81),

$$\frac{v^2}{2} - \frac{\mu}{r_\theta} = -\frac{\mu}{2a}$$

$$\frac{v^2}{2} - \frac{398,600}{8387} = -\frac{398,600}{2 \cdot 8578}$$

$$\boxed{v = 6.970\text{km/s}}$$

(j) Eq. (2.52) gives the flight path angle in terms of the true anomaly of the average radius $\bar{r}_\theta$. Substituting the smaller of the two angles found in part (h) above yields

$$\tan\gamma = \frac{e\sin\theta}{1 + e\cos\theta} = \frac{0.2098 \cdot \sin 96.09°}{1 + 0.2098 \cdot \cos 96.09°} = 0.2134$$

This means that $\boxed{\gamma = 12.05°}$ when the satellite passes through $\bar{r}_\theta$ on its way from the perigee.

(k) To find where γ is a maximum, we must take the derivative of

$$\gamma = \tan^{-1}\frac{e\sin\theta}{1 + e\cos\theta} \tag{a}$$

with respect to θ and set the result equal to zero. Using the rules of calculus,

$$\frac{d\gamma}{d\theta} = \frac{1}{1 + \left(\dfrac{e\sin\theta}{1 + e\cos\theta}\right)^2} \frac{d}{d\theta}\left(\frac{e\sin\theta}{1 + e\cos\theta}\right) = \frac{e(e + \cos\theta)}{(1 + e\cos\theta)^2 + e^2\sin^2\theta}$$

For $e < 1$, the denominator is positive for all values of θ. Therefore, $d\gamma/d\theta = 0$ only if the numerator vanishes (i.e., if $\cos\theta = -e$). Recall from Eq. (2.75) that this true anomaly locates the end point of the minor axis of the ellipse. The maximum positive flight path angle therefore occurs at the true anomaly,

$$\theta = \cos^{-1}(-0.2098)$$

$$\boxed{\theta = 102.1°}$$

Substituting this into Eq. (a), we find the value of the flight path angle to be

$$\gamma_{max} = \tan^{-1}\frac{0.2098 \cdot \sin(102.1°)}{1 + 0.2098 \cdot \cos(102.1°)}$$

$$\boxed{\gamma_{max} = 12.11°}$$

After attaining this greatest magnitude, the flight path angle starts to decrease steadily toward its value of zero at the apogee.

EXAMPLE 2.8

At two points on a geocentric orbit, the altitude and true anomaly are $z_1 = 1545\,\text{km}$, $\theta_1 = 126°$ and $z_2 = 852\,\text{km}$, $\theta_2 = 58°$, respectively. Find (a) the eccentricity, (b) the altitude of perigee, (c) the semimajor axis, and (d) the period.

Solution

The first objective is to find the primary orbital parameters e and h, since all other orbital data can be deduced from them.

(a) Before proceeding, we must remember to add the earth's radius to the given altitudes so that we are dealing with orbital radii. The radii of the two points are

$$r_1 = R_E + z_1 = 6378 + 1545 = 7923\text{km}$$
$$r_2 = R_E + z_2 = 6378 + 852 = 7230\text{km}$$

The only formula we have that relates the orbital position to the orbital parameters e and h is the orbit formula, Eq. (2.45). Writing that equation down for each of the two given points on the orbit yields two equations for e and h. For point 1, we obtain

$$r_1 = \frac{h^2}{\mu}\frac{1}{1 + e\cos\theta_1}$$
$$7923 = \frac{h^2}{398{,}600} \cdot \frac{1}{1 + e\cos 126°} \tag{a}$$
$$h^2 = 3.158(10^9) - 1.856(10^9)e$$

For point 2,

$$r_2 = \frac{h^2}{\mu}\frac{1}{1 + e\cos\theta_2}$$
$$7230 = \frac{h^2}{398{,}600} \cdot \frac{1}{1 + e\cos 58°} \tag{b}$$
$$h^2 = 2.882(10^9) + 1.527(10^9)e$$

Equating Eqs. (a) and (b), the two expressions for h^2, yields a single equation for the eccentricity e,

$$3.158(10^9) - 1.856(10^9)e = 2.882(10^9) + 1.527(10^9)e$$

or

$$3.384(10^9)e = 276.2(10^6)$$

Therefore,

$$\boxed{e = 0.08164} \text{ (an ellipse)} \tag{c}$$

By substituting the eccentricity back into Eq. (a) (or Eq. (b)), we find the angular momentum,

$$h^2 = 3.158(10^9) - 1.856(10^9) \cdot 0.08164 \Rightarrow h = 54{,}830\text{km}^2/\text{s} \tag{d}$$

(b) With the eccentricity and angular momentum available, we can use the orbit equation to obtain the perigee radius (Eq. 2.50),

$$r_p = \frac{h^2}{\mu}\frac{1}{1+e} = \frac{54{,}830^2}{398{,}600}\frac{1}{1+0.08164} = 6974\,\text{km} \tag{e}$$

From this we find the perigee altitude,

$$z_p = r_p - R_E = 6974 - 6378$$

$$\boxed{z_p = 595.5\,\text{km}}$$

(c) The semimajor axis is the average of the perigee and apogee radii. We just found the perigee radius above in Eq. (e). Thus, we need only to compute the apogee radius and that is accomplished by using Eq. (2.70), which is the orbit formula evaluated at the apogee.

$$r_a = \frac{h^2}{\mu}\frac{1}{1-e} = \frac{54{,}830^2}{398{,}600}\frac{1}{1-0.08164} = 8213\,\text{km} \tag{f}$$

From Eqs. (e) and (f) it follows that

$$a = \frac{r_p + r_a}{2} = \frac{8213 + 6974}{2}$$

$$\boxed{a = 7593\,\text{km}}$$

(d) Since the semimajor axis has been determined, it is convenient to use Eq. (2.83) to find the period.

$$T = \frac{2\pi}{\sqrt{\mu}}a^{\frac{3}{2}} = \frac{2\pi}{\sqrt{398{,}600}}7593^{\frac{3}{2}} = 6585\,\text{s}$$

$$\boxed{T = 1.829\,\text{h}}$$

2.8 PARABOLIC TRAJECTORIES ($e = 1$)

If the eccentricity equals 1, then the orbit formula (Eq. 2.45) becomes

$$r = \frac{h^2}{\mu}\frac{1}{1+\cos\theta} \tag{2.89}$$

As the true anomaly θ approaches 180 degrees, the denominator approaches zero, so that r tends toward infinity. According to Eq. (2.60), the energy of a trajectory for which $e = 1$ is zero, so that for a parabolic trajectory the conservation of energy (Eq. 2.57) is

$$\frac{v^2}{2} - \frac{\mu}{r} = 0$$

In other words, the speed anywhere on a parabolic path is

$$v = \sqrt{\frac{2\mu}{r}} \tag{2.90}$$

If the body m_2 is launched on a parabolic trajectory, it will coast to infinity, arriving there with zero velocity relative to m_1. It will not return. Parabolic paths are therefore called escape trajectories. At a given distance r from m_1, the escape velocity is given by Eq. (2.90),

$$v_{esc} = \sqrt{\frac{2\mu}{r}} \tag{2.91}$$

Let v_c be the speed of a satellite in a circular orbit of radius r. Then, from Eqs. (2.63) and (2.91), we have

$$v_{esc} = \sqrt{2}v_c \tag{2.92}$$

That is, to escape from a circular orbit requires a velocity boost of 41.4%. However, remember our assumption is that m_1 and m_2 are the only objects in the universe. A spacecraft launched from the earth with a velocity v_{esc} (relative to the earth) will not coast to infinity (i.e., leave the solar system) because it will eventually succumb to the gravitational influence of the sun and, in fact, end up in the same orbit as the earth. This will be discussed in more detail in Chapter 8.

For the parabola, Eq. (2.52) for the flight path angle takes the form

$$\tan\gamma = \frac{\sin\theta}{1+\cos\theta}$$

Using the trigonometric identities

$$\sin\theta = 2\sin\frac{\theta}{2}\cos\frac{\theta}{2} \qquad \cos\theta = 2\cos^2\frac{\theta}{2} - 1$$

we can write

$$\tan\gamma = \frac{2\sin\frac{\theta}{2}\cos\frac{\theta}{2}}{2\cos^2\frac{\theta}{2}} = \frac{\sin\frac{\theta}{2}}{\cos\frac{\theta}{2}} = \tan\frac{\theta}{2}$$

It follows that

$$\gamma = \frac{\theta}{2} \tag{2.93}$$

That is, on parabolic trajectories the flight path angle is always one-half of the true anomaly (Fig. 2.22).

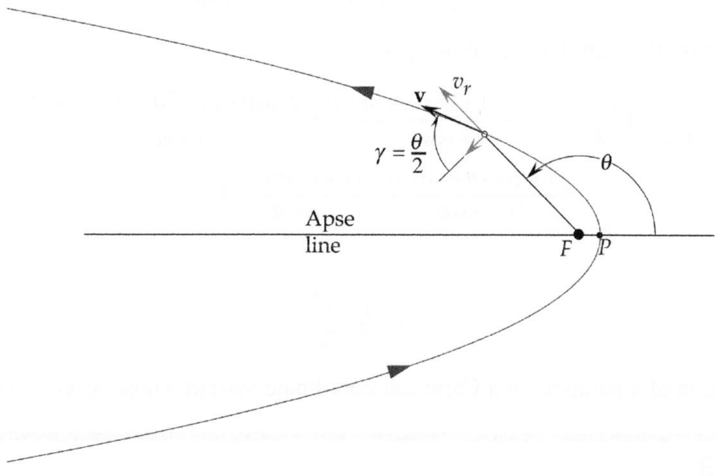

FIG. 2.22

Parabolic trajectory around the focus *F*.

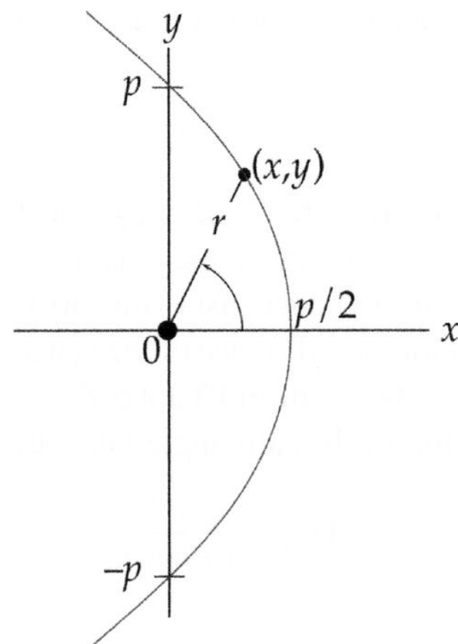

FIG. 2.23

Parabola with focus at the origin of the Cartesian coordinate system.

Eq. (2.53) gives the parameter p of an orbit. Let us substitute that expression into Eq. (2.89) and then plot $r = p/(1+\cos\theta)$ in a Cartesian coordinate system centered at the focus, as illustrated in Fig. 2.23. From the figure, it is clear that

$$x = r\cos\theta = p\frac{\cos\theta}{1+\cos\theta} \quad y = r\sin\theta = p\frac{\sin\theta}{1+\cos\theta} \tag{2.94}$$

Therefore,

$$\frac{x}{p/2} + \left(\frac{y}{p}\right)^2 = 2\frac{\cos\theta}{1+\cos\theta} + \frac{\sin^2\theta}{(1+\cos\theta)^2}$$

Working to simplify the right-hand side, we get

$$\frac{x}{p/2} + \left(\frac{y}{p}\right)^2 = \frac{2\cos\theta(1+\cos\theta)+\sin^2\theta}{(1+\cos\theta)^2} = \frac{2\cos\theta + 2\cos^2\theta + (1-\cos^2\theta)}{(1+\cos\theta)^2}$$
$$= \frac{1+2\cos\theta+\cos^2\theta}{(1+\cos\theta)^2} = \frac{(1+\cos\theta)^2}{(1+\cos\theta)^2} = 1$$

It follows that

$$x = \frac{p}{2} - \frac{y^2}{2p} \tag{2.95}$$

This is the equation of a parabola in a Cartesian coordinate system whose origin serves as the focus.

EXAMPLE 2.9

The perigee radius of a satellite in a parabolic geocentric trajectory of Fig. 2.24 is 7000 km. Find the distance d between points P_1 and P_2 on the orbit, which are 8000km and 16, 000km, respectively, from the center of the earth.

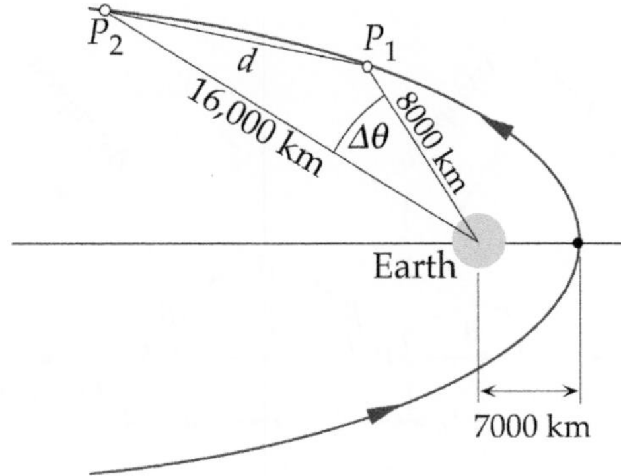

FIG. 2.24

Parabolic geocentric trajectory.

Solution

This would be a simple trigonometry problem if we knew the angle $\Delta\theta$ between the radials to P_1 and P_2. We can find that angle by first determining the true anomalies of the two points. The true anomalies are obtained from the orbit formula, Eq. (2.89), once we have determined the angular momentum h.

We calculate the angular momentum of the satellite by evaluating the orbit equation at perigee,

$$r_p = \frac{h^2}{\mu}\frac{1}{1+\cos(0)} = \frac{h^2}{2\mu}$$

from which

$$h = \sqrt{2\mu r_p} = \sqrt{2 \cdot 398{,}600 \cdot 7000} = 74{,}700\,\text{km}^2/\text{s} \tag{a}$$

Substituting the radii and the true anomalies of points P_1 and P_2 into Eq. (2.89), we get

$$8000 = \frac{74{,}700^2}{398{,}600}\frac{1}{1+\cos\theta_1} \Rightarrow \cos\theta_1 = 0.75 \Rightarrow \theta_1 = 41.41°$$

$$16{,}000 = \frac{74{,}700^2}{398{,}600}\frac{1}{1+\cos\theta_2} \Rightarrow \cos\theta_2 = -0.125 \Rightarrow \theta_2 = 97.18°$$

The difference between the two angles θ_1 and θ_2 is $\Delta\theta = 97.18 - 41.41 = 55.78°$.

The length of the chord $\overline{P_1P_2}$ can now be found by using the law of cosines from trigonometry,

$$d^2 = 8000^2 + 16{,}000^2 - 2 \cdot 8000 \cdot 16{,}000 \cos\Delta\theta$$

$$\boxed{d = 13{,}270\text{km}}$$

2.9 HYPERBOLIC TRAJECTORIES ($e > 1$)

If $e > 1$, the orbit formula,

$$r = \frac{h^2}{\mu}\frac{1}{1+e\cos\theta} \tag{2.96}$$

describes the geometry of the hyperbola shown in Fig. 2.25. The system consists of two symmetric curves. The orbiting body occupies one of them. The other one is its empty mathematical image.

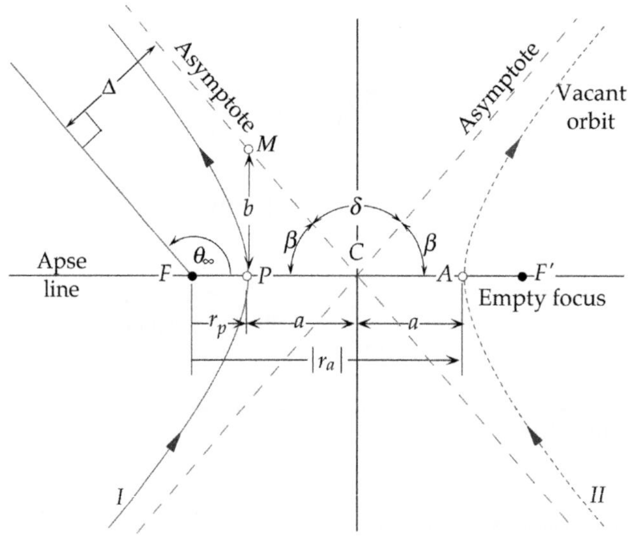

FIG. 2.25

Hyperbolic trajectory.

Clearly, the denominator of Eq. (2.96) goes to zero when $\cos\theta = -1/e$. We denote this value of true anomaly as

$$\theta_\infty = \cos^{-1}(-1/e) \tag{2.97}$$

since the radial distance approaches infinity as the true anomaly approaches θ_∞. θ_∞ is known as the true anomaly of the asymptote. Observe that θ_∞ lies between 90 and 180 degrees. From the trig identity $\sin^2\theta_\infty + \cos^2\theta_\infty = 1$ it follows that

$$\sin\theta_\infty = \sqrt{\frac{e^2-1}{e}} \tag{2.98}$$

For $-\theta_\infty < \theta < \theta_\infty$, the physical trajectory is the occupied hyperbola I shown on the left in Fig. 2.25. For $\theta_\infty < \theta < (360° - \theta_\infty)$, hyperbola II, the vacant orbit around the empty focus F', is traced out. (The vacant orbit is physically impossible, because it would require a repulsive gravitational force.) Periapsis P lies on the apse line on the physical hyperbola I, whereas apoapsis A lies on the apse line on the vacant orbit. The point halfway between periapsis and apoapsis is the center C of the hyperbola. The asymptotes of the hyperbola are the straight lines toward which the curves tend as they approach infinity. The asymptotes intersect at C, making an acute angle β with the apse line, where $\beta = 180° - \theta_\infty$. Therefore, $\cos\beta = -\cos\theta_\infty$, which means

$$\beta = \cos^{-1}(1/e) \tag{2.99}$$

The angle δ between the asymptotes is called the turn angle. This is the angle through which the velocity vector of the orbiting body is rotated as it rounds the attracting body at F and heads back toward infinity. From the figure, we see that $\delta = 180° - 2\beta$, so that

$$\sin\frac{\delta}{2} = \sin\left(\frac{180° - 2\beta}{2}\right) = \sin(90° - \beta) = \cos\beta \overbrace{=}^{\text{Eq.(2.99)}} \frac{1}{e}$$

or

$$\delta = 2\sin^{-1}(1/e) \tag{2.100}$$

Eq. (2.50) gives the distance r_p from the focus F to the periapsis,

$$r_p = \frac{h^2}{\mu}\frac{1}{1+e} \tag{2.101}$$

Just as for an ellipse, the radial coordinate r_a of the apoapsis is found by setting $\theta = 180°$ in Eq. (2.45),

$$r_a = \frac{h^2}{\mu}\frac{1}{1-e} \tag{2.102}$$

Observe that r_a is negative, since $e > 1$ for the hyperbola. This means the apoapsis lies to the right of the focus F. From Fig. 2.25 we see that the distance $2a$ from periapsis P to apoapsis A is

$$2a = |r_a| - r_p = -r_a - r_p$$

Substituting Eqs. (2.101) and (2.102) yields

$$2a = -\frac{h^2}{\mu}\left(\frac{1}{1-e} + \frac{1}{1+e}\right)$$

From this it follows that a, the semimajor axis of the hyperbola, is given by an expression that is nearly identical to that for an ellipse (Eq. 2.72),

$$a = \frac{h^2}{\mu}\frac{1}{e^2 - 1} \tag{2.103}$$

Therefore, Eq. (2.96) may be written for the hyperbola

$$r = a\frac{e^2 - 1}{1 + e\cos\theta} \tag{2.104}$$

This formula is analogous to Eq. (2.72) for the elliptical orbit. Furthermore, from Eq. (2.104) it follows that

$$r_p = a(e - 1) \tag{2.105a}$$

$$r_a = -a(e + 1) \tag{2.105b}$$

The distance b from periapsis to an asymptote, measured perpendicular to the apse line, is the semiminor axis of the hyperbola. From Fig. 2.25, we see that the length b of the semiminor axis $\overline{PM}$ is

$$b = a\tan\beta = a\frac{\sin\beta}{\cos\beta} = a\frac{\sin(180° - \theta_\infty)}{\cos(180° - \theta_\infty)} = a\frac{\sin\theta_\infty}{-\cos\theta_\infty} = a\frac{\frac{\sqrt{e^2-1}}{e}}{-\left(-\frac{1}{e}\right)}$$

so that for the hyperbola,

$$b = a\sqrt{e^2 - 1} \tag{2.106}$$

This relation is analogous to Eq. (2.76) for the semiminor axis of an ellipse.

The distance Δ between the asymptote and a parallel line through the focus is called the aiming radius, which is illustrated in Fig. 2.25. From this figure we see that

$$\begin{aligned}
\Delta &= (r_p + a)\sin\beta \\
&= ae\sin\beta && \text{(Eq.2.105a)} \\
&= ae\frac{\sqrt{e^2-1}}{e} && \text{(Eq.2.99)} \\
&= ae\sin\theta_\infty && \text{(Eq.2.98)} \\
&= ae\sqrt{1-\cos^2\theta_\infty} && \text{(trig identity)} \\
&= ae\sqrt{1-\frac{1}{e^2}} && \text{(Eq.2.97)}
\end{aligned}$$

or

$$\Delta = a\sqrt{e^2 - 1} \tag{2.107}$$

Comparing this result with Eq. (2.106), it is clear that the aiming radius equals the length of the semi-minor axis of the hyperbola.

As with the ellipse and the parabola, we can express the polar form of the equation of the hyperbola in a Cartesian coordinate system whose origin is in this case midway between the two foci, as illustrated in Fig. 2.26. From the figure, it is apparent that

$$x = -a - r_p + r\cos\theta \tag{2.108a}$$

$$y = r\sin\theta \tag{2.108b}$$

Using Eqs. (2.104) and (2.105a) in Eq. (2.108a), we obtain

$$x = -a - a(e-1) + a\frac{e^2-1}{1+e\cos\theta}\cos\theta = -a\frac{e+\cos\theta}{1+e\cos\theta}$$

Substituting Eqs. (2.104) and (2.106) into Eq. (2.108b) yields

$$y = \frac{b}{\sqrt{e^2-1}}\frac{e^2-1}{1+e\cos\theta}\sin\theta = b\frac{\sqrt{e^2-1}\sin\theta}{1+e\cos\theta}$$

It follows that

$$\begin{aligned}
\frac{x^2}{a^2} - \frac{y^2}{b^2} &= \left(\frac{e+\cos\theta}{1+e\cos\theta}\right)^2 - \left(\frac{\sqrt{e^2-1}\sin\theta}{1+e\cos\theta}\right)^2 \\
&= \frac{e^2 + 2e\cos\theta + \cos^2\theta - (e^2-1)(1-\cos^2\theta)}{(1+e\cos\theta)^2} \\
&= \frac{1 + 2e\cos\theta + e^2\cos^2\theta}{(1+e\cos\theta)^2} = \frac{(1+e\cos\theta)^2}{(1+e\cos\theta)^2}
\end{aligned}$$

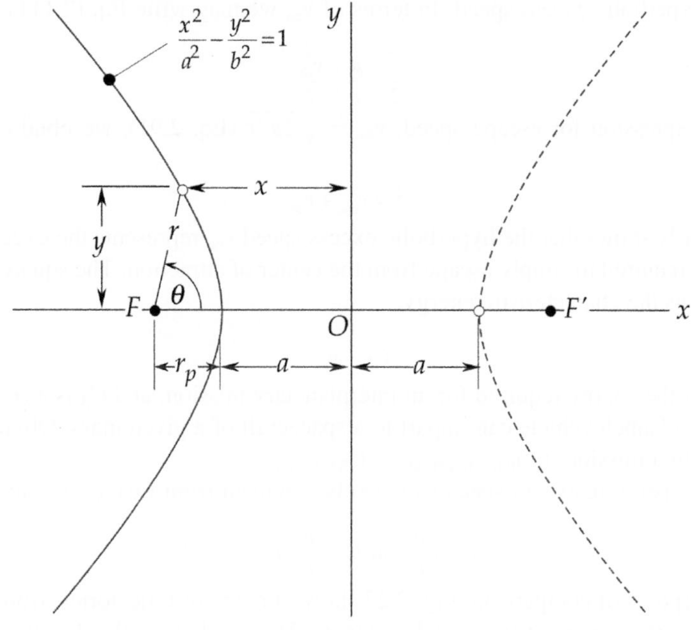

FIG. 2.26

Plot of Eq. (2.104) in a Cartesian coordinate system with origin *O* midway between the two foci.

That is,

$$\frac{x^2}{a^2} - \frac{y^2}{b^2} = 1 \tag{2.109}$$

This is the familiar equation of a hyperbola that is symmetric about the x and y axes, with intercepts on the x axis.

Eq. (2.60) gives the specific energy of the hyperbolic trajectory. Substituting Eq. (2.103) into that expression yields

$$\varepsilon = \frac{\mu}{2a} \tag{2.110}$$

The specific energy of a hyperbolic orbit is clearly positive and independent of the eccentricity. The conservation of energy for a hyperbolic trajectory is

$$\frac{v^2}{2} - \frac{\mu}{r} = \frac{\mu}{2a} \tag{2.111}$$

Let v_∞ denote the speed at which a body on a hyperbolic path arrives at infinity. According to Eq. (2.111)

$$v_\infty = \sqrt{\frac{\mu}{a}} \tag{2.112}$$

v_∞ is called the hyperbolic excess speed. In terms of v_∞ we may write Eq. (2.111) as

$$\frac{v^2}{2} - \frac{\mu}{r} = \frac{v_\infty^2}{2}$$

Substituting the expression for escape speed, $v_{esc} = \sqrt{2\mu/r}$ (Eq. 2.91), we obtain for a hyperbolic trajectory

$$v^2 = v_{esc}^2 + v_\infty^2 \tag{2.113}$$

This equation clearly shows that the hyperbolic excess speed v_∞ represents the excess kinetic energy over that which is required to simply escape from the center of attraction. The square of v_∞ is denoted C_3, and is known as the characteristic energy,

$$C_3 = v_\infty^2 \tag{2.114}$$

C_3 is a measure of the energy required for an interplanetary mission, and C_3 is also a measure of the maximum energy a launch vehicle can impart to a spacecraft of a given mass. Obviously, to match a launch vehicle with a mission, $C_3)_{\text{launch vehicle}} > C_3)_{\text{mission}}$.

Note that the hyperbolic excess speed can also be obtained from Eqs. (2.49) and (2.98),

$$v_\infty = \frac{\mu}{h} e \sin\theta_\infty = \frac{\mu}{h}\sqrt{e^2 - 1} \tag{2.115}$$

Finally, for purposes of comparison, Fig. 2.27 shows a range of trajectories, from a circle through hyperbolas, all having a common focus and periapsis. The parabola is the demarcation between the closed, negative energy orbits (ellipses) and open, positive energy orbits (hyperbolas).

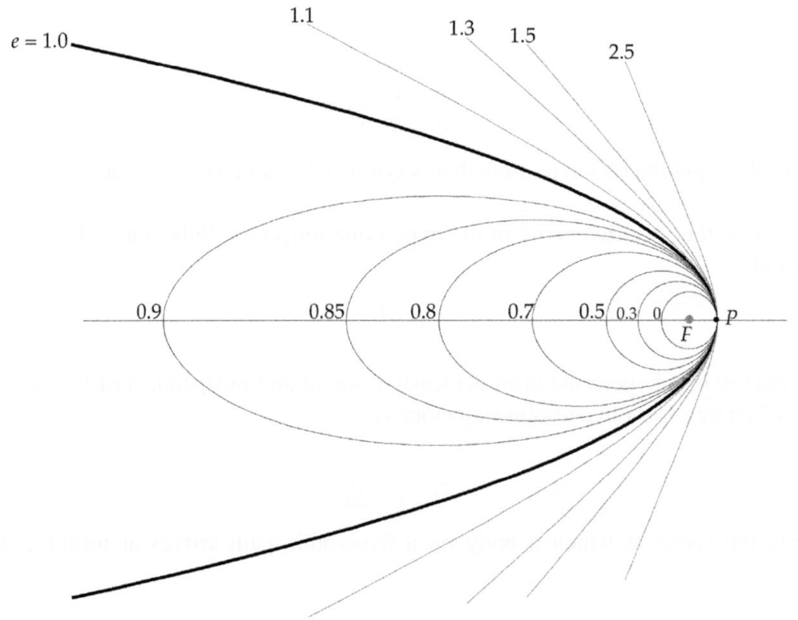

FIG. 2.27

Orbits of various eccentricities, having a common focus F and periapsis P.

At this point, the reader may be understandably overwhelmed by the number of formulas for Keplerian orbits (conic sections) that have been presented thus far in this chapter. As summarized in the Road Map in Appendix B, there is just a small set of equations from which all the others are derived.

Here is a "toolbox" of the only equations necessary for solving two-dimensional curvilinear orbital problems that do not involve time, which is the subject of Chapter 3.

All orbits:

$$h = rv_\perp \quad \text{Eq.(2.31)}$$

$$r = \frac{h^2}{\mu}\frac{1}{1+e\cos\theta} \quad \text{Eq.(2.45)}$$

$$v_r = \frac{\mu}{h}e\sin\theta \quad \text{Eq.(2.49)}$$

$$\tan\gamma = \frac{v_r}{v_\perp} \quad \text{Eq.(2.51)}$$

$$v = \sqrt{v_r^2 + v_\perp^2}$$

Ellipses ($0 \le e < 1$):

$$a = \frac{r_p + r_a}{2} = \frac{h^2}{\mu}\frac{1}{1-e^2} \quad \text{Eq.(2.71)}$$

$$\frac{v^2}{2} - \frac{\mu}{r} = -\frac{\mu}{2a} \quad \text{Eq.(2.81)}$$

$$T = \frac{2\pi}{\sqrt{\mu}}a^{3/2} \quad \text{Eq.(2.83)}$$

$$e = \frac{r_a - r_p}{r_a + r_p} \quad \text{Eq.(2.84)}$$

Parabolas ($e = 1$):

$$\frac{v^2}{2} - \frac{\mu}{r} = 0 \quad \text{Eq.(2.90)}$$

Hyperbolas ($e > 1$):

$$\theta_\infty = \cos^{-1}\left(-\frac{1}{e}\right) \quad \text{Eq.(2.97)}$$

$$\delta = 2\sin^{-1}\left(\frac{1}{e}\right) \quad \text{Eq.(2.100)}$$

$$a = \frac{h^2}{\mu}\frac{1}{e^2-1} \quad \text{Eq. (2.103)}$$

$$\Delta = a\sqrt{e^2-1} \quad \text{Eq. (2.107)}$$

$$\frac{v^2}{2} - \frac{\mu}{r} = \frac{\mu}{2a} \quad \text{Eq. (2.111)}$$

Note that we can rewrite Eqs. (2.103) and (2.111) as follows (where a is positive),

$$-a = \frac{h^2}{\mu}\frac{1}{1-e^2} \qquad \frac{v^2}{2} - \frac{\mu}{r} = -\frac{\mu}{2(-a)}$$

That is, if we assume that the semimajor axis of a hyperbola has a negative value, then the semimajor axis formula and the vis viva equation become identical for ellipses and hyperbolas. There is no advantage at this point in requiring hyperbolas to have negative semimajor axes. However, doing so will be necessary for the universal variable formulation to be presented in the next chapter.

EXAMPLE 2.10

At a given point of a spacecraft's geocentric trajectory, the radius is 14,600 km, the speed is 8.6 km/s, and the flight path angle is 50°. Show that the path is a hyperbola and calculate the following:

(a) angular momentum
(b) eccentricity
(c) true anomaly
(d) radius of the perigee
(e) semimajor axis
(f) C_3
(g) turn angle
(h) aiming radius

This problem is illustrated in Fig. 2.28.

Solution

Since both the radius and the speed are given, we can determine the type of trajectory by comparing the speed with the escape speed (of a parabolic trajectory) at the given radius:

$$v_{esc} = \sqrt{\frac{2\mu}{r}} = \sqrt{\frac{2 \cdot 398,600}{14,600}} = 7.389 \text{km/s}$$

The escape speed is less than the spacecraft's speed of 8.6 km/s, which means the path is a hyperbola.

(a) Before embarking on a quest for the required orbital data, remember that everything depends on the primary orbital parameters, angular momentum h, and eccentricity e. These are among the list of five unknowns for this problem: h, e, θ, v_r, and $v_\perp$. From the "toolbox" we have five equations involving these five quantities and the given data:

$$r = \frac{h^2}{\mu} \frac{1}{1 + e\cos\theta} \tag{a}$$

$$v_r = \frac{\mu}{h} e \sin\theta \tag{b}$$

$$v_\perp = \frac{h}{r} \tag{c}$$

$$v = \sqrt{v_r^2 + v_\perp^2} \tag{d}$$

$$\tan\gamma = \frac{v_r}{v_\perp} \tag{e}$$

From Eq. (e)

$$v_r = v_\perp \tan 50^\circ = 1.1918 v_\perp \tag{f}$$

Substituting this and the given speed into Eq. (d) yields

$$8.6^2 = (1.11918 v_\perp)^2 + v_\perp^2 \Rightarrow v_\perp = 5.528 \text{km/s} \tag{g}$$

The angular momentum may now be found from Eq. (c),

$$h = 14,600 \cdot 5.528 = \boxed{80,708 \text{km}^2/\text{s}}$$

(b) Substituting $v_\perp$ into Eq. (f) we get the radial velocity component,

$$v_r = 1.1918 \cdot 5.528 = 6.588 \text{km/s}$$

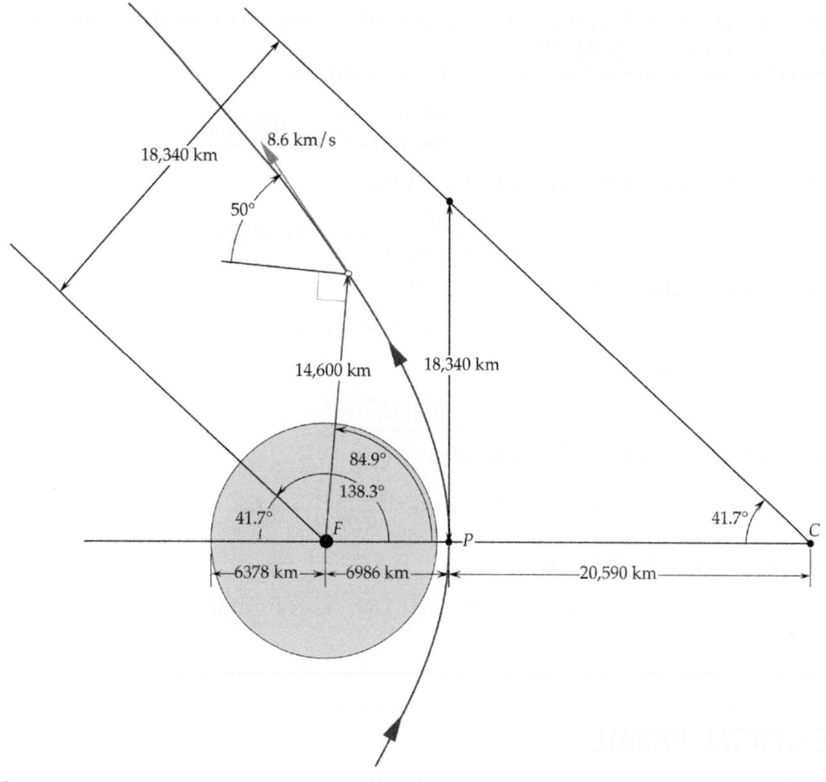

FIG. 2.28

Solution of Example 2.10.

Substituting h and v_r into Eq. (b) yields an expression involving the eccentricity and the true anomaly,

$$6.588 = \frac{398{,}600}{80{,}708} e\sin\theta \Rightarrow e\sin\theta = 1.3339 \tag{h}$$

Similarly, substituting h and r into Eq. (a) we find

$$14{,}600 = \frac{80{,}708^2}{398{,}600} \frac{1}{1 + e\cos\theta} \Rightarrow e\cos\theta = 0.1193 \tag{i}$$

By squaring the expressions in Eqs. (h) and (i) and then summing them, we obtain the eccentricity,

$$e^2 \overbrace{\left(\sin^2\theta + \cos^2\theta\right)}^{=1} = 1.7936$$

$$\boxed{e = 1.3393}$$

(c) To find the true anomaly, substitute the value of e into Eq. (i),

$$1.3393\cos\theta = 0.1193 \Rightarrow \theta = 84.889° \text{ or } \theta = 275.11°$$

We choose the smaller of the angles because Eqs. (h) and (i) imply that both $\sin\theta$ and $\cos\theta$ are positive, which means that θ lies in the first quadrant ($\theta \leq 90°$). Alternatively, we may note that the given flight path angle (50°) is positive, which means the spacecraft is flying away from the perigee, so that the true anomaly must be less than 180°. In any case, the true anomaly is given by $\boxed{\theta = 84.889°}$.

(d) The radius of perigee can now be found from the orbit equation (Eq. a)

$$r_p = \frac{h^2}{\mu}\frac{1}{1+e\cos(0)} = \frac{80{,}710^2}{398{,}600}\frac{1}{1+1.339} = \boxed{6986\,\text{km}}$$

(e) The semimajor axis of the hyperbola is found in Eq. (2.103),

$$a = \frac{h^2}{\mu}\frac{1}{e^2-1} = \frac{80{,}710^2}{398{,}600}\frac{1}{1.339^2-1} = \boxed{20{,}590\,\text{km}}$$

(f) The hyperbolic excess velocity is found using Eq. (2.113),

$$v_\infty^2 = v^2 - v_{\text{esc}}^2 = 8.6^2 - 7.389^2 = 19.36\,\text{km}^2/\text{s}^2$$

From Eq. (2.114) it follows that

$$\boxed{C_3 = 19.36\,\text{km}^2/\text{s}^2}$$

(g) The formula for turn angle is Eq. (2.100), from which

$$\delta = 2\sin^{-1}\left(\frac{1}{e}\right) = 2\sin^{-1}\left(\frac{1}{1.339}\right) = 96.60°$$

(h) According to Eq. (2.107), the aiming radius is

$$\Delta = a\sqrt{e^2-1} = 20{,}590\sqrt{1.339^2-1} = \boxed{18{,}340\,\text{km}}$$

2.10 PERIFOCAL FRAME

The perifocal frame is the "natural frame" for an orbit. It is a Cartesian coordinate system fixed in space and centered at the focus of the orbit. Its $\bar{x}\bar{y}$ plane is the plane of the orbit, and its $\bar{x}$ axis is directed from the focus through the periapsis, as illustrated in Fig. 2.29. The unit vector along the $\bar{x}$ axis (the apse line) is denoted $\hat{\mathbf{p}}$. The $\bar{y}$ axis, with unit vector $\hat{\mathbf{q}}$, lies at 90° true anomaly to the $\bar{x}$ axis. The $\bar{z}$ axis is normal to the plane of the orbit in the direction of the angular momentum vector $\mathbf{h}$. The $\bar{z}$ unit vector is $\hat{\mathbf{w}}$,

$$\hat{\mathbf{w}} = \frac{\mathbf{h}}{h} \tag{2.116}$$

In the perifocal frame, the position vector $\mathbf{r}$ is written (Fig. 2.30)

$$\mathbf{r} = \bar{x}\hat{\mathbf{p}} + \bar{y}\hat{\mathbf{q}} \tag{2.117}$$

where

$$\bar{x} = r\cos\theta \quad \bar{y} = r\sin\theta \tag{2.118}$$

and r, the magnitude of $\mathbf{r}$, is given by the orbit equation, $r = (h^2/\mu)[1/(1 + e\cos\theta)]$. Thus, we may write Eq. (2.117) as

$$\mathbf{r} = \frac{h^2}{\mu}\frac{1}{1+e\cos\theta}(\cos\theta\hat{\mathbf{p}} + \sin\theta\hat{\mathbf{q}}) \tag{2.119}$$

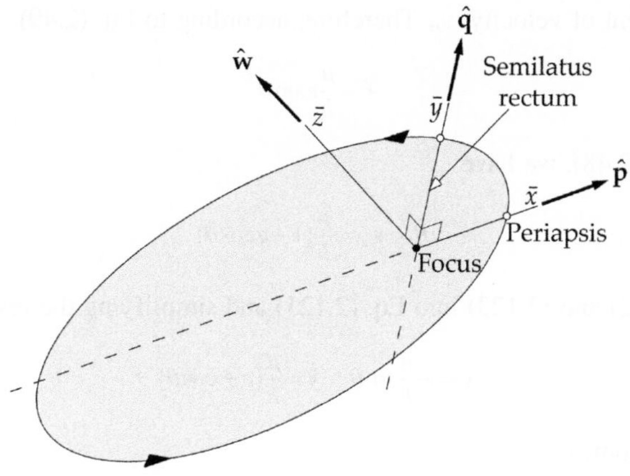

FIG. 2.29

Perifocal frame $\hat{\mathbf{p}}\hat{\mathbf{q}}\hat{\mathbf{w}}$.

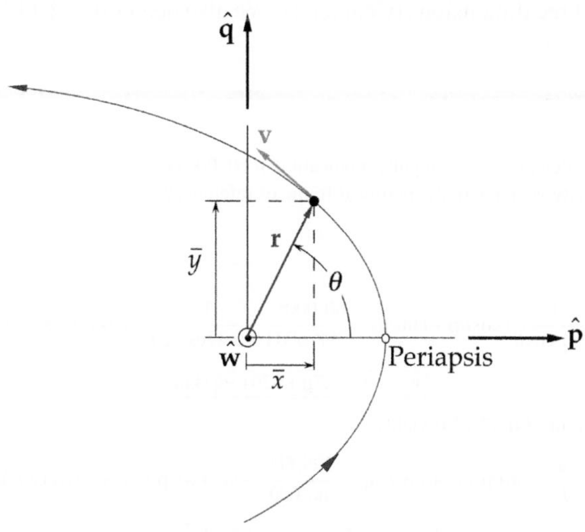

FIG. 2.30

Position and velocity relative to the perifocal frame.

The velocity is found by taking the time derivative of **r**,

$$\mathbf{v} = \dot{\mathbf{r}} = \dot{\bar{x}}\hat{\mathbf{p}} + \dot{\bar{y}}\hat{\mathbf{q}} \tag{2.120}$$

From Eq. (2.118) we obtain

$$\dot{\bar{x}} = \dot{r}\cos\theta - r\dot{\theta}\sin\theta \quad \dot{\bar{y}} = \dot{r}\sin\theta + r\dot{\theta}\cos\theta \tag{2.121}$$

$\dot{r}$ is the radial component of velocity, v_r. Therefore, according to Eq. (2.49),

$$\dot{r} = \frac{\mu}{h} e \sin\theta \tag{2.122}$$

From Eqs. (2.46) and (2.48), we have

$$r\dot{\theta} = v_\perp = \frac{\mu}{h}(1 + e\cos\theta) \tag{2.123}$$

Substituting Eqs. (2.122) and (2.123) into Eq. (2.121) and simplifying the results yields

$$\dot{\bar{x}} = -\frac{\mu}{h}\sin\theta \quad \dot{\bar{y}} = \frac{\mu}{h}(e + \cos\theta) \tag{2.124}$$

Hence, Eq. (2.120) becomes

$$\mathbf{v} = \frac{\mu}{h}[-\sin\theta\hat{\mathbf{p}} + (e + \cos\theta)\hat{\mathbf{q}}] \tag{2.125}$$

Formulating the kinematics of orbital motion in the perifocal frame, as we have done here, is a prelude to the study of orbits in three dimensions (Chapter 4). We also need Eqs. (2.117) and (2.120) in the next section.

EXAMPLE 2.11

An earth orbit has an eccentricity of 0.3, an angular momentum of 60, 000 km^2/s, and a true anomaly of 120°. What are the position vector **r** and velocity vector **v** in the perifocal frame of reference?

Solution

From Eq. (2.119) we have

$$\mathbf{r} = \frac{h^2}{\mu}\frac{1}{1 + e\cos\theta}(\cos\theta\hat{\mathbf{p}} + \sin\theta\hat{\mathbf{q}}) = \frac{60{,}000^2}{398{,}600}\frac{1}{1 + 0.3\cos 120^\circ}(\cos 120^\circ\hat{\mathbf{p}} + \sin 120^\circ\hat{\mathbf{q}})$$

$$\boxed{\mathbf{r} = -5312.7\hat{\mathbf{p}} + 9201.9\hat{\mathbf{q}}\ (\text{km})}$$

Substituting the given data into Eq. (2.125) yields

$$\mathbf{v} = \frac{\mu}{h}[-\sin\theta\hat{\mathbf{p}} + (e + \cos\theta)\hat{\mathbf{q}}] = \frac{398{,}600}{60{,}000}[-\sin 120^\circ\hat{\mathbf{p}} + (0.3 + \cos 120^\circ)\hat{\mathbf{q}}]$$

$$\boxed{\mathbf{v} = -5.7533\hat{\mathbf{p}} - 1.3287\hat{\mathbf{q}}\ (\text{km/s})}$$

EXAMPLE 2.12

An earth satellite has the following position and velocity vectors at a given instant:

$$\mathbf{r} = 7000\hat{\mathbf{p}} + 9000\hat{\mathbf{q}}\ (\text{km})$$
$$\mathbf{v} = -3.3472\hat{\mathbf{p}} + 9.1251\hat{\mathbf{q}}\ (\text{km/s})$$

Calculate the specific angular momentum h, the true anomaly θ, and the eccentricity e.

Solution

This problem is obviously the reverse of the situation presented in the previous example. From Eq. (2.28) the angular momentum is

$$\mathbf{h} = \mathbf{r} \times \mathbf{v} = \begin{vmatrix} \hat{\mathbf{p}} & \hat{\mathbf{q}} & \hat{\mathbf{w}} \\ 7000 & 9000 & 0 \\ -3.3472 & 9.1251 & 0 \end{vmatrix} = 94{,}000\hat{\mathbf{w}}\ (\text{km}^2/\text{s})$$

Hence, the magnitude of the angular momentum is

$$\boxed{h = 94{,}000\ \text{km}^2/\text{s}}$$

The true anomaly is measured from the positive $\bar{x}$ axis. By definition of the dot product, $\mathbf{r} \cdot \hat{\mathbf{p}} = r\cos\theta$. Thus,

$$\cos\theta = \frac{\mathbf{r}}{r} \cdot \hat{\mathbf{p}} = \frac{7000\hat{\mathbf{p}} + 9000\hat{\mathbf{q}}}{\sqrt{7000^2 + 9000^2}} \cdot \hat{\mathbf{p}} = \frac{7000}{11{,}402} = 0.61394$$

which means $\theta = 52.125°$ or $\theta = -52.125°$. Since the $\bar{y}$ component of $\mathbf{r}$ is positive, the true anomaly must lie between 0° and 180°. It follows that

$$\boxed{\theta = 52.125°}$$

Finally, the eccentricity may be found from the orbit formula, $r = (h^2/\mu)/(1 + e\cos\theta)$:

$$\sqrt{7000^2 + 9000^2} = \frac{94{,}000^2}{398{,}600} \frac{1}{1 + e\cos 52.125°}$$

$$\boxed{e = 1.538}$$

The trajectory is a hyperbola.

2.11 THE LAGRANGE COEFFICIENTS

In this section, we will establish what may seem intuitively obvious: if the position and velocity of an orbiting body are known at a given instant, then the position and velocity at any later time are found in terms of the initial values. Let us start with Eqs. (2.117) and (2.120),

$$\mathbf{r} = \bar{x}\hat{\mathbf{p}} + \bar{y}\hat{\mathbf{q}} \tag{2.126}$$

$$\mathbf{v} = \dot{\mathbf{r}} = \dot{\bar{x}}\hat{\mathbf{p}} + \dot{\bar{y}}\hat{\mathbf{q}} \tag{2.127}$$

Attach a subscript "zero" to quantities evaluated at time $t = t_0$. Then the expressions for $\mathbf{r}$ and $\mathbf{v}$ evaluated at $t = t_0$ are

$$\mathbf{r}_0 = \bar{x}_0\hat{\mathbf{p}} + \bar{y}_0\hat{\mathbf{q}} \tag{2.128}$$

$$\mathbf{v}_0 = \dot{\bar{x}}_0\hat{\mathbf{p}} + \dot{\bar{y}}_0\hat{\mathbf{q}} \tag{2.129}$$

The angular momentum $\mathbf{h}$ is constant, so let us calculate it using the initial conditions. Substituting Eqs. (2.128) and (2.129) into Eq. (2.28) yields

$$\mathbf{h} = \mathbf{r}_0 \times \mathbf{v}_0 = \begin{vmatrix} \hat{\mathbf{p}} & \hat{\mathbf{q}} & \hat{\mathbf{w}} \\ \bar{x}_0 & \bar{y}_0 & 0 \\ \dot{\bar{x}}_0 & \dot{\bar{y}}_0 & 0 \end{vmatrix} = \hat{\mathbf{w}}\left(\bar{x}_0\dot{\bar{y}}_0 - \bar{y}_0\dot{\bar{x}}_0\right) \tag{2.130}$$

Recall that $\hat{\mathbf{w}}$ is the unit vector in the direction of $\mathbf{h}$ (Eq. 2.116). Therefore, the coefficient of $\hat{\mathbf{w}}$ on the right-hand side of Eq. (2.130) must be the magnitude of the angular momentum. That is,

$$h=\bar{x}_0\dot{\bar{y}}_0-\bar{y}_0\dot{\bar{x}}_0 \tag{2.131}$$

Now let us solve Eqs. (2.128) and (2.129) for the unit vectors $\hat{\mathbf{p}}$ and $\hat{\mathbf{q}}$ in terms of $\mathbf{r}_0$ and $\mathbf{v}_0$. From Eq. (2.128) we get

$$\hat{\mathbf{q}}=\frac{1}{\bar{y}_0}\mathbf{r}_0-\frac{\bar{x}_0}{\bar{y}_0}\hat{\mathbf{p}} \tag{2.132}$$

Substituting this into Eq. (2.129), combining terms, and using Eq. (2.131) yields

$$\mathbf{v}_0=\dot{\bar{x}}_0\hat{\mathbf{p}}+\dot{\bar{y}}_0\left(\frac{1}{\bar{y}_0}\mathbf{r}_0-\frac{\bar{x}_0}{\bar{y}_0}\hat{\mathbf{p}}\right)=\frac{\bar{y}_0\dot{\bar{x}}_0-\bar{x}_0\dot{\bar{y}}_0}{\bar{y}_0}\hat{\mathbf{p}}+\frac{\dot{\bar{y}}_0}{\bar{y}_0}\mathbf{r}_0=-\frac{h}{\bar{y}_0}\hat{\mathbf{p}}+\frac{\dot{\bar{y}}_0}{\bar{y}_0}\mathbf{r}_0$$

Solve this for $\hat{\mathbf{p}}$ to obtain

$$\hat{\mathbf{p}}=\frac{\dot{\bar{y}}_0}{h}\mathbf{r}_0-\frac{\bar{y}_0}{h}\mathbf{v}_0 \tag{2.133}$$

Putting this result back into Eq. (2.132) gives

$$\hat{\mathbf{q}}=\frac{1}{\bar{y}_0}\mathbf{r}_0-\frac{\bar{x}_0}{\bar{y}_0}\left(\frac{\dot{\bar{y}}_0}{h}\mathbf{r}_0-\frac{\bar{y}_0}{h}\mathbf{v}_0\right)=\frac{h-\bar{x}_0\dot{\bar{y}}_0}{\bar{y}_0}\mathbf{r}_0+\frac{\bar{x}_0}{h}\mathbf{v}_0$$

Upon replacing h with the right-hand side of Eq. (2.131) we get

$$\hat{\mathbf{q}}=-\frac{\dot{\bar{x}}_0}{h}\mathbf{r}_0+\frac{\bar{x}_0}{h}\mathbf{v}_0 \tag{2.134}$$

Eqs. (2.133) and (2.134) give $\hat{\mathbf{p}}$ and $\hat{\mathbf{q}}$ in terms of the initial state vector. Substituting those two expressions back into Eqs. (2.126) and (2.127) yields, respectively

$$\mathbf{r}=\bar{x}\left(\frac{\dot{\bar{y}}_0}{h}\mathbf{r}_0-\frac{\bar{y}_0}{h}\mathbf{v}_0\right)+\bar{y}\left(-\frac{\dot{\bar{x}}_0}{h}\mathbf{r}_0+\frac{\bar{x}_0}{h}\mathbf{v}_0\right)=\frac{\bar{x}\dot{\bar{y}}_0-\bar{y}\dot{\bar{x}}_0}{h}\mathbf{r}_0+\frac{-\bar{x}\bar{y}_0+\bar{y}\bar{x}_0}{h}\mathbf{v}_0$$

$$\mathbf{v}=\dot{\bar{x}}\left(\frac{\dot{\bar{y}}_0}{h}\mathbf{r}_0-\frac{\bar{y}_0}{h}\mathbf{v}_0\right)+\dot{\bar{y}}\left(-\frac{\dot{\bar{x}}_0}{h}\mathbf{r}_0+\frac{\bar{x}_0}{h}\mathbf{v}_0\right)=\frac{\dot{\bar{x}}\dot{\bar{y}}_0-\dot{\bar{y}}\dot{\bar{x}}_0}{h}\mathbf{r}_0+\frac{-\dot{\bar{x}}\bar{y}_0+\dot{\bar{y}}\bar{x}_0}{h}\mathbf{v}_0$$

Therefore,

$$\mathbf{r}=f\mathbf{r}_0+g\mathbf{v}_0 \tag{2.135}$$

$$\mathbf{v}=\dot{f}\mathbf{r}_0+\dot{g}\mathbf{v}_0 \tag{2.136}$$

where f and g are given by

$$f=\frac{\bar{x}\dot{\bar{y}}_0-\bar{y}\dot{\bar{x}}_0}{h} \tag{2.137a}$$

$$g=\frac{-\bar{x}\bar{y}_0+\bar{y}\bar{x}_0}{h} \tag{2.137b}$$

together with their time derivatives

$$\dot{f}=\frac{\dot{\bar{x}}\dot{\bar{y}}_0-\dot{\bar{y}}\dot{\bar{x}}_0}{h} \tag{2.138a}$$

$$\dot{g} = \frac{-\dot{\bar{x}}\bar{y}_0 + \dot{\bar{y}}\bar{x}_0}{h} \tag{2.138b}$$

The f and g functions are referred to as the Lagrange coefficients after Joseph-Louis Lagrange (1736–1813), an Italian mathematical physicist whose numerous contributions include calculations of planetary motion.

From Eqs. (2.135) and (2.136) we see that the position and velocity vectors **r** and **v** are indeed linear combinations of the initial position and velocity vectors. The Lagrange coefficients and their time derivatives in these expressions are themselves functions of time and the initial conditions.

Before proceeding, let us show that the conservation of angular momentum **h** imposes a condition on f and g and their time derivatives $\dot{f}$ and $\dot{g}$. Calculate **h** using Eqs. (2.135) and (2.136),

$$\mathbf{h} = \mathbf{r} \times \mathbf{v} = (f\mathbf{r}_0 + g\mathbf{v}_0) \times (\dot{f}\mathbf{r}_0 + \dot{g}\mathbf{v}_0)$$

Expanding the right-hand side yields

$$\mathbf{h} = (f\mathbf{r}_0 \times \dot{f}\mathbf{r}_0) + (f\mathbf{r}_0 \times \dot{g}\mathbf{v}_0) + (g\mathbf{v}_0 \times \dot{f}\mathbf{r}_0) + (g\mathbf{v}_0 \times \dot{g}\mathbf{v}_0)$$

Factoring out the scalars f, g, $\dot{f}$, and $\dot{g}$, we get

$$\mathbf{h} = f\dot{f}(\mathbf{r}_0 \times \mathbf{r}_0) + f\dot{g}(\mathbf{r}_0 \times \mathbf{v}_0) + \dot{f}g(\mathbf{v}_0 \times \mathbf{r}_0) + g\dot{g}(\mathbf{v}_0 \times \mathbf{v}_0)$$

But $\mathbf{r}_0 \times \mathbf{r}_0 = \mathbf{v}_0 \times \mathbf{v}_0 = \mathbf{0}$, so

$$\mathbf{h} = f\dot{g}(\mathbf{r}_0 \times \mathbf{v}_0) + \dot{f}g(\mathbf{v}_0 \times \mathbf{r}_0)$$

Since

$$\mathbf{v}_0 \times \mathbf{r}_0 = -(\mathbf{r}_0 \times \mathbf{v}_0)$$

this reduces to

$$\mathbf{h} = (f\dot{g} - \dot{f}g)(\mathbf{r}_0 \times \mathbf{v}_0)$$

or

$$\mathbf{h} = (f\dot{g} - \dot{f}g)\mathbf{h}_0$$

where $\mathbf{h}_0 = \mathbf{r}_0 \times \mathbf{v}_0$, which is the angular momentum at $t = t_0$. But the angular momentum is constant (recall Eq. 2.29), which means $\mathbf{h} = \mathbf{h}_0$, so that

$$\mathbf{h} = (f\dot{g} - \dot{f}g)\mathbf{h}$$

Since **h** cannot be zero (unless the body is traveling in a straight line toward the center of attraction), it follows that

$$f\dot{g} - \dot{f}g = 1 \quad \text{(Conservation of angular momentum)} \tag{2.139}$$

Thus, if any three of the functions f, g, $\dot{f}$, and $\dot{g}$ are known, the fourth may be found from Eq. (2.139).

Let us use Eqs. (2.137) and (2.138) to evaluate the Lagrange coefficients and their time derivative in terms of the true anomaly. First of all, note that evaluating Eq. (2.118) at time $t = t_0$ yields

$$\begin{aligned} \bar{x}_0 &= r_0 \cos\theta_0 \\ \bar{y}_0 &= r_0 \sin\theta_0 \end{aligned} \tag{2.140}$$

Likewise, from Eq. (2.124) we get

$$\begin{aligned}\dot{x}_0 &= -\frac{\mu}{h}\sin\theta_0 \\ \dot{y}_0 &= \frac{\mu}{h}(e+\cos\theta_0)\end{aligned} \tag{2.141}$$

To evaluate the function, f we substitute Eqs. (2.118) and (2.141) into Eq. (2.137a),

$$\begin{aligned}f &= \frac{x\dot{y}_0 - y\dot{x}_0}{h} = \frac{1}{h}\left\{r\cos\theta\left[\frac{\mu}{h}(e+\cos\theta_0)\right] - r\sin\theta\left(-\frac{\mu}{h}\sin\theta_0\right)\right\} \\ &= \frac{\mu r}{h^2}[e\cos\theta + (\cos\theta\cos\theta_0 + \sin\theta\sin\theta_0)]\end{aligned} \tag{2.142}$$

If we invoke the trig identity

$$\cos(\theta-\theta_0) = \cos\theta\cos\theta_0 + \sin\theta\sin\theta_0 \tag{2.143}$$

and let $\Delta\theta$ represent the difference between the current and initial true anomalies,

$$\Delta\theta = \theta - \theta_0 \tag{2.144}$$

then Eq. (2.142) reduces to

$$f = \frac{\mu r}{h^2}(e\cos\theta + \cos\Delta\theta) \tag{2.145}$$

Finally, from Eq. (2.45), we have

$$e\cos\theta = \frac{h^2}{\mu r} - 1 \tag{2.146}$$

Substituting this into Eq. (2.145) leads to

$$f = 1 - \frac{\mu r}{h^2}(1-\cos\Delta\theta) \tag{2.147}$$

We obtain r from the orbit formula (Eq. 2.45) in which the true anomaly θ appears, whereas the difference in the true anomalies occurs on the right-hand side of Eq. (2.147). However, we can express the orbit equation in terms of the difference in true anomalies as follows. From Eq. (2.144), we have $\theta = \theta_0 + \Delta\theta$, which means we can write the orbit equation as

$$r = \frac{h^2}{\mu}\frac{1}{1+e\cos(\theta_0+\Delta\theta)} \tag{2.148}$$

By replacing θ_0 with $-\Delta\theta$ in Eq. (2.143), Eq. (2.148) becomes

$$r = \frac{h^2}{\mu}\frac{1}{1+e\cos\theta_0\cos\Delta\theta - e\sin\theta_0\sin\Delta\theta} \tag{2.149}$$

To remove θ_0 from this expression, observe first of all that Eq. (2.146) implies that, at $t = t_0$,

$$e\cos\theta_0 = \frac{h^2}{\mu r_0} - 1 \tag{2.150}$$

Furthermore, from Eq. (2.49) for the radial velocity we obtain

$$e\sin\theta_0 = \frac{h v_r)_0}{\mu} \tag{2.151}$$

Substituting Eqs. (2.150) and (2.151) into Eq. (2.149) yields

$$r = \frac{h^2}{\mu}\frac{1}{1+\left(\dfrac{h^2}{\mu r_0}-1\right)\cos\Delta\theta - \dfrac{h v_r)_0}{\mu}\sin\Delta\theta} \tag{2.152}$$

Using this form of the orbit equation, we can find r in terms of the initial conditions and the change in the true anomaly. Thus f in Eq. (2.147) depends only on $\Delta\theta$.

The Lagrange coefficient g is found by substituting Eqs. (2.118) and (2.140) into Eq. (2.137b),

$$\begin{aligned} g &= \frac{-\bar{x}\bar{y}_0 + \bar{y}\bar{x}_0}{h} \\ &= \frac{1}{h}[(-r\cos\theta)(r_0\sin\theta_0) + (r\sin\theta)(r\cos\theta_0)] \\ &= \frac{r r_0}{h}(\sin\theta\cos\theta_0 - \cos\theta\sin\theta_0) \end{aligned} \tag{2.153}$$

Making use of the trig identity

$$\sin(\theta-\theta_0) = \sin\theta\cos\theta_0 - \cos\theta\sin\theta_0$$

together with Eq. (2.144), we find

$$g = \frac{r r_0}{h}\sin(\Delta\theta) \tag{2.154}$$

To obtain $\dot{g}$, substitute Eqs. (2.124) and (2.140) into Eq. (2.138b),

$$\begin{aligned} \dot{g} = \frac{-\dot{\bar{x}}\bar{y}_0 + \dot{\bar{y}}\bar{x}_0}{h} &= \frac{1}{h}\left\{-\left[-\frac{\mu}{h}\sin\theta\right][r_0\sin\theta_0] + \left[\frac{\mu}{h}(e+\cos\theta)\right](r_0\cos\theta_0)\right\} \\ &= \frac{\mu r_0}{h^2}[e\cos\theta_0 + (\cos\theta\cos\theta_0 + \sin\theta\sin\theta_0)] \end{aligned}$$

With the aid of Eqs. (2.143) and (2.150), this reduces to

$$\dot{g} = 1 - \frac{\mu r_0}{h^2}(1-\cos\Delta\theta) \tag{2.155}$$

$\dot{f}$ can be found using Eq. (2.139). Thus,

$$\dot{f} = \frac{1}{g}(f\dot{g}-1) \tag{2.156}$$

Substituting Eqs. (2.147), (2.153), and (2.155) results in

$$\begin{aligned} \dot{f} &= \frac{1}{\dfrac{r r_0}{h}\sin\Delta\theta}\left\{\left[1-\frac{\mu r}{h^2}(1-\cos\Delta\theta)\right]\left[1-\frac{\mu r_0}{h^2}(1-\cos\Delta\theta)\right]-1\right\} \\ &= \frac{1}{\dfrac{r r_0}{h}\sin\Delta\theta}\frac{h^2\mu r r_0}{h^4}\left[(1-\cos\Delta\theta)^2\frac{\mu}{h^2} - (1-\cos\Delta\theta)\left(\frac{1}{r_0}+\frac{1}{r}\right)\right] \end{aligned}$$

or

$$\dot{f} = \frac{\mu}{h}\frac{1-\cos\Delta\theta}{\sin\Delta\theta}\left[\frac{\mu}{h^2}(1-\cos\Delta\theta) - \frac{1}{r_0} - \frac{1}{r}\right] \tag{2.157}$$

To summarize, the Lagrange coefficients in terms of the change in true anomaly are:

$$f = 1 - \frac{\mu r}{h^2}(1-\cos\Delta\theta) \tag{2.158a}$$

$$g = \frac{r r_0}{h}\sin\Delta\theta \tag{2.158b}$$

$$\dot{f} = \frac{\mu}{h}\frac{1-\cos\Delta\theta}{\sin\Delta\theta}\left[\frac{\mu}{h^2}(1-\cos\Delta\theta) - \frac{1}{r_0} - \frac{1}{r}\right] \tag{2.158c}$$

$$\dot{g} = 1 - \frac{\mu r_0}{h^2}(1-\cos\Delta\theta) \tag{2.158d}$$

where r is given by Eq. (2.152).

The implementation of these four functions in MATLAB is presented in Appendix D.7.

Observe that using the Lagrange coefficients to determine the position and velocity from the initial conditions does not require knowing the type of orbit we are dealing with (ellipse, parabola, or hyperbola), since the eccentricity does not appear in Eqs. (2.152) and (2.158). However, the initial position and velocity give us that information. From $\mathbf{r}_0$ and $\mathbf{v}_0$ we obtain the angular momentum $h = \|\mathbf{r}_0 \times \mathbf{v}_0\|$. The initial radius r_0 is just the magnitude of the vector $\mathbf{r}_0$. The initial radial velocity $v_r)_0$ is the projection of $\mathbf{v}_0$ onto the direction of $\mathbf{r}_0$,

$$v_r)_0 = \mathbf{v}_0 \cdot \frac{\mathbf{r}_0}{r_0}$$

From Eqs. (2.45) and (2.49) we have

$$r_0 = \frac{h^2}{\mu}\frac{1}{1+e\cos\theta_0} \qquad v_r)_0 = \frac{\mu}{h}e\sin\theta_0$$

These two equations can be solved for the eccentricity e and for the true anomaly of the initial point θ_0.

ALGORITHM 2.3

Given $\mathbf{r}_0$ and $\mathbf{v}_0$, find $\mathbf{r}$ and $\mathbf{v}$ after the true anomaly changes by $\Delta\theta$. See Appendix D.8 for an implementation of this procedure in MATLAB.

1. Compute the f and g functions and their derivatives by the following steps:
   (a) Calculate the magnitude of $\mathbf{r}_0$ and $\mathbf{v}_0$:

$$r_0 = \sqrt{\mathbf{r}_0 \cdot \mathbf{r}_0} \qquad v_0 = \sqrt{\mathbf{v}_0 \cdot \mathbf{v}_0}$$

   (b) Calculate the radial component of $\mathbf{v}_0$ by projecting it onto the direction of $\mathbf{r}_0$:

$$v_r)_0 = \frac{\mathbf{r}_0 \cdot \mathbf{v}_0}{r_0}$$

   (c) Calculate the magnitude of the constant angular momentum:

$$h = r_0 v_\perp)_0 = r_0\sqrt{v_0^2 - v_r)_0^{\,2}}$$

(d) Substitute r_0, $v_r)_0$, h, and $\Delta\theta$ in Eq. (2.152) to calculate r.
(e) Substitute r, r_0, h, and $\Delta\theta$ into Eqs. (2.158a) and (2.158b) to find f, g, $\dot{f}$, and $\dot{g}$.

2. Use Eqs. (2.135) and (2.136) to calculate **r** and **v**.

EXAMPLE 2.13

An earth satellite moves in the xy plane of an inertial frame with the origin at the earth's center. Relative to that frame, the position and velocity of the satellite at time t_0 are

$$\begin{aligned}\mathbf{r}_0 &= 8182.4\hat{\mathbf{i}} - 6865.9\hat{\mathbf{j}}\,(\text{km})\\ \mathbf{v}_0 &= 0.47572\hat{\mathbf{i}} + 8.8116\hat{\mathbf{j}}\ (\text{km/s})\end{aligned} \tag{a}$$

Use Algorithm 2.3 to compute the position and velocity vectors after the satellite has traveled through a true anomaly of 120°.

Solution

Step 1:

(a)

$$r_0 = \sqrt{\mathbf{r}_0 \cdot \mathbf{r}_0} = 10{,}861\,\text{km} \quad v_0 = \sqrt{\mathbf{v}_0 \cdot \mathbf{v}_0} = 8.8244\,\text{km/s}$$

(b)

$$v_r)_0 = \mathbf{v}_0 \cdot \frac{\mathbf{r}_0}{r_0} = \frac{\left(0.47572\hat{\mathbf{i}} + 8.8116\hat{\mathbf{j}}\right) \cdot \left(8182.4\hat{\mathbf{i}} - 6865.9\hat{\mathbf{j}}\right)}{10{,}681} = -5.2996\,\text{km/s}$$

(c)

$$h = r_0\sqrt{v_0^2 - v_{r0}^2} = 10{,}861\sqrt{8.8244^2 - (-5.2996)^2} = 75{,}366\ \text{km}^2/\text{s}$$

(d)

$$\begin{aligned} r &= \frac{h^2}{\mu}\frac{1}{1 + \left(\dfrac{h^2}{\mu r_0} - 1\right)\cos\Delta\theta - \dfrac{h v_r)_0}{\mu}\sin\Delta\theta}\\ &= \frac{75{,}366^2}{398{,}600}\frac{1}{1 + \left(\dfrac{75{,}366^2}{398{,}600 \cdot 10{,}681} - 1\right)\cos 120^\circ - \dfrac{75{,}366 \cdot (-5.2996)}{398{,}600}\sin 120^\circ}\\ &= 8378.8\,\text{km}\end{aligned}$$

(e)

$$\begin{aligned} f &= 1 - \frac{\mu r}{h^2}(1 - \cos\Delta\theta)\\ &= 1 - \frac{398{,}600 \cdot 8378.8}{75{,}366^2}(1 - \cos 120^\circ) = 0.11802\ (\text{dimensionless})\\ g &= \frac{r r_0}{h}\sin(\Delta\theta) = \frac{8378.8 \cdot 10{,}681}{75{,}366}\sin 120^\circ = 1028.4\,\text{s}\\ \dot{f} &= \frac{\mu}{h}\frac{1 - \cos\Delta\theta}{\sin\Delta\theta}\left[\frac{\mu}{h^2}(1 - \cos\Delta\theta) - \frac{1}{r_0} - \frac{1}{r}\right]\\ &= \frac{398{,}600}{75{,}366}\frac{1 - \cos 120^\circ}{\sin 120^\circ}\left[\frac{398{,}600}{75{,}366^2}(1 - \cos 120^\circ) - \frac{1}{10{,}681} - \frac{1}{8378.9}\right]\\ &= -9.8666\left(10^{-4}\right)\text{s}^{-1}\end{aligned}$$

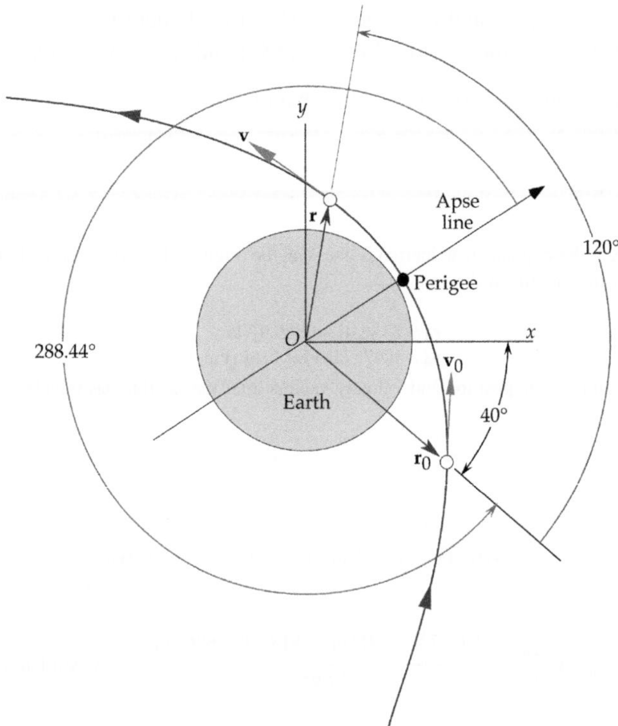

FIG. 2.31

The initial and final position and velocity vectors and the perigee location for Examples 2.13 and 2.14.

$$\dot{g} = \left(1 + \dot{f}g\right)/f = \left[1 + \left(-9.8666 \cdot 10^{-4}\right) \cdot 1028.4\right]/0.11803 = -0.12435 \text{ (dimensionless)}$$

Step 2:

$$\begin{aligned}\mathbf{r} &= f\mathbf{r}_0 + g\mathbf{v}_0 \\ &= 0.118802\left(8182.4\hat{\mathbf{i}} - 6865.9\hat{\mathbf{j}}\right) + 1028.4\left(0.47572\hat{\mathbf{i}} + 8.8116\hat{\mathbf{j}}\right) \\ &= \boxed{1454.9\hat{\mathbf{i}} + 8251.6\hat{\mathbf{j}} \text{ (km)}}\end{aligned}$$

$$\begin{aligned}\mathbf{v} &= \dot{f}\mathbf{r}_0 + \dot{g}\mathbf{v}_0 \\ &= \left(-9.8666 \times 10^{-4}\right)\left(8182.4\hat{\mathbf{i}} - 6865.9\hat{\mathbf{j}}\right) + (-0.12435)\left(0.47572\hat{\mathbf{i}} + 8.8116\hat{\mathbf{j}}\right) \\ &= \boxed{-8.1323\hat{\mathbf{i}} + 5.6785\hat{\mathbf{j}} \text{ (km/s)}}\end{aligned}$$

These results are shown in Fig. 2.31.

EXAMPLE 2.14

Find the eccentricity of the orbit in Example 2.13 as well as the true anomaly at the initial time t_0 and, hence, the location of the perigee for this orbit.

Solution

In Example 2.13, we found

$$r_0 = 10.861\,\text{km} \quad v_r)_0 = -5.2996\,\text{km/s} \quad h = 75{,}366\,\text{km}^2/\text{s} \tag{a}$$

Since $v_r)_0$ is negative, we know that the spacecraft is approaching the perigee, which means that

$$180° < \theta_0 < 360° \tag{b}$$

The orbit formula and the radial velocity formula (Eqs. 2.45 and 2.49) evaluated at t_0 are

$$r_0 = \frac{h^2}{\mu}\frac{1}{1 + e\cos\theta_0} \qquad v_r)_0 = \frac{\mu}{h} e \sin\theta_0$$

Substituting the numerical values from Eqs. (a) into these formulas yields

$$10{,}861 = \frac{75{,}366^3}{398{,}600}\frac{1}{1 + e\cos\theta_0} \qquad -5.2996 = \frac{398{,}600}{75{,}366} e \sin\theta_0$$

From these, we obtain two equations for the two unknowns e and θ_0:

$$e\cos\theta_0 = 0.3341 \qquad e\sin\theta_0 = -1.002 \tag{c}$$

Summing the squares of these two expressions gives

$$e^2\left(\sin^2\theta_0 + \cos^2\theta_0\right) = 1.1157$$

Recalling the trig identity $\sin^2 x + \cos^2 x = 1$, we get

$$\boxed{e = 1.0563} \ \text{(hyperbola)}$$

The eccentricity may be substituted back into either of the two expressions in Eq. (c) to find the true anomaly θ_0. Choosing Eq. $(c)_1$, we find

$$\cos\theta_0 = \frac{0.3341}{1.0563} = 0.3163$$

This means either $\theta_0 = 71.56°$ (moving away from the perigee) or $\theta_0 = 288.44°$ (moving toward the perigee). From Eq. (a) we know that the motion is toward perigee, so that

$$\boxed{\theta_0 = 288.44°}$$

Fig. 2.31 shows the computed location of the perigee relative to the initial and final position vectors.

To use the Lagrange coefficients to find the position and velocity as a function of time instead of true anomaly, we need to come up with a relation between $\Delta\theta$ and time. We deal with that complex problem in Chapter 3. Meanwhile, for times t that are close to the initial time t_0, we can obtain polynomial expressions for f and g in which the variable $\Delta\theta$ is replaced by the time interval $\Delta t = t - t_0$.

To do so, we expand the position vector $\mathbf{r}(t)$, considered to be a function of time, in a Taylor series about $t = t_0$. As pointed out previously (Eqs. 1.97 and 1.98), the Taylor series is given by

$$\mathbf{r}(t) = \sum_{n=0}^{\infty} \frac{1}{n!}\mathbf{r}^{(n)}(t_0)(t - t_0)^n \tag{2.159}$$

where $\mathbf{r}^{(n)}(t_0)$ is the nth time derivative of $\mathbf{r}(t)$, evaluated at t_0,

$$\mathbf{r}^{(n)}(t_0) = \left(\frac{d^n r}{dt^n}\right)_{t=t_0} \tag{2.160}$$

Let us truncate this infinite series at four terms. Then, to that degree of approximation,

$$\mathbf{r}(t)=\mathbf{r}(t_0)+\left(\frac{d\mathbf{r}}{dt}\right)_{t=t_0}\Delta t+\frac{1}{2}\left(\frac{d^2\mathbf{r}}{dt^2}\right)_{t=t_0}\Delta t^2+\frac{1}{6}\left(\frac{d^3\mathbf{r}}{dt^3}\right)_{t=t_0}\Delta t^3+\frac{1}{24}\left(\frac{d^4\mathbf{r}}{dt^4}\right)_{t=t_0}\Delta t^4 \tag{2.161}$$

where $\Delta t = t - t_0$. To evaluate the four derivatives, we note first that $(d\mathbf{r}/dt)_{t=t_0}$ is just the velocity $\mathbf{v}$ at $t = t_0$,

$$\left(\frac{d\mathbf{r}}{dt}\right)_{t=t_0}=\mathbf{v}_0 \tag{2.162}$$

$(d^2\mathbf{r}/dt^2)_{t=t_0}$ is evaluated using Eq. (2.22),

$$\ddot{\mathbf{r}}=-\frac{\mu}{r^3}\mathbf{r} \tag{2.163}$$

Thus,

$$\left(\frac{d^2\mathbf{r}}{dt^2}\right)_{t=t_0}=-\frac{\mu}{r_0^3}\mathbf{r}_0 \tag{2.164}$$

$(d^3\mathbf{r}/dt^3)_{t=t_0}$ is evaluated by differentiating Eq. (2.163),

$$\frac{d^3\mathbf{r}}{dt^3}=-\mu\frac{d}{dt}\left(\frac{\mathbf{r}}{r^3}\right)=-\mu\left(\frac{r^3\mathbf{v}-3\mathbf{r}r^2\dot{r}}{r^6}\right)=-\mu\frac{\mathbf{v}}{r^3}+3\mu\frac{\dot{r}\mathbf{r}}{r^4} \tag{2.165}$$

From Eq. (2.35a) we have

$$\dot{r}=\frac{\mathbf{r}\cdot\mathbf{v}}{r} \tag{2.166}$$

Hence, Eq. (2.165), evaluated at $t = t_0$, is

$$\left(\frac{d^3\mathbf{r}}{dt^3}\right)_{t=t_0}=-\mu\frac{\mathbf{v}_0}{r_0^3}+3\mu\frac{\mathbf{r}_0\cdot\mathbf{v}_0}{r_0^5}\mathbf{r}_0 \tag{2.167}$$

Finally, $(d^4\mathbf{r}/dt^4)_{t=t_0}$ is found by first differentiating Eq. (2.165),

$$\frac{d^4\mathbf{r}}{dt^4}=\frac{d}{dt}\left(-\mu\frac{\dot{\mathbf{r}}}{r^3}+3\mu\frac{\dot{r}\mathbf{r}}{r^4}\right)=-\mu\left(\frac{r^3\ddot{\mathbf{r}}-3r^2\dot{r}\dot{\mathbf{r}}}{r^6}\right)+3\mu\left[\frac{r^4(\ddot{r}\mathbf{r}+\dot{r}\dot{\mathbf{r}})-4r^3\dot{r}^2\mathbf{r}}{r^8}\right] \tag{2.168}$$

$\ddot{r}$ is found in terms of $\mathbf{r}$ and $\mathbf{v}$ by differentiating Eq. (2.166) and making use of Eq. (2.163). This leads to the expression

$$\ddot{r}=\frac{d}{dt}\left(\frac{\mathbf{r}\cdot\dot{\mathbf{r}}}{r}\right)=\frac{v^2}{r}-\frac{\mu}{r^2}-\frac{(\mathbf{r}\cdot\mathbf{v})^2}{r^3} \tag{2.169}$$

Substituting Eqs. (2.163), (2.166), and (2.169) into Eq. (2.168), combining terms, and evaluating the result at $t = t_0$ yields

$$\left(\frac{d^4\mathbf{r}}{dt^4}\right)_{t=t_0}=\left[-2\frac{\mu^2}{r_0^6}+3\mu\frac{v_0^2}{r_0^5}-15\mu\frac{(\mathbf{r}_0\cdot\mathbf{v}_0)^2}{r_0^7}\right]\mathbf{r}_0+6\mu\frac{(\mathbf{r}_0\cdot\mathbf{v}_0)}{r_0^5}\mathbf{v}_0 \tag{2.170}$$

After substituting Eqs. (2.162), (2.164), (2.167), and (2.170) into Eq. (2.161), rearranging, and collecting terms, we obtain

$$\begin{aligned}\mathbf{r}(t)=&\left\{1-\frac{\mu}{2r_0^3}\Delta t^2+\frac{\mu}{2}\frac{\mathbf{r_0}\cdot\mathbf{v_0}}{r_0^5}\Delta t^3+\frac{\mu}{24}\left[-2\frac{\mu}{r_0^6}+3\frac{v_0^2}{r_0^5}-15\frac{(\mathbf{r_0}\cdot\mathbf{v_0})^2}{r_0^7}\right]\Delta t^4\right\}\mathbf{r_0}\\&+\left[\Delta t-\frac{1}{6}\frac{\mu}{r_0^3}\Delta t^3+\frac{\mu}{4}\frac{(\mathbf{r_0}\cdot\mathbf{v_0})}{r_0^5}\Delta t^4\right]\mathbf{v_0}\end{aligned} \tag{2.171}$$

Comparing this expression with Eq. (2.135), we see that, to the fourth order in Δt,

$$\begin{aligned}f&=1-\frac{\mu}{2r_0^3}\Delta t^2+\frac{\mu}{2}\frac{\mathbf{r_0}\cdot\mathbf{v_0}}{r_0^5}\Delta t^3+\frac{\mu}{24}\left[-2\frac{\mu}{r_0^6}+3\frac{v_0^2}{r_0^5}-15\frac{(\mathbf{r_0}\cdot\mathbf{v_0})^2}{r_0^7}\right]\Delta t^4\\g&=\Delta t-\frac{1}{6}\frac{\mu}{r_0^3}\Delta t^3+\frac{\mu}{4}\frac{\mathbf{r_0}\cdot\mathbf{v_0}}{r_0^5}\Delta t^4\end{aligned} \tag{2.172}$$

For small values of elapsed time Δt these f and g series may be used to calculate the position of an orbiting body from the initial conditions.

EXAMPLE 2.15

The orbit of an earth satellite has an eccentricity $e=0.2$ and a perigee radius of 7000 km. Starting at the perigee, plot the radial distance as a function of time using the f and g series and compare the curve with the exact solution.

Solution

Since the satellite starts at the perigee, $t_0=0$, and we have, using the perifocal frame,

$$\mathbf{r}_0=7000\hat{\mathbf{p}}\ (\text{km}) \tag{a}$$

The orbit equation evaluated at the perigee is Eq. (2.50), which in the present case becomes

$$7000=\frac{h^2}{398{,}600}\frac{1}{1+0.2}$$

Solving for the angular momentum, we get $h=57{,}864\ \text{km}^2/\text{s}$. Then, using the angular momentum formula, Eq. (2.31), we find that the speed at the perigee is $v_0=8.2663$ km/s, so that

$$\mathbf{v}_0=8.2663\hat{\mathbf{q}}\ (\text{km/s}) \tag{b}$$

Clearly, $\mathbf{r}_0\cdot\mathbf{v}_0=0$. Hence, with $\mu=398{,}600\ \text{km}^3/\text{s}^2$, the two Lagrange series in Eq. (2.172) become (setting $\Delta t=t$)

$$\begin{aligned}f&=1-5.8105(10^{-7})t^2+9.0032(10^{-14})t^4\\g&=t-1.9368(10^{-7})t^3\end{aligned}$$

where the units of t are seconds. Substituting f and g into Eq. (2.135) yields

$$\mathbf{r}=\left[1-5.8105(10^{-7})t^2+9.0032(10^{-14})t^4\right](7000\hat{\mathbf{p}})+\left[t-1.9368(10^{-7})t^3\right](8.2663\hat{\mathbf{q}})$$

From this we obtain

$$\boxed{r=\|\mathbf{r}\|=\sqrt{49(10^6)+11.389t^2-1.103(10^{-6})t^4-2.5633(10^{-12})t^6+3.9718(10^{-19})t^8}} \tag{c}$$

For the exact solution of r versus time we must appeal to the methods presented in Chapter 3. The exact solution and the series solution (Eq. (c)) are plotted in Fig. 2.32. As can be seen, the series solution begins to seriously diverge from the exact solution after about 10 min.

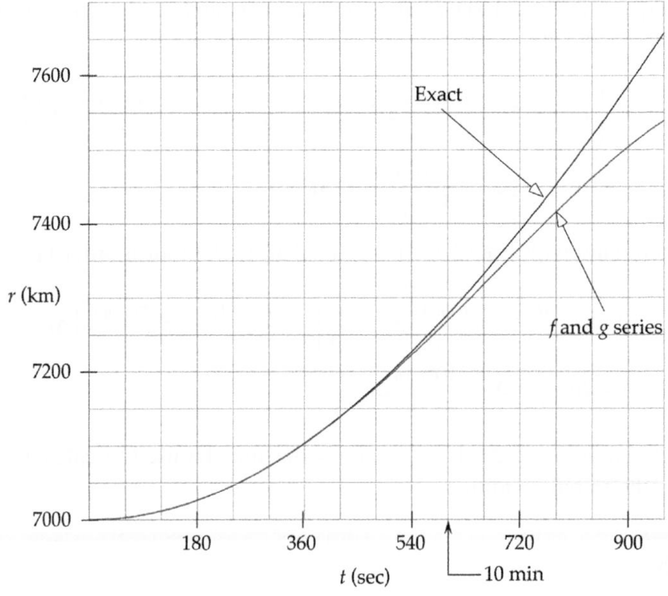

FIG. 2.32

Exact and series solutions for the radial position of the satellite.

If we include terms of fifth and higher orders in the f and g series (Eq. 2.172), then the approximate solution in the above example will agree with the exact solution for a longer time interval than that indicated in Fig. 2.32. However, there is a time interval beyond which the series solution will diverge from the exact one no matter how many terms we include. This time interval is called the radius of convergence. According to Bond and Allman (1996), for the elliptical orbit of Example 2.15, the radius of convergence is 1700 s (not quite half an hour), which is one-fifth of the period of that orbit. This further illustrates the fact that the series forms of the Lagrange coefficients are applicable only over small time intervals. For arbitrary time intervals, the closed form of these functions, presented in Chapter 3, must be employed.

2.12 CIRCULAR RESTRICTED THREE-BODY PROBLEM

Consider two bodies m_1 and m_2 moving under the action of just their mutual gravitation, and let their orbit around each other be a circle of radius r_{12}. Consider as well a noninertial, comoving frame of reference xyz whose origin lies at the center of mass G of the two-body system, with the x axis directed toward m_2, as shown in Fig. 2.33. The y axis lies in the orbital plane, to which the z axis is perpendicular. In this rotating frame of reference, m_1 and m_2 appear to be at rest, the force of gravity on each one

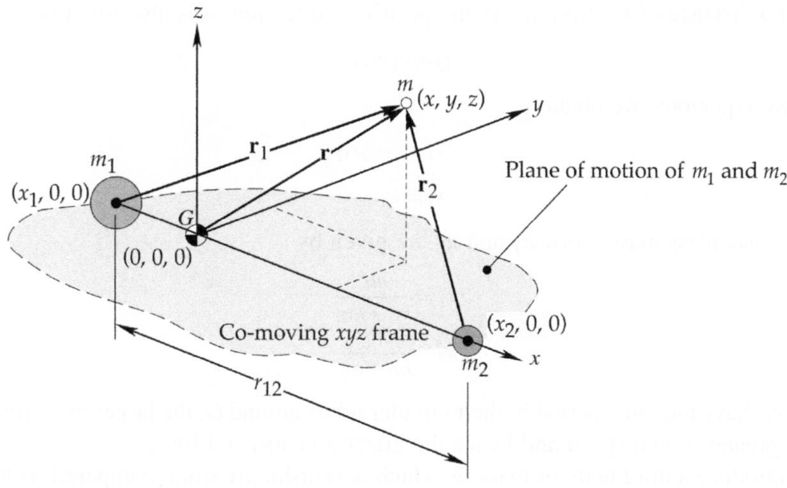

FIG. 2.33

Primary bodies m_1 and m_2 in circular orbit around each other, plus a secondary mass m.

seemingly balanced by the fictitious centripetal force required to hold it in its circular path around the system center of mass. We shall henceforth assume that $m_1 > m_2$, so that body 1 might be the earth and body 2 its moon.

The constant, inertial angular velocity $\mathbf{\Omega}$ is given by

$$\mathbf{\Omega} = \Omega \hat{\mathbf{k}} \tag{2.173}$$

where

$$\Omega = \frac{2\pi}{T}$$

and T is the period of the orbit (Eq. 2.64),

$$T = \frac{2\pi}{\sqrt{\mu}} {r_{12}}^{3/2}$$

Thus,

$$\Omega = \sqrt{\frac{\mu}{{r_{12}}^3}} \tag{2.174}$$

Recall that if M is the total mass of the system,

$$M = m_1 + m_2 \tag{2.175}$$

then

$$\mu = GM \tag{2.176}$$

m_1 and m_2 lie in the orbital plane, so that their y and z coordinates are zero. To determine their locations on the x axis, we use the definition of the center of mass (Eq. 2.2) to write

$$m_1 x_1 + m_2 x_2 = 0$$

Since m_2 is at a distance of r_{12} from m_1 in the positive x direction, it is also true that

$$x_2 = x_1 + r_{12}$$

From these two equations, we obtain

$$x_1 = -\pi_2 r_{12} \tag{2.177a}$$

$$x_2 = \pi_1 r_{12} \tag{2.177b}$$

where the dimensionless mass ratios π_1 and π_2 are given by

$$\begin{aligned} \pi_1 &= \frac{m_1}{m_1 + m_2} \\ \pi_2 &= \frac{m_2}{m_1 + m_2} \end{aligned} \tag{2.178}$$

Since m_1 and m_2 have the same period in their circular orbits around G, the larger mass (the one closest to G) has the greater orbital speed and hence the greatest centripetal force.

We now introduce a third body of mass m, which is vanishingly small compared with the primary masses m_1 and m_2, like the mass of a spacecraft compared with that of a planet or moon of the solar system. This is called the circular restricted three-body problem (CRTBP), because the secondary mass m is assumed to be so small that it has no effect on the circular motion of the primary bodies around each other. We are interested in the motion of m due to the gravitational fields of m_1 and m_2. Unlike the two-body problem, there is no general, closed-form solution for this motion. However, we can set up the equations of motion and draw some general conclusions from them.

In the comoving coordinate system, the position vector of the secondary mass m relative to m_1 is given by

$$\mathbf{r}_1 = (x - x_1)\hat{\mathbf{i}} + y\hat{\mathbf{j}} + z\hat{\mathbf{k}} = (x + \pi_2 r_{12})\hat{\mathbf{i}} + y\hat{\mathbf{j}} + z\hat{\mathbf{k}} \tag{2.179}$$

Relative to m_2 the position of m is

$$\mathbf{r}_2 = (x - \pi_1 r_{12})\hat{\mathbf{i}} + y\hat{\mathbf{j}} + z\hat{\mathbf{k}} \tag{2.180}$$

Finally, the position vector of the secondary body relative to the center of mass is

$$\mathbf{r} = x\hat{\mathbf{i}} + y\hat{\mathbf{j}} + z\hat{\mathbf{k}} \tag{2.181}$$

The inertial velocity of m is found by taking the time derivative of Eq. (2.181). However, relative to inertial space, the xyz coordinate system is rotating with the angular velocity $\boldsymbol{\Omega}$, so that the time derivatives of the unit vectors $\hat{\mathbf{i}}$ and $\hat{\mathbf{j}}$ are not zero. To account for the rotating frame, we use Eq. (1.66) to obtain

$$\dot{\mathbf{r}} = \mathbf{v}_G + \boldsymbol{\Omega} \times \mathbf{r} + \mathbf{v}_{\text{rel}} \tag{2.182}$$

where $\mathbf{v}_G$ is the inertial velocity of the center of mass (the origin of the xyz frame), and $\mathbf{v}_{\text{rel}}$ is the velocity of m as measured in the moving xyz frame. That is

$$\mathbf{v}_{\text{rel}} = \dot{x}\hat{\mathbf{i}} + \dot{y}\hat{\mathbf{j}} + \dot{z}\hat{\mathbf{k}} \tag{2.183}$$

The absolute acceleration of m is found using the "five-term" relative acceleration formula (Eq. 1.70)

$$\ddot{\mathbf{r}} = \mathbf{a}_G + \dot{\boldsymbol{\Omega}} \times \mathbf{r} + \boldsymbol{\Omega} \times (\boldsymbol{\Omega} \times \mathbf{r}) + 2\boldsymbol{\Omega} \times \mathbf{v}_{\text{rel}} + \mathbf{a}_{\text{rel}} \tag{2.184}$$

Recall from Section 2.2 that the velocity $\mathbf{v}_G$ of the center of mass is constant, so that $\mathbf{a}_G = \mathbf{0}$. Furthermore, $\dot{\boldsymbol{\Omega}} = \mathbf{0}$ since the angular velocity of the circular orbit is constant. Therefore, Eq. (2.184) reduces to

$$\ddot{\mathbf{r}} = \boldsymbol{\Omega} \times (\boldsymbol{\Omega} \times \mathbf{r}) + 2\boldsymbol{\Omega} \times \mathbf{v}_{\text{rel}} + \mathbf{a}_{\text{rel}} \tag{2.185}$$

where

$$\mathbf{a}_{\text{rel}} = \ddot{x}\hat{\mathbf{i}} + \ddot{y}\hat{\mathbf{j}} + \ddot{z}\hat{\mathbf{k}} \tag{2.186}$$

Substituting Eqs. (2.173), (2.181), (2.183), and (2.186) into Eq. (2.185) yields

$$\begin{aligned}\ddot{\mathbf{r}} &= (\Omega\hat{\mathbf{k}}) \times \left[(\Omega\hat{\mathbf{k}}) \times \left(x\hat{\mathbf{i}} + y\hat{\mathbf{j}} + z\hat{\mathbf{k}}\right)\right] + 2(\Omega\hat{\mathbf{k}}) \times \left(\dot{x}\hat{\mathbf{i}} + \dot{y}\hat{\mathbf{j}} + \dot{z}\hat{\mathbf{k}}\right) + \ddot{x}\hat{\mathbf{i}} + \ddot{y}\hat{\mathbf{j}} + \ddot{z}\hat{\mathbf{k}} \\ &= -\Omega^2\left(x\hat{\mathbf{i}} + y\hat{\mathbf{j}}\right) + 2\Omega\dot{x}\hat{\mathbf{j}} - 2\Omega\dot{y}\hat{\mathbf{i}} + \ddot{x}\hat{\mathbf{i}} + \ddot{y}\hat{\mathbf{j}} + \ddot{z}\hat{\mathbf{k}}\end{aligned}$$

Collecting terms we find

$$\ddot{\mathbf{r}} = (\ddot{x} - 2\Omega\dot{y} - \Omega^2 x)\hat{\mathbf{i}} + (\ddot{y} + 2\Omega\dot{x} - \Omega^2 y)\hat{\mathbf{j}} + \ddot{z}\hat{\mathbf{k}} \tag{2.187}$$

Now that we have an expression for the inertial acceleration in terms of quantities measured in the rotating frame, let us observe that Newton's second law for the secondary body is

$$m\ddot{\mathbf{r}} = \mathbf{F}_1 + \mathbf{F}_2 \tag{2.188}$$

$\mathbf{F}_1$ and $\mathbf{F}_2$ are the gravitational forces exerted on m by m_1 and m_2, respectively. Recalling Eq. (2.10), we have

$$\begin{aligned}\mathbf{F}_1 &= -\frac{Gm_1 m}{r_1{}^2}\hat{\mathbf{u}}_r)_1 = -\frac{\mu_1 m}{r_1{}^3}\mathbf{r}_1 \\ \mathbf{F}_2 &= -\frac{Gm_2 m}{r_2{}^2}\hat{\mathbf{u}}_r)_2 = -\frac{\mu_2 m}{r_2{}^3}\mathbf{r}_2\end{aligned} \tag{2.189}$$

where

$$\mu_1 = Gm_1 \qquad \mu_2 = Gm_2 \tag{2.190}$$

Substituting Eq. (2.189) into Eq. (2.188) and canceling out m yields

$$\ddot{\mathbf{r}} = -\frac{\mu_1}{r_1{}^3}\mathbf{r}_1 - \frac{\mu_2}{r_2{}^3}\mathbf{r}_2 \tag{2.191}$$

Finally, we substitute Eq. (2.187) on the left and Eqs. (2.179) and (2.180) on the right to obtain

$$\begin{aligned}(\ddot{x} - 2\Omega\dot{y} - \Omega^2 x)\hat{\mathbf{i}} + (\ddot{y} + 2\Omega\dot{x} - \Omega^2 y)\hat{\mathbf{j}} + \ddot{z}\hat{\mathbf{k}} = &-\frac{\mu_1}{r_1{}^3}\left[(x + \pi_2 r_{12})\hat{\mathbf{i}} + y\hat{\mathbf{j}} + z\hat{\mathbf{k}}\right] \\ &-\frac{\mu_2}{r_2{}^3}\left[(x - \pi_1 r_{12})\hat{\mathbf{i}} + y\hat{\mathbf{j}} + z\hat{\mathbf{k}}\right]\end{aligned}$$

Equating the coefficients of $\hat{\mathbf{i}}$, $\hat{\mathbf{j}}$, and $\hat{\mathbf{k}}$ on each side of this equation yields the three scalar equations of motion for the circular restricted three-body problem:

$$\ddot{x} - 2\Omega\dot{y} - \Omega^2 x = -\frac{\mu_1}{r_1{}^3}(x + \pi_2 r_{12}) - \frac{\mu_2}{r_2{}^3}(x - \pi_1 r_{12}) \tag{2.192a}$$

$$\ddot{y}+2\Omega\dot{x}-\Omega^2 y=-\frac{\mu_1}{r_1{}^3}y-\frac{\mu_2}{r_2{}^3}y \tag{2.192b}$$

$$\ddot{z}=-\frac{\mu_1}{r_1{}^3}z-\frac{\mu_2}{r_2{}^3}z \tag{2.192c}$$

2.12.1 LAGRANGE POINTS

Although Eqs. (2.192a), (2.192b), and (2.192c) have no closed-form analytical solution, we can use them to determine the location of the equilibrium points. These are the locations in space where the secondary mass m would have zero velocity and zero acceleration (i.e., where m would appear permanently at rest relative to m_1 and m_2 and therefore appear to an inertial observer to move in circular orbits around m_1 and m_2). Once placed at an equilibrium point (also called libration point or Lagrange point), a body will presumably stay there. The equilibrium points are therefore defined by the conditions

$$\dot{x}=\dot{y}=\dot{z}=0 \text{ and } \boxed{\ddot{x}=\ddot{y}=\ddot{z}=0}$$

Substituting these conditions into Eqs. (2.192a), (2.192b), and (2.192c) yield

$$-\Omega^2 x=-\frac{\mu_1}{r_1{}^3}(x+\pi_2 r_{12})-\frac{\mu_2}{r_2{}^3}(x-\pi_1 r_{12}) \tag{2.193a}$$

$$-\Omega^2 y=-\frac{\mu_1}{r_1{}^3}y-\frac{\mu_2}{r_2{}^3}y \tag{2.193b}$$

$$0=-\frac{\mu_1}{r_1{}^3}z-\frac{\mu_2}{r_2{}^3}z \tag{2.193c}$$

From Eq. (2.193c), we have

$$\left(\frac{\mu_1}{r_1{}^3}+\frac{\mu_2}{r_2{}^3}\right)z=0 \tag{2.194}$$

Since $\mu_1/r_1{}^3>0$ and $\mu_2/r_2{}^3>0$, it must therefore be true that $z=0$. That is, the equilibrium points lie in the orbital plane.

From Eq. (2.178), it is clear that

$$\pi_1=1-\pi_2 \tag{2.195}$$

Using this, along with Eq. (2.174), and assuming $y\neq 0$, we can write Eqs. (2.193a) and (2.193b) as

$$\begin{aligned}(1-\pi_2)(x+\pi_2 r_{12})\frac{1}{r_1{}^3}+\pi_2(x+\pi_2 r_{12}-r_{12})\frac{1}{r_2{}^3}&=\frac{x}{r_{12}{}^3}\\(1-\pi_2)\frac{1}{r_1{}^3}+\pi_2\frac{1}{r_2{}^3}&=\frac{1}{r_{12}{}^3}\end{aligned} \tag{2.196}$$

where we made use of the fact that

$$\pi_1=\mu_1/\mu \quad \pi_2=\mu_2/\mu \tag{2.197}$$

Treating Eq. (2.196) as two linear equations in $1/r_1{}^3$ and $1/r_2{}^3$ we, solve them simultaneously to find that

$$\frac{1}{r_1{}^3}=\frac{1}{r_2{}^3}=\frac{1}{r_{12}{}^3}$$

or

$$r_1 = r_2 = r_{12} \tag{2.198}$$

Using this result, together with $z = 0$ and Eq. (2.195), we obtain from Eqs. (2.179) and (2.180), respectively,

$$r_{12}^2 = (x + \pi_2 r_{12})^2 + y^2 \tag{2.199}$$

$$r_{12}^2 = (x + \pi_2 r_{12} - r_{12})^2 + y^2 \tag{2.200}$$

Equating the right-hand sides of these two equations leads at once to the conclusion that

$$x = \frac{r_{12}}{2} - \pi_2 r_{12} \tag{2.201}$$

Substituting this result into Eq. (2.199) or Eq. (2.200) and solving for y yields

$$y = \pm\frac{\sqrt{3}}{2} r_{12}$$

We have thus found two of the equilibrium points, the Lagrange points L_4 and L_5. As Eq. (2.198) shows, these points are at the same distance r_{12} from the primary bodies m_1 and m_2 that the primary bodies are from each other, and in the comoving coordinate system, their coordinates are

$$L_4, L_5: \quad x = \frac{r_{12}}{2} - \pi_2 r_{12} \quad y = \pm\frac{\sqrt{3}}{2} r_{12} \quad z = 0 \tag{2.202}$$

Therefore, the two primary bodies and these two Lagrange points lie at the vertices of equilateral triangles, as illustrated in Fig. 2.36.

The remaining three equilibrium points L_1, L_2, and L_3, are found by setting $y = 0$ as well as $z = 0$, which satisfy both Eqs. (2.193b) and (2.193c). For these values, Eqs. (2.179) and (2.180) become

$$\mathbf{r}_1 = (x + \pi_2 r_{12})\hat{\mathbf{i}}$$
$$\mathbf{r}_2 = (x - \pi_1 r_{12})\hat{\mathbf{i}} = (x + \pi_2 r_{12} - r_{12})\hat{\mathbf{i}}$$

Therefore

$$r_1 = |x + \pi_2 r_{12}|$$
$$r_2 = |x + \pi_2 r_{12} - r_{12}|$$

Substituting these together with Eqs. (2.174), (2.195), and (2.197) into Eq. (2.193a) yields

$$(1 - \pi_2)\frac{x + \pi_2 r_{12}}{|x + \pi_2 r_{12}|^3} + \pi_2 \frac{x + \pi_2 r_{12} - r_{12}}{|x + \pi_2 r_{12} - r_{12}|^3} - \frac{1}{r_{12}^3} x = 0 \tag{2.203}$$

Further simplification is obtained by nondimensionalizing x,

$$\xi = \frac{x}{r_{12}}$$

In terms of ξ, Eq. (2.203) becomes $f(\pi_2, \xi) = 0$, where

$$f(\pi_2, \xi) = (1 - \pi_2)\frac{\xi + \pi_2}{|\xi + \pi_2|^3} + \pi_2 \frac{\xi + \pi_2 - 1}{|\xi + \pi_2 - 1|^3} - \xi \tag{2.204}$$

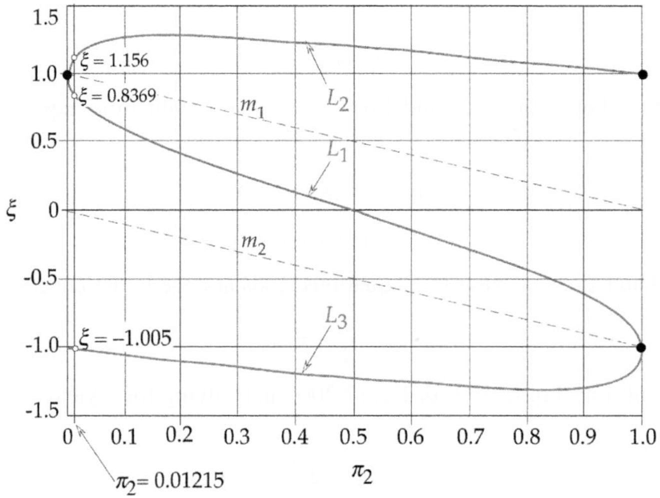

FIG. 2.34

Contour plot of $f(\pi_2, \xi) = 0$ for the collinear equilibrium points of the restricted three-body problem. $\pi_2 = 0.01215$ for the earth-moon system.

Fig. 2.34 shows the plot of an S-shaped contour, which is the locus of points (π_2, ξ) at which f is zero. The horizontal lines $\xi = -1$ and $\xi = 1$ divide the contour into three curves labeled L_1, L_2, and L_3. For a given value of the mass ratio π_2 $(0 < \pi_2 < 1)$, the figure reveals that there are three values for the Lagrange point coordinate ξ, one for each of the three subregions L_1, L_2, and L_3. The two straight lines labeled m_1 and m_2 in Fig. 2.34 are graphs of the nondimensional forms of Eq. (2.177),

$$m_1 : \xi + \pi_2 = 0$$

$$m_2 : \xi + \pi_2 - 1 = 0$$

These relate the nondimensional coordinates ξ_1 and ξ_2 of the primary masses m_1 and m_2 to the mass ratio π_2. Clearly, the curve labeled L_3 in Fig. 2.34 lies below that for m_1; L_1 lies between m_1 and m_2; and L_2 lies above m_2. That is, assuming as in Fig. 2.33 that mass m_2 is positioned to the right of m_1, one of the collinear Lagrange points (L_3) lies to the left of m_1, another (L_1) lies between m_1 and m_2, and the third (L_2) lies beyond m_2 to the right (see Fig. 2.36).

For a given π_2, we cannot read the three values of the Lagrange point coordinates precisely from Fig. 2.34, but we can use the approximate values as starting points of an iterative solution for the roots of the function $f(\pi_2, \xi)$ in Eq. (2.204). The bisection method is a simple, though not very efficient, procedure that we can employ here as well as in other problems that require the root of a nonlinear function.

If r is a root of the function $f(x)$, then $f(r) = 0$. To find r by the bisection method, we first select two values of x that we know lie close to and on each side of the root. Label these values x_l and x_u, where

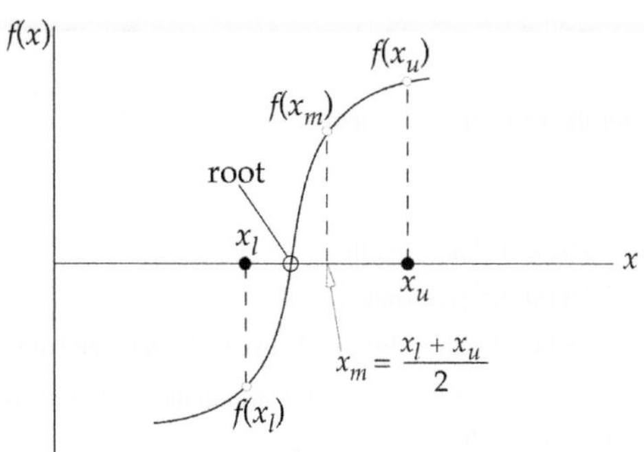

FIG. 2.35

Determining a root by the bisection method.

$x_l < r$ and $x_u > r$. Since the function f changes sign at a root, it follows that $f(x_l)$ and $f(x_u)$ must be of opposite sign, which means $f(x_l) \cdot f(x_u) < 0$. For the sake of argument, suppose $f(x_l) < 0$ and $f(x_u) > 0$, as in Fig. 2.35. Bisect the interval from x_l to x_u by computing $x_m = (x_l + x_u)/2$. If $f(x_m)$ is positive, then the root r lies between x_l and x_m, so (x_l, x_m) becomes our new search interval. If instead $f(x_m)$ is negative, then (x_m, x_u) becomes our search interval. In either case, we bisect the new search interval and repeat the process over and over again, the search interval becoming smaller and smaller, until we eventually converge to r within a desired accuracy E. To achieve that accuracy from the starting values of x_l and x_u requires no more than n iterations, where n is the smallest integer such that (Hahn, 2002)

$$n > \frac{1}{\ln 2} \ln\left(\frac{|x_u - x_l|}{E}\right)$$

Let us summarize the procedure as follows:

ALGORITHM 2.4

Find a root r of the function $f(x)$ using the bisection method. See Appendix D.9 for a MATLAB implementation of this procedure in the script named *bisect.m.*

1. Select values x_l and x_u that are known to be fairly close to r and such that $x_l < r$ and $x_u > r$.
2. Choose a tolerance E and determine the number of iterations n from the above formula.
3. Repeat the following steps n times:
   (a) Compute $x_m = (x_l + x_u)/2$.
   (b) If $f(x_l) \cdot f(x_u) > 0$, then $x_l \leftarrow x_m$; otherwise, $x_r \leftarrow x_m$.
   (c) Return to **a**.
4. $r = x_m$.

EXAMPLE 2.16

Locate the five Lagrange points for the earth–moon system.

Solution

From Table A.1 we find

$$\begin{aligned} m_1 &= 5.974\left(10^{24}\right)\text{kg} \quad \text{(earth)} \\ m_2 &= 7.348\left(10^{22}\right)\text{kg} \quad \text{(moon)} \\ r_{12} &= 3.844\left(10^{5}\right)\text{km} \quad \text{(distance between the earth and moon)} \end{aligned} \tag{2.205}$$

We know that Lagrange points L_4 and L_5 lie on the moon's orbit around the earth, L_4 is 60° ahead of the moon, and L_5 lies 60° behind the moon, as illustrated in Fig. 2.36.

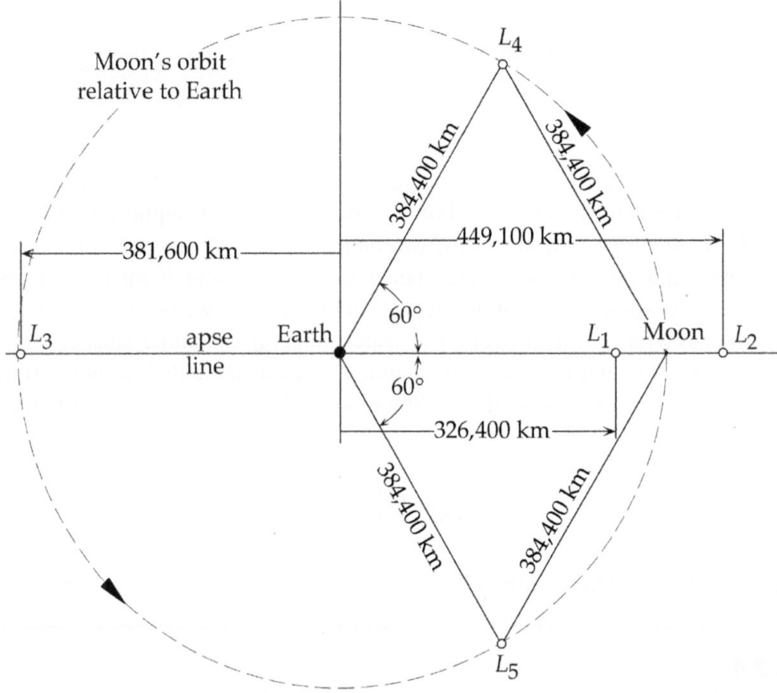

FIG. 2.36

Location of the five Lagrange points of the earth-moon system. These points orbit the Earth with the same period as the moon.

To find L_1, L_2, and L_3 requires finding the roots of Eq. (2.204), in which, for the case at hand, the mass ratio is

$$\pi_2 = \frac{m_2}{m_1 + m_2} = 0.01215$$

Using Algorithm 2.4, we proceed as follows.

Step 1:

For the above value of π_2, Fig. 2.34 shows that L_3 lies near $\xi = -1$, whereas L_1 and L_2 lie on the low and high side, respectively, of $\xi = +1$. We cannot read these values precisely off the graph, but we can use them to select the starting values for the bisection method. For L_3, we choose $\xi_l = -1.1$ and $\xi_u = -0.9$.

Step 2:

Choose an error tolerance of $E = 10^{-6}$, which sets the number of iterations,

$$n > \frac{1}{\ln 2} \ln\left(\frac{|\xi_u - \xi_l|}{E}\right) = \frac{1}{\ln 2} \ln\left(\frac{|-0.9 - (-1.1)|}{10^{-6}}\right) = 17.61$$

Table 2.1 Steps of the bisection method leading to $\xi = -1.0050$ for L_3

n	ξ_l	ξ_u	ξ_m	Sign of $f(\pi_2,\xi_l)\cdot f(\pi_2,\xi_u)$
1	−1.1	−0.9	−1	<0
2	−1.1	−1	−1.05	>0
3	−1.05	−1	−1.025	>0
4	−1.025	−1	−1.0125	>0
5	−1.0125	−1	−1.00625	>0
6	−1.00625	−1	−1.003125	<0
7	−1.00625	−1.003125	−1.0046875	<0
8	−1.00625	−1.0046875	−1.00546875	>0
9	−1.00546875	−1.0046875	−1.005078125	>0
10	−1.005078125	−1.0046875	−1.0049882812	<0
11	−1.005078125	−1.0049882812	−1.004980469	<0
12	−1.004980469	−1.0049882812	−1.005029297	>0
13	−1.005029297	−1.0049882812	−1.005004883	>0
14	−1.005004883	−1.0049882812	−1.004992676	<0
15	−1.005004883	−1.004992676	−1.004998779	>0
16	−1.004998779	−1.004992676	−1.004995728	>0
17	−1.004995728	−1.004992676	−1.004994202	<0
18	−1.004995728	−1.004994202	−1.004994965	>0

That is, $n = 18$.

Step 3:

This is summarized in Table 2.1.

We conclude that, to five significant figures, $\xi_3 = -1.0050$.

The values of ξ for the Lagrange points L_1 and L_2 are found the same way using Algorithm 2.4, starting with the estimates obtained from Fig. 2.34. Rather than repeating the lengthy hand computations, see instead Appendix D.9 for the MATLAB program *Example_2_16.m*, which carries out the calculations of all the three roots. It uses the program *bisect.m* to do the iterations, leading to $\xi_1 = 0.8369$ and $\xi_2 = 1.156$, as well as $\xi_3 = -1.005$ computed in Table 2.1.

Multiplying each dimensionless root by r_{12} yields the x coordinates (relative to the center of mass) of the collinear Lagrange points in kilometers.

$$\boxed{\begin{array}{l} L_1: x = 0.8369 r_{12} = 3.217(10^5)\text{km} \\ L_2: x = 1.156 r_{12} = 4.444(10^5)\text{km} \\ L_3: x = -1.005 r_{12} = -3.863(10^5)\text{km} \end{array}} \tag{2.206}$$

The locations of the five Lagrange points for the earth–moon system are shown in Fig. 2.36. For convenience, all their positions are shown relative to the center of the earth, instead of the center of mass. As can be seen from Eq. (2.177a), the center of mass of the earth–moon system is only 4670 km from the center of the earth. That is, it lies within the earth at 73% of its radius. Since the Lagrange points are fixed relative to the earth and the moon, they follow circular orbits around the earth with the same period as the moon.

If an equilibrium point is stable, then a small mass occupying that point will tend to return to that point if nudged out of position. The perturbation results in a small oscillation (orbit) about the equilibrium point. Thus, objects can be placed in small orbits (called halo orbits) around stable equilibrium points without requiring much in the way of station keeping. On the other hand, if a body located at an unstable equilibrium point is only slightly perturbed, it will oscillate in a divergent fashion, drifting eventually completely away from that point. It turns out (Battin, 1987) that the collinear Lagrange points L_1, L_2, and L_3 are unstable, whereas L_4 and L_5, which lie 60° ahead of m_2 and 60° behind m_2 in its orbit, are stable if

$$\frac{m_1}{m_2}+\frac{m_2}{m_1}\geq 25$$

This will be true as long as the ratio m_1/m_2 exceeds 24.96. For the earth–moon system that ratio is 81.3. However, L_4 and L_5 are destabilized by the influence of the sun's gravity, so that in actuality station keeping would be required to maintain position in the neighborhood of those points of the earth–moon system.

Solar observation spacecraft have been placed in halo orbits around the L_1 point of the sun–earth system. L_1 lies about 1.5 million km from the earth (1/100 the distance to the sun) and well outside the earth's magnetosphere. Three such missions were the International Sun–Earth Explorer 3 launched in August 1978, the Solar and Heliocentric Observatory launched in December 1995, and the Advanced Composition Explorer launched in August 1997.

In June 2001, the 830-kg Wilkinson Microwave Anisotropy Probe (WMAP) was launched aboard a Delta II rocket on a 3-month journey to sun–earth Lagrange point L_2, which lies 1.5 million km from the earth in the opposite direction from L_1. WMAP's several-year mission was to measure the cosmic microwave background radiation. The 6500-kg James Webb Space Telescope is currently scheduled for a 2020 launch aboard an Ariane 5 to an orbit around L_2. This successor to the Hubble Space Telescope, which is in low earth orbit, will use a 6.5-m mirror to gather data in the infrared spectrum over a period of 5–10 years.

2.12.2 JACOBI CONSTANT

Multiply Eq. (2.192a) by $\dot{x}$, Eq. (2.192b) by $\dot{y}$, and Eq. (2.192c) by $\dot{z}$ to obtain

$$\ddot{x}\dot{x}-2\Omega\dot{x}\dot{y}-\Omega^2x\dot{x}=-\frac{\mu_1}{r_1^3}(x\dot{x}+\pi_2r_{12}\dot{x})-\frac{\mu_2}{r_2^3}(x\dot{x}-\pi_1r_{12}\dot{x})$$

$$\ddot{y}\dot{y}+2\Omega\dot{x}\dot{y}-\Omega^2y\dot{y}=-\frac{\mu_1}{r_1^3}y\dot{y}-\frac{\mu_2}{r_2^3}y\dot{y}$$

$$\ddot{z}\dot{z}=-\frac{\mu_1}{r_1^3}z\dot{z}-\frac{\mu_2}{r_2^3}z\dot{z}$$

Sum up the left-hand and right-hand sides of these equations to get

$$\ddot{x}\dot{x}+\ddot{y}\dot{y}+\ddot{z}\dot{z}-\Omega^2(x\dot{x}+y\dot{y})=-\left(\frac{\mu_1}{r_1^3}+\frac{\mu_2}{r_2^3}\right)(x\dot{x}+y\dot{y}+z\dot{z})+r_{12}\left(\frac{\pi_1\mu_2}{r_2^3}-\frac{\pi_2\mu_1}{r_1^3}\right)\dot{x}$$

or, by rearranging terms,

$$\ddot{x}\dot{x}+\ddot{y}\dot{y}+\ddot{z}\dot{z}-\Omega^2(x\dot{x}+y\dot{y})=-\frac{\mu_1}{r_1^3}(x\dot{x}+y\dot{y}+z\dot{z}+\pi_2r_{12}\dot{x})-\frac{\mu_2}{r_2^3}(x\dot{x}+y\dot{y}+z\dot{z}-\pi_1r_{12}\dot{x}) \tag{2.207}$$

Note that

$$\ddot{x}\dot{x}+\ddot{y}\dot{y}+\ddot{z}\dot{z}=\frac{1}{2}\frac{d}{dt}\left(\dot{x}^2+\dot{y}^2+\dot{z}^2\right)=\frac{1}{2}v^2 \tag{2.208}$$

where v is the speed of the secondary mass relative to the rotating frame. Similarly,

$$x\dot{x}+y\dot{y}=\frac{1}{2}\frac{d}{dt}\left(x^2+y^2\right) \tag{2.209}$$

From Eq. (2.179) we obtain

$$r_1{}^2=(x+\pi_2 r_{12})^2+y^2+z^2$$

Therefore

$$2r_1\frac{dr_1}{dt}=2(x+\pi_2 r_{12})\dot{x}+2y\dot{y}+2z\dot{z}$$

or

$$\frac{dr_1}{dt}=\frac{1}{r_1}(\pi_2 r_{12}\dot{x}+x\dot{x}+y\dot{y}+z\dot{z})$$

It follows that

$$\frac{d}{dt}\frac{1}{r_1}=-\frac{1}{r_1{}^2}\frac{dr_1}{dt}=-\frac{1}{r_1{}^3}(x\dot{x}+y\dot{y}+z\dot{z}+\pi_2 r_{12}\dot{x}) \tag{2.210}$$

In a similar fashion, starting with Eq. (2.180), we find

$$\frac{d}{dt}\frac{1}{r_2}=-\frac{1}{r_2{}^3}(x\dot{x}+y\dot{y}+z\dot{z}+\pi_1 r_{12}\dot{x}) \tag{2.211}$$

Substituting Eqs. (2.208)–(2.211) into Eq. (2.207) yields

$$\frac{1}{2}\frac{d}{dt}v^2-\frac{1}{2}\Omega^2\frac{d}{dt}\left(x^2+y^2\right)=\mu_1\frac{d}{dt}\frac{1}{r_1}+\mu_2\frac{d}{dt}\frac{1}{r_2}$$

Alternatively, upon rearranging terms

$$\frac{d}{dt}\left[\frac{1}{2}v^2-\frac{1}{2}\Omega^2\left(x^2+y^2\right)-\frac{\mu_1}{r_1}-\frac{\mu_2}{r_2}\right]=0$$

which means the bracketed expression is a constant

$$\frac{1}{2}v^2-\frac{1}{2}\Omega^2\left(x^2+y^2\right)-\frac{\mu_1}{r_1}-\frac{\mu_2}{r_2}=C \tag{2.212}$$

$v^2/2$ is the kinetic energy per unit mass relative to the rotating frame. $-\mu_1/r_1$ and $-\mu_2/r_2$ are the gravitational potential energies of the two primary masses. $-\Omega^2(x^2+y^2)/2$ may be interpreted as the potential energy of the centrifugal force per unit mass $\Omega^2(x\hat{\mathbf{i}}+y\hat{\mathbf{j}})$ induced by the rotation of the reference frame. The constant C (which is frequently written as $-C/2$ in the literature) is known as the Jacobi constant, after the German mathematician Carl Gustav Jacobi (1804–1851), who discovered it in 1836. Jacobi's constant may be interpreted as the total energy of the secondary particle relative to the rotating frame. C is a constant of the motion of the secondary mass just like energy and angular momentum are constants of the relative motion in the two-body problem.

Solving Eq. (2.212) for v^2 yields

$$v^2 = \Omega^2\left(x^2+y^2\right) + \frac{2\mu_1}{r_1} + \frac{2\mu_2}{r_2} + 2C \tag{2.213}$$

If we restrict the motion of the secondary mass to lie in the plane of motion of the primary masses, then

$$r_1 = \sqrt{\left(x+\pi_2 r_{12}\right)^2 + y^2} \quad r_2 = \sqrt{\left(x-\pi_1 r_{12}\right)^2 + y^2} \tag{2.214}$$

For a given value of the Jacobi constant, v^2 is a function only of position in the rotating frame. Since v^2 cannot be negative, it must be true that

$$\Omega^2\left(x^2+y^2\right) + \frac{2\mu_1}{r_1} + \frac{2\mu_2}{r_2} + 2C \geq 0 \tag{2.215}$$

Trajectories of the secondary body in regions where this inequality is violated are not allowed. The boundaries between forbidden and allowed regions of motion are found by setting $v^2 = 0$. That is

$$\Omega^2\left(x^2+y^2\right) + \frac{2\mu_1}{r_1} + \frac{2\mu_2}{r_2} + 2C = 0 \tag{2.216}$$

For a given value of the Jacobi constant the curves of zero velocity are determined by this equation. These boundaries cannot be crossed by a secondary mass (spacecraft) moving within an allowed region.

Since the first three terms on the left of Eq. (2.216) are all positive, it follows that the zero velocity curves correspond to negative values of the Jacobi constant. Large negative values of C mean that the secondary body is far from the system center of mass ($x^2 + y^2$ is large) or that the body is close to one of the primary bodies (r_1 is small or r_2 is small).

Let us consider again the earth–moon system. From Eqs. (2.174–2.176), (2.190), and (2.205) we have

$$\begin{aligned}
\Omega &= \sqrt{\frac{G(m_1+m_2)}{{r_{12}}^3}} = \sqrt{\frac{6.67259\left(10^{-20}\right)\cdot 6.04748\left(10^{24}\right)}{384{,}400^3}} = 2.66538\left(10^{-6}\right)\text{rad/s} \\
\mu_1 &= Gm_1 = 6.67259\left(10^{-20}\right)\cdot 5.9742\left(10^{24}\right) = 398{,}620\,\text{km}^3/\text{s}^2 \\
\mu_2 &= Gm_2 = 6.67259\left(10^{-20}\right)\cdot 7.348\left(10^{22}\right) = 4903.02\,\text{km}^3/\text{s}^2
\end{aligned} \tag{2.217}$$

Substituting these values into Eq. (2.216), we can plot the zero velocity curves for different values of Jacobi's constant. The curves bound regions in which the motion of a spacecraft is not allowed.

For $C = -1.8\,\text{km}^2/\text{s}^2$, the allowable regions are circles surrounding the earth and the moon, as shown in Fig. 2.37(a). A spacecraft launched from the earth with this value of C cannot reach the moon, to say nothing of escaping the earth–moon system.

Substituting the coordinates of the Lagrange points L_1, L_2, and L_3 into Eq. (2.216), we obtain the successively larger values (smaller negative values) of the Jacobi constants C_1, C_2, and C_3 that are

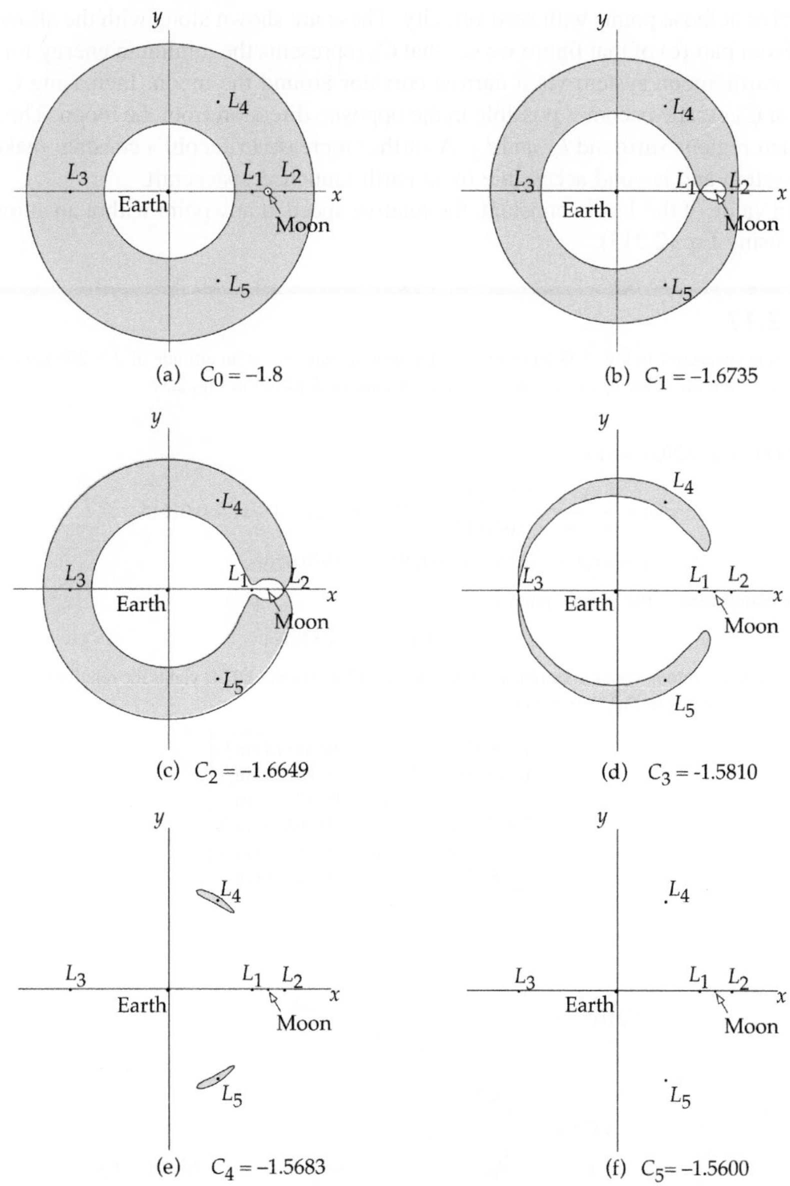

FIG. 2.37

Forbidden regions (*shaded*) within the earth-moon system for increasing values of Jacobi's constant (km^2/s^2).

required to arrive at those points with zero velocity. These are shown along with the allowable regions in Fig. 2.37. From part (c) of that figure we see that C_2 represents the minimum energy for a spacecraft to escape the earth–moon system via a narrow corridor around the moon. Increasing C widens that corridor, and at C_3 escape becomes possible in the opposite direction from the moon. The last vestiges of the forbidden regions surround L_4 and L_5. A further increase in Jacobi's constant makes the entire earth–moon system and beyond accessible to an earth-launched spacecraft.

For a given value of the Jacobi constant, the relative speed at any point within an allowable region can be found using Eq. (2.213).

EXAMPLE 2.17

The earth-orbiting spacecraft in Fig. 2.38 has a relative burnout velocity v_{bo} at an altitude of $d = 200$ km on a radial for which $\phi = -90°$. Find the value of v_{bo} for each of the six scenarios depicted in Fig. 2.37.

Solution

From Eqs. (2.177) and (2.205), we have

$$\pi_1 = \frac{m_1}{m_1 + m_2} = \frac{5.947(10^{24})}{6.047(10^{24})} = 0.9878 \quad \pi_2 = 1 - \pi_1 = 0.01215$$

$$x_1 = -\pi_1 r_{12} = -0.9878 \cdot 384{,}400 = -4670.6\text{km}$$

Therefore, the coordinates of the burnout point are

$$x = -4670.6\text{km} \quad y = -6578\text{km}$$

Substituting these values along with the Jacobi constant into Eqs. (2.213) and (2.214) yields the relative burnout speed v_{bo}. For the six Jacobi constants in Fig. 2.38 we obtain

$$\boxed{\begin{array}{l} C = -1.8000\,\text{km}^2/\text{s}^2: \ v_{bo} = 10.84518\,\text{km/s} \\ C = -1.6735\,\text{km}^2/\text{s}^2: \ v_{bo} = 10.85683\,\text{km/s} \\ C = -1.6649\,\text{km}^2/\text{s}^2: \ v_{bo} = 10.85762\,\text{km/s} \\ C = -1.5810\,\text{km}^2/\text{s}^2: \ v_{bo} = 10.86535\,\text{km/s} \\ C = -1.5683\,\text{km}^2/\text{s}^2: \ v_{bo} = 10.86652\,\text{km/s} \\ C = -1.5600\,\text{km}^2/\text{s}^2: \ v_{bo} = 10.86728\,\text{km/s} \end{array}}$$

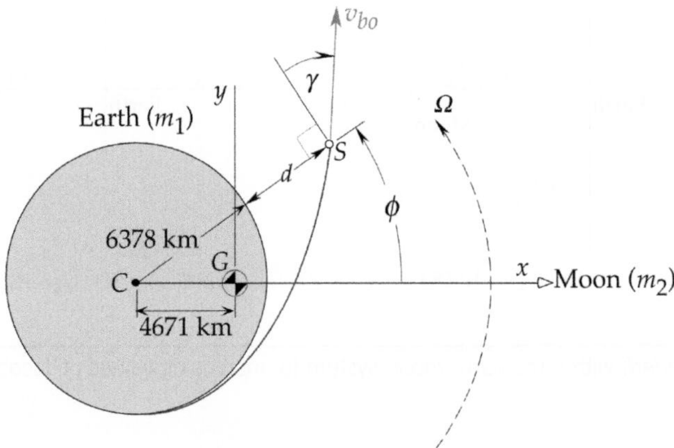

FIG. 2.38

Spacecraft S burnout position and velocity relative to the rotating earth–moon frame.

These burnout velocities all differ less than 1.5% from the escape velocity (Eq. 2.91) at 200 km altitude,

$$v_{esc} = \sqrt{\frac{2\mu}{r}} = \sqrt{\frac{2 \cdot 398,600}{6578}} = 11.01\,\text{km/s}$$

Observe that a change in v_{bo} of less than 10 m/s can have a significant influence on the regions of the earth–moon space accessible to the spacecraft.

EXAMPLE 2.18

For the spacecraft in Fig. 2.38 the initial conditions ($t = 0$) are $d = 200\,\text{km}$, $\phi = -90°$, $\gamma = 20°$, and $v_{bo} = 10.9148$ km/s. Use Eqs. (2.192a), (2.192b), and (2.192c), the circular restricted three-body equations of motion, to determine the trajectory and locate its position at $t = 3.16689$ days.

Solution

Since z and $\dot{z}$ are initially zero, Eq. (2.192c) implies that z remains zero. The motion is therefore confined to the xy plane and is governed by Eqs. (2.192a) and (2.192b). These have no analytical solution, so we must use a numerical approach.

To get Eqs. (2.192a) and (2.192b) into the standard form for numerical solution (Section 1.8), we introduce the auxiliary variables

$$y_1 = x \quad y_2 = y \quad y_3 = \dot{x} \quad y_4 = \dot{y} \tag{a}$$

The time derivatives of these variables are

$$\begin{aligned}
\dot{y}_1 &= y_3 \\
\dot{y}_2 &= y_4 \\
\dot{y}_3 &= 2\Omega y_4 + \Omega^2 y_1 - \frac{\mu_1}{r_1{}^3}(y_1 + \pi_2 r_{12}) - \frac{\mu_2}{r_2{}^3}(y_1 - \pi_1 r_{12}) \qquad \text{(Eq. 2.192a)} \\
\dot{y}_4 &= -2\Omega y_3 + \Omega^2 y_2 - \frac{\mu_1}{r_1{}^3} y_2 - \frac{\mu_2}{r_2{}^3} y_2 \qquad \text{(Eq.2.192b)}
\end{aligned} \tag{b}$$

where, from Eqs. (2.179) and (2.180),

$$r_1 = \sqrt{(y_1 + \pi_2 r_{12})^2 + y_2{}^2} \quad r_2 = \sqrt{(y_1 - \pi_1 r_{12})^2 + y_2{}^2} \tag{c}$$

Eqs. (b) are of the form $\dot{\mathbf{y}} = \mathbf{f}(t, \mathbf{y})$ given by Eq. (1.105).

To solve this system let us use the Runge–Kutta–Fehlberg 4(5) method and Algorithm 1.3, which is implemented in MATLAB as the program *rkf45.m* in Appendix D.4. The MATLAB function named *Example_2_18.m* in Appendix D.10 contains the data for this problem, the given initial conditions, and the time range. To perform the numerical integration, *Example_2_18.m* calls *rkf45.m*, which uses the subfunction *rates*, which is embedded within *Example_2_18.m*, to compute the derivatives in Eq. (b) above. Running *Example_2_18.m* yields the plot of the trajectory shown in Fig. 2.39. After coasting 3.16689 days as specified in the problem statement,

$$\boxed{\begin{array}{l}\text{The spacecraft arrives at the far side of the moon}\\ \text{on the earth} - \text{moon line at an altitude of 256km}\end{array}}$$

For comparison, the 1969 Apollo 11 translunar trajectory, which differed from this one in many details (including the use of midcourse corrections), required 3.04861 days to arrive at the lunar orbit insertion point.

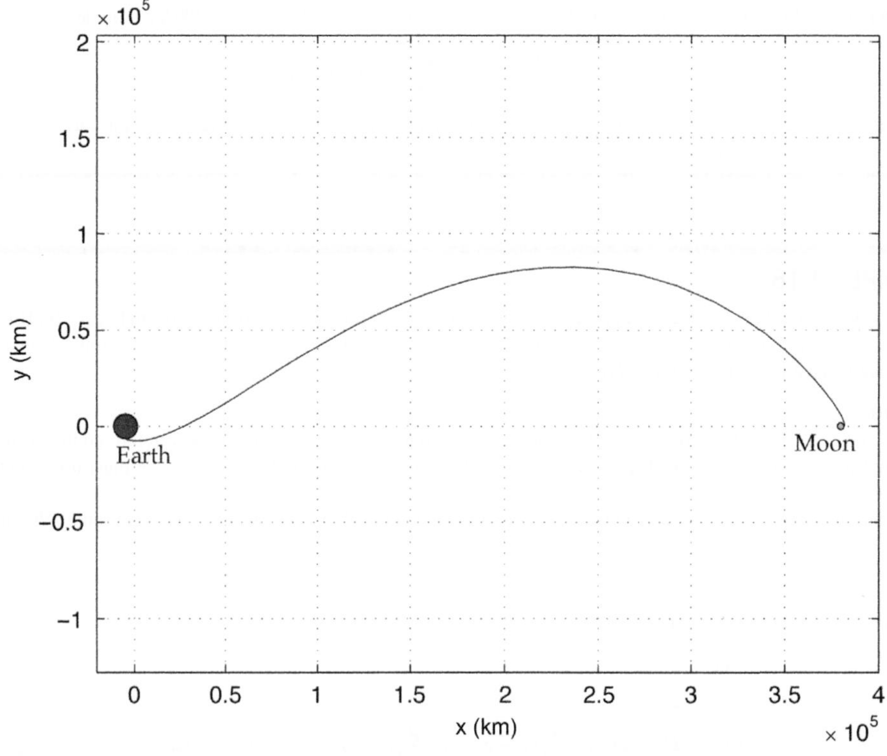

FIG. 2.39

Translunar coast trajectory computed numerically from the restricted three-body differential equations using the *RKF4(5)* method.

PROBLEMS

For man-made earth satellites use $\mu = 398{,}600\,\text{km}^3/\text{s}^2$ and $R_E = 6378\,\text{km}$ (Tables A.1 and A.2).

Section 2.2

2.1 Two particles of identical mass m are acted on only by the gravitational force of one upon the other. If the distance d between the particles is constant, what is the angular velocity of the line joining them? Use Newton's second law with the center of mass of the system as the origin of the inertial frame.

{Ans.: $\omega = \sqrt{2Gm/d^3}$}

2.2 Three particles of identical mass m are acted on only by their mutual gravitational attraction. They are located at the vertices of an equilateral triangle with sides of length d. Consider the motion of any one of the particles about the system center of mass G and, using G as the origin of the inertial frame, employ Newton's second law to determine the angular velocity ω required for d to remain constant.

{Ans.: $\omega = \sqrt{3Gm/d^3}$}

Section 2.3

2.3 Consider the two-body problem illustrated in Fig. 2.1. If a force **T** (such as rocket thrust) acts on m_2 in addition to the mutual force of gravitation $\mathbf{F}_{21}$, show that

(a) Eq. (2.22) becomes

$$\ddot{\mathbf{r}} = -\frac{\mu}{r^3}\mathbf{r} + \frac{\mathbf{T}}{m_2}$$

(b) If the thrust vector **T** has a magnitude T and is aligned with the velocity vector **v**, then

$$\mathbf{T} = T\frac{\mathbf{v}}{v}$$

2.4 At a given instant t_0, a 1000-kg earth-orbiting satellite has the inertial position and velocity vectors $\mathbf{r}_0 = 3207\hat{\mathbf{i}} + 5459\hat{\mathbf{j}} + 2714\hat{\mathbf{k}}$ (km) and $\mathbf{v}_0 = -6.532\hat{\mathbf{i}} + 0.7835\hat{\mathbf{j}} + 6.142\hat{\mathbf{k}}$ (km/s). Solve Eq. (2.22) numerically to find the maximum altitude reached by the satellite and the time at which it occurs.

{Ans.: Using MATLAB's *ode45*, the maximum altitude = 9670 km at 1.66 h after t_0}

2.5 At a given instant, a 1000-kg earth-orbiting satellite has the inertial position and velocity vectors $\mathbf{r}_0 = 6600\hat{\mathbf{i}}$ (km) and $\mathbf{v}_0 = 12\hat{\mathbf{j}}$ (km/s). Solve Eq. (2.22) numerically to find the distance of the spacecraft from the center of the earth and its speed 24 h later.

{Ans.: Using MATLAB's *ode45*, distance = 456,500 km, speed = 5 km/s}

Section 2.4

2.6 If **r**, in meters, is given by $\mathbf{r} = t\sin t\hat{\mathbf{I}} + t^2\cos t\hat{\mathbf{J}} + t^3\sin^2 t\hat{\mathbf{K}}$, where t is the time in seconds, calculate (a) $\dot{r}$ (where $r = \|\dot{\mathbf{r}}\|$) and (b) $\|\dot{\mathbf{r}}\|$ at $t = 2$ s.

{Ans.: (a) $\dot{r} = 4.894$ m/s; (b) $\|\dot{\mathbf{r}}\| = 6.563$ m/s}

2.7 Starting with Eq. (2.35a), prove that $\dot{r} = \mathbf{v} \cdot \hat{\mathbf{u}}_r$ and interpret this result.

2.8 Show that $\hat{\mathbf{u}}_r \cdot d\hat{\mathbf{u}}_r/dt = 0$, where $\hat{\mathbf{u}}_r = \mathbf{r}/r$. Use only the fact that $\hat{\mathbf{u}}_r$ is a unit vector. Interpret this result.

2.9 Show that $v = (\mu/h)\sqrt{1 + 2e\cos\theta + e^2}$ for any orbit.

2.10 Relative to a nonrotating, earth-centered Cartesian coordinate system, the position and velocity vectors of a spacecraft are $\mathbf{r} = 7000\hat{\mathbf{i}} - 2000\hat{\mathbf{j}} - 4000\hat{\mathbf{k}}$ (km) and $\mathbf{v} = 3\hat{\mathbf{i}} - 6\hat{\mathbf{j}} + 5\hat{\mathbf{k}}$ (km/s). Calculate the orbit's (a) eccentricity vector and (b) the true anomaly.

{Ans.: (a) $\mathbf{e} = 0.2888\hat{\mathbf{i}} + 0.08523\hat{\mathbf{j}} - 0.3840\hat{\mathbf{k}}$; (b) $\theta = 33.32°$}

2.11 Show that the eccentricity is 1 for rectilinear orbits ($\mathbf{h} = \mathbf{0}$).

2.12 Relative to a nonrotating, earth-centered Cartesian coordinate system, the velocity of a spacecraft is $\mathbf{v} = -4\hat{\mathbf{i}} + 3\hat{\mathbf{j}} - 5\hat{\mathbf{k}}$ (km/s) and the unit vector in the direction of the radius $\mathbf{r}$ is $\hat{\mathbf{u}}_r = 0.26726\hat{\mathbf{i}} + 0.53452\hat{\mathbf{j}} + 0.80178\hat{\mathbf{k}}$. Calculate (a) the radial component of velocity v_r, (b) the azimuth component of velocity $v_\perp$, and (c) the flight path angle γ.

{Ans.: (a) −3.474 km/s; (b) 6.159 km/s; (c) −29.43°}

Section 2.5

2.13 If the specific energy ε of the two-body problem is negative, show that m_2 cannot move outside a sphere of radius $\mu/|\varepsilon|$ centered at m_1.

2.14 Relative to a nonrotating Cartesian coordinate frame with the origin at the center O of the earth, a spacecraft in a rectilinear trajectory has the velocity $\mathbf{v} = 2\hat{\mathbf{i}} + 3\hat{\mathbf{j}} + 4\hat{\mathbf{k}}$ (km/s) when its distance from O is 10,000 km. Find the position vector $\mathbf{r}$ when the spacecraft comes to rest.

{Ans.: $\mathbf{r} = 5837.4\hat{\mathbf{i}} + 8756.1\hat{\mathbf{j}} + 11{,}675\hat{\mathbf{k}}$ (km)}

Section 2.6

2.15 The specific angular momentum of a satellite in circular earth orbit is 60,000 km^2/s. Calculate the period.

{Ans.: 2.372 h}

2.16 A spacecraft is in a circular orbit of Mars at an altitude of 200 km. Calculate its speed and its period.

{Ans.: 3.451 km/s; 1 h 49 min}

Section 2.7

2.17 Calculate the area A swept out during the time $t = T/4$ since periapsis, where T is the period of the elliptical orbit. See the figure below.

{Ans.: $0.7854ab$}

2.18 Determine the true anomaly θ of the point(s) on an elliptical orbit at which the speed equals the speed of a circular orbit with the same radius (i.e., $v_{\text{ellipse}} = v_{\text{circle}}$). See the figure on the next page.

{Ans.: $\theta = \cos^{-1}(-e)$, where e is the eccentricity of the ellipse}

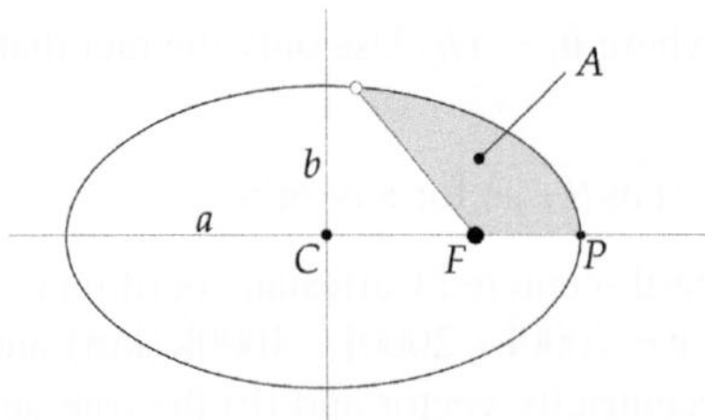

2.19 Calculate the flight path angle at the locations found in Problem 2.18.

{Ans.: $\gamma = \tan^{-1}\left(e/\sqrt{1-e^2}\right)$}

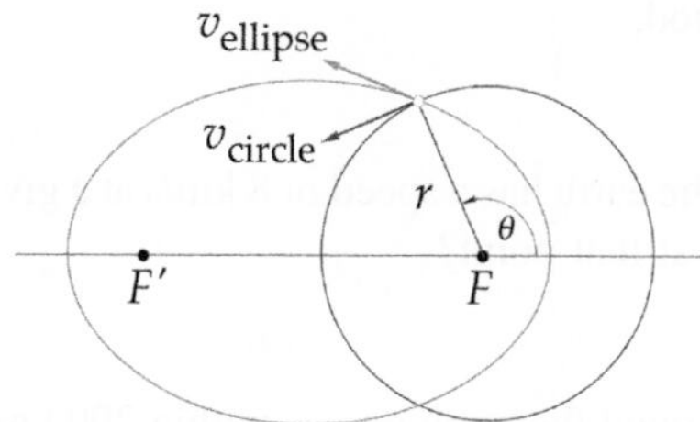

2.20 An unmanned satellite orbits the earth with a perigee radius of 10,000 km and an apogee radius of 100,000 km. Calculate:
(a) the eccentricity of the orbit;
(b) the semimajor axis of the orbit (km);
(c) the period of the orbit (h);
(d) the specific energy of the orbit (km^2/s^2);
(e) the true anomaly (degrees) at which the altitude is 10,000 km;
(f) v_r and $v_\perp$ (km/s) at the points found in part (e);
(g) the speed at perigee and apogee (km/s).

{Partial Ans.: (c) 35.66 h; (e) 82.26°; (g) 8.513 km/s, 0.8513 km/s}

2.21 A spacecraft is in a 400-km-by-600-km low earth orbit. How long (in minutes) does it take to coast from the perigee to the apogee?

{Ans.: 48.34 min}

2.22 The altitude of a satellite in an elliptical orbit around the earth is 2000 km at apogee and 500 km at perigee. Determine:
(a) the eccentricity of the orbit;
(b) the orbital speeds at perigee and apogee;
(c) the period of the orbit.

{Ans.: (a) 0.09832; (b) $v_p = 7.978$ km/s, $v_a = 6.550$ km/s; (c) $T = 110.5$ min}

2.23 A satellite is placed into an earth orbit at perigee at an altitude of 500 km with a speed of 10 km/s. Calculate the flight path angle γ and the altitude of the satellite at a true anomaly of 120°.

{Ans.: $\gamma = 44.60°$, $z = 12,247$ km}

2.24 A satellite is launched into earth orbit at an altitude of 1000 km with a speed of 10 km/s and a flight path angle of 15°. Calculate the true anomaly of the launch point and the period of the orbit.

{Ans.: $\theta = 32.48°$; $T = 30.45$ h}

2.25 A satellite has perigee and apogee altitudes of 500 and 21,000 km. Calculate the orbit period, eccentricity, and the maximum speed.

{Ans.: 6.20 h, 0.5984, 9.625 km/s}

2.26 A satellite is launched parallel to the earth's surface with a speed of 7.6 km/s at an altitude of 500 km. Calculate the period.

{Ans.: 1.61 h}

2.27 A satellite in orbit around the earth has a speed of 8 km/s at a given point of its orbit. If the period is 2 h, what is the altitude at that point?

{Ans.: 648 km}

2.28 A satellite in polar orbit around the earth comes within 200 km of the north pole at its point of closest approach. If the satellite passes over the pole once every 100 min, calculate the eccentricity of its orbit.

{Ans.: 0.07828}

2.29 For an earth orbiter, the altitude is 1000 km at a true anomaly of 40° and 2000 km at a true anomaly of 150°. Calculate
(a) the eccentricity;
(b) the perigee altitude (km);
(c) the semimajor axis (km).

{Partial Ans.: (c) 7863 km}

2.30 An earth satellite has a speed of 7.5 km/s and a flight path angle of 10° when its radius is 8000 km. Calculate
(a) the true anomaly (degrees);
(b) the eccentricity of the orbit.

{Ans.: (a) 63.82°; (b) 0.2151}

2.31 For an earth satellite, the specific angular momentum is 70, 000 km^2/s and the specific energy is $-10\ \text{km}^2/\text{s}^2$. Calculate the apogee and perigee altitudes.

{Ans.: 25,889 and 1214.9 km}

2.32 A rocket launched from the surface of the earth has a speed of 7 km/s when the powered flight ends at an altitude of 1000 km. The flight path angle at this time is 10°. Determine the eccentricity and the period of the orbit.

{Ans.: 0.1963 and 92.0 min}

2.33 If the perigee velocity is c times the apogee velocity, calculate the eccentricity of the orbit in terms of c.

{Ans.: $e = (c - 1)/(c + 1)$}

Section 2.8

2.34 At what true anomaly does the speed on a parabolic trajectory equal α times the speed at the periapsis, where $\alpha \leq 1$?
{Ans.: $\cos^{-1}(2\alpha^2 - 1)$}

2.35 What velocity increase is required for the earth to escape the solar system on a parabolic path?

{Ans.: 12.34 km/s}

Section 2.9

2.36 A hyperbolic earth departure trajectory has a perigee altitude of 250 km and a perigee speed of 11 km/s. Calculate:

(a) the hyperbolic excess speed (km/s);
(b) the radius (km) when the true anomaly is 100°;
(c) v_r and $v_\perp$ (km/s) when the true anomaly is 100°.

{Partial Ans.: (b) 16,179 km}

2.37 A meteoroid is first observed approaching the earth when it is 402,000 km from the center of the earth with a true anomaly of 150°, as shown in the figure below. If the speed of the meteoroid at that time is 2.23 km/s, calculate:

(a) the eccentricity of the trajectory;
(b) the altitude at closest approach;
(c) the speed at the closest approach.

{Ans.: (a) 1.086; (b) 5088 km; (c) 8.516 km/s}

2.38 If α is a number between 1 and $\sqrt{(1+e)/(1-e)}$, calculate the true anomaly at which the speed on a hyperbolic trajectory is α times the hyperbolic excess speed.

$\left\{\text{Ans.: } \cos^{-1}\left[\frac{(\alpha^2-1)(e^2-1)-2}{2e}\right]\right\}$

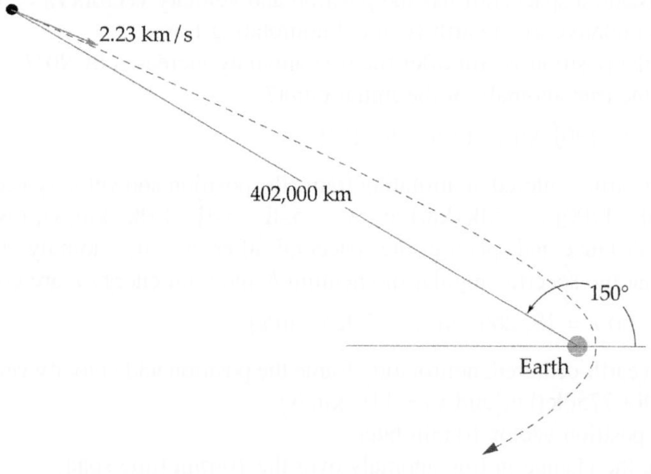

2.23 km/s

402,000 km

150°

Earth

2.39 For a hyperbolic orbit, find the eccentricity in terms of the radius at periapsis r_p and the hyperbolic excess speed v_∞.

{Ans.: $e = 1 + r_p v_\infty^2/\mu$}

2.40 A space vehicle has a velocity of 10 km/s in the direction shown when it is 10,000 km from the center of the earth. Calculate its true anomaly.

{Ans.: 51°}

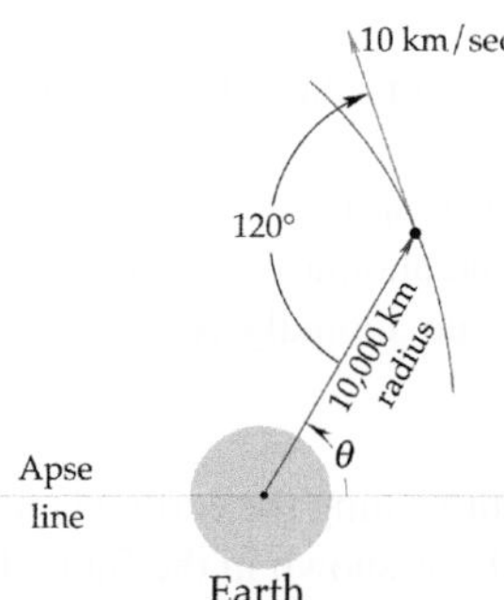

2.41 A spacecraft at a radius r has a speed v and a flight path angle γ. Find an expression for the eccentricity of its orbit in terms of r, v, and γ.

{Ans.: $e=\sqrt{1+\sigma(\sigma-2)\cos^2\gamma}$, where $\sigma = rv^2/\mu$}

2.42 For an orbiting spacecraft, $r = r_1$ when $\theta = \theta_1$, and $r = r_2$ when $\theta = \theta_2$. What is the eccentricity?

{Ans.: $e = (r_1 - r_2)/(r_2 \cos\theta_2 - r_1 \cos\theta_1)$}

Section 2.11

2.43 At a given instant, a spacecraft has the position and velocity vectors $\mathbf{r}_0 = 7000\hat{\mathbf{i}}$ (km) and $\mathbf{v}_0 = 7\hat{\mathbf{i}} + 7\hat{\mathbf{j}}$ (km/s) relative to an earth-centered nonrotating frame.
(a) What is the position vector after the true anomaly increases by 90°?
(b) What is the true anomaly of the initial point?

{Ans.: (a) $\mathbf{r} = 43{,}180\hat{\mathbf{j}}$ (km); (b) $\theta = 99.21°$}

2.44 Relative to an earth-centered, nonrotating frame the position and velocity vectors of a spacecraft are $\mathbf{r}_0 = 3450\hat{\mathbf{i}} - 1700\hat{\mathbf{j}} + 7750\hat{\mathbf{k}}$ (km) and $\mathbf{v}_0 = 5.4\hat{\mathbf{i}} - 5.4\hat{\mathbf{j}} + 1.0\hat{\mathbf{k}}$ (km/s), respectively.
(a) Find the distance and speed of the spacecraft after the true anomaly changes by 82°.
(b) Verify that the specific angular momentum h and total energy ε are conserved.

{Partial Ans.: (a) $r = 19{,}266$ km, $v = 2.925$ km/s}

2.45 Relative to an earth-centered, nonrotating frame the position and velocity vectors of a spacecraft are $\mathbf{r}_0 = 6320\hat{\mathbf{i}} + 7750\hat{\mathbf{k}}$ (km)and $\mathbf{v}_0 = 11\hat{\mathbf{j}}$ (km/s).
(a) Find the position vector 10 min later.
(b) Calculate the change in true anomaly over the 10-min time span.

{Ans.: (a) $\mathbf{r} = 5320\hat{\mathbf{i}} - 6194\hat{\mathbf{j}} + 3073\hat{\mathbf{k}}$ (km); (b) 45°}

Section 2.12

2.46 For the sun–earth system, find the distance of the collinear Lagrange points L_1, L_2, and L_3 from the barycenter.

{Ans.: $x_1 = 148.108(10^6)$ km, $x_2 = 151.101(10^6)$ km, and $x_3 = -149.600(10^6)$ km (opposite side of the sun)}

2.47 Write a program, like that for Example 2.18, to compute the trajectory of a spacecraft using the restricted three-body equations of motion. Use the program to design a trajectory from the earth to the earth–moon Lagrange point L_4, starting at a 200-km altitude burnout point. The path should take the coasting spacecraft to within 500 km of L_4 with a relative speed of not more than 1 km/s.

REFERENCES

Battin, R.H., 1987. An Introduction to the Mathematics and Methods of Astrodynamics. AIAA Education Series, New York.

Bond, V.R., Allman, M.C., 1996. Modern Astrodynamics: Fundamentals and Perturbation Methods. Princeton University Press.

Hahn, B.D., 2002. Essential MATLAB® for Scientists and Engineers, second ed. Butterworth-Heinemann, Oxford.

Zwillinger, D., 2018. Standard Mathematical Tables and Formulae, 33rd ed. CRC Press, New York.

CHAPTER

ORBITAL POSITION AS A FUNCTION OF TIME

3

3.1 INTRODUCTION

In Chapter 2, we found the relationship between position and true anomaly for the two-body problem. The only place time appeared explicitly was in the expression for the period of an ellipse. Obtaining position as a function of time is a simple matter for circular orbits. For elliptical, parabolic, and hyperbolic paths, we are led to the various forms of Kepler's equation relating position to time. These transcendental equations must be solved iteratively using a procedure like Newton's method, which is presented and illustrated in this chapter.

The different forms of Kepler's equation are combined into a single universal Kepler's equation by introducing universal variables. Implementation of this appealing notion is accompanied by the introduction of an unfamiliar class of functions known as Stumpff functions. The universal variable formulation is required for the Lambert and Gauss orbit determination algorithms in Chapter 5.

The road map of Appendix B may aid in grasping how the material presented here depends on that of Chapter 2.

3.2 TIME SINCE PERIAPSIS

The orbit formula, $r=(h^2/\mu)/(1+e\cos\theta)$, gives the position of body m_2 in its orbit around m_1 as a function of the true anomaly. For many practical reasons, we need to be able to determine the position of m_2 as a function of time. For elliptical orbits, we have a formula for the period T (Eq. 2.82), but we cannot yet calculate the time required to fly between any two true anomalies. The purpose of this section is to come up with the formulas that allow us to do that calculation.

The one equation we have that relates true anomaly directly to time is Eq. (2.47), $h=r^2\dot{\theta}$, which can be written as

$$\frac{\mathrm{d}\theta}{\mathrm{d}t}=\frac{h}{r^2}$$

Substituting $r=(h^2/\mu)/(1+e\cos\theta)$ we find, after separating variables,

$$\frac{\mu^2}{h^3}\mathrm{d}t=\frac{\mathrm{d}\theta}{(1+e\cos\theta)^2}$$

Orbital Mechanics for Engineering Students. https://doi.org/10.1016/B978-0-12-824025-0.00003-9

Integrating both sides of this equation yields

$$\frac{\mu^2}{h^3}(t-t_p)=\int_0^\theta \frac{d\vartheta}{(1+e\cos\vartheta)^2} \tag{3.1}$$

where the constant of integration t_p is the time at periapsis passage, where by definition $\theta = 0$. t_p is the sixth constant of the motion that was missing in Chapter 2. The origin of time is arbitrary. It is convenient to measure time from periapsis passage, so we will usually set $t_p = 0$. In that case we have

$$\frac{\mu^2}{h^3}t=\int_0^\theta \frac{d\vartheta}{(1+e\cos\vartheta)^2} \tag{3.2}$$

The integral on the right may be found in any standard mathematical handbook, such as Zwillinger (2018), in which we find

$$\int\frac{dx}{(a+b\cos x)^2}=\frac{1}{(a^2-b^2)^{3/2}}\left[2a\tan^{-1}\left(\sqrt{\frac{a-b}{a+b}}\tan\frac{x}{2}\right)-\frac{b\sqrt{a^2-b^2}\sin x}{a+b\cos x}\right] \tag{3.3}$$

$$\int\frac{dx}{(a+b\cos x)^2}=\frac{1}{a^2}\left(\frac{1}{2}\tan\frac{x}{2}+\frac{1}{6}\tan^3\frac{x}{2}\right)\quad(b=a) \tag{3.4}$$

$$\int\frac{dx}{(a+b\cos x)^2}=\frac{1}{(b^2-a^2)^{3/2}}\left[\frac{b\sqrt{b^2-a^2}\sin x}{a+b\cos x}-a\ln\left(\frac{\sqrt{b+a}+\sqrt{b-a}\tan(x/2)}{\sqrt{b+a}-\sqrt{b-a}\tan(x/2)}\right)\right] \tag{3.5}$$

3.3 CIRCULAR ORBITS ($e = 0$)

If $e = 0$, the integral in Eq. (3.2) is simply $\int_0^\theta d\theta$, which means

$$t=\frac{h^3}{\mu^2}\theta$$

Recall that for a circle (Eq. 2.62), $r = h^2/\mu$. Therefore $h^3 = r^{3/2}\mu^{3/2}$, so that

$$t=\frac{r^{3/2}}{\sqrt{\mu^2}}\theta$$

Finally, substituting the formula (Eq. 2.64) for the period T of a circular orbit, $T=2\pi r^{3/2}/\sqrt{\mu}$, yields

$$t=\frac{\theta}{2\pi}T$$

or

$$\theta=\frac{2\pi}{T}t$$

The reason that t is directly proportional to θ in a circular orbit is simply that the angular velocity $2\pi/T$ is constant. Therefore, the time Δt to fly through a true anomaly of $\Delta\theta$ is $(\Delta\theta/2\pi)T$.

Because the circle is symmetric about any diameter, the apse line—and therefore the periapsis—can be chosen arbitrarily (Fig. 3.1).

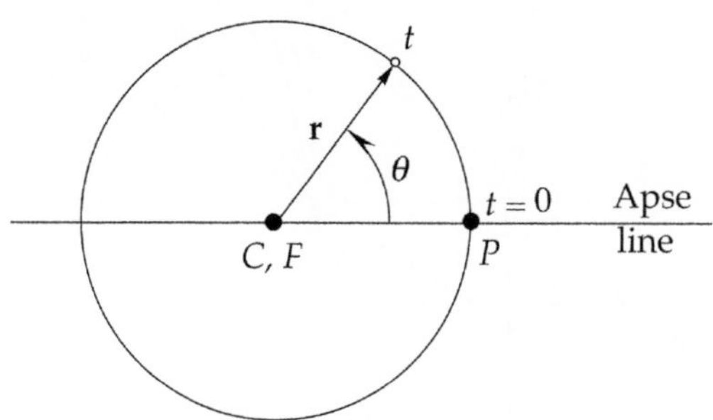

FIG. 3.1

Time since periapsis is directly proportional to true anomaly in a circular orbit.

3.4 ELLIPTICAL ORBITS ($e < 1$)

Set $a = 1$ and $b = e$ in Eq. (3.3) to obtain

$$\int_0^\theta \frac{d\vartheta}{(1+e\cos\vartheta)^2} = \frac{1}{(1-e^2)^{3/2}}\left[2\tan^{-1}\left(\sqrt{\frac{1-e}{1+e}}\tan\frac{\theta}{2}\right) - \frac{e\sqrt{1-e^2}\sin\theta}{1+e\cos\theta}\right]$$

Therefore, Eq. (3.2) in this case becomes

$$\frac{\mu^2}{h^3}t = \frac{1}{(1-e^2)^{3/2}}\left[2\tan^{-1}\left(\sqrt{\frac{1-e}{1+e}}\tan\frac{\theta}{2}\right) - \frac{e\sqrt{1-e^2}\sin\theta}{1+e\cos\theta}\right]$$

or

$$M_e = 2\tan^{-1}\left(\sqrt{\frac{1-e}{1+e}}\tan\frac{\theta}{2}\right) - \frac{e\sqrt{1-e^2}\sin\theta}{1+e\cos\theta} \tag{3.6}$$

where

$$M_e = \frac{\mu^2}{h^3}(1-e^2)^{3/2}t \tag{3.7}$$

M_e is called the mean anomaly. The subscript e reminds us that this is for an ellipse and not for parabolas and hyperbolas, which have their own "mean anomaly" formulas. Eq. (3.6) is plotted in Fig. 3.2. Observe that for all values of the eccentricity e, M_e is a monotonically increasing function of the true anomaly θ.

From Eq. (2.82), the formula for the period T of an elliptical orbit, we have $\mu^2(1 - e^2)^{3/2}/h^3 = 2\pi/T$, so that the mean anomaly can be written much more simply as

$$M_e = \frac{2\pi}{T}t \tag{3.8}$$

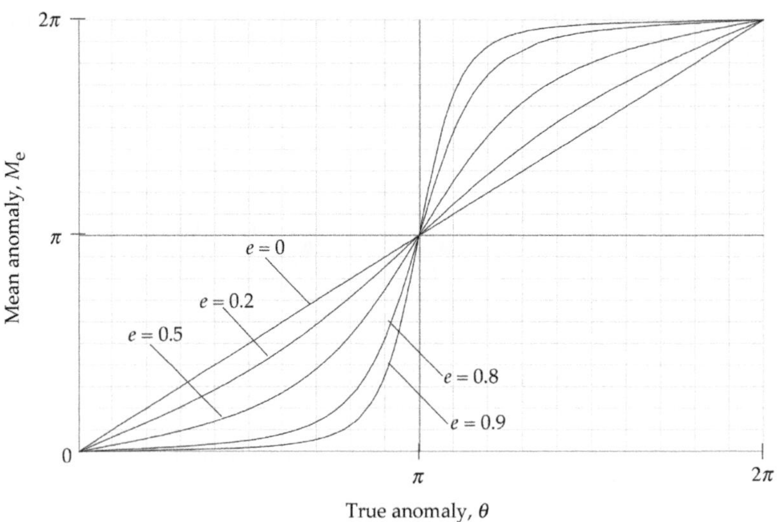

FIG. 3.2

Mean anomaly vs. true anomaly for ellipses of various eccentricities.

The angular velocity of the position vector of an elliptical orbit is not constant, but since 2π radians are swept out per period T, the ratio $2\pi/T$ is the average angular velocity, which is given the symbol n and called the mean motion,

$$n = \frac{2\pi}{T} \tag{3.9}$$

In terms of the mean motion, Eq. (3.8) can be written simpler still,

$$M_e = nt$$

The mean anomaly is the azimuth position (in radians) of a fictitious body moving around the ellipse at the constant angular speed n. For a circular orbit, the mean anomaly M_e and the true anomaly θ are identical.

It is convenient to simplify Eq. (3.6) by introducing an auxiliary angle E called the eccentric anomaly, which is shown in Fig. 3.3. This is done by circumscribing the ellipse with a concentric auxiliary circle having a radius equal to the semimajor axis a of the ellipse. Let S be that point on the ellipse whose true anomaly is θ. Through point S we pass a perpendicular to the apse line, intersecting the auxiliary circle at point Q and the apse line at point V. The angle between the apse line and the radius drawn from the center of the circle to Q on its circumference is the eccentric anomaly E. Observe that E lags the true anomaly θ from periapsis P to apoapsis A ($0 \leq \theta < 180°$), whereas it leads θ from A to P ($180° \leq \theta < 360°$).

To find E as a function of θ, we first observe from Fig. 3.3 that, in terms of the eccentric anomaly, $\overline{OV} = a\cos E$, whereas in terms of the true anomaly, $\overline{OV} = ae + r\cos\theta$. Thus,

$$a\cos E = ae + r\cos\theta$$

Using Eq. (2.72), $r = a(1 - e^2)/(1 + e\cos\theta)$, we can write this as

$$a\cos E = ae + \frac{a(1-e^2)\cos\theta}{1 - e\cos\theta}$$

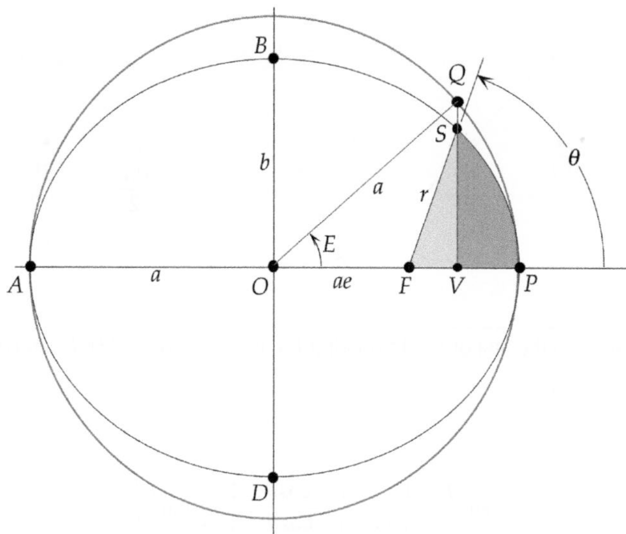

FIG. 3.3

Ellipse and the circumscribed auxiliary circle.

Simplifying the right-hand side, we get

$$\cos E = \frac{e + \cos\theta}{1 + e\cos\theta} \tag{3.10a}$$

Solving this for $\cos\theta$ we obtain the inverse relation,

$$\cos\theta = \frac{e - \cos E}{e\cos E - 1} \tag{3.10b}$$

Substituting Eq. (3.10a) into the trigonometric identity $\sin^2 E + \cos^2 E = 1$ and solving for $\sin E$ yields

$$\sin E = \frac{\sqrt{1 - e^2}\sin\theta}{1 + e\cos\theta} \tag{3.11}$$

Eq. (3.10a) would be fine for obtaining E from θ, except that, given a value of $\cos E$ between -1 and 1, there are two values of E between 0° and 360°, as illustrated in Fig. 3.4. The same comments hold for Eq. (3.11). To resolve this quadrant ambiguity, we use the following trigonometric identity:

$$\tan^2\frac{E}{2} = \frac{\sin^2 E/2}{\cos^2 E/2} = \frac{\dfrac{1 - \cos E}{2}}{\dfrac{1 + \cos E}{2}} = \frac{1 - \cos E}{1 + \cos E} \tag{3.12}$$

From Eq. (3.10a)

$$1 - \cos E = \frac{1 - \cos\theta}{1 + e\cos\theta}(1 - e) \quad \text{and} \quad 1 + \cos E = \frac{1 + \cos\theta}{1 + e\cos\theta}(1 + e)$$

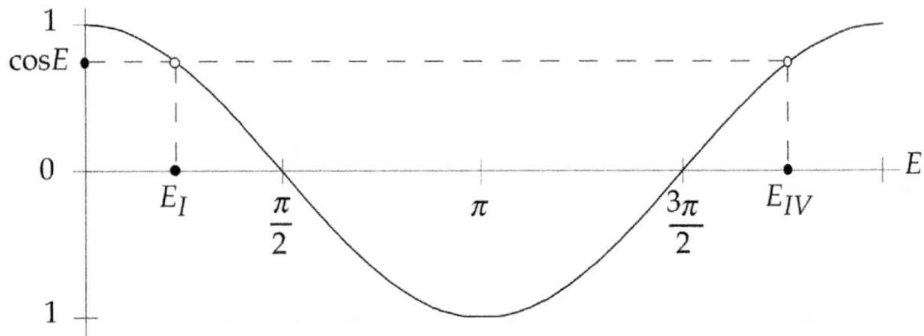

FIG. 3.4

For $0 < \cos E < 1$, E can lie in the first or fourth quadrant. For $-1 < \cos E < 0$, E can lie in the second or third quadrant.

Therefore,

$$\tan^2\frac{E}{2} = \frac{1-e}{1+e} \cdot \frac{1-\cos\theta}{1+\cos\theta} = \frac{1-e}{1+e}\tan^2\frac{\theta}{2}$$

where the last step required applying the trig identity in Eq. (3.12) to the term $(1 - \cos\theta)/(1 + \cos\theta)$. Finally, therefore, we obtain

$$\tan\frac{E}{2} = \sqrt{\frac{1-e}{1+e}}\tan\frac{\theta}{2} \tag{3.13a}$$

or

$$E = 2\tan^{-1}\left(\sqrt{\frac{1-e}{1+e}}\tan\frac{\theta}{2}\right) \tag{3.13b}$$

Observe from Fig. 3.5 that for any value of $\tan(E/2)$, there is only one value of E between 0° and 360°. There is no quadrant ambiguity.

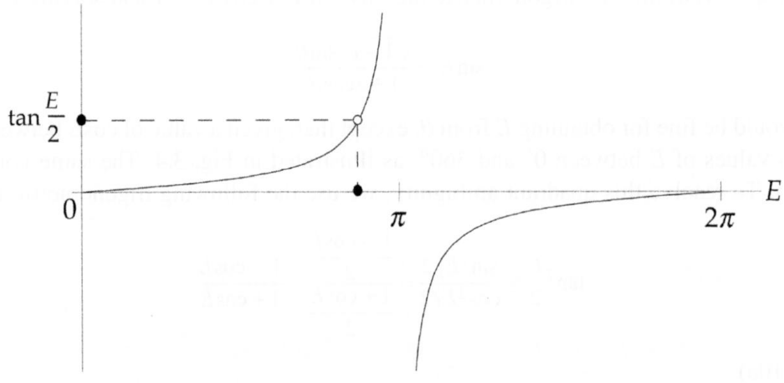

FIG. 3.5

To any value of $\tan(E/2)$, there corresponds a unique value of E in the range 0 to 2π.

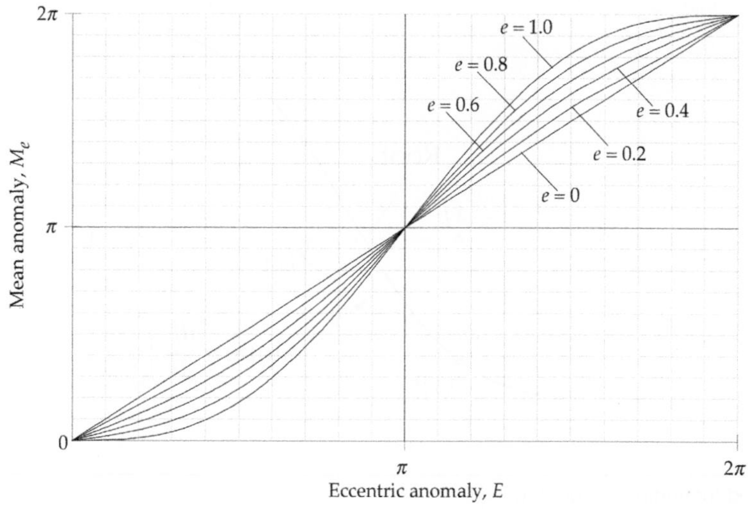

FIG. 3.6

Plot of Kepler's equation for an elliptical orbit.

Substituting Eqs. (3.11) and (3.13b) into Eq. (3.6) yields Kepler's equation,

$$M_e = E - e\sin E \tag{3.14}$$

This monotonically increasing relationship between mean anomaly and eccentric anomaly is plotted for several values of eccentricity in Fig. 3.6.

Given the true anomaly θ, we calculate the eccentric anomaly E using Eq. (3.13). Substituting E into Kepler's formula, Eq. (3.14) yields the mean anomaly directly. From the mean anomaly and the period T we find the time (since periapsis) from Eq. (3.8),

$$t = \frac{M_e}{2\pi}T \tag{3.15}$$

On the other hand, if we are given the time, then Eq. (3.15), yields the mean anomaly M_e. Substituting M_e into Kepler's equation, we get the following expression for the eccentric anomaly:

$$E - e\sin E = M_e$$

We cannot solve this transcendental equation directly for E. A rough value of E might be read from Fig. 3.6. However, an accurate solution requires an iterative, trial-and-error procedure.

Newton's method, or one of its variants, is one of the more common and efficient ways of finding the root of a well-behaved function. To find a root of the equation $f(x) = 0$ in Fig. 3.7, we estimate it to be x_i and evaluate the function $f(x)$ and its first derivative $f'(x)$ at that point. We then extend the tangent to the curve at $f(x_i)$ until it intersects the x axis at x_{i+1}, which becomes our updated estimate of the root. The intercept x_{i+1} is found by setting the slope of the tangent line equal to the slope of the curve at x_i; that is,

$$f'(x_i) = \frac{0 - f(x_i)}{x_{i+1} - x_i}$$

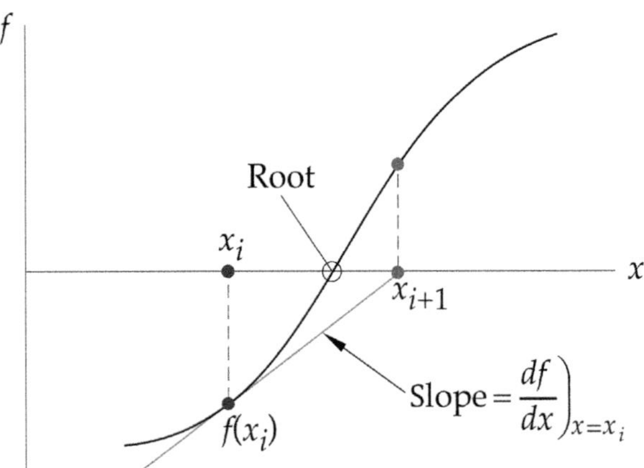

FIG. 3.7

Newton's method for finding a root of $f(x) = 0$.

from which we obtain

$$x_{i+1} = x_i - \frac{f(x_i)}{f'(x_i)} \tag{3.16}$$

The process is repeated, using x_{i+1} to estimate x_{i+2}, and so on, until the root has been found to the desired level of precision.

To apply Newton's method to the solution of Kepler's equation, we form the function

$$f(E) = E - e\sin E - M_e$$

and seek the value of eccentric anomaly that makes $f(E) = 0$. Since

$$f'(E) = 1 - e\cos E$$

for this problem, Eq. (3.16) becomes

$$E_{i+1} = E_i - \frac{E_i - e\sin E_i - M_e}{1 - e\cos E_i} \tag{3.17}$$

ALGORITHM 3.1

Solve Kepler's equation for the eccentric anomaly E given the eccentricity e and the mean anomaly M_e. See Appendix D.11 for the implementation of this algorithm in MATLAB.

1. Choose an initial estimate of the root E as follows (Prussing and Conway, 2013). If $M_e < \pi$, then $E = M_e + e/2$. If $M_e > \pi$, then $E = M_e - e/2$. Remember that the angles E and M_e are in radians. (When using a handheld calculator, be sure that it is in the radian mode.)
2. At any given step, having obtained E_i from the previous step, calculate $f(E_i) = E_i - e \sin E_i - M_e$ and $f'(E_i) = 1 - e \cos E_i$.
3. Calculate $\text{ratio}_i = f(E_i)/f'(E_i)$.

4. If $|\text{ratio}_i|$ exceeds the chosen tolerance (e.g., 10^{-8}), then calculate an updated value of E,

$$E_{i+1} = E_i - \text{ratio}_i$$

Return to Step 2.

5. If $|\text{ratio}_i|$ is less than the tolerance, then accept E_i as the solution to within the chosen accuracy.

EXAMPLE 3.1

A geocentric elliptical orbit has a perigee radius of 9600 km and an apogee radius of 21,000 km, as shown in Fig. 3.8. Calculate the time to fly from perigee P to a true anomaly of 120°.

Solution

Before anything else, let us find the primary orbital parameters e and h. The eccentricity is readily obtained from the perigee and apogee radii by means of Eq. (2.84),

$$e = \frac{r_a - r_p}{r_a + r_p} = \frac{21{,}000 - 9600}{21{,}000 + 9600} = 0.37255 \tag{a}$$

We find the angular momentum using the orbit equation,

$$9600 = \frac{h^2}{398{,}600} \frac{1}{1 + 0.37255\cos(0)} \Rightarrow h = 72{,}472\,\text{km}^2/\text{s}$$

With h and e, the period of the orbit is obtained from Eq. (2.82),

$$T = \frac{2\pi}{\mu^2}\left(\frac{h}{\sqrt{1-e^2}}\right)^3 = \frac{2\pi}{398{,}600^2}\left(\frac{72{,}472}{\sqrt{1-0.37255^2}}\right)^3 = 18{,}834\,\text{s} \tag{b}$$

Eq. (3.13b) yields the eccentric anomaly from the true anomaly,

$$E = 2\tan^{-1}\left(\sqrt{\frac{1-e}{1+e}}\tan\frac{\theta}{2}\right) = 2\tan^{-1}\left(\sqrt{\frac{1-0.37255}{1+0.37255}}\tan\frac{120°}{2}\right) = 1.7281\,\text{rad}$$

Then Kepler's equation (Eq. 3.14) is used to find the mean anomaly,

$$M_e = 1.7281 - 0.37255\sin 1.7281 = 1.3601\,\text{rad}$$

Finally, the time follows from Eq. (3.15),

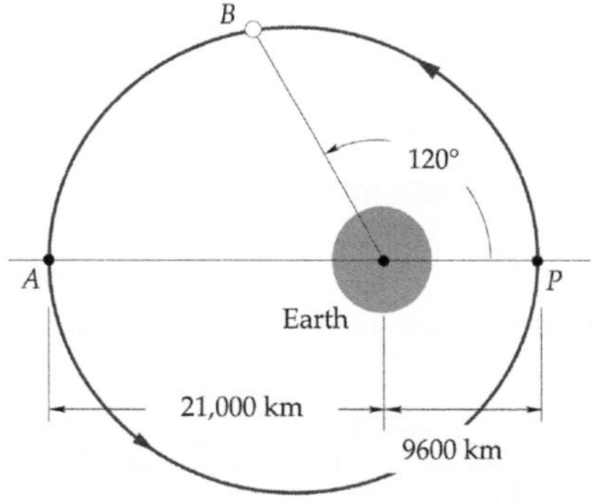

FIG. 3.8

Geocentric elliptical orbit.

$$t=\frac{M_e}{2\pi}T=\frac{1.3601}{2\pi}\cdot 18{,}834=\boxed{4077\text{s }(1.132\text{h})}$$

EXAMPLE 3.2

In the previous example, find the true anomaly at 3 h after perigee passage.

Solution

Since the time (10,800 s) is greater than one-half the period, the true anomaly must be greater than 180°.

First, we use Eq. (3.8) to calculate the mean anomaly for $t=10{,}800\,\text{s}$.

$$M_e=2\pi\frac{t}{T}=2\pi\frac{10{,}800}{18{,}830}=3.6029\,\text{rad} \tag{a}$$

Kepler's equation, $E-e\sin(E)=M_e$ (with all angles in radians) is then employed to find the eccentric anomaly. This transcendental equation will be solved using Algorithm 3.1 with an error tolerance of 10^{-6}. Since $M_e>\pi$, a good starting value for the iteration is $E_0=M_e-e/2=3.4166$. Executing the algorithm yields the following steps:

Step 1:

$$\begin{aligned}
E_0 &= 3.4166\\
f(E_0) &= -0.085124\\
f'(E_0) &= 1.3585\\
\text{ratio} &= \frac{-0.085124}{1.3585} = -0.062658\\
|\text{ratio}| &> 10^{-6} \qquad \therefore \text{ repeat}
\end{aligned}$$

Step 2:

$$\begin{aligned}
E_1 &= 3.4166-(-0.062658) = 3.4793\\
f(E_1) &= -0.0002134\\
f'(E_1) &= 1.3515\\
\text{ratio} &= \frac{-0.0002134}{1.3515} = -1.5778\times 10^{-4}\\
|\text{ratio}| &> 10^{-6} \qquad \therefore \text{repeat}
\end{aligned}$$

Step 3:

$$\begin{aligned}
E_2 &= 3.4793-\left(-1.5778\times 10^{-4}\right) = 3.4794\\
f(E_2) &= -1.5366\times 10^{-9}\\
f'(E_2) &= 1.3515\\
\text{ratio} &= \frac{-1.5366\times 10^{-9}}{1.3515} = -1.137\times 10^{-9}\\
|\text{ratio}| &< 10^{-6} \qquad \therefore \text{stop}
\end{aligned}$$

Convergence to even more than the desired accuracy occurred after just two iterations. With $E=3.4794$, the true anomaly is found from Eq. (3.13a) to be

$$\tan\frac{\theta}{2}=\sqrt{\frac{1+e}{1-e}}\tan\frac{E}{2}=\sqrt{\frac{1+0.37255}{1-0.37255}}\tan\frac{3.4794}{2}=-8.6721 \;\Rightarrow\; \boxed{\theta=193.2°}$$

EXAMPLE 3.3

Let a satellite be in a 500 km by 5000 km orbit with its apse line parallel to the line from the earth to the sun, as shown in Fig. 3.9. Find the time that the satellite is in the earth's shadow if
(a) the apogee is toward the sun
(b) the perigee is toward the sun.

Solution

We start by using the given data to find the primary orbital parameters, e and h. The eccentricity is obtained from Eq. (2.84),

$$e = \frac{r_a - r_p}{r_a + r_p} = \frac{(6378 + 5000) - (6378 + 500)}{(6378 + 5000) + (6378 + 500)} = 0.24649 \quad \text{(a)}$$

The orbit equation can then be used to find the angular momentum

$$r_p = \frac{h^2}{\mu}\frac{1}{1 + e\cos(0)} \Rightarrow 6878 = \frac{h^2}{398{,}600}\frac{1}{1 + 0.24649} \Rightarrow h = 58{,}458\,\text{km}^2/\text{s}$$

The semimajor axis may be found from Eq. (2.71),

$$a = \frac{h^2}{\mu}\frac{1}{1 - e^2} = \frac{58{,}458^2}{398{,}600}\frac{1}{1 - 0.24649^2} = 9128\,\text{km} \quad \text{(b)}$$

or from the fact that $a = (r_p + r_a)/2$. The period of the orbit follows from Eq. (2.83),

$$T = \frac{2\pi}{\sqrt{\mu}}a^{3/2} = \frac{2\pi}{\sqrt{398{,}600}}9128^{3/2} = 86791.1\,\text{s}\,(2.4109\,\text{h})$$

(a) If the apogee is toward the sun, as in Fig. 3.9, then the satellite is in earth's shadow between points a and b on its orbit. These are two of the four points of intersection of the orbit with the lines that are parallel to the earth-sun line, and lie at a distance R_E from the center of the earth. The true anomaly of b is therefore given by $\sin\theta = R_E/r$, where r is the radial position of the satellite. It follows that the radius of b is

$$r = \frac{R_E}{\sin\theta} \quad \text{(c)}$$

From Eq. (2.72), we also have

$$r = \frac{a(1 - e^2)}{1 + e\cos\theta} \quad \text{(d)}$$

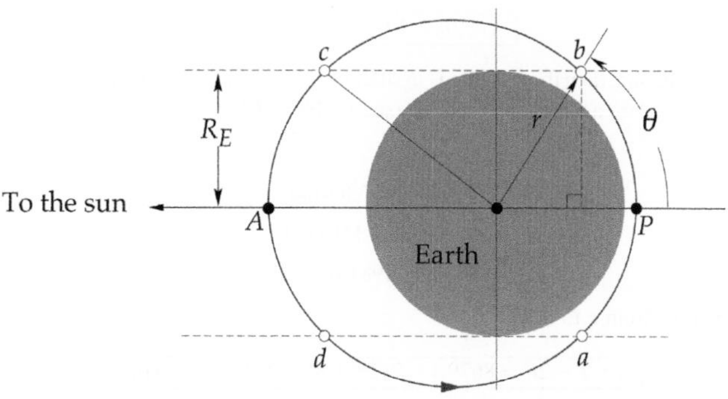

FIG. 3.9

Satellite passing in and out of the earth's shadow.

Substituting Eq. (c) into Eq. (d), collecting terms, and simplifying yields an equation in θ,

$$e\cos\theta - (1-e^2)\frac{a}{R_E}\sin\theta + 1 = 0 \tag{e}$$

Substituting Eqs. (a) and (b) together with $R_E = 6378\,\text{km}$ into Eq. (e) yields

$$0.24649\cos\theta - 1.3442\sin\theta = -1$$

This equation is of the form

$$A\cos\theta + B\sin\theta = C$$

It has two roots, which are given by (see Problem 3.12):

$$\theta = \tan^{-1}\frac{B}{A} \pm \cos^{-1}\left[\frac{C}{A}\cos\left(\tan^{-1}\frac{B}{A}\right)\right]$$

For the case at hand,

$$\theta = \tan^{-1}\frac{-1.3442}{0.24649} \pm \cos^{-1}\left[\frac{-1}{0.24649}\cos\left(\tan^{-1}\frac{-1.3442}{0.24649}\right)\right] = -79.607^\circ \pm 137.03^\circ$$

That is

$$\theta_b = 57.423^\circ$$
$$\theta_c = -216.64^\circ\,(+143.36^\circ)$$

For apogee toward the sun, the flight from perigee to point b will be in shadow. To find the time of flight from perigee to point b, we first compute the eccentric anomaly of b using Eq. (3.13b):

$$E_b = 2\tan^{-1}\left(\sqrt{\frac{1-e}{1+e}}\tan\frac{\theta_b}{2}\right) = 2\tan^{-1}\left(\sqrt{\frac{1-0.24649}{1+0.24649}}\tan\frac{57.423^\circ}{2}\right) = 0.80521\,\text{rad}$$

From this we find the mean anomaly using Kepler's equation,

$$M_e = E - e\sin E = 0.80521 - 0.24649\sin 0.80521 = 0.62749\,\text{rad}$$

Finally, Eq. (3.5) yields the time at b,

$$t_b = \frac{M_e}{2\pi}T = \frac{0.62749}{2\pi}8679.1 = 866.77\,\text{s}$$

The total time in shadow, from a to b, during which the satellite passes through perigee, is

$$\boxed{t = 2t_b = 1734\,\text{s}\ (28.98\,\text{min})} \tag{f}$$

(b) If the perigee is toward the sun, then the satellite is in shadow near apogee, from point c ($\theta_c = 143.36^\circ$) to d on the orbit. Following the same procedure as above, we obtain (see Problem 3.13),

$$E_c = 2.3364\,\text{rad}$$
$$M_c = 2.1587\,\text{rad}$$
$$t_c = 2981.8\,\text{s}$$

The total time in shadow, from c to d, is

$$\boxed{t = T - 2t_c = 8679.1 - 2(2981.8) = 2716\,\text{s}\,(45.26\,\text{min})}$$

The time is longer than that given by Eq. (f) since the satellite travels slower near apogee.

We have observed that there is no closed-form solution for the eccentric anomaly E in Kepler's equation, $E - e\sin E = M_e$. However, there exist infinite series solutions. One of these, due to Lagrange (Battin, 1987), is a power series in the eccentricity e,

$$E = M_e + \sum_{n=1}^{\infty} a_n e^n \tag{3.18}$$

where the coefficients a_n are given by the somewhat intimidating expression

$$a_n = \frac{1}{2^{n-1}} \sum_{k=0}^{floor(n/2)} (-1)^k \frac{1}{(n-k)!k!}(n-2k)^{n-1}\sin[(n-2k)M_e] \tag{3.19}$$

where $floor(x)$ means x rounded to the next lowest integer (e.g., $floor(0.5) = 0$, $floor(\pi) = 3$). If e is sufficiently small, then the Lagrange series converges. This means that by including enough terms in the summation, we can obtain E to any desired degree of precision. Unfortunately, if e exceeds 0.6627434193, the series diverges, which means taking more and more terms yields worse and worse results for some values of M.

The limiting value for the eccentricity was discovered by the French mathematician Pierre-Simone Laplace (1749–1827) and is called the Laplace limit.

In practice, we must truncate the Lagrange series to a finite number of terms N, so that

$$E = M_e + \sum_{n=1}^{N} a_n e^n \tag{3.20}$$

For example, setting $N = 3$ and calculating each a_n by means of Eq. (3.19) leads to

$$E = M_e + e\sin M_e + \frac{e^2}{2}\sin 2M_e + \frac{e^3}{8}(3\sin 3M_e - \sin M_e) \tag{3.21}$$

For small values of the eccentricity e, this yields good agreement with the exact solution of Kepler's equation (plotted in Fig. 3.6). However, as we approach the Laplace limit, the accuracy degrades unless more terms of the series are included. Fig. (3.10) shows that for an eccentricity of 0.65, just below the Laplace limit, Eq. (3.21) ($N = 3$) yields a solution that oscillates around the exact solution but is fairly close to it everywhere. Setting $N = 10$ in Eq. (3.20) produces a curve that, at the given scale, is indistinguishable from the exact solution. On the other hand, for an eccentricity of 0.90, far above the Laplace limit, Fig. 3.11 reveals that Eq. (3.21) is a poor approximation to the exact solution, and using $N = 10$ makes matters even worse.

Another infinite series for E (Battin, 1987) is given by

$$E = M_e + \sum_{n=1}^{\infty} \frac{2}{n} J_n(ne)\sin nM_e \tag{3.22}$$

where the coefficients J_n are Bessel functions of the first kind, attributable to German astronomer Friedrich Bessel (1784–1846). They are defined by the infinite series

$$J_n(x) = \sum_{k=0}^{\infty} \frac{(-1)^k}{k!(n+k)!}\left(\frac{x}{2}\right)^{n+k} \tag{3.23}$$

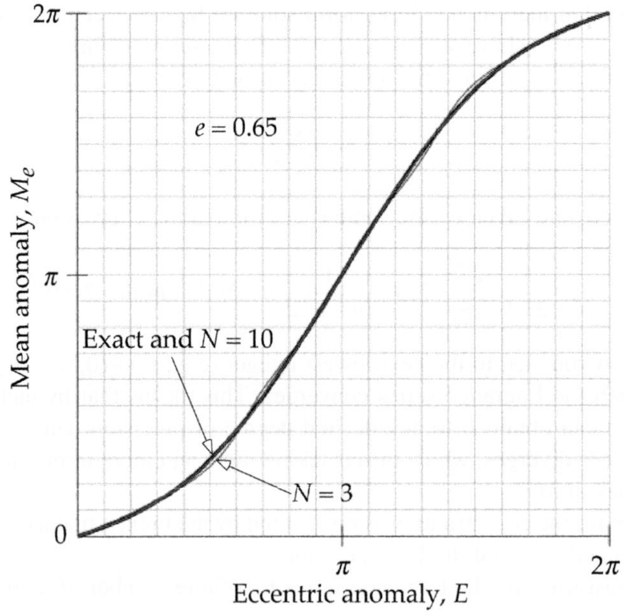

FIG. 3.10

Comparison of the exact solution of Kepler's equation with the truncated Lagrange series solution ($N = 3$ and $N = 10$) for an eccentricity of 0.65.

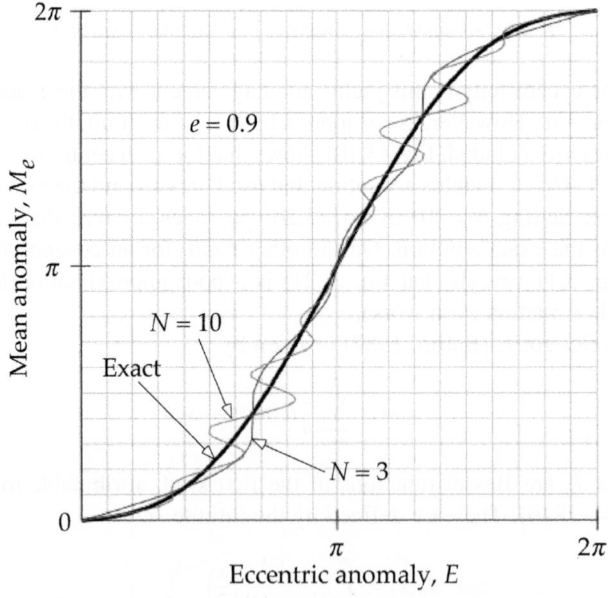

FIG. 3.11

Comparison of the exact solution of Kepler's equation with the truncated Lagrange series solution ($N = 3$ and $N = 10$) for an eccentricity of 0.90.

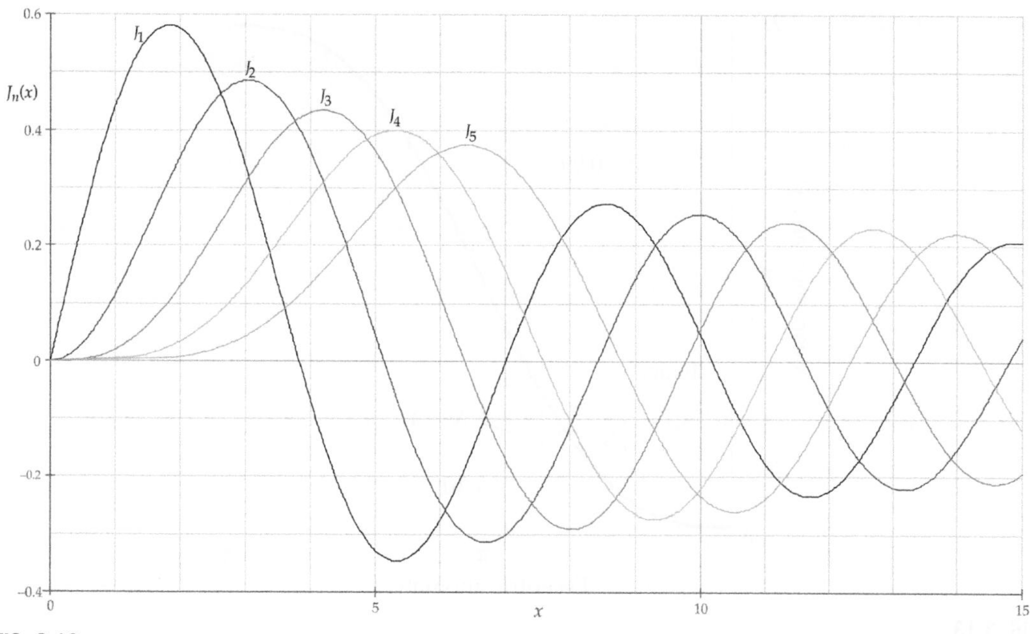

FIG. 3.12

Bessel functions of the first kind.

J_1 through J_5 are plotted in Fig. 3.12. Clearly, they are oscillatory in appearance and tend toward zero with increasing x.

It turns out that, unlike the Lagrange series, the Bessel function series solution converges for all values of the eccentricity <1. Fig. 3.13 shows how the truncated Bessel series solutions

$$E = M_e + \sum_{n=1}^{N} \frac{2}{n} J_n(ne) \sin nM_e \tag{3.24}$$

for $N = 3$ and $N = 10$ compare with the exact solution of Kepler's equation for the very large elliptical eccentricity of $e = 0.99$. It can be seen that the case $N = 3$ yields a poor approximation for all but a few values of M_e. Increasing the number of terms in the series to $N = 10$ obviously improves the approximation, and adding even more terms will make the truncated series solution indistinguishable from the exact solution at the given scale.

Observe that we can combine Eqs. (3.10) and (2.72) as follows to obtain the orbit equation for the ellipse in terms of the eccentric anomaly:

$$r = \frac{a(1-e^2)}{1+e\cos\theta} = \frac{a(1-e^2)}{1+e\left(\dfrac{e-\cos E}{e\cos E - 1}\right)}$$

From this it is easy to see that

$$r = a(1 - e\cos E) \tag{3.25}$$

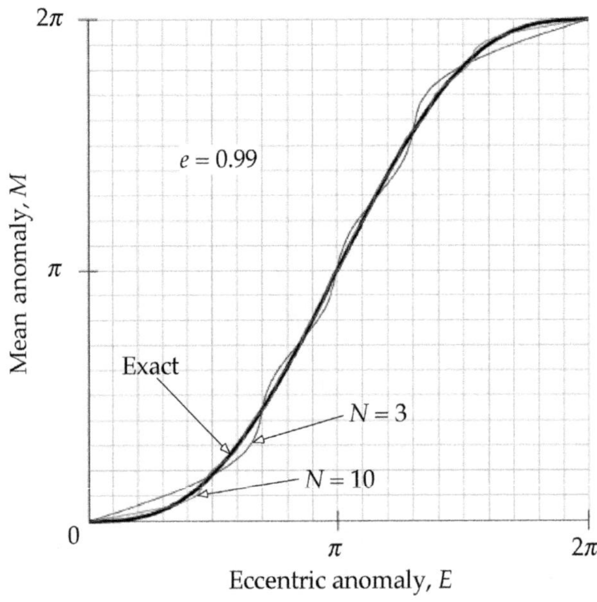

FIG. 3.13

Comparison of the exact solution of Kepler's equation with the truncated Bessel series solution ($N=3$ and $N=10$) for an eccentricity of 0.99.

In Eq. (2.86), we defined the true anomaly-averaged radius $\bar{r}_\theta$ of an elliptical orbit. Alternatively, the time-averaged radius $\bar{r}_t$ of an elliptical orbit is defined as

$$\bar{r}_t = \frac{1}{T}\int_0^T r\,\mathrm{d}t \tag{3.26}$$

According to Eqs. (3.14) and (3.15),

$$t = \frac{T}{2\pi}(E - e\sin E)$$

Therefore,

$$\mathrm{d}t = \frac{T}{2\pi}(1 - e\cos E)\mathrm{d}E$$

Upon using this relationship to change the variable of integration from t to E and substituting Eq. (3.25), Eq. (3.26) becomes

$$\begin{aligned}\bar{r}_t &= \frac{1}{T}\int_0^{2\pi}[a(1-e\cos E)]\left[\frac{T}{2\pi}(1-e\cos E)\right]dE \\ &= \frac{a}{2\pi}\int_0^{2\pi}(1-e\cos E)^2 dE \\ &= \frac{a}{2\pi}\int_0^{2\pi}(1-2e\cos E+e^2\cos^2 E)dE \\ &= \frac{a}{2\pi}(2\pi-0+e^2\pi)\end{aligned}$$

so that

$$\bar{r}_t = a\left(1+\frac{e^2}{2}\right) \quad \text{Time-averaged radius of an elliptical orbit} \tag{3.27}$$

Comparing this result with Eq. (2.87) reveals, as we should have expected, that $\bar{r}_t > \bar{r}_\theta$. In fact, combining Eqs. (2.87) and (3.27) yields

$$\bar{r}_\theta = a\sqrt{3-2\frac{\bar{r}_t}{a}} \tag{3.28}$$

3.5 PARABOLIC TRAJECTORIES ($e = 1$)

For the parabola, Eq. (3.2) becomes

$$\frac{\mu^2}{h^3}t = \int_0^\theta \frac{d\vartheta}{(1+\cos\vartheta)^2} \tag{3.29}$$

Setting $a = b = 1$ in Eq. (3.4) yields

$$\int_0^\theta \frac{d\vartheta}{(1+\cos\vartheta)^2} = \frac{1}{2}\tan\frac{\theta}{2} + \frac{1}{6}\tan^3\frac{\theta}{2}$$

Therefore, Eq. (3.29) may be written as

$$M_p = \frac{1}{2}\tan\frac{\theta}{2} + \frac{1}{6}\tan^3\frac{\theta}{2} \tag{3.30}$$

where

$$M_p = \frac{\mu^2 t}{h^3} \tag{3.31}$$

M_p is dimensionless, and it may be thought of as the "mean anomaly" for the parabola. Eq. (3.30) is plotted in Fig. 3.14. Eq. (3.30) is also known as Barker's equation, after Thomas Barker (1722–1809), a British meteorologist.

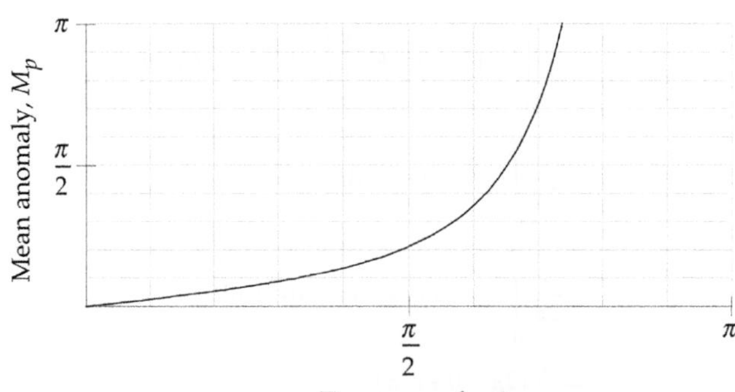

FIG. 3.14

Graph of Barker's equation.

Given the true anomaly θ, we find the time directly from Eqs. (3.30) and (3.31). If time is the given variable, then we must solve the cubic equation

$$\frac{1}{6}\left(\tan\frac{\theta}{2}\right)^3+\frac{1}{2}\tan\frac{\theta}{2}-M_p=0$$

which has but one real root, namely,

$$\tan\frac{\theta}{2}=z-\frac{1}{z} \tag{3.32a}$$

where

$$z=\left(3M_p+\sqrt{1+(3M_p)^2}\right)^{1/3} \tag{3.32b}$$

EXAMPLE 3.4

A geocentric parabola has a perigee velocity of 10 km/s. How far is the satellite from the center of the earth 6 h after perigee passage?

Solution

The first step is to find the orbital parameters e and h. Of course, we know that $e=1$. To get the angular momentum, we can use the given perigee speed and Eq. (2.90) (the energy equation) to find the perigee radius,

$$r_p=\frac{2\mu}{v_p^2}=\frac{2\cdot 398{,}600}{10^2}=7972\,\text{km}$$

It follows from Eq. (2.31) that the angular momentum is

$$h=r_p v_p=7972\cdot 10=79{,}720\,\text{km}^2/\text{s}$$

We can now calculate the parabolic mean anomaly by means of Eq. (3.31),

$$M_p=\frac{\mu^2 t}{h^3}=\frac{398{,}600^2\cdot(6\cdot 3600)}{79{,}720^3}=6.7737\,\text{rad}$$

Therefore, $3M_p = 20.321$ rad, which, when substituted into Eqs. (3.32), yields the true anomaly,

$$z = \left(20.321 + \sqrt{1 + 20.321^2}\right)^{1/3} = 3.4388$$

$$\tan\frac{\theta}{2} = 3.4388 - \frac{1}{3.4388} = 3.1480 \Rightarrow \theta = 144.75°$$

Finally, we substitute the true anomaly into the orbit equation to find the radius,

$$r = \frac{79{,}720^2}{398{,}600} \frac{1}{1 + \cos(144.75°)} = \boxed{86{,}899\text{km}}$$

3.6 HYPERBOLIC TRAJECTORIES ($e > 1$)

Setting $a = 1$ and $b = e$ in Eq. (3.5) yields

$$\int_0^\theta \frac{d\vartheta}{(1 + e\cos\vartheta)^2} = \frac{1}{e^2 - 1}\left\{\frac{e\sin\theta}{1 + e\cos\theta} - \frac{1}{\sqrt{e^2 - 1}} \ln\left[\frac{\sqrt{e+1} + \sqrt{e-1}\tan(\theta/2)}{\sqrt{e+1} - \sqrt{e-1}\tan(\theta/2)}\right]\right\}$$

Therefore, for the hyperbola, Eq. (3.1) becomes

$$\frac{\mu^2}{h^3}t = \frac{1}{e^2 - 1}\frac{e\sin\theta}{1 + e\cos\theta} - \frac{1}{(e^2 - 1)^{3/2}} \ln\left[\frac{\sqrt{e+1} + \sqrt{e-1}\tan(\theta/2)}{\sqrt{e+1} - \sqrt{e-1}\tan(\theta/2)}\right]$$

Multiplying both sides by $(e^2 - 1)^{3/2}$, we get

$$M_h = \frac{e\sqrt{e^2 - 1}\sin\theta}{1 + e\cos\theta} - \ln\left[\frac{\sqrt{e+1} + \sqrt{e-1}\tan(\theta/2)}{\sqrt{e+1} - \sqrt{e-1}\tan(\theta/2)}\right] \quad (3.33)$$

where

$$M_h = \frac{\mu^2}{h^3}(e^2 - 1)^{3/2} t \quad (3.34)$$

M_h is the hyperbolic mean anomaly. Eq. (3.33) is plotted in Fig. 3.15. Recall that θ cannot exceed θ_∞ (Eq. 2.97).

We can simplify Eq. (3.33) by introducing an auxiliary angle analogous to the eccentric anomaly E for the ellipse. Consider a point on a hyperbola whose polar coordinates are r and θ. Referring to Fig. 3.16, let x be the horizontal distance of the point from the center C of the hyperbola, and let y be its distance above the apse line. The ratio y/b defines the hyperbolic sine of the dimensionless variable F that we will use as the hyperbolic eccentric anomaly. That is, we define F to be such that

$$\sinh F = \frac{y}{b} \quad (3.35)$$

In view of the equation of a hyperbola,

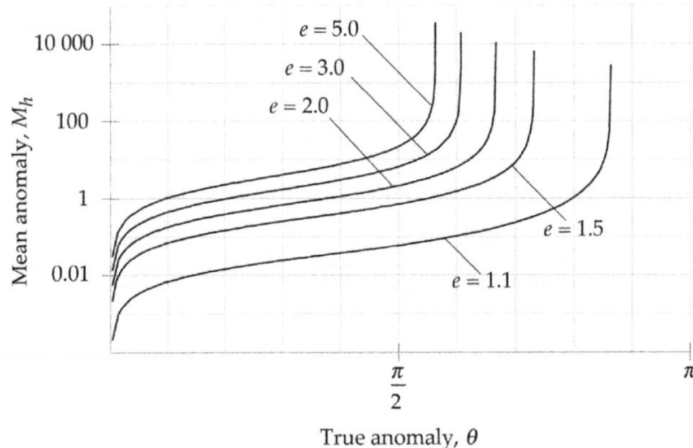

FIG. 3.15

Plots of Eq. (3.33) for several different eccentricities.

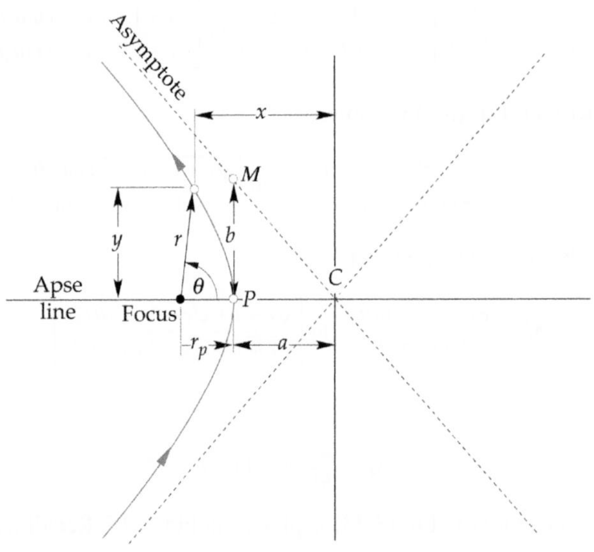

FIG. 3.16

Hyperbola parameters.

$$\frac{x^2}{a^2}-\frac{y^2}{b^2}=1$$

it is consistent with the definition of sinh$F$ to define the hyperbolic cosine as

$$\cosh F=\frac{x}{a} \tag{3.36}$$

(It should be recalled that $\sinh x=(e^x-e^{-x})/2$ and $\cosh x=(e^x+e^{-x})/2$, and that, therefore, $\cosh^2 x-\sinh^2 x=1$.)

From Fig. 3.16 we see that $y = r\sin\theta$. Substituting this into Eq. (3.35), along with $r = a(e^2 - 1)/(1 + e\cos\theta)$ (Eq. 2.104) and $b = a\sqrt{e^2-1}$ (Eq. 2.106), we get

$$\sinh F = \frac{1}{b} r\sin\theta = \frac{1}{a\sqrt{e^2-1}}\frac{a(e^2-1)}{1+e\cos\theta}\sin\theta$$

so that

$$\sinh F = \frac{\sin\theta\sqrt{e^2-1}}{1+e\cos\theta} \tag{3.37}$$

This can be used to obtain F in terms of the true anomaly,

$$F = \sinh^{-1}\left(\frac{\sin\theta\sqrt{e^2-1}}{1+e\cos\theta}\right) \tag{3.38}$$

Using the formula $\sinh^{-1}x = \ln\left(x + \sqrt{x^2+1}\right)$ we can, after simplifying the algebra, write Eq. (3.38) as

$$F = \ln\left(\frac{\sin\theta\sqrt{e^2-1} + \cos\theta + e}{1+e\cos\theta}\right)$$

Substituting the trigonometric identities,

$$\sin\theta = \frac{2\tan(\theta/2)}{1+\tan^2(\theta/2)} \qquad \cos\theta = \frac{1-\tan^2(\theta/2)}{1+\tan^2(\theta/2)}$$

and doing some more algebra yields

$$F = \ln\left[\frac{1+e+(e-1)\tan^2(\theta/2) + 2\tan(\theta/2)\sqrt{e^2-1}}{1+e+(1-e)\tan^2(\theta/2)}\right]$$

Fortunately, but not too obviously, the numerator and the denominator in the brackets have a common factor, so that this expression for the hyperbolic eccentric anomaly reduces to

$$F = \ln\left[\frac{\sqrt{e+1}+\sqrt{e-1}\tan(\theta/2)}{\sqrt{e+1}-\sqrt{e-1}\tan(\theta/2)}\right] \tag{3.39}$$

Substituting Eqs. (3.37) and (3.39) into Eq. (3.33) yields Kepler's equation for the hyperbola,

$$M_h = e\sinh F - F \tag{3.40}$$

This equation is plotted for several different eccentricities in Fig. 3.17.

If we substitute the expression for $\sinh F$ (Eq. 3.37) into the hyperbolic trig identity

$$\cosh^2 F - \sinh^2 F = 1$$

we get

$$\cosh^2 F = 1 + \left(\frac{\sin\theta\sqrt{e^2-1}}{1+e\cos\theta}\right)^2$$

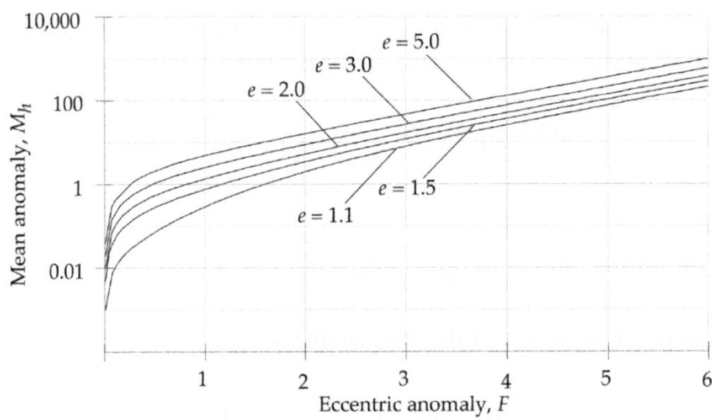

FIG. 3.17

Plot of Kepler's equation for the hyperbola.

A few steps of algebra lead to

$$\cosh^2 F = \left(\frac{\cos\theta + e}{1 + e\cos\theta}\right)^2$$

so that

$$\cosh F = \frac{\cos\theta + e}{1 + e\cos\theta} \tag{3.41a}$$

Solving this for $\cos\theta$, we obtain the inverse relation,

$$\cos\theta = \frac{\cosh F - e}{1 - e\cosh F} \tag{3.41b}$$

The hyperbolic tangent is found in terms of the hyperbolic sine and cosine by the formula

$$\tanh F = \frac{\sinh F}{\cosh F}$$

In mathematical handbooks, we can find the hyperbolic trig identity,

$$\tanh\frac{F}{2} = \frac{\sinh F}{1 + \cosh F} \tag{3.42}$$

Substituting Eqs. (3.37) and (3.41a) into this formula and simplifying yields

$$\tanh\frac{F}{2} = \sqrt{\frac{e-1}{e+1}}\frac{\sin\theta}{1 + \cos\theta} \tag{3.43}$$

Interestingly enough, Eq. (3.42) holds for ordinary trig functions, too; that is,

$$\tan\frac{\theta}{2} = \frac{\sin\theta}{1 + \cos\theta}$$

Therefore, Eq. (3.43) can be written

$$\tanh\frac{F}{2} = \sqrt{\frac{e-1}{e+1}}\tan\frac{\theta}{2} \tag{3.44a}$$

This is a somewhat simpler alternative to Eq. (3.39) for computing eccentric anomaly from true anomaly, and it is a whole lot simpler to invert:

$$\tan\frac{\theta}{2} = \sqrt{\frac{e+1}{e-1}}\tanh\frac{F}{2} \tag{3.44b}$$

If time is the given quantity, then Eq. (3.40)—a transcendental equation—must be solved for F by an iterative procedure, as was the case for the ellipse. To apply Newton's procedure to the solution of Kepler's equation for the hyperbola, we form the function

$$f(F) = e\sinh F - F - M_h$$

and seek the value of F that makes $f(F) = 0$. Since

$$f'(F) = e\cosh F - 1$$

Eq. (3.16) becomes

$$F_{i+1} = F_i - \frac{e\sinh F_i - F_i - M_h}{e\cosh F_i - 1} \tag{3.45}$$

All quantities in this formula are dimensionless (radians, not degrees).

ALGORITHM 3.2

Solve Kepler's equation for the hyperbola for the hyperbolic eccentric anomaly F given the eccentricity e and the hyperbolic mean anomaly M_h. See Appendix D.12 for the implementation of this algorithm in MATLAB.

1. Choose an initial estimate of the eccentric anomaly F.
   a. For hand computations, read a rough value of F_0 (no more than two significant figures) from Fig. 3.17 to keep the number of iterations to a minimum.
   b. In computer software, let $F_0 = M_h$, an inelegant choice that may result in many iterations but will nevertheless rapidly converge on today's high-speed desktops and laptops.
2. At any given step, having obtained F_i from the previous step, calculate $f(F_i) = e \sinh F_i - F_i - M_h$ and $f'(F_i) = e \cosh F_i - 1$.
3. Calculate $\text{ratio}_i = f(F_i)/f'(F_i)$.
4. If $|\text{ratio}_i|$ exceeds the chosen tolerance (e.g., 10^{-8}), then calculate an updated value of F_i. Return to Step 2.
5. If $|\text{ratio}_i|$ is less than the tolerance, then accept F_i as the solution to within the desired accuracy.

EXAMPLE 3.5

A geocentric trajectory has a perigee velocity of 15 km/s and a perigee altitude of 300 km. Find
(a) the radius and the time when the true anomaly is 100°;
(b) the position and speed 3 h later.

Solution

We first calculate the primary orbital parameters e and h. The angular momentum is calculated from Eq. (2.31) and the given perigee data:

$$h = r_p v_p = (6378 + 300) \cdot 15 = 100,170\text{km}^2/\text{s}$$

The eccentricity is found by evaluating Eq. (2.50), the orbit equation, $r = (h^2/\mu)/(1 + e\cos\theta)$, at perigee ($\theta = 0°$):

$$6378 + 300 = \frac{100,170^2}{398,600}\frac{1}{1+e} \Rightarrow e = 2.7696$$

(a) Since $e > 1$, the trajectory is a hyperbola. Note that the true anomaly of the asymptote of the hyperbola is, according to Eq. (2.97),

$$\theta_\infty = \cos^{-1}\left(-\frac{1}{2.7696}\right) = 111.17°$$

Evaluating the orbit equation at the given true anomaly, $\theta = 100°$, yields

$$r = \frac{100,170^2}{398,600}\frac{1}{1 + 2.7696\cos 100°} = \boxed{48,497\ \text{km}}$$

To find the time since perigee passage at $\theta = 100°$, we first use Eq. (3.44a) to calculate the hyperbolic eccentric anomaly,

$$\tanh\frac{F}{2} = \sqrt{\frac{2.7696 - 1}{2.7696 + 1}}\tan\frac{100°}{2} = 0.81653 \Rightarrow F = 2.2927\text{rad}$$

Kepler's equation for the hyperbola then yields the mean anomaly,

$$M_h = e\sinh F - F = 2.7696\sinh 2.2927 - 2.2927 = 11.279\text{rad}$$

The time since perigee passage is found from Eq. (3.34),

$$t = \frac{h^3}{\mu^2}\frac{1}{\left(e^2 - 1\right)^{3/2}}M_h = \frac{100,170^3}{398,600^2}\frac{1}{\left(2.7696^2 - 1\right)^{3/2}}11.279 = \boxed{4141.4\,\text{s}}$$

(b) After 3 h, the time since perigee passage is

$$t = 4141.4 + 3 \cdot 3600 = 14,941\,\text{s}\ (4.15\,\text{h})$$

The corresponding mean anomaly, from Eq. (3.34), is

$$M_h = \frac{398,600^2}{100,170^3}\left(2.7696^2 - 1\right)^{3/2}(14,941) = 40.690\text{rad}$$

We will use Algorithm 3.2 with an error tolerance of 10^{-6} to find the hyperbolic eccentric anomaly F. Referring to Fig. 3.17, we see that for $M_h = 40.69$ and $e = 2.7696$, F lies between 3 and 4. Let us arbitrarily choose $F_0 = 3$ as our initial estimate of F. Executing the algorithm yields the following steps:

$$F_0 = 3$$

Step 1:

$$\begin{aligned} f(F_0) &= -15.944494 \\ f'(F_0) &= 26.883397 \\ \text{ratio} &= -0.59309818 \\ F_1 &= 3-(-0.59309818) = 3.5930982 \\ |\text{ratio}| &> 10^{-6} \quad \therefore \text{repeat.} \end{aligned}$$

Step 2:

$$\begin{aligned} f(F_1) &= 6.0114484 \\ f'(F_1) &= 49.370747 \\ \text{ratio} &= -0.12176134 \\ F_2 &= 3.5930982-(-0.12176134) = 3.4713368 \\ |\text{ratio}| &> 10^{-6} \quad \therefore \text{repeat.} \end{aligned}$$

Step 3:

$$\begin{aligned} f(F_2) &= 0.35812370 \\ f'(F_2) &= 43.605527 \\ \text{ratio} &= 8.2128052 \times 10^{-3} \\ F_3 &= 3.4713368-\left(8.2128052 \times 10^{-3}\right) = 3.4631240 \\ |\text{ratio}| &> 10^{-6} \quad \therefore \text{repeat.} \end{aligned}$$

Step 4:

$$\begin{aligned} f(F_3) &= 1.4973128 \times 10^{-3} \\ f'(F_3) &= 43.241398 \\ \text{ratio} &= 3.4626836 \times 10^{-5} \\ F_4 &= 3.4631240-\left(3.4626836 \times 10^{-5}\right) = 3.4630894 \\ |\text{ratio}| &> 10^{-6} \quad \therefore \text{repeat.} \end{aligned}$$

Step 5:

$$\begin{aligned} f(F_4) &= 2.6470781 \times 10^{-3} \\ f'(F_4) &= 43.239869 \\ \text{ratio} &= 6.1218459 \times 10^{-10} \\ F_5 &= 3.4630894-\left(6.1218459 \times 10^{-10}\right) = 3.4630894 \\ |\text{ratio}| &< 10^{-6} \quad \therefore \text{accept } F = 3.4631 \text{ as the solution.} \end{aligned}$$

We substitute this final value of F into Eq. (3.44b) to find the true anomaly,

$$\tan\frac{\theta}{2} = \sqrt{\frac{e+1}{e-1}}\tanh\frac{F}{2} = \sqrt{\frac{2.7696+1}{2.7696-1}}\tanh\frac{3.4631}{2} = 1.3708 \Rightarrow \theta = 107.78^\circ$$

With the true anomaly, the orbital equation yields the radial coordinate at the final time

$$r = \frac{h^2}{\mu}\frac{1}{1+e\cos\theta} = \frac{100{,}170^2}{398{,}600}\frac{1}{1+2.7696\cos 107.78^\circ} = \boxed{163{,}180\,\text{km}}$$

The velocity components are obtained from Eq. (2.31),

$$v_\perp = \frac{h}{r} = \frac{100{,}170}{163{,}180} = 0.61386\text{km/s}$$

and Eq. (2.49),

$$v_r = \frac{\mu}{h}e\sin\theta = \frac{398{,}600}{100{,}170}2.7696\sin 107.78^\circ = 10.496\text{km/s}$$

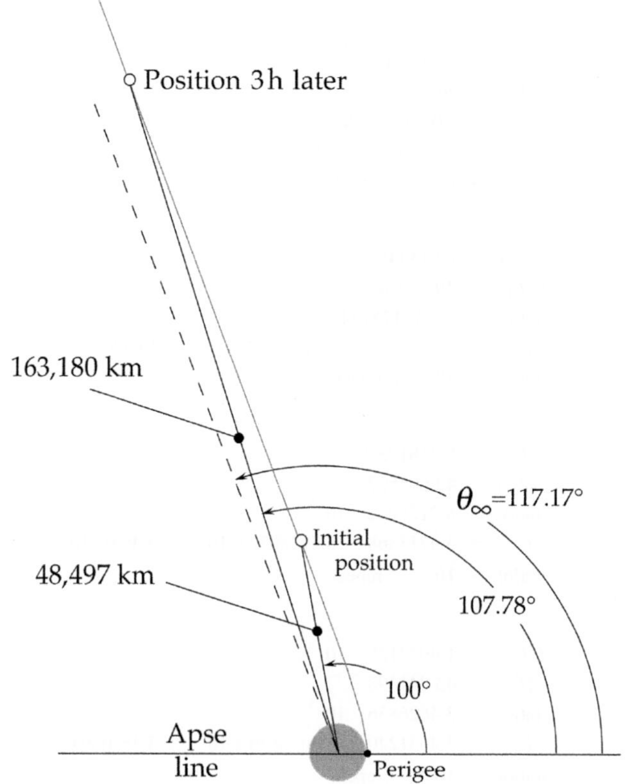

FIG. 3.18

Given and computed data for Example 3.5.

Therefore, the speed of the spacecraft is

$$v=\sqrt{v_r^2+v_\perp^2}=\sqrt{10.494^2+0.61386^2}=\boxed{10.51\text{km/ s}}$$

Note that the hyperbolic excess speed for this orbit is

$$v_\infty=\frac{\mu}{h}e\sin\theta_\infty=\frac{398,600}{100,170}\cdot 2.7696\cdot\sin 111.7°=10.277\text{km/s}$$

The results of this analysis are shown in Fig. 3.18.

When determining orbital position as a function of time with the aid of Kepler's equation, it is convenient to have position r as a function of eccentric anomaly F. This is obtained by substituting Eq. (3.41b) into Eq. (2.104),

$$r=\frac{a(e^2-1)}{1+e\cos\theta}=\frac{a(e^2-1)}{1+e\left(\dfrac{\cos F-e}{1-e\cos F}\right)}$$

This reduces to

$$r = a(e\cosh F - 1) \tag{3.46}$$

3.7 UNIVERSAL VARIABLES

The equations for elliptical and hyperbolic trajectories are very similar, as can be seen from Table 3.1. Observe, for example, that the hyperbolic mean anomaly is obtained from that of the ellipse as follows:

$$\begin{aligned}
M_h &= \frac{\mu^2}{h^3}\left(e^2-1\right)^{3/2}t \\
&= \frac{\mu^2}{h^3}\left[(-1)\left(1-e^2\right)\right]^{3/2}t \\
&= \frac{\mu^2}{h^3}(-1)^{3/2}\left(1-e^2\right)^{3/2}t \\
&= \frac{\mu^2}{h^3}t(-i)\left(1-e^2\right)^{3/2}t \\
&= -i\left[\frac{\mu^2}{h^3}\left(1-e^2\right)^{3/2}t\right] \\
&= -iM_e
\end{aligned}$$

In fact, the formulas for the hyperbola can all be obtained from those of the ellipse by replacing the variables in the ellipse equations according to the following scheme, wherein "←" means "replace by" and $i = \sqrt{-1}$:

$$\begin{aligned}
a &\leftarrow -a \\
b &\leftarrow ib \\
M_e &\leftarrow -iM_h \\
E &\leftarrow iF
\end{aligned}$$

Table 3.1 Comparison of some of the orbital formulas for the ellipse and hyperbola

	Equation	Ellipse ($e < 1$)	Hyperbola ($e > 1$)
1.	Orbit equation vs. true anomaly	$r = \dfrac{h^2}{\mu}\dfrac{1}{1+e\cos\theta}$	Same
2.	Conic equation in Cartesian coordinates	$\dfrac{x^2}{a^2}+\dfrac{y^2}{b^2}=1$	$\dfrac{x^2}{a^2}-\dfrac{y^2}{b^2}=1$
3.	Semimajor axis	$a = \dfrac{h^2}{\mu}\dfrac{1}{1-e^2}$	$a = \dfrac{h^2}{\mu}\dfrac{1}{e^2-1}$
4.	Semiminor axis	$b = \sqrt{1-e^2}$	$b = \sqrt{e^2-1}$
5.	Energy equation	$\dfrac{v^2}{2}-\dfrac{\mu}{r}=-\dfrac{\mu}{2a}$	$\dfrac{v^2}{2}-\dfrac{\mu}{r}=\dfrac{\mu}{2a}$
6.	Mean anomaly	$M_e = \dfrac{\mu^2}{h^3}\left(1-e^2\right)^{3/2}t$	$M_h = \dfrac{\mu^2}{h^3}\left(e^2-1\right)^{3/2}t$
7.	Kepler's equation	$M_e = E - e\sin E$	$M_h = e\sinh F - F$
8.	Orbit equation vs. eccentric anomaly	$r = a(1 - \cos E)$	$r = a(e\cosh F - 1)$

Note in this regard that $\sin(iF) = i\sinh F$ and $\cos(iF) = \cosh F$. Relations among the circular and hyperbolic trig functions are found in mathematics handbooks, such as Zwillinger (2018).

In the universal variable approach, the semimajor axis of the hyperbola is considered to have a negative value, so that the energy equation (row 5 of Table 3.1) has the same form for any type of orbit, including the parabola, for which $a = \infty$. In this formulation, the semimajor axis of any orbit is found using (row 3),

$$a = \frac{h^2}{\mu}\frac{1}{1-e^2} \tag{3.47}$$

If the position r and velocity v are known at a given point on the path, then the energy equation (row 5) is convenient for finding the semimajor axis of any orbit,

$$a = \frac{1}{\dfrac{2}{r} - \dfrac{v^2}{\mu}} \tag{3.48}$$

Kepler's equation may also be written in terms of a universal variable, or universal anomaly χ, that is valid for all orbits. See, for example, Prussing and Conway, 2013, Battin, 1987, and Bond and Allman, 1996. If t_0 is the time when the universal variable is zero, then the value of χ at time $t_0 + \Delta t$ is found by iterative solution of the universal Kepler's equation

$$\sqrt{\mu}\Delta t = \frac{r_0 v_r)_0}{\sqrt{\mu}}\chi^2 C(\alpha\chi^2) + (1 - \alpha r_0)\chi^3 S(\alpha\chi^2) + r_0\chi \tag{3.49}$$

where r_0 and $v_r)_0$ are the radius and radial velocity, respectively, at $t = t_0$, and α is the reciprocal of the semimajor axis

$$\alpha = \frac{1}{a} \tag{3.50}$$

$\alpha < 0$, $\alpha = 0$, and $\alpha > 0$ for hyperbolas, parabolas, and ellipses, respectively. The units of χ are $\text{km}^{1/2}$ (so $\alpha\chi^2$ is dimensionless). The functions $C(z)$ and $S(z)$ belong to the class known as Stumpff functions, named for the German astronomer Karl Stumpff (1895–1970). They are defined by the infinite series,

$$S(z) = \sum_{k=0}^{\infty}(-1)^k\frac{z^k}{(2k+3)!} = \frac{1}{6} - \frac{z}{120} + \frac{z^2}{5040} - \frac{z^3}{362{,}880} + \cdots \tag{3.51a}$$

$$C(z) = \sum_{k=0}^{\infty}(-1)^k\frac{z^k}{(2k+2)!} = \frac{1}{2} - \frac{z}{24} + \frac{z^2}{720} - \frac{z^3}{40{,}320} + \cdots \tag{3.51b}$$

$C(z)$ and $S(z)$ are related to the circular and hyperbolic trig functions as follows:

$$S(z) = \begin{cases} \dfrac{\sqrt{z} - \sin\sqrt{z}}{\left(\sqrt{z}\right)^3} & (z > 0) \\[2ex] \dfrac{\sinh\sqrt{-z} - \sqrt{-z}}{\left(\sqrt{z}\right)^3} & (z < 0) \qquad (z = \alpha\chi^2) \\[2ex] \dfrac{1}{6} & (z = 0) \end{cases} \tag{3.52}$$

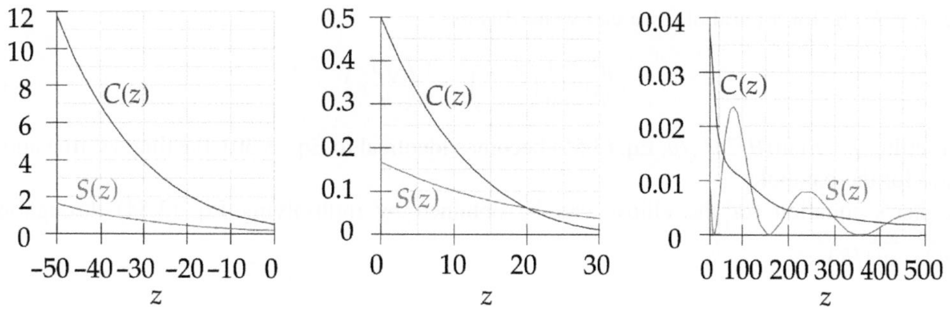

FIG. 3.19

A plot of the Stumpff functions $C(z)$ and $S(z)$.

$$C(z)=\begin{cases}\dfrac{1-\cos\sqrt{z}}{z} & (z>0)\\[2mm] \dfrac{\cosh\sqrt{-z}-1}{-z} & (z<0)\qquad (z=\alpha\chi^2)\\[2mm] \dfrac{1}{2} & (z=0)\end{cases}\tag{3.53}$$

Clearly, $z<0$, $z=0$, and $z>0$ for hyperbolas, parabolas, and ellipses, respectively. It should be pointed out that if $C(z)$ and $S(z)$ are computed by the series expansions (Eqs. 3.51), then the forms of $C(z)$ and $S(z)$, depending on the sign of z, are selected, so to speak, automatically. $C(z)$ and $S(z)$ behave as shown in Fig. 3.19. Both $C(z)$ and $S(z)$ are nonnegative functions of z. They increase without bound as z approaches $-\infty$ and tend toward zero for large positive values of z. As can be seen from Eq. (3.53), for $z>0$, $C(z)=0$ when $\cos\sqrt{z}=1$; that is, when $z=(2\pi)^2, (4\pi)^2, (6\pi)^2, \cdots$.

The price we pay for using the universal variable formulation is having to deal with the relatively unknown Stumpff functions. However, Eqs. (3.52) and (3.53) are easy to implement in both computer programs and programmable calculators. See Appendix D.13 for the implementation of these expressions in MATLAB.

To gain some insight into how Eq. (3.49) represents the Kepler equations for all the conic sections, let t_0 be the time at periapse passage and let us set $t_0=0$, as we have assumed previously. Then $\Delta t=t$, $v_r)_0=0$, and $r_0=r_p$, the periapsis radius. In that case Eq. (3.49) reduces to

$$\sqrt{\mu}t=(1-\alpha r_p)\chi^3 S(\alpha\chi^2)+r_p\chi \quad \text{(time = 0 at periapse passage)}\tag{3.54}$$

Consider first the parabola. In this case $\alpha=0$, and $S=S(0)=1/6$, so that Eq. (3.54) becomes a cubic polynomial in χ,

$$\sqrt{\mu}t=\frac{1}{6}\chi^3+r_p\chi$$

Multiply this equation through by $\left(\sqrt{\mu}/h\right)^3$ to obtain

$$\frac{\mu^2}{h^3}t=\frac{1}{6}\left(\frac{\chi\sqrt{\mu}}{h}\right)^3+r_p\chi\left(\frac{\sqrt{\mu}}{h}\right)^3$$

Since $r_p = h^2/2\mu$ for a parabola, we can write this as

$$\frac{\mu^2}{h^3}t = \frac{1}{6}\left(\frac{\chi\sqrt{\mu}}{h}\right)^3 + \frac{1}{2}\left(\frac{\sqrt{\mu}}{h}\chi\right) \tag{3.55}$$

Upon setting $\chi = h\tan(\theta/2)/\sqrt{\mu}$, Eq. (3.55) becomes identical to Eq. (3.30), the time vs. true anomaly relation for the parabola.

Kepler's equation for the ellipse can be obtained by multiplying Eq. (3.54) throughout by $\left(\sqrt{\mu(1-e^2)}/h\right)^3$:

$$\frac{\mu^2}{h^3}(1-e^2)^{3/2}t = \left(\chi\frac{\sqrt{\mu}}{h}\sqrt{1-e^2}\right)^3(1-\alpha r_p)S(z) + r_p\chi\left(\frac{\sqrt{\mu}}{h}\sqrt{1-e^2}\right)^3 \quad (z=\alpha\chi^2) \tag{3.56}$$

Recall that for the ellipse, $r_p = h^2/[\mu(1+e)]$ and $\alpha = 1/a = \mu(1-e^2)/h^2$. Using these two expressions in Eq. (3.56), along with $S(z) = [\sqrt{\alpha}\chi - \sin(\sqrt{\alpha}\chi)]/(\sqrt{\alpha}\chi)^3$ (from Eq. 3.52), and working through the algebra ultimately leads to

$$M_e = \frac{\chi}{\sqrt{a}} - e\sin\left(\frac{\chi}{\sqrt{a}}\right)$$

Comparing this with Kepler's equation for an ellipse (Eq. 3.14) reveals that the relationship between the universal variable χ and the eccentric anomaly E is $\chi = \sqrt{a}E$. Similarly, it can be shown for hyperbolic orbits that $\chi = \sqrt{-a}F$. In summary, the universal anomaly χ is related to the previously encountered anomalies as follows:

$$\chi = \begin{cases} \dfrac{h}{\sqrt{\mu}}\tan\dfrac{\theta}{2} & \text{parabola} \\ \sqrt{a}E & \text{ellipse} \quad (t=0 \text{ at periapse}) \\ \sqrt{-a}F & \text{hyperbola} \end{cases} \tag{3.57}$$

When t_0 is the time at a point other than periapsis, such that Eq. (3.49) applies, then Eq. (3.57) becomes

$$\chi = \begin{cases} \dfrac{h}{\sqrt{\mu}}\left(\tan\dfrac{\theta}{2} - \tan\dfrac{\theta_0}{2}\right) & \text{parabola} \\ \sqrt{a}(E-E_0) & \text{ellipse} \\ \sqrt{-a}(F-F_0) & \text{hyperbola} \end{cases} \tag{3.58}$$

As before, we can use Newton's method to solve Eq. (3.49) for the universal anomaly χ, given the time interval Δt. To do so, we form the function

$$f(\chi) = \frac{r_0 v_r)_0}{\sqrt{\mu}}\chi^2 C(\alpha\chi^2) + (1-\alpha r_0)\chi^3 S(\alpha\chi^2) + r_0\chi - \sqrt{\mu}\Delta t \tag{3.59}$$

and its derivative

$$\begin{aligned}\frac{df(\chi)}{d\chi} &= 2\frac{r_0 v_r)_0}{\sqrt{\mu}}\chi C(z) + \frac{r_0 v_r)_0}{\sqrt{\mu}}\chi^2\frac{dC(z)}{dz}\frac{dz}{d\chi} \\ &\quad + 3(1-\alpha r_0)\chi^2 S(z) + (1-\alpha r_0)\chi^3\frac{dS(z)}{dz}\frac{dz}{d\chi} + r_0\end{aligned} \tag{3.60}$$

where it is to be recalled that

$$z = \alpha\chi^2 \tag{3.61}$$

which means of course that

$$\frac{dz}{d\chi} = 2\alpha\chi \tag{3.62}$$

It turns out that

$$\begin{aligned}\frac{dS(z)}{dz} &= \frac{1}{2z}[C(z) - 3S(z)] \\ \frac{dC(z)}{dz} &= \frac{1}{2z}[1 - zS(z) - 2C(z)]\end{aligned} \tag{3.63}$$

Substituting Eqs. (3.61–3.63) into Eq. (3.60) and simplifying the result yields

$$\frac{df(\chi)}{d\chi} = \frac{r_0 v_r)_0}{\sqrt{\mu}}\chi\left[1 - \alpha\chi^2 S(z)\right] + (1 - \alpha r_0)\chi^2 C(z) + r_0 \tag{3.64}$$

With Eqs. (3.59) and (3.64), Newton's algorithm (Eq. 3.16) for the universal Kepler's equation becomes

$$\chi_{i+1} = \chi_i - \frac{\dfrac{r_0 v_r)_0}{\sqrt{\mu}}\chi_i{}^2 C(z_i) + (1 - \alpha r_0)\chi_i{}^3 S(z_i) + r_0\chi - \sqrt{\mu}\Delta t}{\dfrac{r_0 v_r)_0}{\sqrt{\mu}}\chi_i\left[1 - \alpha\chi_i{}^2 S(z_i)\right] + (1 - \alpha r_0)\chi_i{}^2 C(z_i) + r_0} \quad (z_i = \alpha\chi_i{}^2) \tag{3.65}$$

According to Chobotov (2002), a reasonable estimate for the starting value χ_0 is

$$\chi_0 = \sqrt{\mu}|\alpha|\Delta t \tag{3.66}$$

ALGORITHM 3.3

Solve the universal Kepler's equation for the universal anomaly χ given Δt, r_0, $v_r)_0$, and α. See Appendix D.14 for an implementation of this procedure in MATLAB.

1. Use Eq. (3.66) for an initial estimate of χ_0.
2. At any given step, having obtained χ_i from the previous step, calculate

$$f(\chi_i) = \frac{r_0 v_r)_0}{\sqrt{\mu}}\chi_i{}^2 C(z_i) + (1 - \alpha r_0)\chi_i{}^3 S(z_i) + r_0\chi_i - \sqrt{\mu}\Delta t$$

and

$$f'(\chi_i) = \frac{r_0 v_r)_0}{\sqrt{\mu}}\chi_i\left[1 - \alpha\chi_i{}^2 S(z_i)\right] + (1 - \alpha r_0)\chi_i{}^2 C(z_i) + r_0$$

where $z_i = \alpha\chi_i{}^2$.

3. Calculate $\text{ratio}_i = f(\chi_i)/f'(\chi_i)$.

4. If $|\text{ratio}_i|$ exceeds the chosen tolerance (e.g., 10^{-8}), then calculate an updated value of χ,

$$\chi_{i+1} = \chi_i - \text{ratio}_i$$

Return to Step 2.

5. If $|\text{ratio}_i|$ is less than the tolerance, then accept χ_i as the solution to within the desired accuracy.

EXAMPLE 3.6

An earth satellite has an initial true anomaly of $\theta = 30°$, a radius of $r_0 = 10,000\,\text{km}$, and a speed of $v_0 = 10\,\text{km/s}$. Use the universal Kepler's equation to find the change in universal anomaly χ after 1 h and use that information to determine the true anomaly θ at that time.

Solution

Using the initial conditions, let us first determine the angular momentum and the eccentricity of the trajectory. From the orbit formula (Eq. 2.45) we have

$$h = \sqrt{\mu r_0(1 + e\cos\theta_0)} = \sqrt{398{,}600 \cdot 10{,}000 \cdot (1 + e\cos 30°)} = 63{,}135\sqrt{1 + 0.86602e} \tag{a}$$

This, together with the angular momentum formula (Eq. 2.31), yields

$$v_\perp)_0 = \frac{h}{r_0} = \frac{63{,}135\sqrt{1 + 0.86602e}}{10{,}000} = 6.3135\sqrt{1 + 0.86602e}$$

Using the radial velocity relation (Eq. 2.49) we find

$$v_r)_0 = \frac{\mu}{h}e\sin\theta_0 = \frac{398{,}600}{63{,}135\sqrt{1 + 0.86602e}}e\sin 30° = 3.1567\frac{e}{\sqrt{1 + 0.86602e}} \tag{b}$$

Since $v_r)_0{}^2 + v_\perp)_0{}^2 = v_0{}^2$, it follows that

$$\left(3.1567\frac{e}{\sqrt{1 + 0.86602e}}\right)^2 + \left(6.3135\sqrt{1 + 0.86602e}\right)^2 = 10^2$$

which simplifies to become $39.86e^2 - 17.563e - 60.14 = 0$. The only positive root of this quadratic equation is

$$e = 1.4682$$

Since e is greater than 1, the orbit is a hyperbola. Substituting this value of the eccentricity back into Eqs. (a) and (b) yields the angular momentum

$$h = 95{,}154\,\text{km}^2/\text{s}$$

as well as the initial radial speed

$$v_r)_0 = 3.0752\,\text{km/s}$$

The hyperbolic eccentric anomaly F_0 for the initial conditions may now be found from Eq. (3.44a),

$$\tanh\frac{F_0}{2} = \sqrt{\frac{e-1}{e+1}}\tan\frac{\theta_0}{2} = \sqrt{\frac{1.4682 - 1}{1.4682 + 1}}\tan\frac{30°}{2} = 0.16670$$

Solving for F_0 yields

$$F_0 = 0.23448\,\text{rad} \tag{c}$$

In the universal variable formulation, we calculate the semimajor axis of the orbit by means of Eq. (3.47),

$$a=\frac{h^2}{\mu}\frac{1}{1-e^2}=\frac{95,154^2}{398,600}\frac{1}{1-1.4682^2}=-19,655\,\text{km} \tag{d}$$

The negative value is consistent with the fact that the orbit is a hyperbola. From Eq. (3.50) we get

$$\alpha=\frac{1}{a}=\frac{1}{-19,655}=-5.0878\left(10^{-5}\right)\text{km}^{-1}$$

which appears throughout the universal Kepler's equation.

We will use Algorithm 3.3 with an error tolerance of 10^{-6} to find the universal anomaly. From Eq. (3.66), our initial estimate is

$$\chi_0=\sqrt{398,600}\cdot\left|-5.0878\left(10^{-5}\right)\right|\cdot 3600=115.6$$

Executing the algorithm yields the following steps:

$$\chi_0=115.6$$

Step 1:

$$\begin{aligned}
f(\chi_0) &= -370,650.01\\
f'(\chi_0) &= 26,956.300\\
\text{ratio} &= -13.750033\\
\chi_1 &= 115.6-(-13.750033)=129.35003\\
|\text{ratio}| &> 10^{-6}\ \therefore\text{repeat}
\end{aligned}$$

Step 2:

$$\begin{aligned}
f(\chi_1) &= 25,729.002\\
f'(\chi_1) &= 30,776.401\\
\text{ratio} &= 0.83599669\\
\chi_2 &= 129.35003-0.83599669=128.51404\\
|\text{ratio}| &> 10^{-6}\ \therefore\text{repeat}
\end{aligned}$$

Step 3:

$$\begin{aligned}
f(\chi_2) &= 102.83891\\
f'(\chi_2) &= 30,530.672\\
\text{ratio} &= 3.3683800\left(10^{-3}\right)\\
\chi_3 &= 128.51404-3.3683800\left(10^{-3}\right)=128.51067\\
|\text{ratio}| &> 10^{-6}\ \therefore\text{repeat}
\end{aligned}$$

Step 4:

$$\begin{aligned}
f(\chi_3) &= 1.6614116\left(10^{-3}\right)\\
f'(\chi_3) &= 30,529.686\\
\text{ratio} &= 5.4419545\left(10^{-8}\right)\\
\chi_4 &= 128.51067-5.4419545\left(10^{-8}\right)=128.51067\\
|\text{ratio}| &< 10^{-6}\ \therefore\text{stop}
\end{aligned}$$

So we accept

$$\chi=128.51\,\text{km}^{1/2}$$

as the solution after four iterations. Substituting this value of χ together with the semimajor axis (Eq. d) into Eq. (3.58) yields

$$F-F_0=\frac{\chi}{\sqrt{-a}}=\frac{128.51}{\sqrt{-(-19,655)}}=0.91664$$

It follows from Eq. (c) that the hyperbolic eccentric anomaly after 1 h is

$$F = 0.23448 + 0.91664 = 1.1511$$

Finally, we calculate the corresponding true anomaly using Eq. (3.44b),

$$\tan\frac{\theta}{2} = \sqrt{\frac{e+1}{e-1}}\tanh\frac{F}{2} = \sqrt{\frac{1.4682+1}{1.4682-1}}\tanh\frac{1.1511}{2} = 1.1926$$

which means that after 1 h

$$\boxed{\theta = 100.04^\circ}$$

Recall from Section 2.11 that the position $\mathbf{r}$ and velocity $\mathbf{v}$ on a trajectory at any time t can be found in terms of the position $\mathbf{r}_0$ and velocity $\mathbf{v}_0$ at time t_0 by means of the Lagrange f and g coefficients and their first derivatives,

$$\mathbf{r} = f\mathbf{r}_0 + g\mathbf{v}_0 \tag{3.67}$$

$$\mathbf{v} = \dot{f}\mathbf{r}_0 + \dot{g}\mathbf{v}_0 \tag{3.68}$$

Eqs. (2.158) give f, g, $\dot{f}$, and $\dot{g}$ explicitly in terms of the change in true anomaly $\Delta\theta$ over the time interval $\Delta t = t - t_0$. The Lagrange coefficients can also be derived in terms of changes in the eccentric anomaly ΔE for elliptical orbits, ΔF for hyperbolas, or $\Delta\tan(\theta/2)$ for parabolas. However, if we take advantage of the universal variable formulation, we can cover all these cases with the same set of Lagrange coefficients. In terms of the universal anomaly χ and the Stumpff functions $C(z)$ and $S(z)$, the Lagrange coefficients are (Bond and Allman, 1996)

$$f = 1 - \frac{\chi^2}{r_0}C(\alpha\chi^2) \tag{3.69a}$$

$$g = \Delta t - \frac{1}{\sqrt{\mu}}\chi^3 S(\alpha\chi^2) \tag{3.69b}$$

$$\dot{f} = \frac{\sqrt{\mu}}{rr_0}\left[\alpha\chi^3 S(\alpha\chi^2) - \chi\right] \tag{3.69c}$$

$$\dot{g} = 1 - \frac{\chi^2}{r}C(\alpha\chi^2) \tag{3.69d}$$

The implementation of these four functions in MATLAB is found in Appendix D.15.

ALGORITHM 3.4

Given $\mathbf{r}_0$ and $\mathbf{v}_0$, find $\mathbf{r}$ and $\mathbf{v}$ at a time Δt later. See Appendix D.16 for an implementation of this procedure in MATLAB.

1. Use the initial conditions to find:
   a. The magnitude of $\mathbf{r}_0$ and $\mathbf{v}_0$,

$$r_0 = \sqrt{\mathbf{r}_0 \cdot \mathbf{r}_0} \qquad v_0 = \sqrt{\mathbf{v}_0 \cdot \mathbf{v}_0}$$

   b. The radial component of velocity $v_r)_0$ by projecting $\mathbf{v}_0$ onto the direction of $\mathbf{r}_0$,

$$v_r)_0 = \frac{\mathbf{v}_0 \cdot \mathbf{r}_0}{r_0}$$

c. The reciprocal α of the semimajor axis, using Eq. (3.48),

$$\alpha = \frac{2}{r_0} - \frac{v_0^2}{\mu}$$

The sign of α determines whether the trajectory is an ellipse ($\alpha > 0$), parabola ($\alpha = 0$), or hyperbola ($\alpha < 0$).

2. With r_0, $v_r)_0$, α, and Δt, use Algorithm 3.3 to find the universal anomaly χ.
3. Substitute α, r_0, Δt, and χ into Eqs. (3.69a) and (3.69b) to obtain f and g.
4. Use Eq. (3.67) to compute $\mathbf{r}$ followed by its magnitude r.
5. Substitute α, r_0, r, and χ into Eqs. (3.69c) and (3.69d) to obtain $\dot{f}$ and $\dot{g}$.
6. Use Eq. (3.68) to compute $\mathbf{v}$.

EXAMPLE 3.7

An earth satellite moves in the xy plane of an inertial frame with origin at the earth's center. Relative to that frame, the position and velocity of the satellite at time t_0 are

$$\begin{aligned} \mathbf{r}_0 &= 7000.0\hat{\mathbf{i}} - 12{,}124\hat{\mathbf{j}}\,(\text{km}) \\ \mathbf{v}_0 &= 2.6679\hat{\mathbf{i}} + 4.6210\hat{\mathbf{j}}\,(\text{km/s}) \end{aligned} \tag{a}$$

Compute the position and velocity vectors of the satellite 60 min later using Algorithm 3.4.

Solution

Step 1:

$$\begin{aligned} r_0 &= \sqrt{7000.0^2 + (-12{,}124)^2} &&= 14{,}000\,\text{km} \\ v_r)_0 &= \frac{2.6679 \cdot 7000.0 + 4.6210 \cdot (-12{,}124)}{14{,}000} &&= -2.6679\,\text{km/s} \\ \alpha &= \frac{2}{14{,}000} - \frac{5.3359^2}{398{,}600} &&= 7.1429\left(10^{-5}\right)\text{km}^{-1} \end{aligned}$$

The trajectory is an ellipse, because α is positive.

Step 2:

Using the results of Step 1, Algorithm 3.3 yields

$$\chi = 253.53\,\text{km}^{1/2}$$

which means

$$z = \alpha\chi^2 = 7.1429\left(10^{-5}\right) \cdot 253.53^2 = 4.5911$$

Step 3:

Substituting the above values of χ and z into Eqs. (3.69a) and (3.69b), we find

$$f = 1 - \frac{\chi^2}{r_0} C(\alpha\chi^2) = 1 - \frac{253.53^2}{14{,}000} \overbrace{C(4.5911)}^{0.3357} = -0.54123$$

$$g = \Delta t - \frac{1}{\sqrt{\mu}} \chi^3 S(\alpha\chi^2) = 3600 - \frac{253.53^3}{\sqrt{398{,}600}} \overbrace{S(4.5911)}^{0.13233} = 184.35 \text{ s}$$

Step 4:

$$\begin{aligned} \mathbf{r} &= f\mathbf{r}_0 + g\mathbf{v}_0 \\ &= (-0.54123)(7000.0\hat{\mathbf{i}} - 12.124\hat{\mathbf{j}}) + 184.35(2.6679\hat{\mathbf{i}} + 4.6210\hat{\mathbf{j}}) \\ &= \boxed{-3296.8\hat{\mathbf{i}} + 7413.9\hat{\mathbf{j}}\ (\text{km})} \end{aligned}$$

Therefore, the magnitude of **r** is

$$r = \sqrt{(-3296.8)^2 + (7413.9)^2} = 8113.9 \text{km}$$

Step 5:

$$\begin{aligned} \dot{f} &= \frac{\sqrt{\mu}}{rr_0}\left[\alpha\chi^3 S(\alpha\chi^2) - \chi\right] \\ &= \frac{\sqrt{398{,}600}}{8113.9 \cdot 14{,}000}\left[7.1429(10^5) \cdot 253.53^3 \cdot \overbrace{S(4.5911)}^{0.13233} - 253.53\right] \\ &= -0.00055298 \text{ s}^{-1} \end{aligned}$$

$$\dot{g} = 1 - \frac{\chi^2}{r} C(\alpha\chi^2) = 1 - \frac{253.53^2}{8113.9} \overbrace{C(4.5911)}^{0.3357} = -1.6593$$

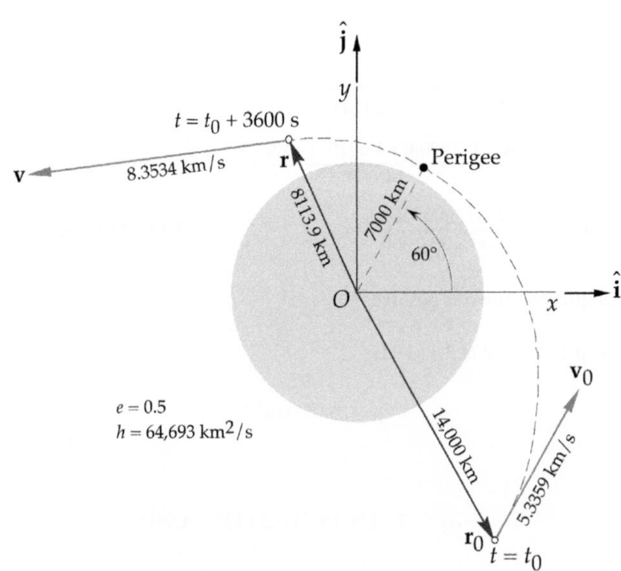

FIG. 3.20

Initial and final points on the geocentric trajectory of Example 3.7.

Step 6:

$$\begin{aligned}\mathbf{v} &= \dot{f}\mathbf{r}_0 + \dot{g}\mathbf{v}_0 \\ &= (-0.00055298)(7000.0\hat{\mathbf{i}} - 12.124\hat{\mathbf{j}}) + (-1.6593)(2.6679\hat{\mathbf{i}} + 4.6210\hat{\mathbf{j}}) \\ &= \boxed{-8.2977\hat{\mathbf{i}} - 0.96309\hat{\mathbf{j}}\ (\text{km/s})}\end{aligned}$$

The initial and final position and velocity vectors, as well as the trajectory, are accurately illustrated in Fig. 3.20.

PROBLEMS

Section 3.2

3.1 If $f = \frac{1}{2}\tan\frac{x}{2} + \frac{1}{6}\tan^3\frac{x}{2}$, then show that $df/dx = 1/(1 + \cos x)^2$, thereby verifying the integral in Eq. (3.4).

Section 3.4

3.2 Find the three positive roots of the equation $10e^{\sin x} = x^2 - 5x + 4$ to eight significant figures. Use
(a) Newton's method.
(b) Bisection method.

3.3 Find the first four nonnegative roots of the equation $\tan(x) = \tanh(x)$ to eight significant figures. Use
(a) Newton's method.
(b) Bisection method.

3.4 In terms of the eccentricity e, the period T, and the angles α and β (in radians), find the time t required to fly from point 1 to point 2 on the ellipse. C is the center of the ellipse.

$$\left\{\text{Ans.: } t = \frac{T}{2\pi}\left(\beta - \alpha - 2e\cos\frac{\beta+\alpha}{2}\sin\frac{\beta-\alpha}{2}\right)\right\}$$

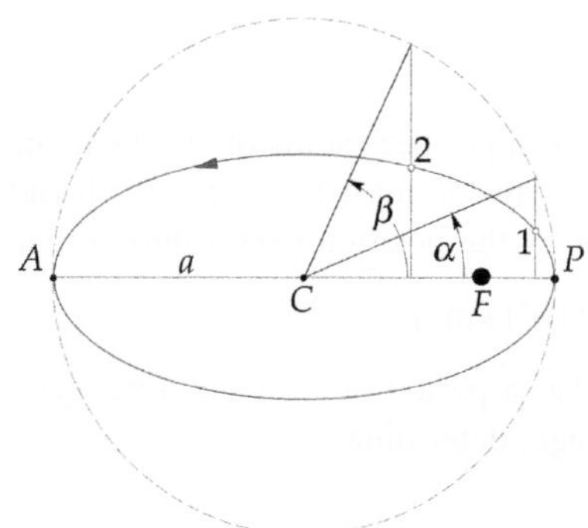

3.5 Calculate the time required to fly from P to B, in terms of the eccentricity e and the period T. B lies on the minor axis.

$$\left\{\text{Ans.: } \left(\frac{1}{4} - \frac{e}{2\pi}\right)T\right\}$$

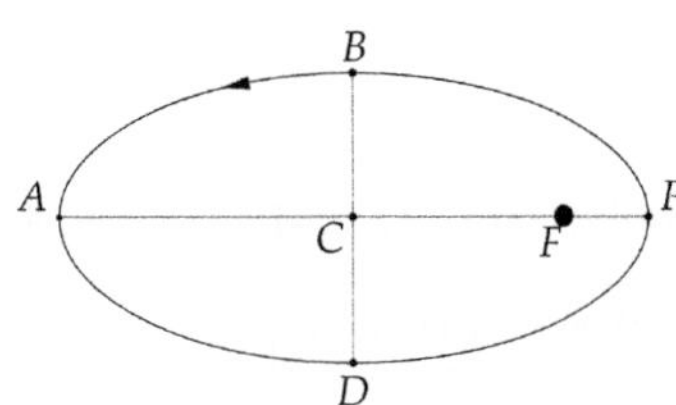

3.6 If the eccentricity of the elliptical orbit is 0.3, calculate, in terms of the period T, the time required to fly from P to B.

{Ans.: 0.156 T}

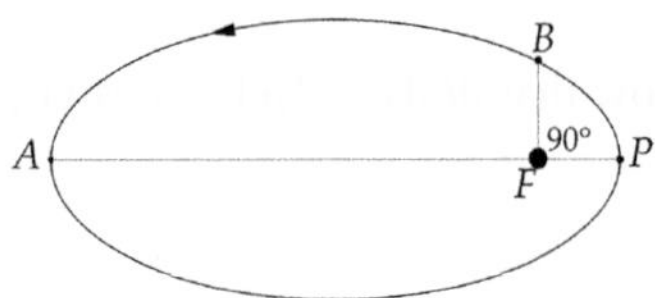

3.7 If the eccentricity of the elliptical orbit is 0.5, calculate, in terms of the period T, the time required to fly from P to B.

{Ans.: 0.170 T}

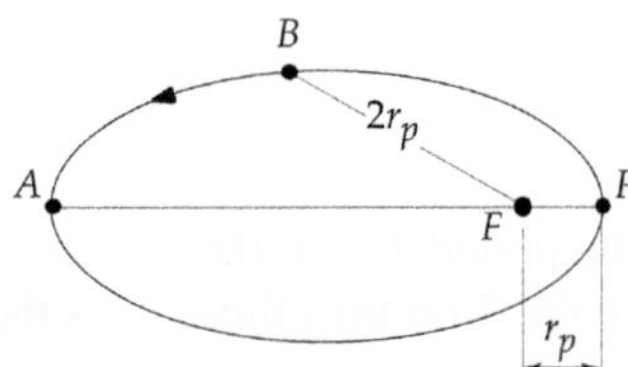

3.8 A satellite is in earth orbit for which the perigee altitude is 200 km and the apogee altitude is 600 km. Find the time interval during which the satellite remains above an altitude of 400 km.

{Ans.: 47.15 min}

3.9 An earth-orbiting satellite has a perigee radius of 7000 km and an apogee radius of 10,000 km.
(a) What true anomaly $\Delta\theta$ is swept out between $t = 0.5$ h and $t = 1.5$ h after perigee passage?
(b) What area is swept out by the position vector during that time interval?

{Ans.: (a) 128.7°; (b) $1.03(10^8)$ km^2}

3.10 An earth-orbiting satellite has a period of 14 h and a perigee radius of 10,000 km. At time $t = 10$ h after perigee passage, determine
(a) The radial position.
(b) The speed.
(c) The radial component of the velocity.

{Ans.: (a) 42,356 km; (b) 2.303 km/s; (c) −1.271 km/s}

3.11 A satellite in earth orbit has perigee and apogee radii of $r_p = 7500$ km and $r_a = 16{,}000$ km, respectively. Find its true anomaly 40 min after passing the true anomaly of 80°.

{Ans.: 174.7°}

3.12 Show that the solution to $a\cos\theta + b\sin\theta = c$, where a, b, and c are given, is

$$\theta = \phi \pm \cos^{-1}\left(\frac{c}{a}\cos\phi\right)$$

where $\tan\phi = b/a$.

3.13 Verify the results of part (b) of Example 3.3.

Section 3.5

3.14 Calculate the time required for a spacecraft launched into a parabolic trajectory at a perigee altitude of 200 km to leave the earth's sphere of influence (see Table A.2).

{Ans.: 7.77 days}

3.15 A spacecraft on a parabolic trajectory around the earth has a perigee radius of 6600 km.

(a) How long does it take to coast from $\theta = -90°$ to $\theta = +90°$?

(b) How far is the spacecraft from the center of the earth 36 h after passing through perigee?

{Ans.: (a) 0.8897 h; (b) 304,700 km}

Section 3.6

3.16 A spacecraft on a hyperbolic trajectory around the earth has a perigee radius of 6600 km and a perigee speed of $1.2v_{esc}$.

(a) How long does it take to coast from $\theta = -90°$ to $\theta = +90°$?

(b) How far is the spacecraft from the center of the earth 24 h after passing through perigee?

{Ans.: (a) 0.9992 h; (b) 656,610 km}

3.17 A trajectory has a perigee velocity $1.1v_{esc}$ and a perigee altitude of 200 km. If at 10 a.m. the satellite is traveling toward the earth with a speed of 8 km/s, how far will it be from the earth's surface at 5 p.m. the same day?

{Ans.: 136,250 km}

3.18 An incoming object is sighted at an altitude of 100,000 km with a speed of 6 km/s and a flight path angle of −80°.

(a) Will it impact the earth or fly by?

(b) What is the time to impact or to closest approach?

{Partial Ans.: (b) 4 h 29 min}

Section 3.7

3.19 At a given instant, the radial position of an earth-orbiting satellite is 7200 km and its radial speed is 1 km/s. If the semimajor axis is 10,000 km, use Algorithm 3.3 to find the universal anomaly 60 min later. Check your result using Eq. (3.58).

3.20 At a given instant, a space object has the following position and velocity vectors relative to an earth-centered inertial frame of reference:

$$\mathbf{r}_0 = 20{,}000\hat{\mathbf{i}} - 105{,}000\hat{\mathbf{j}} - 19{,}000\hat{\mathbf{k}}\,(\text{km})$$
$$\mathbf{v}_0 = 0.9000\hat{\mathbf{i}} - 3.4000\hat{\mathbf{j}} - 1.5000\hat{\mathbf{k}}\,(\text{km/s})$$

Use Algorithm 3.4 to find $\mathbf{r}$ and $\mathbf{v}$ 2 h later.

{Ans.: $\mathbf{r} = 26{,}338\hat{\mathbf{i}} - 128{,}750\hat{\mathbf{j}} - 29{,}656\hat{\mathbf{k}}\,(\text{km})$,
$\mathbf{v} = 0.86280\hat{\mathbf{i}} - 3.2116\hat{\mathbf{j}} - 1.4613\hat{\mathbf{k}}\,(\text{km/s})$}

REFERENCES

Battin, R.H., 1987. An Introduction to the Mathematics and Methods of Astrodynamics. AIAA Education Series, New York.

Bond, V.R., Allman, M.C., 1996. Modern Astrodynamics: Fundamentals and Perturbation Methods. Princeton University Press.

Chobotov, V.A., 2002. Orbital Mechanics, third ed. AIAA Education Series.

Prussing, J.E., Conway, B.A., 2013. Orbital Mechanics, second ed. Oxford University Press, New York.

Zwillinger, D., 2018. Standard Mathematical Tables and Formulae, thirty third ed. CRC Press, New York.

CHAPTER

ORBITS IN THREE DIMENSIONS

4

4.1 INTRODUCTION

The discussion of orbital mechanics up to now has been confined to two dimensions (i.e., to the plane of the orbits themselves). This chapter explores the means of describing orbits in three-dimensional space, which, of course, is the setting for real missions and orbital maneuvers. Our focus will be on the orbits of earth satellites, but the applications are to any two-body trajectories, including interplanetary missions to be discussed in Chapters 8 and 9.

We begin with a discussion of the ancient concept of the celestial sphere and the use of right ascension and declination to define the location of stars, planets, and other celestial objects on the sphere. This leads to the establishment of the inertial geocentric equatorial frame of reference and the concept of state vector. The six components of this vector give the instantaneous position and velocity of an object relative to the inertial frame and define the characteristics of the orbit. Following this discussion is a presentation of the six classical orbital elements, which also uniquely define the shape and orientation of an orbit and the location of a body on it. We then show how to transform the state vector into orbital elements, and vice versa, taking advantage of the perifocal frame introduced in Chapter 2.

We go on to summarize two of the major perturbations of earth orbits due to the earth's nonspherical shape. These perturbations are exploited to place satellites in sun-synchronous and Molniya orbits. The various perturbations of spacecraft trajectories are presented in more detail in Chapter 10.

The chapter concludes with a discussion of ground tracks and how to compute them.

4.2 GEOCENTRIC RIGHT ASCENSION-DECLINATION FRAME

The coordinate system used to describe earth orbits in three dimensions is defined in terms of earth's equatorial plane, the ecliptic plane, and the earth's axis of rotation. The ecliptic is the plane of the earth's orbit around the sun, as illustrated in Fig. 4.1. The earth's axis of rotation, which passes through the north and south poles, is not perpendicular to the ecliptic. It is tilted away by an angle known as the obliquity of the ecliptic, ε. For the earth, ε is approximately 23.4°. Therefore, the earth's equatorial plane and the ecliptic intersect along a line, which is known as the vernal equinox line. On the calendar, "vernal equinox" is the first day of spring in the northern hemisphere, when the noontime sun crosses the equator from south to north. The position of the sun at that instant defines the location of a point in the sky called the vernal equinox, for which the symbol γ is used. On the day of the vernal equinox, the

Orbital Mechanics for Engineering Students. https://doi.org/10.1016/B978-0-12-824025-0.00004-0

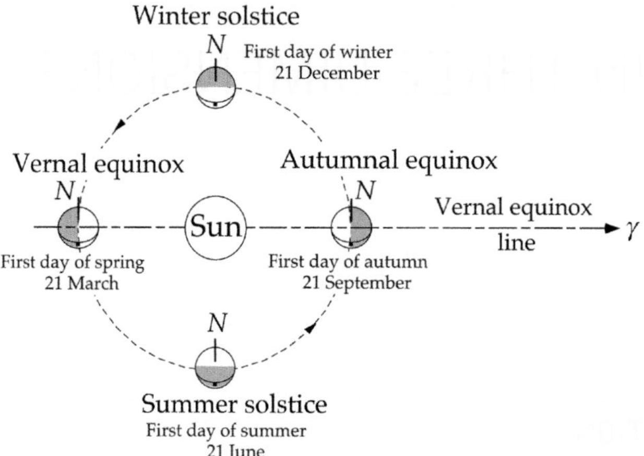

FIG. 4.1

The earth's orbit around the sun, viewed from above the ecliptic plane, showing the change of seasons in the northern hemisphere.

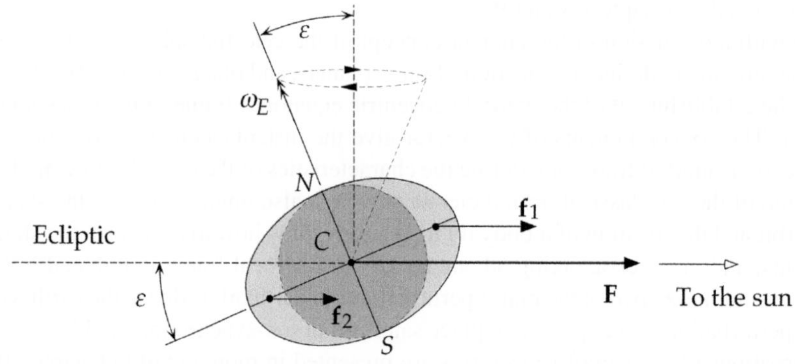

FIG. 4.2

Secondary (perturbing) gravitational forces on the earth.

number of hours of daylight and darkness are equal; hence, the word equinox. The other equinox occurs precisely half a year later, when the sun crosses back over the equator from north to south, thereby defining the first day of autumn. The vernal equinox lies today in the constellation Pisces, which is visible in the night sky during the fall. The direction of the vernal equinox line is from the earth toward γ, as shown in Fig. 4.1.

For many practical purposes, the vernal equinox line may be considered fixed in space. However, it actually rotates slowly because the earth's tilted spin axis precesses westward around the normal to the ecliptic at the rate of about 1.4° per century. This slow precession is due primarily to the action of the sun and the moon on the nonspherical distribution of mass within the earth. Due to the centrifugal force of rotation about its own axis, the earth bulges very slightly outward at its equator. This effect is shown highly exaggerated in Fig. 4.2. One of the bulging sides is closer to the sun than the other, so the force of the sun's gravity $\mathbf{f}_1$ on its mass is slightly larger than the force $\mathbf{f}_2$ on the opposite side, farthest from the sun. The forces $\mathbf{f}_1$ and $\mathbf{f}_2$, along with the dominant force $\mathbf{F}$ on the spherical mass, comprise the total force of the sun on the earth, holding it in its solar orbit. Taken together, $\mathbf{f}_1$ and $\mathbf{f}_2$ produce a net clockwise moment (a vector into the page) about the center of the earth. That moment would rotate the earth's equator into alignment with the ecliptic if it were not for the fact that the earth has an angular momentum directed along its south-to-north polar axis due to its spin around that axis at an angular velocity ω_E of about 360° per day. The effect of the moment is to rotate the angular momentum vector in the direction of the moment (into the page). The result is that the spin axis is forced to precess in a counterclockwise direction around the normal to the ecliptic, sweeping out a cone as illustrated in the figure. The moon exerts a torque on the earth for the same reason, and the combined effect of the sun and the moon is a precession of the spin axis, and hence γ, with a period of about 26,000 years. The moon's action also superimposes a small nutation on the precession. This causes the obliquity ε to vary with a maximum amplitude of 0.0025° over a period of 18.6 years.

About 4000 years ago, when the first recorded astronomical observations were being made, γ was located in the constellation Aries, the ram. The Greek letter γ resembles the ancient symbol representing the head of a ram (♈).

To the human eye, objects in the night sky appear as points on a celestial sphere surrounding the earth, as illustrated in Fig. 4.3. The north and south poles of this fixed sphere correspond to those of the earth rotating within it. Coordinates of latitude and longitude are used to locate points on the celestial

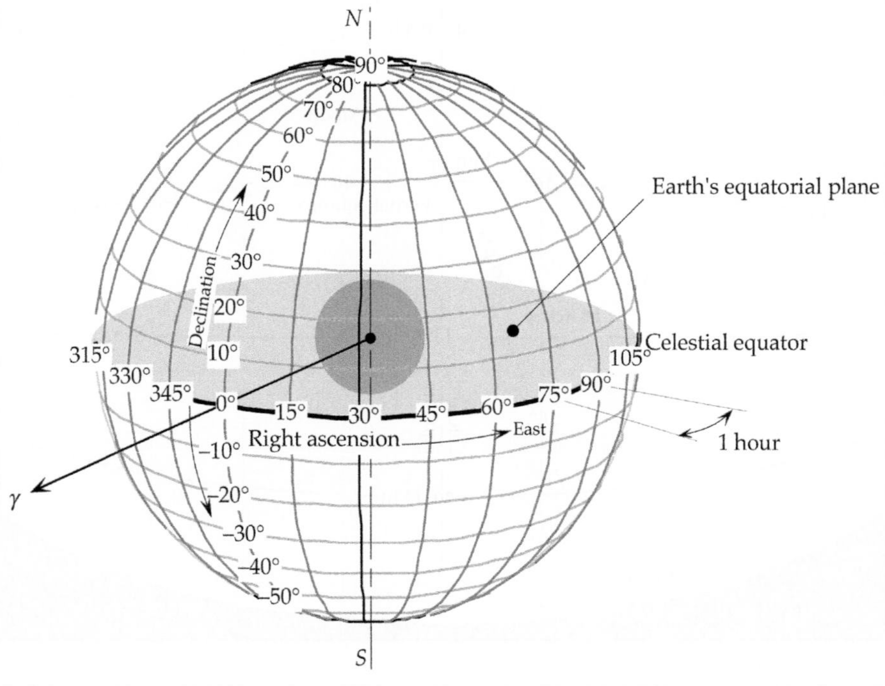

FIG. 4.3

The celestial sphere, with grid lines of right ascension and declination]

sphere in much the same way as on the surface of the earth. The projection of the earth's equatorial plane outward onto the celestial sphere defines the celestial equator. The vernal equinox γ, which lies on the celestial equator, is the origin for measurement of longitude, which in astronomical parlance is called right ascension. Right ascension (RA or α) is measured along the celestial equator in degrees east from the vernal equinox (astronomers measure right ascension in hours instead of degrees, where 24 h equals 360°). Latitude on the celestial sphere is called declination. Declination (Dec or δ) is measured along a meridian in degrees, positive to the north of the equator and negative to the south. Fig. 4.4 is a sky chart showing how the heavenly grid appears from a given point on the earth. Notice that the sun is located at the intersection of the equatorial and ecliptic planes, so this must be the first day of spring.

Stars are so far away from the earth that their positions relative to each other appear stationary on the celestial sphere. Planets, comets, satellites, etc., move on the fixed backdrop of the stars. A table of the coordinates of celestial bodies as a function of time is called an ephemeris [e.g., the *Astronomical Almanac* (Department of the Navy, 2018)]. Table 4.1 is an abbreviated ephemeris for the moon and for Venus. An ephemeris depends on the location of the vernal equinox at a given time or epoch, for we know that even the positions of the stars relative to the equinox change slowly with time. For example, Table 4.2 shows the celestial coordinates of the star Regulus at five epochs since AD 1700. Currently, the position of the vernal equinox in the year 2000 is used to define the standard grid of the celestial sphere. In 2025, the position will be updated to that of the year 2050, and so on at 25-year intervals.

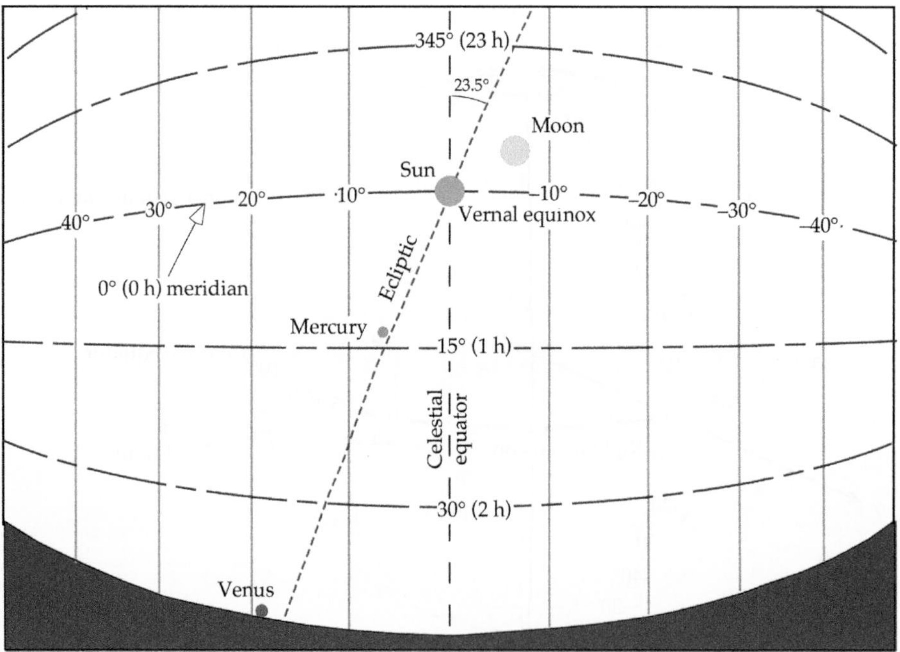

FIG. 4.4

A view of the sky above the eastern horizon from 0° longitude on the equator at 9 a.m. local time, March 20, 2004 (precession epoch AD 2000).

Table 4.1 Venus and moon ephemeris for 0 h universal time (precession epoch: 2000 AD)

	Venus		Moon	
Date	RA	Dec	RA	Dec
1 Jan 2004	21h 05.0min	−18° 36′	1h 44.9min	+8° 47′
1 Feb 2004	23h 28.0min	−04° 30′	4h 37.0min	+24° 11′
1 Mar 2004	01h 30.0min	+10° 26′	6h 04.0min	+08° 32′
1 Apr 2004	03h 37.6min	+22° 51′	9h 18.7min	+21° 08′
1 May 2004	05h 20.3min	+27° 44′	11h 28.8min	+07° 53′
1 Jun 2004	05h 25.9min	+24° 43′	14h 31.3min	−14° 48′
1 Jul 2004	04h 34.5min	+17° 48′	17h 09.0min	−26° 08′
1 Aug 2004	05h 37.4min	+19° 04′	21h 05.9min	−21° 49′
1 Sep 2004	07h 40.9min	+19° 16′	00h 17.0min	−00° 56′
1 Oct 2004	09h 56.5min	+12° 42′	02h 20.9min	+14° 35′
1 Nov 2004	12h 15.8min	+00° 01′	05h 26.7min	+27° 18′
1 Dec 2004	14h 34.3min	−13° 21′	07h 50.3min	+26° 14′
1 Jan 2005	17h 12.9min	−22° 15′	10h 49.4min	+11° 39′

Table 4.2 Variation of the coordinates of the star Regulus due to precession of the equinox

Precession epoch	RA	Dec
1700 AD	9h 52.2min (148.05°)	+13° 25′
1800 AD	9h 57.6min (149.40°)	+12° 56′
1900 AD	10h 3.0min (150.75°)	+12° 27′
1950 AD	10h 5.7min (151.42°)	+12° 13′
2000 AD	10h 8.4min (152.10°)	+11° 58′

Since observations are made relative to the actual orientation of the earth, these measurements must be transformed into the standardized celestial frame of reference. As Table 4.2 suggests, the adjustments will be small if the current epoch is within 25 years of the standard precession epoch.

4.3 STATE VECTOR AND THE GEOCENTRIC EQUATORIAL FRAME

At any given time, the state vector of a satellite comprises its position **r** and orbital velocity **v**. Orbital mechanics is concerned with specifying or predicting state vectors over intervals of time. From Chapter 2, we know that the equation governing the state vector of a satellite traveling around the earth is, under the familiar assumptions,

$$\ddot{\mathbf{r}} = -\frac{\mu}{r^3}\mathbf{r} \tag{4.1}$$

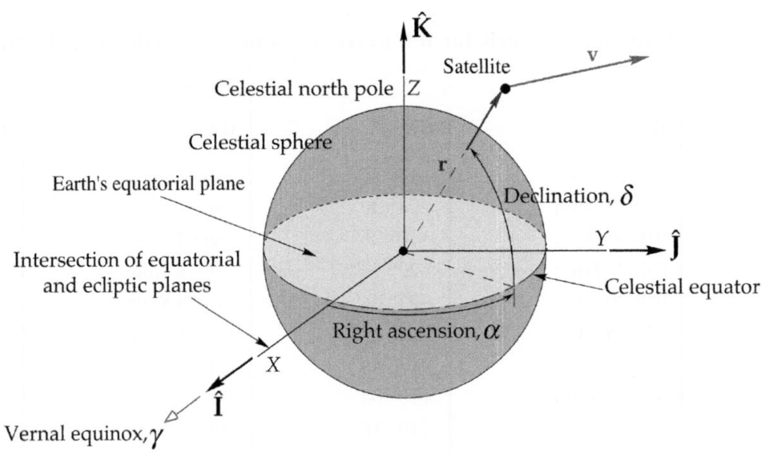

FIG. 4.5

The geocentric equatorial frame.

where **r** is the position vector of the satellite relative to the center of the earth. The components of **r**, and especially, those of its time derivatives $\dot{\mathbf{r}} = \mathbf{v}$ and $\ddot{\mathbf{r}} = \mathbf{a}$, must be measured in a nonrotating frame attached to the earth. A commonly used nonrotating right-handed Cartesian coordinate system is the geocentric equatorial frame shown in Fig. 4.5. The X axis points in the vernal equinox direction. The XY plane is the earth's equatorial plane, and the Z axis coincides with the earth's axis of rotation and points northward. The unit vectors $\hat{\mathbf{I}}$, $\hat{\mathbf{J}}$, and $\hat{\mathbf{K}}$ form a right-handed triad. The nonrotating geocentric equatorial frame serves as an inertial frame for the two-body earth satellite problem, as embodied in Eq. (4.1). It is not truly an inertial frame, however, since the center of the earth is always accelerating toward a third body, the sun (to say nothing of the moon), a fact that we ignore in the two-body formulation.

In the geocentric equatorial frame, the state vector is given in component form by

$$\mathbf{r} = X\hat{\mathbf{I}} + Y\hat{\mathbf{J}} + Z\hat{\mathbf{K}} \tag{4.2}$$

$$\mathbf{v} = v_X\hat{\mathbf{I}} + v_Y\hat{\mathbf{J}} + v_Z\hat{\mathbf{K}} \tag{4.3}$$

If r is the magnitude of the position vector, then

$$\mathbf{r} = r\hat{\mathbf{u}}_r \tag{4.4}$$

Fig. 4.5 shows that the components of $\hat{\mathbf{u}}_r$ (the direction cosines l, m, and n of $\hat{\mathbf{u}}_r$) are found in terms of the right ascension α and declination δ as follows:

$$\hat{\mathbf{u}}_r = l\hat{\mathbf{I}} + m\hat{\mathbf{J}} + n\hat{\mathbf{K}} = \cos\delta\cos\alpha\hat{\mathbf{I}} + \cos\delta\sin\alpha\hat{\mathbf{J}} + \sin\delta\hat{\mathbf{K}} \tag{4.5}$$

From this we see that the declination is obtained as $\delta = \sin^{-1}n$. There is no quadrant ambiguity since, by definition, the declination lies between −90° and +90°, which is precisely the range of the principal values of the arcsine function. It follows that $\cos\delta$ cannot be negative. Eq. (4.5) also reveals that

$l = \cos\delta\cos\alpha$. Hence, we find the right ascension from $\alpha = \cos^{-1}(l/\cos\delta)$, which yields two values of α between 0° and 360°. To determine the correct quadrant for α, we check the sign of the direction cosine, $m=\cos\delta\sin\alpha$. Since $\cos\delta$ cannot be negative, the sign of m is the same as the sign of $\sin\alpha$. If $\sin\alpha > 0$, then α lies in the range 0° to 180°, whereas $\sin\alpha < 0$ means that α lies between 180° and 360°.

ALGORITHM 4.1

Given the position vector $\mathbf{r} = X\hat{\mathbf{I}} + Y\hat{\mathbf{J}} + Z\hat{\mathbf{K}}$, calculate the right ascension α and declination δ. This procedure is implemented in MATLAB as *ra_and_dec_from_r.m*, which appears in Appendix D.17.

1. Calculate the magnitude of **r**: $r = \sqrt{X^2 + Y^2 + Z^2}$.
2. Calculate the direction cosines of **r**: $l = X/r \quad m = Y/r \quad n = Z/r$
3. Calculate the declination: $\delta = \sin^{-1} n$
4. Calculate the right ascension: $\alpha = \begin{cases} \cos^{-1}(l/\cos\delta) & (m > 0) \\ 360° - \cos^{-1}(l/\cos\delta) & (m \le 0) \end{cases}$

Although the position vector furnishes the right ascension and declination, the right ascension and declination alone do not furnish **r**. For that we need the distance r to obtain the position vector from Eq. (4.4).

EXAMPLE 4.1

If the position vector of the International Space Station in the geocentric equatorial frame is $\mathbf{r} = -5368\hat{\mathbf{I}} - 1784\hat{\mathbf{J}} + 3691\hat{\mathbf{K}}$ (km), what are its right ascension and declination?

Solution

We employ Algorithm 4.1.

Step 1:

$$r = \sqrt{(-5368)^2 + (-1784)^2 + 3691^2} = 6754\,\text{km}$$

Step 2:

$$l = \frac{-5368}{6754} = -0.7947 \qquad m = \frac{-1784}{6754} = -0.2642 \qquad n = \frac{3691}{6754} = 0.5462$$

Step 3:

$$\delta = \sin^{-1} 0.5464 = \boxed{33.12°}$$

Step 4:

Since the direction cosine m is negative,

$$\alpha = 360° - \cos^{-1}\left(\frac{l}{\cos\delta}\right) = 360° - \cos^{-1}\left(\frac{-0.7947}{\cos 33.12°}\right) = 360° - 161.6° = \boxed{198.4°}$$

From Section 2.11 we know that if we are provided the state vector $(\mathbf{r}_0, \mathbf{v}_0)$ at a given instant, then we can determine the state vector at any other time in terms of the initial vector by means of the expressions

$$\begin{aligned}\mathbf{r} &= f\mathbf{r}_0 + g\mathbf{v}_0 \\ \mathbf{v} &= \dot{f}\mathbf{r}_0 + \dot{g}\mathbf{v}_0\end{aligned} \tag{4.6}$$

where the Lagrange coefficients f and g and their time derivatives are given in Eq. (3.69). Specifying the total of six components of $\mathbf{r}_0$ and $\mathbf{v}_0$ therefore completely determines the size, shape, and orientation of the orbit.

EXAMPLE 4.2

At time t_0, the state vector of an earth satellite is

$$\mathbf{r}_0 = 1600\hat{\mathbf{I}} + 5310\hat{\mathbf{J}} + 3800\hat{\mathbf{K}}\,(\text{km}) \tag{a}$$

$$\mathbf{v}_0 = -7.350\hat{\mathbf{I}} + 0.4600\hat{\mathbf{J}} + 2.470\hat{\mathbf{K}}\,(\text{km/s}) \tag{b}$$

Determine the position and velocity vectors 3200 s later and plot the orbit in three dimensions.

Solution

We will use the universal variable formulation and Algorithm 3.4, which was illustrated in detail in Example 3.7. Therefore, only the results of each step are presented here.

Step 1: (α here is not to be confused with right ascension)

$$\alpha = 1.4613\left(10^{-4}\right)\text{km}^{-1}$$

Since this is positive, the orbit is an ellipse.

Step 2:

$$\chi = 294.42\,\text{km}^{1/2}$$

Step 3:

$$f = -0.94843 \quad \text{and} \quad g = -354.89\,\text{s}^{-1}$$

Step 4:

$$\boxed{\mathbf{r} = 1090.9\hat{\mathbf{I}} - 5199.4\hat{\mathbf{J}} - 4480.6\hat{\mathbf{K}}\,(\text{km})} \;\Rightarrow\; r = 6949.8\,\text{km}$$

Step 5:

$$\dot{f} = 0.00045324\,\text{s}^{-1} \quad \text{and} \quad \dot{g} = -0.88479$$

Step 6:

$$\boxed{\mathbf{v} = 7.2284\hat{\mathbf{I}} + 1.9997\hat{\mathbf{J}} - 0.46311\hat{\mathbf{K}}\,(\text{km/s})}$$

To plot the elliptical orbit, we observe that one complete revolution means a change in the eccentric anomaly E of 2π radians. According to Eq. (3.57), the corresponding change in the universal anomaly is

$$\chi = \sqrt{a}E = \sqrt{\frac{1}{\alpha}}E = \sqrt{\frac{1}{0.00014613}} \cdot 2\pi = 519.77\,\text{km}^{1/2}$$

Letting χ vary from 0 to 519.77 in small increments, we employ the Lagrange coefficient formulation (Eq. 3.67 plus Eqs. 3.69a and 3.69b) to compute

$$\mathbf{r} = \left[1 - \frac{\chi^2}{r_0}C\left(\alpha\chi^2\right)\right]\mathbf{r}_0 + \left[\Delta t - \frac{1}{\sqrt{\mu}}\chi^3 S\left(\alpha\chi^2\right)\right]\mathbf{v}_0$$

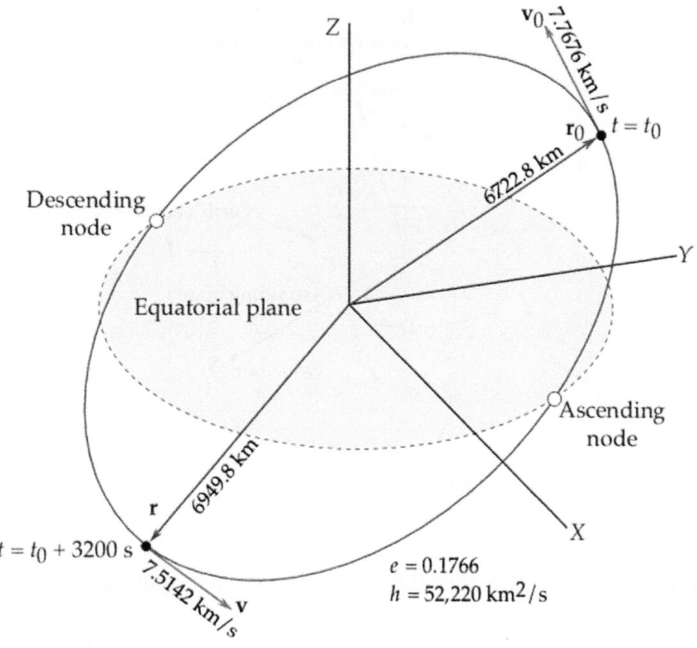

FIG. 4.6

The orbit corresponding to the initial conditions given in Eqs. (a) and (b) of Example 4.2.

where Δt for a given value of χ is given by Eq. (3.49). Using a computer to plot the points obtained in this fashion yields Fig. 4.6, which also shows the state vectors at t_0 and $t_0 + 3200$ s.

The previous example illustrates the fact that the six quantities or orbital elements comprising the state vector **r** and **v** completely determine the orbit. Other elements may be chosen. The classical orbital elements are introduced and related to the state vector in the next section.

4.4 ORBITAL ELEMENTS AND THE STATE VECTOR

To define an orbit in the plane requires two parameters: eccentricity and angular momentum. Other parameters, such as the semimajor axis, the specific energy, and (for an ellipse) the period are obtained from these two. To locate a point on the orbit requires a third parameter, the true anomaly, which leads us to the time since perigee. Describing the orientation of an orbit in three dimensions requires three additional parameters, called the Euler angles, which are illustrated in Fig. 4.7.

First, we locate the intersection of the orbital plane with the equatorial (*XY*) plane. This line is called the node line. The point on the node line where the orbit passes above the equatorial plane from below it is called the ascending node. The node line vector **N** extends outward from the origin through the ascending node. At the other end of the node line, where the orbit dives below the equatorial plane, is the descending node. The angle between the positive *X* axis and the node line is the first Euler angle Ω, the

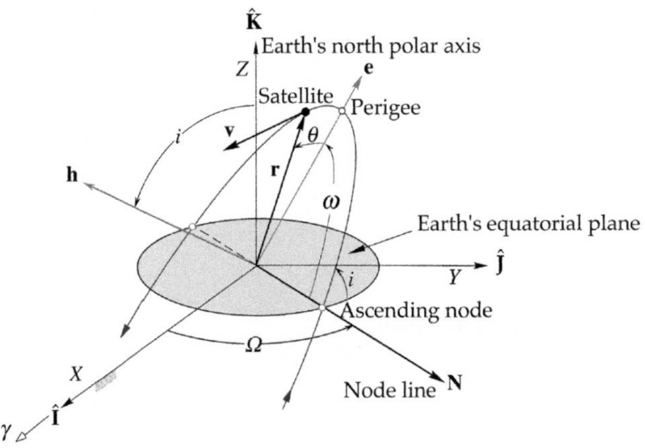

FIG. 4.7

Geocentric equatorial frame and the orbital elements.

right ascension of the ascending node. Recall from Section 4.2 that right ascension is a positive number lying between 0° and 360°.

The dihedral angle between the orbital plane and the equatorial plane is the inclination i, measured according to the right-hand rule (i.e., counterclockwise around the node line vector from the equator to the orbit). The inclination is also the angle between the positive Z axis and the normal to the plane of the orbit. The two equivalent means of measuring i are indicated in Fig. 4.7. Recall from Chapter 2 that the angular momentum vector $\mathbf{h}$ is normal to the plane of the orbit. Therefore, the inclination i is the angle between the positive Z axis and $\mathbf{h}$. The inclination is a positive number between 0° and 180°.

It remains to locate the perigee of the orbit. Recall that perigee lies at the intersection of the eccentricity vector $\mathbf{e}$ with the orbital path. The third Euler angle ω, the argument of perigee, is the angle between the node line vector $\mathbf{N}$ and the eccentricity vector $\mathbf{e}$, measured in the plane of the orbit. The argument of perigee is a positive number between 0° and 360°.

In summary, the six orbital elements are

h specific angular momentum
i inclination
Ω right ascension of the ascending node
e eccentricity
ω argument of perigee
θ true anomaly.

The angular momentum h and true anomaly θ are frequently replaced by the semimajor axis a and the mean anomaly M, respectively.

Given the position $\mathbf{r}$ and velocity $\mathbf{v}$ of a spacecraft in the geocentric equatorial frame, how do we obtain the orbital elements? The step-by-step procedure is outlined next in Algorithm 4.2. Note that each step incorporates results obtained in the previous steps. Several steps require resolving the quadrant ambiguity that arises in calculating the arccosine (recall Fig. 3.4).

ALGORITHM 4.2

Obtain orbital elements from the state vector. A MATLAB version of this procedure appears in Appendix D.18. Applying this algorithm to orbits around other planets or the sun amounts to defining the frame of reference and substituting the appropriate gravitational parameter μ.

1. Calculate the distance, $r = \sqrt{\mathbf{r}\cdot\mathbf{r}} = \sqrt{X^2 + Y^2 + Z^2}$.
2. Calculate the speed, $v = \sqrt{\mathbf{v}\cdot\mathbf{v}} = \sqrt{v_X^2 + v_Y^2 + v_Z^2}$.
3. Calculate the radial velocity, $v_r = \mathbf{r}\cdot\mathbf{v}/r = (Xv_X + Yv_Y + Zv_Z)/r$.
   Note that if $v_r > 0$, the spacecraft is flying away from perigee. If $v_r < 0$, it is flying toward perigee.
4. Calculate the specific angular momentum,

$$\mathbf{h} = \mathbf{r}\times\mathbf{v} = \begin{vmatrix} \hat{\mathbf{I}} & \hat{\mathbf{J}} & \hat{\mathbf{K}} \\ X & Y & Z \\ v_X & v_Y & v_Z \end{vmatrix}$$

5. Calculate the magnitude of the specific angular momentum, $h = \sqrt{\mathbf{h}\cdot\mathbf{h}}$.
   This is the first orbital element.
6. Calculate the inclination,

$$i = \cos^{-1}(h_Z/h) \tag{4.7}$$

   This is the second orbital element. Recall that i must lie between 0° and 180°, which is precisely the range (principal values) of the arccosine function. Hence, there is no quadrant ambiguity to contend with here. If $90° < i \le 180°$, the angular momentum **h** points in a southerly direction. In that case, the orbit is retrograde, which means that the motion of the satellite around the earth is opposite to earth's rotation.
7. Calculate

$$\mathbf{N} = \hat{\mathbf{K}}\times\mathbf{h} = \begin{vmatrix} \hat{\mathbf{I}} & \hat{\mathbf{J}} & \hat{\mathbf{K}} \\ 0 & 0 & 1 \\ h_X & h_Y & h_Z \end{vmatrix} \tag{4.8}$$

   This vector defines the node line.
8. Calculate the magnitude of **N**, $N = \sqrt{\mathbf{N}\cdot\mathbf{N}}$.
9. Calculate the right ascension of the ascending node, $\Omega = \cos^{-1}(N_X/N)$. This is the third orbital element. If $N_X > 0$, then Ω lies in either the first or fourth quadrant. If $N_X < 0$, then Ω lies in either the second or third quadrant. To place Ω in the proper quadrant, observe that the ascending node lies on the positive side of the vertical XZ plane ($0 \le \Omega < 180°$) if $N_Y > 0$. On the other hand, the ascending node lies on the negative side of the XZ plane ($180° \le \Omega < 360°$) if $N_Y < 0$. Therefore, $N_Y > 0$ implies that $0 \le \Omega < 180°$, whereas $N_Y < 0$ implies that $180° \le \Omega < 360°$. In summary,

$$\Omega = \begin{cases} \cos^{-1}\left(\dfrac{N_X}{N}\right) & (N_Y \ge 0) \\ 360° - \cos^{-1}\left(\dfrac{N_X}{N}\right) & (N_Y < 0) \end{cases} \tag{4.9}$$

10. Calculate the eccentricity vector. Starting with Eq. (2.40),

$$\mathbf{e}=\frac{1}{\mu}\left[\mathbf{v}\times\mathbf{h}-\mu\frac{\mathbf{r}}{r}\right]=\frac{1}{\mu}\left[\mathbf{v}\times(\mathbf{r}\times\mathbf{v})-\mu\frac{\mathbf{r}}{r}\right]=\frac{1}{\mu}\left[\overbrace{\mathbf{r}v^2-\mathbf{v}(\mathbf{r}\cdot\mathbf{v})}^{bac-cab\ rule}-\mu\frac{\mathbf{r}}{r}\right]$$

so that

$$\mathbf{e}=\frac{1}{\mu}\left[\left(v^2-\frac{\mu}{r}\right)\mathbf{r}-rv_r\mathbf{v}\right] \tag{4.10}$$

11. Calculate the eccentricity, $e=\sqrt{\mathbf{e}\cdot\mathbf{e}}$, which is the fourth orbital element. Substituting Eq. (4.10) leads to a form depending only on the scalars obtained thus far,

$$e=\sqrt{1+\frac{h^2}{\mu^2}\left(v^2-\frac{2\mu}{r}\right)} \tag{4.11}$$

12. Calculate the argument of perigee,

$$\omega=\cos^{-1}\left(\frac{\mathbf{N}}{N}\cdot\frac{\mathbf{e}}{e}\right)$$

This is the fifth orbital element. If $\mathbf{N}\cdot\mathbf{e}>0$, then ω lies in either the first or fourth quadrant. If $\mathbf{N}\cdot\mathbf{e}<0$, then ω lies in either the second or third quadrant. To place ω in the proper quadrant, observe that perigee lies above the equatorial plane ($0°\le\omega<180°$) if $\mathbf{e}$ points up (in the positive Z direction) and that perigee lies below the plane ($180°\le\omega<360°$) if $\mathbf{e}$ points down. Therefore, $e_Z\ge0$ implies that $0°<\omega<180°$, whereas $e_Z<0$ implies that $180°<\omega<360°$. To summarize,

$$\omega=\begin{cases}\cos^{-1}\left(\dfrac{\mathbf{N}\cdot\mathbf{e}}{Ne}\right) & (e_Z\ge0)\\ 360°-\cos^{-1}\left(\dfrac{\mathbf{N}\cdot\mathbf{e}}{Ne}\right) & (e_Z<0)\end{cases} \tag{4.12}$$

13. Calculate the true anomaly,

$$\theta=\cos^{-1}\left(\frac{\mathbf{e}}{e}\cdot\frac{\mathbf{r}}{r}\right)$$

This is the sixth and final orbital element. If $\mathbf{e}\cdot\mathbf{r}>0$, then θ lies in the first or fourth quadrant. If $\mathbf{e}\cdot\mathbf{r}<0$, then θ lies in the second or third quadrant. To place θ in the proper quadrant, note that if the satellite is flying away from perigee ($\mathbf{r}\cdot\mathbf{v}\ge0$), then $0\le\theta<180°$, whereas if the satellite is flying toward perigee ($\mathbf{r}\cdot\mathbf{v}<0$), then $180°\le\theta<360°$. Therefore, using the results of Step 3 above

$$\theta = \begin{cases} \cos^{-1}\left(\dfrac{\mathbf{e}}{e}\cdot\dfrac{\mathbf{r}}{r}\right) & (v_r \geq 0) \\ 360^\circ - \cos^{-1}\left(\dfrac{\mathbf{e}}{e}\cdot\dfrac{\mathbf{r}}{r}\right) & (v_r < 0) \end{cases} \tag{4.13a}$$

Substituting Eq. (4.10) yields an alternative form of this expression,

$$\theta = \begin{cases} \cos^{-1}\left[\dfrac{1}{e}\left(\dfrac{h^2}{\mu r} - 1\right)\right] & (v_r \geq 0) \\ 360^\circ - \cos^{-1}\left[\dfrac{1}{e}\left(\dfrac{h^2}{\mu r} - 1\right)\right] & (v_r < 0) \end{cases} \tag{4.13b}$$

The procedure described above for calculating the orbital elements is not unique.

EXAMPLE 4.3

Given the state vector,

$$\begin{aligned} \mathbf{r} &= -6045\hat{\mathbf{I}} - 3490\hat{\mathbf{J}} + 2500\hat{\mathbf{K}}\,(\text{km}) \\ \mathbf{v} &= -3.457\hat{\mathbf{I}} + 6.618\hat{\mathbf{J}} + 2.533\hat{\mathbf{K}}\,(\text{km/s}) \end{aligned}$$

find the orbital elements h, i, Ω, e, ω, and θ using Algorithm 4.2.

Solution

Step 1:

$$r = \sqrt{\mathbf{r}\cdot\mathbf{r}} = \sqrt{(-6045)^2 + (-3490)^2 + 2500^2} = 7414\,\text{km}$$

Step 2:

$$v = \sqrt{\mathbf{v}\cdot\mathbf{v}} = \sqrt{(-3.457)^2 + 6.618^2 + 2.533^2} = 7.884\,\text{km/s}$$

Step 3:

$$v_r = \frac{\mathbf{v}\cdot\mathbf{r}}{r} = \frac{(-3.457)\cdot(-6045) + 6.618\cdot(-3490) + 2.533\cdot 2500}{7414} = 0.5575\,\text{km/s} \tag{a}$$

Since $v_r > 0$, the satellite is flying away from perigee.

Step 4:

$$\mathbf{h} = \mathbf{r}\times\mathbf{v} = \begin{vmatrix} \hat{\mathbf{I}} & \hat{\mathbf{J}} & \hat{\mathbf{K}} \\ -6045 & -3490 & 2500 \\ -3.457 & 6.618 & 2.533 \end{vmatrix} = -25{,}380\hat{\mathbf{I}} + 6670\hat{\mathbf{J}} - 52{,}070\hat{\mathbf{K}}\,(\text{km}^2/\text{s})$$

Step 5:

$$h = \sqrt{\mathbf{h}\cdot\mathbf{h}} = \sqrt{(-25,380)^2 + 6670^2 + (-52,070)^2} \Rightarrow \boxed{h = 58{,}310\,\text{km}^2/\text{s}}$$

Step 6:

$$i = \cos^{-1}\frac{h_Z}{h} = \cos^{-1}\left(\frac{-52{,}070}{58{,}310}\right) \Rightarrow \boxed{i = 153.2^\circ}$$

Since i is greater than 90°, this is a retrograde orbit.

Step 7:

$$\mathbf{N}=\hat{\mathbf{K}}\times\mathbf{h}=\begin{vmatrix}\hat{\mathbf{I}} & \hat{\mathbf{J}} & \hat{\mathbf{K}}\\ 0 & 0 & 1\\ -25{,}380 & 6670 & -52{,}070\end{vmatrix}=-6670\hat{\mathbf{I}}-25{,}380\hat{\mathbf{J}}\,(\mathrm{km^2/s}) \tag{b}$$

Step 8:

$$N=\sqrt{\mathbf{N}\cdot\mathbf{N}}=\sqrt{(-6670)^2+(-25,380)^2}=26{,}250\mathrm{km^2/s}$$

Step 9:

$$\Omega=\cos^{-1}\frac{N_X}{N}=\cos^{-1}\left(\frac{-6670}{26{,}250}\right)=104.7^\circ \text{ or } 255.3^\circ$$

From Eq. (b) we know that $N_Y < 0$; therefore, Ω must lie in the third quadrant,

$$\boxed{\Omega=255.3^\circ}$$

Step 10:

$$\begin{aligned}\mathbf{e}&=\frac{1}{\mu}\left[\left(v^2-\frac{\mu}{r}\right)\mathbf{r}-rv_r\mathbf{v}\right]\\ &=\frac{1}{398{,}600}\left[\left(7.884^2-\frac{398{,}600}{7414}\right)(-6045\hat{\mathbf{I}}-3490\hat{\mathbf{J}}+2500\hat{\mathbf{K}})\right.\\ &\left.-(7414)(0.5575)(-3.457\hat{\mathbf{I}}+6.618\hat{\mathbf{J}}+2.533\hat{\mathbf{K}})\right]\\ \mathbf{e}&=-0.09160\hat{\mathbf{I}}-0.1422\hat{\mathbf{J}}+0.02644\hat{\mathbf{K}}\end{aligned} \tag{c}$$

Step 11:

$$e=\sqrt{\mathbf{e}\cdot\mathbf{e}}=\sqrt{(-0.09160)^2+(-0.1422)^2+(0.02644)^2}\Rightarrow\boxed{e=0.1712}$$

Clearly, the orbit is an ellipse.
Step 12:

$$\omega=\cos^{-1}\frac{\mathbf{N}\cdot\mathbf{e}}{Ne}=\cos^{-1}\left[\frac{(-6670)(-0.09160)+(-25,380)(-0.1422)+(0)(0.02644)}{(26,250)(0.1712)}\right]=20.07^\circ \text{ or } 339.9^\circ$$

ω lies in the first quadrant if $e_Z > 0$, which is true in this case, as we see from Eq. (c). Therefore,

$$\boxed{\omega=20.07^\circ}$$

Step 13:

$$\theta=\cos^{-1}\left(\frac{\mathbf{e}\cdot\mathbf{r}}{er}\right)=\cos^{-1}\left[\frac{(-0.09160)(-6045)+(-0.1422)\cdot(-3490)+(0.02644)(2500)}{(0.1712)(7414)}\right]=28.45^\circ \text{ or } 331.6^\circ$$

From Eq. (a) we know that $v_r > 0$, which means $0^\circ \le \theta < 180^\circ$. Therefore,

$$\boxed{\theta=28.45^\circ}$$

Having found the six orbital elements, we can go on to compute other parameters. The perigee and apogee radii are

$$\begin{aligned}r_p&=\frac{h^2}{\mu}\frac{1}{1+e\cos(0)}=\frac{58{,}310^2}{398{,}600}\frac{1}{1+0.1712}=7284\mathrm{km}\\ r_a&=\frac{h^2}{\mu}\frac{1}{1+e\cos(180^\circ)}=\frac{58{,}310^2}{398{,}600}\frac{1}{1-0.1712}=10{,}290\mathrm{km}\end{aligned}$$

From these it follows that the semimajor axis of the ellipse is

$$a=\frac{1}{2}(r_p+r_a)=8788\mathrm{km}$$

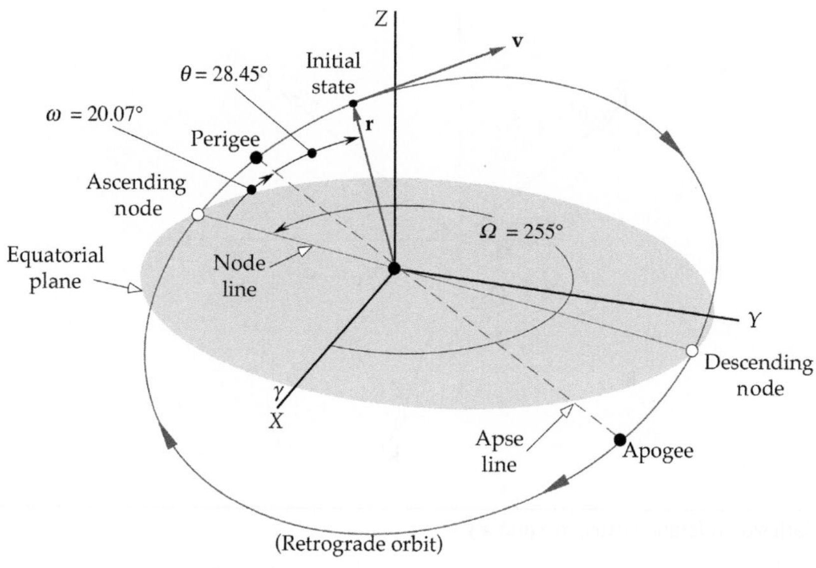

FIG. 4.8

A plot of the orbit identified in Example 4.3.

This leads to the period,

$$T = \frac{2\pi}{\sqrt{\mu}} a^{3/2} = 2.278\,\text{h}$$

The orbit is illustrated in Fig. 4.8.

We have seen how to obtain the orbital elements from the state vector. To arrive at the state vector, given the orbital elements, requires performing coordinate transformations, which are discussed in the next section.

4.5 COORDINATE TRANSFORMATION

The Cartesian coordinate system was introduced in Section 1.2. Fig. 4.9 shows two such coordinate systems: the unprimed system with axes xyz, and the primed system with axes $x'y'z'$. The orthogonal unit basis vectors for the unprimed system are $\hat{\mathbf{i}}$, $\hat{\mathbf{j}}$, and $\hat{\mathbf{k}}$. The fact they are unit vectors means

$$\hat{\mathbf{i}} \cdot \hat{\mathbf{i}} = \hat{\mathbf{j}} \cdot \hat{\mathbf{j}} = \hat{\mathbf{k}} \cdot \hat{\mathbf{k}} = 1 \tag{4.14}$$

Since they are orthogonal,

$$\hat{\mathbf{i}} \cdot \hat{\mathbf{j}} = \hat{\mathbf{i}} \cdot \hat{\mathbf{k}} = \hat{\mathbf{j}} \cdot \hat{\mathbf{k}} = 0 \tag{4.15}$$

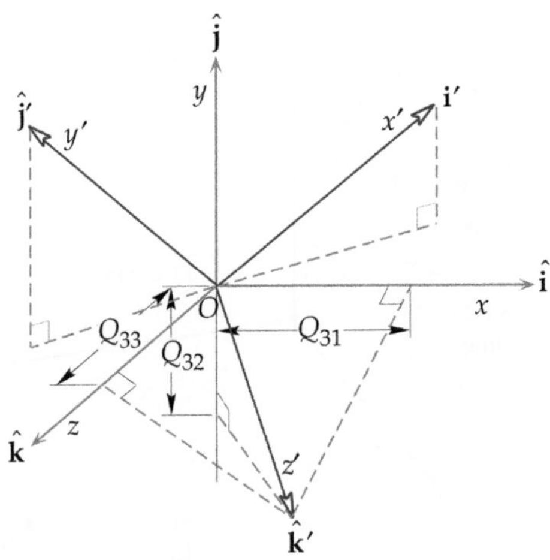

FIG. 4.9

Two sets of Cartesian reference axes, *xyz* and *x′y′z′*.

The orthonormal basis vectors $\hat{\mathbf{i}}'$, $\hat{\mathbf{j}}'$, and $\hat{\mathbf{k}}'$ of the primed system share these same properties. That is,

$$\hat{\mathbf{i}}'\cdot\hat{\mathbf{i}}'=\hat{\mathbf{j}}'\cdot\hat{\mathbf{j}}'=\hat{\mathbf{k}}'\cdot\hat{\mathbf{k}}'=1 \tag{4.16}$$

and

$$\hat{\mathbf{i}}'\cdot\hat{\mathbf{j}}'=\hat{\mathbf{i}}'\cdot\hat{\mathbf{k}}'=\hat{\mathbf{j}}'\cdot\hat{\mathbf{k}}'=0 \tag{4.17}$$

We can express the unit vectors of the primed system in terms of their components in the unprimed system as follows:

$$\begin{aligned}\hat{\mathbf{i}}'&=Q_{11}\hat{\mathbf{i}}+Q_{12}\hat{\mathbf{j}}+Q_{13}\hat{\mathbf{k}}\\ \hat{\mathbf{j}}'&=Q_{21}\hat{\mathbf{i}}+Q_{22}\hat{\mathbf{j}}+Q_{23}\hat{\mathbf{k}}\\ \hat{\mathbf{k}}'&=Q_{31}\hat{\mathbf{i}}+Q_{32}\hat{\mathbf{j}}+Q_{33}\hat{\mathbf{k}}\end{aligned} \tag{4.18}$$

The Q′s in these expressions are just the direction cosines of $\hat{\mathbf{i}}', \hat{\mathbf{j}}'$, and $\hat{\mathbf{k}}'$. Fig. 4.9 illustrates the components of $\hat{\mathbf{k}}'$, which are, of course, the projections of $\hat{\mathbf{k}}'$ onto the x, y, and z axes. The unprimed unit vectors may be resolved into components along the primed system to obtain a set of equations similar to Eq. (4.18).

$$\begin{aligned}\hat{\mathbf{i}}&=Q'_{11}\hat{\mathbf{i}}'+Q'_{12}\hat{\mathbf{j}}'+Q'_{13}\hat{\mathbf{k}}'\\ \hat{\mathbf{j}}&=Q'_{21}\hat{\mathbf{i}}'+Q'_{22}\hat{\mathbf{j}}'+Q'_{23}\hat{\mathbf{k}}'\\ \hat{\mathbf{k}}&=Q'_{31}\hat{\mathbf{i}}'+Q'_{32}\hat{\mathbf{j}}'+Q'_{33}\hat{\mathbf{k}}'\end{aligned} \tag{4.19}$$

However, $\hat{\mathbf{i}}'\cdot\hat{\mathbf{i}}=\hat{\mathbf{i}}\cdot\hat{\mathbf{i}}'$, so that, from Eqs. (4.18) and (4.19), we find $Q_{11}=Q_{11}'$. Likewise, $\hat{\mathbf{i}}'\cdot\hat{\mathbf{j}}=\hat{\mathbf{j}}\cdot\hat{\mathbf{i}}'$, which, according to Eqs. (4.18) and (4.19), means $Q_{12}=Q_{21}'$. Proceeding in this fashion, it is clear that the direction cosines in Eq. (4.18) may be expressed in terms of those in Eq. (4.19). That is, Eq. (4.19) may be written

$$\begin{aligned}\hat{\mathbf{i}} &= Q_{11}\hat{\mathbf{i}}' + Q_{21}\hat{\mathbf{j}}' + Q_{31}\hat{\mathbf{k}}' \\ \hat{\mathbf{j}} &= Q_{12}\hat{\mathbf{i}}' + Q_{22}\hat{\mathbf{j}}' + Q_{32}\hat{\mathbf{k}}' \\ \hat{\mathbf{k}} &= Q_{13}\hat{\mathbf{i}}' + Q_{23}\hat{\mathbf{j}}' + Q_{33}\hat{\mathbf{k}}'\end{aligned} \tag{4.20}$$

Substituting Eq. (4.20) into Eq. (4.14) and making use of Eqs. (4.16) and (4.17), we get the three relations

$$\begin{aligned}\hat{\mathbf{i}}\cdot\hat{\mathbf{i}} = 1 &\Rightarrow Q_{11}^2 + Q_{21}^2 + Q_{31}^2 = 1 \\ \hat{\mathbf{j}}\cdot\hat{\mathbf{j}} = 1 &\Rightarrow Q_{12}^2 + Q_{22}^2 + Q_{32}^2 = 1 \\ \hat{\mathbf{k}}\cdot\hat{\mathbf{k}} = 1 &\Rightarrow Q_{13}^2 + Q_{23}^2 + Q_{33}^2 = 1\end{aligned} \tag{4.21}$$

Substituting Eq. (4.20) into Eq. (4.15) and, again, making use of Eqs. (4.16) and (4.17), we obtain the three equations

$$\begin{aligned}\hat{\mathbf{i}}\cdot\hat{\mathbf{j}} = 0 &\Rightarrow Q_{11}Q_{12} + Q_{21}Q_{22} + Q_{31}Q_{32} = 0 \\ \hat{\mathbf{i}}\cdot\hat{\mathbf{k}} = 0 &\Rightarrow Q_{11}Q_{13} + Q_{21}Q_{23} + Q_{31}Q_{33} = 0 \\ \hat{\mathbf{j}}\cdot\hat{\mathbf{k}} = 0 &\Rightarrow Q_{12}Q_{13} + Q_{22}Q_{23} + Q_{32}Q_{33} = 0\end{aligned} \tag{4.22}$$

Let $[\mathbf{Q}]$ represent the matrix of direction cosines of $\hat{\mathbf{i}}'$, $\hat{\mathbf{j}}'$, and $\hat{\mathbf{k}}'$ relative to $\hat{\mathbf{i}}$, $\hat{\mathbf{j}}$, and $\hat{\mathbf{k}}$, as given by Eq. (4.18). $[\mathbf{Q}]$ is referred to as the direction cosine matrix (DCM).

$$[\mathbf{Q}] = \begin{bmatrix} Q_{11} & Q_{12} & Q_{13} \\ Q_{21} & Q_{22} & Q_{23} \\ Q_{31} & Q_{32} & Q_{33} \end{bmatrix} = \begin{bmatrix} \hat{\mathbf{i}}'\cdot\hat{\mathbf{i}} & \hat{\mathbf{i}}'\cdot\hat{\mathbf{j}} & \hat{\mathbf{i}}'\cdot\hat{\mathbf{k}} \\ \hat{\mathbf{j}}'\cdot\hat{\mathbf{i}} & \hat{\mathbf{j}}'\cdot\hat{\mathbf{j}} & \hat{\mathbf{j}}'\cdot\hat{\mathbf{k}} \\ \hat{\mathbf{k}}'\cdot\hat{\mathbf{i}} & \hat{\mathbf{k}}'\cdot\hat{\mathbf{j}} & \hat{\mathbf{k}}'\cdot\hat{\mathbf{k}} \end{bmatrix} \tag{4.23}$$

The transpose of the matrix $[\mathbf{Q}]$, denoted $[\mathbf{Q}]^T$, is obtained by interchanging the rows and columns of $[\mathbf{Q}]$. Thus,

$$[\mathbf{Q}]^T = \begin{bmatrix} Q_{11} & Q_{21} & Q_{31} \\ Q_{12} & Q_{22} & Q_{32} \\ Q_{13} & Q_{23} & Q_{33} \end{bmatrix} = \begin{bmatrix} \hat{\mathbf{i}}\cdot\hat{\mathbf{i}}' & \hat{\mathbf{i}}\cdot\hat{\mathbf{j}}' & \hat{\mathbf{i}}\cdot\hat{\mathbf{k}}' \\ \hat{\mathbf{j}}\cdot\hat{\mathbf{i}}' & \hat{\mathbf{j}}\cdot\hat{\mathbf{j}}' & \hat{\mathbf{j}}\cdot\hat{\mathbf{k}}' \\ \hat{\mathbf{k}}\cdot\hat{\mathbf{i}}' & \hat{\mathbf{k}}\cdot\hat{\mathbf{j}}' & \hat{\mathbf{k}}\cdot\hat{\mathbf{k}}' \end{bmatrix} \tag{4.24}$$

Forming the product $[\mathbf{Q}]^T[\mathbf{Q}]$, we get

$$\begin{aligned}[\mathbf{Q}]^T[\mathbf{Q}] &= \begin{bmatrix} Q_{11} & Q_{21} & Q_{31} \\ Q_{12} & Q_{22} & Q_{32} \\ Q_{13} & Q_{23} & Q_{33} \end{bmatrix}\begin{bmatrix} Q_{11} & Q_{12} & Q_{13} \\ Q_{21} & Q_{22} & Q_{23} \\ Q_{31} & Q_{32} & Q_{33} \end{bmatrix} \\ &= \begin{bmatrix} Q_{11}^2 + Q_{21}^2 + Q_{31}^2 & Q_{11}Q_{12} + Q_{21}Q_{22} + Q_{31}Q_{32} & Q_{11}Q_{13} + Q_{21}Q_{23} + Q_{31}Q_{33} \\ Q_{12}Q_{11} + Q_{22}Q_{21} + Q_{32}Q_{31} & Q_{12}^2 + Q_{22}^2 + Q_{31}^2 & Q_{12}Q_{13} + Q_{22}Q_{23} + Q_{32}Q_{33} \\ Q_{13}Q_{11} + Q_{23}Q_{21} + Q_{33}Q_{31} & Q_{13}Q_{12} + Q_{23}Q_{22} + Q_{33}Q_{32} & Q_{13}^2 + Q_{23}^2 + Q_{33}^2 \end{bmatrix}\end{aligned}$$

From this we obtain, with the aid of Eqs. (4.21) and (4.22),

$$[\mathbf{Q}]^T[\mathbf{Q}] = [\mathbf{1}] \tag{4.25}$$

where

$$[\mathbf{1}] = \begin{bmatrix} 1 & 0 & 0 \\ 0 & 1 & 0 \\ 0 & 0 & 1 \end{bmatrix}$$

[**1**] stands for the identity matrix or unit matrix.

In a similar fashion, we can substitute Eq. (4.18) into Eqs. (4.16) and (4.17) and make use of Eqs. (4.14) and (4.15) to finally obtain

$$[\mathbf{Q}][\mathbf{Q}]^T = [\mathbf{1}] \tag{4.26}$$

Since [**Q**] satisfies Eqs. (4.25) and (4.26), it is called an orthogonal matrix.

Let **v** be a vector. It can be expressed in terms of its components along the unprimed system

$$\mathbf{v} = v_x\hat{\mathbf{i}} + v_y\hat{\mathbf{j}} + v_z\hat{\mathbf{k}}$$

or along the primed system

$$\mathbf{v} = v'_x\hat{\mathbf{i}}' + v'_y\hat{\mathbf{j}}' + v'_z\hat{\mathbf{k}}'$$

These two expressions for **v** are equivalent (**v** = **v**) since a vector is independent of the coordinate system used to describe it. Thus,

$$v'_x\hat{\mathbf{i}}' + v'_y\hat{\mathbf{j}}' + v'_z\hat{\mathbf{k}}' = v_x\hat{\mathbf{i}} + v_y\hat{\mathbf{j}} + v_z\hat{\mathbf{k}} \tag{4.27}$$

Substituting Eq. (4.20) into the right-hand side of Eq. (4.27) yields

$$\begin{aligned} v'_x\hat{\mathbf{i}}' + v'_y\hat{\mathbf{j}}' + v'_z\hat{\mathbf{k}}' &= \left(Q_{11}\hat{\mathbf{i}}' + Q_{21}\hat{\mathbf{j}}' + Q_{31}\hat{\mathbf{k}}'\right)v_x \\ &+ \left(Q_{12}\hat{\mathbf{i}}' + Q_{22}\hat{\mathbf{j}}' + Q_{32}\hat{\mathbf{k}}'\right)v_y + \left(Q_{13}\hat{\mathbf{i}}' + Q_{23}\hat{\mathbf{j}}' + Q_{33}\hat{\mathbf{k}}'\right)v_z \end{aligned}$$

On collecting terms on the right, we get

$$\begin{aligned} v'_x\hat{\mathbf{i}}' + v'_y\hat{\mathbf{j}}' + v'_z\hat{\mathbf{k}}' &= (Q_{11}v_x + Q_{12}v_y + Q_{13}v_z)\hat{\mathbf{i}}' \\ &+ (Q_{21}v_x + Q_{22}v_y + Q_{23}v_z)\hat{\mathbf{j}}' + (Q_{31}v_x + Q_{32}v_y + Q_{33}v_z)\hat{\mathbf{k}}' \end{aligned}$$

Equating the components of like unit vectors on each side of the equals sign yields

$$\begin{aligned} v'_x &= Q_{11}v_x + Q_{12}v_y + Q_{13}v_z \\ v'_y &= Q_{21}v_x + Q_{22}v_y + Q_{23}v_z \\ v'_z &= Q_{31}v_x + Q_{32}v_y + Q_{33}v_z \end{aligned} \tag{4.28}$$

In matrix notation, this may be written

$$\{\mathbf{v}'\} = [\mathbf{Q}]\{\mathbf{v}\} \tag{4.29}$$

where

$$\{\mathbf{v}'\} = \begin{Bmatrix} v'_x \\ v'_y \\ v'_z \end{Bmatrix} \qquad \{\mathbf{v}\} = \begin{Bmatrix} v_x \\ v_y \\ v_z \end{Bmatrix} \tag{4.30}$$

and $[\mathbf{Q}]$ is given by Eq. (4.23). Eq. (4.28) (or Eq. 4.29) shows how to transform the components of the vector **v** in the unprimed system into its components in the primed system. The inverse transformation, from primed to unprimed, is found by multiplying Eq. (4.29) throughout by $[\mathbf{Q}]^T$:

$$[\mathbf{Q}]^T\{\mathbf{v}'\} = [\mathbf{Q}]^T[\mathbf{Q}]\{\mathbf{v}\}$$

But, according to Eq. (4.25), $[\mathbf{Q}][\mathbf{Q}]^T = [\mathbf{1}]$, so that

$$[\mathbf{Q}]^T\{\mathbf{v}'\} = [\mathbf{1}]\{\mathbf{v}\}$$

Since $[\mathbf{1}]\{\mathbf{v}\} = \{\mathbf{v}\}$, we obtain

$$\{\mathbf{v}\} = [\mathbf{Q}]^T\{\mathbf{v}'\} \tag{4.31}$$

Therefore, to go from the unprimed system to the primed system we use $[\mathbf{Q}]$, and in the reverse direction—from primed to unprimed—we use $[\mathbf{Q}]^T$.

EXAMPLE 4.4

In Fig. 4.10, the x' axis is defined by the line segment $O'P$. The $x'y'$ plane is defined by the intersecting line segments $O'P$ and $O'Q$. The z' axis is normal to the plane of $O'P$ and $O'Q$ and obtained by rotating $O'P$ toward $O'Q$ and using the right-hand rule.

(a) Find the direction cosine matrix $[\mathbf{Q}]$.
(b) If $\{\mathbf{v}\} = \lfloor 2 \;\; 4 \;\; 6 \rfloor^T$, find $\{\mathbf{v}'\}$.
(c) If $\{\mathbf{v}'\} = \lfloor 2 \;\; 4 \;\; 6 \rfloor^T$, find $\{\mathbf{v}\}$.

Solution

(a) Resolve the directed line segments $\overrightarrow{O'P}$ and $\overrightarrow{O'Q}$ into components along the unprimed system:

$$\overrightarrow{O'P} = (-5-3)\hat{\mathbf{i}} + (5-1)\hat{\mathbf{j}} + (4-2)\hat{\mathbf{k}} = -8\hat{\mathbf{i}} + 4\hat{\mathbf{j}} + 2\hat{\mathbf{k}}$$
$$\overrightarrow{O'Q} = (-6-3)\hat{\mathbf{i}} + (3-1)\hat{\mathbf{j}} + (5-2)\hat{\mathbf{k}} = -9\hat{\mathbf{i}} + 2\hat{\mathbf{j}} + 3\hat{\mathbf{k}}$$

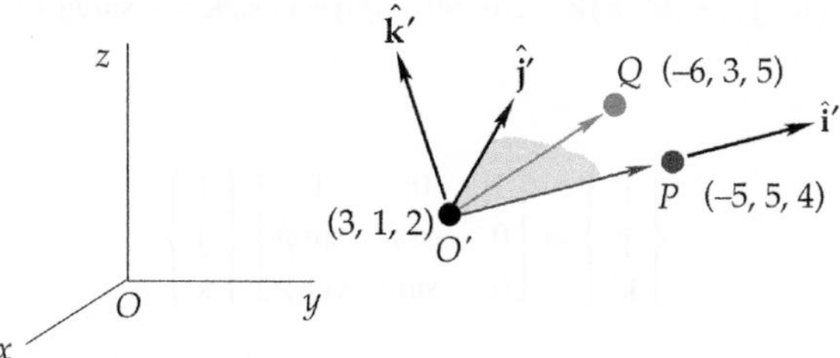

FIG. 4.10

Defining a unit triad from the coordinates of three noncollinear points, O', P, and Q.

Taking the cross product of $\overrightarrow{O'P}$ into $\overrightarrow{O'Q}$ yields a vector $\mathbf{Z}'$, which lies in the direction of the desired positive z' axis:

$$\mathbf{Z}' = \overrightarrow{O'P} \times \overrightarrow{O'Q} = 8\hat{\mathbf{i}} + 6\hat{\mathbf{j}} + 20\hat{\mathbf{k}}$$

Taking the cross product of $\mathbf{Z}'$ into $\overrightarrow{O'P}$ then yields a vector $\mathbf{Y}'$, which points in the positive y' direction:

$$\mathbf{Y}' = \mathbf{Z}' \times \overrightarrow{O'P} = -68\hat{\mathbf{i}} + 176\hat{\mathbf{j}} + 80\hat{\mathbf{k}}$$

Normalizing the vectors $\overrightarrow{O'P}$, $\mathbf{Y}'$, and $\mathbf{Z}'$ produces the $\hat{\mathbf{i}}'$, $\hat{\mathbf{j}}'$, and $\hat{\mathbf{k}}'$ unit vectors, respectively. Thus

$$\begin{aligned}
\hat{\mathbf{i}}' &= \overrightarrow{O'P} \Big/ \left\|\overrightarrow{O'P}\right\| = -0.8729\hat{\mathbf{i}} + 0.4364\hat{\mathbf{j}} + 0.2182\hat{\mathbf{k}} \\
\hat{\mathbf{j}}' &= \mathbf{Y}'/\|\mathbf{Y}'\| = -0.3318\hat{\mathbf{i}} - 0.8588\hat{\mathbf{j}} + 0.3904\hat{\mathbf{k}} \\
\hat{\mathbf{k}}' &= \mathbf{Z}'/\|\mathbf{Z}'\| = 0.3578\hat{\mathbf{i}} + 0.2683\hat{\mathbf{j}} + 0.8944\hat{\mathbf{k}}
\end{aligned}$$

The components of $\hat{\mathbf{i}}'$, $\hat{\mathbf{j}}'$, and $\hat{\mathbf{k}}'$ are the rows of the direction cosine matrix $[\mathbf{Q}]$. Thus,

$$[\mathbf{Q}] = \begin{bmatrix} -0.8729 & 0.4364 & 0.2182 \\ -0.3318 & -0.8588 & 0.3904 \\ 0.3578 & 0.2683 & 0.8944 \end{bmatrix}$$

(b)

$$\{\mathbf{v}'\} = [\mathbf{Q}]\{\mathbf{v}\} = \begin{bmatrix} -0.8729 & 0.4364 & 0.2182 \\ -0.3318 & -0.8588 & 0.3904 \\ 0.3578 & 0.2683 & 0.8944 \end{bmatrix} \begin{Bmatrix} 2 \\ 4 \\ 6 \end{Bmatrix} \Rightarrow \boxed{\{\mathbf{v}'\} = \begin{Bmatrix} 1.309 \\ -1.756 \\ 7.155 \end{Bmatrix}}$$

(c)

$$\{\mathbf{v}\} = [\mathbf{Q}]^T\{\mathbf{v}'\} = \begin{bmatrix} -0.8729 & -0.3318 & 0.3578 \\ 0.4364 & -0.8588 & 0.2683 \\ 0.2182 & 0.3904 & 0.8944 \end{bmatrix} \begin{Bmatrix} 2 \\ 4 \\ 6 \end{Bmatrix} \Rightarrow \boxed{\{\mathbf{v}\} = \begin{Bmatrix} -0.9263 \\ -0.9523 \\ 7.364 \end{Bmatrix}}$$

Let us consider the special case in which the coordinate transformation involves a rotation about only one of the coordinate axes, as shown in Fig. 4.11. If the rotation is about the x axis, then according to Eqs. (4.18) and (4.23),

$$\begin{aligned}
\hat{\mathbf{i}}' &= \hat{\mathbf{i}} \\
\hat{\mathbf{j}}' &= \left(\hat{\mathbf{j}}' \cdot \hat{\mathbf{j}}\right)\hat{\mathbf{j}} + \left(\hat{\mathbf{j}}' \cdot \hat{\mathbf{k}}\right)\hat{\mathbf{k}} = \cos\phi\hat{\mathbf{j}} + \cos(90° - \phi)\hat{\mathbf{k}} = \cos\phi\hat{\mathbf{j}} + \sin\phi\hat{\mathbf{k}} \\
\hat{\mathbf{k}}' &= \left(\hat{\mathbf{k}}' \cdot \hat{\mathbf{j}}\right)\hat{\mathbf{j}} + \left(\hat{\mathbf{k}}' \cdot \hat{\mathbf{k}}\right)\hat{\mathbf{k}} = \cos(90° + \phi)\hat{\mathbf{j}} + \cos\phi\hat{\mathbf{k}} = -\sin\phi\hat{\mathbf{j}} + \cos\phi\hat{\mathbf{k}}
\end{aligned}$$

or

$$\begin{Bmatrix} \hat{\mathbf{i}}' \\ \hat{\mathbf{j}}' \\ \hat{\mathbf{k}}' \end{Bmatrix} = \begin{bmatrix} 1 & 0 & 0 \\ 0 & \cos\phi & \sin\phi \\ 0 & -\sin\phi & \cos\phi \end{bmatrix} \begin{Bmatrix} \hat{\mathbf{i}} \\ \hat{\mathbf{j}} \\ \hat{\mathbf{k}} \end{Bmatrix}$$

The transformation from the xyz coordinate system to the $x'y'z'$ system having a common x axis is given by the direction cosine matrix on the right. Since this is a rotation through the angle ϕ about the x axis, we denote this matrix by $[\mathbf{R}_1(\phi)]$, in which the subscript 1 stands for axis 1 (the x axis). Thus,

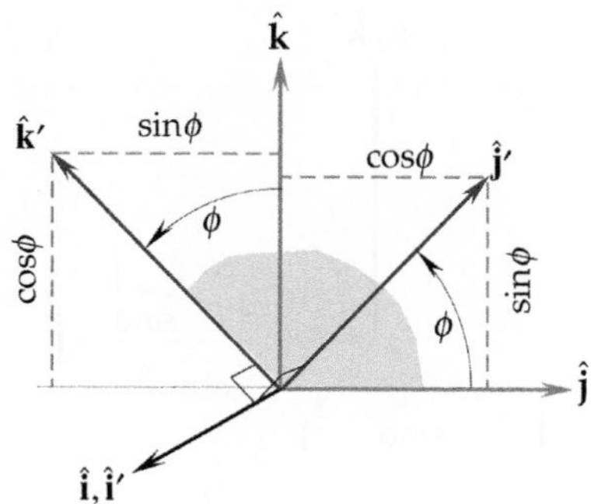

FIG. 4.11

Rotation about the *x* axis.

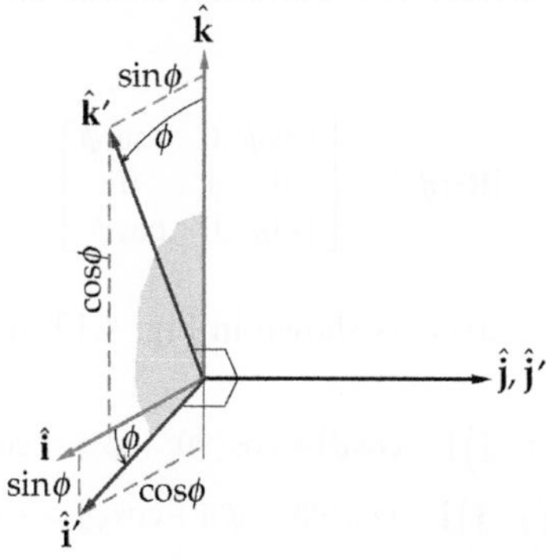

FIG. 4.12

Rotation about the *y* axis.

$$[\mathbf{R}_1(\phi)] = \begin{bmatrix} 1 & 0 & 0 \\ 0 & \cos\phi & \sin\phi \\ 0 & -\sin\phi & \cos\phi \end{bmatrix} \tag{4.32}$$

If the rotation is about the *y* axis, as shown in Fig. 4.12, then Eq. (4.18) yields

$$\begin{aligned}
\hat{\mathbf{i}}' &= \left(\hat{\mathbf{i}}' \cdot \hat{\mathbf{i}}\right)\hat{\mathbf{i}} + \left(\hat{\mathbf{i}}' \cdot \hat{\mathbf{k}}\right)\hat{\mathbf{k}} = \cos\phi\hat{\mathbf{i}} + \cos(\phi + 90°)\hat{\mathbf{k}} = \cos\phi\hat{\mathbf{i}} - \sin\phi\hat{\mathbf{k}} \\
\hat{\mathbf{j}}' &= \hat{\mathbf{j}} \\
\hat{\mathbf{k}}' &= \left(\hat{\mathbf{k}}' \cdot \hat{\mathbf{i}}\right)\hat{\mathbf{i}} + \left(\hat{\mathbf{k}}' \cdot \hat{\mathbf{k}}\right)\hat{\mathbf{k}} = \cos(90° - \phi)\hat{\mathbf{i}} + \cos\phi\hat{\mathbf{k}} = \sin\phi\hat{\mathbf{i}} + \cos\phi\hat{\mathbf{k}}
\end{aligned}$$

or, more compactly,

$$\begin{Bmatrix} \hat{\mathbf{i}}' \\ \hat{\mathbf{j}}' \\ \hat{\mathbf{k}}' \end{Bmatrix} = \begin{bmatrix} \cos\phi & 0 & -\sin\phi \\ 0 & 1 & 0 \\ \sin\phi & 0 & \cos\phi \end{bmatrix} \begin{Bmatrix} \hat{\mathbf{i}} \\ \hat{\mathbf{j}} \\ \hat{\mathbf{k}} \end{Bmatrix}$$

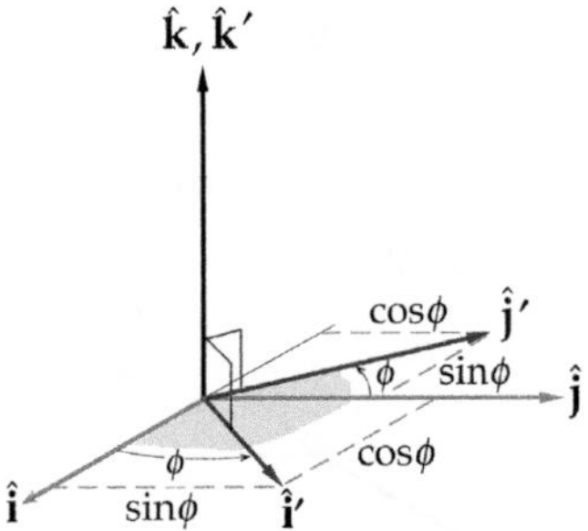

FIG. 4.13

Rotation about the z axis.

We represent this transformation between two Cartesian coordinate systems having a common y axis (axis 2) as $[\mathbf{R}_2(\phi)]$. Therefore,

$$[\mathbf{R}_2(\phi)] = \begin{bmatrix} \cos\phi & 0 & -\sin\phi \\ 0 & 1 & 0 \\ \sin\phi & 0 & \cos\phi \end{bmatrix} \tag{4.33}$$

Finally, if the rotation is about the z axis, as shown in Fig. 4.13, then we have from Eq. (4.18) that

$$\begin{aligned}
\hat{\mathbf{i}}' &= \left(\hat{\mathbf{i}}'\cdot\hat{\mathbf{i}}\right)\hat{\mathbf{i}} + \left(\hat{\mathbf{i}}'\cdot\hat{\mathbf{j}}\right)\hat{\mathbf{j}} = \cos\phi\hat{\mathbf{i}} + \cos(90^\circ - \phi)\hat{\mathbf{j}} = \cos\phi\hat{\mathbf{i}} + \sin\phi\hat{\mathbf{j}} \\
\hat{\mathbf{j}}' &= \left(\hat{\mathbf{j}}'\cdot\hat{\mathbf{i}}\right)\hat{\mathbf{i}} + \left(\hat{\mathbf{j}}'\cdot\hat{\mathbf{j}}\right)\hat{\mathbf{j}} = \cos(90^\circ + \phi)\hat{\mathbf{i}} + \cos\phi\hat{\mathbf{j}} = -\sin\phi\hat{\mathbf{i}} + \cos\phi\hat{\mathbf{j}} \\
\hat{\mathbf{k}}' &= \hat{\mathbf{k}}
\end{aligned}$$

or

$$\begin{Bmatrix} \hat{\mathbf{i}}' \\ \hat{\mathbf{j}}' \\ \hat{\mathbf{k}}' \end{Bmatrix} = \begin{bmatrix} \cos\phi & \sin\phi & 0 \\ -\sin\phi & \cos\phi & 0 \\ 0 & 0 & 1 \end{bmatrix} \begin{Bmatrix} \hat{\mathbf{i}} \\ \hat{\mathbf{j}} \\ \hat{\mathbf{k}} \end{Bmatrix}$$

In this case, the rotation is around axis 3, the z axis, so

$$[\mathbf{R}_3(\phi)] = \begin{bmatrix} \cos\phi & \sin\phi & 0 \\ -\sin\phi & \cos\phi & 0 \\ 0 & 0 & 1 \end{bmatrix} \tag{4.34}$$

The single transformation between the xyz and $x'y'z'$ Cartesian coordinate frames can be viewed as a sequence of three coordinate transformations, starting with xyz:

$$xyz \overbrace{\rightarrow}^{\alpha} x_1y_1z_1 \overbrace{\rightarrow}^{\beta} x_2y_2z_2 \overbrace{\rightarrow}^{\gamma} x'y'z'$$

Each coordinate system is obtained from the previous one by means of a rotation about one of the axes of the previous frame. Two successive rotations cannot be about the same axis. The first rotation angle is α, the second one is β and the final one is γ. In specific applications, the Greek letters that are traditionally used to represent the three rotations are not α, β, and γ. For those new to the subject, however, it might initially be easier to remember that the first, second, and third rotation angles are represented by the first, second, and third letters of the Greek alphabet ($\alpha\beta\gamma$). Each one of the three transformations has the direction cosine matrix $[\mathbf{R}_i(\phi)]$, where $i = 1, 2,$ or 3 and $\phi = \alpha, \beta$ or γ. The sequence of three such elementary rotations relating two different Cartesian frames of reference is called an Euler angle sequence. Each of the 12 possible Euler angle sequences has a direction cosine matrix $[\mathbf{Q}]$, which is the product of three elementary rotation matrices. The six symmetric Euler sequences are those that begin and end with rotation about the same axis:

$$\begin{matrix} [\mathbf{R}_1(\gamma)][\mathbf{R}_2(\beta)][\mathbf{R}_1(\alpha)] & [\mathbf{R}_1(\gamma)][\mathbf{R}_3(\beta)][\mathbf{R}_1(\alpha)] \\ [\mathbf{R}_2(\gamma)][\mathbf{R}_1(\beta)][\mathbf{R}_2(\alpha)] & [\mathbf{R}_2(\gamma)][\mathbf{R}_3(\beta)][\mathbf{R}_2(\alpha)] \\ \underline{[\mathbf{R}_3(\gamma)][\mathbf{R}_1(\beta)][\mathbf{R}_3(\alpha)]} & [\mathbf{R}_3(\gamma)][\mathbf{R}_2(\beta)][\mathbf{R}_3(\alpha)] \end{matrix} \tag{4.35}$$

The asymmetric Euler sequences involve rotations about all three axes:

$$\begin{matrix} \underline{[\mathbf{R}_1(\gamma)][\mathbf{R}_2(\beta)][\mathbf{R}_3(\alpha)]} & [\mathbf{R}_1(\gamma)][\mathbf{R}_3(\beta)][\mathbf{R}_2(\alpha)] \\ \underline{[\mathbf{R}_2(\gamma)][\mathbf{R}_3(\beta)][\mathbf{R}_1(\alpha)]} & [\mathbf{R}_2(\gamma)][\mathbf{R}_1(\beta)][\mathbf{R}_3(\alpha)] \\ [\mathbf{R}_3(\gamma)][\mathbf{R}_1(\beta)][\mathbf{R}_2(\alpha)] & [\mathbf{R}_3(\gamma)][\mathbf{R}_2(\beta)][\mathbf{R}_1(\alpha)] \end{matrix} \tag{4.36}$$

One of the symmetric sequences that has frequent application in space mechanics is the "classical" Euler angle sequence,

$$[\mathbf{Q}] = [\mathbf{R}_3(\gamma)][\mathbf{R}_1(\beta)][\mathbf{R}_3(\alpha)] \quad (0 \le \alpha < 360° \quad 0 \le \beta \le 180° \quad 0 \le \gamma < 360°) \tag{4.37}$$

which is is underlined in Eq. (4.35) and illustrated in Fig. 4.14. The first rotation is around the z axis, through the angle α. It rotates the x and y axes into the x_1 and y_1 directions. Viewed down the z axis, this rotation appears as shown in the insert at the top of the figure. The direction cosine matrix associated with this rotation is $[\mathbf{R}_3(\alpha)]$. The subscript means that the rotation is around the current (3) direction, which was the z axis (and is now the z_1 axis). The second rotation, represented by $[\mathbf{R}_1(\beta)]$, is around the x_1 axis through the angle β required to rotate the z_1 axis into the z_2 axis, which coincides with the target z' axis. Simultaneously, y_1 rotates into y_2. The insert in the lower right of Fig. 4.14 shows how this rotation appears when viewed from the x_1 direction. $[\mathbf{R}_3(\gamma)]$ represents the third and final rotation, which rotates the x_2 axis (formerly the x_1 axis) and the y_2 axis through the angle γ around the z' axis, so that they become aligned with the target x' and y' axes, respectively. This rotation appears from the z' direction as shown in the insert on the left of Fig. 4.14.

Applying the transformation in Eq. (4.37) to the xyz components $\{\mathbf{b}\}_x$ of the vector $\mathbf{b} = b_x\hat{\mathbf{i}} + b_y\hat{\mathbf{j}} + b_z\hat{\mathbf{k}}$ yields the components of the same vector in the $x'y'z'$ frame

$$\{\mathbf{b}\}_{x'} = [\mathbf{Q}]\{\mathbf{b}\}_x \quad \left(\mathbf{b} = b_{x'}\hat{\mathbf{i}}' + b_{y'}\hat{\mathbf{j}}' + b_{z'}\hat{\mathbf{k}}'\right)$$

That is

$$[\mathbf{Q}]\{\mathbf{b}\}_x = [\mathbf{R}_3(\gamma)][\mathbf{R}_1(\beta)]\overbrace{[\mathbf{R}_3(\alpha)]\{\mathbf{b}\}_x}^{\{\mathbf{b}\}_{x_1}} = [\mathbf{R}_3(\gamma)]\overbrace{[\mathbf{R}_1(\beta)]\{\mathbf{b}\}_{x_1}}^{\{\mathbf{b}\}_{x_2}} = \overbrace{[\mathbf{R}_3(\gamma)]\{\mathbf{b}\}_{x_2}}^{\{\mathbf{b}\}_{x'}}$$

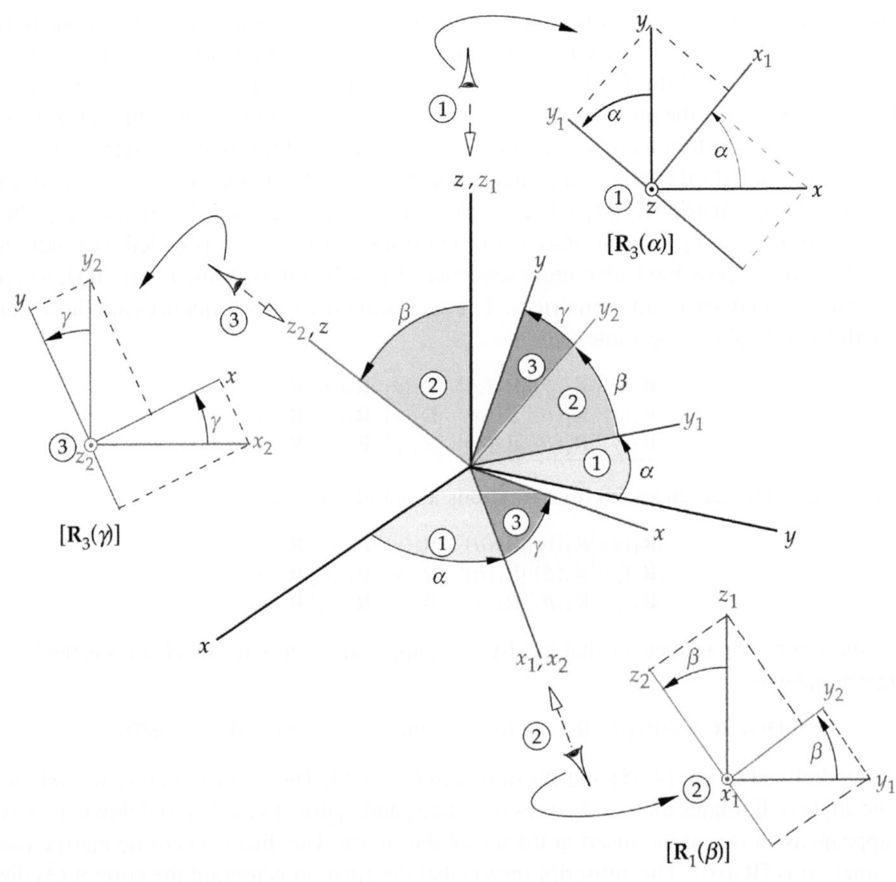

FIG. 4.14

Classical Euler sequence of three rotations transforming *xyz* into *x′y′z′*. The "eye" viewing down an axis sees the illustrated rotation about that axis.

The column vector $\{\mathbf{b}\}_{x_1}$ contains the components of the vector $\mathbf{b}$ $\left(\mathbf{b}=b_{x_1}\hat{\mathbf{i}}_1+b_{y_1}\hat{\mathbf{j}}_1+b_{z_1}\hat{\mathbf{k}}_1\right)$ in the first intermediate frame $x_1y_1z_1$. The column vector $\{\mathbf{b}\}_{x_2}$ contains the components of the vector $\mathbf{b}$ $\left(\mathbf{b}=b_{x_2}\hat{\mathbf{i}}_2+b_{y_2}\hat{\mathbf{j}}_2+b_{z_2}\hat{\mathbf{k}}_2\right)$ in the second intermediate frame $x_2y_2z_2$. Finally, the column vector $\{\mathbf{b}\}_{x'}$ contains the components in the target $x'y'z'$ frame.

Substituting Eqs. (4.32) and (4.34) into Eq. (4.37) yields the direction cosine matrix of the classical Euler sequence $[\mathbf{R}_3(\gamma)][\mathbf{R}_1(\beta)][\mathbf{R}_3(\alpha)]$,

$$[\mathbf{Q}]=\begin{bmatrix}-\sin\alpha\cos\beta\sin\gamma+\cos\alpha\cos\gamma & \cos\alpha\cos\beta\sin\gamma+\sin\alpha\cos\gamma & \sin\beta\sin\gamma\\ -\sin\alpha\cos\beta\sin\gamma-\cos\alpha\cos\gamma & \cos\alpha\cos\beta\sin\gamma-\sin\alpha\cos\gamma & \sin\beta\cos\gamma\\ \sin\alpha\sin\beta & -\cos\alpha\sin\beta & \cos\beta\end{bmatrix} \tag{4.38}$$

From this we can see that, given a direction cosine matrix [**Q**], the angles of the classical Euler sequence may be found as follows:

$$\tan\alpha = \frac{Q_{31}}{-Q_{32}} \qquad \cos\beta = Q_{33} \qquad \tan\gamma = \frac{Q_{13}}{Q_{23}} \qquad \text{(Classical Euler angle sequence)} \tag{4.39}$$

We see that $\beta = \cos^{-1}Q_{33}$. There is no quadrant uncertainty because the principal values of the arc-cosine function coincide with the range of the angle β given in Eq. (4.37) (0° to 180°). Finding α and γ involves computing the inverse tangent (arctan), whose principal values lie in the range −90° to +90°, whereas the range of both α and γ is 0° to 360°. Placing $\tan^{-1}(y/x)$ in the correct quadrant is accomplished by taking into consideration the signs of x and y. The MATLAB function *atan2d_0_360.m* in Appendix D.19 does just that.

ALGORITHM 4.3

Given the direction cosine matrix

$$[\mathbf{Q}] = \begin{bmatrix} Q_{11} & Q_{12} & Q_{13} \\ Q_{21} & Q_{22} & Q_{23} \\ Q_{31} & Q_{32} & Q_{33} \end{bmatrix}$$

find the angles $\alpha\beta\gamma$ of the classical Euler rotation sequence. This algorithm is implemented by the MATLAB function *dcm_to_euler.m* in Appendix D.20.

1. $\alpha = \tan^{-1}(-Q_{31}/Q_{32}) \quad (0 \le \alpha < 360°)$
2. $\beta = \cos^{-1}Q_{33} \quad (0 \le \beta \le 180°)$
3. $\gamma = \tan^{-1}(Q_{13}/Q_{23}) \quad (0 \le \gamma < 360°)$

EXAMPLE 4.5

If the direction cosine matrix for the transformation from xyz to $x'y'z'$ is

$$[\mathbf{Q}] = \begin{bmatrix} 0.64050 & 0.75309 & -0.15038 \\ 0.76737 & -0.63530 & 0.086823 \\ -0.30152 & -0.17101 & -0.98481 \end{bmatrix}$$

find the angles α, β, and γ of the classical Euler sequence.

Solution

We should first verify that the matrix is indeed orthogonal by checking Eq. (4.25) (or Eq. 4.26).

$$[\mathbf{Q}]^T[\mathbf{Q}] = \begin{bmatrix} 1 & -3.5042(10^{-7}) & 4.2773(10^{-7}) \\ -3.5042(10^{-7}) & 1 & 3.4866(10^{-6}) \\ 4.2773(10^{-7}) & 3.4866(10^{-6}) & 1 \end{bmatrix}$$

The off-diagonal elements should all be zero, but they are not due to truncation of the numbers in the original matrix. Since their magnitudes are very much smaller than unity, we may deem [**Q**] as very close to being orthogonal.

Use Algorithm 4.3.

Step 1:

$$\alpha = \tan^{-1}\left(\frac{Q_{31}}{-Q_{32}}\right) = \tan^{-1}\left(\frac{-0.030152}{-[-0.17101]}\right)$$

Since the numerator is negative and the denominator is positive, α must lie in the fourth quadrant. Thus

$$\tan^{-1}\left(\frac{-0.030152}{-[-0.17101]}\right) = \tan^{-1}(-0.17632) = -10^\circ \Rightarrow \boxed{\alpha = 350^\circ}$$

Step 2:

$$\beta = \cos^{-1}Q_{33} = \cos^{-1}(-0.98481) = \boxed{170.0^\circ}$$

Step 3:

$$\gamma = \tan^{-1}\frac{Q_{13}}{Q_{23}} = \tan^{-1}\left(\frac{-0.15038}{0.086823}\right)$$

The numerator is negative and the denominator is positive, so γ lies in the fourth quadrant,

$$\tan^{-1}\left(\frac{-0.15038}{0.086824}\right) = \tan^{-1}(-0.17320) = -60^\circ \Rightarrow \boxed{\gamma = 300^\circ}$$

Another commonly used set of Euler angles for rotating xyz into alignment with $x'y'z'$ is the asymmetric "yaw, pitch, and roll" sequence underlined in Eq. 4.36,

$$[\mathbf{Q}] = [\mathbf{R}_1(\gamma)][\mathbf{R}_2(\beta)][\mathbf{R}_3(\alpha)] \quad (0 \le \alpha < 360^\circ \quad -90^\circ < \beta < 90^\circ \quad 0 \le \gamma < 360^\circ) \tag{4.40}$$

It is illustrated in Fig. 4.15.

The first rotation $[\mathbf{R}_3(\alpha)]$ is about the z axis through the angle α. It carries the y axis into the y_1 axis normal to the plane of z and x', while rotating the x axis into x_1. This rotation appears as shown in the insert at the top right of Fig. 4.15. The second rotation $[\mathbf{R}_2(\beta)]$, shown in the auxiliary view at the bottom right of the figure, is a pitch around y_1 through the angle β. This carries the x_1 axis into x_2, lined up with the target x' direction, and rotates the original z axis (now z_1) into z_2. The final rotation $[\mathbf{R}_1(\gamma)]$ is a roll through the angle γ around the x_2 axis so as to carry y_2 (originally y_1) and z_2 into alignment with the target y' and z' axes.

Substituting Eqs. (4.32)–(4.34) into Eq. (4.40) yields the direction cosine matrix for the yaw, pitch, and roll sequence,

$$[\mathbf{Q}] = \begin{bmatrix} \cos\alpha\cos\beta & \sin\alpha\cos\beta & -\sin\beta \\ \cos\alpha\sin\beta\sin\gamma - \sin\alpha\cos\beta & \sin\alpha\sin\beta\sin\gamma + \cos\alpha\cos\gamma & \cos\beta\sin\gamma \\ \cos\alpha\sin\beta\cos\gamma + \sin\alpha\sin\gamma & \sin\alpha\sin\beta\cos\gamma - \cos\alpha\sin\gamma & \cos\beta\cos\gamma \end{bmatrix} \tag{4.41}$$

From this it is apparent that

$$\tan\alpha = \frac{Q_{12}}{Q_{11}} \qquad \sin\beta = -Q_{13} \qquad \tan\gamma = \frac{Q_{23}}{Q_{33}} \qquad \text{(Yaw, pitch and roll sequence)} \tag{4.42}$$

For β we simply compute $\sin^{-1}(-Q_{13})$. There is no quadrant uncertainty because the principal values of the arcsine function coincide with the range of the pitch angle ($-90^\circ < \beta < 90^\circ$). Finding α and γ involves computing the inverse tangent, so we must once again be careful to place the results of these calculations in the range 0°–360°. As pointed out previously, the MATLAB function *atan2d_0_360.m* in Appendix D.19 takes care of that.

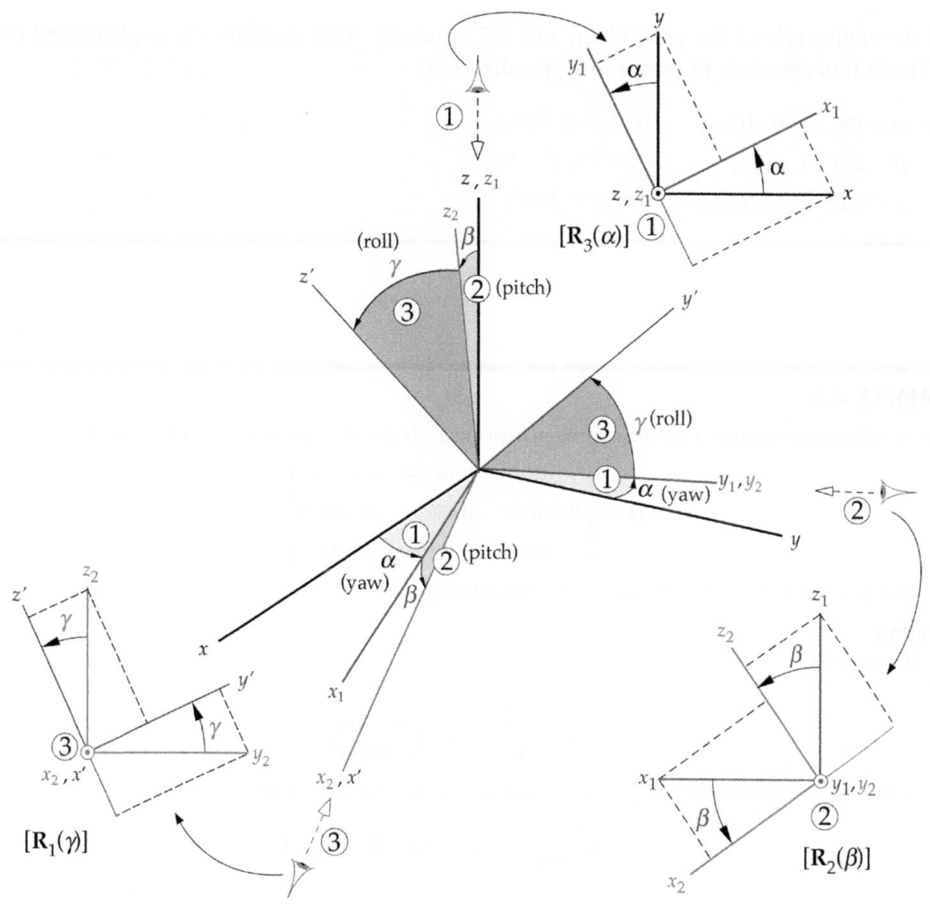

FIG. 4.15

Yaw, pitch, and roll sequence transforming *xyz* into *x′y′z′*.

ALGORITHM 4.4

Given the direction cosine matrix

$$[\mathbf{Q}] = \begin{bmatrix} Q_{11} & Q_{12} & Q_{13} \\ Q_{21} & Q_{22} & Q_{23} \\ Q_{31} & Q_{32} & Q_{33} \end{bmatrix}$$

Find the angles $\alpha\beta\gamma$ of the yaw, pitch, and roll sequence. This algorithm is implemented by the MATLAB function *dcm_to_ypr.m* in Appendix D.21.

1. $\alpha = \tan^{-1}(Q_{12}/Q_{11}) \quad (0 \le \alpha < 360°)$
2. $\beta = \sin^{-1}(-Q_{13}) \quad (-90° < \beta < 90°)$
3. $\gamma = \tan^{-1}(Q_{23}/Q_{33}) \quad (0 \le \gamma < 360°)$

EXAMPLE 4.6

If the direction cosine matrix for the transformation from xyz to $x'y'z'$ is the same as it was in Example 4.5,

$$[\mathbf{Q}] = \begin{bmatrix} 0.64050 & 0.75309 & -0.15038 \\ 0.76737 & -0.63530 & 0.086823 \\ -0.30152 & -0.17101 & -0.98481 \end{bmatrix}$$

find the angles α, β, and γ of the yaw, pitch, and roll sequence.

Solution

Use Algorithm 4.4.

Step 1:

$$\alpha = \tan^{-1}\frac{Q_{12}}{Q_{11}} = \tan^{-1}\left(\frac{0.75309}{0.64050}\right)$$

Since both the numerator and the denominator are positive, α must lie in the first quadrant. Thus,

$$\tan^{-1}\left(\frac{0.75309}{0.64050}\right) = \tan^{-1}1.1758 = \boxed{49.62°}$$

Step 2:

$$\beta = \sin^{-1}(-Q_{13}) = \sin^{-1}[-(-0.15038)] = \sin^{-1}(0.15038) = \boxed{8.649°}$$

Step 3:

$$\gamma = \tan^{-1}\frac{Q_{23}}{Q_{33}} = \tan^{-1}\left(\frac{0.086823}{-0.98481}\right)$$

The numerator is positive and the denominator is negative, so γ lies in the second quadrant,

$$\tan^{-1}\left(\frac{0.086823}{-0.98481}\right) = \tan^{-1}(-0.088162) = -5.0383° \Rightarrow \boxed{\gamma = 174.96°}$$

4.6 TRANSFORMATION BETWEEN GEOCENTRIC EQUATORIAL AND PERIFOCAL FRAMES

The perifocal frame of reference for a given orbit was introduced in Section 2.10. Fig. 4.16 illustrates the relationship between the perifocal and geocentric equatorial frames. Since the orbit lies in the $\bar{x}\bar{y}$ plane, the components of the state vector of a body relative to its perifocal reference are, according to Eqs. (2.119) and (2.125),

$$\mathbf{r} = \bar{x}\hat{\mathbf{p}} + \bar{y}\hat{\mathbf{q}} = \frac{h^2}{\mu}\frac{1}{1+e\cos\theta}(\cos\theta\hat{\mathbf{p}} + \sin\theta\hat{\mathbf{q}}) \tag{4.43}$$

$$\mathbf{v} = \dot{\bar{x}}\hat{\mathbf{p}} + \dot{\bar{y}}\hat{\mathbf{q}} = \frac{\mu}{h}[-\sin\theta\hat{\mathbf{p}} + (e+\cos\theta)\hat{\mathbf{q}}] \tag{4.44}$$

In matrix notation, these may be written

$$\{\mathbf{r}\}_{\bar{x}} = \frac{h^2}{\mu}\frac{1}{1+e\cos\theta}\begin{Bmatrix}\cos\theta\\ \sin\theta\\ 0\end{Bmatrix} \tag{4.45}$$

$$\{\mathbf{v}\}_{\bar{x}} = \frac{\mu}{h}\begin{Bmatrix}-\sin\theta\\ e+\cos\theta\\ 0\end{Bmatrix} \tag{4.46}$$

The subscript $\bar{x}$ is shorthand for "the $\bar{x}\bar{y}\bar{z}$ coordinate system" and is used to indicate that the components of these vectors are given in the perifocal frame, as opposed to, say, the geocentric equatorial frame (Eqs. 4.2 and 4.3).

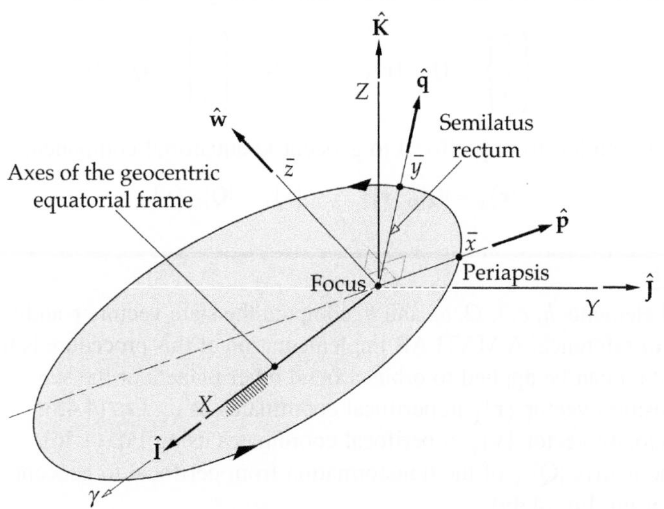

FIG. 4.16

Perifocal ($\bar{x}\bar{y}\bar{z}$) and geocenric equatorial (*XYZ*) frames.

The transformation from the geocentric equatorial frame into the perifocal frame may be accomplished by the classical Euler angle sequence $[\mathbf{R}_3(\gamma)][\mathbf{R}_1(\beta)][\mathbf{R}_3(\alpha)]$ in Eq. (4.37) (see Fig. 4.7). In this case, the first rotation angle is Ω, the right ascension of the ascending node. The second rotation is i, the orbital inclination angle, and the third rotation angle is ω, the argument of perigee. Ω is measured around the Z axis of the geocentric equatorial frame, i is measured around the node line, and ω is measured around the $\bar{z}$ axis of the perifocal frame. Therefore, the direct cosine matrix $[\mathbf{Q}]_{X\bar{x}}$ of the transformation from XYZ to $\bar{x}\,\bar{y}\,\bar{z}$ is

$$[\mathbf{Q}]_{X\bar{x}} = [\mathbf{R}_3(\omega)][\mathbf{R}_1(i)][\mathbf{R}_3(\Omega)] \tag{4.47}$$

From Eq. (4.38) we get

$$[\mathbf{Q}]_{X\bar{x}} = \begin{bmatrix} -\sin\Omega\cos i\sin\omega + \cos\Omega\cos\omega & \cos\Omega\cos i\sin\omega + \sin\Omega\cos\omega & \sin i\sin\omega \\ -\sin\Omega\cos i\cos\omega - \cos\Omega\sin\omega & \cos\Omega\cos i\cos\omega - \sin\Omega\sin\omega & \sin i\cos\omega \\ \sin\Omega\sin i & -\cos\Omega\sin i & \cos i \end{bmatrix} \tag{4.48}$$

Remember that this is an orthogonal matrix, which means that the inverse transformation $[\mathbf{Q}]_{\bar{x}X}$ from $\bar{x}\,\bar{y}\,\bar{z}$ to XYZ, is given by $[\mathbf{Q}]_{\bar{x}X} = \left([\mathbf{Q}]_{X\bar{x}}\right)^T$, or

$$[\mathbf{Q}]_{\bar{x}X} = \begin{bmatrix} -\sin\Omega\cos i\sin\omega + \cos\Omega\cos\omega & -\sin\Omega\cos i\cos\omega - \cos\Omega\sin\omega & \sin\Omega\sin i \\ \cos\Omega\cos i\sin\omega + \sin\Omega co\omega & \cos\cos i\cos\omega - \sin\Omega\sin\omega & -\cos\Omega\sin i \\ \sin i\sin\omega & \sin i\cos\omega & \cos i \end{bmatrix} \tag{4.49}$$

If the components of the state vector are given in the geocentric equatorial frame,

$$\{\mathbf{r}\}_X = \begin{Bmatrix} X \\ Y \\ Z \end{Bmatrix} \qquad \{\mathbf{v}\}_X = \begin{Bmatrix} v_X \\ v_Y \\ v_Z \end{Bmatrix}$$

then the components in the perifocal frame are found by carrying out the matrix multiplications

$$\{\mathbf{r}\}_{\bar{x}} = \begin{Bmatrix} \bar{x} \\ \bar{y} \\ 0 \end{Bmatrix} = [\mathbf{Q}]_{X\bar{x}}\{\mathbf{r}\}_X \qquad \{\mathbf{v}\}_{\bar{x}} = \begin{Bmatrix} \dot{\bar{x}} \\ \dot{\bar{y}} \\ 0 \end{Bmatrix} = [\mathbf{Q}]_{X\bar{x}}\{\mathbf{v}\}_X \tag{4.50}$$

Likewise, the transformation from perifocal to geocentric equatorial components is

$$\{\mathbf{r}\}_X = [\mathbf{Q}]_{\bar{x}X}\{\mathbf{r}\}_{\bar{x}} \qquad \{\mathbf{v}\}_X = [\mathbf{Q}]_{\bar{x}X}\{\mathbf{v}\}_{\bar{x}} \tag{4.51}$$

ALGORITHM 4.5

Given the orbital elements h, e, i, Ω, ω, and θ, compute the state vectors **r** and **v** in the geocentric equatorial frame of reference. A MATLAB implementation of this procedure is listed in Appendix D.22. This algorithm can be applied to orbits around other planets or the sun.

1. Calculate position vector $\{\mathbf{r}\}_{\bar{x}}$ in perifocal coordinates using Eq. (4.45).
2. Calculate velocity vector $\{\mathbf{v}\}_{\bar{x}}$ in perifocal coordinates using Eq. (4.46).
3. Calculate the matrix $[\mathbf{Q}]_{\bar{x}X}$ of the transformation from perifocal to geocentric equatorial coordinates using Eq. (4.49).
4. Transform $\{\mathbf{r}\}_{\bar{x}}$ and $\{\mathbf{v}\}_{\bar{x}}$ into the geocentric frame by means of Eq. (4.51).

EXAMPLE 4.7

For a given earth orbit, the elements are $h = 80,000\,\mathrm{km^2/s}$, $e = 1.4$, $i = 30°$, $\Omega = 40°$, $\omega = 60°$, and $\theta = 30°$. Using Algorithm 4.5, find the state vectors **r** and **v** in the geocentric equatorial frame.

Solution

Step 1:

$$\{\mathbf{r}\}_{\bar{x}} = \frac{h^2}{\mu}\frac{1}{1+e\cos\theta}\begin{Bmatrix}\cos\theta\\ \sin\theta\\ 0\end{Bmatrix} = \frac{80,000^2}{398,600}\frac{1}{1+1.4\cos 30°}\begin{Bmatrix}\cos 30°\\ \sin 30°\\ 0\end{Bmatrix} = \begin{Bmatrix}6285.0\\ 3628.6\\ 0\end{Bmatrix}\ (\mathrm{km})$$

Step 2:

$$\{\mathbf{v}\}_{\bar{x}} = \frac{\mu}{h}\begin{Bmatrix}-\sin\theta\\ e+\cos\theta\\ 0\end{Bmatrix} = \frac{398,600}{80,000}\begin{Bmatrix}-\sin 30°\\ 1.4+\cos 30°\\ 0\end{Bmatrix} = \begin{Bmatrix}-2.4913\\ 11.290\\ 0\end{Bmatrix}\ (\mathrm{km/s})$$

Step 3:

$$[\mathbf{Q}]_{X\bar{x}} = \begin{bmatrix}\cos\omega & \sin\omega & 0\\ -\sin\omega & \cos\omega & 0\\ 0 & 0 & 1\end{bmatrix}\begin{bmatrix}1 & 0 & 0\\ 0 & \cos i & \sin i\\ 0 & -\sin i & \cos i\end{bmatrix}\begin{bmatrix}\cos\Omega & \sin\Omega & 0\\ -\sin\Omega & \cos\Omega & 0\\ 0 & 0 & 1\end{bmatrix}$$

$$= \begin{bmatrix}\cos 60° & \sin 60° & 0\\ -\sin 60° & \cos 60° & 0\\ 0 & 0 & 1\end{bmatrix}\begin{bmatrix}1 & 0 & 0\\ 0 & \cos 30° & \sin 30°\\ 0 & -\sin 30° & \cos 30°\end{bmatrix}\begin{bmatrix}\cos 40° & \sin 40° & 0\\ -\sin 40° & \cos 40° & 0\\ 0 & 0 & 1\end{bmatrix}$$

$$= \begin{bmatrix}-0.099068 & 0.89593 & 0.43301\\ -0.94175 & -0.22496 & 0.25000\\ 0.32139 & -0.38302 & 0.86603\end{bmatrix}$$

This is the direction cosine matrix for $XYZ \rightarrow \bar{x}\,\bar{y}\,\bar{z}$. The transformation matrix for $\bar{x}\,\bar{y}\,\bar{z} \rightarrow XYZ$ is the transpose,

$$[\mathbf{Q}]_{\bar{x}X} = \begin{bmatrix}-0.099068 & -0.94175 & 0.32139\\ 0.89593 & -0.22496 & -0.38302\\ 0.43301 & 0.25000 & 0.86603\end{bmatrix}$$

Step 4:
The geocentric equatorial position vector is

$$\{\mathbf{r}\}_X = [\mathbf{Q}]_{\bar{x}X}\{\mathbf{r}\}_{\bar{x}}$$

$$= \begin{bmatrix}-0.099068 & -0.94175 & 0.32139\\ 0.89593 & -0.22496 & -0.38302\\ 0.43301 & 0.25 & 0.86603\end{bmatrix}\begin{Bmatrix}6285.0\\ 3628.6\\ 0\end{Bmatrix} = \boxed{\begin{Bmatrix}-4040\\ 4815\\ 3629\end{Bmatrix}\ (\mathrm{km})} \tag{a}$$

whereas the geocentric equatorial velocity vector is

$$\{\mathbf{v}\}_X = [\mathbf{Q}]_{\bar{x}X}\{\mathbf{v}\}_{\bar{x}}$$

$$= \begin{bmatrix}-0.099068 & -0.94175 & 0.32139\\ 0.89593 & -0.22496 & -0.38302\\ 0.43301 & 0.25 & 0.86603\end{bmatrix}\begin{Bmatrix}-2.4913\\ 11.290\\ 0\end{Bmatrix} = \boxed{\begin{Bmatrix}-10.39\\ -4.772\\ 1.744\end{Bmatrix}\ (\mathrm{km/s})}$$

The state vectors **r** and **v** are shown in Fig. 4.17. By holding all the orbital parameters except the true anomaly fixed and allowing θ to take on a range of values, we generate a sequence of position vectors $\{\mathbf{r}\}_{\bar{x}}$ from Eq. (4.45). Each of these is

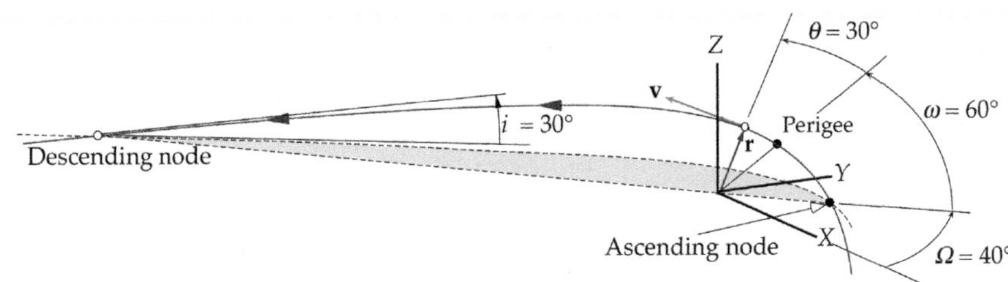

FIG. 4.17

A portion of the hyperbolic trajectory of Example 4.7.

projected into the geocentric equatorial frame as in Eq. (a), using repeatedly the same transformation matrix $[\mathbf{Q}]_{\bar{x}X}$. By connecting the end points of all the position vectors $\{\mathbf{r}\}_X$, we trace out the trajectory illustrated in Fig. 4.17.

4.7 EFFECTS OF THE EARTH'S OBLATENESS

The earth, like all planets with comparable or higher rotational rates, bulges out at the equator because of centrifugal force. The earth's equatorial radius is 21 km (13 miles) larger than the polar radius. This flattening at the poles is called oblateness, which is defined as follows:

$$\text{Oblateness} = \frac{\text{Equatorial radius} - \text{Polar radius}}{\text{Equatorial radius}}$$

The earth is an oblate spheroid, lacking the perfect symmetry of a sphere (a basketball can be made an oblate spheroid by sitting on it). This lack of symmetry means that the force of gravity on an orbiting body is not directed toward the center of the earth. Although the gravitational field of a perfectly spherical planet depends only on the distance from its center, oblateness causes a variation also with latitude (i.e., the angular distance from the equator (or pole)). This is called a zonal variation. The dimensionless parameter that quantifies the major effects of oblateness on orbits is J_2, the second zonal harmonic. J_2 is not a universal constant. Each planet has its own value, as illustrated in Table 4.3, which lists variations of J_2 as well as oblateness.

Table 4.3 Oblateness and second zonal harmonics

Planet	Oblateness	J_2
Mercury	0.000	$60(10^{-6})$
Venus	0.000	$4.458(10^{-6})$
Earth	0.003353	$1.08263(10^{-3})$
Mars	0.00648	$1.96045(10^{-3})$
Jupiter	0.06487	$14.736(10^{-3})$
Saturn	0.09796	$16.298(10^{-3})$
Uranus	0.02293	$3.34343(10^{-3})$
Neptune	0.01708	$3.411(10^{-3})$
(Moon)	0.0012	$202.7(10^{-6})$

Oblateness causes the right ascension Ω and the argument of periapsis ω to vary significantly with time. In Chapter 10, we will show that the average rates of change of these two angles are

$$\dot{\Omega} = -\left[\frac{3}{2}\frac{\sqrt{\mu}J_2R^2}{(1-e^2)^2a^{7/2}}\right]\cos i \tag{4.52}$$

and

$$\dot{\omega} = -\left[\frac{3}{2}\frac{\sqrt{\mu}J_2R^2}{(1-e^2)^2a^{7/2}}\right]\left(\frac{5}{2}\sin^2 i - 2\right) \tag{4.53}$$

where R and μ are the radius and gravitational parameter of the planet, respectively; a and e are the semimajor axis and eccentricity of the orbit, respectively; and i is the orbit's inclination.

In Eq. (4.52), observe that if $0 \le i < 90°$, then $\dot{\Omega} < 0$. That is, for prograde orbits, the node line drifts westward. Since the right ascension of the node continuously decreases, this phenomenon is called regression of the nodes. If $90° < i \le 180°$, we see that $\dot{\Omega} > 0$. The node line of retrograde orbits therefore advances eastward. For polar orbits ($i = 90°$), the node line is stationary.

In Eq. (4.53), we see that if $0° \le i < 63.4°$ or $116.6° < i \le 180°$, then $\dot{\omega}$ is positive, which means the perigee advances in the direction of the motion of the satellite (hence, the name "advance of perigee" for this phenomenon). If $63.4° < i \le 116.6°$, the perigee regresses, moving opposite to the direction of motion. $i = 63.4°$ and $i = 116.6°$ are the critical inclinations at which the apse line does not move.

Observe that the coefficient of the trigonometric terms in Eqs. (4.52) and (4.53) are identical, so that

$$\dot{\omega} = \dot{\Omega}\frac{(5/2)\sin^2 i - 2}{\cos i} \tag{4.54}$$

Fig. 4.18 is a plot of Eqs. (4.52) and (4.53) for several circular low earth orbits. The effect of oblateness on both $\dot{\Omega}$ and $\dot{\omega}$ is greatest at low inclinations, for which the orbit is near the equatorial bulge for longer portions of each revolution. The effect decreases with increasing semimajor axis because the satellite

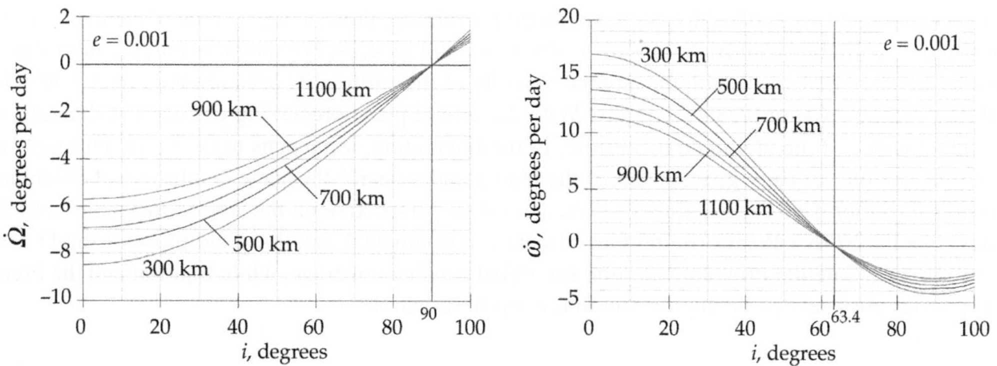

FIG. 4.18

Regression of the node and advance of perigee for nearly circular orbits of altitudes 300–1100 km.

becomes farther from the bulge and its gravitational influence. Obviously, $\dot{\Omega} = \dot{\omega} = 0$ if $J_2 = 0$ (no equatorial bulge).

It turns out (Chapter 10) that the J_2 effect produces zero time-averaged variations of the inclination, eccentricity, angular momentum, and semimajor axis.

EXAMPLE 4.8

A spacecraft is in a 280 km by 400 km orbit with an inclination of 51.43°. Find the rates of node regression and perigee advance.

Solution

The perigee and apogee radii are

$$r_p = 6378 + 280 = 6658\,\text{km} \qquad r_a = 6378 + 400 = 6778\,\text{km}$$

Therefore, the eccentricity and semimajor axis are

$$e = \frac{r_a - r_p}{r_a + r_p} = 0.008931 \qquad a = \frac{r_a + r_p}{2} = 6718\,\text{km}$$

From Eq. (4.52), we obtain the rate of node line regression.

$$\dot{\Omega} = -\left[\frac{3}{2}\frac{\sqrt{398{,}600} \times 0.0010826 \times 6378^2}{\left(1 - 0.0089312^2\right)^2 \times 6718^{7/2}}\right] \cos 51.43° = -1.0465\left(10^{-6}\right)\text{rad}/s$$

or

$$\boxed{\dot{\Omega} = 5.181\,\text{degrees per day to the west}}$$

From Eq. 4.54,

$$\dot{\omega} = -1.0465\left(10^{-6}\right) \cdot \frac{(5/2)\sin^2 51.43° - 2}{\cos 51.43°} = +7.9193\left(10^{-7}\right)\text{rad/s}$$

or

$$\boxed{\dot{\omega} = 3.920\,\text{degrees per day in the flight direction}}$$

The effect of orbit inclination on node regression and advance of perigee is taken advantage of for two very important types of orbits. Sun-synchronous orbits are those whose orbital plane makes a constant angle α with the radial from the sun, as illustrated in Fig. 4.19. For that to occur, the orbital plane must rotate in inertial space with the angular velocity of the earth in its orbit around the sun, which is 360° per 365.26 days, or 0.9856° per day. With the orbital plane precessing eastward at this rate, the ascending node will lie at a fixed local time. In the illustration, it happens to be 3 p.m. During every orbit, the satellite sees any given swath of the planet under nearly the same conditions of daylight or darkness day after day. The satellite also has a constant perspective on the sun. Sun-synchronous satellites, like the NOAA Polar-orbiting Operational Environmental Satellites and those of the Defense Meteorological Satellite Program, are used for global weather coverage, while Landsat and the French SPOT series are intended for high-resolution earth observation.

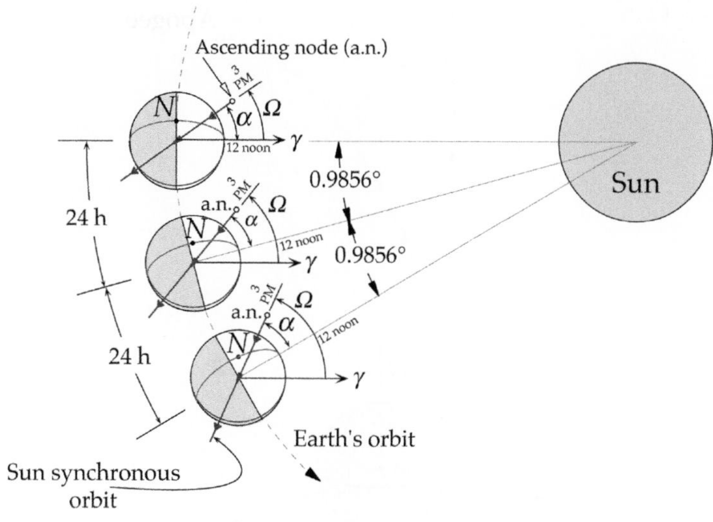

FIG. 4.19

Sun-synchronous orbit.

EXAMPLE 4.9

A satellite is to be launched into a sun-synchronous circular orbit with a period of 100 min. Determine the required altitude and inclination of its orbit.

Solution

We find the altitude z from the period relation for a circular orbit, Eq. (2.64):

$$T=\frac{2\pi}{\sqrt{\mu}}(R_E+z)^{3/2} \Rightarrow 100\times 60=\frac{2\pi}{\sqrt{398{,}600}}(6378+z)^{3/2} \Rightarrow \boxed{z=758.63\,\text{km}}$$

For a sun-synchronous orbit, the ascending node must advance at the rate

$$\dot{\Omega}=\frac{2\pi\,\text{rad}}{365.26\times 24\times 3600\text{s}}=1.991\left(10^{-7}\right)\text{rad/s}$$

Substituting this and the altitude into Eq. (4.47), we obtain,

$$1.991\times 10^{-7}=-\left[\frac{3}{2}\frac{\sqrt{398{,}600}\times 0.00108263\times 6378^2}{\left(1-0^2\right)^2(6378+758.63)^{7/2}}\right]\cos i \Rightarrow \cos i=-0.14658$$

Thus, the inclination of the orbit is

$$\boxed{i=98.43^\circ}$$

This illustrates the fact that sun-synchronous orbits are very nearly polar orbits ($i=90^\circ$).

If a satellite is launched into an orbit with an inclination of 63.4° (prograde) or 116.6° (retrograde), then Eq. (4.53) shows that the apse line will remain stationary. The Russian space program made this a

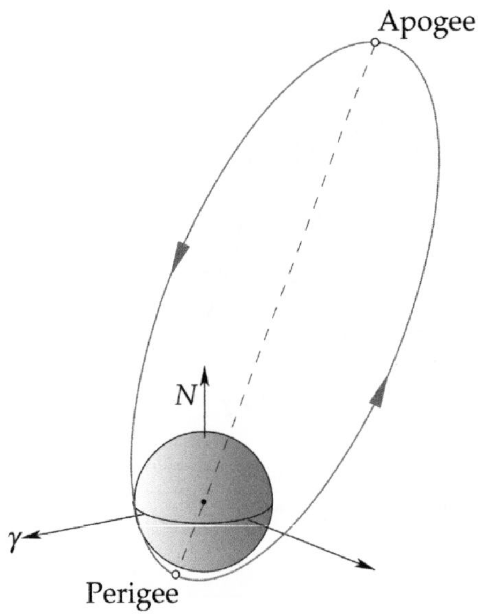

FIG. 4.20

A typical Molniya orbit (to scale).

key element in the design of the system of Molniya ("lightning") communications satellites. All the Russian launch sites are above 45° latitude, with the northernmost, Plesetsk, being located at 62.8°N. As we shall see in Chapter 6, launching a satellite into a geostationary orbit would involve a costly plane change maneuver. Furthermore, recall from Example 2.4 that a geostationary satellite cannot view effectively the far northern latitudes into which Russian territory extends.

The Molniya telecommunications satellites are launched from Plesetsk into 63° inclination orbits having a period of 12 h. From Eq. (2.83), we conclude that the major axis of these orbits is 53,000 km long. Perigee (typically 500 km altitude) lies in the southern hemisphere, while apogee is at an altitude of 40,000 km (25,000 miles) above the northern latitudes, farther out than the geostationary satellites. Fig. 4.20 illustrates a typical Molniya orbit. A Molniya "constellation" consists of eight satellites in planes separated by 45°. Each satellite is above 30°N latitude for over 8 h, coasting toward and away from apogee.

EXAMPLE 4.10

Determine the perigee and apogee for an earth satellite whose orbit satisfies all the following conditions: it is sun synchronous, its argument of perigee is constant, and its period is 3 h.

Solution

The period determines the semimajor axis,

$$T = \frac{2\pi}{\sqrt{\mu}} a^{3/2} \Rightarrow 3 \cdot 3600 = \frac{2\pi}{\sqrt{398,600}} a^{3/2} \Rightarrow a = 10,560\text{km}$$

For the apse line to be stationary, we know from Eq. (4.53) that $i = 64.435°$ or $i = 116.57°$. However, an inclination of less than 90° causes a westward regression of the node, whereas a sun-synchronous orbit requires an eastward advance, which $i = 116.57°$ provides. Substituting this, the semimajor axis and the $\dot{\Omega}$ in radians per second for a sun-synchronous orbit (cf. Example 4.9) into Eq. (4.52), we get

$$1.991 \times 10^{-7} = -\frac{3}{2}\frac{\sqrt{398{,}600} \times 0.0010826 \times 6378^2}{(1-e^2)^2 \times 10{,}560^{7/2}}\cos 116.57° \Rightarrow e = 0.3466$$

Now we can find the angular momentum from the period expression (Eq. 2.82)

$$T = \frac{2\pi}{\mu^2}\left(\frac{h}{\sqrt{1-e^2}}\right)^3 \Rightarrow 3 \cdot 3600 = \frac{2\pi}{398{,}600^2}\left(\frac{h}{\sqrt{1-0.34655^2}}\right)^3 \Rightarrow h = 60{,}850\text{km}^2/\text{s}$$

Finally, to obtain the perigee and apogee radii, we use the orbit formula:

$$z_p + 6378 = \frac{h^2}{\mu}\frac{1}{1+e} = \frac{60{,}860^2}{398{,}600}\frac{1}{1+0.34655} \Rightarrow z_p = \boxed{522.6\text{km}}$$

$$z_a + 6378 = \frac{h^2}{\mu}\frac{1}{1+e} \Rightarrow z_a = \boxed{7842\text{km}}$$

EXAMPLE 4.11

Given the following state vector of a satellite in geocentric equatorial coordinates,

$$\mathbf{r} = -3670\hat{\mathbf{I}} - 3870\hat{\mathbf{J}} + 4400\hat{\mathbf{K}}\,\text{km}$$
$$\mathbf{v} = 4.7\hat{\mathbf{I}} - 7.4\hat{\mathbf{J}} + 1\hat{\mathbf{K}}\,\text{km/s}$$

find the state vector after 4 days (96 h) of coasting flight, assuming that there are no perturbations other than the influence of the earth's oblateness on Ω and ω.

Solution

A time interval of 4 days is long enough for us to take into consideration not only the change in true anomaly, but also the regression of the ascending node and the advance of perigee. First, we must determine the orbital elements at the initial time using Algorithm 4.2, which yields

$$h = 58{,}930\text{km}^2/\text{s}$$
$$i = 39.687°$$
$$e = 0.42607\,(\text{The orbit is an ellipse})$$
$$\Omega_0 = 130.32°$$
$$\omega_0 = 42.373°$$
$$\theta_0 = 52.404°$$

We use Eq. (2.71) to determine the semimajor axis,

$$a = \frac{h^2}{\mu}\frac{1}{1-e^2} = \frac{58{,}930^2}{398{,}600}\frac{1}{1-0.4261^2} = 10{,}640\text{km}$$

so that, according to Eq. (2.83), the period is

$$T = \frac{2\pi}{\sqrt{\mu}}a^{3/2} = 10{,}928\,\text{s}$$

From this we obtain the mean motion

$$n=\frac{2\pi}{T}=0.00057495\,\text{rad/s}$$

The initial value E_0 of eccentric anomaly is found from the true anomaly θ_0 using Eq. (3.13a),

$$\tan\frac{E_0}{2}=\sqrt{\frac{1-e}{1+e}}\tan\frac{\theta_0}{2}=\sqrt{\frac{1-0.42607}{1+0.42607}}\tan\frac{52.404^\circ}{2}\quad\Rightarrow\quad E_0=0.60520\,\text{rad}$$

With E_0, we use Kepler's equation to calculate the time t_0 since perigee at the initial epoch,

$$nt_0=E_0-e\sin E_0\quad\Rightarrow\quad 0.00057495t_0=0.60520-0.42607\sin 0.60520\quad\Rightarrow\quad t_0=631.00\,\text{s}$$

Now we advance the time to t_f, that of the final epoch, given as 96 h later. That is, Δt = 345,600 s, so that

$$t_f=t_0+\Delta t=631.00+345{,}600=346{,}230\,\text{s}$$

The number of periods n_P since passing perigee in the first orbit is

$$n_P=\frac{t_f}{T}=\frac{346{,}230}{10{,}928}=31.682$$

From this we see that the final epoch occurs in the 32nd orbit, whereas t_0 was in orbit 1. Time since passing perigee in the 32nd orbit, which we will denote t_{32}, is

$$t_{32}=(31.682-31)T\quad\Rightarrow\quad t_{32}=7455.7\,\text{s}$$

The mean anomaly corresponding to that time in the 32nd orbit is

$$M_{32}=nt_{32}=0.00057495\times 7455.7=4.2866\,\text{rad}$$

Kepler's equation yields the eccentric anomaly:

$$\begin{aligned}
&E_{32}-e\sin E_{32}=M_{32}\\
&E_{32}-0.42607\sin E_{32}=4.2866\\
&E_{32}=3.9721\,\text{rad}\,(\text{Algorithm 3.1})
\end{aligned}$$

The true anomaly follows in the usual way,

$$\tan\frac{\theta_{32}}{2}=\sqrt{\frac{1+e}{1-e}}\tan\frac{E_{32}}{2}\quad\Rightarrow\quad\theta_{32}=211.25^\circ$$

At this point, we use the newly found true anomaly to calculate the state vector of the satellite in perifocal coordinates. Thus, from Eq. (4.43)

$$\mathbf{r}=r\cos\theta_{32}\hat{\mathbf{p}}+r\sin\theta_{32}\hat{\mathbf{q}}=-11{,}714\hat{\mathbf{p}}-7108.8\hat{\mathbf{q}}\,(\text{km})$$

or, in matrix notation,

$$\{\mathbf{r}\}_{\bar{x}}=\begin{Bmatrix}-11{,}714\\-7108.8\\0\end{Bmatrix}(\text{km})$$

Likewise, from Eq. (4.44),

$$\mathbf{v}=\frac{\mu}{h}\sin\theta_{32}\hat{\mathbf{p}}+\frac{\mu}{h}(e+\cos\theta_{32})\hat{\mathbf{q}}=3.5093\hat{\mathbf{p}}-2.9007\hat{\mathbf{q}}\ (\text{km/s})$$

or

$$\{\mathbf{v}\}_{\bar{x}}=\begin{Bmatrix}3.5093\\-2.9007\\0\end{Bmatrix}(\text{km/s})$$

Before we can project **r** and **v** into the geocentric equatorial frame, we must update the right ascension of the node and the argument of perigee. The regression rate of the ascending node is

$$\dot{\Omega} = -\left[\frac{3}{2}\frac{\sqrt{\mu}J_2R^2}{(1-e^2)^2a^{7/2}}\right]\cos i = -\frac{3}{2}\frac{\sqrt{398{,}600}\times 0.00108263\times 6378^2}{\left(1-0.42607^2\right)^2\times 10{,}644^{7/2}}\cos 39.69°$$
$$= -3.8514\times 10^{-7}\,\text{rad/s}$$

or

$$\dot{\Omega} = -2.2067\times 10^{-5}\,\text{deg/s}$$

Therefore, right ascension at epoch in the 32nd orbit is

$$\Omega_{32} = \Omega_0 + \dot{\Omega}\Delta t = 130.32 + \left(-2.2067\times 10^{-5}\right)\times 345{,}600 = 122.70°$$

Likewise, the perigee advance rate is

$$\dot{\omega} = -\left[\frac{3}{2}\frac{\sqrt{\mu}J_2R^2}{(1-e^2)^2a^{7/2}}\right]\left(\frac{5}{2}\sin^2 i - 2\right) = 4.9072\times 10^{-7}\,\text{rad/s} = 2.8116\times 10^{-5}\,\text{deg/s}$$

which means the argument of perigee at epoch in the 32nd orbit is

$$\omega_{32} = \omega_0 + \omega\Delta t = 42.373 + 2.8116\times 10^{-5}\times 345{,}600 = 52.090°$$

Substituting the updated values of Ω and ω, together with the inclination i, into Eq. (4.47) yields the updated direction cosine matrix from geocentric equatorial to the perifocal frame,

$$[\mathbf{Q}]_{X\bar{x}} = \begin{bmatrix} \cos\omega_{32} & \sin\omega_{32} & 0 \\ -\sin\omega_{32} & \cos\omega_{32} & 0 \\ 0 & 0 & 1 \end{bmatrix}\begin{bmatrix} 1 & 0 & 0 \\ 0 & \cos i & \sin i \\ 0 & -\sin i & \cos i \end{bmatrix}\begin{bmatrix} \cos\Omega_{32} & \sin\Omega_{32} & 0 \\ -\sin\Omega_{32} & \cos\Omega_{32} & 0 \\ 0 & 0 & 1 \end{bmatrix}$$
$$= \begin{bmatrix} \cos 52.09° & \sin 52.09° & 0 \\ -\sin 52.09° & \cos 52.09° & 0 \\ 0 & 0 & 1 \end{bmatrix}\cdot\begin{bmatrix} 1 & 0 & 0 \\ 0 & \cos 39.687° & \sin 39.687° \\ 0 & -\sin 39.687° & \cos 39.687° \end{bmatrix}\cdot\begin{bmatrix} \cos 122.70° & \sin 122.70° & 0 \\ -\sin 122.70° & \cos 122.70° & 0 \\ 0 & 0 & 1 \end{bmatrix}$$

or

$$[\mathbf{Q}]_{X\bar{x}} = \begin{bmatrix} -0.84285 & 0.18910 & 0.50383 \\ 0.028276 & -0.91937 & 0.39237 \\ 0.53741 & 0.34495 & 0.76955 \end{bmatrix}$$

For the inverse transformation, from perifocal to geocentric equatorial, we need the transpose of this matrix,

$$[\mathbf{Q}]_{\bar{x}X} = \begin{bmatrix} -0.84285 & 0.18910 & 0.50383 \\ 0.028276 & -0.91937 & 0.39237 \\ 0.53741 & 0.34495 & 0.76955 \end{bmatrix}^T = \begin{bmatrix} -0.84285 & 0.028276 & 0.53741 \\ 0.18910 & -0.91937 & 0.34495 \\ 0.50383 & 0.39237 & 0.76955 \end{bmatrix}$$

Thus, according to Eq. 4.51, the final state vector in the geocentric equatorial frame is

$$\{\mathbf{r}\}_X = [\mathbf{Q}]_{\bar{x}X}\{\mathbf{r}\}_{\bar{x}} = \begin{bmatrix} -0.84285 & 0.028276 & 0.53741 \\ 0.18910 & -0.91937 & 0.34495 \\ 0.50383 & 0.39237 & 0.76955 \end{bmatrix}\begin{Bmatrix} -11{,}714 \\ -7108.8 \\ 0 \end{Bmatrix} = \begin{Bmatrix} 9672 \\ 4320 \\ -8691 \end{Bmatrix}(\text{km})$$
$$\{\mathbf{v}\}_X = [\mathbf{Q}]_{\bar{x}X}\{\mathbf{v}\}_{\bar{x}} = \begin{bmatrix} -0.84285 & 0.028276 & 0.53741 \\ 0.18910 & -0.91937 & 0.34495 \\ 0.50383 & 0.39237 & 0.76955 \end{bmatrix}\begin{Bmatrix} 3.5093 \\ -2.9007 \\ 0 \end{Bmatrix} = \begin{Bmatrix} -3.040 \\ 3.330 \\ 0.6299 \end{Bmatrix}(\text{km/s})$$

or, in vector notation,

$$\boxed{\mathbf{r} = 9672\hat{\mathbf{I}} + 4320\hat{\mathbf{J}} - 8691\hat{\mathbf{K}}\,(\text{km}) \qquad \mathbf{v} = -3.040\hat{\mathbf{I}} + 3.330\hat{\mathbf{J}} + 0.6299\hat{\mathbf{K}}\,(\text{km/s})}$$

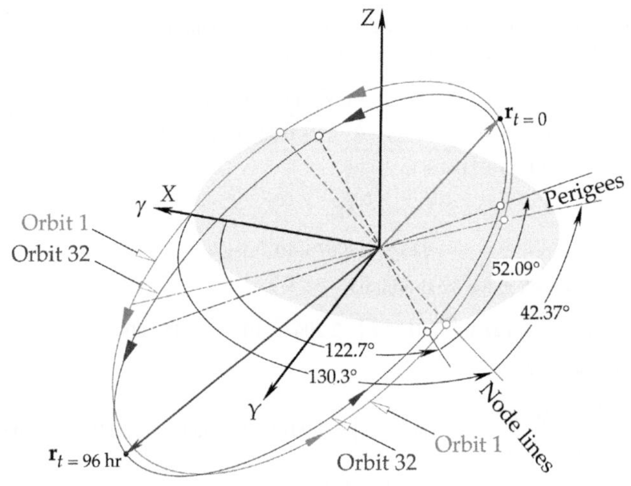

FIG. 4.21

The initial and final position vectors.

The two orbits are plotted in Fig. 4.21.

4.7.1 GROUND TRACKS

The projection of a satellite's orbit onto the earth's surface is called its ground track. At a given instant, we can imagine a radial line drawn outward from the center of the earth to the satellite. Where this line pierces the earth's spherical surface is a point on the ground track. We locate this point by giving its latitude and longitude relative to the earth. As the satellite moves around the earth, the trace of these points is its ground track.

Because the satellite reaches a maximum and minimum latitude ("amplitude") during each orbit, while passing over the equator twice, on a Mercator projection, the ground track of a satellite in low earth orbit often resembles a sine curve. If the earth did not rotate, there would be just one sinusoid-like track, traced repeatedly as the satellite orbits the earth. However, the earth rotates eastward beneath the satellite orbit at 15.04 deg/h, so the ground track advances westward at that rate. Fig. 4.22 shows about two and a half orbits of a satellite, with the beginning and end of this portion of the ground track labeled. The distance between two successive crossings of the equator is measured to be 23.2°, which is the amount of earth rotation in one orbit of the spacecraft. Therefore, the ground track reveals that the period of the satellite is

$$T = \frac{23.2^\circ}{15.04^\circ/\text{h}} = 1.54\,\text{h} = 92.6\,\text{min}$$

This is a typical low earth orbital period.

Given a satellite's position vector $\mathbf{r}$, we can use Algorithm 4.1 to find its right ascension and declination relative to the geocentric equatorial XYZ frame, which is fixed in space. The earth rotates at an angular velocity ω_E relative to this system. Let us attach an $x'y'z'$ Cartesian coordinate system to the

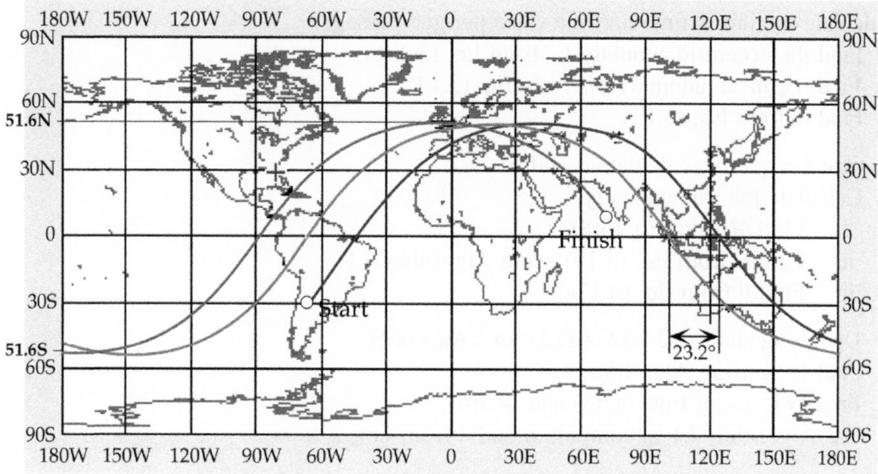

FIG. 4.22

Ground track of a satellite.

earth with its origin located at the earth's center, as illustrated in Fig. 1.18. The $x'y'$ axes lie in the equatorial plane and the z' axis points north. (In Fig. 1.18 the x' axis is directed toward the prime meridian, which passes through Greenwich, England.) The XYZ and $x'y'z'$ axes differ only by the angle θ between the stationary X axis and the rotating x' axis. If the X and x' axes line up at time t_0, then at any time t thereafter the angle θ will be given by $\omega_E(t - t_0)$. The transformation from XYZ to $x'y'z'$ is represented by the elementary rotation matrix (recall Eq. 4.34),

$$[\mathbf{R}_3(\theta)] = \begin{bmatrix} \cos\theta & \sin\theta & 0 \\ -\sin\theta & \cos\theta & 0 \\ 0 & 0 & 1 \end{bmatrix} \quad \theta = \omega_E(t - t_0) \tag{4.55}$$

Thus, if the components of the position vector $\mathbf{r}$ in the inertial XYZ frame are given by $\{\mathbf{r}\}_X$, its components $\{\mathbf{r}\}_{x'}$ in the rotating, earth-fixed $x'y'z'$ frame are

$$\{\mathbf{r}\}_{x'} = [\mathbf{R}_3(\theta)]\{\mathbf{r}\}_X \tag{4.56}$$

Knowing $\{\mathbf{r}\}_{x'}$, we use Algorithm 4.1 to determine the right ascension (longitude east of x') and declination (latitude) in the earth-fixed system. These points are usually plotted on a rectangular Mercator projection of the earth's surface, as in Fig. 4.22.

ALGORITHM 4.6

Given the initial orbital elements (h, e, a, T, i, ω_0, Ω_0, and θ_0) of a satellite relative to the geocentric equatorial frame, compute the right ascension α and declination δ relative to the rotating earth after a time interval Δt. This algorithm is implemented in MATLAB as the script *ground_track.m* in Appendix D.23.

1. Compute $\dot{\Omega}$ and $\dot{\omega}$ from Eqs. (4.52) and (4.53).

2. Calculate the initial time t_0 (time since perigee passage):
   a. Find the eccentric anomaly E_0 from Eq. (3.13b).
   b. Find the mean anomaly M_0 from Eq. (3.14).
   c. Find t_0 from Eq. (3.15).

3. At time $t = t_0 + \Delta t$, calculate α and δ.
   a. Calculate the true anomaly:
      i. Find M from Eq. (3.8).
      ii. Find E from Eq. (3.14) using Algorithm 3.1.
      iii. Find θ from Eq. (3.13a).

   b. Update Ω and ω: $\Omega = \Omega_0 + \dot{\Omega}\Delta t \quad \omega = \omega_0 + \dot{\omega}\Delta t$.
   c. Find $\{\mathbf{r}\}_X$ using Algorithm 4.5.
   d. Find $\{\mathbf{r}\}_{x'}$ using Eqs. (4.55) and (4.56).
   e. Use Algorithm 4.1 to compute α and δ from $\{\mathbf{r}\}_{x'}$.

4. Repeat Step 3 for additional times ($t = t_0 + 2\Delta t$, $t = t_0 + 3\Delta t$, etc.).

EXAMPLE 4.12

An earth satellite has the following orbital parameters:

$r_p = 6700\text{km}$	Perigee
$r_a = 10{,}000\text{km}$	Apogee
$\theta_0 = 230°$	True anomaly
$\Omega_0 = 270°$	Right ascension of the ascending node
$i_0 = 60°$	Inclination
$\omega_0 = 45°$	Argument of perigee

Calculate the right ascension (longitude east of x') and declination (latitude) relative to the rotating earth 45 min later.

Solution

First, we compute the semimajor axis a, eccentricity e, the angular momentum h, the semimajor axis a, and the period T. For the semimajor axis, we recall that

$$a = \frac{r_p + r_a}{2} = \frac{6700 + 10{,}000}{2} = 8350\text{km}$$

From Eq. (2.84) we get the eccentricity,

$$e = \frac{r_a - r_p}{r_a + r_p} = \frac{10{,}000 - 6700}{10{,}000 + 6700} = 0.19760$$

Eq. (2.50) yields the angular momentum

$$h = \sqrt{\mu r_p(1+e)} = \sqrt{398{,}600 \cdot 6700 \cdot (1 + 0.19760)} = 56{,}554\text{km}^2/\text{s}$$

Finally, we obtain the period from Eq. (2.83),

$$T = \frac{2\pi}{\sqrt{\mu}} a^{3/2} = \frac{2\pi}{\sqrt{398{,}600}} 8350^{3/2} = 7593.5\text{s}$$

Now we can proceed with Algorithm 4.6.
Step 1:

$$\dot{\Omega} = -\left[\frac{3}{2}\frac{\sqrt{\mu}J_2R^2}{(1-e^2)^2a^{7/2}}\right]\cos i_0 = -\left[\frac{3}{2}\frac{\sqrt{398,600}\cdot 0.0010836\cdot 6378^2}{(1-0.19760^2)^2\cdot 8350^{7/2}}\right]\cos 60° = -2.3394(10^{-5})°/\text{s}$$

$$\dot{\omega} = \dot{\Omega}\frac{(5/2)\sin^2 i_0 - 2}{\cos i_0} = -2.3394\times 10^{-5}\left[\frac{(5/2)\sin^2 60° - 2}{\cos 60°}\right] = 5.8484(10^{-6})°/\text{s}$$

Step 2:
(a)

$$E = 2\tan^{-1}\left(\tan\frac{\theta}{2}\sqrt{\frac{1-e}{1+e}}\right) = 2\tan^{-1}\left(\tan\frac{230°}{2}\sqrt{\frac{1-0.19760}{1+0.19760}}\right) = -2.1059\,\text{rad}$$

(b)

$$M = E - e\sin E = -2.1059 - 0.19760\sin(-2.1059) = -1.9360\,\text{rad}$$

(c)

$$t_0 = \frac{M}{2\pi}T = \frac{-1.9360}{2\pi}\cdot 7593.5 = -2339.7\,\text{s}\;(\therefore 2339.7\text{ seconds until perigee})$$

Step 3:
(a)

$$t = t_0 + 45\text{ min} = -2339.7 + (45\times 60) = 360.33\text{ s}\;\;(360.33\text{ s after perigee})$$

$$M = 2\pi\frac{t}{T} = 2\pi\frac{360.33}{7593.5} = 0.29815\,\text{rad}$$

$$E - 0.19760\sin E = 0.29815 \overset{\text{Algorithm 3.1}}{\Rightarrow} E = 0.36952\,\text{rad}$$

$$\theta = 2\tan^{-1}\left(\tan\frac{E}{2}\sqrt{\frac{1+e}{1-e}}\right) = 2\tan^{-1}\left(\tan\frac{0.36952}{2}\sqrt{\frac{1+0.19760}{1-0.19760}}\right) = 25.723°$$

(b)

$$\Omega = \Omega_0 + \dot{\Omega}\Delta t = 270° + (-2.3394\times 10^{-5})(2700\,\text{s})$$

$$\omega = \omega_0 + \dot{\omega}\Delta t = 45° + (5.8484\times 10^{-6°}/\text{s})(2700\text{s}) = 45.016°$$

(c)

$$\{\mathbf{r}\}_X \overset{\text{Algorithm 4.5}}{=} \begin{Bmatrix} 3212.6 \\ -2250.5 \\ 5568.6 \end{Bmatrix}(\text{km})$$

(d)

$$\theta = \omega_E\Delta t = \frac{360°\left(1+\dfrac{1}{365.26}\right)}{24\cdot 3600\text{s}}\cdot 2700\text{s} = 11.28°$$

$$[\mathbf{R}_3(\theta)] = \begin{bmatrix} \cos 11.281° & \sin 11.281° & 0 \\ -\sin 11.281° & \cos 11.281° & 0 \\ 0 & 0 & 1 \end{bmatrix} = \begin{bmatrix} 0.98068 & 0.19562 & 0 \\ -0.19562 & 0.98068 & 0 \\ 0 & 0 & 1 \end{bmatrix}$$

$$\{\mathbf{r}\}_{x'} = [\mathbf{R}_3(\theta)]\{\mathbf{r}\}_X = \begin{bmatrix} 0.98068 & 0.19562 & 0 \\ -0.19562 & 0.98068 & 0 \\ 0 & 0 & 1 \end{bmatrix}\begin{Bmatrix} 3212.6 \\ -2250.5 \\ 5568.6 \end{Bmatrix} = \begin{Bmatrix} 2710.3 \\ -2835.4 \\ 5568.6 \end{Bmatrix}(\text{km})$$

(e)

$$\mathbf{r} = 2710.3\hat{\mathbf{i}}' - 2835.4\hat{\mathbf{j}}' + 5568.6\hat{\mathbf{k}}' \overset{\text{Algorithm 4.1}}{\Rightarrow} \boxed{\alpha = 313.7°\;\;\delta = 54.84°}$$

The script *ground_track.m* in Appendix D.23 can be used to plot ground tracks. For the data of Example 4.12, the ground track for 3.25 periods appears in Fig. 4.23. The ground track for one orbit of a Molniya satellite is featured more elegantly in Fig. 4.24.

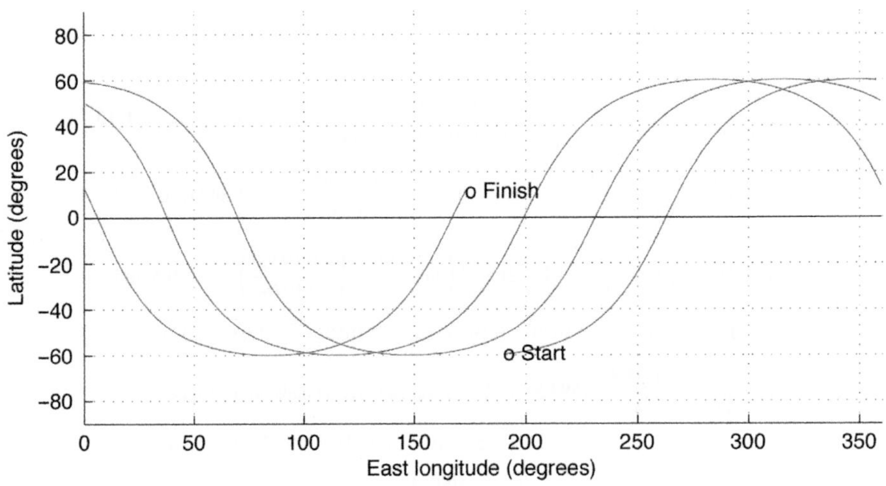

FIG. 4.23

Ground track for 3.25 orbits of the satellite in Example 4.12.

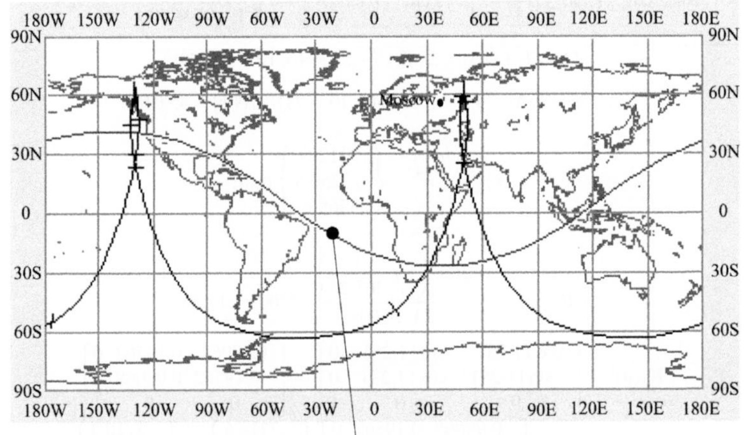

FIG. 4.24

Ground track for two orbits of a Molniya satellite with a 12-h period. Tick marks are 1 h apart.

PROBLEMS

Section 4.3

4.1 For each of the following geocentric equatorial position vectors (in kilometers) find the right ascension and declination.

$$\mathbf{r} = -3000\hat{\mathbf{I}} - 6000\hat{\mathbf{J}} - 9000\hat{\mathbf{K}} \tag{a}$$

$$\mathbf{r} = -3000\hat{\mathbf{I}} - 6000\hat{\mathbf{J}} - 9000\hat{\mathbf{K}} \tag{b}$$

$$\mathbf{r} = -9000\hat{\mathbf{I}} - 3000\hat{\mathbf{J}} - 6000\hat{\mathbf{K}} \tag{c}$$

$$\mathbf{r} = 6000\hat{\mathbf{I}} - 9000\hat{\mathbf{J}} - 3000\hat{\mathbf{K}} \tag{d}$$

{Partial Ans.: (b) $\alpha = 243.4°$, $\delta = -53.30°$}

4.2 At a given instant, a spacecraft is 500 km above the earth, with a right ascension of 300° and a declination of −20° relative to the geocentric equatorial frame. Its velocity is 10 km/s directly north, normal to the equatorial plane. Find α and δ 30 min later.
{Ans.: $\alpha = 120°$, $\delta = -29.98°$}

Section 4.4

4.3 Find the orbital elements of a geocentric satellite whose inertial position and velocity vectors in a geocentric equatorial frame are

$$\mathbf{r} = 2500\hat{\mathbf{I}} + 16{,}000\hat{\mathbf{J}} + 4000\hat{\mathbf{K}}\,(\text{km})$$
$$\mathbf{v} = -3\hat{\mathbf{I}} - \hat{\mathbf{J}} + 5\hat{\mathbf{K}}\,(\text{km/s})$$

{Ans.: $e = 0.4658$, $h = 98,623\,\text{km}^2/\text{s}$, $i = 62.52°$, $\Omega = 73.74°$, $\omega = 22.08°$, $\theta = 353.6°$}

4.4 At a given instant, the position **r** and velocity **v** of a satellite in the geocentric equatorial frame are

$$\mathbf{r} = -13{,}000\hat{\mathbf{K}}\,(\text{km})$$
$$\mathbf{v} = 4\hat{\mathbf{I}} + 5\hat{\mathbf{j}} + 6\hat{\mathbf{K}}\,(\text{km/s})$$

Find the orbital elements.
{Ans.: $e = 1.298$, $h = 83,240\ \text{km}^2/\text{s}$, $i = 90°$, $\Omega = 51.34°$, $\omega = 344.9°$, $\theta = 285.1°$}

4.5 At time t_0 (relative to perigee passage) the position **r** and velocity **v** of a satellite in the geocentric equatorial frame are

$$\mathbf{r} = 6500\hat{\mathbf{I}} - 7500\hat{\mathbf{J}} - 2500\hat{\mathbf{K}}\,(\text{km})$$
$$\mathbf{v} = 4\hat{\mathbf{I}} + 3\hat{\mathbf{J}} - 3\hat{\mathbf{K}}\ (\text{km/s})$$

Find the orbital elements.
{Ans.: $e = 0.2226$, $h = 58,656\ \text{km}^2/\text{s}$, $i = 32.44°$, $\Omega = 107.6°$, $\omega = 72.36°$, $\theta = 134.7°$}

4.6 With respect to the geocentric equatorial frame, the position vector of a spacecraft is $\mathbf{r} = -6000\hat{\mathbf{I}} - 1000\hat{\mathbf{J}} - 5000\hat{\mathbf{K}}\,(\text{km})$ and the orbit's eccentricity vector is $\mathbf{e} = 0.4\hat{\mathbf{I}} + 0.5\hat{\mathbf{j}} + 0.6\hat{\mathbf{K}}$. Calculate the true anomaly θ if the satellite is approaching perigee.
{Ans.: 328.6°}

4.7 Given that, relative to the geocentric equatorial frame, $\mathbf{r} = -6600\hat{\mathbf{I}} - 1300\hat{\mathbf{J}} - 5200\hat{\mathbf{K}}\,(\text{km})$, the eccentricity vector is $\mathbf{e} = -0.4\hat{\mathbf{I}} - 0.5\hat{\mathbf{J}} - 0.6\hat{\mathbf{K}}$, and the satellite is flying toward perigee, calculate the inclination of the orbit.
{Ans.: 43.3°}

Section 4.5

4.8 The right-handed, Cartesian coordinate system $x'y'z'$ is defined by the three points A, B, and C. The $x'y'$ plane is defined by the plane ABC. The x' axis runs from A through B. The z' axis is defined by the cross product of the vector $\overrightarrow{AB}$ into the vector $\overrightarrow{AC}$, so that the $+y'$ axis lies on the same side of the x' axis as point C.

(a) Find the direction cosine matrix $[\mathbf{Q}]$ relating the two coordinate bases.

(b) If the components of a vector $\mathbf{v}$ in the primed system are $\lfloor 2 \quad -1 \quad 3 \rfloor^T$, find the components of $\mathbf{v}$ in the unprimed system.

{Partial Ans.: (b) $\lfloor -1.307 \quad 2.390 \quad 2.565 \rfloor^T$}

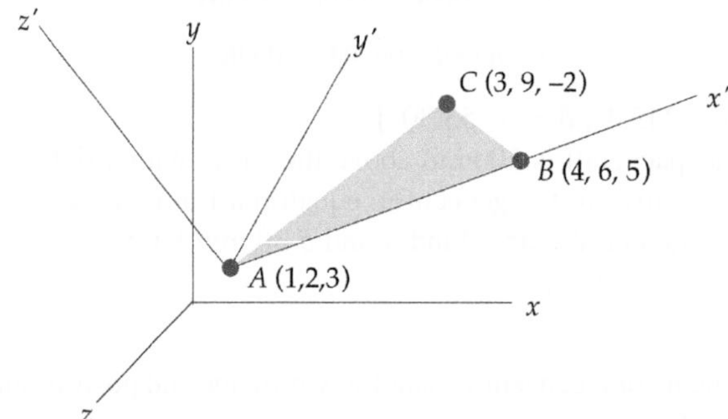

4.9 The unit vectors in a uvw Cartesian coordinate frame have the following components in the xyz frame:

$$\hat{\mathbf{u}} = 0.26726\hat{\mathbf{i}} + 0.53452\hat{\mathbf{j}} + 0.80178\hat{\mathbf{k}}$$
$$\hat{\mathbf{v}} = -0.44376\hat{\mathbf{i}} + 0.80684\hat{\mathbf{j}} + 0.38997\hat{\mathbf{k}}$$
$$\hat{\mathbf{w}} = -0.85536\hat{\mathbf{i}} + 0.25158\hat{\mathbf{j}} + 0.45284\hat{\mathbf{k}}$$

If, in the xyz frame $\mathbf{V} = -50\hat{\mathbf{i}} + 100\hat{\mathbf{j}} + 75\hat{\mathbf{k}}$, find the components of the vector $\mathbf{V}$ in the uvw frame.

{Ans.: $\mathbf{V} = 100.2\hat{\mathbf{u}} + 73.62\hat{\mathbf{v}} + 51.57\hat{\mathbf{w}}$}

4.10 Calculate the direction cosine matrix $[\mathbf{Q}]$ for the sequence of two rotations: $\alpha = 40°$ about the positive X axis, followed by $\beta = 25°$ about the positive y' axis. The result is that the XYZ axes are rotated into the $x''y'z''$ axes.

{Partial Ans.: $Q_{11} = 0.9063 \quad Q_{12} = 0.2716 \quad Q_{13} = -0.3237$ }

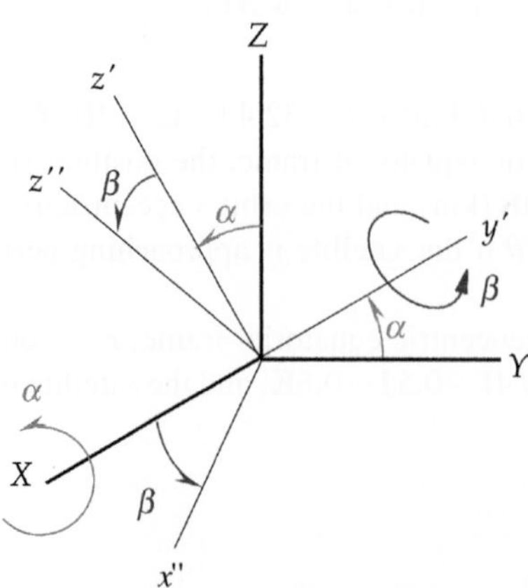

4.11 For the direction cosine matrix

$$[\mathbf{Q}] = \begin{bmatrix} 0.086824 & -0.77768 & 0.62264 \\ -0.49240 & -0.57682 & -0.65178 \\ 0.86603 & -0.25000 & -0.43301 \end{bmatrix}$$

calculate:

(a) The classical Euler angle sequence.

(b) The yaw, pitch, and roll angle sequence.

{Ans.: (a) $\alpha = 73.90°$ $\beta = 115.7°$ $\gamma = 136.31°$
(b) $\alpha = 276.37°$ $\beta = -38.51°$ $\gamma = 236.40°$}

4.12 What yaw, pitch, and roll sequence yields the same direction cosine matrix as the classical Euler sequence $\alpha = 350°$, $\beta = 170°$, $\gamma = 300°$?
{Ans.: $\alpha = 49.62°$, $\beta = 8.649°$, $\gamma = 175.0°$}

4.13 What classical Euler angle sequence yields the same direction cosine matrix as the yaw, pitch, and roll sequence $\alpha = 300°$, $\beta = -80°$, $\gamma = 30°$?
{Ans.: $\alpha = 240.4°$, $\beta = 81.35°$, $\gamma = 84.96°$}

Section 4.6

4.14 At time t_0 (relative to perigee passage), the position **r** and velocity **v** of a satellite in the geocentric equatorial frame are

$$\mathbf{r} = -5000\hat{\mathbf{I}} - 8000\hat{\mathbf{J}} - 2100\hat{\mathbf{K}}\,(\text{km})$$
$$\mathbf{v} = -4\hat{\mathbf{I}} + 3.5\hat{\mathbf{J}} - 3\hat{\mathbf{K}}\,(\text{km/s})$$

Find **r** and **v** at time $t_0 + 50\,\text{min}$.
{Ans.: $\mathbf{r} = -1717\hat{\mathbf{I}} + 7604\hat{\mathbf{J}} - 2101\hat{\mathbf{K}}\,(\text{km})$; $\mathbf{v} = 6.075\hat{\mathbf{I}} + 1.925\hat{\mathbf{J}} + 3.591\hat{\mathbf{K}}\,(\text{km/s})$}

4.15 At time t_0 (relative to perigee passage), a spacecraft has the following orbital parameters: $e = 1.5$; perigee altitude = 300 km; $i = 35°$; $\Omega = 130°$; and $\omega = 115°$. Calculate **r** and **v** at perigee relative to

(a) The perifocal reference frame.

(b) The geocentric equatorial frame.

{Ans.: (a) $\mathbf{r} = 6678\hat{\mathbf{p}}\,(\text{km})$, $\mathbf{v} = 12.22\hat{\mathbf{q}}\,(\text{km/s})$
(b) $\mathbf{r} = -1984\hat{\mathbf{I}} - 5348\hat{\mathbf{J}} + 3471\hat{\mathbf{K}}\,(\text{km})$, $\mathbf{v} = 10.36\hat{\mathbf{I}} - 5.763\hat{\mathbf{J}} - 2.961\hat{\mathbf{K}}\,(\text{km/s})$ }

4.16 For the spacecraft of Problem 4.15, calculate **r** and **v** at two hours past perigee relative to

(a) The perifocal reference frame.

(b) The geocentric equatorial frame.

{Ans.: (a) $\mathbf{r} = -25{,}010\hat{\mathbf{p}} + 48{,}090\hat{\mathbf{q}}\,(\text{km})$, $\mathbf{v} = -4.335\hat{\mathbf{p}} + 5.075\hat{\mathbf{q}}\,(\text{km/s})$
(b) $\mathbf{r} = 48{,}200\hat{\mathbf{I}} - 2658\hat{\mathbf{J}} - 24{,}660\hat{\mathbf{K}}\,(\text{km})$, $\mathbf{v} = 5.590\hat{\mathbf{I}} + 1.078\hat{\mathbf{J}} - 3.484\hat{\mathbf{K}}\,(\text{km/s})$}

4.17 Calculate **r** and **v** relative to the geocentric equatorial frame for the satellite in Problem 4.15 at time $t_0 + 50\,\text{min}$.
{Ans.: $\mathbf{r} = 23{,}047\hat{\mathbf{I}} - 6972.4\hat{\mathbf{J}} - 9219.6\hat{\mathbf{K}}\,(\text{km})$
$\mathbf{v} = 6.6563\hat{\mathbf{I}} + 0.88638\hat{\mathbf{J}} - 3.9680\,(\text{km/s})$}

4.18 For a spacecraft, the following orbital parameters are given: $e = 1.2$; perigee altitude = 200 km; $i = 50°$; $\Omega = 75°$; and $\omega = 80°$. Calculate **r** and **v** at perigee relative to

(a) The perifocal reference frame.
(b) The geocentric equatorial frame.
{Ans. (a) $\mathbf{r} = 6578\hat{\mathbf{p}}$ (km), $\mathbf{v} = 11.55\hat{\mathbf{q}}$ (km/s)
(b) $\mathbf{r} = -3726\hat{\mathbf{I}} + 2181\hat{\mathbf{J}} + 4962\hat{\mathbf{K}}$ (km)
$\mathbf{v} = -4.188\hat{\mathbf{I}} - 10.65\hat{\mathbf{J}} + 1.536\hat{\mathbf{K}}$ (km/s)}

4.19 For the spacecraft of Problem 4.18, calculate **r** and **v** at 2 h past perigee relative to
(a) The perifocal reference frame.
(b) The geocentric equatorial frame.
{Ans.: (a) $\mathbf{r} = -26{,}340\hat{\mathbf{p}} + 37{,}810\hat{\mathbf{q}}$ (km), $\mathbf{v} = -4.306\hat{\mathbf{p}} + 3.298\hat{\mathbf{q}}$ (km/s)
(b) $\mathbf{r} = 1207\hat{\mathbf{I}} - 43{,}600\hat{\mathbf{J}} - 14{,}840\hat{\mathbf{K}}$ (km)
$\mathbf{v} = 1.243\hat{\mathbf{I}} - 4.4700\hat{\mathbf{J}} - 2.810$ (km/s)}

4.20 Given that $e = 0.7$, $h = 75{,}000\,\text{km}^2/\text{s}$, and $\theta = 25°$, calculate the components of velocity in the geocentric equatorial frame if

$$[\mathbf{Q}]_{X\bar{x}} = \begin{bmatrix} -.83204 & -.13114 & 0.53899 \\ 0.02741 & -.98019 & -.19617 \\ 0.55403 & -.14845 & 0.81915 \end{bmatrix}$$

{Ans.: $\mathbf{v} = 2.103\hat{\mathbf{I}} - 8.073\hat{\mathbf{J}} - 2.885\hat{\mathbf{K}}$ (km/s)}

4.21 The apse line of the elliptical orbit lies in the *XY* plane of the geocentric equatorial frame, whose *Z* axis lies in the plane of the orbit. At *B* (for which $\theta = 140°$), the perifocal velocity vector is $\{\mathbf{v}\}_{\bar{x}} = \lfloor -3.208 \quad -0.8288 \quad 0 \rfloor^T$ (km/s). Calculate the geocentric equatorial components of the velocity at *B*.
{Ans.: $\{\mathbf{v}\}_X = \lfloor -1.604 \quad -2.778 \quad -0.8288 \rfloor^T$ (km/s)}

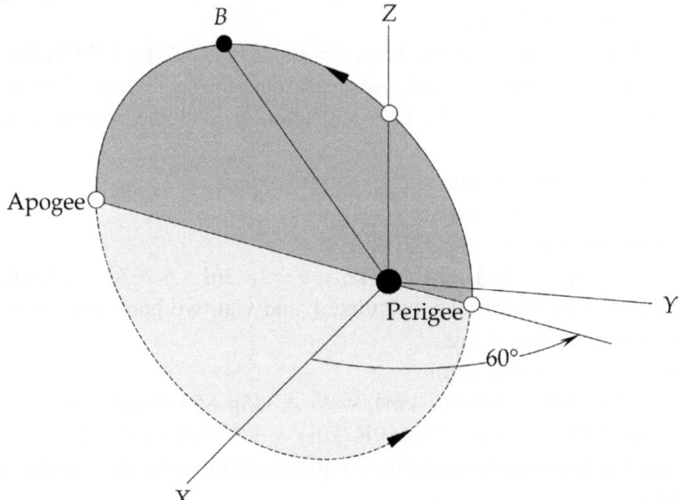

4.22 A satellite in earth orbit has the following orbital parameters: $a = 7016$ km, $e = 0.05$, $i = 45°$, $\Omega = 0°$, $\omega = 20°$, and $\theta = 10°$. Find the position vector in the geocentric equatorial frame.
{Ans.: $\mathbf{r} = 5776.4\hat{\mathbf{I}} + 2358.2\hat{\mathbf{J}} + 2358.2\hat{\mathbf{K}}$ (km)}

Section 4.7

4.23 Calculate the orbital inclination required to place an earth satellite in a 300 km by 600 km sun-synchronous orbit.
{Ans.: 97.21°}.

4.24 A satellite in a circular, sun-synchronous low earth orbit passes over the same point on the equator once each day, at 12 o'clock noon. Calculate the inclination, altitude, and period of the orbit.
{Ans.: This problem has more than one solution.}

4.25 The orbit of a satellite around an unspecified planet has an inclination of 45°, and its perigee advances at the rate of 6° per day. At what rate does the node line regress?
{Ans.: $\dot{\Omega} = 5.656°/\text{day}$}

4.26 At a given time, the position and velocity of an earth satellite in the geocentric equatorial frame are $\mathbf{r} = -2429.1\hat{\mathbf{I}} + 4555.1\hat{\mathbf{J}} + 4577.0\hat{\mathbf{K}}$ (km) and $\mathbf{v} = -4.7689\hat{\mathbf{I}} - 5.6113\hat{\mathbf{J}} + 3.0535\hat{\mathbf{K}}$ (km/s). Find **r** and **v** precisely 72 h later, taking into consideration the node line regression and the advance of perigee.
{Ans.: $\mathbf{r} = 4596\hat{\mathbf{I}} + 5759\hat{\mathbf{J}} - 1266\hat{\mathbf{K}}$(km), $\mathbf{v} = -3.601\hat{\mathbf{I}} + 3.179\hat{\mathbf{J}} + 5.617\hat{\mathbf{K}}$ (km/s)}

Section 4.8

4.27 A spacecraft is in a circular orbit of 180 km altitude and inclination 30°. What is the spacing, in kilometers, between successive ground tracks at the equator, including the effect of earth's oblateness?
{Ans.: 2511 km}

REFERENCE

Department of the Navy, 2018. Nautical Almanac Office: The Astronomical Almanac for the Year 2019. GPO.

CHAPTER

PRELIMINARY ORBIT DETERMINATION

5

5.1 INTRODUCTION

In this chapter, we will consider some (by no means all) of the classical ways in which the orbit of a satellite can be determined from earth-bound observations. All the methods presented here are based on the two-body equations of motion. As such, they must be considered preliminary orbit determination techniques because the actual orbit is influenced over time by other phenomena (perturbations), such as the gravitational force of the moon and sun, atmospheric drag, solar wind, and the nonspherical shape and nonuniform mass distribution of the earth. We took a brief look at the dominant effects of the earth's oblateness in Section 4.7. To accurately propagate an orbit into the future from a set of initial observations requires taking the various perturbations, as well as instrumentation errors themselves, into account. More detailed considerations, including the means of updating the orbit based on additional observations, are beyond our scope. Introductory discussions may be found elsewhere. See Bate et al. (1971), Boulet (1991), Prussing and Conway (2013), and Wiesel (2010), to name but a few.

We begin with the Gibbs method of predicting an orbit using three geocentric position vectors. This is followed by a presentation of Lambert's problem, in which an orbit is determined from two position vectors and the time between them. Both the Gibbs and Lambert procedures are based on the fact that two-body orbits lie in a plane. The Lambert problem is more complex and requires using the Lagrange f and g functions introduced in Chapter 2 as well as the universal variable formulation introduced in Chapter 3. The Lambert algorithm is employed in Chapter 8 to analyze interplanetary missions.

In preparation for explaining how satellites are tracked, the Julian day (JD) numbering scheme is introduced along with the notion of sidereal time. This is followed by a description of the topocentric coordinate systems and the relationships among topocentric right ascension/declination angles and azimuth/elevation angles. We then describe how orbits are determined from measuring the range and the angular orientation of the line of sight, together with their rates. The chapter concludes with a presentation of the Gauss method of angles-only orbit determination.

Orbital Mechanics for Engineering Students. https://doi.org/10.1016/B978-0-12-824025-0.00005-2

5.2 GIBBS METHOD OF ORBIT DETERMINATION FROM THREE POSITION VECTORS

Suppose that from the observations of a space object at the three successive times t_1, t_2, and t_3 ($t_1 < t_2 < t_3$) we have obtained the geocentric position vectors $\mathbf{r}_1$, $\mathbf{r}_2$, and $\mathbf{r}_3$. The problem is to determine the velocities $\mathbf{v}_1$, $\mathbf{v}_2$, and $\mathbf{v}_3$ at t_1, t_2, and t_3 assuming that the object is in a two-body orbit. The solution using purely vector analysis is due to J.W. Gibbs (1839–1903), an American scholar who is known primarily for his contributions to thermodynamics. Our explanation is based on that in Bate et al. (1971).

We know that the conservation of angular momentum requires that the position vectors of an orbiting body must all lie in the same plane. In other words, the unit vector normal to the plane of $\mathbf{r}_2$ and $\mathbf{r}_3$ must be perpendicular to the unit vector in the direction of $\mathbf{r}_1$. Thus, if $\hat{\mathbf{u}}_{r_1} = \mathbf{r}_1/r_1$ and $\hat{\mathbf{C}}_{23} = (\mathbf{r}_2 \times \mathbf{r}_3)/\|\mathbf{r}_2 \times \mathbf{r}_3\|$, then the dot product of these two unit vectors must vanish:

$$\hat{\mathbf{u}}_{r_1} \cdot \hat{\mathbf{C}}_{23} = 0$$

Furthermore, as illustrated in Fig. 5.1, the fact that $\mathbf{r}_1$, $\mathbf{r}_2$, and $\mathbf{r}_3$ lie in the same plane means we can apply scalar factors c_1 and c_3 to $\mathbf{r}_1$ and $\mathbf{r}_3$ so that $\mathbf{r}_2$ is the vector sum of $c_1\mathbf{r}_1$ and $c_3\mathbf{r}_3$:

$$\mathbf{r}_2 = c_1\mathbf{r}_1 + c_3\mathbf{r}_3 \tag{5.1}$$

The coefficients c_1 and c_3 are readily obtained from $\mathbf{r}_1$, $\mathbf{r}_2$, and $\mathbf{r}_3$, as we shall see in Section 5.10 (Eqs. 5.89 and 5.90).

To find the velocity $\mathbf{v}$ corresponding to any of the three given position vectors $\mathbf{r}$ we start with Eq. (2.40), which may be written as

$$\mathbf{v} \times \mathbf{h} = \mu\left(\frac{\mathbf{r}}{r} + \mathbf{e}\right)$$

where $\mathbf{h}$ is the angular momentum, and $\mathbf{e}$ is the eccentricity vector. To isolate the velocity, take the cross product of this equation with the angular momentum,

$$\mathbf{h} \times (\mathbf{v} \times \mathbf{h}) = \mu\left(\frac{\mathbf{h} \times \mathbf{r}}{r} + \mathbf{h} \times \mathbf{e}\right) \tag{5.2}$$

By means of the bac–cab rule (Eq. 2.33), the left-hand side becomes

$$\mathbf{h} \times (\mathbf{v} \times \mathbf{h}) = \mathbf{v}(\mathbf{h} \cdot \mathbf{h}) - \mathbf{h}(\mathbf{h} \cdot \mathbf{v})$$

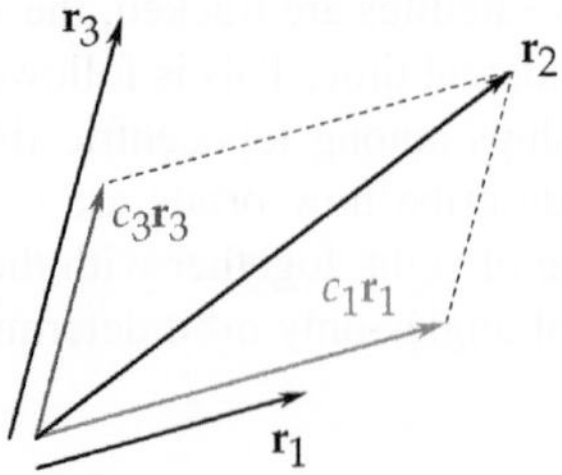

FIG. 5.1

Any one of a set of three coplanar vectors ($\mathbf{r}_1$, $\mathbf{r}_2$, $\mathbf{r}_3$) can be expressed as the vector sum of the other two.

But $\mathbf{h}\cdot\mathbf{h}=h^2$, and $\mathbf{v}\cdot\mathbf{h}=0$, since $\mathbf{v}$ is perpendicular to $\mathbf{h}$. Therefore,

$$\mathbf{h}\times(\mathbf{v}\times\mathbf{h})=h^2\mathbf{v}$$

which means that Eq. (5.2) may be written as

$$\mathbf{v}=\frac{\mu}{h^2}\left(\frac{\mathbf{h}\times\mathbf{r}}{r}+\mathbf{h}\times\mathbf{e}\right) \tag{5.3}$$

In Section 2.10, we introduced the perifocal coordinate system, in which the unit vector $\hat{\mathbf{p}}$ lies in the direction of the eccentricity vector $\mathbf{e}$, and $\hat{\mathbf{w}}$ is the unit vector normal to the orbital plane, in the direction of the angular momentum vector $\mathbf{h}$. Thus, we can write

$$\mathbf{e}=e\hat{\mathbf{p}} \tag{5.4a}$$

$$\mathbf{h}=h\hat{\mathbf{w}} \tag{5.4b}$$

so that Eq. (5.3) becomes

$$\mathbf{v}=\frac{\mu}{h^2}\left(\frac{h\hat{\mathbf{w}}\times\mathbf{r}}{r}+h\hat{\mathbf{w}}\times e\hat{\mathbf{p}}\right)=\frac{\mu}{h}\left[\frac{\hat{\mathbf{w}}\times\mathbf{r}}{r}+e(\hat{\mathbf{w}}\times\hat{\mathbf{p}})\right] \tag{5.5}$$

Since $\hat{\mathbf{p}}$, $\hat{\mathbf{q}}$, and $\hat{\mathbf{w}}$ form a right-handed triad of unit vectors, it follows that $\hat{\mathbf{p}}\times\hat{\mathbf{q}}=\hat{\mathbf{w}}$, $\hat{\mathbf{q}}\times\hat{\mathbf{w}}=\hat{\mathbf{p}}$, and

$$\hat{\mathbf{w}}\times\hat{\mathbf{p}}=\hat{\mathbf{q}} \tag{5.6}$$

Therefore, Eq. (5.5) reduces to

$$\mathbf{v}=\frac{\mu}{h}\left(\frac{\hat{\mathbf{w}}\times\mathbf{r}}{r}+e\hat{\mathbf{q}}\right) \tag{5.7}$$

This is an important result, because if we can somehow use the position vectors $\mathbf{r}_1$, $\mathbf{r}_2$, and $\mathbf{r}_3$ to calculate $\hat{\mathbf{q}}$, $\hat{\mathbf{w}}$, h, and e, then the velocities $\mathbf{v}_1$, $\mathbf{v}_2$, and $\mathbf{v}_3$ will each be determined by this formula.

So far, the only condition we have imposed on the three position vectors is that they are coplanar (Eq. 5.1). To bring in the fact that they describe an orbit, let us take the dot product of Eq. (5.1) with the eccentricity vector $\mathbf{e}$ to obtain the scalar equation

$$\mathbf{r}_2\cdot\mathbf{e}=c_1\mathbf{r}_1\cdot\mathbf{e}+c_3\mathbf{r}_3\cdot\mathbf{e} \tag{5.8}$$

According to Eq. (2.44), the orbit equation, we have the following relations among h, e, and each of the position vectors:

$$\mathbf{r}_1\cdot\mathbf{e}=\frac{h^2}{\mu}-r_1 \quad \mathbf{r}_2\cdot\mathbf{e}=\frac{h^2}{\mu}-r_2 \quad \mathbf{r}_3\cdot\mathbf{e}=\frac{h^2}{\mu}-r_3 \tag{5.9}$$

Substituting these equations into Eq. (5.8) yields

$$\frac{h^2}{\mu}-r_2=c_1\left(\frac{h^2}{\mu}-r_1\right)+c_3\left(\frac{h^3}{\mu}-r_3\right) \tag{5.10}$$

To eliminate the unknown coefficients c_1 and c_3 from this expression, let us take the cross product of Eq. (5.1) first with $\mathbf{r}_1$ and then with $\mathbf{r}_3$. This results in two equations, both having $\mathbf{r}_3\times\mathbf{r}_1$ on the right,

$$\mathbf{r}_2\times\mathbf{r}_1=c_3(\mathbf{r}_3\times\mathbf{r}_1) \quad \mathbf{r}_2\times\mathbf{r}_3=-c_1(\mathbf{r}_3\times\mathbf{r}_1) \tag{5.11}$$

Now multiply Eq. (5.10) through by the vector $\mathbf{r}_3 \times \mathbf{r}_1$ to obtain

$$\frac{h^2}{\mu}(\mathbf{r}_3 \times \mathbf{r}_1) - r_2(\mathbf{r}_3 \times \mathbf{r}_1) = c_1(\mathbf{r}_3 \times \mathbf{r}_1)\left(\frac{h^2}{\mu} - r_1\right) + c_3(\mathbf{r}_3 \times \mathbf{r}_1)\left(\frac{h^2}{\mu} - r_3\right)$$

Using Eq. (5.11), this becomes

$$\frac{h^2}{\mu}(\mathbf{r}_3 \times \mathbf{r}_1) - r_2(\mathbf{r}_3 \times \mathbf{r}_1) = -(\mathbf{r}_2 \times \mathbf{r}_3)\left(\frac{h^2}{\mu} - r_1\right) + (\mathbf{r}_2 \times \mathbf{r}_1)\left(\frac{h^2}{\mu} - r_3\right)$$

Observe that c_1 and c_3 have been eliminated. Rearranging the terms, we get

$$\frac{h^2}{\mu}(\mathbf{r}_1 \times \mathbf{r}_2 + \mathbf{r}_2 \times \mathbf{r}_3 + \mathbf{r}_3 \times \mathbf{r}_1) = r_1(\mathbf{r}_2 \times \mathbf{r}_3) + r_2(\mathbf{r}_3 \times \mathbf{r}_1) + r_3(\mathbf{r}_1 \times \mathbf{r}_2) \tag{5.12}$$

This is an equation involving the given position vectors and the unknown angular momentum h. Let us introduce the following notation for the vectors on each side of Eq. (5.12),

$$\mathbf{N} = r_1(\mathbf{r}_2 \times \mathbf{r}_3) + r_2(\mathbf{r}_3 \times \mathbf{r}_1) + r_3(\mathbf{r}_1 \times \mathbf{r}_2) \tag{5.13}$$

and

$$\mathbf{D} = \mathbf{r}_1 \times \mathbf{r}_2 + \mathbf{r}_2 \times \mathbf{r}_3 + \mathbf{r}_3 \times \mathbf{r}_1 \tag{5.14}$$

Then, Eq. (5.12) may be written more simply as

$$\mathbf{N} = \frac{h^2}{\mu}\mathbf{D}$$

from which we obtain

$$N = \frac{h^2}{\mu}D \tag{5.15}$$

where $N = \|\mathbf{N}\|$ and $D = \|\mathbf{D}\|$. It follows from Eq. (5.15) that the angular momentum h is determined from $\mathbf{r}_1$, $\mathbf{r}_2$, and $\mathbf{r}_3$ by the formula

$$h = \sqrt{\mu \frac{N}{D}} \tag{5.16}$$

Since $\mathbf{r}_1$, $\mathbf{r}_2$, and $\mathbf{r}_3$ are coplanar, all of the cross products $\mathbf{r}_1 \times \mathbf{r}_2$, $\mathbf{r}_2 \times \mathbf{r}_3$, and $\mathbf{r}_3 \times \mathbf{r}_1$ lie in the same direction (namely, normal to the orbital plane). Therefore, it is clear from Eq. (5.14) that $\mathbf{D}$ must be normal to the orbital plane. In the context of the perifocal frame, we use $\hat{\mathbf{w}}$ to denote the orbit unit normal. Therefore,

$$\hat{\mathbf{w}} = \frac{\mathbf{D}}{D} \tag{5.17}$$

So far, we have found h and $\hat{\mathbf{w}}$ in terms of $\mathbf{r}_1$, $\mathbf{r}_2$, and $\mathbf{r}_3$. We need likewise to find an expression for $\hat{\mathbf{q}}$ to use in Eq. (5.7). From Eqs. (5.4a), (5.6), and (5.17), it follows that

$$\hat{\mathbf{q}} = \hat{\mathbf{w}} \times \hat{\mathbf{p}} = \frac{1}{De}(\mathbf{D} \times \mathbf{e}) \tag{5.18}$$

Substituting Eq. (5.14), we get

$$\hat{\mathbf{q}} = \frac{1}{De}[(\mathbf{r}_1 \times \mathbf{r}_2) \times \mathbf{e} + (\mathbf{r}_2 \times \mathbf{r}_3) \times \mathbf{e} + (\mathbf{r}_3 \times \mathbf{r}_1) \times \mathbf{e}] \tag{5.19}$$

We can apply the bac–cab rule to the right-hand side by noting

$$(\mathbf{A} \times \mathbf{B}) \times \mathbf{C} = -\mathbf{C} \times (\mathbf{A} \times \mathbf{B}) = \mathbf{B}(\mathbf{A} \cdot \mathbf{C}) - \mathbf{A}(\mathbf{B} \cdot \mathbf{C})$$

Using this vector identity, we obtain

$$(\mathbf{r}_2 \times \mathbf{r}_3) \times \mathbf{e} = \mathbf{r}_3(\mathbf{r}_2 \cdot \mathbf{e}) - \mathbf{r}_2(\mathbf{r}_3 \cdot \mathbf{e})$$
$$(\mathbf{r}_3 \times \mathbf{r}_1) \times \mathbf{e} = \mathbf{r}_1(\mathbf{r}_3 \cdot \mathbf{e}) - \mathbf{r}_3(\mathbf{r}_1 \cdot \mathbf{e})$$
$$(\mathbf{r}_1 \times \mathbf{r}_2) \times \mathbf{e} = \mathbf{r}_2(\mathbf{r}_1 \cdot \mathbf{e}) - \mathbf{r}_1(\mathbf{r}_2 \cdot \mathbf{e})$$

Once again employing Eq. (5.9), these become

$$(\mathbf{r}_2 \times \mathbf{r}_3) \times \mathbf{e} = \mathbf{r}_3\left(\frac{h^2}{\mu} - r_2\right) - \mathbf{r}_2\left(\frac{h^2}{\mu} - r_3\right) = \frac{h^2}{\mu}(\mathbf{r}_3 - \mathbf{r}_2) + r_3\mathbf{r}_2 - r_2\mathbf{r}_3$$
$$(\mathbf{r}_3 \times \mathbf{r}_1) \times \mathbf{e} = \mathbf{r}_1\left(\frac{h^2}{\mu} - r_3\right) - \mathbf{r}_3\left(\frac{h^2}{\mu} - r_1\right) = \frac{h^2}{\mu}(\mathbf{r}_1 - \mathbf{r}_3) + r_1\mathbf{r}_3 - r_3\mathbf{r}_1$$
$$(\mathbf{r}_1 \times \mathbf{r}_2) \times \mathbf{e} = \mathbf{r}_2\left(\frac{h^2}{\mu} - r_1\right) - \mathbf{r}_1\left(\frac{h^2}{\mu} - r_2\right) = \frac{h^2}{\mu}(\mathbf{r}_2 - \mathbf{r}_1) + r_2\mathbf{r}_1 - r_1\mathbf{r}_2$$

Summing up these three equations, collecting the terms, and substituting the result into Eq. (5.19) yields

$$\hat{\mathbf{q}} = \frac{1}{De}\mathbf{S} \tag{5.20}$$

where

$$\mathbf{S} = \mathbf{r}_1(r_2 - r_3) + \mathbf{r}_2(r_3 - r_1) + \mathbf{r}_3(r_1 - r_2) \tag{5.21}$$

Finally, we substitute Eqs. (5.16), (5.17), and (5.20) into Eq. (5.7) to obtain

$$\mathbf{v} = \frac{\mu}{h}\left(\frac{\hat{\mathbf{w}} \times \mathbf{r}}{r} + e\hat{\mathbf{q}}\right) = \frac{\mu}{\sqrt{\mu\frac{N}{D}}}\left[\frac{\frac{\mathbf{D}}{D} \times \mathbf{r}}{r} + e\left(\frac{1}{De}\mathbf{S}\right)\right]$$

Simplifying this expression for the velocity yields

$$\mathbf{v} = \sqrt{\frac{\mu}{ND}}\left(\frac{\mathbf{D} \times \mathbf{r}}{r} + \mathbf{S}\right) \tag{5.22}$$

All the terms on the right depend only on the given position vectors $\mathbf{r}_1$, $\mathbf{r}_2$, and $\mathbf{r}_3$.

The Gibbs method may be summarized as outlined in the following algorithm.

ALGORITHM 5.1

Given the spacecraft position vectors $\mathbf{r}_1$, $\mathbf{r}_2$, and $\mathbf{r}_3$, determine the orbital elements. A MATLAB implementation of this procedure is found in Appendix D.24.

1. Calculate r_1, r_2, and r_3.
2. Calculate $\mathbf{C}_{12} = \mathbf{r}_1 \times \mathbf{r}_2$, $\mathbf{C}_{23} = \mathbf{r}_2 \times \mathbf{r}_3$, and $\mathbf{C}_{31} = \mathbf{r}_3 \times \mathbf{r}_1$.
3. Verify that $\hat{\mathbf{u}}_{r_1} \cdot \hat{\mathbf{C}}_{23} = 0$.
4. Calculate $\mathbf{N}$, $\mathbf{D}$, and $\mathbf{S}$ using Eqs. (5.13), (5.14), and (5.21), respectively.
5. Calculate $\mathbf{v}_2$ using Eq. (5.22): $\mathbf{v}_2 = (\mathbf{D} \times \mathbf{r}_2/r_2 + \mathbf{S})\sqrt{\mu/(N \cdot D)}$.
6. Use $\mathbf{r}_2$ and $\mathbf{v}_2$ to compute the orbital elements by means of Algorithm 4.2.

EXAMPLE 5.1

The geocentric position vectors of a space object at three successive times are

$$\mathbf{r}_1 = -294.32\hat{\mathbf{I}} + 4265.1\hat{\mathbf{J}} + 5986.7\hat{\mathbf{K}} \text{ (km)}$$

$$\mathbf{r}_2 = -1365.5\hat{\mathbf{I}} + 3637.6\hat{\mathbf{J}} + 6346.8\hat{\mathbf{K}} \text{ (km)}$$

$$\mathbf{r}_3 = -2940.3\hat{\mathbf{I}} + 2473.7\hat{\mathbf{J}} + 6555.8\hat{\mathbf{K}} \text{ (km)}$$

Determine the classical orbital elements using Gibbs method.

Solution

We employ Algorithm 5.1.

Step 1:

$$r_1 = \sqrt{(-294.32)^2 + 4265.1^2 + 5986.7^2} = 7356.5 \text{km}$$

$$r_2 = \sqrt{(-1365.5)^2 + 3637.6^2 + 6346.8^2} = 7441.7 \text{km}$$

$$r_3 = \sqrt{(-2940.3)^2 + 2473.7^2 + 6555.8^2} = 7598.9 \text{km}$$

Step 2:

$$\mathbf{C}_{12} = \begin{vmatrix} \hat{\mathbf{I}} & \hat{\mathbf{J}} & \hat{\mathbf{K}} \\ -294.32 & 4265.1 & 5986.7 \\ -1365.5 & 3637.6 & 6346.8 \end{vmatrix} = (5.2925\hat{\mathbf{I}} - 6.3068\hat{\mathbf{J}} + 4.7534\hat{\mathbf{K}})(10^6) \text{ (km}^2\text{)}$$

$$\mathbf{C}_{23} = \begin{vmatrix} \hat{\mathbf{I}} & \hat{\mathbf{J}} & \hat{\mathbf{K}} \\ -1365.5 & 3637.6 & 6346.8 \\ -294.32 & 2473.7 & 6555.8 \end{vmatrix} = (8.1473\hat{\mathbf{I}} - 9.7096\hat{\mathbf{J}} + 7.3178\hat{\mathbf{K}})(10^6) \text{ (km}^2\text{)}$$

$$\mathbf{C}_{31} = \begin{vmatrix} \hat{\mathbf{I}} & \hat{\mathbf{J}} & \hat{\mathbf{K}} \\ -2940.3 & 2473.7 & 6555.8 \\ -294.32 & 4265.1 & 5986.7 \end{vmatrix} = (-1.3152\hat{\mathbf{I}} + 1.5673\hat{\mathbf{J}} - 1.1813\hat{\mathbf{K}})(10^7) \text{ (km}^2\text{)}$$

Step 3:

$$\hat{\mathbf{C}}_{23} = \frac{\mathbf{C}_{23}}{\|\mathbf{C}_{23}\|} = \frac{8.1473\hat{\mathbf{I}} - 9.7096\hat{\mathbf{J}} + 7.3178\hat{\mathbf{K}}}{\sqrt{8.1473^2 + (-9.7096)^2 + 7.3178^2}} = 0.55667\hat{\mathbf{I}} - 0.66342\hat{\mathbf{J}} + 0.5000\hat{\mathbf{K}}$$

Therefore,

$$\hat{\mathbf{u}}_{r_1}\cdot\hat{\mathbf{C}}_{23}=\left(\frac{-294.32\hat{\mathbf{I}}+4265.1\hat{\mathbf{J}}+5986\hat{\mathbf{K}}}{7356.5}\right).(0.55667\hat{\mathbf{I}}-0.66342\hat{\mathbf{J}}+0.5000\hat{\mathbf{K}})$$
$$=-6.1181(10^{-6})$$

This is close enough to zero for our purposes. The three vectors $\mathbf{r}_1$, $\mathbf{r}_2$, and $\mathbf{r}_3$ are coplanar.

Step 4:

$$\begin{aligned}\mathbf{N}&=r_1\mathbf{C}_{23}+r_2\mathbf{C}_{31}+r_3\mathbf{C}_{12}\\&=7356.5(8.1473\hat{\mathbf{I}}-9.7096\hat{\mathbf{J}}+7.3178\hat{\mathbf{K}})(10^6)\\&\quad+7441.7(-1.3152\hat{\mathbf{I}}+1.5673\hat{\mathbf{J}}-1.1813\hat{\mathbf{K}})(10^6)\\&\quad+7598.9(5.2925\hat{\mathbf{I}}-6.3068\hat{\mathbf{J}}+4.7534\hat{\mathbf{K}})(10^6)\end{aligned}$$

or

$$\mathbf{N}=(2.2811\hat{\mathbf{I}}-2.7186\hat{\mathbf{J}}+2.0481\hat{\mathbf{K}})(10^9)\,(\text{km}^3)$$

so that

$$N=\sqrt{\left[2.2811^2+(-2.7186)^2+2.0481^2\right](10^{18})}=4.0975(10^9)\,(\text{km}^3)$$

$$\begin{aligned}\mathbf{D}&=\mathbf{C}_{12}+\mathbf{C}_{23}+\mathbf{C}_{31}\\&=(5.295\hat{\mathbf{I}}-6.3068\hat{\mathbf{J}}+4.7534\hat{\mathbf{K}})(10^6)+(8.1473\hat{\mathbf{I}}-9.7096\hat{\mathbf{J}}+7.3178\hat{\mathbf{K}})(10^6)\\&\quad+(-1.3152\hat{\mathbf{I}}+1.5673\hat{\mathbf{J}}-1.1813\hat{\mathbf{K}})(10^6)\end{aligned}$$

or

$$\mathbf{D}=(2.8797\hat{\mathbf{I}}-3.4321\hat{\mathbf{J}}+2.5866\hat{\mathbf{K}})(10^5)\,(\text{km}^2)$$

so that

$$D=\sqrt{\left[2.8797^2+(-3.4321)^2+2.5856^2\right](10^{10})}=5.1728(10^5)\,(\text{km}^2)$$

Lastly,

$$\begin{aligned}\mathbf{S}&=\mathbf{r}_1(r_2-r_3)+\mathbf{r}_2(r_3-r_1)+\mathbf{r}_3(r_1-r_2)\\&=(-294.32\hat{\mathbf{I}}+4265.1\hat{\mathbf{J}}+5986.7\hat{\mathbf{K}})(7441.7-7598.9)\\&\quad+(-1365.5\hat{\mathbf{I}}+3637.6\hat{\mathbf{J}}+6346.8\hat{\mathbf{K}})(7598.9-7356.5)\\&\quad+(-2940.3\hat{\mathbf{I}}+2473.7\hat{\mathbf{J}}+6555.8\hat{\mathbf{K}})(7356.5-7441.7)\end{aligned}$$

or

$$\mathbf{S}=-34{,}276\hat{\mathbf{I}}+478.57\hat{\mathbf{J}}+38{,}810\hat{\mathbf{K}}\,(\text{km}^2)$$

Step 5:

$$\begin{aligned}\mathbf{v}_2&=\sqrt{\frac{\mu}{ND}}\left(\frac{\mathbf{D}\times\mathbf{r}_2}{r_2}+\mathbf{S}\right)\\&=\sqrt{\frac{398{,}600}{(4.0971(10^9))(5.1728(10^5))}}\\&\quad\times\left[\frac{\begin{vmatrix}\hat{\mathbf{I}}&\hat{\mathbf{J}}&\hat{\mathbf{K}}\\2.8797(10^6)&-3.4321(10^6)&2.5856(10^6)\\-1365.5&3637.6&6346.8\end{vmatrix}}{7441.7}+(-34,276\hat{\mathbf{I}}+478.57\hat{\mathbf{J}}+38,810\hat{\mathbf{K}})\right]\end{aligned}$$

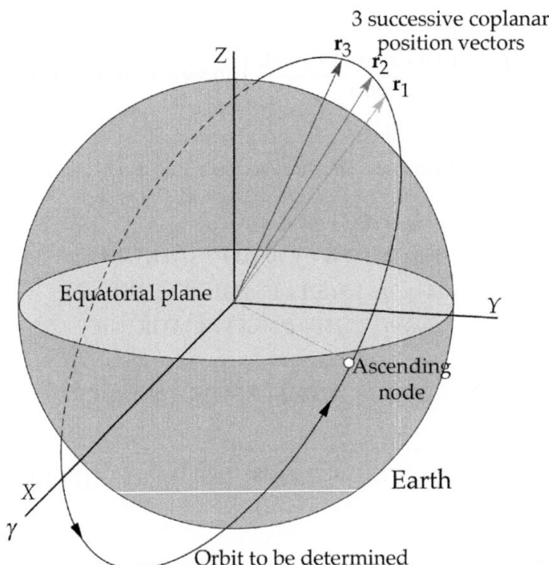

FIG. 5.2

Sketch of the orbit of Example 5.1.

or

$$\mathbf{v}_2 = -6.2174\hat{\mathbf{I}} - 4.0122\hat{\mathbf{J}} + 1.5990\hat{\mathbf{K}}\,(\text{km/s})$$

Step 6:

Using $\mathbf{r}_2$ and $\mathbf{v}_2$, Algorithm 4.2 yields the orbital elements:

$a = 8000\text{km}$
$e = 0.1$
$i = 60°$
$\Omega = 40°$
$\omega = 30°$
$\theta = 50°\,(\text{for position vector}\,\mathbf{r}_2)$

The orbit is sketched in Fig. 5.2.

5.3 LAMBERT'S PROBLEM

Suppose we know the position vectors $\mathbf{r}_1$ and $\mathbf{r}_2$ of two points P_1 and P_2 on the path of mass m around mass M, as illustrated in Fig. 5.3. $\mathbf{r}_1$ and $\mathbf{r}_2$ determine the change in the true anomaly $\Delta\theta$, since

$$\cos\Delta\theta = \frac{\mathbf{r}_1 \cdot \mathbf{r}_2}{r_1 r_2} \tag{5.23}$$

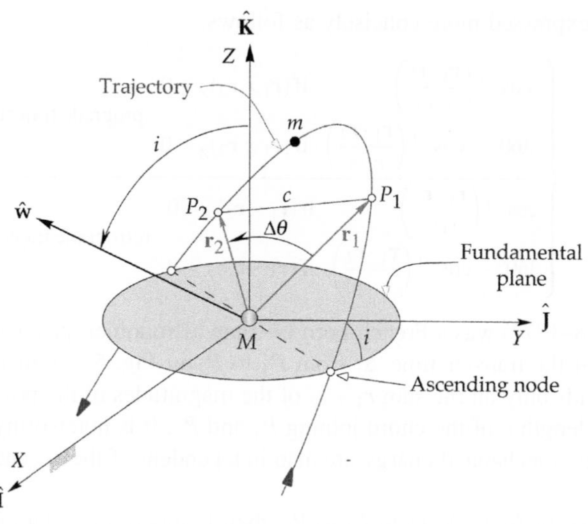

FIG. 5.3

Lambert's problem.

where

$$r_1 = \sqrt{\mathbf{r}_1 \cdot \mathbf{r}_1} \qquad r_2 = \sqrt{\mathbf{r}_2 \cdot \mathbf{r}_2} \tag{5.24}$$

However, if $\cos\Delta\theta > 0$, then $\Delta\theta$ lies in either the first or fourth quadrant; whereas if $\cos\Delta\theta < 0$, then $\Delta\theta$ lies in the second or third quadrant (recall Fig. 3.4). The first step in resolving this quadrant ambiguity is to calculate the Z component of $\mathbf{r}_1 \times \mathbf{r}_2$,

$$(\mathbf{r}_1 \times \mathbf{r}_2)_Z = \hat{\mathbf{K}} \cdot (\mathbf{r}_1 \times \mathbf{r}_2) = \hat{\mathbf{K}} \cdot (r_1 r_2 \sin\Delta\theta \hat{\mathbf{w}}) = r_1 r_2 \sin\Delta\theta (\hat{\mathbf{K}} \cdot \hat{\mathbf{w}})$$

where $\hat{\mathbf{w}}$ is the unit normal to the orbital plane. Therefore, $\hat{\mathbf{K}} \cdot \hat{\mathbf{w}} = \cos i$, where i is the inclination of the orbit, so that

$$(\mathbf{r}_1 \times \mathbf{r}_2)_Z = r_1 r_2 \sin\Delta\theta \cos i \tag{5.25}$$

We use the sign of the scalar $(\mathbf{r}_1 \times \mathbf{r}_2)_Z$ to determine the correct quadrant for $\Delta\theta$.

There are two cases to consider: prograde trajectories ($0 < i < 90°$) and retrograde trajectories ($90° < i < 180°$).

For prograde trajectories (like the one illustrated in Fig. 5.3), $\cos i > 0$, so that if $(\mathbf{r}_1 \times \mathbf{r}_2)_Z > 0$, then Eq. (5.25) implies that $\sin\Delta\theta > 0$, which means $0° < \Delta\theta < 180°$. Since $\Delta\theta$ therefore lies in the first or second quadrant, it follows that $\Delta\theta$ is given by $\cos^{-1}(\mathbf{r}_1 \cdot \mathbf{r}_2 / r_1 r_2)$. On the other hand, if $(\mathbf{r}_1 \times \mathbf{r}_2)_Z < 0$, Eq. (5.25) implies that $\sin\Delta\theta < 0$, which means $180° < \Delta\theta < 360°$. In this case, $\Delta\theta$ lies in the third or fourth quadrant and is given by $360° - \cos^{-1}(\mathbf{r}_1 \cdot \mathbf{r}_2 / r_1 r_2)$. For retrograde trajectories, $\cos i < 0$. Thus, if $(\mathbf{r}_1 \times \mathbf{r}_2)_Z > 0$, then $\sin\Delta\theta < 0$, which places $\Delta\theta$ in the third or fourth quadrant. Similarly, if $(\mathbf{r}_1 \times \mathbf{r}_2)_Z > 0$, $\Delta\theta$ must lie in the first or second quadrant.

This logic can be expressed more concisely as follows:

$$\Delta\theta = \begin{cases} \cos^{-1}\left(\dfrac{\mathbf{r}_1\cdot\mathbf{r}_2}{r_1r_2}\right) & \text{if } (\mathbf{r}_1\times\mathbf{r}_2)_Z \geq 0 \\ 360^\circ - \cos^{-1}\left(\dfrac{\mathbf{r}_1\cdot\mathbf{r}_2}{r_1r_2}\right) & \text{if } (\mathbf{r}_1\times\mathbf{r}_2)_Z < 0 \end{cases} \text{ prograde trajectory} \\ \begin{cases} \cos^{-1}\left(\dfrac{\mathbf{r}_1\cdot\mathbf{r}_2}{r_1r_2}\right) & \text{if } (\mathbf{r}_1\times\mathbf{r}_2)_Z < 0 \\ 360^\circ - \cos^{-1}\left(\dfrac{\mathbf{r}_1\cdot\mathbf{r}_2}{r_1r_2}\right) & \text{if } (\mathbf{r}_1\times\mathbf{r}_2)_Z \geq 0 \end{cases} \text{ retrograde trajectory} \tag{5.26}$$

J.H. Lambert (1728–1777) was a French-born German astronomer, physicist, and mathematician. Lambert proposed that the transfer time Δt from P_1 to P_2 in Fig. 5.3 is independent of the orbit's eccentricity and depends only on the sum $r_1 + r_2$ of the magnitudes of the position vectors, the semimajor axis a, and the length c of the chord joining P_1 and P_2. It is noteworthy that the period (of an ellipse) and the specific mechanical energy are also independent of the eccentricity (Eqs. 2.83, 2.80, and 2.110).

If we know the time of flight Δt from P_1 to P_2, then Lambert's problem is to find the trajectory joining P_1 and P_2. The trajectory is determined once we find $\mathbf{v}_1$, because, according to Eqs. (2.135) and (2.136), the position and velocity of any point on the path are determined by $\mathbf{r}_1$ and $\mathbf{v}_1$. That is, in terms of the notation in Fig. 5.3,

$$\mathbf{r}_2 = f\mathbf{r}_1 + g\mathbf{v}_1 \tag{5.27a}$$

$$\mathbf{v}_2 = \dot{f}\mathbf{r}_1 + \dot{g}\mathbf{v}_1 \tag{5.27b}$$

Solving the first of these for $\mathbf{v}_1$ yields

$$\mathbf{v}_1 = \frac{1}{g}(\mathbf{r}_2 - f\mathbf{r}_1) \tag{5.28}$$

Substitute this result into Eq. (5.27b) to get

$$\mathbf{v}_2 = \dot{f}\mathbf{r}_1 + \frac{\dot{g}}{g}(\mathbf{r}_2 - f\mathbf{r}_1) = \frac{\dot{g}}{g}\mathbf{r}_2 - \frac{f\dot{g} - \dot{f}g}{g}\mathbf{r}_1$$

However, according to Eq. (2.139), $f\dot{g} - \dot{f}g = 1$. Hence,

$$\mathbf{v}_2 = \frac{1}{g}(\dot{g}\mathbf{r}_2 - \mathbf{r}_1) \tag{5.29}$$

By means of Algorithm 4.2, we can find the orbital elements from either $\mathbf{r}_1$ and $\mathbf{v}_1$ or $\mathbf{r}_2$ and $\mathbf{v}_2$. Clearly, Lambert's problem is solved once we determine the Lagrange coefficients f, g, and $\dot{g}$.

The Lagrange f and g coefficients and their time derivatives are listed as functions of the change in true anomaly $\Delta\theta$ in Eq. (2.158),

$$f = 1 - \frac{\mu r_2}{h^2}(1 - \cos\Delta\theta) \qquad g - \frac{r_1r_2}{h}(1 - \sin\Delta\theta) \tag{5.30a}$$

$$\dot{f} = \frac{\mu}{h}\frac{1-\cos\Delta\theta}{\sin\Delta\theta}\left[\frac{\mu}{h^2}(1-\cos\Delta\theta) - \frac{1}{r_1} - \frac{1}{r_2}\right] \qquad \dot{g} = 1 - \frac{\mu r_1}{h_2}(1 - \cos\Delta\theta) \tag{5.30b}$$

Eq. (3.69) express these quantities in terms of the universal anomaly χ,

$$f=1-\frac{\chi^2}{r_1}C(z) \qquad g=\Delta t-\frac{1}{\sqrt{\mu}}\chi^3 S(z) \tag{5.31a}$$

$$\dot{f}=\frac{\sqrt{\mu}}{r_1 r_2}\chi[zS(z)-1] \quad \dot{g}=1-\frac{\chi^2}{r_2}C(z) \tag{5.31b}$$

where $z=\alpha\chi^2$. The f and g functions do not depend on the eccentricity, which would seem to make them an obvious choice for the solution of Lambert's problem.

The unknowns on the right of the above sets of equations are h, χ, and z, whereas $\Delta\theta$, Δt, r_1, and r_2 are given. Equating the four pairs of expressions for f, g, $\dot{f}$, and $\dot{g}$ in Eqs. (5.30) and (5.31) yields four equations in the three unknowns h, χ, and z. However, because of the fact that $f\dot{g}-\dot{f}g=1$, only three of these equations are independent. We must solve them for h, χ, and z to evaluate the Lagrange coefficients and thereby obtain the solution to Lambert's problem. We will follow the procedure presented by Bate et al. (1971) and Bond and Allman (1996).

While $\Delta\theta$ appears throughout Eqs. (5.30a) and (5.30b), the time interval Δt does not. However, Δt does appear in Eqs. (5.31a) and (5.31b). A relationship between $\Delta\theta$ and Δt can therefore be found by equating the two expressions for g,

$$\frac{r_1 r_2}{h}\sin\Delta\theta=\Delta t-\frac{1}{\sqrt{\mu}}\chi^3 S(z) \tag{5.32}$$

To eliminate the unknown angular momentum h, equate the expressions for f in Eqs. (5.30a) and (5.31a),

$$1-\frac{\mu r_2}{h^2}(1-\cos\Delta\theta)=1-\frac{\chi^2}{r_1}C(z)$$

Upon solving this for h, we obtain

$$h=\sqrt{\frac{\mu r_1 r_2(1-\cos\Delta\theta)}{\chi^2 C(z)}} \tag{5.33}$$

(Equating the two expressions for $\dot{g}$ leads to the same result.) Substituting Eq. (5.33) into Eq. (5.32), simplifying, and rearranging the terms yields

$$\sqrt{\mu}\Delta t=\chi^3 S(z)+\chi\sqrt{C(z)}\left(\sin\Delta\theta\sqrt{\frac{r_1 r_2}{1-\cos\Delta\theta}}\right) \tag{5.34}$$

The term in parentheses on the right is a constant that comprises solely the given data. Let us assign it the symbol A,

$$A=\sin\Delta\theta\sqrt{\frac{r_1 r_2}{1-\cos\Delta\theta}} \tag{5.35}$$

Then, Eq. (5.34) assumes the simpler form

$$\sqrt{\mu}\Delta t=\chi^3 S(z)+A\chi\sqrt{C(z)} \tag{5.36}$$

The right-hand side of this equation contains both of the unknown variables χ and z. We cannot use the fact that $z = \alpha\chi^2$ to reduce the unknowns to one since α is the reciprocal of the semimajor axis of the as yet unknown orbit.

To find a relationship between z and χ that does not involve orbital parameters, we equate the expressions for $\dot{f}$ (Eqs. 5.30b and 5.31b) to obtain

$$\frac{\mu}{h}\frac{1-\cos\Delta\theta}{\sin\Delta\theta}\left[\frac{\mu}{h^2}(1-\cos\Delta\theta)-\frac{1}{r_1}-\frac{1}{r_2}\right]=\frac{\sqrt{\mu}}{r_1r_2}\chi[zS(z)-1]$$

Multiplying through by r_1r_2 and substituting for the angular momentum using Eq. (5.33) yields

$$\frac{\mu}{\sqrt{\dfrac{\mu r_1r_2(1-\cos\Delta\theta)}{\chi^2C(z)}}}\frac{1-\cos\Delta\theta}{\sin\Delta\theta}\left[\frac{\mu}{\dfrac{\mu r_1r_2(1-\cos\Delta\theta)}{\chi^2C(z)}}(1-\cos\Delta\theta)-r_1-r_2\right]=\sqrt{\mu}\chi[zS(z)-1]$$

Simplifying and dividing out the common factors leads to

$$\frac{\sqrt{1-\cos\Delta\theta}}{\sqrt{r_1r_2}\sin\Delta\theta}\sqrt{C(z)}\left[\chi^2C(z)-r_1-r_2\right]=zS(z)-1$$

We recognize the reciprocal of A on the left, so we can rearrange this expression to read as follows:

$$\chi^2C(z)=r_1+r_2+A\frac{zS(z)-1}{\sqrt{C(z)}}$$

The right-hand side depends exclusively on z. Let us call that function $y(z)$, so that

$$\chi=\sqrt{\frac{y(z)}{C(z)}} \tag{5.37}$$

where

$$y(z)=r_1+r_2+A\frac{zS(z)-1}{\sqrt{C(z)}} \tag{5.38}$$

Eq. (5.37) is the relation between χ and z that we were seeking. Substituting it back into Eq. (5.36) yields

$$\sqrt{\mu}\Delta t=\left[\frac{y(z)}{C(z)}\right]^{3/2}S(z)+A\sqrt{y(z)} \tag{5.39}$$

We can use this equation to solve for z, given the time interval Δt. It must be done iteratively.

Using Newton's method, we form the function

$$F(z)=\left[\frac{y(z)}{C(z)}\right]^{3/2}S(z)+A\sqrt{y(z)}-\sqrt{\mu}\Delta t \tag{5.40}$$

and its derivative

$$F'(z)=\frac{1}{2\sqrt{y(z)C^5(z)}}\left\{[2C(z)S'(z)-3C'(z)S(z)]y^2(z)+\left[AC^{5/2}(z)+3C(z)S(z)y(z)\right]y'(z)\right\} \tag{5.41}$$

in which $C'(z)$ and $S'(z)$ are the derivatives of the Stumpff functions, which are given by Eq. (3.63). $y'(z)$ is obtained by differentiating $y(z)$ in Eq. (5.38),

$$y'(z)=\frac{A}{2C(z)^{3/2}}\{[1-zS(z)]C'(z)+2[S(z)+zS'(z)]C(z)\}$$

If we substitute Eq. (3.63) into this expression, a much simpler form is obtained; namely

$$y'(z)=\frac{A}{4}\sqrt{C(z)} \tag{5.42}$$

This result can be worked out by using Eqs. (3.52) and (3.53) to express $C(z)$ and $S(z)$ in terms of the more familiar trig functions. Substituting Eq. (5.42) along with Eq. (3.63) into Eq. (5.41) yields

$$F'(z)=\begin{cases}\left[\dfrac{y(z)}{C(z)}\right]^{3/2}\left\{\dfrac{1}{2z}\left[C(z)-\dfrac{3}{2}\dfrac{S(z)}{C(z)}\right]+\dfrac{3}{4}\dfrac{S(z)^2}{C(z)}\right\}+\dfrac{A}{8}\left[3\dfrac{S(z)}{C(z)}\sqrt{y(z)}+A\sqrt{\dfrac{C(z)}{y(z)}}\right] & (z\neq 0)\\[2ex] \dfrac{\sqrt{2}}{40}y(0)^{3/2}+\dfrac{A}{8}\left[\sqrt{y(0)}+A\sqrt{\dfrac{1}{2y(0)}}\right] & (z=0)\end{cases} \tag{5.43}$$

Evaluating $F'(z)$ at $z=0$ must be done carefully (and is therefore shown as a special case) because of the z in the denominator within the curly brackets. To handle $z=0$, we assume that z is very small (almost but not quite zero), so that we can retain just the first two terms in the series expansions of $C(z)$ and $S(z)$ (Eq. 3.51),

$$C(z)=\frac{1}{2}-\frac{z}{24}+\cdots \qquad S(z)=\frac{1}{6}-\frac{z}{120}+\cdots$$

Then, we evaluate the term within the curly brackets as follows:

$$\begin{aligned}\frac{1}{2z}\left[C(z)-\frac{3}{2}\frac{S(z)}{C(z)}\right]&\approx\frac{1}{2z}\left[\left(\frac{1}{2}-\frac{z}{24}\right)-\frac{3}{2}\frac{\left(\frac{1}{6}-\frac{z}{120}\right)}{\left(\frac{1}{2}-\frac{z}{24}\right)}\right]\\ &=\frac{1}{2z}\left[\left(\frac{1}{2}-\frac{z}{24}\right)-3\left(\frac{1}{6}-\frac{z}{120}\right)\left(1-\frac{z}{12}\right)^{-1}\right]\\ &\approx\frac{1}{2z}\left[\left(\frac{1}{2}-\frac{z}{24}\right)-3\left(\frac{1}{6}-\frac{z}{120}\right)\left(1+\frac{z}{12}\right)\right]\\ &=\frac{1}{2z}\left(-\frac{7z}{120}+\frac{z^2}{480}\right)\\ &=-\frac{7}{240}+\frac{z}{960}\end{aligned}$$

In the third step, we used the familiar binomial expansion theorem,

$$(a+b)^n=a^n+na^{n-1}b+\frac{n(n-1)}{2!}a^{n-2}b^2+\frac{n(n-1)(n-2)}{3!}a^{n-3}b^3+\cdots \tag{5.44}$$

to set $(1 - z/12)^{-1} \approx 1 + z/12$, which is true if z is close to zero. Thus, when z is actually zero,

$$\frac{1}{2z}\left[C(z) - \frac{3}{2}\frac{S(z)}{C(z)}\right] = -\frac{7}{240}$$

Evaluating the other terms in $F'(z)$ presents no difficulties.

$F(z)$ in Eq. (5.40) and $F'(z)$ in Eq. (5.43) are used in Newton's formula (Eq. 3.16) for the iterative procedure,

$$z_{i+1} = z_i - \frac{F(z_i)}{F'(z_i)} \tag{5.45}$$

For choice of a starting value for z, recall that $z = (1/a)\chi^2$. According to Eq. (3.57), $z = E^2$ for an ellipse and $z = -F^2$ for a hyperbola. Since we do not know what the orbit is, setting $z_0 = 0$ seems a reasonable, simple choice. Alternatively, we can plot or tabulate $F(z)$ and choose z_0 to be a point near where $F(z)$ changes sign.

Substituting Eqs. (5.37) and (5.39) into Eqs. (5.31a) and (5.31b) yields the Lagrange coefficients as functions of z alone.

$$f = 1 - \frac{\left[\sqrt{\frac{y(z)}{C(z)}}\right]^2}{r_1}C(z) = 1 - \frac{y(z)}{r_1} \tag{5.46a}$$

$$g = \frac{1}{\sqrt{\mu}}\left\{\left[\frac{y(z)}{C(z)}\right]^{3/2} S(z) + A\sqrt{y(z)}\right\} - \frac{1}{\sqrt{\mu}}\left[\frac{y(z)}{C(z)}\right]^{3/2} S(z) = A\sqrt{\frac{y(z)}{\mu}} \tag{5.46b}$$

$$\dot{f} = \frac{\sqrt{\mu}}{r_1 r_2}\sqrt{\frac{y(z)}{C(z)}}[zS(z) - 1] \tag{5.46c}$$

$$\dot{g} = 1 - \frac{\left[\sqrt{\frac{y(z)}{C(z)}}\right]^2}{r_2}C(z) = 1 - \frac{y(z)}{r_2} \tag{5.46d}$$

We are now in a position to present the solution of Lambert's problem in universal variables, following Bond and Allman (1996).

ALGORITHM 5.2

To solve Lambert's problem use the MATLAB implementation that appears in Appendix D.25. Given $\mathbf{r}_1$, $\mathbf{r}_2$, and Δt, the steps are as follows:

1. Calculate r_1 and r_2 using Eq. (5.24).
2. Choose either a prograde or a retrograde trajectory and calculate $\Delta\theta$ using Eq. (5.26).
3. Calculate A in Eq. (5.35).
4. By iteration, using Eqs. (5.40), (5.43), and (5.45), solve Eq. (5.39) for z. The sign of z tells us whether the orbit is a hyperbola ($z < 0$), parabola ($z = 0$), or ellipse ($z = 0$).
5. Calculate y using Eq. (5.38).
6. Calculate the Lagrange f, g, and $\dot{g}$ functions using Eqs. (5.46a), (5.46b), (5.46c), and (5.46d).

7. Calculate $\mathbf{v}_1$ and $\mathbf{v}_2$ from Eqs. (5.28) and (5.29).
8. Use $\mathbf{r}_1$ and $\mathbf{v}_1$ (or $\mathbf{r}_2$ and $\mathbf{v}_2$) in Algorithm 4.2 to obtain the orbital elements.

EXAMPLE 5.2

The position of an earth satellite is first determined to be

$$\mathbf{r}_1 = 5000\hat{\mathbf{I}} + 10{,}000\hat{\mathbf{J}} + 2100\hat{\mathbf{K}}\,(\text{km})$$

After 1 h the position vector is

$$\mathbf{r}_2 = -14{,}600\hat{\mathbf{I}} + 2500\hat{\mathbf{J}} + 7000\hat{\mathbf{K}}\,(\text{km})$$

Determine the orbital elements and find the perigee altitude and the time since perigee passage of the first sighting.

Solution

We must first execute the steps of Algorithm 5.2 to find $\mathbf{v}_1$ and $\mathbf{v}_2$.
Step 1:

$$r_1 = \sqrt{5000^2 + 10{,}000^2 + 2100^2} = 11{,}375 \text{ km}$$

$$r_2 = \sqrt{(-14,600)^2 + 2500^2 + 7000^2} = 16{,}383 \text{ km}$$

Step 2: Assume a prograde trajectory.

$$\mathbf{r}_1 \times \mathbf{r}_2 = \left(64.75\hat{\mathbf{I}} - 65.66\hat{\mathbf{J}} + 158.5\hat{\mathbf{K}}\right)\left(10^6\right)\left(\text{km}^2\right)$$

$$\cos^{-1}\frac{r_1 \cdot r_2}{r_1 r_2} = 100.29° \quad \text{or} \quad 259.71°$$

Since the trajectory is prograde and the z component $\mathbf{r}_1 \times \mathbf{r}_2$ is positive, it follows from Eq. (5.26) that

$$\Delta\theta = 100.29°$$

Step 3:

$$A = \sin\Delta\theta\sqrt{\frac{r_1 r_2}{1-\cos\Delta\theta}} = \sin 100.29°\sqrt{\frac{11{,}375 \times 16{,}383}{1-\cos 100.29°}} = 12{,}372\text{km}$$

Step 4:

Using this value of A and $\Delta t = 3600$ s, we can evaluate the functions $F(z)$ and $F'(z)$ given by Eqs. (5.40) and (5.43), respectively. Let us first plot $F(z)$ to estimate where it crosses the z axis. As can be seen from Fig. 5.4, $F(z) = 0$ near $z = 1.5$. With $z_0 = 1.5$ as our initial estimate, we execute Newton's procedure (Eq. 5.45), $z_{i+1} = z_i - F(z_i)/F'(z_i)$:

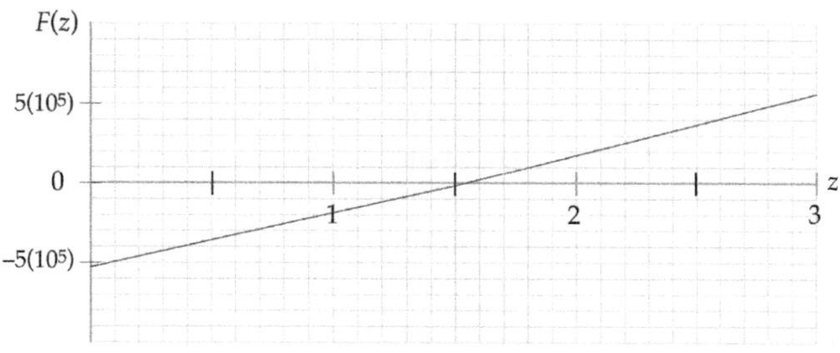

FIG. 5.4

Graph of $F(z)$.

$$z_1 = 1.5 - \frac{-14{,}476.4}{362{,}642} = 1.53991$$

$$z_2 = 1.53991 - \frac{23.6274}{363{,}828} = 1.53985$$

$$z_3 = 1.53985 - \frac{6.29457 \times 10^{-5}}{363{,}826} = 1.53985$$

Thus, to five significant figures $z = 1.5398$. The fact that z is positive means the orbit is an ellipse.

Step 5:

$$y = r_1 + r_2 + A\frac{zS(z) - 1}{\sqrt{C(z)}} = 11{,}375 + 16{,}383 + 12{,}372\frac{1.5398\overbrace{S(1.5398)}^{0.154296}}{\sqrt{\underbrace{C(1.5398)}_{0.439046}}} = 13{,}523\,\text{km}$$

Step 6:
Eqs. (5.46a)–(5.46d) yield the Lagrange functions

$$f = 1 - \frac{y}{r_1} = 1 - \frac{13{,}523}{11{,}375} = -0.18877$$

$$g = A\sqrt{\frac{y}{\mu}} = 12{,}372\sqrt{\frac{13{,}523}{398{,}600}} = 2278.9\,\text{s}$$

$$\dot{g} = 1 - \frac{y}{r_2} = 1 - \frac{13{,}523}{16{,}383} = 0.17457$$

Step 7:

$$\begin{aligned}\mathbf{v}_1 &= \frac{1}{g}(\mathbf{r}_2 - f\mathbf{r}_1)\\ &= \frac{1}{2278.9}\left[(-14{,}600\hat{\mathbf{I}} + 2500\hat{\mathbf{J}} + 7000\hat{\mathbf{K}}) - (-0.18877)(5000\hat{\mathbf{I}} + 10{,}000\hat{\mathbf{J}} + 2100\hat{\mathbf{K}})\right]\\ &= -5.9925\hat{\mathbf{I}} + 1.9254\hat{\mathbf{J}} + 3.2456\hat{\mathbf{K}}\ (\text{km/s})\end{aligned}$$

$$\begin{aligned}\mathbf{v}_2 &= \frac{1}{g}(\dot{g}\mathbf{r}_2 - \mathbf{r}_1)\\ &= \frac{1}{2278.9}\left[(0.17457)(-14{,}600\hat{\mathbf{I}} + 2500\hat{\mathbf{J}} + 7000\hat{\mathbf{K}}) - (5000\hat{\mathbf{I}} + 10{,}000\hat{\mathbf{J}} + 2100\hat{\mathbf{K}})\right]\\ &= -3.3125\hat{\mathbf{I}} - 4.1966\hat{\mathbf{J}} - 0.38529\hat{\mathbf{K}}\ (\text{km/s})\end{aligned}$$

Step 8:
Using $\mathbf{r}_1$ and $\mathbf{v}_1$, Algorithm 4.2 yields the orbital elements:

$$\boxed{\begin{aligned}h &= 80{,}470\,\text{km}^2/\text{s}\\ a &= 20{,}000\,\text{km}\\ e &= 0.4335\\ \Omega &= 44.60^\circ\\ i &= 30.19^\circ\\ \omega &= 30.71^\circ\\ \theta_1 &= 350.8^\circ\end{aligned}}$$

This elliptical orbit is plotted in Fig. 5.5. The perigee of the orbit is

$$r_p = \frac{h^2}{\mu}\frac{1}{1 + e\cos(0)} = \frac{80{,}470^2}{398{,}600}\frac{1}{1 + 0.4335} = 11{,}330\ \text{km}$$

Therefore, the perigee altitude is $11{,}330 - 6378 = \boxed{4952\,\text{km}}$.

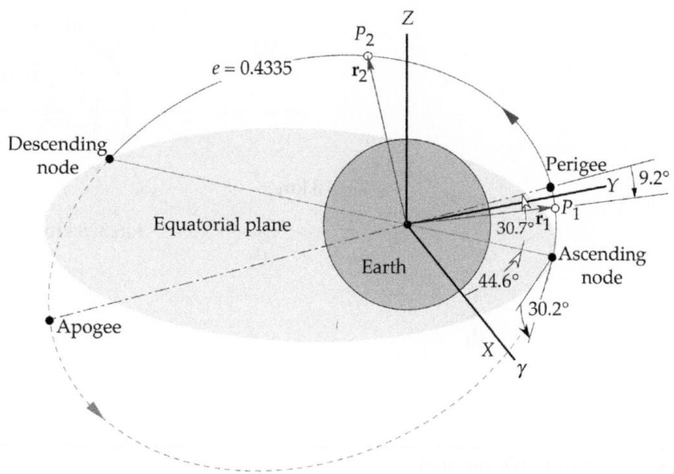

FIG. 5.5

The solution of Example 5.2 (Lambert's problem).

To find the time of the first sighting, we first calculate the eccentric anomaly by means of Eq. (3.13b),

$$E_1 = 2\tan^{-1}\left(\sqrt{\frac{1-e}{1+e}}\tan\frac{\theta}{2}\right) = 2\tan^{-1}\left(\sqrt{\frac{1-0.4335}{1+0.4335}}\tan\frac{350.8°}{2}\right) = 2\tan^{-1}(-0.05041)$$
$$= -0.1007\,\text{rad}$$

Then using Kepler's equation for the ellipse (Eq. 3.14), we find the mean anomaly,

$$M_{e_1} = E_1 - e\sin E_1 = -0.1007 - 0.4335\sin(-0.1007) = -0.05715\,\text{rad}$$

so that from Eq. (3.7), the time since perigee passage is

$$t_1 = \frac{h^3}{\mu^2}\frac{1}{(1-e^2)^{3/2}}M_{e_1} = \frac{80{,}470^3}{398{,}600^2}\frac{1}{(1-.4335^2)^{3/2}}(-0.05715) = -256.1\,\text{s}$$

The minus sign means that, after the initial sighting, there are 256.1 s until perigee encounter.

EXAMPLE 5.3

A meteoroid is sighted at an altitude of 267,000 km. After 13.5 h and a change in true anomaly of 5°, the altitude is observed to be 140,000 km. Calculate the perigee altitude and the time to perigee after the second sighting.

Solution

We have

$$P_1: \quad r_1 = 6378 + 267{,}000 = 273{,}378\,\text{km}$$
$$P_2: \quad r_2 = 6378 + 140{,}000 = 146{,}378\,\text{km}$$
$$\Delta t = 13.5 \times 3600 = 48{,}600\,\text{s}$$
$$\Delta\theta = 5°$$

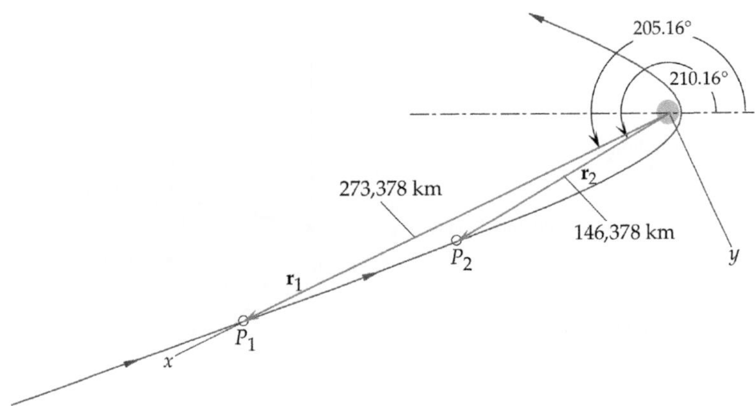

FIG. 5.6

Solution of Example 5.3 (Lambert's problem).

Since r_1, r_2, and $\Delta\theta$ are given, we can skip to Step 3 of Algorithm 5.2 and compute

$$A = 2.8263(10^5)\,\text{km}$$

Then, solving for z as in the previous example, we obtain

$$z = -0.17344$$

Since z is negative, the path of the meteoroid is a hyperbola.

With z available, we evaluate the Lagrange functions,

$$\begin{aligned} f &= 0.95846 \\ g &= 47{,}708\,\text{s} \\ \dot{g} &= 0.92241 \end{aligned} \tag{a}$$

Step 7 requires the initial and final position vectors. Therefore, for the purposes of this problem, let us define a geocentric coordinate system with the x axis aligned with $\mathbf{r}_1$ and the y axis at 90° thereto in the direction of the motion (Fig. 5.6). The z axis is therefore normal to the plane of the orbit. Then,

$$\begin{aligned} \mathbf{r}_1 &= r_1\hat{\mathbf{i}} = 273{,}378\hat{\mathbf{i}}\,(\text{km}) \\ \mathbf{r}_2 &= r_2\cos\Delta\theta\hat{\mathbf{i}} + r_2\sin\Delta\theta\hat{\mathbf{j}} = 145{,}820\hat{\mathbf{i}} + 12{,}758\hat{\mathbf{j}}\,(\text{km}) \end{aligned} \tag{b}$$

With Eqs. (a) and (b), we obtain the velocity at P_1,

$$\begin{aligned} \mathbf{v}_1 &= \frac{1}{g}(\mathbf{r}_2 - f\mathbf{r}_1) \\ &= \frac{1}{47{,}708}\left[\left(145{,}820\hat{\mathbf{i}} + 12{,}758\hat{\mathbf{j}}\right) - 0.95846\left(273{,}378\hat{\mathbf{i}}\right)\right] \\ &= -2.4356\hat{\mathbf{i}} + 0.26741\hat{\mathbf{j}}\,(\text{km/s}) \end{aligned}$$

Using $\mathbf{r}_1$ and $\mathbf{v}_1$, Algorithm 4.2 yields

$$h = 73{,}105\,\text{km}^2/\text{s} \qquad e = 1.0506 \qquad \theta_1 = 205.06°$$

The orbit is now determined except for its orientation in space, for which no information was provided. In the plane of the orbit, the trajectory is as shown in Fig. 5.6.

The perigee radius is

$$r_p = \frac{h^2}{\mu}\frac{1}{1+e\cos(0)} = 6538.2\text{km}$$

which means the perigee altitude is dangerously low for a large meteoroid,

$$z_p = 6538.2 - 6378 = \boxed{160.2\text{km}\,(100\text{miles})}$$

To find the time of flight from P_2 to perigee, we note that the true anomaly of P_2 is

$$\theta_2 = \theta_1 + 5° = 210.16°$$

The hyperbolic eccentric anomaly F_2 follows from Eq. (3.44a),

$$F_2 = 2\tanh^{-1}\left(\sqrt{\frac{e-1}{e+1}}\tan\frac{\theta_2}{2}\right) = -1.3347\text{rad}$$

Substituting this value into Kepler's equation (Eq. 3.40) yields the mean anomaly,

$$M_{h_2} = e\sin h(F_2) - F_2 = -0.52265\text{rad}$$

Finally, Eq. (3.34) yields the time

$$t_2 = \frac{M_{h_2}h^3}{\mu^2(e^2-1)^{3/2}} = -38{,}396\text{s}$$

The minus sign means that 38,396 s (a scant 10.6 h) remain until the meteoroid passes through perigee.

5.4 SIDEREAL TIME

To deduce the orbit of a satellite or celestial body from observations requires, among other things, recording the time of each observation. The time we use in everyday life, the time we set our clocks by, is the solar time. It is reckoned by the motion of the sun across the sky. A solar day is the time required for the sun to return to the same position overhead (i.e., to lie on the same meridian). A solar day—from high noon to high noon—comprises 24 h. Universal time (UT) is determined by the sun's passage across the Greenwich meridian, which is 0° terrestrial longitude (see Fig. 1.18). At noon UT, the sun lies on the Greenwich meridian. Local standard time, or civil time, is obtained from UT by adding 1 h for each time zone between Greenwich and the site, measured westward.

Sidereal time is measured by the rotation of the earth relative to the fixed stars (i.e., the celestial sphere, Fig. 4.3). The time it takes for a distant star to return to its same position overhead (i.e., to lie on the same meridian) is one sidereal day (24 sidereal hours). As illustrated in Fig. 4.20, the earth's orbit around the sun results in the sidereal day being slightly shorter than the solar day. One sidereal day is 23 h and 56 min. To put it another way, the earth rotates 360° in one sidereal day, whereas it rotates 360.986° in a solar day.

Local sidereal time θ (not to be confused with true anomaly θ) of a site is the time elapsed since the local meridian of the site passed through the vernal equinox. The number of degrees (measured eastward) between the vernal equinox and the local meridian is the sidereal time multiplied by 15. To know the location of a point on the earth at any given instant relative to the geocentric equatorial frame requires knowing its local sidereal time. The local sidereal time of a site is found by first determining the Greenwich sidereal time θ_G (the sidereal time of the Greenwich meridian) and then adding the east longitude (or subtracting the west longitude) of the site. Algorithms for determining sidereal time rely on the notion of Julian day.

The Julian day number is the number of days since noon UT on January 1, 4713 BCE. The origin of this timescale is placed in antiquity so that, except for prehistoric events, we do not have to deal with positive and negative dates. The Julian day count is uniform and continuous and does not involve leap years or different numbers of days in different months. The number of days between two events is found by simply subtracting the Julian day of one from that of the other. The Julian day begins at noon rather than at midnight so that astronomers observing the heavens at night would not have to deal with a change of date during their watch.

The Julian day numbering system is not to be confused with the Julian calendar, which the Roman emperor Julius Caesar introduced in 46 BCE. The Gregorian calendar, introduced in 1583, has largely supplanted the Julian calendar and is in common civil use today throughout much of the world.

J_0 is the symbol for the Julian day number at 0 h UT (which is halfway into the Julian day). At any other UT, the Julian day is given by

$$JD = J_0 + \frac{UT}{24} \tag{5.47}$$

Algorithms and tables for obtaining J_0 from the ordinary year (y), month (m), and day (d) exist in the literature and on the World Wide Web. One of the simplest formulas is found in Boulet (1991),

$$J_0 = 367y - \text{INT}\left\{\frac{7\left[y + \text{INT}\left(\frac{m+9}{12}\right)\right]}{4}\right\} + \text{INT}\left(\frac{275m}{9}\right) + d + 1{,}721{,}013.5 \tag{5.48}$$

where y, m, and d are integers lying in the following ranges:

$$1901 \le y \le 2099$$
$$1 \le m \le 12$$
$$1 \le d \le 31$$

INT(x) means retaining only the integer portion of x, without rounding (or, in other words, round toward zero). For example, INT(-3.9) $= -3$ and INT(3.9) $= 3$. Appendix D.26 lists a MATLAB implementation of Eq. (5.48).

EXAMPLE 5.4

What is the Julian day number for May 12, 2004, at 14:45:30 UT?

Solution

In this case $y = 2004$, $m = 5$, and $d = 12$. Therefore, Eq. (5.48) yields the Julian day number at 0 h UT,

$$J_0 = 367 \times 2004 - \text{INT}\left\{\frac{7\left[2004 + \left(\frac{5+9}{12}\right)\right]}{4}\right\} + \text{INT}\left(\frac{275 \times 5}{9}\right) + 12 + 1{,}721{,}013.5$$

$$= 735{,}468 - \text{INT}\left\{\frac{7[2004 + 1]}{4}\right\} + 152 + 12 + 1{,}721{,}013.5$$

$$= 735{,}468 - 3508 + 152 + 12 + 1{,}721{,}013.5$$

or

$$J_0 = 2{,}453{,}137.5\,\text{days}$$

The universal time, in hours, is

$$\text{UT} = 14 + \frac{45}{60} + \frac{30}{3600} = 14.758\,\text{h}$$

Therefore, from Eq. (5.47), we obtain the Julian day number at the desired universal time,

$$JD = 2{,}453{,}137.5 + \frac{14.758}{24} = \boxed{2{,}453{,}138.115\,\text{days}}$$

EXAMPLE 5.5

Find the elapsed time between October 4, 1957 UT 19:26:24, and the date of the previous example.

Solution

Proceeding as in Example 5.4 we find that the Julian day number of the given event (the launch of the first man-made satellite, Sputnik I) is

$$JD_1 = 2{,}436{,}116.3100\,\text{days}$$

The Julian day of the previous example is

$$JD_2 = 2{,}453{,}138.1149\,\text{days}$$

Hence, the elapsed time is

$$\Delta JD = 2{,}453{,}138.1149 - 2{,}436{,}116.3100 = 17{,}021.805\,\text{days}\ \boxed{(46\,\text{years}, 220\,\text{days})}$$

The current Julian epoch is defined to have been noon on January 1, 2000. This epoch is denoted J2000 and has the exact Julian day number 2,451,545.0. Since there are 365.25 days in a Julian year, a Julian century has 36,525 days. It follows that the time T_0 in Julian centuries between the Julian day J_0 and J2000 is

$$T_0 = \frac{J_0 - 2{,}451{,}545}{36{,}525} \tag{5.49}$$

The Greenwich sidereal time θ_{G_0} at 0 h UT may be found in terms of this dimensionless time (Seidelmann, 1992, Section 2.24). θ_{G_0} is in degrees and is given by the series

$$\theta_{G_0} = 100.4606184 + 36{,}000.77004T_0 + 0.000387933T_0^2 - 2.583\left(10^{-8}\right)T_0^3\ (\text{degrees}) \tag{5.50}$$

This formula can yield a value outside the range $0 \le \theta_{G_0} \le 360°$. If so, then the appropriate integer multiple of 360° must be added or subtracted to bring θ_{G_0} into that range.

Once θ_{G_0} has been determined, the Greenwich sidereal time θ_G at any other UT is found using the relation

$$\theta_G = \theta_{G_0} + 360.98564724\frac{\text{UT}}{24} \tag{5.51}$$

where UT is in hours. The coefficient of the second term on the right is the number of degrees the earth rotates in 24 h (solar time).

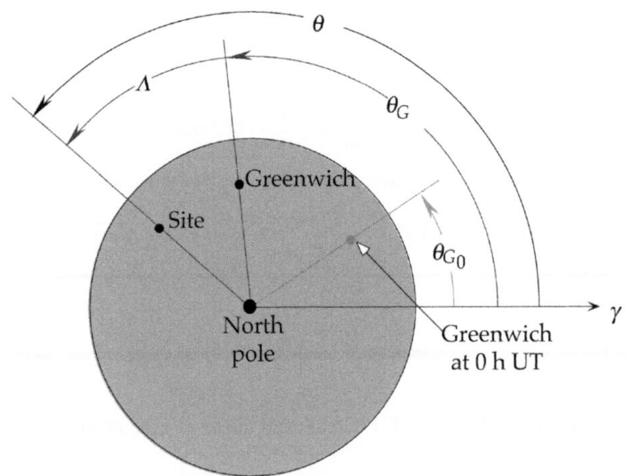

FIG. 5.7

Schematic of the relationship among θ_{G_0}, θ_G, Λ, and θ.

Finally, the local sidereal time θ of a site is obtained by adding its east longitude Λ to the Greenwich sidereal time,

$$\theta = \theta_G + \Lambda \tag{5.52}$$

Here again, it is possible for the computed value of θ to exceed 360°. If so, it must be reduced to within that limit by subtracting the appropriate integer multiple of 360°. Fig. 5.7 illustrates the relationship among θ_{G_0}, θ_G, Λ, and θ.

ALGORITHM 5.3

Calculate the local sidereal time, given the date, the local time, and the east longitude of the site. This is implemented in MATLAB in Appendix D.27.

1. Using the year, month, and day, calculate J_0 using Eq. (5.48).
2. Calculate T_0 by means of Eq. (5.49).
3. Compute θ_{G_0} from Eq. (5.50). If θ_{G_0} lies outside the range $0° \le \theta_{G_0} \le 360°$, then subtract the multiple of 360° required to place θ_{G_0} in that range.
4. Calculate θ_G using Eq. (5.51).
5. Calculate the local sidereal time θ by means of Eq. (5.52), adjusting the final value so it lies between 0° and 360°.

EXAMPLE 5.6

Use Algorithm 5.3 to find the local sidereal time (in degrees) of Tokyo, Japan, on March 3, 2004, at 4:30:00 UT. The east longitude of Tokyo is 139.80°. (This places Tokyo nine time zones ahead of Greenwich, so the local time is 1:30 in the afternoon.)

Step 1:

$$J_0 = 367 \times 2004 - \text{INT}\left\{\frac{7\left[2004 + \text{INT}\left(\frac{3+9}{12}\right)\right]}{4}\right\} + \text{INT}\left(\frac{275 \times 3}{9}\right) + 3 + 1{,}721{,}013.5$$

$$= 2{,}453{,}067.5\,\text{days}$$

Recall that the 0.5 means that we are halfway into the Julian day, which began at noon UT of the previous day.

Step 2:

$$T_0 = \frac{2{,}453{,}067.5 - 2{,}451{,}545}{36{,}525} = 0.041683778$$

Step 3:

$$\begin{aligned}\theta_{G_0} &= 100.4606184 + 36{,}000.77004(0.041683778) \\ &\quad + 0.000387933(0.041683778)^2 - 2.583\left(10^{-8}\right)(0.041683778)^3 \\ &= 1601.1087^\circ\end{aligned}$$

The right-hand side is too large. We must reduce θ_{G_0} to an angle that does not exceed 360°. To that end, observe that

$$\text{INT}(1601.1087/360) = 4$$

Hence,

$$\theta_{G_0} = 1601.1087 - 4 \times 360 = 161.10873^\circ \tag{a}$$

Step 4:

The UT of interest in this problem is

$$UT = 4 + \frac{30}{60} + \frac{0}{3600} = 4.5\text{h}$$

Substitute this and Eq. (a) into Eq. (5.51) to get the Greenwich sidereal time.

$$\theta_G = 161.10873 + 360.98564724\frac{4.5}{24} = 228.79354^\circ$$

Step 5:

Add the east longitude of Tokyo to this value to obtain the local sidereal time,

$$\theta = 228.79354 + 139.80 = 368.59^\circ$$

To reduce this result into the range $0 \le \theta \le 360^\circ$, we must subtract 360° to get

$$\theta = 368.59 - 360 = \boxed{8.59^\circ \;\; 0.573\text{h}}$$

Note that the right ascension of a celestial body lying on Tokyo's meridian is 8.59°.

5.5 TOPOCENTRIC COORDINATE SYSTEM

A topocentric coordinate system is one that is centered at the observer's location on the surface of the earth. Consider an object B, a satellite or celestial body, and an observer O on the earth's surface, as illustrated in Fig. 5.8. $\mathbf{r}$ is the position of the body B relative to the center of attraction C; $\mathbf{R}$ is the position vector of the observer relative to C; and $\boldsymbol{\rho}$ is the position vector of the body B relative to the observer. $\mathbf{r}$, $\mathbf{R}$, and $\boldsymbol{\rho}$ comprise the fundamental vector triangle. The relationship among these three vectors is

$$\mathbf{r} = \mathbf{R} + \boldsymbol{\rho} \tag{5.53}$$

As we know, the earth is not a sphere but a slightly oblate spheroid. This ellipsoidal shape is exaggerated in Fig. 5.8. The location of the observation site O is determined by specifying its east longitude Λ and latitude ϕ. East longitude Λ is measured positive eastward from the Greenwich meridian to the meridian through O. The angle between the vernal equinox direction (XZ plane) and the meridian of

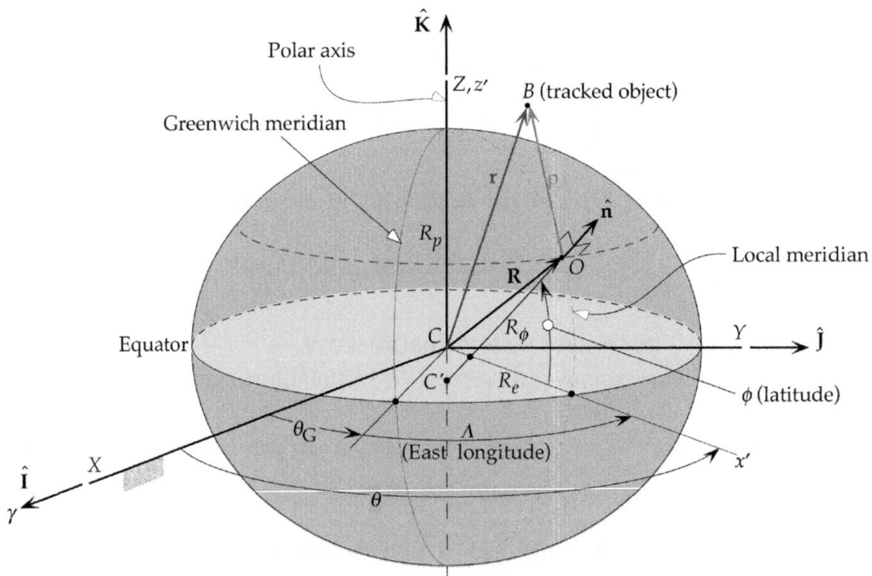

FIG. 5.8

Oblate spheroidal earth (exaggerated).

O is the local sidereal time θ. Likewise, θ_G is the Greenwich sidereal time. Once we know θ_G, then the local sidereal time is given by Eq. (5.52).

Latitude ϕ is the angle between the equator and the normal $\hat{\mathbf{n}}$ to the earth's surface at O. Since the earth is not a perfect sphere, the position vector $\mathbf{R}$, directed from the center C of the earth to O, does not point in the direction of the normal except at the equator and the poles.

The oblateness, or flattening f, was defined in Section 4.7,

$$f = \frac{R_e - R_p}{R_e}$$

where R_e is the equatorial radius, and R_p is the polar radius. (Review from Table 4.3 that $f = 0.003353$ for the earth.) Fig. 5.9 shows the ellipse of the meridian through O. Obviously, R_e and R_p are, respectively, the semimajor and semiminor axes of the ellipse. According to Eq. (2.76),

$$R_p = R_e\sqrt{1-e^2}$$

It is easy to show from the above two relations that flattening and eccentricity are related as follows:

$$e = \sqrt{2f - f^2} \quad f = 1 - \sqrt{1-e^2}$$

As illustrated in Fig. 5.8 and again in Fig. 5.9, the normal $\hat{\mathbf{n}}$ to the earth's surface at O intersects the polar axis at a point C' that lies below the center C of the earth (if O is in the northern hemisphere). The angle ϕ between the normal and the equator is called the geodetic latitude, as opposed to geocentric latitude ϕ', which is the angle between the equatorial plane and the line joining O to the center of the

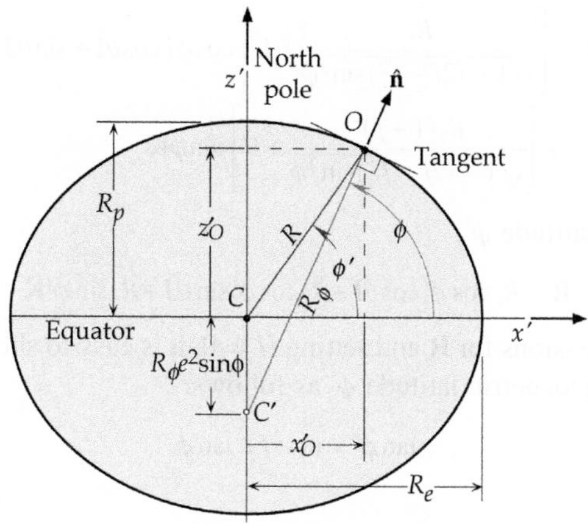

FIG. 5.9

The relationship between geocentric latitude (ϕ') and geodetic latitude (ϕ).

earth. The distance from C to C' is $R_\phi e^2 \sin\phi$, where R_ϕ, the distance from C' to O, is a function of latitude (Seidelmann, 1992, Section 5.2.4)

$$R_\phi = \frac{R_e}{\sqrt{1-e^2\sin^2\phi}} = \frac{R_e}{\sqrt{1-(2f-f^2)\sin^2\phi}} \tag{5.54}$$

Thus, the meridional coordinates of O are

$$x'_O = R_\phi \cos\phi$$
$$z'_O = (1-e^2)R_\phi \sin\phi = (1-f)^2 R_\phi \sin\phi$$

If the observation point O is at an elevation H above the ellipsoidal surface, then we must add $H\cos\phi$ to x'_O and $H\sin\phi$ to z'_O to obtain

$$x'_O = R_c\cos\phi \qquad z'_O = R_s\sin\phi \tag{5.55a}$$

where

$$R_c = R_\phi + H \qquad R_s = (1-f)^2R_\phi + H \tag{5.55b}$$

Observe that, whereas R_c is the distance of O from point C' on the earth's axis, R_s is the distance from O to the intersection of the line OC' with the equatorial plane.

The geocentric equatorial coordinates of O are

$$X = x'_O\cos\theta \qquad Y = x'_O\sin\theta \qquad Z = z'_O$$

where θ is the local sidereal time given in Eq. (5.52). Hence, the position vector $\mathbf{R}$ shown in Fig. 5.8 is

$$\mathbf{R} = R_c\cos\phi\cos\theta\hat{\mathbf{I}} + R_c\cos\phi\sin\theta\hat{\mathbf{J}} + R_s\sin\phi\hat{\mathbf{K}}$$

Substituting Eqs. (5.54) and (5.55b) yields

$$\mathbf{R} = \left[\frac{R_e}{\sqrt{1-(2f-f^2)\sin^2\phi}} + H\right]\cos\phi\left(\cos\theta\hat{\mathbf{I}} + \sin\theta\hat{\mathbf{J}}\right) + \left[\frac{R_e(1-f)^2}{\sqrt{1-(2f-f^2)\sin^2\phi}} + H\right]\sin\phi\hat{\mathbf{K}} \tag{5.56}$$

In terms of the geocentric latitude ϕ',

$$\mathbf{R} = R_e\cos\phi'\cos\theta\hat{\mathbf{I}} + R_e\cos\phi'\sin\theta\hat{\mathbf{J}} + R_e\sin\phi'\hat{\mathbf{K}}$$

By equating these two expressions for $\mathbf{R}$ and setting $H = 0$, it is easy to show that at sea level the geodetic latitude is related to geocentric latitude ϕ' as follows:

$$\tan\phi' = (1-f)^2\tan\phi$$

5.6 TOPOCENTRIC EQUATORIAL COORDINATE SYSTEM

The topocentric equatorial coordinate system with the origin at point O on the surface of the earth uses a nonrotating set of xyz axes through O that coincide in direction with the XYZ axes of the geocentric equatorial frame, as illustrated in Fig. 5.10. As can be inferred from the figure, the relative position vector $\boldsymbol{\rho}$ in terms of the topocentric right ascension and declination is

$$\boldsymbol{\rho} = \rho\cos\delta\cos\alpha\hat{\mathbf{I}} + \rho\cos\delta\sin\alpha\hat{\mathbf{J}} + \rho\sin\delta\hat{\mathbf{K}}$$

since at all times $\hat{\mathbf{i}} = \hat{\mathbf{I}}$, $\hat{\mathbf{j}} = \hat{\mathbf{J}}$, and $\hat{\mathbf{k}} = \hat{\mathbf{K}}$ for this frame of reference. We can write $\boldsymbol{\rho}$ as

$$\boldsymbol{\rho} = \rho\hat{\boldsymbol{\rho}}$$

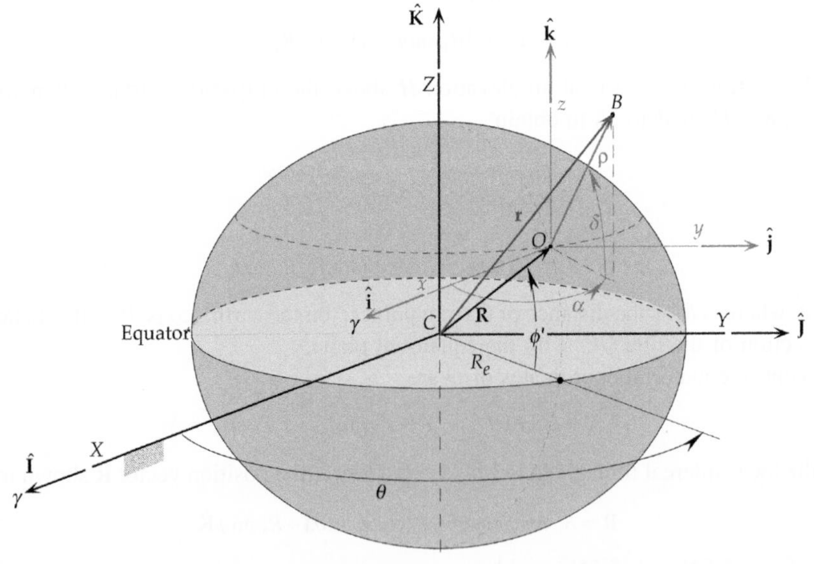

FIG. 5.10

Topocentric equatorial coordinate system.

where ρ is the slant range and $\hat{\boldsymbol{\rho}}$ is the unit vector in the direction of the position vector $\boldsymbol{\rho}$,

$$\hat{\boldsymbol{\rho}} = \cos\delta\cos\alpha\hat{\mathbf{I}} + \cos\delta\sin\alpha\hat{\mathbf{J}} + \sin\delta\hat{\mathbf{K}} \tag{5.57}$$

Since the origins of the geocentric and topocentric systems do not coincide, the direction cosines of the position vectors $\mathbf{r}$ and $\boldsymbol{\rho}$ will in general differ. In particular, the topocentric right ascension and declination of an earth-orbiting body B will not be the same as the geocentric right ascension and declination. This is an example of parallax. On the other hand, if $\|\mathbf{r}\| \gg \|\mathbf{R}\|$, then the difference between the geocentric and topocentric position vectors, and hence, the right ascension and declination, is negligible. This is true for distant planets and stars.

EXAMPLE 5.7

At the instant the Greenwich sidereal time is $\theta_G = 126.7°$, the geocentric equatorial position vector of the International Space Station is

$$\mathbf{r} = -5368\hat{\mathbf{I}} - 1784\hat{\mathbf{J}} + 3691\hat{\mathbf{K}}\,(\text{km})$$

Find its topocentric right ascension and declination at sea level ($H = 0$), latitude $\phi = 20°$, and east longitude $\Lambda = 60°$.

Solution

According to Eq. (5.52), the local sidereal time at the observation site is

$$\theta = \theta_G + \Lambda = 126.7° + 60° = 186.7°$$

Substituting $R_e = 6378\,\text{km}$, $f = 0.003353$ (Table 4.3), $\theta = 186.7°$, and $\phi = 20°$ into Eq. (5.56) yields the geocentric position vector of the site,

$$\mathbf{R} = -5955\hat{\mathbf{I}} - 699.5\hat{\mathbf{J}} + 2168\hat{\mathbf{K}}\,(\text{km})$$

Having found $\mathbf{R}$, we obtain the position vector of the space station relative to the site from Eq. (5.53),

$$\begin{aligned}\boldsymbol{\rho} &= \mathbf{r} - \mathbf{R}\\ &= \left(-5368\hat{\mathbf{I}} - 1784\hat{\mathbf{J}} + 3691\hat{\mathbf{K}}\right) - \left(-5955\hat{\mathbf{I}} - 699.5\hat{\mathbf{J}} + 2168\hat{\mathbf{K}}\right)\\ &= 586.8\hat{\mathbf{I}} - 1084\hat{\mathbf{J}} + 1523\hat{\mathbf{K}}\,(\text{km})\end{aligned}$$

Applying Algorithm 4.1 to this vector yields

$$\boxed{\alpha = 298.4°}\quad \boxed{\delta = 51.01°}$$

Compare these with the geocentric right ascension α_0 and declination δ_0, which were computed in Example 4.1,

$$\alpha_0 = 198.4° \quad \delta_0 = 33.12°$$

5.7 TOPOCENTRIC HORIZON COORDINATE SYSTEM

The topocentric horizon coordinate system was introduced in Section 1.7 and is illustrated again in Fig. 5.11. It is centered at the observation point O whose position vector is $\mathbf{R}$. The xy plane is the local horizon, which is the plane tangent to the ellipsoid at point O. The z axis is normal to this plane and is directed outward toward the zenith. The x axis is directed eastward and the y axis points north. Because the x axis points east, this may be referred to as an *ENZ* (east–north–zenith) frame. In the *SEZ* topocentric reference frame the x axis points toward the south and the y axis toward the east. The *SEZ* frame is obtained from the *ENZ* frame by rotating it 90° clockwise around the zenith. Therefore, the matrix of the transformation from *NEZ* to *SEZ* is $[\mathbf{R}_3(-90°)]$, where $[\mathbf{R}_3(\phi)]$ is found in Eq. (4.34).

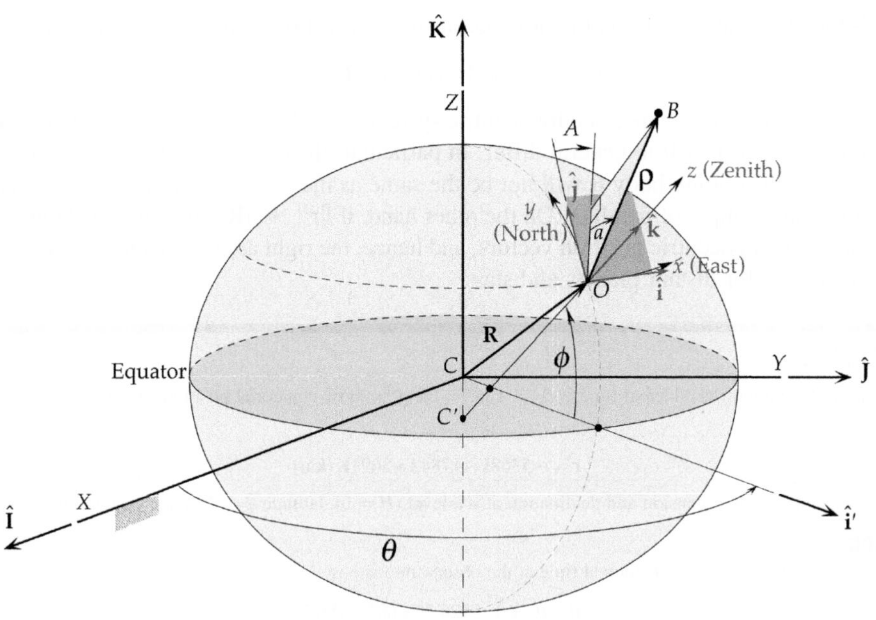

FIG. 5.11

Topocentric horizon (*xyz*) coordinate system on the surface of the oblate earth.

The position vector $\boldsymbol{\rho}$ of a body B relative to the topocentric horizon system in Fig. 5.11 is

$$\boldsymbol{\rho} = \rho\cos a\sin A\hat{\mathbf{i}} + \rho\cos a\cos A\hat{\mathbf{j}} + \rho\sin a\hat{\mathbf{k}}$$

where ρ is the range, A is the azimuth measured positive clockwise from due north ($0° \leq A \leq 360°$), and a is the elevation angle or altitude measured from the horizontal to the line of sight of the body B ($-90° \leq a \leq 90°$). The unit vector $\hat{\boldsymbol{\rho}}$ in the line-of-sight direction is

$$\hat{\boldsymbol{\rho}} = \cos a\sin A\hat{\mathbf{i}} + \cos a\cos A\hat{\mathbf{j}} + \sin a\hat{\mathbf{k}} \tag{5.58}$$

The transformation between geocentric equatorial and topocentric horizon systems is found by first determining the projections of the topocentric base vectors $\hat{\mathbf{i}}\hat{\mathbf{j}}\hat{\mathbf{k}}$ onto those of the geocentric equatorial frame. From Fig. 5.11, it is apparent that

$$\hat{\mathbf{k}} = \cos\phi\hat{\mathbf{i}}' + \sin\phi\hat{\mathbf{K}}$$

and

$$\hat{\mathbf{i}}' = \cos\theta\hat{\mathbf{I}} + \sin\theta\hat{\mathbf{J}}$$

where $\hat{\mathbf{i}}'$ lies in the local meridional plane and is normal to the Z axis. Hence,

$$\hat{\mathbf{k}} = \cos\phi\cos\theta\hat{\mathbf{I}} + \cos\phi\sin\theta\hat{\mathbf{J}} + \sin\phi\hat{\mathbf{K}} \tag{5.59}$$

The eastward-directed unit vector $\hat{\mathbf{i}}$ may be found by taking the cross product of $\hat{\mathbf{K}}$ into the unit normal $\hat{\mathbf{k}}$,

$$\hat{\mathbf{i}} = \frac{\hat{\mathbf{K}} \times \hat{\mathbf{k}}}{\|\hat{\mathbf{K}} \times \hat{\mathbf{k}}\|} = \frac{-\cos\phi\sin\theta\hat{\mathbf{I}} + \cos\phi\cos\theta\hat{\mathbf{J}}}{\sqrt{\cos^2\phi(\sin^2\theta + \cos^2\theta)}} = -\sin\theta\hat{\mathbf{I}} + \cos\theta\hat{\mathbf{J}} \tag{5.60}$$

Finally, crossing $\hat{\mathbf{k}}$ into $\hat{\mathbf{i}}$ yields $\hat{\mathbf{j}}$,

$$\hat{\mathbf{j}} = \hat{\mathbf{k}} \times \hat{\mathbf{i}} = \begin{vmatrix} \hat{\mathbf{I}} & \hat{\mathbf{J}} & \hat{\mathbf{K}} \\ \cos\phi\cos\theta & \cos\phi\sin\theta & \sin\phi \\ -\sin\theta & \cos\theta & 0 \end{vmatrix} = -\sin\phi\cos\theta\hat{\mathbf{I}} - \sin\phi\sin\theta\hat{\mathbf{J}} + \cos\phi\hat{\mathbf{K}} \tag{5.61}$$

Let us denote the matrix of the transformation from the geocentric equatorial to the topocentric horizon as $[\mathbf{Q}]_{Xx}$. Recall from Section 4.5 that the rows of this matrix comprise the direction cosines of $\hat{\mathbf{i}}$, $\hat{\mathbf{j}}$, and $\hat{\mathbf{k}}$, respectively. It follows from Eqs. (5.59)–(5.61) that

$$[\mathbf{Q}]_{Xx} = \begin{bmatrix} -\sin\theta & \cos\theta & 0 \\ -\sin\phi\cos\theta & -\sin\phi\sin\theta & \cos\phi \\ \cos\phi\cos\theta & \cos\phi\sin\theta & \sin\phi \end{bmatrix} \tag{5.62a}$$

The reverse transformation, from the topocentric horizon to the geocentric equatorial, is represented by the transpose of this matrix,

$$[\mathbf{Q}]_{xX} = \begin{bmatrix} -\sin\theta & -\sin\phi\cos\theta & \cos\phi\cos\theta \\ \cos\theta & -\sin\phi\sin\theta & \cos\phi\sin\theta \\ 0 & \cos\phi & \sin\phi \end{bmatrix} \tag{5.62b}$$

Observe that these matrices also represent the transformation between the topocentric horizon and the topocentric equatorial frames, because the unit basis vectors of the latter coincide with those of the geocentric equatorial coordinate system.

EXAMPLE 5.8

The east longitude and latitude of an observer near San Francisco are $\Lambda = 238°$ and $\phi = 38°$, respectively. The local sidereal time is $\theta = 215.1°$ (14 h 20 min). At that time, the planet Jupiter is observed by means of a telescope to be located at azimuth $A = 214.3°$ and angular elevation $a = 43°$. What are Jupiter's right ascension and declination in the topocentric equatorial system?

Solution

The given information allows us to formulate the matrix of the transformation from the topocentric horizon to the topocentric equatorial using Eq. (5.62b),

$$[\mathbf{Q}]_{xX} = \begin{bmatrix} -\sin 215.1° & -\sin 38\cos 215.1° & \cos 38\cos 215.1° \\ \cos 215.1° & -\sin 38\sin 215.1° & \cos 38\sin 215.1° \\ 0 & \cos 38° & \sin 38° \end{bmatrix}$$
$$= \begin{bmatrix} 0.5750 & 0.5037 & -0.6447 \\ -0.8182 & 0.3540 & -0.4531 \\ 0 & 0.7880 & 0.6157 \end{bmatrix}$$

From Eq. (5.58), we have

$$\hat{\boldsymbol{\rho}} = \cos a\sin A\hat{\mathbf{i}} + \cos a\cos A\hat{\mathbf{j}} + \sin a\hat{\mathbf{k}}$$
$$= \cos 43°\sin 214.3°\hat{\mathbf{i}} + \cos 43°\cos 214.3°\hat{\mathbf{j}} + \sin 43°\hat{\mathbf{k}}$$
$$= -0.4121\hat{\mathbf{i}} - 0.6042\hat{\mathbf{j}} + 0.6820\hat{\mathbf{k}}$$

Therefore, in matrix notation, the topocentric horizon components of $\hat{\boldsymbol{\rho}}$ are

$$\{\hat{\boldsymbol{\rho}}\}_x = \begin{Bmatrix} -0.4121 \\ -0.6042 \\ 0.6820 \end{Bmatrix}$$

We obtain the topocentric equatorial components $\{\hat{\boldsymbol{\rho}}\}_X$ by the matrix operation

$$\{\hat{\boldsymbol{\rho}}\}_X = [\mathbf{Q}]_{xX}\{\hat{\boldsymbol{\rho}}\}_x = \begin{bmatrix} 0.5750 & 0.5037 & -0.6447 \\ -0.8182 & 0.3540 & -0.4531 \\ 0 & 0.7880 & 0.6157 \end{bmatrix} \begin{Bmatrix} -0.4121 \\ 0.6042 \\ 0.6820 \end{Bmatrix} = \begin{Bmatrix} -0.9810 \\ -0.1857 \\ -0.05621 \end{Bmatrix}$$

so that the topocentric equatorial line-of-sight unit vector is

$$\hat{\boldsymbol{\rho}} = -0.9810\hat{\mathbf{I}} - 0.1857\hat{\mathbf{J}} - 0.05621\hat{\mathbf{K}}$$

Using this vector in Algorithm 4.1 yields the topocentric equatorial right ascension and declination,

$$\boxed{\alpha = 190.7^\circ \quad \delta = -3.222^\circ}$$

Jupiter is sufficiently far away that we can ignore the radius of the earth in Eq. (5.53). That is, to our level of precision, there is no distinction between the topocentric equatorial and geocentric equatorial systems:

$$\mathbf{r} \approx \boldsymbol{\rho}$$

Therefore, the topocentric right ascension and declination computed above are the same as the geocentric equatorial values.

EXAMPLE 5.9

At a given time, the geocentric equatorial position vector of the International Space Station is

$$\mathbf{r} = -2032.4\hat{\mathbf{I}} + 4591.2\hat{\mathbf{J}} - 4544.8\hat{\mathbf{K}}\,(\text{km})$$

Determine the azimuth and elevation angle relative to a sea level ($H = 0$) observer whose latitude is $\phi = -40^\circ$ and local sidereal time is $\theta = 110^\circ$.

Solution

Using Eq. (5.56), we find the position vector of the observer to be

$$\mathbf{R} = -1673\hat{\mathbf{I}} + 4598\hat{\mathbf{J}} - 4078\hat{\mathbf{K}}\,(\text{km})$$

For the position vector of the space station relative to the observer, we have (Eq. 5.53)

$$\begin{aligned} \boldsymbol{\rho} &= \mathbf{r} - \mathbf{R} \\ &= \left(-2032\hat{\mathbf{I}} + 4591\hat{\mathbf{J}} - 4545\hat{\mathbf{K}}\right) - \left(-1673\hat{\mathbf{I}} + 4598\hat{\mathbf{J}} - 4078\hat{\mathbf{K}}\right) \\ &= -359.0\hat{\mathbf{I}} - 6.342\hat{\mathbf{J}} - 466.9\hat{\mathbf{K}}\,(\text{km}) \end{aligned}$$

or, in matrix notation,

$$\{\boldsymbol{\rho}\}_X = \begin{Bmatrix} -359.0 \\ -6.342 \\ -466.9 \end{Bmatrix} (\text{km})$$

To transform these geocentric equatorial components into the topocentric horizon system, we need the transformation matrix $[\mathbf{Q}]_{Xx}$, which is given by Eq. (5.62a),

$$[\mathbf{Q}]_{Xx} = \begin{bmatrix} -\sin\theta & \cos\theta & 0 \\ -\sin\phi\cos\theta & -\sin\phi\sin\theta & \cos\phi \\ \cos\phi\cos\theta & \cos\phi\sin\theta & \sin\phi \end{bmatrix}$$

$$= \begin{bmatrix} -\sin 110^\circ & \cos 110^\circ & 0 \\ -\sin(-40^\circ)\cos 110^\circ & -\sin(-40^\circ)\sin 110^\circ & \cos(-40^\circ) \\ \cos(-40^\circ)\cos 110^\circ & \cos(-40^\circ)\sin 110^\circ & \sin(-40^\circ) \end{bmatrix}$$

Thus,

$$\{\boldsymbol{\rho}\}_x = [\mathbf{Q}]_{Xx}\{\boldsymbol{\rho}\}_X = \begin{bmatrix} -0.9397 & -0.3420 & 0 \\ -0.2198 & 0.6040 & 0.7660 \\ -0.2620 & 0.7198 & -0.6428 \end{bmatrix} \begin{Bmatrix} -359.0 \\ -6.342 \\ -466.9 \end{Bmatrix} = \begin{Bmatrix} 339.5 \\ -282.6 \\ 389.6 \end{Bmatrix} \text{(km)}$$

or, reverting to vector notation,

$$\boldsymbol{\rho} = 339.5\hat{\mathbf{i}} - 282.6\hat{\mathbf{j}} + 389.6\hat{\mathbf{k}}\,(\text{km})$$

The magnitude of this vector is $\rho = 589.0$ km. Hence, the unit vector in the direction of $\boldsymbol{\rho}$ is

$$\hat{\boldsymbol{\rho}} = \frac{\boldsymbol{\rho}}{\rho} = 0.5765\hat{\mathbf{i}} - 0.4787\hat{\mathbf{j}} + 0.6615\hat{\mathbf{k}}$$

Comparing this with Eq. (5.58), we see that $\sin a = 0.6615$, so that the angular elevation is

$$a = \sin^{-1}0.6615 = \boxed{41.41^\circ}$$

Furthermore,

$$\sin A = \frac{0.5765}{\cos a} = 0.7687$$

$$\cos A = \frac{-0.4787}{\cos a} = -0.6397$$

It follows that

$$A = \cos^{-1}(-0.6397) = 129.8^\circ\,(\text{second quadrant}) \quad \text{or} \quad 230.2^\circ\,(\text{third quadrant})$$

A must lie in the second quadrant because $\sin A > 0$. Thus, the azimuth is

$$\boxed{A = 129.8^\circ}$$

5.8 ORBIT DETERMINATION FROM ANGLE AND RANGE MEASUREMENTS

We know that an orbit around the earth is determined once the state vectors **r** and **v** in the inertial geocentric equatorial frame are provided at a given instant of time (epoch). Satellites are of course observed from the earth's surface and not from its center. Let us briefly consider how the state vector is determined from measurements by an earth-based tracking station.

The fundamental vector triangle formed by the topocentric position vector $\boldsymbol{\rho}$ of a satellite relative to a tracking station, the position vector **R** of the station relative to the center of attraction C, and the geocentric position vector **r** was illustrated in Fig. 5.8 and is shown again schematically in Fig. 5.12. The relationship among these three vectors is given by Eq. (5.53), which can be written as

$$\mathbf{r} = \mathbf{R} + \rho\hat{\boldsymbol{\rho}} \tag{5.63}$$

where the range ρ is the distance of the body B from the tracking site, and $\hat{\boldsymbol{\rho}}$ is the unit vector containing the directional information about B. By differentiating Eq. (5.63) with respect to time, we obtain the velocity **v** and acceleration **a**,

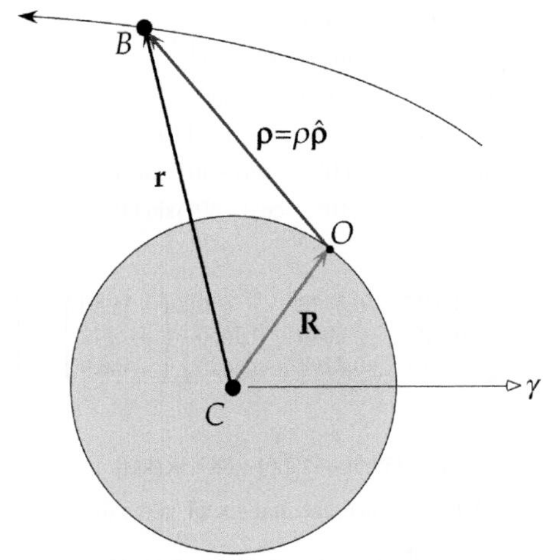

FIG. 5.12

Earth-orbiting body B tracked by an observer O.

$$\mathbf{v}=\dot{\mathbf{r}}=\dot{\mathbf{R}}+\rho\dot{\hat{\boldsymbol{\rho}}}+\dot{\rho}\hat{\boldsymbol{\rho}} \tag{5.64}$$

$$\mathbf{a}=\ddot{\mathbf{r}}=\ddot{\mathbf{R}}+\ddot{\rho}\hat{\boldsymbol{\rho}}+2\dot{\rho}\dot{\hat{\boldsymbol{\rho}}}+\rho\ddot{\hat{\boldsymbol{\rho}}} \tag{5.65}$$

The vectors in these equations must all be expressed in the common basis $(\hat{\mathbf{I}}\hat{\mathbf{J}}\hat{\mathbf{K}})$ of the inertial (non-rotating) geocentric equatorial frame.

Since $\mathbf{R}$ is a vector fixed in the earth, whose constant angular velocity is $\boldsymbol{\Omega}=\omega_E\hat{\mathbf{K}}$ (Eq. 2.67), it follows from Eqs. (1.52) and (1.53) that

$$\dot{\mathbf{R}}=\boldsymbol{\Omega}\times\mathbf{R} \tag{5.66}$$

$$\ddot{\mathbf{R}}=\boldsymbol{\Omega}\times(\boldsymbol{\Omega}\times\mathbf{R}) \tag{5.67}$$

If L_X, L_Y, and L_Z are the topocentric equatorial direction cosines, then the direction cosine vector $\hat{\boldsymbol{\rho}}$ is

$$\hat{\boldsymbol{\rho}}=L_X\hat{\mathbf{I}}+L_Y\hat{\mathbf{J}}+L_Z\hat{\mathbf{K}} \tag{5.68}$$

and its first and second derivatives are

$$\dot{\hat{\boldsymbol{\rho}}}=\dot{L}_X\hat{\mathbf{I}}+\dot{L}_Y\hat{\mathbf{J}}+\dot{L}_Z\hat{\mathbf{K}} \tag{5.69}$$

and

$$\ddot{\hat{\boldsymbol{\rho}}}=\ddot{L}_X\hat{\mathbf{I}}+\ddot{L}_Y\hat{\mathbf{J}}+\ddot{L}_Z\hat{\mathbf{K}} \tag{5.70}$$

Comparing Eqs. (5.57) and (5.68) reveals that the topocentric equatorial direction cosines in terms of the topocentric right ascension α and declination δ are

$$\begin{Bmatrix} L_X \\ L_Y \\ L_Z \end{Bmatrix} = \begin{Bmatrix} \cos\alpha\cos\delta \\ \sin\alpha\cos\delta \\ \sin\delta \end{Bmatrix} \tag{5.71}$$

Differentiating this equation twice yields

$$\begin{Bmatrix} \dot{L}_X \\ \dot{L}_Y \\ \dot{L}_Z \end{Bmatrix} = \begin{Bmatrix} -\dot{\alpha}\sin\alpha\cos\delta - \dot{\delta}\cos\alpha\sin\delta \\ \dot{\alpha}\cos\alpha\cos\delta - \dot{\delta}\sin\alpha\sin\delta \\ \dot{\delta}\cos\delta \end{Bmatrix} \tag{5.72}$$

and

$$\begin{Bmatrix} \ddot{L}_X \\ \ddot{L}_Y \\ \ddot{L}_Z \end{Bmatrix} = \begin{Bmatrix} -\ddot{\alpha}\sin\alpha\cos\delta - \ddot{\delta}\cos\alpha\sin\delta - \left(\dot{\alpha}^2 + \dot{\delta}^2\right)\cos\alpha\cos\delta + 2\dot{\alpha}\dot{\delta}\sin\alpha\sin\delta \\ \ddot{\alpha}\cos\alpha\cos\delta - \ddot{\delta}\sin\alpha\sin\delta - \left(\dot{\alpha}^2 + \dot{\delta}^2\right)\sin\alpha\cos\delta - 2\dot{\alpha}\dot{\delta}\cos\alpha\sin\delta \\ \ddot{\delta}\cos\delta - \dot{\delta}^2\sin\delta \end{Bmatrix} \tag{5.73}$$

Eqs. (5.71)–(5.73) show how the direction cosines and their rates are obtained from right ascension and declination and their rates.

In the topocentric horizon system, the relative position vector is written as

$$\hat{\boldsymbol{\rho}} = l_x\hat{\mathbf{i}} + l_y\hat{\mathbf{j}} + l_z\hat{\mathbf{k}} \tag{5.74}$$

where, according to Eq. (5.58), the direction cosines l_x, l_y, and l_z are found in terms of the azimuth A and elevation a as

$$\begin{Bmatrix} l_x \\ l_y \\ l_z \end{Bmatrix} = \begin{Bmatrix} \sin A\cos a \\ \cos A\cos a \\ \sin a \end{Bmatrix} \tag{5.75}$$

L_X, L_Y, and L_Z are obtained from l_x, l_y, and l_z by the coordinate transformation

$$\begin{Bmatrix} L_X \\ L_Y \\ L_Z \end{Bmatrix} = [\mathbf{Q}]_{xX}\begin{Bmatrix} l_x \\ l_y \\ l_z \end{Bmatrix} \tag{5.76}$$

where $[\mathbf{Q}]_{xX}$ is given by Eq. (5.62b). Thus,

$$\begin{Bmatrix} L_X \\ L_Y \\ L_Z \end{Bmatrix} = \begin{bmatrix} -\sin\theta & -\cos\theta\sin\phi & \cos\theta\cos\phi \\ \cos\theta & -\sin\theta\sin\phi & \sin\theta\cos\phi \\ 0 & \cos\phi & \sin\phi \end{bmatrix}\begin{Bmatrix} \sin A\cos a \\ \cos A\cos a \\ \sin a \end{Bmatrix} \tag{5.77}$$

Substituting Eq. (5.71) we see that topocentric right ascension/declination and azimuth/elevation are related by

$$\begin{Bmatrix} \cos\alpha\cos\delta \\ \sin\alpha\cos\delta \\ \sin\delta \end{Bmatrix} = \begin{bmatrix} -\sin\theta & -\cos\theta\sin\phi & \cos\theta\cos\phi \\ \cos\theta & -\sin\theta\sin\phi & \sin\theta\cos\phi \\ 0 & \cos\phi & \sin\phi \end{bmatrix}\begin{Bmatrix} \sin A\cos a \\ \cos A\cos a \\ \sin a \end{Bmatrix}$$

Expanding the right-hand side and solving for sin δ, sin α, and cos α, we get

$$\sin\delta = \cos\phi\cos A\cos a + \sin\phi\sin a \tag{5.78a}$$

$$\sin\alpha = \frac{(\cos\phi\sin a - \cos A\cos a\sin\phi)\sin\theta + \cos\theta\sin A\cos a}{\cos\delta} \tag{5.78b}$$

$$\cos\alpha = \frac{(\cos\phi\sin a - \cos A\cos a\sin\phi)\cos\theta - \sin\theta\sin A\cos a}{\cos\delta} \tag{5.78c}$$

We can simplify Eqs. (5.78b) and (5.78c) by introducing the hour angle h,

$$h = \theta - \alpha \tag{5.79}$$

where h is the angular distance between the object and the local meridian. If h is positive, the object is west of the meridian; if h is negative, the object is east of the meridian.

Using well-known trig identities, we have

$$\sin(\theta - \alpha) = \sin\theta\cos\alpha - \cos\theta\sin\alpha \tag{5.80a}$$

$$\cos(\theta - \alpha) = \cos\theta\cos\alpha + \sin\theta\sin\alpha \tag{5.80b}$$

Substituting Eqs. (5.78b) and (5.78c) on the right of Eq. (5.80a) and simplifying yields

$$\sin(h) = -\frac{\sin A\cos a}{\cos\delta} \tag{5.81}$$

Likewise, Eq. (5.80b) leads to

$$\cos(h) = \frac{\cos\phi\sin a - \sin\phi\cos A\cos a}{\cos\delta} \tag{5.82}$$

We calculate h from this equation, resolving quadrant ambiguity by checking the sign of sin(h). That is,

$$h = \cos^{-1}\left(\frac{\cos\phi\sin a - \sin\phi\cos A\cos a}{\cos\delta}\right)$$

if sin(h) is positive. Otherwise, we must subtract h from 360°. Since both the elevation angle a and the declination δ lie between −90° and +90°, neither cos$a$ nor cosδ can be negative. It follows from Eq. (5.81) that the sign of sin(h) depends only on that of sin$A$.

To summarize, given the topocentric azimuth A and altitude a of the target together with the sidereal time θ and latitude ϕ of the tracking station, we compute the topocentric declination δ and right ascension α as follows:

$$\delta = \sin^{-1}(\cos\phi\cos A\cos a + \sin\phi\sin a) \tag{5.83a}$$

$$h = \begin{cases} 360^\circ - \cos^{-1}\left(\dfrac{\cos\phi\sin a - \sin\phi\cos A\cos a}{\cos\delta}\right) & 0^\circ < A < 180^\circ \\ \cos^{-1}\left(\dfrac{\cos\phi\sin a - \sin\phi\cos A\cos a}{\cos\delta}\right) & 180^\circ \le A \le 360^\circ \end{cases} \tag{5.83b}$$

$$\alpha = \theta - h \tag{5.83c}$$

If A and a are provided as functions of time, then α and δ are found as functions of time by means of Eqs. (5.83a)–(5.83c). The rates $\dot{\alpha}$, $\ddot{\alpha}$, $\dot{\delta}$, and $\ddot{\delta}$ are determined by differentiating $\alpha(t)$ and $\delta(t)$ and substituting the results into Eqs. (5.68)–(5.73) to calculate the direction cosine vector $\hat{\boldsymbol{\rho}}$ and its rates $\dot{\hat{\boldsymbol{\rho}}}$ and $\ddot{\hat{\boldsymbol{\rho}}}$.

It is a relatively simple matter to find $\dot{\alpha}$ and $\dot{\delta}$ in terms of $\dot{A}$ and $\dot{a}$. Differentiating Eq. (5.78a) with respect to time yields

$$\dot{\delta} = \frac{1}{\cos\delta}\left[-\dot{A}\cos\phi\sin A\cos a + \dot{a}(\sin\phi\cos a - \cos\phi\cos A\sin a)\right] \tag{5.84}$$

Differentiating Eq. (5.81), we get

$$\dot{h}\cos(h) = -\frac{1}{\cos^2\delta}\left[(\dot{A}\cos A\cos a - \dot{a}\sin A\sin a)\cos\delta + \dot{\delta}\sin A\cos a\sin\delta\right]$$

Substituting Eq. (5.82) and simplifying leads to

$$\dot{h} = -\frac{\dot{A}\cos A\cos a - \dot{a}\sin A\sin a + \dot{\delta}\sin A\cos a\tan\delta}{\cos\phi\sin a - \sin\phi\cos A\cos a}$$

But $\dot{h} = \dot{\theta} - \dot{\alpha} = \omega_E - \dot{\alpha}$, so that finally,

$$\dot{\alpha} = \omega_E + \frac{\dot{A}\cos A\cos a - \dot{a}\sin A\sin a + \dot{\delta}\sin A\cos a\tan\delta}{\cos\phi\sin a - \sin\phi\cos A\cos a} \tag{5.85}$$

ALGORITHM 5.4

Given the range ρ, azimuth A, and angular elevation a together with the rates $\dot{\rho}$, $\dot{A}$, and $\dot{a}$ relative to an earth-based tracking station (for which the altitude H, latitude ϕ, and local sidereal time are known), calculate the state vectors **r** and **v** in the geocentric equatorial frame. A MATLAB script of this procedure appears in Appendix D.28.

1. Using the altitude H, latitude ϕ, and local sidereal time θ of the site, calculate its geocentric position vector **R** from Eq. (5.56).
2. Calculate the topocentric declination δ using Eq. (5.83a).
3. Calculate the topocentric right ascension α from Eqs. (5.83b) and (5.83c).
4. Calculate the direction cosine unit vector $\hat{\boldsymbol{\rho}}$ from Eqs. (5.68) and (5.71),

$$\hat{\boldsymbol{\rho}} = \cos\delta(\cos\alpha\hat{\mathbf{I}} + \sin\alpha\hat{\mathbf{J}}) + \sin\delta\hat{\mathbf{K}}$$

5. Calculate the geocentric position vector **r** from Eq. (5.63).
6. Calculate the inertial velocity $\dot{\mathbf{R}}$ of the site from Eq. (5.66).
7. Calculate the declination rate $\dot{\delta}$ using Eq. (5.84).
8. Calculate the right ascension rate $\dot{\alpha}$ by means of Eq. (5.85).
9. Calculate the direction cosine rate vector $\dot{\hat{\boldsymbol{\rho}}}$ from Eqs. (5.69) and (5.72).
10. Calculate the geocentric velocity vector **v** from Eq. (5.64).

EXAMPLE 5.10

At $\theta = 300°$ local sidereal time a sea level ($H = 0$) tracking station at a latitude of $\phi = 60°$ detects a space object and obtains the following data:

Slant range:	$\rho = 2551\,\text{km}$
Azimuth:	$A = 90°$
Elevation:	$a = 30°$
Range rate:	$\dot{\rho} = 0$
Azimuth rate:	$\dot{A} = 1.973\left(10^{-3}\right)\text{rad/s}\,(0.1130°/\text{s})$
Elevation rate:	$\dot{a} = 9.864\left(10^{-4}\right)\text{rad/s}\,(0.05651°/\text{s})$

What are the orbital elements of the object?

Solution

We must first employ Algorithm 5.4 to obtain the state vectors **r** and **v** to compute the orbital elements by means of Algorithm 4.2.

Step 1:

The equatorial radius of the earth is $R_e = 6378$ km and the flattening factor is $f = 0.003353$. It follows from Eq. (5.56) that the position vector of the observer is

$$\mathbf{R} = 1598\hat{\mathbf{I}} - 2769\hat{\mathbf{J}} + 5500\hat{\mathbf{K}}\,(\text{km})$$

Step 2:

$$\begin{aligned}\delta &= \sin^{-1}(\cos\phi\cos A\cos a + \sin\phi\sin a)\\ &= \sin^{-1}(\cos 60°\cos 90°\cos 30° + \sin 60°\sin 30°)\\ &= 25.66°\end{aligned}$$

Step 3:

Since the given azimuth lies between 0° and 180°, Eq. (5.83b) yields

$$\begin{aligned}h &= 360° - \cos^{-1}\left(\frac{\cos\phi\sin a - \sin\phi\cos A\cos a}{\cos\delta}\right)\\ &= 360° - \cos^{-1}\left(\frac{\cos 60°\sin 30° - \sin 60°\cos 90°\cos 30°}{\cos 25.66°}\right)\\ &= 360° - 73.90° = 286.1°\end{aligned}$$

Therefore, the right ascension is

$$\alpha = \theta - h = 300° - 286.1° = 13.90°$$

Step 4:

$$\begin{aligned}\hat{\boldsymbol{\rho}} &= \cos\delta(\cos\alpha\hat{\mathbf{I}} + \sin\alpha\hat{\mathbf{J}}) + \sin\delta\hat{\mathbf{K}}\\ &= \cos 25.66°(\cos 13.90°\hat{\mathbf{I}} + \sin 13.90°\hat{\mathbf{J}}) + \sin 25.66°\hat{\mathbf{K}}\\ &= 0.8750\hat{\mathbf{I}} + 0.2165\hat{\mathbf{J}} + 0.4330\hat{\mathbf{K}}\end{aligned}$$

Step 5:

$$\begin{aligned}\mathbf{r} &= \mathbf{R} + \rho\hat{\boldsymbol{\rho}}\\ &= (1598\hat{\mathbf{I}} - 2769\hat{\mathbf{J}} + 5500\hat{\mathbf{K}}) + 2551(0.8750\hat{\mathbf{I}} + 0.2165\hat{\mathbf{J}} + 0.4330\hat{\mathbf{K}})\\ &= 3831\hat{\mathbf{I}} - 2216\hat{\mathbf{J}} + 6605\hat{\mathbf{K}}\,(\text{km})\end{aligned}$$

Step 6:

Recalling from Eq. (2.67) that the angular velocity ω_E of the earth is $72.92(10^{-6})$ rad/s,

$$\begin{aligned}\dot{\mathbf{R}} &= \boldsymbol{\Omega} \times \mathbf{R}\\ &= 72.92(10^{-6})\hat{\mathbf{K}} \times (1598\hat{\mathbf{I}} - 2769\hat{\mathbf{J}} + 5500\hat{\mathbf{K}})\\ &= 0.2019\hat{\mathbf{I}} + 0.1166\hat{\mathbf{J}}\,(\text{km/s})\end{aligned}$$

Step 7:

$$\dot{\delta} = \frac{1}{\cos\delta}\left[-\dot{A}\cos\phi\sin A\cos a + \dot{a}(\sin\phi\cos a - \cos\phi\cos A\sin a)\right]$$
$$= \frac{1}{\cos 25.66^\circ}\left[-1.973(10^{-3})\cos 60^\circ\sin 90^\circ\cos 30^\circ + 9.864(10^{-4})(\sin 60^\circ\cos 30^\circ - \cos 60^\circ\cos 90^\circ\sin 30^\circ)\right]$$
$$= -1.2696(10^{-4})\ (\text{rad/s})$$

Step 8:

$$\dot{\alpha} - \omega_E = \frac{\dot{A}\cos A\cos a - \dot{a}\sin A\sin a + \dot{\delta}\sin A\cos a\tan\delta}{\cos\phi\sin a - \sin\phi\cos A\cos a}$$
$$= \frac{1.973(10^{-3})\cos 90^\circ\cos 30^\circ - 9.864(10^{-4})\sin 90^\circ\sin 30^\circ + \left[-1.2696(10^{-4})\right]\sin 90^\circ\cos 30^\circ\tan 25.66^\circ}{\cos 60^\circ\sin 30^\circ - \sin 60^\circ\cos 90^\circ\cos 30^\circ}$$
$$= -0.002184$$
$$\therefore \dot{\alpha} = 72.92(10^{-6}) - 0.002184 = -0.002111\ (\text{rad/s})$$

Step 9:

$$\dot{\hat{\boldsymbol{\rho}}} = (-\dot{\alpha}\sin\alpha\cos\delta - \dot{\delta}\cos\alpha\sin\delta)\hat{\mathbf{I}} + (\dot{\alpha}\cos\alpha\cos\delta - \dot{\delta}\sin\alpha\sin\delta)\hat{\mathbf{J}} + \dot{\delta}\cos\delta\hat{\mathbf{K}}$$
$$= [-(-0.002111)\sin 13.90^\circ\cos 25.66^\circ - (-0.1270)\cos 13.90^\circ\sin 25.66^\circ]\hat{\mathbf{I}}$$
$$+ [(-0.002111)\cos 13.90^\circ\cos 25.66^\circ - (-0.1270)\sin 13.90^\circ\sin 25.66^\circ]\hat{\mathbf{J}}$$
$$+ [-0.1270\cos 25.66^\circ]\hat{\mathbf{K}}$$
$$\therefore \dot{\hat{\boldsymbol{\rho}}} = (0.5104\hat{\mathbf{I}} - 1.834\hat{\mathbf{J}} - 0.1144\hat{\mathbf{K}})(10^{-3})\ (\text{rad/s})$$

Step 10:

$$\mathbf{v} = \dot{\mathbf{R}} + \dot{\rho}\hat{\boldsymbol{\rho}} + \rho\dot{\hat{\boldsymbol{\rho}}}$$
$$= (0.2019\hat{\mathbf{I}} + 0.1166\hat{\mathbf{J}}) + 0.(0.8750\hat{\mathbf{I}} + 0.2165\hat{\mathbf{J}} + 0.4330\hat{\mathbf{K}})$$
$$+ 2551\left[0.5104(10^{-3})\hat{\mathbf{I}} - 1.834\times 10^{-3}\hat{\mathbf{J}} - 0.1144\times 10^{-3}\hat{\mathbf{K}}\right]$$
$$\therefore \mathbf{v} = 1.504\hat{\mathbf{I}} - 4.562\hat{\mathbf{J}} - 0.2920\hat{\mathbf{K}}\ (\text{km/s})$$

Using the position and velocity vectors from Steps 5 and 10, the reader can verify that Algorithm 4.2 yields the following orbital elements of the tracked object:

$$a = 5170\text{km}$$
$$i = 113.4^\circ$$
$$\Omega = 109.8^\circ$$
$$e = 0.619$$
$$\omega = 309.8^\circ$$
$$\theta = 165.3^\circ$$

This is a highly elliptical orbit with a semimajor axis less than the earth's radius, so the object will impact the earth (at a true anomaly of 216°).

For objects orbiting the sun (planets, asteroids, comets, and man-made interplanetary probes), the fundamental vector triangle is as illustrated in Fig. 5.13. The tracking station is on the earth, but, of course, the sun rather than the earth is the center of attraction. The procedure for finding the heliocentric state vector **r** and **v** is similar to that outlined above. Because of the vast distances involved, the observer can usually be imagined to reside at the center of the earth. Dealing with **R** is different in this

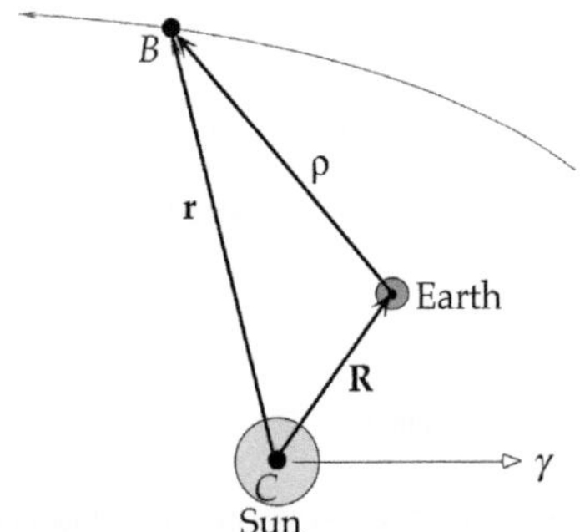

FIG. 5.13

An object B orbiting the sun and tracked from earth.

case. The daily position of the sun relative to the earth ($-\mathbf{R}$ in Fig. 5.13) may be found in ephemerides, such as the *Astronomical Almanac* (Department of the Navy, 2018). A discussion of interplanetary trajectories appears in Chapter 8 of this text.

5.9 ANGLES-ONLY PRELIMINARY ORBIT DETERMINATION

To determine an orbit requires specifying six independent quantities. These can be the six classical orbital elements or all six components of the state vector, $\mathbf{r}$ and $\mathbf{v}$, at a given instant. To determine an orbit solely from observations therefore requires six independent measurements. In the previous section, we assumed the tracking station was able to measure simultaneously the six quantities: range and range rate, azimuth and azimuth rate, plus elevation and elevation rate. These data led directly to the state vector and, hence, to a complete determination of the orbit. In the absence of the capability to measure range and range rate, as with a telescope, we must rely on measurements of just the two angles, azimuth and elevation, to determine the orbit. A minimum of three observations of azimuth and elevation is therefore required to accumulate the six quantities we need to predict the orbit. We shall henceforth assume that the angular measurements are converted to topocentric right ascension α and declination δ, as described in the previous section.

We shall consider the classical method of angles-only orbit determination due to Carl Friedrich Gauss (1777–1855), a German mathematician who many consider was one of the greatest mathematicians ever. This method requires gathering angular information over closely spaced intervals of time and yields a preliminary orbit determination based on those initial observations.

5.10 GAUSS METHOD OF PRELIMINARY ORBIT DETERMINATION

Suppose we have three observations of an orbiting body at times t_1, t_2, and t_3, as shown in Fig. 5.14. At each time, the geocentric position vector $\mathbf{r}$ is related to the observer's position vector $\mathbf{R}$, the slant range ρ, and the topocentric direction cosine vector $\hat{\boldsymbol{\rho}}$ by Eq. (5.63),

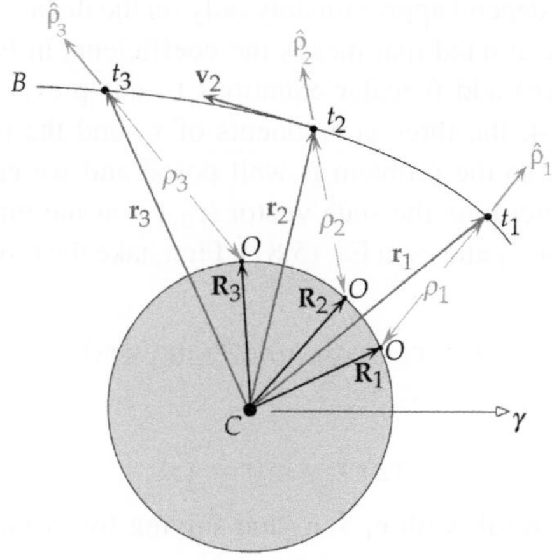

FIG. 5.14

Center of attraction C, observer O, and the tracked body B.

$$\mathbf{r}_1 = \mathbf{R}_1 + \rho_1 \hat{\boldsymbol{\rho}}_1 \tag{5.86a}$$

$$\mathbf{r}_2 = \mathbf{R}_2 + \rho_2 \hat{\boldsymbol{\rho}}_2 \tag{5.86b}$$

$$\mathbf{r}_3 = \mathbf{R}_3 + \rho_3 \hat{\boldsymbol{\rho}}_3 \tag{5.86c}$$

The positions $\mathbf{R}_1$, $\mathbf{R}_2$, and $\mathbf{R}_3$ of the observer O are known from the location of the tracking station and the time of the observations. $\hat{\boldsymbol{\rho}}_1$, $\hat{\boldsymbol{\rho}}_2$, and $\hat{\boldsymbol{\rho}}_3$ are obtained by measuring the right ascension α and declination δ of the body at each of the three times (recall Eq. 5.57). Eqs. (5.86a), (5.86b), and (5.86c) are three vector equations, and therefore there are nine scalar equations, in 12 unknowns: the three components of each of the three vectors $\mathbf{r}_1$, $\mathbf{r}_2$, and $\mathbf{r}_3$, plus the three slant ranges ρ_1, ρ_2, and ρ_3.

An additional three equations are obtained by recalling from Chapter 2 that the conservation of angular momentum requires the vectors $\mathbf{r}_1$, $\mathbf{r}_2$, and $\mathbf{r}_3$ to lie in the same plane. As in our discussion of the Gibbs method in Section 5.2, this means that $\mathbf{r}_2$ is a linear combination $\mathbf{r}_1$ and $\mathbf{r}_3$.

$$\mathbf{r}_2 = c_1 \mathbf{r}_1 + c_3 \mathbf{r}_3 \tag{5.87}$$

Adding this equation to those in Eqs. (5.86) introduces two new unknowns, c_1 and c_3. At this point, we therefore have 12 scalar equations in 14 unknowns.

Another consequence of the two-body equation of motion (Eq. 2.22) is that the state vectors $\mathbf{r}$ and $\mathbf{v}$ of the orbiting body can be expressed in terms of the state vectors at any given time by means of the Lagrange coefficients, Eqs. (2.135) and (2.136). For the case at hand, this means we can express the position vectors $\mathbf{r}_1$ and $\mathbf{r}_3$ in terms of the position $\mathbf{r}_2$ and velocity $\mathbf{v}_2$ at the intermediate time t_2 as follows:

$$\mathbf{r}_1 = f_1 \mathbf{r}_2 + g_1 \mathbf{v}_2 \tag{5.88a}$$

$$\mathbf{r}_3 = f_3 \mathbf{r}_2 + g_3 \mathbf{v}_2 \tag{5.88b}$$

where f_1 and g_1 are the Lagrange coefficients evaluated at t_1, and f_3 and g_3 are those same functions evaluated at time t_3. If the time intervals between the three observations are sufficiently small, then

Eqs. (2.172) reveal that f and g depend approximately only on the distance from the center of attraction at the initial time. For the case at hand that means the coefficients in Eqs. (5.88) depend only on r_2. Hence, Eqs. (5.88a) and (5.88b) add 6 scalar equations to our previous list of 12 while adding to the list of 14 unknowns only 4: the three components of $\mathbf{v}_2$ and the radius r_2. We have arrived at 18 equations in 18 unknowns, so the problem is well posed and we can proceed with the solution. The ultimate objective is to determine the state vector $(\mathbf{r}_2, \mathbf{v}_2)$ at the intermediate time t_2.

Let us start out by solving for c_1 and c_3 in Eq. (5.87). First, take the cross product of each term in that equation with $\mathbf{r}_3$,

$$\mathbf{r}_2 \times \mathbf{r}_3 = c_1(\mathbf{r}_1 \times \mathbf{r}_3) + c_3(\mathbf{r}_3 \times \mathbf{r}_3)$$

Since $\mathbf{r}_3 \times \mathbf{r}_3 = \mathbf{0}$, this reduces to

$$\mathbf{r}_2 \times \mathbf{r}_3 = c_1(\mathbf{r}_1 \times \mathbf{r}_3)$$

Taking the dot product of this result with $\mathbf{r}_1 \times \mathbf{r}_3$ and solving for c_1 yields

$$c_1 = \frac{(\mathbf{r}_2 \times \mathbf{r}_3) \cdot (\mathbf{r}_1 \times \mathbf{r}_3)}{\|\mathbf{r}_1 \times \mathbf{r}_3\|^2} \tag{5.89}$$

In a similar fashion, by forming the dot product of Eq. (5.87) with $\mathbf{r}_1$, we are led to

$$c_3 = \frac{(\mathbf{r}_2 \times \mathbf{r}_1) \cdot (\mathbf{r}_3 \times \mathbf{r}_1)}{\|\mathbf{r}_1 \times \mathbf{r}_3\|^2} \tag{5.90}$$

Let us next use Eqs. (5.88a) and (5.88b) to eliminate $\mathbf{r}_1$ and $\mathbf{r}_3$ from the expressions for c_1 and c_3. First of all,

$$\mathbf{r}_1 \times \mathbf{r}_3 = (f_1\mathbf{r}_2 + g_1\mathbf{v}_2) \times (f_3\mathbf{r}_2 + g_3\mathbf{v}_2) = f_1g_3(\mathbf{r}_2 \times \mathbf{v}_2) + f_3g_1(\mathbf{v}_2 \times \mathbf{r}_2)$$

But $\mathbf{r}_2 \times \mathbf{v}_2 = \mathbf{h}$, where $\mathbf{h}$ is the constant angular momentum of the orbit (Eq. 2.28). It follows that

$$\mathbf{r}_1 \times \mathbf{r}_3 = (f_1g_3 - f_3g_1)\mathbf{h} \tag{5.91}$$

and, of course,

$$\mathbf{r}_3 \times \mathbf{r}_1 = -(f_1g_3 - f_3g_1)\mathbf{h} \tag{5.92}$$

Therefore

$$\|\mathbf{r}_1 \times \mathbf{r}_3\|^2 = (f_1g_3 - f_3g_1)^2h^2 \tag{5.93}$$

Similarly

$$\mathbf{r}_2 \times \mathbf{r}_3 = \mathbf{r}_2 \times (f_3\mathbf{r}_2 + g_3\mathbf{v}_2) = g_3\mathbf{h} \tag{5.94}$$

and

$$\mathbf{r}_2 \times \mathbf{r}_1 = \mathbf{r}_2 \times (f_1\mathbf{r}_2 + g_1\mathbf{v}_2) = g_1\mathbf{h} \tag{5.95}$$

Substituting Eqs. (5.91), (5.93) and (5.94) into Eq. (5.89) yields

$$c_1 = \frac{g_3\mathbf{h} \cdot (f_1g_3 - f_3g_1)\mathbf{h}}{(f_1g_3 - f_3g_1)^2h^2} = \frac{g_3(f_1g_3 - f_3g_1)h^2}{(f_1g_3 - f_3g_1)^2h^2}$$

or

$$c_1 = \frac{g_3}{f_1 g_3 - f_3 g_1} \tag{5.96}$$

Likewise, substituting Eqs. (5.92), (5.93), and (5.95) into Eq. (5.90) leads to

$$c_3 = -\frac{g_1}{f_1 g_3 - f_3 g_1} \tag{5.97}$$

The coefficients in Eq. (5.87) are now expressed solely in terms of the Lagrange functions, and so far no approximations have been made. However, we will have to make some approximations to proceed.

We must approximate c_1 and c_2 under the assumption that the times between observations of the orbiting body are small. To that end, let us introduce the notation

$$\tau_1 = t_1 - t_2 \qquad \tau_3 = t_3 - t_2 \tag{5.98}$$

where τ_1 and τ_3 are the time intervals between the successive measurements of $\hat{\boldsymbol{\rho}}_1$, $\hat{\boldsymbol{\rho}}_2$, and $\hat{\boldsymbol{\rho}}_3$. If the time intervals τ_1 and τ_3 are small enough, we can retain just the first two terms of the series expressions for the Lagrange coefficients f and g in Eq. (2.172), thereby obtaining the approximations

$$f_1 \approx 1 - \frac{1}{2}\frac{\mu}{r_2^3}\tau_1^2 \qquad f_3 \approx 1 - \frac{1}{2}\frac{\mu}{r_2^3}\tau_3^2 \tag{5.99}$$

and

$$g_1 \approx \tau_1 - \frac{1}{6}\frac{\mu}{r_2^3}\tau_1^3 \qquad g_3 \approx \tau_3 - \frac{1}{6}\frac{\mu}{r_2^3}\tau_3^3 \tag{5.100}$$

We want to exclude all terms in f and g beyond the first two, so that only the unknown r_2 appears in Eqs. (5.99) and (5.100). We can see from Eq. (2.172) that the higher order terms include the unknown $\mathbf{v}_2$ as well.

Using Eqs. (5.99) and (5.100), we can calculate the denominator in Eqs. (5.96) and (5.97),

$$f_1 g_3 - f_3 g_1 = \left(1 - \frac{1}{2}\frac{\mu}{r_2^3}\tau_1^2\right)\left(\tau_3 - \frac{1}{6}\frac{\mu}{r_2^3}\tau_3^3\right) - \left(1 - \frac{1}{2}\frac{\mu}{r_2^3}\tau_3^2\right)\left(\tau_1 - \frac{1}{6}\frac{\mu}{r_2^3}\tau_1^3\right)$$

Expanding the right-hand side and collecting the terms yields

$$f_1 g_3 - f_3 g_1 = (\tau_3 - \tau_1) - \frac{1}{6}\frac{\mu}{r_2^3}(\tau_3 - \tau_1)^3 + \frac{1}{12}\frac{\mu^2}{r_2^6}\left(\tau_1^2\tau_3^3 - \tau_1^3\tau_3^2\right)$$

Retaining terms of at most the third order in the time intervals τ_1 and τ_3, and setting

$$\tau = \tau_3 - \tau_1 \tag{5.101}$$

reduces this expression to

$$f_1 g_3 - f_3 g_1 \approx \tau - \frac{1}{6}\frac{\mu}{r_2^3}\tau^3 \tag{5.102}$$

From Eq. (5.98) observe that τ is just the time interval between the first and last observations. Substituting Eqs. (5.100) and (5.102) into Eq. (5.96), we get

$$c_1 \approx \frac{\tau_3 - \dfrac{1}{6}\dfrac{\mu}{r_2^3}\tau_3^3}{\tau - \dfrac{1}{6}\dfrac{\mu}{r_2^3}\tau^3} = \frac{\tau_3}{\tau}\left(1 - \frac{1}{6}\frac{\mu}{r_2^3}\tau_3^2\right)\cdot\left(1 - \frac{1}{6}\frac{\mu}{r_2^3}\tau^2\right)^{-1} \tag{5.103}$$

We can use the binomial theorem to simplify (linearize) the last term on the right. Setting $a=1$, $b=-\mu\tau^2/(6r_2{}^3)$, and $n=-1$ in Eq. (5.44), and neglecting terms of higher order than 2 in τ, yields

$$\left(1-\frac{1}{6}\frac{\mu}{r_2{}^3}\tau^2\right)^{-1}\approx 1+\frac{1}{6}\frac{\mu}{r_2{}^3}\tau^2$$

Hence, Eq. (5.103) becomes

$$c_1\approx\frac{\tau_3}{\tau}\left[1+\frac{1}{6}\frac{\mu}{r_2{}^3}\left(\tau^2-\tau_3{}^2\right)\right] \tag{5.104}$$

where only second-order terms in the time have been retained. In precisely the same way we can show that

$$c_3\approx-\frac{\tau_1}{\tau}\left[1+\frac{1}{6}\frac{\mu}{r_2{}^3}\left(\tau^2-\tau_1{}^2\right)\right] \tag{5.105}$$

Finally, we have managed to obtain approximate formulas for the coefficients in Eq. (5.87) in terms of just the time intervals between observations and the as yet unknown distance r_2 from the center of attraction at the central time t_2.

The next stage of the solution for $\mathbf{r}_2$ and $\mathbf{v}_2$ is to seek formulas for the slant ranges ρ_1, ρ_2, and ρ_3 in terms of c_1 and c_2. To that end substitute Eqs. (5.86) into Eq. (5.87) to get

$$\mathbf{R}_2+\rho_2\hat{\boldsymbol{\rho}}_2=c_1(\mathbf{R}_1+\rho_1\hat{\boldsymbol{\rho}}_1)+c_3(\mathbf{R}_3+\rho_3\hat{\boldsymbol{\rho}}_3)$$

which we rearrange into the form

$$c_1\rho_1\hat{\boldsymbol{\rho}}_1-\rho_2\hat{\boldsymbol{\rho}}_2+c_3\rho_3\hat{\boldsymbol{\rho}}_3=-c_1\mathbf{R}_1+\mathbf{R}_2-c_3\mathbf{R}_3 \tag{5.106}$$

Let us isolate the slant ranges ρ_1, ρ_2, and ρ_3 in turn by taking the dot product of this equation with appropriate vectors. To isolate ρ_1, take the dot product of each term in this equation with $\hat{\boldsymbol{\rho}}_2\times\hat{\boldsymbol{\rho}}_3$, which gives

$$\begin{aligned}&c_1\rho_1\hat{\boldsymbol{\rho}}_1\cdot(\hat{\boldsymbol{\rho}}_2\times\hat{\boldsymbol{\rho}}_3)-\rho_2\hat{\boldsymbol{\rho}}_2\cdot(\hat{\boldsymbol{\rho}}_2\times\hat{\boldsymbol{\rho}}_3)+c_3\rho_3\hat{\boldsymbol{\rho}}_3\cdot(\hat{\boldsymbol{\rho}}_2\times\hat{\boldsymbol{\rho}}_3)\\&=-c_1\mathbf{R}_1\cdot(\hat{\boldsymbol{\rho}}_2\times\hat{\boldsymbol{\rho}}_3)+\mathbf{R}_2\cdot(\hat{\boldsymbol{\rho}}_2\times\hat{\boldsymbol{\rho}}_3)-c_3\mathbf{R}_3\cdot(\hat{\boldsymbol{\rho}}_2\times\hat{\boldsymbol{\rho}}_3)\end{aligned}$$

Since $\hat{\boldsymbol{\rho}}_2\cdot(\hat{\boldsymbol{\rho}}_2\times\hat{\boldsymbol{\rho}}_3)=\hat{\boldsymbol{\rho}}_3\cdot(\hat{\boldsymbol{\rho}}_2\times\hat{\boldsymbol{\rho}}_3)=0$, this reduces to

$$c_1\rho_1\hat{\boldsymbol{\rho}}_1\cdot(\hat{\boldsymbol{\rho}}_2\times\hat{\boldsymbol{\rho}}_3)=(-c_1\mathbf{R}_1+\mathbf{R}_2-c_3\mathbf{R}_3)\cdot(\hat{\boldsymbol{\rho}}_2\times\hat{\boldsymbol{\rho}}_3) \tag{5.107}$$

Let D_0 represent the scalar triple product of $\hat{\boldsymbol{\rho}}_1$, $\hat{\boldsymbol{\rho}}_2$, and $\hat{\boldsymbol{\rho}}_3$,

$$D_0=\hat{\boldsymbol{\rho}}_1\cdot(\hat{\boldsymbol{\rho}}_2\times\hat{\boldsymbol{\rho}}_3) \tag{5.108}$$

We will assume that D_0 is not zero, which means that $\hat{\boldsymbol{\rho}}_1$, $\hat{\boldsymbol{\rho}}_2$, and $\hat{\boldsymbol{\rho}}_3$ do not lie in the same plane. Then, we can solve Eq. (5.107) for ρ_1 to get

$$\rho_1=\frac{1}{D_0}\left(-D_{11}+\frac{1}{c_1}D_{21}-\frac{c_3}{c_1}D_{31}\right) \tag{5.109a}$$

where the Ds stand for the scalar triple products

$$D_{11}=\mathbf{R}_1\cdot(\hat{\boldsymbol{\rho}}_2\times\hat{\boldsymbol{\rho}}_3)\quad D_{21}=\mathbf{R}_2\cdot(\hat{\boldsymbol{\rho}}_2\times\hat{\boldsymbol{\rho}}_3)\quad D_{31}=\mathbf{R}_3\cdot(\hat{\boldsymbol{\rho}}_2\times\hat{\boldsymbol{\rho}}_3) \tag{5.109b}$$

In a similar fashion, by taking the dot product of Eq. (5.106) with $\hat{\boldsymbol{\rho}}_1 \times \hat{\boldsymbol{\rho}}_3$ and then $\hat{\boldsymbol{\rho}}_1 \times \hat{\boldsymbol{\rho}}_2$, we obtain ρ_2 and ρ_3,

$$\rho_2 = \frac{1}{D_0}(-c_1 D_{12} + D_{22} - c_3 D_{32}) \tag{5.110a}$$

where

$$D_{12} = \mathbf{R}_1 \cdot (\hat{\boldsymbol{\rho}}_1 \times \hat{\boldsymbol{\rho}}_3) \quad D_{22} = \mathbf{R}_2 \cdot (\hat{\boldsymbol{\rho}}_1 \times \hat{\boldsymbol{\rho}}_3) \quad D_{32} = \mathbf{R}_3 \cdot (\hat{\boldsymbol{\rho}}_1 \times \hat{\boldsymbol{\rho}}_3) \tag{5.110b}$$

and

$$\rho_3 = \frac{1}{D_0}\left(-\frac{c_1}{c_3} D_{13} + \frac{1}{c_3} D_{23} - D_{33}\right) \tag{5.111a}$$

where

$$D_{13} = \mathbf{R}_1 \cdot (\hat{\boldsymbol{\rho}}_1 \times \hat{\boldsymbol{\rho}}_2) \quad D_{23} = \mathbf{R}_2 \cdot (\hat{\boldsymbol{\rho}}_1 \times \hat{\boldsymbol{\rho}}_2) \quad D_{33} = \mathbf{R}_3 \cdot (\hat{\boldsymbol{\rho}}_1 \times \hat{\boldsymbol{\rho}}_2) \tag{5.111b}$$

To obtain these results, we used the fact that $\hat{\boldsymbol{\rho}}_2 \cdot (\hat{\boldsymbol{\rho}}_1 \times \hat{\boldsymbol{\rho}}_3) = -D_0$ and $\hat{\boldsymbol{\rho}}_3 \cdot (\hat{\boldsymbol{\rho}}_1 \times \hat{\boldsymbol{\rho}}_2) = D_0$ (Eq. 1.21).

Substituting Eqs. (5.104) and (5.105) into Eq. (5.110a) yields the slant range ρ_2,

$$\rho_2 = A + \frac{\mu B}{r_2{}^3} \tag{5.112a}$$

where

$$A = \frac{1}{D_0}\left(-D_{12}\frac{\tau_3}{\tau} + D_{22} + D_{32}\frac{\tau_1}{\tau}\right) \tag{5.112b}$$

$$B = \frac{1}{6D_0}\left[D_{12}\left(\tau_3{}^2 - \tau^2\right)\frac{\tau_3}{\tau} + D_{32}\left(\tau^2 - \tau_1{}^2\right)\frac{\tau_1}{\tau}\right] \tag{5.112c}$$

On the other hand, making the same substitutions into Eqs. (5.109a), (5.109b), (5.111a), and (5.111b) leads to the following expressions for the slant ranges ρ_1 and ρ_3:

$$\rho_1 = \frac{1}{D_0}\left[\frac{6\left(D_{31}\dfrac{\tau_1}{\tau_3} + D_{21}\dfrac{\tau}{\tau_3}\right)r_2{}^3 + \mu D_{31}\left(\tau^2 - \tau_1{}^2\right)\dfrac{\tau_1}{\tau_3}}{6r_2{}^3 + \mu\left(\tau^2 - \tau_3{}^2\right)} - D_{11}\right] \tag{5.113}$$

$$\rho_3 = \frac{1}{D_0}\left[\frac{6\left(D_{13}\dfrac{\tau_3}{\tau_1} - D_{23}\dfrac{\tau}{\tau_1}\right)r_2{}^3 + \mu D_{13}\left(\tau^2 - \tau_3{}^2\right)\dfrac{\tau_3}{\tau_1}}{6r_2{}^3 + \mu\left(\tau^2 - \tau_1{}^2\right)} - D_{33}\right] \tag{5.114}$$

Eq. (5.112a) is a relation between the slant range ρ_2 and the geocentric radius r_2. Another expression relating these two variables is obtained from Eq. (5.86b),

$$\mathbf{r}_2 \cdot \mathbf{r}_2 = (\mathbf{R}_2 + \rho_2\hat{\boldsymbol{\rho}}_2) \cdot (\mathbf{R}_2 + \rho_2\hat{\boldsymbol{\rho}}_2)$$

or

$$r_2{}^2 = \rho_2{}^2 + 2E\rho_2 + R_2{}^2 \tag{5.115a}$$

where

$$E = \mathbf{R}_2 \cdot \hat{\boldsymbol{\rho}}_2 \tag{5.115b}$$

Substituting Eq. (5.112a) into Eq. (5.115a) gives

$$r_2{}^2 = \left(A + \frac{\mu B}{r_2{}^2}\right)^2 + 2E\left(A + \frac{\mu B}{r_2{}^2}\right) + R_2{}^2$$

Expanding and rearranging terms leads to an eighth-degree polynomial,

$$x^8 + ax^6 + bx^3 + c = 0 \tag{5.116}$$

where $x = r_2$ and the coefficients are

$$a = -\left(A^2 + 2AE + R_2{}^2\right) \quad b = -2\mu B(A+E) \quad c = -\mu^2 B^2 \tag{5.117}$$

We solve Eq. (5.116) for r_2 and substitute the result into Eqs. (5.112)–(5.114) to obtain the slant ranges ρ_1, ρ_2, and ρ_3. Then Eqs. (5.86) yield the position vectors $\mathbf{r}_1$, $\mathbf{r}_2$, and $\mathbf{r}_3$. Recall that finding $\mathbf{r}_2$ was one of our objectives.

To attain the other objective, the velocity $\mathbf{v}_2$, we first solve Eq. (5.88a) for $\mathbf{r}_2$,

$$\mathbf{r}_2 = \frac{1}{f_1}\mathbf{r}_1 - \frac{g_1}{f_1}\mathbf{v}_2$$

Substitute this result into Eq. (5.88b) to get

$$\mathbf{r}_3 = \frac{f_3}{f_1}\mathbf{r}_1 + \left(\frac{f_1 g_3 - f_3 g_1}{f_1}\right)\mathbf{v}_2$$

Solving this for $\mathbf{v}_2$ yields

$$\mathbf{v}_2 = \frac{1}{f_1 g_3 - f_3 g_1}\left(-f_3\mathbf{r}_1 + f_1\mathbf{r}_3\right) \tag{5.118}$$

in which we employ the approximate Lagrange functions appearing in Eqs. (5.99) and (5.100).

The approximate values we have found for $\mathbf{r}_2$ and $\mathbf{v}_2$ are used as the starting point for iteratively improving the accuracy of the computed $\mathbf{r}_2$ and $\mathbf{v}_2$ until convergence is achieved. The entire step-by-step procedure is summarized in Algorithms 5.5 and 5.6 (see also Appendix D.29).

ALGORITHM 5.5

The Gauss method of preliminary orbit determination

Given the direction cosine vectors $\hat{\boldsymbol{\rho}}_1$, $\hat{\boldsymbol{\rho}}_2$, and $\hat{\boldsymbol{\rho}}_3$ and the observer's position vectors $\mathbf{R}_1$, $\mathbf{R}_2$, and $\mathbf{R}_3$ at the times t_1, t_2, and t_3, compute the orbital elements.

1. Calculate the time intervals τ_1, τ_3, and τ using Eqs. (5.98) and (5.101).
2. Calculate the cross products $\mathbf{p}_1 = \hat{\boldsymbol{\rho}}_2 \times \hat{\boldsymbol{\rho}}_3$, $\mathbf{p}_2 = \hat{\boldsymbol{\rho}}_1 \times \hat{\boldsymbol{\rho}}_3$, and $\mathbf{p}_3 = \hat{\boldsymbol{\rho}}_1 \times \hat{\boldsymbol{\rho}}_2$.
3. Calculate $D_0 = \hat{\boldsymbol{\rho}}_1 \cdot \mathbf{p}_1$ (Eq. 5.108).
4. From Eqs. (5.109b), (5.110b), and (5.111b) compute the nine scalar quantities

$$\begin{aligned} D_{11} &= \mathbf{R}_1 \cdot \mathbf{p}_1 & D_{12} &= \mathbf{R}_1 \cdot \mathbf{p}_2 & D_{13} &= \mathbf{R}_1 \cdot \mathbf{p}_3 \\ D_{21} &= \mathbf{R}_2 \cdot \mathbf{p}_1 & D_{22} &= \mathbf{R}_2 \cdot \mathbf{p}_2 & D_{23} &= \mathbf{R}_2 \cdot \mathbf{p}_3 \\ D_{31} &= \mathbf{R}_3 \cdot \mathbf{p}_1 & D_{32} &= \mathbf{R}_3 \cdot \mathbf{p}_2 & D_{33} &= \mathbf{R}_3 \cdot \mathbf{p}_3 \end{aligned}$$

5. Calculate A and B using Eqs. (5.112b) and (5.112c).
6. Calculate E using Eq. (5.115b), and calculate $R_2{}^2 = \mathbf{R}_2 \cdot \mathbf{R}_2$.
7. Calculate a, b, and c from Eq. (5.117).
8. Find the roots of Eq. (5.116) and select the most reasonable one as r_2. Newton's method can be used, in which case Eq. (3.16) becomes

$$x_{i+1} = x_i - \frac{x_i{}^8 + ax_i{}^6 + bx_i{}^3 + c}{8x_i{}^7 + 6ax_i{}^5 + 3bx_i{}^2} \tag{5.119}$$

We must first print or graph the function $F = x^8 + ax^6 + bx^3 + c$ for $x > 0$ and choose as an initial estimate a value of x near the point where F changes sign. If there is more than one physically reasonable root, then each one must be used and the resulting orbit checked against the knowledge that may already be available about the general nature of the orbit. Alternatively, the analysis can be repeated using additional sets of observations.
9. Calculate ρ_1, ρ_2, and ρ_3 using Eqs. (5.113), (5.112a), and (5.114).
10. Use Eq. (5.86) to calculate $\mathbf{r}_1$, $\mathbf{r}_2$, and $\mathbf{r}_3$.
11. Calculate the Lagrange coefficients f_1, g_1, f_3, and g_3 from Eqs. (5.99) and (5.100).
12. Calculate $\mathbf{v}_2$ using Eq. (5.118).
13. (a) Use $\mathbf{r}_2$ and $\mathbf{v}_2$ from Steps 10 and 12 to obtain the orbital elements from Algorithm 4.2. (b) Alternatively, proceed to Algorithm 5.6 to improve the preliminary estimate of the orbit.

ALGORITHM 5.6

Iterative improvement of the orbit determined by Algorithm 5.5

Use the values of $\mathbf{r}_2$ and $\mathbf{v}_2$ obtained from Algorithm 5.5 to compute the "exact" values of the f and g functions from their universal formulation as follows.

1. Calculate the magnitude of $\mathbf{r}_2$ $\left(r_2 = \sqrt{\mathbf{r}_2 \cdot \mathbf{r}_2}\right)$ and $\mathbf{v}_2$ $\left(v_2 = \sqrt{\mathbf{v}_2 \cdot \mathbf{v}_2}\right)$.
2. Calculate α, the reciprocal of the semimajor axis: $\alpha = 2/r_2 - v_2^2/\mu$.
3. Calculate the radial component of $\mathbf{v}_2$: $v_r)_2 = \mathbf{v}_2 \cdot \mathbf{r}_2/r_2$.
4. Use Algorithm 3.3 to solve the universal Kepler equation (Eq. 3.49) for the universal variables χ_1 and χ_3 at times t_1 and t_3, respectively:

$$\sqrt{\mu}\tau_1 = \frac{r_2 v_r)_2}{\sqrt{\mu}}\chi_1{}^2 C\left(\alpha\chi_1{}^2\right) + (1 - \alpha r_2)\chi_1{}^3 S\left(\alpha\chi_1{}^2\right) + r_2\chi_1$$

$$\sqrt{\mu}\tau_3 = \frac{r_2 v_r)_2}{\sqrt{\mu}}\chi_3{}^2 C\left(\alpha\chi_3{}^2\right) + (1 - \alpha r_2)\chi_3{}^3 S\left(\alpha\chi_3{}^2\right) + r_2\chi_3$$

5. Use χ_1 and χ_3 to calculate f_1, g_1, f_3, and g_3 from Eqs. (3.69):

$$f_1 = 1 - \frac{\chi_1^2}{r_2}C\left(\alpha\chi_1^2\right) \quad g_1 = \tau_1 - \frac{1}{\sqrt{\mu}}\chi_1^3 S\left(\alpha\chi_1^2\right)$$

$$f_3 = 1 - \frac{\chi_3^2}{r_2}C\left(\alpha\chi_3^2\right) \quad g_3 = \tau_3 - \frac{1}{\sqrt{\mu}}\chi_3^3 S\left(\alpha\chi_3^2\right)$$

6. Use these values of f_1, g_1, f_3, and g_3 to calculate c_1 and c_3 from Eqs. (5.96) and (5.97).
7. Use c_1 and c_3 to calculate updated values of ρ_1, ρ_2, and ρ_3 from Eqs. (5.109), (5.110), and (5.111).

8. Calculate updated $\mathbf{r}_1$, $\mathbf{r}_2$, and $\mathbf{r}_3$ from Eqs. (5.86).
9. Calculate updated $\mathbf{v}_2$ using Eq. (5.118) and the f and g values computed in Step 5.
10. Go back to Step 1 and repeat until, to the desired degree of precision, there is no further change in ρ_1, ρ_2, and ρ_3.
11. Use $\mathbf{r}_2$ and $\mathbf{v}_2$ to compute the orbital elements by means of Algorithm 4.2.

EXAMPLE 5.11

A tracking station is located at $\phi = 40°$ N latitude at an altitude of $H = 1$ km. Three observations of an earth satellite yield the values for the topocentric right ascension and declination listed in Table 5.1, which also shows the local sidereal time θ of the observation site.

Use the Gauss Algorithm 5.5 to estimate the state vector at the second observation time. Recall that $\mu = 398{,}600\,\text{km}^3/\text{s}^2$.

Solution

Recalling that the equatorial radius of the earth is $R_e = 6378\,\text{km}$ and the flattening factor is $f = 0.003353$, we substitute $\phi = 40°$, $H = 1$ km, and the given values of θ into Eq. (5.56) to obtain the inertial position vector of the tracking station at each of the three observation times.

$$\mathbf{R}_1 = 3489.8\hat{\mathbf{I}} + 3430.2\hat{\mathbf{J}} + 4078.5\hat{\mathbf{K}}\,(\text{km})$$
$$\mathbf{R}_2 = 3460.1\hat{\mathbf{I}} + 3460.1\hat{\mathbf{J}} + 4078.5\hat{\mathbf{K}}\,(\text{km})$$
$$\mathbf{R}_3 = 3429.9\hat{\mathbf{I}} + 3490.1\hat{\mathbf{J}} + 4078.5\hat{\mathbf{K}}\,(\text{km})$$

Using Eq. (5.57), we compute the direction cosine vectors at each of the three observation times from the right ascension and declination data

$$\begin{aligned}\hat{\boldsymbol{\rho}}_1 &= \cos(-8.7833°)\cos 43.537°\hat{\mathbf{I}} + \cos(-8.7833°)\sin 43.537°\hat{\mathbf{J}} + \sin(-8.7833°)\hat{\mathbf{K}}\\ &= 0.71643\hat{\mathbf{I}} + 0.68074\hat{\mathbf{J}} - 0.15270\hat{\mathbf{K}}\\ \hat{\boldsymbol{\rho}}_2 &= \cos(-12.074°)\cos 54.420°\hat{\mathbf{I}} + \cos(-12.074°)\sin 54.420°\hat{\mathbf{J}} + \sin(-12.074°)\hat{\mathbf{K}}\\ &= 0.56897\hat{\mathbf{I}} + 0.79531\hat{\mathbf{J}} - 0.20917\hat{\mathbf{K}}\\ \hat{\boldsymbol{\rho}}_3 &= \cos(-15.105°)\cos 64.318°\hat{\mathbf{I}} + \cos(-15.105°)\sin 64.318°\hat{\mathbf{J}} + \sin(-15.105°)\hat{\mathbf{K}}\\ &= 0.41841\hat{\mathbf{I}} + 0.87007\hat{\mathbf{J}} - 0.26059\hat{\mathbf{K}}\end{aligned}$$

We can now proceed with Algorithm 5.5.
Step 1:

$$\tau_1 = 0 - 118.10 = -118.10\,\text{s}$$
$$\tau_3 = 237.58 - 118.10 = 119.47\,\text{s}$$
$$\tau = 119.47 - (-118.1) = 237.58\,\text{s}$$

Table 5.1 Data for Example 5.11

Observation	Time (s)	Right ascension, α (°)	Declination, δ (°)	Local sidereal time, θ (°)
1	0	43.537	−8.7833	44.506
2	118.10	54.420	−12.074	45.000
3	237.58	64.318	−15.105	45.499

Step 2:

$$\mathbf{p}_1 = \hat{\boldsymbol{\rho}}_2 \times \hat{\boldsymbol{\rho}}_3 = -0.025258\hat{\mathbf{I}} + 0.060753\hat{\mathbf{J}} + 0.16229\hat{\mathbf{K}}$$
$$\mathbf{p}_2 = \hat{\boldsymbol{\rho}}_1 \times \hat{\boldsymbol{\rho}}_3 = -0.044538\hat{\mathbf{I}} + 0.12281\hat{\mathbf{J}} + 0.33853\hat{\mathbf{K}}$$
$$\mathbf{p}_3 = \hat{\boldsymbol{\rho}}_1 \times \hat{\boldsymbol{\rho}}_2 = -0.020950\hat{\mathbf{I}} + 0.062977\hat{\mathbf{J}} + 0.18246\hat{\mathbf{K}}$$

Step 3:

$$D_0 = \hat{\boldsymbol{\rho}}_1 \cdot \mathbf{p}_1 = -0.0015198$$

Step 4:

$$\begin{array}{lll} D_{11} = \mathbf{R}_1 \cdot \mathbf{p}_1 = 782.15\,\text{km} & D_{12} = \mathbf{R}_1 \cdot \mathbf{p}_2 = 1646.5\,\text{km} & D_{13} = \mathbf{R}_1 \cdot \mathbf{p}_3 = 887.10\,\text{km} \\ D_{21} = \mathbf{R}_2 \cdot \mathbf{p}_1 = 784.72\,\text{km} & D_{22} = \mathbf{R}_2 \cdot \mathbf{p}_2 = 1651.5\,\text{km} & D_{23} = \mathbf{R}_2 \cdot \mathbf{p}_3 = 889.60\,\text{km} \\ D_{31} = \mathbf{R}_3 \cdot \mathbf{p}_1 = 787.31\,\text{km} & D_{32} = \mathbf{R}_3 \cdot \mathbf{p}_2 = 1656.6\,\text{km} & D_{33} = \mathbf{R}_3 \cdot \mathbf{p}_3 = 892.13\,\text{km} \end{array}$$

Step 5:

$$A = \frac{1}{-0.0015198}\left[-1646.5\frac{119.47}{237.58} + 1651.5 + 1656.6\frac{(-118.10)}{237.58}\right] = -6.6858\,\text{km}$$

$$B = \frac{1}{6(-0.0015198)}\left\{1646.5\left(119.47^2 - 237.58^2\right)\frac{119.47}{237.58}\right.$$
$$\left. +1656.6\left[237.58^2 - (-118.10)^2\right]\frac{(-118.10)}{237.58}\right\} = 7.6667\left(10^9\right)\text{km}\cdot\text{s}^2$$

Step 6:

$$E = \mathbf{R}_2 \cdot \hat{\boldsymbol{\rho}}_2 = 3867.5\,\text{km}$$
$$R_2{}^2 = \mathbf{R}_2 \cdot \mathbf{R}_2 = 4.058\left(10^7\right)\text{km}^2$$

Step 7:

$$a = -\left[(6.6858)^2 + 2(-6.6858)(3875.8) + 4.058 \times 10^7\right] = -4.0528 \times 10^7\,\text{km}^2$$
$$b = -2(389,600)\left(7.6667 \times 10^9\right)(-6.6858 + 3875.8) = -2.3597 \times 10^{19}\,\text{km}^5$$
$$c = -(398,600)^2\left(7.6667 \times 10^9\right)^2 = -9.3387 \times 10^{30}\,\text{km}^8$$

Step 8:

$$F(x) = x^8 - 4.0528 \times 10^7 x^6 - 2.3597 \times 10^{19} x^3 - 9.3387 \times 10^{30} = 0$$

The graph of $F(x)$ in Fig. 5.15 shows that it changes sign near $x = 9000$ km. Let us use that as the starting value in Newton's method for finding the roots of $F(x)$. For the case at hand, Eq. (5.119) is

$$x_{i+1} = x_i - \frac{x_i^8 - 4.0528\left(10^7\right)x_i^6 - 2.3622\left(10^{19}\right)x_i^3 - 9.3186\left(10^{30}\right)}{8x_i^7 - 2.4317\left(10^8\right)x_i^5 - 7.0866\left(10^{19}\right)x_i^2}$$

Stepping through Newton's iterative procedure yields

$$x_0 = 9000$$
$$x_1 = 9000 - (-276.93) = 9276.9$$
$$x_2 = 9276.9 - 34.526 = 9242.4$$
$$x_3 = 9242.4 - 0.63428 = 9241.8$$
$$x_4 = 9241.8 - 0.00021048 = 9241.8$$

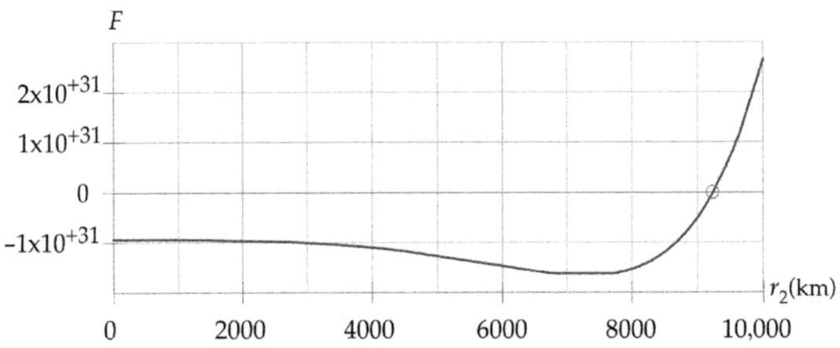

FIG. 5.15

Graph of the polynomial $F(x)$ in Step 8.

Thus, after four steps we converge to

$$r_2 = 9241.8\,\text{km}$$

The other roots are either negative or complex and are therefore physically unacceptable.

Step 9:

$$\rho_1 = \frac{1}{-0.0015198}\times\left\{\frac{6\left(787.31\dfrac{-118.10}{119.47}+784.72\dfrac{237.58}{119.47}\right)9241.8^3+398{,}600\cdot 787.31\cdot\left[237.58^2-(-118.10)^2\right]\dfrac{-118.10}{19.47}}{6\cdot 9241.8^3+398{,}600\left(237.58^2-119.47^2\right)}-782.15\right\}$$
$$= 3639.1\,\text{km}$$

$$\rho_2 = -6.6858 + \frac{398{,}600\cdot 7.6667\left(10^9\right)}{9241.8^3} = 3864.8\,\text{km}$$

$$\rho_3 = \frac{1}{-0.0015198}\times\left[\frac{6\left(887.10\dfrac{119.47}{-118.10}-889.60\dfrac{237.58}{-118.10}\right)9241.8^3+398{,}600\cdot 887.10\left(237.58^2-119.47^2\right)\dfrac{119.47}{-118.10}}{6\cdot 9241.8^3+398{,}600\left[237.58^2-(-118.10)^2\right]}-892.13\right]$$
$$= 4172.8\,\text{km}$$

Step 10:

$$\mathbf{r}_1 = \left(3489.8\hat{\mathbf{I}}+3430.2\hat{\mathbf{J}}+4078.5\hat{\mathbf{K}}\right)+3639.1\left(0.71643\hat{\mathbf{I}}+0.68074\hat{\mathbf{J}}-0.15270\hat{\mathbf{K}}\right)$$
$$= 6096.9\hat{\mathbf{I}}+5907.5\hat{\mathbf{J}}+3522.9\hat{\mathbf{K}}\,(\text{km})$$
$$\mathbf{r}_2 = \left(3460.1\hat{\mathbf{I}}+3460.1\hat{\mathbf{J}}+4078.5\hat{\mathbf{K}}\right)+3864.8\left(0.56897\hat{\mathbf{I}}+0.79531\hat{\mathbf{J}}-0.20917\hat{\mathbf{K}}\right)$$
$$= 5659.1\hat{\mathbf{I}}+6533.8\hat{\mathbf{J}}+3270.1\hat{\mathbf{K}}\,(\text{km})$$
$$\mathbf{r}_3 = \left(3429.9\hat{\mathbf{I}}+3490.1\hat{\mathbf{J}}+4078.5\hat{\mathbf{K}}\right)+4172.8\left(0.41841\hat{\mathbf{I}}+0.87007\hat{\mathbf{J}}-0.26059\hat{\mathbf{K}}\right)$$
$$= 5175.8\hat{\mathbf{I}}+7120.8\hat{\mathbf{J}}+2991.1\hat{\mathbf{K}}\,(\text{km})$$

Step 11:

$$f_1 \approx 1 - \frac{1398,600}{29241.8^3}(-118.10)^2 = 0.99648$$
$$f_3 \approx 1 - \frac{1398,600}{29241.8^3}(119.47)^2 = 0.99640$$
$$g_1 \approx -118.10 - \frac{1}{6}\cdot\frac{398,600}{9241.8^3}(-118.10)^3 = -117.97$$
$$g_3 \approx 119.47 - \frac{1}{6}\cdot\frac{398,600}{69241.8^3}(119.47)^3 = 119.33$$

Step 12:

$$\mathbf{v}_2 = \frac{-0.99640(6096.9\hat{\mathbf{I}} + 5907.5\hat{\mathbf{J}} + 3522.9\hat{\mathbf{K}}) + 0.99648(5175.8\hat{\mathbf{I}} + 7120.8\hat{\mathbf{J}} + 2991.1\hat{\mathbf{K}})}{0.99648 \cdot 119.33 - 0.99640(-117.97)}$$
$$= -3.8800\hat{\mathbf{I}} + 5.1156\hat{\mathbf{J}} - 2.2397\hat{\mathbf{K}}\,(\text{km/s})$$

In summary, the state vector at time t_2 is, approximately,

$$\boxed{\begin{aligned}\mathbf{r}_2 &= 5659.1\hat{\mathbf{I}} + 6533.8\hat{\mathbf{J}} + 3270.1\hat{\mathbf{K}}\,(\text{km})\\ \mathbf{v}_2 &= -3.8800\hat{\mathbf{I}} + 5.1156\hat{\mathbf{J}} - 2.2387\hat{\mathbf{K}}\,(\text{km/s})\end{aligned}}$$

EXAMPLE 5.12

Starting with the state vector determined in Example 5.11, use Algorithm 5.6 to improve the vector to five significant figures.

Step 1:

$$r_2 = \|\mathbf{r}_2\| = \sqrt{5659.1^2 + 6533.8^2 + 3270.1^2} = 9241.8\,\text{km}$$
$$v_2 = \|\mathbf{v}_2\| = \sqrt{(-3.8800)^2 + 5.1156 + (-2.2397)^2} = 6.7999\,\text{km/s}$$

Step 2:

$$\alpha = \frac{2}{r_2} - \frac{v_2^2}{\mu} = \frac{2}{9241.8} - \frac{6.7999^2}{398,600} = 1.0154(10^{-4})\,\text{km}^{-1}$$

Step 3:

$$v_r)_2 = \frac{\mathbf{v}_2 \cdot \mathbf{r}_2}{r_2} = \frac{(-3.8800)\cdot 5659.1 + 5.1156 \cdot 6533.8 + (-2.2397)\cdot 3270.1}{9241.8} = 0.44829\,\text{km/s}$$

Step 4:

The universal Kepler equation at times t_1 and t_3, respectively, becomes

$$\sqrt{398,600}\tau_1 = \frac{9241.8.0\cdot 44829}{\sqrt{398,600}}\chi_1^2 C(1.0040\times 10^{-4}\chi_1^2)$$
$$+ (1 - 1.0040\times 10^{-4}\cdot 9241.8)\chi_1^3 S(1.0040\times 10^{-4}\chi_1^2) + 9241.8\chi_1$$
$$\sqrt{398,600}\tau_3 = \frac{9241.8\cdot 0.44829}{\sqrt{398,600}}\chi_3^2 C(1.0040\times 10^{-4}\chi_3^2)$$
$$+ (1 - 1.0040\times 10^{-4}\cdot 9241.8)\chi_3^3 S(1.0040\times 10^{-4}\chi_3^2) + 9241.8\chi_3$$

or

$$631.35\tau_1 = 6.5622\chi_1^2 C\left(1.0040\times10^{-4}\chi_1^2\right) + 0.072085\chi_1^3 S\left(1.0040\times10^{-4}\chi_1^2 + 9241.8\chi_1\right)$$
$$631.35\tau_3 = 6.5622\chi_3^2 C\left(1.0040\times10^{-4}\chi_3^2\right) + 0.072085\chi_1^3 S\left(1.0040\times10^{-4}\chi_3^2 + 9241.8\chi_3\right)$$

Applying Algorithm 3.3 to each of these equations yields

$$\chi_1 = -8.0908\sqrt{\text{km}}$$
$$\chi_3 = 8.1375\sqrt{\text{km}}$$

Step 5:

$$f_1 = 1 - \frac{\chi_1^2}{r_2}C\left(\alpha\chi_1^2\right) = 1 - \frac{(-8.0908)^2}{9241.8}\cdot\overbrace{C\left(1.0040\times10^{-4}[-8.0908]^2\right)}^{0.49973} = 0.99646$$
$$g_1 = \tau_1 - \frac{1}{\sqrt{\mu}}\chi_1^3 S\left(\alpha\chi_1^2\right)$$
$$= -118.1 - \frac{1}{\sqrt{398,600}}(-8.0908)^3\cdot\overbrace{S\left(1.0040\times10^{-4}[-8.0908]^2\right)}^{0.16661} = -117.96\text{s}$$

and

$$f_3 = 1 - \frac{\chi_3^2}{r_2}C\left(\alpha\chi_3^2\right) = 1 - \frac{8.1375^2}{9241.8}\cdot\overbrace{C\left(1.0040\times10^{-4}\cdot8.1375^2\right)}^{0.49972} = 0.99642$$
$$g_3 = \tau_3 - \frac{1}{\sqrt{\mu}}\chi_3^3 S\left(\alpha\chi_3^2\right)$$
$$= -118.1 - \frac{1}{\sqrt{398,600}}8.1375^3\cdot\overbrace{S\left(1.0040\times10^{-4}\cdot8.1375^2\right)}^{0.16661} = 119.33$$

It turns out that the procedure converges more rapidly if the Lagrange coefficients are set equal to the average of those computed for the current step and those computed for the previous step. Thus, we set

$$f_1 = \frac{0.99648 + 0.99646}{2} = 0.99647$$
$$g_1 = \frac{-117.97 + (-117.96)}{2} = -117.96\text{s}$$
$$f_3 = \frac{0.99642 + 0.99641}{2} = 0.99641$$
$$g_3 = \frac{119.33 + 119.33}{2} = 119.34\text{s}$$

Step 6:

$$c_1 = \frac{119.33}{(0.99647)(119.33) - (0.99641)(-117.96)} = 0.50467$$
$$c_3 = \frac{-117.96}{(0.99647)(119.33) - (0.99641)(-117.96)} = 0.49890$$

Step 7:

$$\rho_1 = \frac{1}{-0.0015198}\left(-782.15 + \frac{1}{0.50467}787.72 - \frac{0.49890}{0.50467}787.31\right) = 3650.6\text{km}$$
$$\rho_2 = \frac{1}{-0.0015198}(-0.50467\cdot1646.5 + 1651.5 - 0.49890\cdot1656.6) = 3877.2\text{km}$$
$$\rho_3 = \frac{1}{-0.0015198}\left(-\frac{0.50467}{0.49890}887.10 + \frac{1}{0.49890}889.60 - 892.13\right) = 4186.2\text{km}$$

Table 5.2 Key results at each step of the iterative procedure

Step	χ_1	χ_3	f_1	g_1	f_3	g_3	ρ_1	ρ_2	ρ_3
1	−8.0908	8.1375	0.99647	−117.97	0.99641	119.33	3650.6	3877.2	4186.2
2	−8.0818	8.1282	0.99647	−117.96	0.996 42	119.33	3643.8	3869.9	4178.3
3	−8.0871	8.1337	0.99647	−117.96	0.996 42	119.33	3644.0	3870.1	4178.6
4	−8.0869	8.1336	0.99647	−117.96	0.996 42	119.33	3644.0	3870.1	4178.6

Step 8:

$$\begin{aligned}\mathbf{r}_1 &= \left(3489.8\hat{\mathbf{I}} + 3430.2\hat{\mathbf{J}} + 4078.5\hat{\mathbf{K}}\right) + 3650.6\left(0.71643\hat{\mathbf{I}} + 0.68074\hat{\mathbf{J}} - 0.15270\hat{\mathbf{K}}\right)\\ &= 6105.2\hat{\mathbf{I}} + 5915.3\hat{\mathbf{J}} + 3521.1\hat{\mathbf{K}}\ (\text{km})\\ \mathbf{r}_2 &= \left(3460.1\hat{\mathbf{I}} + 3460.1\hat{\mathbf{J}} + 4078.5\hat{\mathbf{K}}\right) + 3877.2\left(0.56897\hat{\mathbf{I}} + 0.79531\hat{\mathbf{J}} - 0.20917\hat{\mathbf{K}}\right)\\ &= 5666.6\hat{\mathbf{I}} + 6543.7\hat{\mathbf{J}} + 3267.5\hat{\mathbf{K}}\ (\text{km})\\ \mathbf{r}_3 &= \left(3429.9\hat{\mathbf{I}} + 3490.1\hat{\mathbf{J}} + 4078.5\hat{\mathbf{K}}\right) + 4186.2\left(0.41841\hat{\mathbf{I}} + 0.87007\hat{\mathbf{J}} - 0.26059\hat{\mathbf{K}}\right)\\ &= 5181.4\hat{\mathbf{I}} + 7132.4\hat{\mathbf{J}} + 2987.6\hat{\mathbf{K}}\ (\text{km})\end{aligned}$$

Step 9:

$$\begin{aligned}\mathbf{v}_2 &= \frac{1}{0.99647\cdot 119.33 - 0.99641(-117.96)}\\ &\quad \times \left[-0.99641\left(6105.2\hat{\mathbf{I}} + 5915.3\hat{\mathbf{J}} + 3521.1\hat{\mathbf{K}}\right) + 0.99647\left(5181.4\hat{\mathbf{I}} + 7132.4\hat{\mathbf{J}} + 2987.6\hat{\mathbf{K}}\right)\right]\\ &= -3.8856\hat{\mathbf{I}} + 5.1214\hat{\mathbf{J}} - 2.2434\hat{\mathbf{K}}\ (\text{km/s})\end{aligned}$$

This completes the first iteration.

The updated position $\mathbf{r}_2$ and velocity $\mathbf{v}_2$ are used to repeat the procedure beginning at Step 1. The results of the first and subsequent iterations are shown in Table 5.2. Convergence to five significant figures in the slant ranges ρ_1, ρ_2, and ρ_3 occurs in four steps, at which point the state vector is

$$\boxed{\begin{aligned}\mathbf{r}_2 &= 5662.1\hat{\mathbf{I}} + 6538.0\hat{\mathbf{J}} + 3269.0\hat{\mathbf{K}}\ (\text{km})\\ \mathbf{v}_2 &= -3.8856\hat{\mathbf{I}} + 5.1214\hat{\mathbf{J}} - 2.2433\hat{\mathbf{K}}\ (\text{km/s})\end{aligned}}$$

Using $\mathbf{r}_2$ and $\mathbf{v}_2$ in Algorithm 4.2, we find that the orbital elements are

$$\begin{aligned}a &= 10{,}000\,\text{km}\ \left(h = 62{,}818\,\text{km}^2/\text{s}\right)\\ e &= 0.1000\\ i &= 30^\circ\\ \Omega &= 270^\circ\\ \omega &= 90^\circ\\ \theta &= 45.01^\circ\end{aligned}$$

PROBLEMS

Section 5.2

5.1 The geocentric equatorial position vectors of a satellite at three separate times are

$$\mathbf{r}_1 = 5887\hat{\mathbf{I}} - 3520\hat{\mathbf{J}} - 1204\hat{\mathbf{K}}\ (\text{km})$$
$$\mathbf{r}_2 = 5572\hat{\mathbf{I}} - 3457\hat{\mathbf{J}} - 2376\hat{\mathbf{K}}\ (\text{km})$$
$$\mathbf{r}_3 = 5088\hat{\mathbf{I}} - 3289\hat{\mathbf{J}} - 3480\hat{\mathbf{K}}\ (\text{km})$$

Use Gibbs method to find $\mathbf{v}_2$.
{Partial Ans.: $v_2 = 7.59\ \text{km/s}$}

5.2 Calculate the orbital elements and perigee altitude of the space object in the previous problem.
{Partial Ans.: $z_p = 567$ km}

Section 5.3

5.3 At a given instant, the altitude of an earth satellite is 400 km. Some 30 min later, the altitude is 1000 km, and the true anomaly has increased by 120°. Find the perigee altitude.
{Ans.: 270.4 km}

5.4 At a given instant, the geocentric equatorial position vector of an earth satellite is

$$\mathbf{r}_1 = 3600\hat{\mathbf{I}} + 4600\hat{\mathbf{J}} + 3600\hat{\mathbf{K}}\ (\text{km})$$

Some 30 min later, the position is

$$\mathbf{r}_2 = -5500\hat{\mathbf{I}} + 6240\hat{\mathbf{J}} + 5200\hat{\mathbf{K}}\ (\text{km})$$

Find the specific energy of the orbit.
{Ans.: $-19.871\ (\text{km/s})^2$}

5.5 Compute the perigee altitude and the inclination of the orbit in the previous problem.
{Ans.: 483.59 km, 44.17°}

5.6 At a given instant, the geocentric equatorial position vector of an earth satellite is

$$\mathbf{r}_1 = 5644\hat{\mathbf{I}} + 2830\hat{\mathbf{J}} + 4170\hat{\mathbf{K}}\ (\text{km})$$

Some 20 min later, the position is

$$\mathbf{r}_2 = -2240\hat{\mathbf{I}} + 7320\hat{\mathbf{J}} + 4980\hat{\mathbf{K}}\ (\text{km})$$

Calculate $\mathbf{v}_1$ and $\mathbf{v}_2$.
{Partial Ans.: $v_1 = 10.84\ \text{km/s}$, $v_2 = 9.970\ \text{km/s}$}

5.7 Compute the orbital elements and perigee altitude for the previous problem.
{Partial Ans.: $z_p = 224\ \text{km}$}

Section 5.4

5.8 Calculate the Julian day number for the following epochs:
(a) 5:30 UT on August 14, 1914.
(b) 14:00 UT on April 18, 1946.
(c) 0:00 UT on September 1, 2010.

(d) 12:00 UT on October 16, 2007.
(e) Noon today, your local time.
{Ans.: (a) 2,420,358.729, (b) 2,431,929.083, (c) 2,455,440.500, (d) 2,454,390.000}

5.9 Calculate the number of days from 12:00 UT on your date of birth to 12:00 UT on today's date.

5.10 Calculate the local sidereal time (in degrees) at
(a) Stockholm, Sweden (east longitude 18°03′) at 12:00 UT on January 1, 2008.
(b) Melbourne, Australia (east longitude 144°58′) at 10:00 UT on December 21, 2007.
(c) Los Angeles, California (west longitude 118°15′) at 20:00 UT on July 4, 2005.
(d) Rio de Janeiro, Brazil (west longitude 43°06′) at 3:00 UT on February 15, 2006.
(e) Vladivostok, Russia (east longitude 131°56′) at 8:00 UT on March 21, 2006.
(f) At noon today, your local time and place.
{Ans.: (a) 298.6°, (b) 24.6°, (c) 104.7°, (d) 146.9°, (e) 70.6°}

Section 5.8

5.11 Relative to a tracking station whose local sidereal time is 117° and latitude is +51°, the azimuth and elevation angle of a satellite are 27.5156° and 67.5556°, respectively. Calculate the topocentric right ascension and declination of the satellite.
{Ans.: $\alpha = 145.3°$, $\delta = 68.24°$}

5.12 A sea level tracking station whose local sidereal time is 40° and latitude is 35° makes the following observations of a space object:
Azimuth: 36.0°
Azimuth rate: 0.590°/s
Elevation: 36.6°
Elevation rate: −0.263°/s
Range: 988 km
Range rate: 4.86 km/s
What is the state vector of the object?
{Partial Ans.: $r = 7003.3$ km, $v = 10.922$ km/s}

5.13 Calculate the orbital elements of the satellite in the previous problem.
{Partial Ans.: $e = 1.1$, $i = 40°$}

5.14 A tracking station at latitude −20° and elevation 500 m makes the following observations of a satellite at the given times.

Time (min)	Local sidereal time (°)	Azimuth (°)	Elevation angle (°)	Range (km)
0	60.0	165.931	9.53549	1214.89
2	60.5014	145.967	45.7711	421.441
4	61.0027	2.40962	21.8825	732.079

Use the Gibbs method to calculate the state vector of the satellite at the central observation time.
{Partial Ans.: $r_2 = 6684\,\text{km}$, $v_2 = 7.7239\,\text{km/s}$}

5.15 Calculate the orbital elements of the satellite in the previous problem.
{Partial Ans.: $e = 0.001$, $i = 95°$}

Section 5.10

5.16 A sea level tracking station at latitude +29° makes the following observations of a satellite at the given times.

Time (min)	Local sidereal time (°)	Topocentric right ascension (°)	Topocentric declination (°)
0.0	0	0	51.5110
1.0	0.250684	65.9279	27.9911
2.0	0.501369	79.8500	14.6609

Use the Gauss method without iterative improvement to estimate the state vector of the satellite at the middle observation time.
{Partial Ans.: $r = 6700.9$ km, $v = 8.0757$ km/s}

5.17 Refine the estimate in the previous problem using iterative improvement.
{Partial Ans.: $r = 6701.5$ km, $v = 8.0881$ km/s}

5.18 Calculate the orbital elements from the state vector obtained in the previous problem.
{Partial Ans.: $e = 0.10$, $i = 30°$}

5.19 A sea level tracking station at latitude +29° makes the following observations of a satellite at the given times.

Time (min)	Local sidereal time (°)	Topocentric right ascension (°)	Topocentric declination (°)
0.0	90	15.0394	20.7487
1.0	90.2507	25.7539	30.1410
2.0	90.5014	48.6055	43.8910

Use the Gauss method without iterative improvement to estimate the state vector of the satellite.
{Partial Ans.: $r = 6999.1$ km, $v = 7.5541$ km/s}

5.20 Refine the estimate in the previous problem using iterative improvement.
{Partial Ans.: $r = 7000.0$ km, $v = 7.5638$ km/s}

5.21 Calculate the orbital elements from the state vector obtained in the previous problem.
{Partial Ans.: $e = 0.0048$, $i = 31°$}

5.22 The position vector **R** of a tracking station and the direction cosine vector $\hat{\boldsymbol{\rho}}$ of a satellite relative to the tracking station at three times are as follows:

$$t_1 = 0\text{ min}$$

$$\mathbf{R}_1 = -1825.96\hat{\mathbf{I}} + 3583.66\hat{\mathbf{J}} + 4933.54\hat{\mathbf{K}}\text{ (km)}$$

$$\hat{\boldsymbol{\rho}}_1 = -0.301687\hat{\mathbf{I}} + 0.200673\hat{\mathbf{J}} + 0.932049\hat{\mathbf{K}}$$

$$t_2 = 1\text{ min}$$

$$\mathbf{R}_2 = -1816.30\hat{\mathbf{I}} + 3575.63\hat{\mathbf{J}} + 4933.54\hat{\mathbf{K}}\text{ (km)}$$

$$\hat{\boldsymbol{\rho}}_2 = -0.793090\hat{\mathbf{I}} - 0.210324\hat{\mathbf{J}} + 0.571640\hat{\mathbf{K}}$$

$$t_3 = 2\text{ min}$$

$$\mathbf{R}_3 = -1857.25\hat{\mathbf{I}} + 3567.54\hat{\mathbf{J}} + 4933.54\hat{\mathbf{K}}\text{ (km)}$$

$$\hat{\boldsymbol{\rho}}_3 = -0.873085\hat{\mathbf{I}} - 0.362969\hat{\mathbf{J}} + 0.325539\hat{\mathbf{K}}$$

Use the Gauss method without iterative improvement to estimate the state vector of the satellite at the central observation time.
{Partial Ans.: $r = 6742.3$ km, $v = 7.6799$ km/s}

5.23 Refine the estimate in the previous problem using iterative improvement.
{Partial Ans.: $r = 6743.0$ km, $v = 7.6922$ km/s}

5.24 Calculate the orbital elements from the state vector obtained in the previous problem.
{Partial Ans.: $e = 0.001$, $i = 52°$}

5.25 A tracking station at latitude 60°N and 500-m elevation obtains the following data:

Time (min)	Local sidereal time (°)	Topocentric right ascension (°)	Topocentric declination (°)
0.0	150	157.783	24.2403
5.0	151.253	159.221	27.2993
10.0	152.507	160.526	29.8982

Use the Gauss method without iterative improvement to estimate the state vector of the satellite.
{Partial Ans.: $r = 25{,}132$ km, $v = 6.0588$ km/s}

5.26 Refine the estimate in the previous problem using iterative improvement.
{Partial Ans.: $r = 25{,}169$ km, $v = 6.0671$ km/s}

5.27 Calculate the orbital elements from the state vector obtained in the previous problem.
{Partial Ans.: $e = 1.09$, $i = 63°$}

5.28 The position vector **R** of a tracking station and the direction cosine vector $\hat{\boldsymbol{\rho}}$ of a satellite relative to the tracking station at three times are as follows:

$$t_1 = 0\,\text{min}$$
$$\mathbf{R}_1 = 5582.84\hat{\mathbf{I}} + 3073.90\hat{\mathbf{K}}\,(\text{km})$$
$$\hat{\boldsymbol{\rho}}_1 = 0.846428\hat{\mathbf{I}} + 0.532504\hat{\mathbf{K}}$$
$$t_2 = 5\,\text{min}$$
$$\mathbf{R}_2 = 5581.50\hat{\mathbf{I}} + 122.122\hat{\mathbf{J}} + 3073.90\hat{\mathbf{K}}\,(\text{km})$$
$$\hat{\boldsymbol{\rho}}_2 = 0.749290\hat{\mathbf{I}} + 0.463023\hat{\mathbf{J}} + 0.473470\hat{\mathbf{K}}$$
$$t_3 = 10\,\text{min}$$
$$\mathbf{R}_3 = 5577.50\hat{\mathbf{I}} + 244.186\hat{\mathbf{J}} + 3073.90\hat{\mathbf{K}}\,(\text{km})$$
$$\hat{\boldsymbol{\rho}}_3 = 0.529447\hat{\mathbf{I}} - 0.777163\hat{\mathbf{J}} + 0.340152\hat{\mathbf{K}}$$

Use the Gauss method without iterative improvement to estimate the state vector of the satellite.
{Partial Ans.: $r = 9729.6$ km, $v = 6.0234$ km/s}

5.29 Refine the estimate in the previous problem using iterative improvement.
{Partial Ans.: $r = 9759.8$ km, $v = 6.0713$ km/s}

5.30 Calculate the orbital elements from the state vector obtained in the previous problem.
{Partial Ans.: $e = 0.1$, $i = 30°$}

REFERENCES

Bate, R.R., Mueller, D., White, J.E., 1971. Fundamentals of Astrodynamics. Dover Publications, New York.

Bond, V.R., Allman, M.C., 1996. Modern Astrodynamics: Fundamentals and Perturbation Methods. Princeton University Press.

Boulet, D.L., 1991. Methods of Orbit Determination for the Microcomputer. Willmann-Bell, Richmond.

Department of the Navy, 2018. Nautical Almanac Office: The Astronomical Almanac for the Year 2019. GPO, Washington.

Prussing, J.E., Conway, B.A., 2013. Orbital Mechanics, second ed. Oxford University Press, New York.

Seidelmann, P.K., 1992. Explanatory Supplement to the Astronomical Almanac. University Science Books, Sausalito, CA.

Wiesel, W.E., 2010. Spacecraft Dynamics, third ed. Aphelion Press, Beaver Creek, OH.

CHAPTER

ORBITAL MANEUVERS

6

6.1 INTRODUCTION

Orbital maneuvers transfer a spacecraft from one orbit to another. Orbital changes can be dramatic, such as the transfer from a low earth parking orbit to an interplanetary trajectory. They can also be quite small, as in the final stages of the rendezvous of one spacecraft with another. Changing orbits requires the firing of onboard rocket engines. We will be concerned primarily with impulsive maneuvers in which the rockets fire in relatively short bursts to produce the required velocity change (delta-v).

We start with the classical, energy-efficient Hohmann transfer maneuver and generalize it to the bielliptic Hohmann transfer to see if even more efficiency can be obtained. The phasing maneuver, a form of Hohmann transfer, is considered next. This is followed by a study of non-Hohmann transfer maneuvers with and without rotation of the apse line. We then analyze chase maneuvers, which requires solving Lambert's problem as explained in Chapter 5. Energy-demanding chase maneuvers may be impractical for low earth orbits, but they are necessary for interplanetary missions, as we shall see in Chapter 8. After having focused on impulsive transfers between coplanar orbits, we finally turn our attention to plane change maneuvers and their delta-v requirements, which can be very large.

The chapter concludes with a brief consideration of some orbital transfers in which the propulsion system delivers the impulse during a finite (perhaps very long) time interval instead of instantaneously. This makes it difficult to obtain closed-form solutions, so we illustrate the use of the numerical integration techniques presented in Chapter 1 as an alternative.

6.2 IMPULSIVE MANEUVERS

Impulsive maneuvers are those in which brief firings of onboard rocket motors change the magnitude and direction of the velocity vector instantaneously. During an impulsive maneuver, the position of the spacecraft is considered to be fixed; only the velocity changes. The impulsive maneuver is an idealization by means of which we can avoid having to solve the equations of motion (Eq. 2.22) with the rocket thrust included. The idealization is satisfactory for those cases in which the position of the spacecraft changes only slightly during the time that the maneuvering rockets fire. This is true for high-thrust rockets with burn times that are short compared with the coasting time of the vehicle.

Each impulsive maneuver results in a change $\Delta\mathbf{v}$ in the velocity of the spacecraft. $\Delta\mathbf{v}$ can represent a change in the magnitude (pumping maneuver) or the direction (cranking maneuver) of the velocity

Orbital Mechanics for Engineering Students. https://doi.org/10.1016/B978-0-12-824025-0.00006-4

vector, or both. The magnitude Δv of the velocity increment is related to Δm, the mass of propellant consumed, by the ideal rocket equation (see Eq. 13.30).

$$\frac{\Delta m}{m} = 1 - e^{\frac{\Delta v}{I_{sp} g_0}} \tag{6.1}$$

where m is the mass of the spacecraft before the burn, g_0 is the sea level standard acceleration of gravity, and I_{sp} is the specific impulse of the propellants. Specific impulse is defined as follows:

$$I_{sp} = \frac{\text{Thrust}}{\text{Sea-level weight rate of fuel consumption}}$$

Specific impulse has units of seconds, and it is a measure of the performance of a rocket propulsion system. I_{sp} for some common propellant combinations is shown in Table 6.1. Fig. 6.1 is a graph of Eq. (6.1) for a range of specific impulses. Note that for Δv's on the order of 1 km/s or higher, the required propellant exceeds 25% of the spacecraft mass before the burn.

There are presently no refueling stations in space, so a mission's delta-v schedule must be carefully planned to minimize the propellant mass carried aloft in favor of payload.

Table 6.1 Some typical specific impulses

Propellant	I_{sp} (s)
Cold gas	50
Monopropellant hydrazine	230
Solid propellant	290
Nitric acid/monomethylhydrazine	310
Liquid oxygen/liquid hydrogen	455
Ion propulsion	>3000

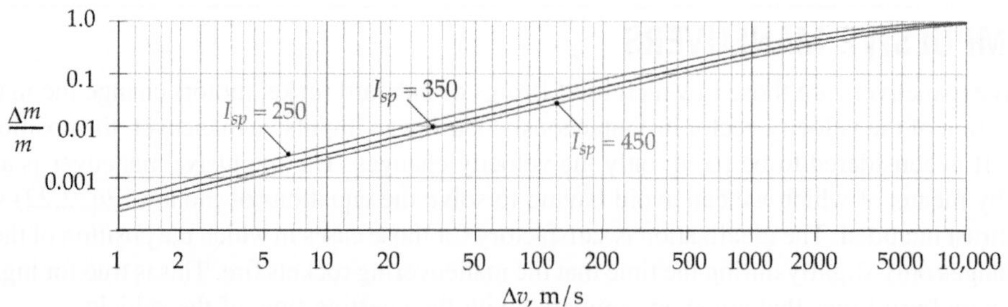

FIG. 6.1

Propellant mass fraction versus Δv for typical specific impulses.

6.3 HOHMANN TRANSFER

Walter Hohmann (1880–1945) was a German engineer whose interest in early rocketry led him to discover the orbital transfer maneuver that now bears his name. In a book published in 1925 (Hohmann, 1925), he showed that the Hohmann transfer is the most energy-efficient two-impulse maneuver for transferring between two coplanar circular orbits sharing a common focus. The Hohmann transfer is an elliptical orbit tangent to both circles on its apse line, as illustrated in Fig. 6.2. The periapsis and apoapsis of the transfer ellipse are the radii of the inner and outer circles, respectively. Obviously, only one-half of the ellipse is flown during the maneuver, which can occur in either direction, from the inner to the outer circle, or vice versa.

It may be helpful in sorting out orbit transfer strategies to use the fact that the energy of an orbit depends only on its semimajor axis a. Recall that for an ellipse (Eq. 2.80), the specific energy is negative,

$$\varepsilon = -\frac{\mu}{2a}$$

Increasing the energy requires reducing its magnitude, to make ε less negative. Therefore, the larger the semimajor axis, the more energy the orbit has. In Fig. 6.2, the energies increase as we move from the inner circle to the outer circle.

Starting at A on the inner circle, a velocity increment Δv_A in the direction of flight is required to boost the vehicle onto the higher energy elliptical trajectory. After coasting from A to B, another forward velocity increment Δv_B places the vehicle on the still higher energy, outer circular orbit. Without the latter delta-v burn, the spacecraft would, of course, remain on the Hohmann transfer ellipse and return to A. The total energy expenditure is reflected in the total delta-v requirement, $\Delta v_{total} = \Delta v_A + \Delta v_B$.

The same total delta-v is required if the transfer begins at B on the outer circular orbit. Since moving to the lower energy inner circle requires lowering the energy of the spacecraft, the Δv's must be accomplished by retrofires. That is, the thrust of the maneuvering rocket is directed opposite to the flight direction to act as a brake on the motion. Since Δv represents the same propellant expenditure regardless of the direction the thruster is aimed, when summing up Δv's, we are concerned only with their magnitudes.

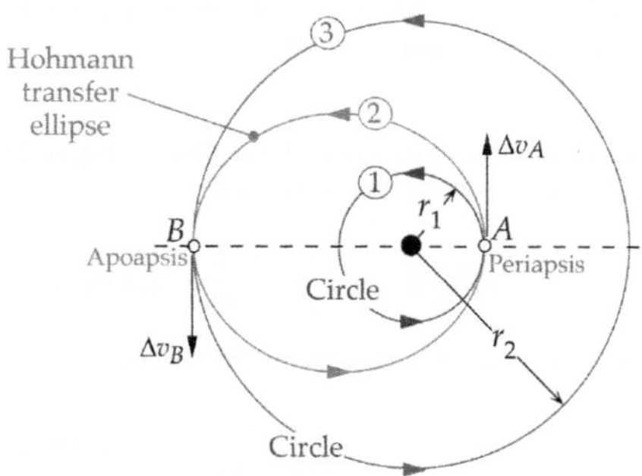

FIG. 6.2

Hohmann transfer.

Recall that the eccentricity of an elliptical orbit is found from its radius to periapsis r_p and its radius to apoapsis r_a by means of Eq. (2.84),

$$e = \frac{r_a - r_p}{r_a + r_p}$$

The radius to periapsis is given by Eq. (2.50),

$$r_p = \frac{h^2}{\mu}\frac{1}{1+e}$$

Combining these last two expressions yields

$$r_p = \frac{h^2}{\mu}\frac{1}{1+\dfrac{r_a - r_p}{r_a + r_p}} = \frac{h^2}{\mu}\frac{r_a + r_p}{2r_a}$$

Solving for the angular momentum h, we get

$$h = \sqrt{2\mu}\sqrt{\frac{r_a r_p}{r_a + r_p}} \tag{6.2}$$

This is a useful formula for analyzing Hohmann transfers, because knowing h we can find the apsidal velocities from Eq. (2.31). Note that for circular orbits ($r_a = r_p$), Eq. (6.2) yields

$$h = \sqrt{\mu r} \quad \text{(circular orbit)}$$

Alternatively, one may prefer to compute the velocities by means of the energy equation (Eq. 2.81) in the form

$$v = \sqrt{2\mu}\sqrt{\frac{1}{r} - \frac{1}{2a}} \tag{6.3}$$

This of course yields Eq. (2.63) for circular orbits.

EXAMPLE 6.1

A 2000-kg spacecraft is in a 480 km by 800 km earth orbit (orbit 1 in Fig. 6.3). Find

(a) The Δv required at perigee A to place the spacecraft in a 480 km by 16,000 km transfer ellipse (orbit 2).

(b) The Δv (apogee kick) required at B of the transfer orbit to establish a circular orbit of 16,000 km altitude (orbit 3).

(c) The total required propellant if the specific impulse is 300 s.

Solution

Since we know the perigee and apogee of all three of the orbits, let us first use Eq. (6.2) to calculate their angular momenta.

Orbit 1: $r_p = 6378 + 480 = 6858\,\text{km}$ $\quad r_a = 6378 + 800 = 7178\,\text{km}$

$$\therefore h_1 = \sqrt{2\cdot 398{,}600}\sqrt{\frac{7178\cdot 6858}{7178 + 6858}} = 52{,}876.5\,\text{km/s}^2 \tag{a}$$

Orbit 2: $r_p = 6378 + 480 = 6858\,\text{km}$ $\quad r_a = 6378 + 16{,}000 = 22{,}378\,\text{km}$

$$\therefore h_2 = \sqrt{2\cdot 398{,}600}\sqrt{\frac{22{,}378\cdot 6858}{22{,}378 + 6858}} = 64{,}689.5\,\text{km/s}^2 \tag{b}$$

Orbit 3: $r_a = r_p = 22{,}378\,\text{km}$

$$\therefore h_3 = \sqrt{398{,}600\cdot 22{,}378} = 94{,}445.1\,\text{km/s}^2 \tag{c}$$

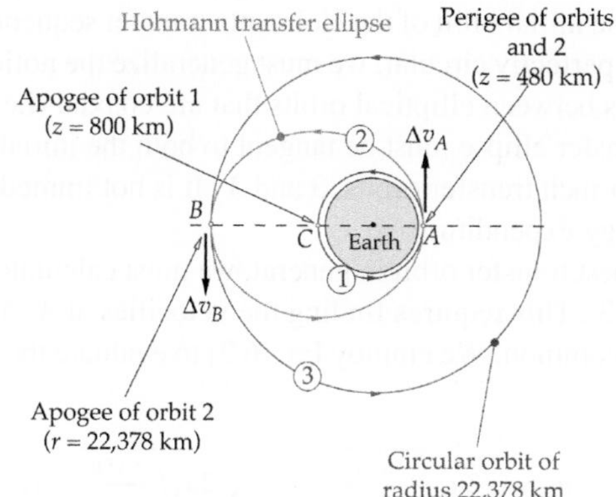

FIG. 6.3

Hohmann transfer between two earth orbits.

(a) The speed on orbit 1 at point A is

$$v_A)_1 = \frac{h_1}{r_A} = \frac{52{,}876}{6858} = 7.71019\,\text{km/s}$$

The speed on orbit 2 at point A is

$$v_A)_2 = \frac{h_2}{r_A} = \frac{64{,}689.5}{6858} = 9.43271\,\text{km/s}$$

Therefore, the delta-v required at point A is

$$\Delta v_A = v_A)_2 - v_A)_1 = \boxed{1.7225\,\text{km/s}}$$

(b) The speed on orbit 2 at point B is

$$v_B)_2 = \frac{h_2}{r_B} = \frac{64{,}689.5}{22{,}378} = 2.89076\,\text{km/s}$$

The speed on orbit 3 at point B is

$$v_B)_3 = \frac{h_3}{r_B} = \frac{94{,}445.1}{22{,}378} = 4.22044\,\text{km/s}$$

Hence, the apogee kick required at point B is

$$\Delta v_B = v_B)_3 - v_B)_2 = \boxed{1.3297\,\text{km/s}}$$

(c) The total delta-v requirement for this Hohmann transfer is

$$\Delta v_{\text{total}} = |\Delta v_A| + |\Delta v_B| = 1.7225 + 1.3297 = 3.0522\,\text{km/s}$$

According to Eq. (6.1) (converting velocity to m/s),

$$\frac{\Delta m}{m} = 1 - e^{-\frac{3052.2}{300\cdot 9.807}} = 0.64563$$

Therefore, the mass of propellant expended is

$$\boxed{\Delta m = 0.64563 \cdot 2000 = 1291.3\,\text{kg}}$$

In the previous example the initial orbit of the Hohmann transfer sequence was an ellipse rather than a circle. Since no real orbit is perfectly circular, we must generalize the notion of a Hohmann transfer to include two-impulse transfers between elliptical orbits that are coaxial (i.e., share the same apse line), as shown in Fig. 6.4. The transfer ellipse must be tangent to both the initial and target ellipses 1 and 2. As can be seen, there are two such transfer orbits, 3 and 3′. It is not immediately obvious which of the two requires the lowest energy expenditure.

To find out which is the best transfer orbit in general, we must calculate the individual total delta-v requirement for orbits 3 and 3′. This requires finding the velocities at A, A', B, and B' for each pair of orbits having those points in common. We employ Eq. (6.2) to evaluate the angular momentum of each of the four orbits in Fig. 6.4.

$$h_1 = \sqrt{2\mu}\sqrt{\frac{r_A r_{A'}}{r_A + r_{A'}}} \quad h_2 = \sqrt{2\mu}\sqrt{\frac{r_B r_{B'}}{r_B + r_{B'}}} \quad h_3 = \sqrt{2\mu}\sqrt{\frac{r_A r_B}{r_A + r_B}} \quad h_{3'} = \sqrt{2\mu}\sqrt{\frac{r_{A'} r_{B'}}{r_{A'} + r_{B'}}}$$

From these we obtain the velocities,

$$\begin{aligned} v_A)_1 &= \frac{h_1}{r_A} & v_A)_3 &= \frac{h_3}{r_A} \\ v_B)_2 &= \frac{h_2}{r_B} & v_B)_3 &= \frac{h_3}{r_B} \\ v_{A'})_1 &= \frac{h_1}{r_{A'}} & v_{A'})_{3'} &= \frac{h_{3'}}{r_{A'}} \\ v_{B'})_2 &= \frac{h_2}{r_{B'}} & v_{B'})_{3'} &= \frac{h_{3'}}{r_{B'}} \end{aligned}$$

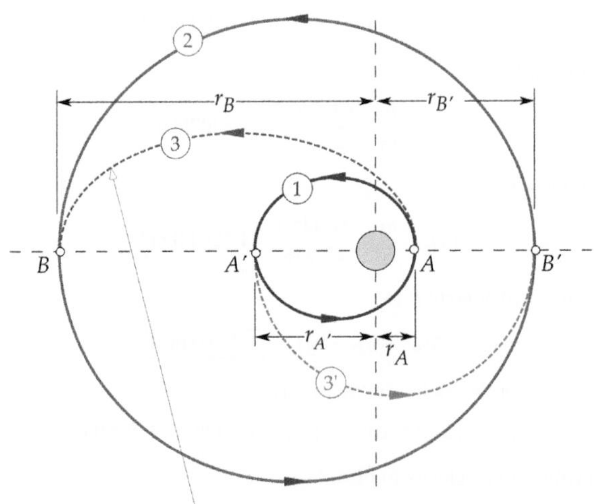

FIG. 6.4

Hohmann transfers between coaxial elliptical orbits. In this illustration, $r_{A'}/r_A = 3$, $r_B/r_A = 8$, and $r_{B'}/r_A = 4$.

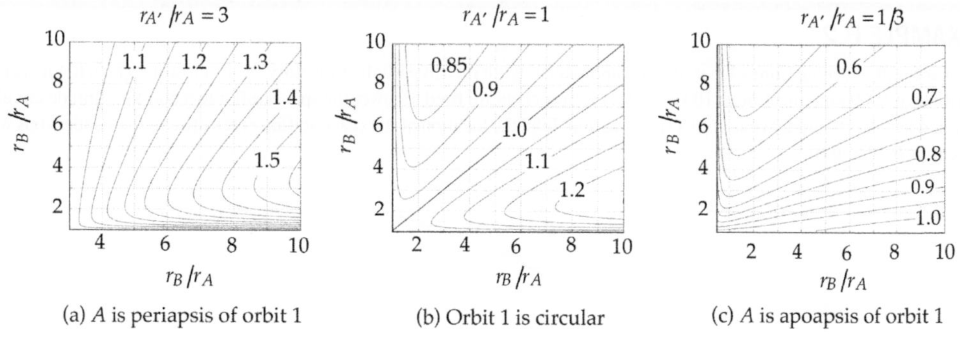

FIG. 6.5

Contour plots of $\Delta v_{\text{total}})_{3'}/\Delta v_{\text{total}})_3$ for different relative sizes of the ellipses in Fig. 6.4. Note that $r_B > r_{A'}$ and $r_{B'} > r_A$.

These lead to the delta-v's

$$\Delta v_A = \left|v_A)_3 - v_A)_1\right| \quad \Delta v_B = \left|v_B)_2 - v_B)_3\right| \quad \Delta v_{A'} = \left|v_{A'})_{3'} - v_{A'})_1\right| \quad \Delta v_{B'} = \left|v_{B'})_2 - v_{B'})_{3'}\right|$$

and, finally, to the total delta-v requirement for the two possible transfer trajectories,

$$\Delta v_{\text{total}})_3 = \Delta v_A + \Delta v_B \qquad \Delta v_{\text{total}})_{3'} = \Delta v_{A'} + \Delta v_{B'}$$

If $\Delta v_{\text{total}})_{3'}/\Delta v_{\text{total}})_3 > 1$, then orbit 3 is the most efficient. On the other hand, if $\Delta v_{\text{total}})_{3'}/\Delta v_{\text{total}})_3 < 1$, then orbit 3′ is more efficient than orbit 3.

Three contour plots of $\Delta v_{\text{total}})_{3'}/\Delta v_{\text{total}})_3$ are shown in Fig. 6.5, for three different shapes of inner orbit 1 of Fig. 6.4. Fig. 6.5a is for $r_{A'}/r_A = 3$, which is the situation represented in Fig. 6.4, in which point A is the periapsis of the initial ellipse. In Fig. 6.5b $r_{A'}/r_A = 1$, which means the starting ellipse is a circle. Finally, in Fig. 6.5c $r_{A'}/r_A = 1/3$, which corresponds to an initial orbit of the same shape as orbit 1 in Fig. 6.4, but with point A being the apoapsis instead of periapsis.

Fig. 6.5a, for which $r_{A'} > r_A$, implies that if point A is the periapsis of orbit 1, then transfer orbit 3 is the most efficient. Fig. 6.5c, for which $r_{A'} < r_A$, shows that if point A' is the periapsis of orbit 1, then transfer orbit 3′ is the most efficient. Together, these results lead us to conclude that it is most efficient for the transfer orbit to begin at the periapsis on inner orbit 1, where its kinetic energy is greatest, regardless of the shape of the outer target orbit. If the starting orbit is a circle, then Fig. 6.5b shows that transfer orbit 3′ is the most efficient if $r_{B'} > r_B$. That is, from an inner circular orbit, the transfer ellipse should terminate at apoapsis of the outer target ellipse, where the speed is slowest.

If Hohmann transfer is in the reverse direction (i.e., to a lower energy inner orbit), the above analysis still applies, since the same total delta-v is required whether the Hohmann transfer runs forward or backward. Thus, from an outer circle or ellipse, to an inner ellipse the most energy-efficient transfer ellipse terminates at periapsis of the inner target orbit. If the inner orbit is a circle, the transfer ellipse should start at apoapsis of the outer ellipse.

We close this section with an illustration of the careful planning required for one spacecraft to rendezvous with another at the end of a Hohmann transfer.

EXAMPLE 6.2

A spacecraft returning from a lunar mission approaches earth on a hyperbolic trajectory. At its closest approach A it is at an altitude of 5000 km, traveling at 10 km/s. At A retrorockets are fired to lower the spacecraft into a 500-km-altitude circular orbit, where it is to rendezvous with a space station. Find the location of the space station at retrofire so that rendezvous will occur at B (Fig. 6.6).

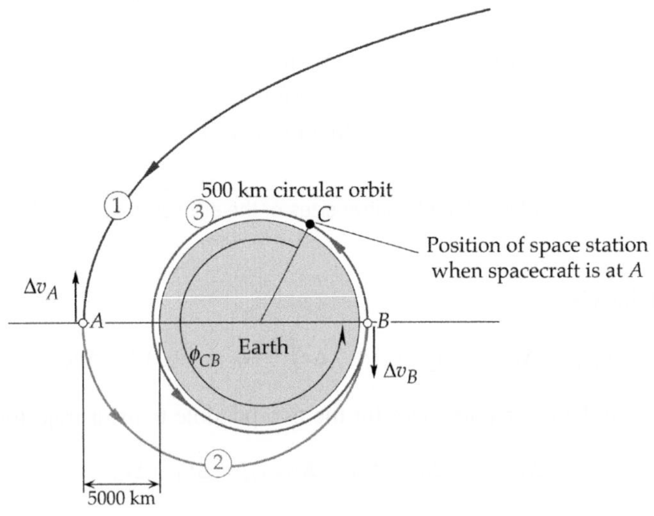

FIG. 6.6

Relative position of spacecraft and space station at beginning of the transfer ellipse.

Solution

The time of flight from A to B is one-half the period T_2 of elliptical transfer orbit 2. While the spacecraft coasts from A to B, the space station coasts through the angle ϕ_{CB} from C to B. Hence, this mission has to be carefully planned and executed, going all the way back to lunar departure, so that the two vehicles meet at B.

According to Eq. (2.83), to find the period T_2 we need to only determine the semimajor axis of orbit 2. The apogee and perigee of orbit 2 are

$$r_A = 5000 + 6378 = 11{,}378 \text{ km} \quad r_B = 500 + 6378 = 6878 \text{ km}$$

Therefore, the semimajor axis is

$$a = \frac{1}{2}(r_A + r_B) = 9128 \text{km}$$

From this we obtain

$$T_2 = \frac{2\pi}{\sqrt{\mu}} a^{3/2} = \frac{2\pi}{\sqrt{398{,}600}} 9128^{3/2} = 8679.1 \text{ s} \tag{a}$$

The period of circular orbit 3 is

$$T_3 = \frac{2\pi}{\sqrt{\mu}} r_B{}^{3/2} = \frac{2\pi}{\sqrt{398{,}600}} 6878^{3/2} = 5676.8 \text{ s} \tag{b}$$

The time of flight from C to B on orbit 3 must equal the time of flight from A to B on orbit 2.

$$\Delta t_{CB} = \frac{1}{2} T_2 = \frac{1}{2} \cdot 8679.1 = 4339.5 \text{ s}$$

Since orbit 3 is a circle, its angular velocity, unlike an ellipse, is constant. Therefore, we can write

$$\frac{\phi_{CB}}{\Delta t_{CB}} = \frac{360°}{T_3} \Rightarrow \phi_{CB} = \frac{4339.5}{5676.8} \cdot 360 = \boxed{275.2°}$$

(The reader should verify that the total delta-v required to lower the spacecraft from the hyperbola into the parking orbit is 5.749 km/s. According to Eq. (6.1), that means over 85% of the returning spacecraft mass must consist of propellant!)

6.4 BIELLIPTIC HOHMANN TRANSFER

A Hohmann transfer from circular orbit 1 to circular orbit 4 in Fig. 6.7 is the dotted ellipse lying inside the outer circle, outside the inner circle, and tangent to both. A bielliptic Hohmann transfer uses two coaxial semiellipses, 2 and 3, which extend beyond the outer target orbit. Each of the two ellipses is tangent to one of the circular orbits, and they are tangent to each other at B, which is the apoapsis of both. The idea is to place B sufficiently far from the focus that the Δv_B will be very small. In fact, as r_B approaches infinity (where the orbital speed is zero), Δv_B approaches zero. For the bielliptic scheme to be more energy efficient than a Hohmann transfer, it must be true that

$$\Delta v_{\text{total}})_{\text{bielliptical}} < \Delta v_{\text{total}})_{\text{Hohmann}}$$

Let v_0 be speed in circular inner orbit 1,

$$v_0 = \sqrt{\frac{\mu}{r_A}}$$

Then calculating the total delta-v requirements of the Hohmann and bielliptic transfers leads to the following two expressions, respectively,

$$\begin{aligned} \Delta\bar{v}_{\text{H}} &= \frac{1}{\sqrt{\alpha}} - \frac{\sqrt{2}(1-\alpha)}{\sqrt{\alpha(1+\alpha)}} - 1 \\ \Delta\bar{v}_{\text{BE}} &= \sqrt{\frac{2(\alpha+\beta)}{\alpha\beta}} - \frac{1+\sqrt{\alpha}}{\sqrt{\alpha}} - \sqrt{\frac{2}{\beta(1+\beta)}}(1-\beta) \end{aligned} \quad (6.4a)$$

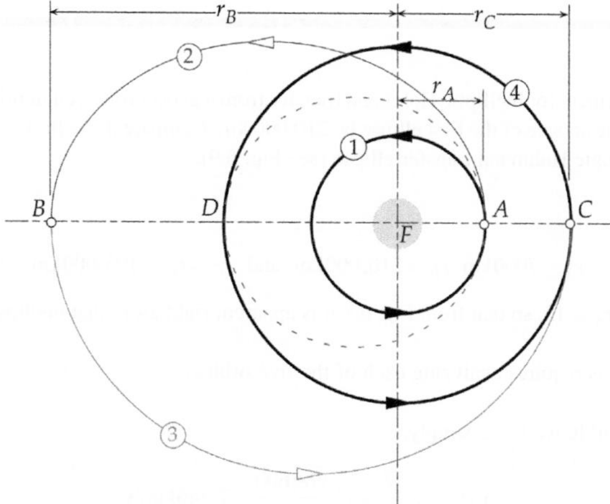

FIG. 6.7

Bielliptic transfer from inner orbit 1 to outer orbit 4.

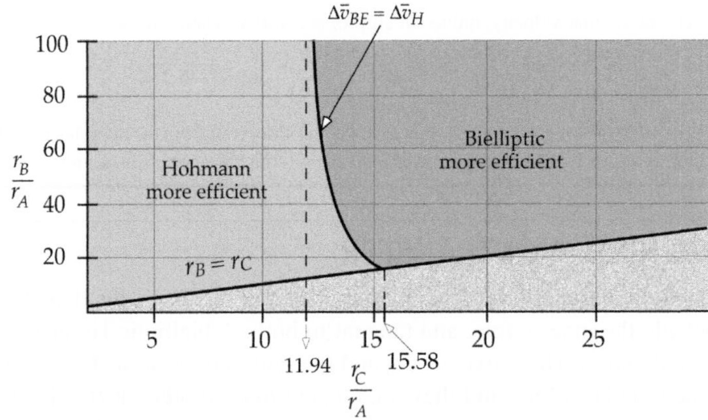

FIG. 6.8

Orbits for which a bielliptic transfer is either less efficient or more efficient than a Hohmann transfer.

where the nondimensional terms are

$$\Delta\bar{v}_{\mathrm{H}} = \left.\frac{\Delta v_{\mathrm{total}}}{v_0}\right)_{\mathrm{Hohmann}} \qquad \Delta\bar{v}_{\mathrm{BE}} = \left.\frac{\Delta v_{\mathrm{total}}}{v_0}\right)_{\mathrm{bielliptical}} \qquad \alpha = \frac{r_C}{r_A} \qquad \beta = \frac{r_B}{r_A} \tag{6.4b}$$

Plotting the difference between $\Delta\bar{v}_{\mathrm{H}}$ and $\Delta\bar{v}_{\mathrm{BE}}$ as a function of α and β reveals the regions in which the difference is positive, negative, or zero. These are shown in Fig. 6.8.

From the figure we see that if the radius of the outer circular target orbit (r_C) is less than 11.94 times that of the inner one (r_A), then the standard Hohmann maneuver is the more energy efficient. If the ratio exceeds 15.58, then the bielliptic strategy is better in that regard. Between those two ratios, large values of the apoapsis radius r_B favor bielliptic transfer, while smaller values favor Hohmann transfer.

Small gains in energy efficiency may be more than offset by the much longer flight times around bielliptic trajectories compared with the time of flight on the single semiellipse of Hohmann transfer.

EXAMPLE 6.3

Find the total delta-v requirement for bielliptic Hohmann transfer from a geocentric circular orbit of 7000 km radius to one of 105,000 km radius. Let the apogee of the first ellipse be 210,000 km. Compare the delta-v schedule and total flight time with that for an ordinary single Hohmann transfer ellipse (see Fig. 6.9).

Solution

Since

$$r_A = 7000\,\mathrm{km} \quad r_B = 210{,}000\,\mathrm{km} \quad \text{and} \quad r_C = r_D = 105{,}000\,\mathrm{km}$$

we have $r_B/r_A = 30$ and $r_C/r_A = 15$, so that from Fig. 6.8 it is apparent right away that bielliptic transfer will be the more energy efficient.

To do the delta-v analysis requires analyzing each of the five orbits.

Orbit 1:

Since this is a circular orbit, we have, simply,

$$v_A)_1 = \sqrt{\frac{\mu}{r_A}} = \sqrt{\frac{398{,}600}{7000}} = 7.546\,\mathrm{km/s} \tag{a}$$

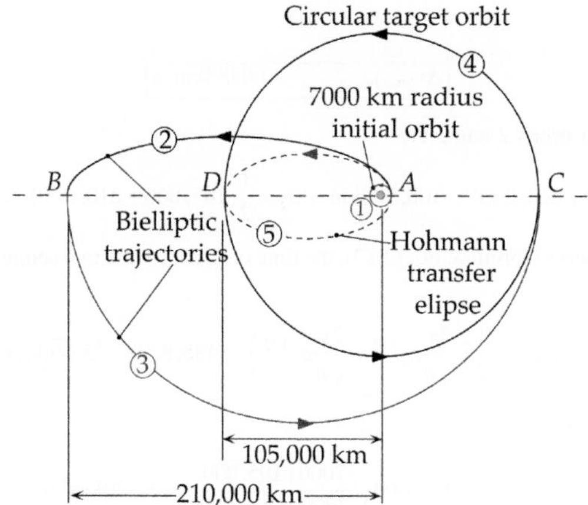

FIG. 6.9

Bielliptic transfer.

Orbit 2:

For this transfer ellipse, Eq. (6.2) yields

$$h_2 = \sqrt{2\mu}\sqrt{\frac{r_A r_B}{r_A + r_B}} = \sqrt{2 \cdot 398{,}600}\sqrt{\frac{7000 \cdot 210{,}000}{7000 + 210{,}000}} = 73{,}487\,\text{km}^2/\text{s}$$

Therefore,

$$v_A)_2 = \frac{h_2}{r_A} = \frac{73{,}487}{7000} = 10.498\,\text{km/s} \tag{b}$$

$$v_B)_2 = \frac{h_2}{r_B} = \frac{73{,}487}{210{,}000} = 0.34994\,\text{km/s} \tag{c}$$

Orbit 3:

For the second transfer ellipse, we have

$$h_3 = \sqrt{2 \cdot 398{,}600}\sqrt{\frac{105{,}000 \cdot 210{,}000}{105{,}000 + 210{,}000}} = 236{,}230\,\text{km}^2/\text{s}$$

From this we obtain

$$v_B)_3 = \frac{h_3}{r_B} = \frac{236{,}230}{210{,}000} = 1.1249\,\text{km/s} \tag{d}$$

$$v_C)_3 = \frac{h_3}{r_C} = \frac{236{,}230}{105{,}000} = 2.2498\,\text{km/s} \tag{e}$$

Orbit 4:

The target orbit, like orbit 1, is a circle, which means

$$v_C)_4 = v_D)_4 = \sqrt{\frac{398{,}600}{105{,}000}} = 1.9484\,\text{km/s} \tag{f}$$

For the bielliptic maneuver, the total delta-v is, therefore,

$$\begin{aligned}\Delta v_{\text{total}})_{\text{bi-elliptical}} &= \Delta v_A + \Delta v_B + \Delta v_C \\ &= \left|v_A)_2 - v_A)_1\right| + \left|v_B)_3 - v_B)_2\right| + \left|v_C)_4 - v_C)_3\right| \\ &= |10.498 - 7.546| + |1.1249 - 0.34994| + |1.9484 - 2.2498|\end{aligned}$$

or

$$\boxed{\Delta v_{\text{total}})_{\text{bielliptical}} = 4.0285\,\text{km/s}} \tag{g}$$

The semimajor axes of transfer orbits 2 and 3 are

$$a_2 = \frac{1}{2}(7000 + 210,000) = 108{,}500\,\text{km} \quad a_3 = \frac{1}{2}(105,000 + 210,000) = 157{,}500\,\text{km}$$

With this information and the period formula, Eq. (2.83), the time of flight for the two semiellipses of the bielliptic transfer is found to be

$$t_{\text{bielliptical}} = \frac{1}{2}\left(\frac{2\pi}{\sqrt{\mu}}a_2{}^{3/2} + \frac{2\pi}{\sqrt{\mu}}a_3{}^{3/2}\right) = 488{,}870\,\text{s} = \boxed{5.66\,\text{days}} \tag{h}$$

For the Hohmann transfer ellipse 5,

$$h_5 = \sqrt{2 \cdot 398{,}600}\sqrt{\frac{7000 \cdot 105{,}000}{7000 + 105{,}000}} = 72{,}330\,\text{km}^2/\text{s}$$

Hence,

$$v_A)_5 = \frac{h_5}{r_A} = \frac{72{,}330}{7000} = 10.333\,\text{km/s} \tag{i}$$

$$v_D)_5 = \frac{h_5}{r_D} = \frac{72{,}330}{105{,}000} = 0.68886\,\text{km/s} \tag{j}$$

It follows that

$$\begin{aligned}\Delta v_{\text{total}})_{\text{Hohmann}} &= \left|v_A)_5 - v_A)_1\right| + \left|v_D)_5 - v_D)_1\right| \\ &= (10.333 - 7.546) + (1.9484 - 0.68886) \\ &= 2.7868 + 1.2595\end{aligned}$$

or

$$\boxed{\Delta v_{\text{total}})_{\text{Hohmann}} = 4.0463\,\text{km/s}} \tag{k}$$

This is only slightly (0.44%) larger than that of the bielliptic transfer.

Since the semimajor axis of the Hohmann semiellipse is

$$\boxed{a_5 = \frac{1}{2}(7000 + 105,000) = 56{,}000\,\text{km}}$$

the time of flight from A to D is

$$t_{\text{Hohmann}} = \frac{1}{2}\left(\frac{2\pi}{\sqrt{\mu}}a_5{}^{3/2}\right) = 65{,}942\,\text{s} = \boxed{0.763\,\text{days}} \tag{l}$$

The time of flight of the bielliptic maneuver is over seven times longer than that of the Hohmann transfer.

6.5 PHASING MANEUVERS

A phasing maneuver is a two-impulse Hohmann transfer from and then back to the same orbit, as illustrated in Fig. 6.10. The Hohmann transfer ellipse is the phasing orbit with a period selected to return the spacecraft to the main orbit within a specified time. Phasing maneuvers are used to change the position of a spacecraft in its orbit. If two spacecraft, destined to rendezvous, are at different locations in the same orbit, then one of them may perform a phasing maneuver to catch the other one. Communications and weather satellites in geostationary earth orbit use phasing maneuvers to move to new

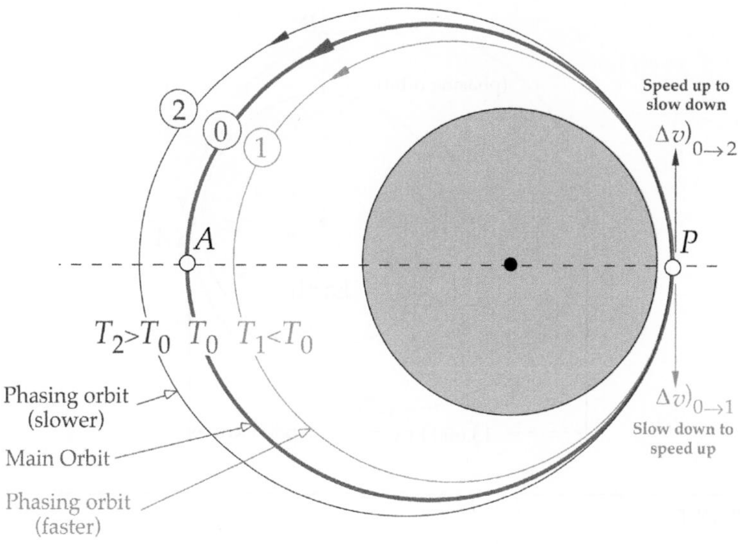

FIG. 6.10

Main orbit (0) and examples of two phasing orbits: faster (1) and slower (2). T_0 is the period of the main orbit.

locations above the equator. In that case, the rendezvous is with an empty point in space rather than with a physical target. In Fig. 6.10, phasing orbit 1 might be used to return to P in less than one period of the main orbit. This would be appropriate if the target is ahead of the chasing vehicle. Note that a retrofire is required to enter orbit 1 at P. That is, it is necessary to slow the spacecraft down to speed it up, relative to the main orbit. If the chaser is ahead of the target, then phasing orbit 2 with its longer period might be appropriate. A forward fire of the thruster boosts the spacecraft's speed to slow it down.

Once the period T of the phasing orbit is established, then Eq. (2.8) should be used to determine the semimajor axis of the phasing ellipse,

$$a = \left(\frac{T\sqrt{\mu}}{2\pi}\right)^{2/3} \tag{6.5}$$

With the semimajor axis established, the radius of point A opposite to P is obtained from the fact that $2a = r_P + r_A$. Eq. (6.2) may then be used to obtain the angular momentum.

EXAMPLE 6.4

Spacecraft at A and B are in the same orbit (1). At the instant shown in Fig. 6.11 the chaser vehicle at A executes a phasing maneuver so as to catch the target spacecraft back at A after just one revolution of the chaser's phasing orbit (2). What is the required total delta-v?

Solution

We must find the angular momenta of orbits 1 and 2 so that we can use Eq. (2.31) to find the velocities on orbits 1 and 2 at point A. (We can alternatively use energy, Eq. (2.81), to find the speeds at A.) These velocities furnish the delta-v required to leave orbit 1 for orbit 2 at the beginning of the phasing maneuver and to return to orbit 1 at the end.

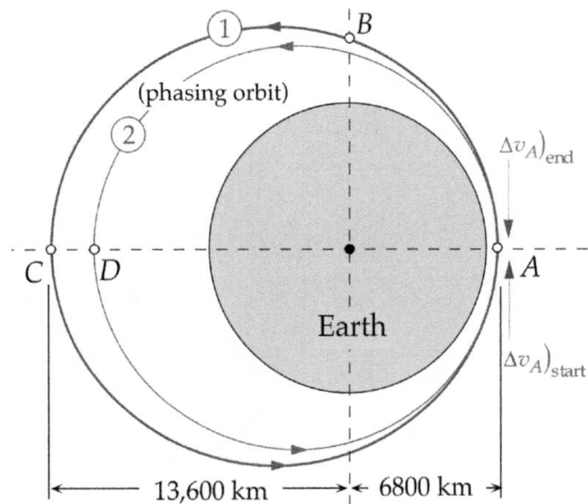

FIG. 6.11

Phasing maneuver.

Angular momentum of orbit 1

From Fig. 6.11 we observe that perigee and apogee radii of orbit 1 are, respectively,

$$r_A = 6800\,\text{km} \quad r_C = 13{,}600\,\text{km}$$

It follows from Eq. (6.2) that the orbit's angular momentum is

$$h_1 = \sqrt{2\mu}\sqrt{\frac{r_A r_C}{r_A + r_C}} = \sqrt{2\cdot 398{,}600}\sqrt{\frac{6800\cdot 13{,}600}{6800 + 13{,}600}} = 60{,}116\,\text{km/s}^2$$

Angular momentum of orbit 2

The phasing orbit must have a period T_2 equal to the time it takes the target vehicle at B to coast around to point A on orbit 1. That flight time equals the period of orbit 1 minus the flight time t_{AB} from A to B. That is,

$$T_2 = T_1 - t_{AB} \tag{a}$$

The period of orbit 1 is found by computing its semimajor axis,

$$a_1 = \frac{1}{2}(r_A + r_C) = 10{,}200\,\text{km}$$

and substituting that result into Eq. (2.83),

$$T_1 = \frac{2\pi}{\sqrt{\mu}}{a_1}^{3/2} = \frac{2\pi}{\sqrt{398{,}600}}10{,}200^{3/2} = 10{,}252\,\text{s} \tag{b}$$

The flight time from the perigee A of orbit 1 to point B is obtained from Kepler's equation (Eqs. 3.8 and 3.14),

$$t_{AB} = \frac{T_1}{2\pi}(E_B - e_1 \sin E_B) \tag{c}$$

Since the eccentricity of orbit 1 is

$$e_1 = \frac{r_C - r_A}{r_C + r_A} = 0.33333 \tag{d}$$

and the true anomaly of B is 90°, it follows from Eq. (3.13b) that the eccentric anomaly of B is

$$E_B = 2\tan^{-1}\left(\sqrt{\frac{1-e_1}{1+e_1}}\tan\frac{\theta_B}{2}\right) = 2\tan^{-1}\left(\sqrt{\frac{1-0.33333}{1+0.33333}}\tan\frac{90^\circ}{2}\right) = 1.2310\,\text{rad} \tag{e}$$

Substituting Eqs. (b), (d), and (e) into Eq. (c) yields

$$t_{AB} = \frac{10{,}252}{2\pi}(1.231 - 0.33333 \cdot \sin 1.231) = 1495.7\,\text{s}$$

It follows from Eq. (a) that

$$T_2 = 10{,}252 - 1495.7 = 8756.3\,\text{s}$$

This, together with the period formula (Eq. 2.83), yields the semimajor axis of orbit 2,

$$a_2 = \left(\frac{\sqrt{\mu}T_2}{2\pi}\right)^{2/3} = \left(\frac{\sqrt{398{,}600} \cdot 8756.2}{2\pi}\right)^{2/3} = 9182.1\,\text{km}$$

Since $2a_2 = r_A + r_D$, we find that the apogee of orbit 2 is

$$r_D = 2a_2 - r_A = 2 \cdot 9182.1 - 6800 = 11{,}564\,\text{km}$$

Finally, Eq. (6.2) yields the angular momentum of orbit 2,

$$h_2 = \sqrt{2\mu}\sqrt{\frac{r_A r_D}{r_A + r_D}} = \sqrt{2 \cdot 398{,}600}\sqrt{\frac{6800 \cdot 11{,}564}{6800 + 11{,}564}} = 58{,}426\,\text{km/s}^2$$

Velocities at A

Since A is the perigee of orbit 1, there is no radial velocity component there. The speed, directed entirely in the transverse direction, is found from the angular momentum formula,

$$v_A)_1 = \frac{h_1}{r_A} = \frac{60{,}116}{6800} = 8.8406\,\text{km/s}$$

Likewise, the speed at the perigee of orbit 2 is

$$v_A)_2 = \frac{h_2}{r_A} = \frac{58{,}426}{6800} = 8.5921\,\text{km/s}$$

At the beginning of the phasing maneuver, the velocity change required to drop into phasing orbit 2 is

$$\Delta v_A = v_A)_2 - v_A)_1 = 8.5921 - 8.8406 = -0.24851\,\text{km/s}$$

At the end of the phasing maneuver, the velocity change required to return to orbit 1 is

$$\Delta v_A = v_A)_1 - v_A)_2 = 8.8406 - 8.5921 = 0.24851\,\text{km/s}$$

The total delta-v required for the chaser to catch up with the target is

$$\Delta v_{\text{total}} = |-0.24851| + |0.24851| = \boxed{0.4970\,\text{km/s}}$$

The delta-v requirement for a phasing maneuver can be lowered by reducing the difference between the period of the main orbit and that of the phasing orbit. In the previous example, we could make Δv_{total} smaller by requiring the chaser to catch the target after n revolutions of the phasing orbit instead of just one. In that case, we would replace Eq. (a) of Example 6.4 by $T_2 = T_1 - t_{AB}/n$.

EXAMPLE 6.5

It is desired to shift the longitude of a GEO satellite 12° westward in three revolutions of its phasing orbit. Calculate the delta-v requirement.

Solution

This problem is illustrated in Fig. 6.12. It may be recalled from Eqs. (2.67)–(2.69) that the angular velocity of the earth, the radius to GEO, and the speed in GEO are, respectively

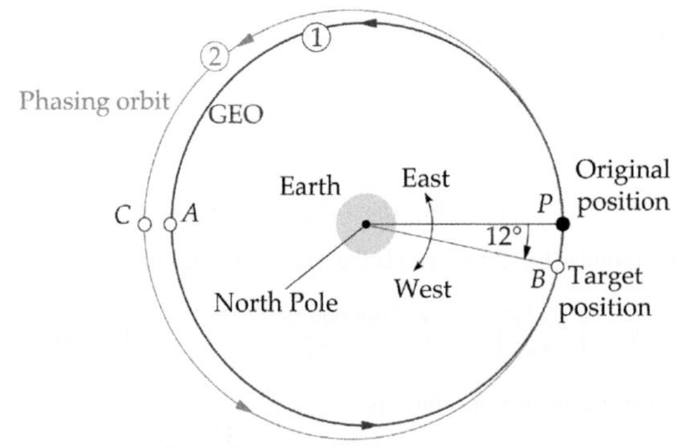

FIG. 6.12

GEO repositioning.

$$\begin{aligned} \omega_E &= \omega_{\text{GEO}} = 72.922(10^{-6})\,\text{rad/s} \\ r_{\text{GEO}} &= 42,164\,\text{km} \\ V_{\text{GEO}} &= 3.0747\,\text{km/s} \end{aligned} \tag{a}$$

Let $\Delta\Lambda$ be the change in longitude in radians. Then the period T_2 of the phasing orbit can be obtained from the following formula,

$$\omega_E(3T_2) = 3 \cdot 2\pi + \Delta\Lambda \tag{b}$$

which states that after three circuits of the phasing orbit, the original position of the satellite will be $\Delta\Lambda$ radians east of P. In other words, the satellite will end up $\Delta\Lambda$ radians west of its original position in GEO, as desired. From Eq. (b) we obtain,

$$T_2 = \frac{1}{3}\frac{\Delta\Lambda + 6\pi}{\omega_E} = \frac{1}{3}\frac{12° \cdot \dfrac{\pi}{180°} + 6\pi}{72.922 \times 10^{-6}} = 87,121\,\text{s}$$

Note that the period of GEO is

$$T_{\text{GEO}} = \frac{2\pi}{\omega_{\text{GEO}}} = 86,163\,\text{s}$$

The satellite in its slower phasing orbit appears to drift westward at the rate

$$\dot{\Lambda} = \frac{\Delta\Lambda}{3T_2} = 8.0133 \times 10^{-7}\,\text{rad/s} = 3.9669\,\text{degrees/day}$$

Having the period, we can use Eq. (6.5) to obtain the semimajor axis of orbit 2,

$$a_2 = \left(\frac{T_2\sqrt{\mu}}{2\pi}\right)^{2/3} = \left(\frac{87,121\sqrt{398,600}}{2\pi}\right)^{2/3} = 42,476\,\text{km}$$

From this we find the radius to the apogee C of the phasing orbit,

$$2a_2 = r_p + r_C \;\Rightarrow\; r_C = 2 \cdot 42,476 - 42,164 = 42,788\,\text{km}$$

The angular momentum of the orbit is given by Eq. (6.2),

$$h_2 = \sqrt{2\mu}\sqrt{\frac{r_B r_C}{r_B + r_C}} = \sqrt{2 \cdot 398,600}\sqrt{\frac{42,164 \cdot 42,748}{42,164 + 42,788}} = 130,120\,\text{km}^2/\text{s}$$

At P the speed in orbit 2 is

$$v_P)_2 = \frac{130,120}{42,164} = 3.0859\,\text{km/s}$$

Therefore, at the beginning of the phasing orbit,

$$\Delta v = v_P)_2 - v_{\text{GEO}} = 3.0859 - 3.0747 = 0.01126\text{km/s}$$

At the end of the phasing maneuver,

$$\Delta v = v_{\text{GEO}} - v_P)_2 = 3.0747 - 3.08597 = -0.01126\text{km/s}$$

It follows that,

$$\Delta v_{\text{total}} = |0.01126| + |-0.01126| = \boxed{0.02252\text{km/s}}$$

6.6 NON-HOHMANN TRANSFERS WITH A COMMON APSE LINE

Fig. 6.13 illustrates a transfer between two coaxial, coplanar elliptical orbits in which the transfer trajectory shares the apse line but is not necessarily tangent to either the initial or target orbit. The problem is to determine whether there exists such a trajectory joining points A and B, and, if so, to find the total delta-v requirement.

The radials r_A and r_B are already known, as are the true anomalies θ_A and θ_B. Because of the common apse line assumption, θ_A and θ_B are the true anomalies of points A and B on the transfer orbit as well. Applying the orbit equation to A and B on the transfer orbit yields

$$r_A = \frac{h_2}{\mu}\frac{1}{1 + e\cos\theta_A} \qquad r_B = \frac{h_2}{\mu}\frac{1}{1 + e\cos\theta_B}$$

Solving these two equations for e and h, we get

$$e = \frac{r_A - r_B}{r_A\cos\theta_A - r_B\cos\theta_B} \tag{6.6a}$$

$$h = \sqrt{\mu r_A r_B}\sqrt{\frac{\cos\theta_A - \cos\theta_B}{r_A\cos\theta_A - r_B\cos\theta_B}} \tag{6.6b}$$

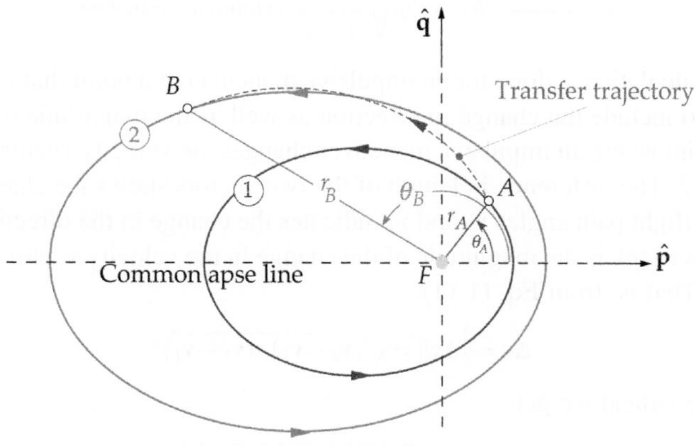

FIG. 6.13

Non-Hohmann transfer between two coaxial elliptical orbits.

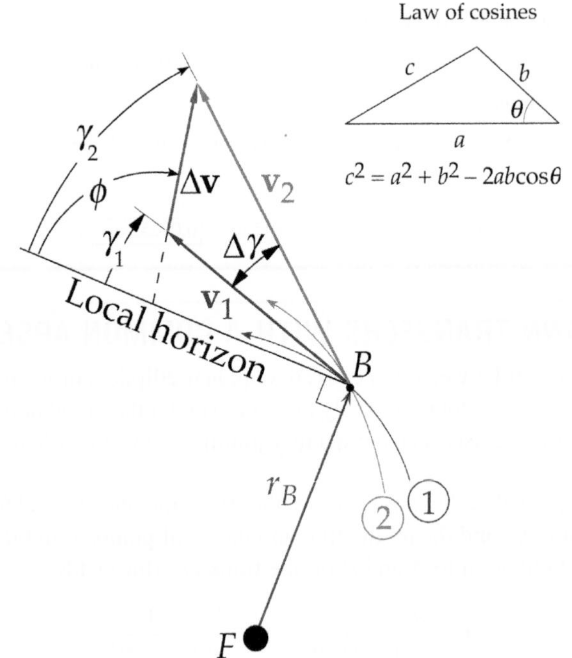

FIG. 6.14

Vector diagram of the change in velocity and flight path angle at the intersection of two orbits (plus a reminder of the law of cosines).

With these orbital elements, the transfer orbit is determined and the velocity may be found at any true anomaly. Note that for a Hohmann transfer, in which $\theta_A = 0$ and $\theta_B = \pi$, Eqs. 6.6 become

$$e = \frac{r_B - r_A}{r_B - r_A} \quad h = \sqrt{2\mu}\sqrt{\frac{r_A r_B}{r_A + r_B}} \quad \text{(Hohmann transfer)} \tag{6.7}$$

When a delta-v calculation is done for an impulsive maneuver at a point that is not on the apse line, care must be taken to include the change in direction as well as the magnitude of the velocity vector. Fig. 6.14 shows a point where an impulsive maneuver changes the velocity vector from $\mathbf{v}_1$ on orbit 1 to $\mathbf{v}_2$ on coplanar orbit 2. The difference in length of the two vectors shows the change in the speed, and the difference in the flight path angles γ_2 and γ_1 indicates the change in the direction. It is important to observe that the Δv we seek is the magnitude of the change in the velocity vector, not the change in its magnitude (speed). That is, from Eq. (1.11),

$$\Delta v = \|\Delta \mathbf{v}\| = \sqrt{(\mathbf{v}_2 - \mathbf{v}_1)\cdot(\mathbf{v}_2 - \mathbf{v}_1)}$$

Expanding under the radical we get

$$\Delta v = \sqrt{\mathbf{v}_1 \cdot \mathbf{v}_1 + \mathbf{v}_2 \cdot \mathbf{v}_2 - 2\mathbf{v}_1 \cdot \mathbf{v}_2}$$

Again, according to Eq. (1.11), $\mathbf{v}_1 \cdot \mathbf{v}_1 = {v_1}^2$ and $\mathbf{v}_2 \cdot \mathbf{v}_2 = {v_2}^2$. Furthermore, since $\gamma_2 - \gamma_1$ is the angle between $\mathbf{v}_1$ and $\mathbf{v}_2$, Eq. (1.7) implies that

$$\mathbf{v}_1 \cdot \mathbf{v}_2 = v_1 v_2 \cos \Delta\gamma$$

where $\Delta\gamma = \gamma_2 - \gamma_1$. Therefore,

$$\Delta v = \sqrt{v_1^2 + v_2^2 - 2v_1 v_2 \cos \Delta\gamma} \quad \text{(impulsive maneuver, coplanar orbits)} \tag{6.8}$$

This is the familiar law of cosines from trigonometry. Only if $\Delta\gamma = 0$, which means that $\mathbf{v}_1$ and $\mathbf{v}_2$ are parallel (as in a Hohmann transfer), is it true that $\Delta v = |v_2 - v_1|$. If $v_2 = v_1 = v$, then Eq. (6.8) yields

$$\Delta v = v\sqrt{2(1 - \cos \Delta\gamma)} \quad \text{(pure rotation of the velocity vector in the orbital plane)} \tag{6.9}$$

Therefore, fuel expenditure is required to change the direction of the velocity even if its magnitude remains the same.

The direction of $\Delta\mathbf{v}$ in Fig. 6.14 shows the required alignment of the thruster that produces the impulse. The orientation of $\Delta\mathbf{v}$ relative to the local horizon is found by replacing v_r and $v_\perp$ in Eq. (2.51) with Δv_r and $\Delta v_\perp$, so that

$$\tan\phi = \frac{\Delta v_r}{\Delta v_\perp} \tag{6.10}$$

where ϕ is the angle from the local horizon to the $\Delta\mathbf{v}$ vector.

Finally, recall from Eq. (2.57) that the specific mechanical energy of a spacecraft is,

$$\varepsilon = \frac{\mathbf{v} \cdot \mathbf{v}}{2} - \frac{\mu}{r}$$

An impulsive maneuver changes the velocity $\mathbf{v}$ but not the position vector $\mathbf{r}$. It follows that

$$\Delta\varepsilon = \frac{(\mathbf{v} + \Delta\mathbf{v}) \cdot (\mathbf{v} + \Delta\mathbf{v})}{2} - \frac{\mathbf{v} \cdot \mathbf{v}}{2} = \mathbf{v} \cdot \Delta\mathbf{v} + \frac{1}{2}\Delta v^2$$

The angle between $\mathbf{v}$ and $\Delta\mathbf{v}$ is $\Delta\gamma$ (Fig. 6.14, with $\mathbf{v}_1 = \mathbf{v}$). Therefore, $\mathbf{v} \cdot \Delta\mathbf{v} = v\Delta v \cos \Delta\gamma$ and we obtain

$$\Delta\varepsilon = v\Delta v \cos \Delta\gamma + \frac{1}{2}\Delta v^2 \tag{6.11}$$

This shows that, for a given Δv, the change in specific energy is larger when the spacecraft is moving fastest and when Δv is aligned with the original velocity ($\Delta\gamma \approx 0$). The larger the $\Delta\varepsilon$ associated with a given Δv, the more efficient the maneuver. As we know, a spacecraft has its greatest speed at periapsis.

EXAMPLE 6.6

A geocentric satellite in orbit 1 of Fig. 6.15 executes a delta-v maneuver at A, which places it on orbit 2, for reentry at D. Calculate Δv at A and its direction relative to the local horizon.

Solution

From the figure we see that

$$r_B = 20{,}000\,\text{km} \quad r_C = 10{,}000\,\text{km} \quad r_D = 6378\,\text{km}$$

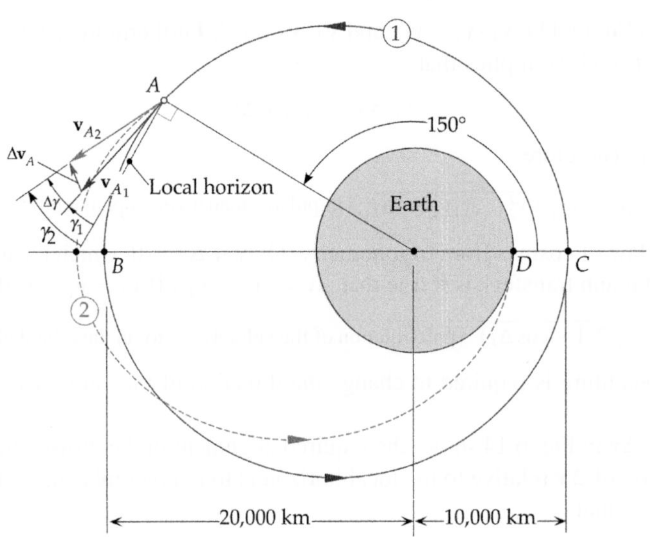

FIG. 6.15

Non-Hohmann transfer with a common apse line.

Orbit 1:

The eccentricity is

$$e_1 = \frac{r_B - r_C}{r_B + r_C} = 0.33333$$

The angular momentum is obtained from Eq. (6.2), noting that point C is perigee:

$$h_1 = \sqrt{2\mu}\sqrt{\frac{r_B r_C}{r_B + r_C}} = \sqrt{2 \cdot 398{,}600}\sqrt{\frac{20{,}000 \cdot 10{,}000}{20{,}000 + 10{,}000}} = 72{,}902\text{km}^2/\text{s}$$

With the angular momentum and the eccentricity, we can use the orbit equation to find the radial coordinate of point A,

$$r_A = \frac{72{,}902^2}{398{,}600}\frac{1}{1 + 0.33333 \cdot \cos 150°} = 18{,}744\text{km}$$

Eqs. (2.31) and (2.49) yield the transverse and radial components of velocity at A on orbit 1,

$$v_{\perp_A})_1 = \frac{h_1}{r_A} = 3.8893\text{km/s}$$

$$v_{r_A})_1 = \frac{\mu}{h_1} e_1 \sin 150° = 0.91127\text{km/s}$$

From these we find the speed at A,

$$v_A)_1 = \sqrt{v_{\perp_A})_1{}^2 + v_{r_A})_1{}^2} = 3.9946\text{km/s}$$

and the flight path angle,

$$\gamma_1 = \tan^{-1}\frac{v_{r_A})_1}{v_{\perp_A})_1} = \tan^{-1}\frac{0.91127}{3.8893} = 13.187°$$

Orbit 2:
The radius and true anomaly of points *A* and *D* on orbit 2 are known. From Eqs. (6.6) we find

$$e_2 = \frac{r_D - r_A}{r_D\cos\theta_D - r_A\cos\theta_A} = \frac{6378 - 18{,}744}{6378\cos 0 - 18{,}744\cos 150°} = 0.5469$$

$$h_2 = \sqrt{\mu r_A r_D}\sqrt{\frac{\cos\theta_D - \cos\theta_A}{r_D\cos\theta_D - r_A\cos\theta_A}} = \sqrt{398{,}600\cdot 18{,}744\cdot 6378}\sqrt{\frac{\cos 0 - \cos 150°}{6378\cos 0 - 18{,}744\cos 150°}}$$
$$= 62{,}711\,\text{km}^2/\text{s}$$

Now we can calculate the radial and perpendicular components of velocity on orbit 2 at point *A*.

$$v_{\perp_A})_2 = \frac{h_2}{r_A} = 3.3456\,\text{km/s}$$
$$v_{r_A})_2 = \frac{\mu}{h_2}e_2\sin 150° = 1.7381\,\text{km/s}$$

Hence, the speed and flight path angle at *A* on orbit 2 are

$$v_A)_2 = \sqrt{v_{\perp_A})_2{}^2 + v_{r_A})_2{}^2} = 3.7702\,\text{km/s}$$
$$\gamma_2 = \tan^{-1}\frac{v_{r_A})_2}{v_{\perp_A})_2} = \tan^{-1}\frac{1.7381}{3.3456} = 27.453°$$

The change in the flight path angle as a result of the impulsive maneuver is

$$\Delta\gamma = \gamma_2 - \gamma_1 = 27.453° - 13.187° = 14.266°$$

With this we can use Eq. (6.8) to finally obtain Δv_A,

$$\Delta v_A = \sqrt{v_A)_1{}^2 + v_A)_2{}^2 - 2v_A)_1 v_A)_2\cos\Delta\gamma} = \sqrt{3.9946^2 + 3.7702^2 - 2\cdot 3.9946\cdot 3.7702\cdot\cos 14.266°}$$

$$\boxed{\Delta v_A = 0.9896\,\text{km/s}}$$

Note that Δv_A is the magnitude of the change in velocity vector $\Delta\mathbf{v}_A$ at *A*. That is not the same as the change in the magnitude of the velocity (i.e., the change in speed), which is

$$v_A)_2 - v_A)_1 = 3.7702 - 3.9946 = -0.2244\,\text{km/s}$$

To find the orientation of $\Delta\mathbf{v}_A$, we use Eq. (6.10),

$$\tan\phi = \frac{\Delta v_{r_A}}{\Delta v_{\perp_A}} = \frac{v_{r_A})_2 - v_{r_A})_1}{v_{\perp_A})_2 - v_{\perp_A})_1} = \frac{1.7381 - 0.9113}{3.3456 - 3.8893} = -1.5207$$

so that

$$\boxed{\phi = 123.3°}$$

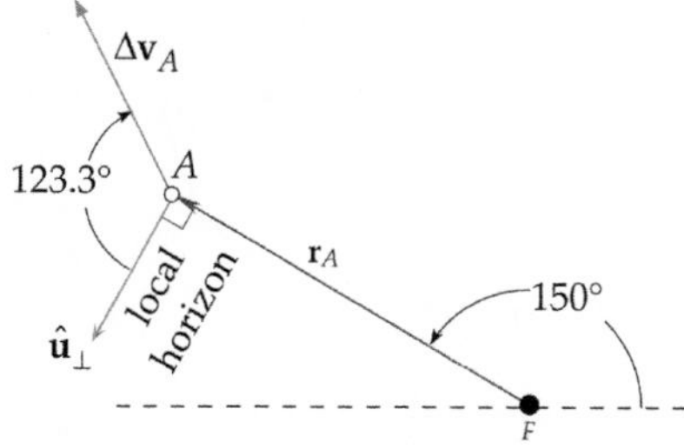

FIG. 6.16

Orientation of $\Delta\mathbf{v}_A$ to the local horizon.

This angle is illustrated in Fig. 6.16. Before firing, the spacecraft would have to be rotated so that the centerline of the rocket motor coincides with the line of action of $\Delta \mathbf{v}_A$, with the nozzle aimed in the opposite direction.

6.7 APSE LINE ROTATION

Fig. 6.17 shows two intersecting orbits that have a common focus, but their apse lines are not collinear. A Hohmann transfer between them is clearly impossible. The opportunity for transfer from one orbit to the other by a single impulsive maneuver occurs where they intersect, at points I and J in this case. As can be seen from the figure, the rotation η of the apse line is the difference between the true anomalies of the point of intersection, measured from periapsis of each orbit. That is,

$$\eta = \theta_1 - \theta_2 \tag{6.12}$$

We will consider two cases of apse line rotation.

The first case is that in which the apse line rotation η is given as well as the orbital parameters e and h of both orbits. The problem is then to find the true anomalies of I and J relative to both orbits. The radius of the point of intersection I is given by either of the following:

$$r_I)_1 = \frac{h_1^2}{\mu}\frac{1}{1+e_1\cos\theta_1} \qquad r_I)_2 = \frac{h_2^2}{\mu}\frac{1}{1+e_2\cos\theta_2}$$

Since $r_I)_1 = r_I)_2$, we can equate these two expressions and rearrange terms to get

$$e_1h_2^2\cos\theta_1 - e_2h_1^2\cos\theta_2 = h_1^2 - h_2^2$$

Setting $\theta_2 = \theta_1 - \eta$ and using the trig identity $\cos(\theta_1 - \eta) = \cos\theta_1\cos\eta + \sin\theta_1\sin\eta$ leads to an equation for θ_1,

$$a\cos\theta_1 + b\sin\theta_1 = c \tag{6.13a}$$

where

$$a = e_1h_2^2 - e_2h_1^2\cos\eta \quad b = -e_2h_1^2\sin\eta \quad c = h_1^2 - h_2^2 \tag{6.13b}$$

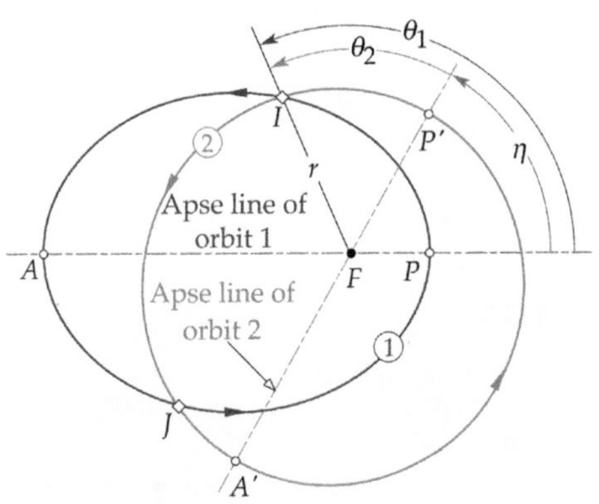

FIG. 6.17

Two intersecting orbits whose apse lines do not coincide.

Eq. (6.13a) has two roots (see Problem 3.12), corresponding to the two points of intersection I and J of the two orbits:

$$\theta_1 = \phi \pm \cos^{-1}\left(\frac{c}{a}\cos\phi\right) \tag{6.14a}$$

where

$$\phi = \tan^{-1}\frac{b}{a} \tag{6.14b}$$

Having found θ_1 we obtain θ_2 from Eq. (6.12). Δv for the impulsive maneuver may then be computed as illustrated in the following example.

EXAMPLE 6.7

An earth satellite is in an 8000 km by 16,000 km radius orbit (orbit 1 of Fig. 6.18). Calculate the delta-v and the true anomaly θ_1 required to obtain a 7000 km by 21,000 km radius orbit (orbit 2) whose apse line is rotated 25° counterclockwise. Indicate the orientation ϕ of $\Delta\mathbf{v}$ to the local horizon.

Solution

The eccentricities of the two orbits are

$$\begin{aligned} e_1 &= \frac{r_{A_1} - r_{P_1}}{r_{A_1} + r_{P_1}} = \frac{16{,}000 - 8000}{16{,}000 + 8000} = 0.33333 \\ e_2 &= \frac{r_{A_2} - r_{P_2}}{r_{A_2} + r_{P_2}} = \frac{21{,}000 - 7000}{21{,}000 + 7000} = 0.5 \end{aligned} \tag{a}$$

The orbit equation yields the angular momenta

$$\begin{aligned} r_{P_1} &= \frac{h_1{}^2}{\mu}\frac{1}{1+e_1\cos(0)} \Rightarrow 8000 = \frac{h_1{}^2}{398{,}600}\frac{1}{1+0.33333} \Rightarrow h_1 = 65{,}205\,\text{km}^2/\text{s} \\ r_{P_2} &= \frac{h_2{}^2}{\mu}\frac{1}{1+e_2\cos(0)} \Rightarrow 7000 = \frac{h_2{}^2}{398{,}600}\frac{1}{1+0.5} \Rightarrow h_2 = 64{,}694\,\text{km}^2/\text{s} \end{aligned} \tag{b}$$

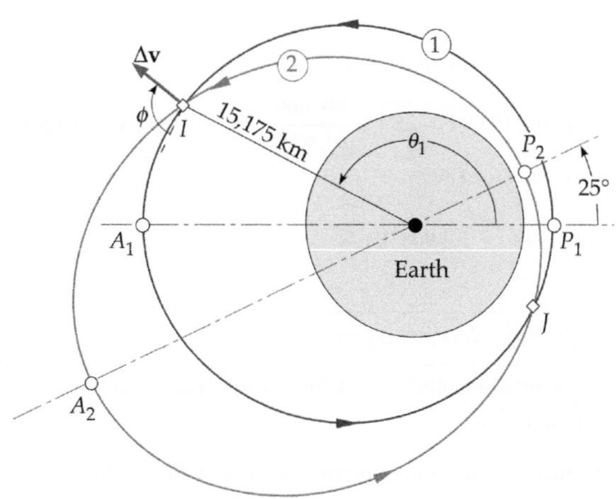

FIG. 6.18

Δv produces a rotation of the apse line.

Using these orbital parameters and the fact that $\eta = 25°$, we calculate the terms in Eq. (6.13b):

$$a = e_1 h_2^2 - e_2 h_1^2 \cos\eta = 0.3333 \cdot 64{,}694^2 - 0.5 \cdot 65{,}205^2 \cdot \cos 25° = -5.3159(10^8)\,\text{km}^4/\text{s}^2$$
$$b = -e_2 h_1^2 \sin\eta = -0.5 \cdot 65{,}205^2 \sin 25° = -8.9843(10^8)\,\text{km}^4/\text{s}^2$$
$$c = h_2^2 - h_2^2 = 65{,}205^2 - 64{,}694^2 = 6.6433(10^7)\,\text{km}^4/\text{s}^2$$

Then Eq. (6.14) yields

$$\phi = \tan^{-1}\frac{-8.9843(10^8)}{-5.3159(10^8)} = 59.388°$$
$$\theta_1 = 59.388° \pm \cos^{-1}\left[\frac{6.6433(10^7)}{-5.3159(10^8)}\cos 59.388°\right] = 59.388° \pm 93.649°$$

Thus, the true anomaly of point I, the point of interest, is

$$\theta_1 = 153.04° \qquad \text{(c)}$$

(For point J, $\theta_1 = 325.74°$.)

With the true anomaly available, we can evaluate the radial coordinate of the maneuver point,

$$r = \frac{h_1{}^2}{\mu}\frac{1}{1 + e_1 \cos 153.04°} = 15{,}175\,\text{km}$$

The velocity components and flight path angle for orbit 1 at point I are

$$v_{\perp_1} = \frac{h_1}{r} = \frac{65{,}205}{15{,}175} = 4.2968\,\text{km/s}$$
$$v_{r_1} = \frac{\mu}{h_1} e_1 \sin 153.04° = \frac{398{,}600}{65{,}205} \cdot 0.33333 \cdot \sin 153.04° = 0.92393\,\text{km/s}$$
$$\gamma_1 = \tan^{-1}\frac{v_{r_1}}{v_{\perp_1}} = 12.135°$$

The speed of the satellite in orbit 1 is, therefore,

$$v_1 = \sqrt{v_{r_1}{}^2 + v_{\perp_1}{}^2} = 4.3950\,\text{km/s}$$

Likewise, for orbit 2,

$$v_{\perp_2} = \frac{h_2}{r} = \frac{64{,}694}{15{,}175} = 4.2631\,\text{km/s}$$
$$v_{r_2} = \frac{\mu}{h_2} e_2 \sin(153.04° - 25°) = \frac{398{,}600}{64{,}694} \cdot 0.5 \cdot \sin 128.04° = 2.4264\,\text{km/s}$$
$$\gamma_2 = \tan^{-1}\frac{v_{r_2}}{v_{\perp_2}} = 29.647°$$
$$v_2 = \sqrt{v_{r_2}{}^2 + v_{\perp_2}{}^2} = 4.9053\,\text{km/s}$$

Eq. (6.8) is used to find Δv,

$$\Delta v = \sqrt{v_1^2 + v_2^2 - 2v_1 v_2 \cos(\gamma_2 - \gamma_1)}$$
$$= \sqrt{4.3950^2 + 4.9053^2 - 2 \cdot 4.3950 \cdot 4.9053 \cos(29.647° - 12.135°)}$$
$$\boxed{\Delta v = 1.503\,\text{km/s}}$$

The angle ϕ that the vector $\mathbf{\Delta v}$ makes with the local horizon is given by Eq. (6.10),

$$\phi = \tan^{-1}\frac{\Delta v_r}{\Delta v_\perp} = \tan^{-1}\frac{v_{r_2} - v_{r_1}}{v_{\perp_2} - v_{\perp_1}} = \tan^{-1}\frac{2.4264 - 0.92393}{4.2631 - 4.2968} = \boxed{91.28°}$$

The second case of apse line rotation is that in which the impulsive maneuver takes place at a given true anomaly θ_1 on orbit 1. The problem is to determine the angle of rotation η and the eccentricity e_2 of the new orbit.

The impulsive maneuver creates a change in the radial and transverse velocity components at point I of orbit 1. From the angular momentum formula, $h = rv_\perp$, we obtain the angular momentum of orbit 2,

$$h_2 = r(v_\perp + \Delta v_\perp) = h_1 + r\Delta v_\perp \tag{6.15}$$

The formula for radial velocity, $v_r = (\mu/h)e \sin \theta$, applied to orbit 2 at point I, where $v_{r_2} = v_{r_1} + \Delta v_r$ and $\theta_2 = \theta_1 - \eta$, yields

$$v_{r_1} + \Delta v_r = \frac{\mu}{h_2} e_2 \sin\theta_2$$

Substituting Eq. (6.15) into this expression and solving for $\sin\theta_2$ leads to

$$\sin\theta_2 = \frac{1}{e_2}\frac{(h_1 + r\Delta v_\perp)(\mu e_1 \sin\theta_1 + h_1 \Delta v_r)}{\mu h_1} \tag{6.16}$$

From the orbit equation, we have at point I

$$r = \frac{h_1{}^2}{\mu}\frac{1}{1 + e_1\cos\theta_1} \quad \text{(orbit 1)}$$
$$r = \frac{h_2{}^2}{\mu}\frac{1}{1 + e_2\cos\theta_2} \quad \text{(orbit 2)}$$

Equating these two expressions for r, substituting Eq. (6.15), and solving for $\cos\theta_2$, yields

$$\cos\theta_2 = \frac{1}{e_2}\frac{(h_1 + r\Delta v_\perp)^2 e_1 \cos\theta_1 + (2h_1 + r\Delta v_\perp) r\Delta v_\perp}{h_1^2} \tag{6.17}$$

Finally, by substituting Eqs. (6.16) and (6.17) into the trigonometric identity $\tan\theta_2 = \sin \theta_2 / \cos \theta_2$ we obtain a formula for θ_2 that does not involve the eccentricity e_2,

$$\tan\theta_2 = \frac{h_1}{\mu}\frac{(h_1 + r\Delta v_\perp)(\mu e_1 \sin\theta_1 + h_1\Delta v_r)}{(h_1 + r\Delta v_\perp)^2 e_1 \cos\theta_1 + (2h_1 + r\Delta v|_\perp) r\Delta v_\perp} \tag{6.18a}$$

Eq. (6.18a) can be simplified a bit by replacing $\mu e_1 \sin \theta_1$ with $h_1 v_{r_1}$ and h_1 with $rv_{\perp_1}$, so that

$$\tan\theta_2 = \frac{(v_{\perp_1} + \Delta v_\perp)(v_{r_1} + \Delta v_r)}{(v_{\perp_1} + \Delta v_\perp)^2 e_1 \cos\theta_1 + (2v_{\perp_1} + \Delta v_\perp)\Delta v_\perp (\mu/r)} \tag{6.18b}$$

Eqs. (6.18) show how the apse line rotation, $\eta = \theta_1 - \theta_2$, is completely determined by the components of $\Delta\mathbf{v}$ imparted at the true anomaly θ_1. Notice that if $\Delta v_r = -v_{r_1}$, then $\theta_2 = 0$, which means that the maneuver point is on the apse line of the new orbit.

After solving Eq. (6.18a) or (6.18b), we substitute θ_2 into either Eq. (6.16) or (6.17) to calculate the eccentricity e_2 of orbit 2. Therefore, with h_2 from Eq. (6.15), the rotated orbit 2 is completely specified.

If the impulsive maneuver takes place at the periapsis of orbit 1, so that $\theta_1 = v_r = 0$, and if it is also true that $\Delta v_\perp = 0$, then Eq. (6.18b) yields

$$\tan\eta = \frac{rv_{\perp 1}}{\mu e_1}\Delta v_r \quad \text{(with radial impulse at periapsis)}$$

Thus, if the velocity vector is given an outward radial component at periapsis, then $\eta < 0$, which means the apse line of the resulting orbit is rotated clockwise relative to the original one. That makes sense, since having acquired $v_r > 0$ means that the spacecraft is now flying away from its new periapsis. Likewise, applying an inward radial velocity component at periapsis rotates the apse line counterclockwise.

EXAMPLE 6.8

An earth satellite in orbit 1 of Fig. 6.19 undergoes the indicated delta-v maneuver at its perigee. Determine the rotation η of its apse line as well as the new perigee and apogee.

Solution

From Fig. 6.19 the apogee and perigee of orbit 1 are

$$r_{A_1} = 17{,}000\,\text{km} \quad r_{P_1} = 7000\,\text{km}$$

Therefore, the eccentricity of orbit 1 is

$$e_1 = \frac{r_{A_1} - r_{P_1}}{r_{A_1} + r_{P_1}} = 0.41667 \tag{a}$$

As usual, we use the orbit equation to find the angular momentum,

$$r_{P_1} = \frac{h_1^2}{\mu}\frac{1}{1 + e_1 \cos(0)} \Rightarrow 7000 = \frac{h_1^2}{398{,}600}\frac{1}{1 + 0.41667} \Rightarrow h_1 = 62{,}871\,\text{km}^2/\text{s}$$

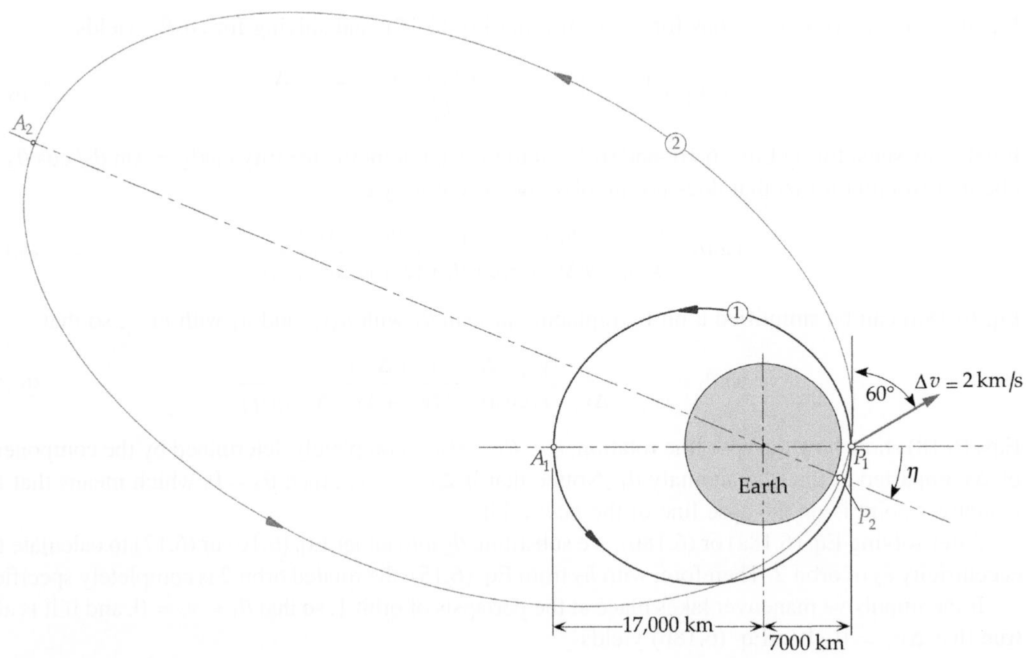

FIG. 6.19

Apse line rotation maneuver.

At the maneuver point P_1, the angular momentum formula and the fact that P_1 is the perigee of orbit 1 ($\theta_1 = 0$) imply that

$$v_{\perp_1} = \frac{h_1}{r_{P_1}} = \frac{62,871}{7000} = 8.9816\,\text{km/s}$$
$$v_{r_1} = 0 \tag{b}$$

From Fig. 6.19 it is clear that

$$\Delta v_{\perp} = \Delta v \cos 60° = 1\,\text{km/s}$$
$$\Delta v_r = \Delta v \sin 60° = 1.7321\,\text{km/s} \tag{c}$$

The angular momentum of orbit 2 is given by Eq. (6.15),

$$h_2 = h_1 + r\Delta v_{\perp} = 62,871 + 7000 \cdot 1 = 69,871\,\text{km}^2/\text{s}$$

To compute θ_2, we use Eq. (6.18b) together with Eqs. (a)–(c):

$$\begin{aligned}
\tan\theta_2 &= \frac{(v_{\perp_1} + \Delta v_{\perp})(v_{r_1} + \Delta v_r)\dfrac{{v_{\perp_1}}^2}{}}{(v_{\perp_1} + \Delta v_{\perp})^2 e_1 \cos\theta_1 + (2v_{\perp_1} + \Delta v_{\perp})\Delta v_{\perp}\,(\mu/r_{P_1})} \\
&= \frac{(8.9816+1)(0+1.7321)\dfrac{8.9816^2}{}}{(8.9816+1)^2 \cdot 0.41667 \cdot \cos(0) + (2 \cdot 8.9816 + 1) \cdot 1\,(398,600/7000)} \\
&= 0.4050
\end{aligned}$$

It follows that $\theta_2 = 22.05°$, so that Eq. (6.12) yields

$$\boxed{\eta = -22.05°}$$

This means that the rotation of the apse line is clockwise, as indicated in Fig. 6.19.

From Eq. (6.17) we obtain the eccentricity of orbit 2,

$$\begin{aligned}
e_2 &= \frac{(h_1 + r_{P_1}\Delta v_{\perp})^2 e_1 \cos\theta_1 + (2h_1 + r_{P_1}\Delta v_{\perp}) r_{P_1}\Delta v_{\perp}}{h_1^2 \cos\theta_2} \\
&= \frac{(62,871 + 7000.1)^2 \cdot 0.41667 \cdot \cos(0) + (2 \cdot 62,871 + 7000 \cdot 1) \cdot 7000 \cdot 1}{62,871^2 \cdot \cos 22.05°} \\
&= 0.80883
\end{aligned}$$

With this and the angular momentum we find using the orbit equation that the perigee and apogee radii of orbit 2 are

$$r_{P_2} = \frac{{h_2}^2}{\mu}\frac{1}{1+e_2} = \frac{69,871^2}{398,600}\frac{1}{1+0.80883} = \boxed{6771.1\,\text{km}}$$
$$r_{A_2} = \frac{69,871^2}{398,600}\frac{1}{1-0.80883} = \boxed{64,069\,\text{km}}$$

6.8 CHASE MANEUVERS

Whereas Hohmann transfers and phasing maneuvers are leisurely, energy-efficient procedures that require some preconditions (e.g., coaxial elliptical, orbits) to work, a chase or intercept trajectory is one that answers the question, "How do I get from point *A* to point *B* in space in a given amount of time?" The nature of the orbit lies in the answer to the question rather than being prescribed at the outset. Intercept trajectories near a planet are likely to require delta-v's beyond the capabilities of today's technology, so they are largely of theoretical rather than practical interest. We might refer to them as "star wars maneuvers." Chase trajectories can be found as solutions to Lambert's problem (Section 5.3), which is useful and practical for interplanetary mission design (Chapter 8).

EXAMPLE 6.9

Spacecraft B and C are both in the geocentric elliptical orbit (1) shown in Fig. 6.20, from which it can be seen that the true anomalies are $\theta_B = 45°$ and $\theta_C = 150°$. At the instant shown, spacecraft B executes a delta-v maneuver, embarking upon a trajectory (2), which will intercept and rendezvous with vehicle C in precisely one hour. Find the orbital parameters (e and h) of the intercept trajectory and the total delta-v required for the chase maneuver.

Solution

First, we must determine the parameters of orbit 1 in the usual way. The eccentricity is found using the orbit's perigee and apogee, shown in Fig. 6.20,

$$e_1 = \frac{18,900 - 8100}{18,900 + 8100} = 0.4000$$

From Eq. (6.2),

$$h_1 = \sqrt{2 \cdot 398,600}\sqrt{\frac{8100 \cdot 18,900}{8100 + 18,900}} = 67,232 \text{km}^2/\text{s}$$

Using Eq. (2.82) yields the period,

$$T_1 = \frac{2\pi}{\mu^2}\left(\frac{h_1}{\sqrt{1-e_1^2}}\right)^3 = \frac{2\pi}{398,600^2}\left(\frac{67,232}{\sqrt{1-0.4^2}}\right)^3 = 15,610\text{s}$$

In perifocal coordinates (Eq. 2.119) the position vector of B is

$$\mathbf{r}_B = \frac{{h_1}^2}{\mu}\frac{1}{1+e_1\cos\theta_B}(\cos\theta_B\hat{\mathbf{p}} + \sin\theta_B\hat{\mathbf{q}}) = \frac{67,232^2}{398,600}\frac{1}{1+0.4\cos 45°}(\cos 45°\hat{\mathbf{p}} + \sin 45°\hat{\mathbf{q}})$$

or

$$\mathbf{r}_B = 6250.6\hat{\mathbf{p}} + 6250.6\hat{\mathbf{q}}\ (\text{km}) \tag{a}$$

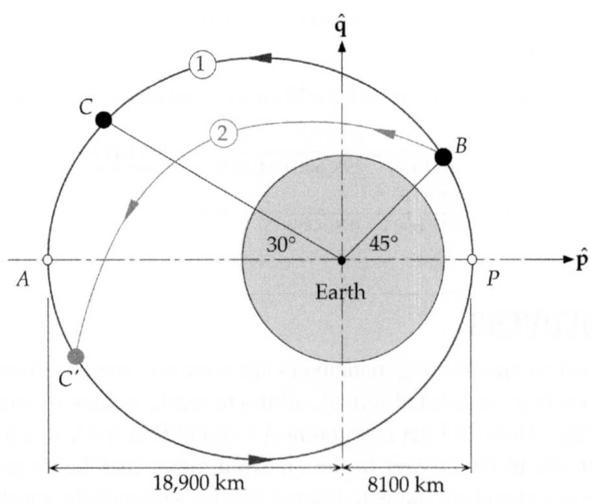

FIG. 6.20

Intercept trajectory (2) required for B to catch C in 1 h.

Likewise, according to Eq. (2.125), the velocity at B on orbit 1 is

$$\mathbf{v}_B)_1 = \frac{\mu}{h}[-\sin\theta_B\hat{\mathbf{p}} + (e + \cos\theta_B)\hat{\mathbf{q}}] = \frac{398{,}600}{67{,}232}[-\sin 45°\hat{\mathbf{p}} + (0.4 + \cos 45°)\hat{\mathbf{q}}]$$

so that

$$\mathbf{v}_B)_1 = -4.1922\hat{\mathbf{p}} + 6.5637\hat{\mathbf{q}}\ (\text{km/s}) \tag{b}$$

Now we need to move spacecraft C along orbit 1 to the position C' that it will occupy one hour later, when it will presumably be met by spacecraft B. To do that, we must first calculate the time since perigee passage at C. Since we know the true anomaly, the eccentric anomaly follows from Eq. (3.13b),

$$E_C = 2\tan^{-1}\left(\sqrt{\frac{1-e_1}{1+e_1}}\tan\frac{\theta_C}{2}\right) = 2\tan^{-1}\left(\sqrt{\frac{1-0.4}{1+0.4}}\tan\frac{150°}{2}\right) = 2.3646\,\text{rad}$$

Substituting this value into Kepler's equation (Eqs. 3.8 and 3.14) yields the time since perigee passage,

$$t_C = \frac{T_1}{2\pi}(E_C - e_1\sin E_C) = \frac{15{,}610}{2\pi}(2.3646 - 0.4\cdot\sin 2.3646) = 5178\,\text{s}$$

An hour later ($\Delta t = 3600\,\text{s}$), the spacecraft will be in intercept position at C',

$$t_{C'} = t_C + \Delta t = 5178 + 3600 = 8778\,\text{s}$$

The corresponding mean anomaly is

$$M_e)_{C'} = 2\pi\frac{t_{C'}}{T_1} = 2\pi\frac{8778}{15{,}610} = 3.5331\,\text{rad}$$

With this value of the mean anomaly, Kepler's equation becomes

$$E_{C'} - e_1\sin E_{C'} = 3.5331$$

Applying Algorithm 3.1 to the solution of this equation we get

$$E_{C'} = 3.4223\,\text{rad}$$

Substituting this result into Eq. (3.13a) yields the true anomaly at C',

$$\tan\frac{\theta_{C'}}{2} = \sqrt{\frac{1+04}{1-04}}\tan\frac{3.4223}{2} = -10.811 \Rightarrow \theta_{C'} = 190.57°$$

We are now able to calculate the perifocal position and velocity vectors at C' on orbit 1

$$\begin{aligned}\mathbf{r}_{C'} &= \frac{67{,}232^2}{398{,}600}\frac{1}{1+0.4\cos 190.57°}(\cos 190.57°\hat{\mathbf{p}} + \sin 190.57°\hat{\mathbf{q}})\\ &= -18{,}372\hat{\mathbf{p}} - 3428.1\hat{\mathbf{q}}\ (\text{km})\end{aligned}$$

$$\begin{aligned}\mathbf{v}_{C'})_1 &= \frac{398{,}600}{67{,}232}[-\sin 190.57°\hat{\mathbf{p}} + (0.4 + \cos 190.57°)\hat{\mathbf{q}}]\\ &= 1.0875\hat{\mathbf{p}} - 3.4566\hat{\mathbf{q}}\ (\text{km/s})\end{aligned} \tag{c}$$

The intercept trajectory connecting points B and C' is found by solving Lambert's problem. Substituting $\mathbf{r}_B$ and $\mathbf{r}_{C'}$ along with $\Delta t = 3600\,\text{s}$, into Algorithm 5.2 yields

$$\mathbf{v}_B)_2 = -8.1349\hat{\mathbf{p}} + 4.0506\hat{\mathbf{q}}\ (\text{km/s}) \tag{d}$$

$$\mathbf{v}_{C'})_2 = -3.4745\hat{\mathbf{p}} - 4.7943\hat{\mathbf{q}}\ (\text{km/s}) \tag{e}$$

These velocities are most easily obtained by running the following MATLAB script, which executes Algorithm 5.2 by means of the function M-file *lambert.m* (Appendix D.25).

```
clear
global mu
deg       = pi/180;
mu        = 398600;
e         = 0.4;
h         = 67232;
theta1    = 45*deg;
theta2    = 190.57*deg;
delta_t   = 3600;
rB        = h^2/mu/(1 + e * cos(theta1))...
            * [cos(theta1),sin(theta1),0];
rC_prime  = h^2/mu/(1 + e * cos(theta2))...
            * [cos(theta2),sin(theta2),0];
string    = 'pro';
[vB2 vC_prime_2] = lambert(rB, rC_prime, delta_t, string)
```

From Eqs. (b) and (d) we find

$$\Delta \mathbf{v}_B = \mathbf{v}_B)_2 - \mathbf{v}_B)_1 = -3.9326\hat{\mathbf{p}} - 2.5132\hat{\mathbf{q}}\,(\text{km/s})$$

whereas Eqs. (c) and (e) yield

$$\Delta \mathbf{v}_{C'} = \mathbf{v}_{C'})_1 - \mathbf{v}_{C'})_2 = -4.5620\hat{\mathbf{p}} + 1.3376\hat{\mathbf{q}}\,(\text{km/s})$$

The anticipated, extremely large delta-v requirement for this chase maneuver is the sum of the magnitudes of these two vectors,

$$\Delta v = \|\Delta \mathbf{v}_B\| + \|\Delta \mathbf{v}_{C'}\| = 4.6755 + 4.7540 = \boxed{9.430\text{km/s}}$$

We know that orbit 2 is an ellipse, because the magnitude of $\mathbf{v}_B)_2$ (9.088 km/s) is less than the escape speed ($\sqrt{2\mu/r_B} = 9.496\text{km/s}$) at B. To pin it down a bit more, we can use $\mathbf{r}_B$ and $\mathbf{v}_B)_2$ to obtain the orbital elements from Algorithm 4.2, which yields

$$\boxed{h_2 = 76{,}167\text{km}^2/\text{s}}$$
$$\boxed{e_2 = 0.8500}$$
$$a_2 = 52{,}449\text{km}$$
$$\theta_B)_2 = 319.52^\circ$$

These may be found quickly by running the following MATLAB script, in which the M-function *coe_from_sv.m* implements Algorithm 4.2 (see Appendix D.18):

```
clear
global mu
mu  = 398600;
rB  = [6250.6,6250.6,0];
vB2 = [-8.1349,4.0506,0];
orbital_elements = coe_from_sv(rB, vB2);
```

The details of the intercept trajectory and the delta-v maneuvers are shown in Fig. 6.21. A far less dramatic though more leisurely (and realistic) way for B to catch up with C would be to use a phasing maneuver.

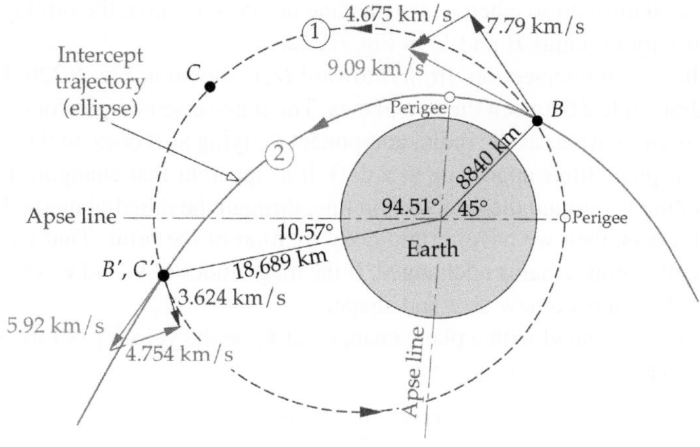

FIG. 6.21

Details of the large elliptical orbit, a portion of which serves as the intercept trajectory.

6.9 PLANE CHANGE MANEUVERS

Orbits having a common focus F need not, and generally do not, lie in a common plane. Fig. 6.22 shows two such orbits and their line of intersection BD. A and P denote the apoapses and periapses. Since the common focus lies in every orbital plane, it must lie on the line of intersection of any two orbits. For a spacecraft in orbit 1 to change its plane to that of orbit 2 by means of a single delta-v maneuver

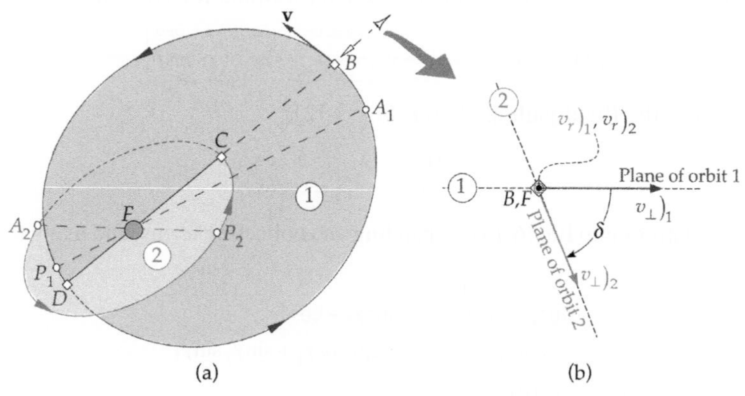

FIG. 6.22

(a) Two noncoplanar orbits about F. (b) A view down the line of intersection of the two orbital planes.

(cranking maneuver), it must do so when it is on the line of intersection of the orbital planes. Those two opportunities occur only at points B and D in Fig. 6.22a.

A view down the line of intersection, from B toward D, is shown in Fig. 6.22b. Here we can see in true view the dihedral angle δ between the two planes. The transverse component of velocity $\mathbf{v}_\perp$ at B is evident in this perspective, whereas the radial component $\mathbf{v}_r$, lying as it does on the line of intersection, is normal to the view plane (thus appearing as a dot). It is apparent that changing the plane of orbit 1 requires simply rotating $\mathbf{v}_\perp$ around the intersection line, through the dihedral angle. If $v_\perp$ and v_r remain unchanged in the process, then we have a rigid body rotation of the orbit. That is, except for its new orientation in space, the orbit remains unchanged. If the magnitudes of $\mathbf{v}_r$ and $\mathbf{v}_\perp$ change in the process, then the rotated orbit acquires a new size and shape.

To find the delta-v associated with a plane change, let $\mathbf{v}_1$ be the velocity before and $\mathbf{v}_2$ the velocity after the impulsive maneuver. Then

$$\mathbf{v}_1 = v_{r_1}\hat{\mathbf{u}}_r + v_{\perp_1}\hat{\mathbf{u}}_{\perp_1}$$
$$\mathbf{v}_2 = v_{r_2}\hat{\mathbf{u}}_r + v_{\perp_2}\hat{\mathbf{u}}_{\perp_2}$$

where $\hat{\mathbf{u}}_r$ is the radial unit vector directed along the line of intersection of the two orbital planes. $\hat{\mathbf{u}}_r$ does not change during the maneuver. As we know, the transverse unit vector $\hat{\mathbf{u}}_\perp$ is perpendicular to $\hat{\mathbf{u}}_r$ and lies in the orbital plane. Therefore, it rotates through the dihedral angle δ from its initial orientation $\hat{\mathbf{u}}_{\perp_1}$ to its final orientation $\hat{\mathbf{u}}_{\perp_2}$.

The change $\Delta\mathbf{v}$ in the velocity vector is

$$\Delta\mathbf{v} = \mathbf{v}_2 - \mathbf{v}_1 = (v_{r_2} - v_{r_1})\hat{\mathbf{u}}_r + v_{\perp_2}\hat{\mathbf{u}}_{\perp_2} - v_{\perp_1}\hat{\mathbf{u}}_{\perp_1}$$

The magnitude Δv is found by taking the dot product of $\Delta\mathbf{v}$ with itself,

$$\Delta v^2 = \Delta\mathbf{v}\cdot\Delta\mathbf{v} = \left[(v_{r_2} - v_{r_1})\hat{\mathbf{u}}_r + v_{\perp_2}\hat{\mathbf{u}}_{\perp_2} - v_{\perp_1}\hat{\mathbf{u}}_{\perp_1}\right]\cdot\left[(v_{r_2} - v_{r_1})\hat{\mathbf{u}}_r + v_{\perp_2}\hat{\mathbf{u}}_{\perp_2} - v_{\perp_1}\hat{\mathbf{u}}_{\perp_1}\right]$$

Carrying out the dot products while noting that $\hat{\mathbf{u}}_r\cdot\hat{\mathbf{u}}_r = \hat{\mathbf{u}}_{\perp_1}\cdot\hat{\mathbf{u}}_{\perp_1} = \hat{\mathbf{u}}_{\perp_2}\cdot\hat{\mathbf{u}}_{\perp_2} = 1$ and $\hat{\mathbf{u}}_r\cdot\hat{\mathbf{u}}_{\perp_1} = \hat{\mathbf{u}}_{\perp_1}\cdot\hat{\mathbf{u}}_{\perp_2} = 0$ yields

$$\Delta v^2 = (v_{r_2} - v_{r_1})^2 + {v_{\perp_1}}^2 + {v_{\perp_2}}^2 - 2v_{\perp_1}v_{\perp_2}(\hat{\mathbf{u}}_{\perp_1}\cdot\hat{\mathbf{u}}_{\perp_2})$$

But $\hat{\mathbf{u}}_{\perp_1}\cdot\hat{\mathbf{u}}_{\perp_2} = \cos\delta$, so that we finally obtain a general formula for Δv with plane change,

$$\boxed{\Delta v = \sqrt{(v_{r_2} - v_{r_1})^2 + {v_{\perp_1}}^2 + {v_{\perp_2}}^2 - 2v_{\perp_1}v_{\perp_2}\cos\delta}} \tag{6.19}$$

From the definition of the flight path angle (cf. Fig. 2.12),

$$v_{r_1} = v_1\sin\gamma_1 \qquad v_{\perp_1} = v_1\cos\gamma_1$$
$$v_{r_2} = v_2\sin\gamma_2 \qquad v_{\perp_2} = v_2\cos\gamma_2$$

Substituting these relations into Eq. (6.19), expanding and collecting terms, and using the trigonometric identities,

$$\sin^2\gamma_1 + \cos^2\gamma_1 = \sin^2\gamma_2 + \cos^2\gamma_2 = 1$$
$$\cos(\gamma_2 - \gamma_1) = \cos\gamma_2\cos\gamma_1 + \sin\gamma_2\sin\gamma_1$$

leads to another version of Eq. (6.19),

$$\Delta v = \sqrt{v_1^2 + v_2^2 - 2v_1v_2[\cos\Delta\gamma - \cos\gamma_2\cos\gamma_1(1 - \cos\delta)]} \tag{6.20}$$

where $\Delta\gamma = \gamma_2 - \gamma_1$. If there is no plane change ($\delta = 0$), then $\cos\delta = 1$ and Eq. (6.20) reduces to

$$\Delta v = \sqrt{v_1^2 + v_2^2 - 2v_1 v_2 \cos\Delta\gamma} \quad \text{(No plane change)}$$

which is the cosine law we have been using to compute Δv in coplanar maneuvers. Therefore, not surprisingly, Eq. (6.19) contains Eq. (6.8) as a special case.

To keep Δv at a minimum, it is clear from Eq. (6.19) that the radial velocity should remain unchanged during a plane change maneuver. For the same reason, it is apparent that the maneuver should occur where $v_\perp$ is smallest, which is at apoapsis. Fig. 6.23 illustrates a plane change maneuver at the apoapsis of both orbits. In this case $v_{r_1} = v_{r_2} = 0$, so that $v_{\perp_1} = v_1$ and $v_{\perp_2} = v_2$, thereby reducing Eq. (6.19) to

$$\Delta v = \sqrt{v_1^2 + v_2^2 - 2v_1 v_2 \cos\delta} \tag{6.21}$$

Eq. (6.21) is for a speed change accompanied by a plane change, as illustrated in Fig. 6.24a. Using the trigonometric identity

$$\cos\delta = 1 - 2\sin^2\frac{\delta}{2}$$

we can rewrite Eq. (6.21) as follows,

$$\Delta v_{\text{I}} = \sqrt{(v_2 - v_1)^2 + 4v_1 v_2 \sin^2\frac{\delta}{2}} \quad \text{(Rotation about the common apse line)} \tag{6.22}$$

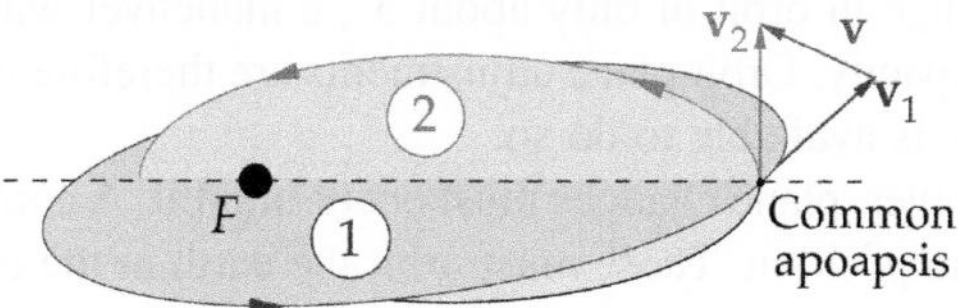

FIG. 6.23

Impulsive plane change maneuver at apoapsis.

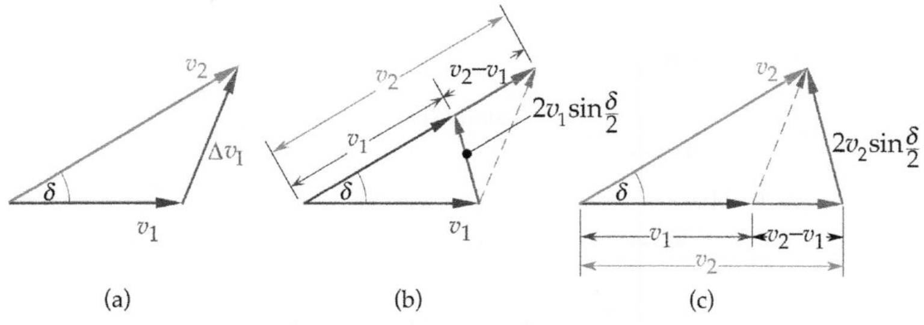

FIG. 6.24

The orbital plane rotates about the common apse line. (a) Speed change accompanied by plane change. (b) Plane change followed by speed change. (c) Speed change followed by plane change.

If there is no change in the speed, so that $v_2 = v_1$, then Eq. (6.22) yields

$$\Delta v_\delta = 2v\sin\frac{\delta}{2} \quad \text{(Pure rotation of the velocity vector)} \tag{6.23}$$

The subscript δ reminds us that this is the delta-v for a pure rotation of the velocity vector through the angle δ.

Another plane change strategy, illustrated in Fig. 6.24b, is to rotate the velocity vector and then change its magnitude. In that case, the delta-v is

$$\Delta v_{\text{II}} = 2v_1\sin\frac{\delta}{2} + |v_2 - v_1|$$

Yet another possibility is to change the speed first, and then rotate the velocity vector (Fig. 6.24c). Then

$$\Delta v_{\text{III}} = |v_2 - v_1| + 2v_2\sin\frac{\delta}{2}$$

Since the sum of the lengths of any two sides of a triangle must be greater than the length of the third side, it is evident from Fig. 6.24 that both Δv_{II} and Δv_{III} are greater than Δv_{I}. It follows that plane change accompanied by speed change is the most efficient of the above three maneuvers.

Eq. (6.23), the delta-v formula for pure rotation of the velocity vector, is plotted in Fig. 6.25, which shows why significant plane changes are so costly in terms of propellant expenditure. For example, a plane change of just 24° requires a delta-v equal to that needed for an escape trajectory (41.4% velocity boost). A 60° plane change requires a delta-v equal to the speed of the spacecraft itself, which in earth orbit operations is about 7.5 km/s. For such a maneuver in LEO, the most efficient chemical propulsion system would require that well over 80% of the spacecraft mass consist of propellant. The space shuttle was capable of a plane change in orbit of only about 3°, a maneuver which would exhaust its entire orbital-maneuvering fuel capacity. Orbit plane adjustments are therefore made during the powered ascent phase when the energy is available to do so.

For some missions, however, plane changes must occur in orbit. A common example is the maneuvering of GEO satellites into position. These must orbit the earth in the equatorial plane, but it is impossible to throw a satellite directly into an equatorial orbit from a launch site that is not on the equator. That is not difficult to understand when we realize that the plane of the orbit must contain the center of the earth (the focus) as well as the point at which the satellite is inserted into orbit, as illustrated in Fig. 6.26. So if the insertion point is anywhere but on the equator, the plane of the orbit will be tilted

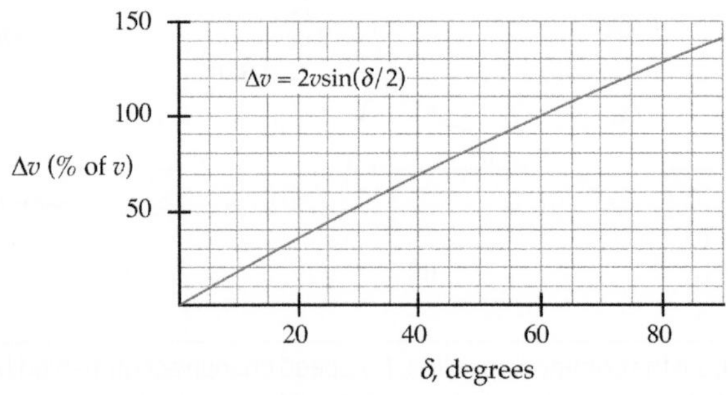

FIG. 6.25

Δv required to rotate the velocity vector through an angle δ.

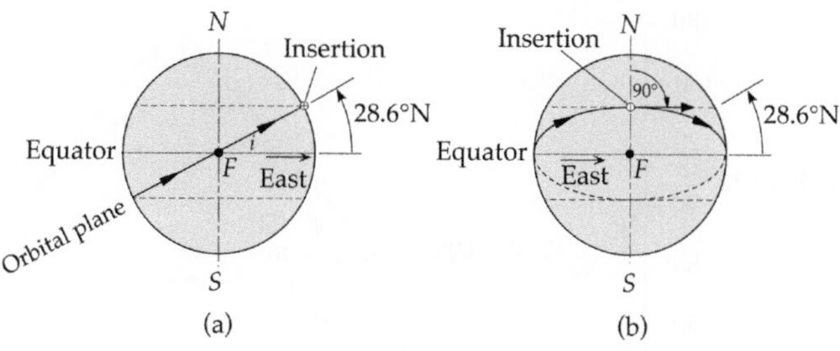

FIG. 6.26

Two views of the orbit of a satellite launched directly east at 28.6°N latitude. (a) Edge-on view of the orbital plane. (b) View toward insertion point meridian.

away from the earth's equator. As we know from Chapter 4, the angle between the equatorial plane and the plane of the orbiting satellite is called the inclination i.

Launching a satellite due east takes full advantage of the earth's rotational velocity, which is 0.46 km/s (about 1000 miles per hour) at the equator and diminishes toward the poles according to the formula

$$v_{\text{rotational}} = v_{\text{equatorial}} \cos\phi$$

where ϕ is the latitude. Fig. 6.26 shows a spacecraft launched due east into low earth orbit at a latitude of 28.6°N, which is the latitude of the Kennedy Space Flight Center (KSC). As can be seen from the figure, the inclination of the orbit will be 28.6°. One-fourth of the way around the earth the satellite will cross the equator. Halfway around the earth it reaches its southernmost latitude, $\phi = 28.6°$ S. It then heads north, crossing over the equator at the three-quarters point, and returning after one complete revolution to $\phi = 28.6°$ N .

Launch azimuth A is the flight direction at insertion, measured clockwise from north on the local meridian. Thus $A = 90°$ is due east. If the launch direction is not directly eastward, then the orbit will have an inclination greater than the launch latitude, as illustrated in Fig. 6.27 for $\phi = 28.6°$ N. Northeasterly ($0 < A < 90°$) or southeasterly ($90° < A < 180°$) launches take only partial advantage of the earth's rotational speed and both produce an inclination i greater than the launch latitude but less than 90°. Since these orbits have an eastward velocity component, they are called prograde orbits. Launches

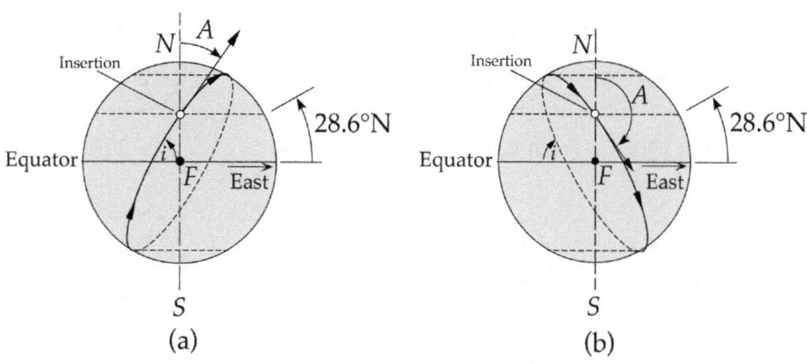

FIG. 6.27

(a) Northeasterly launch ($0 < A < 90°$) from a latitude of 28.6°N. (b) Southeasterly launch ($90° < A < 270°$).

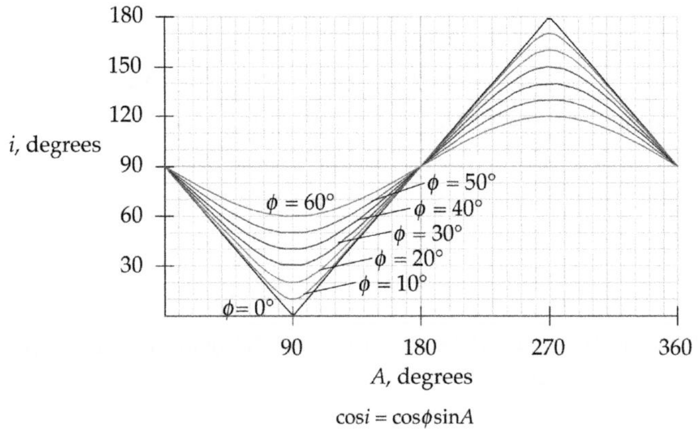

FIG. 6.28

Orbit inclination i versus launch azimuth A for several latitudes ϕ.

to the west produce retrograde orbits with inclinations between 90° and 180°. Launches directly north or directly south result in polar orbits.

Spherical trigonometry is required to obtain the relationship between orbital inclination i, launch platform latitude ϕ, and launch azimuth A. It turns out that

$$\cos i = \cos\phi \sin A \tag{6.24}$$

From this, we verify, for example, that $i = \phi$ when $A = 90°$, as pointed out above. A plot of this relation is presented in Fig. 6.28, while Fig. 6.29 illustrates the orientation of orbits for a range of launch azimuths at $\phi = 28°$.

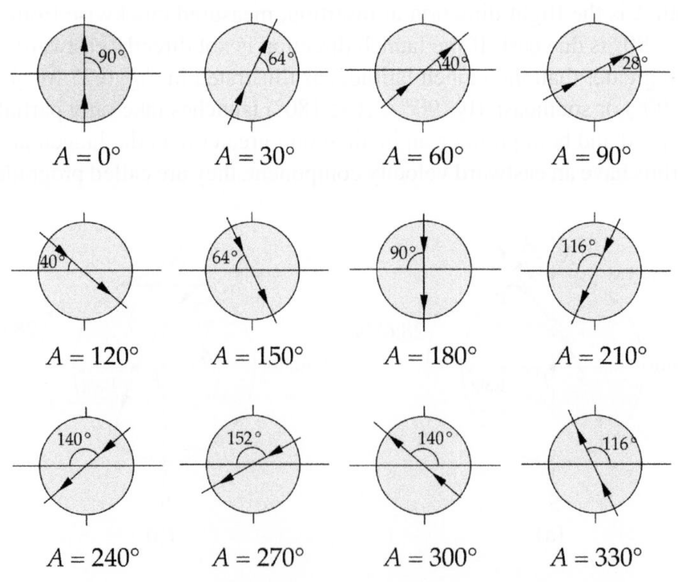

FIG. 6.29

Variation of orbit inclinations with launch azimuth at $\phi = 28^{\circ}$. Note the retrograde orbits for $A > 180^{\circ}$.

EXAMPLE 6.10

Determine the required launch azimuth for the sun-synchronous satellite of Example 4.9 if it is launched from Vandenburgh AFB on the California coast (latitude = 34.5°N).

Solution

In Example 4.9 the inclination of the sun-synchronous orbit was determined to be 98.43°. Eq. (6.24) is used to calculate the launch azimuth,

$$\sin A = \frac{\cos i}{\cos l} = \frac{\cos 98.43^\circ}{\cos 34.5^\circ} = -0.1779$$

From this $A = 190.2°$, a launch to the south, or $A = 349.8°$, a launch to the north.

Fig. 6.30 shows the effect that the choice of launch azimuth has on the sun-synchronous orbit of Example 6.10. It does not change the fact that the orbit is retrograde; it simply determines whether the ascending node will be in the same hemisphere as the launch site or on the opposite side of the earth. Actually, a launch to the north from Vandenburgh is not an option because of the safety hazard to the populated land lying below the ascent trajectory. Launches to the south, over open water, are not a hazard. Working this problem for Kennedy Space Center (latitude 28.6°N) yields nearly the same values of A. Since safety considerations on the Florida east coast limit launch azimuths to between 35° and 120°, polar and sun-synchronous satellites cannot be launched from the eastern test range.

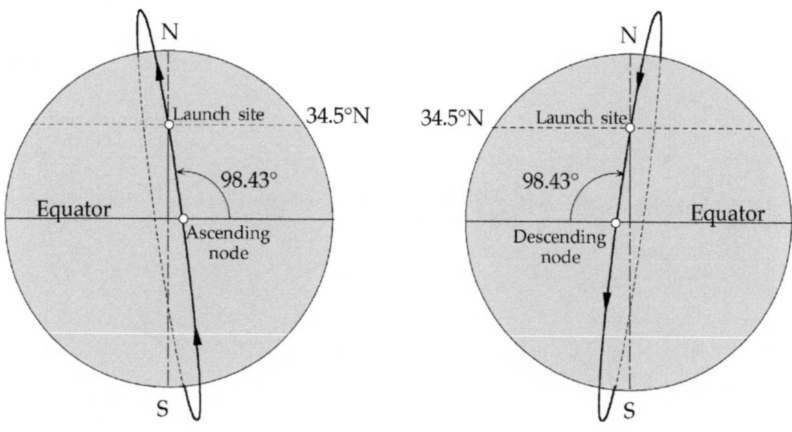

FIG. 6.30

Effect of launch azimuth on the position of the orbit.

EXAMPLE 6.11

Find the delta-v required to transfer a satellite from a circular, 300-km-altitude low earth orbit of 28° inclination to a geostationary equatorial orbit. Circularize and change the inclination at altitude. Compare that delta-v requirement with the one in which the plane change is done in the low earth orbit.

Solution

Fig. 6.31 shows the 28° inclined low earth parking orbit (1), the coplanar Hohmann transfer ellipse (2), and the coplanar GEO orbit (3). From the figure we see that

$$r_B = 6678\,\text{km} \quad r_C = 42,164\,\text{km}$$

Orbit 1:

For this circular orbit the speed at B is

$$v_B)_1 = \sqrt{\frac{\mu}{r_B}} = \sqrt{\frac{398,600}{6678}} = 7.7258\,\text{km/s}$$

Orbit 2:

We first obtain the angular momentum by means of Eq. (6.2),

$$h_2 = \sqrt{2\mu}\sqrt{\frac{r_B r_C}{r_B + r_C}} = 67,792\,\text{km/s}$$

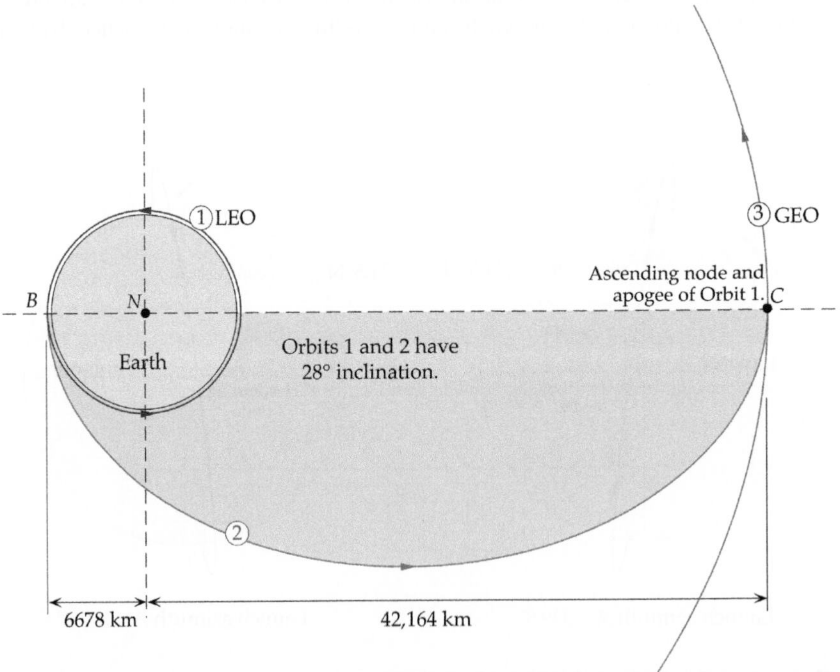

FIG. 6.31

Transfer from LEO to GEO in an orbit of 28° inclination.

The velocities at perigee and apogee of orbit 2 are, from the angular momentum formula,

$$v_B)_2 = \frac{h_2}{r_B} = 10.152\text{km/s} \quad v_C)_2 = \frac{h_2}{r_C} = 1.6078\text{km/s}$$

At this point we can calculate Δv_B,

$$\Delta v_B = v_B)_2 - v_B)_1 = 10.152 - 7.7258 = 2.4258\text{km/s}$$

Orbit 3:
For this GEO orbit, which is circular, the speed at C is

$$v_C)_3 = \sqrt{\frac{\mu}{r_C}} = 3.0747\text{km/s}$$

The spacecraft in orbit 2 arrives at C with a velocity of 1.6078 km/s inclined at 28° to orbit 3. Therefore, both its orbital speed and inclination must be changed at C (Fig. 6.32). The most efficient strategy is to combine the plane change with the speed change (Eq. 6.21), so that

$$\begin{aligned}\Delta v_C &= \sqrt{v_C)_2{}^2 + v_C)_3{}^2 - 2v_C)_2 v_C)_3 \cos\Delta i} \\ &= \sqrt{1.6078^2 + 3.0747^2 - 2\cdot 1.6078\cdot 3.0747\cdot \cos 28°} = 1.8191\text{km/s}\end{aligned}$$

Therefore, the total delta-v requirement is

$$\Delta v_{\text{total}} = \Delta v_B + \Delta v_C = 2.4258 + 1.819 = \boxed{4.2449\text{km/s}} \quad (\text{Plane change at } C)$$

Suppose we make the plane change at LEO instead of at GEO. In that case, Eq. (6.21) provides the initial delta-v,

$$\begin{aligned}\Delta v_B &= \sqrt{v_B)_1{}^2 + v_B)_2{}^2 - 2v_B)_1 v_B)_2 \cos\Delta i} \\ &= \sqrt{7.7258^2 + 10.152^2 - 2\cdot 7.7258\cdot 10.152\cdot \cos 28°} = 4.9242\text{km/s}\end{aligned}$$

The spacecraft travels to C in the equatorial plane, so that when it arrives, the delta-v requirement at C is simply

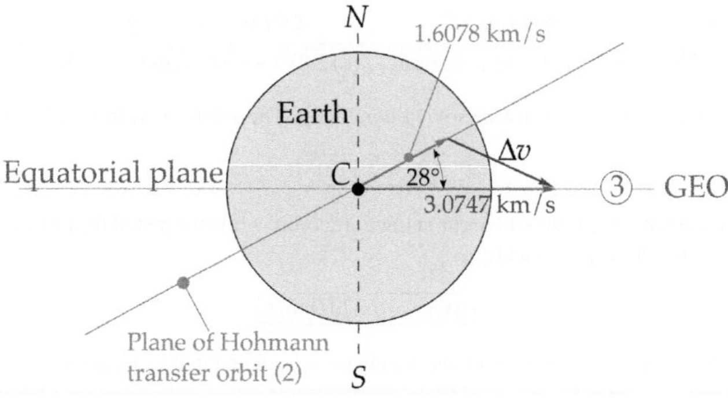

FIG. 6.32

Plane change maneuver required after the Hohmann transfer.

$$\Delta v_C = v_C)_3 - v_C)_2 = 3.0747 - 1.6078 = 1.4668\text{km/s}$$

Therefore, the total delta-v is

$$\Delta v_{\text{total}} = \Delta v_B + \Delta v_C = 4.9242 + 1.4668 = \boxed{6.3910\text{km/s}} \quad \text{(Plane change at } B\text{)}$$

This is a 50% increase over the total delta-v with plane change at GEO. Clearly, it is best to do plane change maneuvers at the largest possible distance (apoapsis) from the primary attractor, where the velocities are smallest.

EXAMPLE 6.12

Suppose in the previous example that part of the plane change, Δi, takes place at B, the perigee of the Hohmann transfer ellipse, and the remainder, $28° - \Delta i$, occurs at the apogee C. What is the value of Δi that results in the minimum Δv_{total}?

Solution

We found in Example 6.11 that if $\Delta i = 0$, then $\Delta v_{\text{total}} = 4.2449\text{km/s}$, whereas $\Delta i = 28°$ made $\Delta v_{\text{total}} = 6.3910\text{km/s}$. Here we are to determine if there is a value of Δi between 0° and 28° that yields a Δv_{total} that is smaller than either of those two.

In this case a plane change occurs at both B and C. The most efficient strategy is to combine the plane change with the speed change, so that the delta-v's at those points are (Eq. 6.21)

$$\begin{aligned}\Delta v_B &= \sqrt{v_B)_1{}^2 + v_B)_2{}^2 - 2v_B)_1 v_B)_2 \cos\Delta i} \\ &= \sqrt{7.7258^2 + 10.152^2 - 2 \cdot 7.7258 \cdot 10.152 \cdot \cos\Delta i} \\ &= \sqrt{162.74 - 156.86\cos\Delta i}\end{aligned}$$

and

$$\begin{aligned}\Delta v_C &= \sqrt{v_C)_2{}^2 + v_C)_3{}^2 - 2v_C)_2 v_C)_3 \cos(28° - \Delta i)} \\ &= \sqrt{1.6078^2 + 3.0747^2 - 2 \cdot 1.1078 \cdot 3.0747 \cdot \cos(28° - \Delta i)} \\ &= \sqrt{12.039 - 9.8874\cos(28° - \Delta i)}\end{aligned}$$

Thus,

$$\Delta v_{\text{total}} = \Delta v_B + \Delta v_C = \sqrt{162.74 - 156.86\cos\Delta i} + \sqrt{12.039 - 9.8874\cos(28° - \Delta i)} \qquad \text{(a)}$$

To determine if there is a Δi that minimizes Δv_{total}, we take its derivative with respect to Δi and set it equal to zero:

$$\frac{d\Delta v_{\text{total}}}{d\Delta i} = \frac{78.43\sin\Delta i}{\sqrt{162.74 - 156.86\cos\Delta i}} - \frac{4.9435\sin(28° - \Delta i)}{\sqrt{12.039 - 9.8871\cos(28° - \Delta i)}} = 0$$

This is a transcendental equation, which must be solved iteratively. The solution, as the reader may verify, is

$$\boxed{\Delta i = 2.1751°} \qquad \text{(b)}$$

That is, an inclination change of 2.1751° should occur in low earth orbit, while the rest of the plane change, 25.825°, is done at GEO. Substituting Eq. (b) into Eq. (a) yields

$$\boxed{\Delta v_{\text{total}} = 4.2207\text{km/s}}$$

This is only very slightly smaller (less than 1%) than the lowest v_{total} computed in Example 6.11.

EXAMPLE 6.13

A spacecraft is in a 500 km by 10,000 km altitude geocentric orbit that intersects the equatorial plane at a true anomaly of 120° (see Fig. 6.33). If the orbit's inclination to the equatorial plane is 15°, what is the minimum velocity increment required to make this an equatorial orbit?

Solution

The orbital parameters are

$$e=\frac{r_A-r_P}{r_A+r_P}=\frac{(6378+10,000)-(6378+500)}{(6378+10,000)+(6378+500)}=0.4085$$

$$h=\sqrt{2\mu}\sqrt{\frac{r_A r_P}{r_A+r_P}}=\sqrt{2\cdot 398,600}\sqrt{\frac{16,378\cdot 6878}{16,378+6878}}=62,141\,\text{km/s}$$

The radial position and velocity components at points B and C, on the line of intersection with the equatorial plane, are

$$r_B=\frac{h^2}{\mu}\frac{1}{1+e\cos\theta_B}=\frac{62,141^2}{398,600}\frac{1}{1+0.4085\cdot\cos 120°}=12,174\,\text{km}$$

$$v_{\perp_B}=\frac{h}{r_B}=\frac{62,141}{12,174}=5.1043\,\text{km/s}$$

$$v_{r_B}=\frac{\mu}{h}e\sin\theta=\frac{398,600}{62,141}\cdot 0.4085\cdot\sin 120°=2.2692\,\text{km/s}$$

and

$$r_C=\frac{h^2}{\mu}\frac{1}{1+e\cos\theta_C}=\frac{62,141^2}{398,600}\frac{1}{1+0.4085\cdot\cos 300°}=8044.6\,\text{km}$$

$$v_{\perp_C}=\frac{h}{r_C}=\frac{62,141}{8044.6}=7.7246\,\text{km/s}$$

$$v_{r_C}=\frac{\mu}{h}e\sin\theta_C=\frac{398,600}{62,141}\cdot 0.4085\cdot\sin 300°=-2.2692\,\text{km/s}$$

These are all shown in Fig. 6.33.

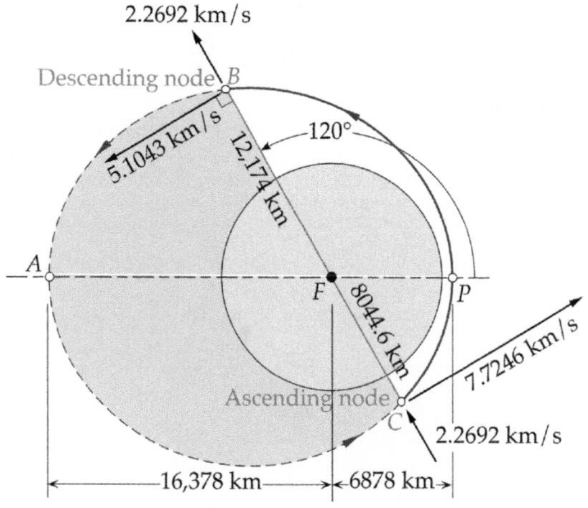

FIG. 6.33

An orbit that intersects the equatorial plane along line *BC*. The equatorial plane makes an angle of 15° with the plane of the page.

All we wish to do here is to rotate the plane of the orbit rigidly around the node line BC. The impulsive maneuver must occur at either B or C. Eq. (6.19) applies, and since the radial and transverse velocity components remain fixed, it reduces to

$$\Delta v = v_{\perp}\sqrt{2(1-\cos\delta)} = 2v_{\perp}\sin\frac{\delta}{2}$$

where $\delta = 15°$. For the minimum Δv, the maneuver must be done where $v_{\perp}$ is smallest, which is at B, the point farthest from the center of attraction F. Thus,

$$\Delta v = 2\cdot 5.1043\cdot \sin\frac{15°}{2} = \boxed{1.3325\,\text{km/s}}$$

EXAMPLE 6.14

Orbit 1 in Fig. 6.34 has angular momentum h and eccentricity e. The direction of motion is shown. Calculate the Δv required to rotate the orbit 90° about its latus rectum BC without changing h and e. The required direction of motion in orbit 2 is shown.

Solution

By symmetry, the required maneuver may occur at either B or C, and it involves a rigid body rotation of the ellipse, so that v_r and $v_{\perp}$ remain unaltered. Because of the directions of motion shown, the true anomalies of B on the two orbits are

$$\theta_B)_1 = -90° \quad \theta_B)_2 = +90°$$

The radial coordinate of B is

$$r_B = \frac{h^2}{\mu}\frac{1}{1+e\cos(\pm 90)} = \frac{h^2}{\mu}$$

For the velocity components at B, we have

$$v_{\perp_B})_1 = v_{\perp_B})_2 = \frac{h}{r_B} = \frac{\mu}{h}$$

$$v_{r_B})_1 = \frac{\mu}{h}e\sin\theta_B)_1 = -\frac{\mu e}{h}$$

$$v_{r_B})_2 = \frac{\mu}{h}e\sin\theta_B)_2 = \frac{\mu e}{h}$$

Substituting these into Eq. (6.19), yields

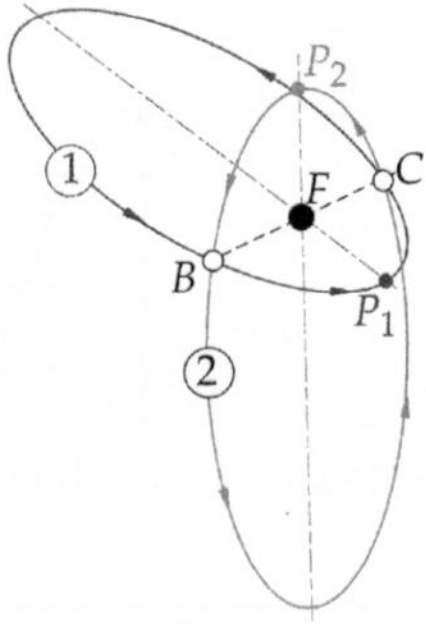

FIG. 6.34

Identical ellipses intersecting at 90° along their common latus rectum, BC.

$$\Delta v_B = \sqrt{\left[v_{r_B})_2 - v_{r_B})_1\right]^2 + v_{\perp_B})_1{}^2 + v_{\perp_B})_2{}^2 - 2v_{\perp_B})_1 v_{\perp_B})_2 \cos 90^\circ}$$
$$= \sqrt{\left[\frac{\mu e}{h} - \left(-\frac{\mu e}{h}\right)\right]^2 + \left(\frac{\mu}{h}\right)^2 + \left(\frac{\mu}{h}\right)^2 - 2\left(\frac{\mu}{h}\right)\left(\frac{\mu}{h}\right)(0)}$$
$$= \sqrt{4\frac{\mu^2}{h^2}e^2 + 2\frac{\mu^2}{h^2}}$$

so that

$$\boxed{\Delta v_B = \frac{\sqrt{2}\mu}{h}\sqrt{1+2e^2}} \tag{a}$$

If the motion on ellipse 2 were opposite to that shown in Fig. 6.34, then the radial velocity components at B (and C) would be in the same rather than in the opposite direction on both ellipses, so that instead of Eq. (a) we would find a smaller velocity increment,

$$\Delta v_B = \frac{\sqrt{2}\mu}{h}$$

6.10 NONIMPULSIVE ORBITAL MANEUVERS

Up to this point we have assumed that delta-v maneuvers take place in zero time, altering the velocity vector but leaving the position vector unchanged. In nonimpulsive maneuvers the thrust acts over a significant time interval and must be included in the equations of motion. According to Problem 2.3, adding an external force $\mathbf{F}$ to the spacecraft yields the following equation of relative motion:

$$\ddot{\mathbf{r}} = -\mu\frac{\mathbf{r}}{r^3} + \frac{\mathbf{F}}{m} \tag{6.25}$$

where m is the mass of the spacecraft. This of course reduces to Eq. (2.22) when $\mathbf{F} = 0$. If the external force is a thrust T in the direction of the velocity vector $\mathbf{v}$, then $\mathbf{F} = T(\mathbf{v}/v)$ and Eq. (6.25) becomes

$$\ddot{\mathbf{r}} = -\mu\frac{\mathbf{r}}{r^3} + \frac{T}{m}\frac{\mathbf{v}}{v} \quad (\mathbf{v} = \dot{\mathbf{r}}) \tag{6.26}$$

(Drag forces act opposite to the velocity vector, and so does thrust during a retrofire maneuver.) The Cartesian component form of Eq. (6.26) is

$$\ddot{x} = -\mu\frac{x}{r^3} + \frac{T}{m}\frac{\dot{x}}{v} \quad \ddot{y} = -\mu\frac{y}{r^3} + \frac{T}{m}\frac{\dot{y}}{v} \quad \ddot{z} = -\mu\frac{z}{r^3} + \frac{T}{m}\frac{\dot{z}}{v} \tag{6.27a}$$

where

$$r = \sqrt{x^2+y^2+z^2} \quad v = \sqrt{\dot{x}^2+\dot{y}^2+\dot{z}^2} \tag{6.27b}$$

While the rocket motor is firing, the spacecraft mass decreases, because propellant combustion products are being discharged into space through the nozzle. According to elementary rocket dynamics (cf. Section 13.3), the mass decreases at a rate given by the formula

$$\frac{dm}{dt} = -\frac{T}{I_{sp}g_0} \tag{6.28}$$

where T and I_{sp} are the thrust and the specific impulse of the propulsion system, and g_0 is the sea level acceleration of gravity.

If the thrust is not zero, then Eqs. (6.27a) may not have a straightforward analytical solution. In any case, they can be solved numerically using methods such as those discussed in Section 1.8. For that purpose, Eqs. (6.27a), (6.27b), and (6.28) must be rewritten as a system of linear differential equations in the form

$$\dot{\mathbf{y}} = \mathbf{f}(t, \mathbf{y}) \tag{6.29}$$

For the case at hand, the vector $\mathbf{y}$ consists of the six components of the state vector (position and velocity vectors) plus the mass. Therefore, with the aid of Eqs. (6.27a), (6.27b), and (6.28), we have

$$\mathbf{y} = \begin{Bmatrix} x \\ y \\ z \\ \dot{x} \\ \dot{y} \\ \dot{z} \\ m \end{Bmatrix} \quad \dot{\mathbf{y}} = \begin{Bmatrix} \dot{x} \\ \dot{y} \\ \dot{z} \\ \ddot{x} \\ \ddot{y} \\ \ddot{z} \\ \dot{m} \end{Bmatrix} \quad \mathbf{f}(t, y) = \begin{Bmatrix} y_4 \\ y_5 \\ y_6 \\ -\mu\dfrac{y_1}{r^3} + \dfrac{T}{m}\dfrac{y_4}{v} \\ -\mu\dfrac{y_2}{r^3} + \dfrac{T}{m}\dfrac{y_5}{v} \\ -\mu\dfrac{y_3}{r^3} + \dfrac{T}{m}\dfrac{y_6}{v} \\ -\dfrac{T}{I_{sp}g_o} \end{Bmatrix} \tag{6.30}$$

The numerical solution of Eq. (6.30) is illustrated in the following examples.

EXAMPLE 6.15

Suppose the spacecraft in Example 6.1 (see Fig. 6.3) has a restartable onboard propulsion system with a thrust of 10 kN and specific impulse of 300 s. Assuming that the thrust vector remains aligned with the velocity vector, solve Example 6.1 without using impulsive (zero time) delta-v burns. Compare the propellant expenditures for the two solutions.

Solution

Refer to Fig. 6.3 as an aid to visualizing the solution procedure described below. Let us assume that the plane of Fig. 6.3 is the xy plane of an earth-centered inertial frame with the z axis directed out of the page. The apse line of orbit 1 is the x axis, which is directed to the right, and y points upward toward the top of the page.

Transfer from perigee of orbit 1 to apogee of orbit 2

According to Example 6.1, the state vector just before the first delta-v maneuver is

$$\mathbf{y}_0 = \begin{Bmatrix} x \\ y \\ z \\ \dot{x} \\ \dot{y} \\ \dot{z} \\ m \end{Bmatrix}_{t=0} = \begin{Bmatrix} 6858\,\text{km} \\ 0 \\ 0 \\ 0 \\ 7.7102\,\text{km/s} \\ 0 \\ 2000 \end{Bmatrix} \tag{a}$$

Using this together with an assumed burn time t_{burn}, we numerically integrate Eq. (6.29) from $t = 0$ to $t = t_{\text{burn}}$. This yields $\mathbf{r}$, $\mathbf{v}$, and the mass m at the start of the coasting trajectory (orbit 2). We can find the true anomaly θ at the start of orbit 2 by substituting these values of $\mathbf{r}$ and $\mathbf{v}$ into Algorithm 4.2. The spacecraft must coast through a true anomaly of $\Delta\theta = 180° - \theta$ to reach apogee. Substituting $\mathbf{r}$, $\mathbf{v}$, and $\Delta\theta$ into Algorithm 2.3 yields the state vector ($\mathbf{r}_a$ and $\mathbf{v}_a$) at apogee.

The apogee radius r_a is the magnitude of $\mathbf{r}_a$. If r_a does not equal the target value of 22,378 km, then we assume a new burn time and repeat the above steps to calculate a new r_a. This trial-and-error process is repeated until r_a is acceptably close to 22,378 km.

The calculations are done in the MATLAB M-function *integrate_thrust.m*, which is listed in Appendix D.30. *rkf45.m* (see Appendix D.4) was chosen as the numerical integrator. The initial conditions $\mathbf{y}_0$ in Eq. (a) are passed to *rkf45*, which solves the system of Eq. (6.29) at discrete times between 0 and t_{burn}. *rkf45.m* employs the subfunction *rates*, embedded in *integrate_thrust.m*, to calculate the vector of derivatives **f** in Eq. (6.30). Output is to the command window, and a revised burn time was entered into the code in the MATLAB editor after each calculation of r_a.

The following output of *integrate_thrust.m* shows that a burn time of 261.1127 s (4.352 min), with a propellant expenditure of 887.5 kg, is required to produce a coasting trajectory with an apogee of 22,378 km. Due to the finite burn time, the apse line in this case is rotated 8.336° counterclockwise from that in Example 6.1 (line *BCA* in Fig. 6.3). Notice that the speed boost Δv imparted by the burn is $9.38984 - 7.71020 = 1.6796$ km/s, compared with the impulsive $\Delta v_A = 1.7725$ km/s in Example 6.1.

```
Before ignition:
  Mass = 2000 kg
  State vector:
    r = [     6858,          0,         0] (km)
      Radius = 6858
    v = [        0,     7.7102,         0] (km/s)
      Speed = 7.7102

Thrust           =            10 kN
Burn time        =     261.112700 s
Mass after burn  = 1.112495E+03 kg

End-of-burn state vector:
    r = [    6551.56, 2185.85,     0] (km)
      Radius = 6906.58
    v = [ -2.42229,    9.07202,    0] (km/s)
      Speed = 9.38984

Postburn trajectory:
  Eccentricity = 0.530257
  Semimajor axis = 14623.7 km
  Apogee state vector:
    r = [-2.21416E+04, -3.24453E+03,  0.00000E+00] (km)
      Radius = 22378
    v = [ 4.19390E-01, -2.86203E+00, -0.00000E+00] (km/s)
      Speed = 2.8926
```

Transfer from apogee of orbit 2 to the circular target orbit 3

The spacecraft mass and state vector at apogee, given by the above MATLAB output (under "`Postburn trajectory`"), are entered as new initial conditions in *integrate_thrust.m*, and the manual trial-and-error process described above is carried out. It is not possible to transfer from the 22,378-km apogee of orbit 2 to a circular orbit of radius 22,378 km using a single finite-time burn. Therefore, the objective in this case is to make the semimajor axis of the final orbit equal to 22,378 km. This was achieved with a burn time of 118.88 s and a propellant expenditure of 404.05 kg, and it yields a nearly circular orbit having an eccentricity of 0.00867 and an apse line rotated 80.85° clockwise from the *x* axis.

The computed spacecraft mass at the end of the second delta-v maneuver is 708.44 kg. Therefore, the total propellant expenditure is $2000 - 708.44 = 1291.6$ kg. This is essentially the same as the propellant requirement (1291.3 kg) calculated in Example 6.1, in which the two delta-v maneuvers were impulsive.

Let us take the dot product of both sides of Eq. (6.26) with the velocity **v**, to obtain

$$\ddot{\mathbf{r}} \cdot \mathbf{v} = -\frac{\mu}{r^3}\mathbf{r} \cdot \mathbf{v} + \frac{T}{m}\frac{\mathbf{v} \cdot \mathbf{v}}{v} \tag{6.31}$$

In Section 2.5, we showed that

$$\ddot{\mathbf{r}}\cdot\mathbf{v}=\frac{1}{2}\frac{dv^2}{dt}\quad\text{and}\quad\frac{\mu}{r^3}\mathbf{r}\cdot\mathbf{v}=-\frac{d}{dt}\left(\frac{\mu}{r}\right)$$

Substituting these together with $\mathbf{v}\cdot\mathbf{v}=v^2$ into Eq. (6.31) yields the energy equation,

$$\frac{d}{dt}\left(\frac{v^2}{2}-\frac{\mu}{r}\right)=\frac{T}{m}v \tag{6.32}$$

This equation may be applied to the approximate solution of a constant tangential thrust orbit transfer problem. If the spacecraft is in a circular orbit, then applying a very low constant thrust T in the forward direction will cause its total energy $\varepsilon=v^2/2-\mu/r$ to slowly increase over time according to Eq. (6.32). This will raise the height after each revolution, resulting in a slow outward spiral (or inward spiral if the thrust is directed aft). If we assume that the speed at any radius of the closely spaced spiral trajectory is essentially that of a circular orbit of that radius (Wiesel, 2010), then we can replace v by $\sqrt{\mu/r}$ to obtain an approximate version of Eq. (6.32),

$$\frac{d}{dt}\left(\frac{1}{2}\frac{\mu}{r}-\frac{\mu}{r}\right)=\frac{T}{m}\sqrt{\frac{\mu}{r}}$$

Simplifying and separating variables leads to

$$\frac{d(\mu/r)}{\sqrt{\mu/r}}=-2\frac{T}{m}dt \tag{6.33}$$

The spacecraft mass is a function of time

$$m=m_0-\dot{m}_e t \tag{6.34}$$

where m_0 is the mass at the start of the orbit transfer ($t=0$), and $\dot{m}_e$ is the constant rate at which propellant is expended. Thus

$$\frac{d(\mu/r)}{\sqrt{\mu/r}}=-2\frac{T}{m_0-\dot{m}_e t}dt \tag{6.35}$$

Integrating both sides of this equation and setting $r=r_0$ when $t=0$ results in

$$\sqrt{\frac{\mu}{r}}-\sqrt{\frac{\mu}{r_0}}=\frac{T}{\dot{m}_e}\ln\left(1-\frac{\dot{m}_e}{m_0}t\right) \tag{6.36}$$

Finally, since $\dot{m}_e=-dm/dt$, Eq. (6.28) implies that we can replace $\dot{m}_e$ with $T/(I_{sp}g_0)$, so that

$$\sqrt{\frac{\mu}{r}}-\sqrt{\frac{\mu}{r_0}}=I_{sp}g_0\ln\left(1-\frac{T}{m_0 g_0 I_{sp}}t\right) \tag{6.37}$$

We may solve this equation for either r or t to get

$$r=\frac{\mu}{\left[\sqrt{\frac{\mu}{r_0}}+I_{sp}g_0\ln\left(1-\frac{T}{m_0 g_0 I_{sp}}t\right)\right]^2} \tag{6.38}$$

$$t=\frac{m_0 g_0 I_{sp}}{T}\left\{1-\exp\left[\frac{\sqrt{\mu}}{I_{sp}g_0}\left(\sqrt{\frac{1}{r}}-\sqrt{\frac{1}{r_0}}\right)\right]\right\} \tag{6.39}$$

where $\exp(x)=e^x$. Although this scalar analysis yields the radius in terms of the elapsed time, it does not provide us the state vector components $\mathbf{r}$ and $\mathbf{v}$.

EXAMPLE 6.16

A 1000-kg spacecraft is in a 6678-km (300-km-altitude) circular equatorial earth orbit. Its ion propulsion system, which has a specific impulse of 10,000 s, exerts a constant tangential thrust of $2500(10^{-6})$ kN.

(a) How long will it take the spacecraft to reach GEO (42,164 km)?
(b) How much fuel will be expended?

Solution

(a) Using Eq. (6.39), and remembering to express the acceleration of gravity in km/s^2, the flight time is

$$t = \frac{1000 \cdot 0.009807 \cdot 10{,}000}{2500(10^{-6})}\left\{1 - \exp\left[\frac{\sqrt{398{,}600}}{10{,}000 \cdot 0.009807}\left(\sqrt{\frac{1}{42{,}164}} - \sqrt{\frac{1}{6678}}\right)\right]\right\}$$

$$\boxed{t = 1{,}817{,}000\,\text{s} = 21.03\,\text{days}}$$

(b) The propellant mass m_p used is

$$m_p = \dot{m}_e t = \frac{T}{I_{sp} g_0} t = \frac{2500(10^{-6})}{10{,}000 \cdot 0.009807} \cdot 1{,}817{,}000$$

$$\boxed{m_p = 46.32\,\text{kg}}$$

In Example 6.11, we found that the total delta-v for a Hohmann transfer from 6678 km to GEO radius, with no plane change, is 3.893 km/s. Assuming a typical chemical rocket specific impulse of 300 s, Eq. (6.1) reveals that the propellant requirement would be 734 kg if the initial mass is 1000 kg. This is almost 16 times that required for the hypothetical ion-propelled spacecraft of Example 6.12. Because of their efficiency (high specific impulse), ion engines—typically using xenon as the propellant—will play an increasing role in deep-space missions and satellite station keeping. However, these extremely low-thrust devices cannot replace chemical rockets in high-acceleration applications, such as launch vehicles.

EXAMPLE 6.17

What will be the orbit after the ion engine in Example 6.16 shuts down upon reaching GEO radius?

Solution

This requires a numerical solution using the MATLAB M-function *integrate_thrust.m*, listed in Appendix D.30. According to the data of Example 6.16, the initial state vector in geocentric equatorial coordinates can be written

$$\mathbf{r}_0 = 6678\hat{\mathbf{I}}\,(\text{km}) \quad \mathbf{v}_0 = \sqrt{\frac{\mu}{r_0}}\hat{\mathbf{J}} = 7.72584\hat{\mathbf{J}}\,(\text{km/s})$$

Using these as the initial conditions, we start by assuming that the elapsed time is 21.03 days, as calculated in Example 6.16. *integrate_thrust.m* computes the final radius for that burn time and outputs the results to the command window. Depending on whether the radius is smaller or greater than 42,164 km, we reenter a slightly larger or slightly smaller time in the MATLAB editor and run the program again. Several of these manual trial-and-error steps yield the following MATLAB output:

```
Before ignition:
  Mass = 1000 kg
  State vector:
    r = [      6678,          0,    0] (km)
      Radius = 6678
```

```
      v = [          0,   7.72584,    0] (km/s)
       Speed = 7.72584

Thrust =                   0.0025 kN
Burn time =              21.037600 days
Mass after burn = 9.536645E+02 kg
End-of-burn state vector:
   r = [     -19028,     -37625.9,     0] (km)
     Radius = 42163.6
   v = [ 2.71001,       -1.45129,     0] (km/s)
     Speed = 3.07415

Postburn trajectory:
   Eccentricity    = 0.0234559
   Semimajor axis = 42149.2 km
   Apogee state vector:
      r = [ 3.77273E+04, -2.09172E+04, 0.00000E+00] (km)
        Radius = 43137.9
      v = [ 1.45656E+00,  2.62713E+00, 0.00000E+00] (km/s)
        Speed = 3.0039
```

From the printout it is evident that to reach GEO radius requires the following time and propellant expenditure:

(a) $\boxed{t = 21.0376\,\text{days}}$

(b) $\boxed{m_p = 46.34\,\text{kg}}$

These are very nearly the same as the values found in the previous example. However, this numerical solution in addition furnishes the end-of-burn state vector, which shows that the postburn orbit is slightly elliptical, having an eccentricity of 0.02346 and a semimajor axis that is only 15 km less than GEO radius.

PROBLEMS

Section 6.2

6.1 A large spacecraft has a mass of 125,000 kg. Its orbital-maneuvering engines produce a thrust of 50 kN. The orbiter is in a 400-km circular earth orbit. A delta-v maneuver transfers the spacecraft to a coplanar 300 km by 400 km elliptical orbit. Neglecting propellant loss and using elementary physics (linear impulse equals change in linear momentum, distance equals speed times time):

(a) Estimate the time required for the Δv burn.

(b) Estimate the distance traveled by the spacecraft during the burn.

(c) Calculate the ratio of your answer for (b) to the circumference of the initial circular orbit.

(d) What percent of the initial mass was expelled as combustion products?

{Ans.: (a) $\Delta t = 71$ s; (b) 548 km; (c) 1.3%; (d) 1%}

6.2 A satellite traveling 8 km/s at a perigee altitude of 500 km fires a retrorocket. What delta-v is necessary to reach a minimum altitude of 200 km during the next orbit?

{Ans.: −473 m/s}

6.3 A spacecraft is in a 500-km-altitude circular earth orbit. Neglecting the atmosphere, find the delta-v required at *A* to impact the earth at (a) point *B* and (b) point *C*.

{Ans.: (a) −0.192 km/s; (b) −7.61 km/s}

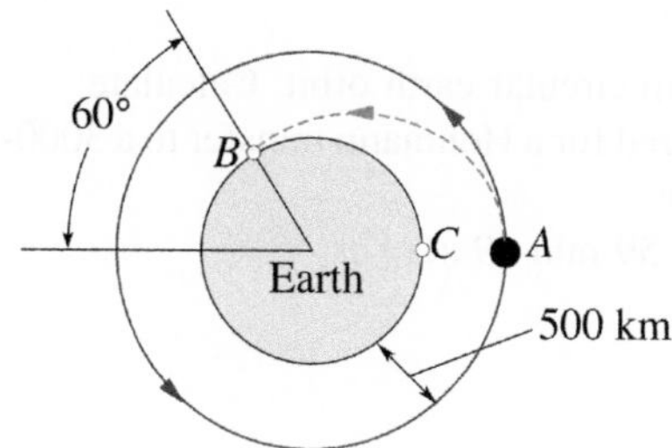

6.4 A satellite is in a circular orbit at an altitude of 250 km above the earth's surface. If an onboard rocket provides a delta-v of 200m/s in the direction of the satellite's motion, calculate the altitude of the new orbit's apogee.
{Ans.: 981 km}

6.5 A spacecraft S is in a geocentric hyperbolic trajectory with a perigee radius of 7000 km and a perigee speed of $1.3v_{esc}$. At perigee, the spacecraft releases a projectile B with a speed of 7.1 km/s parallel to the spacecraft's velocity. How far d from the earth's surface is S at the instant B impacts the earth? Neglect the atmosphere.
{Ans.: $d = 8978$ km}

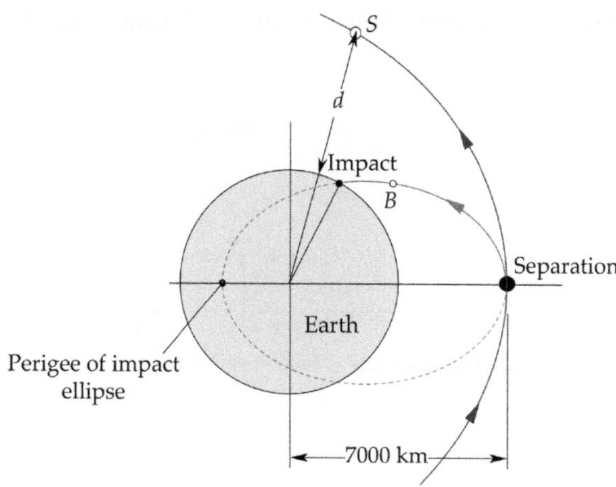

6.6 A spacecraft is in a 200-km circular earth orbit. At $t = 0$, it fires a projectile in the direction opposite to the spacecraft's motion. Some 30 min after leaving the spacecraft, the projectile impacts the earth. What delta-v was imparted to the projectile? Neglect the atmosphere.
{Ans.: $\Delta v = 77.2$ m/s}

6.7 A spacecraft is in a circular orbit of radius r and speed v around an unspecified planet. A rocket on the spacecraft is fired, instantaneously increasing the speed in the direction of motion by the amount $\Delta v = \alpha v$, where $\alpha > 0$. Calculate the eccentricity of the new orbit.
{Ans.: $e = \alpha(\alpha + 2)$}

Section 6.3

6.8 A spacecraft is in a 300-km circular earth orbit. Calculate

(a) The total delta-v required for a Hohmann transfer to a 3000-km coplanar circular earth orbit.

(b) The transfer orbit time.

{Ans.: (a) 1.198 km/s; (b) 59 min 39 s}

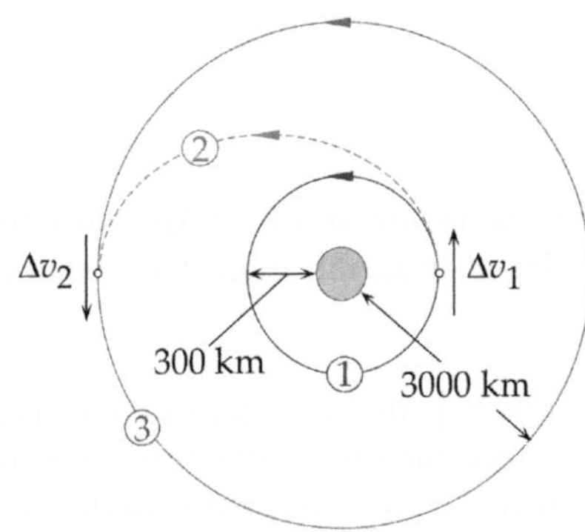

6.9 A space vehicle in a circular orbit at an altitude of 500 km above the earth executes a Hohmann transfer to a 1000-km circular orbit. Calculate the total delta-v requirement.

{Ans.: 0.2624 km/s}

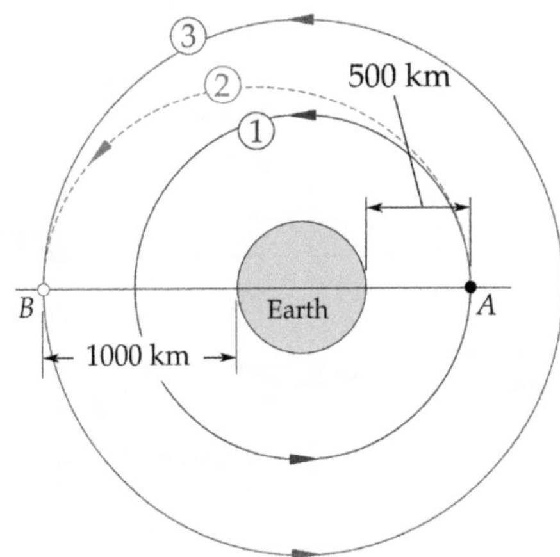

6.10 Assuming the orbits of earth and Mars are circular and coplanar, calculate

(a) The time required for a Hohmann transfer from earth orbit to Mars orbit.

(b) The initial position of Mars (α) in its orbit relative to earth for interception to occur.

Radius of earth orbit = $1.496(10^8)$ km.

Radius of Mars orbit = $2.279(10^8)$ km.

$\mu_{\text{Sun}} = 1.327(10^{11})\,\text{km}^3/\text{s}^2$.

{Ans.: (a) 259 days; (b) $\alpha = 44.3°$}

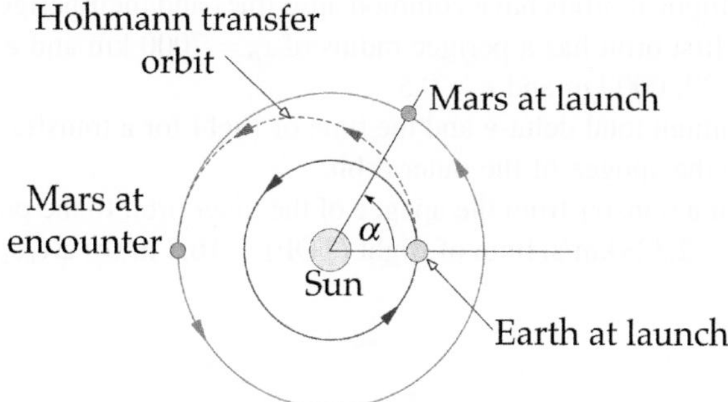

6.11 Calculate the total delta-v required for a Hohmann transfer from the smaller circular orbit to the larger one.
{Ans.: $0.394v_1$, where v_1 is the speed in orbit 1}

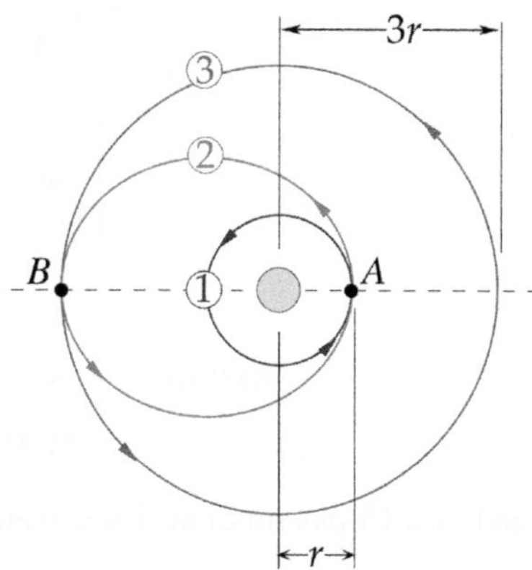

6.12 With a Δv_A of 1.500 km/s, a spacecraft in the circular 6700-km geocentric orbit 1 initiates a Hohmann transfer to the larger circular orbit 3. Calculate Δv_B at apogee of the Hohmann transfer ellipse 2.
{Ans.: $\Delta v_B = 1.874$ km/s}

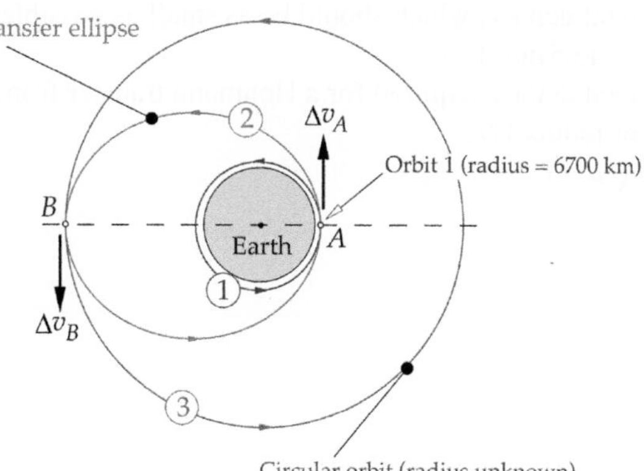

6.13 Two geocentric elliptical orbits have common apse lines and their perigees are on the same side of the earth. The first orbit has a perigee radius of $r_p = 7000$ km and $e = 0.3$, whereas for the second orbit $r_p = 32,000$ km and $e = 0.5$.

(a) Find the minimum total delta-v and the time of flight for a transfer from the perigee of the inner orbit to the apogee of the outer orbit.

(b) Do part (a) for a transfer from the apogee of the inner orbit to the perigee of the outer orbit.

{Ans.: (a) $\Delta v_{\text{total}} = 2.388\,\text{km/s}$, time of flight (TOF) = 16.2 h; (b) $\Delta v_{\text{total}} = 3.611\,\text{km/s}$, TOF = 4.66 h}

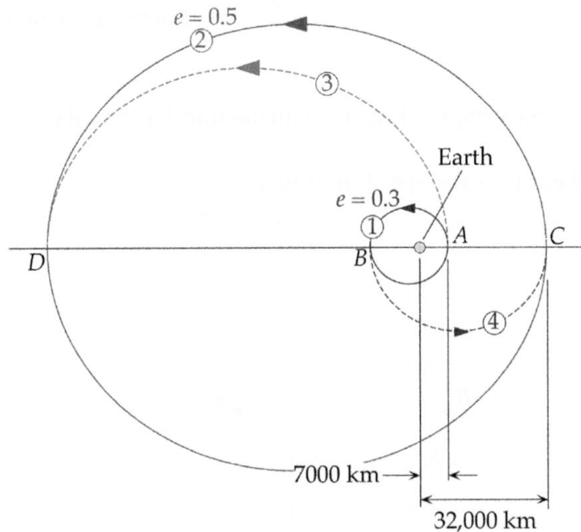

6.14 The space shuttle was launched on a 15-day mission. There were four orbits after injection, all of them at 39° inclination.

Orbit 1: 302 km by 296 km
Orbit 2 (day 11): 291 km by 259 km
Orbit 3 (day 12): 259 km circular
Orbit 4 (day 13): 255 km by 194 km

Calculate the total delta-v, which should be as small as possible, assuming Hohmann transfers.

{Ans.: $\Delta v_{\text{total}} = 43.5\,\text{m/s}$ }

6.15 Calculate the total delta-v required for a Hohmann transfer from a circular orbit of radius r to a circular orbit of radius $12r$.

{Ans.: $0.5342\sqrt{\mu/r}$}

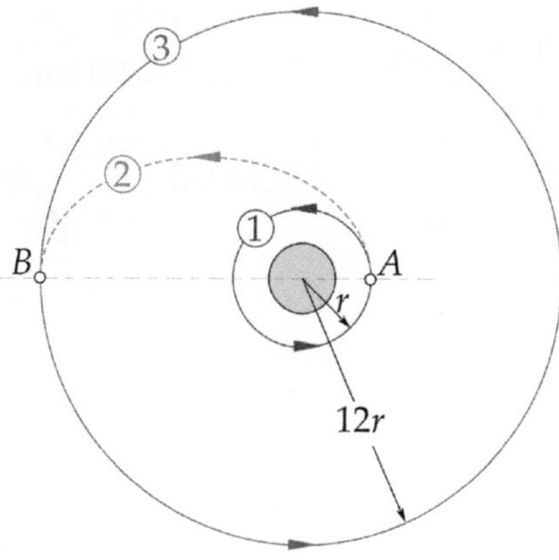

Section 6.4

6.16 A spacecraft in circular orbit 1 of radius r leaves for infinity on parabolic trajectory 2 and returns from infinity on a parabolic trajectory 3 to a circular orbit 4 of radius $12r$. Find the total delta-v required for this non-Hohmann orbit change maneuver.
{Ans.: $0.5338\sqrt{\mu/r}$}

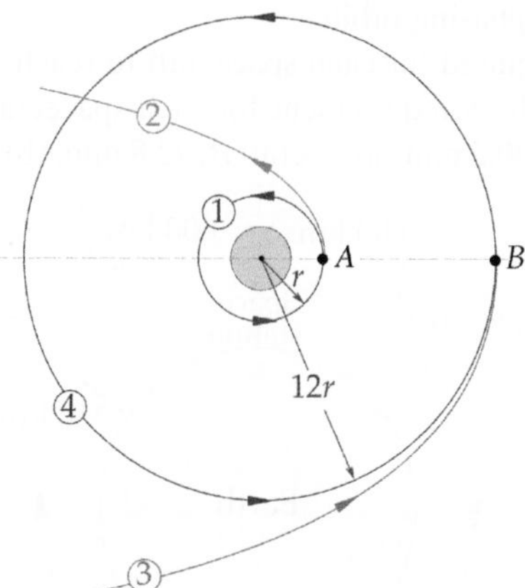

6.17 A spacecraft is in a 300-km circular earth orbit. Calculate

(a) The total delta-v required for the bielliptical transfer to the 3000-km-altitude coplanar circular orbit shown.

(b) The total transfer time.

{Ans.: (a) 2.039 km/s; (b) 2.86 h}

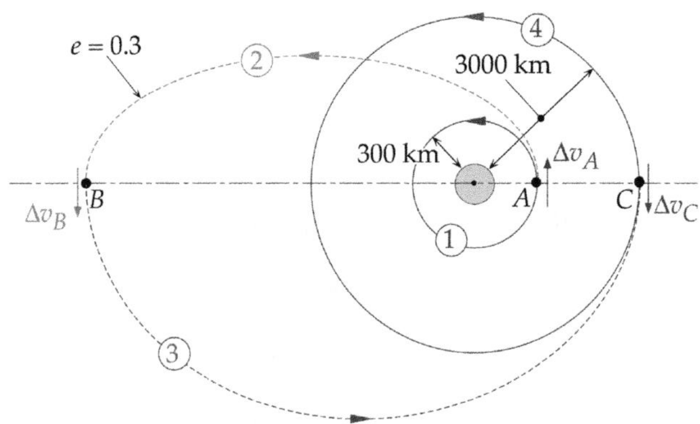

6.18 Verify Eq. (6.4).

Section 6.5

6.19 The space station and spacecraft A and B are all in the same circular earth orbit of 350 km altitude. Spacecraft A is 600 km behind the space station and spacecraft B is 600 km ahead of the space station. At the same instant, both spacecraft apply a $\Delta v_\perp$ so as to arrive at the space station in one revolution of their phasing orbits.

(a) Calculate the time required for each spacecraft to reach the space station.

(b) Calculate the total delta-v requirement for each spacecraft.

{Ans.: (a) Spacecraft A, 90.2 min; spacecraft B, 92.8 min; (b) $\Delta v_A = 73.9\,\text{m/s}$; $\Delta v_B = 71.5\,\text{m/s}$}

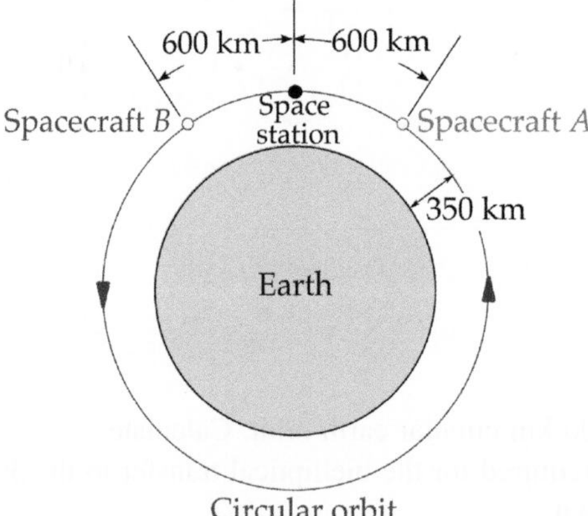

6.20 Satellites A and B are in the same circular orbit of radius r. B is 180° ahead of A. Calculate the semimajor axis of a phasing orbit in which A will rendezvous with B after just one revolution in the phasing orbit.

{Ans.: $a = 0.63r$}

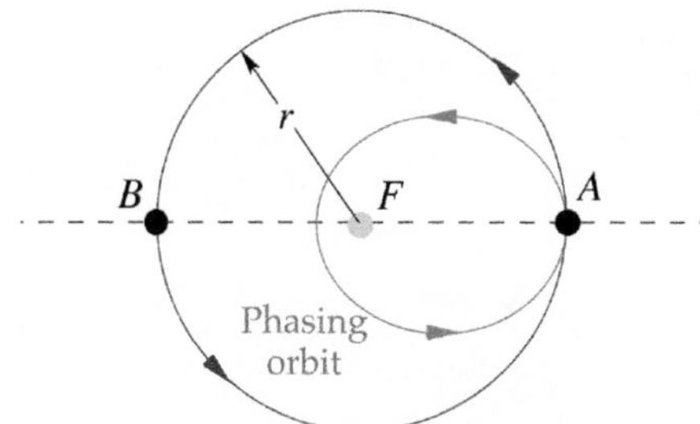

6.21 Two spacecraft are in the same elliptical earth orbit with perigee radius 8000 km and apogee radius 13,000 km. Spacecraft 1 is at perigee and spacecraft 2 is 30° ahead. Calculate the total delta-v required for spacecraft 1 to intercept and rendezvous with spacecraft 2 when spacecraft 2 has traveled 60°.
{Ans.: $\Delta v_{total} = 6.24\,\text{km/s}$}

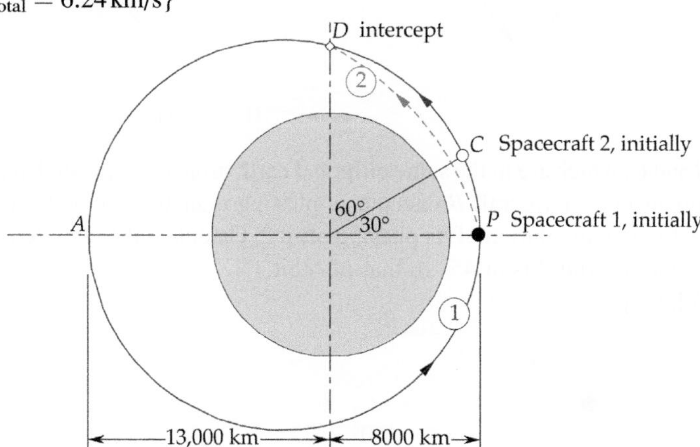

6.22 At the instant shown, spacecraft S_1 is at point A of circular orbit 1 and spacecraft S_2 is at point B of circular orbit 2. At that instant, S_1 executes a Hohmann transfer so as to arrive at point C of orbit 2. After arriving at C, S_1 immediately executes a phasing maneuver to rendezvous with S_2 after one revolution of its phasing orbit. What is the total delta-v requirement?
{Ans.: 2.159 km/s}

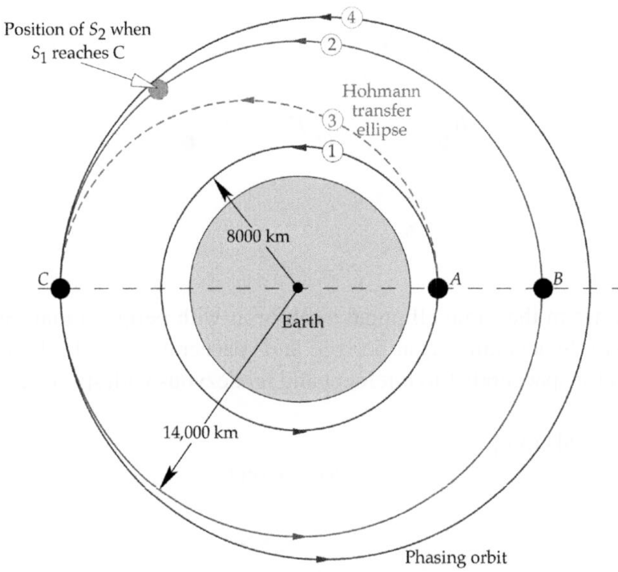

6.23 Spacecraft B and C, which are in the same elliptical earth orbit 1, are located at the true anomalies shown. At this instant, spacecraft B executes a phasing maneuver so as to rendezvous with spacecraft C after one revolution of its phasing orbit 2. Calculate the total delta-v required. Note that the apse line of orbit 2 is at 45° to that of orbit 1.
{Ans.: 3.405 km/s}

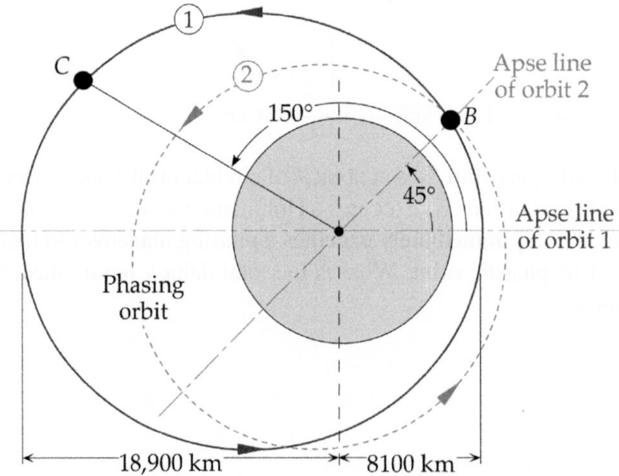

Section 6.6

6.24 **(a)** With a single delta-v maneuver, the earth orbit of a satellite is to be changed from a circle of radius 15,000 km to a collinear ellipse with perigee altitude of 500 km and apogee radius of 22,000 km. Calculate the magnitude of the required delta-v and the change in the flight path angle $\Delta\gamma$.

(b) What is the minimum total delta-v if the orbit change is accomplished instead by a Hohmann transfer?

{Ans.: (a) $\|\Delta\mathbf{v}\| = 2.77\,\text{km/s}$, $\Delta\gamma = 31.51°$; (b) $\Delta v_{\text{Hohmann}} = 1.362$ km/s}

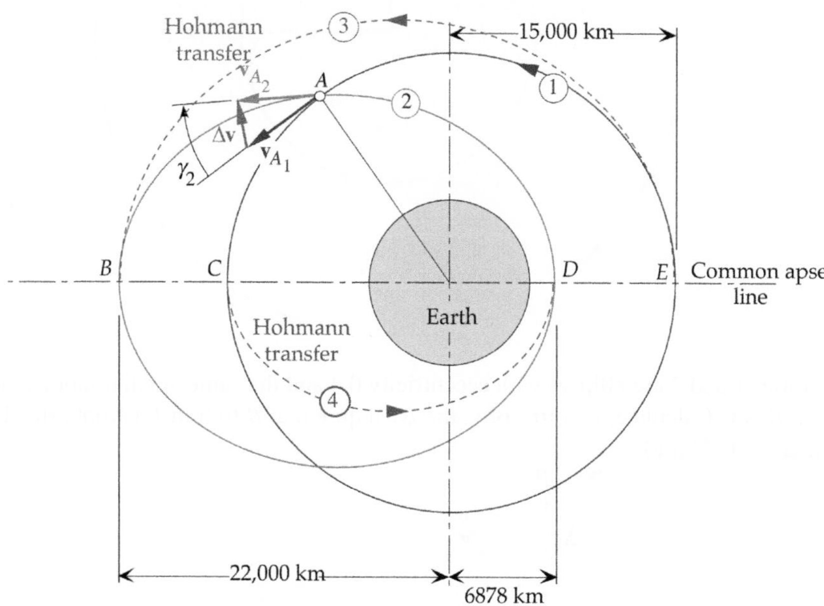

6.25 An earth satellite has a perigee altitude of 1270 km and a perigee speed of 9 km/s. It is required to change its orbital eccentricity to 0.4, without rotating the apse line, by a delta-v maneuver at $\theta = 100°$. Calculate the magnitude of the required $\Delta\mathbf{v}$ and the change in flight path angle $\Delta\gamma$.
{Ans.: $\|\Delta\mathbf{v}\| = 0.915\,\text{km/s}$; $\Delta\gamma = -8.18°$}

6.26 The velocities at points A and B on orbits 1, 2, and 3, respectively, are (relative to the perifocal frame)

$$\begin{aligned}
\mathbf{v}_A)_1 &= -3.7730\hat{\mathbf{p}} + 6.5351\hat{\mathbf{q}}\,(\text{km/s})\\
\mathbf{v}_A)_2 &= -3.2675\hat{\mathbf{p}} + 8.1749\hat{\mathbf{q}}\,(\text{km/s})\\
\mathbf{v}_B)_2 &= -3.2675\hat{\mathbf{p}} - 3.1442\hat{\mathbf{q}}\,(\text{km/s})\\
\mathbf{v}_B)_3 &= -2.6679\hat{\mathbf{p}} - 4.6210\hat{\mathbf{q}}\,(\text{km/s})
\end{aligned}$$

Calculate the total Δv for a transfer from orbit 1 to orbit 3 by means of orbit 2.
{Ans.: 3.310 km/s}

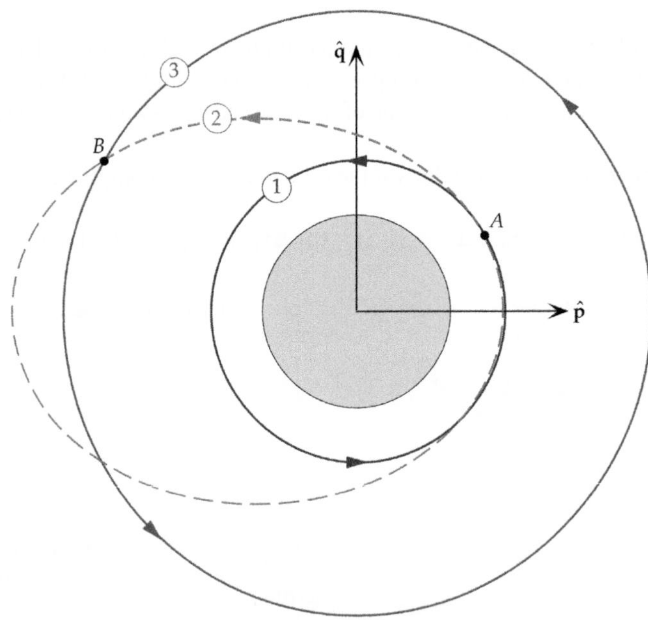

6.27 Trajectories 1 and 2 are ellipses with eccentricity 0.4 and the same angular momentum h. Their speed at B is v. Calculate, in terms of v, the Δv required at B to transfer from orbit 1 to orbit 2.
{Ans.: $\Delta v = 0.7428v$}

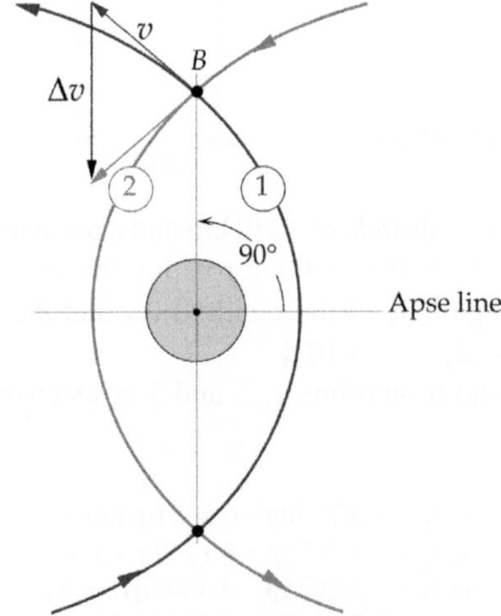

Section 6.7

6.28 A satellite is in a circular earth orbit of altitude 400 km. Determine the new perigee and apogee altitudes if the satellite's onboard rocket

(a) provides a delta-v in the tangential direction of 240 m/s.

(b) provides a delta-v in the radial (outward) direction of 240 m/s.

{Ans.: (a) $z_a = 1320\,\text{km}$, $z_p = 400\,\text{km}$; (b) $z_a = 619\,\text{km}$, $z_p = 194\,\text{km}$}

6.29 At point A on its earth orbit, the radius, speed, and flight path angle of a satellite are $r_A = 12,756\,\text{km}$, $v_A = 6.5992\,\text{km/s}$, and $\gamma_A = 20°$. At point B, at which the true anomaly is 150°, an impulsive maneuver causes $\Delta v_\perp = +0.75820\,\text{km/s}$ and $\Delta v_r = 0$.

(a) What is the time of flight from A to B?

(b) What is the rotation of the apse line as a result of this maneuver?

{Ans.: (a) 2.045 h; (b) 43.39° counterclockwise}

6.30 A satellite is in elliptical orbit 1. Calculate the true anomaly θ (relative to the apse line of orbit 1) of an impulsive maneuver that rotates the apse line an angle η counterclockwise but leaves the eccentricity and the angular momentum unchanged.

{Ans.: $\theta = \eta/2$}

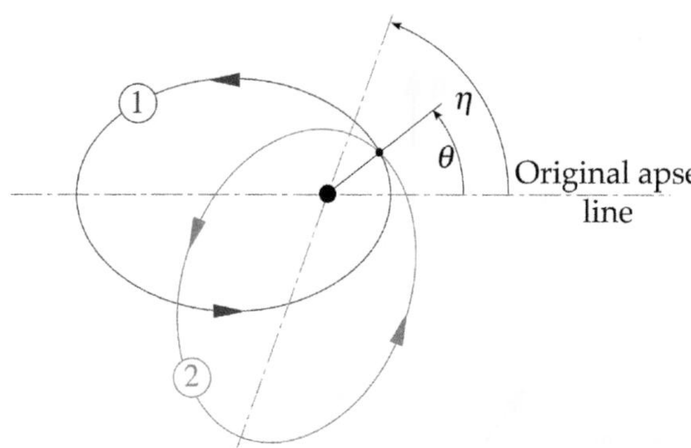

6.31 A satellite in orbit 1 undergoes a delta-v maneuver at perigee P_1 such that the new orbit 2 has the same eccentricity e, but its apse line is rotated 90° clockwise from the original one. Calculate the specific angular momentum of orbit 2 in terms of that of orbit 1 and the eccentricity e.

{Ans.: $h_2 = h_1/\sqrt{1+e}$}

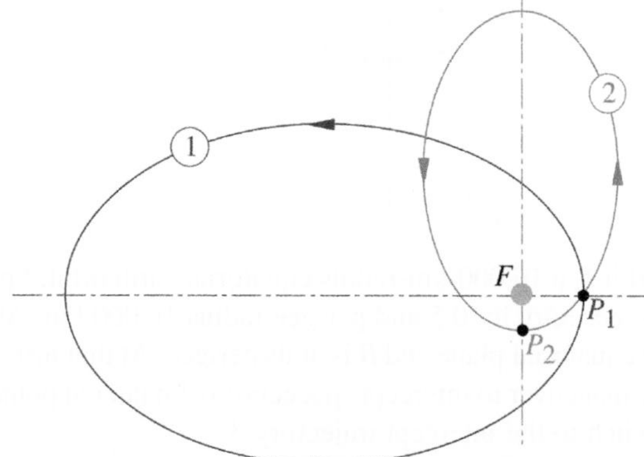

6.32 Calculate the delta-v required at A in orbit 1 for a single impulsive maneuver to rotate the apse line 180° counterclockwise (to become orbit 2), but keep the eccentricity e and the angular momentum h the same.
{Ans.: $\Delta v = 2\mu e/h$}

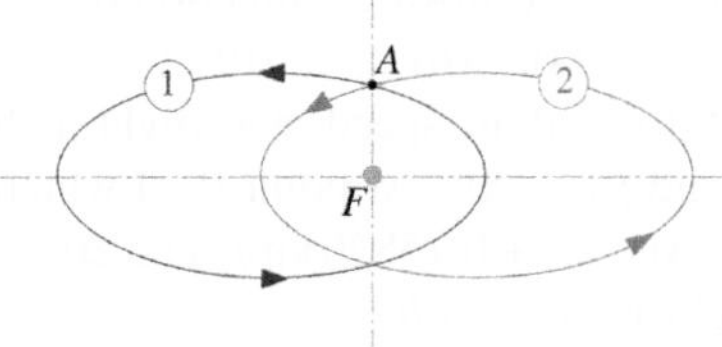

Section 6.8

6.33 Spacecraft A and B are in concentric, coplanar circular orbits 1 and 2, respectively. At the instant shown, spacecraft A executes an impulsive delta-v maneuver to embark on orbit 3 to intercept and rendezvous with spacecraft B in a time equal to the period of orbit 1. Calculate the total delta-v required.
{Ans.: 3.795 km/s}

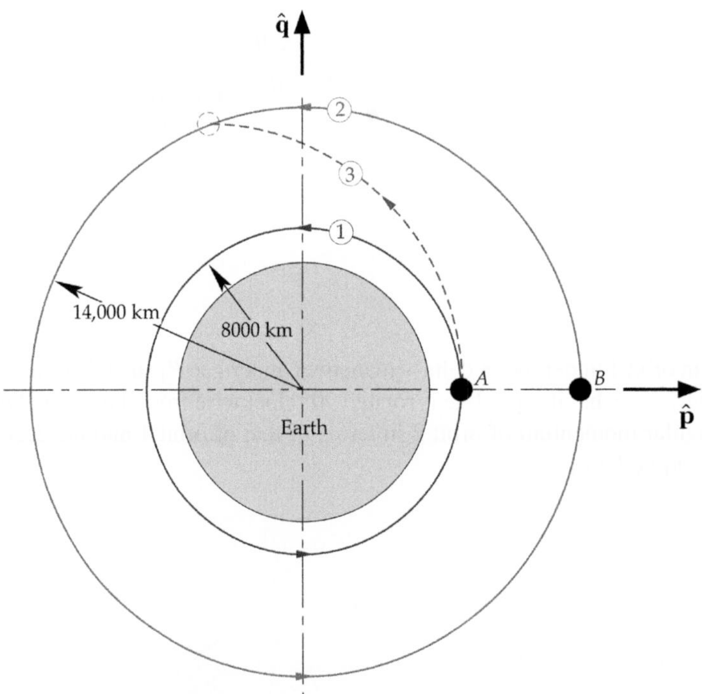

6.34 Spacecraft A is in orbit 1, a 10,000-km-radius equatorial earth orbit. Spacecraft B is in elliptical polar orbit 2, having eccentricity 0.5 and perigee radius 16,000 km. At the instant shown, both spacecraft are in the equatorial plane and B is at its perigee. At that instant, spacecraft A executes an impulsive delta-v maneuver to intercept spacecraft B 1 h later at point C. Calculate the delta-v required for A to switch to the intercept trajectory 3.
{Ans.: 8.117 km/s}

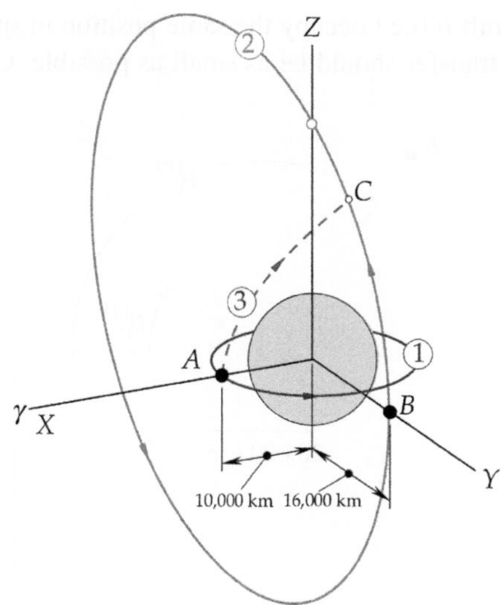

6.35 Spacecraft B and C are in the same elliptical orbit 1, characterized by a perigee radius of 7000 km and an apogee radius of 10,000 km. The spacecraft are in the positions shown when B executes an impulsive transfer to orbit 2 to catch and rendezvous with C when C arrives at apogee A. Find the total delta-v requirement.
{Ans.: 5.066 km/s}

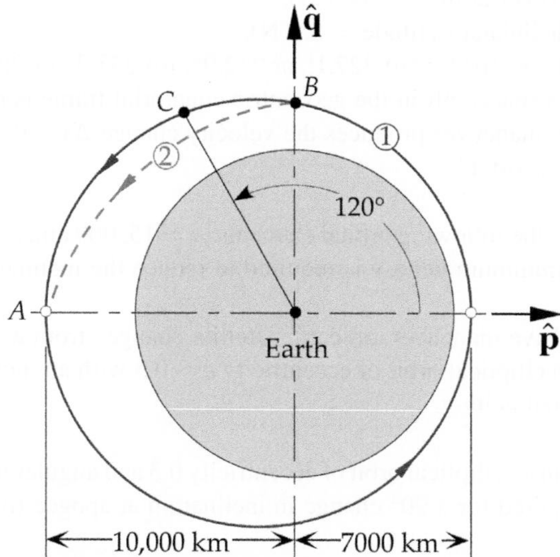

6.36 At time $t = 0$, manned spacecraft a and unmanned spacecraft b are at the positions shown in circular earth orbits 1 and 2, respectively. For assigned values of $\theta_0^{(a)}$ and $\theta_0^{(b)}$, design a series of impulsive maneuvers by means of which spacecraft a transfers from orbit 1 to orbit 2 so as to rendezvous with spacecraft b (i.e., occupy the same position in space). The total time and total delta-v required for the transfer should be as small as possible. Consider earth's gravity only.

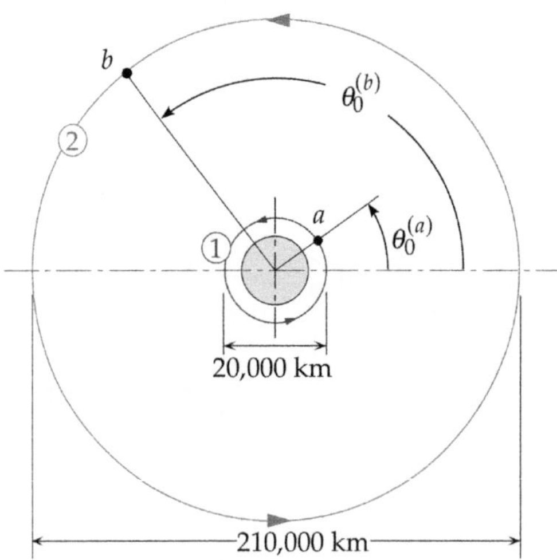

Section 6.9

6.37 What must the launch azimuth be if the satellite in Example 4.8 is launched from

(a) Kennedy Space Center (latitude = 28.5°N);
(b) Vandenburgh AFB (latitude = 34.5°N);
(c) Kourou, French Guiana (latitude = 5.5°N).

{Ans.: (a) 329.4° or 210.6°; (b) 327.1° or 212.9°; (c) 333.3° or 206.7°}

6.38 The state vector of a spacecraft in the geocentric equatorial frame is $\mathbf{r} = r\hat{\mathbf{I}}$ and $\mathbf{v} = v\hat{\mathbf{J}}$. At that instant an impulsive maneuver produces the velocity change $\Delta\mathbf{v} = 0.5v\hat{\mathbf{I}} + 0.5v\hat{\mathbf{K}}$. What is the inclination of the new orbit?

{Ans.: 26.57°}

6.39 An earth satellite has the following orbital elements: $a = 15,000$ km, $e = 0.5$, $\Omega = 45°$, $\omega = 30°$, and $i = 10°$. What minimum delta-v is required to reduce the inclination to zero?

{Ans.: 0.588 km/s}

6.40 With a single impulsive maneuver, an earth satellite changes from a 400-km circular orbit inclined at 60° to an elliptical orbit of eccentricity $e = 0.5$ with an inclination of 40°. Calculate the minimum required delta-v.

{Ans.: 3.41 km/s}

6.41 An earth satellite is in an elliptical orbit of eccentricity 0.3 and angular momentum 60,000 km^2/s. Find the delta-v required for a 90° change in inclination at apogee (no change in speed).

{Ans.: 6.58 km/s}

6.42 A spacecraft is in a circular, equatorial orbit (1) of radius r_0 about a planet. At point B it impulsively transfers to polar orbit (2), whose eccentricity is 0.25 and whose perigee is directly over the north pole. Calculate the minimum delta-v required at B for this maneuver.

{Ans.: $1.436\sqrt{\mu/r_0}$}

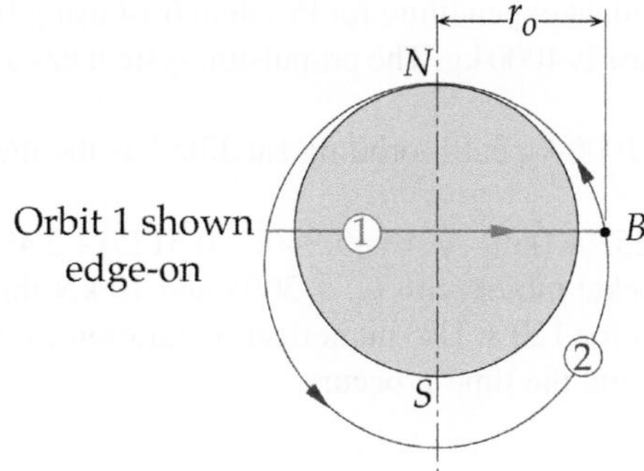

6.43 A spacecraft is in a circular, equatorial orbit (1) of radius r_0 and speed v_0 about an unknown planet ($\mu \neq 398,600\,\text{km}^3/\text{s}^2$). At point C it impulsively transfers to orbit (2), for which the ascending node is point C, the eccentricity is 0.1, the inclination is 30°, and the argument of periapsis is 60°. Calculate, in terms of v_0, the single delta-v required at C for this maneuver.
{Ans.: $\Delta v = 0.5313v_0$}

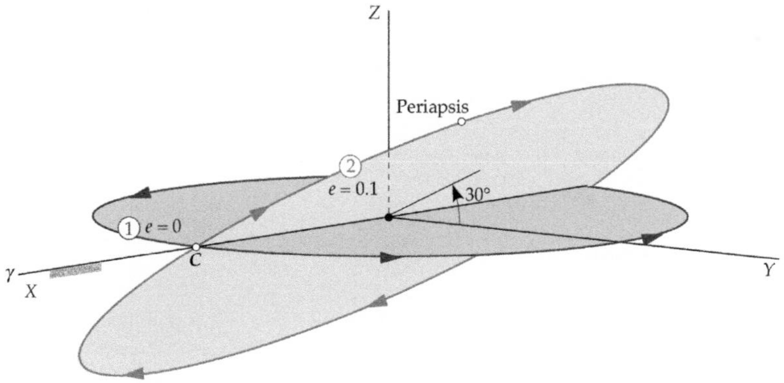

6.44 A spacecraft is in a 300-km circular parking orbit. It is desired to increase the altitude to 600 km and change the inclination by 20°. Find the total delta-v required if

(a) the plane change is made after insertion into the 600-km orbit (so that there are a total of three delta-v burns).

(b) the plane change and insertion into the 600-km orbit are accomplished simultaneously (so that the total number of delta-v burns is two).

(c) the plane change is made upon departing the lower orbit (so that the total number of delta-v burns is two).

{Ans.: (a) 2.793 km/s; (b) 2.696 km/s; (c) 2.783 km/s}

Section 6.10

6.45 Calculate the total propellant expenditure for Problem 6.3 using finite-time delta-v maneuvers. The initial spacecraft mass is 4000 kg. The propulsion system has a thrust of 30 kN and a specific impulse of 280 s.

6.46 Calculate the total propellant expenditure for Problem 6.14 using finite-time delta-v maneuvers. The initial spacecraft mass is 4000 kg. The propulsion system has a thrust of 30 kN and a specific impulse of 280 s.

6.47 At a given instant t_0, a 1000-kg earth-orbiting satellite has the inertial position and velocity vectors

$$\mathbf{r}_0 = 436\hat{\mathbf{i}} + 6083\hat{\mathbf{j}} + 2529\hat{\mathbf{k}}\,(\text{km}) \quad \mathbf{v}_0 = -7.340\hat{\mathbf{i}} - 0.5125\hat{\mathbf{j}} + 2.497\hat{\mathbf{k}}\,(\text{km/s})$$

About 89 min later a rocket motor with $I_{sp} = 300\,\text{s}$ and 10 kN thrust aligned with the velocity vector ignites and burns for 120 s. Use numerical integration to find the maximum altitude reached by the satellite and the time it occurs.

REFERENCES

Hohmann, W., 1925. The Attainability of Celestial Bodies. R. Oldenbourg, Munich (in German).
Wiesel, W.E., 2010. Spacecraft Dynamics, third ed. Aphelion Press, Beaver Creek, OH.

CHAPTER

RELATIVE MOTION AND RENDEZVOUS

7

7.1 INTRODUCTION

Up to now we have mostly referenced the motion of orbiting objects to a nonrotating coordinate system fixed to the center of attraction (e.g., the center of the earth). This platform served as an inertial frame of reference, in which Newton's second law can be written as

$$\mathbf{F}_{\text{net}} = m\mathbf{a}_{\text{absolute}}$$

An exception to this rule was the discussion of the restricted three-body problem at the end of Chapter 2, in which we made use of the relative motion equations developed in Chapter 1. In a rendezvous maneuver two orbiting vehicles observe one another from each of their own free-falling, rotating, clearly noninertial frames of reference. To base impulsive maneuvers on observations made from a moving platform requires transforming relative velocity and acceleration measurements into an inertial frame. Otherwise, the true thrusting forces cannot be sorted out from the fictitious "inertial forces" that appear in Newton's law when it is written incorrectly as

$$\mathbf{F}_{\text{net}} = m\mathbf{a}_{\text{rel}}$$

The purpose of this chapter is to use relative motion analysis to gain some familiarity with the problem of maneuvering one spacecraft relative to another, especially when they are in close proximity.

7.2 RELATIVE MOTION IN ORBIT

A rendezvous maneuver usually involves a target vehicle A, which is passive and nonmaneuvering, and a chase vehicle B, which is active and performs the maneuvers required to bring itself alongside the target. An obvious example was the Space Shuttle, the chaser, rendezvousing with the International Space Station, the target. The position vector of target A in the geocentric equatorial frame is $\mathbf{r}_A$. This radial is sometimes called the "r-bar." The moving frame of reference has its origin at the target, as illustrated in Fig. 7.1. The x axis is directed along the outward radial $\mathbf{r}_A$ to the target. Therefore, the unit vector $\hat{\mathbf{i}}$ along the moving x axis is

$$\hat{\mathbf{i}} = \frac{\mathbf{r}_A}{r_A} \tag{7.1}$$

Orbital Mechanics for Engineering Students. https://doi.org/10.1016/B978-0-12-824025-0.00007-6

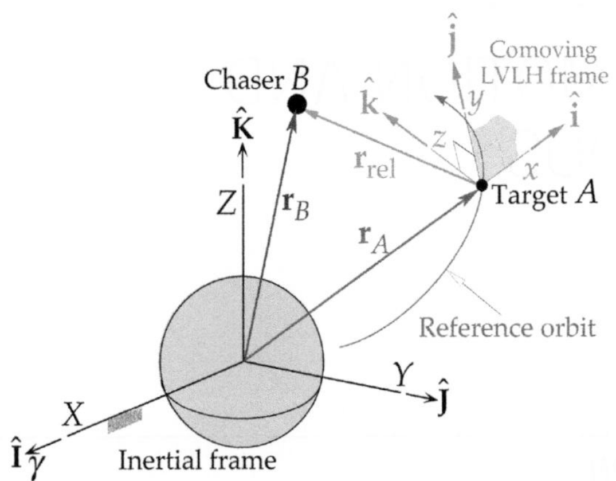

FIG. 7.1

Comoving reference frame attached to A, from which body B is observed.

The z axis is normal to the orbital plane of the target spacecraft and therefore lies in the direction of A's angular momentum vector. It follows that the unit vector along the z axis of the moving frame is given by

$$\hat{\mathbf{k}} = \frac{\mathbf{h}_A}{h_A} \tag{7.2}$$

The y axis is perpendicular to both $\hat{\mathbf{i}}$ and $\hat{\mathbf{k}}$ and points in the direction of the target satellite's local horizon. Therefore, both the x and y axes lie in the target's orbital plane, with the y unit vector completing a right triad; that is,

$$\hat{\mathbf{j}} = \hat{\mathbf{k}} \times \hat{\mathbf{i}} \tag{7.3}$$

We may refer to the comoving xyz frame defined here as a local vertical/local horizontal (LVLH) frame.

The position, velocity, and acceleration of B relative to A, measured in the comoving frame, are given by

$$\mathbf{r}_{\text{rel}} = x\hat{\mathbf{i}} + y\hat{\mathbf{j}} + z\hat{\mathbf{k}} \tag{7.4a}$$

$$\mathbf{v}_{\text{rel}} = \dot{x}\hat{\mathbf{i}} + \dot{y}\hat{\mathbf{j}} + \dot{z}\hat{\mathbf{k}} \tag{7.4b}$$

$$\mathbf{a}_{\text{rel}} = \ddot{x}\hat{\mathbf{i}} + \ddot{y}\hat{\mathbf{j}} + \ddot{z}\hat{\mathbf{k}} \tag{7.4c}$$

The angular velocity vector $\mathbf{\Omega}$ of the xyz axes attached to the target is just the angular velocity of the target's position vector. It is obtained with the aid of Eqs. (2.31) and (2.46) from the fact that

$$\mathbf{h}_A = \mathbf{r}_A \times \mathbf{v}_A = (r_A v_{A_\perp})\hat{\mathbf{k}} = (r_A{}^2 \Omega)\hat{\mathbf{k}} = r_A{}^2 \mathbf{\Omega}$$

from which we obtain

$$\mathbf{\Omega} = \frac{\mathbf{h}_A}{r_A{}^2} = \frac{\mathbf{r}_A \times \mathbf{v}_A}{r_A{}^2} \tag{7.5}$$

To find the angular acceleration vector $\dot{\boldsymbol{\Omega}}$ of the *xyz* frame we take the time derivative of $\boldsymbol{\Omega}$ in Eq. (7.5) and use the fact that the angular momentum $\mathbf{h}_A$ of the passive target is constant,

$$\dot{\boldsymbol{\Omega}} = \mathbf{h}_A \frac{d}{dt}\frac{1}{{r_A}^2} = -2\frac{\mathbf{h}_A}{{r_A}^3}\dot{r}_A$$

Recall from Eq. (2.35a) that $\dot{r}_A = \mathbf{v}_A \cdot \mathbf{r}_A / r_A$, so this may be written as

$$\dot{\boldsymbol{\Omega}} = -2\frac{\mathbf{v}_A \cdot \mathbf{r}_A}{{r_A}^4}\mathbf{h}_A = -2\frac{\mathbf{v}_A \cdot \mathbf{r}_A}{{r_A}^2}\boldsymbol{\Omega} \tag{7.6}$$

After first calculating

$$\mathbf{r}_{\text{rel}} = \mathbf{r}_B - \mathbf{r}_A \tag{7.7}$$

we use Eqs. (7.5) and (7.6) to determine the angular velocity and angular acceleration of the comoving frame, both of which are required in the relative velocity and acceleration formulas (Eqs. 1.66 and 1.70),

$$\mathbf{v}_{\text{rel}} = \mathbf{v}_B - \mathbf{v}_A - \boldsymbol{\Omega} \times \mathbf{r}_{\text{rel}} \tag{7.8}$$

$$\mathbf{a}_{\text{rel}} = \mathbf{a}_B - \mathbf{a}_A - \dot{\boldsymbol{\Omega}} \times \mathbf{r}_{\text{rel}} - \boldsymbol{\Omega} \times (\boldsymbol{\Omega} \times \mathbf{r}_{\text{rel}}) - 2\boldsymbol{\Omega} \times \mathbf{v}_{\text{rel}} \tag{7.9}$$

The vectors in Eqs. (7.7)–(7.9) are all referred to the inertial *XYZ* frame in Fig. 7.1. To find their components in the accelerating *xyz* frame at any instant we must first form the orthogonal direction cosine matrix $[\mathbf{Q}]_{Xx}$, as discussed in Section 4.5. The rows of this matrix comprise the direction cosines of each of the *xyz* axes with respect to the *XYZ* axes. That is, from Eqs. (7.1)–(7.3) we find

$$\begin{aligned} \hat{\mathbf{i}} &= l_x\hat{\mathbf{I}} + m_x\hat{\mathbf{J}} + n_x\hat{\mathbf{K}} \\ \hat{\mathbf{j}} &= l_y\hat{\mathbf{I}} + m_y\hat{\mathbf{J}} + n_y\hat{\mathbf{K}} \\ \hat{\mathbf{k}} &= l_z\hat{\mathbf{I}} + m_z\hat{\mathbf{J}} + n_z\hat{\mathbf{K}} \end{aligned} \tag{7.10}$$

where the *l*s, *m*s, and *n*s are the direction cosines. Then,

$$[\mathbf{Q}]_{Xx} = \begin{bmatrix} l_x & m_x & n_x \\ l_y & m_y & n_y \\ l_z & m_z & n_z \end{bmatrix} \begin{matrix} \leftarrow \text{components of } \hat{\mathbf{i}} \\ \leftarrow \text{components of } \hat{\mathbf{j}} \\ \leftarrow \text{components of } \hat{\mathbf{k}} \end{matrix} \tag{7.11}$$

The components of the relative position, velocity, and acceleration are computed as follows:

$$\{\mathbf{r}_{\text{rel}}\}_x = [\mathbf{Q}]_{Xx}\{\mathbf{r}_{\text{rel}}\}_X \tag{7.12a}$$

$$\{\mathbf{v}_{\text{rel}}\}_x = [\mathbf{Q}]_{Xx}\{\mathbf{v}_{\text{rel}}\}_X \tag{7.12b}$$

$$\{\mathbf{a}_{\text{rel}}\}_x = [\mathbf{Q}]_{Xx}\{\mathbf{a}_{\text{rel}}\}_X \tag{7.12c}$$

in which

$$\{\mathbf{r}_{\text{rel}}\}_X = \begin{Bmatrix} X_B - X_A \\ Y_B - Y_A \\ Z_B - Z_A \end{Bmatrix} \tag{7.13a}$$

$$\{\mathbf{v}_{\text{rel}}\}_X = \begin{Bmatrix} \dot{X}_B - \dot{X}_A + \Omega_Z(Y_B - Y_A) - \Omega_Y(Z_B - Z_A) \\ \dot{Y}_B - \dot{Y}_A - \Omega_Z(X_B - X_A) + \Omega_X(Z_B - Z_A) \\ \dot{Z}_B - \dot{Z}_A + \Omega_Y(X_B - X_A) - \Omega_X(Y_B - Y_A) \end{Bmatrix} \tag{7.13b}$$

$$\{\mathbf{a}_{\text{rel}}\}_X = \begin{Bmatrix} \ddot{X}_B - \ddot{X}_A + 2\Omega_Z(\dot{Y}_B - \dot{Y}_A) - 2\Omega_Y(\dot{Z}_B - \dot{Z}_A)\cdots \\ -({\Omega_Y}^2 + {\Omega_Z}^2)(X_B - X_A) + (\Omega_X\Omega_Y + a\Omega_Z)(Y_B - Y_A) + (\Omega_X\Omega_Z - a\Omega_Y)(Z_B - Z_A) \\ \ddot{Y}_B - \ddot{Y}_A - 2\Omega_Z(\dot{X}_B - \dot{X}_A) + 2\Omega_X(\dot{Z}_B - \dot{Z}_A)\cdots \\ +(\Omega_X\Omega_Y - a\Omega_Z)(X_B - X_A) - ({\Omega_X}^2 + {\Omega_Z}^2)(Y_B - Y_A) + (\Omega_Y\Omega_Z + a\Omega_X)(Z_B - Z_A) \\ \ddot{Z}_B - \ddot{Z}_A + 2\Omega_Y(\dot{X}_B - \dot{X}_A) - 2\Omega_X(\dot{Y}_B - \dot{Y}_A)\cdots \\ +(\Omega_X\Omega_Z + a\Omega_Y)(X_B - X_A) + (\Omega_Y\Omega_Z - a\Omega_X)(Y_B - Y_A) - ({\Omega_X}^2 + {\Omega_Y}^2)(Z_B - Z_A) \end{Bmatrix} \tag{7.13c}$$

The components of $\boldsymbol{\Omega}$ are obtained from Eq. (7.5), and $\dot{\boldsymbol{\Omega}} = a\boldsymbol{\Omega}$, where, according to Eq. (7.6), $a = -2\mathbf{v}_A \cdot \mathbf{r}_A / {r_A}^2$.

ALGORITHM 7.1

Given the state vectors $(\mathbf{r}_A, \mathbf{v}_A)$ of target spacecraft A and $(\mathbf{r}_B, \mathbf{v}_B)$ of chaser spacecraft B, find the position $\{\mathbf{r}_{\text{rel}}\}_x$, velocity $\{\mathbf{v}_{\text{rel}}\}_x$, and acceleration $\{\mathbf{a}_{\text{rel}}\}_x$ of B relative to A along the LVLH axes attached to A. See Appendix D.31 for an implementation of this procedure in MATLAB.

1. Calculate the angular momentum of A, $\mathbf{h}_A = \mathbf{r}_A \times \mathbf{v}_A$.
2. Calculate the unit vectors $\hat{\mathbf{i}}$, $\hat{\mathbf{j}}$, and $\hat{\mathbf{k}}$ of the comoving frame by means of Eqs. (7.1)–(7.3).
3. Calculate the orthogonal direction cosine matrix $[\mathbf{Q}]_{Xx}$ using Eq. (7.11).
4. Calculate $\boldsymbol{\Omega}$ and $\dot{\boldsymbol{\Omega}}$ from Eqs. (7.5) and (7.6).
5. Calculate the absolute accelerations of A and B using Eq. (2.22).

$$\mathbf{a}_A = -\frac{\mu}{{r_A}^3}\mathbf{r}_A \quad \mathbf{a}_B = -\frac{\mu}{{r_B}^3}\mathbf{r}_B$$

6. Calculate $\mathbf{r}_{\text{rel}}$ using Eq. (7.7).
7. Calculate $\mathbf{v}_{\text{rel}}$ using Eq. (7.8).
8. Calculate $\mathbf{a}_{\text{rel}}$ using Eq. (7.9).
9. Calculate $\{\mathbf{r}_{\text{rel}}\}_x$, $\{\mathbf{v}_{\text{rel}}\}_x$, and $\{\mathbf{a}_{\text{rel}}\}_x$ using Eqs. (7.12).

EXAMPLE 7.1

In Fig. 7.2, spacecraft A is in an elliptical earth orbit having the following parameters:

$$h = 52{,}059\ \text{km}^2/\text{s} \quad e = 0.025724 \quad i = 60^\circ \quad \Omega = 40^\circ \quad \omega = 30^\circ \quad \theta = 40^\circ \tag{a}$$

Spacecraft B is likewise in an earth orbit with these parameters:

$$h = 52{,}362\ \text{km}^2/\text{s} \quad e = 0.0072696 \quad i = 50^\circ \quad \Omega = 40^\circ \quad \omega = 120^\circ \quad \theta = 40^\circ \tag{b}$$

Calculate the position $\mathbf{r}_{\text{rel}})_x$, velocity $\mathbf{v}_{\text{rel}})_x$, and acceleration $\mathbf{a}_{\text{rel}})_x$ of spacecraft B relative to spacecraft A, measured along the xyz axes of the comoving coordinate system of spacecraft A, as defined in Fig. 7.1.

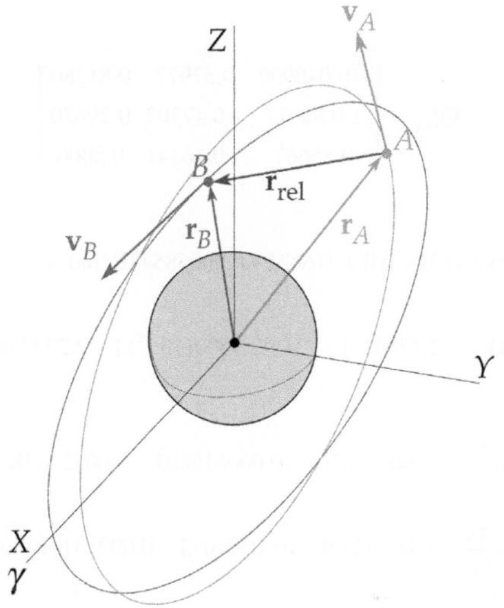

FIG. 7.2

Spacecraft *A* and *B* in slightly different orbits.

Solution

From the orbital elements in Eqs. (a) and (b) we can use Algorithm 4.5 to find the position and the velocity of both spacecraft relative to the geocentric equatorial reference frame. Omitting those familiar calculations here, the reader can verify that for spacecraft *A*

$$\mathbf{r}_A = -266.77\hat{\mathbf{I}} + 3865.8\hat{\mathbf{J}} + 5426.2\hat{\mathbf{K}}\,(\text{km}) \quad (r_A = 6667.8\,\text{km}) \tag{c}$$

$$\mathbf{v}_A = -6.4836\hat{\mathbf{I}} - 3.6198\hat{\mathbf{J}} + 2.4156\hat{\mathbf{K}}\,(\text{km/s}) \quad (v_A = 7.8087\,\text{km/s}) \tag{d}$$

and for spacecraft *B*

$$\mathbf{r}_B = -5890.7\hat{\mathbf{I}} - 2979.8\hat{\mathbf{J}} + 1792.2\hat{\mathbf{K}}\,(\text{km}) \quad (r_B = 6840.4\,\text{km}) \tag{e}$$

$$\mathbf{v}_B = 0.93583\hat{\mathbf{I}} - 5.2403\hat{\mathbf{J}} - 5.5009\hat{\mathbf{K}}\,(\text{km/s}) \quad (v_B = 7.6548\,\text{km/s}) \tag{f}$$

Having found the state vectors we can proceed with Algorithm 7.1.

Step 1:

$$\mathbf{h}_A = \mathbf{r}_A \times \mathbf{v}_A = \begin{vmatrix} \hat{\mathbf{I}} & \hat{\mathbf{J}} & \hat{\mathbf{K}} \\ -266.77 & 3865.8 & 5426.2 \\ -6.4836 & -3.6198 & 2.4156 \end{vmatrix}$$

$$= 28{,}980\hat{\mathbf{I}} - 34{,}537\hat{\mathbf{J}} + 26{,}029\hat{\mathbf{K}}\left(\text{km}^2/\text{s}\right)$$

$$\left(h_A = 52{,}059\,\text{km}^2/\text{s}\right)$$

Step 2:

$$\hat{\mathbf{i}} = \frac{\mathbf{r}_A}{r_A} = -0.040009\hat{\mathbf{I}} + 0.57977\hat{\mathbf{J}} + 0.81380\hat{\mathbf{K}}$$

$$\hat{\mathbf{k}} = \frac{\mathbf{h}_A}{h_A} = 0.55667\hat{\mathbf{I}} - 0.66341\hat{\mathbf{J}} + 0.5000\hat{\mathbf{K}}$$

$$\hat{\mathbf{j}} = \begin{vmatrix} \hat{\mathbf{I}} & \hat{\mathbf{J}} & \hat{\mathbf{K}} \\ 0.55667 & -0.66341 & 0.5000 \\ -0.040008 & 0.57977 & 0.81380 \end{vmatrix} = -0.82977\hat{\mathbf{I}} - 0.47302\hat{\mathbf{J}} + 0.29620\hat{\mathbf{K}}$$

Step 3:

$$[\mathbf{Q}]_{Xx} = \begin{bmatrix} -0.040009 & 0.57977 & 0.81380 \\ -0.82977 & -0.47302 & 0.29620 \\ 0.55667 & -0.66341 & 0.5000 \end{bmatrix}$$

Step 4:

$$\boldsymbol{\Omega} = \frac{\mathbf{h}_A}{r_A{}^2} = 0.00065183\hat{\mathbf{I}} - 0.00077682\hat{\mathbf{J}} + 0.00058547\hat{\mathbf{K}}\,(\text{rad/s})$$

$$\dot{\boldsymbol{\Omega}} = -2\frac{\mathbf{v}_A \cdot \mathbf{r}_A}{r_A{}^2}\boldsymbol{\Omega} = -2.47533\left(10^{-8}\right)\hat{\mathbf{I}} + 2.9500\left(10^{-8}\right)\hat{\mathbf{J}} - 2.2233\left(10^{-8}\right)\hat{\mathbf{K}}\left(\text{rad/s}^2\right)$$

Step 5:

$$\mathbf{a}_A = -\mu\frac{\mathbf{r}_A}{r_A{}^3} = 0.00035870\hat{\mathbf{I}} - 0.00051980\hat{\mathbf{J}} - 0.0072962\hat{\mathbf{K}}\left(\text{km/s}^2\right)$$

$$\mathbf{a}_B = -\mu\frac{\mathbf{r}_B}{r_B{}^3} = 0.0073359\hat{\mathbf{I}} - 0.0037108\hat{\mathbf{J}} - 0.0022319\hat{\mathbf{K}}\left(\text{km/s}^2\right)$$

Step 6:

$$\mathbf{r}_{\text{rel}} = \mathbf{r}_B - \mathbf{r}_A = -5623.9\hat{\mathbf{I}} - 6845.5\hat{\mathbf{J}} - 3634.0\hat{\mathbf{K}}\,(\text{km})$$

Step 7:

$$\begin{aligned} \mathbf{v}_{\text{rel}} &= \mathbf{v}_B - \mathbf{v}_A - \boldsymbol{\Omega} \times \mathbf{r}_{\text{rel}} \\ &= \left(0.93583\hat{\mathbf{I}} - 5.2403\hat{\mathbf{J}} - 5.5009\hat{\mathbf{K}}\right) - \left(-6.4836\hat{\mathbf{I}} - 3.6198\hat{\mathbf{J}} + 2.4156\hat{\mathbf{K}}\right) \\ &\quad - \begin{vmatrix} \hat{\mathbf{I}} & \hat{\mathbf{J}} & \hat{\mathbf{K}} \\ 0.00065183 & -0.00077682 & 0.00058547 \\ -5623.9 & -6845.5 & -3634.0 \end{vmatrix} \\ \mathbf{v}_{\text{rel}} &= 0.58855\hat{\mathbf{I}} - 0.69663\hat{\mathbf{J}} + 0.91436\hat{\mathbf{K}}\,(\text{km/s}) \end{aligned}$$

Step 8:

$$\begin{aligned} \mathbf{a}_{\text{rel}} &= \mathbf{a}_B - \mathbf{a}_A - \dot{\boldsymbol{\Omega}} \times \mathbf{r}_{\text{rel}} - \boldsymbol{\Omega} \times (\boldsymbol{\Omega} \times \mathbf{r}_{\text{rel}}) - 2\boldsymbol{\Omega} \times \mathbf{v}_{\text{rel}} \\ &= \left(0.0073359\hat{\mathbf{I}} + 0.0037108\hat{\mathbf{J}} - 0.0022319\hat{\mathbf{K}}\right) - \left(0.00035870\hat{\mathbf{I}} + 0.0051980\hat{\mathbf{J}} - 0.0072962\mathbf{K}\right) \\ &\quad - \begin{vmatrix} \hat{\mathbf{I}} & \hat{\mathbf{J}} & \hat{\mathbf{K}} \\ -2.4753\left(10^{-8}\right) & 2.9500\left(10^{-8}\right) & -2.2233\left(10^{-8}\right) \\ -5623.9 & -6845.5 & -3634.0 \end{vmatrix} \\ &\quad - \left(0.00065183\hat{\mathbf{I}} - 0.00077682\hat{\mathbf{J}} + 0.0005854\hat{\mathbf{K}}\right) \times \begin{vmatrix} \hat{\mathbf{I}} & \hat{\mathbf{J}} & \hat{\mathbf{K}} \\ 0.00065183 & -0.00077682 & 0.00058547 \\ -5623.9 & -6845.5 & -3634.0 \end{vmatrix} \\ &\quad -2\begin{vmatrix} \hat{\mathbf{I}} & \hat{\mathbf{J}} & \hat{\mathbf{K}} \\ 0.00065183 & -0.00077682 & 0.00058547 \\ 0.58855 & -0.69663 & 0.91436 \end{vmatrix} \\ \mathbf{a}_{\text{rel}} &= 0.00044050\hat{\mathbf{I}} - 0.00037900\hat{\mathbf{J}} + 0.00001858\hat{\mathbf{K}}\left(\text{km/s}^2\right) \end{aligned}$$

Step 9:

$$\mathbf{r}_{\text{rel}}\big)_x = \begin{bmatrix} -0.040008 & 0.57977 & 0.81380 \\ -0.82977 & -0.47302 & 0.29620 \\ 0.55667 & -0.66341 & 0.5000 \end{bmatrix} \begin{Bmatrix} -5623.9 \\ -6845.5 \\ -3634.0 \end{Bmatrix}$$

$$\boxed{\mathbf{r}_{\text{rel}})_x = \begin{Bmatrix} -6701.2 \\ 6828.3 \\ -406.26 \end{Bmatrix} \text{(km)}}$$

$$\mathbf{v}_{\text{rel}}\big)_x = \begin{bmatrix} -0.040008 & 0.57977 & 0.81380 \\ -0.82977 & -0.47302 & 0.29620 \\ 0.55667 & -0.66341 & 0.5000 \end{bmatrix} \begin{Bmatrix} 0.58855 \\ -0.69663 \\ 0.91436 \end{Bmatrix}$$

$$\boxed{\mathbf{v}_{\text{rel}})_x = \begin{Bmatrix} 0.31667 \\ 0.11199 \\ 1.2470 \end{Bmatrix} \text{(km/ s)}}$$

$$\mathbf{a}_{\text{rel}}\big)_x = \begin{bmatrix} -0.040008 & 0.57977 & 0.81380 \\ -0.82977 & -0.47302 & 0.29620 \\ 0.55667 & -0.66341 & 0.5000 \end{bmatrix} \begin{Bmatrix} 0.00044050 \\ -0.00037900 \\ 0.000018581 \end{Bmatrix}$$

$$\boxed{\mathbf{a}_{\text{rel}})_x = \begin{Bmatrix} -0.00022222 \\ -0.00018074 \\ 0.00050593 \end{Bmatrix} \text{(km/ s}^2\text{)}}$$

See Appendix D.31 for the MATLAB solution to this problem.

The motion of one spacecraft relative to another in orbit may be hard to visualize at first. Fig. 7.3 is offered as an assist. Orbit 1 is circular, and orbit 2 is elliptical with an eccentricity of 0.125. Both coplanar orbits were chosen to have the same semimajor axis length, so they both have the same period. A comoving frame is shown attached to the observers *A* in circular orbit 1. At the initial time *I* the spacecraft *B* in elliptical orbit 2 is directly below the observers. In other words, *A* must draw an arrow in the negative local *x* direction to determine the position vector of *B* in the lower orbit. The figure shows eight different instants (*I*, *II*, *III*, ..., *VIII*), equally spaced around the circular orbit, at which observers *A* construct the position vector pointing from them toward *B* in the elliptical orbit. Of course, *A*'s frame is rotating, because its *x* axis must always be directed away from the earth. Observers *A* cannot sense this rotation and record the set of observations in their (to them) fixed *xy* coordinate system, as shown at the right in the figure. Coasting at a uniform speed along this circular orbit, observers *A* see the other vehicle orbiting them clockwise in a sort of bean-shaped path. The distance between the two spacecraft in this case never becomes so great that the earth intervenes.

If observers *A* declared theirs to be an inertial frame of reference, they would be faced with the task of explaining the physical origin of the force holding *B* in its bean-shaped orbit. Of course, there is no such force. The apparent path is due to the actual, combined motion of both spacecraft in their free fall around the earth. When *B* is below *A* (having a negative *x* coordinate), conservation of angular momentum demands that *B* move faster than *A*, thereby speeding up in *A*'s positive *y* direction until the orbits cross ($x = 0$) between *III* and *IV*. When *B*'s *x* coordinate becomes positive (i.e., *B* is above *A*) the laws of momentum dictate that *B* slow down, which it does, progressing in *A*'s negative *y* direction until the

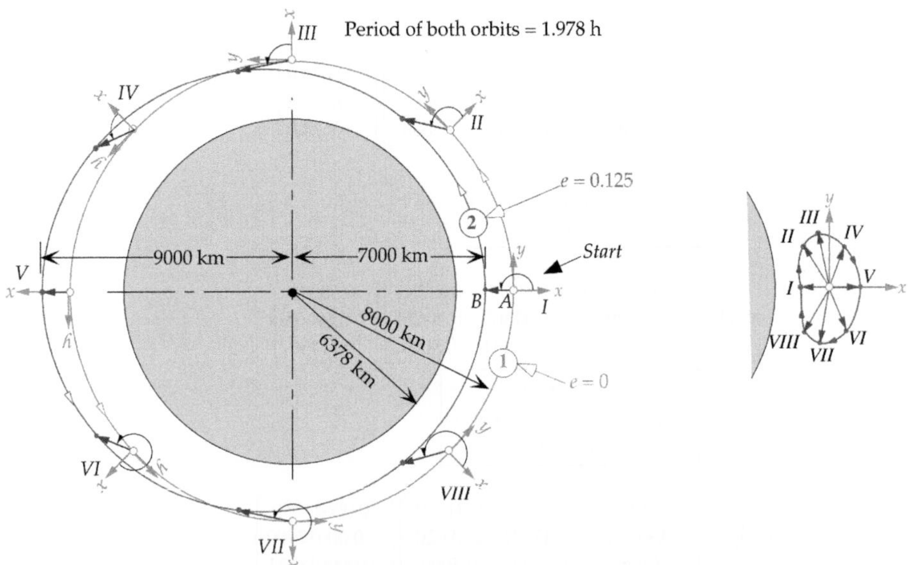

As viewed in the inertial frame. As viewed from the LVLH frame in reference orbit **1**.

FIG. 7.3

Spacecraft B in elliptical orbit 2 appears to orbit the observer A in circular orbit 1.

next crossing of the orbits between *VI* and *VII*. B then falls below A and begins to pick up speed. The process repeats over and over again. From inertial space the process is the motion of two satellites on intersecting orbits, appearing not at all like the orbiting motion seen by the moving observers A.

EXAMPLE 7.2

Plot the motion of spacecraft B relative to spacecraft A in Example 7.1.

Solution

In Example 7.1 we found $\mathbf{r}_{\text{rel}})_x$ at a single time. To plot the path of B relative to A we must find $\mathbf{r}_{\text{rel}})_x$ at a large number of times, so that when we "connect the dots" in three-dimensional space a smooth curve results. Let us outline an algorithm and implement it in MATLAB.

1. Given the orbital elements of spacecraft A and B, calculate their state vectors $(\mathbf{r}_{A_0}, \mathbf{v}_{A_0})$ and $(\mathbf{r}_{B_0}, \mathbf{v}_{B_0})$ at the initial time t_0 using Algorithm 4.5 (as we did in Example 7.1).
2. Calculate the period T_A of A's orbit from Eq. (2.82). (For the data of Example 7.1, $T_A = 5585$ s.)
3. Let the final time t_f for the plot be $t_0 + mT_A$, where m is an arbitrary integer.
4. Let n be the number of points to be plotted, so that the time step is $\Delta t = (t_f - t)/n$.
5. At time $t \geq t_0$:
   a. Calculate the state vectors $(\mathbf{r}_A, \mathbf{v}_A)$ and $(\mathbf{r}_B, \mathbf{v}_B)$ using Algorithm 3.4.
   b. Calculate $\mathbf{r}_{\text{rel}})_x$ using Algorithm 7.1.
   c. Plot the point $(x_{\text{rel}}, y_{\text{rel}}, z_{\text{rel}})$.
6. Let $t \leftarrow t + \Delta t$ and repeat Step 5 until $t = t_f$.

This algorithm is implemented in the MATLAB script *Example_7_02.m* listed in Appendix D.32. The resulting plot of the relative motion for a time interval of 60 periods of spacecraft A is shown in Fig. 7.4. The arrow drawn from A to B is the initial position vector $\mathbf{r}_{\text{rel}})_x$ found in Example 7.1. As can be seen the trajectory of B is a looping, clockwise motion around a circular path about 14,000 km in diameter. The closest approach of B to A is 105.5 km at an elapsed time of 25.75 h.

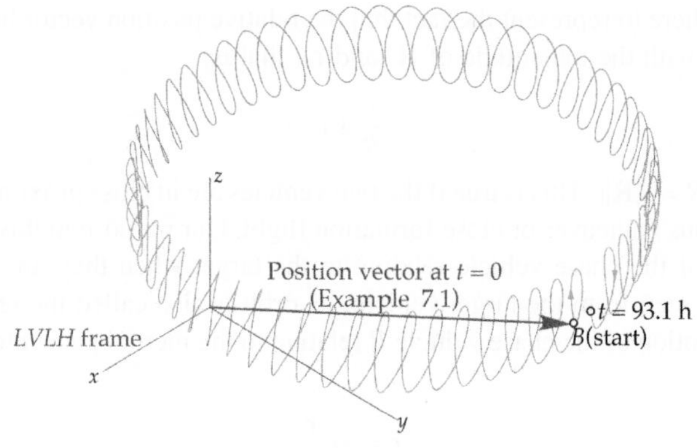

FIG. 7.4

Trajectory of spacecraft B relative to spacecraft A for the data in Example 7.1. The total time is 60 periods of A's orbit (93.1 h).

7.3 LINEARIZATION OF THE EQUATIONS OF RELATIVE MOTION IN ORBIT

Fig. 7.5, similar to Fig. 7.1, shows two spacecraft in earth orbit. Let the inertial position vector of the target vehicle A be denoted $\mathbf{R}$ and that of the chase vehicle B be denoted $\mathbf{r}$. The position vector of the chase vehicle relative to the target is $\delta\mathbf{r}$, so that

$$\mathbf{r} = \mathbf{R} + \delta\mathbf{r} \tag{7.14}$$

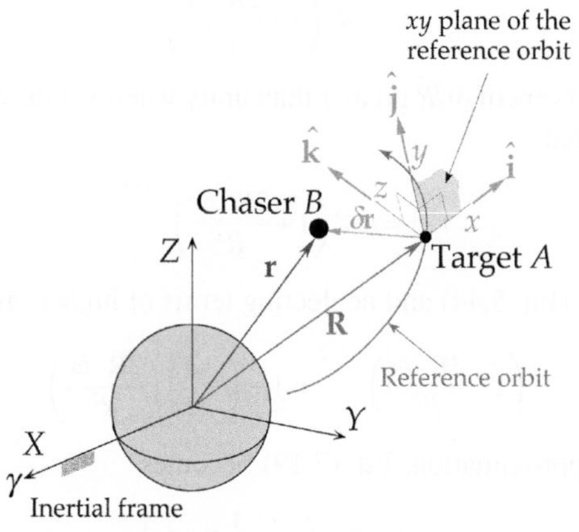

FIG. 7.5

Position of chaser B relative to the target A.

The symbol δ is used here to represent the fact that the relative position vector has a magnitude that is very small compared with the magnitude of $\mathbf{R}$ (and $\mathbf{r}$); that is,

$$\frac{\delta r}{R} \ll 1 \tag{7.15}$$

where $\delta r = \|\delta\mathbf{r}\|$ and $R = \|\mathbf{R}\|$. This is true if the two vehicles are in close proximity to each other, as is the case in a rendezvous maneuver or close formation flight. Our purpose in this section is to seek the equations of motion of the chase vehicle relative to the target when they are close together. Since the relative motion is seen from the target vehicle, its orbit is also called the reference orbit.

The equation of motion of the chase vehicle B relative to the inertial geocentric equatorial frame is Eq. (2.22),

$$\ddot{\mathbf{r}} = -\mu\frac{\mathbf{r}}{r^3} \tag{7.16}$$

where $r = \|\mathbf{r}\|$. Substituting Eq. (7.14) into Eq. (7.16) and writing $\delta\ddot{\mathbf{r}} = \left(\mathrm{d}^2/\mathrm{d}t^2\right)\delta\mathbf{r}$ yields the equation of motion of the chaser relative to the target,

$$\delta\ddot{\mathbf{r}} = -\ddot{\mathbf{R}} - \mu\frac{\mathbf{R}+\delta\mathbf{r}}{r^3} \quad (\text{where } r = \|\mathbf{R}+\delta\mathbf{r}\|) \tag{7.17}$$

We will simplify this equation by making use of the fact that $\|\delta\mathbf{r}\|$ is very small, as expressed in Eq. (7.15).

First, note that

$$r^2 = \mathbf{r}\cdot\mathbf{r} = (\mathbf{R}+\delta\mathbf{r})\cdot(\mathbf{R}+\delta\mathbf{r}) = \mathbf{R}\cdot\mathbf{R} + 2\mathbf{R}\cdot\delta\mathbf{r} + \delta\mathbf{r}\cdot\delta\mathbf{r}$$

Since $\mathbf{R}\cdot\mathbf{R} = R^2$ and $\delta\mathbf{r}\cdot\delta\mathbf{r} = \delta r^2$, we can factor out R^2 on the right to obtain

$$r^2 = R^2\left[1 + \frac{2\mathbf{R}\cdot\delta\mathbf{r}}{R^2} + \left(\frac{\delta r}{R}\right)^2\right]$$

By virtue of Eq. (7.15) we can neglect the last term in the brackets, so that

$$r^2 = R^2\left(1 + \frac{2\mathbf{R}\cdot\delta\mathbf{r}}{R^2}\right) \tag{7.18}$$

In fact, we will neglect all powers of $\delta r/R$ greater than unity wherever they appear. Since $r^{-3} = (r^2)^{-3/2}$ it follows from Eq. (7.18) that

$$r^{-3} = R^{-3}\left(1 + \frac{2\mathbf{R}\cdot\delta\mathbf{r}}{R^2}\right)^{-3/2} \tag{7.19}$$

Using the binomial theorem (Eq. 5.44) and neglecting terms of higher order than 1 in $\delta r/R$, we obtain

$$\left(1 + \frac{2\mathbf{R}\cdot\delta\mathbf{r}}{R^2}\right)^{-3/2} = 1 + \left(-\frac{3}{2}\right)\left(\frac{2\mathbf{R}\cdot\delta\mathbf{r}}{R^2}\right)$$

Therefore, to our level of approximation, Eq. (7.19) becomes

$$r^{-3} = R^{-3}\left(1 - \frac{3}{R^2}\mathbf{R}\cdot\delta\mathbf{r}\right)$$

which can be written as

$$\frac{1}{r^3}=\frac{1}{R^3}-\frac{3}{R^5}\mathbf{R}\cdot\delta\mathbf{r} \tag{7.20}$$

Substituting Eq. (7.20) into Eq. (7.17) (the equation of motion), we get

$$\begin{aligned}\delta\ddot{\mathbf{r}}&=-\ddot{\mathbf{R}}-\mu\left(\frac{1}{R^3}-\frac{3}{R^5}\mathbf{R}\cdot\delta\mathbf{r}\right)(\mathbf{R}+\delta\mathbf{r})\\&=-\ddot{\mathbf{R}}-\mu\left[\frac{\mathbf{R}+\delta\mathbf{r}}{R^3}-\frac{3}{R^5}(\mathbf{R}\cdot\delta\mathbf{r})(\mathbf{R}+\delta\mathbf{r})\right]\\&=-\ddot{\mathbf{R}}-\mu\left[\frac{\mathbf{R}}{R^3}+\frac{\delta\mathbf{r}}{R^3}-\frac{3}{R^5}(\mathbf{R}\cdot\delta\mathbf{r})\mathbf{R}+\overbrace{\text{terms of higher order than 1 in }\delta\mathbf{r}}^{\text{neglect}}\right]\end{aligned}$$

That is, to our degree of approximation,

$$\delta\ddot{\mathbf{r}}=-\ddot{\mathbf{R}}-\mu\frac{\mathbf{R}}{R^3}-\frac{\mu}{R^3}\left[\delta\mathbf{r}-\frac{3}{R^2}(\mathbf{R}\cdot\delta r)\mathbf{R}\right] \tag{7.21}$$

But the equation of motion of the reference orbit is

$$\ddot{\mathbf{R}}=-\mu\frac{\mathbf{R}}{R^3} \tag{7.22}$$

Substituting this into Eq. (7.21) finally yields

$$\delta\ddot{\mathbf{r}}=-\frac{\mu}{R^3}\left[\delta\mathbf{r}-\frac{3}{R^2}(\mathbf{R}\cdot\delta\mathbf{r})\mathbf{R}\right] \tag{7.23}$$

This is the linearized version of Eq. (7.17), the equation that governs the motion of the chaser with respect to the target. The expression is linear because the unknown $\delta\mathbf{r}$ appears only in the numerator and only to the first power throughout. We achieved this by dropping a lot of terms that are insignificant when Eq. (7.15) is valid. Eq. (7.23) is nonlinear in $\mathbf{R}$, which is not an unknown because it is determined independently by solving Eq. (7.22).

In the comoving frame of Fig. 7.5 the x axis lies along the radial $\mathbf{R}$, so that

$$\mathbf{R}=R\hat{\mathbf{i}} \tag{7.24}$$

In terms of its components in the comoving frame the relative position vector $\delta\mathbf{r}$ in Fig. 7.5 is (cf. Eq. 7.4a)

$$\delta\mathbf{r}=\delta x\hat{\mathbf{i}}+\delta y\hat{\mathbf{j}}+\delta z\hat{\mathbf{k}} \tag{7.25}$$

Substituting Eqs. (7.24) and (7.25) into Eq. (7.23) yields

$$\delta\ddot{\mathbf{r}}=-\frac{\mu}{R^3}\left[\left(\delta x\hat{\mathbf{i}}+\delta y\hat{\mathbf{j}}+\delta z\hat{\mathbf{k}}\right)-\frac{3}{R^2}\left[\left(R\hat{\mathbf{i}}\right)\cdot\left(\delta x\hat{\mathbf{i}}+\delta y\hat{\mathbf{j}}+\delta z\hat{\mathbf{k}}\right)\right]\left(R\hat{\mathbf{i}}\right)\right]$$

After expanding the dot product on the right and collecting terms, we find that the linearized equation of relative motion takes a rather simple form when the components of $\mathbf{R}$ and $\delta\mathbf{r}$ are given in the comoving frame,

$$\delta\ddot{\mathbf{r}}=-\frac{\mu}{R^3}\left(-2\delta x\hat{\mathbf{i}}+\delta y\hat{\mathbf{j}}+\delta z\hat{\mathbf{k}}\right) \tag{7.26}$$

Recall that $\delta\ddot{\mathbf{r}}$ is the acceleration of chaser B relative to target A as measured in the inertial frame. That is,

$$\delta\ddot{\mathbf{r}} = \frac{d^2}{dt^2}\delta\mathbf{r} = \frac{d^2}{dt^2}(\mathbf{r}_B - \mathbf{r}_A) = \ddot{\mathbf{r}}_B - \ddot{\mathbf{r}}_A = \mathbf{a}_B - \mathbf{a}_A$$

$\delta\ddot{\mathbf{r}}$ is not to be confused with $\delta\mathbf{a}_{\text{rel}}$, which is the relative acceleration measured in the comoving frame. These two quantities are related by Eq. (7.9),

$$\delta\mathbf{a}_{\text{rel}} = \delta\ddot{\mathbf{r}} - \dot{\mathbf{\Omega}} \times \delta\mathbf{r} - \mathbf{\Omega} \times (\mathbf{\Omega} \times \delta\mathbf{r}) - 2\mathbf{\Omega} \times \delta\mathbf{v}_{\text{rel}} \tag{7.27}$$

Since we arrived at an expression for $\delta\ddot{\mathbf{r}}$ in Eq. (7.26), let us proceed to evaluate each of the three terms on the right that involve $\mathbf{\Omega}$ and $\dot{\mathbf{\Omega}}$. First, recall that the angular momentum of A $(\mathbf{h} = \mathbf{R} \times \dot{\mathbf{R}})$ is normal to A's orbital plane, and so is the z axis of the comoving frame. Therefore, $\mathbf{h} = h\hat{\mathbf{k}}$. It follows that Eqs. (7.5) and (7.6) may be written as

$$\mathbf{\Omega} = \frac{h}{R^2}\hat{\mathbf{k}} \tag{7.28}$$

and

$$\dot{\mathbf{\Omega}} = -\frac{2(\mathbf{V} \cdot \mathbf{R})h}{R^4}\hat{\mathbf{k}} \tag{7.29}$$

where $\mathbf{V} = \dot{\mathbf{R}}$.

From Eqs. (7.25)–(7.29), we find

$$\dot{\mathbf{\Omega}} \times \delta\mathbf{r} = \left[-\frac{2(\mathbf{V} \cdot \mathbf{R})h}{R^4}\hat{\mathbf{k}}\right] \times \left(\delta x\hat{\mathbf{i}} + \delta y\hat{\mathbf{j}} + \delta z\hat{\mathbf{k}}\right) = \frac{2(\mathbf{V} \cdot \mathbf{R})h}{R^4}\left(\delta y\hat{\mathbf{i}} - \delta x\hat{\mathbf{j}}\right) \tag{7.30}$$

and

$$\mathbf{\Omega} \times (\mathbf{\Omega} \times \delta\mathbf{r}) = \frac{h}{R^2}\hat{\mathbf{k}} \times \left[\frac{h}{R^2}\hat{\mathbf{k}} \times \left(\delta x\hat{\mathbf{i}} + \delta y\hat{\mathbf{j}} + \delta z\hat{\mathbf{k}}\right)\right] = -\frac{h^2}{R^4}\left(\delta x\hat{\mathbf{i}} + \delta y\hat{\mathbf{j}}\right) \tag{7.31}$$

According to Eq. (7.4b), $\delta\mathbf{v}_{\text{rel}} = \delta\dot{x}\hat{\mathbf{i}} + \delta\dot{y}\hat{\mathbf{j}} + \delta\dot{z}\hat{\mathbf{k}}$ where $\delta\dot{x} = (d/dt)\delta x$, etc. It follows that

$$2\mathbf{\Omega} \times \delta\mathbf{v}_{\text{rel}} = 2\frac{h}{R^2}\hat{\mathbf{k}} \times \left(\delta\dot{x}\hat{\mathbf{i}} + \delta\dot{y}\hat{\mathbf{j}} + \delta\dot{z}\hat{\mathbf{k}}\right) = 2\frac{h}{R^2}\left(\delta\dot{x}\hat{\mathbf{j}} - \delta\dot{y}\hat{\mathbf{k}}\right) \tag{7.32}$$

Substituting Eq. (7.26) along with Eqs. (7.30)–(7.32) into Eq. (7.27) yields

$$\delta\mathbf{a}_{\text{rel}} = \overbrace{-\frac{\mu}{R^3}\left(-2\delta x\hat{\mathbf{i}} + \delta y\hat{\mathbf{j}} + \delta z\hat{\mathbf{k}}\right)}^{\delta\ddot{\mathbf{r}}} - \overbrace{\frac{2(\mathbf{V} \cdot \mathbf{R})h}{R^4}\left(\delta y\hat{\mathbf{i}} - \delta x\hat{\mathbf{j}}\right)}^{\dot{\mathbf{\Omega}}\times\delta\mathbf{r}} - \overbrace{\left[-\frac{h^2}{R^4}\left(\delta x\hat{\mathbf{i}} + \delta y\hat{\mathbf{j}}\right)\right]}^{\mathbf{\Omega}\times(\mathbf{\Omega}\times\delta\mathbf{r})}$$
$$\overbrace{-2\frac{h}{R^2}\left(\delta\dot{x}\hat{\mathbf{j}} + \delta\dot{y}\hat{\mathbf{i}}\right)}^{2\mathbf{\Omega}\times\delta\mathbf{v}_{\text{rel}}}$$

Referring to Eq. (7.4c) we set $\delta\mathbf{a}_{\text{rel}} = \delta\ddot{x}\hat{\mathbf{i}} + \delta\ddot{y}\hat{\mathbf{j}} + \delta\ddot{z}\hat{\mathbf{k}}$ (where $\delta\ddot{x} = (\mathrm{d}^2/\mathrm{d}t^2)\delta x$, etc.) and collect the terms on the right to obtain

$$\begin{aligned}\delta\ddot{x}\hat{\mathbf{i}} + \delta\ddot{y}\hat{\mathbf{j}} + \delta\ddot{z}\hat{\mathbf{k}} = &\left[\left(\frac{2\mu}{R^3} + \frac{h^2}{R^4}\right)\delta x - \frac{2(\mathbf{V}\cdot\mathbf{R})h}{R^4}\delta y + 2\frac{h}{R^2}\delta\dot{y}\right]\hat{\mathbf{i}}\\ &+\left[\left(\frac{h^2}{R^4} - \frac{\mu}{R^3}\right)\delta y + \frac{2(\mathbf{V}\cdot\mathbf{R})h}{R^4}\delta x - 2\frac{h}{R^2}\delta\dot{x}\right]\hat{\mathbf{j}}\\ &-\frac{\mu}{R^3}\delta z\hat{\mathbf{k}}\end{aligned} \tag{7.33}$$

Finally, by equating the coefficients of the three unit vectors $\hat{\mathbf{i}}$, $\hat{\mathbf{j}}$, and $\hat{\mathbf{k}}$, this vector equation yields the three scalar equations,

$$\delta\ddot{x} - \left(\frac{2\mu}{R^3} + \frac{h^2}{R^4}\right)\delta x + \frac{2(\mathbf{V}\cdot\mathbf{R})h}{R^4}\delta y - 2\frac{h}{R^2}\delta\dot{y} = 0 \tag{7.34a}$$

$$\delta\ddot{y} + \left(\frac{\mu}{R^3} - \frac{h^2}{R^4}\right)\delta y - \frac{2(\mathbf{V}\cdot\mathbf{R})h}{R^4}\delta x + 2\frac{h}{R^2}\delta\dot{x} = 0 \tag{7.34b}$$

$$\delta\ddot{z} + \frac{\mu}{R^3}\delta z = 0 \tag{7.34c}$$

This set of linear second-order differential equations must be solved to obtain the relative position coordinates δx, δy, and δz as a function of time. Eqs. (7.34a) and (7.34b) are coupled since δx and δy appear in each one of them. δz appears by itself in Eq. (7.34c) and nowhere else, which means the relative motion in the z direction is independent of that in the other two directions. If the reference orbit is an ellipse, then **R** and **V** vary with time (although the angular momentum h of the reference orbit is constant). In that case the coefficients in Eq. (7.34) are time dependent, so there is no easy analytical solution. However, we can solve Eq. (7.34) numerically using the methods in Section 1.8.

To that end we recast Eq. (7.34) as a set of first-order differential equations in the standard form

$$\dot{\mathbf{y}} = \mathbf{f}(t, \mathbf{y}) \tag{7.35}$$

where

$$\mathbf{y} = \begin{Bmatrix}\delta x\\ \delta y\\ \delta z\\ \delta\dot{x}\\ \delta\dot{y}\\ \delta\dot{z}\end{Bmatrix} \quad \dot{\mathbf{y}} = \begin{Bmatrix}\delta\dot{x}\\ \delta\dot{y}\\ \delta\dot{z}\\ \delta\ddot{x}\\ \delta\ddot{y}\\ \delta\ddot{z}\end{Bmatrix} \quad \mathbf{f}(t, \mathbf{y}) = \begin{Bmatrix} y_4\\ y_5\\ y_6\\ \left(\dfrac{2\mu}{R^3} + \dfrac{h^2}{R^4}\right)y_1 - \dfrac{2(\mathbf{V}\cdot\mathbf{R})h}{R^4}y_2 + 2\dfrac{h}{R^2}y_5\\ \left(\dfrac{h^2}{R^4} - \dfrac{\mu}{R^3}\right)y_2 + \dfrac{2(\mathbf{V}\cdot\mathbf{R})h}{R^4}y_1 - 2\dfrac{h}{R^2}y_4\\ -\dfrac{\mu}{R^3}y_3\end{Bmatrix} \tag{7.36}$$

These can be solved by Algorithm 1.1 (Runge-Kutta), Algorithm 1.2 (Heun), or Algorithm 1.3 (Runge-Kutta-Fehlberg). In any case, the state vector of the target orbit must be updated at each time step to provide the current values of **R** and **V**. This is done with the aid of Algorithm 3.4. (Alternatively, Eq. (7.22), the equations of motion of the target, can be integrated along with Eq. (7.36) to provide **R** and **V** as a function of time.)

EXAMPLE 7.3

At time $t=0$ the orbital parameters of target vehicle A in an equatorial earth orbit are

$$r_p = 6678\,\text{km} \quad e = 0.1 \quad i = \Omega = \omega = \theta = 0^\circ \tag{a}$$

where r_p is the perigee radius. At that same instant the state vector of the chaser vehicle B relative to A is

$$\delta\mathbf{r}_0 = -1\hat{\mathbf{i}}(\text{km}) \quad \delta\mathbf{v}_{\text{rel}})_0 = 2n\hat{\mathbf{j}}(\text{km/s}) \tag{b}$$

where n is the mean motion of A. Plot the path of B relative to A in the comoving frame for five periods of the reference orbit.

Solution

1. Use Algorithm 4.5 to obtain the initial state vector $(\mathbf{R}_0, \mathbf{V}_0)$ of the target vehicle from the orbital parameters given in Eq. (a).
2. Starting with the initial conditions given in Eq. (b), use Algorithm 1.3 to integrate Eq. (7.36) over the specified time interval. Use Algorithm 3.4 to obtain the reference orbit state vector (**R**, **V**) at each time step in order to evaluate the coefficients in Eq. (7.36).
3. Graph the trajectory $\delta y(t)$ vs. $\delta x(t)$.
   This procedure is implemented in the MATLAB function *Example_7_03.m* listed in Appendix D.33. The output of the program is shown in Fig. 7.6. Observe that since $\delta z_0 = \delta \dot{z}_0 = 0$, no movement develops in the z direction. The motion of the chaser therefore lies in the plane of the target vehicle's orbit. Fig. 7.6 shows that B rapidly moves away from A along the y direction and that the amplitude of its looping motion about the x axis continuously increases. The accuracy of this solution degrades over time because eventually the criterion in Eq. (7.15) is no longer satisfied.

   It is interesting to note that if we change the eccentricity of A to zero, so that the reference orbit is a circle, then Fig. 7.7 results. That is, for the same initial conditions, B orbits the target vehicle instead of drifting away from it.

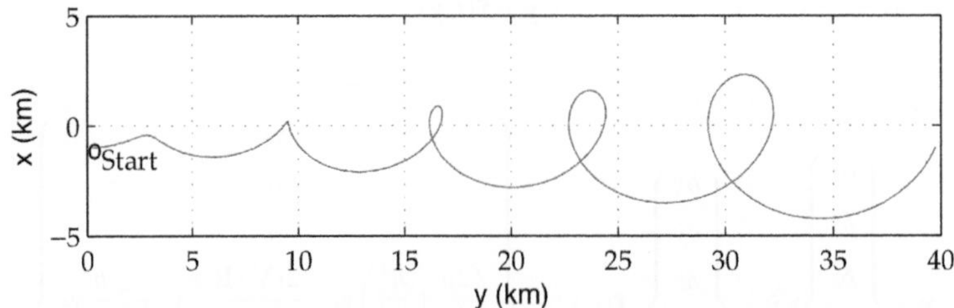

FIG. 7.6

Trajectory of B relative to A in the comoving frame during five of the target's orbits. Eccentricity of the target orbit is 0.1.

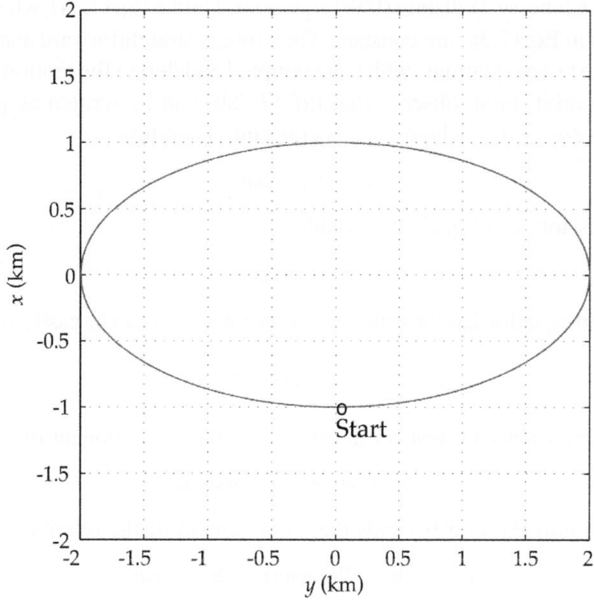

FIG. 7.7

Trajectory of B relative to A in the comoving frame during five of the target's orbits. Eccentricity of the target orbit is 0.

7.4 CLOHESSY-WILTSHIRE EQUATIONS

If the orbit of the target vehicle A in Fig. 7.5 is a circle, then our LVLH frame is called a Clohessy-Wiltshire (CW) frame (Clohessy) (Clohessy and Wiltshire, 1960). In such a frame Eq. (7.34) simplifies considerably. For a circular target orbit $\mathbf{V}\cdot\mathbf{R}=0$ and $h=\sqrt{\mu R}$. Substituting these into Eqs. (7.34) yields

$$\begin{aligned}\delta\ddot{x}-3\frac{\mu}{R^3}\delta x-2\sqrt{\frac{\mu}{R^3}}\delta\dot{y} &=0\\ \delta\ddot{y}+2\sqrt{\frac{\mu}{R^3}}\delta\dot{x} &=0\\ \delta\ddot{z}+\frac{\mu}{R^3}\delta z &=0\end{aligned}\tag{7.37}$$

It is furthermore true for circular orbits that the angular velocity (mean motion) is

$$n=\frac{V}{R}=\frac{\sqrt{\mu/R}}{R}=\sqrt{\frac{\mu}{R^3}}$$

Therefore, Eq. (7.37) may be written as

$$\delta\ddot{x}-3n^2\delta x-2n\delta\dot{y}=0\tag{7.38a}$$

$$\delta\ddot{y}+2n\delta\dot{x}=0\tag{7.38b}$$

$$\delta\ddot{y}+2n\delta\dot{x}=0\tag{7.38c}$$

These are known as the Clohessy-Wiltshire (CW) equations. Unlike Eq. (7.34), where the target orbit is an ellipse, the coefficients in Eq. (7.38) are constant. Therefore, a straightforward analytical solution exists.

We start with the first two equations, which are coupled and define the motion of the chaser in the xy plane of the reference orbit. First, observe that Eq. (7.38b) can be written as $(\mathrm{d}/\mathrm{d}t)(\delta\dot{y}+2n\delta x)=0$, which means that $\delta\dot{y}+2n\delta x=C_1$, where C_1 is a constant. Therefore,

$$\delta\dot{y}=C_1-2n\delta x \tag{7.39}$$

Substituting this expression into Eq. (7.38a) yields

$$\delta\ddot{x}+n^2\delta x=2nC_1 \tag{7.40}$$

This familiar differential equation has the following solution, which can be easily verified by substitution:

$$\delta x=\frac{2}{n}C_1+C_2\sin nt+C_3\cos nt \tag{7.41}$$

Differentiating this expression with respect to time gives the x component of the relative velocity,

$$\delta\dot{x}=C_2n\cos nt-C_3n\sin nt \tag{7.42}$$

Substituting Eq. (7.41) into Eq. (7.39) yields the y component of the relative velocity

$$\delta\dot{y}=-3C_1-2C_2n\sin nt-2C_3n\cos nt \tag{7.43}$$

Integrating this equation with respect to time yields

$$\delta y=-3C_1t+2C_2\cos nt-2C_3\sin nt+C_4 \tag{7.44}$$

The constants C_1 through C_4 are found by applying the initial conditions; namely,

$$\text{At } t=0\quad \delta x=\delta x_0\quad \delta y=\delta y_0\quad \delta\dot{x}=\delta\dot{x}_0\quad \delta\dot{y}=\delta\dot{y}_0$$

Evaluating Eqs. (7.41)–(7.44), respectively, at $t=0$ we get

$$\begin{aligned}\frac{2}{n}C_1+C_3&=\delta x_0\\ C_2n&=\delta\dot{x}_0\\ -3C_1-2C_3n&=\delta\dot{y}_0\\ 2C_2+C_4&=\delta y_0\end{aligned}$$

Solving for C_1 through C_4 yields

$$C_1=2n\delta x_0+\delta\dot{y}_0\quad C_2=\frac{1}{n}\delta\dot{x}_0\quad C_3=-3\delta x_0-\frac{2}{n}\delta\dot{y}_0\quad C_4=-\frac{2}{n}\delta\dot{x}_0+\delta y_0 \tag{7.45}$$

Finally, we turn our attention to Eq. (7.38c), which governs the relative motion normal to the plane of the circular reference orbit. Eq. (7.38c) has the same form as Eq. (7.40) with $C_1=0$. Therefore, its solution is

$$\delta z=C_5\sin nt+C_6\cos nt \tag{7.46}$$

It follows that the relative velocity normal to the reference orbit is

$$\delta\dot{z}=C_5n\cos nt-C_6n\sin nt \tag{7.47}$$

The initial conditions are $\delta z = \delta z_0$ and $\delta \dot{z} = \delta \dot{z}_0$ at $t = 0$, which means

$$C_5 = \frac{\delta \dot{z}_0}{n} \quad C_6 = \delta z_0 \tag{7.48}$$

Substituting Eqs. (7.45) and (7.48) into Eqs. (7.41), (7.44), and (7.46) yields the trajectory of the chaser in the CW frame,

$$\delta x = 4\delta x_0 + \frac{2}{n}\delta \dot{y}_0 + \frac{\delta \dot{x}_0}{n}\sin nt - \left(3\delta x_0 + \frac{2}{n}\delta \dot{y}_0\right)\cos nt \tag{7.49a}$$

$$\delta y = \delta y_0 - \frac{2}{n}\delta \dot{x}_0 - 3(2n\delta x_0 + \delta \dot{y}_0)t + 2\left(3\delta x_0 + \frac{2}{n}\delta \dot{y}_0\right)\sin nt + \frac{2}{n}\delta \dot{x}_0 \cos nt \tag{7.49b}$$

$$\delta z = \frac{1}{n}\delta \dot{z}_0 \sin nt + \delta z_0 \cos nt \tag{7.49c}$$

Observe that all the three components of $\delta \mathbf{r}$ oscillate with a frequency equal to the frequency of revolution (mean motion n) of the CW frame. Only δy has a secular term, which grows linearly with time. Therefore, unless $2n\delta x_0 + \delta \dot{y}_0 = 0$, the chaser will drift away from the target and the distance δr will increase without bound. The accuracy of Eqs. (7.49) will consequently degrade as the criterion (Eq. 7.15) on which this solution is based eventually ceases to be valid. Fig. 7.8 shows the motion of a particle relative to a CW frame with an orbital radius of 6678 km. The particle started at the origin with a velocity of 0.01 km/s in the negative y direction. This delta-v dropped the particle into a lower energy, a slightly elliptical orbit. The subsequent actual relative motion of the particle in the CW frame is graphed in Fig. 7.8 as is the motion given by Eqs. (7.49), the linearized CW solution. Clearly, the two solutions diverge markedly after one orbit of the reference frame, when the distance of the particle from the origin exceeds 150 km.

Now that we have finished solving the CW equations, let us simplify our notation a bit and denote the x, y, and z components of relative velocity in the moving frame as δu, δv, and δw, respectively. That is, let

$$\delta u = \delta \dot{x} \quad \delta v = \delta \dot{y} \quad \delta w = \delta \dot{z} \tag{7.50a}$$

The initial conditions on the relative velocity components are then written as

$$\delta u_0 = \delta \dot{x}_0 \quad \delta v_0 = \delta \dot{y}_0 \quad \delta w_0 = \delta \dot{z}_0 \tag{7.50b}$$

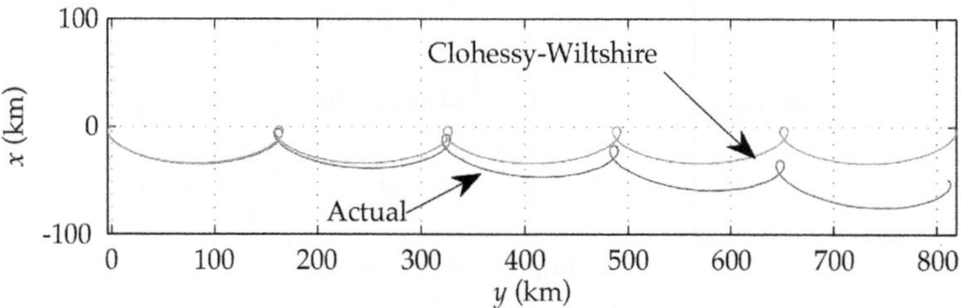

FIG. 7.8

Relative motion of a particle and its Clohessy-Wiltshire approximation.

Using this notation in Eq. (7.49) and rearranging the terms we get

$$\begin{aligned}
\delta x &= (4-3\cos nt)\delta x_0 + \frac{\sin nt}{n}\delta u_0 + \frac{2}{n}(1-\cos nt)\delta v_0 \\
\delta y &= 6(\sin nt - nt)\delta x_0 + \delta y_0 + \frac{2}{n}(\cos nt - 1)\delta u_0 + \frac{1}{n}(4\sin nt - 3nt)\delta v_0 \\
\delta z &= \cos nt\delta z_0 + \frac{1}{n}\sin nt\delta w_0
\end{aligned} \tag{7.51a}$$

Differentiating each of these with respect to time and using Eq. (7.50a) yields

$$\begin{aligned}
\delta u &= 3n\sin nt\delta x_0 + \cos nt\delta u_0 + 2\sin nt\delta v_0 \\
\delta v &= 6n(\cos nt - 1)\delta x_0 - 2\sin nt\delta u_0 + (4\cos nt - 3)\delta v_0 \\
\delta w &= -n\sin nt\delta z_0 + \cos nt\delta w_0
\end{aligned} \tag{7.51b}$$

Let us introduce matrix notation to define the relative position and velocity vectors

$$\{\delta \mathbf{r}(t)\} = \begin{Bmatrix} \delta x(t) \\ \delta y(t) \\ \delta z(t) \end{Bmatrix} \qquad \{\delta \mathbf{v}(t)\} = \begin{Bmatrix} \delta u(t) \\ \delta v(t) \\ \delta w(t) \end{Bmatrix}$$

and their initial values (at $t = 0$)

$$\{\delta \mathbf{r}_0\} = \begin{Bmatrix} \delta x_0 \\ \delta y_0 \\ \delta z_0 \end{Bmatrix} \qquad \{\delta \mathbf{v}_0\} = \begin{Bmatrix} \delta u_0 \\ \delta v_0 \\ \delta w_0 \end{Bmatrix}$$

In matrix notation, Eqs. (7.51) appear more compactly as

$$\{\delta \mathbf{r}(t)\} = [\Phi_{rr}(t)]\{\delta \mathbf{r}_0\} + [\Phi_{rv}(t)]\{\delta \mathbf{v}_0\} \tag{7.52a}$$

$$\{\delta \mathbf{v}(t)\} = [\Phi_{vr}(t)]\{\delta \mathbf{r}_0\}[\Phi_{vv}(t)]\{\delta \mathbf{v}_0\} \tag{7.52b}$$

where the "Clohessy-Wiltshire matrices" comprise the coefficients in Eqs. (7.51):

$$[\boldsymbol{\Phi}_{rr}(t)] = \left[\begin{array}{cc|c} 4-3\cos nt & 0 & 0 \\ 6(\sin nt - nt) & 1 & 0 \\ \hline 0 & 0 & \cos nt \end{array}\right] \tag{7.53a}$$

$$[\boldsymbol{\Phi}_{rv}(t)] = \left[\begin{array}{cc|c} \frac{1}{n}\sin nt & \frac{2}{n}(1-\cos nt) & 0 \\ \frac{2}{n}(\cos nt - 1) & \frac{1}{n}(4\sin nt - 3nt) & 0 \\ \hline 0 & 0 & \frac{1}{n}\sin nt \end{array}\right] \tag{7.53b}$$

$$[\boldsymbol{\Phi}_{vr}(t)] = \left[\begin{array}{cc|c} 3n\sin nt & 0 & 0 \\ 6n(\cos nt - 1) & 0 & 0 \\ \hline 0 & 0 & -n\sin nt \end{array}\right] \tag{7.53c}$$

$$[\boldsymbol{\Phi}_{vv}(t)] = \left[\begin{array}{cc|c} \cos nt & 2\sin nt & 0 \\ -2\sin nt & 4\cos nt - 3 & 0 \\ \hline 0 & 0 & \cos nt \end{array}\right] \tag{7.53d}$$

The subscripts on $\boldsymbol{\Phi}$ remind us which of the vectors $\delta\mathbf{r}$ and $\delta\mathbf{v}$ is related by that matrix to which of the initial conditions $\delta\mathbf{r}_0$ and $\delta\mathbf{v}_0$. For example, $[\boldsymbol{\Phi}_{rv}]$ relates $\delta\mathbf{r}$ to $\delta\mathbf{v}_0$. The partition lines remind us that motion in the xy plane is independent of that in the z direction normal to the target's orbit. In problems where there is no motion in the z direction ($\delta z_0 = \delta w_0 = 0$), we need only use the upper left 2 by 2 corners of CW matrices. Finally, note also that

$$[\boldsymbol{\Phi}_{vr}(t)] = \frac{d}{dt}[\boldsymbol{\Phi}_{rr}(t)] \text{ and } [\boldsymbol{\Phi}_{vv}(t)] = \frac{d}{dt}[\boldsymbol{\Phi}_{rv}(t)]$$

7.5 TWO-IMPULSE RENDEZVOUS MANEUVERS

Fig. 7.9 illustrates the rendezvous problem. At time $t = 0^-$ (the instant preceding $t = 0$) the position $\delta\mathbf{r}_0$ and velocity $\delta\mathbf{v}_0^-$ of the chase vehicle B relative to target A are known. At $t = 0$ an impulsive maneuver instantaneously changes the relative velocity to $\delta\mathbf{v}_0^+$ at $t = 0^+$ (the instant after $t = 0$). The components of $\delta\mathbf{v}_0^+$ are shown in Fig. 7.9. We must determine the values of δu_0^+, δv_0^+, and δw_0^+ at the beginning of the

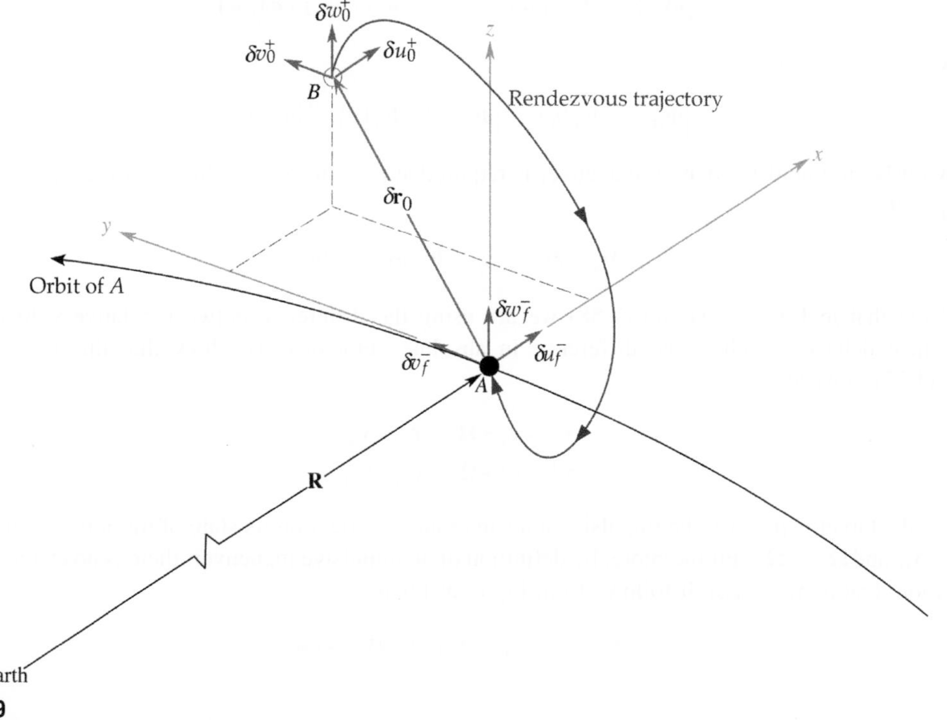

FIG. 7.9

Rendezvous with a target A in the neighborhood of the chase vehicle B.

rendezvous trajectory, so that B will arrive at the target in a specified time t_f. The delta-v required to place B on the rendezvous trajectory is

$$\Delta\mathbf{v}_0 = \delta\mathbf{v}_0^+ - \delta\mathbf{v}_0^- = (\delta u_0^+ - \delta u_0^-)\hat{\mathbf{i}} + (\delta v_0^+ - \delta v_0^-)\hat{\mathbf{j}} + (\delta w_0^+ - \delta w_0^-)\hat{\mathbf{k}} \tag{7.54}$$

At time t_f, B arrives at A, at the origin of the CW frame, which means $\delta\mathbf{r}_f = \delta\mathbf{r}(t_f) = \mathbf{0}$. Evaluating Eq. (7.52a) at t_f we find

$$\{\mathbf{0}\} = [\Phi_{rr}(t_f)]\{\delta\mathbf{r}_0\} + [\Phi_{rv}(t_f)]\{\delta\mathbf{v}_0^+\} \tag{7.55}$$

Solving this for $\{\delta\mathbf{v}_0^+\}$ yields

$$\{\delta\mathbf{v}_0^+\} = -[\Phi_{rv}(t_f)]^{-1}[\Phi_{rr}(t_f)]\{\delta\mathbf{r}_0\} \quad \left(\delta\mathbf{v}_0^+ = \delta u_0^+\hat{\mathbf{i}} + \delta v_0^+\hat{\mathbf{j}} + \delta w_0^+\hat{\mathbf{k}}\right) \tag{7.56}$$

where $[\Phi_{rv}(t_f)]^{-1}$ is the matrix inverse of $[\Phi_{rv}(t_f)]$. Thus, we now have the velocity $\delta\mathbf{v}_0^+$ at the beginning of the rendezvous path. We substitute Eq. (7.56) into Eq. (7.52b) to obtain the velocity $\delta\mathbf{v}_f^-$ at $t = t_f^-$, when B arrives at target A:

$$\begin{aligned}\{\delta\mathbf{v}_f^-\} &= [\Phi_{vr}(t_f)]\{\delta\mathbf{r}_0\} + [\Phi_{vv}(t_f)]\{\delta\mathbf{v}_0^+\} \\ &= [\Phi_{vr}(t_f)]\{\delta\mathbf{r}_0\} + [\Phi_{vv}(t_f)]\left(-[\Phi_{rv}(t_f)]^{-1}[\Phi_{rr}(t_f)]\{\delta\mathbf{r}_0\}\right)\end{aligned}$$

Collecting terms we get

$$\{\delta\mathbf{v}_f^-\} = \left[\widetilde{\Phi}_{vr}\right]\{\delta\mathbf{r}_0\} \quad \left(\delta\mathbf{v}_f^- = \delta u_f^-\hat{\mathbf{i}} + \delta v_f^-\hat{\mathbf{j}} + \delta w_f^-\hat{\mathbf{k}}\right) \tag{7.57a}$$

where

$$\left[\widetilde{\Phi}_{vr}\right] = [\Phi_{vr}(t_f)] - [\Phi_{vv}(t_f)][\Phi_{rv}(t_f)]^{-1}[\Phi_{rr}(t_f)] \tag{7.57b}$$

Obviously, an impulsive delta-v maneuver is required at $t = t_f$ to bring vehicle B to rest relative to A ($\delta\mathbf{v}_f^+ = \mathbf{0}$):

$$\Delta\mathbf{v}_f = \delta\mathbf{v}_f^+ - \delta\mathbf{v}_f^- = \mathbf{0} - \delta\mathbf{v}_f^- = -\delta\mathbf{v}_f^- \tag{7.58}$$

Note that in Eqs. (7.54) and (7.58) we are using the difference between relative velocities to calculate delta-v, which is the difference in absolute velocities. To show that this is valid use Eq. (1.75) to write

$$\begin{aligned}\mathbf{v}^- &= \mathbf{v}_0^- + \boldsymbol{\Omega}^- \times \mathbf{r}_{rel}^- + \mathbf{v}_{rel}^- \\ \mathbf{v}^+ &= \mathbf{v}_0^+ + \boldsymbol{\Omega}^+ \times \mathbf{r}_{rel}^+ + \mathbf{v}_{rel}^+\end{aligned} \tag{7.59}$$

Since the target is passive, the impulsive maneuver has no effect on its state of motion, which means $\mathbf{v}_0^+ = \mathbf{v}_0^-$ and $\boldsymbol{\Omega}^+ = \boldsymbol{\Omega}^-$. Furthermore, by definition of an impulsive maneuver, there is no change in the position; that is, $\mathbf{r}_{rel}^+ = \mathbf{r}_{rel}^-$. It follows from Eq. (7.59) that

$$\mathbf{v}^+ - \mathbf{v}^- = \mathbf{v}_{rel}^+ - \mathbf{v}_{rel}^- \quad \text{or} \quad \Delta\mathbf{v} = \Delta\mathbf{v}_{rel}$$

EXAMPLE 7.4

A space station and another spacecraft are in earth orbits with the following parameters:

	Space station	Spacecraft
Perigee × apogee (altitude)	300 km circular	320.06 km × 513.86 km
Period (computed using above data)	1.5086 h	1.5484 h
True anomaly, θ	60°	349.65°
Inclination, i	40°	40.130°
RA of ascending node, Ω	20°	19.819°
Argument of perigee, ω	0° (arbitrary)	70.662°

Compute the total delta-v required for an 8-h, two-impulse rendezvous trajectory.

Solution

We substitute the given data into Algorithm 4.5 to obtain the state vectors of the two spacecraft in the geocentric equatorial frame.

Space station:

$$\mathbf{R} = 1622.39\hat{\mathbf{I}} + 5305.10\hat{\mathbf{J}} + 3717.44\hat{\mathbf{K}}\ (\text{km})$$
$$\mathbf{V} = -7.29936\hat{\mathbf{I}} + 0.492329\hat{\mathbf{J}} + 2.48304\hat{\mathbf{K}}\ (\text{km/s})$$

Spacecraft:

$$\mathbf{r} = 1612.75\hat{\mathbf{I}} + 5310.19\hat{\mathbf{J}} + 3750.33\hat{\mathbf{K}}\ (\text{km})$$
$$\mathbf{v} = -7.35170\hat{\mathbf{I}} + 0.463828\hat{\mathbf{J}} + 2.46906\hat{\mathbf{K}}\ (\text{km/s})$$

The space station reference frame unit vectors (at this instant) are, by definition,

$$\hat{\mathbf{i}} = \frac{\mathbf{R}}{\|\mathbf{R}\|} = 0.242945\hat{\mathbf{I}} + 0.794415\hat{\mathbf{J}} + 0.556670\hat{\mathbf{K}}$$
$$\hat{\mathbf{j}} = \frac{\mathbf{V}}{\|\mathbf{V}\|} = -0.944799\hat{\mathbf{I}} + 0.063725\hat{\mathbf{J}} + 0.321394\hat{\mathbf{K}}$$
$$\hat{\mathbf{k}} = \hat{\mathbf{i}} \times \hat{\mathbf{j}} = 0.219846\hat{\mathbf{I}} - 0.604023\hat{\mathbf{J}} + 0.766044\hat{\mathbf{K}}$$

Therefore, the direction cosine matrix of the transformation from the geocentric equatorial frame into the space station frame is (at this instant)

$$[\mathbf{Q}]_{Xx} = \begin{bmatrix} 0.242945 & 0.794415 & 0.556670 \\ -0.944799 & 0.063725 & 0.321394 \\ 0.219846 & -0.604023 & 0.766044 \end{bmatrix}$$

The position vector of the spacecraft relative to the space station (in the geocentric equatorial frame) is

$$\delta\mathbf{r} = \mathbf{r} - \mathbf{R} = -9.64015\hat{\mathbf{I}} + 5.08235\hat{\mathbf{J}} + 32.8822\hat{\mathbf{K}}(\text{km})$$

The relative velocity is given by the formula (Eq. 7.8)

$$\delta\mathbf{v} = \mathbf{v} - \mathbf{V} - \mathbf{\Omega}_{\text{space station}} \times \delta\mathbf{r}$$

where $\mathbf{\Omega}_{\text{space station}} = n\hat{\mathbf{k}}$ and n, the mean motion of the space station, is

$$n = \frac{V}{R} = \frac{7.7258}{6678} = 0.00115691\,\text{rad/s} \tag{a}$$

Thus,

$$\delta\mathbf{v} = \overbrace{-7.35170\hat{\mathbf{I}} + 0.463828\hat{\mathbf{J}} + 2.46906\hat{\mathbf{K}}}^{\mathbf{v}} - \overbrace{(-7.29936\hat{\mathbf{I}} + 0.492329\hat{\mathbf{J}} + 2.48304\hat{\mathbf{K}})}^{\mathbf{V}}$$

$$\overbrace{-(0.00115691)\begin{vmatrix} \hat{\mathbf{I}} & \hat{\mathbf{J}} & \hat{\mathbf{K}} \\ 0.219846 & -0.604023 & 0.766044 \\ -9.64015 & 5.08235 & 32.8822 \end{vmatrix}}^{\boldsymbol{\Omega}_{\text{space station}} \times \delta\mathbf{r}}$$

so that

$$\delta\mathbf{v} = -0.024854\hat{\mathbf{I}} - 0.01159370\hat{\mathbf{J}} - 0.00853575\hat{\mathbf{K}}\,(\text{km/s})$$

In space station coordinates the relative position vector $\delta\mathbf{r}_0$ at the beginning of the rendezvous maneuver is

$$\{\delta\mathbf{r}_0\} = [\mathbf{Q}]_{Xx}\{\delta\mathbf{r}\} = \begin{bmatrix} 0.242945 & 0.794415 & 0.556670 \\ -0.944799 & 0.063725 & 0.321394 \\ 0.219846 & -0.604023 & 0.766044 \end{bmatrix} \begin{Bmatrix} -9.64015 \\ 5.08235 \\ 32.8822 \end{Bmatrix} = \begin{Bmatrix} 20 \\ 20 \\ 20 \end{Bmatrix} (\text{km}) \tag{b}$$

Likewise, the relative velocity $\delta\mathbf{v}_0^-$ just before launch into the rendezvous trajectory is

$$\{\delta\mathbf{v}_0^-\} = [\mathbf{Q}]_{Xx}\{\delta\mathbf{v}\} = \begin{bmatrix} 0.242945 & 0.794415 & 0.556670 \\ -0.944799 & 0.063725 & 0.321394 \\ 0.219846 & -0.604023 & 0.766044 \end{bmatrix} \begin{Bmatrix} -0.024854 \\ -0.0115937 \\ -0.00853575 \end{Bmatrix}$$

$$= \begin{Bmatrix} -0.02000 \\ 0.02000 \\ -0.005000 \end{Bmatrix} (\text{km/s})$$

The Clohessy-Wiltshire matrices, for $t = t_f = 8\,\text{h} = 28{,}800\,\text{s}$ and $n = 0.00115691\,\text{rad/s}$ (from Eq. (a)), are

$$[\Phi_{\text{rr}}] = \begin{bmatrix} 4 - 3\cos nt & 0 & 0 \\ 6(\sin nt - nt) & 1 & 0 \\ 0 & 0 & \cos nt \end{bmatrix} = \begin{bmatrix} 4.97849 & 0 & 0 \\ -194.242 & 1.000 & 0 \\ 0 & 0 & -0.326163 \end{bmatrix}$$

$$[\Phi_{\text{rv}}] = \begin{bmatrix} \frac{1}{n}\sin nt & \frac{2}{n}(1 - \cos nt) & 0 \\ \frac{2}{n}(\cos nt - 1) & \frac{1}{n}(4\sin nt - 3nt) & 0 \\ 0 & 0 & \frac{1}{n}\sin nt \end{bmatrix} = \begin{bmatrix} 817.102 & 2292.60 & 0 \\ -2292.60 & -83131.6 & 0 \\ 0 & 0 & 817.103 \end{bmatrix}$$

$$[\Phi_{\text{vr}}] = \begin{bmatrix} 3n\sin nt & 0 & 0 \\ 6n(\cos nt - 1) & 0 & 0 \\ 0 & 0 & -n\sin nt \end{bmatrix} = \begin{bmatrix} 0.00328092 & 0 & 0 \\ -0.00920550 & 0 & 0 \\ 0 & 0 & -0.00109364 \end{bmatrix}$$

$$[\Phi_{\text{vv}}] = \begin{bmatrix} \cos nt & 2\sin nt & 0 \\ -2\sin nt & 4\cos nt - 3 & 0 \\ 0 & 0 & \cos nt \end{bmatrix} = \begin{bmatrix} -0.326164 & 1.89063 & 0 \\ -1.89063 & -4.30466 & 0 \\ 0 & 0 & -0.326164 \end{bmatrix}$$

From Eqs. (7.56) and (b) we find $\delta\mathbf{v}_0^+$:

$$\begin{Bmatrix} \delta u_0^+ \\ \delta v_0^+ \\ \delta w_0^+ \end{Bmatrix} = -\begin{bmatrix} 817.102 & 2292.60 & 0 \\ -2292.60 & -83131.6 & 0 \\ 0 & 0 & 817.103 \end{bmatrix}^{-1} \begin{bmatrix} 4.97849 & 0 & 0 \\ -194.242 & 1.000 & 0 \\ 0 & 0 & -0.326163 \end{bmatrix} \begin{Bmatrix} 20 \\ 20 \\ 20 \end{Bmatrix}$$

$$= -\begin{bmatrix} 817.102 & 2292.60 & 0 \\ -2292.60 & -83131.6 & 0 \\ 0 & 0 & 817.103 \end{bmatrix}^{-1} \begin{bmatrix} 99.5698 \\ -3864.84 \\ -6.52386 \end{bmatrix} = \begin{Bmatrix} 0.00930458 \\ -0.0467472 \\ 0.00798343 \end{Bmatrix} (\text{km/s}) \tag{c}$$

From Eq. (7.52b), evaluated at $t = t_f$, we have

$$\{\delta\mathbf{v}_f^-\} = [\Phi_{vr}(t_f)]\{\delta\mathbf{r}_0\} + [\Phi_{vv}(t_f)]\{\delta\mathbf{v}_0^+\}$$

Substituting Eqs. (b) and (c),

$$\begin{Bmatrix} \delta u_f^- \\ \delta v_f^- \\ \delta w_f^- \end{Bmatrix} = \begin{bmatrix} 0.00328092 & 0 & 0 \\ -0.00920550 & 0 & 0 \\ 0 & 0 & -0.00109364 \end{bmatrix} \begin{Bmatrix} 20 \\ 20 \\ 20 \end{Bmatrix}$$

$$+ \begin{bmatrix} -0.326164 & 1.89063 & 0 \\ -1.89063 & -4.30466 & 0 \\ 0 & 0 & -0.326164 \end{bmatrix} \begin{Bmatrix} 0.00930458 \\ -0.0467472 \\ 0.00798343 \end{Bmatrix}$$

$$\begin{Bmatrix} \delta u_f^- \\ \delta v_f^- \\ \delta w_f^- \end{Bmatrix} = \begin{Bmatrix} -0.0257978 \\ -0.000470870 \\ -0.0244767 \end{Bmatrix} (\text{km/s}) \tag{d}$$

The delta-v at the beginning of the rendezvous maneuver is found as

$$\{\Delta\mathbf{v}_0\} = \{\delta\mathbf{v}_0^+\} - \{\delta\mathbf{v}_0^-\} = \begin{Bmatrix} 0.00930458 \\ -0.0467472 \\ 0.00798343 \end{Bmatrix} - \begin{Bmatrix} -0.02 \\ 0.02 \\ -0.005 \end{Bmatrix} = \begin{Bmatrix} 0.0293046 \\ -0.0667472 \\ 0.0129834 \end{Bmatrix} (\text{km/s})$$

The delta-v at the conclusion of the maneuver is

$$\{\Delta\mathbf{v}_f\} = \{\delta\mathbf{v}_f^+\} - \{\delta\mathbf{v}_f^-\} = \begin{Bmatrix} 0 \\ 0 \\ 0 \end{Bmatrix} - \begin{Bmatrix} -0.0257978 \\ -0.000470870 \\ -0.0244767 \end{Bmatrix} = \begin{Bmatrix} 0.0257978 \\ 0.000470870 \\ 0.0244767 \end{Bmatrix} (\text{km/s})$$

The total delta-v requirement is

$$\Delta v_{\text{total}} = \|\Delta\mathbf{v}_0\| + \|\Delta\mathbf{v}_f\| = 0.0740440 + 0.0355649 = 0.109609\,\text{km/s} = \boxed{109.6\,\text{m/s}}$$

From Eq. (7.52a) we have, for $0 < t < t_f$,

$$\begin{Bmatrix} \delta x(t) \\ \delta y(t) \\ \delta z(t) \end{Bmatrix} = \begin{bmatrix} 4 - 3\cos nt & 0 & 0 \\ 6(\sin nt - nt) & 1 & 0 \\ 0 & 0 & \cos nt \end{bmatrix} \begin{Bmatrix} 20 \\ 20 \\ 20 \end{Bmatrix}$$

$$+ \begin{bmatrix} \frac{1}{n}\sin nt & \frac{2}{n}(1 - \cos nt) & 0 \\ \frac{2}{n}(\cos nt - 1) & \frac{1}{n}(4\sin nt - 3nt) & 0 \\ 0 & 0 & \frac{1}{n}\sin nt \end{bmatrix} \begin{Bmatrix} 0.00930458 \\ -0.0467472 \\ 0.00798343 \end{Bmatrix}$$

Substituting n from Eq. (a) we obtain the relative position vector as a function of time. It is plotted in Fig. 7.10.

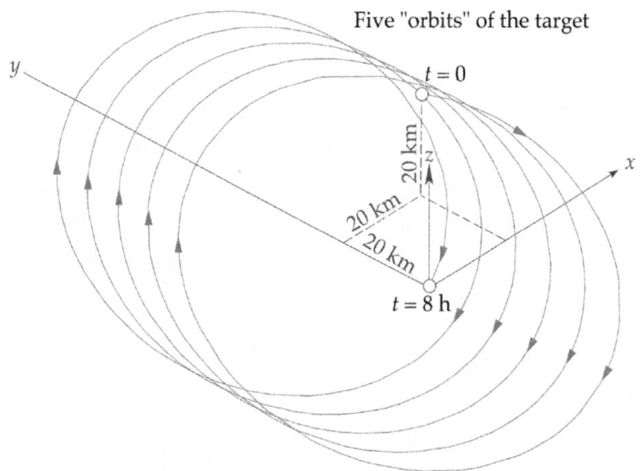

FIG. 7.10

Rendezvous trajectory of the chase vehicle relative to the target.

EXAMPLE 7.5

A target and a chase vehicle are in the same 300-km circular earth orbit. The chaser is 2 km behind the target when the chaser initiates a two-impulse rendezvous maneuver so as to rendezvous with the target in 1.49 h. Find the total delta-v requirement.

Solution

For the circular reference orbit

$$v = \sqrt{\frac{\mu}{r}} = \sqrt{\frac{398{,}600}{6378 + 300}} = 7.7258\,\text{km/s} \tag{a}$$

so that the mean motion is

$$n = \frac{v}{r} = \frac{7.7258}{6678} = 0.0011569\,\text{rad/s} \tag{b}$$

For this mean motion and the rendezvous trajectory time $t = 1.49\text{ h} = 5364\text{ s}$, the Clohessy-Wiltshire matrices are

$$[\boldsymbol{\Phi}_{rr}] = \left[\begin{array}{cc|c} 1.0090 & 0 & 0 \\ -37.699 & 1 & 0 \\ \hline 0 & 0 & 0.99700 \end{array}\right] \qquad [\boldsymbol{\Phi}_{rv}] = \left[\begin{array}{cc|c} -66.946 & 5.1928 & 0 \\ -5.1928 & -16360 & 0 \\ \hline 0 & 0 & -66.946 \end{array}\right]$$

$$[\boldsymbol{\Phi}_{vr}] = \left[\begin{array}{cc|c} -2.6881(10^{-4}) & 0 & 0 \\ -2.0851(10^{-5}) & 0 & 0 \\ \hline 0 & 0 & 8.9603(10^{-5}) \end{array}\right] \qquad [\boldsymbol{\Phi}_{vv}] = \left[\begin{array}{cc|c} 0.99700 & -0.15490 & 0 \\ 0.15490 & 0.98798 & 0 \\ \hline 0 & 0 & 0.99700 \end{array}\right] \tag{c}$$

The initial and final positions of the chaser in the CW frame are

$$\{\delta\mathbf{r}_0\} = \begin{Bmatrix} 0 \\ -2 \\ 0 \end{Bmatrix}(\text{km}) \quad \{\delta\mathbf{r}_f\} = \begin{Bmatrix} 0 \\ 0 \\ 0 \end{Bmatrix}(\text{km}) \tag{d}$$

Since $\delta z_0 = \delta w_0 = 0$ there is no motion in the z direction $[\delta z(t) = 0]$, so we need employ only the upper left 2 by 2 corners of the Clohessy-Wiltshire matrices and treat this as a two-dimensional problem in the plane of the reference orbit. Thus solving the first CW equation, $\{\delta\mathbf{r}_f\} = [\Phi_{rr}]\{\delta\mathbf{r}_0\} + [\Phi_{rv}]\{\delta\mathbf{v}_0^+\}$, for $\{\delta\mathbf{v}_0^+\}$ we get

$$\{\delta\mathbf{v}_0^+\} = -[\Phi_{rv}]^{-1}[\Phi_{rr}]\{\delta\mathbf{r}_0\} = -\begin{bmatrix} -0.014937 & -4.7412(10^{-6}) \\ 4.7412(10^{-6}) & -6.1124(10^{-5}) \end{bmatrix}\begin{bmatrix} 1.0090 & 0 \\ -37.699 & 1 \end{bmatrix}\begin{Bmatrix} 0 \\ -2 \end{Bmatrix}$$
$$= \begin{Bmatrix} 9.4824(10^{-6}) \\ -1.2225(10^{-4}) \end{Bmatrix}$$

or

$$\delta\mathbf{v}_0^+ = -9.4824(10^{-6})\hat{\mathbf{i}} + 1.2225(10^{-4})\hat{\mathbf{j}}(\text{km/s}) \tag{e}$$

Therefore, the second CW equation, $\{\delta\mathbf{v}_f^-\} = [\Phi_{vr}]\{\delta\mathbf{r}_0\} + [\Phi_{vv}]\{\delta\mathbf{v}_0^+\}$, yields

$$\{\delta\mathbf{v}_f^-\} = \begin{bmatrix} -2.6881(10^{-4}) & 0 \\ -2.0851(10^{-5}) & 0 \end{bmatrix}\begin{Bmatrix} 0 \\ -2 \end{Bmatrix} + \begin{bmatrix} 0.99700 & -0.15490 \\ 0.15490 & 0.98798 \end{bmatrix}\begin{Bmatrix} -9.4824(10^{-6}) \\ -1.2225(10^{-4}) \end{Bmatrix}$$
$$= \begin{Bmatrix} 9.4824(10^{-6}) \\ -1.2225(10^{-4}) \end{Bmatrix}$$

or

$$\delta\mathbf{v}_f^- = 9.4824(10^{-6})\hat{\mathbf{i}} - 1.2225(10^{-4})\hat{\mathbf{j}}(\text{km/s}) \tag{f}$$

Since the chaser is in the same circular orbit as the target, its relative velocity is initially zero; that is, $\delta\mathbf{v}_0^- = \mathbf{0}$. (See also Eq. 7.68 at the end of the next section.) Thus,

$$\Delta\mathbf{v}_0 = \delta\mathbf{v}_0^+ - \delta\mathbf{v}_0^- = \left(-9.4824\times10^{-6}\hat{\mathbf{i}} - 1.2225\times10^{-4}\hat{\mathbf{j}}\right) - \mathbf{0}$$
$$= -9.4824\times10^{-6}\hat{\mathbf{i}} + 1.2225\times10^{-4}\hat{\mathbf{j}}(\text{km/s})$$

which implies

$$\|\Delta\mathbf{v}_0\| = 0.1226\,\text{m/s} \tag{g}$$

At the end of the rendezvous maneuver, $\delta\mathbf{v}_f^+ = \mathbf{0}$, so that

$$\Delta\mathbf{v}_f = \delta\mathbf{v}_f^+ - \delta\mathbf{v}_f^- = \mathbf{0} - \left(9.4824\times10^{-6}\hat{\mathbf{i}} - 1.2225\times10^{-4}\hat{\mathbf{j}}\right)$$
$$= -9.4824\times10^{-6}\hat{\mathbf{i}} + 1.2225\times10^{-4}\hat{\mathbf{j}}(\text{km/s})$$

Therefore,

$$\|\Delta\mathbf{v}_f\| = 0.1226\text{m/s} \tag{h}$$

The total delta-v required is

$$\Delta v_{\text{total}} = \|\Delta\mathbf{v}_0\| + \|\Delta\mathbf{v}_f\| = \boxed{0.2452\ \text{m/s}} \tag{i}$$

The coplanar rendezvous trajectory relative to the CW frame is sketched in Fig. 7.11. Notice that in the CW frame circular orbits appear as straight lines parallel to the y axis. This is due to the linearization we did, based on Eq. (7.15).

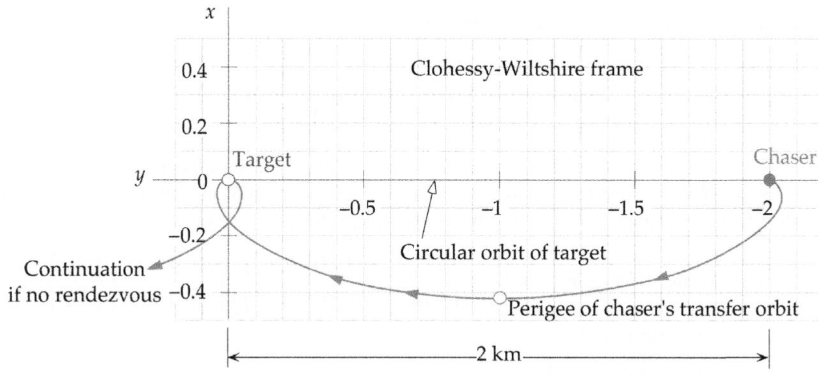

FIG. 7.11

Motion of the chaser relative to the target.

7.6 RELATIVE MOTION IN CLOSE-PROXIMITY CIRCULAR ORBITS

Fig. 7.12 shows two spacecraft in coplanar circular orbits. Let us calculate the velocity $\delta\mathbf{v}$ of the chase vehicle B relative to target A when they are in close proximity. "Close proximity" means that

$$\frac{\delta r}{R} << 1$$

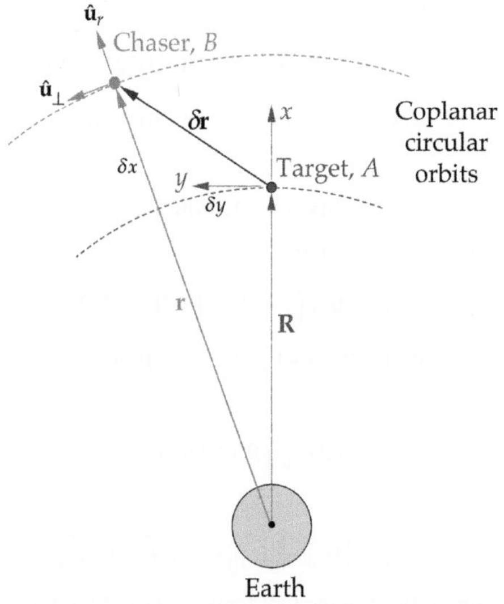

FIG. 7.12

Two spacecraft in close proximity.

To solve this problem we must use the relative velocity equation,

$$\mathbf{v}_B = \mathbf{v}_A + \boldsymbol{\Omega} \times \delta\mathbf{r} + \delta\mathbf{v} \tag{7.60}$$

where $\boldsymbol{\Omega}$ is the angular velocity of the Clohessy-Wiltshire frame attached to A,

$$\boldsymbol{\Omega} = n\hat{\mathbf{k}}$$

n is the mean motion of the target vehicle,

$$n = \frac{v_A}{R} \tag{7.61}$$

where, by virtue of the circular orbit,

$$v_A = \sqrt{\frac{\mu}{R}} \tag{7.62}$$

Solving Eq. (7.60) for the relative velocity $\delta\mathbf{v}$ yields

$$\delta\mathbf{v} = \mathbf{v}_B - \mathbf{v}_A - \left(n\hat{\mathbf{k}}\right) \times \delta\mathbf{r} \tag{7.63}$$

Since the chase orbit is circular, we have for the first term on the right-hand side of Eq. (7.63)

$$\mathbf{v}_B = \sqrt{\frac{\mu}{r}}\hat{\mathbf{u}}_\perp = \sqrt{\frac{\mu}{r}}\left(\hat{\mathbf{k}} \times \hat{\mathbf{u}}_r\right) = \sqrt{\mu}\hat{\mathbf{k}} \times \left(\frac{1}{\sqrt{r}}\frac{\mathbf{r}}{r}\right) \tag{7.64}$$

Since, as is apparent from Fig. 7.12, $\mathbf{r} = \mathbf{R} + \delta\mathbf{r}$, we can write this expression for $\mathbf{v}_B$ as follows:

$$\mathbf{v}_B = \sqrt{\mu}\hat{\mathbf{k}} \times r^{-3/2}(\mathbf{R} + \delta\mathbf{r}) \tag{7.65}$$

Now

$$r^{-3/2} = \left(r^2\right)^{-3/4} = \left[\overbrace{R^2\left(1 + \frac{2\mathbf{R}\cdot\delta\mathbf{r}}{R^2}\right)}^{\text{See Eq.7.18}}\right]^{-3/4} = R^{-3/2}\left(1 + \frac{2\mathbf{R}\cdot\delta\mathbf{r}}{R^2}\right)^{-3/4} \tag{7.66}$$

Using the binomial theorem (Eq. 5.44), and retaining terms at most linear in $\delta\mathbf{r}$, we find

$$\left(1 + \frac{2\mathbf{R}\cdot\delta\mathbf{r}}{R^2}\right)^{-3/4} = 1 - \frac{3}{2}\frac{\mathbf{R}\cdot\delta\mathbf{r}}{R^2}$$

Substituting this into Eq. (7.66) leads to

$$r^{-3/2} = R^{-3/2} - \frac{3}{2}\frac{\mathbf{R}\cdot\delta\mathbf{r}}{R^{7/2}}$$

Upon substituting this result into Eq. (7.65) we get

$$\mathbf{v}_B = \sqrt{\mu}\hat{\mathbf{k}} \times (\mathbf{R} + \delta\mathbf{r})\left(R^{-3/2} - \frac{3}{2}\frac{\mathbf{R}\cdot\delta\mathbf{r}}{R^{7/2}}\right)$$

Retaining terms at most linear in $\delta\mathbf{r}$ we can write this as

$$\mathbf{v}_B = \hat{\mathbf{k}} \times \left\{\sqrt{\frac{\mu}{R}}\frac{\mathbf{R}}{R} + \frac{\sqrt{\mu/R}}{R}\delta\mathbf{r} - \frac{3}{2}\frac{\sqrt{\mu/R}}{R}\left[\left(\frac{\mathbf{R}}{R}\right)\cdot\delta\mathbf{r}\right]\frac{\mathbf{R}}{R}\right\}$$

Using Eqs. (7.61) and (7.62), together with the facts that $\delta\mathbf{r} = \delta x\hat{\mathbf{i}} + \delta y\hat{\mathbf{j}}$ and $\mathbf{R}/R = \hat{\mathbf{i}}$, this reduces to

$$\begin{aligned}\mathbf{v}_B &= \hat{\mathbf{k}} \times \left\{ v_A\hat{\mathbf{i}} + \frac{v_A}{R}\left(\delta x\hat{\mathbf{i}} + \delta y\hat{\mathbf{j}}\right) - \frac{3}{2}\frac{v_A}{R}\left[\hat{\mathbf{i}}\cdot\left(\delta x\hat{\mathbf{i}} + \delta y\hat{\mathbf{j}}\right)\hat{\mathbf{i}}\right]\right\} \\ &= v_A\hat{\mathbf{j}} + \left(-n\delta y\hat{\mathbf{i}} + n\delta x\hat{\mathbf{j}}\right) - \frac{3}{2}n\delta x\hat{\mathbf{j}}\end{aligned}$$

so that

$$\mathbf{v}_B = -n\delta y\hat{\mathbf{i}} + \left(v_A - \frac{1}{2}n\delta x\right)\hat{\mathbf{j}} \tag{7.67}$$

This is the absolute velocity of the chaser resolved into components in the target's CW frame.

Substituting Eq. (7.67) into Eq. (7.63) and using the fact that $\mathbf{v}_A = v_A\hat{\mathbf{j}}$ yields

$$\begin{aligned}\delta\mathbf{v} &= \left[-n\delta y\hat{\mathbf{i}} + \left(v_A - \frac{1}{2}n\delta x\right)\hat{\mathbf{j}}\right] - \left(v_A\hat{\mathbf{j}}\right) - \left(n_A\hat{\mathbf{k}}\right) \times \left(\delta x\hat{\mathbf{i}} + \delta y\hat{\mathbf{j}}\right) \\ &= -n\delta y\hat{\mathbf{i}} + v_A\hat{\mathbf{j}} - \frac{1}{2}n\delta x\hat{\mathbf{j}} - v_A\hat{\mathbf{j}} - n\delta x\hat{\mathbf{j}} + n\delta y\hat{\mathbf{i}}\end{aligned}$$

so that

$$\delta\mathbf{v} = -\frac{3}{2}n\delta x\hat{\mathbf{j}} \tag{7.68}$$

This is the velocity of the chaser as measured in the moving reference frame of the neighboring target. Keep in mind that circular orbits were assumed at the outset.

In the Clohessy-Wiltshire frame, neighboring coplanar circular orbits appear to be straight lines parallel to the y axis, which is the orbit of the origin. Fig. 7.13 illustrates this point, showing also the linear velocity variation according to Eq. (7.68).

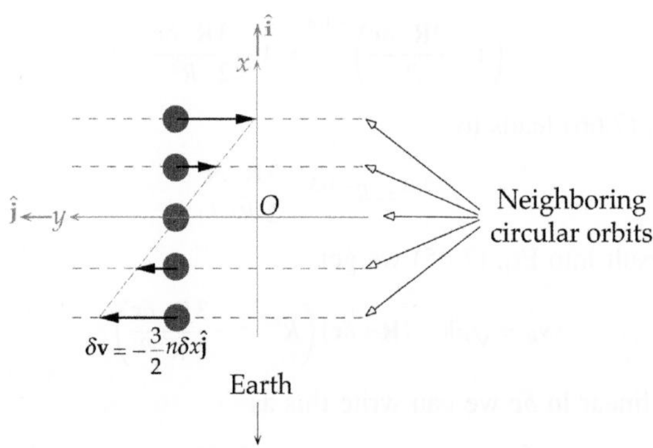

FIG. 7.13

Circular orbits, with relative velocity directions, in the vicinity of the Clohessy-Wiltshire frame.

PROBLEMS

Section 7.3

7.1 Two manned spacecraft, *A* and *B* (see the figure), are in circular polar ($i = 90°$) orbits around the earth. *A*'s orbital altitude is 300 km, *B*'s is 250 km. At the instant shown (*A* over the equator, *B* over the North Pole), calculate (a) the position, (b) velocity, and (c) the acceleration of *B* relative to *A*. *A*'s *y* axis points always in the flight direction and its *x* axis is directed radially outward at all times.
{Ans.: (a) $\mathbf{r}_{\text{rel}})_{xyz} = -6678\hat{\mathbf{i}} + 6628\hat{\mathbf{j}}(\text{km})$; (b) $\mathbf{v}_{\text{rel}})_{xyz} = -0.08693\hat{\mathbf{i}}(\text{km/s})$;
(c) $\mathbf{a}_{\text{rel}})_{xyz} = -1.140\left(10^{-6}\right)\hat{\mathbf{j}}(\text{km/s}^2)$}

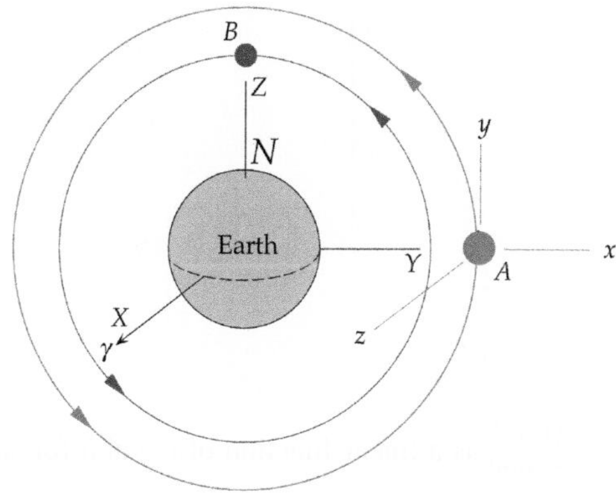

7.2 Spacecraft *A* and *B* are in coplanar, circular geocentric orbits. The orbital radii are shown in the figure. When *B* is directly below *A*, as shown, calculate *B*'s acceleration $\mathbf{a}_{\text{rel}})_{xyz}$ relative to *A*.
{Ans.: $\mathbf{a}_{\text{rel}})_{xyz} = -0.268\hat{\mathbf{i}}(\text{m/s}^2)$}

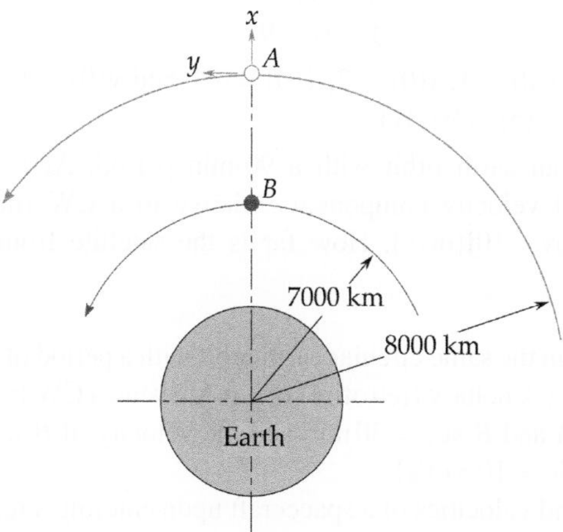

Section 7.3

7.3 Use the order-of-magnitude analysis in this chapter as a guide to answer the following questions:

(a) If $\mathbf{r} = \mathbf{R} + \delta\mathbf{r}$, express $\sqrt{r}$ (where $r = \sqrt{\mathbf{r}\cdot\mathbf{r}}$) to the first order in $\delta\mathbf{r}$ (i.e., to the first order in the components of $\delta\mathbf{r} = \delta x\hat{\mathbf{i}} + \delta y\hat{\mathbf{j}} + \delta z\hat{\mathbf{k}}$). In other words, find $O(\delta\mathbf{r})$, such that $\sqrt{r} = \sqrt{R} + O(\delta\mathbf{r})$, where $O(\delta\mathbf{r})$ is linear in $\delta\mathbf{r}$.

(b) For the special case $\mathbf{R} = 3\hat{\mathbf{i}} + 4\hat{\mathbf{j}} + 5\hat{\mathbf{k}}$ and $\delta\mathbf{r} = 0.01\hat{\mathbf{i}} - 0.01\hat{\mathbf{j}} + 0.03\hat{\mathbf{k}}$, calculate $\sqrt{r} - \sqrt{R}$ and compare that result with $O(\delta\mathbf{r})$.

(c) Repeat Part (b) using $\delta\mathbf{r} = \hat{\mathbf{i}} - \hat{\mathbf{j}} + 3\hat{\mathbf{k}}$ and compare the results.

{Ans.: (a) $O(\delta\mathbf{r}) = \mathbf{R}\cdot\delta\mathbf{r}/(2R^{3/2})$; (b) $O(\delta\mathbf{r})/\left(\sqrt{r} - \sqrt{R}\right) = 0.998$;
(c) $O(\delta\mathbf{r})/\left(\sqrt{r} - \sqrt{R}\right) = 0.903$}

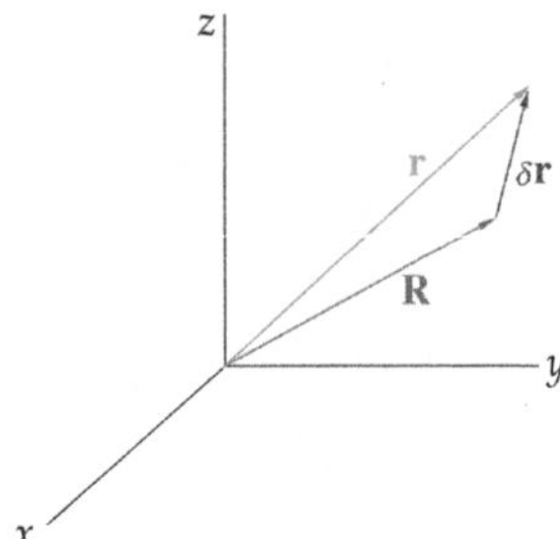

7.4 Write the expression $r = \frac{a\left(1-e^2\right)}{1+e\cos\theta}$ as a linear function of e, valid for small values of e ($e \ll 1$).

Section 7.4

7.5 Given $\ddot{x} + 9x = 10$, with the initial conditions $x = 5$ and $\dot{x} = -3$ at $t = 0$, find x and $\dot{x}$ at $t = 1.2$.
{Ans.: $x(1.2) = -1.934$, $\dot{x}(1.2) = 7.853$}

7.6 Given that

$$\ddot{x} + 10x + 2\dot{y} = 0$$
$$\ddot{y} + 3\dot{x} = 0$$

with initial conditions $x(0) = 1$, $y(0) = 2$, $\dot{x}(0) = -3$, and $\dot{y}(0) = 4$, find x and y at $t = 5$.
{Ans.: $x(5) = -6.460$, $y(5) = 97.31$}

7.7 A space station is in an earth orbit with a 90-min period. At $t = 0$ a satellite has the following position and velocity components relative to a CW frame attached to the space station: $\delta\mathbf{r} = 1\hat{\mathbf{i}}$(km), $\delta\mathbf{v} = 10\hat{\mathbf{j}}$(m/s). How far is the satellite from the space station 15 min later?
{Ans.: 11.2 km}

7.8 Spacecraft A and B are in the same circular earth orbit with a period of 2 h. B is 6 km ahead of A. At $t = 0$, B applies an in-track delta-v (retrofire) of 3 m/s. Using a CW frame attached to A, determine the distance between A and B at $t = 30$ min and the velocity of B relative to A at that instant.
{Ans.: $\delta r = 10.9$ km, $\delta v = 10.8$ m/s}

7.9 The CW coordinates and velocities of a spacecraft upon entering a rendezvous trajectory with the target vehicle are shown. The spacecraft orbits are coplanar. Calculate the distance d of the

spacecraft from the target when $t = \pi/(2n)$, where n is the mean motion of the target's circular orbit.
{Ans.: $0.900\delta r$}

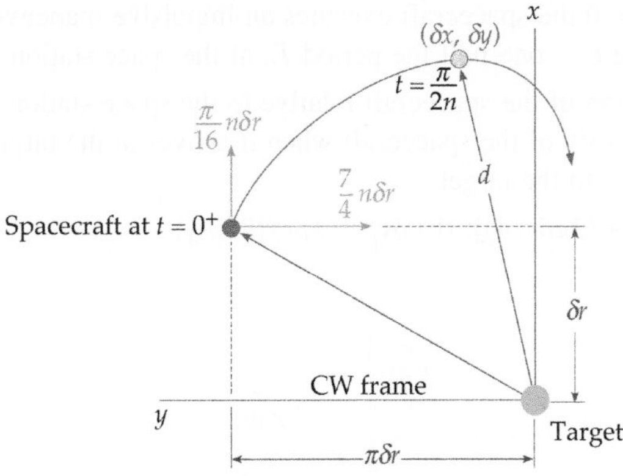

7.10 At time $t = 0$ a particle is at the origin of a CW frame with a relative velocity $\delta \mathbf{v}_0 = v\hat{\mathbf{j}}$. What will be the relative speed of the particle after a time equal to one-half the orbital period of the CW frame?
{Ans.: $7v$}

7.11 The chaser and the target are in close-proximity, coplanar circular orbits. At $t = 0$ the position of the chaser relative to the target is $\delta \mathbf{r}_0 = r\hat{\mathbf{i}} + a\hat{\mathbf{j}}$, where a is given and r is unknown. The relative velocity at $t = 0^+$ is $\delta \mathbf{v}_0^+ = v_0\hat{\mathbf{j}}$ (v_0 is unknown), and the chaser ends up at $\delta \mathbf{r}_f = -a\hat{\mathbf{i}}$ when $t = \pi/n$, where n is the mean motion of the target. Use the Clohessy-Wiltshire equations to find the required value of the orbital spacing r.
{Ans.: $1.424a$}

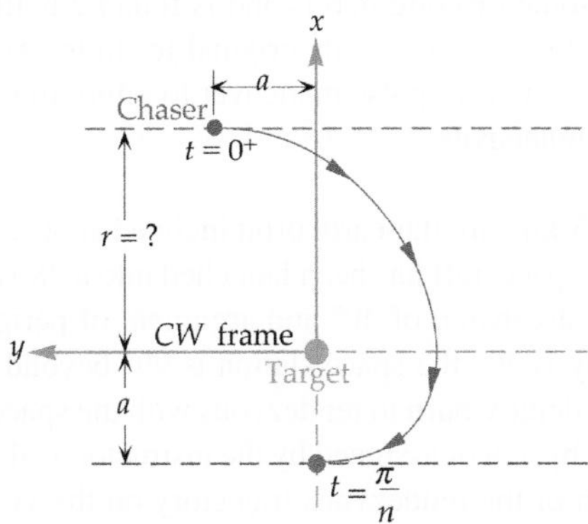

Section 7.5

7.12 A space station is in a circular earth orbit of radius 6600 km. An approaching spacecraft executes a delta-v burn when its position vector relative to the space station is $\delta \mathbf{r}_0 = \hat{\mathbf{i}} + \hat{\mathbf{j}} + \hat{\mathbf{k}}(\text{km})$. Just before the burn the relative velocity of the spacecraft was $\delta \mathbf{v}_0^- = 5(\text{m/s})$. Calculate the total

delta-v required for the spacecraft to rendezvous with the station in a time equal to one-third period of the space station's orbit.
{Ans.: 6.21 m/s}

7.13 A space station is in circular orbit 2 of radius R. A spacecraft is in coplanar circular orbit 1 of radius $R + \delta r$. At $t = 0$ the spacecraft executes an impulsive maneuver to rendezvous with the space station at time $t_f =$ one-half the period T_0 of the space station. If $\delta u_0^+ = 0$ find

(a) The initial position of the spacecraft relative to the space station.

(b) The relative velocity of the spacecraft when it arrives at the target. Sketch the rendezvous trajectory relative to the target.

{Ans.: (a) $\delta \mathbf{r}_0 = \delta r \hat{\mathbf{i}} + (3\pi\delta r/4)\hat{\mathbf{j}}$; (b) $\delta \mathbf{v}_f^- = \pi\delta r/(2T_0)\hat{\mathbf{j}}$}

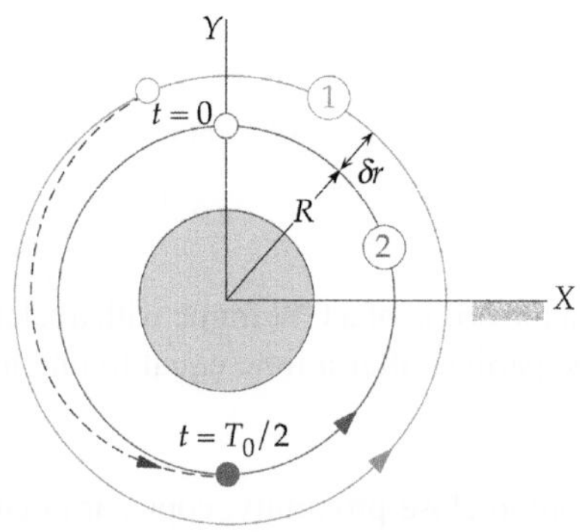

7.14 If $\delta u_0^+ = 0$, calculate the total delta-v required for rendezvous if $\delta \mathbf{r}_0 = \delta y_0 \hat{\mathbf{j}}$, $\delta \mathbf{v}_0^- = \mathbf{0}$, and $t_f =$ the period T of the circular target orbit. Sketch the rendezvous trajectory relative to the target.
{Ans.: $\Delta v_{\text{tot}} = 2\delta y_o/(3T)$}

7.15 A GEO satellite strikes some orbiting debris and is found 2 h afterward to have drifted to the position $\delta \mathbf{r} = -10\hat{\mathbf{i}} + 10\hat{\mathbf{j}}$(km) relative to its original location. At that time the only slightly damaged satellite initiates a two-impulse maneuver to return to its original location in 6 h. Find the total delta-v for this maneuver.
{Ans.: 3.5 m/s}

7.16 A space station is in a 245-km circular earth orbit inclined at 30°. The right ascension of its node line is 40°. Meanwhile, a spacecraft has been launched into a 280 km by 250 km orbit inclined at 30.1°, with a nodal right ascension of 40° and argument of perigee equal to 60°. When the spacecraft's true anomaly is 40° the space station is 99° beyond its node line. At that instant the spacecraft executes a delta-v burn to rendezvous with the space station in (precisely) t_f hours, where t_f is either selected by you or assigned by the instructor. Calculate the total delta-v required and sketch the projection of the rendezvous trajectory on the xy plane of the space station coordinates.

7.17 The target A is in a circular earth orbit with mean motion n. The chaser B is directly above A in a slightly larger circular orbit having the same plane as A. What relative initial velocity $\delta \mathbf{v}_0^+$ is required so that B arrives at target A at time $t_f =$ one-half the target's period?
{Ans.: $\delta \mathbf{v}_0^+ = -0.589n\delta x_0 \hat{\mathbf{i}} - 1.75n\delta x_0$}

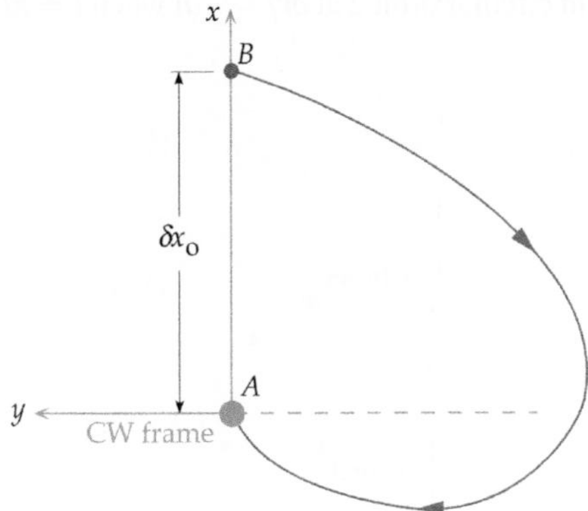

Section 7.6

7.18 The space station is in a circular earth orbit of radius 6600 km. Another spacecraft is also in a circular orbit in the same plane as the space station. At the instant that the position of the spacecraft relative to the space station, in Clohessy-Wiltshire coordinates, is $\delta\mathbf{r} = 5\hat{\mathbf{i}}$(km), what is the relative velocity $\delta\mathbf{v}$ of the spacecraft in meters/s?

{Ans.: 8.83 m/s}

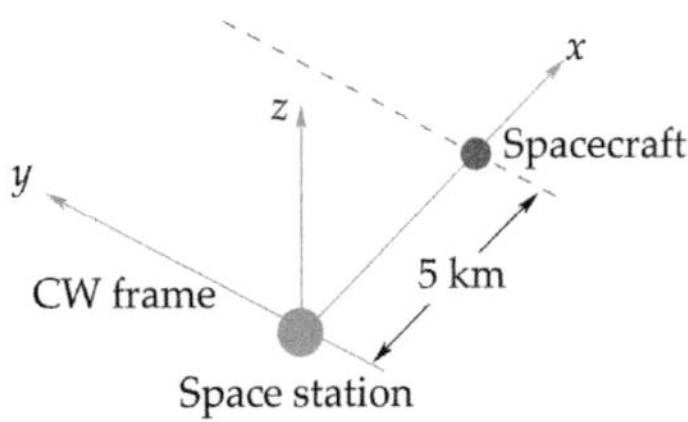

7.19 A spacecraft and the space station are in coplanar circular orbits. The space station has an orbital radius R and a mean motion n. The spacecraft's radius is $R - d$ ($d \ll R$). If a two-impulse rendezvous maneuver with $t_f = \pi/(4n)$ is initiated with zero relative velocity in the x direction ($\delta u_0^+ = 0$), calculate the total delta-v.

{Ans.: $4.406nd$}

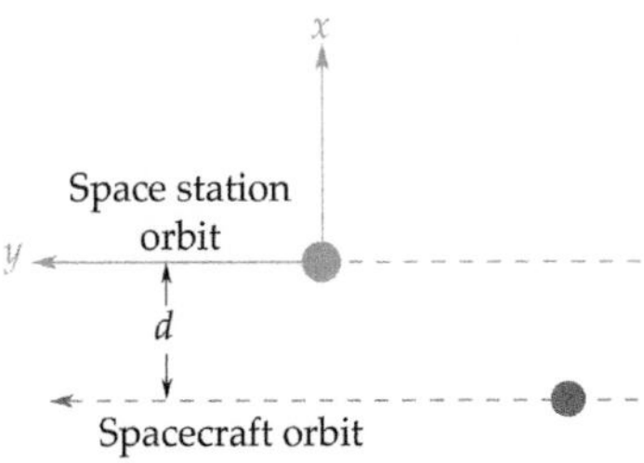

7.20 The chaser and the target are in close-proximity, coplanar circular orbits. At $t = 0$ the position of the chaser relative to the target is $\delta\mathbf{r}_0 = a\hat{\mathbf{i}}$. Use the CW equations to find the total delta-v required

for the chaser to end up in circular orbit 2 at $\delta\mathbf{r}_f = -a\hat{\mathbf{i}}$ when $t = \pi/n$, where n is the mean motion of the target.
{Ans.: na}

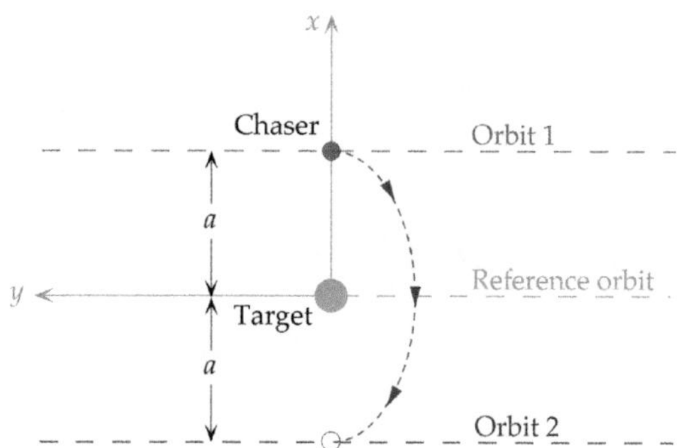

REFERENCE

Clohessy, W.H., Wiltshire, R.S., 1960. Terminal guidance system for satellite rendezvous. J. Aerosol Sci. 27, 653–658.

CHAPTER

INTERPLANETARY TRAJECTORIES 8

8.1 INTRODUCTION

In this chapter, we consider some basic aspects of planning interplanetary missions. We begin by considering Hohmann transfers, which are the easiest to analyze and the most energy efficient. The orbits of the planets involved must lie in the same plane and the planets must be positioned just right for a Hohmann transfer to be used. The time between such opportunities is derived. The method of patched conics is employed to divide the mission up into three parts: the hyperbolic departure trajectory relative to the home planet, the cruise ellipse relative to the sun, and the hyperbolic arrival trajectory relative to the target planet.

The use of patched conics is justified by calculating the radius of a planet's sphere of influence and showing how small it is on the scale of the solar system. Matching the velocity of the spacecraft at the home planet's sphere of influence to that required to initiate the outbound cruise phase and then specifying the periapse radius of the departure hyperbola determines the delta-v requirement at departure. The sensitivity of the target radius to the burnout conditions is discussed. Matching the velocities at the target planet's sphere of influence and specifying the periapse of the arrival hyperbola yields the delta-v at the target for a planetary rendezvous or the direction of the outbound hyperbola for a planetary flyby. Flyby maneuvers are discussed, including the effect of leading- and trailing-side flybys, and some noteworthy examples of the use of gravity assist maneuvers are presented.

The chapter concludes with an analysis of the situation in which the planets' orbits are not coplanar and the transfer ellipse is tangent to neither orbit. This is akin to the chase maneuver in Chapter 6 and requires the solution of Lambert's problem using Algorithm 5.2.

8.2 INTERPLANETARY HOHMANN TRANSFERS

As can be seen from Table A.1, the orbits of most of the planets in the solar system lie very close to the earth's orbital plane (the ecliptic plane). The innermost planet, Mercury, and the formerly outermost planet, Pluto (which was reclassified by the International Astronomical Union as a *dwarf planet* in 2006), differ most in inclination (7° and 17°, respectively). The orbital planes of the other planets lie within 3.5° degrees of the ecliptic. It is also evident from Table A.1 that most of the planetary orbits have small eccentricities, the exceptions once again being Mercury and Pluto.

Orbital Mechanics for Engineering Students. https://doi.org/10.1016/B978-0-12-824025-0.00008-8

Besides Pluto, there are currently four other IAU-recognized dwarf planets orbiting the Sun (namely, Ceres (the smallest), Haumea, Makemake, and Eris (the largest)). Ceres lies in the asteroid belt between Mars and Jupiter. Reclassified as a dwarf planet in 2006, it was visited by NASA's *Dawn* spacecraft in 2015. The three other dwarfs lie at the outer reaches of the solar system, beyond Neptune's orbit. Eris is roughly the size of Pluto, but its highly elliptical orbit takes it far beyond that of Pluto. All of the dwarf planets have orbits that are significantly inclined to the ecliptic, from 10.6° for Ceres to 44.2° for Eris.

To simplify the beginning of our study of interplanetary trajectories, we will focus on the eight major planets and assume that all of their orbits are circular and coplanar. In Section 8.10, we will relax this assumption.

The most energy-efficient way for a spacecraft to transfer from one planet's orbit to another is to use a Hohmann transfer ellipse (Section 6.3). Consider Fig. 8.1, which shows a Hohmann transfer from an inner planet 1 to an outer planet 2. The departure point D is at periapsis (perihelion) of the transfer ellipse and the arrival point is at apoapsis (aphelion). The circular orbital speed of planet 1 relative to the sun is given by Eq. (2.63),

$$V_1 = \sqrt{\frac{\mu_{\text{sun}}}{R_1}} \tag{8.1}$$

The specific angular momentum h of the transfer ellipse relative to the sun is found from Eq. (6.2), so that the heliocentric speed $V_D^{(v)}$ of the space vehicle on the transfer ellipse at the departure point D is

$$V_D^{(v)} = \frac{h}{R_1} = \sqrt{2\mu_{\text{sun}}}\sqrt{\frac{R_2}{R_1(R_1+R_2)}} \tag{8.2}$$

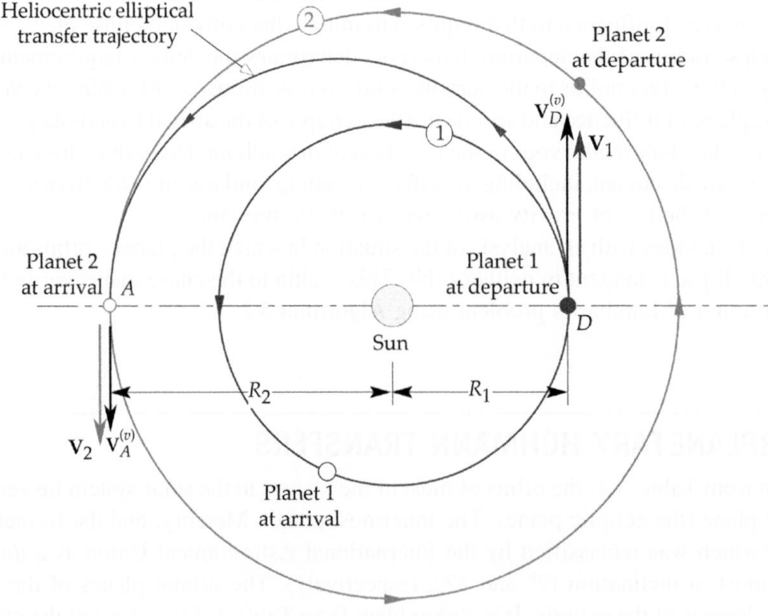

FIG. 8.1

Hohmann transfer from inner planet 1 to outer planet 2. $V_D^{(v)} > V_1$ and $V_A^{(v)} < V_2$.

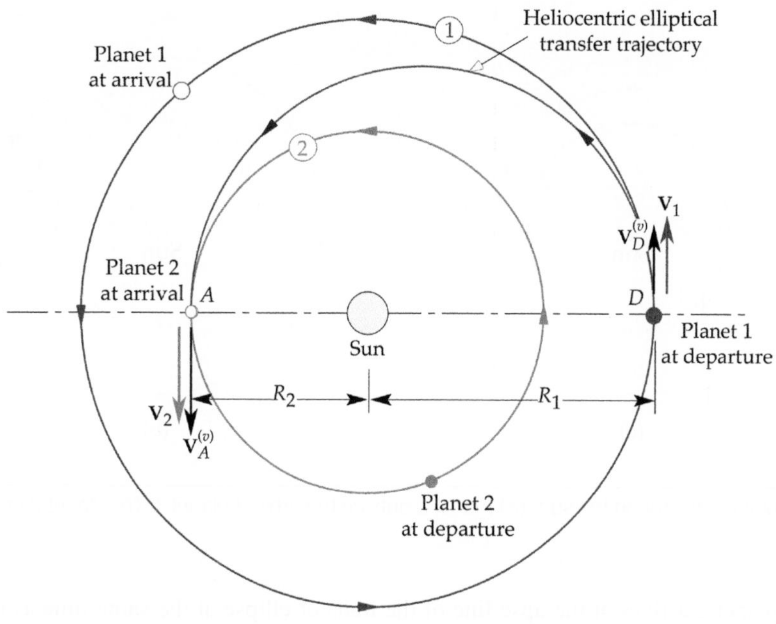

FIG. 8.2

Hohmann transfer from outer planet 1 to inner planet 2. $V_D^{(v)} < V_1$ and $V_A^{(v)} > V_2$.

This is greater than the speed V_1 of the planet. Therefore, the required spacecraft delta-v at D is

$$\Delta V_D = V_D^{(v)} - V_1 = \sqrt{\frac{\mu_{sun}}{R_1}}\left(\sqrt{\frac{2R_2}{R_1+R_2}} - 1\right) \tag{8.3}$$

Likewise, the delta-v at the arrival point A is

$$\Delta V_A = V_2 - V_A^{(v)} = \sqrt{\frac{\mu_{sun}}{R_2}}\left(1 - \sqrt{\frac{2R_1}{R_1+R_2}}\right) \tag{8.4}$$

This velocity increment, like that at point D, is positive since planet 2 is traveling faster than the spacecraft at point A.

For a mission from an outer planet to an inner planet, as illustrated in Fig. 8.2, the delta-v's computed using Eqs. (8.3) and (8.4) will both be negative instead of positive. This is because the departure point and the arrival point are now at aphelion and perihelion, respectively, of the transfer ellipse. The speed of the spacecraft must be reduced for it to drop into the lower energy transfer ellipse at the departure point D, and it must be reduced again at point A to arrive in the lower energy circular orbit of planet 2.

8.3 RENDEZVOUS OPPORTUNITIES

The purpose of an interplanetary mission is for the spacecraft to not only intercept a planet's orbit but also to rendezvous with the planet when it gets there. For rendezvous to occur at the end of a Hohmann transfer, the location of planet 2 in its orbit at the time of the spacecraft's departure from planet 1 must

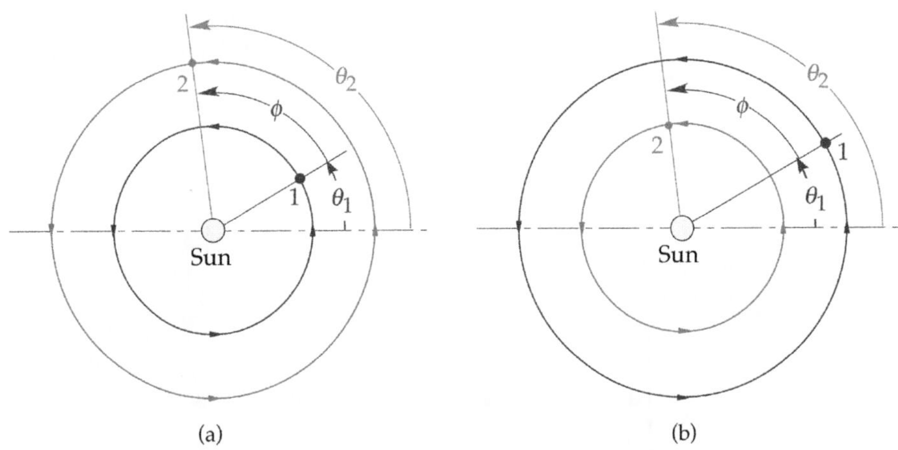

FIG. 8.3

Planets in circular orbits around the sun. (a) Planet 2 outside the orbit of planet 1. (b) Planet 2 inside the orbit of planet 1.

be such that planet 2 arrives at the apse line of the transfer ellipse at the same time as the spacecraft does. Phasing maneuvers (Section 6.5) are clearly not practical, especially for manned missions, due to the large periods of the heliocentric orbits.

Consider planet 1 and planet 2 in circular orbits around the sun, as shown in Fig. 8.3. Since the orbits are circular, we can choose a common horizontal apse line from which to measure the true anomaly θ. The true anomalies of planets 1 and 2, respectively, are

$$\theta_1 = \theta_1)_0 + n_1 t \tag{8.5}$$

$$\theta_2 = \theta_2)_0 + n_2 t \tag{8.6}$$

where n_1 and n_2 are the mean motions (angular velocities) of the planets and $\theta_1)_0$, and $\theta_2)_0$ are their true anomalies at time $t = 0$. The phase angle between the position vectors of the two planets is defined as

$$\phi = \theta_2 - \theta_1 \tag{8.7}$$

ϕ is the angular position of planet 2 relative to planet 1. Substituting Eqs. (8.5) and (8.6) into Eq. (8.7) we get

$$\phi = \phi_0 + (n_2 - n_1)t \tag{8.8}$$

where ϕ_0 is the phase angle at time zero, and $n_2 - n_1$ is the orbital angular velocity of planet 2 relative to planet 1. If the orbit of planet 1 lies inside that of planet 2, as in Fig. 8.3a, then $n_1 > n_2$. Therefore, the relative angular velocity $n_2 - n_1$ is negative, which means planet 2 moves clockwise relative to planet 1. On the other hand, if planet 1 is outside planet 2, then $n_2 - n_1$ is positive, so that the relative motion is counterclockwise.

The phase angle obviously varies linearly with time according to Eq. (8.8). If the phase angle is ϕ_0 at $t = 0$, how long will it take to become ϕ_0 again? The answer: when the position vector of planet 2 rotates through 2π radians relative to planet 1. The time required for the phase angle to return to its initial value is called the synodic period, which is denoted T_{syn}. For the case shown in Fig. 8.3a, in which the relative motion is clockwise, T_{syn} is the time required for ϕ to change from ϕ_0 to $\phi_0 - 2\pi$. From Eq. (8.8) we have

$$\phi_0 - 2\pi = \phi_0 + (n_2 - n_1)T_{syn}$$

so that

$$T_{syn} = \frac{2\pi}{n_1 - n_2} \quad (n_1 > n_2)$$

For the situation illustrated in Fig. 8.3b ($n_2 > n_1$), T_{syn} is the time required for ϕ to go from ϕ_0 to $\phi_0 + 2\pi$, in which case Eq. (8.8) yields

$$T_{syn} = \frac{2\pi}{n_2 - n_1} \quad (n_2 > n_1)$$

Both cases are covered by writing

$$T_{syn} = \frac{2\pi}{|n_1 - n_2|} \tag{8.9}$$

Recalling Eq. (3.9), we can write $n_1 = 2\pi/T_1$ and $n_2 = 2\pi/T_2$. Thus, in terms of the orbital periods of the two planets,

$$T_{syn} = \frac{T_1 T_2}{|T_1 - T_2|} \tag{8.10}$$

Observe that T_{syn} is the orbital period of planet 2 relative to planet 1.

EXAMPLE 8.1

Calculate the synodic period of Mars relative to that of the earth.

Solution

In Table A.1 we find the orbital periods of earth and Mars:

$$T_{earth} = 365.26 \text{days} (1 \text{year})$$
$$T_{Mars} = 1 \text{ year plus } 321.73 \text{ days} = 687.99 \text{ days}$$

Hence,

$$T_{syn} = \frac{T_{earth} T_{Mars}}{|T_{earth} - T_{Mars}|} = \frac{365.26 \times 687.99}{|365.26 - 687.99|} = \boxed{777.9 \text{days}}$$

These are earth days (1 day = 24 h). Therefore, it takes 2.13 years for a given configuration of Mars relative to the earth to occur again.

Fig. 8.4 depicts a mission from planet 1 to planet 2. Following a heliocentric Hohmann transfer, the spacecraft intercepts and undergoes rendezvous with planet 2. Later it returns to planet 1 by means of another Hohmann transfer. The major axis of the heliocentric transfer ellipse is the sum of the radii of

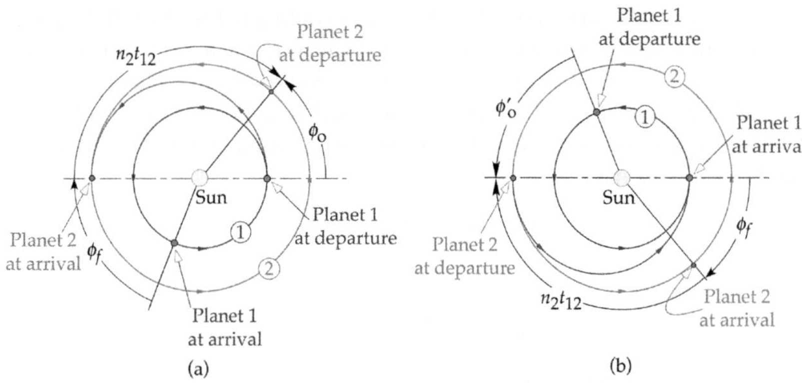

FIG. 8.4

Round trip mission, with layover, to planet 2. (a) Departure and rendezvous with planet 2. (b) Return and rendezvous with planet 1.

the two planets' orbits, $R_1 + R_2$. The time t_{12} required for the transfer is one-half the period of the ellipse. Hence, according to the period formula (Eq. 2.83),

$$t_{12} = \frac{\pi}{\sqrt{\mu_{sun}}}\left(\frac{R_1 + R_2}{2}\right)^{3/2} \tag{8.11}$$

During the time it takes the spacecraft to fly from orbit 1 to orbit 2, through an angle of π radians, planet 2 must move around its circular orbit and end up at a point directly opposite planet 1's position when the spacecraft departed. Since planet 2's angular velocity is n_2, the angular distance traveled by the planet during the spacecraft's trip is n_2t_{12}. Hence, as can be seen from Fig. 8.4a, the initial phase angle ϕ_0 between the two planets is

$$\phi_0 = \pi - n_2t_{12} \tag{8.12}$$

When the spacecraft arrives at planet 2, the phase angle will be ϕ_f, which is found using Eqs. (8.8) and (8.12).

$$\begin{aligned}\phi_f &= \phi_0 + (n_2 - n_1)t_{12} = (\pi - n_2t_{12}) + (n_2 - n_1)t_{12} \\ \phi_f &= \pi - n_1t_{12}\end{aligned} \tag{8.13}$$

For the situation illustrated in Fig. 8.4, planet 2 ends up being behind planet 1 by an amount equal to the magnitude of ϕ_f.

At the start of the return trip, illustrated in Fig. 8.4b, planet 2 must be ϕ_0' radians ahead of planet 1. Since the spacecraft flies the same Hohmann transfer trajectory back to planet 1, the time of flight is t_{12}, the same as the outbound leg. Therefore, the distance traveled by planet 1 during the return trip is the same as the outbound leg, which means

$$\phi_0' = -\phi_f \tag{8.14}$$

In any case, the phase angle at the beginning of the return trip must be the negative of the phase angle at arrival from planet 1. The time required for the phase angle to reach its proper value is called the wait time, t_{wait}. Setting time equal to zero at the instant we arrive at planet 2, Eq. (8.8) becomes

$$\phi = \phi_f + (n_2 - n_1)t$$

ϕ becomes $-\phi_f$ after the time t_{wait}. That is,

$$-\phi_f = \phi_f + (n_2 - n_1)t_{wait}$$

or

$$t_{wait} = \frac{-2\phi_f}{n_2 - n_1} \tag{8.15}$$

where ϕ_f is given by Eq. (8.13). Eq. (8.15) may yield a negative result, which means the desired phase relation occurred in the past. Therefore, we must add or subtract an integral multiple of 2π to the numerator to get a positive value for t_{wait}. Specifically, if $N = 0, 1, 2, \ldots$, then

$$t_{wait} = \frac{-2\phi_f - 2\pi N}{n_2 - n_1} \quad (n_1 > n_2) \tag{8.16}$$

$$t_{wait} = \frac{-2\phi_f + 2\pi N}{n_2 - n_1} \quad (n_1 < n_2) \tag{8.17}$$

where N is chosen to make t_{wait} positive. t_{wait} would probably be the smallest positive number thus obtained.

EXAMPLE 8.2

Calculate the minimum wait time for initiating a return trip from Mars to earth.

Solution

From Tables A.1 and A.2 we have

$$R_{earth} = 149.6 \times 10^6 \text{ km}$$
$$R_{Mars} = 227.9 \times 10^6 \text{ km}$$
$$\mu_{sun} = 132.71 \times 10^9 \text{ km}^3/\text{s}^2$$

According to Eq. (8.11), the time of flight from earth to Mars is

$$t_{12} = \frac{\pi}{\sqrt{\mu_{sun}}}\left(\frac{R_{earth} + R_{Mars}}{2}\right)^{3/2}$$
$$= \frac{\pi}{\sqrt{132.71 \times 10^9}}\left(\frac{149.6 \times 10^6 + 227.9 \times 10^6}{2}\right)^{3/2} = 2.2362 \times 10^7 \text{ s}$$

or

$$t_{12} = 258.82 \text{days}$$

From Eq. (3.9) and the orbital periods of earth and Mars (see Example 8.1 above) we obtain the mean motions of the earth and Mars

$$n_{earth} = \frac{2\pi}{365.26} = 0.017202 \text{rad/day}$$
$$n_{Mars} = \frac{2\pi}{687.99} = 0.0091327 \text{rad/day}$$

The phase angle between earth and Mars when the spacecraft reaches Mars is given by Eq. (8.13).

$$\phi_f = \pi - n_{\text{earth}} t_{12} = \pi - 0.017202 \cdot 258.82 = -1.3107\,\text{rad}$$

Since $n_{\text{earth}} > n_{\text{Mars}}$, we choose Eq. (8.16) to find the wait time

$$t_{\text{wait}} = \frac{-2\phi_f - 2\pi N}{n_{\text{Mars}} - n_{\text{earth}}} = \frac{-2(-1.3107) - 2\pi N}{0.0091327 - 0.017202} = 778.65N - 324.85\,\text{days}$$

$N = 0$ yields a negative value, which we cannot accept. Setting $N = 1$, we find

$$\boxed{t_{\text{wait}} = 453.8\,\text{days}}$$

This is the minimum wait time (1.24 years). Obviously, we could set $N = 2, 3, \ldots$ to obtain longer wait times.

For a spacecraft to depart on a mission to Mars by means of a Hohmann (minimum energy) transfer, the phase angle between earth and Mars must be that given by Eq. (8.12). Using the results of Example 8.2, we find it to be

$$\phi_0 = \pi - n_{\text{Mars}} t_{12} = \pi - 0.0091327 \cdot 258.82 = 0.7778\,\text{rad} = 44.57°$$

This opportunity occurs once every synodic period, which we found to be 2.13 years in Example 8.1. In Example 8.2, we found that the time to fly to Mars is 258.8 days, followed by a wait time of 453.8 days, followed by a return trip time of 258.8 days. Hence, the minimum total time for a manned Mars mission, using Hohmann transfers is

$$t_{\text{total}} = 258.8 + 453.8 + 258.8 = 971.4\,\text{days} = 2.66\,\text{years}$$

8.4 SPHERE OF INFLUENCE

The sun, of course, is the dominant celestial body in the solar system. It is over 1000 times more massive than the largest planet, Jupiter, and has a mass of over 300,000 earths. The sun's gravitational pull holds all the planets in its grasp according to Newton's law of gravity (Eq. 1.31). However, near a given planet, the influence of its own gravity exceeds that of the sun. For example, at its surface the earth's gravitational force is over 1600 times greater than the sun's. The inverse-square nature of the law of gravity means that the force of gravity F_g drops off rapidly with distance r from the center of attraction. If F_{g0} is the gravitational force at the surface of a planet with radius r_0, then Fig. 8.5 shows how rapidly the force diminishes with distance. At 10 body radii, the force is 1% of its value at the surface. Eventually, the force of the sun's gravitational field overwhelms that of the planet.

To estimate the radius of a planet's gravitational sphere of influence, consider the three-body system comprising a planet p of mass m_p, the sun s of mass m_s, and a space vehicle v of mass m_v illustrated in Fig. 8.6. The position vectors of the planet and spacecraft relative to an inertial frame centered at the sun are $\mathbf{R}$ and $\mathbf{R}_v$, respectively. The position vector of the space vehicle relative to the planet is $\mathbf{r}$. (Throughout this chapter we will use uppercase letters to represent position, velocity, and acceleration measured relative to the sun and lowercase letters when they are measured relative to a planet.) The gravitational force exerted on the vehicle by the planet is denoted $\mathbf{F}_p^{(v)}$, and that exerted by the

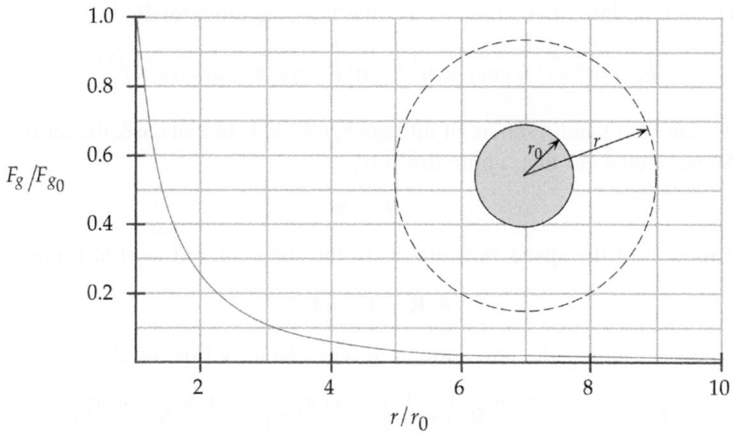

FIG. 8.5

Decrease of gravitational force with distance from a planet's surface.

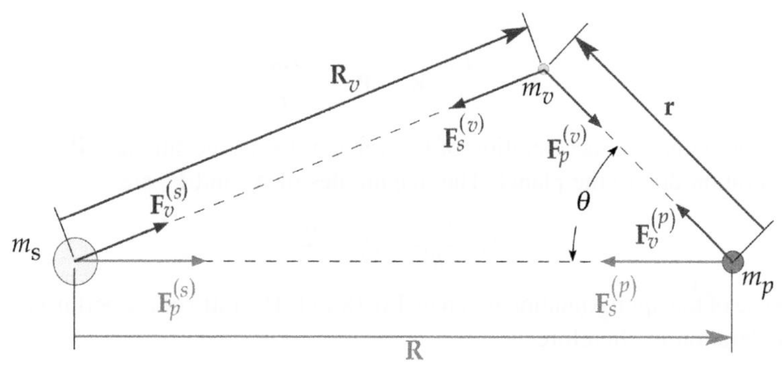

FIG. 8.6

Relative position and gravitational force vectors among the three bodies.

sun is $\mathbf{F}_s^{(v)}$. Likewise, the forces on the planet are $\mathbf{F}_s^{(p)}$ and $\mathbf{F}_v^{(p)}$, whereas on the sun we have $\mathbf{F}_v^{(s)}$ and $\mathbf{F}_p^{(s)}$. According to Newton's law of gravitation (Eq. 2.10), these forces are

$$\mathbf{F}_p^{(v)} = -\frac{Gm_v m_p}{r^3}\mathbf{r} \tag{8.18a}$$

$$\mathbf{F}_s^{(v)} = -\frac{Gm_v m_s}{R_v{}^3}\mathbf{R}_v \tag{8.18b}$$

$$\mathbf{F}_s^{(p)} = -\frac{Gm_p m_s}{R^3}\mathbf{R} \tag{8.18c}$$

Observe that

$$\mathbf{R}_v = \mathbf{R} + \mathbf{r} \tag{8.19}$$

From Fig. 8.6 and the law of cosines we see that the magnitude of $\mathbf{R}_v$ is

$$R_v = \left(R^2 + r^2 - 2Rr\cos\theta\right)^{1/2} = R\left[1 - 2(r/R)\cos\theta + (r/R)^2\right]^{1/2} \tag{8.20}$$

We expect that within the planet's sphere of influence, $r/R \ll 1$. In that case, the terms involving r/R in Eq. (8.20) can be neglected, so that, approximately,

$$R_v = R \tag{8.21}$$

The equation of motion of the spacecraft relative to the sun-centered inertial frame is

$$m_v\ddot{\mathbf{R}}_v = \mathbf{F}_s^{(v)} + \mathbf{F}_p^{(v)}$$

Solving for $\ddot{\mathbf{R}}_v$ and substituting the gravitational forces given by Eqs. (8.18a) and (8.18b), we get

$$\ddot{\mathbf{R}}_v = \frac{1}{m_v}\left(-\frac{Gm_vm_s}{R_v{}^3}\mathbf{R}_v\right) + \frac{1}{m_v}\left(-\frac{Gm_vm_p}{r^3}\mathbf{r}\right) = -\frac{Gm_s}{R_v{}^3}\mathbf{R}_v - \frac{Gm_p}{r^3}\mathbf{r} \tag{8.22}$$

Let us write this as

$$\ddot{\mathbf{R}}_v = \mathbf{A}_s + \mathbf{P}_p \tag{8.23}$$

where

$$\mathbf{A}_s = -\frac{Gm_s}{R_v{}^3}\mathbf{R}_v \quad \mathbf{P}_p = -\frac{Gm_p}{r^3}\mathbf{r} \tag{8.24}$$

$\mathbf{A}_s$ is the primary gravitational acceleration of the vehicle due to the sun, and $\mathbf{P}_p$ is the secondary or perturbing acceleration due to the planet. The magnitudes of $\mathbf{A}_s$ and $\mathbf{P}_p$ are

$$A_s = \frac{Gm_s}{R^2} \quad P_p = \frac{Gm_p}{r^2} \tag{8.25}$$

where we made use of the approximation given by Eq. (8.21). The ratio of the perturbing acceleration to the primary acceleration is, therefore,

$$\frac{P_p}{A_s} = \frac{\dfrac{Gm_p}{r^2}}{\dfrac{Gm_s}{R^2}} = \frac{m_p}{m_s}\left(\frac{R}{r}\right)^2 \tag{8.26}$$

The equation of motion of the planet relative to the inertial frame is

$$m_p\ddot{\mathbf{R}} = \mathbf{F}_v^{(p)} + \mathbf{F}_s^{(p)}$$

Solving for $\ddot{\mathbf{R}}$, noting that $\mathbf{F}_v^{(p)} = -\mathbf{F}_p^{(v)}$, and using Eqs. (8.18b) and (8.18c) yields

$$\ddot{\mathbf{R}} = \frac{1}{m_p}\left(\frac{Gm_vm_p}{r^3}\mathbf{r}\right) + \frac{1}{m_p}\left(-\frac{Gm_pm_s}{R^3}\mathbf{R}\right) = \frac{Gm_v}{r^3}\mathbf{r} - \frac{Gm_s}{R^3}\mathbf{R} \tag{8.27}$$

Subtracting Eq. (8.27) from Eq. (8.22) and collecting terms, we find

$$\ddot{\mathbf{R}}_v - \ddot{\mathbf{R}} = -\frac{Gm_p}{r^3}\mathbf{r}\left(1 + \frac{m_v}{m_p}\right) - \frac{Gm_s}{R_v{}^3}\left[\mathbf{R}_v - \left(\frac{R_v}{R}\right)^3\mathbf{R}\right]$$

Recalling Eq. (8.19), we can write this as

$$\ddot{\mathbf{r}} = -\frac{Gm_p}{r^3}\mathbf{r}\left(1+\frac{m_v}{m_p}\right) - \frac{Gm_s}{R_v{}^3}\left\{\mathbf{r} + \left[1-\left(\frac{R_v}{R}\right)^3\right]\mathbf{R}\right\} \tag{8.28}$$

This is the equation of motion of the vehicle relative to the planet. By using Eq. (8.21) and the fact that $m_v \ll m_p$, we can write this in approximate form as

$$\ddot{\mathbf{r}} = \mathbf{a}_p + \mathbf{p}_s \tag{8.29}$$

where

$$\mathbf{a}_p = -\frac{Gm_p}{r^3}\mathbf{r} \quad \mathbf{p}_s = -\frac{Gm_s}{R^3}\mathbf{r} \tag{8.30}$$

In this case $\mathbf{a}_p$ is the primary gravitational acceleration of the vehicle due to the planet and $\mathbf{p}_s$ is the perturbation caused by the sun. The magnitudes of these vectors are

$$a_p = \frac{Gm_p}{r^2} \quad p_s = \frac{Gm_s}{R^3}r \tag{8.31}$$

The ratio of the perturbing acceleration to the primary acceleration is

$$\frac{p_s}{a_p} = \frac{Gm_s\dfrac{r}{R^3}}{\dfrac{Gm_p}{r^2}} = \frac{m_s}{m_p}\left(\frac{r}{R}\right)^3 \tag{8.32}$$

For motion relative to the planet, the ratio p_s/a_p is a measure of the deviation of the vehicle's orbit from the Keplerian orbit arising from the planet acting by itself ($p_s/a_p = 0$). Likewise, P_p/A_s is a measure of the planet's influence on the orbit of the vehicle relative to the sun. If

$$\frac{p_s}{a_p} < \frac{P_p}{A_s} \tag{8.33}$$

then the perturbing effect of the sun on the vehicle's orbit around the planet is less than the perturbing effect of the planet on the vehicle's orbit around the sun. We say that the vehicle is therefore within the planet's sphere of influence. Substituting Eqs. (8.26) and (8.32) into Eq. (8.33) yields

$$\frac{m_s}{m_p}\left(\frac{r}{R}\right)^3 < \frac{m_p}{m_s}\left(\frac{R}{r}\right)^2$$

which means

$$\left(\frac{r}{R}\right)^5 < \left(\frac{m_p}{m_s}\right)^2$$

or

$$\frac{r}{R} < \left(\frac{m_p}{m_s}\right)^{2/5}$$

Let r_{SOI} be the radius of the sphere of influence. Within the planet's sphere of influence, defined by

$$\frac{r_{\mathrm{SOI}}}{R} = \left(\frac{m_p}{m_s}\right)^{2/5} \tag{8.34}$$

the motion of the spacecraft is determined by its equations of motion relative to the planet (Eq. 8.28). Outside the sphere of influence, the path of the spacecraft is computed relative to the sun (Eq. 8.22).

The sphere of influence radius presented in Eq. (8.34) is not an exact quantity. It is simply a reasonable estimate of the distance beyond which the sun's gravitational attraction dominates that of a planet. The spheres of influence of all the planets and the earth's moon are listed in Table A.2.

EXAMPLE 8.3

Calculate the radius of the earth's sphere of influence.

Solution

In Table A.1 we find

$$m_{\text{earth}} = 5.974\left(10^{24}\right)\text{kg}$$

$$m_{\text{sun}} = 1.989\left(10^{30}\right)\text{kg}$$

$$R_{\text{earth}} = 149.6\left(10^{6}\right)\text{km}$$

Substituting these data into Eq. (8.34) yields

$$r_{\text{SOI}} = 149.6 \times 10^{6}\left(\frac{5.974 \times 10^{24}}{1.989 \times 10^{24}}\right)^{2/5} = \boxed{925 \times 10^{6}\,\text{km}}$$

Since the radius of the earth is 6378 km,

$$r_{\text{SOI}} = 145\,\text{earth radii}$$

Relative to the earth, its sphere of influence is very large. However, relative to the sun it is tiny, as illustrated in Fig. 8.7.

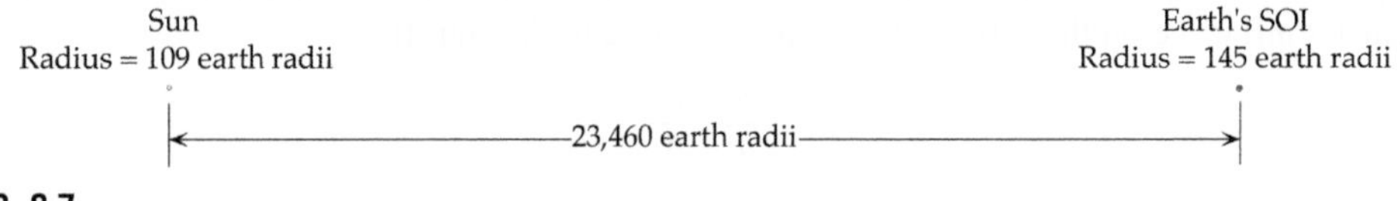

FIG. 8.7

The earth's sphere of influence and the sun, drawn to scale.

8.5 METHOD OF PATCHED CONICS

"Conics" refers to the fact that two-body, or Keplerian, orbits are conic sections with the focus at the attracting body. To study an interplanetary trajectory, we assume that when the spacecraft is outside the sphere of influence of a planet it follows an unperturbed Keplerian orbit around the sun. Because interplanetary distances are so vast, for heliocentric orbits we may neglect the size of the spheres of influence and consider them, like the planets they surround, to be just points in space coinciding with the planetary centers. Within each planetary sphere of influence, the spacecraft travels an unperturbed Keplerian path about the planet. While the sphere of influence appears as a mere speck on the scale of the solar system, from the point of view of the planet it is very large indeed and may be considered to lie at infinity.

To analyze a mission from planet 1 to planet 2 using the method of patched conics, we first determine the heliocentric trajectory, such as the Hohmann transfer ellipse discussed in Section 8.2, that will intersect the desired positions of the two planets in their orbits. This trajectory takes the spacecraft from the sphere of influence of planet 1 to that of planet 2. At the spheres of influence, the heliocentric velocities of the transfer orbit are computed relative to the planet to establish the velocities "at infinity," which are then used to determine planetocentric departure trajectory at planet 1 and arrival trajectory at planet 2. In this way, we "patch" together the three conics, one centered at the sun and the other two centered at the planets in question.

Whereas the method of patched conics is remarkably accurate for interplanetary trajectories, such is not the case for lunar rendezvous and return trajectories. The orbit of the moon is determined primarily by the earth, whose sphere of influence extends well beyond the moon's 384,400-km orbital radius. To apply patched conics to lunar trajectories we ignore the sun and consider the motion of a spacecraft as influenced by just the earth and moon, as in the restricted three-body problem discussed in Section 2.12. The size of the moon's sphere of influence is found using Eq. (8.34), with the earth playing the role of the sun:

$$r_{SOI} = R\left(\frac{m_{moon}}{m_{earth}}\right)^{2/5}$$

where R is the radius of the moon's orbit. Thus, using Table A.1,

$$r_{SOI} = 384,400\left[\frac{73.48\left(10^{21}\right)}{5974\left(10^{21}\right)}\right]^{2/5} = 66,200\text{km}$$

as recorded in Table A.2. The moon's sphere of influence extends out to over one-sixth of the distance to the earth. We can hardly consider it to be a mere speck relative to the earth. Another complication is the fact that the earth and the moon are somewhat comparable in mass, so that their center of mass lies almost three-quarters of an earth radius from the center of the earth. The motion of the moon cannot be accurately described as rotating around the center of the earth.

Complications such as these place the analysis of cislunar trajectories beyond the scope of this chapter. (In Example 2.18, we did a lunar trajectory calculation not by using patched conics but by integrating the equations of motion of a spacecraft within the context of the restricted three-body problem.) We extend the patched conic technique to lunar trajectories in Chapter 9.

8.6 PLANETARY DEPARTURE

To escape the gravitational pull of a planet, the spacecraft must travel a hyperbolic trajectory relative to the planet, arriving at its sphere of influence with a relative speed v_∞ (hyperbolic excess speed) greater than zero. On a parabolic trajectory, according to Eq. (2.91), the spacecraft will arrive at the sphere of influence ($r = \infty$) with a relative speed of zero. In that case, the spacecraft remains in the same orbit as the planet and does not embark upon a heliocentric elliptical path.

Fig. 8.8 shows a spacecraft departing on a Hohmann trajectory from planet 1 toward a target planet 2, which is farther away from the sun (as in Fig. 8.1). On crossing the sphere of influence, the heliocentric velocity $\mathbf{V}_D^{(v)}$ of the spacecraft is parallel to the asymptote of the departure hyperbola as well as to the planet's heliocentric velocity vector $\mathbf{V}_1$. $\mathbf{V}_D^{(v)}$ and $\mathbf{V}_1$ must be parallel and in the same direction for

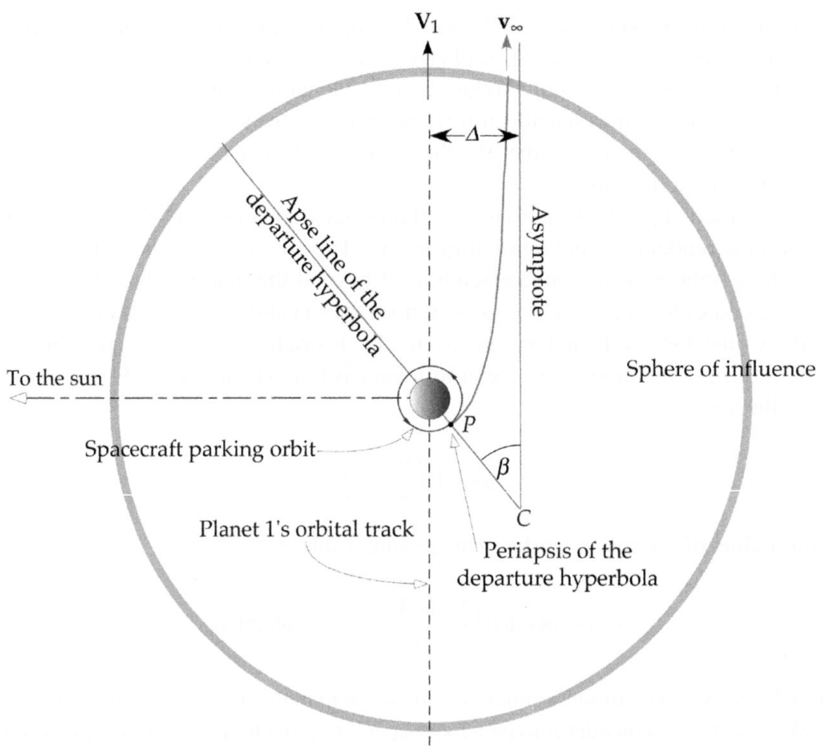

FIG. 8.8

Departure of a spacecraft on a mission from an inner planet to an outer planet.

a Hohmann transfer such that ΔV_D in Eq. (8.3) is positive. Clearly, ΔV_D is the hyperbolic excess speed of the departure hyperbola,

$$v_\infty = \sqrt{\frac{\mu_{\text{sun}}}{R_1}}\left(\sqrt{\frac{2R_2}{R_1+R_2}}-1\right) \tag{8.35}$$

It would be well at this point for the reader to review Section 2.9 on hyperbolic trajectories and compare Fig. 8.8 and Fig. 2.25. Recall that point C is the center of the hyperbola.

A space vehicle is ordinarily launched into an interplanetary trajectory from a circular parking orbit. The radius of this parking orbit equals the periapse radius r_p of the departure hyperbola. According to Eq. (2.50), the periapsis radius is given by

$$r_p = \frac{h^2}{\mu_1}\frac{1}{1+e} \tag{8.36}$$

where h is the angular momentum of the departure hyperbola (relative to the planet), e is the eccentricity of the hyperbola, and μ_1 is the planet's gravitational parameter. The hyperbolic excess speed is found in Eq. (2.115), from which we obtain

$$h = \frac{\mu_1\sqrt{e^2-1}}{v_\infty} \tag{8.37}$$

Substituting this expression for the angular momentum into Eq. (8.36) and solving for eccentricity yields

$$e = 1 + \frac{r_p {v_\infty}^2}{\mu_1} \tag{8.38}$$

We place this result back into Eq. (8.37) to obtain the following expression for the angular momentum:

$$h = r_p \sqrt{{v_\infty}^2 + \frac{2\mu_1}{r_p}} \tag{8.39}$$

Since the hyperbolic excess speed is specified by the mission requirements (Eq. 8.35), choosing a departure periapsis r_p yields the parameters e and h of the departure hyperbola. From the angular momentum, we get the periapsis speed,

$$v_p = \frac{h}{r_p} = \sqrt{{v_\infty}^2 + \frac{2\mu_1}{r_p}} \tag{8.40}$$

which can also be found from an energy approach using Eq. (2.113). With Eq. (8.40) and the speed of the circular parking orbit (Eq. 2.63),

$$v_c = \sqrt{\frac{\mu_1}{r_p}} \tag{8.41}$$

we can calculate the delta-v required to put the vehicle onto the hyperbolic departure trajectory,

$$\Delta v = v_p - v_c = v_c \left(\sqrt{2 + \left(\frac{v_\infty}{v_c} \right)^2} - 1 \right) \tag{8.42}$$

The location of periapsis, where the delta-v maneuver must occur, is found using Eq. (2.99) and Eq. (8.38),

$$\beta = \cos^{-1}\left(\frac{1}{e} \right) = \cos^{-1}\left(\frac{1}{1 + \frac{r_p {v_\infty}^2}{\mu_1}} \right) \tag{8.43}$$

β gives the orientation of the apse line of the hyperbola to the planet's heliocentric velocity vector.

It should be pointed out that the only requirement on the orientation of the plane of the departure hyperbola is that it must contain the center of mass of the planet as well as the relative velocity vector $\mathbf{v}_\infty$. Therefore, as shown in Fig. 8.9, the hyperbola can be rotated about a line A–A, which passes through the planet's center of mass and is parallel to $\mathbf{v}_\infty$ (or $\mathbf{V}_1$, which of course is parallel to $\mathbf{v}_\infty$ for Hohmann transfers). Rotating the hyperbola in this way sweeps out a surface of revolution on which all possible departure hyperbolas lie. The periapsis of the hyperbola traces out a circle which, for the specified periapsis radius r_p, is the locus of all possible points of injection into a departure trajectory toward the target planet. This circle is the base of a cone with its vertex at the center of the planet. From Fig. 2.25 we can determine that its radius is $r_p \sin\beta$, where β is given just above in Eq. (8.43).

The plane of the parking orbit, or direct ascent trajectory, must contain the line A–A and the launch site at the time of launch. The possible inclinations of a prograde orbit range from a minimum of i_{min}, where i_{min} is the latitude of the launch site, to i_{max}, which cannot exceed 90°. Launch site safety

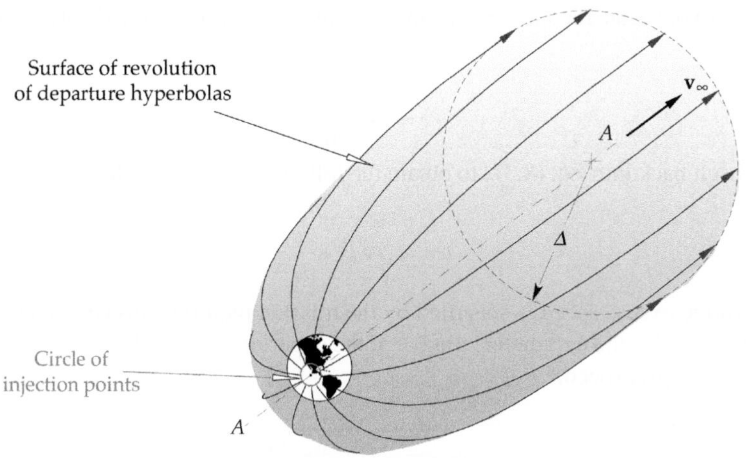

FIG. 8.9

Locus of possible departure trajectories for a given v_∞ and r_p.

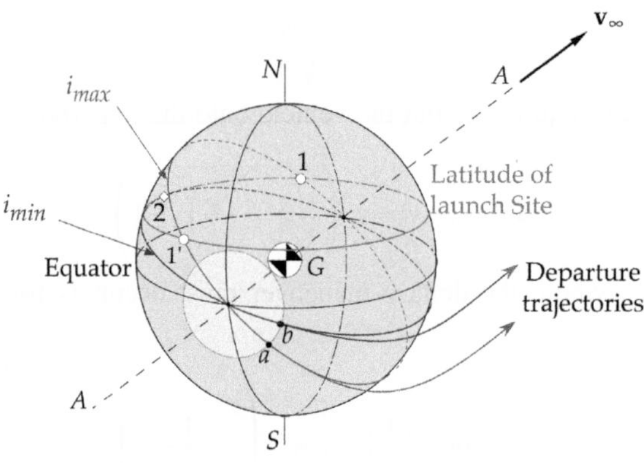

FIG. 8.10

Parking orbits and departure trajectories for a launch site at a given latitude.

considerations may place additional limits on this range. For example, orbits originating from the Kennedy Space Center in Florida (latitude 28.5°), are limited to inclinations between 28.5° and 52.5°. For the scenario illustrated in Fig. 8.10, the location of the launch site limits access to just the departure trajectories having periapses lying between *a* and *b*. The figure shows that there are two times per day—when the planet rotates the launch site through positions 1 and 1′—that a spacecraft can be launched into a parking orbit. These times are closer together (the launch window is smaller), the lower the inclination of the parking orbit.

Once a spacecraft is established in its parking orbit, then an opportunity for launch into the departure trajectory occurs at each orbital circuit.

If the mission is to send a spacecraft from an outer planet to an inner planet, as in Fig. 8.2, then the spacecraft's heliocentric speed $V_D^{(v)}$ at departure must be less than that of the planet. That means the

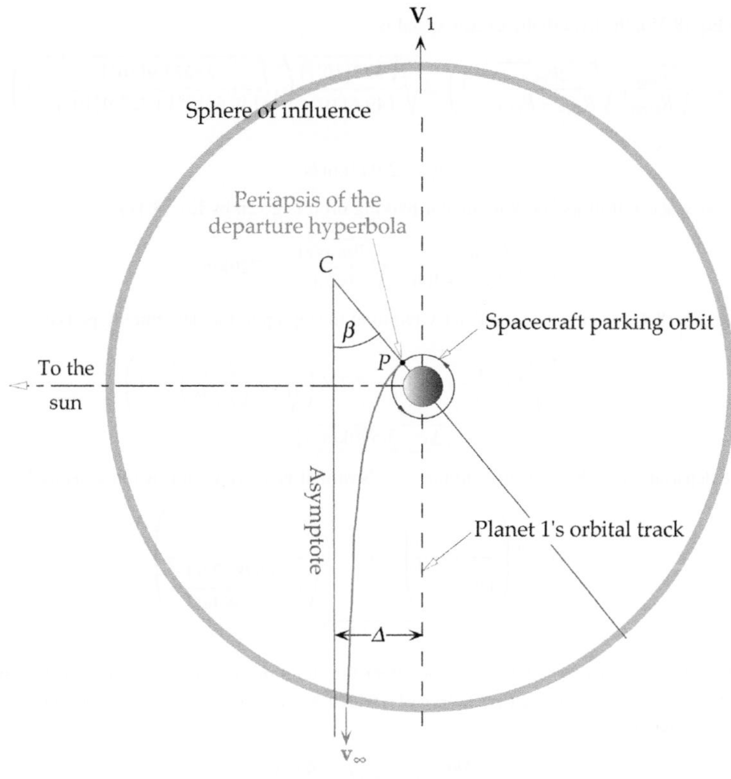

FIG. 8.11

Departure of a spacecraft on a trajectory from an outer planet to an inner planet.

spacecraft must emerge from the backside of the sphere of influence with its relative velocity vector $\mathbf{v}_\infty$ directed opposite to $\mathbf{V}_1$, as shown in Fig. 8.11. Figs. 8.9 and 8.10 apply to this situation as well.

EXAMPLE 8.4

A spacecraft is launched on a mission to Mars starting from a 300-km circular parking orbit. Calculate (a) the delta-v required, (b) the location of perigee of the departure hyperbola, and (c) the amount of propellant required as a percentage of the spacecraft mass before the delta-v burn, assuming a specific impulse of 300 s.

Solution

From Tables A.1 and A.2, we obtain the gravitational parameters for the sun and the earth,

$$\mu_{\text{sun}} = 1.327\left(10^{11}\right)\text{km}^3/\text{s}^2$$
$$\mu_{\text{earth}} = 398{,}600\text{km}^3/\text{s}^2$$

and the orbital radii of the earth and Mars,

$$R_{\text{earth}} = 149.6\left(10^6\right)\text{km}$$
$$R_{\text{Mars}} = 227.9\left(10^6\right)\text{km}$$

(a) According to Eq. (8.35), the hyperbolic excess speed is

$$v_\infty = \sqrt{\frac{\mu_{sun}}{R_{earth}}}\left(\sqrt{\frac{2R_{Mars}}{R_{earth}+R_{Mars}}}-1\right) = \sqrt{\frac{1.327(10^{11})}{149.6(10^6)}}\left(\sqrt{\frac{2\cdot 227.9(10^6)}{149.6(10^6)+227.9(10^6)}}-1\right)$$

from which

$$v_\infty = 2.943\text{km/s}$$

The speed of the spacecraft in its 300-km circular parking orbit is given by Eq. (8.41),

$$v_c = \sqrt{\frac{\mu_{earth}}{R_{earth}+300}} = \sqrt{\frac{398,600}{6678}} = 7.726\text{km/s}$$

Finally, we use Eq. (8.42) to calculate the delta-v required to step up to the departure hyperbola

$$\Delta v = v_c\left(\sqrt{2+\left(\frac{v_\infty}{v_c}\right)^2}-1\right) = 7.726\left(\sqrt{2+\left(\frac{2.943}{7.726}\right)^2}-1\right)$$

$$\boxed{\Delta v = 3.590\text{km/s}}$$

(b) Perigee of the departure hyperbola, relative to the earth's orbital velocity vector, is found using Eq. (8.43),

$$\beta = \cos^{-1}\left(\frac{1}{1+\dfrac{r_p v_\infty{}^2}{\mu_{earth}}}\right) = \cos^{-1}\left(\frac{1}{1+\dfrac{6678\cdot 2.943^2}{368,600}}\right)$$

$$\boxed{\beta = 29.16°}$$

Fig. 8.12 shows that the perigee can be located on either the sunlit or the dark side of the earth. It is likely that the parking orbit would be a prograde orbit (west to east), which would place the burnout point on the dark side.

(c) From Eq. (6.1), we have

$$\frac{\Delta m}{m} = 1 - \exp\left(-\frac{\Delta v}{I_{sp}g_0}\right)$$

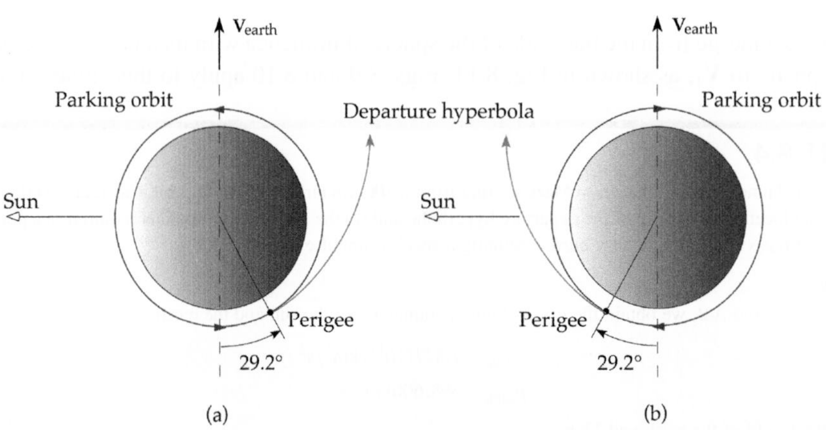

FIG. 8.12

Departure trajectory to Mars initiated from (a) the dark side and (b) the sunlit side of the earth (Example 8.4).

Substituting $\Delta v = 3.590\,\text{km/s}$, $I_{sp} = 300\,\text{s}$, and $g_0 = 9.81(10^{-3})\,\text{km/s}^2$, this yields

$$\boxed{\frac{\Delta m}{m} = 0.705}$$

That is, prior to the delta-v maneuver over, 70% of the spacecraft mass must be propellant.

8.7 SENSITIVITY ANALYSIS

The initial maneuvers required to place a spacecraft on an interplanetary trajectory occur well within the sphere of influence of the departure planet. Since the sphere of influence is just a point on the scale of the solar system, we may ask what effects small errors in position and velocity at the maneuver point have on the trajectory. Assuming the mission is from an inner to an outer planet, let us consider the effect that small changes in the burnout velocity v_p and radius r_p have on the target radius R_2 of the heliocentric Hohmann transfer ellipse (see Figs. 8.1 and 8.8).

R_2 is the radius of aphelion, so we use Eq. (2.70) to obtain

$$R_2 = \frac{h^2}{\mu_{sun}} \frac{1}{1-e}$$

Substituting $h = R_1 V_D^{(v)}$ and $e = (R_2 - R_1)/(R_2 + R_1)$, and solving for R_2, yields

$$R_2 = \frac{R_1^2 \left(V_D^{(v)}\right)^2}{2\mu_{sun} - R_1 \left(V_D^{(v)}\right)^2} \tag{8.44}$$

(This expression holds as well for a mission from an outer to an inner planet.) The change δR_2 in R_2 due to a small variation $\delta V_D^{(v)}$ of $V_D^{(v)}$ is

$$\delta R_2 = \frac{dR_2}{dV_D^{(v)}} \delta V_D^{(v)} = \frac{4R_1^2 \mu_{sun}}{\left[2\mu_{sun} - R_1 \left(V_D^{(v)}\right)^2\right]^2} V_D^{(v)} \delta V_D^{(v)}$$

Dividing this equation by Eq. (8.44) leads to

$$\frac{\delta R_2}{R_2} = \frac{2}{1 - \dfrac{R_1 \left(V_D^{(v)}\right)^2}{2\mu_{sun}}} \frac{\delta V_D^{(v)}}{V_D^{(v)}} \tag{8.45}$$

The departure speed $V_D^{(v)}$ of the space vehicle is the sum of the planet's speed V_1 and excess speed v_∞

$$V_D^{(v)} = V_1 + v_\infty$$

We can solve Eq. (8.40) for v_∞,

$$v_\infty = \sqrt{v_p{}^2 - \frac{2\mu_1}{r_p}}$$

Hence

$$V_D^{(v)} = V_1 + \sqrt{v_p{}^2 - \frac{2\mu_1}{r_p}} \tag{8.46}$$

The change in $V_D^{(v)}$ due to variations δr_p and δv_p of the burnout position (periapse) r_p and speed v_p is given by

$$\delta V_D^{(v)} = \frac{\partial V_D^{(v)}}{\partial r_p}\delta r_p + \frac{\partial V_D^{(v)}}{\partial v_p}\delta v_p \tag{8.47}$$

From Eq. (8.46), we obtain

$$\frac{\partial V_D^{(v)}}{\partial r_p} = \frac{\mu_1}{v_\infty r_p^2} \qquad \frac{\partial V_D^{(v)}}{\partial v_p} = \frac{v_p}{v_\infty}$$

Therefore,

$$\delta V_D^{(v)} = \frac{\mu_1}{v_\infty r_p^2}\delta r_p + \frac{v_p}{v_\infty}\delta v$$

Once again making use of Eq. (8.40), this can be written as follows:

$$\frac{\delta V_D^{(v)}}{V_D^{(v)}} = \frac{\mu_1}{V_D^{(v)} v_\infty r_p}\frac{\delta r_p}{r_p} + \frac{v_\infty + \dfrac{2\mu_1}{r_p v_\infty}}{V_D^{(v)}}\frac{\delta v_p}{v_p} \tag{8.48}$$

Substituting this into Eq. (8.45) finally yields the desired result: an expression for the variation of R_2 due to variations in r_p and v_p

$$\frac{\delta R_2}{R_2} = \frac{2}{1 - \dfrac{R_1\left(V_D^{(v)}\right)^2}{2\mu_{sun}}}\left(\frac{\mu_1}{V_D^{(v)} v_\infty r_p}\frac{\delta r_p}{r_p} + \frac{v_\infty + \dfrac{2\mu_1}{r_p v_\infty}}{V_D^{(v)}}\frac{\delta v_p}{v_p}\right) \tag{8.49}$$

Consider a mission from earth to Mars, starting from a 300-km parking orbit. We have

$$\begin{aligned}
\mu_{sun} &= 1.327\left(10^{11}\right)\text{km}^3/\text{s}^2\\
\mu_1 = \mu_{earth} &= 398{,}600\text{km}^3/\text{s}^2\\
R_1 &= 149.6\left(10^6\right)\text{km}\\
R_2 &= 227.9\left(10^6\right)\text{km}\\
r_p &= 6678\text{km}
\end{aligned}$$

In addition, from Eqs. (8.1) and (8.2),

$$V_1 = V_{earth} = \sqrt{\frac{\mu_{sun}}{R_1}} = \sqrt{\frac{1.327\left(10^{11}\right)}{149.6\left(10^6\right)}} = 29.78\text{km/s}$$

$$\begin{aligned}
V_D^{(v)} &= \sqrt{2\mu_{sun}}\sqrt{\frac{R_2}{R_1(R_1+R_2)}} = \sqrt{2\cdot 1.327\left(10^{11}\right)}\sqrt{\frac{227.9\left(10^6\right)}{149.6\left(10^6\right)\left[149.6\left(10^6\right)+227.9\left(10^6\right)\right]}}\\
&= 32.73\text{km/s}
\end{aligned}$$

Therefore,

$$v_\infty = V_D^{(v)} - V_{\text{earth}} = 2.943\,\text{km/s}$$

and, from Eq. (8.40),

$$v_p = \sqrt{{v_\infty}^2 + \frac{2\mu_{\text{earth}}}{r_p}} = \sqrt{2.943^2 + \frac{2 \cdot 398{,}600}{6678}} = 11.32\,\text{km/s}$$

Substituting these values into Eq. (8.49) yields

$$\frac{\delta R_2}{R_2} = 3.127\frac{\delta r_p}{r_p} + 6.708\frac{\delta v_p}{v_p}$$

This expression shows that a 0.01% variation (1.1 m/s) in the burnout speed v_p changes the target radius R_2 by 0.067% or 153,000 km! Likewise, an error of 0.01% (0.67 km) in burnout radius r_p produces an error of over 70,000 km. Thus, small errors that are likely to occur in the launch phase of the mission must be corrected by midcourse maneuvers during the coasting flight along the elliptical transfer trajectory.

8.8 PLANETARY RENDEZVOUS

A spacecraft arrives at the sphere of influence of the target planet with a hyperbolic excess velocity $\mathbf{v}_\infty$ relative to the planet. In the case illustrated in Fig. 8.1, a mission from an inner planet 1 to an outer planet 2 (e.g., earth to Mars), the spacecraft's heliocentric approach velocity $\mathbf{V}_A^{(v)}$ is smaller in magnitude than that of the planet, $\mathbf{V}_2$. Therefore, it crosses the forward portion of the sphere of influence, as shown in Fig. 8.13. For a Hohmann transfer, $\mathbf{V}_A^{(v)}$ and $\mathbf{V}_2$ are parallel, so the magnitude of the hyperbolic excess velocity is, simply,

$$v_\infty = V_2 - V_A^{(v)} \tag{8.50}$$

If the mission is as illustrated in Fig. 8.2, from an outer planet to an inner one (e.g., earth to Venus), then $V_A^{(v)}$ is greater than V_2, and the spacecraft must cross the rear portion of the sphere of influence, as shown in Fig. 8.14. In that case

$$v_\infty = V_A^{(v)} - V_2 \tag{8.51}$$

What happens after crossing the sphere of influence depends on the nature of the mission. If the goal is to impact the planet (or its atmosphere), the aiming radius Δ of the approach hyperbola must be such that the hyperbola's periapsis radius r_p equals essentially the radius of the planet. If the intent is to go into orbit around the planet, then Δ must be chosen so that the delta-v burn at periapsis will occur at the correct altitude above the planet. If there is no impact with the planet and no drop into a capture orbit around the planet, then the spacecraft will simply continue past periapsis on a flyby trajectory, exiting the sphere of influence with the same relative speed v_∞ as it entered, but with the velocity vector rotated through the turn angle δ, given by Eq. (2.100),

$$\delta = 2\sin^{-1}\left(\frac{1}{e}\right) \tag{8.52}$$

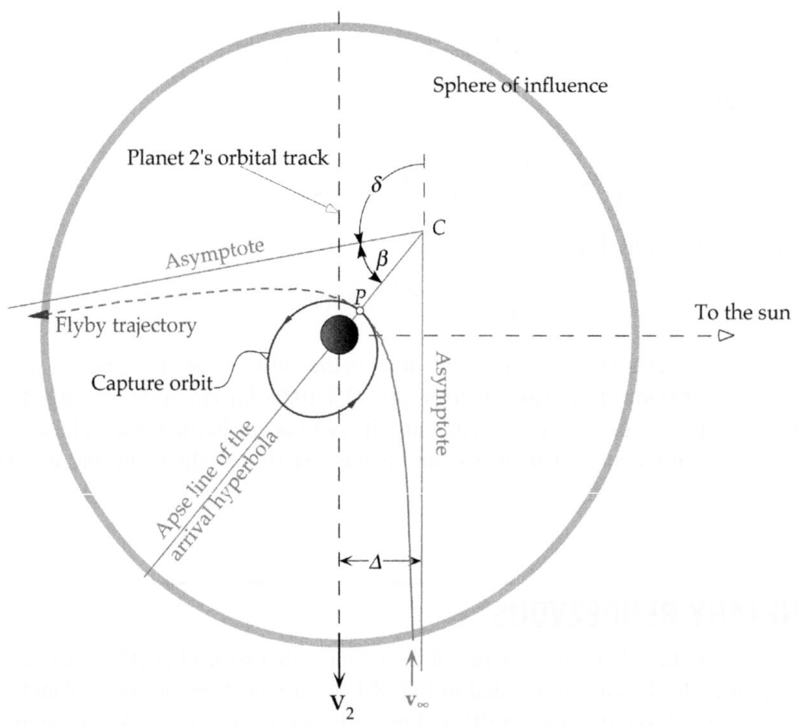

FIG. 8.13

Spacecraft approach trajectory for a Hohmann transfer to an outer planet from an inner one. *P* is the periapse of the approach hyperbola.

With the hyperbolic excess speed v_∞ and the periapse radius r_p specified, the eccentricity of the approach hyperbola is found from Eq. (8.38),

$$e = 1 + \frac{r_p {v_\infty}^2}{\mu_2} \tag{8.53}$$

where μ_2 is the gravitational parameter of planet 2. Hence, the turn angle is

$$\delta = 2\sin^{-1}\left(\frac{1}{1 + \dfrac{r_p {v_\infty}^2}{\mu_2}}\right) \tag{8.54}$$

We can combine Eqs. (2.103) and (2.107) to obtain the following expression for the aiming radius:

$$\Delta = \frac{h^2}{\mu_2}\frac{1}{\sqrt{e^2 - 1}} \tag{8.55}$$

The angular momentum of the approach hyperbola relative to the planet is found using Eq. (8.39),

$$h = r_p\sqrt{{v_\infty}^2 + \frac{2\mu_2}{r_p}} \tag{8.56}$$

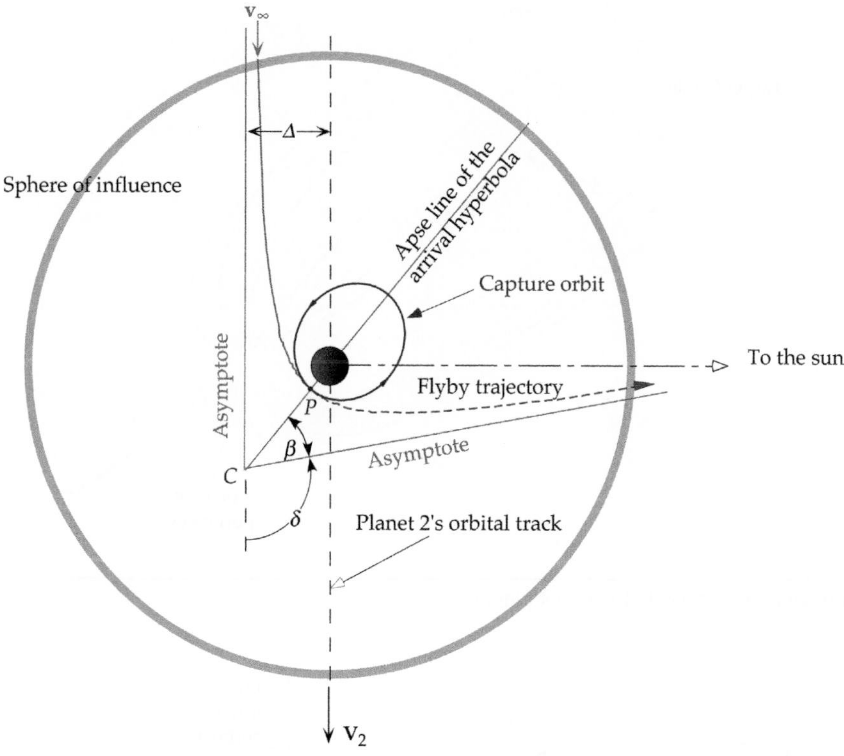

FIG. 8.14

Spacecraft approach trajectory for a Hohmann transfer to an inner planet from an outer one. *P* is the periapse of the approach hyperbola.

Substituting Eqs. (8.53) and (8.56) into Eq. (8.55) yields the aiming radius in terms of the periapse radius and the hyperbolic excess speed,

$$\Delta = r_p \sqrt{1 + \frac{2\mu_2}{r_p v_\infty{}^2}} \tag{8.57}$$

Just as we observed when discussing departure trajectories, the approach hyperbola does not lie in a unique plane. We can rotate the hyperbolas illustrated in Figs. 8.11 and 8.12 about a line *A–A* parallel to $\mathbf{v}_\infty$ and passing through the target planet's center of mass, as shown in Fig. 8.15. The approach hyperbolas in that figure terminate at the circle of periapses. Fig. 8.16 is a plane through the solid of revolution revealing the shape of hyperbolas having a common v_∞ but varying Δ.

Let us suppose that the purpose of the mission is to enter an elliptical orbit of eccentricity *e* around the planet. This will require a delta-v maneuver at periapsis *P* (Figs. 8.13 and 8.14), which is also periapsis of the ellipse. The speed in the hyperbolic trajectory at periapsis is given by Eq. (8.40),

$$v_p)_{\text{hyp}} = \sqrt{v_\infty{}^2 + \frac{2\mu_2}{r_p}} \tag{8.58}$$

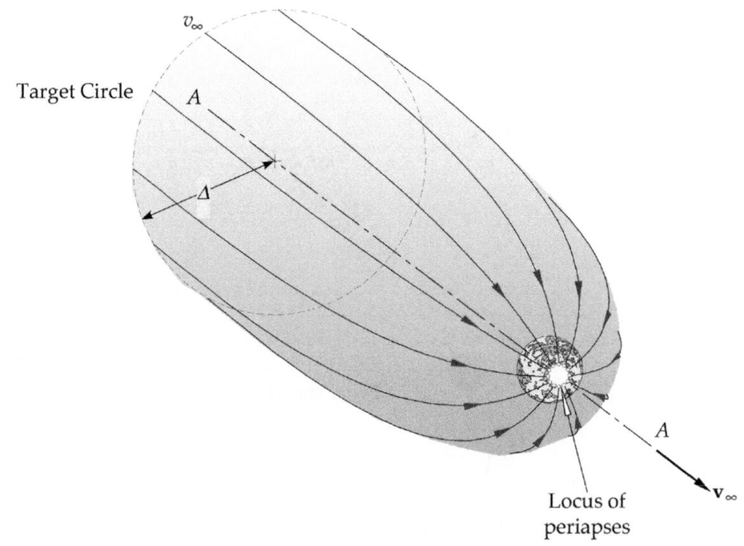

FIG. 8.15

Locus of approach hyperbolas to the target planet.

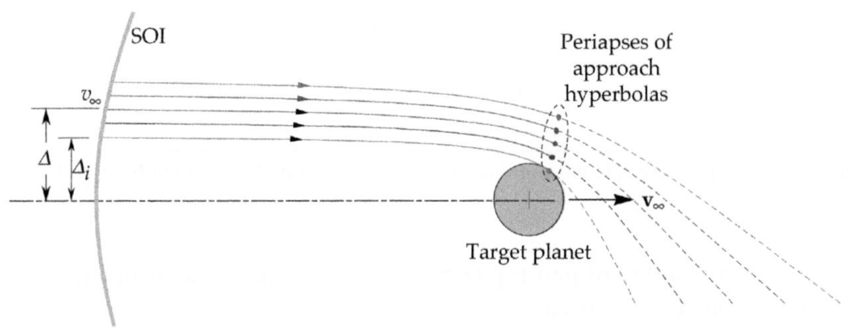

FIG. 8.16

Family of approach hyperbolas having the same v_∞ but different Δ.

The velocity at periapsis of the capture orbit is found by setting $h = r_p v_p$ in Eq. (2.50) and solving for v_p

$$v_p)_{\text{capture}} = \sqrt{\frac{\mu_2(1+e)}{r_p}} \tag{8.59}$$

Hence, the required delta-v is

$$\Delta v = v_p)_{\text{hyp}} - v_p)_{\text{capture}} = \sqrt{v_\infty{}^2 + \frac{2\mu_2}{r_p}} - \sqrt{\frac{\mu_2(1+e)}{r_p}} \tag{8.60}$$

For a given v_∞, Δv clearly depends on the choice of periapse radius r_p and capture orbit eccentricity e. Requiring the maneuver point to remain the periapsis of the capture orbit means that Δv is maximum for a circular capture orbit and decreases with increasing eccentricity until $\Delta v = 0$, which, of course, means no capture (flyby).

To determine optimal capture radius, let us write Eq. (8.60) in nondimensional form as

$$\frac{\Delta v}{v_\infty} = \sqrt{1+\frac{2}{\xi}} - \sqrt{\frac{1+e}{\xi}} \tag{8.61}$$

where

$$\xi = \frac{r_p {v_\infty}^2}{\mu_2} \tag{8.62}$$

The first and second derivatives of $\Delta v/v_\infty$ with respect to ξ are

$$\frac{d}{d\xi}\frac{\Delta v}{v_\infty} = \left(-\frac{1}{\sqrt{\xi+2}} + \frac{\sqrt{1+e}}{2}\right)\frac{1}{\xi^{3/2}} \tag{8.63}$$

$$\frac{d^2}{d\xi^2}\frac{\Delta v}{v_\infty} = \left(\frac{2\xi+3}{(\xi+2)^{3/2}} - \frac{3}{4}\sqrt{1+e}\right)\frac{1}{\xi^{5/2}} \tag{8.64}$$

Setting the first derivative equal to zero and solving for ξ yields

$$\xi = 2\frac{1-e}{1+e} \tag{8.65}$$

Substituting this value of ξ into Eq. (8.64), we get

$$\frac{d^2}{d\xi^2}\frac{\Delta v}{v_\infty} = \frac{\sqrt{2}}{64}\frac{(1+e)^3}{(1-e)^{3/2}} \tag{8.66}$$

This expression is positive for elliptical orbits ($0 \le e < 1$), which means that when ξ is given by Eq. (8.65), Δv is a minimum. Therefore, from Eq. (8.62), the optimal periapse radius as far as fuel expenditure is concerned is

$$r_p = \frac{2\mu_2}{{v_\infty}^2}\frac{1-e}{1+e} \tag{8.67}$$

We can combine Eqs. (2.50) and (2.70) to get

$$\frac{1-e}{1+e} = \frac{r_p}{r_a} \tag{8.68}$$

where r_a is the apoapsis radius. Thus, Eq. (8.67) implies

$$r_a = \frac{2\mu_2}{{v_\infty}^2} \tag{8.69}$$

That is, the apoapsis of this capture ellipse is independent of the eccentricity and equals the radius of the optimal circular orbit.

Substituting Eq. (8.65) back into Eq. (8.61) yields the minimum Δv,

$$\Delta v = v_\infty \sqrt{\frac{1-e}{2}} \tag{8.70}$$

Finally, placing the optimal r_p into Eq. (8.57) leads to an expression for the aiming radius required for minimum Δv,

$$\Delta = 2\sqrt{2}\frac{\sqrt{1-e}}{1+e}\frac{\mu_2}{v_\infty{}^2} = \sqrt{\frac{2}{1-e}}r_p \tag{8.71}$$

Clearly, the optimal Δv (and periapsis height) are reduced for highly eccentric elliptical capture orbits ($e \to 1$). However, it should be pointed out that the use of optimal Δv may have to be sacrificed in favor of a variety of other mission requirements.

EXAMPLE 8.5

After a Hohmann transfer from earth to Mars, calculate

(a) the minimum delta-v required to place a spacecraft in orbit with a period of 7 h
(b) the periapsis radius
(c) the aiming radius
(d) the angle between periapsis and Mars' velocity vector.

Solution

The following data are required from Tables A.1 and A.2:

$$\begin{aligned}
\mu_{\text{sun}} &= 1.327\left(10^{11}\right)\text{km}^3/\text{s}^2 \\
\mu_{\text{Mars}} &= 42{,}830\text{km}^3/\text{s}^2 \\
R_{\text{earth}} &= 149.6\left(10^6\right)\text{km} \\
R_{\text{Mars}} &= 227.9\left(10^6\right)\text{km} \\
r_{\text{Mars}} &= 3396\text{km}
\end{aligned}$$

(a) The hyperbolic excess speed is found using Eq. (8.4),

$$v_\infty = \Delta V_A = \sqrt{\frac{\mu_{\text{sun}}}{R_{\text{Mars}}}}\left(1-\sqrt{\frac{2R_{\text{earth}}}{R_{\text{earth}}+R_{\text{Mars}}}}\right) = \sqrt{\frac{1.327\left(10^{11}\right)}{227.9\left(10^6\right)}}\left(1-\sqrt{\frac{2\cdot 149.6\left(10^6\right)}{149.6\left(10^6\right)+227.9\left(10^6\right)}}\right)$$

$$\therefore v_\infty = 2.648\text{km/s}$$

We can use Eq. (2.83) to express the semimajor axis a of the capture orbit in terms of its period T,

$$a = \left(\frac{T\sqrt{\mu_{\text{Mars}}}}{2\pi}\right)^{2/3}$$

Substituting $T = 7\cdot 3600\text{s}$ yields

$$a = \left(\frac{25{,}200\sqrt{42{,}830}}{2\pi}\right)^{2/3} = 8832\text{km}$$

From Eq. (2.73) we obtain

$$a = \frac{r_p}{1-e}$$

On substituting the optimal periapsis radius (Eq. 8.67) this becomes

$$a = \frac{2\mu_{\text{Mars}}}{v_\infty{}^2}\frac{1}{1+e}$$

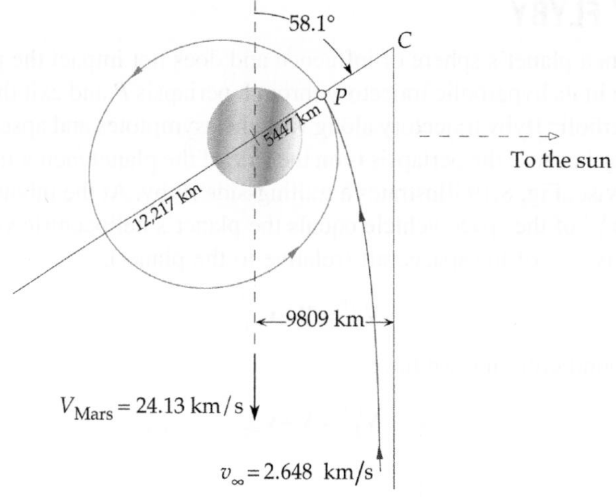

FIG. 8.17

An optimal approach to a Mars capture orbit with a 7-h period ($r_{Mars} = 3396$ km).

from which

$$e = \frac{2\mu_{Mars}}{av_\infty^2} - 1 = \frac{2 \cdot 42{,}830}{8832 \cdot 2.648^2} - 1 = 0.3833$$

Thus, using Eq. (8.70), we find

$$\Delta v = v_\infty\sqrt{\frac{1-e}{2}} = 2.648\sqrt{\frac{1-03833}{2}} = \boxed{1.470\text{km/s}}$$

(b) From Eq. (8.66), we obtain the periapse radius

$$r_p = \frac{2\mu_{Mars}}{v_\infty^2}\frac{1-e}{1+e} = \frac{2 \cdot 42{,}830}{2.648^2}\frac{1-0.3833}{1+0.3833} = \boxed{5447\text{km}}$$

(c) The aiming radius is given by Eq. (8.71),

$$\Delta = r_p\sqrt{\frac{2}{1-e}} = 5447\sqrt{\frac{2}{1-0.3833}} = \boxed{9809\text{km}}$$

(d) Using Eq. (8.43), we get the angle to periapsis

$$\beta = \cos^{-1}\left(\frac{1}{1+\dfrac{r_p v_\infty^2}{\mu_{Mars}}}\right) = \cos^{-1}\left(\frac{1}{1+\dfrac{5447 \cdot 2.648^2}{42{,}830}}\right) = \boxed{58.09°}$$

Mars, the approach hyperbola, and the capture orbit are shown to scale in Fig. 8.17. The approach could also be made from the dark side of the planet instead of the sunlit side. The approach hyperbola and capture ellipse would be the mirror image of that shown, as is the case in Fig. 8.12.

8.9 PLANETARY FLYBY

A spacecraft that enters a planet's sphere of influence and does not impact the planet or go into orbit around it will continue in its hyperbolic trajectory through periapsis *P* and exit the sphere of influence. Fig. 8.18 shows a hyperbolic flyby trajectory along with the asymptotes and apse line of the hyperbola. It is a leading-side flyby because the periapsis is on the side of the planet facing into the direction of the planet's motion. Likewise, Fig. 8.19 illustrates a trailing-side flyby. At the inbound crossing point, the heliocentric velocity $\mathbf{V}_1^{(v)}$ of the space vehicle equals the planet's heliocentric velocity $\mathbf{V}$ plus the hyperbolic excess velocity $\mathbf{v}_{\infty_1}$ of the spacecraft (relative to the planet),

$$\mathbf{V}_1^{(v)} = \mathbf{V} + \mathbf{v}_{\infty_1} \tag{8.72}$$

Similarly, at the outbound crossing, we have

$$\mathbf{V}_2^{(v)} = \mathbf{V} + \mathbf{v}_{\infty_2} \tag{8.73}$$

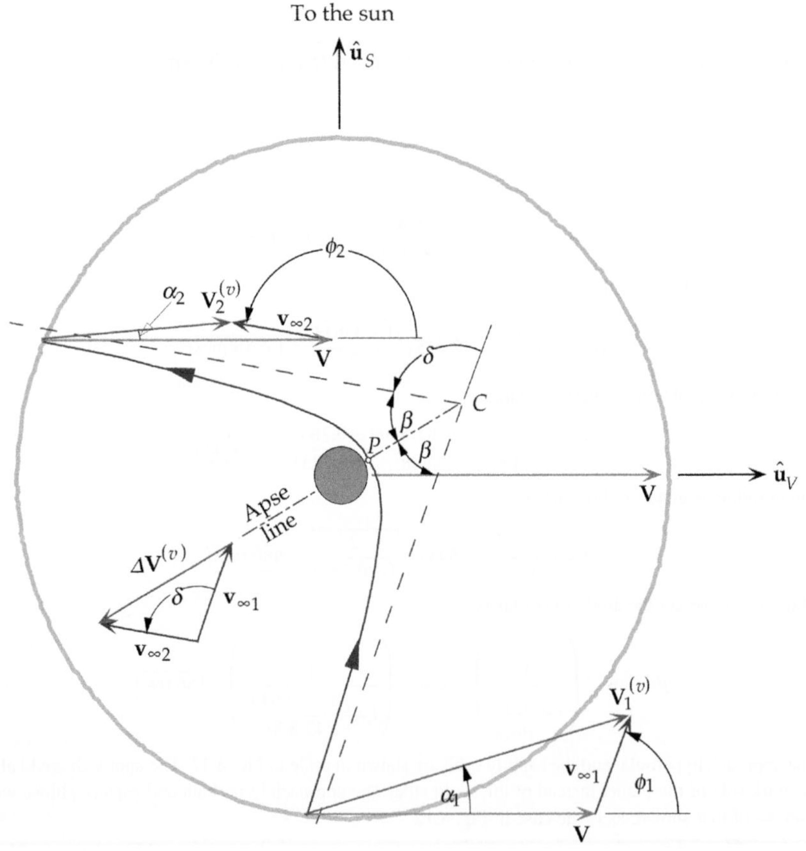

FIG. 8.18

Leading-side planetary flyby.

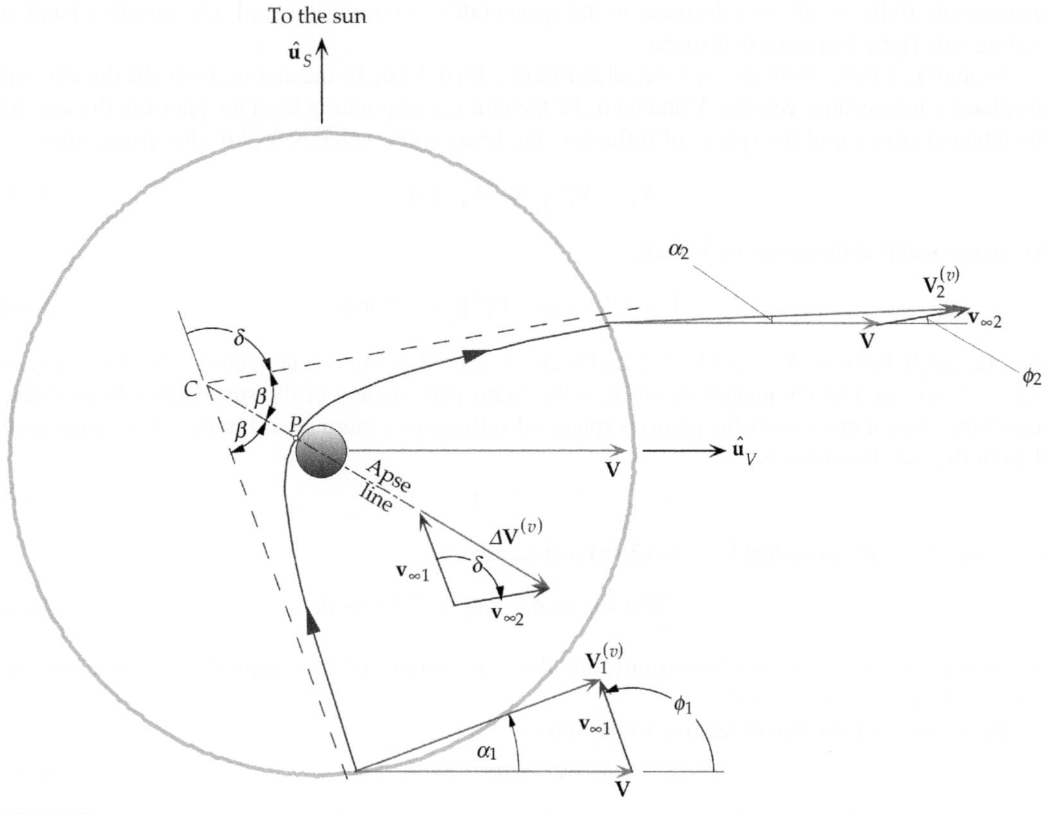

FIG. 8.19

Trailing-side planetary flyby.

The change $\Delta\mathbf{V}^{(v)}$ in the spacecraft's heliocentric velocity is

$$\Delta\mathbf{V}^{(v)} = \mathbf{V}_2^{(v)} - \mathbf{V}_1^{(v)} = (\mathbf{V} + \mathbf{v}_{\infty_2}) - (\mathbf{V} + \mathbf{v}_{\infty_1})$$

which means

$$\Delta\mathbf{V}^{(v)} = \mathbf{v}_{\infty_2} - \mathbf{v}_{\infty_1} = \Delta\mathbf{v}_{\infty} \tag{8.74}$$

The hyperbolic excess velocities $\mathbf{v}_{\infty_1}$ and $\mathbf{v}_{\infty_2}$ lie along the asymptotes of the hyperbola and are therefore inclined at the same angle β to the apse line (see Fig. 2.25), with $\mathbf{v}_{\infty_1}$ pointing toward and $\mathbf{v}_{\infty_2}$ pointing away from the center C. They both have the same magnitude v_{∞}, with $\mathbf{v}_{\infty_2}$ having simply rotated relative to $\mathbf{v}_{\infty_1}$ by the turn angle δ. Hence, $\Delta\mathbf{v}_{\infty}$—and therefore $\Delta\mathbf{V}^{(v)}$—is a vector that lies along the apse line and always points away from periapsis, as illustrated in Figs. 8.18 and 8.19. From these figures it can be seen that, in a leading-side flyby, the component of $\Delta\mathbf{V}^{(v)}$ in the direction of the planet's velocity is negative, whereas for the trailing-side flyby, it is positive. This means that a leading-side flyby results in a decrease in the spacecraft's heliocentric speed. On the other hand, a trailing-side flyby increases that speed.

To analyze a flyby problem, we proceed as follows. First, let $\hat{\mathbf{u}}_V$ be the unit vector in the direction of the planet's heliocentric velocity $\mathbf{V}$ and let $\hat{\mathbf{u}}_S$ be the unit vector pointing from the planet to the sun. At the inbound crossing of the sphere of influence, the heliocentric velocity $\mathbf{V}_1^{(v)}$ of the spacecraft is

$$\mathbf{V}_1^{(v)} = V_1^{(v)}\Big)_V \hat{\mathbf{u}}_V + V_1^{(v)}\Big)_S \hat{\mathbf{u}}_S \tag{8.75}$$

where the scalar components of $\mathbf{V}_1^{(v)}$ are

$$V_1^{(v)}\Big)_V = V_1^{(v)} \cos\alpha_1 \quad V_1^{(v)}\Big)_S = V_1^{(v)} \sin\alpha_1 \tag{8.76}$$

α_1 is the angle between $\mathbf{V}_1^{(v)}$ and $\mathbf{V}$. All angles are measured positive counterclockwise. Referring to Fig. 2.12, we see that the magnitude of α_1 is the flight path angle γ of the spacecraft's heliocentric trajectory when it encounters the planet's sphere of influence (a mere speck) at the planet's distance R from the sun. Furthermore,

$$V_1^{(v)}\Big)_V = V_{\perp_1}^{(v)} \quad V_1^{(v)}\Big)_S = -V_{r_1}^{(v)} \tag{8.77}$$

${V_\perp}_1^{(v)}$ and ${V_r}_1^{(v)}$ are furnished by Eqs. (2.48) and (2.49),

$$V_{\perp_1}^{(v)} = \frac{\mu_{\text{sun}}}{h_1}(1 + e_1\cos\theta_1) \quad V_{r_1}^{(v)} = \frac{\mu_{\text{sun}}}{h_1} e_1 \sin\theta_1 \tag{8.78}$$

in which e_1, h_1, and θ_1 are the eccentricity, angular momentum, and true anomaly of the heliocentric approach trajectory, respectively.

The velocity of the planet relative to the sun is

$$\mathbf{V} = V\hat{\mathbf{u}}_V \tag{8.79}$$

where $V = \sqrt{\mu_{\text{sun}}/R}$. At the inbound crossing of the planet's sphere of influence, the hyperbolic excess velocity of the spacecraft is obtained from Eq. (8.72),

$$\mathbf{v}_{\infty_1} = \mathbf{V}_1^{(v)} - \mathbf{V}$$

Using this we find

$$\mathbf{v}_{\infty_1} = v_{\infty_1})_V \hat{\mathbf{u}}_V + v_{\infty_1})_S \hat{\mathbf{u}}_S \tag{8.80}$$

where the scalar components of $\mathbf{v}_{\infty_1}$ are

$$v_{\infty_1})_V = V_1^{(v)}\cos\alpha_1 - V \quad v_{\infty_1})_S = V_1^{(v)}\sin\alpha_1 \tag{8.81}$$

v_∞ is the magnitude of $\mathbf{v}_{\infty_1}$,

$$v_\infty = \sqrt{\mathbf{v}_{\infty_1}\cdot\mathbf{v}_{\infty_1}} = \sqrt{\left(V_1^{(v)}\right)^2 + V^2 - 2V_1^{(v)}V\cos\alpha_1} \tag{8.82}$$

At this point, v_∞ is known, so that upon specifying the periapsis radius r_p we can compute the angular momentum and eccentricity of the flyby hyperbola (relative to the planet), using Eqs. (8.38) and (8.39).

$$h = r_p\sqrt{{v_\infty}^2 + \frac{2\mu}{r_p}} \quad e = 1 + \frac{r_p {v_\infty}^2}{\mu} \tag{8.83}$$

where μ is the gravitational parameter of the planet.

The angle between $\mathbf{v}_{\infty_1}$ and the planet's heliocentric velocity $\mathbf{V}$ is ϕ_1. It is found using the components of $\mathbf{v}_{\infty_1}$ shown in Eq. (8.81),

$$\phi_1 = \tan^{-1}\frac{\left.v_{\infty_1}\right)_S}{\left.v_{\infty_1}\right)_V} = \tan^{-1}\frac{V_1^{(v)}\sin\alpha_1}{V_1^{(v)}\cos\alpha_1 - V} \tag{8.84}$$

At the outbound crossing, the angle between $\mathbf{v}_{\infty_2}$ and $\mathbf{V}$ is ϕ_2, where

$$\phi_2 = \phi_1 + \delta \tag{8.85}$$

For the leading-side flyby in Fig. 8.18, the turn angle δ is positive (counterclockwise), whereas in Fig. 8.19 it is negative. Since the magnitude of $\mathbf{v}_{\infty_2}$ is v_∞, we can express $\mathbf{v}_{\infty_2}$ in components as

$$\mathbf{v}_{\infty_2} = v_\infty\cos\phi_2\hat{\mathbf{u}}_V + v_\infty\sin\phi_2\hat{\mathbf{u}}_S \tag{8.86}$$

Therefore, the heliocentric velocity of the spacecraft at the outbound crossing is

$$\mathbf{V}_2^{(v)} = \mathbf{V} + \mathbf{v}_{\infty_2} = \left.V_2^{(v)}\right)_V\hat{\mathbf{u}}_V + \left.V_2^{(v)}\right)_S\hat{\mathbf{u}}_S \tag{8.87}$$

where the components of $\mathbf{V}_2^{(v)}$ are

$$\left.V_2^{(v)}\right)_V = V + v_\infty\cos\phi_2 \quad \left.V_2^{(v)}\right)_S = v_\infty\sin\phi_2 \tag{8.88}$$

From this we obtain the spacecraft's radial and transverse heliocentric velocity components,

$$V_{\perp_2}^{(v)} = \left.V_2^{(v)}\right)_V \quad V_{r_2}^{(v)} = -\left.V_2^{(v)}\right)_S \tag{8.89}$$

From these, we finally obtain the three elements e_2, h_2, and θ_2 of the new heliocentric departure trajectory by means of Eq. (2.21),

$$h_2 = RV_{\perp_2}^{(v)} \tag{8.90}$$

Eq. (2.45),

$$R = \frac{h_2^2}{\mu_{\text{sun}}}\frac{1}{1 + e_2\cos\theta_2} \tag{8.91}$$

and Eq. (2.49),

$$V_{r_2}^{(v)} = \frac{\mu_{\text{sun}}}{h_2}e_2\sin\theta_2 \tag{8.92}$$

Notice that the flyby is considered to be an impulsive maneuver during which the heliocentric radius of the spacecraft, which is confined within the planet's sphere of influence, remains fixed at R. The heliocentric velocity analysis is similar to that described in Section 6.7.

EXAMPLE 8.6

A spacecraft departs earth with a velocity perpendicular to the sun line on a flyby mission to Venus. Encounter occurs at a true anomaly in the approach trajectory of −30°. Periapsis altitude is to be 300 km.

(a) For an approach from the dark side of the planet, show that the postflyby orbit is as illustrated in Fig. 8.20.

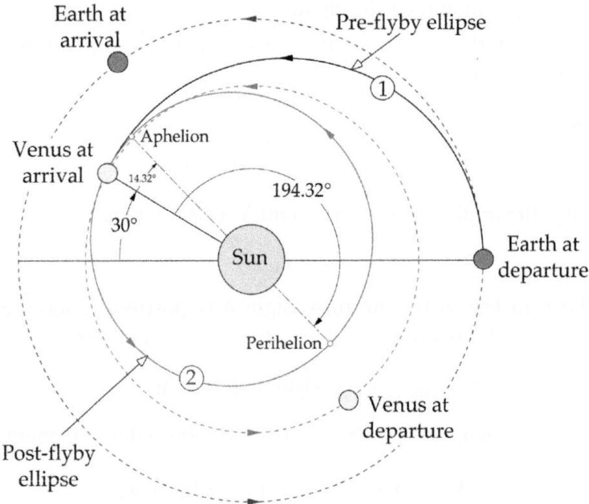

FIG. 8.20

Spacecraft orbits before and after a flyby of Venus, approaching from the dark side.

(b) For an approach from the sunlit side of the planet, show that the postflyby orbit is as illustrated in Fig. 8.21.

Solution

The following data are found in Tables A.1 and A.2:

$$\mu_{\text{sun}} = 1.3271(10^{11})\,\text{km}^3/\text{s}^2$$
$$\mu_{\text{Venus}} = 324{,}900\,\text{km}^3/\text{s}^2$$
$$R_{\text{earth}} = 149.6(10^6)\,\text{km}$$
$$R_{\text{Venus}} = 108.2(10^6)\,\text{km}$$
$$r_{\text{Venus}} = 6052\,\text{km}$$

Preflyby ellipse (orbit 1)

Evaluating the orbit formula (Eq. 2.45) at aphelion of orbit 1 yields

$$R_{\text{earth}} = \frac{{h_1}^2}{\mu_{\text{sun}}}\frac{1}{1-e_1}$$

Thus,

$$ {h_1}^2 = \mu_{\text{sun}} R_{\text{earth}}(1-e_1) \tag{a}$$

At intercept,

$$R_{\text{Venus}} = \frac{{h_1}^2}{\mu_{\text{sun}}}\frac{1}{1+e_1\cos(\theta_1)}$$

Substituting Eq. (a) and $\theta_1 = -30°$ into this expression and solving the resulting expression for e_1 leads to

$$e_1 = \frac{R_{\text{earth}} - R_{\text{Venus}}}{R_{\text{earth}} + R_{\text{Venus}}\cos(\theta_1)} = \frac{149.6(10^6) - 108.2(10^6)}{149.6(10^6) + 108.2(10^6)\cos(-30°)} = 0.1702$$

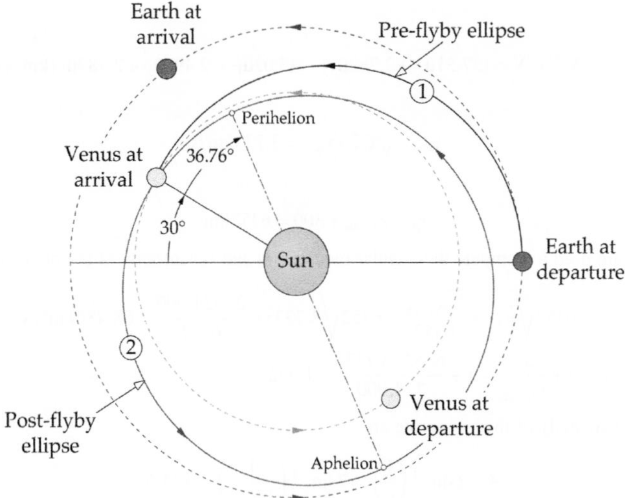

FIG. 8.21

Spacecraft orbits before and after a flyby of Venus, approaching from the sunlit side.

With this result, and Eq. (a) yields

$$h_1 = \sqrt{1.327(10^{11}) \cdot 149.6(10^6)(1-0.1702)} = 4.059(10^9)\,\text{km}^2/\text{s}$$

Now we can use Eqs. (2.31) and (2.49) to calculate the radial and transverse components of the spacecraft's heliocentric velocity at the inbound crossing of Venus' sphere of influence

$$V_{\perp_1}^{(v)} = \frac{h_1}{R_{\text{Venus}}} = \frac{4.059(10^9)}{108.2(10^6)} = 37.51\,\text{km/s}$$

$$V_{r_1}^{(v)} = \frac{\mu_{\text{sun}}}{h_1} e_1 \sin(\theta_1) = \frac{1.327(10^{11})}{4.059(10^9)} \cdot 0.1702 \cdot \sin(-30°) = -2.782\text{km/s}$$

The flight path angle, from Eq. (2.51), is

$$\gamma_1 = \tan^{-1}\frac{V_{r_1}^{(v)}}{V_{\perp_1}^{(v)}} = \tan^{-1}\left(\frac{-2.782}{37.51}\right) = -4.241°$$

The negative sign is consistent with the fact that the spacecraft is flying toward perihelion of the preflyby elliptical trajectory (orbit 1).

The speed of the space vehicle at the inbound crossing is

$$V_1^{(v)} = \sqrt{\left(V_{r_1}^{(v)}\right)^2 + \left(V_{\perp_1}^{(v)}\right)^2} = \sqrt{(-2.782)^2 + 37.51^2} = 37.62\text{km/s} \tag{b}$$

Flyby hyperbola
From Eqs. (8.75) and (8.77), we obtain

$$\mathbf{V}_1^{(v)} = 37.51\hat{\mathbf{u}}_V + 2.782\hat{\mathbf{u}}_S\,(\text{km/s})$$

The velocity of Venus in its presumed circular orbit around the sun is

$$\mathbf{V} = \sqrt{\frac{\mu_{\text{sun}}}{R_{\text{Venus}}}}\hat{\mathbf{u}}_V = \sqrt{\frac{1.327(10)^{11}}{108.2(10)^6}}\hat{\mathbf{u}}_V = 35.02\hat{\mathbf{u}}_V\,(\text{km/s}) \tag{c}$$

Hence

$$\mathbf{v}_{\infty_1} = \mathbf{V}_1^{(v)} - \mathbf{V} = (37.51\hat{\mathbf{u}}_V + 2.782\hat{\mathbf{u}}_S) - 35.02\hat{\mathbf{u}}_V = 2.490\hat{\mathbf{u}}_V + 2.782\hat{\mathbf{u}}_S\ (\text{km/s}) \tag{d}$$

It follows that

$$v_\infty = \sqrt{\mathbf{v}_{\infty_1} \cdot \mathbf{v}_{\infty_1}} = 3.733\ \text{km/s}$$

The periapsis radius is

$$r_p = r_{\text{Venus}} + 300 = 6352\ \text{km}$$

Eqs. (8.38) and (8.39) are used to compute the angular momentum and eccentricity of the planetocentric hyperbola.

$$h = 6352\sqrt{v_\infty{}^2 + \frac{2\mu_{\text{Venus}}}{6352}} = 6352\sqrt{3.733^2 + \frac{2 \cdot 324{,}900}{6352}} = 68{,}480\ \text{km}^2/\text{s}$$

$$e = 1 + \frac{r_p v_\infty{}^2}{\mu_{\text{Venus}}} = 1 + \frac{6352 \cdot 3.733^2}{324{,}900} = 1.272$$

The turn angle and true anomaly of the asymptote are

$$\delta = 2\sin^{-1}\left(\frac{1}{e}\right) = 2\sin^{-1}\left(\frac{1}{1.272}\right) = 103.6^\circ$$

$$\theta_\infty = \cos^{-1}\left(-\frac{1}{e}\right) = \cos^{-1}\left(-\frac{1}{1.272}\right) = 141.8^\circ$$

From Eqs. (2.50), (2.103), and (2.107), the aiming radius is

$$\Delta = r_p\sqrt{\frac{e+1}{e-1}} = 6352\sqrt{\frac{1.272+1}{1.272-1}} = 18{,}340\ \text{km} \tag{e}$$

Finally, from Eqs. (8.84) and (d) we obtain the angle between $\mathbf{v}_{\infty_1}$ and $\mathbf{V}$,

$$\phi_1 = \tan^{-1}\frac{v_{\infty_1})_S}{v_{\infty_1})_V} = \tan^{-1}\frac{2.782}{2.490} = 48.17^\circ \tag{f}$$

There are two flyby approaches, as shown in Fig. 8.22. In the dark-side approach, the turn angle is counterclockwise (+103.6°), whereas for the sunlit side approach, it is clockwise (−103.6°).

(a) *Dark-side approach*
According to Eq. (8.85), the angle between $\mathbf{v}_\infty$ and $\mathbf{V}_{\text{Venus}}$ at the outbound crossing is

$$\phi_2 = \phi_1 + \delta = 48.17^\circ + 103.6^\circ = 151.8^\circ$$

Hence, by Eq. (8.86),

$$\mathbf{v}_{\infty_2} = 3.733(\cos 151.8^\circ \hat{\mathbf{u}}_V + \sin 151.8^\circ \hat{\mathbf{u}}_S) = -3.289\hat{\mathbf{u}}_V + 1.766\hat{\mathbf{u}}_S\ (\text{km/s})$$

Using this and Eq. (c), we compute the spacecraft's heliocentric velocity at the outbound crossing.

$$\mathbf{V}_2^{(v)} = \mathbf{V} + \mathbf{v}_{\infty_2} = 31.73\hat{\mathbf{u}}_V + 1.766\hat{\mathbf{u}}_S\ (\text{km/s})$$

It follows from Eq. (8.89) that

$$V_{\perp_2}^{(v)} = 31.73\ \text{km/s} \quad V_{r_2}^{(v)} = -1.766\ \text{km/s} \tag{g}$$

The speed of the spacecraft at the outbound crossing is

$$V_2^{(v)} = \sqrt{\left(V_{r_2}^{(v)}\right)^2 + \left(V_{\perp_2}^{(v)}\right)^2} = \sqrt{(-1.766)^2 + 31.73^2} = 31.78\ \text{km/s}$$

This is 5.83 km/s less than the inbound speed.

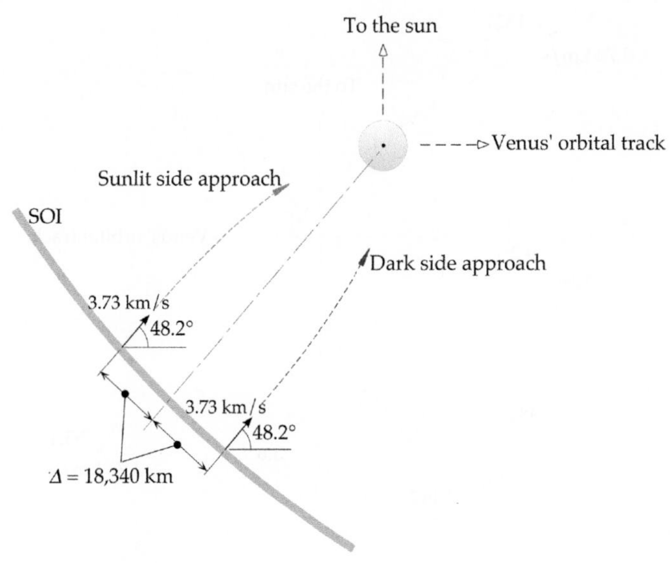

FIG. 8.22

Initiation of a sunlit-side approach and dark-side approach at the inbound crossing.

Postflyby ellipse (orbit 2) for the dark-side approach

For the heliocentric postflyby trajectory, labeled orbit 2 in Fig. 8.20, the angular momentum is found using Eq. (8.90)

$$h_2 = R_{\text{Venus}} V_{\perp_2}^{(v)} = 108.2(10)^6 \cdot 31.73 = 3.434\left(10^9\right)\left(\text{km}^2/\text{s}\right) \tag{h}$$

From Eq. (8.91),

$$e\cos\theta_2 = \frac{h_2{}^2}{\mu_{\text{sun}} R_{\text{Venus}}} - 1 = \frac{\left[3.434\left(10^9\right)\right]^2}{1.327\left(10^{11}\right)\cdot 108.2\left(10^6\right)} - 1 = -0.1790 \tag{i}$$

and from Eq. (8.92)

$$e\sin\theta_2 = \frac{V_{r_2}^{(v)} h_2}{\mu_{\text{sun}}} = \frac{-1.766\cdot 3.434(10)^9}{1.327\left(10^{11}\right)} = -0.04569 \tag{j}$$

Thus

$$\tan\theta_2 = \frac{e\sin\theta_2}{e\cos\theta_2} = \frac{-0.04569}{-0.1790} = 0.2553 \tag{k}$$

which means

$$\theta_2 = 14.32° \text{ or } 194.32° \tag{l}$$

But θ_2 must lie in the third quadrant since, according to Eqs. (i) and (j), both the sine and cosine are negative. Hence,

$$\theta_2 = 194.32° \tag{m}$$

With this value of θ_2, we can use either Eq. (i) or Eq. (j) to calculate the eccentricity,

$$e_2 = 0.1847 \tag{n}$$

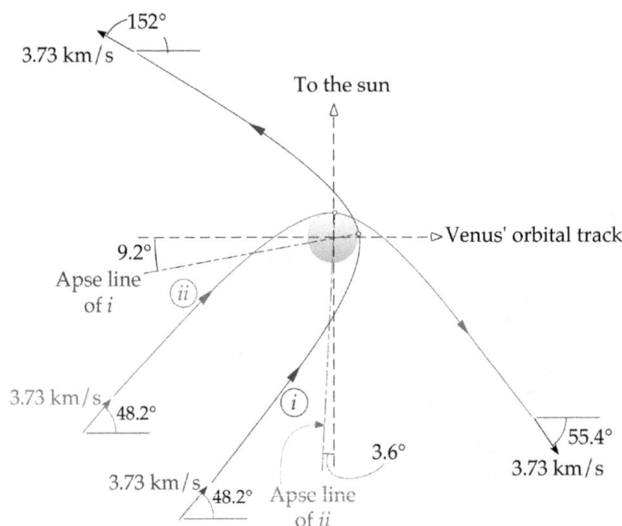

FIG. 8.23

Hyperbolic flyby trajectories for (i) the dark-side approach and (ii) the sunlit-side approach.

Perihelion of the departure orbit lies 194.32° clockwise from the encounter point (so that aphelion is 14.32° therefrom), as illustrated in Fig. 8.20. The perihelion radius is given by Eq. (2.50),

$$R_{\text{perihelion}} = \frac{h_2{}^2}{\mu_{\text{sun}}} \frac{1}{1+e_2} = \frac{\left[3.434\left(10^9\right)\right]^2}{1.327\left(10^{11}\right)} \frac{1}{1+0.1847} = 74.98\left(10^6\right)\text{km}$$

which is well within the orbit of Venus.

(b) *Sunlit-side approach*

In this case, the angle between $\mathbf{v}_\infty$ and $\mathbf{V}_{\text{Venus}}$ at the outbound crossing is

$$\phi_2 = \phi_1 - \delta = 48.17° - 103.6° = -55.44°$$

Therefore,

$$\mathbf{v}_{\infty_2} = 3.733[\cos(-55.44°)\hat{\mathbf{u}}_V + \sin(-55.44°)\hat{\mathbf{u}}_S] = 2.118\hat{\mathbf{u}}_V - 3.074\hat{\mathbf{u}}_S\,(\text{km/s})$$

The spacecraft's heliocentric velocity at the outbound crossing is

$$\mathbf{V}_2^{(v)} = \mathbf{V}_{\text{Venus}} + \mathbf{v}_{\infty_2} = 37.14\hat{\mathbf{u}}_V - 3.074\hat{\mathbf{u}}_S\,(\text{km/s})$$

which means

$$V_{\perp_2}^{(v)} = 37.14\text{km/s} \quad V_{r_2}^{(v)} = 3.074\text{km/s}$$

The speed of the spacecraft at the outbound crossing is

$$V_2^{(v)} = \sqrt{\left(V_{r_2}^{(v)}\right)^2 + \left(V_{\perp_2}^{(v)}\right)^2} = \sqrt{3.050^2 + 37.14^2} = 37.27\text{km/s}$$

This speed is just 0.348 km/s less than the inbound crossing speed. The relatively small speed change is due to the fact that the apse line of this hyperbola is nearly perpendicular to Venus' orbital track, as shown in Fig. 8.23. Nevertheless, the periapses of both hyperbolas are on the leading side of the planet.

Postflyby ellipse (orbit 2) for the sunlit-side approach

To determine the heliocentric postflyby trajectory, labeled orbit 2 in Fig. 8.21, we repeat Steps (h) through (n) above.

$$h_2 = R_{\text{Venus}} V_{\perp_2}^{(v)} = 108.2(10^6) \cdot 37.14 = 4.019(10^9)\ (\text{km}^2/\text{s})$$

$$e\cos\theta_2 = \frac{h_2^2}{\mu_{\text{sun}} R_{\text{Venus}}} - 1 = \frac{[4.019(10^9)]^2}{1.327(10^{11}) \cdot 108.2(10^6)} - 1 = 0.1246 \tag{o}$$

$$e\sin\theta_2 = \frac{V_{r_2}^{(v)} h_2}{\mu_{\text{sun}}} = \frac{3.074 \cdot 4.019(10^9)}{1.327(10^{11})} = 0.09309 \tag{p}$$

$$\tan\theta_2 = \frac{e\sin\theta_2}{e\cos\theta_2} = \frac{0.09309}{0.1246} = 0.7469 \Rightarrow \theta_2 = 36.76° \text{ or } 216.76°$$

θ_2 must lie in the first quadrant since both the sine and cosine are positive. Hence,

$$\theta_2 = 36.76° \tag{q}$$

With this value of θ_2, we can use either Eq. (o) or (p) to calculate the eccentricity,

$$e_2 = 0.1556$$

Perihelion of the departure orbit lies 36.76° clockwise from the encounter point as illustrated in Fig. 8.21. The perihelion radius is

$$R_{\text{perihelion}} = \frac{h_2^2}{\mu_{\text{sun}}} \frac{1}{1+e_2} = \frac{[4.019(10^9)]^2}{1.327(10^{11})} \frac{1}{1+0.1556} = 105.3(10^6)\ \text{km}$$

which is just within the orbit of Venus. Aphelion lies between the orbits of earth and Venus.

Gravity assist maneuvers are used to add momentum to a spacecraft over and above that available from a spacecraft's onboard propulsion system. A sequence of flybys of planets can impart the delta-v needed to reach regions of the solar system that would be inaccessible using only existing propulsion technology. The technique can also reduce the flight time. Interplanetary missions using gravity assist flybys must be carefully designed to take advantage of the relative positions of planets.

Pioneer 11, a 260-kg spacecraft launched from Cape Canaveral, Florida, in April 1973, used a December 1974 flyby of Jupiter to gain the momentum required to carry it to the first ever flyby encounter with Saturn on September 1, 1979. Contact with Pioneer 11 was finally lost in September 1995.

Mariner 10 was a 503-kg spacecraft launched from Cape Canaveral on a mission to Mercury on November 3, 1973. It flew by Venus once and Mercury three times. It was deactivated on March 29, 1974, and is probably still in orbit around the sun.

Following its September 1977 launch from Cape Canaveral, the 826-kg Voyager 1, like Pioneer 11 before it, used a flyby of Jupiter (March 1979) to reach Saturn in November 1980. In August 1977, Voyager 2 was launched on its "grand tour" of the outer planets and beyond. This involved gravity assist flybys of Jupiter (July 1979), Saturn (August 1981), Uranus (January 1986), and Neptune (August 1989), after which the spacecraft departed the solar system at an angle of 30° to the ecliptic.

With a mass nine times that of Pioneer 11, the dual-spin Galileo spacecraft was launched from the space shuttle *Atlantis*, departing on October 18, 1989 for an extensive international exploration of Jupiter and its satellites that lasted until the spacecraft was deorbited on September 21, 2003. Galileo used gravity assist flybys of Venus (February 1990), earth (December 1990), and earth again (December 1992) before arriving at Jupiter in December 1995.

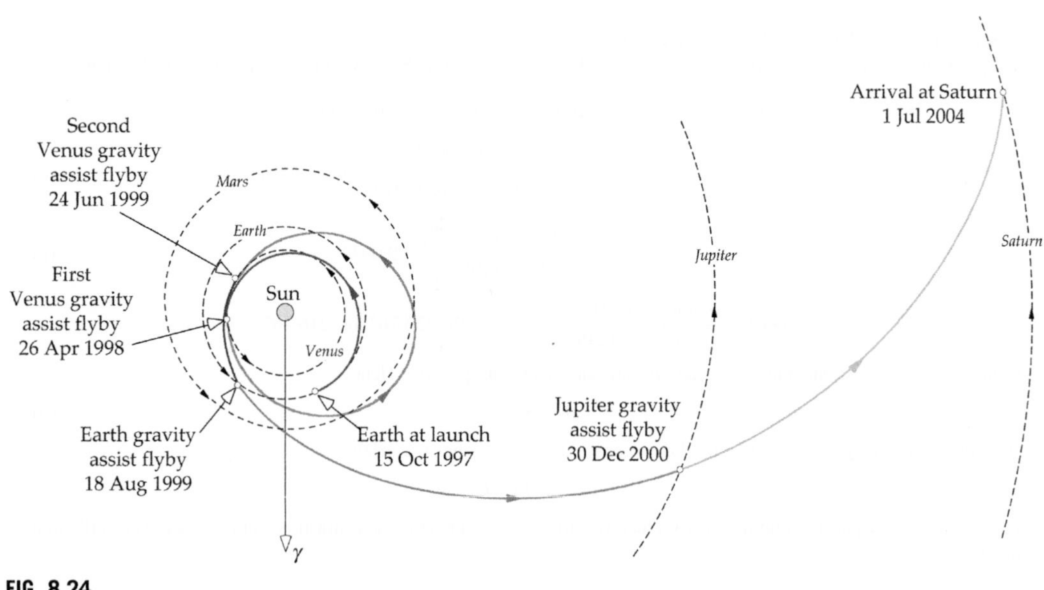

FIG. 8.24

Cassini's 7-year mission to Saturn.

Ulysses was a 370-kg international spacecraft launched from the space shuttle *Discovery* in 1990. Its flyby of Jupiter in early February 1992 increased its heliocentric orbital inclination to over 80°. With a period of about six years, this orbit allowed Ulysses to expore the polar regions of the sun until it was decommisioned in June 2009.

The international Cassini mission to Saturn made extensive use of gravity assist flyby maneuvers. The 5712-kg Cassini spacecraft was launched on October 15, 1997, from Cape Canaveral, Florida, and arrived at Saturn nearly 7 years later, on July 1, 2004. The mission involved four flybys, as illustrated in Fig. 8.24. A little over 8 months after launch, on April 26, 1998, Cassini flew by Venus at a periapsis altitude of 284 km and received a speed boost of about 7 km/s. This placed the spacecraft in an orbit that sent it just outside the orbit of Mars (but well away from the planet) and returned it to Venus on June 24, 1999, for a second flyby, this time at an altitude of 600 km. The result was a trajectory that vectored Cassini toward the earth for an August 18, 1999, flyby at an altitude of 1171 km. The 5.5-km/s speed boost at earth sent the spacecraft toward Jupiter for its next flyby maneuver. This occurred on December 30, 2000, at a distance of 9.7 million km from Jupiter, boosting Cassini's speed by about 2 km/s and adjusting its trajectory so as to rendezvous with Saturn about three and a half years later. After 13 years in orbit, on September 12, 2015, its fuel exhausted, the Cassini mission ended with the spacecraft plunging into Saturn's upper atmosphere.

Messenger was a 1108-kg spacecraft launched from Cape Canaveral, Florida, on August 3, 2004. It flew by earth, Venus, and Mercury on its way finally to orbit insertion around Mercury on March 2011. Messenger deorbited four years later.

8.10 PLANETARY EPHEMERIS

The state vector (**R**, **V**) of a planet is defined relative to the heliocentric ecliptic frame of reference, as illustrated in Fig. 8.25. This is very similar to the geocentric equatorial frame of Fig. 4.7. The sun replaces the earth as the center of attraction, and the plane of the ecliptic replaces the earth's equatorial plane. The vernal equinox continues to define the inertial X axis.

To design realistic interplanetary missions, we must be able to determine the state vector of a planet at any given time. Table 8.1 provides the orbital elements of the planets and their rates of change per century (Cy) with respect to the J2000 epoch (January 1, 2000, 12 h UT). The table, covering the years 1800–2050, is sufficiently accurate for our needs. Alternatively, one can use JPL's online HORIZONS system (JPL Horizons Web-Interface, 2018) or, within MATLAB, the function `planetEphemeris`. From the orbital elements, we can infer the state vector using Algorithm 4.5.

To interpret Table 8.1, observe the following:

- One astronomical unit (1 AU) is $1.49597871(10^8)$ km, the average distance between the earth and the sun.
- One arcsecond ($1''$) is 1/3600 of a degree.
- a is the semimajor axis.
- e is the eccentricity.
- i is the inclination to the ecliptic plane.
- Ω is the right ascension of the ascending node (relative to the J2000 vernal equinox).
- ϖ, the longitude of perihelion, is defined as $\varpi = \omega + \Omega$, where ω is the argument of perihelion.

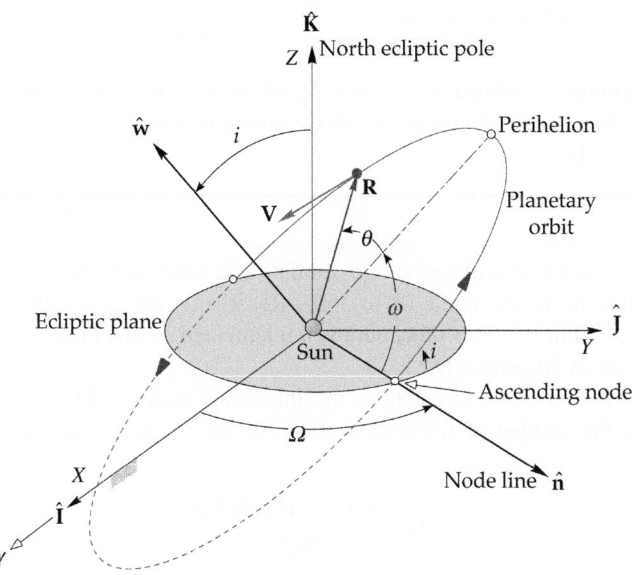

FIG. 8.25

Planetary orbit in the heliocentric ecliptic frame.

Table 8.1 Planetary orbital elements and their centennial rates

	a (AU) $\dot{a}$ (AU/Cy)	e $\dot{e}$ (1/Cy)	i (°) $\dot{i}$ (°/Cy)	Ω (°) $\dot{\Omega}$ (°/Cy)	ϖ (°) $\dot{\varpi}$ (°/Cy)	L (°) $\dot{L}$ (°/Cy)
Mercury	0.38709927	0.20563593	7.00497902	48.33076593	77.45779628	252.25032350
	0.00000037	0.00001906	−0.00594749	−0.12534081	0.16047689	149,472.67411175
Venus	0.72333566	0.00677672	3.39467605	76.67984255	131.60246718	181.97909950
	0.00000390	−0.00004107	−0.00078890	−0.27769418	0.00268329	58,517.81538729
Earth	1.00000261	0.01671123	−0.00001531	0.0	102.93768193	100.46457166
	0.00000562	−0.00004392	−0.01294668	0.0	0.32327364	35,999.37244981
Mars	1.52371034	0.09339410	1.84969142	49.55953891	−23.94362959	−4.55343205
	0.0001847	0.00007882	−0.00813131	−0.29257343	0.44441088	19,140.30268499
Jupiter	5.20288700	0.04838624	1.30439695	100.47390909	14.72847983	34.39644501
	−0.00011607	−0.00013253	−0.00183714	0.20469106	0.21252668	3034.74612775
Saturn	9.53667594	0.05386179	2.48599187	113.66242448	92.59887831	49.95424423
	−0.00125060	−0.00050991	0.00193609	−0.28867794	−0.41897216	1222.49362201
Uranus	19.18916464	0.04725744	0.77263783	74.01692503	170.95427630	313.23810451
	−0.00196176	−0.00004397	−0.00242939	0.04240589	0.40805281	428.48202785
Neptune	30.06992276	0.00859048	1.77004347	131.78422574	44.96476227	−55.12002969
	0.00026291	0.00005105	0.00035372	−0.00508664	−0.32241464	218.45945325
(Pluto)	39.48211675	0.24882730	17.14001206	110.30393684	224.06891629	238.92903833
	−0.00031596	0.00005170	0.00004818	−0.01183482	−0.04062942	145.20780515

Reproduced with permission from Standish et al. (2013).

- L, the mean longitude, is defined as $L = \varpi + M$, where M is the mean anomaly.
- $\dot{a}$, $\dot{e}$, $\dot{\Omega}$, etc., are the rates of change of the above orbital elements per Julian century. One century (Cy) equals 36,525 days.

ALGORITHM 8.1

Determine the state vector of a planet at a given date and time. All angular calculations must be adjusted so that they lie in the range 0° to 360°. Recall that the gravitational parameter of the sun is $\mu = 1.327(10^{11})\ \text{km}^3/\text{s}^2$. This procedure is implemented in MATLAB as the function *planet_elements_and_sv.m* in Appendix D.35.

1. Use Eqs. (5.47) and (5.48) to calculate the Julian day number JD.
2. Calculate T_0, the number of Julian centuries between J2000 and the date in question (Eq. 5.49).

$$T_0 = \frac{JD - 2{,}451{,}545.0}{36{,}525}$$

3. If Q is any one of the six planetary orbital elements listed in Table 8.1, then calculate its value at JD by means of the formula

$$Q = Q_0 + \dot{Q}T_0 \tag{8.93b}$$

where Q_0 is the value listed for J2000, and $\dot{Q}$ is the tabulated rate. All angular quantities must be adjusted to lie in the range 0°–360°.

4. Use the semimajor axis a and the eccentricity e to calculate the angular momentum h at JD from Eq. (2.71)

$$h = \sqrt{\mu a(1-e^2)}$$

5. Obtain the argument of perihelion ω and mean anomaly M at JD from the results of Step 3 by means of the definitions

$$\omega = \varpi - \Omega$$
$$M = L - \varpi$$

6. Substitute the eccentricity e and the mean anomaly M at JD into Kepler's equation (Eq. 3.14) and calculate the eccentric anomaly E.
7. Calculate the true anomaly θ using Eq. (3.13).
8. Use h, e, Ω, i, ω, and θ to obtain the heliocentric position vector **R** and velocity **V** by means of Algorithm 4.5, with the heliocentric ecliptic frame replacing the geocentric equatorial frame.

EXAMPLE 8.7

Find the distance between earth and Mars at 12 h UT on August 27, 2003. Use Algorithm 8.1.

Step 1:

According to Eq. (5.48), the Julian day number J_0 for midnight (0h UT) of this date is

$$\begin{aligned} J_0 &= 367 \cdot 2003 - \text{INT}\left\{\frac{7\left[2003 + \text{INT}\left(\frac{8+9}{12}\right)\right]}{4}\right\} + \text{INT}\left(\frac{275 \cdot 8}{9}\right) + 27 + 1{,}721{,}013.5 \\ &= 735{,}101 - 3507 + 244 + 27 + 1{,}721{,}013.5 \\ &= 2{,}452{,}878.5 \end{aligned}$$

At $UT = 12$, the Julian day number is

$$JD = 2{,}452{,}878.5 + \frac{12}{24} = 2{,}452{,}879.0$$

Step 2:

The number of Julian centuries between J2000 and this date is

$$T_0 = \frac{JD - 2{,}451{,}545}{36{,}525} = \frac{2{,}452{,}879 - 2{,}451{,}545}{36{,}525} = 0.036523\text{Cy}$$

Step 3:

Table 8.1 and Eq. (8.93b) yield the orbital elements of earth and Mars at 12 h UT on August 27, 2003:

	a (km)	e	i (°)	Ω (°)	ϖ (°)	L (°)
Earth	$1.4960(10^8)$	0.016710	−0.00048816	0.0	102.95	335.27
Mars	$2.2794(10^8)$	0.093397	1.8494	49.549	336.07	334.51

Step 4:

$$h_{\text{earth}} = 4.4451(10^9)\,\text{km}^2/\text{s}$$
$$h_{\text{Mars}} = 5.4760(10^9)\,\text{km}^2/\text{s}$$

Step 5:

$$\omega_{\text{earth}} = (\varpi - \Omega)_{\text{earth}} = 102.95 - 0 = 102.95°$$
$$\omega_{\text{Mars}} = (\varpi - \Omega)_{\text{Mars}} = 336.07 - 49.549 = 286.52°$$
$$M_{\text{earth}} = (L - \varpi)_{\text{earth}} = 335.27 - 102.95 = 232.32°$$
$$M_{\text{Mars}} = (L - \varpi)_{\text{Mars}} = 334.51 - 336.07 = -1.56°(358.43°)$$

Step 6:

$$E_{\text{earth}} - 0.016710\sin E_{\text{earth}} = 232.32°(\pi/180) \Rightarrow E_{\text{earth}} = 231.57°$$
$$E_{\text{Mars}} - 0.093397\sin E_{\text{Mars}} = 358.43°(\pi/180) \Rightarrow E_{\text{Mars}} = 358.27°$$

Step 7:

$$\theta_{\text{earth}} = 2\tan^{-1}\left(\sqrt{\frac{1+0.016710}{1-0.016710}}\tan\frac{231.57°}{2}\right) = -129.18 \Rightarrow \theta_{\text{earth}} = 230.8°$$

$$\theta_{\text{Mars}} = 2\tan^{-1}\left(\sqrt{\frac{1+0.093397}{1-0.093397}}\tan\frac{358.27°}{2}\right) = -1.8998° \Rightarrow \theta_{\text{Mars}} = 358.10°$$

Step 8:
From Algorithm 4.5,

$$\mathbf{R}_{\text{earth}} = \left(135.59\hat{\mathbf{I}} - 66.803\hat{\mathbf{J}} - 0.00056916\hat{\mathbf{K}}\right)(10^6)\,(\text{km})$$
$$\mathbf{V}_{\text{earth}} = \left(12.680\hat{\mathbf{I}} - 26.610\hat{\mathbf{J}} - 0.00022672\hat{\mathbf{K}}\right)(\text{km/s})$$
$$\mathbf{R}_{\text{Mars}} = \left(185.95\hat{\mathbf{I}} - 89.959\hat{\mathbf{J}} - 6.4534\hat{\mathbf{K}}\right)(10^6)\,(\text{km})$$
$$\mathbf{V}_{\text{Mars}} = 11.478\hat{\mathbf{I}} - 23.881\hat{\mathbf{J}} - 0.21828\hat{\mathbf{K}}\,(\text{km})$$

The distance d between the two planets is therefore

$$\begin{aligned} d &= \|\mathbf{R}_{\text{Mars}} - \mathbf{R}_{\text{earth}}\| \\ &= \sqrt{(185.95 - 135.59)^2 + [-89.959 - (-66.803)]^2 + (-6.4534 - 0.00056916)^2}\,(10^6) \end{aligned}$$

or

$$\boxed{d = 55.80(10^6)\,\text{km}}$$

The positions of earth and Mars are illustrated in Fig. 8.26. It is a rare event for Mars to be in opposition (lined up with earth on the same side of the sun) when Mars is at or near perihelion. The two planets had not been this close in recorded history.

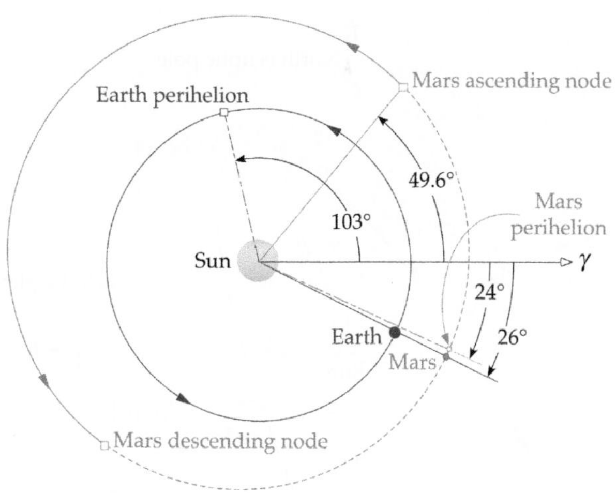

FIG. 8.26

Earth and Mars on August 27, 2003. Angles shown are heliocentric latitude, measured in the plane of the ecliptic counterclockwise from the vernal equinox of J2000.

8.11 NON-HOHMANN INTERPLANETARY TRAJECTORIES

To implement a systematic patched conic procedure for three-dimensional trajectories, we will use vector notation and the procedures described in Sections 4.4 and 4.6 (Algorithms 4.2 and 4.5), together with the solution of Lambert's problem presented in Section 5.3 (Algorithm 5.2). The mission is to send a spacecraft from planet 1 to planet 2 in a specified time t_{12}. As previously discussed in this chapter, we break the mission down into three parts: the departure phase, the cruise phase, and the arrival phase. We start with the cruise phase.

The frame of reference that we use is the heliocentric ecliptic frame, as shown in Fig. 8.27. The first step is to obtain the state vector of planet 1 at departure (time t) and the state vector of planet 2 at arrival (time $t + t_{12}$). This is accomplished by means of Algorithm 8.1.

The next step is to determine the spacecraft's transfer trajectory from planet 1 to planet 2. We first observe that, according to the patched conic procedure, the heliocentric position vector of the spacecraft at time t is that of planet 1 ($\mathbf{R}_1$) and at time $t + t_{12}$ its position vector is that of planet 2 ($\mathbf{R}_2$). With $\mathbf{R}_1$, $\mathbf{R}_2$, and the time of flight t_{12} we can use Algorithm 5.2 (Lambert's problem) to obtain the spacecraft's departure and arrival velocities $\mathbf{V}_D^{(v)}$ and $\mathbf{V}_A^{(v)}$ relative to the sun. Either of the state vectors ($\mathbf{R}_1$, $\mathbf{V}_D^{(v)}$) or ($\mathbf{R}_2$, $\mathbf{V}_A^{(v)}$) can be used to obtain the transfer trajectory's six orbital elements by means of Algorithm 4.2.

The spacecraft's hyperbolic excess velocity on exiting the sphere of influence of planet 1 is

$$\mathbf{v}_\infty)_{\text{Departure}} = \mathbf{V}_D^{(v)} - \mathbf{V}_1 \tag{8.94a}$$

and its excess speed is

$$v_\infty)_{\text{Departure}} = \left\|\mathbf{V}_D^{(v)} - \mathbf{V}_1\right\| \tag{8.94b}$$

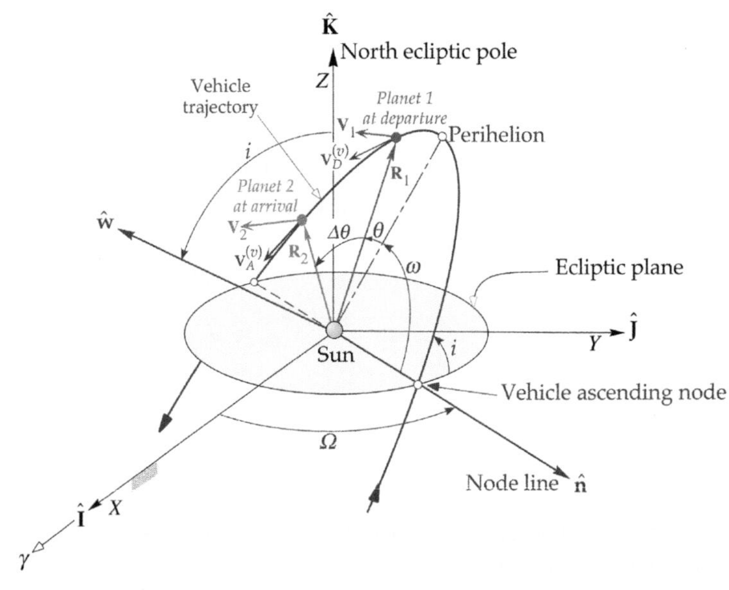

FIG. 8.27

Heliocentric orbital elements of a three-dimensional transfer trajectory from planet 1 to planet 2.

Likewise, at the sphere of influence crossing at planet 2,

$$\mathbf{v}_\infty)_{\text{Arrival}} = \mathbf{V}_A^{(v)} - \mathbf{V}_2 \tag{8.95a}$$

$$v_\infty)_{\text{Arrival}} = \left\| \mathbf{V}_A^{(v)} - \mathbf{V}_2 \right\| \tag{8.95b}$$

ALGORITHM 8.2

Given the departure and arrival dates (and, therefore, the time of flight), determine the trajectory for a mission from planet 1 to planet 2. This procedure is implemented as the MATLAB function *interplanetary.m* in Appendix D.36.

1. Use Algorithm 8.1 to determine the state vector $(\mathbf{R}_1, \mathbf{V}_1)$ of planet 1 at departure and the state vector $(\mathbf{R}_2, \mathbf{V}_2)$ of planet 2 at arrival.
2. Use $\mathbf{R}_1$, $\mathbf{R}_2$, and the time of flight in Algorithm 5.2 to find the spacecraft velocity $\mathbf{V}_D^{(v)}$ at departure from planet 1's sphere of influence and its velocity $\mathbf{V}_A^{(v)}$ upon arrival at planet 2's sphere of influence.
3. Calculate the hyperbolic excess velocities at departure and arrival using Eqs. (8.94) and (8.95).

EXAMPLE 8.8

A spacecraft departs earth's sphere of influence on November 7, 1996 (0 h UT), on a prograde coasting flight to Mars, arriving at Mars' sphere of influence on September 12, 1997 (0 h UT). Use Algorithm 8.2 to determine the trajectory and then compute the hyperbolic excess velocities at departure and arrival.

Solution

Step 1:

Algorithm 8.1 yields the state vectors for earth and Mars.

$$\mathbf{R}_{\text{earth}} = 1.0499(10^8)\hat{\mathbf{I}} + 1.0465(10^8)\hat{\mathbf{J}} + 716.93\hat{\mathbf{K}}\,(\text{km}) \qquad (R_{\text{earth}} = 1.4824(10^8)\,\text{km})$$
$$\mathbf{V}_{\text{earth}} = -21.515\hat{\mathbf{I}} + 20.958\hat{\mathbf{J}} + 0.00014376\hat{\mathbf{K}}\,(\text{km/s}) \qquad (V_{\text{earth}} = 30.055\,\text{km/s})$$
$$\mathbf{R}_{\text{Mars}} = -2.0858(10^7)\hat{\mathbf{I}} - 2.1842(10^8)\hat{\mathbf{J}} + 4.06244(10^6)\hat{\mathbf{K}}\,(\text{km}) \qquad (R_{\text{Mars}} = 2.1945(10^8)\,\text{km})$$
$$\mathbf{V}_{\text{Mars}} = 25.037\hat{\mathbf{I}} + 0.22311\hat{\mathbf{J}} - 0.62018\hat{\mathbf{K}}\,(\text{km/s}) \qquad (V_{\text{Mars}} = 25.046\,\text{km/s})$$

Step 2:

The position vector $\mathbf{R}_1$ of the spacecraft at crossing the earth's sphere of influence is just that of the earth,

$$\mathbf{R}_1 = \mathbf{R}_{\text{earth}} = 1.0499(10^8)\hat{\mathbf{I}} + 1.0465(10^8)\hat{\mathbf{J}} + 716.93\hat{\mathbf{K}}\,(\text{km})$$

On arrival at Mars' sphere of influence, the spacecraft's position vector is

$$\mathbf{R}_2 = \mathbf{R}_{\text{mars}} = -2.0858(10^7)\hat{\mathbf{I}} - 2.1842(10^8)\hat{\mathbf{J}} - 4.06244(10^6)\hat{\mathbf{K}}\,(\text{km})$$

According to Eqs. (5.47) and (5.48)

$$JD_{\text{Departure}} = 2{,}450{,}394.5$$
$$JD_{\text{Arrival}} = 2{,}450{,}703.5$$

Hence, the time of flight is

$$t_{12} = 2{,}450{,}703.5 - 2{,}450{,}394.5 = 309\,\text{days}$$

Entering $\mathbf{R}_1$, $\mathbf{R}_2$, and t_{12} into Algorithm 5.2 yields

$$\mathbf{V}_D^{(v)} = -24.429\hat{\mathbf{I}} + 21.782\hat{\mathbf{J}} + 0.94810\hat{\mathbf{K}}\,(\text{km/s}) \quad \left(V_D^{(v)} = 32.743\,\text{km/s}\right)$$
$$\mathbf{V}_A^{(v)} = 22.157\hat{\mathbf{I}} + 0.19959\hat{\mathbf{J}} + 0.45793\hat{\mathbf{K}}\,(\text{km/s}) \quad \left(V_A^{(v)} = 22.162\,\text{km/s}\right)$$

Using the state vector $(\mathbf{R}_1, \mathbf{V}_D^{(v)})$, we employ Algorithm 4.2 to find the orbital elements of the heliocentric transfer trajectory.

$$\boxed{\begin{aligned} h &= 4.8456(10^6)\,\text{km}^2/\text{s} \\ e &= 0.20581 \\ \Omega &= 44.898^\circ \\ i &= 1.6622^\circ \\ \omega &= 19.973^\circ \\ \theta_1 &= 340.04^\circ \\ a &= 1.8475(10^8)\,\text{km} \end{aligned}}$$

Step 3:
At departure, the hyperbolic excess velocity is

$$\mathbf{v}_\infty)_{\text{Departure}} = \mathbf{V}_D^{(v)} - \mathbf{V}_{\text{earth}} = -2.9138\hat{\mathbf{I}} + 0.79525\hat{\mathbf{J}} + 0.94796\hat{\mathbf{K}}\,(\text{km/s})$$

Therefore, the hyperbolic excess speed is

$$v_\infty)_{\text{Departure}} = \|\mathbf{v}_\infty)_{\text{Departure}}\| = \boxed{3.1656\text{km/s}} \quad \text{(a)}$$

Likewise, at arrival

$$\mathbf{v}_\infty)_{\text{Arrival}} = \mathbf{V}_A^{(v)} - \mathbf{V}_{\text{Mars}} = -2.58805\hat{\mathbf{I}} + 0.023514\hat{\mathbf{J}} + 0.16254\hat{\mathbf{K}}\,(\text{km/s})$$

so that

$$\mathbf{v}_\infty)_{\text{Arrival}} = \|\mathbf{v}_\infty)_{\text{Arrival}}\| = \boxed{2.8852\text{km/s}} \quad \text{(b)}$$

For the previous example, Fig. 8.28 shows the orbits of earth, Mars, and the spacecraft from directly above the ecliptic plane. Dotted lines indicate the portions of an orbit that are below the plane. λ is the heliocentric longitude measured counterclockwise from the vernal equinox of J2000. Also shown are the position of Mars at departure and the position of the earth at arrival.

The transfer orbit resembles that of the Mars Global Surveyor, which departed earth on November 7, 1996, and arrived at Mars 309 days later, on September 12, 1997.

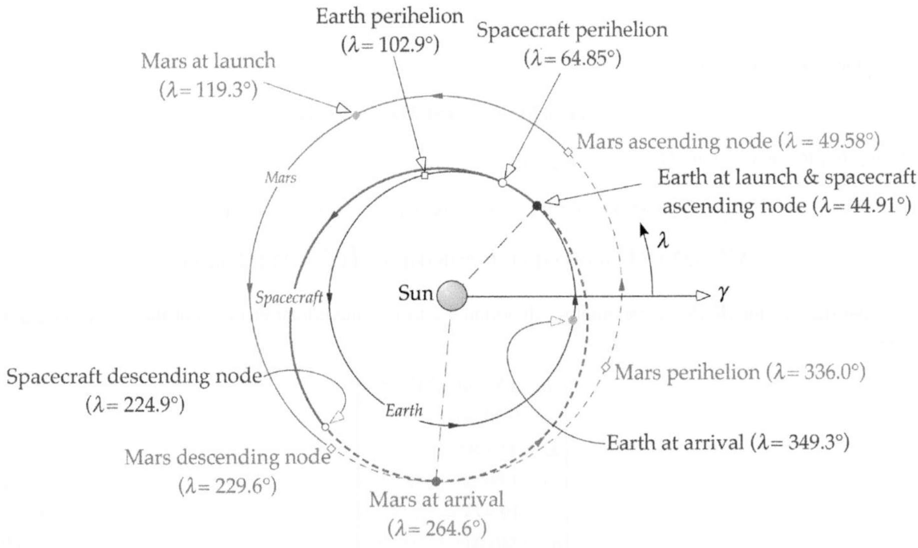

FIG. 8.28

The transfer trajectory of Example 8.8, together with the orbits of earth and Mars, as viewed from directly above the plane of the ecliptic.

EXAMPLE 8.9

In Example 8.8, calculate the delta-v required to launch the spacecraft onto its cruise trajectory from a 180-km circular parking orbit. Sketch the departure trajectory.

Solution

Recall that

$$r_{\text{earth}} = 6378\,\text{km}$$

$$\mu_{\text{earth}} = 398{,}600\,\text{km}^3/\text{s}^2$$

The radius to periapsis of the departure hyperbola is the radius of the earth plus the altitude of the parking orbit,

$$r_p = 6378 + 180 = 6558\,\text{km}$$

Substituting this and Eq. (a) from Example 8.8 into Eq. (8.40) we get the speed of the spacecraft at periapsis of the departure hyperbola,

$$v_p)_{\text{Departure}} = \sqrt{\left[v_\infty)_{\text{Departure}}\right]^2 + \frac{2\mu_{\text{earth}}}{r_p}} = \sqrt{3.1651^2 + \frac{2.398{,}600}{6558}} = 11.47\,\text{km/s}$$

The speed of the spacecraft in its circular parking orbit is

$$v_c = \sqrt{\frac{\mu_{\text{earth}}}{r_p}} = \sqrt{\frac{398{,}600}{6558}} = 7.796\,\text{km/s}$$

Hence, the delta-v requirement is

$$\Delta v = v_p)_{\text{Departure}} - v_c = \boxed{3.674\,\text{km/s}}$$

The eccentricity of the hyperbola is given by Eq. (8.38),

$$e = 1 + \frac{r_p\left[v_\infty)_{\text{Departure}}\right]^2}{\mu_{\text{earth}}} = 1 + \frac{6558 \cdot 3.1656^2}{398{,}600} = 1.165$$

If we assume that the spacecraft is launched from a parking orbit of 28° inclination, then the departure appears as shown in the three-dimensional sketch in Fig. 8.29.

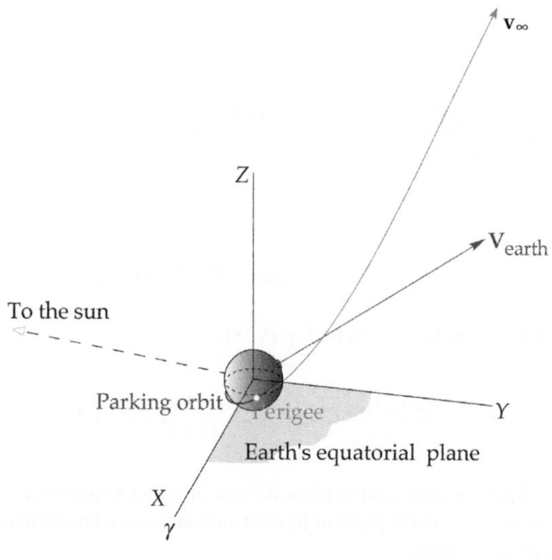

FIG. 8.29

The departure hyperbola, assumed to be at 28° inclination to earth's equator.

EXAMPLE 8.10

In Example 8.8, calculate the delta-v required to place the spacecraft in an elliptical capture orbit around Mars with a periapsis altitude of 300 km and a period of 48 h. Sketch the approach hyperbola.

Solution

From Tables A.1 and A.2, we know that

$$r_{\text{Mars}} = 3380\,\text{km}$$
$$\mu_{\text{Mars}} = 42{,}830\,\text{km}^3/\text{s}^2$$

The radius to periapsis of the arrival hyperbola is the radius of Mars plus the periapsis of the elliptical capture orbit,

$$r_p = 3380 + 300 = 3680\,\text{km}$$

According to Eq. (8.40) and Eq. (b) of Example 8.8, the speed of the spacecraft at periapsis of the arrival hyperbola is

$$v_p)_{\text{Arrival}} = \sqrt{[v_\infty)_{\text{Arrival}}]^2 + \frac{2\mu_{\text{Mars}}}{r_p}} = \sqrt{2.8852^2 + \frac{2 \cdot 42{,}830}{3680}} = 5.621\,\text{km/s}$$

To find the speed $v_p)_{\text{ellipse}}$ at periapsis of the capture ellipse, we use the required period (48 h) to determine the ellipse's semimajor axis, using Eq. (2.83),

$$a_{\text{ellipse}} = \left(\frac{T\sqrt{\mu_{\text{Mars}}}}{2\pi}\right)^{3/2} = \left(\frac{48 \cdot 3600 \cdot \sqrt{42{,}830}}{2\pi}\right)^{3/2} = 31{,}880\,\text{km}$$

From Eq. (2.73), we obtain

$$e_{\text{ellipse}} = 1 - \frac{r_p}{a_{\text{ellipse}}} = 1 - \frac{3680}{31{,}880} = 0.8846$$

Then Eq. (8.59) yields

$$v_p)_{\text{ellipse}} = \sqrt{\frac{\mu_{\text{Mars}}}{r_p}(1 + e_{\text{ellipse}})} = \sqrt{\frac{42{,}830}{3680}}(1 + 0.8846) = 4.683\,\text{km/s}$$

Hence, the delta-v requirement is

$$\Delta v = v_p)_{\text{Arrival}} - v_p)_{\text{ellipse}} = \boxed{0.9382\,\text{km/s}}$$

The eccentricity of the approach hyperbola is given by Eq. (8.38),

$$e = 1 + \frac{r_p (v_\infty)_{\text{Arrival}})^2}{\mu_{\text{Mars}}} = 1 + \frac{3680 \cdot 2.8852^2}{42{,}830} = 1.715$$

Assuming that the capture ellipse is a polar orbit of Mars, then the approach hyperbola is as illustrated in Fig. 8.30. Note that Mars' equatorial plane is inclined 25° to the plane of its orbit around the sun. Furthermore, the vernal equinox of Mars lies at an angle of 85° from that of the earth.

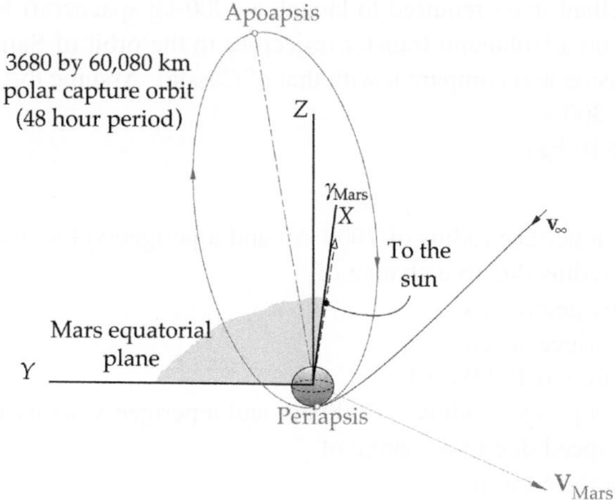

FIG. 8.30

The Mars approach hyperbola and capture ellipse.

PROBLEMS

Section 8.2

8.1 Find the total delta-v required for a Hohmann transfer from earth's orbit to Saturn's orbit.
[Ans.: 15.74 km/s]

8.2 Find the total delta-v required for a Hohmann transfer from Mars' orbit to Jupiter's orbit.
[Ans.: 10.15 km/s]

Section 8.3

8.3 Calculate the synodic period of Venus relative to the earth.
{Ans.: 583.9 days}

8.4 Calculate the synodic period of Jupiter relative to Mars.
{Ans.: 816.6 days}

Section 8.4

8.5 Calculate the radius of the spheres of influence of Mercury, Venus, Mars, and Jupiter.
{Ans.: See Table A.2}

8.6 Calculate the radius of the spheres of influence of Saturn, Uranus, and Neptune.
{Ans.: See Table A.2}

Section 8.6

8.7 On a date when the earth was 147.4(10^6) km from the sun, a spacecraft parked in a 200-km-altitude circular earth orbit was launched directly into an elliptical orbit around the sun with perihelion of 120(10^6) km and aphelion equal to the earth's distance from the sun on the launch date. Calculate the delta-v required and v_∞ of the departure hyperbola.
{Ans.: $\Delta v = 3.34$ km/s, $v_\infty = 1.579$ km/s}

8.8 Calculate the propellant mass required to launch a 2000-kg spacecraft from a 180-km-altitude circular earth orbit on a Hohmann transfer trajectory to the orbit of Saturn. Calculate the time required for the mission and compare it with that of Cassini. Assume the propulsion system has a specific impulse of 300 s.
{Ans.: 6.03 yr; 21,810 kg}

Section 8.7

8.9 An earth orbit has a perigee radius of 7000 km and a perigee velocity of 9 km/s. Calculate the change in apogee radius due to a change of
(a) 1 km in the perigee radius
(b) 1 m/s in the perigee speed.
{Ans.: (a) 13.27 km; (b) 10.99 km}

8.10 An earth orbit has a perigee radius of 7000 km and a perigee velocity of 9 km/s. Calculate the change in apogee speed due to a change of
(a) 1 km in the perigee radius
(b) 1 m/s in the perigee speed.

{Ans.: (a) −1.81 m/s; (b) −0.406 m/s}

Section 8.8

8.11 Estimate the total delta-v requirement for a Hohmann transfer from earth to Mercury, assuming a 150-km-altitude circular parking orbit at earth and a 150-km circular capture orbit at Mercury. Furthermore, assume that the planets have coplanar circular orbits with radii equal to the semimajor axes listed in Table A.1.
{Ans.: 13.08 km/s}

Section 8.9

8.12 Suppose a spacecraft approaches Jupiter on a Hohmann transfer ellipse from earth. If the spacecraft flies by Jupiter at an altitude of 200,000 km on the sunlit side of the planet, determine the orbital elements of the postflyby trajectory and the delta-v imparted to the spacecraft by Jupiter's gravity. Assume that all the orbits lie in the same (ecliptic) plane.
{Ans.: $\Delta V = 10.6$ km/s, $a = 4.79(10^6)$ km, $e = 0.8453$}

Section 8.10

8.13 Use Table 8.1 to verify the orbital elements for earth and Mars presented in Example 8.7.

8.14 Use Table 8.1 to determine the day of the year 2005 when the earth was farthest from the sun.
{Ans.: July 4}

Section 8.11

8.15 On December 1, 2005, a spacecraft left a 180-km-altitude circular orbit around the earth on a mission to Venus. It arrived at Venus 121 days later on April 1, 2006, entering a 300-km-by-9000-km capture ellipse around the planet. Calculate the total delta-v requirement for this mission.
{Ans.: 6.75 km/s}

8.16 On August 15, 2005, a spacecraft in a 190-km, 52°-inclination circular parking orbit around the earth departed on a mission to Mars, arriving at the red planet on March 15, 2006, whereupon retrorockets placed it into a highly elliptic orbit with a periapsis of 300 km and a period of 35 h. Determine the total delta-v required for this mission.
{Ans.: 4.86 km/s}

REFERENCES

JPL Horizons Web-Interface, 2018. https://ssd.jpl.nasa.gov/horizons.cgi. (Accessed 29 June 2018).

Standish, E.M., Williams, J.G., 2013. Orbital ephemerides of the sun, moon and planets. In: Seidelmann, P.K. (Ed.), Explanatory Supplement to the Astronomical Almanac, p. 338. University Science Books.

CHAPTER

LUNAR TRAJECTORIES

9

9.1 INTRODUCTION

The orbit of the moon around the earth is an ellipse having a small eccentricity ($e = 0.0549$) and perigee and apogee radii of $r_p = 363{,}400\,\text{km}$ and $r_a = 405{,}500\,\text{km}$, respectively. Therefore, the semimajor axis a of the moon's orbit is

$$a = \frac{r_a + r_p}{2} = 384{,}400\,\text{km} \tag{9.1}$$

A circular orbit of this radius has the same period as the moon's elliptical orbit (see Eq. 2.83). Therefore, to simplify our analysis, let us assume that the moon's path around the earth is a circle of radius D, where

$$D = 384{,}400\,\text{km} \tag{9.2}$$

Recalling that the earth's gravitational parameter is

$$\mu_e = 398{,}600\,\text{km}^3/\text{s}^2$$

Eq. (2.63) gives the circular orbital speed v_m of the moon as

$$v_m = \sqrt{\frac{\mu_e}{D}} = 1.0183\,\text{km/s} \tag{9.3}$$

Imagine that a spacecraft has been placed in a circular earth orbit of 320 km altitude (radius $r_c = 6698\,\text{km}$) and that its orbit is coplanar with that of the moon. The speed of the vehicle in this circular parking orbit is

$$v_c = \sqrt{\frac{\mu_e}{r_c}} = 7.7143\,\text{km/s}$$

To most efficiently transfer the spacecraft from the low earth orbit out to lunar orbit requires a Hohmann transfer (namely, orbit 2 in Fig. 9.1).

At its perigee orbit 2 is tangent to the circular low earth orbit, so that $r_p = 6698\,\text{km}$, and at apogee it is tangent to the moon's orbit, which means that $r_a = 384{,}400\,\text{km}$. Therefore, the semimajor axis of this transfer ellipse is

$$a = \frac{r_a + r_p}{2} = 195{,}549\,\text{km}$$

Orbital Mechanics for Engineering Students. https://doi.org/10.1016/B978-0-12-824025-0.00009-X

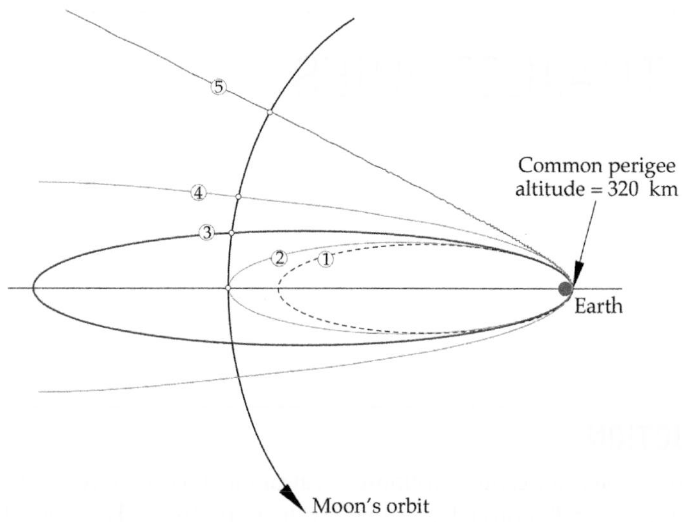

FIG. 9.1

Several strategies for reaching the moon's orbit from low earth orbit. The perigee speeds of the orbits (in kilometers per second) are: (1) 10.8, (2) 10.815, (3) 10.85, (4) 10.9, and (5) 11.2. Clearly, the energy of orbit 1 is too low to reach the moon.

According to Eq. (2.84), the eccentricity e of the orbit is

$$e = \frac{r_a - r_p}{r_a + r_p} = 0.96575$$

and its period T is given by Eq. (2.83),

$$T = \frac{2\pi}{\sqrt{\mu_e}} a^{3/2} = 8.6059(10^5)\,\mathrm{s} = 239.05\mathrm{h}$$

If follows that the time of flight t_F for this Hohmann transfer is

$$t_F = \frac{T}{2} = 119.52\mathrm{h}(4.98\,\mathrm{days})$$

The angular momentum of the transfer ellipse is given by Eq. (6.2),

$$h = \sqrt{2\mu_e}\sqrt{\frac{r_a r_p}{r_a + r_p}} = 72{,}444\mathrm{km}^2/\mathrm{s}$$

We use the angular momentum to find the speeds v_p and v_a at perigee and apogee, respectively,

$$v_p = \frac{h}{r_p} = 10.815\mathrm{km/s} \quad v_a = \frac{h}{r_a} = 0.18846\mathrm{km/s}$$

The delta-v required to transfer from the initial circular parking orbit to the Hohmann transfer trajectory at its perigee is

$$\Delta v_p = v_p - v_c = 10.815 - 7.7143 = 3.1007\mathrm{km/s}$$

The delta-v required to finally transfer from the Hohmann ellipse to the moon's orbit is

$$\Delta v_a = v_m - v_a = 1.0183 - 0.18846 = 0.82984\,\text{km/s}$$

The total delta-v is the sum of these two,

$$\Delta v_{\text{total}} = 3.1007 + 0.82984 = 3.9305\,\text{km/s}$$

To reduce the flight time to lunar orbit from the same departure point and the same departure flight path angle (zero degrees), we must increase the injection speed to a value greater than 10.815 km/s. Let us choose orbit 3 in Fig. 9.1, for which $v_p = 10.85\,\text{km/s}$, which is still below the escape speed of 10.91 km/s (see Eq. 2.91). The angular momentum of orbit 3 is

$$h = r_p v_p = 6698 \cdot 10.85 = 72{,}673\,\text{km}^2/\text{s}$$

Solving Eq. (2.50) for the eccentricity e yields

$$e = \frac{h^2}{\mu_e r_p} - 1 = 0.97819$$

The new semimajor axis a is obtained from Eq. (2.71),

$$a = \frac{h^2}{\mu_e}\frac{1}{1-e^2} = 307{,}104\,\text{km}$$

and from this we find the period

$$T = \frac{2\pi}{\sqrt{\mu_e}} a^{3/2} = 470.48\,\text{h}$$

Next, we set $r = 384{,}400\,\text{km}$ in Eq. (2.45) and solve for the true anomaly at which transfer ellipse 3 first crosses the moon's orbit,

$$\theta = \cos^{-1}\left[\frac{1}{e}\left(\frac{h^2}{\mu_e r} - 1\right)\right] = 170.77°$$

From Eqs. (3.13b), (3.14), and (3.15), the flight time t_F (since perigee) to lunar orbit crossing is

$$t_F = \frac{T}{2\pi}\left\{2\tan^{-1}\left(\sqrt{\frac{1-e}{1+e}}\tan\frac{\theta}{2}\right) - e\sin\left[2\tan^{-1}\left(\sqrt{\frac{1-e}{1+e}}\tan\frac{\theta}{2}\right)\right]\right\}$$

Substituting $\theta = 170.77°$, $e = 0.97819$, and $T = 470.48$ h yields

$$t_F = 66.343\,\text{h}$$

This is a little over half the time required to reach the moon's orbit on the Hohmann transfer ellipse (orbit 2).

The radial (v_r) and transverse ($v_\perp$) components of spacecraft velocity on orbit 3 at the lunar orbit crossing are found using Eqs. (2.49) and (2.31),

$$v_r = \frac{\mu_e}{h} e\sin\theta = \frac{398{,}600}{72{,}673} \cdot 0.97819\sin 170.77° = 0.86025\,\text{km/s}$$

$$v_\perp = \frac{h}{r} = \frac{72{,}673}{384{,}400} = 0.18906\,\text{km/s}$$

Therefore, the speed is

$$v = \sqrt{v_r^2 + v_\perp^2} = 0.88078\,\text{km/s}$$

and, according to Eq. (2.51), the flight path angle γ is

$$\gamma = \tan^{-1}\left(\frac{v_r}{v_\perp}\right) = 77.605°$$

From Eq. (9.3) we know that the speed v_m in the circular lunar orbit is 1.0183 km/s, and the flight path angle is zero. The delta-v required to transfer to the lunar orbit from orbit 3 is given by Eq. (6.8),

$$\Delta v_F = \sqrt{v^2 + v_m^2 - 2vv_m\cos\Delta\gamma} = \sqrt{0.88078^2 + 1.0183^2 - 2\cdot 0.88078\cdot 1.0183\cos(0 - 77.605°)}$$
$$= 1.1949\,\text{km/s}$$

We previously calculated the speed of the initial circular parking orbit as 7.7143 km/s. Therefore, the necessary delta-v for translunar injection on orbit 3 is

$$\Delta v_0 = 10.85 - 7.7143 = 3.1357\,\text{km/s}$$

It follows that the total delta-v required to enter the moon's earth orbit after the 66-h flight is

$$\Delta v_{\text{total}} = \Delta v_0 + \Delta v_F = 3.1357 + 1.1949 = 4.3306\,\text{km/s}$$

This is about ten percent more than the total delta-v that we found for the nearly 120 hour Hohmann transfer strategy (orbit 2).

Increasing the perigee speed to 10.9 and 11.2 km/s yields, respectively, orbits 4 and 5 in Fig. 9.1. These orbits have still smaller flight times but larger delta-v requirements. On the other hand, orbit 1, with a perigee speed of only 10.8 km/s, cannot reach lunar orbit.

An obvious omission from our brief analysis so far is the moon itself. The goal of a lunar mission is not to simply reach the moon's orbit but to go into orbit around the moon or to either impact or land on its surface. Thus, as a spacecraft approaches the moon's orbit, the moon should be nearing the same position. That means lunar gravity will increasingly affect the spacecraft's trajectory, bending it more and more toward the moon. In this chapter we show how the method of patched conics, employed in Chapter 8 for interplanetary trajectories, can be applied to lunar trajectories.

We conclude this chapter with a numerical integration approach to lunar trajectory analysis.

9.2 COPLANAR PATCHED CONIC LUNAR TRAJECTORIES

In this section we shall for simplicity continue to assume that the moon's orbit is a circle and, furthermore, that the spacecraft's translunar trajectory lies in the moon's orbital plane, as illustrated in Fig. 9.2, which is a not-to-scale view looking down on that plane. Our approach is similar to those of Bate et al. (1971), Chobotov (1996), and Brown (1998). The x axis of the earth-centered, nonrotating xyz coordinate system is directed toward the position of the moon at the instant the spacecraft crosses the moon's sphere of influence (SOI), so that the moon's position vector $\mathbf{r}_m$ at that instant is

$$\mathbf{r}_m = D\hat{\mathbf{i}} \tag{9.4}$$

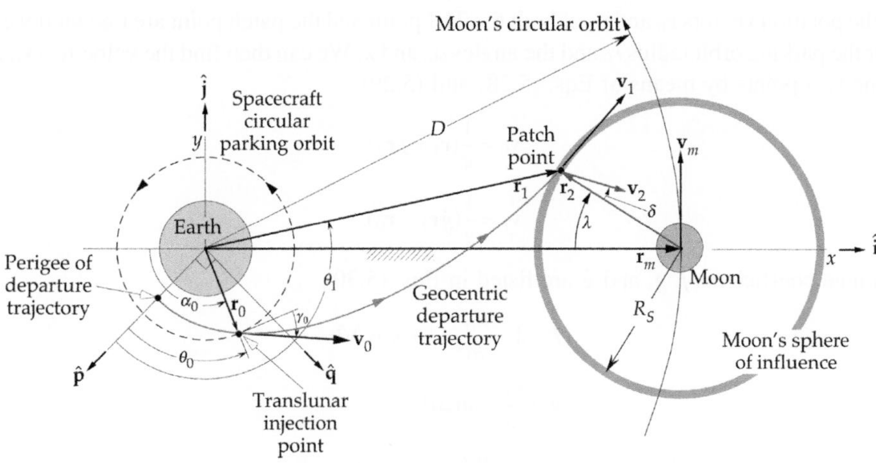

FIG. 9.2

Coplanar translunar trajectory from earth orbit to crossing of the moon's sphere of influence. The earth-centered *xy* axes do not rotate. Not to scale.

The z axis points out of the plane, and the y axis completes the right-handed triad, which means the moon's circular velocity $\mathbf{v}_m$ at this instant is

$$\mathbf{v}_m = v_m \hat{\mathbf{j}} \tag{9.5}$$

We calculated the circular orbital speed v_m of the moon in Eq. (9.3).

As shown in Fig. 9.2, the position vector $\mathbf{r}_0$ of the spacecraft at translunar injection (TLI), when it departs earth orbit, is

$$\mathbf{r}_0 = -r_0 \cos\alpha_0 \hat{\mathbf{i}} - r_0 \sin\alpha_0 \hat{\mathbf{j}} \tag{9.6}$$

where r_0 is the distance from the earth, and α_0 is its angular position relative to the earth-moon line. When the spacecraft arrives at the moon's SOI, its position vector $\mathbf{r}_2$ relative to the moon is

$$\mathbf{r}_2 = -R_S \cos\lambda \hat{\mathbf{i}} + R_S \sin\lambda \hat{\mathbf{j}} \tag{9.7}$$

where λ is the lunar arrival angle. The radius R_S of the SOI is given by Eq. (8.34),

$$R_S = D(m_m/m_e)^{2/5} \tag{9.8}$$

where m_m and m_e are the mass of the moon and the earth, respectively. Since $m_e = 5.974(10^{24})$ kg and $m_m = 7.348(10^{22})$ kg, it follows that

$$R_S = 0.172D = 66{,}183 \text{ km} \tag{9.9}$$

From Fig. 9.2 it is clear that the position vector $\mathbf{r}_1$ of the patch point relative to the earth is

$$\mathbf{r}_1 = \mathbf{r}_m + \mathbf{r}_2 \tag{9.10}$$

Substituting Eqs. (9.3) and (9.7) into this expression yields

$$\mathbf{r}_1 = (D - R_S \cos\lambda)\hat{\mathbf{i}} + R_S \sin\lambda \hat{\mathbf{j}} \tag{9.11}$$

Clearly, the position vectors $\mathbf{r}_0$ and $\mathbf{r}_1$ of both the TLI point and the patch point are known if we provide values for the parking orbit radius r_0 and the angles α_0 and λ. We can then find the velocities $\mathbf{v}_0$ and $\mathbf{v}_1$ at those same two points by means of Eqs. (5.28) and (5.29),

$$\mathbf{v}_0 = \frac{1}{g}(\mathbf{r}_1 - f\mathbf{r}_0) \tag{9.12a}$$

$$\mathbf{v}_1 = \frac{1}{g}(\dot{g}\mathbf{r}_1 - \mathbf{r}_0) \tag{9.12b}$$

The Lagrange coefficients f, g, and $\dot{g}$ are listed in Eqs. (5.30),

$$f = 1 - \frac{\mu_e r_1}{h_1{}^2}(1 - \cos\Delta\theta) \tag{9.13a}$$

$$g = \frac{r_0 r_1}{h_1}\sin\Delta\theta \tag{9.13b}$$

$$\dot{g} = 1 - \frac{\mu_e r_0}{h_1{}^2}(1 - \cos\Delta\theta) \tag{9.13c}$$

where h_1 is the angular momentum of the departure trajectory and $\Delta\theta$ is the difference $\theta_1 - \theta_0$ between the true anomalies of the position vectors $\mathbf{r}_0$ and $\mathbf{r}_1$. We will refer to this difference as the "sweep angle." The sweep angle is found by means of Eq. (5.23),

$$\cos\Delta\theta = \hat{\mathbf{u}}_{r_0} \cdot \hat{\mathbf{u}}_{r_1} \tag{9.14}$$

where the radial unit vectors at each end of the departure trajectory are

$$\hat{\mathbf{u}}_{r_0} = \frac{\mathbf{r}_0}{r_0} \quad \hat{\mathbf{u}}_{r_1} = \frac{\mathbf{r}_1}{r_1} \tag{9.15}$$

It remains to find an expression for the angular momentum h_1 that appears in Eqs. (9.13a)–(9.13c). To that end, we first use Eq. (9.12a) to calculate the radial component of the TLI velocity,

$$v_{r_0} = \hat{\mathbf{u}}_{r_0} \cdot \mathbf{v}_0 = \hat{\mathbf{u}}_{r_0} \cdot \frac{1}{g}(r_1\hat{\mathbf{u}}_{r_1} - fr_0\hat{\mathbf{u}}_{r_0}) = \frac{1}{g}(r_1\hat{\mathbf{u}}_{r_0} \cdot \mathbf{u}_{r_1} - fr_0\hat{\mathbf{u}}_{r_0} \cdot \hat{\mathbf{u}}_{r_0})$$

so that, with the aid of Eq. (9.14), we get

$$v_{r_0} = \frac{1}{g}(r_1\cos\Delta\theta - fr_0) \tag{9.16}$$

Combining Eqs. (2.31) and (2.51), we can calculate the radial component of the velocity $\mathbf{v}_0$ by the formula

$$v_{r_0} = \frac{h_1}{r_0}\tan\gamma_0$$

where γ_0 is the flight path angle at TLI. Substituting this expression into Eq. (9.16) yields

$$\frac{h_1}{r_0}\tan\gamma_0 = \frac{1}{g}(r_1\cos\Delta\theta - fr_0) \tag{9.17}$$

Replacing f and g by their expressions in Eqs. (9.13a)–(9.13c) leads to

$$\tan\gamma_0 = \frac{1}{r_1\sin\Delta\theta}\left\{r_1\cos\Delta\theta - \left[1 - \frac{\mu_e r_1}{h_1{}^2}(1 - \cos\Delta\theta)\right]r_0\right\}$$

Finally, solving this equation for the angular momentum h_1 yields

$$h_1 = \sqrt{\mu_e r_0}\sqrt{\frac{1-\cos\Delta\theta}{\dfrac{r_0}{r_1}+\sin\Delta\theta\tan\gamma_0-\cos\Delta\theta}} \tag{9.18}$$

Clearly, by specifying the initial and final position vectors $\mathbf{r}_0$ and $\mathbf{r}_1$ (and hence, $\Delta\theta$) as well as the initial flight path angle γ_0, we are able to find the angular momentum h_1 of the translunar trajectory. Then Eqs. (9.12a), (9.12b), and (9.13a)–(9.13c) provide the initial and final velocities, $\mathbf{v}_0$ and $\mathbf{v}_1$, so that the orbit is completely determined. Note that the radial components of the velocities at the end points of the translunar trajectory are

$$v_{r_0} = \mathbf{v}_0 \cdot \hat{\mathbf{u}}_{r_0} \qquad v_{r_1} = \mathbf{v}_1 \cdot \hat{\mathbf{u}}_{r_1} \tag{9.19}$$

Since $h_1 = r_0 v_\perp)_0 = v_0 r_0 \cos\gamma_0$ and $v_{esc})_0 = \sqrt{2\mu_e/r_0}$, Eq. (9.18) can be written alternatively as

$$\left.\frac{v}{v_{esc}}\right)_0 = \frac{1}{\sqrt{2}\cos\gamma_0}\sqrt{\frac{1-\cos\Delta\theta}{\dfrac{r_0}{r_1}+\sin\Delta\theta\tan\gamma_0-\cos\Delta\theta}}$$

This equation limits the range of the sweep angle $\Delta\theta$ to those values for which the denominator in the radical is positive. That is,

$$\frac{r_0}{r_1}+\sin\Delta\theta\tan\gamma_0-\cos\Delta\theta > 0 \tag{9.20}$$

Within that range, only those $\Delta\theta$'s for which $v/v_{esc})_0 < 1$ yield elliptical orbits. For example, if $r_0 = 6700\,\text{km}$ and $r_1 = 335{,}000\,\text{km}$ ($r_0/r_1 = 0.02$), the allowable values of γ_0 and $\Delta\theta$ lie within the triangular shaded region shown in Fig. 9.3. Observe that larger sweep angles require smaller initial flight path angles.

If prior to departing for the moon the spacecraft is in a circular parking orbit of radius r_0, then its speed is $v_c = \sqrt{\mu_e/r_0}$ (see Eq. 2.63). The magnitude of the TLI velocity in Eq. (9.12a) is

$$v_0 = \|\mathbf{v}_0\| = \frac{1}{g}\sqrt{f^2 r_0^2 + r_1^2 - 2 f r_0 r_1 \cos\Delta\theta} \tag{9.21}$$

This speed is reached by supplying a delta-v given by Eq. (6.2),

$$\Delta v_0 = \sqrt{v_c^2 + v_0^2 - 2 v_c v_0 \cos\gamma_0} \tag{9.22}$$

The eccentricity vector of the translunar trajectory can be obtained from Eq. (4.10),

$$\mathbf{e}_1 = \frac{1}{\mu_e}\left[\left(v_0^2 - \frac{\mu_e}{r_0}\right)\mathbf{r}_0 - r_0 v_{r_0}\mathbf{v}_0\right] \tag{9.23}$$

The magnitude e_1 of this vector is the orbit's eccentricity, and since the orbit must be an ellipse, e_1 must be less than unity. From the angular momentum h_1 and the eccentricity e_1 we obtain the semimajor axis and the period by means of Eqs. (2.71) and (2.83), respectively,

$$a_1 = \frac{h_1^2}{\mu_e}\frac{1}{1-e_1^2} \tag{9.24}$$

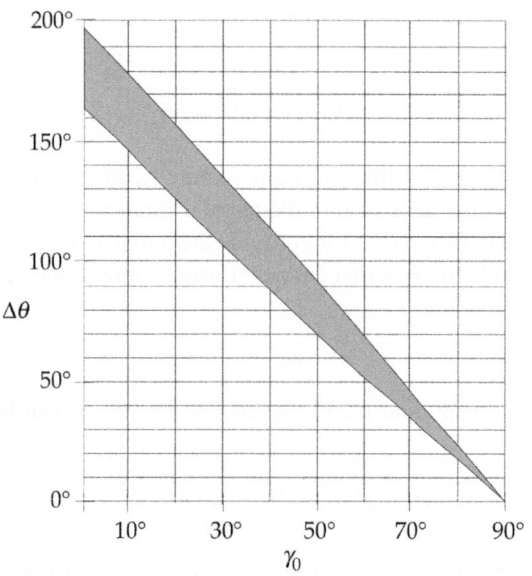

FIG. 9.3

Allowable values of flight path angle γ_0 and sweep angle $\Delta\theta$ if $r_0/r_1 = 0.02$.

$$T_1 = 2\pi\sqrt{\frac{a_1{}^3}{\mu_e}} \tag{9.25}$$

The perifocal unit vectors (see Figs. 2.29 and 2.30) of the translunar trajectory are

$$\hat{\mathbf{p}}_1 = \frac{\mathbf{e}_1}{e_1} \quad \hat{\mathbf{w}}_1 = \frac{\mathbf{r}_1 \times \mathbf{v}_1}{h_1} \quad \mathbf{q}_1 = \hat{\mathbf{w}}_1 \times \hat{\mathbf{p}}_1 \tag{9.26}$$

The true anomaly θ_0 at TLI is the angle between perigee and the radial $\mathbf{r}_0$. Therefore, we obtain θ_0 from the formula

$$\theta_0 = \cos^{-1}(\hat{\mathbf{p}}_1 \cdot \mathbf{r}_0/r_0) \tag{9.27}$$

This true anomaly is less than 180°, because the spacecraft is flying outbound toward apogee, which lies beyond the patch point. Since the trajectory is an ellipse, we find the time t_0 at TLI by means of the sequence of calculations listed in Eqs. (3.13b), (3.14), and (3.15), which yield

$$t_0 = \frac{T_1}{2\pi}\left\{2\tan^{-1}\left(\sqrt{\frac{1-e_1}{1+e_1}}\tan\frac{\theta_0}{2}\right) - e_1\sin\left[2\tan^{-1}\left(\sqrt{\frac{1-e_1}{1+e_1}}\tan\frac{\theta_0}{2}\right)\right]\right\} \tag{9.28}$$

Having found the true anomaly at TLI, the true anomaly θ_1 at the patch point follows from

$$\theta_1 = \theta_0 + \Delta\theta \tag{9.29}$$

where the sweep angle $\Delta\theta$ was obtained from Eq. (9.14). The time t_1 at the patch point follows by replacing θ_0 with θ_1 in Eq. (9.28),

$$t_1 = \frac{T_1}{2\pi}\left\{2\tan^{-1}\left(\sqrt{\frac{1-e_1}{1+e_1}}\tan\frac{\theta_1}{2}\right) - e_1\sin\left[2\tan^{-1}\left(\sqrt{\frac{1-e_1}{1+e_1}}\tan\frac{\theta_1}{2}\right)\right]\right\} \tag{9.30}$$

The total flight time Δt_1 from TLI to the SOI is

$$\Delta t_1 = t_1 - t_0 \tag{9.31}$$

When the spacecraft arrives at the patch point, its velocity $\mathbf{v}_2$ relative to the moon is

$$\mathbf{v}_2 = \mathbf{v}_1 - \mathbf{v}_m \tag{9.32}$$

where the geocentric velocities $\mathbf{v}_m$ and $\mathbf{v}_1$ are found in Eqs. (9.5) and (9.12b), respectively.

Within the SOI, the spacecraft moves under the influence of the moon's gravity exclusively (according to the patched conic approximation). At the patch point, its specific angular momentum $\mathbf{h}_2$ *relative to the moon* is

$$\mathbf{h}_2 = \mathbf{r}_2 \times \mathbf{v}_2 \quad h_2 = \|\mathbf{h}_2\| \tag{9.33}$$

where $\mathbf{r}_2$ is given by Eq. (9.7). We know that all points on this coasting orbit around the moon have the same angular momentum $\mathbf{h}_2$. According to Eq. (2.40) the eccentricity vector of the lunar approach trajectory is

$$\mathbf{e}_2 = \frac{\mathbf{v}_2 \times \mathbf{h}_2}{\mu_m} - \hat{\mathbf{u}}_{r_2} \tag{9.34}$$

where $\mu_m = 4902.8\,\text{km}^3/\text{s}^2$, and $\hat{\mathbf{u}}_{r_2} = \mathbf{r}_2/r_2 = -\cos\lambda\hat{\mathbf{i}} + \sin\lambda\hat{\mathbf{j}}$. The magnitude of $\mathbf{e}_2$ is the orbit's eccentricity e_2, and since the approach trajectory is a hyperbola, e_2 must exceed unity. Observe that if $\hat{\mathbf{h}}_2 \cdot \hat{\mathbf{k}} < 0$, then the motion around the moon is retrograde (clockwise), whereas $\hat{\mathbf{h}}_2 \cdot \hat{\mathbf{k}} > 0$ means the motion is posigrade. For rectilinear motion directly toward the center of the moon, $\hat{\mathbf{h}}_2 \cdot \hat{\mathbf{k}} = 0$. At the patch point, the angle between the probe's relative velocity vector $\mathbf{v}_2$ and the moon's position $(-\mathbf{r}_1)$ relative to the probe is labeled δ in Fig. 9.2. This deviation angle may be computed from the fact that

$$\cos\delta = (-\mathbf{r}_1/r_1)\cdot(\mathbf{v}_2/v_2) \tag{9.35}$$

Clearly, $\delta = 0$ is another indication that the probe is destined for lunar impact.

Eq. (2.50) gives us the perilune radius r_{p_2} of the approach hyperbola,

$$r_{p_2} = \frac{h_2^2}{\mu_m}\frac{1}{1+e_2} \tag{9.36}$$

To get the perilune altitude z_{p_2}, we subtract the moon's radius $R_m = 1737\,\text{km}$,

$$z_{p_2} = r_{p_2} - R_m \tag{9.37}$$

If $z_{p_2} < 0$, then the spacecraft impacts the lunar surface. From Eq. (9.36) and the fact that $h_2 = r_{p_2}v_{p_2}$, we find that the perilune speed of the lunar approach orbit is

$$v_{p_2} = \sqrt{1+e_2}\sqrt{\frac{\mu_m}{r_{p_2}}} \tag{9.38}$$

If the objective is to enter a circular orbit at perilune, then the delta-v required at that point is

$$\Delta v_2 = \sqrt{\frac{\mu_m}{r_{p_2}}} - v_{p_2} = \left(1 - \sqrt{1+e_2}\right)\sqrt{\frac{\mu_m}{r_{p_2}}}$$

The total delta-v for this scenario, starting at TLI, is

$$\Delta v = \Delta v_0 + \Delta v_2 \tag{9.39}$$

The unit vectors of the approach hyperbola's perifocal frame are

$$\hat{\mathbf{p}}_2 = \frac{\mathbf{e}_2}{e_2} \quad \hat{\mathbf{w}}_2 = \frac{\mathbf{h}_2}{h_2} \quad \hat{\mathbf{q}}_2 = \hat{\mathbf{w}}_2 \times \hat{\mathbf{p}}_2 \tag{9.40}$$

According to Eq. (4.13a), the true anomaly θ_2 of the patch point *relative to perilune* is

$$\theta_2 = 360° - \cos^{-1}(\hat{\mathbf{p}}_2 \cdot \hat{\mathbf{u}}_{r_2}) \tag{9.41}$$

because the spacecraft is flying toward perilune ($v_{r_2} < 0$). Given the true anomaly θ_2, we find the time t_2 at the patch point of the hyperbolic lunar orbit by means of the following calculation (see Eqs. 3.34, 3.40, and 3.44a),

$$t_2 = \frac{{h_2}^3}{{\mu_m}^2({e_2}^2-1)^{3/2}}\left\{e_2 \sinh\left[2\tanh^{-1}\left(\sqrt{\frac{e_2-1}{e_2+1}}\tan\frac{\theta_2}{2}\right)\right] - 2\tanh^{-1}\left(\sqrt{\frac{e_2-1}{e_2+1}}\tan\frac{\theta_2}{2}\right)\right\} \tag{9.42}$$

According to Eq. (3.2), time is zero at the periapsis of an orbit. In the present case, as the spacecraft flies from the patch point to perilune, time increases from t_2 to zero. Hence, t_2 must be negative. The flight time from patch point to perilune is

$$\Delta t_2 = 0 - t_2 \quad (\Delta t_2 > 0) \tag{9.43}$$

The total flight time Δt from translunar injection to perilune is

$$\Delta t = \Delta t_1 + \Delta t_2 \tag{9.44}$$

Keep in mind that we describe the motion of the spacecraft within the moon's SOI relative to a rotating frame of reference that is attached to the moon with its origin at the center of the moon. Let us label the axes of this rotating moon-fixed frame $x'y'z'$ to distinguish them from the xyz axes of the earth-fixed system shown in Fig. 9.2. (The z axes of both frames are normal to the moon's orbital plane and therefore coincide.) As illustrated in Fig. 9.4, the x' axis lies on the rotating earth–moon radial and is directed away from the earth. The y' axis is perpendicular to x' and points in the direction of the moon's velocity relative to the earth. The angle ϕ between the x and x' axes increases at a rate equal to the moon's angular velocity, and it is zero at the instant the spacecraft crosses the moon's SOI.

According to Eq. (2.119), the position vector of the spacecraft relative to the rotating moon-fixed $x'y'z'$ frame is

$$\mathbf{r}_{\text{rel}} = r\cos\theta\hat{\mathbf{p}}_2 + r\sin\theta\hat{\mathbf{q}}_2 \quad r = {h_2}^2/[\mu_m(1+e\cos\theta)] \tag{9.45}$$

where $\hat{\mathbf{p}}_2$ and $\hat{\mathbf{q}}_2$ are given by Eq. (9.40). The angle between $\hat{\mathbf{p}}_2$ and the x' axis (the argument of perilune) is ω_p, which is constant. Therefore,

$$\begin{aligned}\hat{\mathbf{p}}_2 &= \cos\omega_p\hat{\mathbf{i}}' + \sin\omega_p\hat{\mathbf{j}}' \\ \hat{\mathbf{q}}_2 &= -\sin\omega_p\hat{\mathbf{i}}' + \cos\omega_p\hat{\mathbf{j}}'\end{aligned} \tag{9.46}$$

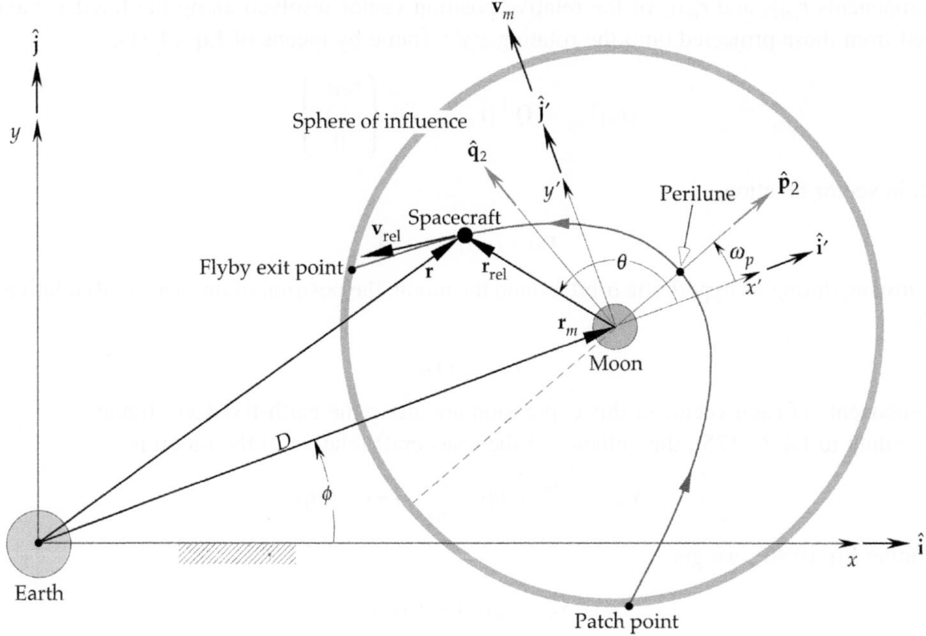

FIG. 9.4

Within the sphere of influence, the orbit and its apse line appear fixed relative to the moon. Not to scale.

Substituting these into Eq. (9.45) yields

$$\mathbf{r}_{rel} = r_{rel})_{x'}\hat{\mathbf{i}}' + r_{rel})_{y'}\hat{\mathbf{j}}'$$

where

$$r_{rel})_{x'} = r\cos\theta\cos\omega_p - r\sin\theta\sin\omega_p$$
$$r_{rel})_{y'} = r\cos\theta\sin\omega_p + r\sin\theta\cos\omega_p$$

In matrix notation

$$\{\mathbf{r}_{rel}\}_{x'y'z'} = \begin{Bmatrix} r_{rel})_{x'} \\ r_{rel})_{y'} \\ r_{rel})_{z'} \end{Bmatrix} = \begin{Bmatrix} r\cos\theta\cos\omega_p - r\sin\theta\sin\omega_p \\ r\cos\theta\sin\omega_p + r\sin\theta\cos\omega_p \\ 0 \end{Bmatrix}$$

At a given instant, the direction cosine matrix $[\mathbf{Q}]$ of the transformation from xyz into $x'y'z'$ is found in Eq. (4.34),

$$[\mathbf{Q}] = [\mathbf{R}_3(\phi)] = \begin{bmatrix} \cos\phi & \sin\phi & 0 \\ -\sin\phi & \cos\phi & 0 \\ 0 & 0 & 1 \end{bmatrix} \tag{9.47}$$

The components $r_{\text{rel}})_x$ and $r_{\text{rel}})_y$ of the relative position vector resolved along the fixed xyz axes are obtained from those projected onto the rotating $x'y'z'$ frame by means of Eq. (4.31),

$$\{\mathbf{r}_{\text{rel}}\}_{xyz} = [\mathbf{Q}]^T \{\mathbf{r}_{\text{rel}}\}_{x'y'z'} = \begin{Bmatrix} r_{\text{rel}})_x \\ r_{\text{rel}})_y \\ 0 \end{Bmatrix} \tag{9.48}$$

so that, in vector notation,

$$\mathbf{r}_{\text{rel}} = r_{\text{rel}})_x \hat{\mathbf{i}} + r_{\text{rel}})_y \hat{\mathbf{j}}$$

At any instant, during its hyperbolic orbit around the moon, the position of the spacecraft relative to the earth is

$$\mathbf{r} = \mathbf{r}_m + \mathbf{r}_{\text{rel}} \tag{9.49}$$

The components of each vector in this expression are along the earth-fixed xyz frame.

According to Eq. (2.125), the velocity of the spacecraft relative to the moon is

$$\mathbf{v}_{\text{rel}} = -\frac{\mu_m}{h_2}\sin\theta\hat{\mathbf{p}}_2 + \frac{\mu_m}{h_2}(e_2 + \cos\theta)\hat{\mathbf{q}}_2$$

Substituting Eq. (9.46), we get

$$\mathbf{v}_{\text{rel}} = v_{\text{rel}})_{x'} \hat{\mathbf{i}}' + v_{\text{rel}})_{y'} \hat{\mathbf{j}}'$$

where

$$v_{\text{rel}})_{x'} = -\frac{\mu_m}{h_2}\sin\theta\cos\omega_p - \frac{\mu_m}{h_2}(e_2 + \cos\theta)\sin\omega_p$$
$$v_{\text{rel}})_{y'} = -\frac{\mu_m}{h_2}\sin\theta\sin\omega_p + \frac{\mu_m}{h_2}(e_2 + \cos\theta)\cos\omega_p$$

In matrix notation

$$\{\mathbf{v}_{\text{rel}}\}_{x'y'z'} = \begin{Bmatrix} v_{\text{rel}})_{x'} \\ v_{\text{rel}})_{y'} \\ v_{\text{rel}})_{z'} \end{Bmatrix} = \begin{Bmatrix} -\frac{\mu_m}{h_2}\sin\theta\cos\omega_p - \frac{\mu_m}{h_2}(e_2 + \cos\theta)\sin\omega_p \\ -\frac{\mu_m}{h_2}\sin\theta\sin\omega_p + \frac{\mu_m}{h_2}(e_2 + \cos\theta)\cos\omega_p \\ 0 \end{Bmatrix}$$

The components $v_{\text{rel}})_x$ and $v_{\text{rel}})_y$ of this relative velocity vector in the fixed xyz system are obtained from those in the rotating $x'y'z'$ frame as in Eq. (9.48), by the operation

$$\{\mathbf{v}_{\text{rel}}\}_{xyz} = [\mathbf{Q}]^T \{\mathbf{v}_{\text{rel}}\}_{x'y'z'} = \begin{Bmatrix} v_{\text{rel}})_x \\ v_{\text{rel}})_y \\ 0 \end{Bmatrix} \tag{9.50}$$

so that

$$\mathbf{v}_{\text{rel}} = v_{\text{rel}})_x \hat{\mathbf{i}} + v_{\text{rel}})_y \hat{\mathbf{j}}$$

According to Eq. (1.66), the absolute velocity of the spacecraft within the SOI is

$$\overbrace{\mathbf{v}}^{\substack{\text{absolute}\\ \text{velocity}}} = \overbrace{\mathbf{v}_m}^{\substack{\text{velocity of origin}\\ \text{of moving frame}}} + \overbrace{\boldsymbol{\omega}_m}^{\substack{\text{angular velocity}\\ \text{of moving frame}}} \times \overbrace{\mathbf{r}_{\text{rel}}}^{\substack{\text{position vector relative}\\ \text{to moving frame}}} + \overbrace{\mathbf{v}_{\text{rel}}}^{\substack{\text{velocity relative}\\ \text{to moving frame}}} \tag{9.51}$$

Since we are assuming that the position vector $\mathbf{r}_m$ of the moon's orbit has constant magnitude, it follows from Eq. (1.52) that $\mathbf{v}_m = \boldsymbol{\omega}_m \times \mathbf{r}_m$. Therefore, $\mathbf{v} = \boldsymbol{\omega}_m \times (\mathbf{r}_m + \mathbf{r}_{rel}) + \mathbf{v}_{rel}$, or

$$\mathbf{v} = \boldsymbol{\omega}_m \times \mathbf{r} + \mathbf{v}_{rel} \tag{9.52}$$

where $\mathbf{r}$ is given by Eq. (9.49). Within the moon's SOI we use this formula to determine the absolute velocity of the spacecraft from its velocity relative to the moon. All vector components in Eq. (9.52) are along the earth-fixed *xyz* frame.

EXAMPLE 9.1

A spacecraft is in a circular earth orbit of 320 km altitude. When $\alpha_0 = 28°$, it is launched into a translunar trajectory with a flight path angle of $\gamma_0 = 6°$ (see Fig. 9.2). The lunar arrival angle is $\lambda = 55°$. Find the perilune altitude and the total flight time from TLI to perilune.

Solution

Since $r_0 = 6378 + 320 = 6698$ km, the position vector of the spacecraft at TLI is, according to Eq. (9.6),

$$\begin{aligned}\mathbf{r}_0 &= -r_0 \cos\alpha_0 \hat{\mathbf{i}} - r_0 \sin\alpha_0 \hat{\mathbf{j}} \\ &= -6698\cos 28°\hat{\mathbf{i}} - 6698\sin 28°\hat{\mathbf{j}} \\ &= -5914.0\hat{\mathbf{i}} - 3144.5\hat{\mathbf{j}}\,(\text{km})\end{aligned}$$

so that

$$\hat{\mathbf{u}}_{r_0} = \frac{\mathbf{r}_0}{r_0} = -0.88295\hat{\mathbf{i}} - 0.46947\hat{\mathbf{j}}$$

Likewise, since $r_2 = R_S = 66{,}183$ km, it follows from Eq. (9.7) that

$$\begin{aligned}\mathbf{r}_2 &= -R_S \cos\lambda \hat{\mathbf{i}} + R_S \sin\lambda \hat{\mathbf{j}} \\ &= -66,183\cos 55°\hat{\mathbf{i}} + 66,183\sin 55°\hat{\mathbf{j}} \\ &= -37,961\hat{\mathbf{i}} + 54,214\hat{\mathbf{j}}\,(\text{km})\end{aligned}$$

$$\therefore \hat{\mathbf{u}}_{r_2} = \frac{\mathbf{r}_2}{r_2} = -0.57358\hat{\mathbf{i}} + 0.81915\hat{\mathbf{j}} \tag{a}$$

From Eq. (9.2) we know that $\mathbf{r}_m = 384{,}400\hat{\mathbf{i}}$ (km) at SOI encounter, so that, according to Eq. (9.10), the position vector $\mathbf{r}_1$ of the patch point relative to the earth is

$$\begin{aligned}\mathbf{r}_1 &= \mathbf{r}_m + \mathbf{r}_2 \\ &= 384,400\hat{\mathbf{i}} + \left(-37,961\hat{\mathbf{i}} + 54,214\hat{\mathbf{j}}\right) \\ &= 346,440\hat{\mathbf{i}} + 54,214\hat{\mathbf{j}}\,(\text{km}) \\ r_1 &= 350,655\,\text{km}\end{aligned}$$

$$\therefore \hat{\mathbf{u}}_{r_1} = \frac{\mathbf{r}_1}{r_1} = 0.98798\hat{\mathbf{i}} + 0.15461\hat{\mathbf{j}}$$

Using Eq. (9.14) we find

$$\cos\Delta\theta = \overbrace{\left(-0.88295\hat{\mathbf{i}} - 0.46947\hat{\mathbf{j}}\right)}^{\hat{\mathbf{u}}_{r_0}} \cdot \overbrace{\left(0.98798\hat{\mathbf{i}} + 0.15461\hat{\mathbf{j}}\right)}^{\hat{\mathbf{u}}_{r_1}} = -0.94491$$

which means the sweep angle is

$$\Delta\theta = 160.89°$$

We can now use Eq. (9.18) to find the angular momentum h_1 of the translunar orbit,

$$h_1 = \sqrt{\mu_e r_0}\sqrt{\frac{1-\cos\Delta\theta}{\dfrac{r_0}{r_1}+\sin\Delta\theta\tan\gamma_0-\cos\Delta\theta}}$$

$$= \sqrt{398{,}600\cdot 6698}\sqrt{\frac{1-(-0.94491)}{\dfrac{6698}{350{,}655}+\sin 160.89°\cdot\tan 6°-(-0.94491)}}$$

$$=72{,}117\,\text{km}^2/\text{s}$$

Eqs. (9.13a)–(9.13c) then yield the values of the Lagrange coefficients,

$$f=1-\frac{\mu_e r_1}{h_1{}^2}(1-\cos\Delta\theta)=1-\frac{398{,}600\cdot 350{,}655}{72{,}117^2}[1-(-0.94491)]=-51.269$$

$$g=\frac{r_0 r_1}{h_1}\sin\Delta\theta=\frac{6698\cdot 350{,}655}{72{,}117}\sin 160.89°=10{,}660\,\text{s}$$

$$\dot{g}=1-\frac{\mu_e r_0}{h_1{}^2}(1-\cos\Delta\theta)=1-\frac{398{,}600\cdot 6698}{72{,}117^2}[1-(-0.94491)]=0.0015816$$

From Eq. (9.12a) and (9.12b) we finally obtain the spacecraft velocities $\mathbf{v}_0$ and $\mathbf{v}_1$ at the beginning and the end of the geocentric departure trajectory:

$$\mathbf{v}_0=\overbrace{\frac{1}{10{,}660}\left[\left(346{,}440\hat{\mathbf{i}}+54{,}214\hat{\mathbf{j}}\right)-(-51.269)\left(-5914.0\hat{\mathbf{i}}-3144.5\hat{\mathbf{j}}\right)\right]}^{\frac{1}{g}(\mathbf{r}_1-f\mathbf{r}_0)} \tag{b}$$

$$=4.05556\hat{\mathbf{i}}-10.0379\hat{\mathbf{j}}\,(\text{km/s})$$

$$v_0=10.826\,\text{km/s}$$

$$v_{r_0}=\mathbf{v}_0\cdot\hat{\mathbf{u}}_{r_0}=1.1316\,\text{km/s}>0$$

$$\mathbf{v}_1=\overbrace{\frac{1}{10{,}660}\left[0.0015816\left(346{,}440\hat{\mathbf{i}}+54{,}214\hat{\mathbf{j}}\right)-\left(-5914.0\hat{\mathbf{i}}-3144.5\hat{\mathbf{j}}\right)\right]}^{\frac{1}{g}(\dot{g}\mathbf{r}_1-\mathbf{r}_0)}$$

$$=0.60618\hat{\mathbf{i}}+0.30302\hat{\mathbf{j}}\,(\text{km/s})$$

$$v_1=0.67770\,\text{km/s}$$

$$v_{r_1}=\mathbf{v}_1\cdot\hat{\mathbf{u}}_{r_1}=0.64574\,\text{km/s}>0$$

It follows from Eq. (9.23) that the eccentricity vector is

$$\mathbf{e}_1=\frac{1}{\mu_e}\left[\left(v_0{}^2-\frac{\mu_e}{r_0}\right)\mathbf{r}_0-r_0 v_{r_0}\mathbf{v}_0\right]$$

$$=\frac{1}{398{,}600}\left[\left(10.826^2-\frac{398{,}600}{6698}\right)\left(-5914.0\hat{\mathbf{i}}-3.1445\hat{\mathbf{j}}\right)-6698\cdot 1.1316\left(4.0556\hat{\mathbf{i}}-10.038\hat{\mathbf{j}}\right)\right]$$

$$=-0.93315\hat{\mathbf{i}}-0.26428\hat{\mathbf{j}}$$

$$\therefore e_1=0.96985$$

From Eq. (9.26), the perifocal unit vectors of the elliptical translunar trajectory are

$$\hat{\mathbf{p}}_1=\frac{\mathbf{e}_1}{e_1}=\frac{-0.93315\hat{\mathbf{i}}-0.26428\hat{\mathbf{j}}}{0.96985}=-0.96216\hat{\mathbf{i}}-0.27249\hat{\mathbf{j}}$$

$$\hat{\mathbf{w}}_1=\frac{\mathbf{r}_0\times\mathbf{v}_0}{h_1}=\frac{72{,}117\hat{\mathbf{k}}}{72{,}117}=\hat{\mathbf{k}}$$

$$\hat{\mathbf{q}}_1=\hat{\mathbf{w}}_1\times\hat{\mathbf{p}}_1=\hat{\mathbf{k}}\times\left(-0.96216\hat{\mathbf{i}}-0.27249\hat{\mathbf{j}}\right)=0.27249\hat{\mathbf{i}}-0.96216\hat{\mathbf{j}}$$

Eqs. (9.24) and (9.25) yield the semimajor axis a_1 and the period T_1,

$$a_1 = \frac{h_1{}^2}{\mu_e}\frac{1}{1-e_1{}^2} = \frac{72{,}117^2}{398{,}600}\frac{1}{1-0.96985^2} = 219{,}714\,\text{km}$$

$$T_1 = 2\pi\sqrt{\frac{a_1{}^3}{\mu_e}} = 2\pi\sqrt{\frac{219{,}714^3}{398{,}600}} = 1.0249\left(10^6\right)\text{s} = 11.863\,\text{days}$$

According to Eq. (9.27), the true anomaly θ_0 of the injection point is

$$\begin{aligned}\theta_0 &= \cos^{-1}\left(\hat{\mathbf{p}}_1 \cdot \hat{\mathbf{u}}_{r_0}\right)\\ &= \cos^{-1}\left[\left(-0.96216\hat{\mathbf{i}} - 0.27249\hat{\mathbf{j}}\right)\cdot\left(-0.88295\hat{\mathbf{i}} - 0.46957\hat{\mathbf{j}}\right)\right]\\ &= \cos^{-1}0.97746\\ \therefore \theta_0 &= 12.187^\circ\end{aligned}$$

From this we find the time t_0 by means of Eq. (9.28),

$$\begin{aligned}t_0 &= \frac{T_1}{2\pi}\left\{2\tan^{-1}\left(\sqrt{\frac{1-e_1}{1+e_1}}\tan\frac{\theta_0}{2}\right) - e_1\sin\left[2\tan^{-1}\left(\sqrt{\frac{1-e_1}{1+e_1}}\tan\frac{\theta_0}{2}\right)\right]\right\}\\ &= \frac{1.0249\left(10^6\right)}{2\pi}\left\{2\tan^{-1}\left(\sqrt{\frac{1-0.96985}{1+0.96985}}\tan\frac{12.187^\circ}{2}\right)\right.\\ &\qquad \left. - 0.96985\sin\left[2\tan^{-1}\left(\sqrt{\frac{1-0.96985}{1+0.96985}}\tan\frac{12.187^\circ}{2}\right)\right]\right\}\\ &= 130.37\,\text{s}\end{aligned}$$

Since we know that the sweep angle is 160.89°, it follows from Eq. (9.29) that the true anomaly θ_1 at the patch point is

$$\theta_1 = \overbrace{12.187^\circ + 160.89^\circ}^{\theta_0 + \Delta\theta} = 173.08^\circ$$

Then Eq. (9.30) yields t_1, the time at the patch point:

$$\begin{aligned}t_1 &= \frac{T_1}{2\pi}\left\{2\tan^{-1}\left(\sqrt{\frac{1-e_1}{1+e_1}}\tan\frac{\theta_1}{2}\right) - e_1\sin\left[2\tan^{-1}\left(\sqrt{\frac{1-e_1}{1+e_1}}\tan\frac{\theta_1}{2}\right)\right]\right\}\\ &= \frac{1.0249\left(10^6\right)}{2\pi}\left\{2\tan^{-1}\left(\sqrt{\frac{1-0.96985}{1+0.96985}}\tan\frac{173.08^\circ}{2}\right)\right.\\ &\qquad \left. -0.96985\sin\left[2\tan^{-1}\left(\sqrt{\frac{1-0.96985}{1+0.96985}}\tan\frac{173.08^\circ}{2}\right)\right]\right\}\\ &= 239{,}370\,\text{s}\end{aligned}$$

From Eq. (9.31) we obtain the flight time Δt_1 from TLI to the patch point,

$$\Delta t_1 = \overbrace{239{,}370 - 130.38}^{t_1 - t_0} = 239{,}236\text{s} = 66.454\text{h}$$

The counterclockwise angular velocity ω_m of the moon in its assumed circular orbit around the earth is

$$\omega_m = \frac{v_m}{D} = \frac{1.0183\,\text{km/s}}{384{,}400\,\text{km}} = 2.6491\left(10^{-6}\right)\text{rad/s} = 0.5464\,\text{deg/h} \tag{c}$$

Multiplying this by the time interval Δt_1 yields the moon's lead angle of 36.31°, as shown in Fig. 9.5. The moon moves through this angle as the spacecraft flies to the patch point along its geocentric departure trajectory.

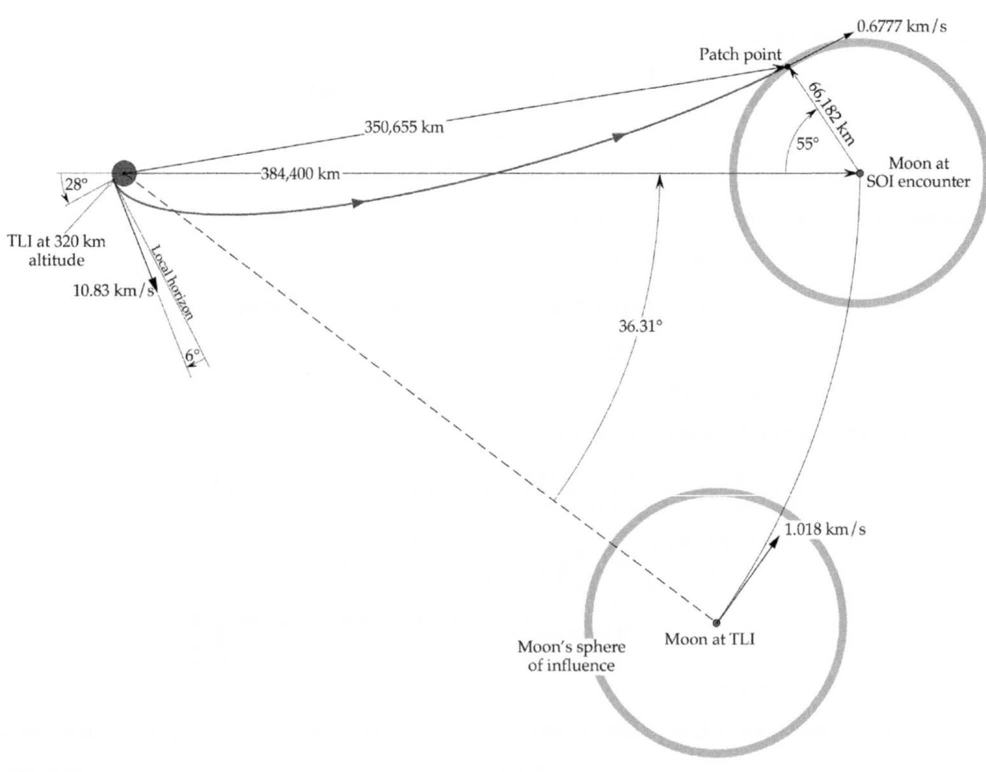

FIG. 9.5

Translunar trajectory of the spacecraft and concomitant motion of the moon. Drawn to scale.

At the patch point, the velocity $\mathbf{v}_2$ of the spacecraft relative to the moon is obtained from Eqs. (9.5) and (9.32)

$$\mathbf{v}_2 = \overbrace{(0.60618\hat{\mathbf{i}} + 0.30302\hat{\mathbf{j}})}^{\mathbf{v}_1} - \overbrace{1.0183\hat{\mathbf{j}}}^{\mathbf{v}_m} = 0.60618\hat{\mathbf{i}} - 0.71528\hat{\mathbf{j}}\,(\text{km/s})$$
$$v_2 = 0.93759\,\text{km/s}$$

We use $\hat{\mathbf{u}}_{r_2}$ from Eq. (a) above to calculate the radial component v_{r_2} of the relative velocity at the patch point,

$$v_{r_2} = \overbrace{\left(0.60618\hat{\mathbf{i}} - 0.71528\hat{\mathbf{j}}\right)}^{\mathbf{v}_2} \cdot \overbrace{\left(-0.57358\hat{\mathbf{i}} + 0.81915\hat{\mathbf{j}}\right)}^{\hat{\mathbf{u}}_{r_2}} = -0.93361\,\text{km/s} \tag{d}$$

The negative sign indicates, as it should, that the spacecraft is flying toward perilune. From Eq. (9.33), the probe's angular momentum relative to the moon is

$$\mathbf{h}_2 = \overbrace{\left(-37{,}961\hat{\mathbf{i}} + 54{,}214\hat{\mathbf{j}}\right)}^{\mathbf{r}_2} \times \overbrace{\left(0.60616\hat{\mathbf{i}} - 0.71528\hat{\mathbf{j}}\right)}^{\mathbf{v}_2} = -5710.78\hat{\mathbf{k}}\,(\text{km}^2/\text{s})$$
$$h_2 = 5710.78\text{km}^2/\text{s}$$

The fact that $\mathbf{h}_2$ points toward the orbital plane (in the negative z direction) means that the motion of the spacecraft is retrograde (i.e., clockwise around the moon).

The eccentricity vector $\mathbf{e}_2$ may now be calculated by using Eq. (9.34)

$$\begin{aligned}
\mathbf{e}_2 &= \frac{\mathbf{v}_2 \times \mathbf{h}_2}{\mu_m} - \hat{\mathbf{u}}_{r_2} \\
&= \frac{\left(0.60618\hat{\mathbf{i}} - 0.71528\hat{\mathbf{j}}\right) \times \left(-5710.78\hat{\mathbf{k}}\right)}{4902.8} - \left(-0.57358\hat{\mathbf{i}} + 0.81915\hat{\mathbf{j}}\right) \\
&= 1.4067\hat{\mathbf{i}} - 0.11307\hat{\mathbf{j}} \\
\therefore e_2 &= 1.44127
\end{aligned}$$

Since the eccentricity exceeds unity, the inbound orbit is indeed a hyperbola, relative to the moon. The perifocal unit vector $\hat{\mathbf{p}}_2$ directed from the center of the moon through perilune of the hyperbolic approach trajectory is, from Eq. (9.40),

$$\hat{\mathbf{p}}_2 = \overbrace{\left(\frac{1.4067\hat{\mathbf{i}} - 0.11307\hat{\mathbf{j}}}{1.44127}\right)}^{\mathbf{e}_2/e_2} = 0.99678\hat{\mathbf{i}} - 0.080122\hat{\mathbf{j}}$$

Substituting this into Eq. (9.41) and using the fact that $v_{r_2} < 0$, we find the true anomaly of the patch point on the lunar approach hyperbola, measured positive clockwise from perilune,

$$\begin{aligned}
\theta_2 &= 360^\circ - \cos^{-1}\overbrace{\left[\left(0.99678\hat{\mathbf{i}} - 0.080122\hat{\mathbf{j}}\right) \cdot \left(-0.57358\hat{\mathbf{i}} + 0.81915\hat{\mathbf{j}}\right)\right]}^{\hat{\mathbf{p}}_2 \cdot \hat{\mathbf{u}}_{r_2}} \\
&= 360^\circ - 129.60^\circ \\
&= 230.40^\circ
\end{aligned} \tag{e}$$

We find the time relative to perilune at this point on the hyperbola by means of Eq. (9.42),

$$\begin{aligned}
t_2 &= \frac{h_2^3}{\mu_m^2 (e_2^2 - 1)^{3/2}} \left\{ e_2 \sinh\left[2\tanh^{-1}\left(\sqrt{\frac{e_2 - 1}{e_2 + 1}}\tan\frac{\theta_2}{2}\right)\right] - 2\tanh^{-1}\left(\sqrt{\frac{e_2 - 1}{e_2 + 1}}\tan\frac{\theta_2}{2}\right)\right\} \\
&= \frac{5710.77^3}{4902.8^2 \left(1.41126^2 - 1\right)^{3/2}} \left\{ 1.41126 \sinh\left[2\tanh^{-1}\left(\sqrt{\frac{1.41127 - 1}{1.41127 + 1}}\tan\frac{230.40^\circ}{2}\right)\right] \right. \\
&\qquad \left. - 2\tanh^{-1}\left(\sqrt{\frac{1.41127 - 1}{1.41127 + 1}}\tan\frac{230.40^\circ}{2}\right)\right\} \\
&= -63{,}116\,\text{s} = -17.532\,\text{h}
\end{aligned} \tag{f}$$

The minus sign means that t_2 is the time *until* perilune. The elapsed time Δt_2 from patch point to perilune is $\Delta t_2 = t_{\text{perilune}} - t_2 = 0 - (-17.532\,\text{h}) = 17.532\,\text{h}$. The total time from translunar injection to perilune passage is

$$\Delta t_{\text{total}} = \overbrace{66.454}^{\Delta t_1} + \overbrace{17.532}^{\Delta t_2} = 83.986\,\text{h} = \boxed{3.4994\ \text{days}}$$

According to Eqs. (9.36) and (9.37), the perilune radius r_{p_2} and altitude are z_{p_2}

$$r_{p_2} = \frac{h_2^2}{\mu_m}\frac{1}{1 + e_2} = \frac{5710.8^2}{4902.8}\frac{1}{1 + 1.41127} = 2758.67\,\text{km}$$

$$z_{p_2} = \overbrace{2758.67 - 1737}^{r_{p_2} - R_m} = \boxed{1021.67\ \text{km}}$$

From Eq. (9.38) we know that the spacecraft's speed at perilune relative to the moon is

$$v_{p_2} = \sqrt{1 + e_2}\sqrt{\frac{\mu_m}{r_{p_2}}} = \sqrt{1 + 1.41127}\sqrt{\frac{4902.8}{2758.67}} = 2.07012\,\text{km/s}$$

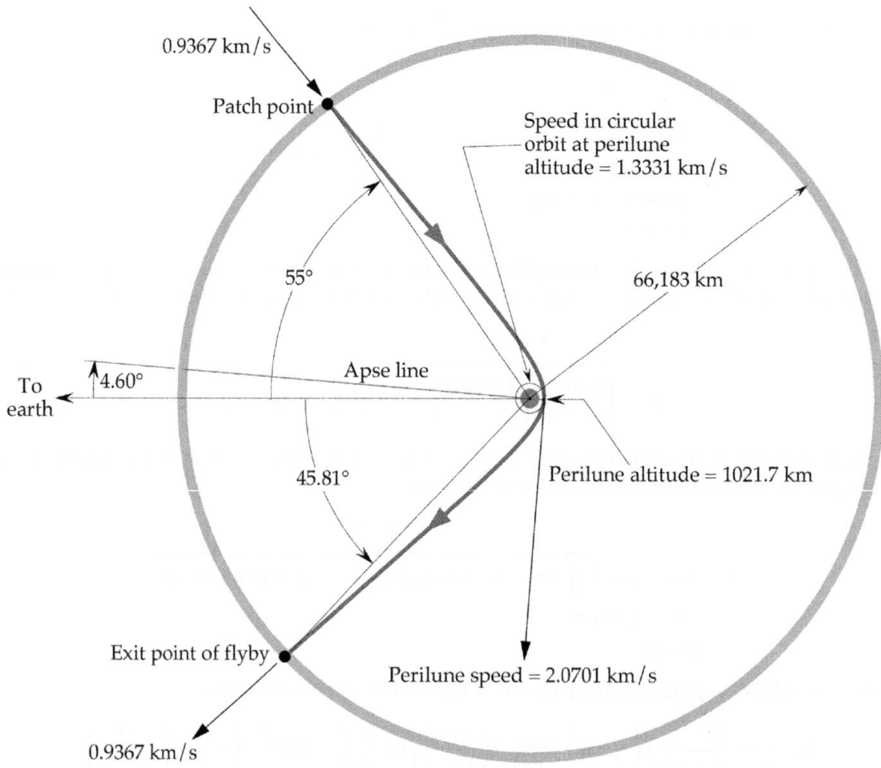

FIG. 9.6

Motion of the spacecraft within the lunar sphere of influence, relative to the moon-fixed frame of reference. Drawn to scale.

Therefore, the delta-v required at perilune to enter a circular lunar orbit of radius r_{p_2} is

$$\Delta v_2 = \sqrt{\frac{\mu_m}{r_{p_2}}} - v_{p_2} = \sqrt{\frac{4902.8}{2758.67}} - 2.07012 = -0.73698\,\text{km/s}$$

Fig. 9.6 shows the hyperbolic path of the spacecraft within the moon's SOI, relative to the moon. Also shown is the small circular capture orbit with a radius equal to that of perilune. If there is no delta-v maneuver at perilune, then the spacecraft continues on its hyperbolic path around the moon and leaves the SOI 17.548 h after perilune passage with the same relative speed at which it entered.

The moon-fixed hyperbolic flyby trajectory pictured in Fig. 9.6 is plotted relative to the stationary, earth-fixed coordinates in Fig. 9.7. The position vector of each point of the hyperbola is transformed into the earth-fixed xy frame using Eqs. (9.48) and (9.49). The result is a flyby trajectory that resembles a partial figure eight (between "SOI entrance" and "SOI exit" in Fig. 9.7) and bears little resemblance to the hyperbola as seen from the moon's perspective in Fig. 9.6.

To determine the trajectory of the spacecraft after leaving the moon's SOI on a flyby, we must first calculate the geocentric state vector at the SOI exit (namely, the position vector $\mathbf{r}_3$ and velocity vector $\mathbf{v}_3$ relative to the earth). The true anomaly on the hyperbola at SOI entrance was found in Eq. (e) to be $\theta_2 = -129.6°$, measured counterclockwise from perilune. The time of that initial SOI crossing was $t_2 = -17.532\,\text{h}$, according to Eq. (f). Therefore, the time and true anomaly of the SOI exit are

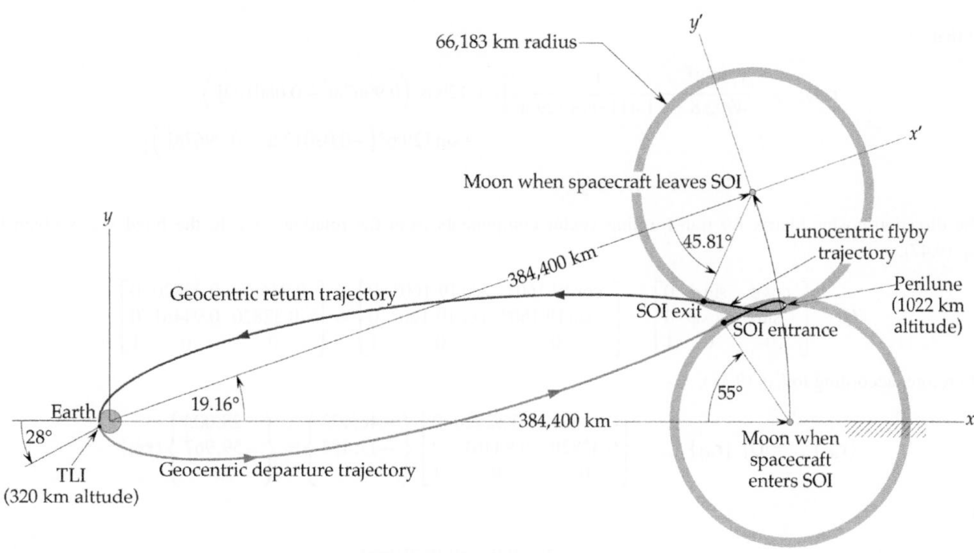

FIG. 9.7

The complete coplanar ballistic trajectory consisting of TLI, coast to moon, flyby, and coasting return to earth, all relative to the earth-fixed reference frame. Drawn to scale.

$$\begin{aligned} t_3 &= +17.532\text{h} = 63,115\text{s} \\ \theta_3 &= +129.6^\circ \end{aligned} \tag{g}$$

Recall that the moon's position angle ϕ is zero at the initial SOI encounter, so that the xyz and $x'y'z'$ axes instantaneously coincide. The angular position of the moon at the end of the hyperbolic fly-around is

$$\phi = \omega_m(t_3 - t_2) = \left[2.6491\left(10^{-6}\right)\frac{\text{rad}}{\text{s}}\right](126,230\text{s}) = 0.33439\text{rad} = 19.159^\circ$$

From Fig. 9.4 we see that the position vector $\mathbf{r}_m$ of the moon is

$$\mathbf{r}_m = D\left(\cos\phi\hat{\mathbf{i}} + \sin\phi\hat{\mathbf{j}}\right)$$

Substituting $D = 384{,}400\text{km}$ and $\phi = 19.159^\circ$ yields

$$\mathbf{r}_m = 363,107\hat{\mathbf{i}} + 126,160\hat{\mathbf{j}}\ (\text{km}) \tag{h}$$

The position of the spacecraft relative to the moon is given by Eq. (9.45),

$$\mathbf{r}_{\text{rel}}\big)_{x'y'z'} = \frac{h_2{}^2}{\mu_m}\frac{1}{1 + e_2\cos\theta_3}\left(\cos\theta_3\hat{\mathbf{p}}_2 + \sin\theta_3\hat{\mathbf{q}}_2\right)$$

where, in terms of the rotating but moon-fixed unit vectors $\hat{\mathbf{i}}'\hat{\mathbf{j}}'\hat{\mathbf{k}}'$,

$$\begin{aligned} \hat{\mathbf{p}}_2 &= \frac{\mathbf{e}_2}{e_2} = \frac{1.4067\hat{\mathbf{i}}' - 0.11307\hat{\mathbf{j}}'}{1.41127} = 0.99678\hat{\mathbf{i}}' - 0.080122\hat{\mathbf{j}}' \\ \hat{\mathbf{w}}_2 &= \frac{\mathbf{r}_2 \times \mathbf{v}_2}{h_2} = \frac{-5710.8\hat{\mathbf{k}}}{5710.8} = -\hat{\mathbf{k}} \\ \hat{\mathbf{q}}_2 &= \hat{\mathbf{w}}_2 \times \hat{\mathbf{p}}_2 = -\hat{\mathbf{k}} \times \left(0.99678\hat{\mathbf{i}}' - 0.080122\hat{\mathbf{j}}'\right) = -0.080122\hat{\mathbf{i}}' - 0.99678\hat{\mathbf{j}}' \end{aligned}$$

so that

$$\mathbf{r}_{\text{rel}}\big)_{x'y'z'} = \frac{5710.8^2}{4902.8}\frac{1}{1+1.4113\cos 129.6°}\left[\cos 129.6°\left(0.99678\hat{\mathbf{i}}' - 0.080122\hat{\mathbf{j}}'\right) + \sin 129.6°\left(-0.080122\hat{\mathbf{i}}' - 0.99678\hat{\mathbf{j}}'\right)\right]$$

$$= -46{,}133\hat{\mathbf{i}}' - 47{,}454\hat{\mathbf{j}}' \text{ (km)}$$

The direction cosine matrix for transforming vector components from the rotating $x'y'z'$ to the fixed xyz is given in Eq. (9.47),

$$[\mathbf{Q}] = \begin{bmatrix} \cos\phi & \sin\phi & 0 \\ -\sin\phi & \cos\phi & 0 \\ 0 & 0 & 1 \end{bmatrix} = \begin{bmatrix} \cos 19.160° & \sin 19.160° & 0 \\ -\sin 19.160° & \cos 19.160° & 0 \\ 0 & 0 & 1 \end{bmatrix} = \begin{bmatrix} 0.94461 & 0.32820 & 0 \\ -0.32820 & 0.94461 & 0 \\ 0 & 0 & 1 \end{bmatrix}$$

Therefore, according to Eq. (9.48),

$$\{\mathbf{r}_{\text{rel}}\}_{xyz} = [\mathbf{Q}]^T\{\mathbf{r}_{\text{rel}}\}_{x'y'z'} = \begin{bmatrix} 0.94461 & -0.32820 & 0 \\ 0.32820 & 0.94461 & 0 \\ 0 & 0 & 1 \end{bmatrix}\begin{Bmatrix} -46{,}133 \\ -47{,}454 \\ 0 \end{Bmatrix} = \begin{Bmatrix} -28{,}003 \\ -59{,}967 \\ 0 \end{Bmatrix} \text{(km)}$$

or

$$\mathbf{r}_{\text{rel}} = -28{,}003\hat{\mathbf{i}} - 59{,}967\hat{\mathbf{j}} \text{(km)}$$

Substituting this along with Eq. (h) into Eq. (9.49) yields the spacecraft position vector at SOI exit, relative to the earth-fixed frame

$$\mathbf{r}_3 = 335{,}104\hat{\mathbf{i}} + 66194\hat{\mathbf{j}} \text{(km)} \tag{i}$$

The velocity of the spacecraft at SOI exit relative to the moon is

$$\mathbf{v}_{\text{rel}}\big)_{x'y'z'} = \frac{\mu_m}{h_2}\left[-\sin\theta_3\hat{\mathbf{p}}_2 + (e_2 + \cos\theta_3)\hat{\mathbf{q}}_2\right]$$

That is,

$$\mathbf{v}_{\text{rel}}\big)_{x'y'z'} = \frac{4902.8}{5710.8}\left[-\sin 129.6°\left(0.99678\hat{\mathbf{i}}' - 0.080122\hat{\mathbf{j}}'\right) + (e_2 + \cos 129.6°)\left(-0.080122\hat{\mathbf{i}}' - 0.99678\hat{\mathbf{j}}'\right)\right]$$

$$= -0.71265\hat{\mathbf{i}}' - 0.60927\hat{\mathbf{j}}' \text{ (km/s)}$$

We use Eq. (9.50) to obtain the components of $\mathbf{v}_{\text{rel}}$ in the xyz frame,

$$\{\mathbf{v}_{\text{rel}}\}_{xyz} = [\mathbf{Q}]^T\{\mathbf{v}_{\text{rel}}\}_{x'y'z'} = \begin{bmatrix} 0.94461 & -0.32820 & 0 \\ 0.32820 & 0.94461 & 0 \\ 0 & 0 & 1 \end{bmatrix}\begin{Bmatrix} -0.71265 \\ -0.60927 \\ 0 \end{Bmatrix} = \begin{Bmatrix} -0.47321 \\ -0.80941 \\ 0 \end{Bmatrix} \text{(km/s)}$$

or

$$\mathbf{v}_{\text{rel}} = -0.47321\hat{\mathbf{i}} - 0.80941\hat{\mathbf{j}} \text{(km/s)}$$

Substituting this expression along with Eq. (i) and the angular velocity of the moon (Eq. c) into Eq. (9.52) yields

$$\mathbf{v}_3 = \boldsymbol{\omega}_m \times \mathbf{r}_3 + \mathbf{v}_{\text{rel}}$$
$$= 2.6491\left(10^{-6}\right)\hat{\mathbf{k}} \times \left(335{,}104\hat{\mathbf{i}} + 66194\hat{\mathbf{j}}\right) + \left(-0.47321\hat{\mathbf{i}} - 0.80941\hat{\mathbf{j}}\right)$$

Therefore, the absolute velocity of the spacecraft at the SOI exit is

$$\mathbf{v}_3 = -0.64856\hat{\mathbf{i}} + 0.078302\hat{\mathbf{j}} \text{(km/s)} \tag{j}$$

From the state vector $(\mathbf{r}_3, \mathbf{v}_3)$, we obtain the transearth trajectory's orbital elements. That is,

$$\mathbf{h}_3 = \mathbf{r}_3 \times \mathbf{v}_3 = 69,170\hat{\mathbf{k}}\ (\mathrm{km}^2/\mathrm{s})$$
$$h_3 = 69,170\mathrm{km}^2/\mathrm{s}$$
$$\mathrm{e}_3 = \frac{1}{\mu_e}(\mathrm{v}_3 \times \mathrm{h}_3) - \frac{\mathrm{r}_3}{r_3} = -0.967456\hat{\mathrm{i}} - 0.0812406\hat{\mathrm{j}}$$
$$e_3 = 0.970860$$

With this information, we can plot the geocentric return trajectory, as shown in Fig. 9.7. Its perigee is

$$r_{p_3} = \frac{h_3{}^2}{\mu_e}\frac{1}{1+e_3} = \frac{69,170^2}{398,600}\frac{1}{1+0.97086} = 6090.4\mathrm{km}$$

Since this is less than the radius of the earth, the spacecraft will impact the atmosphere.

The path followed by the spacecraft in this example is called a free return trajectory because the single delta-v maneuver at TLI yields a lunar flyby followed by a return to earth.

9.3 A SIMPLIFIED LUNAR EPHEMERIS

We will employ the geocentric equatorial *XYZ* frame (Section 4.3) to describe the three-dimensional translunar trajectory of a spacecraft as well as the motion of the moon around the earth. The state vector of the moon at any time is found by means of a lunar ephemeris. High-precision ephemerides are found in the Jet Propulsion Laboratory's authoritative DE (*development ephemeris*) series. These currently may be accessed online at the JPL Horizons ephemeris system website (JPL Horizons Web-Interface, 2018) and within MATLAB by means of the function `planetEphemeris`. Simpson (1999) developed a simplified lunar ephemeris, which is a curve fit of JPL's 1984 DE200 ephemeris model. It is easy to use and the precision is sufficient for our needs.

Simpson's lunar ephemeris yields the geocentric equatorial coordinates *X*, *Y*, and *Z* of the moon in kilometers for any year in the range CE 2000 through CE 2100, according to the formula

$$X_i = \sum_{j=1}^{7} a_{ij} \sin\left(b_{ij}t + c_{ij}\right) \quad (i = 1, 2, 3) \tag{9.54}$$

where $X_1 = X$, $X_2 = Y$, $X_3 = Z$, and t is the time in Julian centuries since J2000 (see Eq. 5.49),

$$t = \frac{JD - 2{,}451{,}545}{36{,}525}\ (\text{centuries}) \tag{9.55}$$

and JD is the Julian date (in days). The components of the 3-by-7 matrices [**a**], [**b**], and [**c**] are

$$[\mathbf{a}] = \begin{bmatrix} 383{,}000 & 31{,}500 & 10{,}600 & 6{,}200 & 3{,}200 & 2{,}300 & 800 \\ 351{,}000 & 28{,}900 & 13{,}700 & 9{,}700 & 5{,}700 & 2{,}900 & 2{,}100 \\ 153{,}200 & 31{,}500 & 12{,}500 & 4{,}200 & 2{,}500 & 3{,}000 & 1{,}800 \end{bmatrix} (\mathrm{km})$$

$$[\mathbf{b}] = \begin{bmatrix} 8399.685 & 70.990 & 16728.377 & 1185.622 & 7143.070 & 15613.745 & 8467.263 \\ 8399.687 & 70.997 & 8433.466 & 16728.380 & 1185.667 & 7143.058 & 15613.755 \\ 8399.672 & 8433.464 & 70.996 & 16728.364 & 1185.645 & 104.881 & 8399.116 \end{bmatrix} (\mathrm{rad/century})$$

$$[\mathbf{c}] = \begin{bmatrix} 5.381 & 6.169 & 1.453 & 0.481 & 5.017 & 0.857 & 1.010 \\ 3.811 & 4.596 & 4.766 & 6.165 & 5.164 & 0.300 & 5.565 \\ 3.807 & 1.629 & 4.595 & 6.162 & 5.167 & 2.555 & 6.248 \end{bmatrix} (\mathrm{rad})$$

The moon's geocentric equatorial velocity components $\dot{X}$, $\dot{Y}$, and $\dot{Z}$ are found by simply differentiating Eq. (9.53) with respect to time,

$$\dot{X}_i = \frac{1}{t_C}\sum_{j=1}^{7} a_{ij} b_{ij} \cos\left(b_{ij} t + c_{ij}\right) \quad (i = 1, 2, 3)$$

The conversion factor t_C is required to convert kilometers per century to kilometers per second

$$t_C = \left(36,525\frac{\text{d}}{\text{cy}}\right)\cdot\left(24\frac{\text{h}}{\text{d}}\right)\cdot\left(3600\frac{\text{s}}{\text{h}}\right) = 3.15576\left(10^9\right)\frac{\text{s}}{\text{cy}}$$

A MATLAB implementation of Simpson's lunar ephemeris is listed in Appendix D.37.

EXAMPLE 9.2

Use Simpson's ephemeris to find the variation of the moon's inclination to the earth's equatorial plane from January 1, 2000, UT 12:00:00 through January 1, 2100, UT 12:00:00.

Solution

The following MATLAB script computes and plots the inclinations.

```
clear all; close all; clc
%
%...Initial and final Julian dates (jd1 & jd2):
jd1  = julian_day(2000,1,1,12,0,0); %January 1,2000 UT 12:00:00
jd2  = julian_day(2100,1,1,12,0,0); %January 1,2100 UT 12:00:00
%
%...The following initially empty column vectors will contain the lunar
%   inclinations and the Julian day for which each one is evaluated:
incl = [];
days = [];
%
%...Loop through the sequence of Julian days, one at a time:
for jd = jd1 : jd2 + 20*365.25
    days     = [days; jd - jd1];                  %Store the elapsed
%                                                    time in the
%                                                    days array.
    [r_,v_] = simpsons_lunar_ephemeris(jd); %Compute the
%                                                    position and
%                                                    velocity vectors,
%                                                    r_ and v_.
    h_       = cross(r_,v_);                      %h_ is the angular
%                                                    momentum vector.
    h        = norm(h_);                          %h is its magnitude.
    incl     = [incl; acosd(h_(3)/h)];            %Compute the
%                                                    inclination from
%                                                    Eq. 4.7 and add
%                                                    it to the
%                                                    incl array.
end
%
%...Smooth out the high frequency content with MATLAB's
%   smoothdata function using a Savitzky-Golay filter:
```

```
incl_smooth = smoothdata(incl, 'sgolay');
%
%...Plot the computed values of incl against the elapsed time
%   in years:
plot(days/365.25,incl_smooth,'b') %(365.25 days per Julian year)
title('Variation of lunar inclination from J2000 to J2100')
axis([0 100 15 30])
xlabel('Elapsed time, years')
ylabel('Inclination, degrees')
grid on
hold on
plot([0 100],[28.5 28.5], '--r')
plot([0 100],[18.4 18.4], '--r')
```

Fig. 9.8 shows the output of the above script. Clearly, the angle between the moon's orbital plane and the earth's equator varies from 18.4° to 28.5° over a period of 18.6 years.

The moon's orbital plane is inclined to the ecliptic plane by an angle (obliquity) of 5.14°, whereas the earth's obliquity is 23.4°. Due primarily to solar gravity, the moon's orbital plane precesses

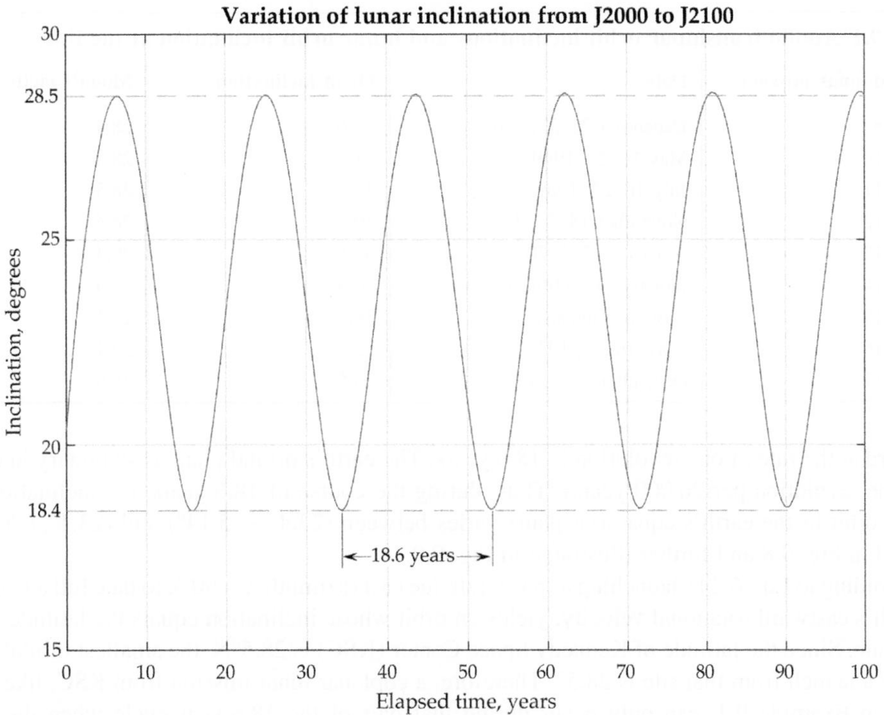

FIG. 9.8

Variation of the moon's orbital inclination with time (Simpson's ephemeris).

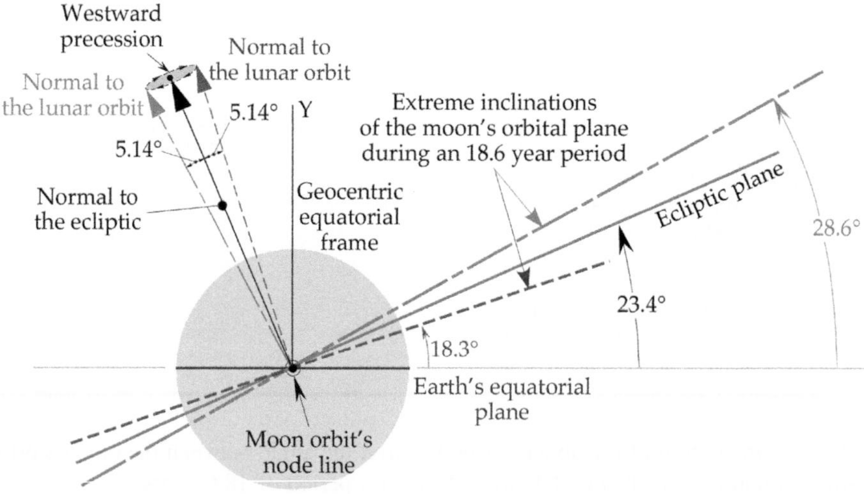

FIG. 9.9

Precession of the lunar orbit around the normal to the ecliptic plane.

Table 9.1 Apollo translunar orbit inclinations and lunar orbit inclination at the time

Manned lunar mission	Date	Orbit inclination (°)	Moon's inclination (°)
Apollo 8	December 21–27, 1968	30.6	28.4
Apollo 10	May 18–26, 1969	31.7	28.5
Apollo 11	July 16–24, 1969	31.4	28.5
Apollo 12	November 14–24, 1969	30.4	28.4
Apollo 13	April 11–17, 1970	31.8	28.4
Apollo 14	January 31–February 9, 1971	30.8	27.9
Apollo 15	July 26–August 7, 1971	29.7	27.4
Apollo 16	April 16–27, 1972	32.5	26.4
Apollo 17	December 7–19, 1972	28.5	25.5

westward at the rate of one revolution in 18.6 years. The earth's orbital plane is stationary in comparison (one revolution per 26,000 years). Thus, during the course of 18.6 years, the inclination of the moon's orbit to the earth's equatorial plane varies between (23.4° − 5.14°) and (23.4° + 5.14°), as revealed in Fig. 9.8 and further illustrated in Fig. 9.9.

According to Eq. (6.24), launching a spacecraft due east (azimuth $A = 90°$), to take full advantage of the earth's eastward rotational velocity, yields an orbit whose inclination equals the latitude ϕ of the launch site. Since the latitude of Kennedy Space Center (KSC) is 28.5°N, the smallest orbital inclination i of a launch from that site is 28.5°. Therefore, a coplanar lunar mission from KSC, like that described in Example 9.1, can only occur during that part of the 18.6-year cycle when the moon's inclination is at or near 28.5°. Table 9.1 compares the orbital inclination of each Apollo lunar mission (Orloff, 2000) with the moon's inclination at the time, which is obtained from the JPL ephemeris.

9.4 PATCHED CONIC LUNAR TRAJECTORIES IN THREE DIMENSIONS

In Section 9.2 we cast the procedure for coplanar patched conic lunar trajectory analysis in vector format. Therefore, the notation and procedure for three dimensions remains essentially the same. We simply drop the assumption that the paths of the spacecraft and the moon lie in the same plane. We obtain the position and velocity of the moon from an accurate lunar ephemeris, instead of assuming that the moon moves around the earth in a circular path. The plane of the translunar trajectory is not that of the moon's orbit, but is determined by the spacecraft's position vector $\mathbf{r}_0$ at TLI and the position vector $\mathbf{r}_m$ of the moon when the spacecraft crosses into the moon's SOI. We continue to assume that the motion of the spacecraft after TLI is ballistic, which means there are no midcourse propulsive maneuvers prior to lunar encounter. As in Section 9.2, let us do the trajectory analysis in four parts.

I. After TLI, travel a ballistic geocentric trajectory until entering the moon's SOI.
II. Determine the spacecraft trajectory inside the moon's SOI, relative to the rotating moon-fixed frame.
III. Transform the trajectory in II into the inertial geocentric equatorial frame.
IV. If flyby occurs, then determine the geocentric trajectory after departing the moon's SOI.

We will use a numerical example and simply outline the procedure, since the details were mostly covered in Section 9.2.

I. *Translunar trajectory up to the moon's SOI* (Fig. 9.10)
Recall that $\mu_e = 398,600\,\text{km}^3/\text{s}^2$.
1. Choose values for the independent variables of the problem.
   a. Select the date for the moon's position at SOI intercept.
   May 4, 2020, 12:00:00 UT. Julian day: 2,458,974.
   b. Select the value for the arrival angle λ (i.e., the angle between the radials from moon to earth and from moon to spacecraft at the moon's SOI) (see Fig. 9.10).

$$\lambda = 50^\circ$$

   c. Select the probe's radius r_0, right ascension α_L, declination δ_L, and flight path angle γ_0 at TLI:

$$r_0 = 6698\,\text{km} \quad \alpha_L = 40^\circ \quad \delta_L = 10^\circ \quad \gamma_0 = 10^\circ$$

2. Use an ephemeris to determine the moon's geocentric equatorial state vector $(\mathbf{r}_m, \mathbf{v}_m)$ on the date in I.1.a.

$$\mathbf{r}_m = -359,984\hat{\mathbf{I}} - 25,810.2\hat{\mathbf{J}} + 22,885.4\hat{\mathbf{K}}\,(\text{km}) \qquad r_m = 361,835\,\text{km}$$
$$\mathbf{v}_m = 0.0805809\hat{\mathbf{I}} - 0.990137\hat{\mathbf{J}} - 0.437526\hat{\mathbf{K}}\,(\text{km/s}) \qquad v_m = 1.08558\,\text{km/s}$$

3. Calculate $\hat{\mathbf{s}}$, the unit vector along the earth-to-moon radial:

$$\hat{\mathbf{s}} = \frac{\mathbf{r}_m}{\|\mathbf{r}_m\|} \qquad \therefore \hat{\mathbf{s}} = -0.994882\hat{\mathbf{I}} - 0.0787934\hat{\mathbf{J}} + 0.0632482\hat{\mathbf{K}}$$

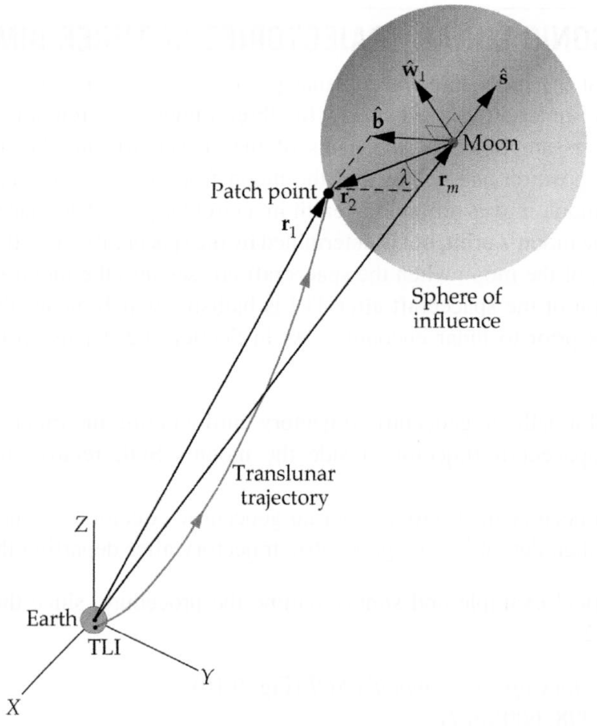

FIG. 9.10

Translunar trajectory up to encounter of moon's sphere of influence.

4. Calculate $\boldsymbol{\omega}_m$, the instantaneous angular velocity of the moon at the time of SOI intercept:

$$\boldsymbol{\omega}_m = \frac{\mathbf{r}_m \times \mathbf{v}_m}{{r_m}^2} \quad \therefore \boldsymbol{\omega}_m = \left(0.268368\hat{\mathbf{I}} - 1.18891\hat{\mathbf{J}} + 2.740245\hat{\mathbf{K}}\right)\left(10^{-6}\right)(\text{rad/s})$$

$$\omega_m = 2.99908\left(10^{-6}\right)\text{rad/s}$$

5. Calculate the geocentric position vector $\mathbf{r}_0$ at TLI using Eqs. (4.4) and (4.5) and the data in I.1.c:

$$\mathbf{r}_0 = r_0\left(\cos\alpha_L \cos\delta_L \hat{\mathbf{I}} + \sin\alpha_L \cos\delta_L \hat{\mathbf{J}} + \sin\delta_L \hat{\mathbf{K}}\right) = 5053.02\hat{\mathbf{I}} + 4239.98\hat{\mathbf{J}} + 1163.10\hat{\mathbf{K}}\,(\text{km})$$
$$(r_0 = 6698\,\text{km})$$

6. Calculate $\hat{\mathbf{w}}_1$, the unit normal to the plane of the translunar trajectory:

$$\hat{\mathbf{w}}_1 = \frac{\mathbf{r}_0 \times \mathbf{r}_m}{\|\mathbf{r}_0 \times \mathbf{r}_m\|} \quad \therefore \hat{\mathbf{w}}_1 = 0.0875163\hat{\mathbf{I}} - 0.359180\hat{\mathbf{J}} + 0.929156\hat{\mathbf{K}}$$

7. Calculate $\hat{\mathbf{b}}$, the unit normal to the plane of $\hat{\mathbf{s}}$ and $\hat{\mathbf{w}}_1$. $\hat{\mathbf{b}}$ lies in the plane of the translunar trajectory (see Fig. 9.10):

$$\hat{\mathbf{b}} = \frac{\hat{\mathbf{w}}_1 \times \hat{\mathbf{s}}}{\|\hat{\mathbf{w}}_1 \times \hat{\mathbf{s}}\|} \quad \therefore \hat{\mathbf{b}} = 0.0504938\hat{\mathbf{I}} - 0.929936\hat{\mathbf{J}} - 0.364238\hat{\mathbf{K}}$$

8. Calculate $\hat{\mathbf{n}}$, the unit vector from the center of the moon to the SOI patch point:

$$\hat{\mathbf{n}} = -\cos\lambda\hat{\mathbf{s}} + \sin\lambda\hat{\mathbf{b}} \quad \therefore \hat{\mathbf{n}} = 0.678179\hat{\mathbf{I}} - 0.661725\hat{\mathbf{J}} - 0.319678\hat{\mathbf{K}}$$

9. Calculate $\mathbf{r}_2$, the position vector of the patch point relative to the moon:

$$\mathbf{r}_2 = R_S\hat{\mathbf{n}} \quad \therefore \mathbf{r}_2 = 44{,}883.7\hat{\mathbf{I}} - 43{,}794.8\hat{\mathbf{J}} - 21{,}157.2\hat{\mathbf{K}}\,(\text{km})$$
$$r_2 = 66{,}183\,\text{km}$$

10. Calculate $\mathbf{r}_1$, the position vector of the patch point relative to the earth:

$$\mathbf{r}_1 = \mathbf{r}_m + \mathbf{r}_2 \quad \therefore \mathbf{r}_1 = -315{,}100\hat{\mathbf{I}} - 72{,}305.0\hat{\mathbf{J}} + 1728.29\hat{\mathbf{K}}\,(\text{km})$$
$$r_1 = 323{,}294\,\text{km}$$

11. Use Eq. (9.14) to calculate the sweep angle $\Delta\theta$:

$$\cos\Delta\theta = (\mathbf{r}_0/r_0)\cdot(\mathbf{r}_1/r_1) \;\Rightarrow\; \Delta\theta = 151.156^\circ$$

12. Calculate the angular momentum h_1 of the translunar trajectory using Eq. (9.18):

$$h_1 = 71{,}426.1\,\text{km}^2/\text{s}$$

13. Calculate the Lagrange coefficients f, g, and $\dot{g}$ from Eqs. (9.13a)–(9.13c):

$$f = -46.3848 \quad g = 14625.9\,\text{s} \quad \dot{g} = 0.0182827$$

14. Calculate the velocity $\mathbf{v}_0$ at TLI and the velocity $\mathbf{v}_1$ at the patch point by means of Eqs. (9.12a) and (9.12b):

$$\mathbf{v}_0 = -5.51878\hat{\mathbf{I}} + 8.503129\hat{\mathbf{J}} + 3.80683\hat{\mathbf{K}}\,(\text{km/s})$$
$$\mathbf{v}_1 = -0.739368\hat{\mathbf{I}} - 0.380279\hat{\mathbf{J}} - 0.0773628\hat{\mathbf{K}}\,(\text{km/s})$$

15. Calculate the radial component of velocity v_{r_0} at TLI:

$$v_{r_0} = \mathbf{v}_0\cdot(\mathbf{r}_0/r_0) \quad \therefore v_{r_0} = 1.88031\,\text{km/s}\,(\text{positive})$$

16. Using the TLI state vector $(\mathbf{r}_0, \mathbf{v}_0)$, calculate the eccentricity vector $\mathbf{e}_1$ of the translunar trajectory from Eq. (2.40). The eccentricity, $e_1 = \|\mathbf{e}_1\|$, must be less than 1:

$$\mathbf{e}_1 = \frac{\mathbf{v}_0\times(\mathbf{r}_0\times\mathbf{v}_0)}{\mu_e} - \frac{\mathbf{r}_0}{r_0} \quad \therefore \mathbf{e}_1 = 0.906360\hat{\mathbf{I}} + 0.345541\hat{\mathbf{J}} + 0.0482052\hat{\mathbf{K}}$$
$$e_1 = 0.971190$$

17. Calculate the semimajor axis a_1 and the period T_1 of the translunar trajectory from Eqs. (9.24) and (9.25), respectively:

$$a_1 = 225{,}375\,\text{km} \quad T_1 = 1{,}064{,}806\,\text{s} = 12.3241\,\text{d}$$

18. Calculate the triad of perifocal unit vectors $\hat{\mathbf{p}}_1$, $\hat{\mathbf{q}}_1$, and $\hat{\mathbf{w}}_1$ for the translunar trajectory:

$$\hat{\mathbf{p}}_1 = \frac{\mathbf{e}_1}{e_1} \quad \therefore \hat{\mathbf{p}}_1 = 0.933246\hat{\mathbf{I}} + 0.355791\hat{\mathbf{J}} + 0.0496352\hat{\mathbf{K}}$$
$$\hat{\mathbf{w}}_1 = 0.0875163\hat{\mathbf{I}} - 0.359180\hat{\mathbf{J}} + 0.929156\hat{\mathbf{K}} \quad \text{(Calculated in Step I.6 above.)}$$
$$\hat{\mathbf{q}}_1 = \hat{\mathbf{w}}_1\times\hat{\mathbf{p}}_1 \quad \therefore \hat{\mathbf{q}}_1 = -0.348413\hat{\mathbf{I}} + 0.862788\hat{\mathbf{J}} + 0.366341\hat{\mathbf{K}}$$

19. Calculate the true anomaly θ_0 at the TLI point using Eq. (9.27) and noting from Step I.16 that $v_{r_0} > 0$ at TLI:

$$\theta_0 = 20.2998°$$

20. Calculate the time t_0 since perigee at the TLI point using Eq. (9.28):

$$t_0 = 213.532\text{s}$$

21. Calculate the true anomaly θ_1 at the patch point, $\theta_1 = \theta_0 + \Delta\theta$, where we found the sweep angle $\Delta\theta$ in Step I.11:

$$\theta_1 = 20.2998° + 151.156° = 171.455°$$

22. Calculate the time t_1 since perigee at the patch point using Eq. (9.30):

$$t_1 = 54.8899\text{h}$$

23. Calculate the flight time Δt_1 from TLI to the patch point, $\Delta t_1 = t_1 - t_0$:

$$\Delta t_1 = 54.8306\text{h}$$

II. *Determine the lunar approach trajectory within the moon's SOI, relative to the moon.* Recall that the gravitational parameter and the radius of the moon are $\mu_m = 4902.8\,\text{km}^3/\text{s}^2$ and $R_m = 1727$ km, respectively.

1. Calculate the velocity $\mathbf{v}_2$ of the spacecraft relative to the moon at the patch point:

$$\mathbf{v}_2 = \mathbf{v}_1 - \mathbf{v}_m \quad \therefore \mathbf{v}_2 = -0.819949\hat{\mathbf{I}} + 0.609957\hat{\mathbf{J}} + 0.360164\hat{\mathbf{K}}\,(\text{km/s})$$
$$v_2 = 1.08355\text{km/s}$$

$\mathbf{v}_m$ was obtained in Step I.2.

2. Calculate the radial speed v_{r_2} relative to the moon at the patch point:

$$v_{r_2} = \mathbf{v}_2 \cdot (\mathbf{r}_2/r_2) \quad \therefore v_{r_2} = -1.07483\text{km/s} \quad (\mathbf{r}_2 \text{ was found in Step I.9})$$

3. Calculate the angular momentum $\mathbf{h}_2$ of the trajectory relative to the moon:

$$\mathbf{h}_2 = \mathbf{r}_2 \times \mathbf{v}_2 \quad \therefore \mathbf{h}_2 = -2868.33\hat{\mathbf{I}} + 1182.29\hat{\mathbf{J}} - 8532.32\hat{\mathbf{J}}\,(\text{km}^2/\text{s})$$
$$h_2 = 9078.86\text{km}^2/\text{s}$$

4. Use Eq. (2.40) to calculate the eccentricity vector $\mathbf{e}_2$ of the trajectory, relative to the moon. e_2 must be greater than 1:

$$\mathbf{e}_2 = \frac{1}{\mu_m}\left(\mathbf{v}_2 \times \mathbf{h}_2 - \frac{\mathbf{r}_2}{r_2}\right) \quad \therefore \mathbf{e}_2 = -1.82654\hat{\mathbf{I}} - 0.975939\hat{\mathbf{J}} + 0.478800\hat{\mathbf{K}}$$
$$e_2 = 2.12554$$

5. Calculate the perilune radius r_{p_2} and altitude z_{p_2} of the hyperbolic lunar approach trajectory:

$$r_{p_2} = \frac{h_2{}^2}{\mu_m}\frac{1}{1+e_2} = 5378.89\text{km} \quad \therefore z_{p_2} = r_{p_2} - R_m = 3641.9\text{km}$$

6. Calculate the perifocal unit vectors $\hat{\mathbf{p}}_2$, $\hat{\mathbf{q}}_2$, and $\hat{\mathbf{w}}_2$ of the hyperbolic lunar approach trajectory:

$$\hat{\mathbf{p}}_2 = \frac{\mathbf{e}_2}{e_2} \qquad \therefore \hat{\mathbf{p}}_2 = -0.859326\hat{\mathbf{I}} - 0.459147\hat{\mathbf{J}} + 0.225260\hat{\mathbf{K}}$$

$$\hat{\mathbf{w}}_2 = \frac{\mathbf{h}_2}{h_2} \qquad \therefore \hat{\mathbf{w}}_2 = -0.315936\hat{\mathbf{I}} + 0.130224\hat{\mathbf{J}} - 0.939801\hat{\mathbf{K}}$$

$$\hat{\mathbf{q}}_2 = \hat{\mathbf{w}}_2 \times \hat{\mathbf{p}}_2 \quad \therefore \hat{\mathbf{q}}_2 = -0.402173\hat{\mathbf{I}} + 0.878763\hat{\mathbf{J}} + 0.256966\hat{\mathbf{K}}$$

7. Calculate the triad of orthogonal unit vectors $\hat{\mathbf{i}}$, $\hat{\mathbf{j}}$, and $\hat{\mathbf{k}}$ directed along the rotating *xyz* moon-fixed Cartesian coordinate axes at the instant the spacecraft crosses the lunar SOI. Note that the *z* axis lies in the direction of the moon's angular velocity vector $\boldsymbol{\omega}_m$, which we computed in Step I.4.

$$\hat{\mathbf{i}} = \frac{\mathbf{r}_m}{r_m} \qquad \therefore \hat{\mathbf{i}} = -0.994882\hat{\mathbf{I}} - 0.0787934\hat{\mathbf{J}} + 0.0632482\hat{\mathbf{K}}$$

$$\hat{\mathbf{k}} = \frac{\boldsymbol{\omega}_m}{\omega_m} \qquad \therefore \hat{\mathbf{k}} = 0.0894832\hat{\mathbf{I}} - 0.396426\hat{\mathbf{J}} + 0.913695\hat{\mathbf{K}}$$

$$\hat{\mathbf{j}} = \hat{\mathbf{k}} \times \hat{\mathbf{i}} \qquad \therefore \hat{\mathbf{j}} = 0.0469199\hat{\mathbf{I}} - 0.914679\hat{\mathbf{J}} - 0.401448\hat{\mathbf{K}}$$

8. Use the above geocentric equatorial components of $\hat{\mathbf{i}}$, $\hat{\mathbf{j}}$, and $\hat{\mathbf{k}}$ to calculate the instantaneous direction cosine matrix $[\mathbf{Q}]$ of the transformation from the geocentric equatorial *XYZ* frame to the moon-fixed *xyz* frame (see Eq. 4.23):

$$[\mathbf{Q}] = \begin{bmatrix} i_X & i_Y & i_Z \\ j_X & j_Y & j_Z \\ k_X & k_Y & k_Z \end{bmatrix} \quad \therefore [\mathbf{Q}] = \begin{bmatrix} -0.994882 & -0.0787934 & 0.0632482 \\ 0.0469199 & -0.914679 & -0.401448 \\ 0.0894832 & -0.396426 & 0.913695 \end{bmatrix}$$

9. Calculate the components of the perifocal unit vectors $\hat{\mathbf{p}}_2$ and $\hat{\mathbf{q}}_2$ (calculated in the *XYZ* frame in II.6) in the rotating *xyz* moon-fixed frame:

$$\{\hat{\mathbf{p}}_2\}_{xyz} = [\mathbf{Q}]\{\hat{\mathbf{p}}_2\}_{XYZ} \quad \therefore \{\hat{\mathbf{p}}_2\}_{xyz} = 0.905354\hat{\mathbf{i}} + 0.289223\hat{\mathbf{j}} + 0.310942\hat{\mathbf{k}}$$

$$\{\hat{\mathbf{q}}_2\}_{xyz} = [\mathbf{Q}]\{\hat{\mathbf{q}}_2\}_{XYZ} \quad \therefore \{\hat{\mathbf{q}}_2\}_{xyz} = 0.347127\hat{\mathbf{i}} - 0.925815\hat{\mathbf{j}} - 0.149563\hat{\mathbf{k}}$$

10. Calculate the time t_2 ($t_2 < 0$) at the patch point on the hyperbolic approach trajectory using Eqs. (9.41) and (9.42):

$$t_2 = -15.8112\text{h}$$

The elapsed time from the patch point to perilune is $\Delta t_2 = t_{perilune} - t_2 = 0 - t_2$. The total time from TLI to perilune is $\Delta t = \Delta t_1 + \Delta t_2$. If the spacecraft proceeds from perilune on a flyby, it will exit the SOI at the time $t_3 = -t_2 = 15.8112\text{h}$ by virtue of the symmetry of the flyby hyperbola, which is illustrated in Fig. 9.11.

III. *At any time t within the moon's SOI ($t_2 \le t \le t_3$), evaluate the spacecraft's geocentric equatorial position vector **r** and velocity vector **v**.* For example, let us choose $t = 0$ (perilune).

1. For the time *t*, use a lunar ephemeris to calculate the moon's position vector $\mathbf{r}_m$ and velocity vector $\mathbf{v}_m$ relative to the geocentric equatorial frame.

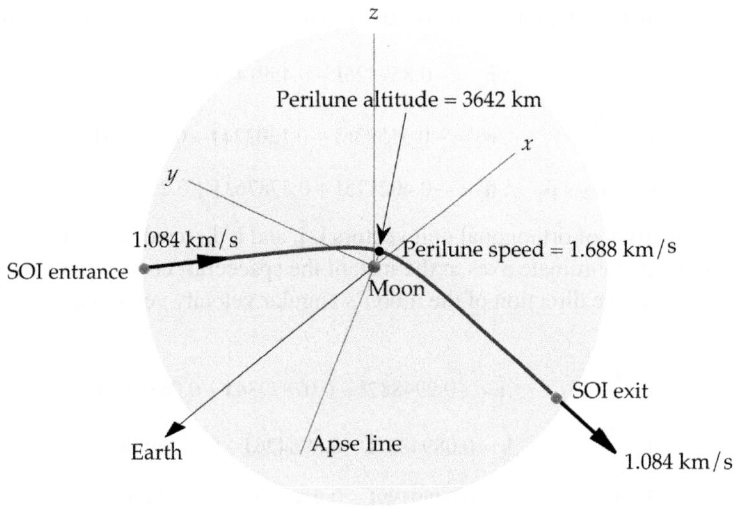

FIG. 9.11

Hyperbolic trajectory within the sphere of influence.

The perilune time $t = 0$ is 15.8112 h after passing the patch point, which means the Julian date at perilune is 2,458,974.66 and for that time the JPL Horizons ephemeris yields:

$$\mathbf{r}_m = -350{,}457\hat{\mathbf{I}} - 84{,}251.0\hat{\mathbf{J}} - 2235.54\hat{\mathbf{K}}\,(\text{km}) \qquad \therefore r_m = 360{,}448\,\text{km}$$
$$\mathbf{v}_m = 0.253698\hat{\mathbf{I}} - 0.963737\hat{\mathbf{J}} - 0.443114\hat{\mathbf{K}}\,(\text{km/s}) \qquad \therefore v_m = 1.09064\,\text{km/s}$$

2. Calculate the angular velocity of the moon, $\boldsymbol{\omega}_m = (\mathbf{r}_m \times \mathbf{v}_m)/{r_m}^2$, at the instant t:

$$\boldsymbol{\omega}_m = \left(0.270763\hat{\mathbf{I}} - 1.19963\hat{\mathbf{J}} + 2.76412\hat{\mathbf{K}}\right)\left(10^{-6}\right)(\text{rad/s})$$
$$\therefore \omega_m = 3.02535\left(10^{-6}\right)\text{rad/s}$$

3. Calculate the triad of orthogonal unit vectors $\hat{\mathbf{i}}$, $\hat{\mathbf{j}}$, and $\hat{\mathbf{k}}$ directed along the rotating *xyz* moon-fixed Cartesian coordinate axes at the instant the spacecraft is at perilune:

$$\hat{\mathbf{i}} = \mathbf{r}_m/r_m \quad \hat{\mathbf{i}} = -0.972280\hat{\mathbf{I}} - 0.233739\hat{\mathbf{J}} - 0.00620221\hat{\mathbf{K}}$$
$$\hat{\mathbf{k}} = \boldsymbol{\omega}_m/\omega_m \quad \hat{\mathbf{k}} = 0.0894979\hat{\mathbf{I}} - 0.396525\hat{\mathbf{J}} + 0.913651\hat{\mathbf{K}}$$
$$\hat{\mathbf{j}} = \hat{\mathbf{k}} \times \hat{\mathbf{i}} \quad \hat{\mathbf{j}} = 0.216015\hat{\mathbf{I}} - 0.887769\hat{\mathbf{J}} - 0.406453\hat{\mathbf{K}}$$

4. Use the geocentric equatorial components of $\hat{\mathbf{i}}$, $\hat{\mathbf{j}}$, and $\hat{\mathbf{k}}$ to calculate the instantaneous direction cosine matrix $[\mathbf{Q}]$ of the transformation from the geocentric equatorial *XYZ* frame to the moon-fixed *xyz* frame (see Eq. 4.23):

$$[\mathbf{Q}] = \begin{bmatrix} i_X & i_Y & i_Z \\ j_X & j_Y & j_Z \\ k_X & k_Y & k_Z \end{bmatrix} \quad \therefore [\mathbf{Q}] = \begin{bmatrix} -0.972280 & -0.233739 & -0.00620212 \\ 0.216015 & -0.887769 & -0.406453 \\ 0.0894979 & -0.396525 & 0.913651 \end{bmatrix}$$

5. Calculate the components of the perifocal unit vectors $\hat{\mathbf{p}}_2$ and $\hat{\mathbf{q}}_2$ (found in the *XYZ* frame in II.6) in the rotating *xyz* moon-fixed frame:

$$\{\hat{\mathbf{p}}_2\}_{xyz} = [\mathbf{Q}]\{\hat{\mathbf{p}}_2\}_{XYZ} \quad \therefore \{\hat{\mathbf{p}}_2\}_{xyz} = 0.905354\hat{\mathbf{i}} + 0.289223\hat{\mathbf{j}} + 0.310942\hat{\mathbf{k}}$$
$$\{\hat{\mathbf{q}}_2\}_{xyz} = [\mathbf{Q}]\{\hat{\mathbf{q}}_2\}_{XYZ} \quad \therefore \{\hat{\mathbf{q}}_2\}_{xyz} = 0.347127\hat{\mathbf{i}} - 0.925815\hat{\mathbf{j}} - 0.149563\hat{\mathbf{k}}$$

6. Determine the true anomaly θ at the time t, which is perilune in this example, so that $t = 0$:

$$M = \frac{\mu_m{}^2}{h_2{}^3}\left(e_2{}^2 - 1\right)^{3/2} t \qquad M = \frac{\mu_m{}^2}{h_2{}^3}\left(e_2{}^2 - 1\right)^{3/2}(0) \qquad \therefore M = 0$$
$$e_2 \sinh F - F = M \qquad e_2 \sinh F - F = 0 \qquad \therefore F = 0$$
$$\theta = 2\tan^{-1}\left(\sqrt{\frac{e_2+1}{e_2-1}}\tanh\frac{F}{2}\right) \quad \theta = 2\tan^{-1}\left(\sqrt{\frac{e_2+1}{e_2-1}}\tanh(0)\right) \quad \therefore \theta = 0$$

7. Calculate the components of the position vector $\mathbf{r}_{\text{rel}} = x\hat{\mathbf{i}} + y\hat{\mathbf{j}} + z\hat{\mathbf{k}}$ and the velocity vector $\mathbf{v}_{\text{rel}} = \dot{x}\hat{\mathbf{i}} + \dot{y}\hat{\mathbf{j}} + \dot{z}\hat{\mathbf{k}}$ of the spacecraft relative to the moon-fixed *xyz* frame:

$$\begin{aligned}
\{\mathbf{r}_{\text{rel}}\}_{xyz} &= \frac{h_2{}^2}{\mu_m}\frac{1}{1+e_2\cos\theta}\left(\cos\theta\{\hat{\mathbf{p}}_2\}_{xyz} + \sin\theta\{\hat{\mathbf{q}}_2\}_{xyz}\right) \\
&= \frac{9078.86^2}{4902.8}\frac{1}{1+2.12554\cos(0)}\left(\cos(0)\{\hat{\mathbf{p}}_2\}_{xyz} + \sin(0)\{\hat{\mathbf{q}}_2\}_{xyz}\right) \\
&= 5378.90\{\hat{\mathbf{p}}_2\}_{xyz}
\end{aligned}$$

$$\begin{aligned}
\therefore \mathbf{r}_{\text{rel}} &= 5378.90\left(0.905354\hat{\mathbf{i}} + 0.289223\hat{\mathbf{j}} + 0.310942\hat{\mathbf{k}}\right) \\
&= 4869.80\hat{\mathbf{i}} + 1555.70\hat{\mathbf{j}} + 1672.52\hat{\mathbf{k}}\ (\text{km})
\end{aligned}$$

$$\begin{aligned}
\{\mathbf{v}_{\text{rel}}\}_{xyz} &= -\frac{\mu_m}{h_2}\sin\theta\{\hat{\mathbf{p}}_2\}_{xyz} + \frac{\mu_m}{h_2}[e_2 + \cos\theta]\{\hat{\mathbf{q}}_2\}_{xyz} \\
&= -\frac{4902.8}{9078.86}\sin(0)\{\hat{\mathbf{p}}_2\}_{xyz} + \frac{4902.8}{9078.86}[2.12554 + \cos(0)]\{\hat{\mathbf{q}}_2\}_{xyz} \\
&= 1.68787\{\hat{\mathbf{q}}_2\}_{xyz}
\end{aligned}$$

$$\begin{aligned}
\therefore \mathbf{v}_{\text{rel}} &= 1.68786\left(0.347127\hat{\mathbf{i}} - 0.925815\hat{\mathbf{j}} - 0.149563\hat{\mathbf{k}}\right) \\
&= 0.585904\hat{\mathbf{i}} - 1.56265\hat{\mathbf{j}} - 0.252443\hat{\mathbf{k}}\ (\text{km/s})
\end{aligned}$$

8. Obtain the components of the relative position vector $\mathbf{r}_{\text{rel}}$ and the relative velocity $\mathbf{v}_{\text{rel}}$ in the geocentric equatorial *XYZ* frame:

$$\{\mathbf{r}_{\text{rel}}\}_{XYZ} = [\mathbf{Q}]^T\{\mathbf{r}_{\text{rel}}\}_{xyz} = \begin{bmatrix} -0.972280 & 0.2160154 & 0.0894979 \\ -0.233739 & -0.887769 & -0.396525 \\ -0.00620212 & -0.4064527 & 0.913651 \end{bmatrix}\begin{Bmatrix} 4869.81 \\ 1555.70 \\ 1672.53 \end{Bmatrix}$$

$$\therefore \mathbf{r}_{\text{rel}} = -4249.06\hat{\mathbf{I}} - 3182.56\hat{\mathbf{J}} + 865.578\hat{\mathbf{K}}\ (\text{km})$$

$$\{\mathbf{v}_{\text{rel}}\}_{XYZ} = [\mathbf{Q}]^T\{\mathbf{v}_{\text{rel}}\}_{xyz} = \begin{bmatrix} -0.972280 & 0.2160154 & 0.0894979 \\ -0.233739 & -0.887769 & -0.396525 \\ -0.00620212 & -0.4064527 & 0.913651 \end{bmatrix}\begin{Bmatrix} 0.585904 \\ -1.56265 \\ -0.252442 \end{Bmatrix}$$

$$\therefore \mathbf{v}_{\text{rel}} = -0.929814\hat{\mathbf{I}} + 1.35043\hat{\mathbf{J}} + 0.400866\hat{\mathbf{K}}\ (\text{km/s})$$

9. Calculate the absolute position vector **r** and the absolute velocity vector **v** of the spacecraft in the geocentric equatorial *XYZ* frame:

$$\begin{aligned}\mathbf{r} &= \mathbf{r}_m + \mathbf{r}_{\text{rel}} \\ &= \left(-350,457\hat{\mathbf{I}} - 84,251.0\hat{\mathbf{J}} - 2235.54\hat{\mathbf{K}}\right) + \left(-4249.06\hat{\mathbf{I}} - 3182.56\hat{\mathbf{J}} + 865.578\hat{\mathbf{K}}\right) \\ &= -354,706\hat{\mathbf{I}} - 87433.6\hat{\mathbf{J}} - 1369.97\hat{\mathbf{K}}\,(\text{km}) \\ r &= 365,325\,\text{km}\end{aligned}$$

For the velocity we must use Eq. (9.51) instead of (9.52), because the radius of the moon's orbit is not constant as we assumed previously (which means $\mathbf{v}_m \neq \boldsymbol{\omega}_m \times \mathbf{r}_m$):

$$\overbrace{\mathbf{v}}^{\substack{\text{absolute}\\\text{velocity of probe}}} = \overbrace{\mathbf{v}_m}^{\substack{\text{velocity of the}\\\text{moon}}} + \overbrace{\boldsymbol{\omega}_m}^{\substack{\text{angular velocity}\\\text{of the moon}}} \times \overbrace{\mathbf{r}_{\text{rel}}}^{\substack{\text{position vector of}\\\text{probe relative to the moon}}} + \overbrace{\mathbf{v}_{\text{rel}}}^{\substack{\text{velocity of the probe}\\\text{relative to the moon}}}$$

$\mathbf{v}_m$ and $\boldsymbol{\omega}_m$ are given in Steps III.1 and III.2, respectively, whereas $\mathbf{r}_{\text{rel}}$ and $\mathbf{v}_{\text{rel}}$ were calculated in Step III.8. Making these substitutions and noting that

$$\begin{aligned}\boldsymbol{\omega}_m \times \mathbf{r}_{\text{rel}} &= \left[\left(0.270763\hat{\mathbf{I}} - 1.19963\hat{\mathbf{J}} + 2.76412\hat{\mathbf{K}}\right)\left(10^{-6}\right)\right] \times \left(-4249.06\hat{\mathbf{I}} - 3182.56\hat{\mathbf{J}} + 865.578\hat{\mathbf{K}}\right) \\ &= 0.00775859\hat{\mathbf{I}} - 0.0119792\hat{\mathbf{J}} - 0.00595901\hat{\mathbf{K}}\,(\text{km/s})\end{aligned}$$

we get

$$\begin{aligned}\mathbf{v} = &\overbrace{\left(0.253698\hat{\mathbf{I}} - 0.963737\hat{\mathbf{J}} - 0.443114\hat{\mathbf{K}}\right)}^{\mathbf{v}_m} + \overbrace{\left(0.00775859\hat{\mathbf{I}} - 0.0119792\hat{\mathbf{J}} - 0.00595901\hat{\mathbf{K}}\right)}^{\boldsymbol{\omega}_m \times \mathbf{r}_{\text{rel}}} \\ &+ \overbrace{\left(-0.929814\hat{\mathbf{I}} + 1.35043\hat{\mathbf{J}} + 0.400866\hat{\mathbf{K}}\right)}^{\mathbf{v}_{\text{rel}}}\end{aligned}$$

or

$$\begin{aligned}\mathbf{v} &= -0.668357\hat{\mathbf{I}} + 0.374711\hat{\mathbf{J}} - 0.0482070\hat{\mathbf{K}}\,(\text{km/s}) \\ v &= 0.767746\,\text{km/s}\end{aligned}$$

Repeating the calculations of **r** in Steps III.1 through III.9 for a sufficient number of times between $t_2 = -\ 15.81$ h and $t_3 = +\ 15.81$ h and "connecting the dots" yields the three-dimensional trace of the position vector **r** within the moon's SOI, relative to the earth. This is the portion of the curve in Fig. 9.12 between "Enter SOI" and "Exit SOI".

IV. *Determine the spacecraft's geocentric orbit upon departing the SOI after a lunar flyby.*

1. Evaluate the geocentric equatorial state vector $(\mathbf{r}_3, \mathbf{v}_3)$ at t_3 (the SOI exit), and use Algorithm 4.2 to determine the geocentric orbital elements of the lunar departure trajectory.
2. If earth is the return target, determine the delta-v required for an acceptable perigee. In this particular case, an in-track delta-v of –0.1225 km/s is required upon exiting the SOI to establish the return trajectory shown in Fig. 9.12. Just after that maneuver the state vector is

$$\begin{aligned}\mathbf{r}_3 &= -357,478\hat{\mathbf{I}} - 77,874.4\hat{\mathbf{J}} - 16,825.7\hat{\mathbf{K}}\,(\text{km}) & r_3 &= 366,249\,\text{km} \\ \mathbf{v}_3 &= 0.0700313\hat{\mathbf{I}} + 0.0736792\hat{\mathbf{J}} - 0.1821388\hat{\mathbf{K}}\,(\text{km/s}) & v_3 &= 0.208585\,\text{km/s}\end{aligned}$$

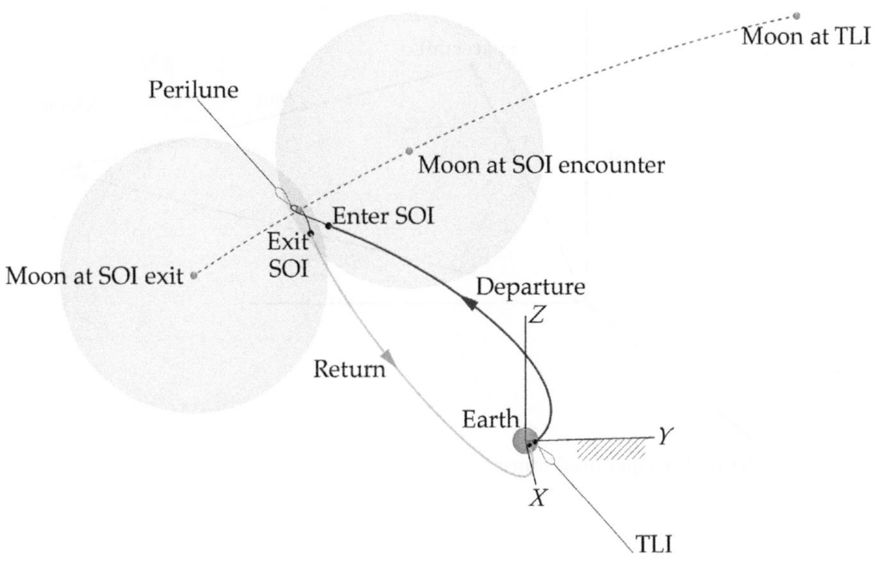

FIG. 9.12

Translunar injection followed by translunar coast, lunar flyaround, and transearth return.

and the orbital elements are

$$\theta_3 = 180.802° \quad e_3 = 0.965378 \quad h_3 = 71,192.1\,\text{km}^2/\text{s}$$
$$i_3 = 107.059° \quad \Omega_3 = 13.0981° \quad \omega_3 = 1.95243$$

Upon return to earth, the perigee altitude is 91.65 km, and its right ascension and declination are $\alpha_p = 12.52°$ and $\delta_p = 1.866°$, respectively.

9.5 LUNAR TRAJECTORIES BY NUMERICAL INTEGRATION

In this section we will put aside the notion of sphere of influence and assume that a spacecraft moving within the earth–moon environment is always attracted to both bodies. For simplicity, we ignore the gravity of the sun as well as that of all other members of the solar system. This leaves us with a three-body system, as illustrated in Fig. 9.13.

The equations of motion of each member of a three-body system are given in Appendix C. As shown in Fig. 9.13, we choose the geocentric equatorial frame as our inertial reference and denote the position vector of the spacecraft by **r**. According to Section 10.10, it follows from Eq. (C.2) that the equation of motion of the spacecraft relative to the earth is Eq. (2.22), with a term **p** added to account for the acceleration due to lunar gravity. Thus,

$$\ddot{\mathbf{r}} = -\mu_e \frac{\mathbf{r}}{r^3} + \mathbf{p} \tag{9.57}$$

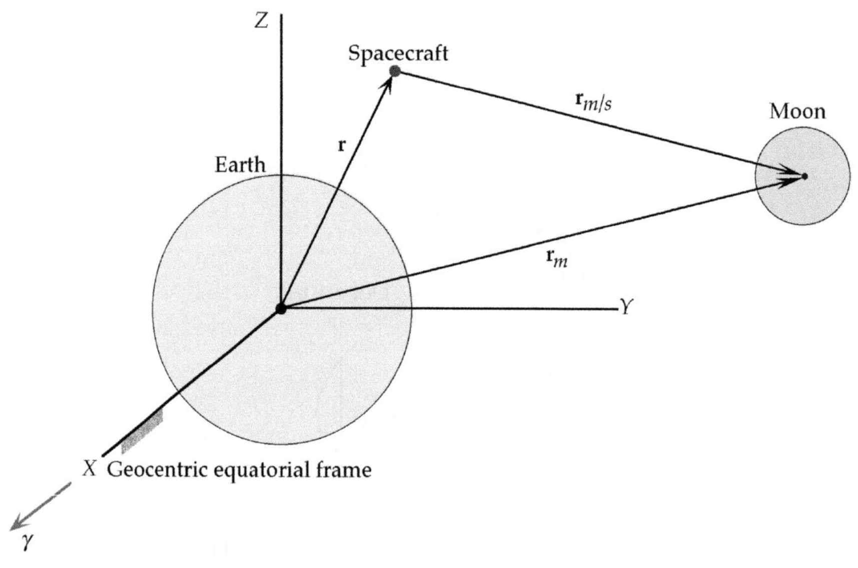

FIG. 9.13

Three-body system of earth, moon, and spacecraft.

where μ_e is earth's gravitational parameter. The moon's gravitational parameter μ_m appears in the term

$$\mathbf{p} = \mu_m \left(\frac{\mathbf{r}_{m/s}}{r_{m/s}{}^3} - \frac{\mathbf{r}_m}{r_m{}^3} \right) \tag{9.58}$$

As shown in Fig. 9.13, $\mathbf{r}_m$ is the position vector of the moon relative to the earth. $\mathbf{r}_{m/s}$ is the position vector of the moon relative to the spacecraft, so that

$$\mathbf{r}_{m/s} = \mathbf{r}_m - \mathbf{r} \tag{9.59}$$

$\mathbf{r}_m$ is obtained from an accurate lunar ephemeris, as discussed in Section 9.3. Therefore, unlike the circular restricted three-body problem in Section 2.12, we assume neither that the moon's orbit around the earth is circular nor that it is coplanar with the spacecraft's.

To numerically integrate Eq. (9.57), we follow the procedure introduced in Section 1.8 and rewrite the second-order differential equation as a system of first-order differential equations. To that end, let

$$\mathbf{y}_1 = \mathbf{r} \tag{9.60a}$$

$$\mathbf{y}_2 = \dot{\mathbf{r}} \tag{9.60b}$$

$\mathbf{y}_1$ and $\mathbf{y}_2$, the position and velocity vectors, form the state vector of the spacecraft,

$$\mathbf{y} = \begin{Bmatrix} \mathbf{y}_1 \\ \mathbf{y}_2 \end{Bmatrix} \tag{9.61}$$

It follows from Eqs. (9.60a) and (9.60b) that

$$\dot{\mathbf{y}}_1 = \mathbf{y}_2$$

Differentiating Eq. (9.60b) and substituting Eq. (9.57) along with $\mathbf{r} = \mathbf{y}_1$, we find

$$\dot{\mathbf{y}}_2 = -\mu_e \frac{\mathbf{r}}{r^3} + \mathbf{p} = -\mu_e \frac{\mathbf{y}_1}{\|\mathbf{y}_1\|^3} + \mathbf{p}$$

Using Eqs. (9.58), (9.59), and (9.60a), the expression for the lunar gravitational acceleration **p** becomes

$$\mathbf{p} = \mu_m \left(\frac{\mathbf{r}_m(t) - \mathbf{y}_1}{\|\mathbf{r}_m(t) - \mathbf{y}_1\|^3} - \frac{\mathbf{r}_m(t)}{r_m(t)^3} \right)$$

In summary, the two first-order, nonlinear differential equations of motion are

$$\dot{\mathbf{y}}_1 = \mathbf{y}_2$$
$$\dot{\mathbf{y}}_2 = -\mu_e \frac{\mathbf{y}_1}{\|\mathbf{y}_1\|^3} + \mu_m \left(\frac{\mathbf{r}_m(t) - \mathbf{y}_1}{\|\mathbf{r}_m(t) - \mathbf{y}_1\|^3} - \frac{\mathbf{r}_m(t)}{r_m(t)^3} \right)$$

This system of equations may be written more compactly in the standard form

$$\dot{\mathbf{y}} = \mathbf{f}(t, \mathbf{y}) \tag{9.62}$$

where the components of the state vector **y** are found in Eq. (9.61) and the rate functions **f** are

$$\mathbf{f}(t, \mathbf{y}) = \left\{ \begin{array}{c} \mathbf{y}_2 \\ -\mu_e \dfrac{\mathbf{y}_1}{\|\mathbf{y}_1\|^3} + \mu_m \left(\dfrac{\mathbf{r}_m(t) - \mathbf{y}_1}{\|\mathbf{r}_m(t) - \mathbf{y}_1\|^3} - \dfrac{\mathbf{r}_m(t)}{r_m(t)^3} \right) \end{array} \right\}$$

Keep in mind that the lunar position vector $\mathbf{r}_m(t)$, obtained from an ephemeris, is a *known* function of time.

Any of the well-known numerical integrators designed to solve Eq. (9.62), such as those described in Section 1.8, can be brought to bear upon the problem of determining a lunar trajectory from the initial conditions in low earth orbit.

EXAMPLE 9.3

A spacecraft in low earth orbit is launched on a ballistic lunar trajectory so as to arrive at the moon on May 4, 2020, at 12:00 UT after a 3-day flight. The conditions at TLI are (see Fig. 4.5)

Altitude:	$z = 320$ km
Right ascension:	$\alpha = 90°$
Declination:	$\delta = 15°$
Flight path angle:	$\gamma = 40°$
Speed:	$v = 10.8267$ km/s

If the objective is to simply fly around the moon, determine the perilune altitude, and find the location of the spacecraft 5.67 days after perilune passage.

Solution

From Eqs. (4.4) and (4.5) the geocentric equatorial position vector of the spacecraft at TLI is

$$\begin{aligned}\mathbf{r}_0 &= r_0\left(\cos\delta\cos\alpha\hat{\mathbf{I}} + \cos\delta\sin\alpha\hat{\mathbf{J}} + \sin\delta\hat{\mathbf{K}}\right) \\ &= (6378 + 320)\left(\cos 15°\cos 90°\hat{\mathbf{I}} + \cos 15°\sin 90°\hat{\mathbf{J}} + \sin 15°\hat{\mathbf{K}}\right) \\ &= 6469.77\hat{\mathbf{J}} + 1733.57\hat{\mathbf{K}}\ (\text{km})\end{aligned} \tag{a}$$

Substituting $y = 2020$, $m = 5$, $d = 4$, and $UT = 12$ into Eqs. (5.47) and (5.48) yields the Julian day of the arrival time,

$$JD = 2{,}458{,}974\ \text{days}$$

This together with Eqs. (9.54) and (9.55) determines the moon's geocentric equatorial position vector $\mathbf{r}_m$ three days after TLI,

$$\mathbf{r}_m = -358{,}887\hat{\mathbf{I}} - 32{,}072.3\hat{\mathbf{J}} + 18{,}358.9\hat{\mathbf{K}}\,(\text{km})$$
$$r_m = 360{,}785\,\text{km}$$

Both of the position vectors $\mathbf{r}_0$ and $\mathbf{r}_m$ define the initial plane of the translunar trajectory, which means that the unit normal $\hat{\mathbf{w}}$ to that orbital plane may be found by normalizing the cross product of $\mathbf{r}_0$ into $\mathbf{r}_m$,

$$\hat{\mathbf{w}} = \frac{\mathbf{r}_0 \times \mathbf{r}_m}{\|\mathbf{r}_0 \times \mathbf{r}_m\|} = 0.0723516\hat{\mathbf{I}} - 0.258141\hat{\mathbf{J}} + 0.963394\hat{\mathbf{K}}$$

The unit vector $\hat{\mathbf{u}}_r$ in the direction of $\mathbf{r}_0$ is

$$\hat{\mathbf{u}}_r = \frac{\mathbf{r}_0}{r_0} = 0.965926\hat{\mathbf{J}} + 0.248819\hat{\mathbf{K}}$$

Let $\hat{\mathbf{u}}_\perp$ be the unit vector that is normal to both of the orthogonal vectors $\hat{\mathbf{u}}_r$ and $\hat{\mathbf{w}}$ and therefore lies in the plane of the translunar trajectory. $\hat{\mathbf{u}}_\perp$ is simply the cross product of the unit vector $\hat{\mathbf{w}}$ into the unit vector $\hat{\mathbf{u}}_r$,

$$\hat{\mathbf{u}}_\perp = \hat{\mathbf{w}} \times \hat{\mathbf{u}}_r = -0.997379\hat{\mathbf{I}} - 0.0187260\hat{\mathbf{J}} + 0.0698863\hat{\mathbf{K}}$$

The radial and transverse components of the TLI velocity $\mathbf{v}_0$ are found from the initial speed v_0 and the flight path angle γ

$$v_{r_0} = v_0 \sin\gamma \qquad v_{\perp_0} = v_0 \cos\gamma$$

Therefore, the velocity vector $\mathbf{v}_0$ may be written

$$\begin{aligned}\mathbf{v}_0 &= v_0 \sin\gamma \hat{\mathbf{u}}_r + v_0 \cos\gamma \hat{\mathbf{u}}_\perp \\ &= 10.8267 \sin 40°\left(0.965926\hat{\mathbf{J}} + 0.248819\hat{\mathbf{K}}\right) \\ &\quad + 10.8267 \cos 40°\left(-0.997379\hat{\mathbf{I}} - 0.0187260\hat{\mathbf{J}} + 0.0698863\hat{\mathbf{K}}\right)\end{aligned}$$

or

$$\mathbf{v}_0 = -8.27203\hat{\mathbf{I}} + 6.56685\hat{\mathbf{J}} + 2.38082\hat{\mathbf{K}}\,(\text{km/s}) \tag{b}$$

The initial value of the state vector $\mathbf{y}$ in Eq. (9.61) comprises $\mathbf{r}_0$ and $\mathbf{v}_0$, as given here in Eqs. (a) and (b).

Starting with the state vector $\mathbf{y}_0$ at time t_0 and using Simpson's ephemeris for the moon (Section 9.3), we numerically integrate Eq. (9.62) to obtain the values $\mathbf{y}_i$ of the state vector at n discrete times t_i between t_0 and the final time t_f. Using MATLAB's *ode45* with $t_0 = 0$ and $t_f = 5.667$ days, and with both the relative and absolute tolerances set to 10^{-10}, we obtain the free return trajectory shown in Fig. 9.14.

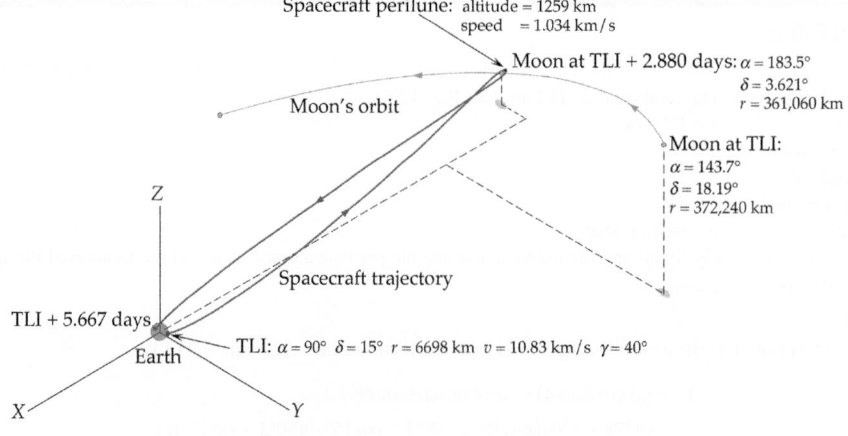

FIG. 9.14

Ballistic lunar flyaround trajectory obtained from numerical integration of the restricted three-body equations of motion.

It may be convenient to view the motion from a noninertial xyz frame with origin at the center of the earth, but having an x axis that always points to the moon. The x axis is therefore defined by the moon's instantaneous position vector $\mathbf{r}_m$. The z axis lies in the direction of the normal to the moon's orbital plane, which is defined by the cross product $\mathbf{r}_m \times \mathbf{v}_m$, where $\mathbf{v}_m$ is the moon's instantaneous velocity. The y axis is normal to the x and the z axes, according to the right-hand rule. The instantaneous unit vectors $\hat{\mathbf{i}}$, $\hat{\mathbf{j}}$, and $\hat{\mathbf{k}}$ of the rotating frame are therefore obtained from $\mathbf{r}_m$ and $\mathbf{v}_m$ as follows:

$$\hat{\mathbf{i}} = \frac{\mathbf{r}_m}{r_m} \quad \hat{\mathbf{k}} = \frac{\mathbf{r}_m \times \mathbf{v}_m}{\|\mathbf{r}_m \times \mathbf{v}_m\|} \quad \hat{\mathbf{j}} = \hat{\mathbf{k}} \times \hat{\mathbf{i}} \tag{c}$$

As we know from Chapter 4, the direction cosine matrix $[\mathbf{Q}]_{Xx}$ of the transformation from the geocentric equatorial XYZ frame to the rotating moon-fixed xyz frame is

$$[\mathbf{Q}]_{Xx} = \begin{bmatrix} i_X & i_Y & i_Z \\ j_X & j_Y & j_Z \\ k_X & k_Y & k_Z \end{bmatrix}$$

where the rows of this matrix comprise the components of each rotating unit vector in Eq. (c) along the axes of the inertial frame. Thus, if the coordinates of a point in the geocentric inertial frame are $\lfloor X \ Y \ Z \rfloor$, then the coordinates of that same point in the moon-fixed frame are $\lfloor x \ y \ z \rfloor$, where

$$\begin{Bmatrix} x \\ y \\ z \end{Bmatrix} = [\mathbf{Q}]_{Xx} \begin{Bmatrix} X \\ Y \\ Z \end{Bmatrix}$$

Applying this transformation to each point of the spacecraft and moon trajectories in Fig. 9.14 yields the trajectory shown in Fig. 9.15, in which the moon simply oscillates between the points on the x axis defined by the perigee and apogee of its noncircular orbit.

The MATLAB listing of the code for this example is found in Appendix D.38.

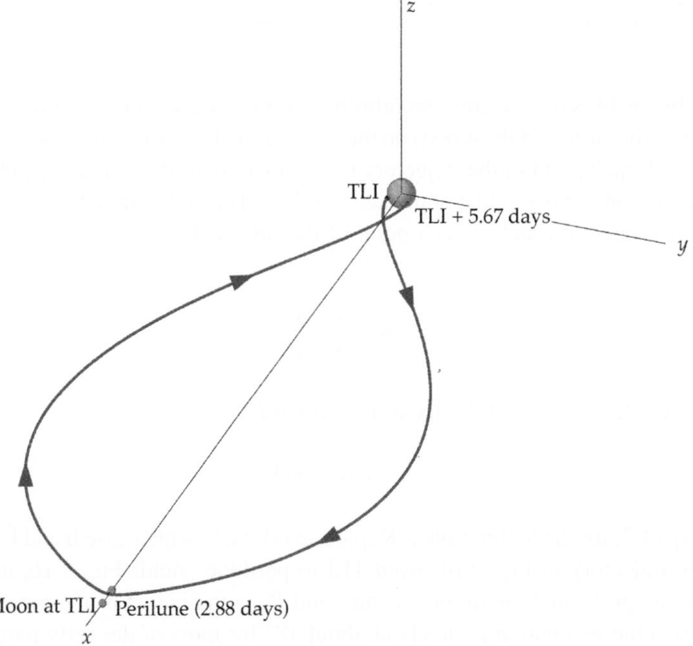

FIG. 9.15

The lunar trajectory of Fig. 9.14 viewed relative to the rotating earth-centered frame whose x axis is the earth-moon line.

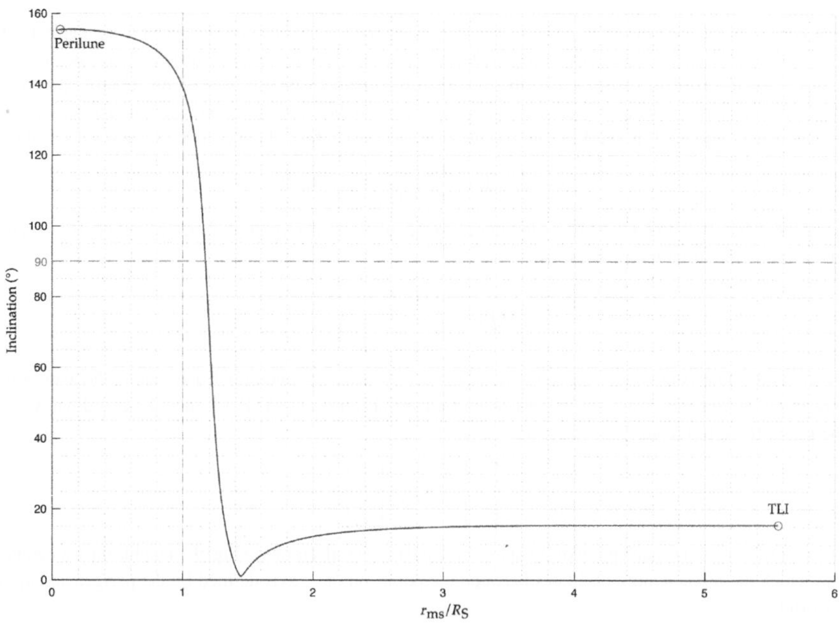

FIG. 9.16

Variation of inclination with distance from the moon for the orbit of Fig. 9.14.

The curve in Fig. 9.14 is not a Keplerian orbit because it results from the gravitational force of not one, but two bodies (the earth and the moon) on the spacecraft. Therefore, the trajectory does not lie in a single plane. Instead, each point of the trajectory has its own *osculating* plane, which is defined by the velocity and acceleration vectors of that point (see Fig. 1.9). The unit normal $\hat{\mathbf{b}}$ to the osculating plane is called the *binormal*. The binormal of each point of the orbit is found by means of the cross product operation,

$$\hat{\mathbf{b}} = \frac{\mathbf{v} \times \mathbf{a}}{\|\mathbf{v} \times \mathbf{a}\|} \tag{9.63}$$

Knowing the binormal, we can find the local inclination i,

$$i = \cos^{-1}\left(\hat{b}_Z\right) \tag{9.64}$$

just as we used Eq. (4.7) for the inclination of Keplerian orbits, in which case $\hat{\mathbf{h}}$ and $\hat{\mathbf{b}}$ coincide. Plotting Eq. (9.64) for the trajectory of Fig. 9.14, from TLI to perilune, yields Fig. 9.16, in which r_{ms} is the distance between the probe and the moon's center and R_S is the radius of the moon's SOI (Eq. 9.9). Fig. 9.16 shows that the inclination is steady at about 18° for most of the early part of the trajectory, so that the orbit is prograde relative to the earth. However, as the probe nears SOI, the inclination increases dramatically, soon exceeding 90°, at which point the orbit becomes retrograde, with the probe beginning to curve to the right, eventually passing behind the moon to reach perilune.

PROBLEMS

Section 9.2

9.1 A spacecraft in a circular 160-km earth orbit, coplanar with that of the moon, is launched at perigee onto an elliptical trajectory whose apogee lies on the moon's SOI. Calculate the Δv required to boost the spacecraft onto this trajectory.
{Ans.: 3.122 km/s}

9.2 If $\delta = 5°$ at the patch point, determine the eccentricity e and the relative speed v_2 required for the perilune altitude to be 100 km.
{Ans.: $e = 1.164$; $v_2 = 0.7653$ km/s}

9.3 At the patch point, $v_2 = 0.7$ km/s, $\lambda = 30°$, and $\delta = -15°$. Calculate the angle λ_p between the earth-moon line and perilune.
{Ans.: 101.4° clockwise}

9.4 Perilune of a lunar approach trajectory is 250 km. If $v_2 = 0.5$ km/s, calculate the required value of δ and the Δv at perilune required to place the probe in a circular prograde lunar orbit.
{Ans.: $\delta = -7.745°$, $\Delta v = -673$ m/s}

9.5 A spacecraft is launched from a circular earth orbit of 300 km altitude onto a Hohmann transfer trajectory to Mars.

(a) What is the burnout speed relative to the earth?
(b) How long does it take the probe to reach the moon's orbit on its way to Mars?
(c) What is the speed of the probe relative to the earth when it crosses the moon's orbit?
(d) Through what angle does the moon move in the time it takes for the probe to coast to lunar orbit?

{Ans.: (a) 11.04 km/s; (b) 40.51 h; (c) 2.15 km/s; (d) 22.1°}

9.6 A lunar probe in a prograde translunar trajectory arrives at the moon's SOI with $\lambda = 0°$. Its speed and flight path angle relative to the earth are 0.7 km/s and 45°, respectively.

(a) Find the perilune radius.
(b) How long does the probe remain within the SOI?

{Ans.: (a) 45,177 km; (b) 39 h}

9.7 In Fig. 9.2, the altitude of TLI is 320 km, its right ascension to the earth-moon line is $\alpha_0 = 37°$, and the flight path angle is $\gamma_0 = 10°$. If $\lambda = 45°$, use the patched conic method to find the perilune altitude for a ballistic, coplanar translunar trajectory. Assume the moon's orbit is circular.
{Ans.: 202.3 km}

9.8 In Fig. 9.2, the following are given: TLI altitude = 185 km, $\alpha_0 = 20°$, $\gamma_0 = 17.18°$, and $\lambda = -60°$. Use the patched conic method to find the perilune altitude for a ballistic, coplanar translunar trajectory, assuming the moon's orbit is circular.
{Ans.: 491.2 km}

Section 9.3

9.9 Use Simpson's lunar ephemeris to find, for November 2034:

(a) The day and the UT of the moon's perigee.
(b) The perigee's distance.
(c) The perigee's right ascension and declination.

{Partial Ans.: (b) 357,400 km}

9.10 Find the radial speed of the moon on April 30, 2025, at 06:00:00 UT.
{Ans.: 56.7 m/s}

9.11 Verify the entries in the last column of Table 9.1.

Section 9.4

9.12 A lunar probe is launched on a ballistic lunar flyaround trajectory from an earth altitude of 180 km with a flight path angle of $\gamma_0 = 13°$. TLI is located at right ascension $\alpha_0 = 42°$ and declination $\delta_0 = 9°$. Upon reaching the moon's SOI the phase angle between earth and the probe is $\lambda = 47°$, and the moon's geocentric state vector is

$$\mathbf{r}_{\text{moon}} = -387{,}639\hat{\mathbf{I}} - 4443.51\hat{\mathbf{J}} + 11{,}750.5\hat{\mathbf{K}}\,(\text{km})$$
$$\mathbf{v}_{\text{moon}} = -0.0603414\hat{\mathbf{I}} - 0.955154\hat{\mathbf{J}} - 0.321928\hat{\mathbf{K}}\,(\text{km/s})$$

(a) Show that the path around the moon is retrograde.
(b) Find the perilune altitude .
(c) Determine the flight time from TLI to perilune.
{Ans.: (b) 71.2 km; (c) 3.20 days}

9.13 At TLI the geocentric state vector of a ballistic lunar probe is

$$\mathbf{r}_0 = 3340.59\hat{\mathbf{I}} + 5346.06\hat{\mathbf{J}} + 1807.63\hat{\mathbf{K}}\,(\text{km})$$
$$\mathbf{v}_0 = -6.97237\hat{\mathbf{I}} + 7.92687\hat{\mathbf{J}} + 3.09093\hat{\mathbf{K}}\,(\text{km/s})$$

From TLI to the moon's SOI requires 44.101 h. If the moon's state vector at SOI encounter is the same as in Problem 9.11, calculate
(a) The lunar arrival angle λ.
(b) The additional time required to reach perilune.
(c) The perilune altitude.
{Ans.: (a) 31°; (b) 12.14 h; (c) 73.5 km}

9.14 At TLI the geocentric state vector of a ballistic lunar probe is

$$\mathbf{r}_0 = -5716.11\hat{\mathbf{I}} - 3168.49\hat{\mathbf{J}} - 571.785\hat{\mathbf{K}}\,(\text{km})$$
$$\mathbf{v}_0 = 3.95652\hat{\mathbf{I}} - 9.76084\hat{\mathbf{J}} - 2.91477\hat{\mathbf{K}}\,(\text{km/s})$$

It arrives at the lunar SOI with a phase angle of $\lambda = 61°$ when the moon's state vector is

$$\mathbf{r}_{\text{moon}} = 359{,}880\hat{\mathbf{I}} - 9215.13\hat{\mathbf{J}} - 21{,}798.5\hat{\mathbf{K}}\,(\text{km})$$
$$\mathbf{v}_{\text{moon}} = 0.068456\hat{\mathbf{I}} + 1.03141\hat{\mathbf{J}} + 0.35269\hat{\mathbf{K}}\,(\text{km/s})$$

Calculate the perilune altitude.
{Ans.: 100.4 km}

Section 9.5

9.15 A lunar probe is launched on a ballistic trajectory to reach the moon on May 4, 2020, at 12:00 UT after a 3-day flight from TLI to perilune. The conditions at TLI are:
Altitude: z = 180 km
Right ascension: α = 70°
Declination: δ = 20°

Flight path angle: $\gamma = 30°$
Speed$v$ = 10.9395 km/s
Determine the perilune altitude and show that the path around the moon is retrograde.
{Ans.: 205 km}

9.16 A lunar probe is launched on a ballistic trajectory to reach the moon on June 13, 2035, at 12:00 UT after a 3.3-day flight from TLI to perilune. The conditions at TLI are:
Altitude: z = 180 km
Right ascension: $\alpha = 65°$
Declination: $\delta = 25°$
Flight path angle: $\gamma = 30°$
Speed v = 10.9472 km/s
Determine the perilune altitude and show that the path around the moon is retrograde.
{Ans.: 174 km. Trajectory becomes retrograde at 2.4 days after TLI}

9.17 Show that Eq. (9.63) becomes $\hat{\mathbf{b}} = \mathbf{h}/h$ for a Keplerian orbit.

REFERENCES

Bate, R.B., Mueller, D.D., White, J.E., 1971. Fundamentals of Astrodynamics. Dover, New York.

Brown, C.D., 1998. Spacecraft Mission Design, second ed. AIAA Education Series.

Chobotov, V.V. (Ed.), 1996. Orbital Mechanics. second ed. AIAA Education Series.

JPL Horizons Web-Interface, 2018. https://ssd.jpl.nasa.gov/horizons.cgi. (Accessed 29 June 2018).

Orloff, R.W., 2000. Apollo by the Numbers: A Statistical Reference, NASA SP-2000-4029.

Simpson, D.G., 1999. An Alternative Lunar Ephemeris Model for On-Board Flight Software Use, Proceedings of the 1999 NASA/GSFC Flight Mechanics Symposium, pp. 175–184.

CHAPTER

INTRODUCTION TO ORBITAL PERTURBATIONS 10

10.1 INTRODUCTION

Keplerian orbits are the closed-form solutions of the two-body equation of relative motion (Eq. 2.22),

$$\ddot{\mathbf{r}} = -\mu \frac{\mathbf{r}}{r^3} \tag{10.1}$$

This equation is based on the assumption that there are only two objects in space, and that their spherically symmetric gravitational fields are the only source of interaction between them. Any effect that causes the motion to deviate from a Keplerian trajectory is known as a perturbation. Common perturbations of two-body motion include a nonspherical central body, atmospheric drag, propulsive thrust, solar radiation pressure, and gravitational interactions with celestial objects like the moon and the sun. To account for perturbations, we add a term $\mathbf{p}$ to the right-hand side of Eq. (10.1) to get

$$\ddot{\mathbf{r}} = -\mu \frac{\mathbf{r}}{r^3} + \mathbf{p} \tag{10.2}$$

The vector $\mathbf{p}$ is the net perturbative acceleration from all sources other than the spherically symmetric gravitational attraction between the two bodies. The magnitude of $\mathbf{p}$ is usually small compared with the primary gravitational acceleration $a_0 = \mu/r^2$. An exception is atmospheric drag which, at an altitude of about 100 km, is large enough to deorbit a satellite. The drag effect decreases rapidly with altitude and becomes negligible ($p_{\text{drag}} < 10^{-10}a_0$) above 1000 km. The other effects depend on the altitude to various extents or, in the case of solar radiation pressure, not at all. At 1000 km altitude, their disturbing accelerations in decreasing order are (Fortescue et al., 2011; Montenbruck and Eberhard, 2000)

$$\begin{aligned} &p_{\text{earth's oblateness}} && \approx 10^{-2}a_0 \\ &p_{\text{lunar gravity}} \approx p_{\text{solar gravity}} && \approx 10^{-7}a_0 \\ &p_{\text{solar radiation}} && \approx 10^{-9}a_0 \end{aligned} \tag{10.3}$$

Starting with a set of initial conditions $(\mathbf{r}_0, \mathbf{v}_0)$ and the functional form of the perturbation $\mathbf{p}$, we can numerically integrate Eq. (10.2) to find the position $\mathbf{r}$ and velocity $\mathbf{v}$ at any time thereafter. The classical orbital elements at any instant are then furnished by Algorithm 4.2. Conversely, we can numerically integrate what are known as Lagrange's planetary equations to obtain the orbital elements instead of the state vector as functions of time. From the orbital elements, we obtain the state vector $(\mathbf{r}, \mathbf{v})$ at any instant by using Algorithm 4.5.

Orbital Mechanics for Engineering Students. https://doi.org/10.1016/B978-0-12-824025-0.00010-6

We start this chapter with a look at the concept of osculating orbits and the two classical techniques for numerically integrating Eq. (10.2) namely, Cowell's method and Encke's method. These methods are then used to carry out special perturbations analyses of the effects of atmospheric drag and the earth's oblateness. Next, we discuss the method of variation of parameters, which is familiar to all students of a first course in differential equations. The method is applied to the solution of Eq. (10.2) to obtain the conditions that are imposed on the osculating elements of a perturbed trajectory. These conditions lead to Lagrange's planetary equations, a set of differential equations, that govern the time variation of the osculating orbital elements. A variant of Lagrange's planetary equations, Gauss' variational equations, will be derived in detail and used for our special perturbations analyses in the rest of this chapter. We will employ Gauss' variational equations to obtain the analytical expressions for the rates of change of the osculating elements. For geopotential perturbations, the method of averaging will be applied to these equations to smooth out the short-period variations, leaving us with simplified formulas for only long-term secular variations. The chapter concludes with applications of Gauss' variational equations to the special perturbations analysis of the effects of solar radiation pressure, lunar gravity, and solar gravity.

10.2 COWELL'S METHOD

Philip H. Cowell (1870–1949) was a British astronomer whose name is attached to the method of direct numerical integration of Eq. (10.2). Cowell's work at the turn of the 20th century (e.g., predicting the time of the closest approach to the sun of Halley's Comet upon its return in 1910) relied entirely on hand calculations to numerically integrate the equations of motion using classical methods dating from Isaac Newton's time. Today, of course, we have high-speed digital computers on which modern, extremely accurate integration algorithms can be easily implemented. A few of these methods were presented in Section 1.8.

We used Cowell's method in Chapter 2 to integrate the three-body equations (derived in Appendix C) for the particular scenario depicted in Figs. 2.4 and 2.5. In Section 6.10, we set $\mathbf{p} = (T/m)(\mathbf{v}/v)$ in Eq. (10.2) to simulate tangential thrust, and then we numerically integrated the ordinary differential equations to obtain the results in Example 6.15 for a high-thrust situation and in Example 6.17 for a low-thrust application.

Upon solving Eq. (10.2) for the state vector of the perturbed path at any time t, the orbital elements may be found by means of Algorithm 4.2. However, these orbital elements describe the *osculating orbit*, not the perturbed orbit. The osculating orbit is the two-body trajectory that would be followed after time t if at that instant the perturbing acceleration $\mathbf{p}$ were to suddenly vanish, thereby making Eq. (10.1) valid. Since the state vectors of both the perturbed orbit and the osculating orbit are identical at time t, the two orbits touch and are tangential at that point. (It is interesting to note that *osculate* has its roots in the Latin word for *kiss*.) Every point of a perturbed trajectory has its own osculating trajectory.

We illustrate the concept of osculating orbits in Fig. 10.1, which shows the earth orbit of a spacecraft containing an onboard rocket engine that exerts a constant tangential thrust T starting at O. The continuous addition of energy causes the trajectory to spiral outward away from the earth. The thrust is the perturbation that forces the orbit to deviate from a Keplerian (elliptical) path. If at time t_1 the engine

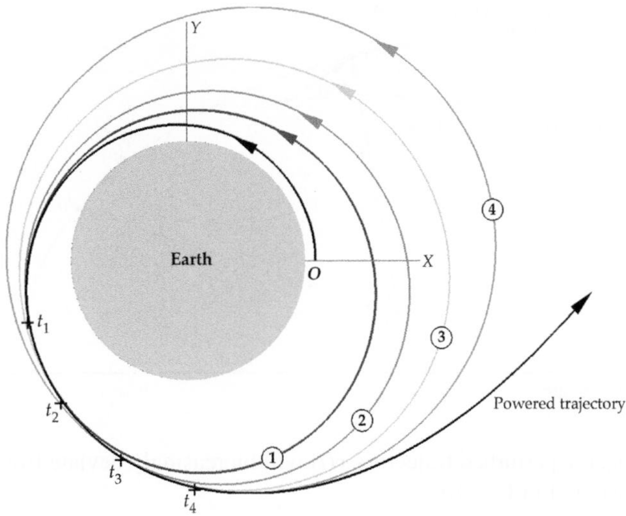

FIG 10.1

Osculating orbits 1 through 4 corresponding to times t_1 through t_4, respectively, on the powered, spiral trajectory that starts at point *O*. The *circled labels* are centered at each orbit's apogee.

were to shut down, then the spacecraft would enter the elliptical osculating orbit 1 shown in the figure. Also shown are the osculating orbits 2, 3, and 4 at times t_2, t_3, and t_4. Observe that the thruster not only increases the semimajor axis but also causes the apse line to rotate counterclockwise, in the direction of the orbital motion.

10.3 ENCKE'S METHOD

In the method developed originally by the German astronomer Johann Franz Encke (1791–1865), the two-body motion due solely to the primary attractor is treated separately from that due to the perturbation. The two-body osculating orbit $\mathbf{r}_{osc}(t)$ is used as a reference orbit upon which the unknown deviation $\delta\mathbf{r}(t)$ due to the perturbation is superimposed to obtain the perturbed orbit $\mathbf{r}(t)$.

Let $(\mathbf{r}_0, \mathbf{v}_0)$ be the state vector of an orbiting object at time t_0. The osculating orbit at that time is governed by Eq. (10.1),

$$\ddot{\mathbf{r}}_{osc} = -\mu\frac{\mathbf{r}_{osc}}{r^3} \tag{10.4}$$

with the initial conditions $\mathbf{r}_{osc}(t_0) = \mathbf{r}_0$ and $\mathbf{v}_{osc}(t_0) = \mathbf{v}_0$. For times $t > t_0$, the state vector $(\mathbf{r}_{osc}, \mathbf{v}_{osc})$ of the osculating, two-body trajectory may be found analytically using Lagrange's coefficients (Eqs. 3.67 and 3.68),

$$\begin{aligned}\mathbf{r}_{osc}(t) &= f(t)\mathbf{r}_0 + g(t)\mathbf{v}_0\\ \mathbf{v}_{osc}(t) &= \dot{f}(t)\mathbf{r}_0 + \dot{g}(t)\mathbf{v}_0\end{aligned} \tag{10.5}$$

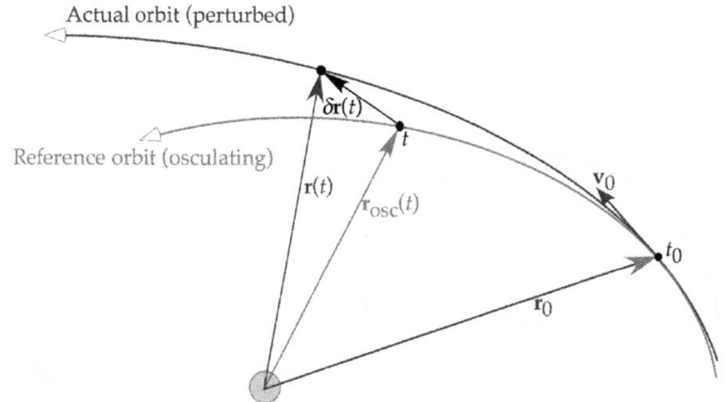

FIG 10.2

Perturbed and osculating orbits.

After the initial time t_0, the perturbed trajectory $\mathbf{r}(t)$ will increasingly deviate from the osculating path $\mathbf{r}_{osc}(t)$, so that, as illustrated in Fig. 10.2,

$$\mathbf{r}(t) = \mathbf{r}_{osc}(t) + \delta\mathbf{r}(t) \tag{10.6}$$

Substituting $\mathbf{r}_{osc} = \mathbf{r} - \delta\mathbf{r}$ into Eq. (10.4) and setting $\delta\mathbf{a} = \delta\ddot{\mathbf{r}}$ yields

$$\delta\mathbf{a} = \ddot{\mathbf{r}} + \mu\frac{\mathbf{r} - \delta\mathbf{r}}{r^3}$$

We may then substitute Eq. (10.2) into this expression to get

$$\delta\mathbf{a} = -\mu\frac{\mathbf{r}}{r^3} + \mu\frac{\mathbf{r} - \delta\mathbf{r}}{r_{osc}{}^3} + \mathbf{p} = -\frac{\mu}{r_{osc}{}^3}\delta\mathbf{r} + \mu\left(\frac{1}{r_{osc}{}^3} - \frac{1}{r^3}\right)\mathbf{r} + \mathbf{p}$$

A final rearrangement of the terms leads to

$$\delta\mathbf{a} = -\frac{\mu}{r_{osc}{}^3}\left[\delta\mathbf{r} - \left(1 - \frac{r_{osc}{}^3}{r^3}\right)\mathbf{r}\right] + \mathbf{p} \tag{10.7}$$

As is evident from Fig. 10.3, r_{osc} and r may become very nearly equal, in which case accurately calculating the difference $F = 1 - (r_{osc}/r)^3$ can be problematic for a digital computer. In that case, we refer to Appendix F to rewrite Eq. (10.7) as

$$\delta\mathbf{a} = -\mu[\delta\mathbf{r} - F(q)]/r_{osc}^3 + \mathbf{p}$$

where $q = \delta\mathbf{r} \cdot (2\mathbf{r} - \delta\mathbf{r})/r^2$, and $F(q)$ is given by Eq. (F.3).

Recall that $\mathbf{p}$ is the perturbing acceleration, which is a known function of time. In Encke's method, we integrate Eq. (10.7) to obtain the deviation $\delta\mathbf{r}(t)$. This is added to the osculating motion $\mathbf{r}_{osc}(t)$ to obtain the perturbed trajectory from Eq. (10.6). If at any time the ratio $\delta r/r$ exceeds a preset tolerance, then the osculating orbit is redefined to be that of the perturbed orbit at time t. This process, which is called rectification, is illustrated in Fig. 10.3.

The following is an implementation of Encke's method in which rectification occurs at the beginning of every time step Δt. Other implementations may be found in textbooks such as Bate et al. (1971) (Section 9.3), Vallado (2007) (Section 8.3), and Schaub and Junkins (2009) (Section 10.2).

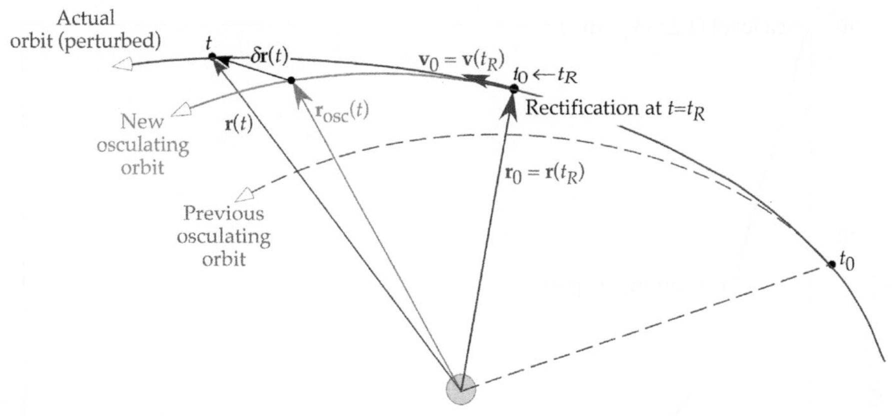

FIG. 10.3

Resetting the reference orbit at time t_R.

ALGORITHM 10.1

Given the functional form of the perturbing acceleration $\mathbf{p}$, and the state vector components $\mathbf{r}_0$ and $\mathbf{v}_0$ at time t, calculate $\mathbf{r}$ and $\mathbf{v}$ at time $t + \Delta t$.

1. Set $\delta\mathbf{r} = \mathbf{0}$ and $\delta\mathbf{v} = \mathbf{0}$.
2. For the time span t to $t + \Delta t$, execute Algorithm 1.3 (or another numerical integration procedure), with $\mathbf{y} = \lfloor \delta\mathbf{r} \;\; \delta\mathbf{v} \rfloor^T$ and $\mathbf{f} = \lfloor \delta\mathbf{v} \;\; \delta\mathbf{a} \rfloor^T$, to obtain $\delta\mathbf{r}(t + \Delta t)$ and $\delta\mathbf{v}(t + \Delta t)$. At each step i of the numerical integration from t to $t + \Delta t$:
   (a) $\delta\mathbf{r}_i$ and $\delta\mathbf{v}_i$ are available from the previous step.
   (b) Compute $\mathbf{r}_{osc_i}$ and $\mathbf{v}_{osc_i}$ from $\mathbf{r}_0$ and $\mathbf{v}_0$ using Algorithm 3.4.
   (c) Compute $\mathbf{r}_i = \mathbf{r}_{osc_i} + \delta\mathbf{r}_i$ from Eq. (10.6).
   (d) Compute $\mathbf{v}_i = \mathbf{v}_{osc_i} + \delta\mathbf{v}_i$.
   (e) Compute $\delta\mathbf{a}_i$ from Eq. (10.7).
   (f) $\delta\mathbf{v}_i$ and $\delta\mathbf{a}_i$ are used to furnish $\delta\mathbf{r}$ and $\delta\mathbf{v}$ for the next step of the numerical integration.
3. $\mathbf{r}_0 \leftarrow \mathbf{r}(t + \Delta t)$ and $\mathbf{v}_0 \leftarrow \mathbf{v}(t + \Delta t)$. (Rectification)

10.4 ATMOSPHERIC DRAG

For the earth, the commonly accepted altitude at which space "begins" is 100 km (60 miles). Although over 99.9999% of the earth's atmosphere lies below 100 km, the air density at that altitude is nevertheless sufficient to exert drag and cause aerodynamic heating on objects moving at orbital speeds. (Recall from Eq. 2.63 that the speed required for a circular orbit at 100 km altitude is 7.8 km/s.) The drag will lower the speed and the height of a spacecraft, and the heating can produce temperatures of 2000°C or more. A spacecraft will likely burn up unless it is protected with a heat shield. Note that the altitude (entry interface) at which the thermally protected space shuttle orbiter entered the atmosphere on its return from space was considered to be 120 km.

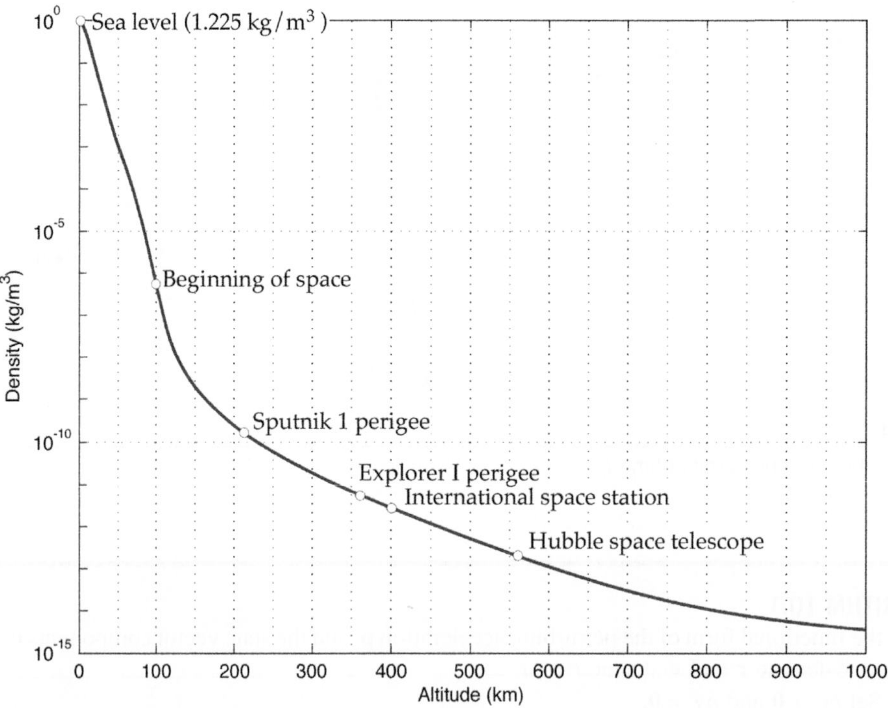

FIG. 10.4

US Standard Atmosphere 1976: density versus attitude.

There are a number of models that describe the variation of atmospheric properties with altitude (AIAA, 2010). One of them is USSA76, the *US Standard Atmosphere 1976* (NOAA/NASA/USAF, 1976). Fig. 10.4 shows the US Standard Atmosphere density profile from sea level to an altitude of 1000 km. This figure was obtained by selecting the density ρ_i at 28 altitudes z_i in the USSA76 table and interpolating between them with the exponential functions $\rho(z) = \rho_i \exp\,[-(z - z_i)/H_i]$, where $z_i \le z < z_{i+1}$, and $H_i = (z_{i+1} - z_i)/\ln\,(\rho_{i+1}/\rho_i)$. The simple procedure is implemented in the MATLAB function *atmosphere.m*, which is listed in Appendix D.39. For several equispaced altitudes (kilometers), *atmosphere.m* yields the following densities (kilograms per cubic meter):

```
>>
z = logspace(0,3,6);
for i = 1:6
    altitude = z(i);
    density = atmosphere(z(i));
    fprintf('%12.3f km %12.3e kg/m^3\n', altitude, density)
end
      1.000 km     1.068e+00 kg/m^3
      3.981 km     7.106e-01 kg/m^3
     15.849 km     1.401e-01 kg/m^3
```

```
     63.096 km     2.059e-04 kg/m^3
    251.189 km     5.909e-11 kg/m^3
   1000.000 km     3.561e-15 kg/m^3
>>
```

According to USSA76, the atmosphere is a spherically symmetric 1000-km-thick gaseous shell surrounding the earth. Its properties throughout are steady state and are consistent with a period of moderate solar activity. The hypothetical variation of properties with altitude approximately represents the year-round conditions at midlatitudes averaged over many years. The model provides realistic values of atmospheric density that, however, may not match the actual values at a given place or time.

Drag affects the life of an orbiting spacecraft. Sputnik 1, the world's first artificial satellite, had a perigee altitude of 228 km, where the air density is about five orders of magnitude greater than at its apogee altitude of 947 km. The drag force associated with repeated passage through the thicker air eventually robbed the spherical, 80-kg Sputnik of the energy needed to stay in orbit. It fell from orbit and burned up on January 4, 1958, almost 3 months to the day after it was launched by the Soviet Union. Soon thereafter, on January 31, the United States launched its first satellite, the cylindrical, 14-kg Explorer 1, into a 358-km-by-2550-km orbit. In this higher orbit, Explorer experienced less drag than Sputnik, and it remained in orbit for 18 years. The current International Space Station's nearly circular orbit is about 400 km above the earth. At that height, drag causes orbital decay that requires frequent reboosts, usually provided by the propulsion systems of visiting supply vehicles. The altitude of the Hubble Space Telescope's circular orbit is 560 km, and it is also degraded by drag. Between 1993 and 2009, space shuttle orbiters visited Hubble five times to service it and boost its orbit. With no propulsion system of its own, the venerable Hubble is expected to deorbit around 2025, 35 years after it was launched.

If the inertial velocity of a spacecraft is $\mathbf{v}$ and that of the atmosphere at that point is $\mathbf{v}_{\text{atm}}$, then the spacecraft velocity relative to the atmosphere is

$$\mathbf{v}_{\text{rel}} = \mathbf{v} - \mathbf{v}_{\text{atm}} \tag{10.8}$$

If the atmosphere rotates with the earth, whose angular velocity is $\boldsymbol{\omega}_E$, then relative to the origin O of the geocentric equatorial frame, $\mathbf{v}_{\text{atm}} = \boldsymbol{\omega}_E \times \mathbf{r}$, where $\mathbf{r}$ is the spacecraft position vector. Thus,

$$\mathbf{v}_{\text{rel}} = \mathbf{v} - \boldsymbol{\omega}_E \times \mathbf{r} \tag{10.9}$$

Since the drag force $\mathbf{D}$ on an object acts in the direction opposite to the relative velocity vector, we can write

$$\mathbf{D} = -D\hat{\mathbf{v}}_{\text{rel}} \tag{10.10}$$

where $\hat{\mathbf{v}}_{\text{rel}} = \mathbf{v}_{\text{rel}}/v_{\text{rel}}$ is the unit vector in the direction of the relative velocity, and

$$D = \frac{1}{2}\rho v_{\text{rel}}^2 C_D A \tag{10.11}$$

where ρ is the atmospheric density, A is the frontal area of the spacecraft (the area normal to the relative velocity vector), and C_D is the dimensionless drag coefficient. If the mass of the spacecraft is m, then the perturbing acceleration due to the drag force is $\mathbf{p} = \mathbf{D}/m$, so that

$$\mathbf{p} = -\frac{1}{2}\rho v_{rel}\left(\frac{C_D A}{m}\right)\mathbf{v}_{\text{rel}} \tag{10.12}$$

There apparently is no universal agreement on the name of the quantity in parentheses. We will call it the ballistic coefficient,

$$B = \frac{C_D A}{m} \tag{10.13}$$

The reader may encounter alternative definitions of the ballistic coefficient, such as the reciprocal, $m/(C_D A)$.

EXAMPLE 10.1

A small spherical earth satellite has a diameter of 1 m and a mass of 100 kg. At a given time t_0, its orbital parameters are

Perigee radius:	$r_p = 6593\text{km}\,(215\text{km altitude})$
Apogee radius:	$r_a = 7317\text{km}\,(939\text{km altitude})$
Right ascension of the ascending node:	$\Omega = 340°$
Inclination:	$i = 65.1°$
Argument of perigee:	$\omega = 58°$
True anomaly:	$\theta = 332°$

Assuming a drag coefficient of $C_D = 2.2$ and using the 1976 US Standard Atmosphere rotating with the earth, employ Cowell's method to find the time for the orbit to decay to 100 km. Recall from Eq. (2.67) that the angular velocity of the earth is $72.9211(10^{-6})$rad/s.

Solution

Let us first use the given data to compute some additional orbital parameters, recalling that for the earth, $\mu = 398,600\ \text{km}^3/\text{s}^2$:

Eccentricity:	$e = (r_a - r_p)/(r_a + r_p)$	$= 0.052049$
Semimajor axis:	$a = (r_a + r_p)/2$	$= 6955\text{km}$
Angular momentum:	$h = \sqrt{\mu a(1-e^2)}$	$= 52{,}580.1\,\text{km}^2/\text{s}$
Period:	$T = 2\pi\sqrt{a^3/\mu}$	$= 96.207\ \text{min}$

Now we can use the six elements h, e, i, Ω, ω, and θ in Algorithm 4.5 to obtain the state vector $(\mathbf{r}_0, \mathbf{v}_0)$ at the initial time, relative to the geocentric equatorial frame,

$$\begin{aligned} \mathbf{r}_0 &= 5873.40\hat{\mathbf{I}} - 658.522\hat{\mathbf{J}} + 3007.49\hat{\mathbf{K}}\,(\text{km}) \\ \mathbf{v}_0 &= -2.89641\hat{\mathbf{I}} + 4.94010\hat{\mathbf{J}} + 6.14446\hat{\mathbf{K}}\,(\text{km/s}) \end{aligned} \tag{a}$$

Referring to Section 1.8, we next write Eq. (10.2), a second-order differential equation in $\mathbf{r}$, as two first-order ordinary differential equations in $\mathbf{r}$ and $\dot{\mathbf{r}}(=\mathbf{v})$,

$$\frac{d}{dt}\begin{Bmatrix} \mathbf{r} \\ \mathbf{v} \end{Bmatrix} = \begin{Bmatrix} \mathbf{v} \\ \mathbf{a} \end{Bmatrix} = \begin{Bmatrix} \mathbf{v} \\ -\mu\dfrac{\mathbf{r}}{r_3} + \mathbf{p} \end{Bmatrix} \tag{b}$$

The drag acceleration $\mathbf{p}$ is given by Eq. (10.12) along with Eq. (10.9). With Eq. (a) as the initial conditions, we can solve for $\mathbf{r}$ and $\mathbf{v}$ on the time interval $[t_0, t_f]$ using, for example, MATLAB's built-in numerical integrator *ode45*. From the solution, we obtain the satellite's altitude, which oscillates with time between the extreme values at perigee and apogee. It is a simple matter to extract just the extrema and plot them, as shown in Fig. 10.5, for $t_0 = 0$ and $t_f = 120$ days.

$$\boxed{\text{The orbit decays to 100km in 108 days}}$$

Observe that the apogee starts to decrease immediately, whereas the perigee remains nearly constant until the very end. As the apogee approaches perigee, the eccentricity approaches zero. We say that atmospheric drag tends to circularize an elliptical orbit.

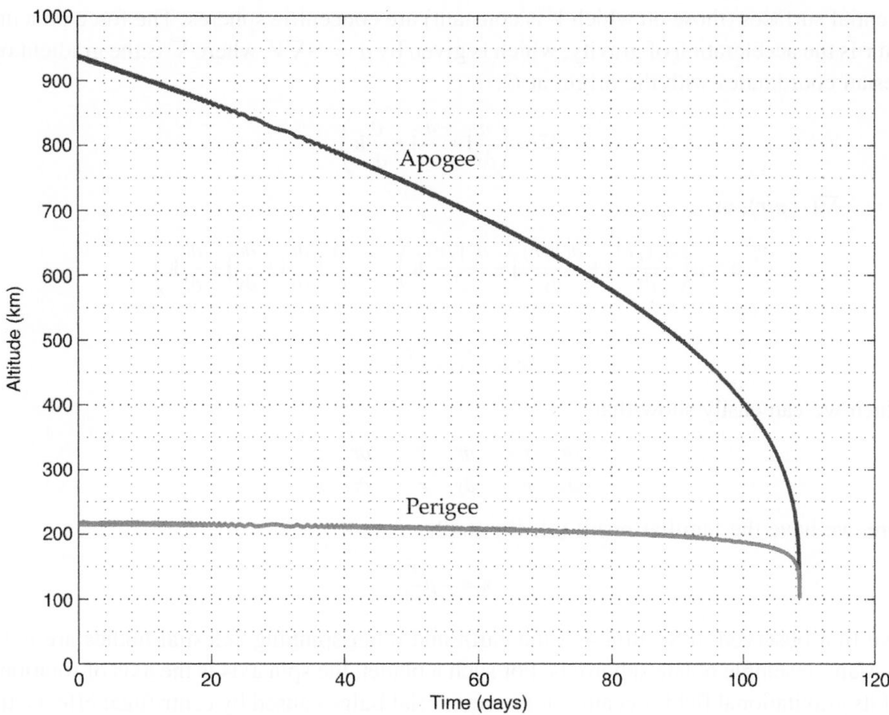

FIG. 10.5

Perigee and apogee versus time.

The MATLAB script *Example_10_01.m* for this problem is located in Appendix D.40.

10.5 GRAVITATIONAL PERTURBATIONS

In Appendix E, we show that if the central attractor (e.g., the earth) is a sphere of radius R with a spherically symmetric mass distribution, then its external gravitational potential field will be spherically symmetric, acting as though all of the mass were concentrated at the center O of the sphere. The gravitational potential energy per unit mass ($m = 1$ in Eq. E.10) is

$$V = -\frac{\mu}{r} \tag{10.14}$$

where $\mu = GM$, G is the universal gravitational constant, M is the sphere's mass, and r is the distance from O to a point outside the sphere ($r > R$). Here r is the magnitude of the position vector $\mathbf{r}$, which, in Cartesian coordinates centered at O, is given by

$$\mathbf{r} = x\hat{\mathbf{i}} + y\hat{\mathbf{j}} + z\hat{\mathbf{k}}$$

Equipotential surfaces (those on which V is constant) are concentric spheres. The force on a unit mass placed at $\mathbf{r}$ is the acceleration of gravity, which is given by $a = -\nabla V$, where ∇ is the gradient operator. In Cartesian coordinates with the origin at O,

$$\nabla = \left(\frac{\partial}{\partial x}\hat{\mathbf{i}} + \frac{\partial}{\partial y}\hat{\mathbf{j}} + \frac{\partial}{\partial z}\hat{\mathbf{k}}\right)$$

Thus, $\mathbf{a} = -\nabla(-\mu/r)$, or

$$\mathbf{a} = \mu\left(\frac{\partial(1/r)}{\partial x}\hat{\mathbf{i}} + \frac{\partial(1/r)}{\partial y}\hat{\mathbf{j}} + \frac{\partial(1/r)}{\partial z}\hat{\mathbf{k}}\right) = -\frac{\mu}{r^2}\left(\frac{\partial r}{\partial x}\hat{\mathbf{i}} + \frac{\partial r}{\partial y}\hat{\mathbf{j}} + \frac{\partial r}{\partial z}\hat{\mathbf{k}}\right)$$

But

$$r = \sqrt{x^2 + y^2 + z^2} \tag{10.15}$$

from which we can easily show that

$$\frac{\partial r}{\partial x} = \frac{x}{r} \qquad \frac{\partial r}{\partial y} = \frac{y}{r} \qquad \frac{\partial r}{\partial z} = \frac{z}{r} \tag{10.16}$$

Therefore, we have the familiar result (cf. Eq. 2.22)

$$\mathbf{a} = -\mu\frac{\mathbf{r}}{r^3} \tag{10.17}$$

As we first observed in Section 4.7, the earth and other spinning celestial bodies are not perfect spheres. Many resemble oblate spheroids. For such a planet, the spin axis is the axis of rotational symmetry of its gravitational field. Because of the equatorial bulge caused by centrifugal effects, the gravitational field varies with the latitude as well as radius. This more complex gravitational potential is dominated by the familiar point mass contribution given by Eq. (10.14), on which is superimposed the perturbation due to oblateness.

It is convenient to use the spherical coordinate system shown in Fig. 10.6. The origin O is at the planet's center of mass, and the z axis of the associated Cartesian coordinate system is the axis of rotational symmetry. (The rotational symmetry means that our discussion is independent of the choice of Cartesian coordinate frame as long as each shares a common z axis.) r is the distance of a point P from O, ϕ is the polar angle measured from the positive z axis to the radial, and θ is the azimuth angle measured from the positive x axis to the projection of the radial onto the xy plane. Observe that

$$\phi = \tan^{-1}\frac{\sqrt{x^2 + y^2}}{z} \tag{10.18}$$

Since the gravitational field is rotationally symmetric, it does not depend on the azimuth angle θ. Therefore, the gravitational potential may be written as

$$V(r,\phi) = -\frac{\mu}{r} + \Phi(r,\phi) \tag{10.19}$$

where Φ is the perturbation of the gravitational potential due to the planet's oblateness.

The rotationally symmetric perturbation $\Phi(r,\phi)$ is given by the infinite series (Battin, 1999)

$$\Phi(r,\phi) = \frac{\mu}{r}\sum_{k=2}^{\infty} J_k\left(\frac{R}{r}\right)^k P_k(\cos\phi) \tag{10.20}$$

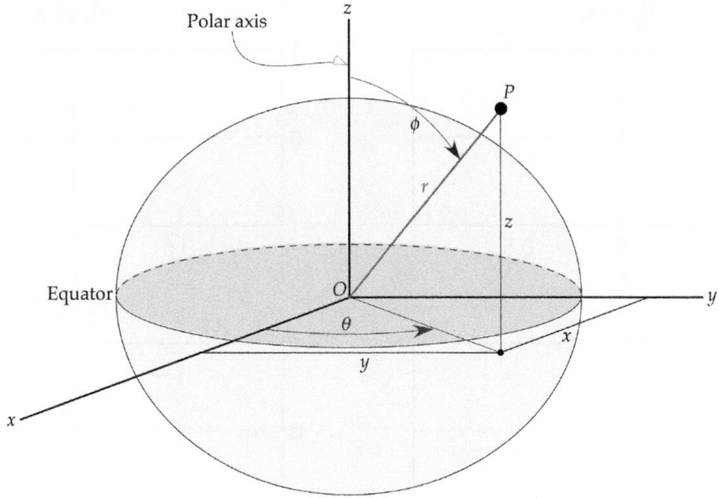

FIG. 10.6

Spherical coordinate system.

where J_k are the zonal harmonics of the planet, R is its equatorial radius ($R/r < 1$), and P_k are the Legendre polynomials (see below). The zonal harmonics are dimensionless numbers that are not derived from mathematics but are inferred from observations of satellite motion around a planet, and they are unique to that planet. The summation starts at $k = 2$ instead of $k = 1$ because $J_1 = 0$, due to the fact that the origin of the spherical coordinate system is at the planet center of mass. For the earth, the next six zonal harmonics are (Vallado, 2007)

$$\begin{array}{ll} J_2 = 0.00108263 & J_3 = -2.33936\left(10^{-3}\right)J_2 \\ J_4 = -1.49601\left(10^{-3}\right)J_2 & J_5 = -0.20995\left(10^{-3}\right)J_2 \\ J_6 = 0.49941\left(10^{-3}\right)J_2 & J_7 = 0.32547\left(10^{-3}\right)J_2 \end{array} \quad \text{Earth zonal harmonics} \tag{10.21}$$

This set of zonal harmonics is clearly dominated by J_2. For $k > 7$ the zonal harmonics all remain more than three orders of magnitude smaller than J_2.

The Legendre polynomials are named for the French mathematician Adrien-Marie Legendre (1752–1833). The polynomial $P_k(x)$ may be obtained from a formula derived by another French mathematician Olinde Rodrigues (1795–1851), as part of his 1816 doctoral thesis,

$$P_k(x) = \frac{1}{2^k k!}\frac{d^k}{dx^k}\left(x^2 - 1\right)^k \quad \text{Rodrigues' formula} \tag{10.22}$$

Using Rodrigues' formula, we can calculate the first few Legendre polynomials that appear in Eq. (10.20),

$$\begin{array}{ll} P_2(x) = \frac{1}{2}\left(3x^2 - 1\right) & P_3(x) = \frac{1}{2}\left(5x^3 - 3x\right) \\ P_4(x) = \frac{1}{8}\left(35x^4 - 30x^2 + 3\right) & P_5(x) = \frac{1}{8}\left(63x^5 - 70x^3 + 15x\right) \\ P_6(x) = \frac{1}{16}\left(231x^6 - 315x^4 + 105x^2 - 5\right) & P_7(x) = \frac{1}{16}\left(429x^7 - 693x^5 + 315x^3 - 35x\right) \end{array} \tag{10.23}$$

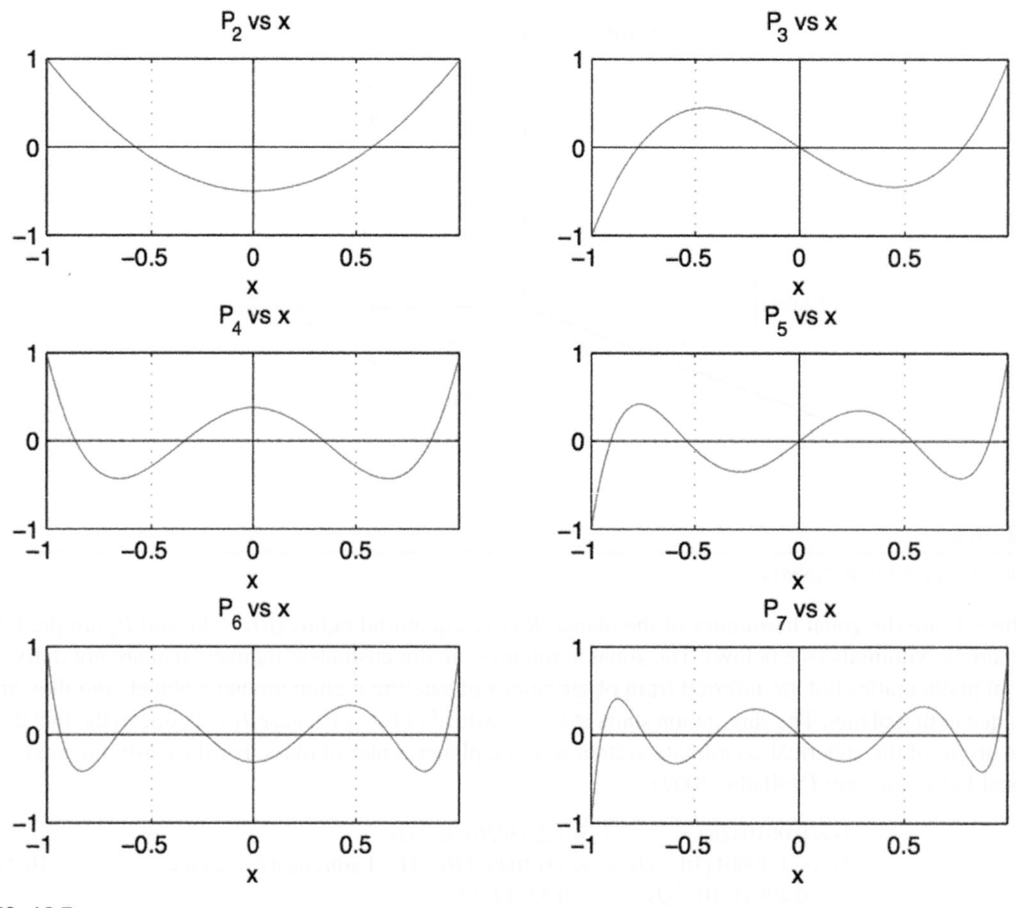

FIG. 10.7

Legendre polynomials P_2 through P_7.

These are plotted in Fig. 10.7.

Since J_2 is by far the largest zonal harmonic, we shall in the interest of simplicity focus only on its contribution to the gravitational perturbation, thereby ignoring all but the J_2 term in the series for $\Phi(r,\phi)$. In that case, Eq. (10.20) yields

$$\Phi(r,\phi) = \frac{J_2 \mu}{2\,r}\left(\frac{R}{r}\right)^2 \left(3\cos^2\phi - 1\right) \tag{10.24}$$

Observe that $\Phi = 0$ when $\cos\phi = \sqrt{1/3}$, which corresponds to about $\pm 35°$ geocentric latitude. This band reflects the earth's equatorial bulge (oblateness). The perturbing acceleration is the negative of the gradient of Φ,

$$\mathbf{p} = -\nabla\Phi = -\frac{\partial\Phi}{\partial x}\hat{\mathbf{i}} - \frac{\partial\Phi}{\Phi y}\hat{\mathbf{j}} - \frac{\partial\Phi}{\partial z}\hat{\mathbf{k}} \tag{10.25}$$

Noting that $\partial\Phi/\partial\theta = 0$, we have from the chain rule of calculus that

$$\frac{\partial\Phi}{\partial x} = \frac{\partial\Phi}{\partial r}\frac{\partial r}{\partial x} + \frac{\partial\Phi}{\partial\phi}\frac{\partial\phi}{\partial x} \quad \frac{\partial\Phi}{\partial y} = \frac{\partial\Phi}{\partial r}\frac{\partial r}{\partial y} + \frac{\partial\Phi}{\partial\phi}\frac{\partial\phi}{\partial y} \quad \frac{\partial\Phi}{\partial z} = \frac{\partial\Phi}{\partial r}\frac{\partial r}{\partial z} + \frac{\partial\Phi}{\partial\phi}\frac{\partial\phi}{\partial z} \tag{10.26}$$

Differentiating Eq. (10.24), we obtain

$$\begin{aligned} \frac{\partial\Phi}{\partial r} &= -\frac{3}{2}J_2\frac{\mu}{r^2}\left(\frac{R}{r}\right)^2\left(3\cos^2\phi - 1\right) \\ \frac{\partial\Phi}{\partial\phi} &= -\frac{3}{2}J_2\frac{\mu}{r}\left(\frac{R}{r}\right)^2\sin\phi\cos\phi \end{aligned} \tag{10.27}$$

Use Eq. (10.18) to find the required partial derivatives of ϕ:

$$\begin{aligned} \frac{\partial\phi}{\partial x} &= \frac{xz}{x^2+y^2+z^2}\frac{1}{\sqrt{x^2+y^2}} = \frac{xz}{r^3\sin\phi} \\ \frac{\partial\phi}{\partial y} &= \frac{yz}{x^2+y^2+z^2}\frac{1}{\sqrt{x^2+y^2}} = \frac{yz}{r^3\sin\phi} \\ \frac{\partial\phi}{\partial z} &= -\frac{\sqrt{x^2+y^2}}{x^2+y^2+z^2} = \frac{\sin\phi}{r} \end{aligned} \tag{10.28}$$

Substituting Eqs. (10.16), (10.27), and (10.28) into Eq. (10.26) and using the fact that $\cos\phi = z/r$ leads to the following expressions for the gradient of perturbing potential Φ:

$$\begin{aligned} \frac{\partial\Phi}{\partial x} &= -\frac{3}{2}J_2\frac{\mu}{r^2}\left(\frac{R}{r}\right)^2\frac{x}{r}\left[5\left(\frac{z}{r}\right)^2 - 1\right] \\ \frac{\partial\Phi}{\partial y} &= -\frac{3}{2}J_2\frac{\mu}{r^2}\left(\frac{R}{r}\right)^2\frac{y}{r}\left[5\left(\frac{z}{r}\right)^2 - 1\right] \\ \frac{\partial\Phi}{\partial z} &= -\frac{3}{2}J_2\frac{\mu}{r^2}\left(\frac{R}{r}\right)^2\frac{z}{r}\left[5\left(\frac{z}{r}\right)^2 - 3\right] \end{aligned} \tag{10.29}$$

The perturbing gravitational acceleration $\mathbf{p}$ due to J_2 is found by substituting these equations into Eq. (10.25), yielding the vector

$$\mathbf{p} = \frac{3}{2}\frac{J_2\mu R^2}{r^4}\left[\frac{x}{r}\left(5\frac{z^2}{r^2} - 1\right)\hat{\mathbf{i}} + \frac{y}{r}\left(5\frac{z^2}{r^2} - 1\right)\hat{\mathbf{j}} + \frac{z}{r}\left(5\frac{z^2}{r^2} - 3\right)\hat{\mathbf{k}}\right] \tag{10.30}$$

The perturbing accelerations for higher zonal harmonics may be evaluated in a similar fashion. Schaub and Junkins (2009) (p. 553) lists the accelerations for J_3 through J_6.

Irregularities in the earth's geometry and its mass distribution cause the gravitational field to vary not only with latitude ϕ but with longitude θ as well. To mathematically account for this increased physical complexity, the series expansion of the potential function Φ in Eq. (10.19) must be generalized to include the azimuth angle θ. As a consequence, it turns out that we can identify *sectorial harmonics*, which account for the longitudinal variation over domains of the earth resembling segments of an orange. We also discover a tile-like patchwork of *tesseral harmonics*, which model how specific regions of the earth deviate locally from a perfect, homogeneous sphere. Incorporating these additional levels of detail into the gravitational model is essential for accurate long-term prediction of satellite orbits (e.g., global positioning satellite constellations). For an in-depth treatment of this subject, including the mathematical details, see Vallado (2007).

EXAMPLE 10.2

At time $t = 0$, an earth satellite has the following orbital parameters:

Perigee radius:	$r_p = 6678\text{km}\,(300\text{km altitude})$
Apogee radius:	$r_a = 9940\text{km}\,(3062\text{km altitude})$
Right ascension of the ascending node:	$\Omega = 45°$
Inclination:	$i = 28°$
Argument of perigee:	$\omega = 30°$
True anomaly:	$\theta = 40°$

Use Encke's method to determine the effect of J_2 perturbation on the variation of right ascension of the node, argument of perigee, angular momentum, eccentricity, and inclination over the next 48 hours.

Solution

Using Fig. 2.18 along with Eqs. (2.83), (2.84), and (2.71), we can find other orbital parameters in addition to the ones given. In particular, recalling that $\mu = 398,600\ \text{km}^3/\text{s}^2$ for the earth,

Eccentricity:	$e = (r_a - r_p)/(r_a + r_p) = 0.17136$
Semimajor axis:	$a = (r_a + r_p)/2 = 8059\text{km}$
Period:	$T = 2\pi\sqrt{a^3/\mu} = 7200\text{s}\ (2\text{h})$
Angular momentum:	$h = \sqrt{\mu a(1-e^2)} = 55{,}839\text{km}^2/\text{s}$

The six elements h, e, i, Ω, ω, and θ together with Algorithm 4.5 yield the state vector $(\mathbf{r}_0, \mathbf{v}_0)$ at the initial time, relative to the geocentric equatorial frame,

$$\begin{aligned}\mathbf{r}_0 &= -2384.46\hat{\mathbf{I}} + 5729.01\hat{\mathbf{J}} + 3050.46\hat{\mathbf{K}}\ (\text{km})\\ \mathbf{v}_0 &= -7.36138\hat{\mathbf{I}} - 2.98997\hat{\mathbf{J}} + 1.64354\hat{\mathbf{K}}\ (\text{km/s})\end{aligned} \tag{a}$$

Using this state vector as a starting point, with $t_0 = 0$, $t_f = 48 \times 3600$ s, and $\Delta t = (t_f - t_0)/1000$, we enter the Encke procedure (Algorithm 10.1) with MATLAB's *ode45* as the numerical integrator to find **r** and **v** at each of the 1001 equally spaced times. At these times, we then use Algorithm 4.2 to compute the right ascension of the node and the argument of perigee, which are plotted in Fig. 10.8.

Fig. 10.8 shows that the J_2 perturbation causes a drift in both Ω and ω over time. For this particular orbit, Ω decreases whereas ω increases. We see that both parameters have a straight line or secular variation on which a small or short-periodic variation is superimposed. Approximate average values of the slopes of the curves for Ω and ω are most

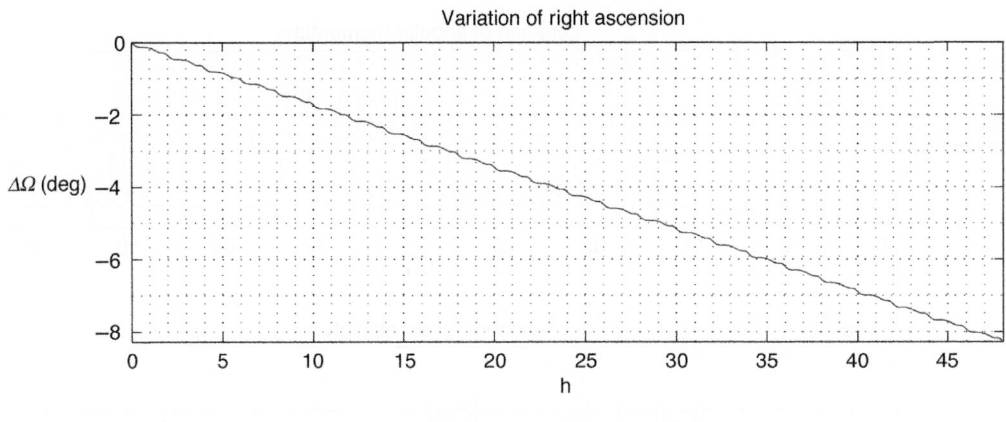

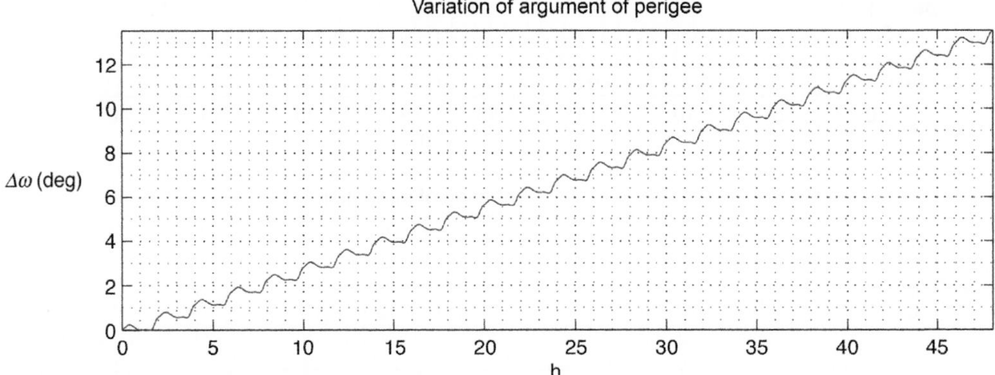

FIG. 10.8

Histories of the right ascension of the ascending node and the argument of perigee for a time span of 48 h. The plotted points were obtained using Encke's method.

simply found by dividing the difference between the computed values at t_f and t_0 by the time span $t_f - t_0$. In this way, we find that

$$\dot{\bar{\Omega}} = -0.172 \text{ deg/h} \tag{b}$$

$$\dot{\bar{\omega}} = 0.282 \text{ deg/h} \tag{c}$$

As pointed out in Section 4.7, the decrease in the node angle Ω with time is called regression of the node, whereas the increase of argument of perigee ω with time is called the advance of perigee.

The computed time histories of h, e, and i are shown in Fig. 10.9. Clearly, none of these osculating elements shows a secular variation due to the earth's oblateness. They are all constant except for the short-periodic perturbations evident in all three plots.

The MATLAB script *Example_10_02.m* for this problem is in Appendix D.41.

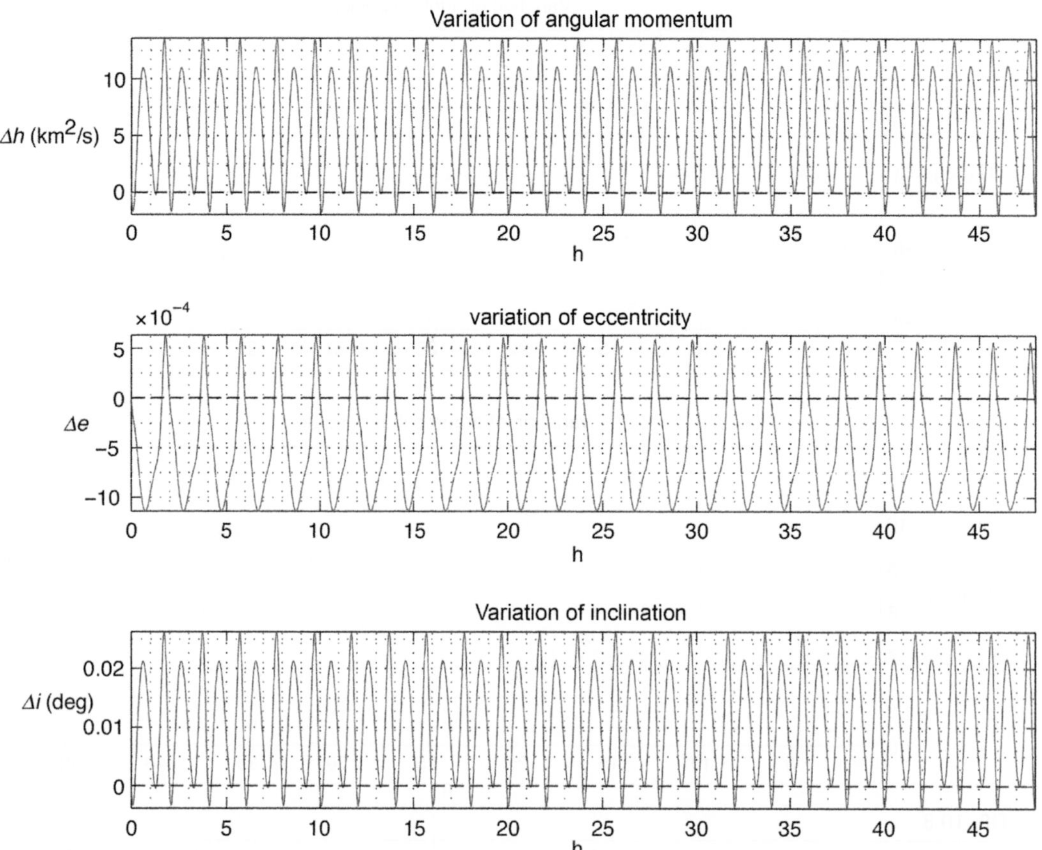

FIG. 10.9

Histories of the J_2 perturbed angular momentum, eccentricity, and inclination for a time span of 48 h.

The results obtained in the above example follow from the specific initial conditions that we supplied to the Encke numerical integration procedure. We cannot generalize the conclusions of this special perturbation analysis to other orbits. In fact, whether or not the node line and the eccentricity vector advance or regress depends on the nature of the particular orbit. We can confirm this by means of repeated numerical analyses with different initial conditions. Obtaining formulas that describe perturbation phenomena for general cases is the object of general perturbation analysis, which is beyond the scope of this book. However, a step in that direction is to derive expressions for the time variations of the osculating elements. These are Lagrange's planetary equations or their variants, Gauss' variational equations. Both are based on the variation-of-parameters approach to solving differential equations.

10.6 VARIATION OF PARAMETERS

For the two-body problem, the equation of motion (Eq. 2.22)

$$\ddot{\mathbf{r}}+\mu\frac{\mathbf{r}}{r^3}=\mathbf{0} \tag{10.31}$$

yields the position vector $\mathbf{r}$ as a function of time t and six parameters or *orbital elements* $\mathbf{c}$,

$$\mathbf{r}=\mathbf{f}(\mathbf{c},t) \tag{10.32}$$

where $\mathbf{c}$ stands for the set of six parameters $(c_1, c_2, \ldots, c_6)$. Keep in mind that the vector $\mathbf{f}$ has three scalar components.

EXAMPLE 10.3

At time t_0, the state vector of a spacecraft in two-body motion is $\lfloor \mathbf{r}_0\ \mathbf{v}_0 \rfloor$. What is $\mathbf{f}(\mathbf{c},t)$ in this case?

Solution

The constant orbital elements $\mathbf{c}$ are the six components of the state vector,

$$\mathbf{c}=\lfloor \mathbf{r}_0\ \mathbf{v}_0 \rfloor$$

By means of Algorithm 3.4, we find

$$\mathbf{r}=f(t)\mathbf{r}_0+g(t)\mathbf{v}_0 \tag{a}$$

where the time-dependent Lagrange f and g functions are given by Eq. (3.69) (Examples 3.7 and 4.2). Thus,

$$\boxed{\mathbf{f}(\mathbf{c},t)=f(t)\mathbf{r}_0+g(t)\mathbf{v}_0}$$

As we know from Section 4.4, the orbital parameters that may alternatively be selected as the classical elements at time t_0, are the longitude of the node Ω, inclination i, argument of periapse ω, eccentricity e, angular momentum h (or semimajor axis a), and true anomaly θ (or the mean anomaly M or the eccentric anomaly E). For $t > t_0$, only the anomalies vary with time, a fact that is reflected by the presence of the argument t in $\mathbf{f}(\mathbf{c},t)$ above. As pointed out in Section 3.2, we could use time of periapse passage t_p as the sixth orbital parameter instead of true anomaly. The advantage of doing so in the present context would be that t_p is a constant for Keplerian orbits. (Up to now we have usually set that constant equal to zero.)

The velocity $\mathbf{v}$ is the time derivative of the position vector $\mathbf{r}$, so that from Eq. (10.32),

$$\mathbf{v}=\frac{\mathrm{d}}{\mathrm{d}t}\mathbf{f}(\mathbf{c},t)=\frac{\partial}{\partial t}\mathbf{f}(\mathbf{c},t)+\sum_{\alpha=1}^{6}\frac{\partial}{\partial c_\alpha}\mathbf{f}(\mathbf{c},t)\frac{\mathrm{d}c_\alpha}{\mathrm{d}t} \tag{10.33}$$

Since the orbital elements are constant in two-body motion, their time derivatives are zero,

$$\frac{\mathrm{d}c_\alpha}{\mathrm{d}t}=0 \quad \alpha=1,\ldots,6$$

Therefore, Eq. (10.33) becomes simply

$$\mathbf{v}=\frac{\partial}{\partial t}\mathbf{f}(\mathbf{c},t) \quad \text{Two}-\text{body motion} \tag{10.34}$$

We find the acceleration by taking the time derivative of $\mathbf{v}$ in Eq. (10.34) while holding $\mathbf{c}$ constant,

$$\mathbf{a}=\frac{\partial^2}{\partial t^2}\mathbf{f}(\mathbf{c},t) \quad \text{Two}-\text{body motion}$$

Substituting this and Eq. (10.32) into Eq. (10.31) yields

$$\frac{\partial^2}{\partial t^2}\mathbf{f}(\mathbf{c},t)+\mu\frac{\mathbf{f}(\mathbf{c},t)}{\|\mathbf{f}(\mathbf{c},t)\|^3}=\mathbf{0} \tag{10.35}$$

A perturbing force produces a perturbing acceleration $\mathbf{p}$ that results in a perturbed motion $\mathbf{r}_p$ for which the equation of motion is

$$\ddot{\mathbf{r}}_p+\mu\frac{\mathbf{r}_p}{r_p{}^3}=\mathbf{p} \tag{10.36}$$

The variation-of-parameters method requires that the solution to Eq. (10.36) have the same mathematical form $\mathbf{f}$ as it does for the two-body problem, except that the six constants $\mathbf{c}$ in Eq. (10.32) are replaced by six functions of time $\mathbf{u}(t)$, so that

$$\mathbf{r}_p=\mathbf{f}(\mathbf{u}(t),t) \tag{10.37}$$

where $\mathbf{u}(t)$ stands for the set of six functions $u_1(t)$, $u_2(t)$, … , $u_6(t)$. The six parameters $\mathbf{u}(t)$ are the orbital elements of the osculating orbit that is tangent to $\mathbf{r}_p$ at time t.

We find the velocity vector $\mathbf{v}_p$ for the perturbed motion by differentiating Eq. (10.37) with respect to time and using the chain rule,

$$\mathbf{v}_p=\frac{\mathrm{d}\mathbf{r}_p}{\mathrm{d}t}=\frac{\partial\mathbf{f}(\mathbf{u},t)}{\partial t}+\sum_{\beta=1}^{6}\frac{\partial\mathbf{f}(\mathbf{u},t)}{\partial u_\beta}\frac{\mathrm{d}u_\beta}{\mathrm{d}t} \tag{10.38}$$

For $\mathbf{v}_p$ to have the same mathematical form as $\mathbf{v}$ for the unperturbed case (Eq. 10.34), we impose the following three conditions on the osculating elements $\mathbf{u}(t)$:

$$\sum_{\beta=1}^{6}\frac{\partial\mathbf{f}(\mathbf{u},t)}{\partial u_\beta}\frac{\mathrm{d}u_\beta}{\mathrm{d}t}=\mathbf{0} \tag{10.39}$$

Eq. (10.38) therefore becomes simply

$$\mathbf{v}_p=\frac{\partial\mathbf{f}(\mathbf{u},t)}{\partial t} \tag{10.40}$$

To find $\ddot{\mathbf{r}}_p$, the acceleration of the perturbed motion, we differentiate the velocity $\mathbf{v}_p$ with respect to time. It follows from Eq. (10.40) that

$$\ddot{\mathbf{r}}_p=\frac{\mathrm{d}\mathbf{v}_p}{\mathrm{d}t}=\frac{\partial^2\mathbf{f}(\mathbf{u},t)}{\partial t^2}+\sum_{\beta=1}^{6}\frac{\partial^2\mathbf{f}(\mathbf{u},t)}{\partial u_\beta\partial t}\frac{\mathrm{d}u_\beta}{\mathrm{d}t} \tag{10.41}$$

Substituting Eqs. (10.37) and (10.41) into Eq. (10.36) yields

$$\frac{\partial^2\mathbf{f}(\mathbf{u},t)}{\partial t^2}+\sum_{\beta=1}^{6}\frac{\partial\mathbf{f}^2(\mathbf{u},t)}{\partial u_\beta\partial t}\frac{\mathrm{d}u_\beta}{\mathrm{d}t}+\mu\frac{\mathbf{f}(\mathbf{u},t)}{\|\mathbf{f}(u,t)\|^3}=\mathbf{p} \tag{10.42}$$

But

$$\frac{\partial^2\mathbf{f}(\mathbf{u},t)}{\partial t^2}+\mu\frac{\mathbf{f}(\mathbf{u},t)}{\|\mathbf{f}(\mathbf{u},t)\|^3}=\mathbf{0} \tag{10.43}$$

because the six osculating orbital elements **u** evaluated at any instant of time define a Keplerian orbit for which Eq. (10.31) is valid. That means Eq. (10.42) reduces to

$$\sum_{\beta=1}^{6} \frac{\partial^2 \mathbf{f}(\mathbf{u}, t)}{\partial u_\beta \partial t} \frac{du_\beta}{dt} = \mathbf{p} \tag{10.44}$$

These are three conditions on the six functions $\mathbf{u}(t)$ in addition to the three conditions listed above in Eq. (10.39). If we simplify our notation by letting $\mathbf{r} = \mathbf{f}(\mathbf{u}, t)$ and $\mathbf{v} = \partial \mathbf{f}(\mathbf{u}, t)/\partial t$, then the six formulas (Eqs. 10.39 and 10.44), respectively, become

$$\sum_{\beta=1}^{6} \frac{\partial \mathbf{r}}{\partial u_\beta} \frac{du_\beta}{dt} = \mathbf{0} \tag{10.45a}$$

$$\sum_{\beta=1}^{6} \frac{\partial \mathbf{v}}{\partial u_\beta} \frac{du_\beta}{dt} = \mathbf{p} \tag{10.45b}$$

These are the six osculating conditions imposed by the variation of parameters to ensure that the osculating orbit at each point of a perturbed trajectory is Keplerian (two body) in nature.

In matrix form, Eqs. (10.45) may be written as

$$[\mathbf{L}]\{\dot{\mathbf{u}}\} = \{\mathbf{P}\} \tag{10.46a}$$

where

$$[\mathbf{L}] = \begin{bmatrix} \partial x_i/\partial u_\alpha \\ \partial v_i/\partial u_\alpha \end{bmatrix} \quad \{\dot{\mathbf{u}}\} = \{\dot{u}_\alpha\} \quad \{\mathbf{P}\} = \begin{Bmatrix} 0 \\ p_i \end{Bmatrix} \quad \begin{matrix} i = 1,2,3 \\ \alpha = 1,2,\ldots,6 \end{matrix} \tag{10.46b}$$

where $[\mathbf{L}]$ is the 6-by-6 Lagrangian matrix, and x_1, x_2, and x_3 are the xyz components of the position vector **r** in a Cartesian inertial reference. Likewise, v_1, v_2, and v_3 are the inertial velocity components, whereas p_1, p_2, and p_3 are the xyz components of the perturbing acceleration. The solution of Eqs. (10.46) yields the time variations of the osculating elements

$$\{\dot{\mathbf{u}}\} = [\mathbf{L}]^{-1}\{\mathbf{P}\} \tag{10.47}$$

An alternative straightforward manipulation of Eqs. (10.45) reveals that the Lagrange matrix $[\mathbf{L}]$ and the acceleration vector $\{\mathbf{P}\}$ may be written as

$$[\mathbf{L}] = \left[\sum_{i=1}^{3} \left(\frac{\partial x_i}{\partial u_\alpha} \frac{\partial v_i}{\partial u_\beta} - \frac{\partial v_i}{\partial u_\alpha} \frac{\partial x_i}{\partial u_\beta} \right) \right] \quad \{\mathbf{P}\} = \left\{ \sum_{i=1}^{3} p_i \frac{\partial x_i}{\partial u_\alpha} \right\} \quad \alpha, \beta = 1,2,\ldots,6 \tag{10.48}$$

These are the forms attributable to Lagrange, and in that case Eqs. (10.48) are known as Lagrange's planetary equations. If the perturbing forces, and hence the perturbing accelerations, are conservative, like gravity, then **p** is the spatial gradient of a scalar potential function R. That is,

$$p_i = \frac{\partial R}{\partial x_i} \quad i = 1,2,3$$

It follows from Eqs. (10.48) that $\{\mathbf{P}\}$ is the gradient of a function of the orbital elements,

$$\{\mathbf{P}\} = \left\{ \sum_{i=1}^{3} \frac{\partial R}{\partial x_i} \frac{\partial x_i}{\partial u_\alpha} \right\} = \left\{ \frac{\partial R}{\partial u_\alpha} \right\} \quad \alpha = 1,2,\ldots,6 \tag{10.49}$$

Lagrange's planetary equations for the variation of the six classical orbital elements a, e, t_p, Ω, i, and ω, as derived by Battin (1999) and others, may be written as follows:

$$\frac{da}{dt} = -\frac{2a^2}{\mu}\frac{\partial R}{\partial t_p} \tag{10.50a}$$

$$\frac{de}{dt} = -\frac{\sqrt{1-e^2}}{\sqrt{\mu a e}}\frac{\partial R}{\partial \omega} - \frac{a(1-e^2)}{\mu e}\frac{\partial R}{\partial t_p} \tag{10.50b}$$

$$\frac{dt_p}{dt} = \frac{2a^2}{\mu}\frac{\partial R}{\partial a} + \frac{a(1-e^2)}{\mu e}\frac{\partial R}{\partial e} \tag{10.50c}$$

$$\frac{d\Omega}{dt} = \frac{1}{\sqrt{\mu a(1-e^2)}\sin i}\frac{\partial R}{\partial i} \tag{10.50d}$$

$$\frac{di}{dt} = \frac{1}{\sqrt{\mu a(1-e^2)}}\left(\frac{1}{\tan i}\frac{\partial R}{\partial \omega} - \frac{1}{\sin i}\frac{\partial R}{\partial \Omega}\right) \tag{10.50e}$$

$$\frac{d\omega}{dt} = -\frac{1}{\sqrt{\mu a(1-e^2)}\tan i}\frac{\partial R}{\partial i} + \frac{\sqrt{1-e^2}}{\sqrt{\mu a e}}\frac{\partial R}{\partial e} \tag{10.50f}$$

Lagrange's planetary equations were obtained by writing the state vector components **r** and **v** in terms of the orbital parameters **u**, then taking partial derivatives to form the Lagrangian matrix [**L**], and finally inverting [**L**] to find the time variations of **u**, as in Eqs. (10.50). The more direct Gauss approach is to obtain the orbital elements **u** from the state vector, as in Algorithm 4.2, and then differentiate those expressions with respect to time to get the equations of variation. Gauss' form of Lagrange's planetary equations relaxes the requirement for perturbations to be conservative and avoids the lengthy though systematic procedure devised by Lagrange for computing the Lagrange matrix [**L**] in Eq. (10.48). We will pursue the Gauss approach in the next section.

10.7 GAUSS' VARIATIONAL EQUATIONS

Let u be an osculating element. Its time derivative is

$$\frac{du}{dt} = \frac{\partial u}{\partial \mathbf{r}} \cdot \frac{d\mathbf{r}}{dt} + \frac{\partial u}{\partial \mathbf{v}} \cdot \frac{d\mathbf{v}}{dt} \tag{10.51}$$

The acceleration $d\mathbf{v}/dt$ consists of the two-body part plus that due to the perturbation,

$$\frac{d\mathbf{v}}{dt} = -\mu\frac{\mathbf{r}}{r^3} + \mathbf{p} \tag{10.52}$$

Therefore, Eq. (10.51) becomes

$$\frac{du}{dt} = \frac{\partial u}{\partial \mathbf{r}} \cdot \frac{d\mathbf{r}}{dt} + \frac{\partial u}{\partial \mathbf{v}} \cdot \left(-\mu\frac{\mathbf{r}}{r^3}\right) + \frac{\partial u}{\partial \mathbf{v}} \cdot \mathbf{p}$$

or

$$\frac{du}{dt} = \left.\frac{du}{dt}\right)_{\text{two-body}} + \frac{\partial u}{\partial \mathbf{v}} \cdot \mathbf{p} \tag{10.53}$$

Except for the true, mean, and eccentric anomalies, the Keplerian elements are constant, so that $\mathrm{d}u/\mathrm{d}t)_{\text{two-body}} = 0$, whereas

$$\left.\frac{\mathrm{d}\theta}{\mathrm{d}t}\right)_{\text{two-body}} = \frac{h}{r^2} \tag{10.54}$$

$$\left.\frac{\mathrm{d}M}{\mathrm{d}t}\right)_{\text{two-body}} = n \tag{10.55}$$

$$\left.\frac{\mathrm{d}E}{\mathrm{d}t}\right)_{\text{two-body}} = \frac{na}{r} \tag{10.56}$$

We usually do orbital mechanics and define the orbital elements relative to a Cartesian inertial frame with origin at the center of the primary attractor. For earth-centered missions, we have employed the geocentric equatorial frame extensively throughout this book. Although not necessary, it may be convenient to imagine it as our inertial frame in what follows. The orthogonal unit vectors of the inertial frame are $\hat{\mathbf{I}}$, $\hat{\mathbf{J}}$, and $\hat{\mathbf{K}}$. As illustrated in Fig. 4.7, they form a right-hand triad, so that $\hat{\mathbf{K}} = \hat{\mathbf{I}} \times \hat{\mathbf{J}}$.

Another Cartesian inertial reference that we have employed for motion around a central attractor is the perifocal frame, as illustrated in Fig. 2.29. The orthogonal unit vectors along its $\overline{x}$, $\overline{y}$, and $\overline{z}$ axes are $\hat{\mathbf{p}}$, $\hat{\mathbf{q}}$, and $\hat{\mathbf{w}}$, respectively. The unit vector $\hat{\mathbf{p}}$ (not to be confused with the perturbing acceleration $\mathbf{p}$) lies in the direction of the eccentricity vector of the orbiting body, $\hat{\mathbf{w}}$ is normal to the orbital plane, and $\hat{\mathbf{q}}$ completes the right-handed triad: $\hat{\mathbf{w}} = \hat{\mathbf{p}} \times \hat{\mathbf{q}}$. Eq. (4.48), repeated here, gives the direction cosine matrix for the transformation from XYZ to $\overline{x}\overline{y}\overline{z}$:

$$[\mathbf{Q}]_{X\overline{x}} = \begin{bmatrix} -\sin\Omega\cos i\sin\omega + \cos\Omega\cos\omega & \cos\Omega\cos i\sin\omega + \sin\Omega\cos\omega & \sin i\sin\omega \\ -\sin\Omega\cos i\sin\omega + \cos\Omega\sin\omega & \cos\Omega\cos i\cos\omega + \sin\Omega\sin\omega & \sin i\cos\omega \\ \sin\Omega\sin i & -\cos\Omega\sin i & \cos i \end{bmatrix} \tag{10.57}$$

Ω, ω, and i are, respectively, the right ascension of the ascending node, argument of periapsis, and inclination.

We have also made use of noninertial local vertical/local horizontal (LVLH) frames in connection with the analysis of relative motion (see Fig. 7.1). It will be convenient to use such a reference for the perturbation analysis in this section. The orthogonal unit vectors of this frame are $\hat{\mathbf{r}}$, $\hat{\mathbf{s}}$, and $\hat{\mathbf{w}}$. $\hat{\mathbf{r}}$ is directed radially outward from the central attractor to the orbiting body and defines the direction of the local vertical. As in the perifocal frame, $\hat{\mathbf{w}}$ is the unit vector normal to the osculating orbital plane of the orbiting body. $\hat{\mathbf{w}}$ lies in the direction of the angular momentum vector $\mathbf{h}$, so that $\hat{\mathbf{w}} = \mathbf{h}/h$. The transverse unit vector $\hat{\mathbf{s}}$ (which we have previously denoted $\hat{\mathbf{u}}_{\perp}$) is normal to both $\hat{\mathbf{r}}$ and $\hat{\mathbf{w}}$ and therefore points in the direction of the orbiting body's local horizon: $\hat{\mathbf{s}} = \hat{\mathbf{w}} \times \hat{\mathbf{r}}$. The *rsw* and *pqr* frames are illustrated in Fig. 10.10.

The transformation from *pqw* to *rsw* is simply a rotation about the normal $\hat{\mathbf{w}}$ through the true anomaly θ. According to Eq. (4.34), the direction cosine matrix for this rotation is

$$[\mathbf{R}_3(\theta)] = \begin{bmatrix} \cos\theta & \sin\theta & 0 \\ -\sin\theta & \cos\theta & 0 \\ 0 & 0 & 1 \end{bmatrix}$$

Therefore, the transformation from *XYZ* to *rsw* is represented by $[\mathbf{Q}]_{Xr} = [\mathbf{R}_3(\theta)][\mathbf{Q}]_{X\overline{x}}$. Carrying out matrix multiplication and using the trig identities

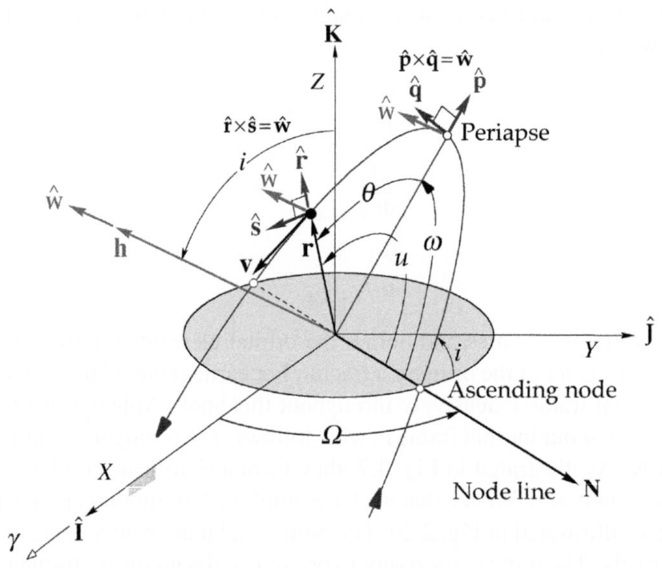

FIG. 10.10

The perifocal *pqw* frame and the local vertical/local horizontal rsw frame. *u* is the argument of latitude.

$$\sin(\omega+\theta) = \sin\omega\cos\theta + \cos\omega\sin\theta$$
$$\cos(\omega+\theta) = \cos\omega\cos\theta - \sin\omega\sin\theta$$

leads to

$$[\mathbf{Q}]_{Xr} = \begin{bmatrix} -\sin\Omega\cos i\sin u + \cos\Omega\cos u & \cos\Omega\cos i\sin u + \sin\Omega\cos u & \sin i\sin u \\ -\sin\Omega\cos i\sin u - \cos\Omega\sin u & \cos\Omega\cos i\cos u - \sin\Omega\sin u & \sin i\cos u \\ \sin\Omega\sin i & -\cos\Omega\sin i & \cos i \end{bmatrix} \tag{10.58}$$

where $u = \omega + \theta$. u is known as the argument of latitude. The direction cosine matrix $[\mathbf{Q}]_{Xr}$ could of course be obtained from Eq. (10.57) by simply replacing the argument of periapsis ω with the argument of latitude u.

In terms of its components in the inertial *XYZ* frame, the perturbing acceleration **p** is expressed analytically as follows:

$$\mathbf{p} = p_X\hat{\mathbf{I}} + p_Y\hat{\mathbf{J}} + p_Z\hat{\mathbf{K}} \tag{10.59}$$

whereas in the noninertial *rsw* frame

$$\mathbf{p} = p_r\hat{\mathbf{r}} + p_s\hat{\mathbf{s}} + p_w\hat{\mathbf{w}} \tag{10.60}$$

The transformation between these two sets of components is

$$\begin{Bmatrix} p_r \\ p_s \\ p_w \end{Bmatrix} = [\mathbf{Q}]_{Xr} \begin{Bmatrix} p_X \\ p_Y \\ p_Z \end{Bmatrix} \tag{10.61}$$

and

$$\begin{Bmatrix} p_X \\ p_Y \\ p_Z \end{Bmatrix} = [\mathbf{Q}]_{Xr}^T \begin{Bmatrix} p_r \\ p_s \\ p_w \end{Bmatrix} \tag{10.62}$$

where $[\mathbf{Q}]_{Xr}^T$ is the transpose of the direction cosine matrix in Eq. (10.58).

Each row of $[\mathbf{Q}]_{Xr}$ comprises the direction cosines of the unit vectors $\hat{\mathbf{r}}$, $\hat{\mathbf{s}}$, and $\hat{\mathbf{w}}$, respectively, relative to the inertial XYZ axes. Therefore, from Eq. (10.58), it is apparent that

$$\hat{\mathbf{r}} = (-\sin\Omega\cos i\sin u + \cos\Omega\cos u)\hat{\mathbf{I}} + (\cos\Omega\cos i\sin u + \sin\Omega\cos u)\hat{\mathbf{J}} + \sin i\sin u\hat{\mathbf{K}} \tag{10.63a}$$

$$\hat{\mathbf{s}} = (-\sin\Omega\cos i\cos u - \cos\Omega\cos u)\hat{\mathbf{I}} + (\cos\Omega\cos i\sin u - \sin\Omega\cos u)\hat{\mathbf{J}} + \sin i\cos u\hat{\mathbf{K}} \tag{10.63b}$$

$$\hat{\mathbf{w}} = \sin\Omega\sin i\hat{\mathbf{I}} - \cos\Omega\sin i\hat{\mathbf{J}} + \cos i\hat{\mathbf{K}} \tag{10.63c}$$

These will prove useful in the following derivation of the time derivatives of the osculating orbital elements h, e, θ, Ω, i, and ω. The formulas that we obtain for these derivatives will be our version of Gauss' planetary equations.

We will employ familiar orbital mechanics formulas from Chapters 1, 2, and 4 to find the time derivatives that we need. The procedure involves the use of basic differential calculus, some vector operations, and a lot of algebra. Those who prefer not to read through the derivations can skip to the summary listing of Gauss' planetary equations in Eq. (10.84).

10.7.1 VARIATION OF THE SPECIFIC ANGULAR MOMENTUM *h*

The time derivative of the angular momentum $\mathbf{h} = \mathbf{r} \times \mathbf{v}$ due to the perturbing acceleration $\mathbf{p}$ is

$$\frac{d\mathbf{h}}{dt} = \frac{d\mathbf{r}}{dt} \times \mathbf{v} + \mathbf{r} \times \frac{d\mathbf{v}}{dt} = \mathbf{v} \times \mathbf{v} + \mathbf{r} \times \left(-\mu\frac{\mathbf{r}}{r^3} + \mathbf{p}\right)$$

But $\mathbf{v} \times \mathbf{v} = \mathbf{r} \times \mathbf{r} = \mathbf{0}$, so this becomes

$$\frac{d\mathbf{h}}{dt} = \mathbf{r} \times \mathbf{p} \tag{10.64}$$

Since the magnitude of the angular momentum is $h = \sqrt{\mathbf{h}\cdot\mathbf{h}}$, its time derivative is

$$\frac{dh}{dt} = \frac{d}{dt}\sqrt{\mathbf{h}\cdot\mathbf{h}} = \frac{1}{2}\frac{1}{\sqrt{\mathbf{h}\cdot\mathbf{h}}}\left(2\mathbf{h}\cdot\frac{d\mathbf{h}}{dt}\right) = \frac{\mathbf{h}}{h}\cdot\frac{d\mathbf{h}}{dt} = \hat{\mathbf{w}}\cdot\frac{d\mathbf{h}}{dt}$$

Substituting Eq. (10.64) yields

$$\frac{dh}{dt} = \hat{\mathbf{w}}\cdot(\mathbf{r} \times \mathbf{p}) \tag{10.65}$$

Using the vector identity in Eq. (1.21) (interchange of the dot and the cross), we can modify this to read

$$\frac{dh}{dt} = (\hat{\mathbf{w}} \times \mathbf{r})\cdot\mathbf{p}$$

Since $\mathbf{r} = r\hat{\mathbf{r}}$ and $\hat{\mathbf{w}} \times \hat{\mathbf{r}} = \hat{\mathbf{s}}$, it follows that

$$\frac{dh}{dt} = rp_s \tag{10.66}$$

where $p_s = \mathbf{p}\cdot\hat{\mathbf{s}}$. Clearly, the variation of the angular momentum with time depends only on perturbation components that lie in the transverse (local horizon) direction.

10.7.2 VARIATION OF THE ECCENTRICITY *e*

The eccentricity may be found from Eq. (4.11),

$$e = \sqrt{1 + \frac{h^2}{\mu^2}\left(v^2 - \frac{2\mu}{r}\right)} \tag{10.67}$$

To find its time derivative, we will use Eq. (10.53),

$$\frac{de}{dt} = \frac{\partial e}{\partial \mathbf{v}} \cdot \mathbf{p} \tag{10.68}$$

since $de/dt)_{\text{two-body}} = 0$. Differentiating Eq. (10.67) with respect to $\mathbf{v}$ and using $\partial v^2/\partial \mathbf{v} = 2\mathbf{v}$ together with $\partial h^2/\partial \mathbf{v} = 2\mathbf{h} \times \mathbf{r}$, yields

$$\frac{\partial e}{\partial \mathbf{v}} = \frac{1}{2e}\frac{\partial}{\partial \mathbf{v}}\left[\frac{h^2}{\mu^2}\left(v^2 - \frac{2\mu}{r}\right)\right] = \frac{1}{\mu^2 e}\left[h^2\mathbf{v} + \left(v^2 - \frac{2\mu}{r}\right)(\mathbf{h} \times \mathbf{r})\right]$$

Substituting $\mathbf{v} = v_r\hat{\mathbf{r}} + v_s\hat{\mathbf{s}}$ and $\mathbf{h} \times \mathbf{r} = h\hat{\mathbf{w}} \times r\hat{\mathbf{r}} = hr\hat{\mathbf{s}}$ we get

$$\frac{\partial e}{\partial \mathbf{v}} = \frac{1}{\mu^2 e}\left\{h^2 v_r \hat{\mathbf{r}} + \left[hr\left(v^2 - \frac{2\mu}{r}\right) + h^2 v_s\right]\hat{\mathbf{s}}\right\}$$

Keeping in mind that v_s is the same as $v_\perp$, we can use Eq. (2.31) ($v_s = h/r$) along with Eq. (2.49) ($v_r = \mu e \sin\theta/h$) to write this as

$$\frac{\partial e}{\partial \mathbf{v}} = \frac{h}{\mu}\sin\theta\hat{\mathbf{r}} + \frac{h}{\mu^2 e}\left[r\left(v^2 - \frac{2\mu}{r}\right) + \frac{h^2}{r}\right]\hat{\mathbf{s}} \tag{10.69}$$

According to Problem 2.10, $v = (\mu/h)\sqrt{1 + 2e\cos\theta + e^2}$. Using the orbit formula (Eq. 2.45) in the form $e\cos\theta = (h^2/\mu r) - 1$, we can write this as $v = \sqrt{2\mu/r - \mu^2(1-e^2)/h^2}$ and substitute it into Eq. (10.69) to get

$$\frac{\partial e}{\partial \mathbf{v}} = \frac{h}{\mu}\sin\theta\hat{\mathbf{r}} + \frac{1}{\mu h}\left[(h^2 + \mu r)\cos\theta + \mu e r\right]\hat{\mathbf{s}}$$

Finally, it follows from Eq. (10.68) that

$$\frac{de}{dt} = \frac{h}{\mu}\sin\theta p_r + \frac{1}{\mu h}\left[(h^2 + \mu r)\cos\theta + \mu e r\right]p_s \tag{10.70}$$

where $p_r = \mathbf{p} \cdot \hat{\mathbf{r}}$ and $p_s = \mathbf{p} \cdot \hat{\mathbf{s}}$. Clearly, the eccentricity is affected only by perturbations that lie in the orbital plane.

10.7.3 VARIATION OF THE TRUE ANOMALY *θ*

According to Eqs. (10.53) and (10.54),

$$\frac{d\theta}{dt} = \frac{h}{r^2} + \frac{\partial \theta}{\partial \mathbf{v}} \cdot \mathbf{p} \tag{10.71}$$

To find $\partial\theta/\partial\mathbf{v}$, we start with the orbit formula (Eq. 2.45), writing it as

$$\mu e r = \frac{h^2 - \mu r}{\cos\theta} \tag{10.72}$$

Another basic equation containing the true anomaly is the radial speed formula (Eq. 2.49), from which we obtain

$$\mu e r = \frac{h\mathbf{r}\cdot\mathbf{v}}{\sin\theta} \tag{10.73}$$

Equating these two expressions for $\mu e r$ yields

$$\left(h^2 - \mu r\right)\sin\theta = h(\mathbf{r}\cdot\mathbf{v})\cos\theta$$

Applying the partial derivative with respect to $\mathbf{v}$ and rearranging the terms leads to

$$\left[\left(h^2 - \mu r\right)\cos\theta + h(\mathbf{r}\cdot\mathbf{v})\sin\theta\right]\frac{\partial\theta}{\partial\mathbf{v}} = h\frac{\partial(\mathbf{r}\cdot\mathbf{v})}{\partial\mathbf{v}}\cos\theta + \left[(\mathbf{r}\cdot\mathbf{v})\cos\theta - 2h\sin\theta\right]\frac{\partial h}{\partial\mathbf{v}}$$

The use of Eqs. (10.72) and (10.73) simplifies the square brackets on each side, so that

$$\mu e r\frac{\partial\theta}{\partial\mathbf{v}} = h\frac{\partial(\mathbf{r}\cdot\mathbf{v})}{\partial\mathbf{v}}\cos\theta - \left(h^2 + \mu r\right)\frac{\sin\theta}{h}\frac{\partial h}{\partial\mathbf{v}}$$

Making use of $\partial(\mathbf{r}\cdot\mathbf{v})/\partial\mathbf{v} = \mathbf{r}$ and $\partial h/\partial\mathbf{v} = (\mathbf{h}\times\mathbf{r})/h$ yields

$$\frac{\partial\theta}{\partial\mathbf{v}} = \frac{h}{\mu e}\cos\theta\hat{\mathbf{r}} - \frac{1}{e}\left(\frac{h^2}{\mu} + r\right)\frac{\sin\theta}{h}\hat{\mathbf{s}} \tag{10.74}$$

where it is to be recalled that $\hat{\mathbf{s}} = \hat{\mathbf{w}}\times\hat{\mathbf{r}}$. Substituting this expression for $\partial\theta/\partial\mathbf{v}$ into Eq. (10.71) finally yields the time variation of true anomaly,

$$\frac{d\theta}{dt} = \frac{h}{r^2} + \frac{1}{eh}\left[\frac{h^2}{\mu}\cos\theta p_r - \left(\frac{h^2}{\mu} + r\right)\sin\theta p_s\right] \tag{10.75}$$

Like the eccentricity, true anomaly is unaffected by perturbations that are normal to the orbital plane.

10.7.4 VARIATION OF RIGHT ASCENSION Ω

As we know, and as can be seen from Fig. 10.10, the right ascension of the ascending node is the angle between the inertial X axis and the node line vector $\mathbf{N}$. Therefore, it may be found by using the vector dot product,

$$\cos\Omega = \frac{\mathbf{N}}{\sqrt{\mathbf{N}\cdot\mathbf{N}}}\cdot\hat{\mathbf{I}}$$

We employed this formula in Algorithm 4.2. Alternatively, it should be evident from Fig. 10.11 that we can use the equation $\tan\Omega = -h_X/h_Y$. Since $h_X = \mathbf{h}\cdot\hat{\mathbf{I}}$ and $h_Y = \mathbf{h}\cdot\hat{\mathbf{J}}$, this can be written as

$$\tan\Omega = -\frac{\mathbf{h}\cdot\hat{\mathbf{I}}}{\mathbf{h}\cdot\hat{\mathbf{J}}} \tag{10.76}$$

The time derivative of Eq. (10.76) is

$$\frac{1}{\cos^2}\frac{d\Omega}{dt} = -\frac{(\mathbf{h}\cdot\hat{\mathbf{J}})\left(\frac{d\mathbf{h}}{dt}\cdot\hat{\mathbf{I}}\right) - (\mathbf{h}\cdot\hat{\mathbf{I}})\left(\frac{d\mathbf{h}}{dt}\cdot\hat{\mathbf{J}}\right)}{(\mathbf{h}\cdot\hat{\mathbf{J}})^2} = -\frac{h_Y\left(\frac{d\mathbf{h}}{dt}\cdot\mathbf{I}\right) - h_X\left(\frac{d\mathbf{h}}{dt}\cdot\hat{\mathbf{J}}\right)}{h_Y^2}$$

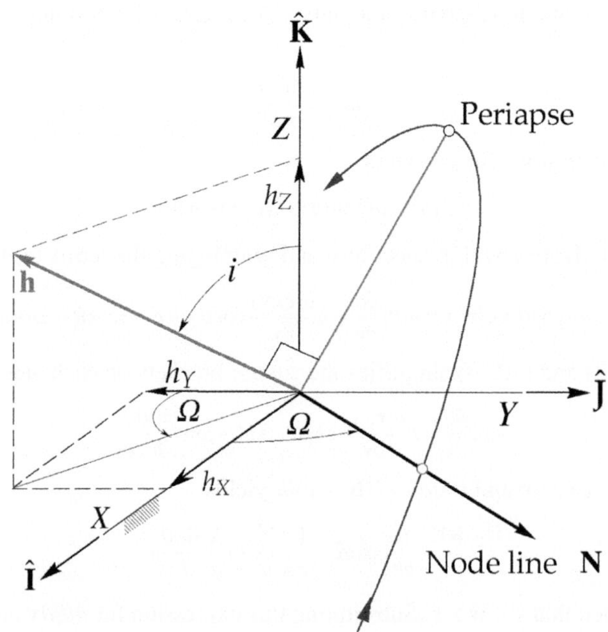

FIG. 10.11

Relation between the components of angular momentum and the node angle Ω.

or, more simply,

$$\frac{d\Omega}{dt} = \cos^2\Omega \frac{d\mathbf{h}}{dt} \cdot \frac{h_X\hat{\mathbf{J}} - h_Y\hat{\mathbf{I}}}{h_Y{}^2} \tag{10.77}$$

Since $\mathbf{h} = h\hat{\mathbf{w}}$ and $\hat{\mathbf{w}}$ is given by Eq. (10.63c), we have

$$\mathbf{h} = h\sin\Omega\sin i\hat{\mathbf{I}} - h\cos\Omega\sin i\hat{\mathbf{J}} + h\cos i\hat{\mathbf{K}}$$

This shows that $h_X = h \sin \Omega \sin i$ and $h_Y = -h \cos \Omega \sin i$, so that Eq. (10.77) becomes

$$\frac{d\Omega}{dt} = \frac{d\mathbf{h}}{dt} \cdot \frac{h\sin i(\cos\Omega\hat{\mathbf{I}} + \sin\Omega\hat{\mathbf{J}})}{(h\sin i)^2} = \frac{1}{h\sin i}\hat{\mathbf{N}} \cdot \frac{d\mathbf{h}}{dt}$$

in which $\hat{\mathbf{N}}$ is the unit vector along the node line. Recalling from Eq. (10.64) that $d\mathbf{h}/dt = \mathbf{r} \times \mathbf{p}$ yields

$$\frac{d\Omega}{dt} = \frac{1}{h\sin i}\hat{\mathbf{N}} \cdot (\mathbf{r} \times \mathbf{p}) = \frac{1}{h\sin i}(\hat{\mathbf{N}} \times \mathbf{r}) \cdot \mathbf{p} \tag{10.78}$$

where we interchanged the dot and the cross by means of the identity in Eq. (1.21). The angle between the node line $\hat{\mathbf{N}}$ and the radial $\mathbf{r}$ is the argument of latitude u, and since $\hat{\mathbf{w}}$ is normal to the plane of $\hat{\mathbf{N}}$ and $\mathbf{r}$, it follows from the definition of the cross product operation that $\hat{\mathbf{N}} \times \mathbf{r} = r\sin u\hat{\mathbf{w}}$. Therefore, Eq. (10.78) may be written as

$$\frac{d\Omega}{dt} = \frac{r\sin u}{h\sin i}\hat{\mathbf{w}} \cdot \mathbf{p} \tag{10.79}$$

or, since $\hat{\mathbf{w}}\cdot\mathbf{p}=p_w$,

$$\frac{d\Omega}{dt}=\frac{r\sin u}{h\sin i}p_w \tag{10.80}$$

Clearly, the variation of right ascension Ω is influenced only by perturbations that are normal to the orbital plane.

10.7.5 VARIATION OF THE INCLINATION *i*

The orbital inclination, which is the angle i between the Z axis and the normal to the orbital plane (Fig. 10.10), may be found from Eq. (4.7),

$$\cos i=\frac{\mathbf{h}\cdot\hat{\mathbf{K}}}{h}$$

Differentiating this expression with respect to time yields

$$-\frac{di}{dt}\sin i=\frac{1}{h}\frac{d\mathbf{h}}{dt}\cdot\hat{\mathbf{K}}-\frac{1}{h^2}\frac{dh}{dt}(\mathbf{h}\cdot\hat{\mathbf{K}})$$

Using Eqs. (10.64) and (10.65), along with $\mathbf{h}\cdot\hat{\mathbf{K}}=h\cos i$, we find that

$$\frac{di}{dt}\sin i=\frac{r}{h}(\hat{\mathbf{w}}\cos i-\hat{\mathbf{K}})\cdot(\hat{\mathbf{r}}\times\mathbf{p})$$

Replacing $\hat{\mathbf{w}}$ by the expression in Eq. (10.63c) yields

$$\begin{aligned}\frac{di}{dt}&=\frac{r}{h}(\sin\Omega\cos i\hat{\mathbf{I}}-\cos\Omega\cos i\mathbf{J}-\sin i\hat{\mathbf{K}})\cdot(\hat{\mathbf{r}}\times\mathbf{p})\\&=\frac{r}{h}\left[(\sin\Omega\cos i\hat{\mathbf{I}}-\cos\Omega\cos i\hat{\mathbf{J}}-\sin i\hat{\mathbf{K}})\times\hat{\mathbf{r}}\right]\cdot\mathbf{p}\end{aligned}$$

where we once again used Eq. (1.21) to interchange the dot and the cross. Replacing the unit vector $\hat{\mathbf{r}}$ by Eq. (10.63a) and using the familiar determinant formula for the cross product, we get

$$\frac{di}{dt}=\frac{r}{h}\begin{vmatrix}\hat{\mathbf{I}} & \hat{\mathbf{J}} & \hat{\mathbf{K}}\\ \sin\Omega\cos i & -\cos\Omega\cos i & -\sin i\\ -\sin\Omega\cos i\sin u+\cos\Omega\cos u & \cos\Omega\cos i\sin u+\sin\Omega\cos u & \sin i\sin u\end{vmatrix}\cdot\mathbf{p}$$

Expanding the determinant and recognizing $\hat{\mathbf{w}}$ from Eq. (10.63c), we find

$$\frac{di}{dt}=\frac{r}{h}\cos u(\sin\Omega\sin i\hat{\mathbf{I}}-\cos\Omega\sin i\hat{\mathbf{J}}+\cos i\hat{\mathbf{K}})\cdot\mathbf{p}=\frac{r}{h}\cos u(\hat{\mathbf{w}}\cdot\mathbf{p})$$

Since $\hat{\mathbf{w}}\cdot\mathbf{p}=p_w$,

$$\frac{di}{dt}=\left(\frac{r}{h}\cos u\right)p_w \tag{10.81}$$

Like Ω, the orbital inclination is affected only by perturbation components that are normal to the orbital plane.

10.7.6 VARIATION OF ARGUMENT OF PERIAPSIS ω

The arguments of periapsis and latitude are related by $\omega = u - \theta$. Let us first seek an expression for du/dt and then obtain the variation dω/dt from the fact that dω/dt = du/dt − dθ/dt, where we found dθ/dt above in Eq. (10.75).

Since the argument of latitude is the angle u between the node line vector **N** and the position vector **r**, it is true that $\cos u = \hat{\mathbf{r}} \cdot \hat{\mathbf{N}}$. Differentiating this expression with respect to the velocity vector **v** and noting that $\hat{\mathbf{N}} = \cos\Omega \hat{\mathbf{I}} + \sin\Omega \hat{\mathbf{J}}$, we get

$$\frac{\partial u}{\partial \mathbf{v}} = -\frac{1}{\sin u}\hat{\mathbf{r}} \cdot \frac{\partial \hat{\mathbf{N}}}{\partial \mathbf{v}} = \frac{1}{\sin u}\left(\hat{\mathbf{r}} \cdot \hat{\mathbf{I}} \sin\Omega - \hat{\mathbf{r}} \cdot \hat{\mathbf{J}} \cos\Omega\right)\frac{\partial \Omega}{\partial \mathbf{v}}$$

Using the expression for the radial unit vector $\hat{\mathbf{r}}$ in Eq. (10.63a), we conclude that

$$\hat{\mathbf{r}} \cdot \hat{\mathbf{I}} \sin\Omega - \hat{\mathbf{r}} \cdot \hat{\mathbf{J}} \cos\Omega = -\cos i \sin u$$

which simplifies our result,

$$\frac{\partial u}{\partial \mathbf{v}} = -\cos i \frac{\partial \Omega}{\partial \mathbf{v}} \tag{10.82}$$

From Eq. (10.53), we know that dΩ/dt = $(\partial\Omega/\partial\mathbf{v}) \cdot \mathbf{p}$, whereas according to Eq. (10.79), $\mathrm{d}\Omega/\mathrm{d}t = (r \sin u / h \sin i)\hat{\mathbf{w}} \cdot \mathbf{p}$. It follows that

$$\frac{\partial \Omega}{\partial \mathbf{v}} = \frac{r \sin u}{h \sin i}\hat{\mathbf{w}}$$

Therefore, Eq. (10.82) yields

$$\frac{\partial u}{\partial \mathbf{v}} = -\frac{r \sin u}{h \tan i}\hat{\mathbf{w}}$$

Finally, since du/dt = $(\partial u/\partial \mathbf{v}) \cdot \mathbf{p}$, we conclude that

$$\frac{\mathrm{d}u}{\mathrm{d}t} = -\frac{r \sin u}{h \tan i} p_w$$

Substituting this result into dω/dt = du/dt − dθ/dt and making use of Eq. (10.74) (the variation of θ due solely to perturbations) yields

$$\frac{\mathrm{d}\omega}{\mathrm{d}t} = -\frac{1}{eh}\left[\frac{h^2}{\mu}\cos\theta p_r - \left(r + \frac{h^2}{\mu}\right)\sin\theta p_s\right] - \frac{r \sin(\omega + \theta)}{h \tan i} p_w \tag{10.83}$$

We see that dω/dt depends on all three components of the perturbing acceleration.

For convenience, let us summarize the Gauss form of the planetary equations that govern variations in the orbital elements: angular momentum h, eccentricity e, true anomaly θ, node angle Ω, inclination i, and argument of periapsis ω.

$$\frac{\mathrm{d}h}{\mathrm{d}t} = r p_s \tag{10.84a}$$

$$\frac{\mathrm{d}e}{\mathrm{d}t} = \frac{h}{\mu}\sin\theta p_r + \frac{1}{\mu h}\left[\left(h^2 + \mu r\right)\cos\theta + \mu e r\right] p_s \tag{10.84b}$$

$$\frac{d\theta}{dt}=\frac{h}{r^2}+\frac{1}{eh}\left[\frac{h^2}{\mu}\cos\theta p_r-\left(r+\frac{h^2}{\mu}\right)\sin\theta p_s\right] \tag{10.84c}$$

$$\frac{d\Omega}{dt}=\frac{r}{h\sin i}\sin(\omega+\theta)p_w \tag{10.84d}$$

$$\frac{di}{dt}=\frac{r}{h}\cos(\omega+\theta)p_w \tag{10.84e}$$

$$\frac{d\omega}{dt}=-\frac{1}{eh}\left[\frac{h^2}{\mu}\cos\theta p_r-\left(r+\frac{h^2}{\mu}\right)\sin\theta p_s\right]-\frac{r\sin(\omega+\theta)}{h\tan i}p_w \tag{10.84f}$$

where, from Eq. (2.45), $r = h^2/[\mu(1 + e\cos\theta)]$.

Given the six orbital elements h_0, e_0, θ_0, Ω_0, i_0, and ω_0 at time t_0 and the functional form of the perturbing acceleration **p**, we can numerically integrate the six planetary equations to obtain the osculating orbital elements and therefore the state vector at subsequent times. Formulas for the variations of alternative orbital elements can be derived, but it is not necessary here because at any instant all other osculating orbital parameters are found in terms of the six listed above. For example,

Semimajor axis (Eq. 2.72): $a=\frac{h^2}{\mu}\frac{1}{1-e^2}$

Eccentric anomaly (Eq. 3.13): $E=2\tan^{-1}\left(\sqrt{\frac{1-e}{1+e}}\tan\frac{\theta}{2}\right)$

Mean anomaly (Eq. 3.6): $M=2\tan^{-1}\left(\sqrt{\frac{1-e}{1+e}}\tan\frac{\theta}{2}\right)-\frac{e\sqrt{1-e^2}\sin\theta}{1+e\cos}$

EXAMPLE 10.4

Obtain the form of Gauss' planetary equations for a perturbing force that is tangent to the orbit.

Solution

If the perturbing force is tangent to the orbit, then the perturbing acceleration **p** lies in the direction of the velocity vector **v**,

$$\mathbf{p}=p_v\frac{\mathbf{v}}{v}$$

The radial and transverse components of the perturbing acceleration are found by projecting **p** onto the radial and transverse directions

$$p_r=\left(p_v\frac{\mathbf{v}}{v}\right)\cdot\hat{\mathbf{r}}=\frac{p_v}{v}(\hat{\mathbf{v}}\cdot\hat{\mathbf{r}})=\frac{p_v}{v}v_r$$
$$p_s=\left(p_v\frac{\mathbf{v}}{v}\right)\cdot\hat{\mathbf{s}}=\frac{p_v}{v}(\mathbf{v}\cdot\hat{\mathbf{s}})=\frac{p_v}{v}v_s$$

The component p_w normal to the orbital plane is zero. From Eqs. (2.31) and (2.49), we have

$$v_s=\frac{h}{r}\quad v_r=\frac{\mu}{h}e\sin\theta$$

Therefore, for a tangential perturbation, the *rsw* components of acceleration are

$$p_r=\frac{\mu}{vh}ep_v\sin\theta\quad p_s=\frac{h}{vr}p_v\quad p_w=0$$

Substituting these expressions into Eq. (10.84) leads to Gauss' planetary equations for tangential perturbations,

$$\boxed{\begin{array}{lll}\dfrac{dh}{dt}=\dfrac{h}{v}p_v & \dfrac{de}{dt}=\dfrac{2}{v}(e+\cos\theta)p_v & \dfrac{d\theta}{dt}=\dfrac{h}{r^2}-\dfrac{2}{ev}p_v\sin\theta \\ \dfrac{d\Omega}{dt}=0 & \dfrac{di}{dt}=0 & \dfrac{d\omega}{dt}=\dfrac{2}{ev}p_s\sin\theta\end{array}}\quad \text{Tangential perturbation} \tag{10.85}$$

Clearly, the right ascension and the inclination are unaffected.

If the perturbing acceleration is due to a tangential thrust T, then $p_v = T/m$, where m is the instantaneous mass of the spacecraft. If the thrust is in the direction of the velocity **v**, then p_v is positive and tends to speed up the spacecraft, producing an outwardly spiraling trajectory like that shown in Fig. 10.1. According to Eq. (10.85), a positive tangential acceleration causes the eccentricity to increase and the perigee to advance (rotating the apse line counterclockwise). Both of these effects are evident in Fig. 10.1. If the thrust is opposite to the direction of **v**, then the effects are opposite: the spacecraft slows down, spiraling inward.

Atmospheric drag acts opposite to the direction of motion. If we neglect the rotation of the atmosphere, then according to Eq. (10.12), the tangential acceleration is

$$p_v = -\frac{1}{2}\rho v^2 B \tag{10.86}$$

where v is the inertial speed, and B is the ballistic coefficient. Substituting this expression into Eq. (10.85) yields Gauss' planetary equations for atmospheric drag,

$$\begin{array}{lll}\dfrac{dh}{dt}=-\dfrac{B}{2}h\rho v & \dfrac{de}{dt}=-B(e+\cos\theta)\rho v & \dfrac{d\theta}{dt}=\dfrac{h}{r^2}+B\dfrac{\rho v}{e}\sin\theta \\ \dfrac{d\Omega}{dt}=0 & \dfrac{di}{dt}=0 & \dfrac{d\omega}{dt}=-B\dfrac{\rho v}{e}\sin\theta\end{array} \tag{10.87}$$

EXAMPLE 10.5

Find the components p_r, p_s, and p_w of the gravitational perturbation of Eq. (10.30) in the rsw frame of Fig. 10.10.

Solution

The transformation from XYZ to rsw is given by Eq. (10.61). Using the direction cosine matrix in Eq. (10.58) we therefore have

$$\begin{Bmatrix}p_r\\p_s\\p_w\end{Bmatrix}=\frac{3}{2}\frac{J_2\mu R^2}{r^4}\begin{bmatrix}-\sin\Omega\cos i\sin u+\cos\Omega\cos u & \cos\Omega\cos i\sin u+\sin\Omega\cos u & \sin i\sin u\\ -\sin\Omega\cos i\cos u-\cos\Omega\sin u & \cos\Omega\cos i\cos u-\sin\Omega\sin u & \sin i\cos u\\ \sin\Omega\sin i & -\cos\Omega\sin i & \cos i\end{bmatrix}\cdot\begin{Bmatrix}\dfrac{X}{r}\left(5\dfrac{Z^2}{r^2}-1\right)\\ \dfrac{Y}{r}\left(5\dfrac{Z^2}{r^2}-1\right)\\ \dfrac{Z}{r}\left(5\dfrac{Z^2}{r^2}-3\right)\end{Bmatrix} \tag{a}$$

Before carrying out the multiplication, note that in the spherical coordinate system of Fig. 10.12,

$$Z=r\sin\delta \quad X=r\cos\delta\cos\alpha \quad Y=r\cos\delta\sin\alpha \tag{b}$$

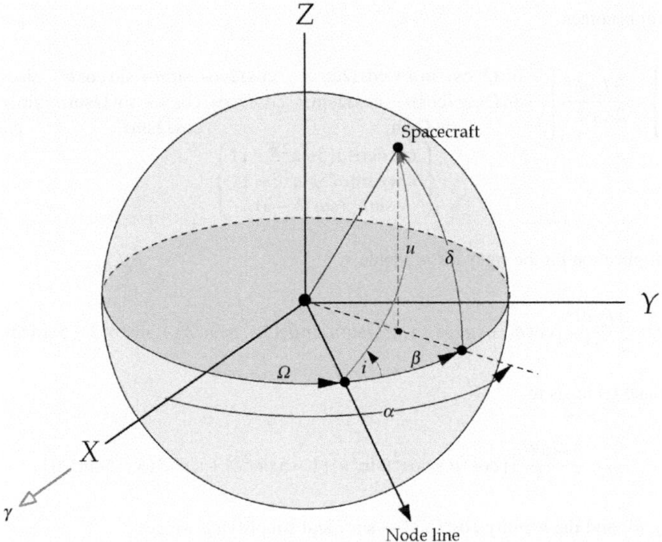

FIG. 10.12

Spherical coordinates and the geocentric equatorial frame.

where α is the azimuth angle measured in the *XY* plane positive from the *X* axis. (We use α instead of the traditional θ to avoid confusion with true anomaly.) The declination δ is the complement of the usual polar angle ϕ, which is measured positive from the polar (*Z*) axis toward the equator. As shown in Fig. 10.12, δ is measured positive northward from the equator. Using δ instead of ϕ makes it easier to take advantage of spherical trigonometry formulas. The angle β in Fig. 10.12 is the difference between the azimuth angle α and the right ascension of the ascending node,

$$\beta = \alpha - \Omega \tag{c}$$

On the unit sphere, the angles i, u, β, and δ appear as shown in Fig. 10.12. Spherical trigonometry (Zwillinger, 2012) yields the following relations among these four angles:

$$\sin\delta = \sin i \sin u \tag{d}$$

$$\cos u = \cos\delta\cos\beta \tag{e}$$

$$\sin\beta = \tan\delta\cot i \tag{f}$$

$$\cos u = \cos\delta\cos\beta \tag{g}$$

From Eqs. (d) and (f), we find

$$\sin\beta = (\sin\delta/\cos\delta)(\cos i/\sin i) = (\sin i \sin u/\cos\delta)(\cos i/\sin i)$$

or

$$\sin\beta = \frac{\sin u \cos i}{\cos\delta} \tag{h}$$

whereas Eq. (g) may be written as

$$\cos\beta = \frac{\cos u}{\cos\delta} \tag{i}$$

Using Eqs. (b), Eq. (a) becomes

$$\begin{Bmatrix} p_r \\ p_s \\ p_w \end{Bmatrix} = \frac{3}{2}\frac{J_2\mu R^2}{r4} \begin{bmatrix} -\sin\Omega\cos i\sin u + \cos\Omega\cos u & \cos\Omega\cos i\sin u + \sin\cos u & \sin i\sin u \\ -\sin\Omega\cos i\cos u - \cos\Omega\sin u & \cos\Omega\cos i\cos u - \sin\Omega\sin u & \sin i\cos u \\ \sin\Omega\sin i & -\cos\Omega\sin i & \cos i \end{bmatrix} \cdot \begin{Bmatrix} \cos\delta\cos\alpha(5\sin^2\delta - 1) \\ \cos\delta\sin\alpha(5\sin^2\delta - 1) \\ \sin\delta(5\sin^2\delta - 3) \end{Bmatrix} \tag{j}$$

Carrying out the multiplication on the right for p_r yields

$$p_r = \frac{3}{2}\frac{J_2\mu R^2}{r^4}\left[\cos\delta(\cos u\cos\beta + \cos i\sin u\sin\beta)(1 - 5\sin^2\delta) + \sin^2\delta(3 - 5\sin^2\delta)\right]$$

Substituting Eqs. (h) and (i) leads to

$$p_r = \frac{3}{2}\frac{J_2\mu R^2}{r^4}\left[(\cos^2 u + \cos^2 i\sin^2 u)(1 - 5\sin^2\delta) + \sin^2\delta(3 - 5\sin^2\delta)\right]$$

After substituting Eq. (d) and the identity $\cos^2 i = 1 - \sin^2 i$ and simplifying we get

$$p_r = -\frac{3}{2}\frac{J_2\mu R^2}{r^2}(1 - 3\sin^2 i\sin^2 u)$$

In a similar fashion we find p_s and p_w, so that in summary

$$\boxed{\begin{aligned} p_r &= -\frac{3}{2}\frac{J_2\mu R^2}{r^4}\left[1 - 3\sin^2 i\cdot\sin^2(\omega + \theta)\right] \\ p_s &= -\frac{3}{2}\frac{J_2\mu R^2}{r^4}\sin^2 i\cdot\sin 2(\omega + \theta) \\ p_w &= -\frac{3}{2}\frac{J_2\mu R^2}{r^4}\sin 2i\cdot\sin(\omega + \theta) \end{aligned}} \tag{10.88}$$

We may now substitute p_r, p_s, and p_w from Eq. (10.88) into Gauss' planetary equations (Eq. 10.84) to obtain the variation of the osculating elements due to the J_2 gravitational perturbation (keeping in mind that $u = \omega + \theta$). After some straightforward algebraic manipulations, we obtain:

$$\frac{dh}{dt} = -\frac{3}{2}\frac{J_2\mu R^2}{r^3}\sin^2 i\sin 2u \tag{10.89a}$$

$$\frac{de}{dt} = \frac{3}{2}\frac{J_2\mu R^2}{hr^3}\left\{\frac{h^2}{\mu r}\sin\theta(3\sin^2 i\sin^2 u - 1) - \sin 2u\sin^2 i[(3 + e\cos\theta)\cos\theta + e]\right\} \tag{10.89b}$$

$$\frac{d\theta}{dt} = \frac{h}{r^2} + \frac{3}{2}\frac{J_2\mu R^2}{ehr^3}\left[\frac{h^2}{\mu r}\cos\theta(3\sin^2 i\sin^2 u - 1) + (2 + e\cos\theta)\sin 2u\sin^2 i\sin\theta\right] \tag{10.89c}$$

$$\frac{d\Omega}{dt} = -3\frac{J_2\mu R^2}{hr^3}\sin^2 u\cos i \tag{10.89d}$$

$$\frac{di}{dt} = -\frac{3}{4}\frac{J_2\mu R^2}{hr^3}\sin 2u\sin 2i \tag{10.89e}$$

$$\frac{d\omega}{dt} = \frac{3}{2}\frac{J_2\mu R^2}{ehr^3}\left[\frac{h^2}{\mu r}\cos\theta(1-3\sin^2 i\sin^2 u) - (2+e\cos\theta)\sin 2u\sin^2 i\sin\theta + 2e\cos^2 i\sin^2 u\right] \tag{10.89f}$$

EXAMPLE 10.6

At time $t=0$, an earth satellite has the following orbital parameters:

Perigee radius: $r_p = 6678\,\text{km}$ (300 km altitude)

Apogee radius: $r_a = 9440\,\text{km}$ (3062 km altitude)

Right ascension of the ascending node: $\Omega = 45°$ (a)

Inclination: $i = 28°$ (b)

Argument of perigee: $\omega = 30°$ (c)

True anomaly: $\theta = 40°$ (d)

Use Gauss' variational equations (Eqs. 10.89) to determine the effect of the J_2 perturbation on the variation of orbital elements h, e, Ω, i, and ω over the next 48 hours.

Solution

From the given information, we find the eccentricity and the angular momentum from Eqs. (2.84) and (2.50) with $\mu = 398{,}600\,\text{km}^3/\text{s}^2$,

$$e = \frac{r_a - r_p}{r_a + r_p} = 0.17136 \tag{e}$$

$$h = \sqrt{\mu(1+e)r_p} = 55{,}839\,\text{km}^2/\text{s} \tag{f}$$

We write the differential equations in Eqs. (10.89) as $\mathbf{dy}/dt = \mathbf{f(y)}$, where

$$\mathbf{y} = \lfloor h \;\; e \;\; \theta \;\; \Omega \;\; i \;\; \omega \rfloor^T$$
$$\mathbf{f} = \lfloor \dot{h} \;\; \dot{e} \;\; \dot{\theta} \;\; \dot{\Omega} \;\; \dot{i} \;\; \dot{\omega} \rfloor^T$$

and the rates $\dot{h}$, $\dot{e}$, etc., are given explicitly in Eqs. (10.89), with $r = h^2/[\mu(1 + e\cos\theta)]$ and $u = \omega + \theta$. With Eqs. (a) through (f) comprising the initial conditions for vector $\mathbf{y}_0$, we can solve for $\mathbf{y}$ on the time interval $[t_0, t_f]$ using MATLAB's numerical integrator *ode45* (or one of those described in Section 1.8). For $t_0 = 0$ and $t_f = 48\,\text{h}$, the solutions are plotted in Fig. 10.13. The MATLAB script *Example_10_06.m* for this example is in Appendix D.42.

The data for this problem are the same as for Example 10.2, wherein the results are identical to what we found here. In Example 10.2, the perturbed state vector (**r**,**v**) was found over a 2-day time interval, at each point of which the osculating elements were then derived from Algorithm 4.2. Here, on the other hand, we directly obtained the osculating elements, from which at any time on the solution interval we can calculate the state vector by means of Algorithm 4.5.

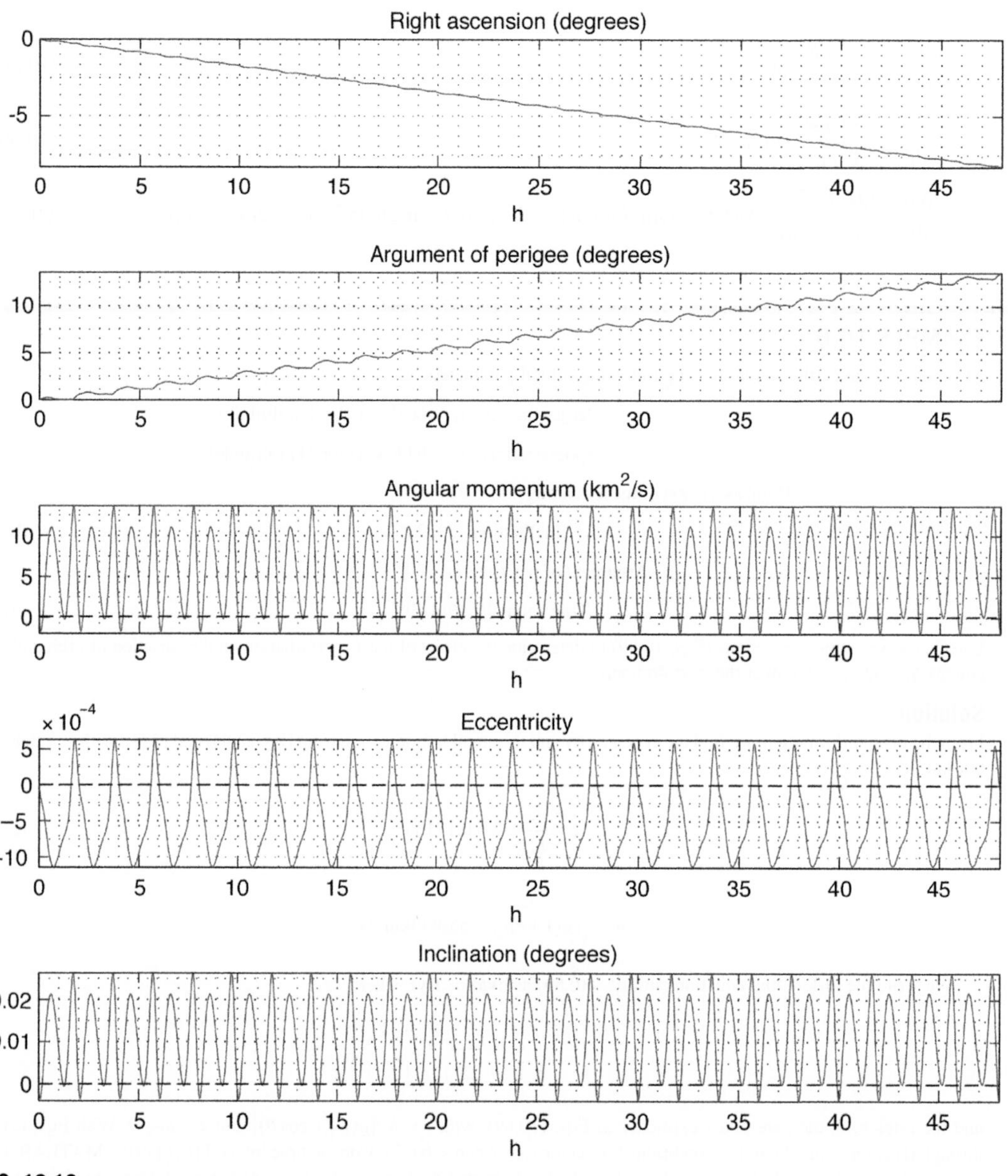

FIG. 10.13

Variation of the osculating elements due to J_2 perturbation.

10.8 METHOD OF AVERAGING

One advantage of Gauss' planetary equations for the J_2 perturbation is that they show explicitly the dependence of the element variations on the elements themselves and the position of the spacecraft in the gravitational field. We see in Eqs. (10.89) that the short-period "ripples" superimposed on the long-term secular trends are due to the presence of the trigonometric terms in the true anomaly θ. To separate the secular terms from the short-period terms the method of averaging is used. Let f be an osculating element and $\dot{f}$ its variation as given in Eqs. (10.89). Since $\dot{f} = (\mathrm{d}f/\mathrm{d}\theta)\dot{\theta}$, the average of $\dot{f}$ over one orbit is

$$\bar{\dot{f}} = \overline{\frac{\mathrm{d}f}{\mathrm{d}\theta}} n \tag{10.90}$$

where n is the mean motion (Eq. 3.9)

$$n = \bar{\dot{\theta}} = \frac{1}{T}\int_0^T \frac{\mathrm{d}\theta}{\mathrm{d}t}\mathrm{d}t = \int_0^{2\pi} \mathrm{d}\theta = \frac{2\pi}{T}$$

Substituting the formula for the period (Eq. 2.83), we can write n as

$$n = \sqrt{\frac{\mu}{a^3}} \tag{10.91}$$

For the average value of $\mathrm{d}f/\mathrm{d}\theta$ we have

$$\overline{\frac{\mathrm{d}f}{\mathrm{d}\theta}} = \frac{1}{2\pi}\int_0^{2\pi} \frac{\mathrm{d}f}{\mathrm{d}\theta}\mathrm{d}\theta = \frac{1}{2\pi}\int_0^{2\pi} \frac{\mathrm{d}f}{\mathrm{d}t}\frac{1}{\dot{\theta}}\mathrm{d}\theta = \frac{1}{2\pi}\int_0^{2\pi} \frac{\mathrm{d}f}{\mathrm{d}t}\frac{r^2}{h}\mathrm{d}\theta$$

where we made use of Eq. (2.47). Substituting this into Eq. (10.90) yields a formula for the time-averaged variation,

$$\bar{\dot{f}} = \frac{n}{2\pi}\int_0^{2\pi} \frac{\mathrm{d}f}{\mathrm{d}t}\frac{r^2}{h}\mathrm{d}\theta \tag{10.92}$$

In doing the integral, the only variable is θ, all the other orbital elements are held fixed.

Let us use Eq. (10.92) to compute the orbital averages of each of the rates in Eqs. (10.89). In doing so, we will make frequent use of the orbit formula $r = h^2/[\mu(1 + e\cos\theta)]$.

10.8.1 ORBITAL-AVERAGED ANGULAR MOMENTUM VARIATION

$$\begin{aligned}
\bar{\dot{h}} &= \frac{n}{2\pi}\int_0^{2\pi} \frac{\mathrm{d}h}{\mathrm{d}t}\frac{r^2}{h}\mathrm{d}\theta = \frac{n}{2\pi}\int_0^{2\pi}\left(-\frac{3}{2}\frac{J_2\mu R^2}{r^3}\sin^2 i \sin 2u\right)\frac{r^2}{h}\mathrm{d}\theta \\
&= \frac{n}{2\pi}\left(-\frac{3}{2}\frac{J_2\mu R^2}{h}\sin^2 i\right)\int_0^{2\pi}\left(\frac{1}{r}\sin 2u\right)\mathrm{d}\theta \\
&= \frac{n}{2\pi}\left(-\frac{3}{2}\frac{J_2\mu^2 R^2}{h^3}\sin^2 i\right)\int_0^{2\pi}(1 + e\cos\theta)\sin 2(\omega + \theta)\mathrm{d}\theta
\end{aligned}$$

Evaluate the integral as follows, remembering to hold the orbital element ω constant:

$$\begin{aligned}\int_0^{2\pi}(1+e\cos\theta)\sin 2(\omega+\theta)d\theta &= \int_0^{2\pi}(1+e\cos\theta)(\sin 2\omega\cos 2\theta+\cos 2\omega\sin 2\theta)d\theta \\ &= \sin 2\omega\int_0^{2\pi}\cos 2\theta d\theta+\cos 2\omega\int_0^{2\pi}\sin 2\theta d\theta \\ &\quad +e\sin 2\omega\int_0^{2\pi}\cos\theta\cos 2\theta d\theta+e\cos 2\omega\int_0^{2\pi}\cos\theta\sin 2\theta d\theta\end{aligned}$$

The four integrals on the right are all zero,

$$\int_0^{2\pi}\cos 2\theta d\theta=\left.\frac{\sin 2\theta}{2}\right|_0^{2\pi}=0 \qquad \int_0^{2\pi}\sin 2\theta d\theta=-\left.\frac{\cos 2\theta}{2}\right|_0^{2\pi}=0$$

$$\int_0^{2\pi}\cos\theta\cos 2\theta d\theta=\left.\left(\frac{\sin\theta}{2}+\frac{\sin 3\theta}{6}\right)\right|_0^{2\pi}=0 \qquad \int_0^{2\pi}\cos\theta\sin 2\theta d\theta=-\left.\left(\frac{\cos\theta}{2}+\frac{\cos 3\theta}{6}\right)\right|_0^{2\pi}=0$$

Therefore, $\bar{\dot{h}}=0$.

10.8.2 ORBITAL-AVERAGED ECCENTRICITY VARIATION

$$\begin{aligned}\bar{\dot{e}} &= \frac{n}{2\pi}\int_0^{2\pi}\frac{de}{dt}\frac{r^2}{h}d\theta \\ &= \frac{n}{2\pi}\int_0^{2\pi}\frac{3J_2\mu R^2}{2\;hr^3}\left\{\frac{h^2}{\mu r}\sin\theta\left[3\sin^2 i\sin^2 u-1\right]-\left[(2+e\cos\theta)\cos\theta+e\right]\sin 2u\sin^2 i\right\}\frac{r^2}{h}d\theta \\ &= \frac{n}{2\pi}\left(\frac{3J_2\mu R^2}{2\;h^2}\right)\int_0^{2\pi}\frac{1}{r}\left\{\frac{h^2}{\mu r}\sin\theta\left[3\sin^2 i\sin^2 u-1\right]-\left[(2+e\cos\theta)\cos\theta+e\right]\sin 2u\sin^2 i\right\}d\theta \\ &= \frac{n}{2\pi}\left(\frac{3J_2\mu^2R^2}{2\;h^4}\right)\int_0^{2\pi}(1+e\cos\theta)\left\{(1+e\cos)\sin\theta\left[3\sin^2 i\sin^2 u-1\right]\right. \\ &\qquad \left.-\left[(2+e\cos\theta)\cos\theta+e\right]\sin 2u\sin^2 i\right\}d\theta\end{aligned}$$

But

$$\int_0^{2\pi}(1+e\cos\theta)\left\{(1+e\cos\theta)\sin\theta\left[3\sin^2 i\sin^2 u-1\right]-\left[(2+e\cos\theta)\cos\theta+e\right]\sin 2u\sin^2 i\right\}d\theta=0$$

A convenient way to evaluate this lengthy integral is to use MATLAB's symbolic math feature, as illustrated in the following Command Window session, in which `w` and `q` represent ω and θ, respectively:

```
syms w q e i positive
f = (1+e*cos(q))*((1+e*cos(q))*sin(q)*(3*sin(i)^2*sin(w+q)^2-1) ...
        -((2+e*cos(q))*cos(q)[Note +e)*sin(2*(w+q))*sin(i)^2);
integral = int(f, q, 0, 2*pi)
integral =
0
```

It follows that $\bar{\dot{e}} = 0$.

10.8.3 ORBITAL-AVERAGED TRUE ANOMALY VARIATION

$$\bar{\dot{\theta}} = \frac{n}{2\pi}\int_0^{2\pi}\frac{d\theta}{dt}\frac{r^2}{h}d\theta$$

$$= \frac{n}{2\pi}\int_0^{2\pi}\left\{\frac{h}{r^2} + \frac{3}{2}\frac{J_2\mu R^2}{ehr^3}\left[\frac{h^2}{\mu r}\cos\theta\left(3\sin^2 i\sin^2 u - 1\right) + (2 + e\cos\theta)\sin 2u\sin^2 i\sin\theta\right]\right\}\frac{r^2}{h}d\theta$$

$$= \frac{n}{2\pi}\int_0^{2\pi}\left\{1 + \frac{3}{2}\frac{J_2\mu R^2}{eh^2 r}\left[\frac{h^2}{\mu r}\cos\theta\left(3\sin^2 i\sin^2 u - 1\right) + (2 + e\cos\theta)\sin 2u\sin^2 i\sin\theta\right]\right\}d\theta$$

$$= n + \frac{n}{2\pi}\frac{3}{2}\frac{J_2\mu^2 R^2}{eh^4}\int_0^{2\pi}\left\{(1 + e\cos\theta)\left[(1 + e\cos\theta)\cos\theta\left(3\sin^2 i\sin^2 u - 1\right)\right]\right.$$
$$\left. + (2 + e\cos\theta)\sin 2u\sin^2 i\sin\theta\right\}d\theta$$

The integral evaluates to $(1 - 3\cos^2 i)\pi e$, as we see in this MATLAB Command Window session:

```
syms w q e i positive
f = (1+e*cos(q))*((1+e*cos(q))*cos(q)*(3*sin(i)^2*sin(w+q)^2-1) ...
    +(2+e*cos(q))*sin(2*(w+q))*sin(i)^2*sin(q));
f = collect(expand(f));
integral = int(f, q, 0, 2*pi)
integral =
pi*e - 3*pi*e*cos(i)^2
```

Thus,

$$\bar{\dot{\theta}} = n\left[1 + \frac{3}{4}\frac{J_2\mu^2 R^2}{h^4}\left(1 - 3\cos^2 i\right)\right]$$

Substituting Eq. (10.91), $h^2 = \mu a(1 - e^2)$, and $\cos^2 i = 1 - \sin^2 i$, we can write this as

$$\bar{\dot{\theta}} = n + \frac{3}{4}\frac{J_2 R^2\sqrt{\mu}}{a^{7/2}(1 - e^2)^2}\left(3\sin^2 i - 2\right)$$

10.8.4 ORBITAL-AVERAGED RIGHT ASCENSION OF ASCENDING NODE VARIATION

$$\begin{aligned}\bar{\dot{\Omega}} &= \frac{n}{2\pi}\int_0^{2\pi}\frac{\mathrm{d}\Omega}{\mathrm{d}t}\frac{r^2}{h}\mathrm{d}\theta \\ &= \frac{n}{2\pi}\int_0^{2\pi}\left(-3\frac{J_2\mu R^2}{hr^3}\sin^2 u\cos i\right)\frac{r^2}{h}\mathrm{d}\theta \\ &= -\frac{3nJ_2\mu R^2}{2\pi\ h^2}\cos i\int_0^{2\pi}\frac{1}{r}\sin^2 u\mathrm{d}\theta \\ &= -\frac{3nJ_2\mu^2R^2}{2\pi\ h^4}\cos i\int_0^{2\pi}(1+e\cos\theta)\sin^2(\omega+\theta)\mathrm{d}\theta\end{aligned}$$

But

$$\int_0^{2\pi}(1+e\cos\theta)\sin^2(\omega+\theta)\mathrm{d}\theta=\pi$$

as is evident from the MATLAB session:

```
syms w q e positive
f = (1 + e*cos(q))*sin(w + q)^2;
integral = int(f, q, 0, 2*pi)
integral =
pi
>>
```

Hence,

$$\bar{\dot{\Omega}} = -\frac{3nJ_2\mu^2R^2}{2\ h^4}\cos i$$

Substituting Eq. (10.91) along with $h^4 = \mu^2a^2(1-e^2)^2$ from Eq. (2.71), we obtain

$$\bar{\dot{\Omega}} = -\left[\frac{3}{2}\frac{J_2\sqrt{\mu}R^2}{a^{7/2}(1-e^2)^2}\right]\cos i$$

10.8.5 ORBITAL-AVERAGED INCLINATION VARIATION

$$\bar{\dot{i}} = \frac{n}{2\pi}\int_0^{2\pi} \frac{di}{dt}\frac{r^2}{h^2}d\theta$$

$$= \frac{n}{2\pi}\int_0^{2\pi}\left(-\frac{3}{4}\frac{J_2\mu R^2}{hr^3}\sin 2u \sin 2i\right)\frac{r^2}{h}d\theta$$

$$= -\frac{n}{2\pi}\frac{3}{4}\frac{J_2\mu R^2}{h^2}\sin 2i\int_0^{2\pi}\frac{1}{r}\sin 2u d\theta$$

$$= -\frac{n}{2\pi}\frac{3}{4}\frac{J_2\mu^2 R^2}{h^4}\sin 2i\int_0^{2\pi}(1+e\cos\theta)\sin 2(\omega+\theta)d\theta$$

Using MATLAB, we see that the integral vanishes:

```
syms w q e positive
f = (1 + e*cos(q))*sin(2*(w + q));
integral = int(f, q, 0, 2*pi)
integral =
0
```

Therefore, $\bar{\dot{i}} = 0$.

10.8.6 ORBITAL-AVERAGED ARGUMENT OF PERIGEE VARIATION

$$\bar{\dot{\omega}} = \frac{n}{2\pi}\int_0^{2\pi}\frac{d\omega}{dt}\frac{r^2}{h}d\theta$$

$$= \frac{n}{2\pi}\int_0^{2\pi}\frac{3}{2}\frac{J_2\mu R^2}{ehr^3}\left[\frac{h^2}{\mu r}\cos\theta(1-3\sin^2 i\sin^2 u) - (2+e\cos\theta)\sin 2u\sin^2 i\sin\theta + 2e\cos^2 i\sin^2 u\right]\frac{r^2}{h}d\theta$$

$$= \frac{n}{2\pi}\frac{3}{2}\frac{J_2\mu R^2}{eh^2}\int_0^{2\pi}\frac{1}{r}\left[\frac{h^2}{\mu r}\cos\theta\left(1-3\sin^2 i\sin^2 u\right) - (2+e\cos\theta)\sin 2u\sin^2 i\sin\theta + 2e\cos^2 i\sin^2 u\right]d\theta$$

$$= \frac{n}{2\pi}\frac{3}{2}\frac{J_2\mu R^2}{eh^4}\int_0^{2\pi}(1+e\cos\theta)\left[\cos\theta(1+e\cos\theta)\left(1-3\sin^2 i\sin^2 u\right)\right.$$
$$\left. -(2+e\cos\theta)\sin 2u\sin^2 i\sin\theta + 2e\cos^2 i\sin^2 u\right]d\theta$$

With the aid of MATLAB, we find that the integral evaluates to $\pi e(5\cos^2 i - 1)$.

```
syms w q e i positive
f = (1+e*cos(q))*...
```

```
        ((1+e*cos(q))*cos(q)*(1-3*sin(i)^2*sin(w+q)^2)...
        -(2+e*cos(q))*sin(2*(w+q))*sin(i)^2*sin(q)...
        + 2*e*cos(i)^2*sin(w+q)^2);
f = collect(expand(f));
integral = int(f, q, 0, 2*pi)
integral =
5*pi*e*cos(i)^2 - pi*e
```

Therefore,

$$\bar{\dot{\omega}} = \frac{3}{2}\frac{J_2\mu^2R^2}{h^4}\frac{n}{2}(5\cos^2 i - 1)$$

Substituting Eq. (10.91), along with $h^4 = \mu^2a^2(1-e^2)^2$ from Eq. (2.71) and the trig identity $\cos^2 i = 1 - \sin^2 i$, we obtain

$$\bar{\dot{\omega}} = -\left[\frac{3}{2}\frac{J_2\sqrt{\mu}R^2}{a^{7/2}(1-e^2)^2}\left(\frac{5}{2}\sin^2 i - 2\right)\right]$$

Let us summarize our calculations of the average rates of variation of the orbital elements due to the J_2 perturbation:

$$\bar{\dot{h}} = \bar{\dot{e}} = \bar{\dot{i}} = 0 \tag{10.93a}$$

$$\bar{\dot{\Omega}} = -\left[\frac{3}{2}\frac{J_2\sqrt{\mu}R^2}{a^{7/2}(1-e^2)^2}\right]\cos i \tag{10.93b}$$

$$\bar{\dot{\omega}} = -\left[\frac{3}{2}\frac{J_2\sqrt{\mu}R^2}{a^{7/2}(1-e^2)^2}\right]\left(\frac{5}{2}\sin^2 i - 2\right) \tag{10.93c}$$

$$\bar{\dot{\theta}} = n - \left[\frac{3}{2}\frac{J_2\sqrt{\mu}R^2}{a^{7/2}(1-e^2)^2}\right]\left(1 - \frac{3}{2}\sin^2 i\right) \tag{10.93d}$$

Formulas such as these for the average rates of variation of the orbital elements are useful for the design of frozen orbits. Frozen orbits are those whose size, shape, and/or orientation remain, on average, constant over long periods of time. Careful selection of the orbital parameters can minimize or eliminate the drift caused by perturbations. For example, the apse line will be frozen in space ($\bar{\dot{\omega}} = 0$) if the orbital inclination i is such that $\sin i = \sqrt{4/5}$. Similarly, the J_2 gravitational perturbation on the mean motion $\bar{\dot{\theta}}$ vanishes if $\sin i = \sqrt{2/3}$. The precession $\bar{\dot{\Omega}}$ of an orbital plane is prevented if $\cos i = 0$. Practical applications of Molniya, sun-synchronous, and polar orbits are discussed in Section 4.7.

Eq. (10.93) agrees with the plotted results of Examples 10.2 and 10.6. We can average out the high-frequency components ("ripples") of the curves in Figs. 10.8, 10.9, and 10.13 with a numerical smoothing technique such as that presented by Garcia (2010), yielding the curves in Fig. 10.14.

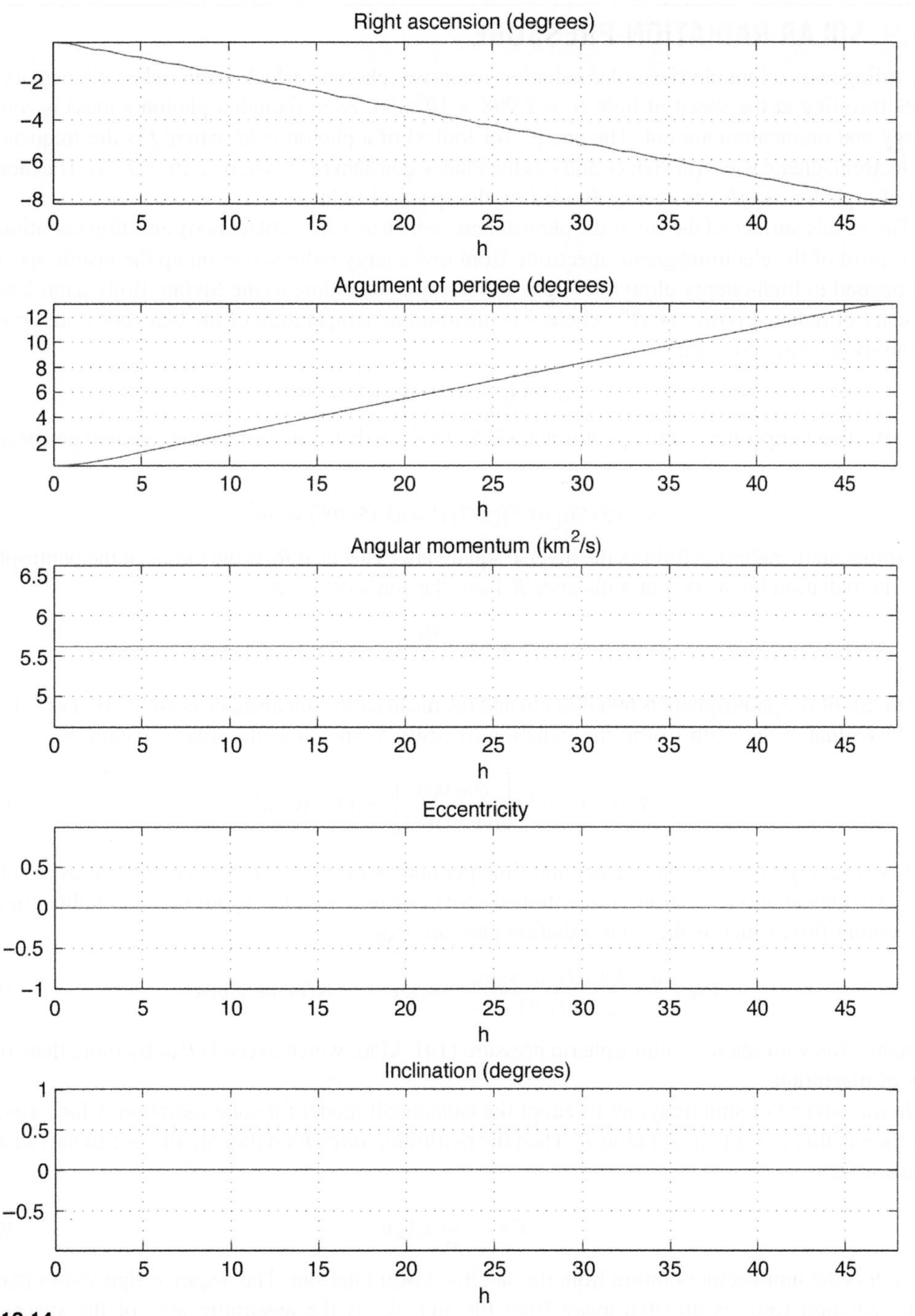

FIG. 10.14

Long term variations of the osculating elements due to J_2.

10.9 SOLAR RADIATION PRESSURE

According to quantum physics, solar radiation comprises photons, which are massless elementary particles traveling at the speed of light ($c = 2.998 \times 10^8$m/s). Even though a photon's mass is zero, its energy and momentum are not. The energy (in Joules) of a photon is hf, where f is the frequency of its electromagnetic wave (in Hertz), and h is the Planck constant ($h = 6.626 \times 10^{-34}$J · s). The momentum of a photon is hf/c, its energy divided by the speed of light.

The visible surface of the sun is the photosphere, which acts like a blackbody emitting radiation that spans most of the electromagnetic spectrum, from low-energy radio waves on up the visible spectrum and beyond to high-energy ultraviolet light and X-rays. According to the Stefan–Boltzmann law, the intensity of radiated power is σT^4, where T is the absolute temperature of the blackbody, and σ is the Stefan–Boltzmann constant,

$$\sigma = 5.670\left(10^{-8}\right)\ \text{W/m}^2\text{K}^4$$

The effective temperature of the photosphere is 5777 K, so that at its surface the radiated power intensity is

$$S_0 = 5.670\left(10^{-8}\right)(5777)^4 = 63.15\left(10^6\right)\ \text{W/m}^2$$

Electromagnetic radiation follows the inverse square law. That is, if R_0 is the radius of the photosphere, then the radiation intensity S at a distance R from the sun's center is

$$S = S_0\left(\frac{R_0}{R}\right)^2$$

The radius of the photosphere is 696,000 km and the mean earth–sun distance is 149.6(10^6) km (1 AU). It follows that at the earth's orbit the radiation intensity S, known as the solar constant, is

$$S = 63.15\left(10^6\right)\left[\frac{696{,}000}{149.6\left(10^6\right)}\right]^2 = 1367\ \text{W/m}^2 \tag{10.94}$$

This is the energy flux (the energy per unit time per unit area) transported by photons across a surface normal to the radiation direction. As mentioned above, we must divide S by the speed of light to find the momentum flux, which is the solar radiation pressure P_{SR},

$$P_{SR} = \frac{S}{c} = \frac{1367(\text{N}\cdot\text{m/s})/\text{m}^2}{2.998\left(10^8\right)\ \text{m/s}} = 4.56\left(10^{-6}\right)\ \text{N/m}^2\ (4.56\mu\text{Pa}) \tag{10.95}$$

Compare this with sea level atmospheric pressure (101 kPa), which exceeds P_{SR} by more than 10 orders of magnitude.

In the interest of simplicity, let us adopt the cannonball model for solar radiation, which assumes that the satellite is a sphere of radius R. Then the perturbing force $\mathbf{F}$ on the satellite due to the radiation pressure S/c is

$$\mathbf{F} = -\nu\frac{S}{c}C_R A_{sc}\hat{\mathbf{u}} \tag{10.96}$$

where $\hat{\mathbf{u}}$ is the unit vector pointing from the satellite toward the sun. The negative sign shows that the solar radiation force is directed away from the sun. A_{sc} is the absorbing area of the spacecraft, which is πR^2 for the cannonball model. ν is the shadow function, which has the value 0 if the satellite is in the earth's shadow; otherwise, $\nu = 1$. C_R is the radiation pressure coefficient, which lies between 1 and 2. C_R equals 1 if the surface is a blackbody, absorbing all of the momentum of the incident photon stream and giving rise to the pressure in Eq. (10.95). When C_R equals 2, all the incident radiation is reflected, so that the incoming photon momentum is reversed in direction, doubling the force on the satellite.

Because the sun is so far from the earth, the angle between the earth—sun line and the satellite—sun line is less than 0.02 degrees, even for geostationary satellites. Therefore, it will be far simpler and sufficiently accurate for our purposes to let $\hat{\mathbf{u}}$ in Eq. (10.96) be the unit vector pointing toward the sun from the earth instead of from the satellite. Then $\hat{\mathbf{u}}$ tracks only the relative motion of the sun around the earth and does not include the motion of the satellite around the earth.

If m is the mass of the satellite, then the perturbing acceleration $\mathbf{p}$ due to solar radiation is $\mathbf{F}/m$, or

$$\mathbf{p} = -p_{SR}\hat{\mathbf{u}} \tag{10.97}$$

where the magnitude of the perturbation is

$$p_{SR} = \nu \frac{SC_R A_{sc}}{c\ \ m} \tag{10.98}$$

The magnitude of the solar radiation pressure perturbation clearly depends on the satellite's area-to-mass ratio A_{sc}/m. Very large spacecraft with a very low mass (like solar sails) are the most affected by solar radiation pressure. Extreme examples of such spacecraft were the Echo 1, Echo 2, and Pageos passive communication balloon satellites launched by the United States in the 1960s. They were very thin–walled, highly reflective spheres about 100 ft (30 m) in diameter, and they had area-to-mass ratios on the order of 10 m^2/kg.

The influence of solar radiation pressure is more pronounced at higher orbital altitudes where atmospheric drag is comparatively negligible. To get an idea of where the tradeoff between the two perturbations occurs, set the magnitude of the drag perturbation equal to that of the solar radiation perturbation, $p_D = p_{SR}$, or

$$\frac{1}{2}\rho v^2\left(\frac{C_D A}{m}\right) = \frac{S C_R A_{sc}}{c\ \ m}$$

Solving for the atmospheric density and assuming that the orbit is circular ($v^2 = \mu/r$), we get

$$\rho = 2\frac{A_{sc}}{A}\frac{C_R}{C_D}\frac{S/c}{\mu}r$$

If $A_{sc}/A = 1, C_R = 1, C_D = 2$, and $r = 6378 + z$, where z is the altitude in kilometers, then this becomes

$$\rho = 2\cdot 1\cdot\frac{1}{2}\cdot\frac{\left[4.56\left(10^{-6}\right)\mathrm{kg}/(\mathrm{m}\cdot\mathrm{s}^2)\right]}{398.6\left(10^{12}\right)\mathrm{m}^3/\mathrm{s}^2}[(6378+z)(\mathrm{km})](1000\mathrm{m}/\mathrm{km})$$

or

$$\rho\left(\mathrm{kg}/\mathrm{m}^3\right) = 1.144\left(10^{-17}\right)(6378+z)(\mathrm{km})$$

When $z = 625\,\mathrm{km}$, this formula gives $\rho = 8.01(10^{-14})\mathrm{kg/m^3}$, whereas according to the US Standard Atmosphere, $\rho = 7.998(10^{-14})\mathrm{kg/m^3}$ at that altitude. So 625 km is a rough estimate of the altitude of the transition from the dominance of the perturbative effect of atmospheric drag to that of solar radiation pressure. This estimate is about 20% lower than the traditionally accepted value of 800 km (Vallado, 2007).

Recall from Section 4.2 that the angle between earth's equatorial plane and the ecliptic plane is the obliquity of the ecliptic ε. The obliquity varies slowly with time and currently is 23.44°. Therefore, the plane of the sun's apparent orbit around the earth is inclined 23.44° to the earth's equator. In the geocentric ecliptic

$(X'Y'Z')$ frame, the Z' axis is normal to the ecliptic and the X' axis lies along the vernal equinox direction. In this frame, the unit vector $\hat{\mathbf{u}}$ along the earth-sun line is provided by the solar ecliptic longitude λ,

$$\hat{\mathbf{u}} = \cos\lambda\hat{\mathbf{I}}' + \sin\lambda\hat{\mathbf{J}}' \tag{10.99}$$

where λ is the angle between the vernal equinox line and the earth-sun line.

The geocentric equatorial frame (XYZ) and the geocentric ecliptic frame $(X'Y'Z')$ share the vernal equinox line as their common X axis. Transformation from one frame to the other is therefore simply a rotation through the obliquity ε around the positive X axis. The transformation from $X'Y'Z'$ to XYZ is represented by the direction cosine matrix found in Eq. (4.32) with $\phi = -\varepsilon$. It follows that the components in XYZ of the unit vector $\hat{\mathbf{u}}$ in Eq. (10.99) are

$$\{\hat{\mathbf{u}}\}_{XYZ} = [\mathbf{R}_1(-\varepsilon)]\{\hat{\mathbf{u}}\}_{X'Y'Z'} = \begin{bmatrix} 1 & 0 & 0 \\ 0 & \cos\varepsilon & -\sin\varepsilon \\ 0 & \sin\varepsilon & \cos\varepsilon \end{bmatrix} \begin{Bmatrix} \cos\lambda \\ \sin\lambda \\ 0 \end{Bmatrix} = \begin{Bmatrix} \cos\lambda \\ \cos\varepsilon\sin\lambda \\ \sin\varepsilon\sin\lambda \end{Bmatrix} \tag{10.100}$$

Substituting this vector expression back into Eq. (10.97) yields the components of the solar radiation perturbation in the geocentric equatorial frame,

$$\{\mathbf{p}\}_{XYZ} = -p_{SR} \begin{Bmatrix} \cos\lambda \\ \cos\varepsilon\sin\lambda \\ \sin\varepsilon\sin\lambda \end{Bmatrix} \tag{10.101}$$

To use Gauss' planetary equations (Eqs. 10.84) to determine the effects of solar radiation pressure on variation of the orbital elements, we must find the components of the perturbation $\mathbf{p}$ in the rsw frame of Fig. (10.10). To do so, we use Eq. (10.61),

$$\begin{Bmatrix} p_r \\ p_s \\ p_w \end{Bmatrix} = [\mathbf{Q}]_{Xr} \begin{Bmatrix} -p_{SR}\cos\lambda \\ -p_{SR}\sin\lambda\cos\varepsilon \\ -p_{SR}\sin\lambda\sin\varepsilon \end{Bmatrix} \tag{10.102}$$

The direction cosine matrix $[\mathbf{Q}]_{Xr}$ of the transformation from XYZ to rsw is given by Eq. (10.58). Thus,

$$\begin{Bmatrix} p_r \\ p_s \\ p_w \end{Bmatrix} = -p_{SR} \begin{bmatrix} -\sin\Omega\cos i\sin u + \cos\Omega\cos u & \cos\Omega\cos i\sin u + \sin\Omega\cos u & \sin i\sin u \\ -\sin\Omega\cos i\cos u - \cos\Omega\sin u & \cos\Omega\cos i\cos u - \sin\Omega\sin u & \sin i\cos u \\ \sin\Omega\sin i & -\cos\Omega\sin i & \cos i \end{bmatrix} \cdot \begin{Bmatrix} \cos\lambda \\ \sin\lambda\cos\varepsilon \\ \sin\lambda\sin\varepsilon \end{Bmatrix} \tag{10.103}$$

Carrying out the matrix multiplications leads to

$$p_r = -p_{SR}u_r \quad p_s = -p_{SR}u_s \quad p_w = -p_{SR}u_w \tag{10.104}$$

where u_r, u_s, and u_w are components of the unit vector $\hat{\mathbf{u}}$ in the rsw frame. Namely,

$$\begin{aligned} u_r = {} & \sin\lambda\cos\varepsilon\cos\Omega\cos i\sin u + \sin\lambda\cos\varepsilon\sin\Omega\cos u \\ & - \cos\lambda\sin\Omega\cos i\sin u + \cos\lambda\cos\Omega\cos u + \sin\lambda\sin\varepsilon\sin i\sin u \end{aligned} \tag{10.105a}$$

$$\begin{aligned} u_s = {} & \sin\lambda\cos\varepsilon\cos\Omega\cos i\cos u - \sin\lambda\cos\varepsilon\sin\Omega\sin u \\ & - \cos\lambda\sin\Omega\cos i\cos u - \cos\lambda\cos\Omega\sin u + \sin\lambda\sin\varepsilon\sin i\cos u \end{aligned} \tag{10.105b}$$

$$u_w = -\sin\lambda\cos\varepsilon\cos\Omega\sin i + \cos\lambda\sin\Omega\sin i + \sin\lambda\sin\varepsilon\cos i \tag{10.105c}$$

Substituting Eq. (10.104) into Eq. (10.84) yields Gauss' planetary equations for solar radiation pressure, where it should be recalled from Eq. (10.98) that $p_{SR} = \nu(S/c)C_R(A_{sc}/m)$:

$$\frac{dh}{dt} = -p_{SR} r u_s \tag{10.106a}$$

$$\frac{de}{dt} = -p_{SR}\left\{\frac{h}{\mu}\sin\theta u_r + \frac{1}{\mu h}\left[\left(h^2+\mu r\right)\cos\theta + \mu e r\right]u_s\right\} \tag{10.106b}$$

$$\frac{d\theta}{dt} = \frac{h}{r^2} - \frac{p_{SR}}{eh}\left[\frac{h^2}{\mu}\cos\theta u_r - \left(r + \frac{h^2}{\mu}\right)\sin\theta u_s\right] \tag{10.106c}$$

$$\frac{d\Omega}{dt} = -p_{SR}\frac{r}{h\sin i}\sin(\omega+\theta)u_w \tag{10.106d}$$

$$\frac{di}{dt} = -p_{SR}\frac{r}{h}\cos(\omega+\theta)u_w \tag{10.106e}$$

$$\frac{d\omega}{dt} = -p_{SR}\left\{-\frac{1}{eh}\left[\frac{h^2}{\mu}\cos\theta u_r - \left(r + \frac{h^2}{\mu}\right)\sin\theta u_s\right] - \frac{r\sin(\omega+\theta)}{h\tan i}u_w\right\} \tag{10.106f}$$

To numerically integrate Eqs. (10.106) requires that we know the time variation of the obliquity ε and solar ecliptic longitude λ, both of which appear throughout the expressions for u_r, u_s, and u_w. We also need the time history of the earth-sun distance $r_\odot$ ($\odot$ is the astronomical symbol for the sun) to compute the geocentric equatorial position vector of the sun (namely, $\mathbf{r}_\odot = r_\odot\hat{\mathbf{u}}$). $\mathbf{r}_\odot$ together with the geocentric position vector of the satellite allow us to determine when the satellite is in the earth's shadow ($\nu = 0$ in Eq. 10.98).

According to *The Astronomical Almanac* (National Almanac Office, 2018), the apparent solar ecliptic longitude (in degrees) is given by the formula

$$\lambda = L + 1.915°\sin M + 0.0200°\sin 2M \ \ (0° \le \lambda \le 360°) \tag{10.107}$$

where L and M are, respectively, the mean longitude and mean anomaly of the sun, both in degrees:

$$L = 280.459° + 0.98564736°n \ \ (0° \le L \le 360°) \tag{10.108}$$

$$M = 357.529° + 0.98560023°n \ \ (0° \le M \le 360°) \tag{10.109}$$

n is the number of days since J2000,

$$n = JD - 2{,}451{,}545.0 \tag{10.110}$$

The concepts of Julian day number JD and the epoch J2000 are explained in Section 5.4. The above formulas for L, M, and λ may deliver angles outside the range 0° to 360°. In those cases, the angle should be reduced by appropriate multiples of 360°, so as to place it in that range. (If the `angle` is a number, then the MATLAB function `mod(angle,360)` yields a number in the range 0° to 360°.)

In terms of n, the obliquity is

$$\varepsilon = 23.439° - 3.56\left(10^{-7}\right)n \tag{10.111}$$

Finally, *The Astronomical Almanac* gives the distance r_S from the earth to the sun in terms of the mean anomaly,

$$r_\odot = (1.00014 - 0.01671\cos M - 0.000140\cos 2M)\ \text{AU} \tag{10.112}$$

where AU is the astronomical unit (1 AU = 149,597,870.691 km).

The following algorithm delivers ε, λ, and $\mathbf{r}_\odot$.

ALGORITHM 10.2

Given the year, month, day, and universal time, calculate the obliquity of the ecliptic ε, ecliptic longitude of the sun λ, and the geocentric position vector of the sun $\mathbf{r}_\odot$:

1. Compute the Julian day number JD using Eqs. (5.47) and (5.48).
2. Calculate n, the number of days since J2000 from Eq. (10.110).
3. Calculate the mean anomaly M using Eq. (10.109).
4. Calculate the mean solar longitude L by means of Eq. (10.108).
5. Calculate the longitude λ using Eq. (10.107).
6. Calculate the obliquity ε from Eq. (10.111).
7. Calculate the unit vector $\hat{\mathbf{u}}$ from the earth to the sun (Eq. 10.100):

$$\hat{\mathbf{u}} = \cos\lambda\hat{\mathbf{I}} + \sin\lambda\cos\varepsilon\hat{\mathbf{J}} + \sin\lambda\sin\varepsilon\hat{\mathbf{K}}$$

8. Calculate the distance $r_\odot$ of the sun from the earth using Eq. (10.112).
9. Calculate the sun's geocentric position vector $\mathbf{r}_\odot = r_\odot\hat{\mathbf{u}}$.

EXAMPLE 10.7

Use Algorithm 10.2 to find the geocentric position vector of the sun at 08:00 UT on July 25, 2013.

Solution

Step 1:

According to Eq. (5.48), with $y = 2013$, $m = 7$, and $d = 25$, the Julian day number at 0 h UT is

$$J_0 = 2{,}456{,}498.5$$

Therefore, from Eq. (5.47), the Julian day number at 08:00 UT is

$$JD = 2{,}456{,}498.5 + \frac{8}{24} = 2{,}456{,}498.8333 \text{ days}$$

Step 2:

$$n = 2{,}456{,}498.8333 - 2{,}451{,}545.0 = 4953.8333$$

Step 3:

$$M = 357.529° + 0.98560023°(4953.8333) = 5240.03° \ (200.028°)$$

Step 4:

$$L = 280.459° + 0.98564736°(4953.8333) = 5163.19° \ (123.192°)$$

Step 5:

$$\lambda = 123.192° + 1.915° \sin(200.028°) + 0.020°(2 \cdot 200.028°) = 122.549°$$

Step 6:

$$\varepsilon = 23.439° - 3.56°(10^{-7})(4953.8333) = 23.4372°$$

Step 7:

$$\begin{aligned}\hat{\mathbf{u}} &= \cos 122.549°\hat{\mathbf{I}} + (\sin 122.549°)(\cos 23.4372°)\hat{\mathbf{J}} + (\sin 122.549°)(\sin 23.4372°)\hat{\mathbf{K}} \\ &= -0.538017\hat{\mathbf{I}} + 0.773390\hat{\mathbf{J}} + 0.335269\hat{\mathbf{K}}\end{aligned}$$

Step 8:

$$r_{\odot} = [1.00014 - 0.01671\cos 200.028° - 0.000140\cos(2 \cdot 200.028°)](149,597,870.691)$$
$$= 151,951,387\,\text{km}$$

Step 9:

$$\mathbf{r}_{\odot} = (151,951,387)(-0.5380167\hat{\mathbf{I}} + 0.7733903\hat{\mathbf{J}} + 0.3352693\hat{\mathbf{K}})$$
$$= \boxed{-81,752,385\hat{\mathbf{I}} + 117,517,729\hat{\mathbf{J}} + 50,944,632\hat{\mathbf{K}}\ (\text{km})}$$

To determine when a satellite is in the earth's shadow (so that the solar radiation pressure perturbation is "off"), we can use the following elementary procedure (Vallado, 2007). First, consider two spacecraft A and B orbiting a central body of radius R. The two position vectors $\mathbf{r}_A$ and $\mathbf{r}_B$ define a plane, which is the plane of Fig. 10.15. That plane contains the circular profile C of the central body. The angle θ between the two position vectors may be found from the dot product operation,

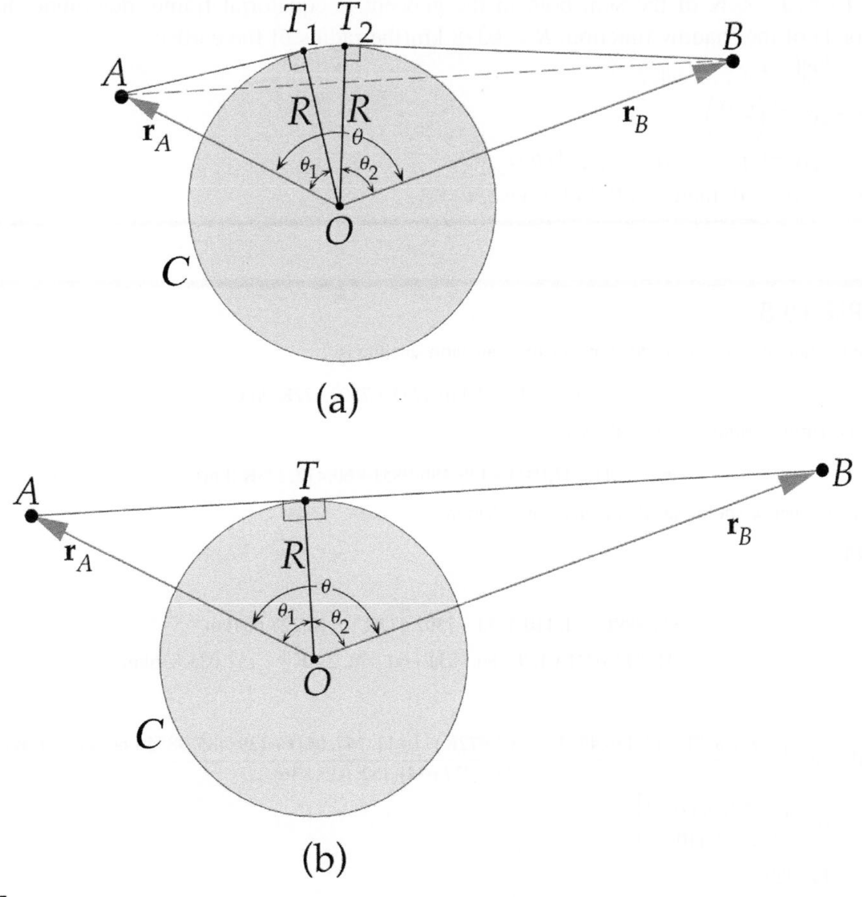

FIG. 10.15

(a) AB intersects the central body ($\theta_1+\theta_2 < \theta$). (b) AB is tangent to the central body ($\theta_1+\theta_2=\theta$).

$$\theta = \cos^{-1}\left(\frac{\mathbf{r}_A \cdot \mathbf{r}_B}{r_A r_B}\right) \tag{10.113}$$

In Fig. 10.15a, T_1 and T_2 are points of tangency to C of lines drawn from A and B, respectively. The radii OT_1 and OT_2 along with the tangent lines AT_1 and BT_2 and the position vectors $\mathbf{r}_A$ and $\mathbf{r}_B$ comprise the two right triangles OAT_1 and OBT_2. The angles at the vertex O of these two triangles are obtained from

$$\theta_1 = \cos^{-1}\frac{R}{r_A} \qquad \theta_2 = \cos^{-1}\frac{R}{r_B} \tag{10.114}$$

If, as in Fig. 10.15a, the line AB intersects the central body, which means there is no line of sight, then $\theta_1 + \theta_2 < \theta$. If the AB is tangent to C (Fig. 10.15b) or lies outside it, then $\theta_1 + \theta_2 \geq \theta$ and there is line of sight.

ALGORITHM 10.3

Given the position vector $\mathbf{r} = X\hat{\mathbf{I}} + Y\hat{\mathbf{J}} + Z\hat{\mathbf{K}}$ of a satellite and the apparent position vector $\mathbf{r}_\odot = X_\odot\hat{\mathbf{I}} + Y_\odot\hat{\mathbf{J}} + Z_\odot\hat{\mathbf{K}}$ of the sun, both in the geocentric equatorial frame, determine the value of ν (0 or 1) of the shadow function. $R = 6378$ km(the radius of the earth):

1. $r = \|\mathbf{r}\| \quad r_\odot = \|\mathbf{r}_\odot\|$
2. $\theta = \cos^{-1}\left(\frac{\mathbf{r}_\odot \cdot \mathbf{r}}{r_\odot r}\right)$
3. $\theta_1 = \cos^{-1}(R/r) \quad \theta_2 = \cos^{-1}(R/r_\odot)$
4. If $\theta_1 + \theta_2 \leq \theta$, then $\nu = 0$. Otherwise, $\nu = 1$.

EXAMPLE 10.8

At a given instant, the geocentric position vector of an earth satellite is

$$\mathbf{r} = 2817.899\hat{\mathbf{I}} - 14{,}110.473\hat{\mathbf{J}} - 7502.672\hat{\mathbf{K}}\ (\text{km})$$

and the geocentric position vector of the sun is

$$\mathbf{r}_\odot = -11{,}747{,}041\hat{\mathbf{I}} + 139{,}486{,}985\hat{\mathbf{J}} + 60{,}472{,}278\hat{\mathbf{K}}\ (\text{km})$$

Determine whether or not the satellite is in earth's shadow.

Solution

Step 1:

$$r = \left\|2817.899\hat{\mathbf{I}} - 14,110.473\hat{\mathbf{J}} - 7502.672\hat{\mathbf{K}}\right\| = 16{,}227.634\text{km}$$

$$r_\odot = \left\|-11,747,041\hat{\mathbf{I}} + 139,486,985\hat{\mathbf{J}} + 60,472,278\hat{\mathbf{K}}\right\| = 152{,}035{,}836\text{km}$$

Step 2:

$$\theta = \cos^{-1}\frac{\left(2817.899\hat{\mathbf{I}} - 14,110.473\hat{\mathbf{J}} - 7502.672\hat{\mathbf{K}}\right)\cdot\left(-11,747,041\hat{\mathbf{I}} + 139,486,985\hat{\mathbf{J}} + 60,472,278\hat{\mathbf{K}}\right)}{(16,227.634)(152,035,836)}$$

$$= \cos^{-1}\left[\frac{-2.4252\left(10^{12}\right)}{2.4672\left(10^{12}\right)}\right]$$

$$= 169.420°$$

Step 3:

$$\theta_1 = \cos^{-1}\frac{6378}{16{,}227.634} = 66.857° \quad \theta_2 = \cos^{-1}\frac{6378}{152{,}035{,}836} = 89.998°$$

Step 4:

$$\theta_1 + \theta_2 = 156.85° < \theta. \ \boxed{\text{Therefore, the spacecraft is in earth's shadow}}$$

EXAMPLE 10.9

A spherical earth satellite has an absorbing area-to-mass ratio (A_{sc}/m) of 2m^2/kg. At time t_0 (Julian date $JD_0 = 2,438,400.5$) its orbital parameters are

$$\text{Angular momentum: } h_0 = 63{,}383.4\,\text{km}^2/\text{s} \quad \text{(a)}$$

$$\text{Eccentricity: } e_0 = 0.025422 \quad \text{(b)}$$

$$\text{Right ascension of the node: } \Omega_0 = 45.3812° \quad \text{(c)}$$

$$\text{Inclination: } i_0 = 88.3924° \quad \text{(d)}$$

$$\text{Argument of perigee: } \omega_0 = 227.493° \quad \text{(e)}$$

$$\text{True anomaly: } \theta_0 = 343.427° \quad \text{(f)}$$

Use numerical integration to plot the variation of these orbital parameters over the next 3 years ($t_f = 1095$ days) due just to solar radiation pressure. Assume that the radiation pressure coefficient is 2.

Solution

We write the system of differential equations in Eq. (10.106) as $\mathbf{dy}/dt = \mathbf{f}(\mathbf{y},t)$, where the components of $\mathbf{y}$ are the orbital elements

$$\mathbf{y} = \lfloor h \ e \ \theta \ \Omega \ i \ \omega \rfloor^T \quad \text{(g)}$$

and

$$\mathbf{f} = \lfloor \dot{h} \ \dot{e} \ \dot{\theta} \ \dot{\Omega} \ \dot{i} \ \dot{\omega} \rfloor^T \quad \text{(h)}$$

The osculating element rates in $\mathbf{f}$ are found on the right-hand side of Eqs. (10.106) with $p_{SR} = \nu(S/c)C_R(A_{sc}/m)$.

The initial conditions vector $\mathbf{y}_0$ comprises Eqs. (a)–(f). We can solve the system $\mathbf{dy}/dt = \mathbf{f}(\mathbf{y},t)$ for $\mathbf{y}$ on the interval $[t_0, t_f]$ by using a numerical integrator such as MATLAB's *ode45*. At each time step, the integrator relies on a subroutine to compute the rates $\mathbf{f}$ from the current values of $\mathbf{y}$ and the time t, as follows:

Update the Julian day number: JD ← $\text{JD}_0 + t$(days).

Use $\mathbf{y}$ to compute the position vector $\mathbf{r}$ of the satellite by means of Algorithm 4.5.

Compute the magnitude of $\mathbf{r}$ ($r = \|\mathbf{r}\|$).

Using JD, compute the sun's apparent ecliptic longitude λ, the obliquity of the ecliptic ε, and the geocentric equatorial position vector $\mathbf{r}_\odot$ of the sun by means of Algorithm 10.2.

Compute the components u_r, u_s, and u_w of the earth–sun unit vector from Eqs. (10.105).
Calculate the value of the shadow function ν from $\mathbf{r}$ and $\mathbf{r}_\odot$ using Algorithm 10.3.
Compute the solar radiation perturbation $p_{SR} = \nu(S/c)C_R(A_{sc}/m)$.
Calculate the components of $\mathbf{f}$ in Eq. (h).

Plots of the orbital parameter variations $h(t) - h_0$, $e(t) - e_0$, etc., are shown in Fig. 10.16. The MATLAB M-file *Example_10_09.m* is listed in Appendix D.45.

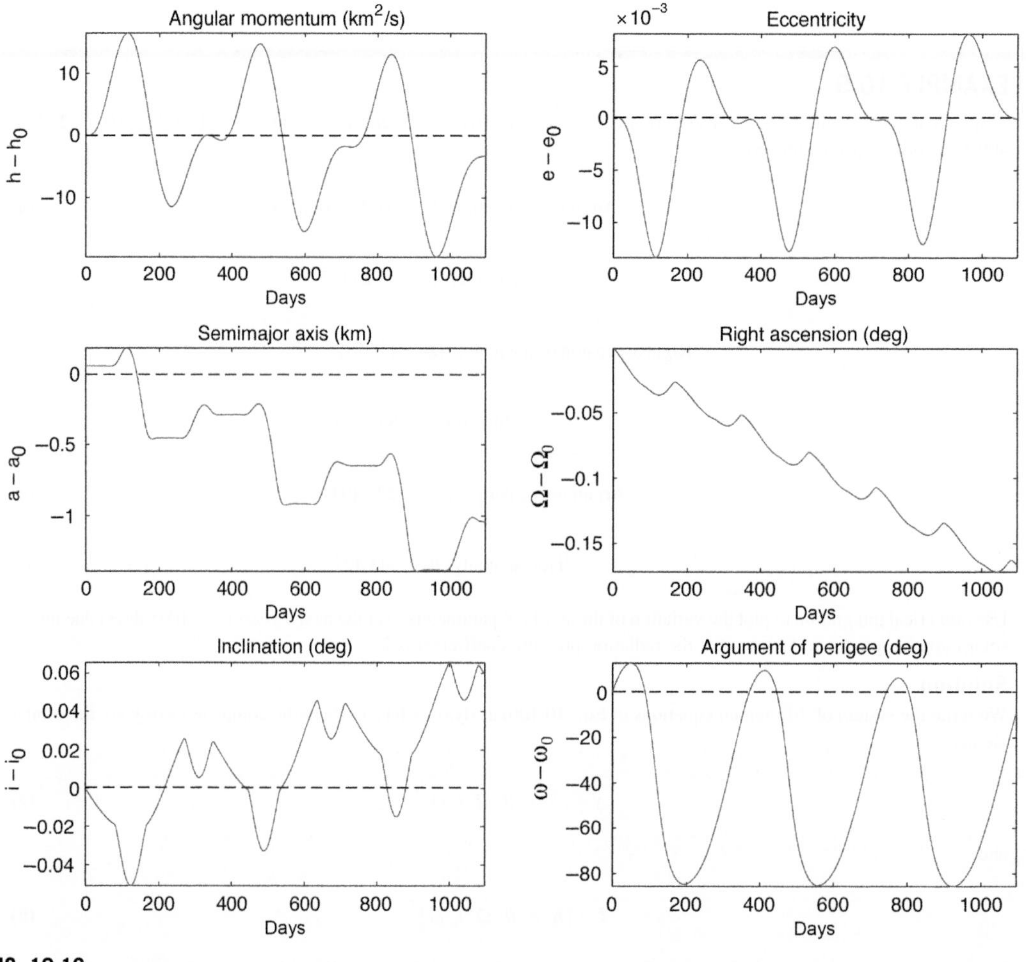

FIG. 10.16

Solar radiation perturbations of h, e, a, Ω, i, and ω during a 3-year interval following Julian date 2438400.5 (1 January 1964).

10.10 LUNAR GRAVITY

Fig. C.1 of Appendix C shows a three-body system comprising masses m_1, m_2, and m_3. The position vectors of the three masses relative to the origin of an inertial *XYZ* frame are $\mathbf{R}_1$, $\mathbf{R}_2$, and $\mathbf{R}_3$, respectively. Eqs. (C.2) give the absolute accelerations $\mathbf{a}_1 = d^2\mathbf{R}_1/dt^2$, $\mathbf{a}_2 = d^2\mathbf{R}_2/dt^2$, and $\mathbf{a}_3 = d^2\mathbf{R}_3/dt^2$ of the three masses due to the mutual gravitational attraction among them. The acceleration $\mathbf{a}_{2/1}$ of body 2 relative to body 1 is $\mathbf{a}_2 - \mathbf{a}_1$. Therefore, from Eqs. (C.2a) and (C.2b), we obtain

$$\mathbf{a}_{2/1} = \overbrace{\left(Gm_1\frac{\mathbf{R}_1-\mathbf{R}_2}{\|\mathbf{R}_1-\mathbf{R}_2\|^3}+Gm_3\frac{\mathbf{R}_3-\mathbf{R}_2}{\|\mathbf{R}_3-\mathbf{R}_2\|^3}\right)}^{\mathbf{a}_2} - \overbrace{\left(Gm_2\frac{\mathbf{R}_2-\mathbf{R}_1}{\|\mathbf{R}_2-\mathbf{R}_1\|^3}+Gm_3\frac{\mathbf{R}_3-\mathbf{R}_1}{\|\mathbf{R}_3-\mathbf{R}_1\|^3}\right)}^{\mathbf{a}_1}$$

Rearranging terms yields

$$\mathbf{a}_{2/1} = -\mu\frac{\mathbf{R}_2-\mathbf{R}_1}{\|\mathbf{R}_2-\mathbf{R}_1\|^3}+\mu_3\left(\frac{\mathbf{R}_3-\mathbf{R}_2}{\|\mathbf{R}_3-\mathbf{R}_2\|^3}-\frac{\mathbf{R}_3-\mathbf{R}_1}{\|\mathbf{R}_3-\mathbf{R}_1\|^3}\right) \tag{10.115}$$

where $\mu = G(m_1 + m_2)$ and $\mu_3 = Gm_3$.

Suppose body 1 is the earth, body 2 is an artificial earth satellite (s), and body 3 is the moon (m). Let us simplify the notation so that, as pictured in Fig. 10.17,

$\mathbf{r} = \mathbf{R}_2 - \mathbf{R}_1$	Position of the spacecraft relative to the earth
$\ddot{\mathbf{r}} = \mathbf{a}_{2/1}$	Acceleration of the spacecraft relative to the earth
$\mathbf{r}_m = \mathbf{R}_3 - \mathbf{R}_1$	Position of the moon relative to the earth
$\mathbf{r}_{m/s} = \mathbf{R}_3 - \mathbf{R}_2 = \mathbf{r}_m - \mathbf{r}$	Position of the moon relative to the spacecraft

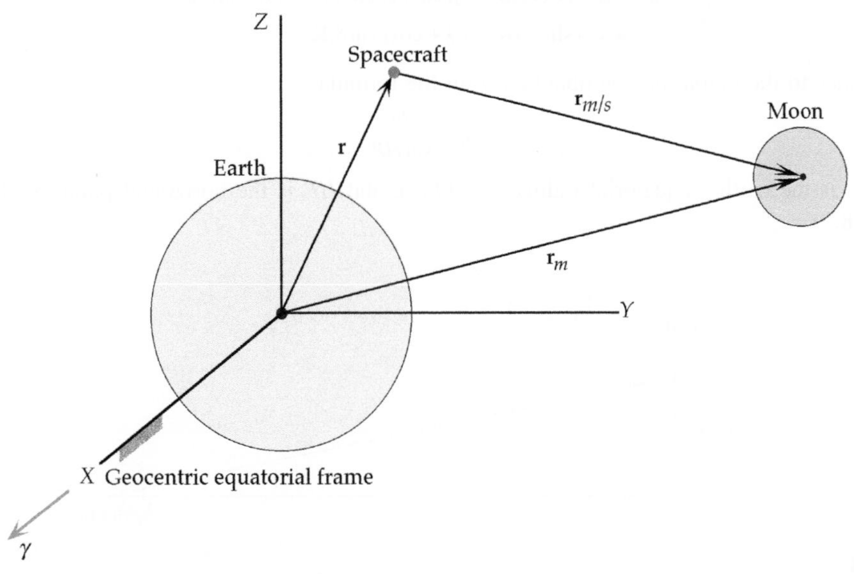

FIG. 10.17

Perturbation of a spacecraft orbit by a third body (the moon).

Then, Eq. (10.115) becomes

$$\ddot{\mathbf{r}} = -\mu\frac{\mathbf{r}}{r^3} + \mu_m\left(\frac{\mathbf{r}_{m/s}}{r_{m/s}{}^3} - \frac{\mathbf{r}_m}{r_m{}^3}\right) \tag{10.116}$$

where $\mu = \mu_{earth} = 398,600\ \text{km}^3/\text{s}^2$ and $\mu_m = \mu_{moon} = 4903\ \text{km}^3/\text{s}^2$. The last term in this equation is the perturbing acceleration due to lunar gravity,

$$\mathbf{p} = \mu_m\left(\frac{\mathbf{r}_{m/s}}{r_{m/s}{}^3} - \frac{\mathbf{r}_m}{r_m{}^3}\right) \tag{10.117}$$

If $\mathbf{p} = 0$, then Eq. (10.116) reduces to Eq. (2.22), the fundamental equation of Keplerian motion.

The unit vector $\hat{\mathbf{u}}$ from the center of the earth to that of the moon is given in the geocentric ecliptic frame by an expression similar to Eq. (10.99),

$$\hat{\mathbf{u}} = \cos\delta\cos\lambda\hat{\mathbf{I}}' + \cos\delta\sin\lambda\hat{\mathbf{J}}' + \sin\delta\hat{\mathbf{K}}' \tag{10.118}$$

where λ is the lunar ecliptic longitude, and δ is the lunar ecliptic latitude. If $\delta = 0$, then this expression reduces to that for the sun, which does not leave the ecliptic plane. The components of $\hat{\mathbf{u}}$ in the geocentric equatorial *(XYZ)* system are found as in Eq. (10.100),

$$\begin{aligned}\{\hat{\mathbf{u}}\}_{XYZ} = [\mathbf{R}_1(-\varepsilon)]\{\hat{\mathbf{u}}\}_{X'Y'Z'} &= \begin{bmatrix} 1 & 0 & 0 \\ 0 & \cos\varepsilon & -\sin\varepsilon \\ 0 & \sin\varepsilon & \cos\varepsilon \end{bmatrix}\begin{Bmatrix} \cos\delta\cos\lambda \\ \cos\delta\sin\lambda \\ \sin\delta \end{Bmatrix} \\ &= \begin{Bmatrix} \cos\delta\cos\lambda \\ \cos\varepsilon\cos\delta\sin\lambda - \sin\varepsilon\sin\delta \\ \sin\varepsilon\cos\delta\sin\lambda + \cos\varepsilon\sin\delta \end{Bmatrix}\end{aligned} \tag{10.119}$$

The geocentric equatorial position of the moon is $\mathbf{r}_m = r_m\hat{\mathbf{u}}$, so that

$$\begin{aligned}\mathbf{r}_m &= r_m\cos\delta\cos\lambda\hat{\mathbf{I}} + r_m(\cos\varepsilon\cos\delta\sin\lambda - \sin\varepsilon\sin\delta)\hat{\mathbf{J}} \\ &\quad + r_m(\sin\varepsilon\cos\delta\sin\lambda + \cos\varepsilon\sin\delta)\hat{\mathbf{K}}\end{aligned} \tag{10.120}$$

The distance to the moon may be obtained from the formula

$$r_m = \frac{R_E}{\sin HP} \tag{10.121}$$

where R_E is the earth's equatorial radius (6378 km), and *HP* is the horizontal parallax, defined in Fig. 10.18.

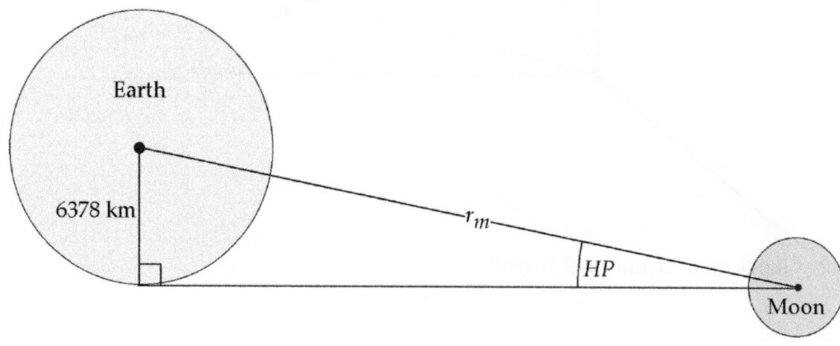

FIG. 10.18

Horizontal parallax *HP*.

The formulas presented in *The Astronomical Almanac* (National Almanac Office, 2018) for the time variation of lunar ecliptic longitude λ, lunar ecliptic latitude δ, and lunar horizontal parallax HP are

$$\lambda = b_0 + c_0 T_0 + \sum_{i=1}^{6} a_i \sin(b_i + c_i T_0) \qquad (0^\circ \le \lambda < 360^\circ) \tag{10.122}$$

$$\delta = \sum_{i=1}^{4} d_i \sin(e_i + f_i T_0) \qquad (0^\circ \le \lambda < 360^\circ) \tag{10.123}$$

$$HP = g_0 + \sum_{i=1}^{4} g_i \cos(h_i + k_i T_0) \qquad (0^\circ \le \lambda < 180^\circ) \tag{10.124}$$

where T_0 is the number of Julian centuries since J2000 for the current Julian day JD,

$$T_0 = \frac{JD - 2{,}451{,}545.0}{36{,}525} \tag{10.125}$$

The coefficients in these formulas are listed in Table 10.1.

Recall from the previous section that for the apparent motion of the sun around the earth *The Astronomical Almanac* uses n, the number of days since J2000, for the time variable instead of the number of centuries T_0 that is employed for the moon's motion. It is obvious from Eqs. (10.110) and (10.125) that the relation between n and T_0 is simply

$$n = 36{,}525 T_0 \tag{10.126}$$

In terms of T_0, the formula for obliquity of the ecliptic (Eq. 10.111) is

$$\varepsilon = 23.439^\circ - 0.0130042 T_0 \tag{10.127}$$

Table 10.1 Coefficients for computing lunar position

	Longitude, λ			Latitude, δ			Horizontal parallax, HP		
i	a_i	b_i	c_i	d_i	e_i	f_i	g_i	h_i	k_i
0	–	218.32	481267.881	–	–	–	0.9508	–	–
1	6.29	135.0	477198.87	5.13	93.3	483202.03	0.0518	135.0	477198.87
2	−1.27	259.3	−413335.36	0.28	220.2	960400.89	0.0095	259.3	−413335.38
3	0.66	235.7	890534.22	−0.28	318.3	6003.15	0.0078	253.7	890534.22
4	0.21	269.9	954397.74	−0.17	217.6	−407332.21	0.0028	269.9	954397.70
5	−0.19	357.5	35999.05	–	–	–	–	–	–
6	−0.11	106.5	966404.03	–	–	–	–	–	–

ALGORITHM 10.4

Given the year, month, day, and universal time, calculate the geocentric position vector of the moon:

1. Compute the Julian day number JD using Eqs. (5.47) and (5.48).
2. Calculate T_0, the number of Julian centuries since J2000, using Eq. (10.125).
3. Calculate the obliquity ε using Eq. (10.127).
4. Calculate the lunar ecliptic longitude λ by means of Eq. (10.122).
5. Calculate the lunar ecliptic latitude δ using Eq. (10.123).
6. Calculate the lunar horizontal parallax HP by means of Eq. (10.124).
7. Calculate the distance r_m from the earth to the moon using Eq. (10.121).
8. Compute the geocentric equatorial position $\mathbf{r}_m$ of the moon from Eq. (10.120).

EXAMPLE 10.10

Use Algorithm 10.4 to find the geocentric equatorial position vector of the moon at 08:00 UT on July 25, 2013. This is the same epoch as used to compute the sun's apparent position in Example 10.7.

Solution

Step 1:

According to Step 1 of Example 10.7, the Julian day number is

$$JD = 2{,}456{,}498.8333 \text{ days}$$

Step 2:

$$T_0 = \frac{2{,}456{,}498.8333 - 2{,}451{,}545.0}{36{,}525} = 0.135629 \text{ Cy}$$

Step 3:

$$\varepsilon = 23.439° - 0.0130042(0.135629) = 23.4375°$$

Step 4:

$$\lambda = b_0 + c_0(0.135629) + \sum_{i=1}^{6} a_i \sin\left[b_i + c_i(0.135629)\right] = 338.155°$$

Step 5:

$$\delta = \sum_{i=1}^{4} d_i \sin\left[e_i + f_i(0.135629)\right] = 4.55400°$$

Step 6:

$$HP = g_0 + \sum_{i=1}^{4} g_i \cos\left[h_i + k_i(0.135629)\right] = 0.991730°$$

Step 7:

$$r_m = \frac{6378}{\sin(0.991730°)} = 368{,}498 \text{ km}$$

Step 8:

$$\mathbf{r}_m = (368,498)\{(\cos 4.55400°)(\cos 338.155°)\hat{\mathbf{I}}\}$$
$$+[(\cos 23.4375°)(\cos 4.55400°)(\sin 338.155°) - (\sin 23.4375°)\ (\sin 4.55400°)]\hat{\mathbf{J}}$$
$$+[(\sin 23.4375°)(\cos 4.55400°)(\sin 338.155°) + (\cos 23.4375°)(\sin 4.55400°)]\hat{\mathbf{K}}$$

$$\boxed{\mathbf{r}_m = 340{,}958\hat{\mathbf{I}} - 137{,}043\hat{\mathbf{J}} - 27{,}521.3\hat{\mathbf{K}}\,(\text{km})}$$

EXAMPLE 10.11

The orbital parameters of three earth satellites at time t_0 (Julian date $JD_0 = 2,454,283.0$) are as follows. For each orbit, find the variation of Ω, ω, and i over the following 60 days due to lunar gravity.

Low earth orbit (LEO)	Highly elliptic earth orbit (HEO)	Geostationary earth orbit (GEO)	
$h_0 = 51{,}591.1\,\text{km}^2/\text{s}$	$h_0 = 69{,}084.1\,\text{km}^2/\text{s}$	$h_0 = 129{,}640\,\text{km}^2/\text{s}$	(a)
$e_0 = 0.01$	$e_0 = 0.741$	$e_0 = 0.0001$	(b)
$\Omega_0 = 0°$	$\Omega_0 = 0°$	$\Omega_0 = 0°$	(c)
$i_0 = 28.5°$	$i_0 = 63.4°$	$i_0 = 1°$	(d)
$\omega_0 = 0°$	$\omega_0 = 270°$	$\omega_0 = 0°$	(e)
$\theta_0 = 0°$	$\theta_0 = 0°$	$\theta_0 = 0°$	(f)
$a_0 = 6678.126\,\text{km}$	$a_0 = 26{,}553.4\,\text{km}$	$a_0 = 42{,}164\,\text{km}$	(g)
$T_0 = 1.50866\,\text{h}$	$T_0 = 11.9616\,\text{h}$	$T_0 = 23.9343\,\text{h}$	(h)

For each orbit, find the variation of Ω, ω, and i over the following 60 days due to lunar gravity.

Solution

For each orbit in turn we write the system of differential equations given by Eq. (10.84) as $d\mathbf{y}/dt = \mathbf{f}(\mathbf{y}, t)$, where the six components of $\mathbf{y}$ are the orbital elements

$$\mathbf{y} = \lfloor h\ \ e\ \ \theta\ \ \Omega\ \ i\ \ \omega \rfloor^T \tag{i}$$

and

$$\mathbf{f} = \lfloor \dot{h}\ \ \dot{e}\ \ \dot{\theta}\ \ \dot{\Omega}\ \ \dot{i}\ \ \dot{\omega} \rfloor^T \tag{j}$$

The six time rates are found on the right-hand side of Gauss' planetary equations (Eqs. 10.84).

The initial conditions vector $\mathbf{y}_0$ for each of the three orbits consists of Eqs. (a) through (f). The system $d\mathbf{y}/dt = \mathbf{f}(\mathbf{y}, t)$ is solved for $\mathbf{y}$ on the interval $[t_0, t_f]$ using a numerical integrator such as MATLAB's *ode45*. At each time step, the integrator relies on a subroutine to compute the rates $\mathbf{f}$ from the current values of $\mathbf{y}$ and the time t, as follows:

Update the Julian day number: $JD \leftarrow JD_0 + t(\text{days})$.

Using JD, compute the geocentric equatorial position vector $\mathbf{r}_m$ of the moon by means of Algorithm 10.4.

Use $\mathbf{y}$ to compute the state vector $(\mathbf{r}, \mathbf{v})$ of the satellite by means of Algorithm 4.5.

Compute the position of the moon relative to the satellite: $\mathbf{r}_{m/s} = \mathbf{r}_m - \mathbf{r}$.

Compute the perturbing acceleration of the moon: $\mathbf{p} = \mu_m \left(\mathbf{r}_{m/s}/r_{m/s}{}^3 - \mathbf{r}_m/r_m{}^3\right)$.

Compute the unit vectors of the *rsw* frame (Fig. 10.10):

$$\hat{\mathbf{r}}=\frac{\mathbf{r}}{\|\mathbf{r}\|}\quad \hat{\mathbf{w}}=\frac{\mathbf{r}\times\mathbf{v}}{\|\mathbf{r}\times\mathbf{v}\|}\quad \hat{\mathbf{s}}=\frac{\hat{\mathbf{w}}\times\hat{\mathbf{r}}}{\|\hat{\mathbf{w}}\times\hat{\mathbf{r}}\|}$$

Compute the components of the perturbing acceleration along the *rsw* axes:

$$p_r=\mathbf{p}\cdot\hat{\mathbf{r}}\quad p_s=\mathbf{p}\cdot\hat{\mathbf{s}}\quad p_w=\mathbf{p}\cdot\hat{\mathbf{w}}$$

Use Eq. (10.84) to calculate the components of $\mathbf{f}$ in Eq. (j).
Plots of the orbital parameter variations $\delta\Omega=\Omega(t)-\Omega_0$, $\delta i=i(t)-i_0$, and $\delta\omega=\omega(t)-\omega_0$ are shown in Fig. 10.19.
The MATLAB M-file *Example_10_11.m* is listed in Appendix D.47.

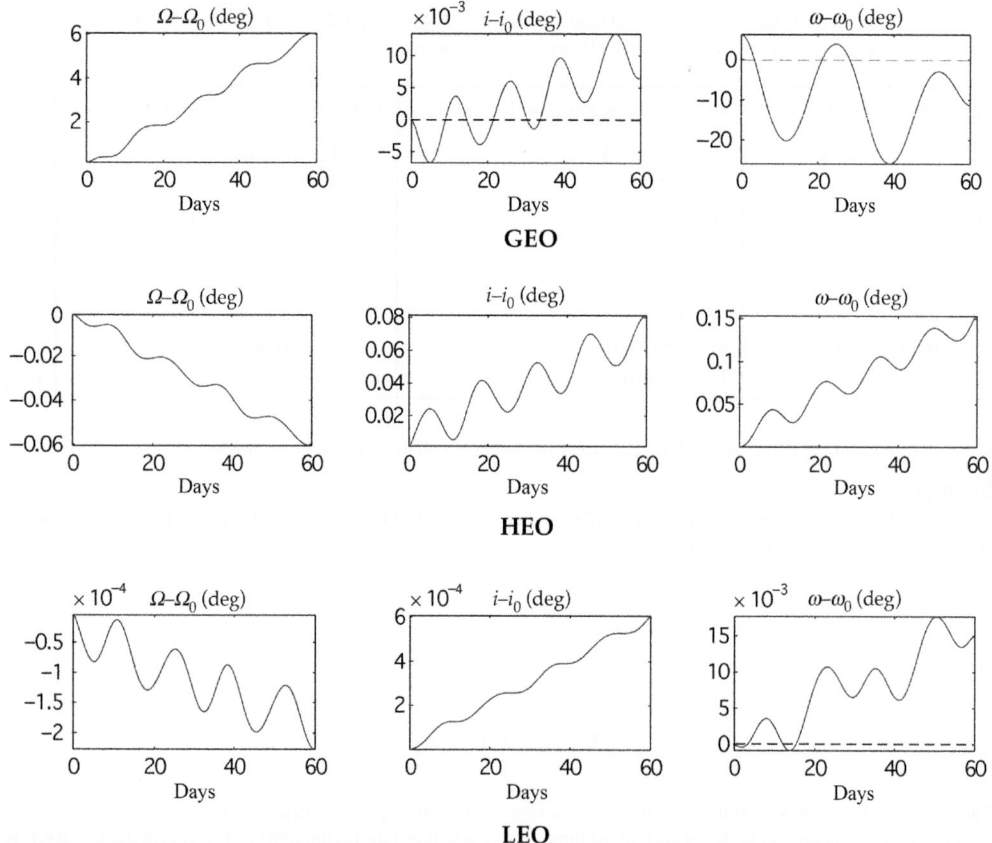

FIG. 10.19

Lunar gravity perturbations of Ω, i, and ω of the three orbits during a 60-day interval following Julian date 2454283.0 (1 July 2007).

10.11 SOLAR GRAVITY

The special perturbations approach to assessing the sun's influence on the orbits of earth satellites proceeds as it did for the moon in the previous section. The sun replaces the moon as the third body, as illustrated in Fig. 10.20, in which ⊙ represents the sun and "s" stands for spacecraft. The perturbing acceleration of the sun may be inferred from that of the moon in Eq. (10.117),

$$\mathbf{p} = \mu_\odot \left(\frac{\mathbf{r}_{\odot/s}}{r_{\odot/s}{}^3} - \frac{\mathbf{r}_\odot}{r_\odot{}^3} \right) \tag{10.128}$$

According to Table A.2, the gravitational parameter $\mu_\odot$ of the sun is 132.712(10^9) km^3/s^2. The sun's geocentric position vector $\mathbf{r}_\odot$ in its apparent motion around the earth is found by using Algorithm 10.2, as we did in our study of solar radiation pressure effects.

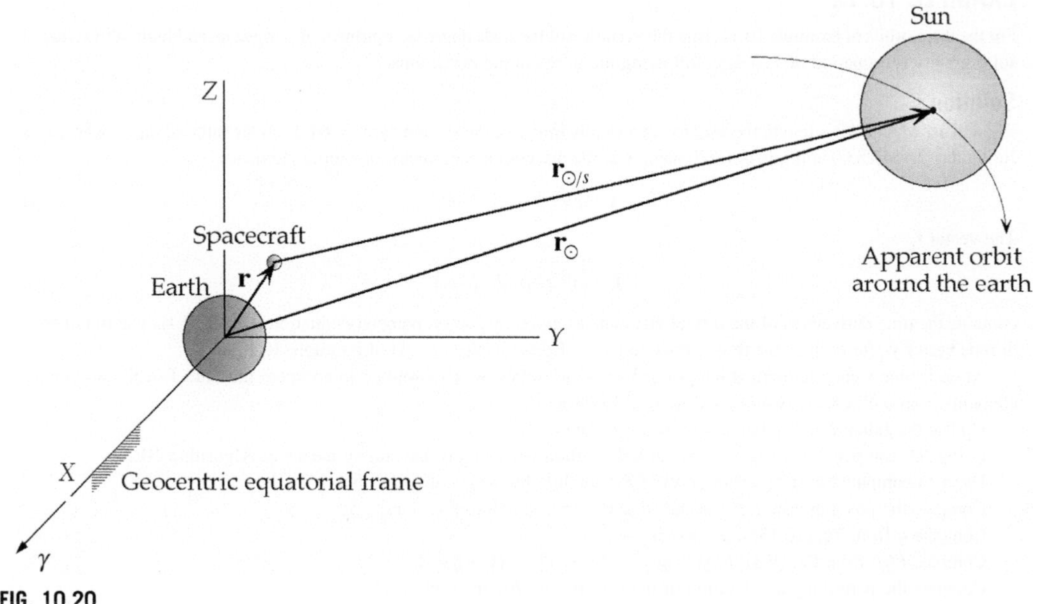

FIG. 10.20

Perturbation of a spacecraft's earth orbit by solar gravity.

Using the fact that $\mathbf{r}_\odot = \mathbf{r} + \mathbf{r}_{\odot/s}$, we may rewrite Eq. (10.128) as

$$\mathbf{p} = \frac{\mu_\odot}{r_{\odot/s}{}^3} \left[\mathbf{r}_\odot \left(1 - \frac{r_{\odot/s}{}^3}{r_\odot{}^3} \right) - \mathbf{r} \right] \tag{10.129}$$

Because the sun is so far from the earth, the ratio $\left(r_{\odot/s}/r_\odot\right)^3$ is very nearly 1. Therefore, evaluating Eq. (10.129) involves subtracting two nearly equal numbers, which should be avoided in a digital computer. We do so by referring to Appendix F to write Eq. (10.129) as

$$\mathbf{p} = \frac{\mu_\odot}{r_{\odot/s}{}^3}\left[F(q)\mathbf{r}_\odot - \mathbf{r}\right] \tag{10.130}$$

where, according to Eq. (F.4),

$$q = \frac{\mathbf{r}\cdot(2\mathbf{r}_\odot - \mathbf{r})}{r_\odot{}^2} \tag{10.131}$$

and $F(q)$ is given by Eq. (F.3).

EXAMPLE 10.12

For the three orbits of Example 10.11, find the variation of the node angle Ω, argument of perigee ω, and inclination i due to solar gravity for a period of 720 days following the given initial conditions.

Solution

We will use MATLAB's function *ode45* to numerically integrate the system $\mathbf{dy}/dt = \mathbf{f}(\mathbf{y}, t)$ on the interval $[t_0, t_f]$, where t_0 is Julian day 2,454,283.0 and t_f is $t_0 + 720$ days. $\mathbf{y}$ is the six-component vector of orbital elements,

$$\mathbf{y} = \lfloor h \;\; e \;\; \theta \;\; \Omega \;\; i \;\; \omega \rfloor^T \tag{a}$$

The vector $\mathbf{f}$,

$$\mathbf{f} = \lfloor \dot{h} \;\; \dot{e} \;\; \dot{\theta} \;\; \dot{\Omega} \;\; \dot{i} \;\; \dot{\omega} \rfloor^T \tag{b}$$

contains the time derivatives of the orbital elements as given by Gauss' planetary equations (Eq. 10.84). The initial conditions vector $\mathbf{y}_0$ for each of the three orbits consists of Eqs. (a) through (f) of Example 10.11.

At each time step, a numerical integrator like *ode45* relies on a subroutine to compute the rates $\mathbf{f}$ of the osculating elements from their current $\mathbf{y}$ and the time t, as follows:

Update the Julian day number: $JD \leftarrow JD_0 + t$ (days).
Using JD, compute the geocentric equatorial position vector $\mathbf{r}_\odot$ of the sun by means of Algorithm 10.2.
Use $\mathbf{y}$ to compute the state vector $(\mathbf{r}, \mathbf{v})$ of the satellite by means of Algorithm 4.5.
Compute the position vector of the sun relative to the satellite: $\mathbf{r}_{\odot/s} = \mathbf{r}_\odot - \mathbf{r}$.
Compute q from Eq. (10.131): $q = \mathbf{r}\cdot(2\mathbf{r}_\odot - \mathbf{r})/r_\odot{}^2$.
Compute $F(q)$ from Eq. (F.3): $F(q) = q(q^2 - 3q + 3)/[1 + (1 - q)^{3/2}]$.
Compute the perturbing acceleration of the sun: $\mathbf{p} = \mu_\odot\left[F(q)\mathbf{r}_\odot - \mathbf{r}\right]/r_{\odot/s}{}^3$.
Compute the unit vectors of the *rsw* frame (Fig. 10.10):

$$\hat{\mathbf{r}} = \frac{\mathbf{r}}{\|\mathbf{r}\|} \qquad \hat{\mathbf{w}} = \frac{\mathbf{r}\times\mathbf{v}}{\|\mathbf{r}\times\mathbf{v}\|} \qquad \hat{\mathbf{s}} = \frac{\hat{\mathbf{w}}\times\hat{\mathbf{r}}}{\|\hat{\mathbf{w}}\times\hat{\mathbf{r}}\|}$$

Compute the components of the perturbing acceleration along the *rsw* axes:

$$p_r = \mathbf{p}\cdot\hat{\mathbf{r}} \qquad p_s = \mathbf{p}\cdot\hat{\mathbf{s}} \qquad p_w = \mathbf{p}\cdot\hat{\mathbf{w}}$$

Use Eq. (10.84) to calculate the components of $\mathbf{f}$.
Plots of the orbital parameter variations $\delta\Omega = \Omega(t) - \Omega_0$, $\delta i = i(t) - i_0$, and $\delta\omega = \omega(t) - \omega_0$ are shown in Fig. 10.21.
The MATLAB script file *Example_10_12.m* is listed in Appendix D.48.

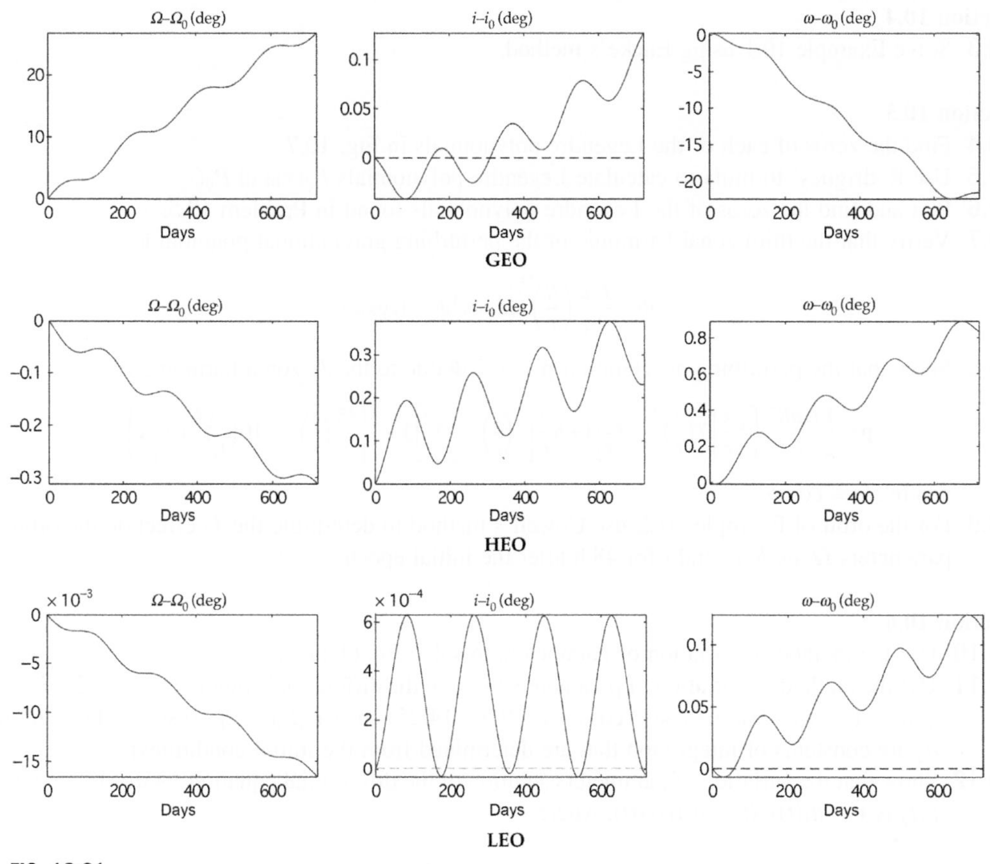

FIG. 10.21

Solar gravity perturbations of Ω, i, and ω of the three orbits in Example 12.11 during a 720-day interval following Julian date 2454283.0 (1 July 2007).

PROBLEMS

Section 10.2

10.1 In Fig. 10.1, the radius at perigee O is 7000 km and the speed is $v_0 = \sqrt{\mu/r_0}$. The initial mass of the spacecraft is $m_0 = 2000\,\text{kg}$ and the thrust of the propulsion system is 0.5 kN. Using Cowell's method and *ode45*, find the perigee and the eccentricity of the osculating orbits at the following times after t_0: (a) 1 h; (b) 1.2 h; (c) 1.4 h; (d) 1.6 h.
{Ans.: (a) 0.1856, 7903 km; (b) 0.2046, 8450 km; (c) 0.2272, 9123 km; (d) 0.2825, 9895 km}

Section 10.3

10.2 Repeat Problem 10.1 using *ode45* and Encke's method.

Section 10.4

10.3 Solve Example 10.1 using Encke's method.

Section 10.5

10.4 Find the zeros of each of the Legendre polynomials in Fig. 10.7.

10.5 Use Rodrigues' formula to calculate Legendre polynomials $P_8(x)$and $P_9(x)$.

10.6 Plot and find the zeros of the Legendre polynomials found in Problem 10.5.

10.7 Verify that the third zonal harmonic of the perturbing gravitational potential is

$$\Phi = \frac{J_3 \mu}{2\, r}\left(\frac{R}{r}\right)^3 \left(5\cos^3\phi - 3\cos\phi\right)$$

10.8 Show that the perturbing acceleration $\mathbf{p} = -\nabla\Phi$ due to the J_3 zonal harmonic is

$$\mathbf{p} = \frac{1}{2}\frac{J_3 \mu R^3}{r^5}\left\{5\frac{x}{r}\left[7\left(\frac{z}{r}\right)^3 - 3\frac{z}{r}\right]\hat{\mathbf{i}} + 5\frac{y}{r}\left[7\left(\frac{z}{r}\right)^3 - 3\frac{z}{r}\right]\hat{\mathbf{j}} + 3\left[\frac{35}{3}\left(\frac{z}{r}\right)^4 - 10\left(\frac{z}{r}\right)^2 + 1\right]\hat{\mathbf{k}}\right\}$$

where $z/r = \cos\phi$.

10.9 For the orbit of Example 10.2, use Cowell's method to determine the J_3 effect on the orbital parameters Ω, ω, h, e, and i for 48 h after the initial epoch.

Section 10.6

10.10 Use the method of variation of parameters to solve Eq. (1.113).

10.11 Use the method of variation of parameters to solve the differential equation $\ddot{x} + 2\dot{x} + 2x = t\sin t$. {Ans.: $x = a_1 e^{-t}\sin t + a_2 e^{-t}\cos t + (-2t/5 + 14/25)\cos t + (t/5 - 2/25)\sin t$, where a_1 and a_2 are constants of integration that are determined from the initial conditions}

10.12 Show that the variation of parameters solution of the differential equation $\ddot{x} + a_1(t)\dot{x} + a_2(t)x = f(t)$ is $x = u_1(t)x_1(t) + u_2(t)x_2(t)$, where

$$u_1 = \int \frac{x_2 f(t)}{x_2 \dot{x}_1 - \dot{x}_2 x_1}dt + C_1 \qquad u_2 = \int \frac{x_1 f(t)}{x_1 \dot{x}_2 - \dot{x}_1 x_2}dt + C_2$$

where x_1 and x_2 are solutions of the reduced homogeneous equation $\ddot{x} + a_1(t)\dot{x} + a_2(t)x = 0$, and C_1 and C_2 are constants.

10.13 Using the definition of the dot product of two vectors ($\mathbf{b}\cdot\mathbf{c} = \sum_{i=1}^{3} b_i c_i$) it is easy to see from Eq. (10.48) that the 36 components of the Lagrange matrix, called *Lagrange brackets*, may be written

$$L_{\alpha\beta} = \frac{\partial \mathbf{r}}{\partial u_\alpha}\cdot\frac{\partial \mathbf{v}}{\partial u_\beta} - \frac{\partial \mathbf{v}}{\partial u_\alpha}\cdot\frac{\partial \mathbf{r}}{\partial u_\beta} \qquad \alpha,\beta = 1,\ldots,6$$

Show that $L_{\alpha\beta} = -L_{\beta\alpha}$ and that $L_{\alpha\beta} = 0$ when $\alpha = \beta$, so that there are only 15 independent Lagrange brackets.

10.14 Using the facts that $\partial\mathbf{r}/\partial t = \mathbf{v}$ and $\partial\mathbf{v}/\partial t = \mathbf{a}$, where $\mathbf{v}$ is the velocity and $\mathbf{a}$ is the acceleration, show that the time derivative of $L_{\alpha\beta}$ in Problem 10.13 is

$$\frac{\partial L_{\alpha\beta}}{\partial t} = \frac{\partial \mathbf{a}}{\partial u_\beta}\cdot\frac{\partial \mathbf{r}}{\partial u_\alpha} - \frac{\partial \mathbf{a}}{\partial u_\alpha}\cdot\frac{\partial \mathbf{r}}{\partial u_\beta}$$

10.15 If **a** is the two-body relative acceleration vector, $\mathbf{a} = -\mu\mathbf{r}/r^3$, show that

$$\frac{\partial \mathbf{a}}{\partial u_\alpha} = -\frac{\mu}{r^3}\left(\frac{\partial \mathbf{r}}{\partial u_\alpha} - 3\frac{\mathbf{r}}{r}\frac{\partial r}{\partial u_\alpha}\right)$$

where u_α ($\alpha = 1, \ldots, 6$) are a set of osculating orbital elements.

10.16 Use the formula for $\partial\mathbf{a}/\partial u_\alpha$ in Problem 10.15 plus the fact that $\mathbf{r}\cdot(\partial\mathbf{r}/\partial u_\beta) = r(\partial r/\partial u_\beta)$ to verify that

$$\frac{\partial \mathbf{a}}{\partial u_\alpha}\cdot\frac{\partial \mathbf{r}}{\partial u_\beta} = -\frac{\mu}{r^3}\left(\frac{\partial \mathbf{r}}{\partial u_\alpha}\cdot\frac{\partial \mathbf{r}}{\partial u_\beta} - 3\frac{\partial r}{\partial u_\alpha}\frac{\partial r}{\partial u_\beta}\right)$$

10.17 Combine the results of Problems 10.14 and 10.16 to show that

$$\frac{\partial L_{\alpha\beta}}{\partial t} = 0 \quad \alpha,\beta = 1,\ldots,6$$

That is, on a given osculating orbit (i.e., for a given set of orbital elements u_α), the Lagrangian brackets are constant. This means that [**L**] may be computed at a point where the orbit formulas have their simplest algebraic form, which usually is at periapsis (as it was for obtaining the specific energy formula in Section 2.6).

10.18 For the orbital elements

$$u_1 = h \quad u_2 = e \quad u_3 = \theta \quad u_4 = \Omega \quad u_5 = i \quad u_6 = \omega$$

it can be shown that the Lagrange matrix is

$$[\mathbf{L}] = \begin{bmatrix} 0 & 0 & -\dfrac{1-e}{1+e} & -\cos i & 0 & -1 \\ 0 & 0 & -\dfrac{eh}{(1+e)^2} & 0 & 0 & 0 \\ \dfrac{1-e}{1+e} & \dfrac{eh}{(1+e)^2} & 0 & 0 & 0 & 0 \\ \cos i & 0 & 0 & 0 & -h\sin i & 0 \\ 0 & 0 & 0 & h\sin i & 0 & 0 \\ 1 & 0 & 0 & 0 & 0 & 0 \end{bmatrix}$$

Solve Eq. (10.47), where

$$\{\mathbf{P}\} = \lfloor \partial R/\partial h \;\; \partial R/\partial e \;\; \partial R/\partial\theta \;\; \partial R/\partial\Omega \;\; \partial R/\partial\omega \;\; \partial R/\partial i \rfloor^T$$

for the element rates, $\dot{h}$, $\dot{e}$, $\dot{\theta}$, $\dot{\Omega}$, and $\dot{\omega}$.

10.19 For the orbital elements

$$u_1 = \Omega \quad u_2 = i \quad u_3 = \omega \quad u_4 = a \quad u_5 = e \quad u_6 = t_p$$

where t_p is the time of periapse passage, it can be shown that the Lagrange matrix is

$$[\mathbf{L}] = \begin{bmatrix} 0 & -nab\sin i & 0 & \frac{1}{2}nb\cos i & -\frac{na^3e}{b}\cos i & 0 \\ nab\sin i & 0 & 0 & 0 & 0 & 0 \\ 0 & 0 & 0 & \frac{1}{2}nb & -\frac{na^3e}{b} & 0 \\ -\frac{1}{2}nb\cos i & 0 & -\frac{1}{2}nb & 0 & 0 & \frac{1}{2}n^2a \\ \frac{na^3e}{b}\cos i & 0 & \frac{na^3e}{b} & 0 & 0 & 0 \\ 0 & 0 & 0 & -\frac{1}{2}n^2a & 0 & 0 \end{bmatrix}$$

where a and b are the semimajor and semiminor axes, and n is the mean motion. Solve Eq. (10.47) with

$$\{\mathbf{P}\} = \lfloor \partial R/\partial\Omega \;\; \partial R/\partial i \;\; \partial R/\partial\omega \;\; \partial R/\partial a \;\; \partial R/\partial\varepsilon \;\; \partial R/\partial t_p \rfloor^T$$

to obtain Lagrange's planetary equations listed in Eq. (10.50).

Section 10.7

10.20 Show that $d\mathbf{r}/dt = \mathbf{0}$ implies that $dr/dt = 0$.

10.21 Verify that for unperturbed two-body motion, $d\theta/dt = h/r^2$.

10.22 Verify that for unperturbed two-body motion, $dM/dt = n$.

10.23 Verify that for unperturbed two-body motion, $dE/dt = na/r$.

10.24 Show that $(da/dt) = (2a^2/h)[e\sin\theta p_r + h^2 p_s/(\mu r)]$.

10.25 Show that there is no time-averaged J_3 perturbation of the semimajor axis $(\overline{da/dt} = 0)$.

10.26 Show that $\partial v^2/\partial\mathbf{v} = 2\mathbf{v}$.

10.27 Show that $\partial(\mathbf{r}\cdot\mathbf{v})/\partial\mathbf{v} = \mathbf{r}$.

10.28 Show that $\partial h/\partial\mathbf{v} = (\mathbf{h}\times\mathbf{r})/h$.

10.29 Show that $\partial h^2/\partial\mathbf{v} = 2\mathbf{h}\times\mathbf{r}$.

10.30 Find Gauss' variational equation for dr_p/dt.

10.31 Find Gauss' variational equation for dr_a/dt.

10.32 Find Gauss' variational equations for a radial acceleration perturbation, $\mathbf{p} = p_r\hat{\mathbf{r}}$.

10.33 Show that for a tangential perturbing acceleration p_v, the variation of the semimajor axis is $da/dt = (2a^2/\mu)(p_v/v)$.

10.34 Numerically integrate Gauss' planetary equations for a given perturbation **p** and set of initial conditions.

10.35 Find the p_r, p_s, and p_w components in the *rsw* frame of the J_3 gravitational perturbation **p** found in Problem 10.8.

10.36 Find the expression for dh/dt due to J_3.

10.37 Find the expression for de/dt due to J_3.

10.38 Find the expression for $d\Omega/dt$ due to J_3.

10.39 Find the expression for di/dt due to J_3.

10.40 Find the expression for $d\omega/dt$ due to J_3.

Section 10.8

10.41–10.45 Find the orbital averages of the J_3 rates found in Problems 10.35–10.39.

Section 10.9

10.46 An earth satellite has the following orbital parameters on Julian date 2,456,793 (May 15, 2014):
$r_p = 10{,}000\,\text{km}$ $r_a = 15{,}000\,\text{km}$ $\theta = 40°$ $\Omega = 300°$ $\omega = 110°$ $i = 55°$
Assuming an absorbing area-to-mass ratio of $2\,\text{m}^2/\text{kg}$ and a radiation pressure coefficient of 2, use Cowell's method and MATLAB's *ode45* (or similar) to plot the variation of these orbital parameters over the next 3 years due just to solar radiation pressure.

Section 10.10

10.47 Solve Problem 10.46, neglecting solar radiation pressure and including only the perturbing effect of lunar gravity.

10.48 Solve Problem 10.46, retaining the effect of solar radiation pressure and adding that of lunar gravity as well.

10.49 For the orbits of Example 10.11, plot the variations a, e, and θ over the same time interval.

10.50 For the orbits of Example 10.11, plot the variations of r_p and r_a over the same time interval.

10.51 For the data of Example 10.11, use Cowell's method to integrate Eq. (10.2) with lunar gravity as the perturbation and then use Algorithm 4.2 to obtain the time histories of the orbital parameters. Compare the results with Fig. 10.19.

10.52 For the data of Example 10.11, use Encke's method to integrate Eq. (10.2) with lunar gravity as the perturbation and then use Algorithm 4.2 to obtain the time histories of the orbital parameters. Compare the results with Fig. 10.19.

Section 10.11

10.53 Solve Problem 10.46, neglecting solar radiation pressure and including only the perturbing effect of solar gravity.

10.54 Solve Problem 10.46, retaining the effect of solar radiation pressure and adding that of solar gravity as well.

10.55 Solve Problem 10.46, retaining the effect of solar radiation pressure and adding those of lunar and solar gravity as well.

10.56 Plot the variation of a, e, and θ in Example 10.12.

10.57 Plot the variations of r_p and r_a in Example 10.12.

10.58 For the data of Example 10.12, use Cowell's method to integrate Eq. (10.2) with solar gravity as the perturbation and then use Algorithm 4.2 to obtain the time histories of the orbital parameters. Compare the results with Fig. 10.21.

10.59 For the data of Example 10.12, use Encke's method to integrate Eq. (10.2) with solar gravity as the perturbation and then use Algorithm 4.2 to obtain the time histories of the orbital parameters. Compare the results with Fig. 10.21.

REFERENCES

AIAA, 2010. Guide to Reference and Standard Atmosphere Models. American Institute of Aeronautics and Astronautics.

Bate, R.R., Mueller, D.D., White, J.E., 1971. Fundamentals of Astrodynamics. Dover.

Battin, R.H., 1999. An Introduction to the Mathematics and Methods of Astrodynamics. Revised Edition, AIAA Education Series.
Fortescue, P.W., Swinerd, G.G., Stark, J.P.W. (Eds.), 2011. Spacecraft Systems Engineering. fourth ed. Wiley.
Garcia, D., 2010. Robust smoothing of gridded data in one and higher dimensions with missing values. Comput. Stat. Data Anal. 54, 1167–1178.
Montenbruck, O., Eberhard, G., 2000. Satellite Orbits: Models, Methods and Applications. Springer.
National Almanac Office, 2018. The Astronomical Almanac for the Year 2019. GPO.
NOAA/NASA/USAF, 1976. U.S. Standard Atmosphere, 1976. GPO.
Schaub, H., Junkins, J.L., 2009. Analytical Mechanics of Space Systems, second ed. AIAA Education Series.
Vallado, D.A., 2007. Fundamentals of Astrodynamics and Applications, third ed. Springer, New York.
Zwillinger, D. (Ed.), 2012. Standard Mathematical Tables and Formulae, 32nd ed. CRC Press, Boca Raton.

CHAPTER

RIGID BODY DYNAMICS

11

11.1 INTRODUCTION

Just as Chapter 1 provides a foundation for development of the equations of orbital mechanics, this chapter serves as a basis for developing the equations of satellite attitude dynamics. Chapter 1 deals with particles, whereas here we are concerned with rigid bodies. Those familiar with rigid body dynamics can move on to the next chapter, perhaps returning from time to time to review concepts.

The kinematics of rigid bodies is presented first. The subject depends on a theorem of the French mathematician Michel Chasles (1793–1880). Chasles' theorem states that the motion of a rigid body can be described by the displacement of any point of the body (the base point) plus a rotation about a unique axis through that point. The magnitude of the rotation does not depend on the base point. Thus, at any instant, a rigid body in a general state of motion has an angular velocity vector whose direction is that of the instantaneous axis of rotation. Describing the rotational component of the motion of a rigid body in three dimensions requires taking advantage of the vector nature of angular velocity and knowing how to take the time derivative of moving vectors, which is explained in Chapter 1. Several examples in the current chapter illustrate how this is done.

We then move on to study the interaction between the motion of a rigid body and the forces acting on it. Describing the translational component of the motion requires simply concentrating all of the mass at a point known as the center of mass and applying the methods of particle mechanics to determine its motion. Indeed, our study of the two-body problem up to this point has focused on the motion of their centers of mass without regard to the rotational aspect. Analyzing the rotational dynamics requires computing the body's angular momentum, and that in turn requires accounting for how the mass is distributed throughout the body. The mass distribution is described by the six components of the moment of inertia tensor.

Writing the equations of rotational motion relative to coordinate axes embedded in the rigid body and aligned with the principal axes of inertia yields the nonlinear Euler equations of motion, which are applied to a study of the dynamics of a spinning top (or one-axis gyro).

The expression for the kinetic energy of a rigid body is derived because it will be needed in the following chapter.

The chapter next describes the two sets of three angles commonly employed to specify the orientation of a body in three-dimensional space. One of these comprises the Euler angles, which are the same as the right ascension of the node (Ω), the argument of periapsis (ω), and the inclination (i). These were introduced in Chapter 4 to orient orbits in space. The other set comprises the yaw, pitch, and roll

Orbital Mechanics for Engineering Students. https://doi.org/10.1016/B978-0-12-824025-0.00011-8

angles, which are suitable for describing the orientation of an airplane. Both the Euler angles and the yaw, pitch, and roll angles will be employed in Chapter 12.

The chapter concludes with a brief discussion of quaternions and an example of how they are used to describe the evolution of the attitude of a rigid body.

11.2 KINEMATICS

Fig. 11.1 shows a moving rigid body and its instantaneous axis of rotation, which defines the direction of the absolute angular velocity vector $\boldsymbol{\omega}$. The *XYZ* axes are a fixed, inertial frame of reference. The position vectors $\mathbf{R}_A$ and $\mathbf{R}_B$ of two points on the rigid body are measured in the inertial frame. The vector $\mathbf{R}_{B/A}$ drawn from point *A* to point *B* is the position vector of *B* relative to *A*. Since the body is rigid, $\mathbf{R}_{B/A}$ has a constant magnitude even though its direction is continuously changing. Clearly,

$$\mathbf{R}_B = \mathbf{R}_A + \mathbf{R}_{B/A}$$

Differentiating this equation through with respect to time, we get

$$\dot{\mathbf{R}}_B = \dot{\mathbf{R}}_A + \frac{d\mathbf{R}_{B/A}}{dt} \tag{11.1}$$

where $\dot{\mathbf{R}}_A$ and $\dot{\mathbf{R}}_B$ are the absolute velocities $\mathbf{v}_A$ and $\mathbf{v}_B$ of points *A* and *B*. Because the magnitude of $\mathbf{R}_{B/A}$ does not change, its time derivative is given by Eq. (1.52). That is,

$$\frac{d\mathbf{R}_{B/A}}{dt} = \boldsymbol{\omega} \times \mathbf{R}_{B/A}$$

Thus, Eq. (11.1) becomes

$$\mathbf{v}_B = \mathbf{v}_A + \boldsymbol{\omega} \times \mathbf{R}_{B/A} \tag{11.2}$$

Taking the time derivative of Eq. (11.1) yields

$$\ddot{\mathbf{R}}_B = \ddot{\mathbf{R}}_A + \frac{d^2\mathbf{R}_{B/A}}{dt^2} \tag{11.3}$$

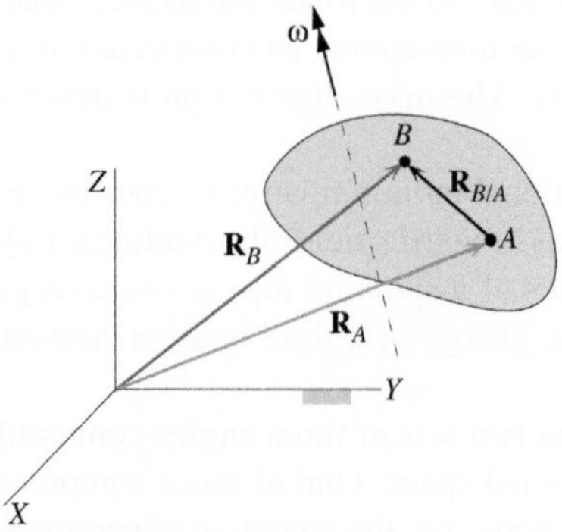

FIG. 11.1

Rigid body and its instantaneous axis of rotation.

where $\ddot{\mathbf{R}}_A$ and $\ddot{\mathbf{R}}_B$ are the absolute accelerations $\mathbf{a}_A$ and $\mathbf{a}_B$ of the two points of the rigid body, while from Eq. (1.53) we have

$$\frac{d^2\mathbf{R}_{B/A}}{dt^2} = \boldsymbol{\alpha} \times \mathbf{R}_{B/A} + \boldsymbol{\omega} \times \left(\boldsymbol{\omega} \times \mathbf{R}_{B/A}\right)$$

where $\boldsymbol{\alpha}$ is the angular acceleration, $\boldsymbol{\alpha} = d\boldsymbol{\omega}/dt$. Therefore, Eq. (11.3) can be written

$$\mathbf{a}_B = \mathbf{a}_A + \boldsymbol{\alpha} \times \mathbf{R}_{B/A} + \boldsymbol{\omega} \times \left(\boldsymbol{\omega} \times \mathbf{R}_{B/A}\right) \tag{11.4}$$

Eqs. (11.2) and (11.4) are the relative velocity and acceleration formulas. Note that all quantities in these expressions are measured in the same inertial frame of reference.

When the rigid body under consideration is connected to and moving relative to another rigid body, computation of its inertial angular velocity $\boldsymbol{\omega}$ and angular acceleration $\boldsymbol{\alpha}$ must be done with care. The key is to remember that angular velocity is a vector. It may be found as the vector sum of a sequence of angular velocities, each measured relative to another, starting with one measured relative to an absolute frame, as illustrated in Fig. 11.2. In that case, the absolute angular velocity $\boldsymbol{\omega}$ of body 4 is

$$\boldsymbol{\omega} = \boldsymbol{\omega}_1 + \boldsymbol{\omega}_{2/1} + \boldsymbol{\omega}_{3/2} + \boldsymbol{\omega}_{4/3} \tag{11.5}$$

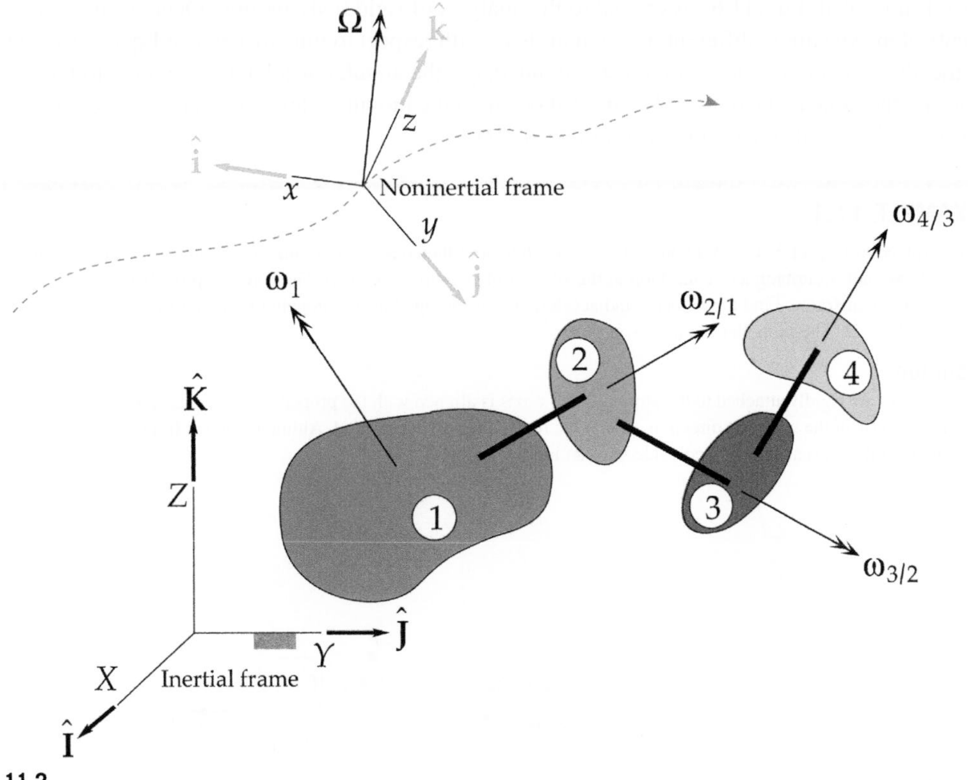

FIG. 11.2

Angular velocity is the vector sum of the relative angular velocities starting with $\boldsymbol{\omega}_1$, measured relative to the inertial frame.

Each of these angular velocities is resolved into components along the axes of the moving frame of reference xyz as shown in Fig. 11.2, so that the vector sum may be written as

$$\boldsymbol{\omega} = \omega_x \hat{\mathbf{i}} + \omega_y \hat{\mathbf{j}} + \omega_z \hat{\mathbf{k}} \tag{11.6}$$

The moving frame is chosen for convenience of analysis, and its own inertial angular velocity vector is denoted as $\boldsymbol{\Omega}$, as discussed in Section 1.6. According to Eq. (1.56), the absolute angular acceleration $\boldsymbol{\alpha} = \mathrm{d}\boldsymbol{\omega}/\mathrm{d}t$ is obtained from Eq. (11.6) by means of the following calculation:

$$\boldsymbol{\alpha} = \left.\frac{\mathrm{d}\boldsymbol{\omega}}{\mathrm{d}t}\right)_{\text{rel}} + \boldsymbol{\Omega} \times \boldsymbol{\omega} \tag{11.7}$$

where

$$\left.\frac{\mathrm{d}\boldsymbol{\omega}}{\mathrm{d}t}\right)_{\text{rel}} = \dot{\omega}_x \hat{\mathbf{i}} + \dot{\omega}_y \hat{\mathbf{j}} + \dot{\omega}_z \hat{\mathbf{k}} \tag{11.8}$$

and $\dot{\omega}_x = \mathrm{d}\omega_x/\mathrm{d}t$.

Being able to express the absolute angular velocity vector in an appropriately chosen moving reference frame, as in Eq. (11.6), is crucial to the analysis of rigid body motion. Once we have the components of $\boldsymbol{\omega}$, we simply differentiate each of them with respect to time to arrive at Eq. (11.8). Observe that the absolute angular acceleration $\boldsymbol{\alpha}$ and $\mathrm{d}\boldsymbol{\omega}/\mathrm{d}t)_{\text{rel}}$, the angular acceleration relative to the moving frame, are the same if and only if $\boldsymbol{\Omega} = \boldsymbol{\omega}$. That occurs if the moving reference is a body-fixed frame (i.e., a set of xyz axes imbedded in the rigid body itself).

EXAMPLE 11.1

The airplane in Fig. 11.3 flies at a constant speed v while simultaneously undergoing a constant yaw rate ω_{yaw} about a vertical axis and describing a circular loop in the vertical plane with a radius ρ. The constant propeller spin rate is ω_{spin} relative to the airframe. Find the velocity and acceleration of the tip P of the propeller relative to the hub H, when P is directly above H. The propeller radius is ℓ.

Solution

The xyz axes are rigidly attached to the airplane. The x axis is aligned with the propeller's spin axis. The y axis is vertical, and the z axis is in the spanwise direction, so that xyz forms a right-handed triad. Although the xyz frame is not inertial, we can imagine it to instantaneously coincide with an inertial frame.

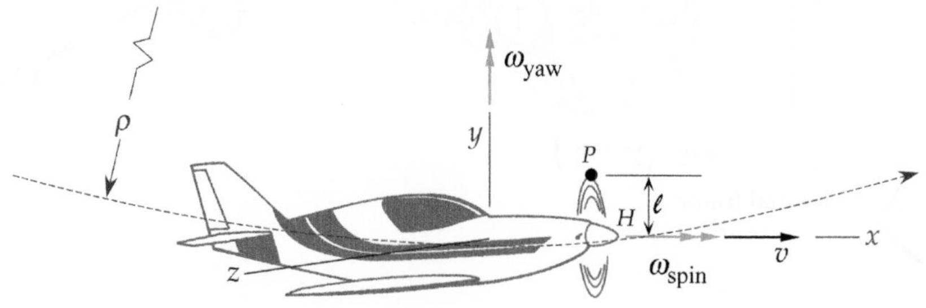

FIG. 11.3

Airplane with attached *xyz* body frame.

The absolute angular velocity of the airplane has two components, the yaw rate and the counterclockwise pitch angular velocity v/ρ of its rotation in the circular loop,

$$\boldsymbol{\omega}_{\text{airplane}} = \omega_{\text{yaw}}\hat{\mathbf{j}} + \omega_{\text{pitch}}\hat{\mathbf{k}} = \omega_{\text{yaw}}\hat{\mathbf{j}} + \frac{v}{\rho}\hat{\mathbf{k}}$$

The angular velocity of the body-fixed moving frame is that of the airplane, $\boldsymbol{\Omega} = \boldsymbol{\omega}_{\text{airplane}}$, so that

$$\boldsymbol{\Omega} = \omega_{\text{yaw}}\hat{\mathbf{j}} + \frac{v}{\rho}\hat{\mathbf{k}} \tag{a}$$

The absolute angular velocity of the propeller is that of the airplane plus the angular velocity of the propeller relative to the airplane

$$\boldsymbol{\omega}_{\text{prop}} = \boldsymbol{\omega}_{\text{airplane}} + \omega_{\text{spin}}\hat{\mathbf{i}} = \boldsymbol{\Omega} + \omega_{\text{spin}}\hat{\mathbf{i}}$$

which means

$$\boldsymbol{\omega}_{\text{prop}} = \omega_{\text{spin}}\hat{\mathbf{i}} + \omega_{\text{yaw}}\hat{\mathbf{j}} + \frac{v}{\rho}\hat{\mathbf{k}} \tag{b}$$

From Eq. (11.2), the velocity of point P on the propeller relative to point H on the hub, $\mathbf{v}_{P/H}$, is given by

$$\mathbf{v}_{P/H} = \mathbf{v}_P - \mathbf{v}_H = \boldsymbol{\omega}_{\text{prop}} \times \mathbf{r}_{P/H}$$

where $\mathbf{r}_{P/H}$ is the position vector of P relative to H at this instant,

$$\mathbf{r}_{P/H} = \ell\hat{\mathbf{j}} \tag{c}$$

Thus, using Eqs. (b) and (c),

$$\mathbf{v}_{P/H} = \left(\omega_{\text{spin}}\hat{\mathbf{i}} + \omega_{\text{yaw}}\hat{\mathbf{j}} + \frac{v}{\rho}\hat{\mathbf{k}}\right) \times \left(\ell\hat{\mathbf{j}}\right)$$

from which

$$\boxed{\mathbf{v}_{P/H} = -\frac{v}{\rho}\ell\hat{\mathbf{i}} + \omega_{\text{spin}}\ell\hat{\mathbf{k}}}$$

The absolute angular acceleration of the propeller is found by substituting Eqs. (a) and (b) into Eq. (11.7),

$$\begin{aligned}
\boldsymbol{\alpha}_{\text{prop}} &= \left.\frac{d\boldsymbol{\omega}_{\text{prop}}}{dt}\right)_{\text{rel}} + \boldsymbol{\Omega} \times \boldsymbol{\omega}_{\text{prop}} \\
&= \left(\frac{d\omega_{\text{spin}}}{dt}\hat{\mathbf{i}} + \frac{d\omega_{\text{yaw}}}{dt}\hat{\mathbf{j}} + \frac{d(v/\rho)}{dt}\hat{\mathbf{k}}\right) + \left(\omega_{\text{yaw}}\hat{\mathbf{j}} + \frac{v}{\rho}\hat{\mathbf{k}}\right) \times \left(\omega_{\text{spin}}\hat{\mathbf{i}} + \omega_{\text{yaw}}\hat{\mathbf{j}} + \frac{v}{\rho}\hat{\mathbf{k}}\right)
\end{aligned}$$

Since ω_{spin}, ω_{yaw}, v, and ρ are all constant, this reduces to

$$\boldsymbol{\alpha}_{\text{prop}} = \left(\omega_{\text{yaw}}\hat{\mathbf{j}} + \frac{v}{\rho}\hat{\mathbf{k}}\right) \times \left(\omega_{\text{spin}}\hat{\mathbf{i}} + \omega_{\text{yaw}}\hat{\mathbf{j}} + \frac{v}{\rho}\hat{\mathbf{k}}\right)$$

Carrying out the cross product yields

$$\boldsymbol{\alpha}_{\text{prop}} = \frac{v}{\rho}\omega_{\text{spin}}\hat{\mathbf{j}} - \omega_{\text{yaw}}\omega_{\text{spin}}\hat{\mathbf{k}} \tag{d}$$

From Eq. (11.4), the acceleration of P relative to H, $\mathbf{a}_{P/H}$, is given by

$$\mathbf{a}_{P/H} = \mathbf{a}_P - \mathbf{a}_H = \boldsymbol{\alpha}_{\text{prop}} \times \mathbf{r}_{P/H} + \boldsymbol{\omega}_{\text{prop}} \times \left(\boldsymbol{\omega}_{\text{prop}} \times \mathbf{r}_{P/H}\right)$$

Substituting Eqs. (b), (c), and (d) into this expression yields

$$\mathbf{a}_{P/H} = \left(\frac{v}{\rho}\omega_{\text{spin}}\hat{\mathbf{j}} - \omega_{\text{yaw}}\omega_{\text{spin}}\hat{\mathbf{k}}\right) \times \left(\ell\hat{\mathbf{j}}\right) + \left(\omega_{\text{spin}}\hat{\mathbf{i}} + \omega_{\text{yaw}}\hat{\mathbf{j}} + \frac{v}{\rho}\hat{\mathbf{k}}\right) \times \left[\left(\omega_{\text{spin}}\hat{\mathbf{i}} + \omega_{\text{yaw}}\hat{\mathbf{j}} + \frac{v}{\rho}\hat{\mathbf{k}}\right) \times \left(\ell\hat{\mathbf{j}}\right)\right]$$

From this we find that

$$\begin{aligned}\mathbf{a}_{P/H} &= \left(\omega_{\text{yaw}}\omega_{\text{spin}}\ell\hat{\mathbf{i}}\right) + \left(\omega_{\text{spin}}\hat{\mathbf{i}} + \omega_{\text{yaw}}\hat{\mathbf{j}} + \frac{v}{\rho}\hat{\mathbf{k}}\right) \times \left[-\frac{v}{\rho}\ell\hat{\mathbf{i}} + \omega_{\text{spin}}\ell\hat{\mathbf{k}}\right] \\ &= \left(\omega_{\text{yaw}}\omega_{\text{spin}}\ell\hat{\mathbf{i}}\right) + \left[\omega_{\text{yaw}}\omega_{\text{spin}}\ell\hat{\mathbf{i}} - \left(\frac{v^2}{\rho^2} + \omega_{\text{spin}}{}^2\right)\ell\hat{\mathbf{j}} + \omega_{\text{yaw}}\frac{v}{\rho}\ell\hat{\mathbf{k}}\right]\end{aligned}$$

so that finally,

$$\boxed{\mathbf{a}_{P/H} = 2\omega_{\text{yaw}}\omega_{\text{spin}}\ell\hat{\mathbf{i}} - \left(\frac{v^2}{\rho^2} + \omega_{\text{spin}}{}^2\right)\ell\hat{\mathbf{j}} + \omega_{\text{yaw}}\frac{v}{\rho}\ell\hat{\mathbf{k}}}$$

EXAMPLE 11.2

The satellite in Fig. 11.4 is rotating about the z axis at a constant rate N. The xyz axes are attached to the spacecraft, and the z axis has a fixed orientation in inertial space. The solar panels rotate at a constant rate $\dot{\theta}$ in the direction shown. Relative to point O, which lies at the center of the spacecraft and on the centerline of the panels, calculate for point A on the panel
(a) its absolute velocity and
(b) its absolute acceleration.

Solution

(a) Since the moving xyz frame is attached to the body of the spacecraft, its angular velocity is

$$\boldsymbol{\Omega} = N\hat{\mathbf{k}} \tag{a}$$

The absolute angular velocity of the panel is the absolute angular velocity of the spacecraft plus the angular velocity of the panel relative to the spacecraft,

$$\boldsymbol{\omega}_{\text{panel}} = -\dot{\theta}\hat{\mathbf{j}} + N\hat{\mathbf{k}} \tag{b}$$

The position vector of A relative to O is

$$\mathbf{r}_{A/O} = -\frac{w}{2}\sin\theta\hat{\mathbf{i}} + d\hat{\mathbf{j}} + \frac{w}{2}\cos\theta\hat{\mathbf{k}} \tag{c}$$

According to Eq. (11.2), the velocity of A relative to O is

$$\mathbf{v}_{A/O} = \mathbf{v}_A - \mathbf{v}_O = \boldsymbol{\omega}_{\text{panel}} \times \mathbf{r}_{A/O} = \begin{vmatrix} \hat{\mathbf{i}} & \hat{\mathbf{j}} & \hat{\mathbf{k}} \\ 0 & -\dot{\theta} & N \\ -\dfrac{w}{2}\sin\theta & d & \dfrac{w}{2}\cos\theta \end{vmatrix}$$

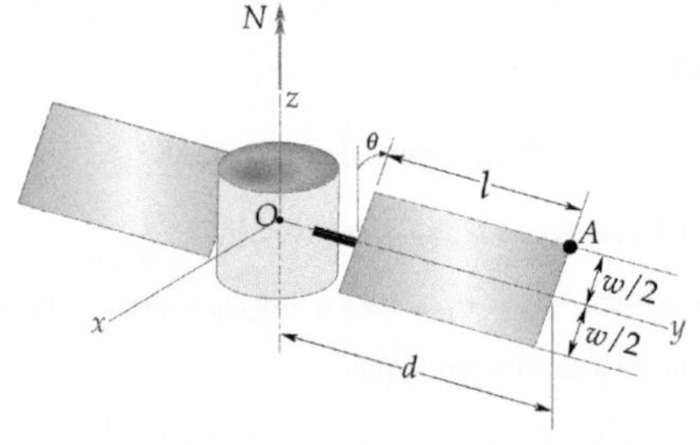

FIG. 11.4

Rotating solar panel on a rotating satellite.

from which we get

$$\boxed{\mathbf{v}_{A/O} = -\left(\frac{w}{2}\dot{\theta}\cos\theta + Nd\right)\hat{\mathbf{i}} - \frac{w}{2}N\sin\theta\hat{\mathbf{j}} - \frac{w}{2}\dot{\theta}\sin\theta\hat{\mathbf{k}}}$$

(b) The absolute angular acceleration of the panel is found by substituting Eqs. (a) and (b) into Eq. (11.7),

$$\begin{aligned}\boldsymbol{\alpha}_{\text{panel}} &= \left.\frac{d\boldsymbol{\omega}_{\text{panel}}}{dt}\right)_{\text{rel}} + \boldsymbol{\Omega}\times\boldsymbol{\omega}_{\text{panel}} \\ &= \left[\frac{d(-\dot{\theta})}{dt}\hat{\mathbf{j}} + \frac{dN}{dt}\hat{\mathbf{k}}\right] + (N\hat{\mathbf{k}})\times\left(-\dot{\theta}\hat{\mathbf{j}} + N\hat{\mathbf{k}}\right)\end{aligned}$$

Since N and $\dot{\theta}$ are constants, this reduces to

$$\boldsymbol{\alpha}_{\text{panel}} = \dot{\theta}N\hat{\mathbf{i}} \tag{d}$$

To find the acceleration of A relative to O, we substitute Eqs. (b) through (d) into Eq. (11.4),

$$\begin{aligned}\mathbf{a}_{A/O} &= \mathbf{a}_A - \mathbf{a}_O = \boldsymbol{\alpha}_{\text{panel}}\times\mathbf{r}_{A/O} + \boldsymbol{\omega}_{\text{panel}}\times\left(\boldsymbol{\omega}_{\text{panel}}\times\mathbf{r}_{A/O}\right) \\ &= \begin{vmatrix}\hat{\mathbf{i}} & \hat{\mathbf{j}} & \hat{\mathbf{k}} \\ \dot{\theta}N & 0 & 0 \\ -\frac{w}{2}\sin\theta & d & \frac{w}{2}\cos\theta\end{vmatrix} + \left(-\dot{\theta}\hat{\mathbf{j}} + N\hat{\mathbf{k}}\right)\times\begin{vmatrix}\hat{\mathbf{i}} & \hat{\mathbf{j}} & \hat{\mathbf{k}} \\ 0 & -\dot{\theta} & N \\ -\frac{w}{2}\sin\theta & d & \frac{w}{2}\cos\theta\end{vmatrix} \\ &= \left(-\frac{w}{2}N\dot{\theta}\cos\theta\hat{\mathbf{j}} + N\dot{\theta}d\hat{\mathbf{k}}\right) + \begin{vmatrix}\hat{\mathbf{i}} & \hat{\mathbf{j}} & \hat{\mathbf{k}} \\ 0 & -\dot{\theta} & N \\ -\frac{w}{2}\dot{\theta}\cos\theta - Nd & -N\frac{w}{2}\sin\theta & -\frac{w}{2}\dot{\theta}\sin\theta\end{vmatrix}\end{aligned}$$

which leads to

$$\boxed{\mathbf{a}_{A/O} = \frac{w}{2}\left(N^2 + \dot{\theta}^2\right)\sin\theta\hat{\mathbf{i}} - N\left(Nd + w\dot{\theta}\cos\theta\right)\hat{\mathbf{j}} - \frac{w}{2}\dot{\theta}^2\cos\theta\hat{\mathbf{k}}}$$

EXAMPLE 11.3

The gyro rotor illustrated in Fig. 11.5 has a constant spin rate ω_{spin} around axis b–a in the direction shown. The XYZ axes are fixed. The xyz axes are attached to the gimbal ring, whose angle θ with the vertical is increasing at a constant rate $\dot{\theta}$ in the direction shown. The assembly is forced to precess at a constant rate N around the vertical. For the rotor in the position shown, calculate

(a) the absolute angular velocity and

(b) the absolute angular acceleration.

Express the results in both the fixed XYZ frame and the moving xyz frame.

Solution

(a) We will need the instantaneous relationship between the unit vectors of the inertial XYZ axes and the comoving xyz frame, which on inspecting Fig. 11.6 can be seen to be

$$\begin{aligned}\hat{\mathbf{I}} &= -\cos\theta\hat{\mathbf{j}} + \sin\theta\hat{\mathbf{k}} \\ \hat{\mathbf{J}} &= \hat{\mathbf{i}} \\ \hat{\mathbf{K}} &= \sin\theta\hat{\mathbf{j}} + \cos\theta\hat{\mathbf{k}}\end{aligned} \tag{a}$$

so that the matrix of the transformation from xyz to XYZ is (Section 4.5)

$$[\mathbf{Q}]_{xX} = \begin{bmatrix}0 & -\cos\theta & \sin\theta \\ 1 & 0 & 0 \\ 0 & \sin\theta & \cos\theta\end{bmatrix} \tag{b}$$

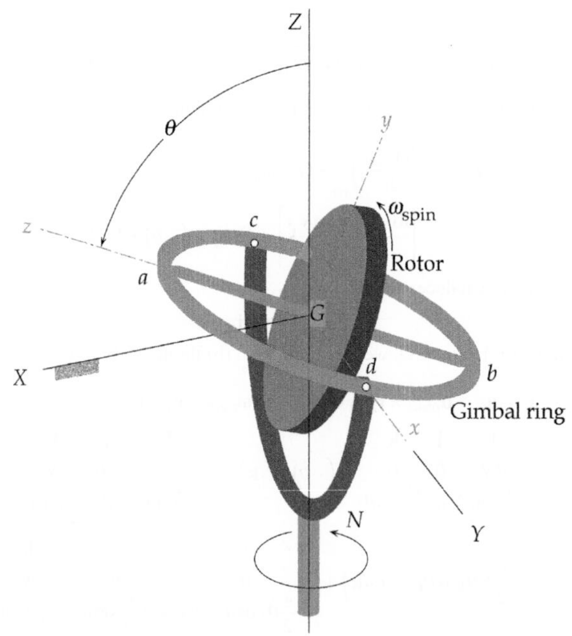

FIG. 11.5

Rotating, precessing, nutating gyro.

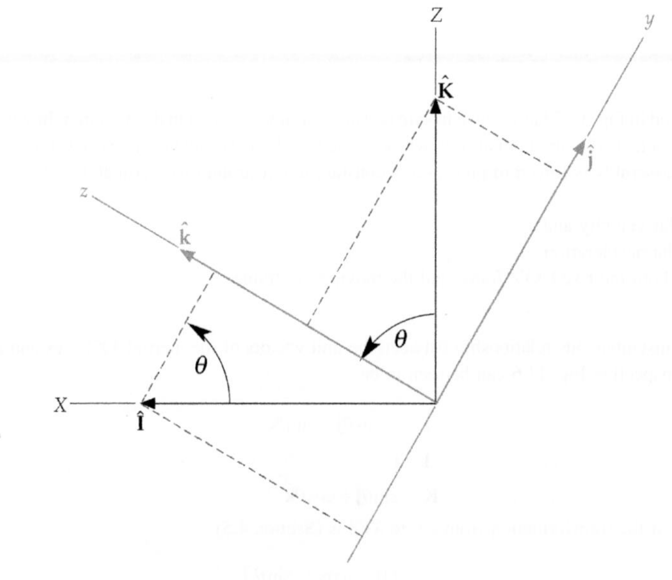

FIG. 11.6

Orientation of the fixed *XZ* axes relative to the rotating *xz* axes.

The absolute angular velocity of the gimbal ring is that of the base ($N\hat{\mathbf{K}}$) plus the angular velocity of the gimbal relative to the base ($\dot{\theta}\hat{\mathbf{i}}$), so that

$$\boldsymbol{\omega}_{\text{gimbal}} = N\hat{\mathbf{K}} + \dot{\theta}\hat{\mathbf{i}} = N\left(\sin\theta\hat{\mathbf{j}} + \cos\theta\hat{\mathbf{k}}\right) + \dot{\theta}\hat{\mathbf{i}} = \dot{\theta}\hat{\mathbf{i}} + N\sin\theta\hat{\mathbf{j}} + N\cos\theta\hat{\mathbf{k}} \tag{c}$$

where we made use of Eq. (a)$_3$ above. Since the moving xyz frame is attached to the gimbal, $\boldsymbol{\Omega} = \boldsymbol{\omega}_{\text{gimbal}}$, so that

$$\boldsymbol{\Omega} = \dot{\theta}\hat{\mathbf{i}} + N\sin\theta\hat{\mathbf{j}} + N\cos\theta\hat{\mathbf{k}} \tag{d}$$

The absolute angular velocity of the rotor is its spin relative to the gimbal, plus the angular velocity of the gimbal,

$$\boldsymbol{\omega}_{\text{rotor}} = \boldsymbol{\omega}_{\text{gimbal}} + \omega_{\text{spin}}\hat{\mathbf{k}} \tag{e}$$

From Eq. (c), it follows that

$$\boxed{\boldsymbol{\omega}_{\text{rotor}} = \dot{\theta}\hat{\mathbf{i}} + N\sin\theta\hat{\mathbf{j}} + \left(N\cos\theta + \omega_{\text{spin}}\right)\hat{\mathbf{k}}} \tag{f}$$

Because $\hat{\mathbf{i}}$, $\hat{\mathbf{j}}$, and $\hat{\mathbf{k}}$ move with the gimbal, expression (f) is valid for any time, not just the instant shown in Fig. 11.5. Alternatively, applying the vector transformation

$$\{\boldsymbol{\omega}_{\text{rotor}}\}_{XYZ} = [\mathbf{Q}]_{xX}\{\boldsymbol{\omega}_{\text{rotor}}\}_{xyz} \tag{g}$$

we obtain the angular velocity of the rotor in the inertial frame, but only at the instant shown in the figure (i.e., when the x axis aligns with the Y axis):

$$\begin{Bmatrix} \omega_X \\ \omega_Y \\ \omega_Z \end{Bmatrix} = \begin{bmatrix} 0 & -\cos\theta & \sin\theta \\ 1 & 0 & 0 \\ 0 & \sin\theta & \cos\theta \end{bmatrix} \begin{Bmatrix} \dot{\theta} \\ N\sin\theta \\ N\cos\theta + \omega_{\text{spin}} \end{Bmatrix} = \begin{Bmatrix} \omega_{\text{spin}}\sin\theta \\ \dot{\theta} \\ N + \omega_{\text{spin}}\cos\theta \end{Bmatrix}$$

or

$$\boxed{\boldsymbol{\omega}_{\text{rotor}} = \omega_{\text{spin}}\sin\theta\hat{\mathbf{I}} + \dot{\theta}\hat{\mathbf{J}} + \left(N + \omega_{\text{spin}}\cos\theta\right)\hat{\mathbf{K}}} \tag{h}$$

(b) The angular acceleration of the rotor can be found by substituting Eqs. (d) and (f) into Eq. (11.7), recalling that N, $\dot{\theta}$, and ω_{spin} are independent of time:

$$\boldsymbol{\alpha}_{\text{rotor}} = \left.\frac{d\boldsymbol{\omega}_{\text{rotor}}}{dt}\right)_{\text{rel}} + \boldsymbol{\Omega} \times \boldsymbol{\omega}_{\text{rotor}}$$

$$= \left[\frac{d(\dot{\theta})}{dt}\hat{\mathbf{i}} + \frac{d(N\sin\theta)}{dt}\hat{\mathbf{j}} + \frac{d\left(N\cos\theta + \omega_{\text{spin}}\right)}{dt}\hat{\mathbf{k}}\right] + \begin{vmatrix} \hat{\mathbf{i}} & \hat{\mathbf{j}} & \hat{\mathbf{k}} \\ \dot{\theta} & N\sin\theta & N\cos\theta \\ \dot{\theta} & N\sin\theta & N\cos\theta + \omega_{\text{spin}} \end{vmatrix}$$

$$= \left(N\dot{\theta}\cos\theta\hat{\mathbf{j}} - N\dot{\theta}\sin\theta\hat{\mathbf{k}}\right) + \left[N\omega_{\text{spin}}\sin\theta\hat{\mathbf{i}} - \omega_{\text{spin}}\dot{\theta}\hat{\mathbf{j}} + (0)\hat{\mathbf{k}}\right]$$

Upon collecting terms, we get

$$\boxed{\boldsymbol{\alpha}_{\text{rotor}} = N\omega_{\text{spin}}\sin\theta\hat{\mathbf{i}} + \dot{\theta}\left(N\cos\theta - \omega_{\text{spin}}\right)\hat{\mathbf{j}} - N\dot{\theta}\sin\theta\hat{\mathbf{k}}} \tag{i}$$

This expression, like Eq. (f), is valid at any time.

The components of $\boldsymbol{\alpha}_{\text{rotor}}$ along the XYZ axes are found in the same way as for $\boldsymbol{\omega}_{\text{rotor}}$,

$$\{\boldsymbol{\alpha}_{\text{rotor}}\}_{XYZ} = [\mathbf{Q}]_{xX}\{\boldsymbol{\alpha}_{\text{rotor}}\}_{xyz}$$

which means

$$\begin{Bmatrix} \alpha_X \\ \alpha_Y \\ \alpha_Z \end{Bmatrix} = \begin{bmatrix} 0 & -\cos\theta & \sin\theta \\ 1 & 0 & 0 \\ 0 & \sin\theta & \cos\theta \end{bmatrix} \begin{Bmatrix} N\omega_{\text{spin}}\sin\theta \\ \dot{\theta}\left(N\cos\theta - \omega_{\text{spin}}\right) \\ -N\dot{\theta}\sin\theta \end{Bmatrix} = \begin{Bmatrix} \dot{\theta}\omega_{\text{spin}}\cos\theta - N\dot{\theta} \\ N\omega_{\text{spin}}\sin\theta \\ -\dot{\theta}\omega_{\text{spin}}\sin\theta \end{Bmatrix}$$

or

$$\boxed{\boldsymbol{\alpha}_{\text{rotor}} = \dot{\theta}\left(\omega_{\text{spin}}\cos\theta - N\right)\hat{\mathbf{I}} + N\omega_{\text{spin}}\sin\theta\hat{\mathbf{J}} - \dot{\theta}\omega_{\text{spin}}\sin\theta\hat{\mathbf{K}}} \tag{j}$$

Note carefully that Eq. (j) is not simply the time derivative of Eq. (h). Eqs. (h) and (j) are valid only at the instant that the *xyz* and *XYZ* axes have the alignments shown in Fig. 11.5.

11.3 EQUATIONS OF TRANSLATIONAL MOTION

Fig. 11.7 again shows an arbitrary, continuous, three-dimensional body of mass m. "Continuous" means that as we zoom in on a point it remains surrounded by a continuous distribution of matter having the infinitesimal mass dm in the limit. The point never ends up in a void. In particular, we ignore the actual atomic and molecular microstructure in favor of this continuum hypothesis, as it is called. Molecular microstructure does bear on the overall dynamics of a finite body. We will use G to denote the center of mass. The position vectors of points relative to the origin of the inertial frame will be designated by capital letters. Thus, the position of the center of mass is $\mathbf{R}_G$, defined as

$$m\mathbf{R}_G = \int_m \mathbf{R}dm \tag{11.9}$$

$\mathbf{R}$ is the position of a mass element dm within the continuum. Each element of mass is acted on by a net external force $\mathbf{dF}_{\text{net}}$ and a net internal force $\mathbf{df}_{\text{net}}$. The external force comes from direct contact with other objects and from action at a distance, such as gravitational attraction. The internal forces are those exerted from within the body by neighboring particles. These are the forces that hold the body together. For each mass element, Newton's second law (Eq. 1.38) is written as

$$\mathbf{dF}_{\text{net}} + \mathbf{df}_{\text{net}} = dm\ddot{\mathbf{R}} \tag{11.10}$$

Writing this equation for the infinite number of mass elements of which the body is composed, and then summing them all together, leads, to the integral

$$\int_m \mathbf{dF}_{\text{net}} + \int_m \mathbf{df}_{\text{net}} = \int_m \ddot{\mathbf{R}}dm$$

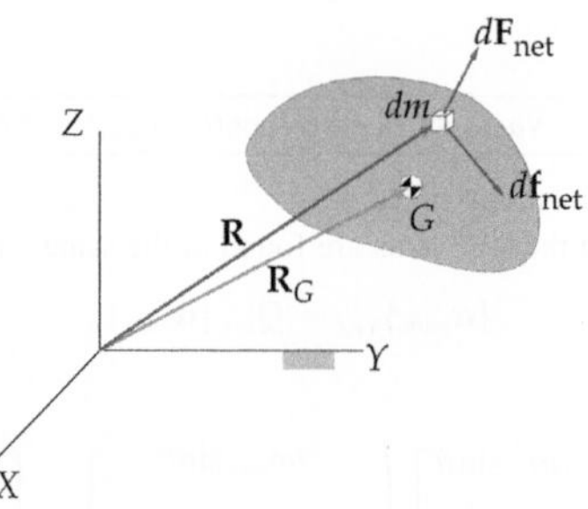

FIG. 11.7

Forces on the mass element dm of a continuous medium.

Because the internal forces occur in action–reaction pairs, $\int_m d\mathbf{f}_{net} = \mathbf{0}$. (External forces on the body are those without an internal reactant; the reactant lies outside the body and, hence, is outside our purview.) Thus,

$$\mathbf{F}_{net} = \int_m \ddot{\mathbf{R}} dm \tag{11.11}$$

where $\mathbf{F}_{net}$ is the resultant external force on the body, $\mathbf{F}_{net} = \int_m d\mathbf{F}_{net}$. From Eq. (11.9),

$$\int_m \ddot{\mathbf{R}} dm = m\ddot{\mathbf{R}}_G$$

where $\ddot{\mathbf{R}}_G = \mathbf{a}_G$, the absolute acceleration of the center of mass. Therefore, Eq. (11.11) can be written as

$$\mathbf{F}_{net} = m\ddot{\mathbf{R}}_G \tag{11.12}$$

We are therefore reminded that the motion of the center of mass of a body is determined solely by the resultant of the external forces acting on it. So far, our study of orbiting bodies has focused exclusively on the motion of their centers of mass. In this chapter, we turn our attention to rotational motion around the center of mass. To simplify things, we ultimately assume that the body is not only continuous but that it is also rigid. This means all points of the body remain at a fixed distance from each other and there is no flexing, bending, or twisting deformation.

11.4 EQUATIONS OF ROTATIONAL MOTION

Our development of the rotational dynamics equations does not require at the outset that the body under consideration be rigid. It may be a solid, fluid, or gas.

Point P in Fig. 11.8 is arbitrary; it need not be fixed in space nor attached to a point on the body. Then the moment about P of the forces on mass element dm (cf. Fig. 11.7) is

$$d\mathbf{M}_P = \mathbf{r} \times d\mathbf{F}_{net} + \mathbf{r} \times d\mathbf{f}_{net}$$

where $\mathbf{r}$ is the position vector of the mass element dm relative to the point P. Writing the right-hand side as $\mathbf{r} \times (d\mathbf{F}_{net} + d\mathbf{f}_{net})$, substituting Eq. (11.10), and integrating over all the mass elements of the body yields

$$\mathbf{M}_P)_{net} = \int_m \mathbf{r} \times \ddot{\mathbf{R}} dm \tag{11.13}$$

where $\ddot{\mathbf{R}}$ is the absolute acceleration of dm relative to the inertial frame and

$$\mathbf{M}_P)_{net} = \int_m \mathbf{r} \times d\mathbf{F}_{net} + \int_m \mathbf{r} \times d\mathbf{f}_{net}$$

But $\int_m \mathbf{r} \times d\mathbf{f}_{net} = \mathbf{0}$ because the internal forces occur in action–reaction pairs. Thus,

$$\mathbf{M}_P)_{net} = \int_m \mathbf{r} \times d\mathbf{F}_{net}$$

which means the net moment includes only the moment of all the external forces on the body.

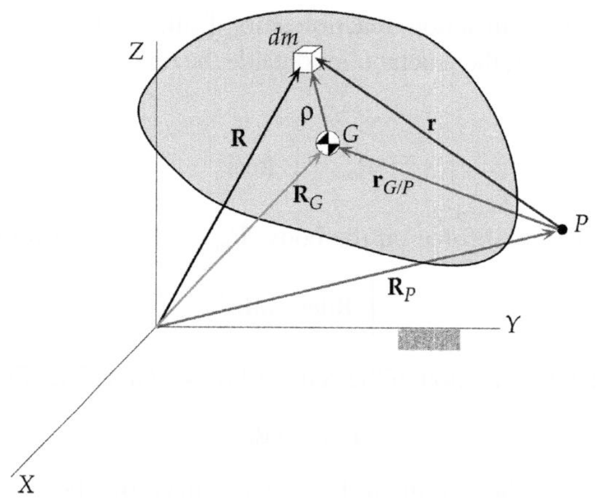

FIG. 11.8

Position vectors of a mass element in a continuum from several key reference points.

From the product rule of calculus, we know that $\mathrm{d}(\mathbf{r}\times\dot{\mathbf{R}})/\mathrm{d}t=\mathbf{r}\times\ddot{\mathbf{R}}+\dot{\mathbf{r}}\times\dot{\mathbf{R}}$, so that the integrand in Eq. (11.13) may be written as

$$\mathbf{r}\times\ddot{\mathbf{R}}=\frac{\mathrm{d}}{\mathrm{d}t}(\mathbf{r}\times\dot{\mathbf{R}})-\dot{\mathbf{r}}\times\dot{\mathbf{R}} \tag{11.14}$$

Furthermore, Fig. 11.8 shows that $\mathbf{r}=\mathbf{R}-\mathbf{R}_P$, where $\mathbf{R}_P$ is the absolute position vector of P. It follows that

$$\dot{\mathbf{r}}\times\dot{\mathbf{R}}=(\dot{\mathbf{R}}-\dot{\mathbf{R}}_P)\times\dot{\mathbf{R}}=-\dot{\mathbf{R}}_P\times\dot{\mathbf{R}} \tag{11.15}$$

Substituting Eq. (11.15) into Eq. (11.14) and then moving that result into Eq. (11.13), yields

$$\mathbf{M}_P)_{\text{net}}=\frac{\mathrm{d}}{\mathrm{d}t}\int_m \mathbf{r}\times\dot{\mathbf{R}}\mathrm{d}m+\dot{\mathbf{R}}_P\times\int_m \dot{\mathbf{R}}\mathrm{d}m \tag{11.16}$$

Now, $\mathbf{r}\times\dot{\mathbf{R}}\mathrm{d}m$ is the moment of the absolute linear momentum of mass element $\mathrm{d}m$ about P. The moment of momentum, or angular momentum, of the entire body is the integral of this cross product over all of its mass elements. That is, the absolute angular momentum of the body relative to point P is

$$\mathbf{H}_P=\int_m \mathbf{r}\times\dot{\mathbf{R}}\mathrm{d}m \tag{11.17}$$

Observing from Fig. 11.8 that $\mathbf{r}=\mathbf{r}_{G/P}+\boldsymbol{\rho}$, we can write Eq. (11.17) as

$$\mathbf{H}_P=\int_m (\mathbf{r}_{G/P}+\boldsymbol{\rho})\times\dot{\mathbf{R}}\mathrm{d}m=\mathbf{r}_{G/P}\times\int_m \dot{\mathbf{R}}\mathrm{d}m+\int_m \boldsymbol{\rho}\times\dot{\mathbf{R}}\mathrm{d}m \tag{11.18}$$

The last term is the absolute angular momentum relative to the center of mass G,

$$\mathbf{H}_G=\int_m \boldsymbol{\rho}\times\dot{\mathbf{R}}\mathrm{d}m \tag{11.19}$$

Furthermore, by the definition of center of mass (Eq. 11.9),

$$\int_m \dot{\mathbf{R}} dm = m\dot{\mathbf{R}}_G \tag{11.20}$$

Eqs. (11.19) and (11.20) allow us to write Eq. (11.18) as

$$\mathbf{H}_P = \mathbf{H}_G + \mathbf{r}_{G/P} \times m\mathbf{v}_G \tag{11.21}$$

This useful relationship shows how to obtain the absolute angular momentum about any point P once $\mathbf{H}_G$ is known.

For calculating the angular momentum about the center of mass, Eq. (11.19) can be cast in a much more useful form by making the substitution (cf. Fig. 11.8) $\mathbf{R} = \mathbf{R}_G + \boldsymbol{\rho}$, so that

$$\mathbf{H}_G = \int_m \boldsymbol{\rho} \times (\dot{\mathbf{R}}_G + \dot{\boldsymbol{\rho}}) dm = \int_m \boldsymbol{\rho} \times \dot{\mathbf{R}}_G dm + \int_m \boldsymbol{\rho} \times \dot{\boldsymbol{\rho}} dm$$

In the two integrals on the right, the variable is $\boldsymbol{\rho}$. $\dot{\mathbf{R}}_G$ is fixed and can therefore be factored out of the first integral to obtain

$$\mathbf{H}_G = \left(\int_m \boldsymbol{\rho} dm\right) \times \dot{\mathbf{R}}_G + \int_m \boldsymbol{\rho} \times \dot{\boldsymbol{\rho}} dm$$

By definition of the center of mass, $\int_m \boldsymbol{\rho} dm = 0$ (the position vector of the center of mass relative to itself is zero), which means

$$\mathbf{H}_G = \int_m \boldsymbol{\rho} \times \dot{\boldsymbol{\rho}} dm \tag{11.22}$$

Since $\boldsymbol{\rho}$ and $\dot{\boldsymbol{\rho}}$ are the position and velocity vectors relative to the center of mass G, $\int_m \boldsymbol{\rho} \times \dot{\boldsymbol{\rho}} dm$ is the total moment about the center of mass of the linear momentum relative to the center of mass, $\mathbf{H}_G)_{\text{rel}}$. In other words,

$$\mathbf{H}_G = \mathbf{H}_G)_{\text{rel}} \tag{11.23}$$

This is a rather surprising fact, hidden in Eq. (11.19), and is true in general for no other point of the body.

Another useful angular momentum formula, similar to Eq. (11.21), may be found by substituting $\mathbf{R} = \mathbf{R}_P + \mathbf{r}$ into Eq. (11.17),

$$\mathbf{H}_P = \int_m \mathbf{r} \times (\dot{\mathbf{R}}_P + \dot{\mathbf{r}}) dm = \left(\int_m \mathbf{r} dm\right) \times \dot{\mathbf{R}}_P + \int_m \mathbf{r} \times \dot{\mathbf{r}} dm \tag{11.24}$$

The term on the far right is the net moment of relative linear momentum about P,

$$\mathbf{H}_P)_{\text{rel}} = \int_m \mathbf{r} \times \dot{\mathbf{r}} dm \tag{11.25}$$

Also, $\int_m \mathbf{r} dm = m\mathbf{r}_{G/P}$, where $\mathbf{r}_{G/P}$ is the position vector of the center of mass relative to P. Thus, Eq. (11.24) can be written as

$$\mathbf{H}_P = \mathbf{H}_P)_{\text{rel}} + \mathbf{r}_{G/P} \times m\mathbf{v}_P \tag{11.26}$$

Finally, substituting this into Eq. (11.21), solving for $\mathbf{H}_{P})_{\text{rel}}$, and noting that $\mathbf{v}_G - \mathbf{v}_P = \mathbf{v}_{G/P}$, yields

$$\mathbf{H}_{P})_{\text{rel}} = \mathbf{H}_G + \mathbf{r}_{G/P} \times m\mathbf{v}_{G/P} \tag{11.27}$$

This expression is useful when the absolute velocity $\mathbf{v}_G$ of the center of mass, which is required in Eq. (11.21), is not available.

So far, we have written down some formulas for calculating the angular momentum about an arbitrary point in space and about the center of mass of the body itself. Let us now return to the problem of relating angular momentum to the applied torque. Substituting Eqs. (11.17) and (11.20) into Eq. (11.16), we obtain

$$\mathbf{M}_{P})_{\text{net}} = \dot{\mathbf{H}}_P + \dot{\mathbf{R}}_P \times m\dot{\mathbf{R}}_G$$

Thus, for an arbitrary point P,

$$\mathbf{M}_{P})_{\text{net}} = \dot{\mathbf{H}}_P + \mathbf{v}_P \times m\mathbf{v}_G \tag{11.28}$$

where $\mathbf{v}_P$ and $\mathbf{v}_G$ are the absolute velocities of points P and G, respectively. This expression is applicable to two important special cases.

If the point P is at rest in inertial space ($\mathbf{v}_P = \mathbf{0}$), then Eq. (11.28) reduces to

$$\mathbf{M}_{P})_{\text{net}} = \dot{\mathbf{H}}_P \tag{11.29}$$

This equation holds as well if $\mathbf{v}_P$ and $\mathbf{v}_G$ are parallel (e.g., if P is the point of contact of a wheel rolling while slipping in the plane). Note that the validity of Eq. (11.29) depends neither on the body being rigid nor on it being in pure rotation about P. If point P is chosen to be the center of mass, then, since $\mathbf{v}_G \times \mathbf{v}_G = \mathbf{0}$, Eq. (11.28) becomes

$$\boxed{\mathbf{M}_{G})_{\text{net}} = \dot{\mathbf{H}}_G} \tag{11.30}$$

This equation is valid for any state of motion.

If Eq. (11.30) is integrated over a time interval, then we obtain the angular impulse–momentum principle,

$$\int_{t_1}^{t_2} \mathbf{M}_{G})_{\text{net}} dt = \mathbf{H}_{G})_2 - \mathbf{H}_{G})_1 \tag{11.31}$$

A similar expression follows from Eq. (11.29). $\int \mathbf{M}dt$ is the angular impulse. If the net angular impulse is zero, then $\Delta\mathbf{H} = \mathbf{0}$, which is a statement of the conservation of angular momentum. Keep in mind that the angular impulse–momentum principle is not valid for just any reference point.

Additional versions of Eqs. (11.29) and (11.30) can be obtained that may prove useful in special circumstances. For example, substituting the expression for $\mathbf{H}_P$ (Eq. 11.21) into Eq. (11.28) yields

$$\begin{aligned}
\mathbf{M}_{P})_{\text{net}} &= \left[\dot{\mathbf{H}}_G + \frac{d}{dt}(\mathbf{r}_{G/P} \times m\mathbf{v}_G)\right] + \mathbf{v}_P \times m\mathbf{v}_G \\
&= \dot{\mathbf{H}}_G + \frac{d}{dt}[(\mathbf{r}_G - \mathbf{r}_P) \times m\mathbf{v}_G] + \mathbf{v}_P \times m\mathbf{v}_G \\
&= \dot{\mathbf{H}}_G + (\mathbf{v}_G - \mathbf{v}_P) \times m\mathbf{v}_G + \mathbf{r}_{G/P} \times m\mathbf{a}_G + \mathbf{v}_P \times m\mathbf{v}_G
\end{aligned}$$

or, finally,

$$\mathbf{M}_P)_{\text{net}} = \dot{\mathbf{H}}_G + \mathbf{r}_{G/P} \times m\mathbf{a}_G \tag{11.32}$$

This expression is useful when it is convenient to compute the net moment about a point other than the center of mass. Alternatively, by simply differentiating Eq. (11.27) we get

$$\dot{\mathbf{H}}_P)_{\text{rel}} = \dot{\mathbf{H}}_G + \overbrace{\mathbf{v}_{G/P} \times m\mathbf{v}_{G/P}}^{=\mathbf{0}} + \mathbf{r}_{G/P} \times m\mathbf{a}_{G/P}$$

Solving for $\dot{\mathbf{H}}_G$, invoking Eq. (11.30), and using the fact that $\mathbf{a}_{P/G} = -\mathbf{a}_{G/P}$ leads to

$$\mathbf{M}_G)_{\text{net}} = \dot{\mathbf{H}}_P)_{\text{rel}} + \mathbf{r}_{G/P} \times m\mathbf{a}_{P/G} \tag{11.33}$$

Finally, if the body is rigid, the magnitude of the position vector $\boldsymbol{\rho}$ of any point relative to the center of mass does not change with time. Therefore, Eq. (1.52) requires that $\dot{\boldsymbol{\rho}} = \boldsymbol{\omega} \times \boldsymbol{\rho}$, leading us to conclude from Eq. (11.22) that

$$\mathbf{H}_G = \int_m \boldsymbol{\rho} \times (\boldsymbol{\omega} \times \boldsymbol{\rho}) dm \quad \text{(Rigid body)} \tag{11.34}$$

Again, the absolute angular momentum about the center of mass depends only on the absolute angular velocity and not on the absolute translational velocity of any point of the body.

No such simplification of Eq. (11.17) exists for an arbitrary reference point P. However, if the point P is fixed in inertial space and the rigid body is rotating about P, then the position vector $\mathbf{r}$ from P to any point of the body is constant. It follows from Eq. (1.52) that $\dot{\mathbf{r}} = \boldsymbol{\omega} \times \mathbf{r}$. According to Fig. 11.8,

$$\mathbf{R} = \mathbf{R}_P + \mathbf{r}$$

Differentiating with respect to time gives

$$\dot{\mathbf{R}} = \dot{\mathbf{R}}_P + \dot{\mathbf{r}} = \mathbf{0} + \boldsymbol{\omega} \times \mathbf{r} = \boldsymbol{\omega} \times \mathbf{r}$$

Substituting this into Eq. (11.17) yields the formula for angular momentum in this special case as

$$\mathbf{H}_P = \int_m \mathbf{r} \times (\boldsymbol{\omega} \times \mathbf{r}) dm \quad \text{(Rigid body rotating about fixed point } P\text{)} \tag{11.35}$$

Although Eqs. (11.34) and (11.35) are mathematically identical, we must keep in mind the notation of Fig. 11.8. Eq. (11.35) applies only if the rigid body is in pure rotation about a stationary point in inertial space, whereas Eq. (11.34) applies unconditionally to any situation.

11.5 MOMENTS OF INERTIA

To use Eq. (11.29) or Eq. (11.30) to solve problems, the vectors within them have to be resolved into components. To find the components of angular momentum, we must appeal to its definition. We focus on the formula for angular momentum of a rigid body about its center of mass (Eq. (11.34)) because the expression for fixed point rotation (Eq. (11.35)) is mathematically the same. The integrand of Eq. (11.34) can be rewritten using the bac–cab vector identity presented in Eq. (2.33),

$$\boldsymbol{\rho} \times (\boldsymbol{\omega} \times \boldsymbol{\rho}) = \boldsymbol{\omega}\rho^2 - \boldsymbol{\rho}(\boldsymbol{\omega} \cdot \boldsymbol{\rho}) \tag{11.36}$$

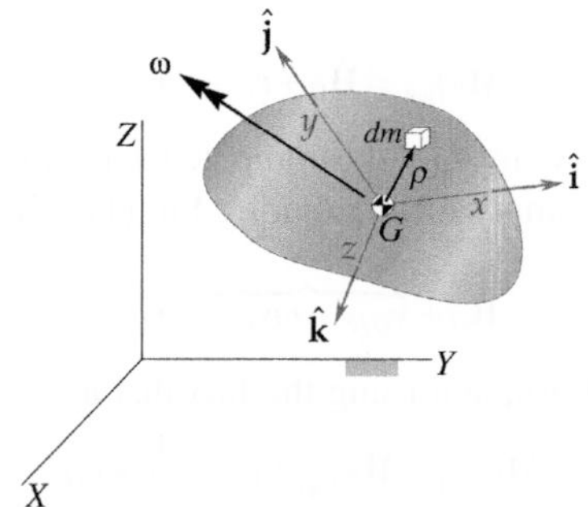

FIG. 11.9

Comoving *xyz* frame used to compute the moments of inertia.

Let the origin of a comoving *xyz* coordinate system be attached to the center of mass G, as shown in Fig. 11.9. The unit vectors of this frame are $\hat{\mathbf{i}}$, $\hat{\mathbf{j}}$, and $\hat{\mathbf{k}}$. The vectors $\boldsymbol{\rho}$ and $\boldsymbol{\omega}$ can be resolved into components in the *xyz* directions to get $\boldsymbol{\rho}=x\hat{\mathbf{i}}+y\hat{\mathbf{j}}+z\hat{\mathbf{k}}$ and $\boldsymbol{\omega}=\omega_x\hat{\mathbf{i}}+\omega_y\hat{\mathbf{j}}+\omega_z\hat{\mathbf{k}}$. Substituting these vector expressions into the right-hand side of Eq. (11.36) yields

$$\boldsymbol{\rho}\times(\boldsymbol{\omega}\times\boldsymbol{\rho})=\left(\omega_x\hat{\mathbf{i}}+\omega_y\hat{\mathbf{j}}+\omega_z\hat{\mathbf{k}}\right)\left(x^2+y^2+z^2\right)-\left(x\hat{\mathbf{i}}+y\hat{\mathbf{j}}+z\hat{\mathbf{k}}\right)\left(\omega_x x+\omega_y y+\omega_z z\right)$$

Expanding the right-hand side and collecting the terms having the unit vectors $\hat{\mathbf{i}}$, $\hat{\mathbf{j}}$, and $\hat{\mathbf{k}}$ in common, we get

$$\begin{aligned}\boldsymbol{\rho}\times(\boldsymbol{\omega}\times\boldsymbol{\rho})=&\left[\left(y^2+z^2\right)\omega_x-xy\omega_y-xz\omega_z\right]\hat{\mathbf{i}}+\left[-yx\omega_x+\left(x^2+z^2\right)\omega_y-yz\omega_z\right]\hat{\mathbf{j}}\\&+\left[-zx\omega_x-zy\omega_y+\left(x^2+y^2\right)\omega_z\right]\hat{\mathbf{k}}\end{aligned}\tag{11.37}$$

We put this result into the integrand of Eq. (11.34) to obtain

$$\mathbf{H}_G=H_x\hat{\mathbf{i}}+H_y\hat{\mathbf{j}}+H_z\hat{\mathbf{k}}\tag{11.38}$$

where

$$\begin{Bmatrix}H_x\\H_y\\H_z\end{Bmatrix}=\begin{bmatrix}I_x & I_{xy} & I_{xz}\\I_{yx} & I_y & I_{yz}\\I_{zx} & I_{zy} & I_z\end{bmatrix}\begin{Bmatrix}\omega_x\\\omega_y\\\omega_z\end{Bmatrix}\tag{11.39a}$$

or, in matrix notation,

$$\{\mathbf{H}\}=[\mathbf{I}]\{\boldsymbol{\omega}\}\tag{11.39b}$$

The nine components of the moment of inertia matrix $[\mathbf{I}]$ about the center of mass are

$$\begin{array}{lll}I_x=\displaystyle\int_m\left(y^2+z^2\right)dm & I_{xy}=-\displaystyle\int_m xydm & I_{xz}=-\displaystyle\int_m xzdm\\I_{yx}=-\displaystyle\int_m yxdm & I_y=\displaystyle\int_m\left(x^2+z^2\right)dm & I_{yz}=-\displaystyle\int_m yzdm\\I_{zx}=-\displaystyle\int_m zxdm & I_{zy}=-\displaystyle\int_m zydm & I_z=\displaystyle\int_m\left(x^2+y^2\right)dm\end{array}\tag{11.40}$$

Since $I_{yx}=I_{xy}$, $I_{zx}=I_{xz}$, and $I_{zy}=I_{yz}$, it follows that $[\mathbf{I}]$ is a symmetric matrix (i.e., $[\mathbf{I}]^T=[\mathbf{I}]$). Therefore, $[\mathbf{I}]$ has just six independent components instead of nine. Observe that, whereas the products of inertia I_{xy}, I_{xz}, and I_{yz} can be positive, negative, or zero, the moments of inertia I_x, I_y, and I_z are always positive (never zero or negative) for bodies of finite dimensions. For this reason, $[\mathbf{I}]$ is a symmetric positive definite matrix. Keep in mind that Eqs. (11.38) and (11.39) are valid as well for axes attached to a fixed point P about which the body is rotating.

The moments of inertia reflect how the mass of a rigid body is distributed. They manifest a body's rotational inertia (i.e., its resistance to being set into rotary motion or stopped once rotation is under way). It is not an object's mass alone but how that mass is distributed that determines how the body will respond to applied torques.

If the xy plane is a plane of symmetry, then for any x and y within the body there are identical mass elements located at $+z$ and $-z$. This means the products of inertia with z in the integrand vanish. Similar statements are true if xz or yz are symmetry planes. In summary, we conclude

If the xy plane is a plane of symmetry of the body, then $I_{xz}=I_{yz}=0$.
If the xz plane is a plane of symmetry of the body, then $I_{xy}=I_{yz}=0$.
If the yz plane is a plane of symmetry of the body, then $I_{xy}=I_{xz}=0$.

It follows that if the body has two planes of symmetry relative to the xyz frame of reference, then all three products of inertia vanish, and $[\mathbf{I}]$ becomes a diagonal matrix such that,

$$[\mathbf{I}]=\begin{bmatrix}A & 0 & 0\\ 0 & B & 0\\ 0 & 0 & C\end{bmatrix} \tag{11.41}$$

where A, B, and C are the principal moments of inertia (all positive), and the xyz axes are the body's principal axes of inertia or principal directions. In this case, relative to either the center of mass or a fixed point of rotation, as appropriate, we have

$$H_x=A\omega_x \quad H_y=B\omega_y \quad H_z=C\omega_z \tag{11.42}$$

In general, the angular velocity vector $\boldsymbol{\omega}$ and the angular momentum vector $\mathbf{H}$ are not parallel ($\boldsymbol{\omega}\times\mathbf{H}\neq 0$). However, if, for example, $\boldsymbol{\omega}=\omega\hat{\mathbf{i}}$, then according to Eq. (11.42), $\mathbf{H}=A\boldsymbol{\omega}$. In other words, if the angular velocity is aligned with a principal direction, so is the angular momentum. In that case, the two vectors $\boldsymbol{\omega}$ and $\mathbf{H}$ are indeed parallel.

Each of the three principal moments of inertia can be expressed as follows:

$$A=mk_x{}^2 \quad B=mk_y{}^2 \quad C=mk_z{}^2 \tag{11.43}$$

where m is the mass of the body and k_x, k_y, and k_z are the three radii of gyration. One may imagine the mass of a body to be concentrated around a principal axis at a distance equal to the radius of gyration.

The moments of inertia for several common shapes are listed in Fig. 11.10. By symmetry, their products of inertia vanish for the coordinate axes used. Formulas for other solid geometries can be found in engineering handbooks and in dynamics textbooks.

For a mass concentrated at a point, the moments of inertia in Eq. (11.40) are just the mass times the integrand evaluated at the point. That is, the moment of inertia matrix $[\mathbf{I}^{(m)}]$ of a point mass m is given by

$$\left[\mathbf{I}^{(m)}\right]=\begin{bmatrix}m(y^2+z^2) & -mxy & -mxz\\ -mxy & m(x^2+z^2) & -myz\\ -mxz & -myz & m(x^2+y^2)\end{bmatrix} \tag{11.44}$$

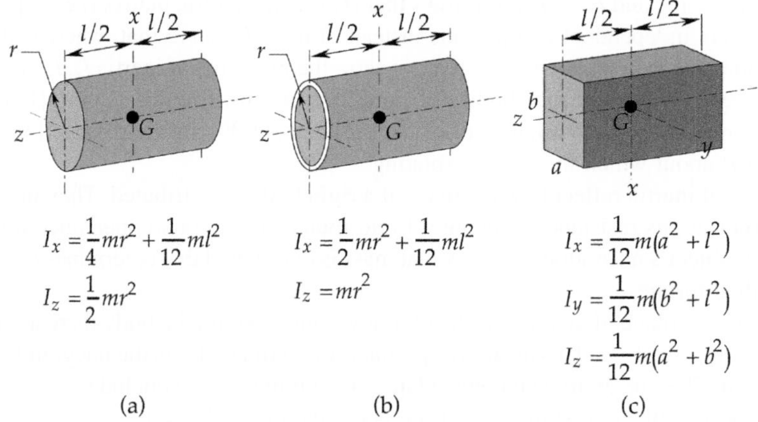

FIG. 11.10

Moments of inertia for three common homogeneous solids of mass m. (a) Solid circular cylinder. (b) Circular cylindrical shell. (c) Rectangular parallelepiped.

EXAMPLE 11.4

The following table lists the mass and coordinates of seven point masses. Find the center of mass of the system and the moments of inertia about the origin.

Point, i	Mass, m_i (kg)	x_i (m)	y_i (m)	z_i (m)
1	3	−0.5	0.2	0.3
2	7	0.2	0.75	−0.4
3	5	1	−0.8	0.9
4	6	1.2	−1.3	1.25
5	2	−1.3	1.4	−0.8
6	4	−0.3	1.35	0.75
7	1	1.5	−1.7	0.85

Solution

The total mass of this system is

$$m = \sum_{i=1}^{7} m_i = 28 \text{ kg}$$

For concentrated masses, the integral in Eq. (11.9) is replaced by the mass times its position vector. Therefore, in this case, the three components of the position vector of the center of mass are $x_G = (1/m)\sum_{i=1}^{7} m_i x_i$, $y_G = (1/m)\sum_{i=1}^{7} m_i y_i$, and $z_G = (1/m)\sum_{i=1}^{7} m_i z_i$, so that

$$\boxed{x_G = 0.35 \text{ m} \quad y_G = 0.01964 \text{ m} \quad z_G = 0.4411 \text{ m}}$$

The total moment of inertia is the sum over all the particles of Eq. (11.44) evaluated at each point. Thus,

$$[\mathbf{I}] = \overbrace{\begin{bmatrix} 0.39 & 0.3 & 0.45 \\ 0.3 & 1.02 & -0.18 \\ 0.45 & -0.18 & 0.87 \end{bmatrix}}^{(1)} + \overbrace{\begin{bmatrix} 5.0575 & -1.05 & 0.56 \\ -1.05 & 1.4 & 2.1 \\ 0.56 & 2.1 & 4.2175 \end{bmatrix}}^{(2)} + \overbrace{\begin{bmatrix} 7.25 & 4 & -4.5 \\ 4 & 9.05 & 3.6 \\ -4.5 & 3.6 & 8.2 \end{bmatrix}}^{(3)}$$

$$+ \overbrace{\begin{bmatrix} 19.515 & 9.36 & -9 \\ 9.36 & 18.015 & 9.75 \\ -9 & 9.75 & 18.78 \end{bmatrix}}^{(4)} + \overbrace{\begin{bmatrix} 5.2 & 3.64 & -2.08 \\ 3.64 & 4.66 & 2.24 \\ -2.08 & 2.24 & 7.3 \end{bmatrix}}^{(5)} + \overbrace{\begin{bmatrix} 9.54 & 1.62 & 0.9 \\ 1.62 & 2.61 & -4.05 \\ 0.9 & -4.05 & 7.65 \end{bmatrix}}^{(6)}$$

$$+ \overbrace{\begin{bmatrix} 3.6125 & 2.55 & -1.275 \\ 2.55 & 2.9725 & 1.445 \\ -1.275 & 1.445 & 5.14 \end{bmatrix}}^{(7)}$$

or

$$\boxed{[\mathbf{I}] = \begin{bmatrix} 50.56 & 20.42 & -14.94 \\ 20.42 & 39.73 & 14.90 \\ -14.94 & 14.90 & 52.16 \end{bmatrix} (\mathrm{kg \cdot m^2})}$$

EXAMPLE 11.5

Calculate the moments of inertia of a slender, homogeneous straight rod of length ℓ and mass m shown in Fig. 11.11. One end of the rod is at the origin and the other has coordinates (a, b, c).

Solution

A slender rod is one whose cross-sectional dimensions are negligible compared with its length. The mass is concentrated along its centerline. Since the rod is homogeneous, the mass per unit length ρ is uniform and given by

$$\rho = \frac{m}{\ell} \tag{a}$$

The length of the rod is

$$\ell = \sqrt{a^2 + b^2 + c^2}$$

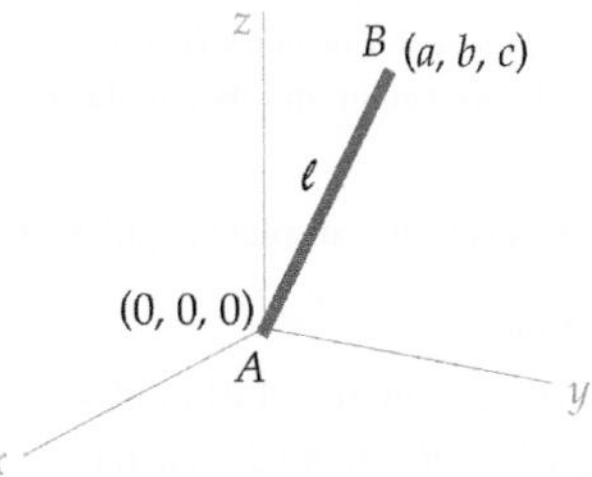

FIG. 11.11

Uniform slender bar of mass m and length ℓ.

Starting with I_x we have from Eq. (11.40),

$$I_x = \int_0^\ell (y^2 + z^2)\rho ds$$

in which we replaced the element of mass dm by ρds, where ds is the element of length along the rod. The distance s is measured from end A of the rod, so that the x, y, and z coordinates of any point along it are found in terms of s by the following relations:

$$x = \frac{s}{\ell}a \quad y = \frac{s}{\ell}b \quad z = \frac{s}{\ell}c$$

Thus,

$$I_x = \int_0^\ell \left(\frac{s}{\ell^2}b^2 + \frac{s}{\ell^2}c^2\right)\rho ds = \rho\frac{b^2 + c^2}{\ell^2}\int_0^\ell s^2 ds = \frac{1}{3}\rho\left(b^2 + c^2\right)\ell$$

Substituting Eq. (a) yields

$$\boxed{I_x = \frac{1}{3}m\left(b^2 + c^2\right)}$$

In precisely the same way, we find

$$\boxed{I_y = \frac{1}{3}m\left(a^2 + c^2\right)} \quad \boxed{I_z = \frac{1}{3}m\left(a^2 + b^2\right)}$$

For I_{xy} we have

$$I_{xy} = -\int_0^\ell xy\rho ds = -\int_0^\ell \left(\frac{s}{\ell}a\right)\left(\frac{s}{\ell}b\right)\rho ds = -\rho\frac{ab}{\ell^2}\int_0^\ell s^2 ds = -\frac{1}{3}\rho ab\ell$$

Once again using Eq. (a),

$$\boxed{I_{xy} = -\frac{1}{3}mab}$$

Likewise,

$$\boxed{I_{xz} = -\frac{1}{3}mac} \quad \boxed{I_{yz} = -\frac{1}{3}mbc}$$

EXAMPLE 11.6

The gyro rotor (Fig. 11.12) in Example 11.3 has a mass m of 5 kg, radius r of 0.08 m, and thickness t of 0.025 m. If $N = 2.1$ rad/s, $\dot{\theta} = 4$ rad/s, $\omega = 10.5$ rad/s, and $\theta = 60°$, calculate

(a) the angular momentum of the rotor about its center of mass G in the body-fixed xyz frame and

(b) the angle between the rotor's angular velocity vector and its angular momentum vector.

Solution

Eq. (f) from Example 11.3 gives the components of the absolute angular velocity of the rotor in the moving xyz frame.

$$\begin{aligned}
\omega_x &= \dot{\theta} = 4 \text{ rad/s} \\
\omega_y &= N\sin\theta = 2.1.\sin 60° = 1.819 \text{ rad/s} \\
\omega_z &= \omega_{\text{spin}} + N\cos\theta = 10.5 + 2.1.\cos 60° = 11.55 \text{ rad/s}
\end{aligned} \tag{a}$$

Therefore,

$$\boldsymbol{\omega} = 4\hat{\mathbf{i}} + 1.819\hat{\mathbf{j}} + 11.55\hat{\mathbf{k}} \text{ (rad/s)} \tag{b}$$

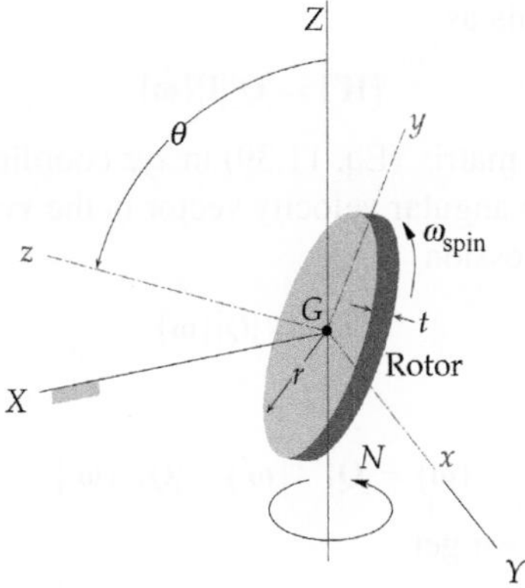

FIG. 11.12

Rotor of the gyroscope in Fig. 11.4.

All three coordinate planes of the body-fixed *xyz* frame contain the center of mass *G* and all are planes of symmetry of the circular cylindrical rotor. Therefore, $I_{xy} = I_{xz} = I_{yz} = 0$.

From Fig. 11.10a, we see that the nonzero diagonal entries in the moment of inertia tensor are

$$\begin{aligned} A = B &= \frac{1}{12}mt^2 + \frac{1}{4}mr^2 = \frac{1}{12}(5)(0.025)^2 + \frac{1}{4}(5)(0.08)^2 = 0.008260 \text{ kg} \cdot \text{m}^2 \\ C &= \frac{1}{2}mr^2 = \frac{1}{2}(5)(0.08)^2 = 0.0160 \text{ kg} \cdot \text{m}^2 \end{aligned} \tag{c}$$

We can use Eq. (11.42) to calculate the angular momentum, because the origin of the *xyz* frame is the rotor's center of mass (which in this case also happens to be a fixed point of rotation, which is another reason why we can use Eq. 11.42). Substituting Eqs. (a) and (c) into Eq. (11.42) yields

$$\begin{aligned} H_x &= A\omega_x = (0.008260)(4) = 0.03304 \text{ kg} \cdot \text{m}^2/\text{s} \\ H_y &= B\omega_y = (0.008260)(1.819) = 0.0150 \text{ kg} \cdot \text{m}^2/\text{s} \\ H_z &= C\omega_z = (0.0160)(11.55) = 0.1848 \text{ kg} \cdot \text{m}^2/\text{s} \end{aligned} \tag{d}$$

so that

$$\boxed{\mathbf{H} = 0.03304\hat{\mathbf{i}} + 0.0150\hat{\mathbf{j}} + 0.1848\hat{\mathbf{k}} \ (\text{kg} \cdot \text{m}^2/\text{s})} \tag{e}$$

The angle ϕ between $\mathbf{H}$ and $\boldsymbol{\omega}$ is found by taking the dot product of the two vectors,

$$\phi = \cos^{-1}\left(\frac{\mathbf{H} \cdot \boldsymbol{\omega}}{H\omega}\right) = \cos^{-1}\left(\frac{2.294}{0.1883 \cdot 12.36}\right) = \boxed{9.717°} \tag{f}$$

As this problem illustrates, the angular momentum and the angular velocity are in general not collinear.

Consider a Cartesian coordinate system $x'y'z'$ with the same origin as *xyz* but a different orientation. Let $[\mathbf{Q}]$ be the orthogonal matrix ($[\mathbf{Q}]^{-1} = [\mathbf{Q}]^T$) that transforms the components of a vector from the *xyz* system to the $x'y'z'$ frame. Recall from Section 4.5 that the rows of $[\mathbf{Q}]$ are the direction cosines of the $x'y'z'$ axes relative to *xyz*. If $\{\mathbf{H}'\}$ comprises the components of the angular momentum vector along the $x'y'z'$ axes, then $\{\mathbf{H}'\}$ is obtained from its components $\{\mathbf{H}\}$ in the *xyz* frame by the relation

$$\{\mathbf{H}'\} = [\mathbf{Q}]\{\mathbf{H}\}$$

From Eq. (11.39), we can write this as

$$\{\mathbf{H}'\} = [\mathbf{Q}][\mathbf{I}]\{\boldsymbol{\omega}\} \tag{11.45}$$

where $[\mathbf{I}]$ is the moment of inertia matrix (Eq. 11.39) in xyz coordinates. Like the angular momentum vector, the components $\{\boldsymbol{\omega}\}$ of the angular velocity vector in the xyz system are related to those in the primed system ($\{\boldsymbol{\omega}'\}$) by the expression

$$\{\boldsymbol{\omega}'\} = [\mathbf{Q}]\{\boldsymbol{\omega}\}$$

The inverse relation is simply

$$\{\boldsymbol{\omega}\} = [\mathbf{Q}]^{-1}\{\boldsymbol{\omega}'\} = [\mathbf{Q}]^T\{\boldsymbol{\omega}'\} \tag{11.46}$$

Substituting this into Eq. (11.45), we get

$$\{\mathbf{H}'\} = [\mathbf{Q}][\mathbf{I}][\mathbf{Q}]^T\{\boldsymbol{\omega}'\} \tag{11.47}$$

But the components of angular momentum and angular velocity in the $x'y'z'$ frame are related by an equation of the same form as Eq. (11.39), so that

$$\{\mathbf{H}'\} = [\mathbf{I}']\{\boldsymbol{\omega}'\} \tag{11.48}$$

where $[\mathbf{I}']$ comprises the components of the inertia matrix in the primed system. Comparing the right-hand sides of Eqs. (11.47) and (11.48), we conclude that

$$[\mathbf{I}'] = [\mathbf{Q}][\mathbf{I}][\mathbf{Q}]^T \tag{11.49a}$$

That is,

$$\begin{bmatrix} I_{x'} & I_{x'y'} & I_{x'z'} \\ I_{y'x'} & I_{y'} & I_{y'z'} \\ I_{z'x'} & I_{z'y'} & I_{z'} \end{bmatrix} = \begin{bmatrix} Q_{11} & Q_{12} & Q_{13} \\ Q_{21} & Q_{22} & Q_{23} \\ Q_{31} & Q_{32} & Q_{33} \end{bmatrix} \begin{bmatrix} I_x & I_{xy} & I_{xz} \\ I_{yx} & I_y & I_{yz} \\ I_{zx} & I_{zy} & I_z \end{bmatrix} \begin{bmatrix} Q_{11} & Q_{21} & Q_{31} \\ Q_{12} & Q_{22} & Q_{32} \\ Q_{13} & Q_{23} & Q_{33} \end{bmatrix} \tag{11.49b}$$

This shows how to transform the components of the inertia matrix from the xyz coordinate system to any other orthogonal system with a common origin. Thus, for example,

$$\begin{aligned} I_{x'} &= \overbrace{\lfloor Q_{11}\ Q_{12}\ Q_{13} \rfloor}^{\lfloor \text{Row 1} \rfloor} \begin{bmatrix} I_x & I_{xy} & I_{xz} \\ I_{yx} & I_y & I_{yz} \\ I_{zx} & I_{zy} & I_z \end{bmatrix} \overbrace{\begin{Bmatrix} Q_{11} \\ Q_{12} \\ Q_{13} \end{Bmatrix}}^{\lfloor \text{Row 1} \rfloor^T} \\ I_{y'z'} &= \overbrace{\lfloor Q_{21}\ Q_{22}\ Q_{23} \rfloor}^{\lfloor \text{Row 2} \rfloor} \begin{bmatrix} I_x & I_{xy} & I_{xz} \\ I_{yx} & I_y & I_{yz} \\ I_{zx} & I_{zy} & I_z \end{bmatrix} \overbrace{\begin{Bmatrix} Q_{31} \\ Q_{32} \\ Q_{33} \end{Bmatrix}}^{\lfloor \text{Row 3} \rfloor^T} \end{aligned} \tag{11.50}$$

etc.

Any object represented by a square matrix whose components transform according to Eq. (11.49) is called a second-order tensor. We may therefore refer to $[\mathbf{I}]$ as the inertia tensor.

EXAMPLE 11.7

Find the mass moment of inertia of the system of point masses in Example 11.4 about an axis from the origin through the point with coordinates (2 m, −3 m, 4 m).

Solution

From Example 11.4, the moment of inertia tensor for the system of point masses is

$$[\mathbf{I}] = \begin{bmatrix} 50.56 & 20.42 & -14.94 \\ 20.42 & 39.73 & 14.90 \\ -14.94 & 14.90 & 52.16 \end{bmatrix} \left(\text{kg}\cdot\text{m}^2\right)$$

The vector $\mathbf{V}$ connecting the origin with (2 m, −3 m, 4 m) is

$$\mathbf{V} = 2\hat{\mathbf{i}} - 3\hat{\mathbf{j}} + 4\hat{\mathbf{k}}$$

The unit vector in the direction of $\mathbf{V}$ is

$$\hat{\mathbf{u}}_V = \frac{\mathbf{V}}{\|\mathbf{V}\|} = 0.3714\hat{\mathbf{i}} - 0.5571\hat{\mathbf{j}} + 0.7428\hat{\mathbf{k}}$$

We may consider $\hat{\mathbf{u}}_V$ as the unit vector along the x' axis of a rotated Cartesian coordinate system. Then, from Eq. (11.50) we get

$$I_{V'} = \lfloor 0.3714 \quad -0.5571 \quad 0.7428 \rfloor \begin{bmatrix} 50.56 & 20.42 & -14.94 \\ 20.42 & 39.73 & 14.90 \\ -14.94 & 14.90 & 52.16 \end{bmatrix} \begin{bmatrix} 0.3714 \\ -0.5571 \\ 0.7428 \end{bmatrix}$$

$$= \lfloor 0.3714 \quad -0.5571 \quad 0.7428 \rfloor \begin{Bmatrix} -3.695 \\ -3.482 \\ 24.90 \end{Bmatrix} = \boxed{19.06 \text{ kg}\cdot\text{m}^2}$$

EXAMPLE 11.8

For the satellite of Example 11.2, which is reproduced in Fig. 11.13, the data are as follows: $N = 0.1$ rad/s and $\dot{\theta} = 0.01$ rad/s, in the directions shown. $\theta = 40°$ and $d_O = 1.5$ m. The length, width, and thickness of the panel are $\ell = 6$ m, $w = 2$ m, and $t = 0.025$ m. The uniformly distributed mass of the panel is 50 kg. Find the angular momentum of the panel relative to the center of mass O of the satellite.

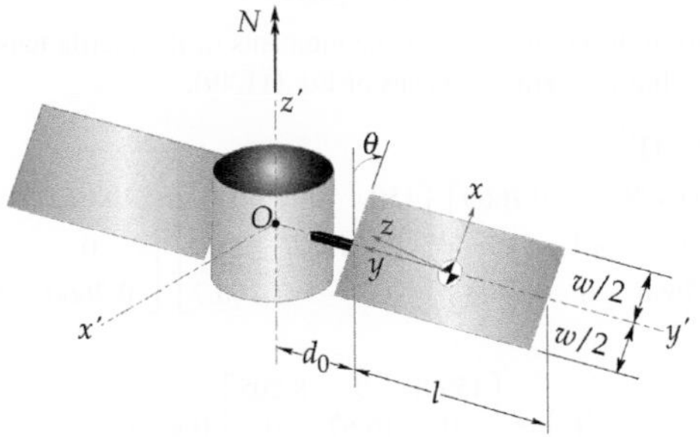

FIG. 11.13

Satellite and solar panel.

Solution

We can treat the panel as a thin parallelepiped. The panel's *xyz* axes have their origin at the center of mass *G* of the panel and are parallel to its three edge directions. According to Fig. 11.10c, the moments of inertia of the panel relative to the *xyz* coordinate system are

$$\begin{aligned}
I_G)_x &= \frac{1}{12}m(\ell^2+t^2) = \frac{1}{12}\cdot 50\cdot(6^2+0.025^2) = 150.0\ \text{kg}\cdot\text{m}^2 \\
I_G)_y &= \frac{1}{12}m(w^2+t^2) = \frac{1}{12}\cdot 50\cdot(2^2+0.025^2) = 16.67\ \text{kg}\cdot\text{m}^2 \\
I_G)_z &= \frac{1}{12}m(w^2+\ell^2) = \frac{1}{12}\cdot 50\cdot(2^2+6^2) = 166.7\ \text{kg}\cdot\text{m}^2 \\
I_G)_{xy} &= I_G)_{xz} = I_G)_{yz} = 0
\end{aligned} \tag{a}$$

In matrix notation,

$$[\mathbf{I}_G] = \begin{bmatrix} 150.0 & 0 & 0 \\ 0 & 16.67 & 0 \\ 0 & 0 & 166.7 \end{bmatrix} (\text{kg}\cdot\text{m}^2) \tag{b}$$

The unit vectors of the satellite's $x'y'z'$ system are related to those of the panel's *xyz* frame by inspection.

$$\begin{aligned}
\hat{\mathbf{i}}' &= -\sin\theta\hat{\mathbf{i}} + \cos\theta\hat{\mathbf{k}} = -0.6428\hat{\mathbf{i}} + 0.7660\hat{\mathbf{k}} \\
\hat{\mathbf{j}}' &= -\hat{\mathbf{j}} \\
\hat{\mathbf{k}}' &= \cos\theta\hat{\mathbf{i}} + \sin\theta\hat{\mathbf{k}} = 0.7660\hat{\mathbf{i}} + 0.6428\hat{\mathbf{k}}
\end{aligned} \tag{c}$$

The matrix $[\mathbf{Q}]$ of the transformation from *xyz* to $x'y'z'$ comprises the direction cosines of $\hat{\mathbf{i}}'$, $\hat{\mathbf{j}}'$, and $\hat{\mathbf{k}}'$, which we infer from Eqs. (c),

$$[\mathbf{Q}] = \begin{bmatrix} -0.6428 & 0 & 0.7660 \\ 0 & -1 & 0 \\ 0.7660 & 0 & 0.6428 \end{bmatrix} \tag{d}$$

In Example 11.2, we found that the absolute angular velocity of the panel, in the satellite's $x'y'z'$ frame of reference, is

$$\boldsymbol{\omega} = -\dot{\theta}\hat{\mathbf{j}}' + N\hat{\mathbf{k}}' = -0.01\hat{\mathbf{j}}' + 0.1\hat{\mathbf{k}}'\ (\text{rad/s})$$

That is,

$$\{\boldsymbol{\omega}'\} = \begin{Bmatrix} 0 \\ -0.01 \\ 0.1 \end{Bmatrix} (\text{rad/s}) \tag{e}$$

To find the absolute angular momentum $\{\mathbf{H}'_G\}$ of the panel in the satellite system requires the use of Eq. (11.39),

$$\{\mathbf{H}'_G\} = [\mathbf{I}'_G]\{\boldsymbol{\omega}'\} \tag{f}$$

Before doing so, we must transform the components of the moments of the inertia tensor in Eq. (b) from the unprimed (panel) system to the primed (satellite) system, by means of Eq. (11.49),

$$\begin{aligned}
[\mathbf{I}'_G] &= [\mathbf{Q}][\mathbf{I}_G][\mathbf{Q}]^T \\
&= \begin{bmatrix} -0.6428 & 0 & 0.7660 \\ 0 & -1 & 0 \\ 0.7660 & 0 & 0.6428 \end{bmatrix} \begin{bmatrix} 150 & 0 & 0 \\ 0 & 16.67 & 0 \\ 0 & 0 & 166.7 \end{bmatrix} \begin{bmatrix} -0.6428 & 0 & 0.7660 \\ 0 & -1 & 0 \\ 0.7660 & 0 & 0.6428 \end{bmatrix}
\end{aligned}$$

so that

$$[\mathbf{I}'_G] = \begin{bmatrix} 159.8 & 0 & 8.205 \\ 0 & 16.67 & 0 \\ 8.205 & 0 & 156.9 \end{bmatrix} (\text{kg}\cdot\text{m}^2) \tag{g}$$

Then Eq. (f) yields

$$\{\mathbf{H}'_G\} = \begin{bmatrix} 159.8 & 0 & 8.205 \\ 0 & 16.67 & 0 \\ 8.205 & 0 & 156.9 \end{bmatrix} \begin{Bmatrix} 0 \\ -0.01 \\ 0.1 \end{Bmatrix} = \begin{Bmatrix} 0.8205 \\ -0.1667 \\ 15.69 \end{Bmatrix} (\text{kg}\cdot\text{m}^2/\text{s})$$

or, in vector notation,

$$\mathbf{H}_G = 0.8205\hat{\mathbf{i}}' - 0.1667\hat{\mathbf{j}}' + 15.69\mathbf{k}' \ (\text{kg}\cdot\text{m}^2/\text{s}) \tag{h}$$

This is the absolute angular momentum of the panel about its own center of mass G, and it is used in Eq. (11.27) to calculate the angular momentum $\mathbf{H}_O)_{\text{rel}}$ relative to the satellite's center of mass O,

$$\mathbf{H}_O)_{\text{rel}} = \mathbf{H}_G + \mathbf{r}_{G/O} \times m\mathbf{v}_{G/O} \tag{i}$$

$\mathbf{r}_{G/O}$ is the position vector from O to G,

$$\mathbf{r}_{G/O} = \left(d_O + \frac{\ell}{2}\right)\hat{\mathbf{j}}' = \left(1.5 + \frac{6}{2}\right)\hat{\mathbf{j}}' = 4.5\hat{\mathbf{j}}' \ (\text{m}) \tag{j}$$

The velocity of G relative to O, $\mathbf{v}_{G/O}$, is found from Eq. (11.2),

$$\mathbf{v}_{G/O} = \boldsymbol{\omega}_{\text{satellite}} \times \mathbf{r}_{G/O} = N\hat{\mathbf{k}}' \times \mathbf{r}_{G/O} = 0.1\hat{\mathbf{k}}' \times 4.5\hat{\mathbf{j}}' = -0.45\hat{\mathbf{i}}' \ (\text{m/s}) \tag{k}$$

Substituting Eqs. (h), (j), and (k) into Eq. (i) finally yields

$$\begin{aligned} \mathbf{H}_O)_{\text{rel}} &= \left(0.8205\hat{\mathbf{i}}' - 0.1667\hat{\mathbf{j}}' + 15.69\hat{\mathbf{k}}'\right) + 4.5\hat{\mathbf{j}}' \times \left[50\left(-0.45\hat{\mathbf{i}}'\right)\right] \\ &= \boxed{0.8205\hat{\mathbf{i}}' - 0.1667\hat{\mathbf{j}}' + 116.9\hat{\mathbf{k}}' \ (\text{kg}\cdot\text{m}^2/\text{s})} \end{aligned} \tag{l}$$

Note that we were unable to use Eq. (11.21) to find the absolute angular momentum $\mathbf{H}_O$ because that requires knowing the absolute velocity $\mathbf{v}_G$, which in turn depends on the absolute velocity of O, which was not provided.

How can we find the direction cosine matrix [**Q**] such that Eq. (11.49) will yield a moment of inertia matrix [**I**′] that is diagonal (i.e., of the form given by Eq. (11.41))? In other words, how do we find the principal directions (eigenvectors) and the corresponding principal values (eigenvalues) of the moment of inertia tensor?

Let the angular velocity vector $\boldsymbol{\omega}$ be parallel to the principal direction defined by the vector **e**, so that $\boldsymbol{\omega} = \beta\mathbf{e}$, where β is a scalar. Since $\boldsymbol{\omega}$ points in the principal direction of the inertia tensor, so must **H**, which means **H** is also parallel to **e**. Therefore, $\mathbf{H} = \alpha\mathbf{e}$, where α is a scalar. From Eq. (11.39), it follows that

$$\alpha\{\mathbf{e}\} = \{\mathbf{I}\}(\beta\{\mathbf{e}\})$$

or

$$[\mathbf{I}]\{\mathbf{e}\} = \lambda\{\mathbf{e}\}$$

where $\lambda = \alpha/\beta$ (a scalar). That is,

$$\begin{bmatrix} I_x & I_{xy} & I_{xz} \\ I_{xy} & I_y & I_{yz} \\ I_{xz} & I_{yz} & I_z \end{bmatrix} \begin{Bmatrix} e_x \\ e_y \\ e_z \end{Bmatrix} = \lambda \begin{Bmatrix} e_x \\ e_y \\ e_z \end{Bmatrix}$$

This can be written

$$\begin{bmatrix} I_x-\lambda & I_{xy} & I_{xz} \\ I_{xy} & I_y-\lambda & I_{yz} \\ I_{xz} & I_{yz} & I_z-\lambda \end{bmatrix} \begin{Bmatrix} e_x \\ e_y \\ e_z \end{Bmatrix} = \begin{Bmatrix} 0 \\ 0 \\ 0 \end{Bmatrix} \tag{11.51}$$

The trivial solution of Eq. (11.51) is $\mathbf{e}=\mathbf{0}$, which is of no interest. The only way that Eq. (11.51) will not yield the trivial solution is if the coefficient matrix on the left is singular. That will occur if its determinant vanishes. That is, if

$$\begin{vmatrix} I_x-\lambda & I_{xy} & I_{xz} \\ I_{xy} & I_y-\lambda & I_{yz} \\ I_{xz} & I_{yz} & I_z-\lambda \end{vmatrix} = 0 \tag{11.52}$$

Expanding the determinant, we find

$$\begin{vmatrix} I_x-\lambda & I_{xy} & I_{xz} \\ I_{xy} & I_y-\lambda & I_{yz} \\ I_{xz} & I_{yz} & I_z-\lambda \end{vmatrix} = -\lambda^3+J_1\lambda^2-J_2\lambda+J_3 \tag{11.53}$$

where

$$J_1=I_x+I_y+I_z \quad J_2=\begin{vmatrix} I_x & I_{xy} \\ I_{xy} & I_y \end{vmatrix}+\begin{vmatrix} I_x & I_{xz} \\ I_{xz} & I_z \end{vmatrix}+\begin{vmatrix} I_y & I_{yz} \\ I_{yz} & I_z \end{vmatrix} \quad J_3=\begin{vmatrix} I_x & I_{xy} & I_{xz} \\ I_{xy} & I_y & I_{yz} \\ I_{xz} & I_{yz} & I_z \end{vmatrix} \tag{11.54}$$

J_1, J_2, and J_3 are invariants (i.e., they have the same value in every Cartesian coordinate system).

Eqs. (11.52) and (11.53) yield the characteristic equation of the tensor [**I**],

$$\lambda^3-J_1\lambda^2+J_2\lambda-J_3=0 \tag{11.55}$$

The three roots λ_p ($p=1,2,3$) of this cubic equation are real, since [**I**] is symmetric; furthermore, they are all positive, since [**I**] is a positive definite matrix. We substitute each root, or eigenvalue, λ_p back into Eq. (11.51) to obtain

$$\begin{bmatrix} I_x-\lambda_p & I_{xy} & I_{xz} \\ I_{xy} & I_y-\lambda_p & I_{yz} \\ I_{xz} & I_{yz} & I_z-\lambda_p \end{bmatrix} \begin{Bmatrix} e_x^{(p)} \\ e_y^{(p)} \\ e_z^{(p)} \end{Bmatrix} = \begin{Bmatrix} 0 \\ 0 \\ 0 \end{Bmatrix} \quad (p=1,2,3) \tag{11.56}$$

Solving this system yields the three eigenvectors $\mathbf{e}^{(p)}$ corresponding to each of the three eigenvalues λ_p. The three eigenvectors are orthogonal, also due to the symmetry of matrix [**I**]. Each eigenvalue is a principal moment of inertia and its corresponding eigenvector is a principal direction.

EXAMPLE 11.9

Find the principal moments of inertia and the principal axes of inertia of the inertia tensor

$$[\mathbf{I}] = \begin{bmatrix} 100 & -20 & -100 \\ -20 & 300 & -50 \\ -100 & -50 & 500 \end{bmatrix} \text{kg}\cdot\text{m}^2$$

Solution

We seek the nontrivial solutions of the system $[\mathbf{I}]\{\mathbf{e}\} = \lambda\{\mathbf{e}\}$. That is,

$$\begin{bmatrix} 100-\lambda & -20 & -100 \\ -20 & 300-\lambda & -50 \\ -100 & -50 & 500-\lambda \end{bmatrix} \begin{Bmatrix} e_x \\ e_y \\ e_z \end{Bmatrix} = \begin{Bmatrix} 0 \\ 0 \\ 0 \end{Bmatrix} \tag{a}$$

From Eq. (11.54),

$$\begin{aligned} J_1 &= 100+300+500 = 900 \\ J_2 &= \begin{vmatrix} 100 & -20 \\ -20 & 300 \end{vmatrix} + \begin{vmatrix} 100 & -100 \\ -100 & 500 \end{vmatrix} + \begin{vmatrix} 300 & -50 \\ -50 & 500 \end{vmatrix} = 217{,}100 \\ J_3 &= \begin{vmatrix} 100 & -20 & -100 \\ -20 & 300 & -50 \\ -100 & -50 & 500 \end{vmatrix} = 11{,}350{,}000 \end{aligned} \tag{b}$$

Thus, the characteristic equation is

$$\lambda^3 - 900\lambda^2 + 217{,}100\lambda - 11{,}350{,}000 = 0 \tag{c}$$

The three roots are the principal moments of inertia, which are found to be

$$\boxed{\lambda_1 = 532.052 \quad \lambda_2 = 295.840 \quad \lambda_3 = 72.1083}\ (\text{kg}\cdot\text{m}^2) \tag{d}$$

We substitute each of these, in turn, back into Eq. (a) to find its corresponding principal direction.

Substituting $\lambda_1 = 532.052\ \text{kg}\cdot\text{m}^2$ into Eq. (a) we obtain

$$\begin{bmatrix} -432.052 & -20.0000 & -100.0000 \\ -20.0000 & -232.052 & -50.0000 \\ -100.0000 & -50.0000 & -32.0519 \end{bmatrix} \begin{Bmatrix} e_x^{(1)} \\ e_y^{(1)} \\ e_z^{(1)} \end{Bmatrix} = \begin{Bmatrix} 0 \\ 0 \\ 0 \end{Bmatrix} \tag{e}$$

Since the determinant of the coefficient matrix is zero, at most two of the three equations in Eq. (e) are independent. Thus, at most, two of the three components of the vector $\mathbf{e}^{(1)}$ can be found in terms of the third. We can therefore arbitrarily set $e_x^{(1)} = 1$ and solve for $e_y^{(1)}$ and $e_z^{(1)}$ using any two of the independent equations in Eq. (e). With $e_x^{(1)} = 1$, the first two of Eq. (e) become

$$\begin{aligned} -20.0000e_y^{(1)} - 100.000e_z^{(1)} &= 432.052 \\ -232.052e_y^{(1)} - 50.000e_z^{(1)} &= 20.0000 \end{aligned} \tag{f}$$

Solving these two equations for $e_y^{(1)}$ and $e_z^{(1)}$ yields, together with the assumption that $e_x^{(1)} = 1$,

$$e_x^{(1)} = 1.00000 \quad e_y^{(1)} = 0.882793 \quad e_z^{(1)} = -4.49708 \tag{g}$$

The unit vector in the direction of $\mathbf{e}^{(1)}$ is

$$\hat{\mathbf{e}}_1 = \frac{\mathbf{e}^{(1)}}{\|\mathbf{e}^{(1)}\|} = \frac{1.00000\hat{\mathbf{i}} + 0.882793\hat{\mathbf{j}} - 4.49708\hat{\mathbf{k}}}{\sqrt{1.00000^2 + 0.882793^2 + (-4.49708)^2}}$$

or

$$\boxed{\hat{\mathbf{e}}_1 = 0.213186\hat{\mathbf{i}} + 0.188199\hat{\mathbf{j}} - 0.958714\hat{\mathbf{k}} \ \ (\lambda_1 = 532.052\ \text{kg}\cdot\text{m}^2)} \tag{h}$$

Substituting $\lambda_2 = 295.840\ \text{kg}\cdot\text{m}^2$ into Eq. (a) and proceeding as above we find that

$$\boxed{\hat{\mathbf{e}}_2 = 0.17632\hat{\mathbf{i}} - 0.972512\hat{\mathbf{j}} - 0.151609\hat{\mathbf{k}} \ \ (\lambda_2 = 295.840\ \text{kg}\cdot\text{m}^2)} \tag{i}$$

The two unit vectors $\hat{\mathbf{e}}_1$ and $\hat{\mathbf{e}}_2$ define two of the three principal directions of the inertia tensor. Observe that $\hat{\mathbf{e}}_1 \cdot \hat{\mathbf{e}}_2 = 0$, as must be the case for symmetric matrices.

To obtain the third principal direction $\hat{\mathbf{e}}_3$, we can substitute $\lambda_3 = 72.1083\ \text{kg}\cdot\text{m}^2$ into Eq. (a) and proceed as above. However, since the inertia tensor is symmetric, we know that the three principal directions are mutually orthogonal, which means $\hat{\mathbf{e}}_3 = \hat{\mathbf{e}}_1 \times \hat{\mathbf{e}}_2$. Substituting Eqs. (h) and (i) into this cross product, we find that

$$\boxed{\hat{\mathbf{e}}_3 = -0.960894\hat{\mathbf{i}} - 0.137114\hat{\mathbf{j}} - 0.240587\hat{\mathbf{k}}\ \ (\lambda_3 = 72.1083\ \text{kg}\cdot\text{m}^2)} \tag{j}$$

We can check our work by substituting λ_3 and $\hat{\mathbf{e}}_3$ into Eq. (a) and verify that it is indeed satisfied:

$$\begin{bmatrix} 100-72.1083 & -20 & -100 \\ -20 & 300-72.1083 & -50 \\ -100 & -50 & 500-72.1083 \end{bmatrix} \begin{Bmatrix} -0.960894 \\ -0.137114 \\ -0.240587 \end{Bmatrix} \overset{\text{verify}}{=} \begin{Bmatrix} 0 \\ 0 \\ 0 \end{Bmatrix} \tag{k}$$

The components of the vectors $\hat{\mathbf{e}}_1$, $\hat{\mathbf{e}}_2$, and $\hat{\mathbf{e}}_3$ define the three rows of the orthogonal transformation $[\mathbf{Q}]$ from the xyz system into the $x'y'z'$ system that is aligned along the three principal directions:

$$[\mathbf{Q}] = \begin{bmatrix} 0.213186 & 0.188199 & -0.958714 \\ 0.176732 & -0.972512 & -0.151609 \\ -0.960894 & -0.137114 & -0.240587 \end{bmatrix} \tag{l}$$

Indeed, if we apply the transformation in Eq. (11.49), $[\mathbf{I}'] = [\mathbf{Q}][\mathbf{I}][\mathbf{Q}]^T$, we find

$$\begin{aligned} [\mathbf{I}'] &= \begin{bmatrix} 0.213186 & 0.188199 & -0.958714 \\ 0.176732 & -0.972512 & -0.151609 \\ -0.960894 & -0.137114 & -0.240587 \end{bmatrix} \begin{bmatrix} 100 & -20 & -100 \\ -20 & 300 & -50 \\ -100 & -50 & 500 \end{bmatrix} \\ &\quad \times \begin{bmatrix} 0.213186 & 0.176732 & -0.960894 \\ 0.188199 & -0.972512 & -0.137114 \\ -0.958714 & -0.151609 & -0.240587 \end{bmatrix} \\ &= \begin{bmatrix} 532.052 & 0 & 0 \\ 0 & 295.840 & 0 \\ 0 & 0 & 72.1083 \end{bmatrix} (\text{kg}\cdot\text{m}^2) \end{aligned}$$

An alternative to the above hand calculations in Example 11.9 is to type the following lines in the MATLAB Command Window:

```
I = [ 100  -20  -100
      -20  300   -50
     -100  -50   500];
[eigenVectors, eigenValues] = eig(I)
```

Hitting the Enter (or Return) key yields the following output to the Command Window:

```
eigenVectors =
    0.9609   0.1767  -0.2132
    0.1371  -0.9725  -0.1882
    0.2406  -0.1516   0.9587
eigenValues =
    72.1083         0         0
          0  295.8398         0
          0         0  532.0519
```

Two of the eigenvectors delivered by MATLAB are opposite in direction to those calculated in Example 11.9. This illustrates the fact that we can determine an eigenvector only to within an arbitrary scalar factor. To show this, suppose **e** is an eigenvector of the tensor [**I**] so that $[\mathbf{I}]\{\mathbf{e}\} = \lambda\{\mathbf{e}\}$. Multiplying this equation through by an arbitrary scalar a yields $([\mathbf{I}]\{\mathbf{e}\})a = (\lambda\{\mathbf{e}\})a$, or $[\mathbf{I}]\{a\mathbf{e}\} = \lambda\{a\mathbf{e}\}$, which means that $\{a\mathbf{e}\}$ is an eigenvector corresponding to the same eigenvalue λ.

11.5.1 PARALLEL AXIS THEOREM

Suppose the rigid body in Fig. 11.14 is in pure rotation about point P. Then, according to Eq. (11.39),

$$\{\mathbf{H}_P)_{\text{rel}}\} = [\mathbf{I}_P]\{\boldsymbol{\omega}\} \tag{11.57}$$

where $[\mathbf{I}_P]$ is the moment of inertia tensor about P, given by Eq. (11.40) with

$$x = x_{G/P} + \xi \quad y = y_{G/P} + \eta \quad z = z_{G/P} + \zeta$$

On the other hand, we have from Eq. (11.27) that

$$\mathbf{H}_P)_{\text{rel}} = \mathbf{H}_G + \mathbf{r}_{G/P} \times m\mathbf{v}_{G/P} \tag{11.58}$$

The vector $\mathbf{r}_{G/P} \times m\mathbf{v}_{G/P}$ is the angular momentum about P of the concentrated mass m located at the center of mass G. Using matrix notation, it is computed as follows:

$$\{\mathbf{r}_{G/P} \times m\mathbf{v}_{G/P}\} \equiv \mathbf{H}_P^{(m)}\Big)_{\text{rel}} = \left[\mathbf{I}_P^{(m)}\right]\{\boldsymbol{\omega}\} \tag{11.59}$$

where $[\mathbf{I}_P^{(m)}]$, the moment of inertia of the point mass m about P, is obtained from Eq. (11.44), with $x = x_{G/P}$, $y = y_{G/P}$, and $z = z_{G/P}$. That is,

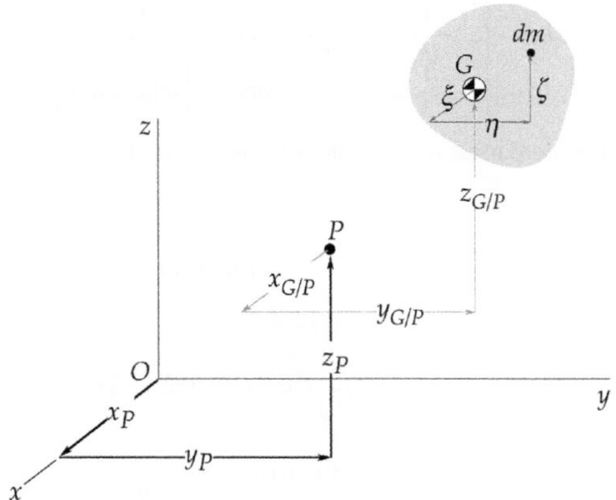

FIG. 11.14

The moments of inertia are to be computed at P, given their values at G.

$$\left[\mathbf{I}_P^{(m)}\right] = \begin{bmatrix} m\left(y_{G/P}^2 + z_{G/P}^2\right) & -mx_{G/P}y_{G/P} & -mx_{G/P}z_{G/P} \\ -mx_{G/P}y_{G/P} & m\left(x_{G/P}^2 + z_{G/P}^2\right) & -my_{G/P}z_{G/P} \\ -mx_{G/P}z_{G/P} & -my_{G/P}z_{G/P} & m\left(x_{G/P}^2 + y_{G/P}^2\right) \end{bmatrix} \tag{11.60}$$

Of course, Eq. (11.39) requires

$$\{\mathbf{H}_G\} = [\mathbf{I}_G]\{\boldsymbol{\omega}\}$$

Substituting this together with Eqs. (11.57) and (11.59) into Eq. (11.58) yields

$$[\mathbf{I}_P]\{\boldsymbol{\omega}\} = [\mathbf{I}_G]\{\boldsymbol{\omega}\} + \left[\mathbf{I}_P^{(m)}\right]\{\boldsymbol{\omega}\} = \left[\mathbf{I}_G + \mathbf{I}_P^{(m)}\right]\{\boldsymbol{\omega}\}$$

From this, we may infer the parallel axis theorem,

$$\mathbf{I}_P = \mathbf{I}_G + \mathbf{I}_P^{(m)} \tag{11.61}$$

The moment of inertia about P is the moment of inertia about the parallel axes through the center of mass plus the moment of inertia of the center of mass about P. That is,

$$\begin{aligned} &I_{P_x} = I_{G_x} + m\left(y_{G/P}^2 + z_{G/P}^2\right) \quad I_{P_y} = I_{G_y} + m\left(y_{G/P}^2 + x_{G/P}^2\right) \quad I_{P_z} = I_{G_z} + m\left(x_{G/P}^2 + y_{G/P}^2\right) \\ &I_{P_{xy}} = I_{G_{xy}} - mx_{G/P}y_{G/P} \qquad I_{P_{xz}} = I_{G_{xz}} - mx_{G/P}z_{G/P} \qquad I_{P_{yz}} = I_{G_{yz}} - my_{G/P}z_{G/P} \end{aligned} \tag{11.62}$$

EXAMPLE 11.10

Find the moments of inertia of the rod in Example 11.5 (Fig. 11.15) about its center of mass G.

Solution

From Example 11.5,

$$[\mathbf{I}_A] = \begin{bmatrix} \frac{1}{3}m(b^2+c^2) & -\frac{1}{3}mab & -\frac{1}{3}mac \\ -\frac{1}{3}mab & \frac{1}{3}m(a^2+c^2) & -\frac{1}{3}mbc \\ -\frac{1}{3}mac & -\frac{1}{3}mbc & \frac{1}{3}m(a^2+b^2) \end{bmatrix}$$

Using Eq. $(11.62)_1$ and noting the coordinates of the center of mass in Fig. 11.15,

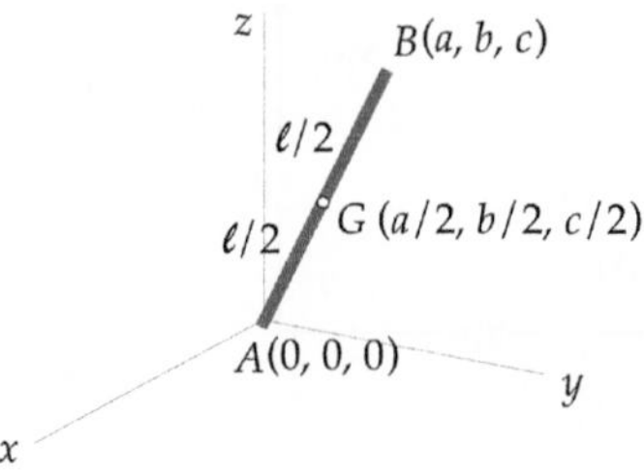

FIG. 11.15

Uniform slender rod.

$$I_{G_x} = I_{A_x} - m\left[(y_G - 0)^2 + (z_G - 0)^2\right] = \frac{1}{3}m\left(b^2 + c^2\right) - m\left[\left(\frac{b}{2}\right)^2 + \left(\frac{c}{2}\right)^2\right] = \frac{1}{12}m\left(b^2 + c^2\right)$$

Eq. $(11.62)_4$ yields

$$I_{G_{xy}} = I_{A_{xy}} + m(x_G - 0)(y_G - 0) = -\frac{1}{3}mab + m \cdot \frac{a}{2} \cdot \frac{b}{2} = -\frac{1}{12}mab$$

The remaining four moments of inertia are found in a similar fashion, so that

$$[\mathbf{I}_G] = \begin{bmatrix} \frac{1}{12}m\left(b^2 + c^2\right) & -\frac{1}{2}mab & -\frac{1}{12}mac \\ -\frac{1}{12}mab & \frac{1}{12}m\left(a^2 + c^2\right) & -\frac{1}{12}mbc \\ -\frac{1}{12}mac & -\frac{1}{12}mbc & \frac{1}{12}m\left(a^2 + b^2\right) \end{bmatrix} \tag{11.63}$$

EXAMPLE 11.11

Calculate the principal moments of inertia about the center of mass and the corresponding principal directions for the bent rod in Fig. 11.16. Its mass is uniformly distributed at 2 kg/m.

Solution

The mass of each of the four slender rod segments is

$$m_1 = 2 \cdot 0.4 = 0.8\text{kg} \quad m_2 = 2 \cdot 0.5 = 1\text{kg} \quad m_3 = 2 \cdot 0.3 = 0.6\text{kg} \quad m_4 = 2 \cdot 0.2 = 0.4\text{kg} \tag{a}$$

The total mass of the system is

$$m = \sum_{i=1}^{4} m_i = 2.8\text{kg} \tag{b}$$

The coordinates of each segment's center of mass are

$$\begin{array}{lll} x_{G_1} = 0 & y_{G_1} = 0 & z_{G_1} = 0.2\text{m} \\ x_{G_2} = 0 & y_{G_2} = 0.25\text{m} & z_{G_2} = 0.2\text{m} \\ x_{G_3} = 0.15\text{m} & y_{G_3} = 0.5\text{m} & z_{G_3} = 0 \\ x_{G_4} = 0.3\text{m} & y_{G_4} = 0.4\text{m} & z_{G_4} = 0 \end{array} \tag{c}$$

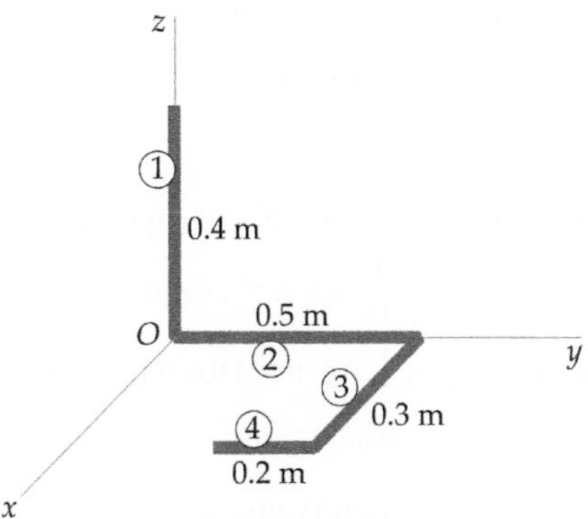

FIG. 11.16

Bent rod for which the principal moments of inertia are to be determined.

If the slender rod in Fig. 11.15 is aligned with, say, the x axis, then $a = \ell$ and $b = c = 0$, so that according to Eq. (11.63),

$$[\mathbf{I}_G] = \begin{bmatrix} 0 & 0 & 0 \\ 0 & \frac{1}{12}m\ell^2 & 0 \\ 0 & 0 & \frac{1}{12}m\ell^2 \end{bmatrix}$$

That is, the moment of inertia of a slender rod about the axes normal to the rod at its center of mass is $m\ell^2/12$, where m and ℓ are the mass and length of the rod, respectively. Since the mass of a slender bar is assumed to be concentrated along the axis of the bar (its cross-sectional dimensions are infinitesimal), the moment of inertia about the centerline is zero. By symmetry, the products of inertia about the axes through the center of mass are all zero. Using this information and the parallel axis theorem, we find the moments and products of inertia of each rod segment about the origin O of the xyz system as follows:

Rod 1:

$$I_x^{(1)} = \left(I_{G_1}^{(1)}\right)_x + m_1\left(y_{G_1}^2 + z_{G_1}^2\right) = \frac{1}{12}\cdot 0.8\cdot 0.4^2 + 0.8\left(0 + 0.2^2\right) = 0.04267 \text{ kg}\cdot\text{m}^2$$

$$I_y^{(1)} = \left(I_{G_1}^{(1)}\right)_y + m_1\left(x_{G_1}^2 + z_{G_1}^2\right) = \frac{1}{12}\cdot 0.8\cdot 0.4^2 + 0.8\left(0 + 0.2^2\right) = 0.04267 \text{ kg}\cdot\text{m}^2$$

$$I_z^{(1)} = \left(I_{G_1}^{(1)}\right)_z + m_1\left(x_{G_1}^2 + y_{G_1}^2\right) = 0 + 0.8(0 + 0) = 0$$

$$I_{xy}^{(1)} = \left(I_{G_1}^{(1)}\right)_{xy} - m_1 x_{G_1} y_{G_1} = 0 - 0.8(0)(0) = 0$$

$$I_{xz}^{(1)} = \left(I_{G_1}^{(1)}\right)_{xz} - m_1 x_{G_1} z_{G_1} = 0 - 0.8(0)(0.2) = 0$$

$$I_{xz}^{(1)} = \left(I_{G_1}^{(1)}\right)_{yz} - m_1 y_{G_1} z_{G_1} = 0 - 0.8(0)(0) = 0$$

Rod 2:

$$I_x^{(2)} = \left(I_{G_2}^{(2)}\right)_x + m_2\left(y_{G_2}^2 + z_{G_2}^2\right) = \frac{1}{12}\cdot 1.0\cdot 0.5^2 + 1.0\left(0 + 0.25^2\right) = 0.08333 \text{ kg}\cdot\text{m}^2$$

$$I_y^{(2)} = \left(I_{G_2}^{(2)}\right)_y + m_2\left(x_{G_2}^2 + z_{G_2}^2\right) = 0 + 1.0(0 + 0) = 0$$

$$I_z^{(2)} = \left(I_{G_2}^{(2)}\right)_z + m_2\left(x_{G_2}^2 + y_{G_2}^2\right) = \frac{1}{12}\cdot 1.0\cdot 0.5^2 + 1.0\left(0 + 0.5^2\right) = 0.08333 \text{ kg}\cdot\text{m}^2$$

$$I_{xy}^{(2)} = \left(I_{G_2}^{(2)}\right)_{xy} - m_2 x_{G_2} y_{G_2} = 0 - 1.0(0)(0.5) = 0$$

$$I_{xz}^{(2)} = \left(I_{G_2}^{(2)}\right)_{xz} - m_2 x_{G_2} z_{G_2} = 0 - 1.0(0)(0) = 0$$

$$I_{yz}^{(2)} = \left(I_{G_2}^{(2)}\right)_{yz} - m_2 y_{G_2} z_{G_2} = 0 - 1.0(0.5)(0) = 0$$

Rod 3:

$$I_x^{(3)} = \left(I_{G_3}^{(3)}\right)_x + m_3\left(y_{G_3}^2 + z_{G_3}^2\right) = 0 + 0.6\left(0.5^2 + 0\right) = 0.15 \text{ kg}\cdot\text{m}^2$$

$$I_y^{(3)} = \left(I_{G_3}^{(2)}\right)_y + m_3\left(x_{G_3}^2 + z_{G_3}^2\right) = \frac{1}{12}\cdot 0.6\cdot 0.3^2 + 0.6\left(0.15^2 + 0\right) = 0.018 \text{ kg}\cdot\text{m}^2$$

$$I_z^{(3)} = \left(I_{G_3}^{(3)}\right)_z + m_3\left(x_{G_3}^2 + y_{G_3}^2\right) = \frac{1}{2}\cdot 0.6\cdot 0.3^2 + 0.6\left(0.15^2 + 0.5^2\right) = 0.1680 \text{ kg}\cdot\text{m}^2$$

$$I_{xy}^{(3)} = \left(I_{G_3}^{(3)}\right)_{xy} - m_3 x_{G_3} y_{G3} = 0 - 0.6(0.15)(0.5) = -0.045 \text{ kg}\cdot\text{m}^2$$

$$I_{xz}^{(3)} = \left(I_{G_3}^{(3)}\right)_{xz} - m_3 x_{G_3} z_{G_3} = 0 - 0.6(0.15)(0) = 0$$

$$I_{yz}^{(3)} = \left(I_{G_3}^{(3)}\right)_{yz} - m_3 y_{G_3} z_{G_3} = 0 - 0.6(0.5)(0) = 0$$

Rod 4:

$$I_x^{(4)} = \left.I_{G_4}^{(4)}\right)_x + m_4\left(y_{G_4}^2 + z_{G_4}^2\right) = \frac{1}{12}\cdot 0.4\cdot 0.2^2 + 0.4\left(0.4^2 + 0\right) = 0.06533\ \text{kg}\cdot\text{m}^2$$
$$I_y^{(4)} = \left.I_{G_4}^{(4)}\right)_y + m_4\left(x_{G_4}^2 + z_{G4}^2\right) = 0 + 0.4\left(0.3^2 + 0\right) = 0.0360\ \text{kg}\cdot\text{m}^2$$
$$I_z^{(4)} = \left.I_{G_4}^{(4)}\right)_z + m_4\left(x_{G_4}^2 + y_{G_4}^2\right) = \frac{1}{12}\cdot 0.4\cdot 0.2^2 + 0.4\left(0.3^2 + 0.4^2\right) = 0.1013\ \text{kg}\cdot\text{m}^2$$
$$I_{xy}^{(4)} = \left.I_{G_4}^{(4)}\right)_{xy} - m_4 x_{G4} y_{G4} = 0 - 0.4(0.3)(0.4) = -0.0480\ \text{kg}\cdot\text{m}^2$$
$$I_{xz}^{(4)} = \left.I_{G_4}^{(4)}\right)_{xz} - m_4 x_{G_4} z_{G_4} = 0 - 0.4(0.3)(0) = 0$$
$$I_{yz}^{(4)} = \left.I_{G_4}^{(4)}\right)_{yz} - m_4 y_{G_4} z_{G_4} = 0 - 0.4(0.4)(0) = 0$$

The total moments of inertia for all the four rods about O are

$$I_x = \sum_{i=1}^{4} I_x^{(i)} = 0.3413\ \text{kg}\cdot\text{m}^2 \quad I_y = \sum_{i=1}^{4} I_y^{(i)} = 0.09667\ \text{kg}\cdot\text{m}^2 \quad I_z = \sum_{i=1}^{4} I_z^{(i)} = 0.3527\ \text{kg}\cdot\text{m}^2$$
$$I_{xy} = \sum_{i=1}^{4} I_{xy}^{(i)} = 0.0930\ \text{kg}\cdot\text{m}^2 \quad I_{xz} = \sum_{i=1}^{4} I_{xz}^{(i)} = 0 \quad I_{yz} = \sum_{i=1}^{4} I_{yz}^{(i)} = 0 \tag{d}$$

The coordinates of the center of mass of the system of four rods are, from Eqs. (a) through (c),

$$x_G = \frac{1}{m}\sum_{i=1}^{4} m_i x_{G_i} = \frac{1}{2.8}\cdot 0.21 = 0.075\text{m}$$
$$y_G = \frac{1}{m}\sum_{i=1}^{4} m_i y_{G_i} = \frac{1}{2.8}\cdot 0.71 = 0.2536\text{m} \tag{e}$$
$$z_G = \frac{1}{m}\sum_{i=1}^{4} m_i z_{G_i} = \frac{1}{2.8}\cdot 0.16 = 0.05714\text{m}$$

We use the parallel axis theorems to shift the moments of inertia in Eq. (d) to the center of mass G of the system

$$I_{G_x} = I_x - m\left(y_G^2 + z_G^2\right) = 0.3413 - 0.1892 = 0.1522\ \text{kg}\cdot\text{m}^2$$
$$I_{G_y} = I_y - m\left(x_G^2 + z_G^2\right) = 0.09667 - 0.02489 = 0.17177\ \text{kg}\cdot\text{m}^2$$
$$I_{G_z} = I_z - m\left(x_G^2 + y_G^2\right) = -0.3527 - 0.1958 = 0.1569\ \text{kg}\cdot\text{m}^2$$
$$I_{G_{xy}} = I_{xy} + m x_G y_G = -0.093 + 0.05325 = -0.03975\ \text{kg}\cdot\text{m}^2$$
$$I_{G_{xz}} = I_{xz} + m x_G z_G = 0 + 0.012 = 0.012\ \text{kg}\cdot\text{m}^2$$
$$I_{G_{yz}} = I_{yz} + m y_G z_G = 0 + 0.04057 = 0.04057\ \text{kg}\cdot\text{m}^2$$

Therefore, the inertia tensor, relative to the center of mass, is

$$[\mathbf{I}] = \begin{bmatrix} I_{G_x} & I_{G_{xy}} & I_{G_{xz}} \\ I_{G_{xy}} & I_{G_y} & I_{G_{yz}} \\ I_{G_{xz}} & I_{G_{yz}} & I_{G_z} \end{bmatrix} = \begin{bmatrix} 0.1522 & -0.03975 & 0.012 \\ -0.03975 & 0.07177 & 0.04057 \\ 0.012 & 0.04057 & 0.1569 \end{bmatrix} \left(\text{kg}\cdot\text{m}^2\right) \tag{f}$$

To find the three principal moments of inertia, we may proceed as in Example 11.9, or simply enter the following lines in the MATLAB Command Window:

```
IG = [ 0.1522  -0.03975  0.012
      -0.03975  0.07177  0.04057
       0.012    0.04057  0.1569];
[eigenVectors, eigenValues] = eig(IG)
```

to obtain

```
eigenVectors =
    0.3469  -0.8482  -0.4003
    0.8742   0.1378   0.4656
   -0.3397  -0.5115   0.7893
eigenValues =
    0.0402        0        0
         0   0.1658        0
         0        0   0.1747
```

Hence, the three principal moments of inertia and their principal directions are

$$\boxed{\begin{array}{ll} \lambda_1 = 0.04023 \text{ kg}\cdot\text{m}^2 & \mathbf{e}^{(1)} = 0.3469\hat{\mathbf{i}} + 0.8742\hat{\mathbf{j}} - 0.3397\hat{\mathbf{k}} \\ \lambda_2 = 0.1658 \text{ kg}\cdot\text{m}^2 & \mathbf{e}^{(2)} = -0.8482\hat{\mathbf{i}} + 0.1378\hat{\mathbf{j}} - 0.5115\hat{\mathbf{k}} \\ \lambda_3 = 0.1747 \text{ kg}\cdot\text{m}^2 & \mathbf{e}^{(3)} = -0.4003\hat{\mathbf{i}} + 0.4656\hat{\mathbf{j}} + 0.7893\hat{\mathbf{k}} \end{array}}$$

11.6 EULER EQUATIONS

For either the center of mass G or for a fixed point P about which the body is in pure rotation, we know from Eqs. (11.29) and (11.30) that

$$\mathbf{M}_{\text{net}} = \dot{\mathbf{H}} \tag{11.64}$$

Using a comoving coordinate system, with angular velocity $\mathbf{\Omega}$ and its origin located at the point (G or P), the angular momentum has the analytical expression

$$\mathbf{H} = H_x\hat{\mathbf{i}} + H_y\hat{\mathbf{j}} + H_z\hat{\mathbf{k}} \tag{11.65}$$

For simplicity, we shall henceforth assume

(a) the moving xyz axes are the principal axes of inertia; (11.66a)

(b) the moments of inertia relative to xyz are constant in time. (11.66b)

Eqs. (11.42) and (11.66a) imply that

$$\mathbf{H} = A\omega_x\hat{\mathbf{i}} + B\omega_y\hat{\mathbf{j}} + C\omega_z\hat{\mathbf{k}} \tag{11.67}$$

where A, B, and C are the principal moments of inertia.

According to Eq. (1.56), the time derivative of $\mathbf{H}$ is $\dot{\mathbf{H}} = \dot{\mathbf{H}})_{\text{rel}} + \mathbf{\Omega} \times \mathbf{H}$, so that Eq. (11.64) can be written as

$$\mathbf{M}_{\text{net}} = \dot{\mathbf{H}})_{\text{rel}} + \mathbf{\Omega} \times \mathbf{H} \tag{11.68}$$

Keep in mind that, whereas $\mathbf{\Omega}$ (the angular velocity of the moving xyz coordinate system) and $\boldsymbol{\omega}$ (the angular velocity of the rigid body itself) are both absolute kinematic quantities, Eq. (11.68) contains their components as projected onto the axes of the noninertial xyz frame given by

$$\boldsymbol{\omega} = \omega_x\hat{\mathbf{i}} + \omega_y\hat{\mathbf{j}} + \omega_z\hat{\mathbf{k}}$$
$$\mathbf{\Omega} = \Omega_x\hat{\mathbf{i}} + \Omega_y\hat{\mathbf{j}} + \Omega_z\hat{\mathbf{k}}$$

The absolute angular acceleration $\boldsymbol{\alpha}$ is obtained using Eq. (1.56) as

$$\boldsymbol{\alpha} = \dot{\boldsymbol{\omega}} = \overbrace{\dot{\omega}_x\hat{\mathbf{i}} + \dot{\omega}_y\hat{\mathbf{j}} + \dot{\omega}_z\hat{\mathbf{k}}}^{\boldsymbol{\alpha}_{\text{rel}}} + \boldsymbol{\Omega}\times\boldsymbol{\omega}$$

That is,

$$\boldsymbol{\alpha} = \left(\dot{\omega}_x + \Omega_y\dot{\omega}_z - \Omega_z\omega_y\right)\hat{\mathbf{i}} + \left(\dot{\omega}_y + \Omega_z\omega_x - \Omega_x\omega_z\right)\hat{\mathbf{j}} + \left(\dot{\omega}_z + \Omega_x\omega_y - \Omega_y\omega_x\right)\hat{\mathbf{k}} \tag{11.69}$$

Clearly, it is generally true that

$$\alpha_x \neq \dot{\omega}_x \quad \alpha_y \neq \dot{\omega}_y \quad \alpha_z \neq \dot{\omega}_z$$

From Eq. (1.57) and Eq. (11.67),

$$\dot{\mathbf{H}})_{\text{rel}} = \frac{\mathrm{d}(A\omega_x)}{\mathrm{d}t}\hat{\mathbf{i}} + \frac{\mathrm{d}(B\omega_y)}{\mathrm{d}t}\hat{\mathbf{j}} + \frac{\mathrm{d}(C\omega_z)}{\mathrm{d}t}\hat{\mathbf{k}}$$

Since A, B, and C are constant, this becomes

$$\dot{\mathbf{H}})_{\text{rel}} = A\dot{\omega}_x\hat{\mathbf{i}} + B\dot{\omega}_y\hat{\mathbf{j}} + C\dot{\omega}_z\hat{\mathbf{k}} \tag{11.70}$$

Substituting Eqs. (11.67) and (11.70) into Eq. (11.68) yields

$$\mathbf{M}_{\text{net}} = A\dot{\omega}_x\hat{\mathbf{i}} + B\dot{\omega}_y\hat{\mathbf{j}} + C\dot{\omega}_z\hat{\mathbf{k}} + \begin{vmatrix} \hat{\mathbf{i}} & \hat{\mathbf{j}} & \hat{\mathbf{k}} \\ \Omega_x & \Omega_y & \Omega_z \\ A\omega_x & B\omega_y & C\omega_z \end{vmatrix}$$

Expanding the cross product and collecting the terms leads to

$$\begin{aligned} M_x)_{\text{net}} &= A\dot{\omega}_x + C\Omega_y\omega_z - B\Omega_z\omega_y \\ M_y)_{\text{net}} &= B\dot{\omega}_y + A\Omega_z\omega_x - C\Omega_z\omega_z \\ M_z)_{\text{net}} &= C\dot{\omega}_z + B\Omega_x\omega_y - A\Omega_y\omega_x \end{aligned} \tag{11.71}$$

If the comoving frame is a body-fixed frame, then its angular velocity vector is the same as that of the body (i.e., $\boldsymbol{\Omega} = \boldsymbol{\omega}$). In that case, Eq. (11.68) reduces to the classical Euler equation of motion, namely,

$$\mathbf{M}_{\text{net}} = \dot{\mathbf{H}})_{\text{rel}} + \boldsymbol{\omega}\times\mathbf{H} \tag{11.72a}$$

the three components of which are obtained from Eq. (11.71) as

$$\begin{aligned} M_x)_{\text{net}} &= A\dot{\omega}_x + (C-B)\omega_y\omega_z \\ M_y)_{\text{net}} &= B\dot{\omega}_y + (A-C)\omega_z\omega_x \\ M_z)_{\text{net}} &= C\dot{\omega}_z + (B-A)\omega_x\omega_y \end{aligned} \tag{11.72b}$$

Eq. (11.68) is sometimes referred to as the modified Euler equation.

When $\boldsymbol{\Omega} = \boldsymbol{\omega}$, it follows from Eq. (11.69) that

$$\dot{\omega}_x = \alpha_x \quad \dot{\omega}_y = \alpha_y \quad \dot{\omega}_z = \alpha_z \tag{11.73}$$

That is, the relative angular acceleration equals the absolute angular acceleration when $\boldsymbol{\Omega} = \boldsymbol{\omega}$. Rather than calculating the time derivatives $\dot{\omega}_x$, $\dot{\omega}_y$, and $\dot{\omega}_z$ for use in Eq. (11.72), we may in this case first compute $\boldsymbol{\alpha}$ in the absolute XYZ frame

$$\boldsymbol{\alpha} = \frac{\mathrm{d}\boldsymbol{\omega}}{\mathrm{d}t} = \frac{\mathrm{d}\omega_X}{\mathrm{d}t}\hat{\mathbf{I}} + \frac{\mathrm{d}\omega_Y}{\mathrm{d}t}\hat{\mathbf{J}} + \frac{\mathrm{d}\omega_z}{\mathrm{d}t}\hat{\mathbf{K}}$$

and then project these components onto the xyz body frame, so that

$$\begin{Bmatrix} \dot{\omega}_x \\ \dot{\omega}_y \\ \dot{\omega}_z \end{Bmatrix} = [\mathbf{Q}]_{Xx} \begin{Bmatrix} d\omega_X/dt \\ d\omega_Y/dt \\ d\omega_Z/dt \end{Bmatrix} \tag{11.74}$$

where $[\mathbf{Q}]_{Xx}$ is the time-dependent orthogonal transformation from the inertial XYZ frame to the non-inertial xyz frame.

EXAMPLE 11.12

Calculate the net moment on the solar panel of Examples 11.2 and 11.8 (Fig. 11.17).

Solution

Since the comoving frame is rigidly attached to the panel, the Euler equation (Eq. 11.72a) applies to this problem. That is

$$\mathbf{M}_G)_{\text{net}} = \dot{\mathbf{H}}_G)_{\text{rel}} + \boldsymbol{\omega} \times \mathbf{H}_G \tag{a}$$

where

$$\mathbf{H}_G = A\omega_x\hat{\mathbf{i}} + B\omega_y\hat{\mathbf{j}} + C\omega_z\hat{\mathbf{k}} \tag{b}$$

and

$$\dot{\mathbf{H}}_G)_{\text{rel}} = A\dot{\omega}_x\hat{\mathbf{i}} + B\dot{\omega}_y\hat{\mathbf{j}} + C\dot{\omega}_z\hat{\mathbf{k}} \tag{c}$$

In Example 11.2, the angular velocity of the panel in the satellite's $x'y'z'$ frame was found to be

$$\boldsymbol{\omega} = -\dot{\theta}\hat{\mathbf{j}}' + N\hat{\mathbf{k}}' \tag{d}$$

In Example 11.8, we showed that the direction cosine matrix for the transformation from the panel's xyz frame to that of the satellite is

$$[\mathbf{Q}] = \begin{bmatrix} -\sin\theta & 0 & \cos\theta \\ 0 & -1 & 0 \\ \cos\theta & 0 & \sin\theta \end{bmatrix} \tag{e}$$

We use the transpose of $[\mathbf{Q}]$ to transform the components of $\boldsymbol{\omega}$ into the panel frame of reference,

$$\{\boldsymbol{\omega}\}_{xyz} = [\mathbf{Q}]^T\{\boldsymbol{\omega}\}_{x'y'z'} = \begin{bmatrix} -\sin\theta & 0 & \cos\theta \\ 0 & -1 & 0 \\ \cos\theta & 0 & \sin\theta \end{bmatrix} \begin{Bmatrix} 0 \\ -\dot{\theta} \\ N \end{Bmatrix} = \begin{Bmatrix} N\cos\theta \\ \dot{\theta} \\ N\sin\theta \end{Bmatrix}$$

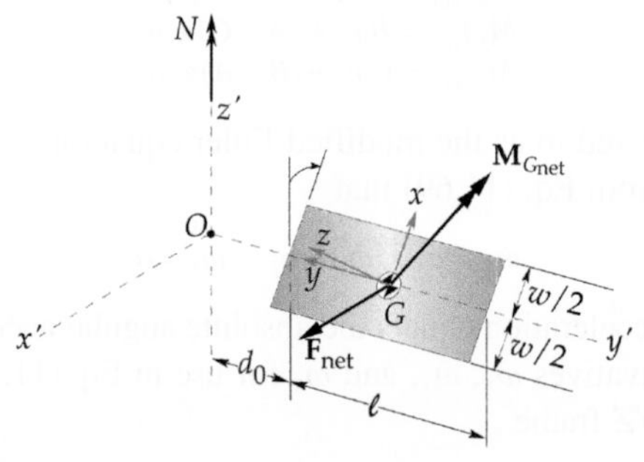

FIG. 11.17

Free body diagram of the solar panel in Examples 11.2 and 11.8.

or

$$\omega_x = N\cos\theta \quad \omega_y = \dot{\theta} \quad \omega_z = N\sin\theta \tag{f}$$

In Example 11.2, N and $\dot{\theta}$ were said to be constant. Therefore, the time derivatives of Eq. (f) are

$$\dot{\omega}_x = \frac{\mathrm{d}(N\cos\theta)}{\mathrm{d}t} = -N\dot{\theta}\sin\theta \quad \dot{\omega}_x = \frac{\mathrm{d}\dot{\theta}}{\mathrm{d}t} = 0 \quad \dot{\omega}_z = \frac{\mathrm{d}(N\sin\theta)}{\mathrm{d}t} = N\dot{\theta}\cos\theta \tag{g}$$

In Example 11.8, the moments of inertia in the panel frame of reference were listed as

$$A = \frac{1}{12}m\left(\ell^2 + t^2\right) \quad B = \frac{1}{12}m\left(w^2 + t^2\right) \quad C = \frac{1}{12}m\left(w^2 + \ell^2\right) \quad I_G)_{xy} = I_G)_{xz} = I_G)_{yz} = 0 \tag{h}$$

Substituting Eqs. (b), (c), (f), (g), and (h) into Eq. (a) yields

$$\mathbf{M}_G)_{\text{net}} = \frac{1}{12}m\left(\ell^2 + t^2\right)\left(-N\dot{\theta}\sin\theta\right)\hat{\mathbf{i}} + \frac{1}{12}m\left(w^2 + t^2\right)(0)\hat{\mathbf{j}} + \frac{1}{12}m\left(w^2 + \ell^2\right)\left(N\dot{\theta}\cos\theta\right)\hat{\mathbf{k}}$$
$$+ \begin{vmatrix} \hat{\mathbf{i}} & \hat{\mathbf{j}} & \hat{\mathbf{k}} \\ N\cos\theta & \dot{\theta} & N\sin\theta \\ \frac{1}{12}m\left(\ell^2 + t^2\right)(N\cos\theta) & \frac{1}{12}m\left(w^2 + t^2\right)\dot{\theta} & \frac{1}{12}m\left(w^2 + \ell^2\right)(N\sin\theta) \end{vmatrix}$$

Upon expanding the cross product determinant and collecting terms, this reduces to

$$\mathbf{M}_G)_{\text{net}} = -\frac{1}{6}mt^2N\dot{\theta}\sin\theta\hat{\mathbf{i}} + \frac{1}{24}m\left(t^2 - w^2\right)N^2\sin 2\theta\hat{\mathbf{j}} + \frac{1}{6}mw^2N\dot{\theta}\cos\theta\hat{\mathbf{k}}$$

Using the numerical data of Example 11.8 ($m = 50$ kg, $N = 0.1$ rad/s, $\theta = 40°$, $\dot{\theta} = 0.01$ rad/s, $\ell = 6$ m, $w = 2$ m, and $t = 0.025$ m), we find

$$\boxed{\mathbf{M}_G)_{\text{net}} = -3.348\left(10^{-6}\right)\hat{\mathbf{i}} - 0.08205\hat{\mathbf{j}} + 0.02554\hat{\mathbf{k}}\,(\mathrm{N\cdot m})}$$

EXAMPLE 11.13

Calculate the net moment on the gyro rotor of Examples 11.3 and 11.6.

Solution

Fig. 11.18 is a free body diagram of the rotor. Since in this case the comoving frame is not rigidly attached to the rotor, we must use Eq. (11.68) to find the net moment about G. That is

$$\mathbf{M}_G)_{\text{net}} = \dot{\mathbf{H}}_G)_{\text{rel}} + \boldsymbol{\Omega} \times \mathbf{H}_G \tag{a}$$

where

$$\mathbf{H}_G = A\omega_x\hat{\mathbf{i}} + B\omega_y\hat{\mathbf{j}} + C\omega_z\hat{\mathbf{k}} \tag{b}$$

and

$$\dot{\mathbf{H}}_G)_{\text{rel}} = A\dot{\omega}_x\hat{\mathbf{i}} + B\dot{\omega}_y\hat{\mathbf{j}} + C\dot{\omega}_z\hat{\mathbf{k}} \tag{c}$$

From Eq. (f) of Example 11.3, we know that the components of the angular velocity of the rotor in the moving reference frame are

$$\omega_x = \dot{\theta} \quad \omega_y = N\sin\theta \quad \omega_z = \omega_{\text{spin}} + N\cos\theta \tag{d}$$

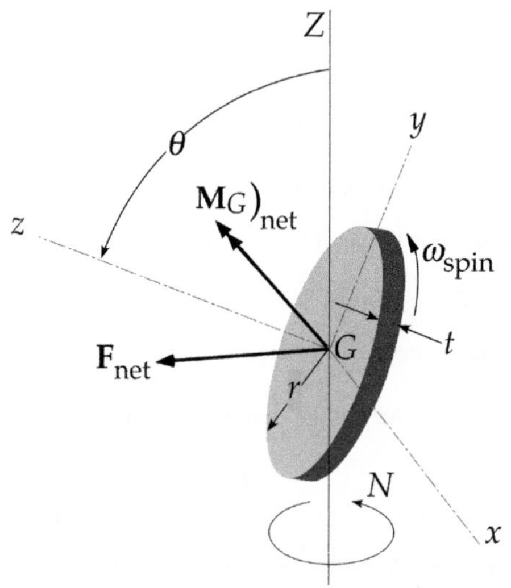

FIG. 11.18

Free body diagram of the gyro rotor of Examples 11.6 and 11.3.

Since, as specified in Example 11.3, $\dot{\theta}$, N, and ω_{spin} are all constant, it follows that

$$\begin{aligned}
\dot{\omega}_x &= \frac{d\dot{\theta}}{dt} = 0 \\
\dot{\omega}_y &= \frac{d(N\sin\theta)}{dt} = N\dot{\theta}\cos\theta \\
\dot{\omega}_z &= \frac{d\left(\omega_{\text{spin}} + N\cos\theta\right)}{dt} = -N\dot{\theta}\sin\theta
\end{aligned} \tag{e}$$

The angular velocity $\boldsymbol{\Omega}$ of the comoving xyz frame is that of the gimbal ring, which equals the angular velocity of the rotor minus its spin. Therefore,

$$\Omega_x = \dot{\theta} \quad \Omega_y = N\sin\theta \quad \Omega_z = N\cos\theta \tag{f}$$

In Example 11.6, we found that

$$A = B = \frac{1}{12}mt^2 + \frac{1}{4}mr^2 \quad C = \frac{1}{2}mr^2 \tag{g}$$

Substituting Eqs. (b) through (g) into Eq. (a), we get

$$\begin{aligned}
\mathbf{M}_G)_{\text{net}} = &\left(\frac{1}{12}mt^2 + \frac{1}{4}mr^2\right)(0)\hat{\mathbf{i}} + \left(\frac{1}{12}mt^2 + \frac{1}{4}mr^2\right)(N\dot{\theta}\cos\theta)\hat{\mathbf{j}} + \frac{1}{2}mr^2(-N\dot{\theta}\sin\theta)\hat{\mathbf{k}} \\
&+ \begin{vmatrix} \hat{\mathbf{i}} & \hat{\mathbf{j}} & \hat{\mathbf{k}} \\ \dot{\theta} & N\sin\theta & N\cos\theta \\ \left(\frac{1}{12}mt^2 + \frac{1}{4}mr^2\right)\dot{\theta} & \left(\frac{1}{12}mt^2 + \frac{1}{4}mr^2\right)N\sin\theta & \frac{1}{2}mr^2\left(\omega_{\text{spin}} + N\cos\theta\right) \end{vmatrix}
\end{aligned}$$

Expanding the cross product, collecting terms, and simplifying leads to

$$\begin{aligned}
\mathbf{M}_G)_{\text{net}} = &\left[\frac{1}{2}\omega_{\text{spin}} + \frac{1}{12}\left(3 - \frac{t^2}{r^2}\right)N\cos\theta\right] mr^2 N\sin\theta\hat{\mathbf{i}} \\
&+ \left(\frac{1}{6}\frac{t^2}{r^2}N\cos\theta - \frac{1}{2}\omega_{\text{spin}}\right) mr^2\dot{\theta}\hat{\mathbf{j}} - \frac{1}{2}N\dot{\theta}\sin\theta mr^2\hat{\mathbf{k}}
\end{aligned} \tag{h}$$

In Example 11.3, the following numerical data were provided: $m = 5$ kg, $r = 0.08$ m, $t = 0.025$ m, $N = 2.1$ rad/s, $\theta = 60°$, $\dot{\theta} = 4$ rad/s, and $\omega_{\text{spin}} = 105$ rad/s. For this set of numbers, Eq. (h) becomes

$$\boxed{\mathbf{M}_G)_{\text{net}} = 0.3203\hat{\mathbf{i}} - 0.6698\hat{\mathbf{j}} - 0.1164\hat{\mathbf{k}}\,(\text{N}\cdot\text{m})}$$

11.7 KINETIC ENERGY

The kinetic energy T of a rigid body is the integral of the kinetic energy $(1/2)v^2\text{d}m$ of its individual mass elements,

$$T = \int_m \frac{1}{2}v^2\text{d}m = \int_m \frac{1}{2}\mathbf{v}.\mathbf{v}\text{d}m \tag{11.75}$$

where $\mathbf{v}$ is the absolute velocity $\dot{\mathbf{R}}$ of the element of mass dm. From Fig. 11.8, we infer that $\dot{\mathbf{R}} = \dot{\mathbf{R}}_G + \dot{\boldsymbol{\rho}}$. Furthermore, Eq. (1.52) requires that $\dot{\boldsymbol{\rho}} = \boldsymbol{\omega} \times \boldsymbol{\rho}$. Thus, $\mathbf{v} = \mathbf{v}_G + \boldsymbol{\omega} \times \dot{\boldsymbol{\rho}}$, which means

$$\mathbf{v}\cdot\mathbf{v} = (\mathbf{v}_G + \boldsymbol{\omega}\times\boldsymbol{\rho})\cdot(\mathbf{v}_G + \boldsymbol{\omega}\times\boldsymbol{\rho}) = v_G{}^2 + 2\mathbf{v}_G\cdot(\boldsymbol{\omega}\times\boldsymbol{\rho}) + (\boldsymbol{\omega}\times\boldsymbol{\rho})\cdot(\boldsymbol{\omega}\times\boldsymbol{\rho})$$

We can apply the vector identity introduced in Eq. (1.21), namely

$$\mathbf{A}\cdot(\mathbf{B}\times\mathbf{C}) = \mathbf{B}\cdot(\mathbf{C}\times\mathbf{A}) \tag{11.76}$$

to the last term to get

$$\mathbf{v}\cdot\mathbf{v} = v_G{}^2 + 2\mathbf{v}_G\cdot(\boldsymbol{\omega}\times\boldsymbol{\rho}) + \boldsymbol{\omega}\cdot[\boldsymbol{\rho}\times(\boldsymbol{\omega}\times\boldsymbol{\rho})]$$

Therefore, Eq. (11.75) becomes

$$T = \int_m \frac{1}{2}v_G{}^2\text{d}m + \mathbf{v}_G\cdot\left(\boldsymbol{\omega}\times\int_m \boldsymbol{\rho}\text{d}m\right) + \frac{1}{2}\boldsymbol{\omega}\cdot\int_m \boldsymbol{\rho}\times(\boldsymbol{\omega}\times\boldsymbol{\rho})\text{d}m$$

Since the position vector $\boldsymbol{\rho}$ is measured from the center of mass, $\int_m \boldsymbol{\rho}\text{d}m = \mathbf{0}$. Recall that, according to Eq. (11.34),

$$\int_m \boldsymbol{\rho}\times(\boldsymbol{\omega}\times\boldsymbol{\rho})\text{d}m = \mathbf{H}_G$$

It follows that the kinetic energy may be written as

$$T = \frac{1}{2}mv_G{}^2 + \frac{1}{2}\boldsymbol{\omega}\cdot\mathbf{H}_G \tag{11.77}$$

The second term is the rotational kinetic energy T_R,

$$T_R = \frac{1}{2}\boldsymbol{\omega}\cdot\mathbf{H}_G \tag{11.78}$$

If the body is rotating about a point P that is at rest in inertial space, we have from Eq. (11.2) and Fig. 11.8 that

$$\mathbf{v}_G = \mathbf{v}_P + \boldsymbol{\omega}\times\mathbf{r}_{G/P} = \mathbf{0} + \boldsymbol{\omega}\times\mathbf{r}_{G/P} = \boldsymbol{\omega}\times\mathbf{r}_{G/P}$$

It follows that

$$v_G{}^2 = \mathbf{v}_G\cdot\mathbf{v}_G = (\boldsymbol{\omega}\times\mathbf{r}_{G/P})\cdot(\boldsymbol{\omega}\times\mathbf{r}_{G/P})$$

Making use once again of the vector identity in Eq. (11.76), we find

$$v_G{}^2 = \boldsymbol{\omega} \cdot \left[\mathbf{r}_{G/P} \times \left(\boldsymbol{\omega} \times \mathbf{r}_{G/P}\right)\right] = \boldsymbol{\omega} \cdot \left(\mathbf{r}_{G/P} \times \mathbf{v}_G\right)$$

Substituting this into Eq. (11.77) yields

$$T = \frac{1}{2}\boldsymbol{\omega} \cdot \left[\mathbf{H}_G + \mathbf{r}_{G/P} \times m\mathbf{v}_G\right]$$

Eq. (11.21) shows that this can be written as

$$T = \frac{1}{2}\boldsymbol{\omega} \cdot \mathbf{H}_P \tag{11.79}$$

In this case, of course, all the kinetic energy is rotational.

In terms of the components of $\boldsymbol{\omega}$ and $\mathbf{H}$, whether it is $\mathbf{H}_P$ or $\mathbf{H}_G$, the rotational kinetic energy expression becomes, with the aid of Eq. (11.39),

$$T_R = \frac{1}{2}\left(\omega_x H_x + \omega_y H_y + \omega_y H_z\right) = \frac{1}{2}\lfloor \omega_x \quad \omega_y \quad \omega_z \rfloor \begin{bmatrix} I_x & I_{xy} & I_{xz} \\ I_{xy} & I_y & I_{yz} \\ I_{xz} & I_{yz} & I_z \end{bmatrix} \begin{Bmatrix} \omega_x \\ \omega_y \\ \omega_z \end{Bmatrix}$$

Expanding, we obtain

$$T_R = \frac{1}{2}I_x\omega_x{}^2 + \frac{1}{2}I_y\omega_y{}^2 + \frac{1}{2}I_z\omega^2{}_z + I_{xy}\omega_x\omega_y + I_{xz}\omega_x\omega_z + I_{yz}\omega_y\omega_z \tag{11.80}$$

Obviously, if the xyz axes are principal axes of inertia, then Eq. (11.80) simplifies considerably,

$$T_R = \frac{1}{2}A\omega_x{}^2 + \frac{1}{2}B\omega_y{}^2 + \frac{1}{2}C\omega_z{}^2 \tag{11.81}$$

EXAMPLE 11.14

A satellite in a circular geocentric orbit of 300 km altitude has a mass of 1500 kg and the moments of inertia relative to a body frame with origin at the center of mass G are

$$[\mathbf{I}] = \begin{bmatrix} 2000 & -1000 & 2500 \\ -1500 & 3000 & -1500 \\ 2500 & -1500 & 4000 \end{bmatrix} (\text{kg} \cdot \text{m}^2)$$

If at a given instant the components of angular velocity in this frame of reference are

$$\boldsymbol{\omega} = 1\mathbf{i} - 0.9\hat{\mathbf{j}} + 1.5\hat{\mathbf{k}}\,(\text{rad/s})$$

calculate the total kinetic energy of the satellite.

Solution

The speed of the satellite in its circular orbit is

$$v = \sqrt{\frac{\mu}{r}} = \sqrt{\frac{398{,}600}{6378 + 300}} = 7.7258\,\text{km/s}$$

The angular momentum of the satellite is

$$\{\mathbf{H}_G\} = [\mathbf{I}_G]\{\boldsymbol{\omega}\} = \begin{bmatrix} 2000 & -1000 & 2500 \\ -1500 & 3000 & -1500 \\ 2500 & -1500 & 4000 \end{bmatrix} \begin{Bmatrix} 1 \\ -0.9 \\ 1.5 \end{Bmatrix} = \begin{Bmatrix} 6650 \\ -5950 \\ 9850 \end{Bmatrix} (\text{kg}\cdot\text{m}^2/\text{s})$$

Therefore, the total kinetic energy is

$$T = \frac{1}{2}mv_G{}^2 + \frac{1}{2}\boldsymbol{\omega}\cdot\mathbf{H}_G = \overbrace{\frac{1}{2}(1500)(7.7258\times 10^3)^2}^{44.766(10^9)\text{J}} + \overbrace{\frac{1}{2}\lfloor 1 \quad -0.9 \quad 1.5 \rfloor \begin{Bmatrix} 6650 \\ -5950 \\ 9850 \end{Bmatrix}}^{13,390\text{J}}$$

$$\boxed{T = 44.766\text{GJ}}$$

Obviously, the kinetic energy is dominated by that due to the orbital motion.

11.8 THE SPINNING TOP

Let us analyze the motion of the simple axisymmetric top in Fig. 11.19. It is constrained to rotate about point O, which is fixed in space.

The moving xyz coordinate system is chosen to have its origin at O. The z axis is aligned with the spin axis of the top (the axis of rotational symmetry). The x axis is the node line, which passes through O and is perpendicular to the plane defined by the inertial Z axis and the spin axis of the top. The y axis is then perpendicular to x and z, such that $\hat{\mathbf{j}} = \hat{\mathbf{k}} \times \hat{\mathbf{i}}$. By symmetry, the moment of inertia matrix of the top relative to the xyz frame is diagonal, with $I_x = I_y = A$ and $I_z = C$. From Eqs. (11.68) and (11.70), we have

$$\mathbf{M}_O)_{\text{net}} = A\dot{\omega}_x\hat{\mathbf{i}} + A\dot{\omega}_y\hat{\mathbf{j}} + C\dot{\omega}_z\hat{\mathbf{k}} + \begin{vmatrix} \hat{\mathbf{i}} & \hat{\mathbf{j}} & \hat{\mathbf{k}} \\ \Omega_x & \Omega_y & \Omega_z \\ A\omega_x & A\omega_y & C\omega_z \end{vmatrix} \tag{11.82}$$

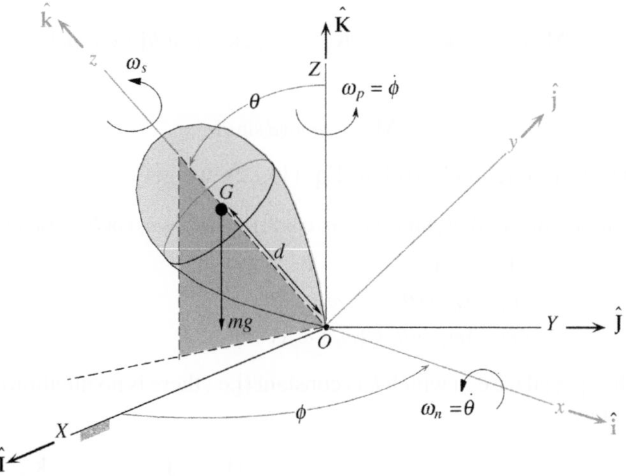

FIG. 11.19

Simple top rotating about the fixed point O.

The angular velocity $\boldsymbol{\omega}$ of the top is the vector sum of the spin rate ω_s and the rates of precession ω_p and nutation ω_n, where

$$\omega_p = \dot{\phi} \quad \omega_n = \dot{\theta} \tag{11.83}$$

Thus,

$$\boldsymbol{\omega} = \omega_n \hat{\mathbf{i}} + \omega_p \hat{\mathbf{K}} + \omega_s \hat{\mathbf{k}}$$

From the geometry, we see that

$$\hat{\mathbf{K}} = \sin\theta \hat{\mathbf{j}} + \cos\theta \hat{\mathbf{k}} \tag{11.84}$$

Therefore, relative to the comoving system,

$$\boldsymbol{\omega} = \omega_n \hat{\mathbf{i}} + \omega_p \sin\theta \hat{\mathbf{j}} + (\omega_s + \omega_p \cos\theta)\hat{\mathbf{k}} \tag{11.85}$$

From Eq. (11.85), we see that

$$\omega_x = \omega_n \quad \omega_y = \omega_p \sin\theta \quad \omega_z = \omega_s + \omega_p \cos\theta \tag{11.86}$$

Computing the time rates of these three expressions yields the components of angular acceleration relative to the *xyz* frame, given by

$$\dot{\omega}_x = \dot{\omega}_n \quad \dot{\omega}_y = \dot{\omega}_p \sin\theta + \omega_p \omega_n \cos\theta \quad \dot{\omega}_z = \dot{\omega}_s + \dot{\omega}_p \cos\theta - \omega_p \omega_n \sin\theta \tag{11.87}$$

The angular velocity $\boldsymbol{\Omega}$ of the *xyz* system is $\boldsymbol{\Omega} = \omega_p \hat{\mathbf{K}} + \omega_n \hat{\mathbf{i}}$, so that, using Eq. (11.84),

$$\boldsymbol{\Omega} = \omega_n \hat{\mathbf{i}} + \omega_p \sin\theta \hat{\mathbf{j}} + \omega_p \cos\theta \hat{\mathbf{k}} \tag{11.88}$$

From Eq. (11.88), we obtain

$$\Omega_x = \omega_n \quad \Omega_y = \omega_p \sin\theta \quad \Omega_z = \omega_p \cos\theta \tag{11.89}$$

In Fig. 11.19, the moment about O is that of the weight vector acting through the center of mass G:

$$\left.\mathbf{M}_O\right)_{\text{net}} = (d\hat{\mathbf{k}}) \times (-mg\hat{\mathbf{K}}) = -mgd\hat{\mathbf{k}} \times \left(\sin\theta \hat{\mathbf{j}} + \cos\theta \hat{\mathbf{k}}\right)$$

or

$$\left.\mathbf{M}_O\right)_{\text{net}} = mgd \sin\theta \hat{\mathbf{i}} \tag{11.90}$$

Substituting Eqs. (11.86) through (11.90) into Eq. (11.82), we get

$$mgd \sin\theta \hat{\mathbf{i}} = A\dot{\omega}_n \hat{\mathbf{i}} + A(\dot{\omega}_p \sin\theta + \dot{\omega}_p \dot{\omega}_n \cos\theta)\hat{\mathbf{j}} + C(\dot{\omega}_s + \dot{\omega}_p \cos\theta - \dot{\omega}_p \dot{\omega}_n \sin\theta)\hat{\mathbf{k}} + \begin{vmatrix} \hat{\mathbf{i}} & \hat{\mathbf{j}} & \hat{\mathbf{k}} \\ \omega_n & \omega_p \sin\theta & \omega_p \cos\theta \\ A\omega_n & A\omega_p \sin\theta & C(\omega_s + \omega_p \cos\theta) \end{vmatrix} \tag{11.91}$$

Let us consider the special case in which θ is constant (i.e., there is no nutation), so that $\omega_n = \dot{\omega}_n = 0$. Then, Eq. (11.91) reduces to

$$mgd \sin\theta \hat{\mathbf{i}} = A\dot{\omega}_p \sin\theta \hat{\mathbf{j}} + C(\dot{\omega}_s + \dot{\omega}_p \cos\theta)\hat{\mathbf{k}} + \begin{vmatrix} \hat{\mathbf{i}} & \hat{\mathbf{j}} & \hat{\mathbf{k}} \\ 0 & \omega_p \sin\theta & \omega_p \cos\theta \\ 0 & A\omega_p \sin\theta & C(\omega_s + \omega_p \cos\theta) \end{vmatrix} \tag{11.92}$$

Expanding the determinant yields

$$mgd\sin\theta\hat{\mathbf{i}} = A\dot{\omega}_p\sin\theta\hat{\mathbf{j}} + C(\dot{\omega}_s + \dot{\omega}_p\cos\theta)\hat{\mathbf{k}} + \left[C\omega_p\omega_s\sin\theta + (C-A)\omega_p{}^2\cos\theta\sin\theta\right]\hat{\mathbf{i}}$$

Equating the coefficients of $\hat{\mathbf{i}}$, $\hat{\mathbf{j}}$, and $\hat{\mathbf{k}}$ on each side of this equation and assuming that $0° < \theta < 180°$ leads to

$$mgd = C\omega_p\omega_s + (C-A)\omega_p{}^2\cos\theta \tag{11.93a}$$

$$A\dot{\omega}_p = 0 \tag{11.93b}$$

$$C(\dot{\omega}_s + \dot{\omega}_p\cos\theta) = 0 \tag{11.93c}$$

Eq. (11.93b) implies $\dot{\omega}_p = 0$, and from Eq. (11.93c) it follows that $\dot{\omega}_s = 0$. Therefore, the rates of spin and precession are both constant. From Eq. (11.93a), we find

$$(A-C)\cos\theta\omega_p{}^2 - C\omega_s\omega_p + mgd = 0 \tag{11.94}$$

If the spin rate is zero, Eq. (11.94) yields

$$\omega_p\Big)_{\omega_s=0} = \pm\sqrt{\frac{mgd}{(C-A)\cos\theta}} \quad \text{if } (C-A)\cos\theta > 0 \tag{11.95}$$

In this case, the top rotates about O at this rate, without spinning. If $A > C$ (prolate), its symmetry axis must make an angle between 90° and 180° to the vertical; otherwise, ω_p is imaginary. On the other hand, if $A < C$ (oblate), the angle lies between 0° and 90°. Thus, in steady rotation without spin, the top's axis sweeps out a cone that lies either below the horizontal plane ($A > C$) or above the plane ($A < C$).

In the special case where $(A - C)\cos\theta = 0$, Eq. (11.94) yields a steady precession rate that is inversely proportional to the spin rate,

$$\omega_p = \frac{mgd}{C\omega_s} \quad \text{if } (A-C)\cos\theta = 0 \tag{11.96}$$

If $A = C$, this precession apparently occurs irrespective of tilt angle θ. If $A \neq C$, this rate of precession occurs at $\theta = 90°$ (i.e., the spin axis is perpendicular to the precession axis).

In general, Eq. (11.94) is a quadratic equation in ω_p, so we can use the quadratic formula to find

$$\omega_p = \frac{C}{2(A-C)\cos\theta}\left(\omega_s \pm \sqrt{\omega_s{}^2 - \frac{4mgd(A-C)\cos\theta}{C^2}}\right) \tag{11.97}$$

Thus, for a given spin rate and tilt angle θ ($\theta \neq 90°$), there are two rates of precession $\dot{\phi}$.

Observe that if $(A - C)\cos\theta > 0$, then ω_p is imaginary when $\omega_s{}^2 < 4mgd(A-C)\cos\theta/C^2$. Therefore, the minimum spin rate required for steady precession at a constant inclination θ is

$$\omega_s\Big)_{\min} = \frac{2}{C}\sqrt{mgd(A-C)\cos\theta} \quad \text{if } (A-C)\cos\theta > 0 \tag{11.98}$$

If $(A - C)\cos\theta < 0$, the radical in Eq. (11.97) is real for all ω_s. In this case, as $\omega_s \to 0$, ω_p approaches the value given above in Eq. (11.95).

EXAMPLE 11.15

Calculate the precession rate ω_p for a toy top like that in Fig. 11.19 if $m = 0.5$ kg, $A(=I_x = I_y) = 12(10^{-4})\text{kg}\cdot\text{m}^2$, $C\,(=I_z) = 4.5(10^{-4})\text{kg}\cdot\text{m}^2$, and $d = 0.05$ m.

Solution

For an inclination of, say, 60°, $(A - C)\cos\theta > 0$, so that Eq. (11.98) requires $\omega_s)_{\text{min}} = 407.01$ rpm. Let us choose the spin rate to be $\omega_s = 1000$ rpm $= 104.7$ rad/s. Then, from Eq. (11.97), the steady precession rate as a function of the inclination θ is given by either one of the following formulas:

$$\boxed{\omega_p = 31.42\frac{1+\sqrt{1-0.3312\cos\theta}}{\cos\theta} \quad \text{and} \quad \omega_p = 31.42\frac{1-\sqrt{1-0.3312\cos\theta}}{\cos\theta}} \tag{a}$$

These are plotted in Fig. 11.20. For $\theta = 60°$, the high-energy precession rate is 1148.1 rpm, which exceeds the spin rate, whereas the low-energy precession rate is a leisurely 51.93 rpm.

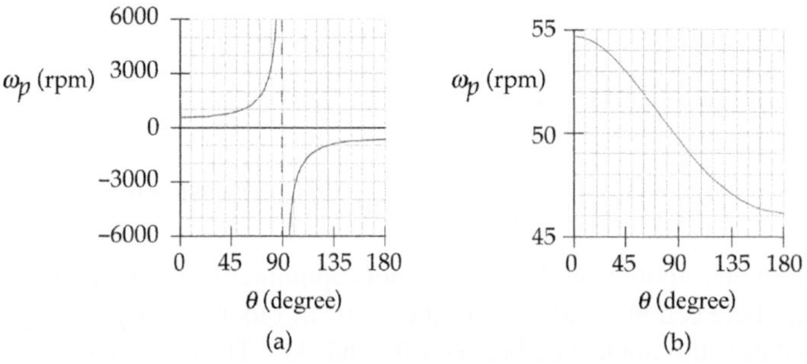

FIG. 11.20

(a) High-energy precession rate (unlikely to be observed). (b) Low-energy precession rate (the one almost always seen).

Fig. 11.21 shows an axisymmetric rotor mounted so that its spin axis (z) remains perpendicular to the precession axis (y). In that case, Eq. (11.85) with $\theta = 90°$ yields

$$\boldsymbol{\omega} = \omega_p\hat{\mathbf{j}} + \omega_s\hat{\mathbf{k}} \tag{11.99}$$

Likewise, from Eq. (11.88), the angular velocity of the comoving xyz system is $\boldsymbol{\Omega} = \omega_p\hat{\mathbf{j}}$. If we assume that the spin rate and precession rate are constant ($d\omega_p/dt = d\omega_s/dt = 0$), then Eq. (11.68), written for the center of mass G, becomes

$$\mathbf{M}_G)_{\text{net}} = \boldsymbol{\Omega}\times\mathbf{H} = \left(\omega_p\hat{\mathbf{j}}\right)\times\left(A\omega_p\hat{\mathbf{j}} + C\omega_s\hat{\mathbf{k}}\right) \tag{11.100}$$

where A and C are the moments of inertia of the rotor about the x and z axes, respectively. Setting $C\omega_s\hat{\mathbf{k}} = \mathbf{H}_s$, the spin angular momentum, and $\omega_p\hat{\mathbf{j}} = \boldsymbol{\omega}_p$, we obtain

$$\mathbf{M}_G)_{\text{net}} = \boldsymbol{\omega}_p\times\mathbf{H}_s \quad (\mathbf{H}_s = C\omega_s\mathbf{k}) \tag{11.101}$$

Since the center of mass is the reference point, there is no restriction on the motion G for which Eq. (11.101) is valid. Observe that the net gyroscopic moment $\mathbf{M}_G)_{\text{net}}$ exerted on the rotor by its supports is perpendicular to the plane of the spin and the precession vectors. If a spinning rotor is forced to

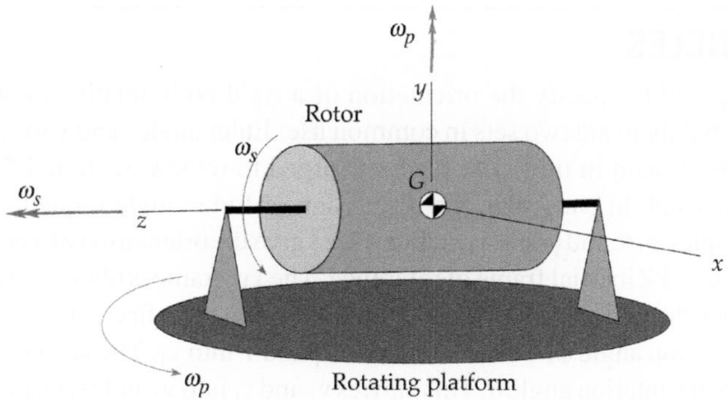

FIG. 11.21

A spinning rotor on a rotating platform.

precess, the gyroscopic moment $\mathbf{M}_G)_{\text{net}}$ develops. Or, if a moment is applied normal to the spin axis of a rotor, it will precess so as to cause the spin axis to turn toward the moment axis.

EXAMPLE 11.16

A uniform cylinder of radius r, length L, and mass m spins at a constant angular velocity ω_s. It rests on simple supports (which cannot exert couples), mounted on a platform that rotates at an angular velocity of ω_p. Find the reactions at A and B. Neglect the weight (i.e., calculate the reactions due just to the gyroscopic effects).

Solution

The net vertical force on the cylinder is zero, so the reactions at each end must be equal and opposite in direction, as shown on the free body diagram insert in Fig. 11.22. Noting that the moment of inertia of a uniform cylinder about its axis of rotational symmetry is $mr^2/2$, Eq. (11.101) yields

$$RL\hat{\mathbf{i}} = \left(\omega_p\hat{\mathbf{j}}\right) \times \left(\frac{mr^2}{2}\omega_s\hat{\mathbf{k}}\right) = \frac{1}{2}mr^2\omega_p\omega_s\hat{\mathbf{i}}$$

so that

$$\boxed{R = \frac{mr^2\omega_p\omega_s}{2L}}$$

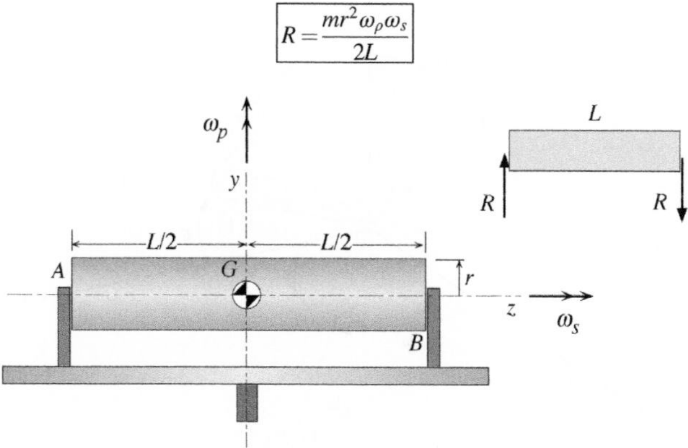

FIG. 11.22

Illustration of the gyroscopic effect.

11.9 EULER ANGLES

Three angles are required to specify the orientation of a rigid body relative to an inertial frame. The choice is not unique, but there are two sets in common use: Euler angles and yaw, pitch, and roll angles. We will discuss each of them in turn. The reader is urged to review Section 4.5 on orthogonal coordinate transformations and, in particular, the discussion of Euler angle sequences.

The three Euler angles ϕ, θ, and ψ shown in Fig. 11.23 give the orientation of a body-fixed *xyz* frame of reference relative to the *XYZ* inertial frame of reference. The *xyz* frame is obtained from the *XYZ* frame by a sequence of rotations through each of the Euler angles in turn. The first rotation is around the $Z\,(=z_1)$ axis through the precession angle ϕ. This takes X into x_1 and Y into y_1. The second rotation is around the $x_2\,(=x_1)$ axis through the nutation angle θ. This carries y_1 and z_1 into y_2 and z_2, respectively. The third and final rotation is around the $z\,(=z_2)$ axis through the spin angle ψ, which takes x_2 into x and y_2 into y.

The matrix $[\mathbf{Q}]_{Xx}$ of the transformation from the inertial frame to the body-fixed frame is given by the classical Euler angle sequence (Eq. 4.37):

$$[\mathbf{Q}]_{Xx} = [\mathbf{R}_3(\psi)][\mathbf{R}_1(\theta)][\mathbf{R}_3(\phi)] \tag{11.102}$$

From Eqs. (4.32) and (4.34), we have

$$[\mathbf{R}_3(\psi)] = \begin{bmatrix} \cos\psi & \sin\psi & 0 \\ -\sin\psi & \cos\psi & 0 \\ 0 & 0 & 1 \end{bmatrix} \quad [\mathbf{R}_1(\theta)] = \begin{bmatrix} 1 & 0 & 0 \\ 0 & \cos\theta & \sin\theta \\ 0 & -\sin\theta & \cos\theta \end{bmatrix} \quad [\mathbf{R}_3(\phi)] = \begin{bmatrix} \cos\phi & \sin\phi & 0 \\ -\sin\phi & \cos\phi & 0 \\ 0 & 0 & 1 \end{bmatrix} \tag{11.103}$$

According to Eq. (4.38), the direction cosine matrix is

$$[\mathbf{Q}]_{Xx} = \begin{bmatrix} -\sin\phi\cos\theta\sin\psi + \cos\phi\cos\psi & \cos\phi\cos\theta\sin\psi + \sin\phi\cos\psi & \sin\theta\sin\psi \\ -\sin\phi\cos\theta\cos\gamma - \cos\phi\sin\psi & \cos\phi\cos\theta\cos\psi - \sin\phi\sin\psi & \sin\theta\cos\psi \\ \sin\psi\sin\theta & -\cos\phi\sin\theta & \cos\theta \end{bmatrix} \tag{11.104}$$

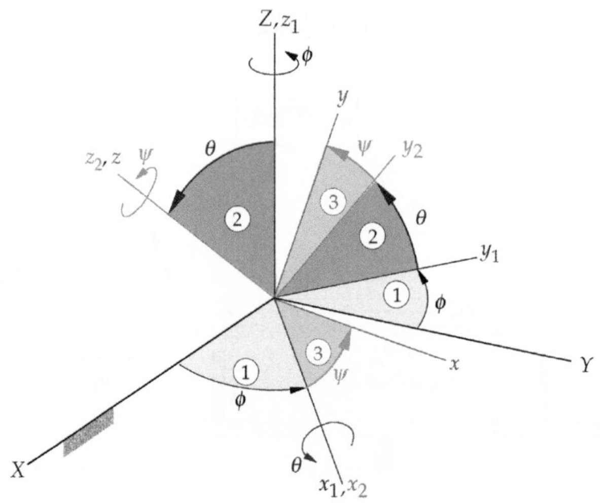

FIG. 11.23

Classical Euler angle sequence (see also fig. 4.14).

Since this is an orthogonal matrix, the inverse transformation from xyz to XYZ is $[\mathbf{Q}]_{xX} = [\mathbf{Q}]_{Xx}^T$,

$$[\mathbf{Q}]_{xX} = \begin{bmatrix} -\sin\phi\cos\theta\sin\psi + \cos\phi\cos\psi & -\sin\phi\cos\theta\cos\gamma - \cos\phi\sin\psi & \sin\psi\sin\theta \\ \cos\phi\cos\theta\sin\psi + \sin\phi\cos\psi & \cos\phi\cos\theta\cos\psi - \sin\phi\sin\psi & -\cos\phi\sin\theta \\ \sin\theta\sin\psi & \sin\theta\cos\psi & \cos\theta \end{bmatrix} \tag{11.105}$$

Algorithm 4.3 is used to find the three Euler angles θ, ϕ, and ψ from a given direction cosine matrix $[\mathbf{Q}]_{Xx}$.

EXAMPLE 11.17

The direction cosine matrix of an orthogonal transformation from XYZ to xyz is

$$[\mathbf{Q}] = \begin{bmatrix} -0.32175 & 0.89930 & -0.29620 \\ 0.57791 & -0.061275 & -0.81380 \\ -0.75000 & -0.43301 & -0.5000 \end{bmatrix}$$

Use Algorithm 4.3 to find the Euler angles ϕ, θ, and ψ for this transformation.

Solution

Step 1 (precession angle, ϕ):

$$\phi = \tan^{-1}\left(\frac{Q_{31}}{-Q_{32}}\right) = \tan^{-1}\left(\frac{-0.75000}{0.43301}\right) \quad (0 \le \phi < 360°)$$

Since the numerator is negative and the denominator is positive, the angle ϕ lies in the fourth quadrant:

$$\boxed{\phi = \tan^{-1}(-1.7320) = 300°}$$

Step 2 (nutation angle, θ):

$$\theta = \cos^{-1} Q_{33} = \cos^{-1}(-0.5000) \quad (0 \le \theta \le 180°)$$

$$\boxed{\theta = 120°}$$

Step 3 (spin angle, ψ):

$$\psi = \tan^{-1}\frac{Q_{13}}{Q_{23}} = \tan^{-1}\left(\frac{-0.29620}{-0.81380}\right) \quad (0 \le \psi < 360°)$$

Since both the numerator and denominator are negative, the angle ψ lies in the third quadrant:

$$\boxed{\psi = \tan^{-1}(0.36397) = 200°}$$

The time rates of change of the Euler angles ϕ, θ, and ψ are, respectively, the precession rate ω_p, the nutation rate ω_n, and the spin ω_s. That is,

$$\omega_p = \dot{\phi} \quad \omega_n = \dot{\theta} \quad \omega_s = \dot{\psi} \tag{11.106}$$

The absolute angular velocity $\boldsymbol{\omega}$ of a rigid body can be resolved into components ω_x, ω_y, and ω_z along the body-fixed xyz axes, so that

$$\boldsymbol{\omega} = \omega_x\hat{\mathbf{i}} + \omega_y\hat{\mathbf{j}} + \omega_z\hat{\mathbf{k}} \tag{11.107}$$

Fig. 11.23 shows that precession is measured around the inertial Z axis (unit vector $\hat{\mathbf{K}}$), nutation is measured around the intermediate x_1 axis (node line) with unit vector $\hat{\mathbf{i}}_1$, and spin is measured around the body-fixed z axis (unit vector $\hat{\mathbf{k}}$). Therefore, the absolute angular velocity can alternatively be written in terms of the nonorthogonal Euler angle rates as

$$\boldsymbol{\omega} = \omega_p \hat{\mathbf{K}} + \omega_n \hat{\mathbf{i}}_1 + \omega_s \hat{\mathbf{k}} \tag{11.108}$$

To find the relationship between the body rates ω_x, ω_y, and ω_z and the Euler angle rates ω_p, ω_n, and ω_s, we must express $\hat{\mathbf{K}}$ and $\hat{\mathbf{i}}_1$ in terms of the unit vectors $\hat{\mathbf{i}}\hat{\mathbf{j}}\hat{\mathbf{k}}$ of the body-fixed frame. To accomplish that, we proceed as follows.

The first rotation $[\mathbf{R}_3(\phi)]$ in Eq. (11.102) rotates the unit vectors $\hat{\mathbf{I}}\hat{\mathbf{J}}\hat{\mathbf{K}}$ of the inertial frame into the unit vectors $\hat{\mathbf{i}}_1\hat{\mathbf{j}}_1\hat{\mathbf{k}}_1$ of the intermediate $x_1y_1z_1$ axes in Fig. 11.23. Hence $\hat{\mathbf{i}}_1\hat{\mathbf{j}}_1\hat{\mathbf{k}}_1$ are rotated into $\hat{\mathbf{I}}\hat{\mathbf{J}}\hat{\mathbf{K}}$ by the inverse transformation given by

$$\begin{Bmatrix} \hat{\mathbf{I}} \\ \hat{\mathbf{J}} \\ \hat{\mathbf{K}} \end{Bmatrix} = [\mathbf{R}_3(\phi)]^T \begin{Bmatrix} \hat{\mathbf{i}}_1 \\ \hat{\mathbf{j}}_1 \\ \hat{\mathbf{k}}_1 \end{Bmatrix} = \begin{bmatrix} \cos\phi & -\sin\phi & 0 \\ \sin\phi & \cos\phi & 0 \\ 0 & 0 & 1 \end{bmatrix} \begin{Bmatrix} \hat{\mathbf{i}}_1 \\ \hat{\mathbf{j}}_1 \\ \hat{\mathbf{k}}_1 \end{Bmatrix} \tag{11.109}$$

The second rotation $[\mathbf{R}_1(\theta)]$ rotates $\hat{\mathbf{i}}_1\hat{\mathbf{j}}_1\hat{\mathbf{k}}_1$ into the unit vectors $\hat{\mathbf{i}}_2\hat{\mathbf{j}}_2\hat{\mathbf{k}}_2$ of the second intermediate frame $x_2y_2z_2$ in Fig. 11.23. The inverse transformation rotates $\hat{\mathbf{i}}_2\hat{\mathbf{j}}_2\hat{\mathbf{k}}_2$ back into $\hat{\mathbf{i}}_1\hat{\mathbf{j}}_1\hat{\mathbf{k}}_1$:

$$\begin{Bmatrix} \hat{\mathbf{i}}_1 \\ \hat{\mathbf{j}}_1 \\ \hat{\mathbf{k}}_1 \end{Bmatrix} = [\mathbf{R}_1(\theta)]^T \begin{Bmatrix} \hat{\mathbf{i}}_2 \\ \hat{\mathbf{j}}_2 \\ \hat{\mathbf{k}}_2 \end{Bmatrix} = \begin{bmatrix} 1 & 0 & 0 \\ 0 & \cos\theta & -\sin\theta \\ 0 & \sin\theta & \cos\theta \end{bmatrix} \begin{Bmatrix} \hat{\mathbf{i}}_2 \\ \hat{\mathbf{j}}_2 \\ \hat{\mathbf{k}}_2 \end{Bmatrix} \tag{11.110}$$

Finally, the third rotation $[\mathbf{R}_3(\psi)]$ rotates $\hat{\mathbf{i}}_2\hat{\mathbf{j}}_2\hat{\mathbf{k}}_2$ into $\hat{\mathbf{i}}\hat{\mathbf{j}}\hat{\mathbf{k}}$, the target unit vectors of the body-fixed xyz frame. $\hat{\mathbf{i}}_2\hat{\mathbf{j}}_2\hat{\mathbf{k}}_2$ are obtained from $\hat{\mathbf{i}}\hat{\mathbf{j}}\hat{\mathbf{k}}$ by the reverse rotation,

$$\begin{Bmatrix} \hat{\mathbf{i}}_2 \\ \hat{\mathbf{j}}_2 \\ \hat{\mathbf{k}}_2 \end{Bmatrix} = [\mathbf{R}_3(\psi)]^T \begin{Bmatrix} \hat{\mathbf{i}} \\ \hat{\mathbf{j}} \\ \hat{\mathbf{k}} \end{Bmatrix} = \begin{bmatrix} \cos\psi & -\sin\psi & 0 \\ \sin\psi & \cos\psi & 0 \\ 0 & 0 & 1 \end{bmatrix} \begin{Bmatrix} \hat{\mathbf{i}} \\ \hat{\mathbf{j}} \\ \hat{\mathbf{k}} \end{Bmatrix} \tag{11.111}$$

From Eqs. (11.109) through (11.111), we observe that

$$\hat{\mathbf{K}} \overset{11.109}{\widehat{=}} \hat{\mathbf{k}}_1 \overset{11.110}{\widehat{=}} \sin\theta\hat{\mathbf{j}}_2 + \cos\theta\hat{\mathbf{k}}_2 \overset{11.111}{\widehat{=}} \sin\theta\left(\sin\psi\hat{\mathbf{i}} + \cos\psi\hat{\mathbf{j}}\right) + \cos\theta\hat{\mathbf{k}}$$

or

$$\hat{\mathbf{K}} = \sin\theta\sin\psi\hat{\mathbf{i}} + \sin\theta\cos\psi\hat{\mathbf{j}} + \cos\theta\hat{\mathbf{k}} \tag{11.112}$$

Similarly, Eqs. (11.110) and (11.111) imply that

$$\hat{\mathbf{i}}_1 = \hat{\mathbf{i}}_2 = \cos\psi\hat{\mathbf{i}} - \sin\psi\hat{\mathbf{j}} \tag{11.113}$$

Substituting Eqs. (11.112) and (11.113) into Eq. (11.108) yields

$$\boldsymbol{\omega} = \omega_p\left(\sin\theta\sin\psi\hat{\mathbf{i}} + \sin\theta\cos\psi\hat{\mathbf{j}} + \cos\theta\hat{\mathbf{k}}\right) + \omega_n\left(\cos\psi\hat{\mathbf{i}} - \sin\psi\hat{\mathbf{j}}\right) + \omega_s\hat{\mathbf{k}}$$

or

$$\boldsymbol{\omega} = (\omega_p\sin\theta\sin\psi + \omega_n\cos\psi)\hat{\mathbf{i}} + (\omega_p\sin\theta\cos\psi - \omega_n\sin\psi)\hat{\mathbf{j}} + (\omega_s + \omega_p\cos\theta)\hat{\mathbf{k}} \tag{11.114}$$

Comparing the coefficients of $\hat{\mathbf{i}}\hat{\mathbf{j}}\hat{\mathbf{k}}$ in this equation with those in Eqs. (11.107), we see that

$$\begin{aligned}\omega_x &= \omega_p \sin\theta \sin\psi + \omega_n \cos\psi \\ \omega_y &= \omega_p \sin\theta \cos\psi - \omega_n \sin\psi \\ \omega_z &= \omega_s + \omega_p \cos\theta\end{aligned} \tag{11.115a}$$

or

$$\begin{Bmatrix}\omega_x \\ \omega_y \\ \omega_z\end{Bmatrix} = \begin{bmatrix}\sin\theta\sin\psi & \cos\psi & 0 \\ \sin\theta\sin\psi & -\sin\psi & 0 \\ \cos\theta & & 1\end{bmatrix}\begin{Bmatrix}\omega_p \\ \omega_n \\ \omega_s\end{Bmatrix} \tag{11.115b}$$

(Note that the precession angle ϕ does not appear.) We solve these three equations to obtain the Euler rates in terms of ω_x, ω_y, and ω_z:

$$\begin{Bmatrix}\omega_p \\ \omega_n \\ \omega_s\end{Bmatrix} = \begin{bmatrix}\sin\psi/\sin\theta & \cos\psi/\sin\theta & 0 \\ \cos\psi & -\sin\psi & 0 \\ -\sin\psi/\tan\theta & -\cos\psi/\tan\theta & 1\end{bmatrix}\begin{Bmatrix}\omega_x \\ \omega_y \\ \omega_z\end{Bmatrix} \tag{11.116a}$$

or

$$\begin{aligned}\omega_p &= \dot{\phi} = \frac{1}{\sin\theta}(\omega_x \sin\psi + \omega_y \cos\psi) \\ \omega_n &= \dot{\theta} = \omega_x \cos\psi - \omega_y \sin\psi \\ \omega_s &= \dot{\psi} = -\frac{1}{\tan\theta}(\omega_x \sin\psi + \omega_y \cos\psi) + \omega_z\end{aligned} \tag{11.116b}$$

Observe that if ω_x, ω_y, and ω_z are given functions of time, found by solving the Euler equations of motion (Eq. 11.72), then Eqs. (11.116b) are three coupled differential equations that may be solved to obtain the three time-dependent Euler angles, namely

$$\phi = \phi(t) \quad \theta = \theta(t) \quad \psi = \psi(t)$$

With this solution, the orientation of the xyz frame, and hence the body to which it is attached, is known for any given time t. Note, however, that Eqs. (11.116) "blow up" when $\theta = 0$ (i.e., when the xy plane is parallel to the XY plane).

EXAMPLE 11.18

At a given instant, the unit vectors of a body frame are

$$\begin{aligned}\hat{\mathbf{i}} &= 0.40825\hat{\mathbf{I}} - 0.40825\hat{\mathbf{J}} + 0.81649\hat{\mathbf{K}} \\ \hat{\mathbf{j}} &= -0.10102\hat{\mathbf{I}} - 0.90914\hat{\mathbf{J}} - 0.40405\hat{\mathbf{K}} \\ \hat{\mathbf{k}} &= 0.90726\hat{\mathbf{I}} + 0.082479\hat{\mathbf{J}} - 0.41240\hat{\mathbf{K}}\end{aligned} \tag{a}$$

and the angular velocity is

$$\boldsymbol{\omega} = -3.1\hat{\mathbf{I}} + 2.5\hat{\mathbf{J}} + 1.7\hat{\mathbf{K}}\,(\text{rad/s}) \tag{b}$$

Calculate ω_p, ω_n, and ω_s (the precession, nutation, and spin rates) at this instant.

Solution

We will ultimately use Eq. (11.116) to find ω_p, ω_n, and ω_s. To do so we must first obtain the Euler angles ϕ, θ, and ψ as well as the components of the angular velocity in the body frame.

The three rows of the direction cosine matrix $[\mathbf{Q}]_{Xx}$ comprise the components of the given unit vectors $\hat{\mathbf{i}}$, $\hat{\mathbf{j}}$, and $\hat{\mathbf{k}}$, respectively,

$$[\mathbf{Q}]_{Xx} = \begin{bmatrix} 0.40825 & -0.40825 & 0.81649 \\ -0.10102 & -0.90914 & -0.40405 \\ 0.90726 & 0.082479 & -0.41240 \end{bmatrix} \tag{c}$$

Therefore, the components of the angular velocity in the body frame are

$$\{\boldsymbol{\omega}\}_x = [\mathbf{Q}]_{Xx}\{\boldsymbol{\omega}\}_X = \begin{bmatrix} 0.40825 & -0.40825 & 0.81649 \\ -0.10102 & -0.90914 & -0.40405 \\ 0.90726 & 0.082479 & -0.41240 \end{bmatrix} \begin{Bmatrix} -3.1 \\ 2.5 \\ 1.7 \end{Bmatrix} = \begin{Bmatrix} -0.89817 \\ -2.6466 \\ -3.3074 \end{Bmatrix}$$

or

$$\omega_x = -0.89817\,\text{rad/s} \quad \omega_y = -2.6466\,\text{rad/s} \quad \omega_z = -3.3074\,\text{rad/s} \tag{d}$$

To obtain the Euler angles ϕ, θ, and ψ from the direction cosine matrix in Eq. (c), we use Algorithm 4.3, as was illustrated in Example 11.17. That algorithm is implemented as the MATLAB function *dcm_to_Euler.m* in Appendix D.20. Typing the following lines in the MATLAB Command Window:

```
Q = [ .40825  -.40825    .81649
     -.10102  -.90914   -.40405
      .90726   .082479  -.41240];
[phi theta psi] = dcm_to_euler(Q)
```

produces the following output:

```
phi =
      95.1945
theta =
      114.3557
psi =
      116.3291
```

Substituting $\theta = 114.36°$ and $\psi = 116.33°$ together with the angular velocities of Eq. (d) into Eqs. (11.116) yields

$$\begin{aligned}\omega_p &= \frac{1}{\sin 114.36°}[-0.89817 \cdot \sin 116.33° + (-2.6466) \cdot \cos 116.33°] \\ &= \boxed{0.40492\,\text{rad/s}}\end{aligned}$$

$$\begin{aligned}\omega_n &= -0.89817 \cdot \cos 116.33° - (-2.6466) \cdot \sin 116.33° \\ &= \boxed{2.7704\,\text{rad/s}}\end{aligned}$$

$$\begin{aligned}\omega_s &= -\frac{1}{\tan 114.36°}[-0.89817 \cdot \sin 116.33° + (-2.6466) \cdot \cos 116.33°] + (-3.3074) \\ &= \boxed{-3.1404\,\text{rad/s}}\end{aligned}$$

EXAMPLE 11.19

The mass moments of inertia of a body about the principal body frame axes with origin at the center of mass G are

$$A = 1000\ \text{kg}\cdot\text{m}^2 \quad B = 2000\ \text{kg}\cdot\text{m}^2 \quad C = 3000\ \text{kg}\cdot\text{m}^2 \tag{a}$$

The Euler angles in radians are given as functions of time in seconds as follows:

$$\begin{aligned}\phi &= 2te^{-0.05t}\\ \theta &= 0.02 + 0.3\sin 0.25t\\ \psi &= 0.6t\end{aligned} \tag{b}$$

At $t = 10$ s, find
(a) the net moment about G and
(b) the components α_X, α_Y, and α_Z of the absolute angular acceleration in the inertial frame.

Solution

(a) We must use Euler equations (Eq. 11.72) to calculate the net moment, which means we must first obtain $\omega_x, \omega_y, \omega_z, \dot{\omega}_x, \dot{\omega}_y$, and $\dot{\omega}_z$. Since we are given the Euler angles as functions of time, we can compute their time derivatives and then use Eq. (11.115) to find the body frame angular velocity components and their derivatives. Starting with the first of Eqs. (b), we get

$$\omega_p = \frac{d\phi}{dt} = \frac{d}{dt}\left(2te^{-0.05t}\right) = 2e^{-0.05t} - 0.1te^{-0.05t}$$

$$\dot{\omega}_p = \frac{d\omega_p}{dt} = \frac{d}{dt}\left(2e^{-0.05t} - 0.1e^{-0.05t}\right) = -0.2e^{-0.05t} + 0.005te^{-0.05t}$$

Proceeding to the remaining two Euler angles leads to

$$\omega_n = \frac{d\theta}{dt} = \frac{d}{dt}(0.02 + 0.3\sin 0.25t) = 0.075\cos 0.25t$$

$$\dot{\omega}_n = \frac{d\omega_n}{dt} = \frac{d}{dt}(0.075\cos 0.25t) = -0.01875\sin 0.25t$$

$$\omega_s = \frac{d\psi}{dt} = \frac{d}{dt}(0.6t) = 0.6$$

$$\dot{\omega}_s = \frac{d\omega_s}{dt} = 0$$

Evaluating all these quantities, including those in Eqs. (b), at $t = 10$ s yields

$$\begin{aligned}&\phi = 335.03^\circ \quad \omega_p = 0.60653\,\text{rad/s} \quad \dot{\omega}_p = -0.09098\,\text{rad/s}^2\\ &\theta = 11.433^\circ \quad \omega_n = -0.060086\,\text{rad/s} \quad \dot{\omega} = -0.011221\,\text{rad/s}^2\\ &\psi = 343.77 \quad \omega_s = 0.6\,\text{rad/s} \quad \dot{\omega}_s = 0\end{aligned} \tag{c}$$

Eq. (11.115) relates the Euler angle rates to the angular velocity components,

$$\begin{aligned}\omega_x &= \omega_p\sin\theta\sin\psi + \omega_n\cos\psi\\ \omega_y &= \omega_p\sin\theta\cos\psi - \omega_n\sin\psi\\ \omega_z &= \omega_s + \omega_p\cos\theta\end{aligned} \tag{d}$$

Taking the time derivative of each of these equations in turn leads to the following three equations:

$$\begin{aligned}\dot{\omega}_x &= \omega_p\omega_n\cos\theta\sin\psi + \omega_p\omega_s\sin\theta\cos\psi - \omega_n\omega_s\sin\psi + \dot{\omega}_p\sin\theta\sin\psi + \dot{\omega}_n\cos\psi\\ \dot{\omega}_y &= \omega_p\omega_n\cos\theta\cos\psi - \omega_p\omega_s\sin\theta\sin\psi - \omega_n\omega_s\cos\psi + \dot{\omega}_p\sin\theta\cos\psi - \dot{\omega}_n\sin\psi\\ \dot{\omega}_z &= -\omega_p\omega_n\sin\theta + \dot{\omega}_p\cos\theta + \dot{\omega}_s\end{aligned} \tag{e}$$

Substituting the data in Eqs. (c) into Eqs. (d) and (e) yields

$$\begin{aligned} &\omega_x = -0.091286\,\text{rad/s} \quad \omega_y = 0.098649\,\text{rad/s} \qquad \omega_z = 1.1945\,\text{rad/s} \\ &\dot{\omega}_x = 0.063435\,\text{rad/s}^2 \quad \dot{\omega}_y = 2.2346\left(10^{-5}\right)\text{rad/s}^2 \quad \dot{\omega}_z = -0.08195\,\text{rad/s}^2 \end{aligned} \tag{f}$$

With Eqs. (a) and (f) we have everything we need for the Euler equations, namely,

$$\begin{aligned} M_x)_{\text{net}} &= A\dot{\omega}_x + (C-B)\omega_y\omega_z \\ M_y)_{\text{net}} &= B\dot{\omega}_y + (A-C)\omega_z\omega_x \\ M_z)_{\text{net}} &= C\dot{\omega}_z + (B-A)\omega_x\omega_y \end{aligned}$$

from which we find

$$\boxed{\begin{aligned} M_x)_{\text{net}} &= 181.27\ \text{N}\cdot\text{m} \\ M_y)_{\text{net}} &= 218.12\ \text{N}\cdot\text{m} \\ M_z)_{\text{net}} &= -254.86\ \text{N}\cdot\text{m} \end{aligned}}$$

(b) Since the comoving *xyz* frame is a body frame, rigidly attached to the solid, we know from Eq. (11.74) that

$$\begin{Bmatrix} \alpha_X \\ \alpha_Y \\ \alpha_Z \end{Bmatrix} = [\mathbf{Q}]_{xX} \begin{Bmatrix} \dot{\omega}_x \\ \dot{\omega}_y \\ \dot{\omega}_z \end{Bmatrix} \tag{g}$$

In other words, the absolute angular acceleration and the relative angular acceleration of the body are the same. All we have to do is project the components of relative acceleration in Eqs. (f), onto the axes of the inertial frame. The required orthogonal transformation matrix is given in Eq. (11.105),

$$[\mathbf{Q}]_{xX} = \begin{bmatrix} -\sin\phi\cos\theta\sin\psi + \cos\phi\cos\psi & -\sin\phi\cos\theta\cos\gamma - \cos\phi\sin\psi & \sin\phi\sin\theta \\ \cos\phi\cos\theta\sin\psi + \sin\phi\cos\psi & \cos\phi\cos\theta\cos\psi - \sin\phi\sin\psi & -\cos\phi\sin\theta \\ \sin\theta\sin\psi & \sin\theta\cos\psi & \cos\theta \end{bmatrix}$$

Upon substituting the numerical values of the Euler angles from Eqs. (c), this becomes

$$[\mathbf{Q}]_{xX} = \begin{bmatrix} 0.75484 & 0.65055 & -0.083668 \\ -0.65356 & 0.73523 & -0.17970 \\ -0.055386 & 0.19033 & 0.98016 \end{bmatrix}$$

Substituting this and the relative angular velocity rates from Eqs. (f) into Eq. (g) yields

$$\begin{aligned} \begin{Bmatrix} \alpha_X \\ \alpha_Y \\ \alpha_Z \end{Bmatrix} &= \begin{bmatrix} 0.75484 & 0.65055 & -0.083668 \\ -0.65356 & 0.73523 & -0.17970 \\ -0.055386 & 0.19033 & 0.98016 \end{bmatrix} \begin{Bmatrix} 0.063435 \\ 2.2345\left(10^{-5}\right) \\ -0.08195 \end{Bmatrix} \\ &= \boxed{\begin{Bmatrix} 0.054755 \\ -0.026716 \\ -0.083833 \end{Bmatrix} (\text{rad/s}^2)} \end{aligned}$$

EXAMPLE 11.20

Fig. 11.24 shows a rotating platform on which is mounted a rectangular parallelepiped shaft (with dimensions b, h, and ℓ) spinning about the inclined axis *DE*. If the mass of the shaft is m, and the angular velocities ω_p and ω_s are constant, calculate the bearing forces at D and E as a function of ϕ and ψ. Neglect gravity, since we are interested only in the gyroscopic forces. (The small extensions shown at each end of the parallelepiped are just for clarity; the distance between the bearings at D and E is ℓ.)

Solution

The inertial *XYZ* frame is centered at O on the platform, and it is right handed $\left(\hat{\mathbf{I}} \times \hat{\mathbf{J}} = \hat{\mathbf{K}}\right)$. The origin of the right-handed comoving body frame *xyz* is at the shaft's center of mass G, and it is aligned with the symmetry axes of the parallelepiped.

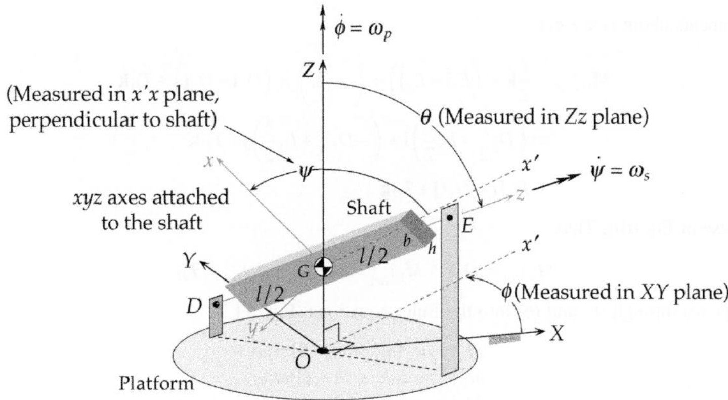

FIG. 11.24

Spinning block mounted on rotating platform.

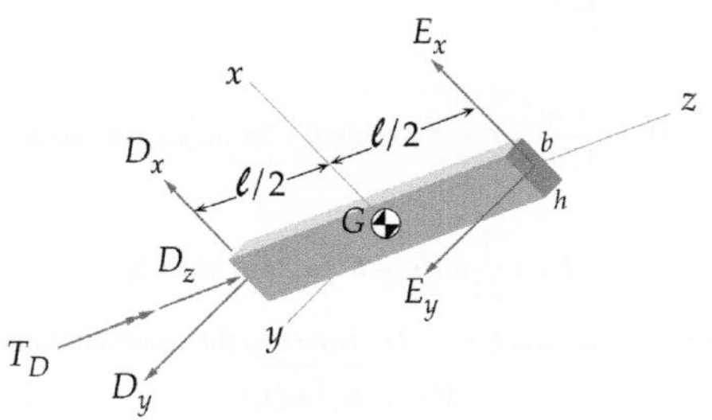

FIG. 11.25

Free body diagram of the block in Fig. 11.24.

The three Euler angles ϕ, θ, and ψ are shown in Fig. 11.24. Since θ is constant, the nutation rate is zero ($\omega_n = 0$). Thus, Eqs. (11.115) reduce to

$$\omega_x = \omega_p \sin\theta \sin\psi \quad \omega_y = \omega_p \sin\theta \cos\psi \quad \omega_z = \omega_p \cos\theta + \omega_s \tag{a}$$

Since ω_p, ω_s, and θ are constant, it follows (recalling Eq. 11.106) that

$$\dot{\omega}_x = \omega_p \omega_s \sin\theta \cos\psi \quad \dot{\omega}_y = -\omega_p \omega_s \sin\theta \sin\psi \quad \dot{\omega}_z = 0 \tag{b}$$

The principal moments of inertia of the parallelepiped are (Fig. 11.10c)

$$A = I_x = \frac{1}{12} m\left(h^2 + \ell^2\right) \quad B = I_y = \frac{1}{12} m\left(b^2 + l^2\right) \quad C = I_z = \frac{1}{12} m\left(b^2 + h^2\right) \tag{c}$$

Fig. 11.25 is a free body diagram of the shaft. Let us assume that the bearings at D and E are such as to exert just the six body frame components of force shown. Thus, D is a thrust bearing to which the axial torque T_D is applied from, say, a motor of some kind. At E, there is a simple journal bearing.

From Newton's laws of motion, we have $\mathbf{F}_{\text{net}} = m\mathbf{a}_G$. But G is fixed in inertial space, so $\mathbf{a}_G = \mathbf{0}$. Thus,

$$\left(D_x\hat{\mathbf{i}} + D_y\hat{\mathbf{j}} + D_z\hat{\mathbf{k}}\right) + \left(E_x\hat{\mathbf{i}} + E_y\hat{\mathbf{j}}\right) = 0$$

It follows that

$$E_x = -D_x \quad E_y = -D_y \quad D_z = 0 \tag{d}$$

Summing moments about G we get

$$\begin{aligned}\mathbf{M}_G)_{\text{net}} &= \frac{\ell}{2}\hat{\mathbf{k}} \times \left(E_x\hat{\mathbf{i}} + E_y\hat{\mathbf{j}}\right) + \left(-\frac{\ell}{2}\hat{\mathbf{k}}\right) \times \left(D_x\hat{\mathbf{i}} + D_y\hat{\mathbf{j}}\right) + T_D\hat{\mathbf{k}} \\ &= \left(D_y\frac{\ell}{2} - E_y\frac{\ell}{2}\right)\hat{\mathbf{i}} + \left(-D_x\frac{\ell}{2} + E_x\frac{\ell}{2}\right)\hat{\mathbf{j}} + T_D\hat{\mathbf{k}} \\ &= D_y\ell\hat{\mathbf{i}} - D_x\ell\hat{\mathbf{j}} + T_D\hat{\mathbf{k}}\end{aligned}$$

where we made use of Eq. (d)$_2$. Thus,

$$M_x)_{\text{net}} = D_y\ell \quad M_y)_{\text{net}} = -D_x\ell \quad M_z)_{\text{net}} = T_D \tag{e}$$

We substitute Eqs. (a) through (c) and (e) into the Euler equations (Eqs. 11.72):

$$\begin{aligned}M_x)_{\text{net}} &= A\dot{\omega}_x + (C-B)\omega_y\omega_z \\ M_y)_{\text{net}} &= B\dot{\omega}_y + (A-C)\omega_x\omega_z \\ M_z)_{\text{net}} &= C\dot{\omega}_z + (B-A)\omega_x\omega_y\end{aligned} \tag{f}$$

After making the substitutions and simplifying, the first Euler equation, Eq. (f)$_1$, becomes

$$D_y = \frac{1}{12}\frac{m}{\ell}\left[\left(h^2 - l^2\right)\omega_p\cos\theta + 2h^2\omega_s\right]\omega_p\sin\theta\cos\psi \tag{g}$$

Likewise, from Eq. (f)$_2$ we obtain

$$D_x = \frac{1}{12}\frac{m}{\ell}\left[\left(b^2 - l^2\right)\omega_p\cos\theta + 2b^2\omega_s\right]\omega_p\sin\theta\sin\psi \tag{h}$$

Finally, Eq. (f)$_3$ yields

$$T_D = \frac{1}{24}m\left(b^2 - h^2\right)\omega_p{}^2\sin^2\theta\sin 2\psi \tag{i}$$

This completes the solution, since $E_y = -D_y$ and $E_z = -D_z$. Note that the resultant transverse bearing load V at D (and E) is

$$V = \sqrt{D_x{}^2 + D_y{}^2} \tag{j}$$

As a numerical example, let

$$\ell = 1\text{m} \quad h = 0.1\text{m} \quad b = 0.025\text{m} \quad \theta = 30° \quad m = 10\text{kg}$$

and

$$\omega_p = 100\text{rpm} = 10.47\text{rad/s} \quad \omega_s = 2000\text{rpm} = 209.4\text{rad/s}$$

For these numbers, the variation of V and T_D with ψ are as shown in Fig. 11.26.

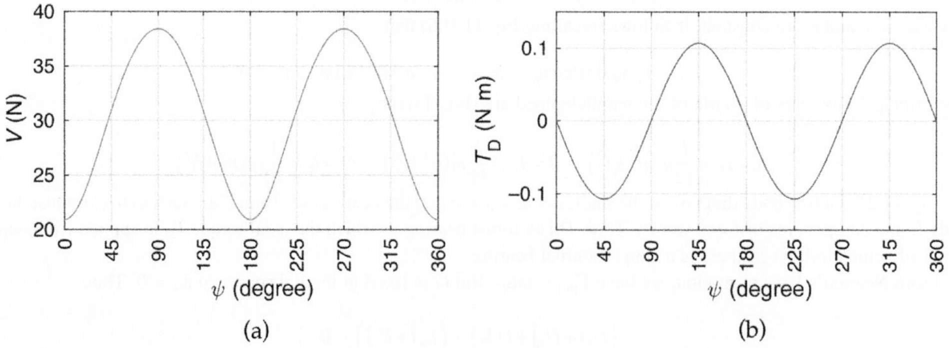

FIG. 11.26

(A) Transverse bearing load. (B) Axial torque at D.

11.10 YAW, PITCH, AND ROLL ANGLES

The problem of the Euler angle relations (Eq. 11.116) becoming singular when the nutation angle θ is zero can be alleviated by using the yaw, pitch, and roll angles discussed in Section 4.5. As in the classical Euler sequence, the yaw–pitch–roll sequence rotates the inertial *XYZ* axes into the triad of body-fixed *xyz* axes by means of a series of three elementary rotations, as illustrated in Fig. 11.27. Like the classical Euler sequence, the first rotation is around the Z $(=z_1)$ axis through the yaw angle ϕ. This takes X into x_1 and Y into y_1. The second rotation is around the y_2 $(=y_1)$ axis through the pitch angle θ. This carries x_1 and z_1 into x_2 and z_2, respectively. The third and final rotation is around the x $(=x_2)$ axis through the roll angle ψ, which takes y_2 into y and z_2 into z.

Eq. (4.40) gives the matrix $[\mathbf{Q}]_{Xx}$ of the transformation from the inertial frame into the body-fixed frame,

$$[\mathbf{Q}]_{Xx} = [\mathbf{R}_1(\psi)][\mathbf{R}_2(\theta)][\mathbf{R}_3(\phi)] \tag{11.117}$$

From Eqs. (4.32) through (4.34), the elementary rotation matrices are

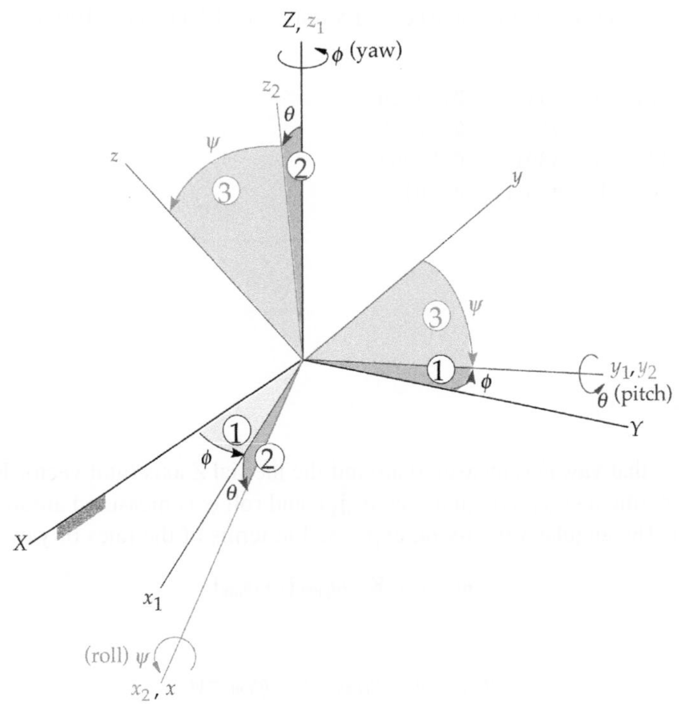

FIG. 11.27

Yaw, pitch, and roll sequence (see also fig. 4.15).

$$[\mathbf{R}_1(\psi)] = \begin{bmatrix} 1 & 0 & 0 \\ 0 & \cos\psi & \sin\psi \\ 0 & -\sin\psi & \cos\psi \end{bmatrix} \quad [\mathbf{R}_2(\theta)] = \begin{bmatrix} \cos\theta & 0 & -\sin\theta \\ 0 & 1 & 0 \\ \sin\theta & 0 & \cos\theta \end{bmatrix}$$
$$[\mathbf{R}_3(\phi)] = \begin{bmatrix} \cos\phi & \sin\phi & 0 \\ -\sin\phi & \cos\phi & 0 \\ 0 & 0 & 1 \end{bmatrix} \tag{11.118}$$

According to Eq. (4.41), the multiplication on the right of Eq. (11.117) yields the following direction cosine matrix for the yaw, pitch, and roll sequence:

$$[\mathbf{Q}]_{Xx} = \begin{bmatrix} \cos\phi\cos\theta & \sin\phi\cos\theta & -\sin\theta \\ \cos\phi\sin\theta\sin\psi - \sin\phi\cos\psi & \sin\phi\sin\theta\sin\psi + \cos\phi\cos\psi & \cos\theta\sin\psi \\ \cos\phi\sin\theta\cos\psi + \sin\phi\sin\psi & \sin\phi\sin\theta\cos\psi - \cos\phi\sin\psi & \cos\theta\cos\psi \end{bmatrix} \tag{11.119}$$

The inverse matrix $[\mathbf{Q}]_{xX}$, which transforms *xyz* into *XYZ*, is just the transpose

$$[\mathbf{Q}]_{xX} = \begin{bmatrix} \cos\phi\cos\theta & \cos\phi\sin\theta\sin\psi - \sin\phi\cos\psi & \cos\phi\sin\theta\cos\psi + \sin\phi\sin\psi \\ \sin\phi\cos\theta & \sin\phi\sin\theta\sin\psi + \cos\phi\cos\psi & \sin\phi\sin\theta\sin\psi - \cos\phi\sin\psi \\ -\sin\theta & \cos\theta\sin\psi & \cos\theta\cos\psi \end{bmatrix} \tag{11.120}$$

Algorithm 4.4 (*dcm_to_ypr.m* in Appendix D.21) is used to determine the yaw, pitch, and roll angles for a given direction cosine matrix. The following brief MATLAB session reveals that the yaw, pitch, and roll angles for the direction cosine matrix in Example 11.17 are $\phi = 109.69°$, $\theta = 17.230°$, and $\psi = 238.43°$.

```
Q = [-0.32175   0.89930   -0.29620
      0.57791  -0.061275  -0.81380
     -0.75000  -0.43301   -0.5000];
[yaw pitch roll] = dcm_to_ypr(Q)

yaw =
      109.6861
pitch =
      17.2295
roll =
      238.4334
```

Fig. 11.27 shows that yaw ϕ is measured around the inertial Z axis (unit vector $\hat{\mathbf{K}}$), pitch θ is measured around the intermediate y_1 axis (unit vector $\hat{\mathbf{j}}_1$), and roll ψ is measured around the body-fixed x axis (unit vector $\hat{\mathbf{i}}$). The angular velocity $\boldsymbol{\omega}$, expressed in terms of the rates of yaw, pitch, and roll, is

$$\boldsymbol{\omega} = \omega_{\text{yaw}}\hat{\mathbf{K}} + \omega_{\text{pitch}}\hat{\mathbf{j}}_2 + \omega_{\text{roll}}\hat{\mathbf{i}} \tag{11.121}$$

in which

$$\omega_{\text{yaw}} = \dot{\phi} \quad \omega_{\text{pitch}} = \dot{\theta} \quad \omega_{\text{roll}} = \dot{\psi} \tag{11.122}$$

The first rotation $[\mathbf{R}_3(\phi)]$ in Eq. (11.117) rotates the unit vectors $\hat{\mathbf{I}}\hat{\mathbf{J}}\hat{\mathbf{K}}$ of the inertial frame into the unit vectors $\hat{\mathbf{i}}_1\hat{\mathbf{j}}_1\hat{\mathbf{k}}_1$ of the intermediate $x_1y_1z_1$ axes in Fig. 11.27. Thus, $\hat{\mathbf{i}}_1\hat{\mathbf{j}}_1\hat{\mathbf{k}}_1$ are rotated into $\hat{\mathbf{I}}\hat{\mathbf{J}}\hat{\mathbf{K}}$ by the inverse transformation

$$\begin{Bmatrix} \hat{\mathbf{I}} \\ \hat{\mathbf{J}} \\ \hat{\mathbf{K}} \end{Bmatrix} = \begin{bmatrix} \cos\phi & -\sin\phi & 0 \\ \sin\phi & \cos\phi & 0 \\ 0 & 0 & 1 \end{bmatrix} \begin{Bmatrix} \hat{\mathbf{i}}_1 \\ \hat{\mathbf{j}}_1 \\ \hat{\mathbf{k}}_1 \end{Bmatrix} \tag{11.123}$$

The second rotation $[\mathbf{R}_2(\theta)]$ rotates $\hat{\mathbf{i}}_1\hat{\mathbf{j}}_1\hat{\mathbf{k}}_1$ into the unit vectors $\hat{\mathbf{i}}_2\hat{\mathbf{j}}_2\hat{\mathbf{k}}_2$ of the second intermediate frame $x_2y_2z_2$ in Fig. 11.27. The inverse transformation rotates $\hat{\mathbf{i}}_2\hat{\mathbf{j}}_2\hat{\mathbf{k}}_2$ back into $\hat{\mathbf{i}}_1\hat{\mathbf{j}}_1\hat{\mathbf{k}}_1$:

$$\begin{Bmatrix} \hat{\mathbf{i}}_1 \\ \hat{\mathbf{j}}_1 \\ \hat{\mathbf{k}}_1 \end{Bmatrix} = \begin{bmatrix} \cos\theta & 0 & \sin\theta \\ 0 & 1 & 0 \\ -\sin\theta & 0 & \cos\theta \end{bmatrix} \begin{Bmatrix} \hat{\mathbf{i}}_2 \\ \hat{\mathbf{j}}_2 \\ \hat{\mathbf{k}}_2 \end{Bmatrix} \tag{11.124}$$

Lastly, the third rotation $[\mathbf{R}_1(\psi)]$ rotates $\hat{\mathbf{i}}_2\hat{\mathbf{j}}_2\hat{\mathbf{k}}_2$ into $\hat{\mathbf{i}}\hat{\mathbf{j}}\hat{\mathbf{k}}$, the unit vectors of the body-fixed xyz frame. $\hat{\mathbf{i}}_2\hat{\mathbf{j}}_2\hat{\mathbf{k}}_2$ are obtained from $\hat{\mathbf{i}}\hat{\mathbf{j}}\hat{\mathbf{k}}$ by the reverse transformation,

$$\begin{Bmatrix} \hat{\mathbf{i}}_2 \\ \hat{\mathbf{j}}_2 \\ \hat{\mathbf{k}}_2 \end{Bmatrix} = \begin{bmatrix} 1 & 0 & 0 \\ 0 & \cos\psi & -\sin\psi \\ 0 & \sin\psi & \cos\psi \end{bmatrix} \begin{Bmatrix} \hat{\mathbf{i}} \\ \hat{\mathbf{j}} \\ \hat{\mathbf{k}} \end{Bmatrix} \tag{11.125}$$

From Eqs. (11.123) through (11.125), we see that

$$\hat{\mathbf{K}} \overset{11.123}{\widehat{=}} \hat{\mathbf{k}}_1 \overset{11.124}{\widehat{=}} -\sin\theta\hat{\mathbf{i}}_2 + \cos\theta\hat{\mathbf{k}}_2 \overset{11.125}{\widehat{=}} -\sin\theta\hat{\mathbf{i}} + \cos\theta\left(\sin\psi\hat{\mathbf{j}} + \cos\psi\hat{\mathbf{k}}\right)$$

or

$$\hat{\mathbf{K}} = -\sin\theta\hat{\mathbf{i}} + \cos\theta\sin\psi\hat{\mathbf{j}} + \cos\theta\cos\psi\hat{\mathbf{k}} \tag{11.126}$$

From Eq. (11.125),

$$\hat{\mathbf{j}}_2 = \cos\psi\hat{\mathbf{j}} - \sin\psi\hat{\mathbf{k}} \tag{11.127}$$

Substituting Eqs. (11.126) and (11.127) into Eq. (11.121) yields

$$\boldsymbol{\omega} = \omega_{\text{yaw}}\left(-\sin\theta\hat{\mathbf{i}} + \cos\theta\sin\psi\hat{\mathbf{j}} + \cos\theta\cos\psi\hat{\mathbf{k}}\right) + \omega_{\text{pitch}}\left(\cos\psi\hat{\mathbf{j}} - \sin\psi\hat{\mathbf{k}}\right) + \omega_{\text{roll}}\hat{\mathbf{i}}$$

or

$$\begin{aligned} \boldsymbol{\omega} = &\left(-\omega_{\text{yaw}}\sin\theta + \omega_{\text{roll}}\right)\hat{\mathbf{i}} + \left(\omega_{\text{yaw}}\cos\theta\sin\psi + \omega_{\text{pitch}}\cos\psi\right)\hat{\mathbf{j}} \\ &+ \left(\omega_{\text{yaw}}\cos\theta\cos\psi - \omega_{\text{pitch}}\sin\psi\right)\hat{\mathbf{k}} \end{aligned} \tag{11.128}$$

Comparing the coefficients of $\hat{\mathbf{i}}\hat{\mathbf{j}}\hat{\mathbf{k}}$ in Eqs. (11.107) and (11.128), we conclude that the body angular velocities are related to the yaw, pitch, and roll rates as follows:

$$\begin{aligned} \omega_x &= \omega_{\text{roll}} - \omega_{\text{yaw}}\sin\theta_{\text{pitch}} \\ \omega_y &= \omega_{\text{yaw}}\cos\theta_{\text{pitch}}\sin\psi_{\text{roll}} + \omega_{\text{pitch}}\cos\psi_{\text{roll}} \\ \omega_z &= \omega_{\text{yaw}}\cos\theta_{\text{pitch}}\cos\psi_{\text{roll}} - \omega_{\text{pitch}}\sin\psi_{\text{roll}} \end{aligned} \tag{11.129a}$$

or

$$\begin{Bmatrix} \omega_x \\ \omega_y \\ \omega_z \end{Bmatrix} = \begin{bmatrix} -\sin\theta_{\text{pitch}} & 0 & 1 \\ \cos\theta_{\text{pitch}}\sin\psi_{\text{roll}} & \cos\psi_{\text{roll}} & 0 \\ \cos\theta_{\text{pitch}}\cos\psi_{\text{roll}} & -\sin\psi_{\text{roll}} & 0 \end{bmatrix} \begin{Bmatrix} \omega_{\text{yaw}} \\ \omega_{\text{pitch}} \\ \omega_{\text{roll}} \end{Bmatrix} \tag{11.129b}$$

wherein the subscript on each symbol helps us remember the physical rotation it describes. Note that ϕ_{yaw} does not appear. The inverse of these equations is

$$\begin{Bmatrix} \omega_{\text{yaw}} \\ \omega_{\text{pitch}} \\ \omega_{\text{roll}} \end{Bmatrix} = \begin{bmatrix} 0 & \sin\psi_{\text{roll}}/\cos\theta_{\text{pitch}} & \cos\psi_{\text{roll}}/\cos\theta_{\text{pitch}} \\ 0 & \cos\psi_{\text{roll}} & -\sin\psi_{\text{roll}} \\ 1 & \sin\psi_{\text{roll}}\tan\theta_{\text{pitch}} & \cos\psi_{\text{roll}}\tan\theta_{\text{pitch}} \end{bmatrix} \begin{Bmatrix} \omega_x \\ \omega_y \\ \omega_z \end{Bmatrix} \tag{11.130a}$$

or

$$\begin{aligned} \omega_{\text{yaw}} &= \frac{1}{\cos\theta_{\text{pitch}}}\left(\omega_y \sin\psi_{\text{roll}} + \omega_z \cos\psi_{\text{roll}}\right) \\ \omega_{\text{pitch}} &= \omega_y \cos\psi_{\text{roll}} - \omega_z \sin\psi_{\text{roll}} \\ \omega_{\text{roll}} &= \omega_x + \omega_y \tan\theta_{\text{pitch}} \sin\psi_{\text{roll}} + \omega_z \tan\theta_{\text{pitch}} \cos\psi_{\text{roll}} \end{aligned} \tag{11.130b}$$

Note that this system becomes singular ($\cos\theta_{\text{pitch}} = 0$) when the pitch angle is $\pm 90°$.

11.11 QUATERNIONS

In Chapter 4, we showed that the transformation from any Cartesian coordinate frame to another having the same origin can be accomplished by three Euler angle sequences, each being an elementary rotation about one of the three coordinate axes. We have focused on the commonly used classical Euler angle sequence $[\mathbf{R}_3(\gamma)][\mathbf{R}_1(\beta)][\mathbf{R}_3(\alpha)]$ and the yaw–pitch–roll sequence $[\mathbf{R}_1(\gamma)][\mathbf{R}_2(\beta)][\mathbf{R}_3(\alpha)]$.

Another of Euler's theorems, which we used in Section 1.6, states that any two Cartesian coordinate frames are related by a unique rotation about a single line through their common origin. This line is called the Euler axis and the angle is referred to as the principal angle.

Let $\hat{\mathbf{u}}$ be the unit vector along the Euler axis. A vector $\mathbf{v}$ can be resolved into orthogonal components $\mathbf{v}_\perp$ normal to $\hat{\mathbf{u}}$ and $\mathbf{v}_\parallel$ parallel to $\hat{\mathbf{u}}$, so that we may write

$$\mathbf{v} = \mathbf{v}_\parallel + \mathbf{v}_\perp \tag{11.131}$$

The component of $\mathbf{v}$ along $\hat{\mathbf{u}}$ is given by $\mathbf{v} \cdot \hat{\mathbf{u}}$. That is,

$$\mathbf{v}_\parallel = (\mathbf{v} \cdot \hat{\mathbf{u}})\hat{\mathbf{u}} \tag{11.132}$$

From Eqs. (11.131) and (11.132), we have

$$\mathbf{v}_\perp = \mathbf{v} - (\mathbf{v} \cdot \hat{\mathbf{u}})\hat{\mathbf{u}} \tag{11.133}$$

Let $\mathbf{v}'$ be the vector obtained by rotating $\mathbf{v}$ through an angle θ around $\hat{\mathbf{u}}$, as illustrated in Fig. 11.28. This rotation leaves the magnitude of $\mathbf{v}_\perp$ and its component along $\hat{\mathbf{u}}$ unchanged. That is

$$\|\mathbf{v}'_\perp\| = \|\mathbf{v}_\perp\| \tag{11.134}$$

$$\mathbf{v}'_\parallel = (\mathbf{v} \cdot \hat{\mathbf{u}})\hat{\mathbf{u}} \tag{11.135}$$

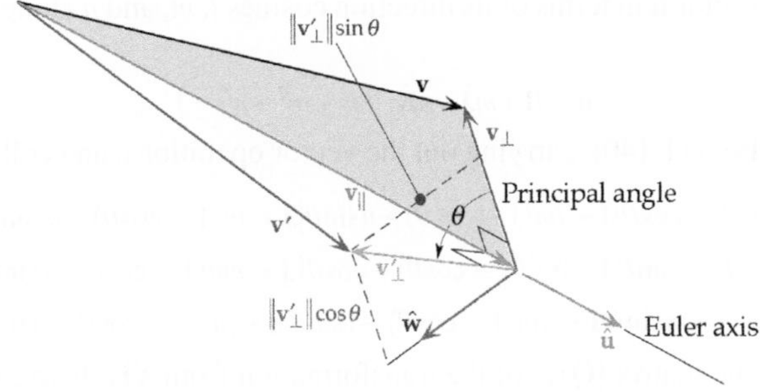

FIG. 11.28

Rotation of a vector through an angle θ about an axis with unit vector $\hat{\mathbf{u}}$.

$\mathbf{v}'_\perp$, having been rotated about $\hat{\mathbf{u}}$, has the component $\|\mathbf{v}'_\perp\| \cos\theta$ along the original vector $\mathbf{v}_\perp$ and the component $\|\mathbf{v}'_\perp\| \sin\theta$ along the vector normal to the plane of $\hat{\mathbf{u}}$ and $\mathbf{v}$. Let $\hat{\mathbf{w}}$ be the unit vector normal to that plane. Then,

$$\hat{\mathbf{w}} = \hat{\mathbf{u}} \times \frac{\mathbf{v}_\perp}{\|\mathbf{v}_\perp\|} \tag{11.136}$$

Thus,

$$\mathbf{v}'_\perp = \|\mathbf{v}'_\perp\| \cos\theta \frac{\mathbf{v}_\perp}{\|\mathbf{v}_\perp\|} + \|\mathbf{v}'_\perp\| \sin\theta \frac{\hat{\mathbf{u}} \times \mathbf{v}_\perp}{\|\mathbf{v}_\perp\|}$$

According to Eq. (11.134), this reduces to

$$\mathbf{v}'_\perp = \cos\theta \mathbf{v}_\perp + \sin\theta \hat{\mathbf{u}} \times \mathbf{v}_\perp \tag{11.137}$$

Observe that

$$\hat{\mathbf{u}} \times \mathbf{v}_\perp = \hat{\mathbf{u}} \times \left(\mathbf{v} - \mathbf{v}_\|\right) = \hat{\mathbf{u}} \times \mathbf{v}$$

since $\mathbf{v}_\|$ is parallel to $\hat{\mathbf{u}}$. This, together with Eq. (11.133), means we can write Eq. (11.137) as

$$\mathbf{v}'_\perp = \cos\theta[\mathbf{v} - (\mathbf{v} \cdot \hat{\mathbf{u}})\hat{\mathbf{u}}] + \sin\theta(\hat{\mathbf{u}} \times \mathbf{v}) \tag{11.138}$$

Since $\mathbf{v}' = \mathbf{v}'_\perp + \mathbf{v}'_\|$, we find, upon substituting Eqs. (11.135) and (11.138) and collecting terms, that

$$\mathbf{v}' = \cos\theta \mathbf{v} + (1 - \cos\theta)(\hat{\mathbf{u}} \cdot \mathbf{v})\hat{\mathbf{u}} + \sin\theta(\hat{\mathbf{u}} \times \mathbf{v}) \tag{11.139}$$

This is known as Rodrigues' *rotation* formula, named for the same French mathematician who gave us the Rodrigues' formula for Legendre polynomials (Eq. 10.22). Eq. (11.139) is useful for determining the result of rotating a vector about a line.

We can obtain the body-fixed *xyz* Cartesian frame from the inertial *XYZ* frame by a single rotation through the principal angle θ about the Euler axis $\hat{\mathbf{u}}$. The unit vectors $\hat{\mathbf{I}}\hat{\mathbf{J}}\hat{\mathbf{K}}$ are thereby rotated into $\hat{\mathbf{i}}\hat{\mathbf{j}}\hat{\mathbf{k}}$. The two sets of unit vectors are related by Eq. (11.139). Thus,

$$\begin{aligned}
\hat{\mathbf{i}} &= \cos\theta \hat{\mathbf{I}} + (1 - \cos\theta)\left(\hat{\mathbf{u}} \cdot \hat{\mathbf{I}}\right)\hat{\mathbf{u}} + \sin\theta \hat{\mathbf{u}} \times \hat{\mathbf{I}} \\
\hat{\mathbf{j}} &= \cos\theta \hat{\mathbf{J}} + (1 - \cos\theta)\left(\hat{\mathbf{u}} \cdot \hat{\mathbf{J}}\right)\hat{\mathbf{u}} + \sin\theta \hat{\mathbf{u}} \times \hat{\mathbf{J}} \\
\hat{\mathbf{k}} &= \cos\theta \hat{\mathbf{K}} + (1 - \cos\theta)\left(\hat{\mathbf{u}} \cdot \hat{\mathbf{K}}\right)\hat{\mathbf{u}} + \sin\theta \hat{\mathbf{u}} \times \hat{\mathbf{K}}
\end{aligned} \tag{11.140}$$

Let us express the unit vector $\hat{\mathbf{u}}$ in terms of its direction cosines l, m, and n along the original XYZ axes. That is

$$\hat{\mathbf{u}} = l\hat{\mathbf{I}} + m\hat{\mathbf{J}} + n\hat{\mathbf{K}} \quad l^2 + m^2 + n^2 = 1 \tag{11.141}$$

Substituting these into Eq. (11.140), carrying out the vector operations, and collecting the terms yields

$$\begin{aligned}
\hat{\mathbf{i}} &= \left[l^2(1-\cos\theta) + \cos\theta\right]\hat{\mathbf{I}} + [lm(1-\cos\theta) + n\sin\theta]\hat{\mathbf{J}} + [ln(1-\cos\theta) - m\sin\theta]\hat{\mathbf{K}} \\
\hat{\mathbf{j}} &= [lm(1-\cos\theta) - \sin\theta]\hat{\mathbf{I}} + \left[m^2(1-\cos\theta) + \cos\theta\right]\hat{\mathbf{J}} + [mn(1-\cos\theta) + l\sin\theta]\hat{\mathbf{K}} \\
\hat{\mathbf{k}} &= [ln(1-\cos\theta) + m\sin\theta]\hat{\mathbf{I}} + [mn(1-\cos\theta) - l\sin\theta]\hat{\mathbf{J}} + \left[n^2(1-\cos\theta) + \cos\theta\right]\hat{\mathbf{K}}
\end{aligned} \tag{11.142}$$

Recall that the rows of the matrix $[\mathbf{Q}]_{Xx}$ of the transformation from XYZ to xyz comprise the direction cosines of the unit vectors $\hat{\mathbf{i}}$, $\hat{\mathbf{j}}$, and $\hat{\mathbf{k}}$, respectively. That is,

$$[\mathbf{Q}]_{Xx} = \begin{bmatrix} l^2(1-\cos\theta) + \cos\theta & lm(1-\cos\theta) + n\sin\theta & ln(1-\cos\theta) - m\sin\theta \\ lm(1-\cos\theta) - n\sin\theta & m^2(1-\cos\theta) + \cos\theta & mn(1-\cos\theta) + l\sin\theta \\ ln(1-\cos\theta) + m\sin\theta & mn(1-\cos\theta) - l\sin\theta & n^2(1-\cos\theta) + \cos\theta \end{bmatrix} \tag{11.143}$$

The direction cosine matrix is thus expressed in terms of the Euler axis direction cosines and the principal angle.

Quaternions (also known as Euler symmetric parameters) were introduced in 1843 by the Irish mathematician Sir William R. Hamilton (1805–65). They provide an alternative to the use of direction cosine matrices for describing the orientation of a body frame in three-dimensional space. Quaternions can be used to avoid encountering the singularities we observed for the classical Euler angle sequence when the nutation angle θ becomes zero (Eqs. 11.116) or for the yaw, pitch, and roll sequence when the pitch angle θ approaches 90° (Eqs. (11.130)).

As the name implies, a quaternion $\widehat{\mathbf{q}}$ comprises four numbers:

$$\widehat{\mathbf{q}} = \begin{Bmatrix} q_1 \\ q_2 \\ q_3 \\ \hline q_4 \end{Bmatrix} = \begin{Bmatrix} \mathbf{q} \\ \hline q_4 \end{Bmatrix} \tag{11.144}$$

where $\mathbf{q}$ is called the vector part $\left(\mathbf{q} = q_1\hat{\mathbf{i}} + q_2\hat{\mathbf{j}} + q_3\hat{\mathbf{k}}\right)$, and q_4 is the scalar part. (It is common to see the scalar part denoted q_0 and listed first, in which case $\widehat{\mathbf{q}} = \begin{Bmatrix} q_0 \\ \hline \mathbf{q} \end{Bmatrix}$.) Regardless, a quaternion whose scalar part is zero is called a pure quaternion.

The norm $\|\widehat{\mathbf{q}}\|$ of the quaternion $\widehat{\mathbf{q}}$ is defined as

$$\|\widehat{\mathbf{q}}\| = \sqrt{\|\mathbf{q}\|^2 + q_4^2} = \sqrt{\mathbf{q}\cdot\mathbf{q} + q_4^2} \tag{11.145}$$

Obviously, the norm of a pure quaternion ($q_4 = 0$) is just the norm of its vector part. A unit quaternion is one whose norm is unity ($\|\widehat{\mathbf{q}}\| = 1$).

Quaternions obey the familiar vector rules of addition and scalar multiplication. That is,

$$\widehat{\mathbf{p}} + \widehat{\mathbf{q}} = \begin{Bmatrix} \mathbf{p} + \mathbf{q} \\ \hline p_4 + q_4 \end{Bmatrix} \qquad a\widehat{\mathbf{p}} = \begin{Bmatrix} a\mathbf{p} \\ \hline ap_4 \end{Bmatrix}$$

Addition is both associative and commutative, so that

$$(\widehat{\mathbf{p}}+\widehat{\mathbf{q}})+\widehat{\mathbf{r}}=\widehat{\mathbf{p}}+(\widehat{\mathbf{q}}+\widehat{\mathbf{r}}) \quad \widehat{\mathbf{p}}+\widehat{\mathbf{q}}=\widehat{\mathbf{q}}+\widehat{\mathbf{p}}$$

We use the special symbol ⊗ to denote the product or "composition" of two quaternions. The somewhat complicated rule for quaternion multiplication involves ordinary scalar multiplication as well as the familiar vector dot product and cross product operations,

$$\widehat{\mathbf{p}}\otimes\widehat{\mathbf{q}}=\left\{\begin{array}{c} p_4\mathbf{q}+q_4\mathbf{p}+\mathbf{p}\times\mathbf{q} \\ \hdashline p_4q_4-\mathbf{p}\cdot\mathbf{q} \end{array}\right\} \tag{11.146}$$

Switching the order of multiplication yields

$$\widehat{\mathbf{q}}\otimes\widehat{\mathbf{p}}=\left\{\begin{array}{c} q_4\mathbf{p}+p_4\mathbf{q}+\mathbf{q}\times\mathbf{p} \\ \hdashline q_4p_4-\mathbf{q}\cdot\mathbf{p} \end{array}\right\}$$

We are familiar with the fact that $\mathbf{q}\times\mathbf{p}=-(\mathbf{p}\times\mathbf{q})$, which means that quaternion multiplication is generally not commutative,

$$\widehat{\mathbf{p}}\otimes\widehat{\mathbf{q}}\neq\widehat{\mathbf{q}}\otimes\widehat{\mathbf{p}}$$

EXAMPLE 11.21

Find the product of the quaternions

$$\widehat{\mathbf{p}}=\left\{\begin{array}{c}\mathbf{p}\\ \hdashline p_4\end{array}\right\}=\left\{\begin{array}{c}\hat{\mathbf{j}}\\ \hdashline 1\end{array}\right\} \quad \widehat{\mathbf{q}}=\left\{\begin{array}{c}\mathbf{q}\\ \hdashline q_4\end{array}\right\}=\left\{\begin{array}{c}0.5\hat{\mathbf{i}}+0.5\hat{\mathbf{j}}+0.75\hat{\mathbf{k}}\\ \hdashline 1\end{array}\right\} \tag{a}$$

Solution

$$\begin{aligned}
\widehat{\mathbf{p}}\otimes\widehat{\mathbf{q}} &=\left\{\begin{array}{c} p_4\mathbf{q}+q_4\mathbf{p}+\mathbf{p}\times\mathbf{q} \\ \hdashline p_4q_4-\mathbf{p}\cdot\mathbf{q} \end{array}\right\} \\
&=\left\{\begin{array}{c} \overbrace{1\cdot(0.5\hat{\mathbf{i}}+0.5\hat{\mathbf{j}}+0.75\hat{\mathbf{k}})}^{0.5\hat{\mathbf{i}}+0.5\hat{\mathbf{j}}+0.75\hat{\mathbf{k}}}+\overbrace{1\cdot\hat{\mathbf{j}}}^{\hat{\mathbf{j}}}+\overbrace{\hat{\mathbf{j}}\times(0.5\hat{\mathbf{i}}+0.5\hat{\mathbf{j}}+0.75\hat{\mathbf{k}})}^{-0.5\hat{\mathbf{k}}+0.75\hat{\mathbf{i}}} \\ \hdashline \underbrace{1\cdot 1}_{1}-\underbrace{\hat{\mathbf{j}}\cdot(0.5\hat{\mathbf{i}}+0.5\hat{\mathbf{j}}+0.75\hat{\mathbf{k}})}_{0.5} \end{array}\right\} \\
&=\left\{\begin{array}{c} (0.5+0.75)\hat{\mathbf{i}}+(0.5+1.0)\hat{\mathbf{j}}+(0.75-0.5)\hat{\mathbf{k}} \\ \hdashline 0.5 \end{array}\right\}
\end{aligned}$$

or

$$\boxed{\widehat{\mathbf{p}}\otimes\widehat{\mathbf{q}} = \begin{Bmatrix} 1.25\hat{\mathbf{i}} + 1.5\hat{\mathbf{j}} + 0.25\hat{\mathbf{k}} \\ \hdashline 0.5 \end{Bmatrix}}$$

The conjugate $\widehat{\mathbf{q}}^*$ of a quaternion $\widehat{\mathbf{q}}$ is found by simply multiplying its vector part by -1, thereby changing the signs of its vector components:

$$\widehat{\mathbf{q}}^* = \begin{Bmatrix} -\mathbf{q} \\ \hdashline q_4 \end{Bmatrix} \tag{11.147}$$

The identity quaternion $\widehat{\mathbf{1}}$ has zero for its vector part and 1 for its scalar part,

$$\widehat{\mathbf{1}} = \begin{Bmatrix} \mathbf{0} \\ \hdashline 1 \end{Bmatrix} \tag{11.148}$$

The product of any quaternion with $\widehat{\mathbf{1}}$ is commutative and yields the original quaternion,

$$\widehat{\mathbf{q}}\otimes\widehat{\mathbf{1}} = \widehat{\mathbf{1}}\otimes\widehat{\mathbf{q}} = \begin{Bmatrix} q_4\cdot\mathbf{0} + 1\cdot\mathbf{q} + \mathbf{q}\times\mathbf{0} \\ \hdashline q_4\cdot 1 - \mathbf{q}\cdot\mathbf{0} \end{Bmatrix} = \begin{Bmatrix} \mathbf{q} \\ \hdashline q_4 \end{Bmatrix} = \widehat{\mathbf{q}}$$

Multiplication of a quaternion by its conjugate is also a commutative operation that yields a quaternion proportional to $\widehat{\mathbf{1}}$,

$$\widehat{\mathbf{q}}\otimes\widehat{\mathbf{q}}^* = \widehat{\mathbf{q}}^*\otimes\widehat{\mathbf{q}} = \begin{Bmatrix} q_4(-\mathbf{q}) + q_4\mathbf{q} + \mathbf{q}\times(-\mathbf{q}) \\ \hdashline q_4q_4 - \mathbf{q}\cdot(-\mathbf{q}) \end{Bmatrix} = \begin{Bmatrix} \mathbf{0} \\ \hdashline \|\widehat{\mathbf{q}}\|^2 \end{Bmatrix} = \|\widehat{\mathbf{q}}\|^2\widehat{\mathbf{1}} \tag{11.149}$$

The inverse $\widehat{\mathbf{q}}^{-1}$ of a quaternion is defined as

$$\widehat{\mathbf{q}}^{-1} = \frac{\widehat{\mathbf{q}}^*}{\|\widehat{\mathbf{q}}\|^2} \tag{11.150}$$

Substituting $\widehat{\mathbf{q}}^* = \|\widehat{\mathbf{q}}\|^2\widehat{\mathbf{q}}^{-1}$ into Eq. (11.149) yields

$$\widehat{\mathbf{q}}\otimes\widehat{\mathbf{q}}^{-1} = \widehat{\mathbf{q}}^{-1}\otimes\widehat{\mathbf{q}} = \widehat{\mathbf{1}} \tag{11.151}$$

Clearly, for unit quaternions the inverse and the conjugate are the same, $\widehat{\mathbf{q}}^* = \widehat{\mathbf{q}}^{-1}$, and

$$\widehat{\mathbf{q}}\otimes\widehat{\mathbf{q}}^* = \widehat{\mathbf{q}}^*\otimes\widehat{\mathbf{q}} = \widehat{\mathbf{1}} \quad \left(\text{if } \|\widehat{\mathbf{q}}\| = 1\right) \tag{11.152}$$

Let us restrict our attention to unit quaternions, in which case

$$\widehat{\mathbf{q}}=\left\{\begin{array}{c}\sin(\theta/2)\hat{\mathbf{u}}\\ \hdashline \cos(\theta/2)\end{array}\right\} \tag{11.153}$$

where $\hat{\mathbf{u}}$ is the unit vector along the Euler axis around which the inertial reference frame is rotated into the body-fixed frame, and θ is the Euler principal rotation angle. Recalling Eq. (11.141), we observe that

$$q_1 = l\sin(\theta/2) \quad q_2 = m\sin(\theta/2) \quad q_3 = n\sin(\theta/2) \quad q_4 = \cos(\theta/2) \tag{11.154}$$

The conjugate quaternion $\widehat{\mathbf{q}}^*$ is found by reversing the sign of the vector part of $\widehat{\mathbf{q}}$, so that

$$\widehat{\mathbf{q}}^*=\left\{\begin{array}{c}-\sin(\theta/2)\hat{\mathbf{u}}\\ \hdashline \cos(\theta/2)\end{array}\right\} \tag{11.155}$$

Employing these and the trigonometric identities

$$\cos\theta = \cos^2(\theta/2) - \sin^2(\theta/2) \quad \sin\theta = 2\cos(\theta/2)\sin(\theta/2) \tag{11.156}$$

we show in Appendix G that the direction cosine matrix $[\mathbf{Q}]_{Xx}$ of the body frame in Eq. (11.143) is obtained from the quaternion $\widehat{\mathbf{q}}$ by means of the following algorithm.

ALGORITHM 11.1

Obtain the direction cosine matrix $[\mathbf{Q}]_{Xx}$ from the unit quaternion $\widehat{\mathbf{q}}$. This procedure is implemented in the MATLAB function *dcm_from_q.m* in Appendix D.49.

1. Write the quaternion as

$$\widehat{\mathbf{q}}=\left\{\begin{array}{c}q_1\\ q_2\\ q_3\\ \hdashline q_4\end{array}\right\}$$

where $\lfloor q_1 \ q_2 \ q_3 \rfloor^T$ is the vector part, q_4 is the scalar part, and $\|\widehat{\mathbf{q}}\| = 1$.

2. Compute the direction cosine matrix of the transformation from *XYZ* to *xyz* as follows:

$$[\mathbf{Q}]_{Xx} = \begin{bmatrix} q_1^2 - q_2^2 - q_3^2 + q_4^2 & 2(q_1q_2 + q_3q_4) & 2(q_1q_3 - q_2q_4) \\ 2(q_2q_1 - q_3q_4) & -q_1^2 + q_2^2 - q_3^2 + q_4^2 & 2(q_2q_3 + q_1q_4) \\ 2(q_3q_1 + q_2q_4) & 2(q_3q_2 - q_1q_4) & -q_1^2 - q_2^2 + q_3^2 + q_4^2 \end{bmatrix} \tag{11.157}$$

Note that every element of the matrix $[\mathbf{Q}]_{Xx}$ in Eq. (11.157) contains products of two components of $\widehat{\mathbf{q}}$. Since $\widehat{\mathbf{q}}$ and $-\widehat{\mathbf{q}}$ therefore yield the same direction cosine matrix, they represent the same rotation. We can verify by carrying out the matrix multiplication and using Eq. (11.145) that $[\mathbf{Q}]_{Xx}$ in Eq. (11.157) exhibits the required orthogonality property,

$$[\mathbf{Q}]_{Xx}[\mathbf{Q}]_{Xx}{}^T = [\mathbf{Q}]_{Xx}{}^T[\mathbf{Q}]_{Xx} = [\mathbf{1}]$$

To find the unit quaternion ($q_1^2 + q_2^2 + q_3^2 + q_4^2 = 1$) for a given direction cosine matrix, we observe from Eq. (11.157) that

$$\begin{gathered} q_4 = \frac{1}{2}\sqrt{1 + Q_{11} + Q_{22} + Q_{33}} \\ q_1 = \frac{Q_{23} - Q_{32}}{4q_4} \quad q_2 = \frac{Q_{31} - Q_{13}}{4q_4} \quad q_3 = \frac{Q_{12} - Q_{21}}{4q_4} \end{gathered} \tag{11.158}$$

This procedure obviously fails for pure quaternions ($q_4 = 0$). The following algorithm (Bar-Itzhack, 2000) avoids having to deal with this situation.

ALGORITHM 11.2

Obtain the (unit) quaternion from the direction cosine matrix $[\mathbf{Q}]_{Xx}$. This procedure is implemented as the MATLAB function *q_from_dcm.m* in Appendix D.50.

1. Form the 4-by-4 *symmetric* matrix

$$[\mathbf{K}] = \frac{1}{3}\begin{bmatrix} Q_{11} - Q_{22} - Q_{33} & Q_{21} + Q_{12} & Q_{31} + Q_{13} & Q_{23} - Q_{32} \\ Q_{21} + Q_{12} & -Q_{11} + Q_{22} - Q_{33} & Q_{32} + Q_{23} & Q_{31} - Q_{13} \\ Q_{31} + Q_{13} & Q_{32} + Q_{23} & -Q_{11} - Q_{22} + Q_{33} & Q_{12} - Q_{21} \\ Q_{23} - Q_{32} & Q_{31} - Q_{13} & Q_{12} - Q_{21} & Q_{11} + Q_{22} + Q_{33} \end{bmatrix} \tag{11.159}$$

2. Solve the eigenvalue problem $[\mathbf{K}]\{\mathbf{e}\} = \lambda\{\mathbf{e}\}$ for the largest eigenvalue λ_{max}. The corresponding eigenvector is the quaternion, $\left\{\widehat{\mathbf{q}}\right\} = \{\mathbf{e}\}$. Since we are interested in only the dominant eigenvalue of $[\mathbf{K}]$, we can use the iterative power method (Jennings, 1977), which converges to λ_{max}. Thus, starting with an estimate $\{\mathbf{e}_0\}$ of the eigenvector, we normalize it,

$$\{\hat{\mathbf{e}}_0\} = \frac{\{\mathbf{e}_0\}}{\|\mathbf{e}_0\|}$$

and use $\{\hat{\mathbf{e}}_0\}$ to compute an updated normalized estimate

$$\{\hat{\mathbf{e}}_1\} = \frac{[\mathbf{K}]\{\hat{\mathbf{e}}_0\}}{\|[\mathbf{K}]\{\hat{\mathbf{e}}_0\}\|} \quad (\|\hat{\mathbf{e}}_1\| = 1)$$

We estimate the corresponding eigenvalue by using the Rayleigh quotient,

$$r_1 = \frac{\{\hat{\mathbf{e}}_1\}^T[\mathbf{K}]\{\hat{\mathbf{e}}_1\}}{\|\hat{\mathbf{e}}_1\|^2} = \{\hat{\mathbf{e}}_1\}^T[\mathbf{K}]\{\hat{\mathbf{e}}_1\}$$

We repeat this process, using $\hat{\mathbf{e}}_1$ to compute an updated normalized estimate $\hat{\mathbf{e}}_2$ followed by using the Rayleigh quotient to find r_2, and so on, over and over again. After n steps we have $\hat{\mathbf{e}}_n$ and r_n. As n increases, r_n converges toward $\lambda_{\max}$, the *maximum* eigenvalue of $[\mathbf{K}]$. When $|(r_n - r_{n-1})/r_{n-1}| < \varepsilon$, where ε is our chosen tolerance, we terminate the iteration and declare that $\lambda_{\max} = r_n$ and that $\hat{\mathbf{e}}_n$ is the corresponding eigenvector.

Of course, instead of the power method we can take advantage of commercial software, such as MATLAB's eigenvalue extraction program `eig`.

EXAMPLE 11.22

(a) Write down the unit quaternion for a rotation about the x axis through an angle θ.
(b) Obtain the corresponding direction cosine matrix.

Solution

(a) According to Eq. (11.151),

$$\mathbf{q} = \sin(\theta/2)\hat{\mathbf{i}} \quad q_4 = \cos(\theta/2) \tag{a}$$

so that

$$\boxed{\hat{\mathbf{q}} = \begin{Bmatrix} \sin(\theta/2) \\ 0 \\ 0 \\ \hline \cos(\theta/2) \end{Bmatrix}} \tag{b}$$

(b) Substituting $q_1 = \sin(\theta/2)$, $q_2 = q_3 = 0$, and $q_4 = \cos(\theta/2)$ into Eq. (11.152) yields

$$[\mathbf{Q}] = \begin{bmatrix} \sin^2(\theta/2) + \cos^2(\theta/2) & 0 & 0 \\ 0 & -\sin^2(\theta/2) + \cos^2(\theta/2) & 2\sin(\theta/2)\cos(\theta/2) \\ 0 & -2\sin(\theta/2)\cos(\theta/2) & -\sin^2(\theta/2) + \cos^2(\theta/2) \end{bmatrix} \tag{c}$$

From trigonometry, we have

$$\sin^2\frac{\theta}{2} + \cos^2\frac{\theta}{2} = 1 \quad 2\sin\frac{\theta}{2}\cos\frac{\theta}{2} = \sin\theta \quad \cos^2\frac{\theta}{2} - \sin^2\frac{\theta}{2} = \cos\theta$$

Therefore, Eq. (c) becomes

$$\boxed{[\mathbf{Q}] = \begin{bmatrix} 1 & 0 & 0 \\ 0 & \cos\theta & \sin\theta \\ 0 & -\sin\theta & \cos\theta \end{bmatrix}} \tag{d}$$

We recognize this as the direction cosine matrix $[\mathbf{R}_1(\theta)]$ for a rotation θ around the x axis (Eq. 4.33).

EXAMPLE 11.23

For the yaw–pitch–roll sequence $\phi_{\text{yaw}} = 50°$, $\theta_{\text{pitch}} = 90°$, and $\psi_{\text{roll}} = 120°$, calculate
(a) the quaternion and
(b) the rotation angle and the axis of rotation.

Solution

(a) Substituting the given angles into Eq. (11.119) yields the direction cosine matrix

$$[\mathbf{Q}]_{Xx} = \begin{bmatrix} 0 & 0 & -1 \\ 0.93969 & 0.34202 & 0 \\ 0.34202 & -0.93969 & 0 \end{bmatrix} \tag{a}$$

Substituting the components of $[\mathbf{Q}]_{Xx}$ into Eq. (11.159), we get

$$[\mathbf{K}] = \begin{bmatrix} -0.11401 & 0.31323 & -0.21933 & 0.31323 \\ 0.31323 & 0.11401 & -0.31323 & 0.44734 \\ -0.21933 & -0.31323 & -0.11401 & -0.31323 \\ 0.31323 & 0.44734 & -0.31323 & 0.11401 \end{bmatrix} \tag{b}$$

The following is a MATLAB script that implements the power method described in Algorithm 11.2.

```
K                  = [-0.11401  0.31323 -0.21933  0.31323
                       0.31323  0.11401 -0.31323  0.44734
                      -0.21933 -0.31323 -0.11401 -0.31323
                       0.31323  0.44734 -0.31323  0.11401];

v0                 = [1 1 1 1]';   %Initial estimate of the eigenvector.
v0                 = v0/norm(v0);  %Normalize it.
lamda_new          = v0'*K*v0;     %Rayleigh quotient (norm(v0) = 1)
                                   % estimate of the eigenvalue.
lamda_old          = 10*lamda_new; %Just to begin the iteration.
no_iterations      = 0;            %Count the number of iterations.
tolerance          = 1.e-10;

while abs((lamda_new - lamda_old)/lamda_old) > tolerance
    no_iterations = no_iterations + 1;
    lamda_old     = lamda_new;
    v             = v0;
    vnew          = K*v/norm(K*v);
    lamda_new     = vnew'*K*vnew; %Rayleigh quotient (norm(vnew) = 1).
    v0            = vnew;
end

no_iterations      = no_iterations
disp(' ')
lamda_max          = lamda_new
eigenvector        = vnew
```

The output of this program to the Command Window is as follows:

```
no_iterations =
    12
lamda_max =
          1
eigenvector =
      0.40558
      0.57923
     -0.40558
      0.57923
```

The quaternion is the eigenvector associated with λ_{max}, so that

$$\boxed{\widehat{\mathbf{q}} = \left\{ \begin{array}{c} 0.40558 \\ 0.57923 \\ -0.40558 \\ \hdashline 0.57923 \end{array} \right\}}$$

Observe that $\|\widehat{\mathbf{q}}\| = 1$. $\widehat{\mathbf{q}}$ must be a unit quaternion.

(b) From Eq. (11.146), we find that the principal angle is

$$\theta = 2\cos^{-1}(q_4) = 2\cos^{-1}(0.57923) = \boxed{54.604°}$$

and the Euler axis is

$$\hat{\mathbf{u}} = \frac{0.40558\hat{\mathbf{I}} + 0.57923\hat{\mathbf{J}} - 0.40558\hat{\mathbf{K}}}{\sin(54.604°/2)} = \boxed{0.4975\hat{\mathbf{I}} + 0.71056\hat{\mathbf{J}} - 0.49754\hat{\mathbf{K}}}$$

We have seen that a unit quaternion of Eq. (11.153) represents a rotation about the unit vector $\hat{\mathbf{u}}$ through the angle θ. Let us show that the Rodrigues' formula (Eq. 11.139) for rotating the vector $\mathbf{v}$ into the vector $\mathbf{v}'$ may be written in terms of quaternions as follows:

$$\widehat{\mathbf{v}}' = \widehat{\mathbf{q}} \otimes \widehat{\mathbf{v}} \otimes \widehat{\mathbf{q}}^* \tag{11.160}$$

where the conjugate quaternion $\widehat{\mathbf{q}}^*$ is given by Eq. (11.155), and $\widehat{\mathbf{v}}$ and $\widehat{\mathbf{v}}'$ are the pure quaternions having $\mathbf{v}$ and $\mathbf{v}'$ as their vector parts,

$$\widehat{\mathbf{v}} = \left\{ \begin{array}{c} \mathbf{v} \\ \hdashline 0 \end{array} \right\} \qquad \widehat{\mathbf{v}}' = \left\{ \begin{array}{c} \mathbf{v}' \\ \hdashline 0 \end{array} \right\} \tag{11.161}$$

Eq. (11.160) is implemented in MATLAB as the function *quat_rotate.m* in Appendix D.51.

Using Eq. (11.146) we first calculate the product of $\widehat{\mathbf{q}}$ and $\widehat{\mathbf{v}}$,

$$\widehat{\mathbf{q}} \otimes \widehat{\mathbf{v}} = \left\{ \begin{array}{c} \cos(\theta/2)\mathbf{v} + \sin(\theta/2)(\hat{\mathbf{u}} \times \mathbf{v}) \\ \hdashline -\sin(\theta/2)(\hat{\mathbf{u}} \cdot \mathbf{v}) \end{array} \right\}$$

Multiply this quaternion on the right by $\widehat{\mathbf{q}}^*$ to get

$$\widehat{\mathbf{v}}' = (\widehat{\mathbf{q}} \otimes \widehat{\mathbf{v}}) \otimes \widehat{\mathbf{q}}^* = \left\{ \begin{array}{c} \mathbf{v}' \\ \hdashline v_4' \end{array} \right\} \tag{11.162}$$

where $\mathbf{v}'$ and v_4' are, respectively, the vector and scalar parts of the quaternion $\widehat{\mathbf{v}}'$.

Once again we employ Eq. (11.146) to obtain

$$\begin{aligned}
\mathbf{v}' &= \overbrace{[-\sin(\theta/2)(\hat{\mathbf{u}}\cdot\mathbf{v})]}^{(\widehat{\mathbf{q}}\otimes\widehat{\mathbf{v}})_4}\overbrace{(-\sin(\theta/2)\hat{\mathbf{u}})}^{\mathbf{q}^*} + \overbrace{\cos(\theta/2)}^{q^*_4}\overbrace{[\cos(\theta/2)\mathbf{v}+\sin(\theta/2)(\hat{\mathbf{u}}\times\mathbf{v})]}^{\mathbf{q}\otimes\mathbf{v}}\\
&\quad+\underbrace{[\cos(\theta/2)\mathbf{v}+\sin(\theta/2)(\hat{\mathbf{u}}\times\mathbf{v})]}_{\mathbf{q}\otimes\mathbf{v}}\times\underbrace{(-\sin(\theta/2)\hat{\mathbf{u}})}_{\mathbf{q}^*}\\
&= \sin^2(\theta/2)\hat{\mathbf{u}}(\hat{\mathbf{u}}\cdot\mathbf{v}) + [\cos^2(\theta/2)\mathbf{v}+\cos(\theta/2)\sin(\theta/2)(\hat{\mathbf{u}}\times\mathbf{v})]\\
&\quad+\cos(\theta/2)\sin(\theta/2)(\hat{\mathbf{u}}\times\mathbf{v}) - \sin^2(\theta/2)(\hat{\mathbf{u}}\times\mathbf{v})\times\hat{\mathbf{u}}
\end{aligned}$$

According to the bac–cab rule (Eq. 1.20), $(\hat{\mathbf{u}}\times\mathbf{v})\times\hat{\mathbf{u}}=\mathbf{v}-\hat{\mathbf{u}}(\hat{\mathbf{u}}\cdot\mathbf{v})$. Substituting this into the above equation and collecting terms we get

$$\mathbf{v}'=\mathbf{v}[\cos^2(\theta/2)-\sin^2(\theta/2)]+\hat{\mathbf{u}}(\hat{\mathbf{u}}\cdot\mathbf{v})[2\sin^2(\theta/2)]+(\hat{\mathbf{u}}\times\mathbf{v})[2\cos(\theta/2)\sin(\theta/2)]$$

But

$$\cos^2(\theta/2)-\sin^2(\theta/2)=\cos\theta \quad 2\sin^2(\theta/2)=1-\cos\theta \quad 2\cos(\theta/2)\sin(\theta/2)=\sin\theta$$

so that finally

$$\mathbf{v}'=\mathbf{v}\cos\theta+\hat{\mathbf{u}}(\hat{\mathbf{u}}\cdot\mathbf{v})(1-\cos\theta)+(\hat{\mathbf{u}}\times\mathbf{v})\sin\theta \tag{11.163}$$

According to Eq. (11.146), the scalar part v_4 of the quaternion product in Eq. (11.162) is

$$\begin{aligned}
v'_4 &= \overbrace{[-\sin(\theta/2)(\hat{\mathbf{u}}\cdot\mathbf{v})]}^{(\widehat{\mathbf{q}}\otimes\widehat{\mathbf{v}})_4}\overbrace{[\cos(\theta/2)]}^{q^*_4}-\overbrace{[\cos(\theta/2)\mathbf{v}+\sin(\theta/2)(\hat{\mathbf{u}}\times\mathbf{v})]}^{\mathbf{q}\otimes\mathbf{v}}\cdot\overbrace{[-\sin(\theta/2)\hat{\mathbf{u}}]}^{\mathbf{q}^*}\\
&=-\sin(\theta/2)\cos(\theta/2)(\hat{\mathbf{u}}\cdot\mathbf{v})+\cos(\theta/2)\sin(\theta/2)(\hat{\mathbf{u}}\cdot\mathbf{v})+\sin^2(\theta/2)(\hat{\mathbf{u}}\times\mathbf{v})\cdot\hat{\mathbf{u}}\\
&=0
\end{aligned}$$

Thus, the scalar part of $\widehat{\mathbf{v}}'$ vanishes, which means that $\widehat{\mathbf{v}}'$ is a pure quaternion whose vector part $\mathbf{v}'$ is identical to Eq. (11.139). We have therefore shown that the quaternion operation $\widehat{\mathbf{q}}\otimes\widehat{\mathbf{v}}\otimes\widehat{\mathbf{q}}^*$ indeed rotates the vector $\mathbf{v}$ around the axis of the quaternion (the Euler axis) through the angle θ. In the same way we can show that the operation $\widehat{\mathbf{q}}^*\otimes\widehat{\mathbf{v}}\otimes\widehat{\mathbf{q}}$ rotates the vector $\mathbf{v}$ through the angle $-\theta$. In fact, if we follow the operation $\widehat{\mathbf{q}}\otimes\widehat{\mathbf{v}}\otimes\widehat{\mathbf{q}}^*$ (rotation through $+\theta$) with the operation $\widehat{\mathbf{q}}^*\otimes\widehat{\mathbf{v}}\otimes\widehat{\mathbf{q}}$ (rotation through $-\theta$) we end up where we started (namely, with the pure quaternion $\widehat{\mathbf{v}}$):

$$\widehat{\mathbf{q}}\otimes\left(\widehat{\mathbf{q}}^*\otimes\widehat{\mathbf{v}}\otimes\widehat{\mathbf{q}}\right)\otimes\widehat{\mathbf{q}}^*=\left(\widehat{\mathbf{q}}\otimes\widehat{\mathbf{q}}^*\right)\otimes\widehat{\mathbf{v}}\otimes\left(\widehat{\mathbf{q}}\otimes\widehat{\mathbf{q}}^*\right)=\widehat{\mathbf{1}}\otimes\widehat{\mathbf{v}}\otimes\widehat{\mathbf{1}}=\left(\widehat{\mathbf{1}}\otimes\widehat{\mathbf{v}}\right)\otimes\widehat{\mathbf{1}}=\widehat{\mathbf{v}}\otimes\widehat{\mathbf{1}}=\widehat{\mathbf{v}}$$

The operation in Eq. (11.160) is a vector rotation. The frame of reference remains fixed while the vector $\mathbf{v}$ is rotated into the vector $\mathbf{v}'$. On the other hand, the familiar operation $\{\mathbf{v}\}_{x'}=[\mathbf{Q}]_{xx'}\{\mathbf{v}\}$ is a frame rotation (a coordinate transformation), in which the vector $\mathbf{v}$ remains fixed while the reference frame is rotated. We can easily illustrate this by revisiting Example 11.22.

EXAMPLE 11.24

Consider the vector $\mathbf{v}=v\hat{\mathbf{j}}$. Using the quaternion and corresponding direction cosine matrix in Example 11.22, carry out the following operations and interpret the results geometrically:

(i) $\widehat{\mathbf{v}}'=\widehat{\mathbf{q}}\otimes\widehat{\mathbf{v}}\otimes\widehat{\mathbf{q}}^*$

(ii) $\{\mathbf{v}'\} = [\mathbf{Q}]\{\mathbf{v}\}$
where

$$\widehat{\mathbf{v}} = \left\{\begin{array}{c} v\hat{\mathbf{j}} \\ \hdashline 0 \end{array}\right\} \qquad \widehat{\mathbf{q}} = \left\{\begin{array}{c} \sin(\theta/2)\hat{\mathbf{i}} \\ \hdashline \cos(\theta/2) \end{array}\right\} \qquad [\mathbf{Q}] = \begin{bmatrix} 1 & 0 & 0 \\ 0 & \cos\theta & \sin\theta \\ 0 & -\sin\theta & \cos\theta \end{bmatrix}$$

Solution

(i) We do the two quaternion products one after the other using Eq. (11.146). The first product is

$$\widehat{\mathbf{q}} \otimes \widehat{\mathbf{v}} = \left\{\begin{array}{c} \cos(\theta/2)(v\hat{\mathbf{j}}) + 0\cdot\sin(\theta/2)\hat{\mathbf{i}} + \sin(\theta/2)\hat{\mathbf{i}} \times v\hat{\mathbf{j}} \\ \hdashline \cos(\theta/2)\cdot 0 - \sin(\theta/2)\hat{\mathbf{i}}\cdot v\hat{\mathbf{j}} \end{array}\right\}$$

$$= \left\{\begin{array}{c} v\cos(\theta/2)\hat{\mathbf{j}} + v\sin(\theta/2)\hat{\mathbf{k}} \\ \hdashline 0 \end{array}\right\}$$

Following this by the second product, we get

$$\widehat{\mathbf{q}} \otimes \widehat{\mathbf{v}} \otimes \widehat{\mathbf{q}}^* = \left\{\begin{array}{c} v\cos(\theta/2)\hat{\mathbf{j}} + v\sin(\theta/2)\hat{\mathbf{k}} \\ \hdashline 0 \end{array}\right\} \otimes \left\{\begin{array}{c} -\sin(\theta/2)\hat{\mathbf{i}} \\ \hdashline \cos(\theta/2) \end{array}\right\}$$

$$= \left\{\begin{array}{c} 0\cdot\left[-\sin(\theta/2)\hat{\mathbf{i}}\right] + \cos(\theta/2)\cdot\left[v\cos(\theta/2)\hat{\mathbf{j}} + v\sin(\theta/2)\hat{\mathbf{k}}\right] + \left[v\cos(\theta/2)\hat{\mathbf{j}} + v\sin(\theta/2)\hat{\mathbf{k}}\right] \times \left[-\sin(\theta/2)\hat{\mathbf{i}}\right] \\ \hdashline 0\cdot\left[\cos(\theta/2)\right] - \left[v\cos(\theta/2)\hat{\mathbf{j}} + v\sin(\theta/2)\hat{\mathbf{k}}\right]\cdot\left[-\sin(\theta/2)\hat{\mathbf{i}}\right] \end{array}\right\}$$

$$= \left\{\begin{array}{c} \overbrace{\left[v\cos^2(\theta/2) - v\sin^2(\theta/2)\right]}^{\cos\theta}\hat{\mathbf{j}} + \overbrace{\left[2v\sin(\theta/2)\cos(\theta/2)\right]}^{\sin\theta}\hat{\mathbf{k}} \\ \hdashline 0 \end{array}\right\} = \left\{\begin{array}{c} v\cos\theta\hat{\mathbf{j}} + v\sin\theta\hat{\mathbf{k}} \\ \hdashline 0 \end{array}\right\}$$

Finally, therefore,

$$\boxed{\mathbf{v}' = v\cos\theta\hat{\mathbf{j}} + v\sin\theta\hat{\mathbf{k}}} \tag{a}$$

(ii)

$$\{\mathbf{v}'\} = \begin{bmatrix} 1 & 0 & 0 \\ 0 & \cos\theta & \sin\theta \\ 0 & -\sin\theta & \cos\theta \end{bmatrix} \begin{Bmatrix} 0 \\ v \\ 0 \end{Bmatrix} = \begin{Bmatrix} 0 \\ v\cos\theta \\ -v\sin\theta \end{Bmatrix}$$

or

$$\boxed{\mathbf{v}' = v\cos\theta\hat{\mathbf{j}} - v\sin\theta\hat{\mathbf{k}}} \tag{b}$$

These two results are illustrated in Fig. 11.29.

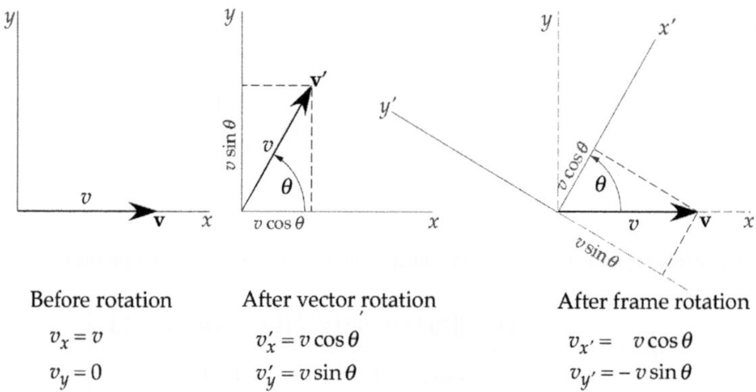

FIG. 11.29

Vector rotation vs frame rotation.

The Euler equations of motion for a rigid body (Eqs. 11.72) provide the angular velocity rates $\dot{\omega}_x$, $\dot{\omega}_y$, and $\dot{\omega}_z$ as functions of time. We can integrate those equations to find the time history of the angular velocities ω_x, ω_y, and ω_z. In addition, to obtain the orientation history of the body, we need the time history of the Euler angles ϕ, θ, and ψ. These are found by integrating Eq. (11.116),

$$\begin{Bmatrix} \dot{\phi} \\ \dot{\theta} \\ \dot{\psi} \end{Bmatrix} = \begin{bmatrix} \dfrac{\sin\psi}{\sin\theta} & \dfrac{\cos\psi}{\sin\theta} & 0 \\ \cos\psi & -\sin\psi & 0 \\ -\dfrac{\sin\psi}{\tan\theta} & -\dfrac{\cos\psi}{\tan\theta} & 1 \end{bmatrix} \begin{Bmatrix} \omega_x \\ \omega_y \\ \omega_z \end{Bmatrix}$$

which provide the Euler angle rates (precession, nutation, and spin) in terms of the angular velocities. If we elect to use quaternions instead of Euler angles to describe the attitude of the body, then we need a formula for the rate of change of $\widehat{\mathbf{q}}$ in terms of the angular velocities.

To find the time derivative of a pure quaternion $\widehat{\mathbf{q}}$, we simply differentiate Eq. (11.153) to get

$$\dot{\widehat{\mathbf{q}}} = \begin{Bmatrix} \dfrac{d}{dt}\left[\hat{\mathbf{u}}\sin(\theta/2)\right] \\ \hdashline \dfrac{d}{dt}\cos(\theta/2) \end{Bmatrix} = \begin{Bmatrix} \dot{\hat{\mathbf{u}}}\sin(\theta/2) + \hat{\mathbf{u}}(\dot{\theta}/2)\cos(\theta/2) \\ \hdashline -(\dot{\theta}/2)\sin(\theta/2) \end{Bmatrix}$$

The Euler axis unit vector $\hat{\mathbf{u}}$ is constant in magnitude, but not in direction. According to Gelman (1971) and Hughes (2004), its time derivative is

$$\dot{\hat{\mathbf{u}}} = \frac{1}{2}\left[\hat{\mathbf{u}} \times \boldsymbol{\omega} - \cot(\theta/2)\hat{\mathbf{u}} \times (\hat{\mathbf{u}} \times \boldsymbol{\omega})\right] \tag{11.164}$$

where $\boldsymbol{\omega}$ is the angular velocity vector. Clearly, if the instantaneous axis of rotation and the Euler axis happen to coincide (i.e., if $\boldsymbol{\omega} = \omega\hat{\mathbf{u}}$), then $\dot{\hat{\mathbf{u}}} = \mathbf{0}$, because in that case $\hat{\mathbf{u}} \times \boldsymbol{\omega} = \mathbf{0}$. However, in general

$\dot{\hat{\mathbf{u}}}$ does not vanish. Substituting Eq. (11.164) into the expression for $\dot{\widehat{\mathbf{q}}}$, we find after expanding and collecting terms that

$$\dot{\widehat{\mathbf{q}}} = \frac{1}{2}\left\{\begin{array}{c} \sin(\theta/2)\hat{\mathbf{u}}\times\boldsymbol{\omega} + \cos(\theta/2)\boldsymbol{\omega} \\ \hline -\dot{\theta}\sin(\theta/2) \end{array}\right\} \tag{11.165}$$

From Eq. (11.153) we know that $\hat{\mathbf{u}}\sin(\theta/2) = \mathbf{q}$ and $\cos(\theta/2) = q_4$. Observing furthermore that $\dot{\theta}$ is the component of the angular velocity $\boldsymbol{\omega}$ along the Euler axis ($\dot{\theta} = \boldsymbol{\omega}\cdot\hat{\mathbf{u}}$), the expression for $\dot{\widehat{\mathbf{q}}}$ becomes

$$\dot{\widehat{\mathbf{q}}} = \frac{1}{2}\left\{\begin{array}{c} \mathbf{q}\times\boldsymbol{\omega} + q_4\boldsymbol{\omega} \\ \hline -\boldsymbol{\omega}\cdot\mathbf{q} \end{array}\right\} \tag{11.166}$$

According to the quaternion composition rule (Eq. 11.146), this may be written

$$\dot{\widehat{\mathbf{q}}} = \frac{1}{2}\widehat{\mathbf{q}}\otimes\widehat{\boldsymbol{\omega}}$$

where $\widehat{\boldsymbol{\omega}} = \left\{\begin{array}{c} \boldsymbol{\omega} \\ \hline 0 \end{array}\right\}$ is the pure quaternion version of the angular velocity vector. Substituting $\mathbf{q} = q_1\hat{\mathbf{i}} + q_2\hat{\mathbf{j}} + q_3\hat{\mathbf{k}}$ and $\boldsymbol{\omega} = \omega_1\hat{\mathbf{i}} + \omega_2\hat{\mathbf{j}} + \omega_3\hat{\mathbf{k}}$ into Eq. (11.166) and expanding the vector and scalar products leads to

$$\dot{\widehat{\mathbf{q}}} = \frac{1}{2}\left\{\begin{array}{c} (q_2\omega_3 - q_3\omega_2) + q_4\omega_1 \\ (q_3\omega_1 - q_1\omega_3) + q_4\omega_2 \\ (q_1\omega_2 - q_2\omega_1) + q_4\omega_3 \\ \hline -\omega_1 q_1 - \omega_2 q_2 - \omega_3 q_3 \end{array}\right\} = \frac{1}{2}\left[\begin{array}{ccc|c} 0 & \omega_3 & -\omega_2 & \omega_1 \\ -\omega_3 & 0 & \omega_1 & \omega_2 \\ \omega_2 & -\omega_1 & 0 & \omega_3 \\ \hline -\omega_1 & -\omega_2 & -\omega_3 & 0 \end{array}\right]\left\{\begin{array}{c} q_1 \\ q_2 \\ q_3 \\ \hline q_4 \end{array}\right\}$$

That is,

$$\frac{d}{dt}\left\{\widehat{\mathbf{q}}\right\} = \frac{1}{2}[\boldsymbol{\Omega}]\left\{\widehat{\mathbf{q}}\right\} \tag{11.167}$$

where

$$[\boldsymbol{\Omega}] = \left[\begin{array}{ccc|c} 0 & \omega_3 & -\omega_2 & \omega_1 \\ -\omega_3 & 0 & \omega_1 & \omega_2 \\ \omega_2 & -\omega_1 & 0 & \omega_3 \\ \hline -\omega_1 & -\omega_2 & -\omega_3 & 0 \end{array}\right] \tag{11.168}$$

ω_1, ω_2, and ω_3 are the x, y, and z components of angular velocity in the body-fixed frame.

If the components of the angular velocity are constant, then the matrix $[\boldsymbol{\Omega}]$ is constant and we can readily integrate Eq. (11.167) to obtain

$$\left\{\widehat{\mathbf{q}}\right\} = \exp\left(\frac{[\boldsymbol{\Omega}]}{2}t\right)\left\{\widehat{\mathbf{q}}_0\right\} \tag{11.169}$$

where $\left\{\widehat{\mathbf{q}}_0\right\}$ is the value of the quaternion at time $t = 0$. This expression may be inferred directly from scalar calculus, in which we know that if c is a constant, then the solution of the differential equation

$dx/dt = cx$ is simply $x = x_0 e^{ct}$. In linear algebra we learn that a 4-by-4 matrix $[\mathbf{A}]$ has four eigenvalues λ_i and four corresponding eigenvectors $\{\mathbf{e}_i\}$, satisfying the equation

$$[\mathbf{A}]\{\mathbf{e}_i\} = \lambda_i\{\mathbf{e}_i\} \quad i = 1,\ldots,4$$

It turns out that

$$\exp([\mathbf{A}]) = [\mathbf{V}]\exp([\boldsymbol{\Lambda}])[\mathbf{V}]^{-1} \tag{11.170}$$

where $[\boldsymbol{\Lambda}]$ is the 4-by-4 diagonal matrix of eigenvalues,

$$[\boldsymbol{\Lambda}] = \begin{bmatrix} \lambda_1 & 0 & 0 & 0 \\ 0 & \lambda_2 & 0 & 0 \\ 0 & 0 & \lambda_3 & 0 \\ 0 & 0 & 0 & \lambda_4 \end{bmatrix}$$

and $[\mathbf{V}]$ is the 4-by-4 matix whose columns comprise the four distinct eigenvectors $\{\mathbf{e}_i\}$,

$$[\mathbf{V}] = [\{\mathbf{e}_1\} \;\; \{\mathbf{e}_2\} \;\; \{\mathbf{e}_3\} \;\; \{\mathbf{e}_4\}]$$

Using, for example, MATLAB's symbolic math feature, we find that the eigenvalues and corresponding eigenvectors of the matrix $[\boldsymbol{\Omega}]$ in Eq. (11.168) are

$$\lambda_1 = \lambda_2 = \omega i: \quad \mathbf{e}_1 = \begin{Bmatrix} (\omega_y\omega i - \omega_x\omega_z)/\omega_{xy}{}^2 \\ -(\omega_x\omega i + \omega_y\omega_z)/\omega_{xy}{}^2 \\ 1 \\ 0 \end{Bmatrix} \quad \mathbf{e}_2 = \begin{Bmatrix} -(\omega_x\omega i + \omega_y\omega_z)/\omega_{xy}{}^2 \\ (-\omega_y\omega i + \omega_x\omega_z)/\omega_{xy}{}^2 \\ 0 \\ 1 \end{Bmatrix}$$

$$\lambda_3 = \lambda_4 = -\omega i: \quad \mathbf{e}_3 = \begin{Bmatrix} -(\omega_y\omega i + \omega_x\omega_z)/\omega_{xy}{}^2 \\ (\omega_x\omega i - \omega_y\omega_z)/\omega_{xy}{}^2 \\ 1 \\ 0 \end{Bmatrix} \quad \mathbf{e}_4 = \begin{Bmatrix} (\omega_x\omega i - \omega_y\omega_z)/\omega_{xy}{}^2 \\ (\omega_y\omega i + \omega_x\omega_z)/\omega_{xy}{}^2 \\ 0 \\ 1 \end{Bmatrix}$$

where $\omega = \sqrt{\omega_x{}^2 + \omega_y{}^2 + \omega_z{}^2}$ (the magnitude of the angular velocity vector), $\omega_{xy} = \sqrt{\omega_x{}^2 + \omega_y{}^2}$, and $i = \sqrt{-1}$. Substituting these results into Eq. (11.170) yields, again with the considerable aid of MATLAB,

$$\exp\left(\frac{[\boldsymbol{\Omega}]}{2}t\right) = \begin{bmatrix} \cos\dfrac{\omega t}{2} & \dfrac{\omega_z}{\omega}\sin\dfrac{\omega t}{2} & -\dfrac{\omega_y}{\omega}\sin\dfrac{\omega t}{2} & \dfrac{\omega_x}{\omega}\sin\dfrac{\omega t}{2} \\ -\dfrac{\omega_z}{\omega}\sin\dfrac{\omega t}{2} & \cos\dfrac{\omega t}{2} & \dfrac{\omega_x}{\omega}\sin\dfrac{\omega t}{2} & \dfrac{\omega_y}{\omega}\sin\dfrac{\omega t}{2} \\ \dfrac{\omega_y}{\omega}\sin\dfrac{\omega t}{2} & -\dfrac{\omega_x}{\omega}\sin\dfrac{\omega t}{2} & \cos\dfrac{\omega t}{2} & \dfrac{\omega_z}{\omega}\sin\dfrac{\omega t}{2} \\ -\dfrac{\omega_x}{\omega}\sin\dfrac{\omega t}{2} & -\dfrac{\omega_y}{\omega}\sin\dfrac{\omega t}{2} & -\dfrac{\omega_z}{\omega}\sin\dfrac{\omega t}{2} & \cos\dfrac{\omega t}{2} \end{bmatrix} \tag{11.171}$$

We know that for Eq. (11.171) to be valid, the angular velocity components must all be constant. The rigid body equations of motion (Eq. 11.72) show that $\dot{\omega}_x = \dot{\omega}_y = \dot{\omega}_z = 0$ if the net torque on the body is zero *and* the principal moments of inertia are all the same. Whereas torque-free motion (Chapter 12) is quite common for space vehicles, spherical symmetry ($A = B = C$) is not. Thus, we cannot make much practical use of Eqs. (11.169) and (11.171). In general, we must instead use numerical integration to obtain the angular velocities from the Euler equations and the quaternions from Eq. (11.167).

EXAMPLE 11.25

At time $t = 0$ the body-fixed axes and inertial angular velocity of a rigid body are those given in Example 11.18, namely

$$\begin{aligned} \hat{\mathbf{i}}_0 &= 0.40825\hat{\mathbf{I}} - 0.40825\hat{\mathbf{J}} + 0.81649\hat{\mathbf{K}} \\ \hat{\mathbf{j}}_0 &= -0.10102\hat{\mathbf{I}} - 0.90914\hat{\mathbf{J}} - 0.40405\hat{\mathbf{K}} \\ \hat{\mathbf{k}}_0 &= 0.90726\hat{\mathbf{I}} + 0.082479\hat{\mathbf{J}} - 0.41240\hat{\mathbf{K}} \end{aligned} \tag{a}$$

and

$$\boldsymbol{\omega}_X = -3.1\hat{\mathbf{I}} + 2.5\hat{\mathbf{J}} + 1.7\hat{\mathbf{K}}\,(\text{rad/s}) \tag{b}$$

If the angular velocity is constant, find the time histories of the Euler angles and the quaternion.

Solution

Because the angular velocity is constant, the motion of the body will be pure rotation about the fixed axis of rotation defined by the angular velocity vector. Once we find the direction cosine matrix as a function of time, we can use Algorithm 4.3 to obtain the Euler angles at each time.

Step 1:

As in Example 11.18, we find that the direction cosine matrix at time $t = 0$ is

$$[\mathbf{Q}_0]_{Xx} = \begin{bmatrix} 0.40825 & -0.40825 & 0.81649 \\ -0.10102 & -0.90914 & -0.40405 \\ 0.90726 & 0.082479 & -0.41240 \end{bmatrix} \tag{c}$$

Step 2:

As in Example 11.18, use $[\mathbf{Q}_0]_{Xx}$ to project the angular velocity $\boldsymbol{\omega}_X$ onto the axes of the body-fixed frame

$$\{\boldsymbol{\omega}\}_x = [\mathbf{Q}_0]_{Xx}\{\boldsymbol{\omega}\}_X = \begin{bmatrix} 0.40825 & -0.40825 & 0.81649 \\ -0.10102 & -0.90914 & -0.40405 \\ 0.90726 & 0.082479 & -0.41240 \end{bmatrix} \begin{Bmatrix} -3.1 \\ 2.5 \\ 1.7 \end{Bmatrix}$$

so that

$$\omega_x = -0.89817\,\text{rad/s} \quad \omega_y = -2.6466\,\text{rad/s} \quad \omega_z = -3.3074\,\text{rad/s} \tag{d}$$

The magnitude of the constant angular velocity is

$$\omega = \sqrt{\omega_x{}^2 + \omega_y{}^2 + \omega_z{}^2} = 4.3301\,\text{rad/s} \tag{e}$$

The period of the rotation is $T = 2\pi/\omega = 1.451$ s.

Step 3:

Use the angular velocities in (d) to form the matrix $[\boldsymbol{\Omega}]$ in Eq. (11.168),

$$[\boldsymbol{\Omega}] = \begin{bmatrix} 0 & -3.3074 & 2.6466 & -0.89817 \\ 3.3074 & 0 & -0.89817 & -2.6466 \\ -2.6466 & 0.89817 & 0 & -3.3074 \\ 0.89817 & 2.6466 & 3.3074 & 0 \end{bmatrix} \tag{f}$$

$[\boldsymbol{\Omega}]$ remains constant.

Step 4:

Use Algorithm 11.2 to obtain the quaternion at $t = 0$ from the direction cosine matrix in (c).

$$\widehat{\mathbf{q}}_0 = \begin{Bmatrix} -0.82610 \\ 0.15412 \\ -0.52165 \\ \hdashline 0.14724 \end{Bmatrix} \tag{g}$$

Step 5

At each time through t_{final}:

Compute the quaternion $\widehat{\mathbf{q}}(t)$ from Eqs. (11.169) and (11.171).

Use $\widehat{\mathbf{q}}(t)$ to update the direction cosine matrix $[\mathbf{Q}(t)]_{Xx}$ using Algorithm 11.1.

Use $[\mathbf{Q}(t)]_{Xx}$ to calculate the Euler angles $\phi(t)$, $\theta(t)$, and $\psi(t)$by means of Algorithm 4.3.

Fig. 11.30 shows the time variation of the four components of $\widehat{\mathbf{q}}$ during one rotation of the body. The variation of the three Euler angles is shown in Fig. 11.31. Observe that their values at $t = 0$ agree with those found in Example 11.18. Fig. 11.32 shows the initial orientation of the orthonormal body-fixed xyz axes given in Eq. (a). The dotted lines trace out the subsequent motion of their end points as they rotate at 4.33 rad/s about the fixed angular velocity vector $\boldsymbol{\omega}$. Finally, Fig. 11.33 shows the initial position of the Euler axis $\hat{\mathbf{u}}$ and its subsequent motion during rotation of the body. The unit vector $\hat{\mathbf{u}}$ is obtained from the unit quaternion $\widehat{\mathbf{q}}(t)$ at any instant by means of Eq. (11.153),

$$\hat{\mathbf{u}}(t) = \frac{\mathbf{q}(t)}{\sin\{\cos^{-1}[q_4(t)]\}}$$

where $\mathbf{q}(t)$ is the vector part of $\widehat{\mathbf{q}}(t)$. Fig. 11.33 amply illustrates the fact that the unit vectors $\hat{\boldsymbol{\omega}}$ and $\hat{\mathbf{u}}$ are not the same.

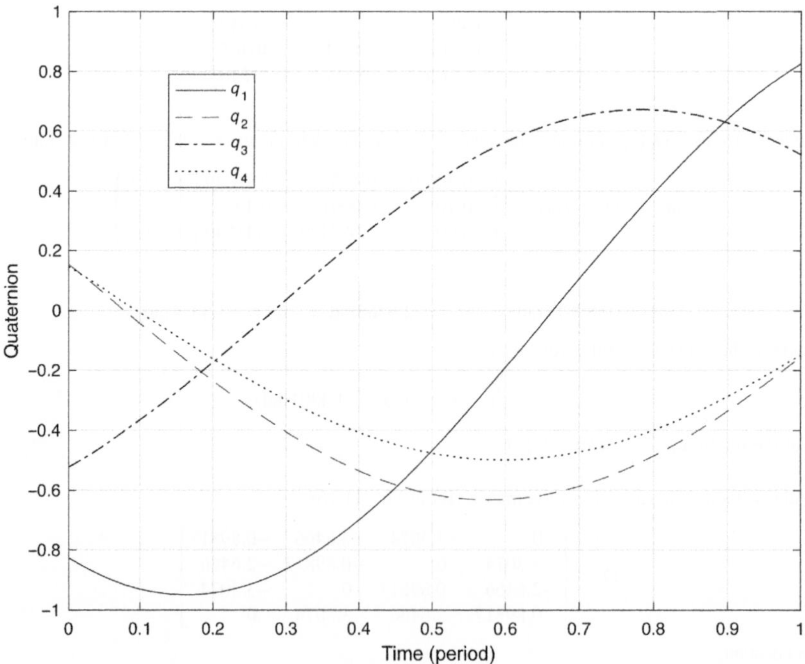

FIG. 11.30

History of the components of $\widehat{\mathbf{q}}$ for one rotation of the body.

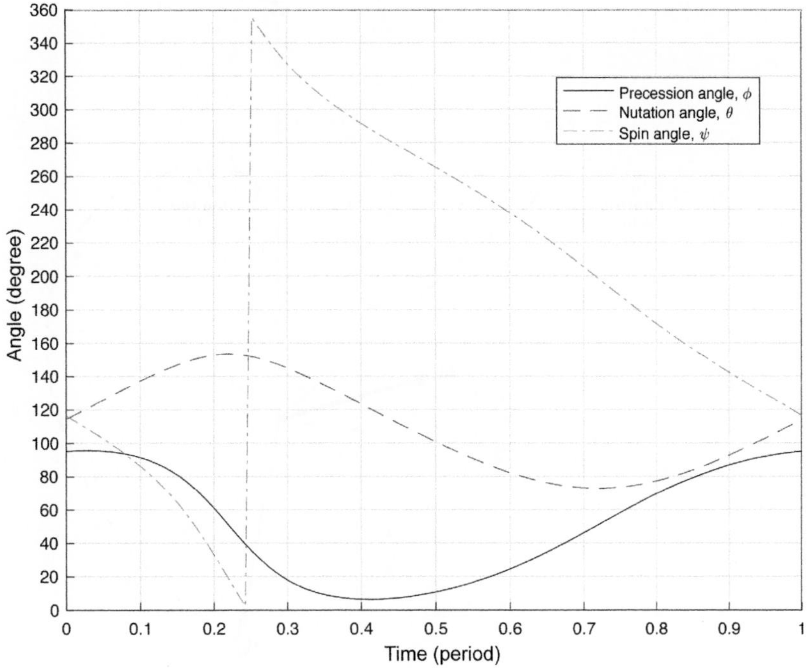

FIG. 11.31

History of the three Euler angles for one rotation of the body.

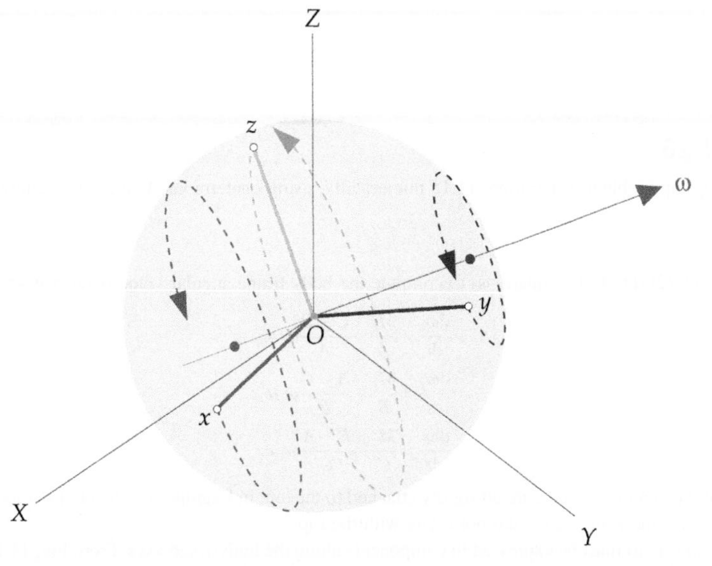

FIG. 11.32

The motion of three orthogonal lines during rotation of the body.

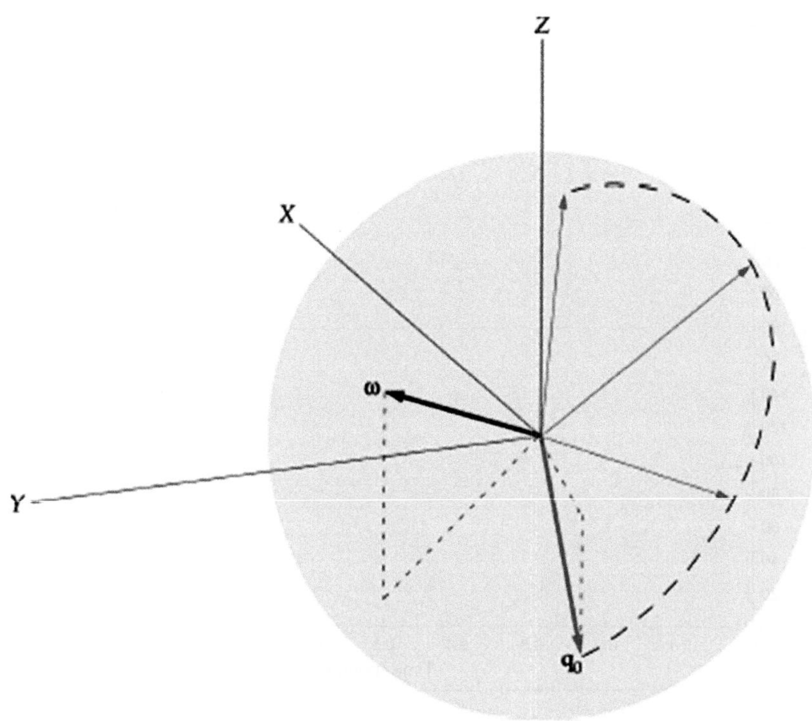

FIG. 11.33

Motion of the Euler axis during one rotation of the body.

EXAMPLE 11.26

Solve the spinning top problem of Example 11.15 numerically, using quaternions. Use the low-energy precession rate, $\omega_p = 51.93\,\text{rpm}$.

Solution

We will use Eq. (11.72) (the Euler equations) to compute the body frame angular velocity derivatives:

$$\begin{aligned}\frac{d\omega_x}{dt} &= \frac{M_x}{A} - \frac{C-B}{A}\omega_y\omega_z \\ \frac{d\omega_y}{dt} &= \frac{M_y}{B} - \frac{A-C}{B}\omega_z\omega_x \\ \frac{d\omega_z}{dt} &= \frac{M_z}{C} - \frac{B-A}{C}\omega_x\omega_y\end{aligned} \tag{a}$$

These require that the moving xyz axes are all rigidly attached to the top. In Example 11.15 only the x axis was fixed to the top along its spin axis; the y and z axes did not rotate with the top.

The moments in Eq. (a) must be expressed in components along the body-fixed axes. From Fig. 11.19, the moment of the weight vector about O is

$$\mathbf{M} = d\hat{\mathbf{k}} \times \left(-mg\hat{\mathbf{K}}\right) = -mgd\left(\hat{\mathbf{k}} \times \hat{\mathbf{K}}\right) \tag{b}$$

where, recalling Eq. $(4.18)_3$

$$\hat{\mathbf{k}} = Q_{31}\hat{\mathbf{I}} + Q_{32}\hat{\mathbf{J}} + Q_{33}\hat{\mathbf{K}} \tag{c}$$

The Qs are the time-dependent components of the direction cosine matrix $[\mathbf{Q}]_{Xx}$ in Eq. (11.157). Carrying out the cross product in Eq. (b) yields the components of $\mathbf{M}$ along the XYZ axes of the fixed space frame,

$$\{\mathbf{M}\}_X = \begin{Bmatrix} -mgdQ_{32} \\ mgdQ_{31} \\ 0 \end{Bmatrix}$$

To obtain the components of $\mathbf{M}$ in the body-fixed frame, we perform the transformation

$$\{\mathbf{M}\}_x = [\mathbf{Q}]_{Xx}\{\mathbf{M}\}_X = \begin{bmatrix} Q_{11} & Q_{12} & Q_{13} \\ Q_{21} & Q_{22} & Q_{23} \\ Q_{31} & Q_{32} & Q_{33} \end{bmatrix} \begin{Bmatrix} -mgdQ_{32} \\ mgdQ_{31} \\ 0 \end{Bmatrix} = mgd \begin{Bmatrix} Q_{12}Q_{31} - Q_{32}Q_{11} \\ Q_{22}Q_{31} - Q_{32}Q_{21} \\ 0 \end{Bmatrix} \tag{d}$$

It can be shown that (Problem 11.27)

$$Q_{12}Q_{31} - Q_{32}Q_{11} = Q_{23} \quad Q_{22}Q_{31} - Q_{32}Q_{21} = -Q_{13}$$

Therefore, at any instant the moments in Eq. (a) are

$$M_x = mgdQ_{23} \quad M_y = -mgdQ_{13} \quad M_z = 0 \tag{e}$$

The MATLAB implementation of the following procedure is listed in Appendix D.51.
Step 1:
Specify the initial orientation of the xyz axes of the body frame, thereby defining the initial value of the direction cosine matrix $[\mathbf{Q}]_{Xx}$.

According to Fig. 11.19, the body z axis is the top's spin axis, and we shall assume here that it initially lies in the global YZ plane, tilted 60° away from the Z axis, as it is in Example 11.15. Let the body x axis be initially aligned with the global X axis. The body y axis is then found from the cross product $\hat{\mathbf{j}} = \hat{\mathbf{k}} \times \hat{\mathbf{i}}$. Thus,

$$\begin{aligned} \hat{\mathbf{k}} &= -\sin 60°\hat{\mathbf{J}} + \cos 60°\hat{\mathbf{K}} \\ \hat{\mathbf{i}} &= \hat{\mathbf{I}} \\ \hat{\mathbf{j}} &= \hat{\mathbf{k}} \times \hat{\mathbf{i}} = \cos 60°\hat{\mathbf{J}} + \sin 60°\hat{\mathbf{K}} \end{aligned}$$

It follows that the direction cosine matrix relating XYZ to xyz at the start of the simulation is

$$[\mathbf{Q}_0]_{Xx} = \begin{bmatrix} 1 & 0 & 0 \\ 0 & \cos 60° & \sin 60° \\ 0 & -\sin 60° & \cos 60° \end{bmatrix} = \begin{bmatrix} 1 & 0 & 0 \\ 0 & 1/2 & \sqrt{3}/2 \\ 0 & -\sqrt{3}/2 & 1/2 \end{bmatrix} \tag{f}$$

Step 2:
Compute the initial quaternion $\widehat{\mathbf{q}}_0$ using Algorithm 11.2. Substituting Eq. (f) into Eq. (11.150) yields

$$[\mathbf{K}] = \begin{bmatrix} 0 & 0 & 0 & 0.57735 \\ 0 & -0.33333 & 0 & 0 \\ 0 & 0 & -0.33333 & 0 \\ 0.57735 & 0 & 0 & 0.66667 \end{bmatrix}$$

Rather than finding the dominant eigenvector by means of the power method, as we did in Example 11.23, we shall here use MATLAB's `eig` function, which obtains all of the eigenpairs. The snippet of MATLAB code for doing so is:

```
[eigvectors, eigvalues] = eig(K);
%Find the dominant eigenvalue and the column of 'eigvectors' that
% contains its eigenvector:
[dominant_eigvalue,column] = max(max(abs(eigvalues)));
dominant_eigenvector = eigvectors(:,column)
```

The output of this code is,

```
dominant_eigenvector =
    0.5
    0
    0
    0.86603
```

Therefore, the initial value of the quaternion is

$$\widehat{\mathbf{q}}_0 = \left\{ \begin{array}{c} 0.5 \\ 0 \\ 0 \\ \hline 0.86603 \end{array} \right\} \tag{g}$$

Step 3:
Specify the initial values of the body frame components of angular velocity $\boldsymbol{\omega}_0 = \lfloor \omega_x)_0 \;\; \omega_y)_0 \;\; \omega_z)_0 \rfloor^T$.

Recall that Eqs. (11.115) relate these body frame angular velocities to the initial values of the top's Euler angles and their rates,

$$\begin{aligned} \omega_x)_0 &= \omega_p)_0 \sin\theta_0 \sin\psi_0 + \omega_n)_0 \cos\psi_0 \\ \omega_y)_0 &= \omega_p)_0 \sin\theta_0 \cos\psi_0 - \omega_n)_0 \sin\psi_0 \\ \omega_z)_0 &= \omega_s)_0 + \omega_p)_0 \cos\theta_0 \end{aligned} \tag{h}$$

The top is released from rest with a given tilt angle θ_0 and spin rate $\omega_s)_0$. According to Example 11.15,

$$\begin{aligned} \theta_0 &= 60° \\ \psi_0 &= 0 \\ \omega_s)_0 &= 1000\,\text{rpm} = 104.72\ \text{rad/s} \\ \omega_p)_0 &= 51.93\,\text{rpm} = 5.438\ \text{rad/s (low energy precession rate)} \\ \omega_n)_0 &= 0 \end{aligned} \tag{i}$$

Substituting these into Eq. (h) we find

$$\boldsymbol{\omega}_0 = \lfloor 0 \;\; 4.7095 \;\; 107.44 \rfloor^T \text{(rad/s)} \tag{j}$$

Step 4:
Supply $\boldsymbol{\omega}_0$ and $\widehat{\mathbf{q}}_0$ as initial conditions to, say, the Runge–Kutta–Fehlberg 4(5) numerical integration procedure (Algorithm 1.3) to solve the system $\{\dot{\mathbf{y}}\} = \{\mathbf{f}\}$, where

$$\begin{aligned}\{\mathbf{y}\} &= \lfloor \omega_x \ \omega_y \ \omega_z \ q_1 \ q_2 \ q_3 \ q_4 \rfloor^T \\ \{\mathbf{f}\} &= \lfloor d\omega_x/dt \ d\omega_y/dt \ d\omega_z/dt \ dq_1/dt \ dq_2/dt \ dq_3/dt \ dq_4/dt \rfloor^T\end{aligned} \tag{k}$$

thereby obtaining the angular velocity $\boldsymbol{\omega}$ and quaternion $\widehat{\mathbf{q}}$ as functions of time. At each step of the numerical integration process:

(i) Use the current value of $\widehat{\mathbf{q}}$ to compute $[\mathbf{Q}]_{Xx}$ from Algorithm 11.1.
(ii) Use the current value of $[\mathbf{Q}]_{Xx}$ and $\boldsymbol{\omega}$ to compute $d\boldsymbol{\omega}/dt$ from (a), (d), and (e).
(iii) Use the current value of $\widehat{\mathbf{q}}$ and $\boldsymbol{\omega}$ to compute $d\widehat{\mathbf{q}}/dt$ from Eqs. (11.164).

Step 5:
At each solution time:

(i) Use Algorithm 11.1 to compute the direction cosine matrix $[\mathbf{Q}]_{Xx}$.
(ii) Use Algorithm 4.3 to compute the Euler angles ϕ (precession), θ (nutation), and ψ(spin).
(iii) Use Eq. (11.116) to compute the Euler angle rates $\dot{\phi}$, $\dot{\theta}$, and $\dot{\psi}$.

Step 6:
Plot the time histories of the Euler angles and their rates.

Fig. 11.34 shows the numerical solution for the precession, nutation, and spin angles as well as their rates as functions of time. We see that the constant precession rate (51.93 rpm) and spin rate (1000 rpm) are in agreement with Example 11.15, as is the nutation angle, which is fixed at 60°. The sawtooth appearance of the spin angle $\psi(t)$ reflects the fact that it is confined to the range 0 to 360°.

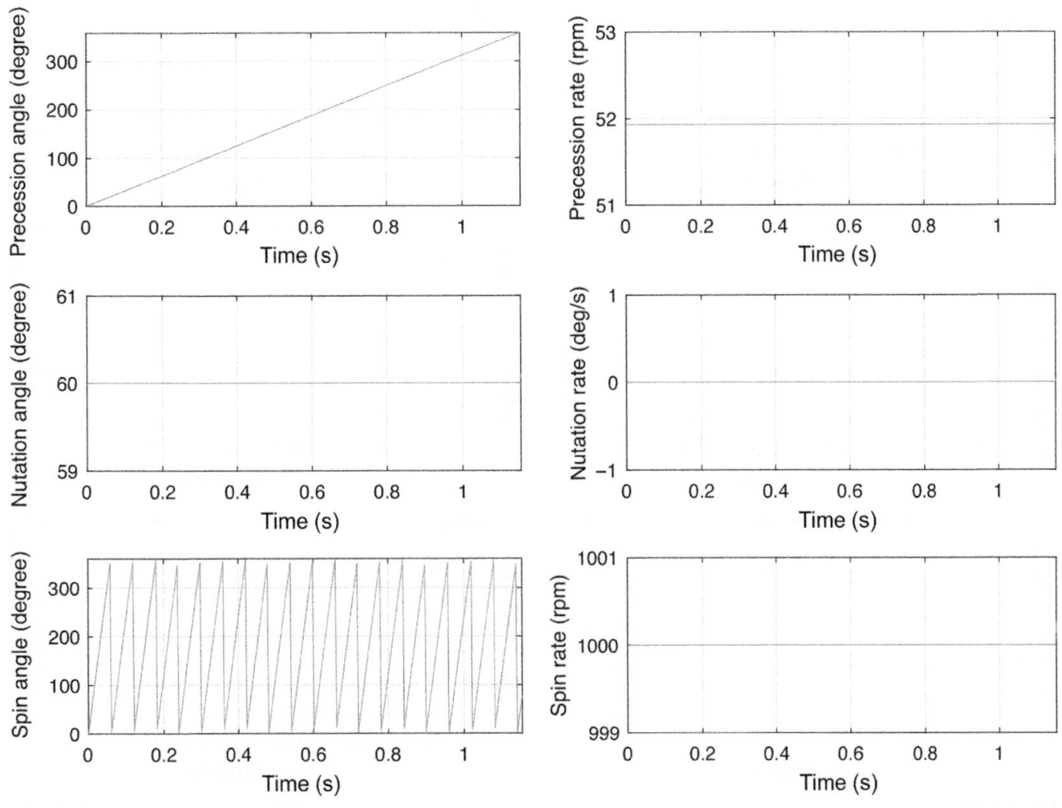

FIG. 11.34

Precession (ϕ), nutation (θ), and spin (ψ) angles and their rates for the top in Example 11.15. $A = B = 0.0012\ \text{kg}\cdot\text{m}^2$, $C = 0.00045\ \text{kg}\cdot\text{m}^2$.

Clearly , the solution of the steady-state spinning top problem by numerical integration yields no new insight into the top's motion and may be deemed a waste of effort. However, suppose we solve the same problem, but release the top from rest with *zero* precession rate, so that instead of Eqs. (i), the initial conditions are

$$\begin{aligned} \theta_0 &= 60^\circ \\ \psi_0 &= 0 \\ \omega_s)_0 &= 1000\text{rpm} = 104.72 \text{ rad/s} \\ \omega_p)_0 &= 0 \\ \omega_n)_0 &= 0 \end{aligned} \tag{l}$$

The initial orientation of the top is unchanged, so the initial quaternion $\widehat{\mathbf{q}}_0$ remains as shown in Eq. (g). On the other hand, the initially zero precession rate yields a different initial angular velocity vector, namely,

$$\boldsymbol{\omega}_0 = \lfloor 0 \;\; 0 \;\; 104.72 \rfloor^T (\text{rad/s}) \tag{m}$$

With only this change, the above numerical integration procedure yields the results shown in Fig. 11.35.

This is an example of unsteady precession, in which we see that the spin axis, instead of making a constant angle of 60° to the vertical, nutates between 60° and 75.4° at a rate of about 5.7 Hz, while the spin rate itself varies between 975 and

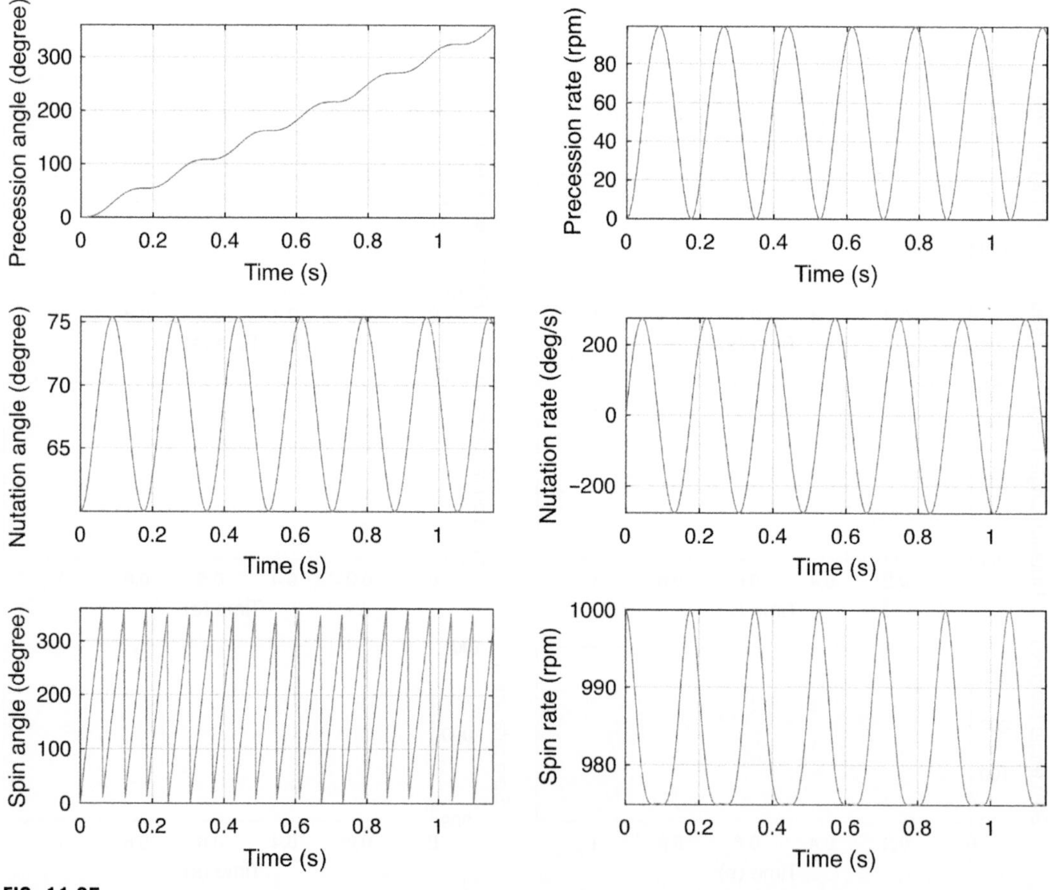

FIG. 11.35

Precession (ϕ), nutation (θ), and spin (ψ) angles and their rates for the top in Fig. 11.19, released from rest with initially zero precession. $A = B = 0.0012\ \text{kg}\cdot\text{m}^2$, $C = 0.00045\ \text{kg}\cdot\text{m}^2$.

1000 rpm at the same frequency. The precession rate oscillates between 0 and 99.4 rpm, also at a frequency of 5.7 Hz, with an average rate of 51.9 rpm, which happens to be the steady-state precession rate (Fig. 11.34).

These numerical results can be compared with formulas from the classical analysis of tops in unsteady precession. For example, it can be shown (Greenwood, 1988) that the relationship between the minimum and maximum nutation angles is

$$\cos\theta_{max} = \lambda - \sqrt{\lambda^2 - 2\lambda\cos\theta_{min} + 1}$$

where $\lambda = C^2\omega_z^{\,2}/(4Amgd)$. For the data of this problem, $\lambda = 1.887$, so that

$$\cos\theta_{max} = 1.887 - \sqrt{1.887^2 - 2\cdot 1.887\cdot\cos 60^\circ + 1} = 0.2518$$
$$\theta_{max} = 75.41^\circ$$

This is precisely what we observe for the nutation angle in Fig. 11.35. By the way, ω_z remains constant at its initial value of 104.72 rad/s because the top is axisymmetric ($A = B$) and $M_z = 0$ (Eq. (e)$_3$), so that $d\omega_z/dt = 0$ (Eq. (a)$_3$).

PROBLEMS

Section 2

11.1 Rigid, bent shaft 1 (ABC) rotates at a constant angular velocity of $2\hat{\mathbf{K}}$ rad/s around the positive Z axis of the inertial frame. Bent shaft 2 (CDE) rotates around BC with a constant angular velocity of $3\hat{\mathbf{j}}$ rad/s, relative to BC. Spinner 3 at E rotates around DE with a constant angular velocity of $4\hat{\mathbf{j}}$ rad/s relative to DE. Calculate the magnitude of the absolute angular acceleration vector $\boldsymbol{\alpha}_3$ of the spinner at the instant shown.

{Ans.: $\|\boldsymbol{\alpha}_3\| = \sqrt{180 + 64\sin^2\theta - 144\cos\theta}(\text{rad/s}^2)$}

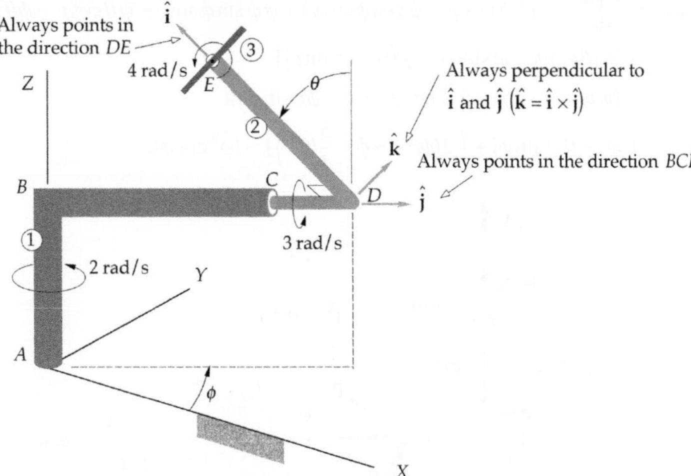

11.2 All the spin rates shown are constant. Calculate the magnitude of the absolute angular acceleration vector $\boldsymbol{\alpha}_3$ of the spinner at the instant shown (i.e., at the instant when the unit vector $\hat{\mathbf{i}}$ is parallel to the X axis and the unit vector $\hat{\mathbf{j}}$ is parallel to the Y axis).

{Ans.: $\|\boldsymbol{\alpha}_3\| = 63$ rad/s^2}

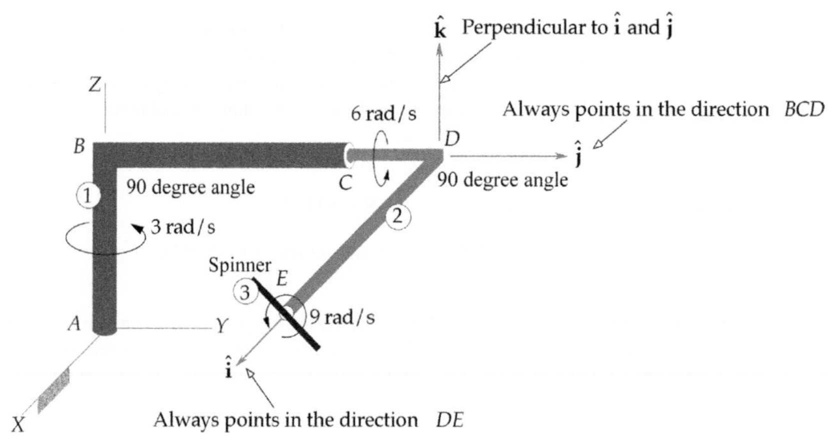

11.3 The body-fixed xyz frame is attached to the cylinder as shown. The cylinder rotates around the inertial Z axis, which is collinear with the z axis, with a constant absolute angular velocity $\dot{\theta}\hat{\mathbf{k}}$. Rod AB is attached to the cylinder and aligned with the y axis. Rod BC is perpendicular to AB and rotates around AB with the constant angular velocity $\dot{\phi}\hat{\mathbf{j}}$ relative to the cylinder. Rod CD is perpendicular to BC and rotates around BC with the constant angular velocity $\dot{\nu}\hat{\mathbf{m}}$ relative to BC, where $\hat{\mathbf{m}}$ is the unit vector in the direction of BC. The plate $abcd$ rotates around CD with a constant angular velocity $\dot{\psi}\hat{\mathbf{n}}$ relative to CD, where the unit vector $\hat{\mathbf{n}}$ points in the direction of CD. Thus, the absolute angular velocity of the plate is $\boldsymbol{\omega}_{\text{plate}} = \dot{\theta}\hat{\mathbf{k}} + \dot{\phi}\hat{\mathbf{j}} + \dot{\nu}\hat{\mathbf{m}} + \dot{\psi}\hat{\mathbf{n}}$. Show that

(a) $$\boldsymbol{\omega}_{\text{plate}} = (\dot{\nu}\sin\phi - \dot{\psi}\cos\phi\sin\nu)\hat{\mathbf{i}} + (\dot{\phi} + \dot{\psi}\cos\nu)\hat{\mathbf{j}} + (\dot{\theta} + \dot{\nu}\cos\phi + \dot{\psi}\sin\phi\sin\nu)\hat{\mathbf{k}}$$

(b) $$\begin{aligned}\boldsymbol{\alpha}_{\text{plate}} = \frac{d\boldsymbol{\omega}_{\text{plate}}}{dt} &= \left[\dot{\nu}\left(\dot{\phi}\cos\phi - \dot{\psi}\cos\phi\cos\nu\right) + \dot{\psi}\dot{\phi}\sin\phi\sin\nu - \dot{\psi}\dot{\theta}\cos\nu - \dot{\phi}\dot{\theta}\right]\hat{\mathbf{i}} \\ &+ \left[\dot{\nu}\left(\dot{\theta}\sin\phi - \dot{\psi}\sin\nu\right) - \dot{\psi}\dot{\theta}\cos\phi\sin\nu\right]\hat{\mathbf{j}} \\ &+ \left(\dot{\psi}\dot{\nu}\cos\nu\sin\phi + \dot{\psi}\dot{\phi}\cos\phi\sin\nu - \dot{\phi}\dot{\nu}\sin\varphi\right)\hat{\mathbf{k}}\end{aligned}$$

(c) $$\mathbf{a}_C = -l(\dot{\phi}^2 + \dot{\theta}^2)\sin\phi\hat{\mathbf{i}} + \left(2l\dot{\phi}\dot{\theta}\cos\phi - \frac{5}{4}l\dot{\theta}^2\right)\hat{\mathbf{j}} - l\dot{\phi}^2\cos\phi\hat{\mathbf{k}}$$

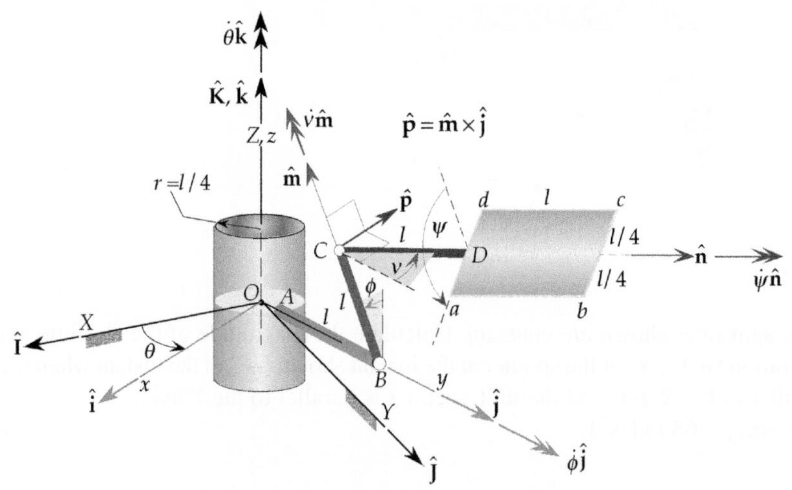

11.4 The mass center G of a rigid body has a velocity $\mathbf{v} = t^3\hat{\mathbf{i}} + 4\hat{\mathbf{j}}$ (m/s) and an angular velocity $\boldsymbol{\omega} = 2t^2\hat{\mathbf{k}}$ (rad/s), where t is time in seconds. The $\hat{\mathbf{i}}$, $\hat{\mathbf{j}}$, $\hat{\mathbf{k}}$ unit vectors are attached to and rotate with the rigid body. Calculate the magnitude of the acceleration $\mathbf{a}_G$ of the center of mass at $t = 2$ s.
{Ans.: $\mathbf{a}_G = -20\hat{\mathbf{i}} + 64\hat{\mathbf{j}}$ (m/s^2)}

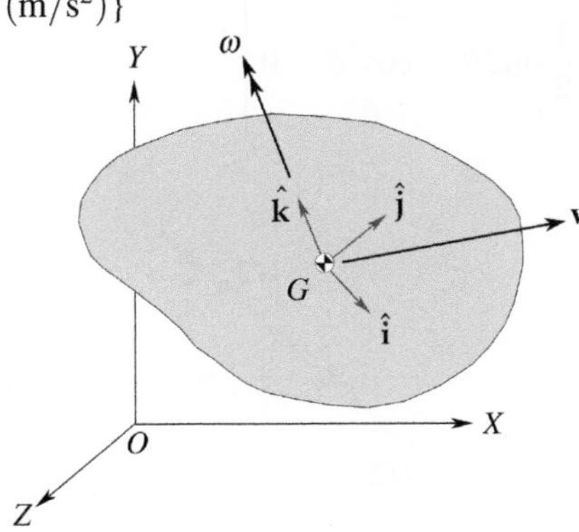

11.5 A rigid body is in pure rotation with angular velocity $\boldsymbol{\omega} = \omega_x\hat{\mathbf{i}} + \omega_y\hat{\mathbf{j}} + \omega_z\hat{\mathbf{k}}$ about the origin of the inertial xyz frame. If point A with position vector $\mathbf{r}_A = 2\hat{\mathbf{i}} + 2\hat{\mathbf{j}} - 2\hat{\mathbf{k}}$ (m) has velocity $\mathbf{v}_A = \hat{\mathbf{i}} + 2\hat{\mathbf{j}} + 3\hat{\mathbf{k}}$ (m/s), what is the magnitude of the velocity of the point B with position vector $\mathbf{r}_B = \hat{\mathbf{i}} + \hat{\mathbf{j}} - \hat{\mathbf{k}}$ (m)?
{Ans.: 1.871 m/s}

11.6 The inertial angular velocity of a rigid body is $\boldsymbol{\omega} = \omega_x\hat{\mathbf{i}} + \omega_y\hat{\mathbf{j}} + \omega_z\hat{\mathbf{k}}$, where $\hat{\mathbf{i}}$, $\hat{\mathbf{j}}$, and $\hat{\mathbf{k}}$ are the unit vectors of a comoving frame whose inertial angular velocity is $\boldsymbol{\omega} = \omega_x\hat{\mathbf{i}} + \omega_y\hat{\mathbf{j}}$. Calculate the components of angular acceleration of the rigid body in the moving frame, assuming that ω_x, ω_y, and ω_z are all constant.
{Ans.: $\boldsymbol{\alpha} = \omega_y\omega_z\hat{\mathbf{i}} - \omega_x\omega_z\hat{\mathbf{j}}$}

Section 5

11.7 Find the moments of inertia about the center of mass of the system of six point masses listed in the table.

Point, i	Mass, m_i (kg)	x_i (m)	y_i (m)	z_i (m)
1	10	1	1	1
2	10	−1	−1	−1
3	8	4	−4	4
4	8	−2	2	−2
5	12	3	−3	−3
6	12	−3	3	3

{Ans.: $[\mathbf{I}_G] = \begin{bmatrix} 783.5 & 351.7 & 40.27 \\ 351.7 & 783.5 & -80.27 \\ 40.27 & -80.27 & 783.5 \end{bmatrix}$ (kg·m^2)}

11.8 Find the mass moment of inertia of the configuration of Problem 11.7 about an axis through the origin and the point with coordinates (1, 2, 2 m).
{Ans.: 898.7 kg·m^2}

11.9 A uniform slender rod of mass m and length l lies in the xy plane inclined to the x axis by an angle θ. Use the results of Example 11.10 to find the mass moments of inertia about the xyz axes passing through the center of mass G.

$$\left\{\text{Ans.: } [\mathbf{I}_G] = \tfrac{1}{12}ml^2 \begin{bmatrix} \sin^2\theta & -\frac{1}{2}\sin 2\theta & 0 \\ -\frac{1}{2}\sin 2\theta & \cos^2\theta & 0 \\ 0 & 0 & 1 \end{bmatrix}\right\}$$

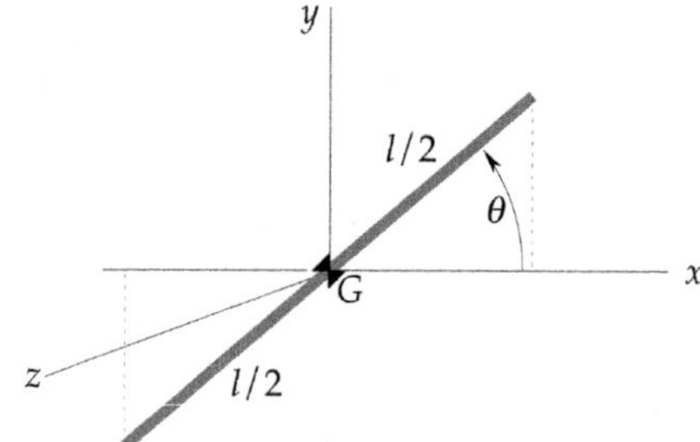

11.10 The uniform rectangular box has a mass of 1000 kg. The dimensions of its edges are shown.

(a) Find the mass moments of inertia about the xyz axes.

$$\left\{\text{Ans.: } [\mathbf{I}_O] = \begin{bmatrix} 1666.7 & -1500 & -750 \\ -1500 & 3333.3 & -500 \\ -750 & -500 & 4333.3 \end{bmatrix} (\text{kg}\cdot\text{m}^2)\right\}$$

(b) Find the principal moments of inertia and the principal directions about the xyz axes through O.
{Partial Ans.: $I_1 = 568.9\ \text{kg}\cdot\text{m}^2$, $\hat{\mathbf{e}}_1 = 0.8366\hat{\mathbf{i}} + 0.4960\hat{\mathbf{j}} + 0.2326\hat{\mathbf{k}}$}

(c) Find the moment of inertia about the line through O and the point with coordinates (3 m, 2 m, 1 m).
{Ans.: 583.3 $\text{kg}\cdot\text{m}^2$}

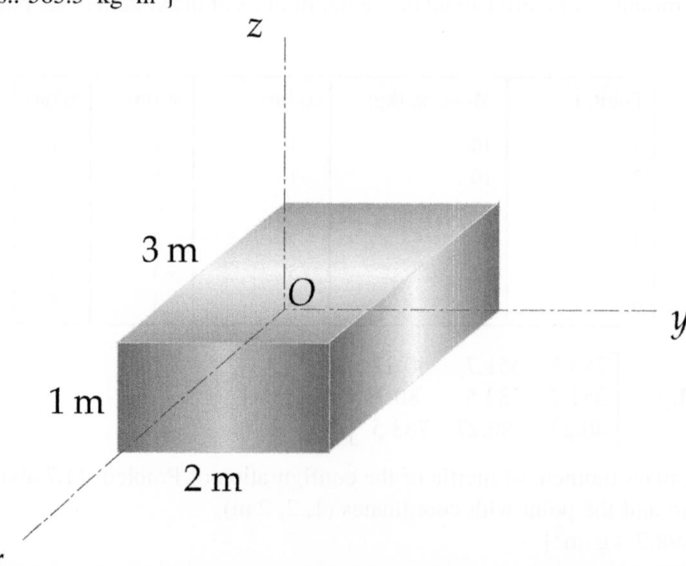

11.11 A taxiing airplane turns about its vertical axis with an angular velocity Ω while its propeller spins at an angular velocity $\omega = \dot{\theta}$. Determine the components of the angular momentum of the propeller about the body-fixed xyz axes centered at P. Treat the propeller as a uniform slender rod of mass m and length l.
{Ans.: $\mathbf{H}_p = \frac{1}{12} m\omega l^2 \hat{\mathbf{i}} - \frac{1}{24} m\Omega l^2 \sin 2\theta \hat{\mathbf{j}} + \left(\frac{1}{12} ml^2 \cos^2\theta + md^2\right)\Omega \hat{\mathbf{k}}$}

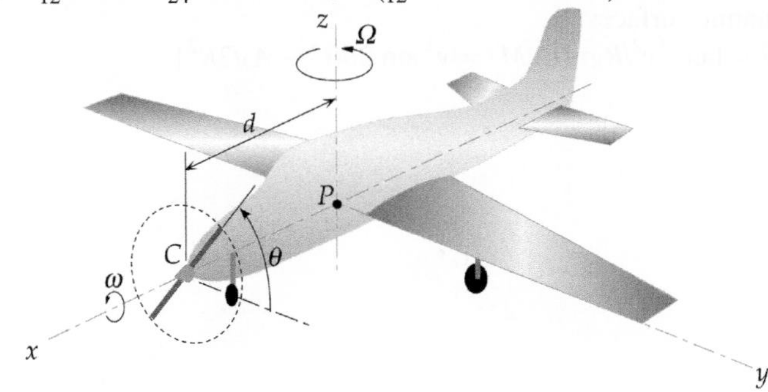

11.12 Relative to an xyz frame of reference the components of angular momentum $\mathbf{H}$ are given by

$$\{\mathbf{H}\} = \begin{bmatrix} 1000 & 0 & -300 \\ 0 & 1000 & 500 \\ -300 & 500 & 1000 \end{bmatrix} \begin{Bmatrix} \omega_x \\ \omega_y \\ \omega_z \end{Bmatrix} \ (\mathrm{kg \cdot m^2/s})$$

where ω_x, ω_y, and ω_z are the components of the angular velocity vector $\boldsymbol{\omega}$. Find the components $\boldsymbol{\omega}$ such that $\{\mathbf{H}\} = 1000\{\boldsymbol{\omega}\}$, where the magnitude of $\boldsymbol{\omega}$ is 20 rad/s.
{Ans.: $\boldsymbol{\omega} = 174.15\hat{\mathbf{i}} + 10.29\hat{\mathbf{j}}$ (rad/s)}

11.13 Relative to a body-fixed xyz frame $[\mathbf{I}_G] = \begin{bmatrix} 10 & 0 & 0 \\ 0 & 20 & 0 \\ 0 & 0 & 30 \end{bmatrix}$ $(\mathrm{kg \cdot m^2})$ and $\boldsymbol{\omega} = 2t^2\hat{\mathbf{i}} + 4\hat{\mathbf{j}} + 3t\hat{\mathbf{k}}$ (rad/s), where t is the time in seconds. Calculate the magnitude of the net moment about the center of mass G at $t = 3$ s.
{Ans.: 3374 N m}

11.14 In Example 11.11, the system is at rest when a 100-N force is applied to point A as shown. Calculate the inertial components of angular acceleration at that instant.
{Ans.: $\alpha_X = 143.9\ \mathrm{rad/s^2}$, $\alpha_Y = 553.1\ \mathrm{rad/s^2}$, $\alpha_Z = 7.61\ \mathrm{rad/s^2}$}

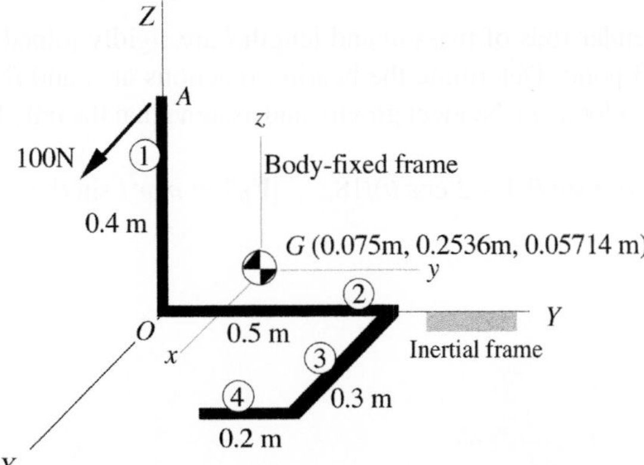

11.15 The body-fixed xyz axes pass through the center of mass G of the airplane and are the principal axes of inertia. The moments of inertia about these axes are A, B, and C, respectively. The airplane is in a level turn of radius R with a speed v.

(a) Calculate the bank angle θ.

(b) Use the Euler equations to calculate the rolling moment M_y that must be applied by the aerodynamic surfaces.

{Ans.: (a) $\theta = \tan^{-1} v^2/Rg$; (b) $M_y = v^2 \sin 2\theta(C - A)/2R^2$}

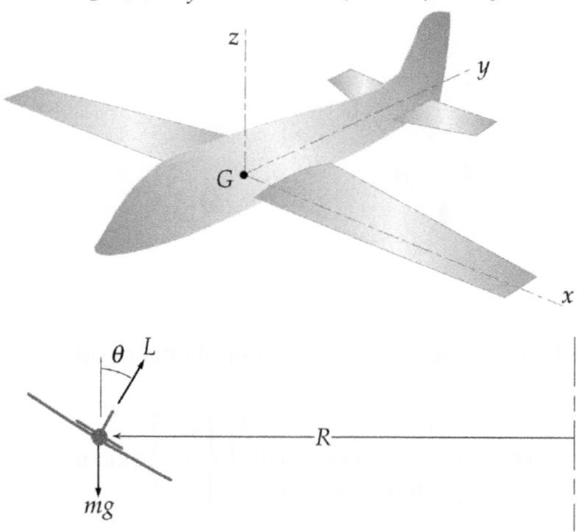

11.16 The airplane in Problem 11.15 is spinning with an angular velocity ω_Z about the vertical Z axis. The nose is pitched down at the angle α. What external moments must accompany this maneuver?

{Ans.: $M_y = M_z = 0$, $M_x = \omega_Z{}^2 \sin 2\alpha (C - B)/2$}

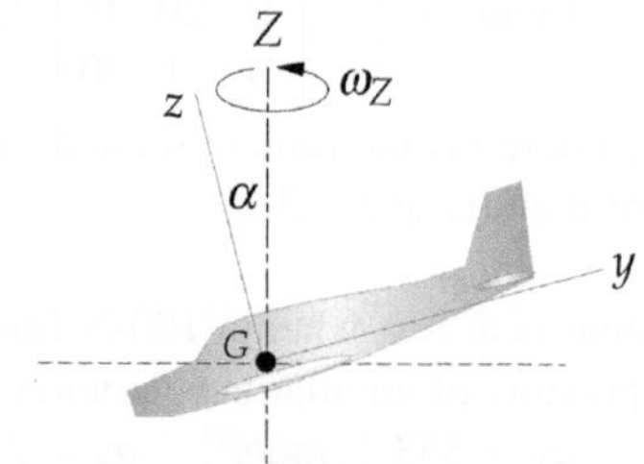

11.17 Two identical slender rods of mass m and length l are rigidly joined together at an angle θ at point C, their 2/3 point. Determine the bearing reactions at A and B if the shaft rotates at a constant angular velocity ω. Neglect gravity and assume that the only bearing forces are normal to rod AB.

{Ans.: $\|\mathbf{F}_A\| = m\omega^2 l \sin\theta(1 + 2\cos\theta)/18$, $\|\mathbf{F}_B\| = m\omega^2 l \sin\theta(1 - \cos\theta)/9$}

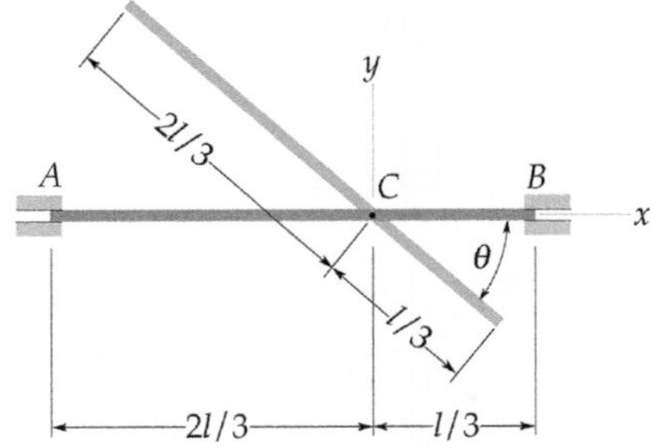

11.18 The flywheel ($A = B = 5\ \text{kg}\cdot\text{m}^2$, $C = 10\ \text{kg}\cdot\text{m}^2$) spins at a constant angular velocity of $\boldsymbol{\omega}_s = 100\hat{\mathbf{k}}$ (rad/s). It is supported by a massless gimbal that is mounted on the platform as shown. The gimbal is initially stationary relative to the platform, which rotates with a constant angular velocity of $\boldsymbol{\omega}_p = 0.5\hat{\mathbf{j}}$ (rad/s). What will be the gimbal's angular acceleration when the torquer applies a torque of $600\hat{\mathbf{i}}$ (N m) to the flywheel?
{Ans.: $70\hat{\mathbf{i}}$ (rad/s^2)}

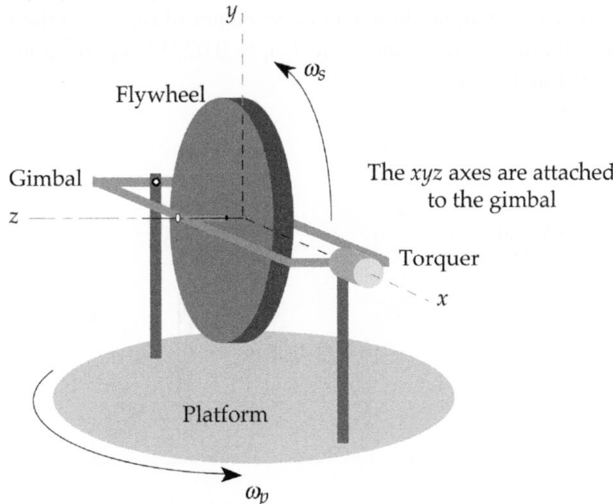

11.19 A uniform slender rod of length L and mass m is attached by a smooth pin at O to a vertical shaft that rotates at constant angular velocity ω. Use the Euler equations and the body frame shown to calculate ω at the instant shown.
{Ans.: $\omega = \sqrt{3g/(2L\cos\theta)}$}

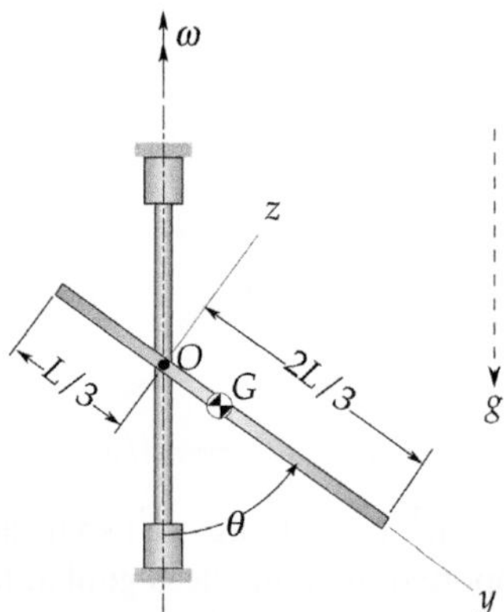

11.20 A uniform, thin circular disk of mass 10 kg spins at a constant angular velocity of 630 rad/s about axis OG, which is normal to the disk and pivots about the frictionless ball joint at O. Neglecting the mass of the shaft OG, determine the rate of precession if OG remains horizontal as shown. Gravity acts down, as shown. G is the center of mass and the y axis remains fixed in space. The moments of inertia about G are $I_{G})_z = 0.02812\ \text{kg}\cdot\text{m}^2$ and $I_{G})_x = I_{G})_y = 0.01406\ \text{kg}\cdot\text{m}^2$
{Ans.: 1.38 rad/s}

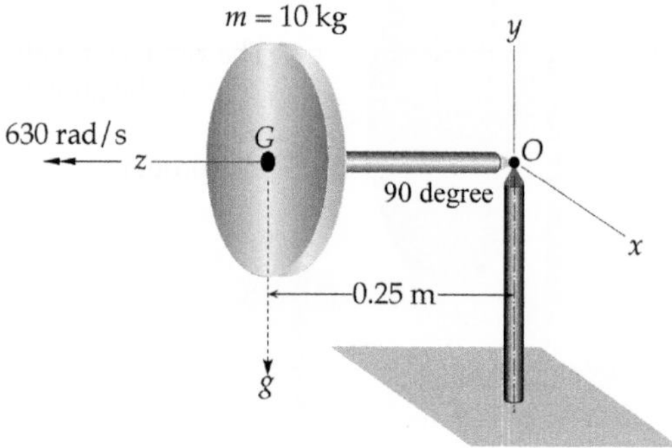

Section 7

11.21 Consider a rigid body experiencing rotational motion associated with an angular velocity vector $\boldsymbol{\omega}$. The inertia tensor (relative to body-fixed axes through the center of mass G) is

$$\begin{bmatrix} 20 & -10 & 0 \\ -10 & 30 & 0 \\ 0 & 0 & 40 \end{bmatrix} \ (\text{kg}\cdot\text{m}^2)$$

and $\boldsymbol{\omega} = 10\hat{\mathbf{i}} + 20\hat{\mathbf{j}} + 30\hat{\mathbf{k}}$ (rad/s). Calculate

(a) the angular momentum $\mathbf{H}_G$ and

(b) the rotational kinetic energy (about G).

{Partial Ans.: (b) $T_R = 23,000$ J}

Section 8

11.22 At the end of its takeoff run, an airplane with retractable landing gear leaves the runway with a speed of 130 km/h. The gear rotates into the wing with an angular velocity of 0.8 rad/s with the wheels still spinning. Calculate the gyroscopic bending moment in the wheel bearing B. The wheels have a diameter of 0.6 m, a mass of 25 kg, and a radius of gyration of 0.2 m.
{Ans.: 96.3 N m}

11.23 The gyro rotor, including shaft AB, has a mass of 4 kg and a radius of gyration 7 cm around AB. The rotor spins at 10,000 rpm while also being forced to rotate around the gimbal axis CC at 2 rad/s. What are the transverse forces exerted on the shaft at A and B? Neglect gravity.
{Ans.: 1.03 kN}

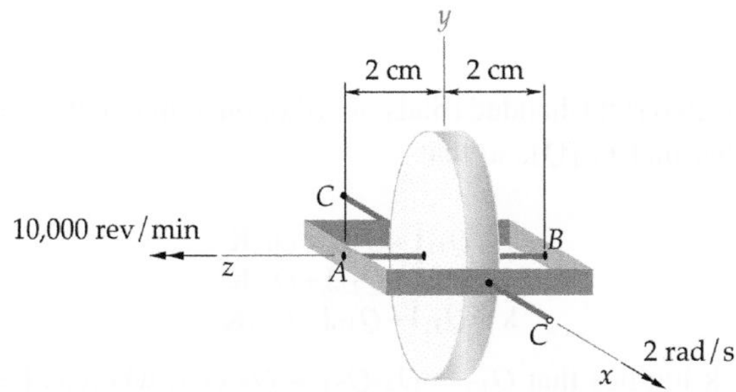

11.24 A jet aircraft is making a level, 2.5-km radius turn to the left at a speed of 650 km/h. The rotor of the turbojet engine has a mass of 200 kg, a radius of gyration of 0.25 m, and rotates at 15,000 rpm clockwise as viewed from the front of the airplane. Calculate the gyroscopic moment that the engine exerts on the airframe and explain why it tends to pitch the nose up or down.
{Ans.: 1.418 kN m; pitch down}

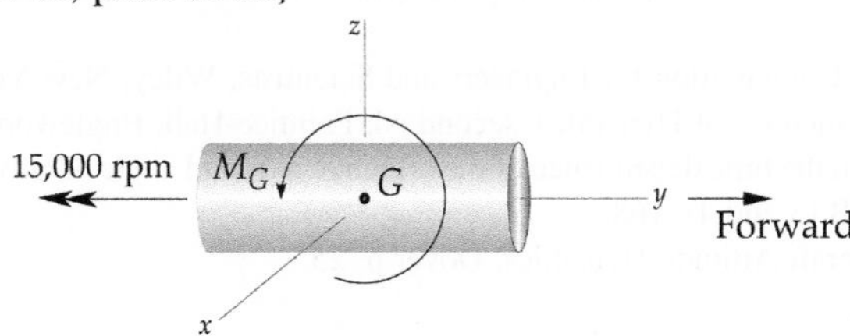

11.25 A cylindrical rotor of mass 10 kg, radius 0.05 m, and length 0.60 m is simply supported at each end in a cradle that rotates at a constant 20 rad/s counterclockwise as viewed from above. Relative to the cradle, the rotor spins at 200 rad/s counterclockwise as viewed from the right

(from B toward A). Assuming that there is no gravity, calculate the bearing reactions R_A and R_B. Use the comoving xyz frame shown, which is attached to the cradle but not to the rotor.
{Ans.: $R_A = -R_B = 83.3$ N}

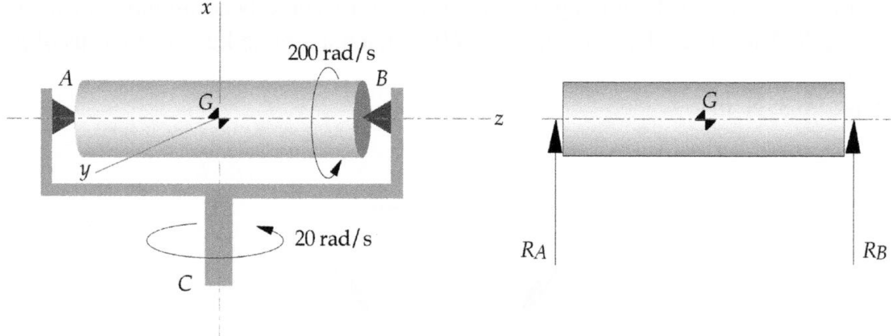

Section 9

11.26 The Euler angles of a rigid body are $\phi = 50°$, $\theta = 25°$, and $\psi = 70°$. Calculate the angle (a positive number) between the body-fixed x axis and the inertial X axis.
{Ans.: 115.6°}

Section 11

11.27 Let $\hat{\mathbf{i}}\hat{\mathbf{j}}\hat{\mathbf{k}}$ and $\hat{\mathbf{I}}\hat{\mathbf{J}}\hat{\mathbf{K}}$ be two right-handed triads of orthogonal unit vectors related as in Eq. (4.18) by the direction cosine matrix $[\mathbf{Q}]$, so that

$$\hat{\mathbf{i}} = Q_{11}\hat{\mathbf{I}} + Q_{12}\hat{\mathbf{J}} + Q_{13}\hat{\mathbf{K}}$$
$$\hat{\mathbf{j}} = Q_{21}\hat{\mathbf{I}} + Q_{22}\hat{\mathbf{J}} + Q_{23}\hat{\mathbf{K}}$$
$$\hat{\mathbf{k}} = Q_{31}\hat{\mathbf{I}} + Q_{32}\hat{\mathbf{J}} + Q_{33}\hat{\mathbf{K}}$$

Show that $\hat{\mathbf{i}} = \hat{\mathbf{j}} \times \hat{\mathbf{k}}$ implies that $Q_{13} = Q_{32}Q_{21} - Q_{22}Q_{31}$, whereas $\hat{\mathbf{j}} = \hat{\mathbf{k}} \times \hat{\mathbf{i}}$ implies that

$$Q_{23} = Q_{12}Q_{31} - Q_{32}Q_{11}.$$

REFERENCES

Bar-Itzhack, I.Y., 2000. New Method for Extracting the Quaternion From the Rotation Matrix. J. Guid. Control Dyn. 23, 1085–1087.
Jennings, A., 1977. Matrix Computation for Engineers and Scientists. Wiley, New York.
Greenwood, D.T., 1988. Principles of Dynamics, second ed. Prentice-Hall, Englewood Cliffs, NJ.
Gelman, H., 1971. A note on the time dependence of the effective axis and angle of a rotation. J. Res. Natl. Bureau Stand. B Math. Sci. 72B (3 and 4), 168.
Hughes, P.C., 2004. Spacecraft Attitude Dynamics. Dover p. 25.

CHAPTER

SPACECRAFT ATTITUDE DYNAMICS

12

12.1 INTRODUCTION

In this chapter, we apply the equations of rigid body motion presented in Chapter 11 to the study of the attitude dynamics of satellites. We begin with spin-stabilized spacecraft. Spinning a satellite around its axis is a very simple way to keep the vehicle pointed in a desired direction. We investigate the stability of a spinning satellite to show that only oblate spinners are stable over long times. Overcoming this restriction on the shape of spin-stabilized spacecraft led to the development of dual-spin vehicles, which consist of two interconnected segments rotating at different rates about a common axis. We consider the stability of that type of configuration as well. The nutation damper and its effect on the stability of spin-stabilized spacecraft are covered next.

The rest of the chapter is devoted to some of the common means of changing the attitude or motion of a spacecraft by applying external or internal forces or torques. The coning maneuver changes the attitude of a spinning spacecraft by using thrusters to apply impulsive torque, which alters the angular momentum and hence the orientation of the spacecraft. The much-used yo-yo despin maneuver reduces or eliminates the spin rate by releasing small masses attached to cords initially wrapped around the cylindrical vehicle.

An alternative to spin stabilization is three-axis stabilization by gyroscopic attitude control. In this case, the vehicle does not continuously rotate. Instead, the desired attitude is maintained by the spin of small wheels within the spacecraft. These are called reaction wheels or momentum wheels. If allowed to pivot relative to the vehicle, they are known as control moment gyros. The attitude of the vehicle can be changed by varying the speed or orientation of these internal gyros. Small thrusters may also be used to supplement gyroscopic attitude control and to hold the spacecraft orientation fixed when it is necessary to despin or reorient the gyros that have become saturated (reached their maximum spin rate or deflection) over time.

The chapter concludes with a discussion of how the earth's gravitational field by itself can stabilize the attitude of large satellites in low earth orbits.

12.2 TORQUE-FREE MOTION

Gravity is the only force acting on a satellite coasting in orbit (if we neglect secondary drag forces and the gravitational influence of bodies other than the planet being orbited). Unless the satellite is

Orbital Mechanics for Engineering Students. https://doi.org/10.1016/B978-0-12-824025-0.00012-X

unusually large, the gravitational force is concentrated at the center of mass G. Since the net moment about the center of mass is zero, the satellite is torque free and according to Eq. (11.30),

$$\dot{\mathbf{H}}_G = \mathbf{0} \tag{12.1}$$

The angular momentum $\mathbf{H}_G$ about the center of mass does not depend on time. It is a vector fixed in inertial space. We will use $\mathbf{H}_G$ to define the Z axis of an inertial frame, as shown in Fig. 12.1. The xyz axes in the figure comprise the principal body frame, centered at G. The angle between the z axis and $\mathbf{H}_G$ is (by definition of the Euler angles) the nutation angle θ. Let us determine the conditions for which θ is constant. From the dot product operation, we know that

$$\cos\theta = \frac{\mathbf{H}_G}{\|\mathbf{H}_G\|} \cdot \hat{\mathbf{k}}$$

Differentiating this expression with respect to time, keeping in mind Eq. (12.1), we get

$$\frac{d\cos\theta}{dt} = \frac{\mathbf{H}_G}{\|H_G\|} \cdot \frac{d\hat{\mathbf{k}}}{dt}$$

But $d\hat{\mathbf{k}}/dt = \boldsymbol{\omega} \times \hat{\mathbf{k}}$, according to Eq. (1.52), so

$$\frac{d\cos\theta}{dt} = \frac{\mathbf{H}_G \cdot \left(\boldsymbol{\omega} \times \hat{\mathbf{k}}\right)}{\|\mathbf{H}_G\|} \tag{12.2}$$

Now,

$$\boldsymbol{\omega} \times \hat{\mathbf{k}} = \left(\omega_x\hat{\mathbf{i}} + \omega_y\hat{\mathbf{j}} + \omega_z\hat{\mathbf{k}}\right) \times \hat{\mathbf{k}} = \omega_y\hat{\mathbf{i}} - \omega_x\hat{\mathbf{j}}$$

Furthermore, we know from Eq. (11.67) that the angular momentum is related to the angular velocity in the principal body frame by the expression

$$\mathbf{H}_G = A\omega_x\hat{\mathbf{i}} + B\omega_y\hat{\mathbf{j}} + C\omega_z\hat{\mathbf{k}}$$

Thus,

$$\mathbf{H}_G \cdot \left(\boldsymbol{\omega} \times \hat{\mathbf{k}}\right) = \left(A\omega_x\hat{\mathbf{i}} + B\omega_y\hat{\mathbf{j}} + C\omega_z\hat{\mathbf{k}}\right) \cdot \left(\omega_y\hat{\mathbf{i}} - \omega_x\hat{\mathbf{j}}\right) = (A - B)\omega_x\omega_y$$

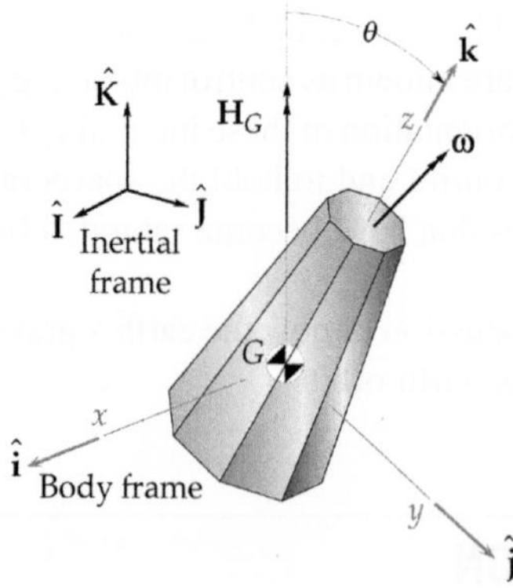

FIG. 12.1

Rotationally symmetric satellite in torque-free motion.

so that Eq. (12.2) can be written as

$$\dot{\theta} = \omega_n = -\frac{(A-B)\omega_x\omega_y}{\|\mathbf{H}_G\|\sin\theta} \tag{12.3}$$

From this, we see that the nutation rate $\dot{\theta}$ vanishes only if $A = B$. If $A \neq B$, the nutation angle θ will not in general be constant.

Relative to the body frame, Eq. (12.1) is written (cf. Eq. 1.56) as

$$\dot{\mathbf{H}}_{G}\Big)_{\text{rel}} + \boldsymbol{\omega} \times \mathbf{H}_G = \mathbf{0}$$

This is the Euler equation with $\mathbf{M}_{G})_{\text{net}} = \mathbf{0}$, and its components are given by Eq. (11.72b),

$$\begin{aligned} A\dot{\omega}_x + (C-B)\omega_z\omega_y &= 0 \\ B\dot{\omega}_y + (A-C)\omega_x\omega_z &= 0 \\ C\dot{\omega}_z + (B-A)\omega_y\omega_x &= 0 \end{aligned} \tag{12.4}$$

In the interest of simplicity, let us consider the special case illustrated in Fig. 12.1 (namely, that in which the z axis is an axis of rotational symmetry), so that $A = B$. Then Eq. (12.4) may be written

$$\begin{aligned} A\dot{\omega}_x + (C-A)\omega_y\omega_z &= 0 \\ A\dot{\omega}_y + (A-C)\omega_z\omega_x &= 0 \\ C\dot{\omega}_z &= 0 \end{aligned} \tag{12.5}$$

From Eq. (12.5_3) we see that the body frame z component of the angular velocity is constant.

$$\omega_z = \omega_o \text{ (constant)} \tag{12.6}$$

The assumption of rotational symmetry therefore reduces the three differential equations in Eq. (12.4) to just the first two in Eq. (12.5). Substituting Eq. (12.6) into Eqs. (12.5_1) and (12.5_2) and introducing the notation

$$\lambda = \frac{A-C}{A}\omega_o \tag{12.7}$$

they can be written as

$$\begin{aligned} \dot{\omega}_x - \lambda\omega_y &= 0 \\ \dot{\omega}_y + \lambda\omega_x &= 0 \end{aligned} \tag{12.8}$$

Note that the sign of λ depends on the relative values of the principal moments of inertia A and C.

To reduce Eq. (12.8) in ω_x and ω_y to just one equation in ω_x, we first differentiate Eq. (12.8_1) with respect to time to get

$$\ddot{\omega}_x - \lambda\dot{\omega}_y = 0 \tag{12.9}$$

We then solve Eq. (12.8_2) for $\dot{\omega}_y$ and substitute the result into Eq. (12.9), which leads to

$$\ddot{\omega}_x + \lambda^2\omega_x = 0 \tag{12.10}$$

The solution of this well-known differential equation is

$$\omega_x = \omega_{xy}\sin\lambda t \tag{12.11}$$

where the constant amplitude ω_{xy} ($\omega_{xy} \neq 0$) has yet to be determined. (Without loss of generality, we have set the phase angle, the other constant of integration, equal to zero.) Substituting Eq. (12.11) back into Eq. (12.8_1) yields the solution for ω_y,

$$\omega_y = \frac{1}{\lambda}\frac{d\omega_x}{dt} = \frac{1}{\lambda}\frac{d}{dt}(\omega_{xy}\sin\lambda t)$$

or

$$\omega_y = \omega_{xy}\cos\lambda t \tag{12.12}$$

Eqs. (12.6), (12.11), and (12.12) give the components of the absolute angular velocity vector $\boldsymbol{\omega}$ along the three principal body axes,

$$\boldsymbol{\omega} = \omega_{xy}\sin\lambda t\hat{\mathbf{i}} + \omega_{xy}\cos\lambda t\hat{\mathbf{j}} + \omega_o\hat{\mathbf{k}}$$

or

$$\boldsymbol{\omega} = \boldsymbol{\omega}_\perp + \omega_o\hat{\mathbf{k}} \tag{12.13}$$

where

$$\boldsymbol{\omega}_\perp = \boldsymbol{\omega}_{xy}\left(\sin\lambda t\hat{\mathbf{i}} + \cos\lambda t\hat{\mathbf{j}}\right) \tag{12.14}$$

$\boldsymbol{\omega}_\perp$ (omega-perp) is the component of $\boldsymbol{\omega}$ normal to the z axis. It sweeps out a circle of radius ω_{xy} in the xy plane at an angular velocity λ. Thus, $\boldsymbol{\omega}$ sweeps out a cone, as illustrated in Fig. 12.2. If ω_o is positive, then the body has an inertial counterclockwise rotation around the positive z axis ($\lambda > 0$) if $A > C$. However, an observer fixed in the body would see the world rotating in the opposite direction, clockwise around positive z, as the figure shows. Of course, the situation is reversed if $A < C$.

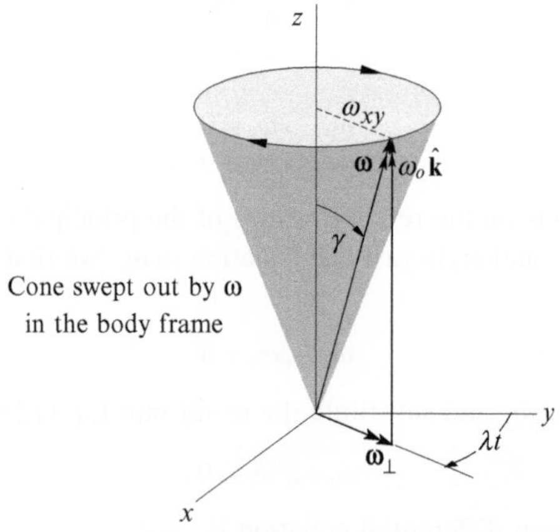

FIG. 12.2

Components of the angular velocity $\boldsymbol{\omega}$ in the body frame.

From Eq. (11.116), the three Euler orientation angles (and their rates) are related to the body frame angular velocity components ω_x, ω_y, and ω_z by

$$\omega_p = \dot{\phi} = \frac{1}{\sin\theta}(\omega_x \sin\psi + \omega_y \cos\psi)$$
$$\omega_n = \dot{\theta} = \omega_x \cos\psi - \omega_y \sin\psi$$
$$\omega_s = \dot{\psi} = -\frac{1}{\tan\theta}(\omega_x \sin\psi + \omega_y \cos\psi) + \omega_z$$

Substituting Eqs. (12.6), (12.11), and (12.12) into these three equations yields

$$\begin{aligned}\omega_p &= \frac{\omega_{xy}}{\sin\theta}\cos(\lambda t - \psi)\\ \omega_n &= \omega_{xy}\sin(\lambda t - \psi)\\ \omega_s &= \omega_o - \frac{\omega_{xy}}{\tan\theta}\cos(\lambda t - \psi)\end{aligned} \tag{12.15}$$

Since $A = B$, we know from Eq. (12.3) that $\omega_n = 0$. It follows from Eq. (12.15_2) that

$$\psi = \lambda t \tag{12.16}$$

(Actually, $\lambda t - \psi = n\pi$, $n = 0, 1, 2, \ldots$. We can set $n = 0$ without loss of generality.) Substituting Eq. (12.16) into Eqs. (12.15_1) and (12.15_3) yields

$$\omega_p = \frac{\omega_{xy}}{\sin\theta} \tag{12.17}$$

and

$$\omega_s = \omega_o - \frac{\omega_{xy}}{\tan\theta} \tag{12.18}$$

We have thus obtained the Euler angle rates ω_p and ω_s in terms of the components of the angular velocity $\boldsymbol{\omega}$ in the body frame.

Differentiating Eq. (12.16) with respect to time shows that

$$\lambda = \dot{\psi} = \omega_s \tag{12.19}$$

That is, the rate λ at which $\boldsymbol{\omega}$ rotates around the body's z axis equals the spin rate. Substituting the spin rate for λ in Eq. (12.7) shows that ω_s is related to ω_o alone,

$$\omega_s = \frac{A - C}{A}\omega_o \tag{12.20}$$

Observe that ω_s and ω_o are opposite in sign if $A < C$.

Eliminating ω_s from Eqs. (12.18) and (12.20) yields the relationship between the magnitudes of the orthogonal components of the angular velocity in Eq. (12.13),

$$\omega_{xy} = \frac{C}{A}\omega_o \tan\theta \tag{12.21}$$

A similar relationship exists between ω_p and ω_s, which generally are *not* orthogonal. Substitute Eq. (12.21) into Eq. (12.17) to obtain

$$\omega_o = \frac{A}{C}\omega_p \cos\theta \tag{12.22}$$

Placing this result in Eq. (12.20) leaves an expression involving only ω_p and ω_s, from which we get a useful formula relating the precession of a torque-free body to its spin,

$$\omega_p = \frac{C}{A-C}\frac{\omega_s}{\cos\theta} \tag{12.23}$$

Observe that if $A > C$ (i.e., the body is prolate, like a soup can or an American football), then ω_p has the same sign as ω_s, which means the precession is prograde. For an oblate body (like a tuna fish can or a frisbee), $A < C$ and the precession is retrograde.

The components of angular momentum along the body frame axes are obtained from the body frame components of $\boldsymbol{\omega}$,

$$\mathbf{H}_G = A\omega_x\hat{\mathbf{i}} + A\omega_y\hat{\mathbf{j}} + C\omega_z\hat{\mathbf{k}}$$

or

$$\mathbf{H}_G = \mathbf{H}_\perp + C\omega_o\hat{\mathbf{k}} \tag{12.24}$$

where

$$\mathbf{H}_\perp = A\omega_{xy}\left(\sin\omega_s t\hat{\mathbf{i}} + \cos\omega_s t\hat{\mathbf{j}}\right) = A\boldsymbol{\omega}_\perp \tag{12.25}$$

Since $\omega_o\hat{\mathbf{k}}$ and $C\omega_o\hat{\mathbf{k}}$ are collinear, as are $\boldsymbol{\omega}_\perp$ and $A\boldsymbol{\omega}_\perp$, it follows that $\hat{\mathbf{k}}$, $\boldsymbol{\omega}$, and $\mathbf{H}_G$ all lie on the same plane. $\mathbf{H}_G$ and $\boldsymbol{\omega}$ both rotate around the z axis at the same rate ω_s. These details are illustrated in Fig. 12.3. See how the precession and spin angular velocities, ω_p and ω_s, add up vectorially to give $\boldsymbol{\omega}$. Note also that from the point of view of inertial space, where $\mathbf{H}_G$ is fixed, $\boldsymbol{\omega}$ and $\hat{\mathbf{k}}$ rotate around $\mathbf{H}_G$ with angular velocity ω_p.

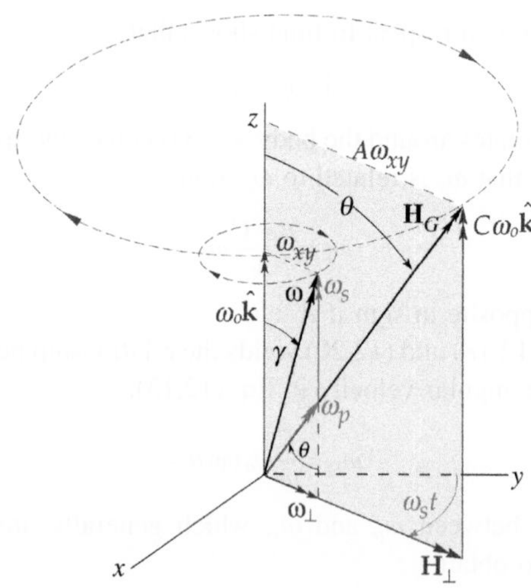

FIG. 12.3

Angular velocity and angular momentum vectors in the body frame ($A > C$).

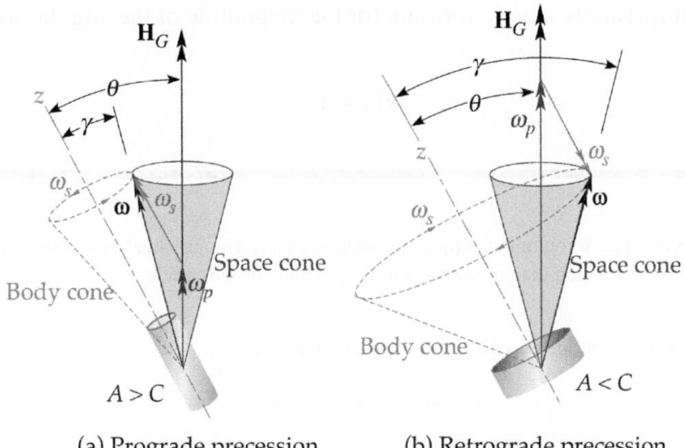

FIG. 12.4

Space and body cones for a rotationally symmetric body in torque-free motion. (a) Prolate body and (b) oblate body.

Let γ be the angle between the angular velocity vector $\boldsymbol{\omega}$ and the spin axis z, as shown in Figs. 12.2 and 12.3. γ is sometimes referred to as the wobble angle. From the figures, it is clear that $\tan\gamma = \omega_{xy}/\omega_0$ and $\tan\theta = A\omega_{xy}/C\omega_0$. It follows that

$$\tan\theta = \frac{A}{C}\tan\gamma \tag{12.26}$$

From this, we conclude that if $A > C$, then $\gamma < \theta$, whereas $C > A$ means $\gamma > \theta$. That is, the angular velocity vector $\boldsymbol{\omega}$ lies between the z axis and the angular momentum vector $\mathbf{H}_G$ when $A > C$ (prolate body). On the other hand, when $C > A$ (oblate body), $\mathbf{H}_G$ lies between the z axis and $\boldsymbol{\omega}$. These two situations are illustrated in Fig. 12.4, which also shows the body cone and space cone. The space cone is swept out in inertial space by the angular velocity vector as it rotates with angular velocity ω_p around $\mathbf{H}_G$, whereas the body cone is the trace of $\boldsymbol{\omega}$ in the body frame as it rotates with angular velocity ω_s about the z axis. From inertial space, the motion may be visualized as the body cone rolling on the space cone, with the line of contact being the angular velocity vector. From the body frame, it appears as though the space cone rolls on the body cone. Fig. 12.4 graphically confirms our deduction from Eq. (12.23) (namely, that precession and spin are in the same direction for prolate bodies and opposite in direction for oblate shapes).

Finally, we know from Eqs. (12.24) and (12.25) that the magnitude $\|\mathbf{H}_G\|$ of the angular momentum is

$$\|\mathbf{H}_G\| = H_G = \sqrt{A^2\omega_{xy}{}^2 + C^2\omega_o{}^2}$$

Using Eqs. (12.17) and (12.22), we can write this as

$$\mathbf{H}_G = \sqrt{A^2(\omega_p \sin\theta)^2 + C^2\left(\frac{A}{C}\omega_p\cos\theta\right)^2} = \sqrt{A^2\omega_p(\sin^2\theta + \cos^2\theta)}$$

so that we obtain a surprisingly simple formula for the magnitude of the angular momentum in torque-free motion,

$$H_G = A\omega_p \tag{12.27}$$

EXAMPLE 12.1

The cylindrical shell in Fig. 12.5 is rotating in torque-free motion about its longitudinal axis. If the axis is wobbling slightly, determine the ratios of l/r for which the precession will be prograde or retrograde.

Solution

Fig. 11.10 shows the moments of inertia of a thin-walled circular cylinder,

$$C = mr^2 \qquad A = \frac{1}{2}mr^2 + \frac{1}{12}ml^2$$

According to Eq. (12.23) and Fig. 12.4, direct or prograde precession exists if $A > C$. That is, if

$$\frac{1}{2}mr^2 + \frac{1}{12}ml^2 > mr^2$$

or

$$\frac{1}{12}ml^2 > \frac{1}{2}mr^2$$

Thus,

$$\boxed{\begin{array}{l} l > 2.45r \Rightarrow \text{Direct precession} \\ l < 2.45r \Rightarrow \text{Rectrograde precession} \end{array}}$$

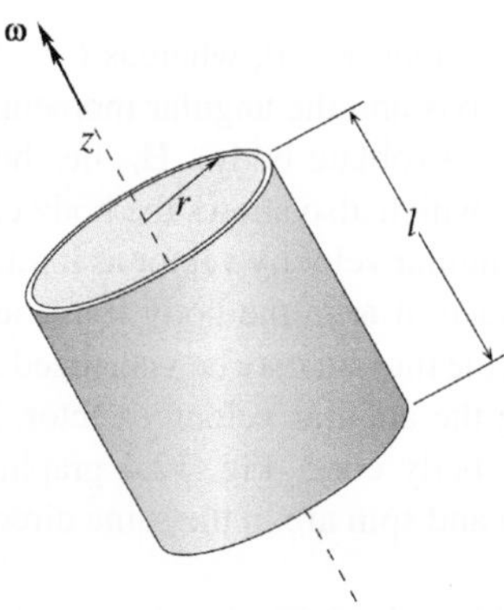

FIG. 12.5

Cylindrical shell in torque-free motion.

EXAMPLE 12.2

In the previous example, let $r = 1$ m, $l = 3$ m, $m = 100$ kg, and let the nutation angle θ be 20°. How long does it take the cylinder to precess through 180° if the spin rate is 2π rad/min?

Solution

Since $l > 2.45r$, the precession is direct. Furthermore,

$$\begin{aligned} C &= mr^2 &&= 100 \cdot 1^2 &&= 100 \text{ kg} \cdot \text{m}^2 \\ A &= \frac{1}{2}mr^2 + \frac{1}{12}ml^2 &&= \frac{1}{2} \cdot 100 \cdot 1^2 + \frac{1}{12}100 \cdot 3^2 &&= 125 \text{ kg} \cdot \text{m}^2 \end{aligned}$$

Thus, Eq. (12.23) yields

$$\omega_p = \frac{C}{A - C\cos\theta}\omega_s = \frac{100}{125 - 100\cos 20°}\frac{2\pi}{} = 26.75 \text{ rad/min}$$

At this rate, the time for the spin axis to precess through an angle of 180° is

$$t = \frac{\pi}{\omega_p} = \boxed{0.1175 \text{ min}}$$

EXAMPLE 12.3

What is the torque-free motion of a spacecraft for which $A = B = C$?

Solution

If $A = B = C$, the spacecraft is spherically symmetric. Any orthogonal triad at the center of mass G is a principal body frame, so $\mathbf{H}_G$ and $\boldsymbol{\omega}$ are collinear,

$$\mathbf{H}_G = C\boldsymbol{\omega}$$

Substituting this and $\mathbf{M}_G)_{\text{net}} = \mathbf{0}$ into the Euler equations (Eq. 11.72) yields

$$C\frac{d\boldsymbol{\omega}}{dt} + \boldsymbol{\omega} \times (C\boldsymbol{\omega}) = \mathbf{0}$$

That is, $\boldsymbol{\omega}$ is constant. The angular velocity vector of a spherically symmetric spacecraft in torque-free motion is fixed in magnitude and direction. We considered this problem in more detail in Example 11.25.

EXAMPLE 12.4

The inertial components of the angular momentum of a torque-free rigid body are

$$\mathbf{H}_G = 320\hat{\mathbf{I}} - 375\hat{\mathbf{J}} + 450\hat{\mathbf{K}} \ (\text{kg} \cdot \text{m}^2/\text{s}) \tag{a}$$

The Euler angles are

$$\phi = 20° \qquad \theta = 50° \qquad \psi = 75° \tag{b}$$

If the inertia tensor in the body-fixed principal frame is

$$[\mathbf{I}_G] = \begin{bmatrix} 1000 & 0 & 0 \\ 0 & 2000 & 0 \\ 0 & 0 & 3000 \end{bmatrix} (\text{kg} \cdot \text{m}^2) \tag{c}$$

calculate the inertial components of the (absolute) angular acceleration.

Solution

Substituting the Euler angles from Eq. (b) into Eq. (11.104), we obtain the direction cosine matrix of the transformation from the inertial frame to the body-fixed frame,

$$[\mathbf{Q}]_{Xx} = \begin{bmatrix} 0.03086 & 0.6720 & 0.7399 \\ -0.9646 & -0.1740 & 0.1983 \\ 0.2620 & -0.7198 & 0.6428 \end{bmatrix} \tag{d}$$

We use this to obtain the components of $\mathbf{H}_G$ in the body frame,

$$\begin{aligned}\{\mathbf{H}_G\}_x = [\mathbf{Q}]_{Xx}\{\mathbf{H}_G\}_X &= \begin{bmatrix} 0.03086 & 0.6720 & 0.7399 \\ -0.9646 & -0.1740 & 0.1983 \\ 0.2620 & -0.7198 & 0.6428 \end{bmatrix} \begin{Bmatrix} 320 \\ -375 \\ 450 \end{Bmatrix} \\ &= \begin{Bmatrix} 90.86 \\ -154.2 \\ 643.0 \end{Bmatrix} (\mathrm{kg \cdot m^2/s}) \end{aligned} \tag{e}$$

In the body frame $\{\mathbf{H}_G\}_x = [\mathbf{I}_G]\{\boldsymbol{\omega}\}_x$, where $\{\boldsymbol{\omega}\}_x$ comprises the components of angular velocity in the body frame. Thus,

$$\begin{Bmatrix} 90.86 \\ -154.2 \\ 643.0 \end{Bmatrix} = \begin{bmatrix} 1000 & 0 & 0 \\ 0 & 2000 & 0 \\ 0 & 0 & 3000 \end{bmatrix} \{\boldsymbol{\omega}\}_x$$

or, solving for $\{\boldsymbol{\omega}\}_x$,

$$\{\boldsymbol{\omega}\}_x = \begin{bmatrix} 1000 & 0 & 0 \\ 0 & 2000 & 0 \\ 0 & 0 & 3000 \end{bmatrix}^{-1} \begin{Bmatrix} 90.86 \\ -154.2 \\ 643.0 \end{Bmatrix} = \begin{Bmatrix} 0.09086 \\ -0.07709 \\ 0.2144 \end{Bmatrix} (\mathrm{rad/s}) \tag{f}$$

Euler's equations of motion (Eq. 11.72a) may be written for the case at hand as

$$[\mathbf{I}_G]\{\boldsymbol{\alpha}\}_x + \{\boldsymbol{\omega}\}_x \times ([\mathbf{I}_G]\{\boldsymbol{\omega}\}_x) = \{\mathbf{0}\} \tag{g}$$

where $\{\boldsymbol{\alpha}\}_x$ is the absolute acceleration in body frame components. Substituting Eqs. (c) and (f) into this expression, we get

$$\begin{bmatrix} 1000 & 0 & 0 \\ 0 & 2000 & 0 \\ 0 & 0 & 3000 \end{bmatrix} \{\boldsymbol{\alpha}\}_x + \begin{Bmatrix} 0.09086 \\ -0.07709 \\ 0.2144 \end{Bmatrix} \times \left(\begin{bmatrix} 1000 & 0 & 0 \\ 0 & 2000 & 0 \\ 0 & 0 & 3000 \end{bmatrix} \begin{Bmatrix} 0.09086 \\ -0.07709 \\ 0.2144 \end{Bmatrix} \right) = \begin{Bmatrix} 0 \\ 0 \\ 0 \end{Bmatrix}$$

$$\begin{bmatrix} 1000 & 0 & 0 \\ 0 & 2000 & 0 \\ 0 & 0 & 3000 \end{bmatrix} \{\boldsymbol{\alpha}\}_x + \begin{Bmatrix} -16.52 \\ -38.95 \\ -7.005 \end{Bmatrix} = \begin{Bmatrix} 0 \\ 0 \\ 0 \end{Bmatrix}$$

so that, finally,

$$\{\boldsymbol{\alpha}\}_x = -\begin{bmatrix} 1000 & 0 & 0 \\ 0 & 2000 & 0 \\ 0 & 0 & 3000 \end{bmatrix}^{-1} \begin{Bmatrix} -16.52 \\ -38.95 \\ -7.005 \end{Bmatrix} = \begin{Bmatrix} 0.01652 \\ 0.01948 \\ 0.002335 \end{Bmatrix} (\mathrm{rad/s^2}) \tag{h}$$

These are the components of the angular acceleration in the body frame (xyz). To transform them into the inertial frame (XYZ) we use

$$\begin{aligned}\{\boldsymbol{\alpha}\}_X = [\mathbf{Q}]_{xX}\{\boldsymbol{\alpha}\}_x = ([\mathbf{Q}]_{Xx})^T\{\boldsymbol{\alpha}\}_x \\ = \begin{bmatrix} 0.03086 & -0.9646 & 0.2620 \\ 0.6720 & -0.1740 & -0.7198 \\ 0.7399 & 0.1983 & 0.6428 \end{bmatrix} \begin{Bmatrix} 0.01652 \\ 0.01948 \\ 0.002335 \end{Bmatrix} = \begin{Bmatrix} -0.01766 \\ 0.006033 \\ 0.01759 \end{Bmatrix} (\mathrm{rad/s^2}) \end{aligned}$$

That is,

$$\boxed{\boldsymbol{\alpha} = -0.01766\hat{\mathbf{i}} + 0.006033\hat{\mathbf{j}} + 0.01759\hat{\mathbf{k}}\ (\mathrm{rad/s^2})}$$

12.3 STABILITY OF TORQUE-FREE MOTION

Let a rigid body be in torque-free motion with its angular velocity vector directed along the principal body z axis, so that $\boldsymbol{\omega} = \omega_o \hat{\mathbf{k}}$, where ω_o is constant. The nutation angle is zero, and there is no precession. Let us perturb the motion slightly, as illustrated in Fig. 12.6, so that

$$\omega_x = \delta\omega_x \quad \omega_y = \delta\omega_y \quad \omega_z = \omega_o + \delta\omega_z \tag{12.28}$$

As in Chapter 7, δ means a very small quantity. In this case, $\delta\omega_x \ll \omega_o$ and $\delta\omega_y \ll \omega_o$. Thus, the angular velocity vector has become slightly inclined to the z axis. For torque-free motion, $M_{G})_x = M_{G})_y = M_{G})_z = 0$, so that the Euler equations (Eq. 11.72b) are

$$\begin{aligned} A\dot{\omega}_x + (C - B)\omega_y\omega_z &= 0 \\ B\dot{\omega}_y + (A - C)\omega_x\omega_z &= 0 \\ C\dot{\omega}_z + (B - A)\omega_x\omega_y &= 0 \end{aligned} \tag{12.29}$$

Observe that we have not assumed $A = B$, as we did in the previous section. Substituting Eq. (12.28) into Eq. (12.29) and keeping in mind our assumption that $\dot{\omega}_o = 0$, we get

$$\begin{aligned} A\delta\dot{\omega}_x + (C - B)\omega_o\delta\omega_y + (C - B)\delta\omega_y\delta\omega_z &= 0 \\ B\delta\dot{\omega}_y + (A - C)\omega_o\delta\omega_x + (C - B)\delta\omega_x\delta\omega_z &= 0 \\ C\delta\dot{\omega}_z + (B - A)\delta\omega_x\delta\omega_y &= 0 \end{aligned} \tag{12.30}$$

Neglecting all the products of the $\delta\omega$s (because they are arbitrarily small), Eq. 12.30 becomes

$$\begin{aligned} A\delta\dot{\omega}_x + (C - B)\omega_o\delta\omega_y &= 0 \\ B\delta\dot{\omega}_y + (A - C)\omega_o\delta\omega_x &= 0 \\ C\delta\dot{\omega}_z &= 0 \end{aligned} \tag{12.31}$$

Eq. (12.31_3) implies that $\delta\omega_z$ is constant.

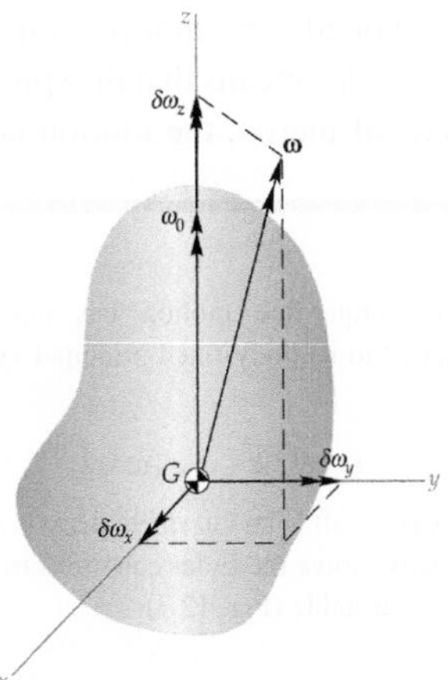

FIG. 12.6

Principal body axes of a rigid body rotating primarily about the body z axis.

Differentiating Eq. (12.31_1) with respect to time, we get

$$A\delta\ddot{\omega}_x + (C-B)\omega_o\delta\dot{\omega}_y = 0 \tag{12.32}$$

Solving Eq. (12.31_2) for $\delta\dot{\omega}_y$ yields $\delta\dot{\omega}_y = -[(A-C)/B]\omega_o\delta\omega_x$, and substituting this into Eq. (12.32) gives

$$\delta\ddot{\omega}_x - \frac{(A-C)(C-B)}{AB}\omega_o^2\delta\omega_x = 0 \tag{12.33}$$

Likewise, differentiating Eq. (12.31_2) and then substituting $\delta\dot{\omega}_x$ from Eq. (12.31_1) yields

$$\delta\ddot{\omega}_y - \frac{(A-C)(C-B)}{AB}\omega_o^2\delta\omega_y = 0 \tag{12.34}$$

If we define

$$k = \frac{(A-C)(B-C)}{AB}\omega_o^2 \tag{12.35}$$

then both Eqs. (12.33) and (12.34) may be written in the form

$$\delta\ddot{\omega} + k\delta\omega = 0 \tag{12.36}$$

If $k > 0$, then $\delta\omega = c_1 e^{i\sqrt{k}t} + c_2 e^{-i\sqrt{k}t}$, which means $\delta\omega_x$ and $\delta\omega_y$ vary sinusoidally with small amplitude. The motion is therefore bounded and neutrally stable. That means the amplitude does not die out with time, but it does not exceed the small amplitude of the perturbation. Observe from Eq. (12.35) that $k > 0$ if C is larger than both A and B or if C is smaller than both A and B. This means that the spin axis (z axis) is either the major axis of inertia or the minor axis of inertia. That is, if the spin axis is either the major or minor axis of inertia, the motion is stable. The stability is neutral for a rigid body, because there is no damping.

On the other hand, if $k < 0$, then $\delta\omega = c_1 e^{\sqrt{k}t} + c_2 e^{-\sqrt{k}t}$, which means that the initially small perturbations $\delta\omega_x$ and $\delta\omega_y$ increase without bound. The motion is unstable. From Eq. (12.35) we see that $k < 0$ if either $A > C > B$ or $A < C < B$. This means that the spin axis is the intermediate axis of inertia. If the spin axis is the intermediate axis of inertia, the motion is unstable.

EXAMPLE 12.5

A homogeneous, box-shaped satellite in torque-free motion has mass moments of inertia $A = 1000\ \text{kg}\cdot\text{m}^2$, $B = 300\ \text{kg}\cdot\text{m}^2$, and $C = 800\ \text{kg}\cdot\text{m}^2$, relative to its body-fixed principal xyz axes. The angular velocity relative to the body-fixed frame is

$$\boldsymbol{\omega} = 1.0\hat{\mathbf{k}} + \boldsymbol{\omega}_{xy}\ (\text{rad/s}) \tag{a}$$

where $\boldsymbol{\omega}_{xy} = \left(\hat{\mathbf{i}} + \hat{\mathbf{j}}\right)(10^{-8})$. Therefore, the spacecraft is an intermediate-axis spinner with an extremely small (essentially zero) transverse component of angular velocity. Solve the Euler equations to verify that the transverse perturbations will grow in time, so that the motion of Eq. (a) is unstable (Fig. 12.7).

Solution

First note that according to Eq. (11.67), the angular momentum vector starts out as

$$\mathbf{H}_G = C\omega_z\hat{\mathbf{k}} = 800\hat{\mathbf{k}}\ \text{kg}\cdot\text{m}^2/\text{s}$$

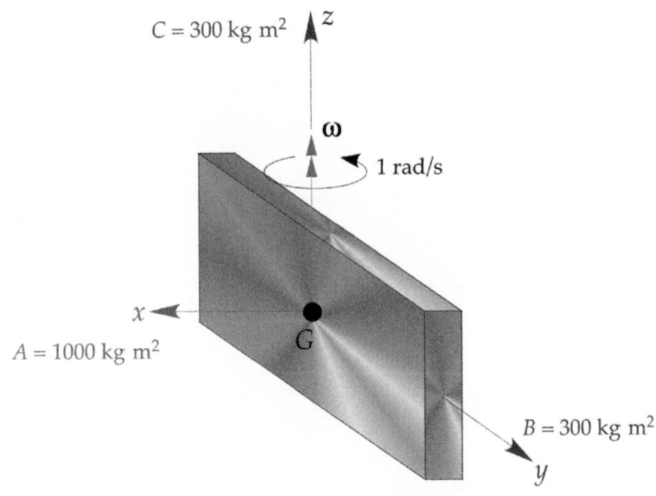

FIG. 12.7

Spacecraft as an intermediate axis spinner.

if we neglect $\boldsymbol{\omega}_{xy}$. Because the net external moment is zero, $\mathbf{H}_G$ must remain constant and aligned with the inertial z direction. The development of a significant transverse component of angular velocity will result in the spin axis (the z axis) precessing around $\mathbf{H}_G$, as shown in Fig. 12.4.

The Euler equations of motion (Eq. 12.4) for this case are

$$\dot{\omega}_x = \frac{B-C}{A}\omega_y\omega_z = -0.5000\omega_y\omega_z \quad \dot{\omega}_y = -0.6667\omega_z\omega_x \quad \dot{\omega}_z = 0.8750\omega_x\omega_y \tag{b}$$

We use the methods of Section 1.8 to numerically integrate this set of coupled, nonlinear, ordinary differential equations. To obtain Eq. (1.95), we set $y_1 = \omega_x$, $y_2 = \omega_y$, and $y_3 = \omega_z$, and write the system (b) as $\dot{\mathbf{y}} = \mathbf{f}(\mathbf{y})$, where

$$\mathbf{y} = \begin{Bmatrix} y_1 \\ y_2 \\ y_3 \end{Bmatrix} \quad \mathbf{f}(\mathbf{y}) = \begin{Bmatrix} -0.5000y_2y_3 \\ -0.6667y_3y_1 \\ 0.875y_1y_2 \end{Bmatrix} \tag{c}$$

Observe that the time t does not appear explicitly here. The initial conditions are

$$\mathbf{y}_0 = \lfloor (10^{-8}) \;\; (10^{-8}) \;\; 1.0 \rfloor^T \tag{d}$$

The following MATLAB code integrates Eq. (b) over a 450-s time interval and plots the results.

```
y0     = [1.e-8; 1.e-8; 1.0000]; % Set the initial conditions
tspan  = linspace(0,450,1000);   % Set the solution time interval
[t,y]  = ode45(@f, tspan, w0);   % Integrate the equations dydt in 'f' below
figure('color','w')
%...Plot the angular velocity histories:
subplot(2,1,1)
  plot(t,y(:,3)); grid on; axis([0 450 -1.2 1.2])
  xlabel('time (s)')
  ylabel('\omega_z (rad/s)')
subplot(2,1,2)
  plot(t,sqrt(y(:,1).^2 + y(:,2).^2)); grid on; axis([0 450 -0.2 1.2])
  xlabel('time (s)')
 ylabel('\omega_{xy} (rad/s)')
```

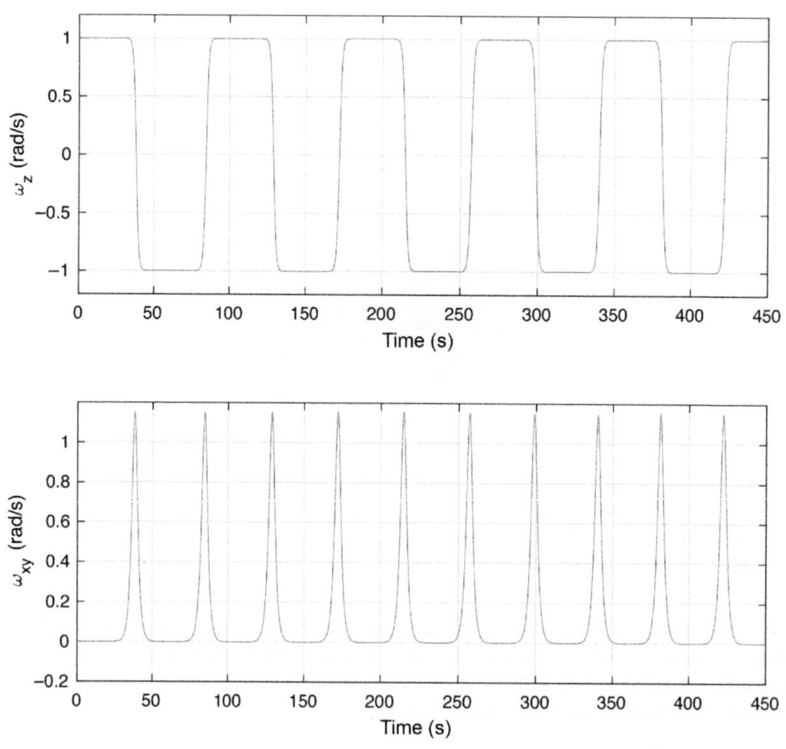

FIG. 12.8

Histories of the axial (*upper*) and transverse (*lower*) components of angular velocity for the spinning body in Fig. 12.7.

```
%~~~~~~~~~~~~~~~~~~~~~~~~~~~~~~~
function dydt = f(t,y)
%-------------------------------
%...Angular velocities in:
  wx = y(1); wy = y(2); wz = y(3);
%...Angular accelerations out:
  wx_dot = -0.5000*wy*wz;
  wy_dot = -0.6667*wz*wx;
  wz_dot =  0.8750*wx*wy;
  dydt   = [wx_dot; wy_dot; wz_dot];
end %f
```

The variations of ω_z and the transverse component $\omega_{xy} = \sqrt{\omega_x^2 + \omega_y^2}$ are shown in Fig. 12.8. Clearly, the spin is not asymptotically stable since ω_z is not constant but changes sign about once every 43 s. That means the body and its embedded spin axis periodically reverse their spatial orientation, as illustrated in Fig. 12.9. Likewise, the amplitude of the transverse angular velocity rises from zero to over 1 rad/s during the time when the body flips upside down. The complex transitional motion is dictated by the fact that the angular momentum vector must retain its original orientation in inertial space.

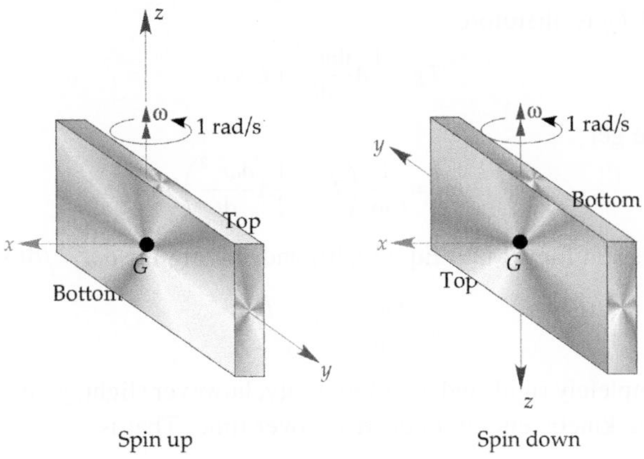

FIG. 12.9

The two semistable states of spin of the body in Fig. 12.7.

If the angular velocity vector of a satellite lies in the direction of its major axis of inertia, the satellite is called a major-axis spinner or oblate spinner. A minor-axis spinner or prolate spinner has its minor axis of inertia aligned with the angular velocity vector. Intermediate-axis spinners are unstable, causing a continual 180° reorientation of the spin axis, if the satellite is a rigid body. However, the flexibility inherent in any real satellite leads to an additional instability, as we shall now see.

Consider again the rotationally symmetric satellite in torque-free motion discussed in Section 12.2. From Eqs. (12.24) and (12.25), we know that angular momentum $\mathbf{H}_G$ is given by

$$\mathbf{H}_G = A\boldsymbol{\omega}_\perp + C\omega_z\hat{\mathbf{k}} \tag{12.37}$$

Hence,

$$\mathrm{H}_G{}^2 = A^2\omega_\perp{}^2 + C^2\omega_z{}^2 \qquad (\omega_\perp = \omega_{xy}) \tag{12.38}$$

Differentiating this equation with respect to time yields

$$\frac{\mathrm{dH}_G{}^2}{\mathrm{d}t} = A^2\frac{\mathrm{d}\omega_\perp{}^2}{\mathrm{d}t} + 2C^2\omega_z\dot{\omega}_z \tag{12.39}$$

But, according to Eq. (12.1), $\mathbf{H}_G$ is constant, so that $\mathrm{dH}_G^2/\mathrm{d}t = 0$ and Eq. (12.39) can be written

$$\frac{\mathrm{d}\omega_\perp{}^2}{\mathrm{d}t} = -2\frac{C^2}{A^2}\omega_z\dot{\omega}_z \tag{12.40}$$

The rotary kinetic energy of a rotationally symmetric body ($A = B$) is found using Eq. (11.81),

$$T_R = \frac{1}{2}A\omega_x{}^2 + \frac{1}{2}A\omega_y{}^2 + \frac{1}{2}C\omega_z{}^2 = \frac{1}{2}A\left(\omega_x{}^2 + \omega_y{}^2\right) + \frac{1}{2}C\omega_z{}^2$$

From Eq. (12.13), we know that $\omega_x{}^2 + \omega_y{}^2 = \omega_\perp{}^2$, which means

$$T_R = \frac{1}{2}A\omega_\perp{}^2 + \frac{1}{2}C\omega_z{}^2 \tag{12.41}$$

The time derivative of T_R is, therefore,

$$\dot{T}_R = \frac{1}{2}A\frac{d\omega_\perp^2}{dt} + C\omega_z\dot{\omega}_z$$

Solving this for $\dot{\omega}_z$, we get

$$\dot{\omega}_z = \frac{1}{C\omega_z}\left(\dot{T}_R - \frac{1}{2}A\frac{d\omega_\perp^2}{dt}\right)$$

Substituting this expression for $\dot{\omega}_z$ into Eq. (12.40) and solving for $d\omega_\perp^2/dt$ yields

$$\frac{d\omega_\perp^2}{dt} = 2\frac{C}{AC - A}\dot{T}_R \tag{12.42}$$

Real bodies are not completely rigid, and their flexibility, however slight, gives rise to small dissipative effects, which cause the kinetic energy to decrease over time. That is,

$$\dot{T}_R < 0 \quad \text{For spacecraft with dissipation} \tag{12.43}$$

Substituting this inequality into Eq. (12.42) leads us to conclude that

$$\begin{aligned} \frac{d\omega_\perp^2}{dt} &< 0 \;\text{ if } C > A \;\text{ (oblate spinner)} \\ \frac{d\omega_\perp^2}{dt} &> 0 \;\text{ if } C < A \;\text{ (prolate spinner)} \end{aligned} \tag{12.44}$$

If $d\omega_\perp^2/dt$ is negative, the spin is asymptotically stable. Should a nonzero value of $\omega_\perp$ develop for some reason, it will drift back to zero over time, so that once again the angular velocity lies completely in the spin direction. On the other hand, if $d\omega_\perp^2/dt$ is positive, the spin is unstable. $\omega_\perp$ does not damp out, and the angular velocity vector drifts away from the spin axis as $\omega_\perp$ increases without bound. We pointed out above that spin about a minor axis of inertia is stable with respect to small disturbances. Now we see that only major-axis spin is stable in the long run if dissipative mechanisms exist.

For some additional insight into this phenomenon, solve Eq. (12.38) for $\omega_\perp^2$,

$$\omega_\perp^2 = \frac{H_G^2 - C^2\omega_z^2}{A^2}$$

and substitute this result into the expression for kinetic energy (Eq. 12.41) to obtain

$$T_R = \frac{1}{2}\frac{H_G^2}{A} + \frac{1}{2}\frac{(A - C)C}{A}\omega_z^2 \tag{12.45}$$

According to Eq. (12.24),

$$\omega_z = \frac{H_G)_z}{C} = \frac{H_G\cos\theta}{C}$$

Substituting this into Eq. (12.45) yields the kinetic energy as a function of just the inclination angle θ,

$$T_R = \frac{1}{2}\frac{H_G^2}{A}\left(1 + \frac{A - C}{C}\cos^2\theta\right) \tag{12.46}$$

The extreme values of T_R occur at $\theta = 0$ or $\theta = \pi$,

$$T_R = \frac{1}{2}\frac{H_G^2}{C} \quad \text{(major axis spinner)}$$

and $\theta = \pi/2$,

$$T_R = \frac{1}{2}\frac{H_G{}^2}{A} \quad \text{(minor axis spinner)}$$

Clearly, the kinetic energy of a torque-free satellite is smallest when the spin is around the major axis of inertia. We may think of a satellite with dissipation ($dT_R/dt < 0$) as seeking the state of minimum kinetic energy, which occurs when it spins about its major axis.

EXAMPLE 12.6

A rigid spacecraft is modeled by the solid cylinder B, which has a mass of 300 kg, and the slender rod R, which passes through the cylinder and has a mass of 30 kg. Which of the principal axes x, y, and z can be an axis about which stable torque-free rotation can occur (Fig. 12.10)?

Solution

For the solid cylinder B, we have

$$r_B = 0.5\ \text{m} \quad l_B = 1.0\ \text{m} \quad m_B = 300\ \text{kg}$$

The principle moments of inertia about the center of mass are found in Fig. 11.10a,

$$I_B)_x = \frac{1}{4}m_B r_B{}^2 + \frac{1}{12}m_B l_B{}^2 = 43.75\ \text{kg}\cdot\text{m}^2$$
$$I_B)_y = I_B)_x = 43.75\ \text{kg}\cdot\text{m}^2$$
$$I_B)_z = \frac{1}{2}m_B r_B^2 = 37.5\ \text{kg}\cdot\text{m}^2$$

The properties of the transverse slender rod are

$$l_R = 1.0\ \text{m} \quad m_R = 30\ \text{kg}$$

Fig. 11.10a, with $r = 0$, yields the moments of inertia,

$$I_R)_y = 0$$

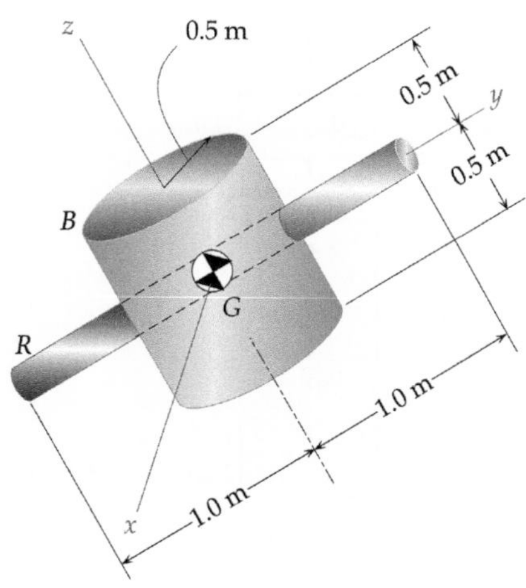

FIG. 12.10

Built-up satellite structure.

$$I_R)_z = I_R)_x = \frac{1}{12} m_A r_A^2 = 10.0\ \text{kg}\cdot\text{m}^2$$

The moments of inertia of the assembly is the sum of the moments of inertia of the cylinder and the rod,

$$I_x = I_B)_x + I_R)_x = 53.75\ \text{kg}\cdot\text{m}^2$$
$$I_y = I_B)_y + I_R)_y = 43.75\ \text{kg}\cdot\text{m}^2$$
$$I_z = I_B)_z + I_R)_z = 47.50\ \text{kg}\cdot\text{m}^2$$

Since I_z is clearly the intermediate mass moment of inertia, rotation about the z axis is unstable. With energy dissipation, rotation is stable in the long term only about the major axis, which in this case is the x axis.

12.4 DUAL-SPIN SPACECRAFT

If a satellite is to be spin-stabilized, it must be an oblate spinner. The diameter of the spacecraft is restricted by the cross-section of the launch vehicle's upper stage, and its length is limited by stability requirements. Therefore, oblate spinners cannot take full advantage of the payload volume available in a given launch vehicle, which after all are slender, prolate shapes for aerodynamic reasons. The dual-spin design permits spin stabilization of a prolate shape.

The axisymmetric, dual-spin configuration, or gyrostat, consists of an axisymmetric rotor and a smaller axisymmetric platform joined together along a common longitudinal spin axis at a bearing, as shown in Fig. 12.11. The platform and rotor have their own components of angular velocity, ω_p and ω_r, respectively, along the spin axis direction $\hat{\mathbf{k}}$. The platform spins at a much slower rate than the rotor. The assembly acts like a rigid body as far as transverse rotations are concerned (i.e., the rotor and the platform have the transverse angular velocity $\omega_\perp$ in common). An electric motor integrated into the axle bearing connecting the two components acts to overcome frictional torque that would otherwise eventually cause the relative angular velocity between the rotor and platform to go to zero. If that

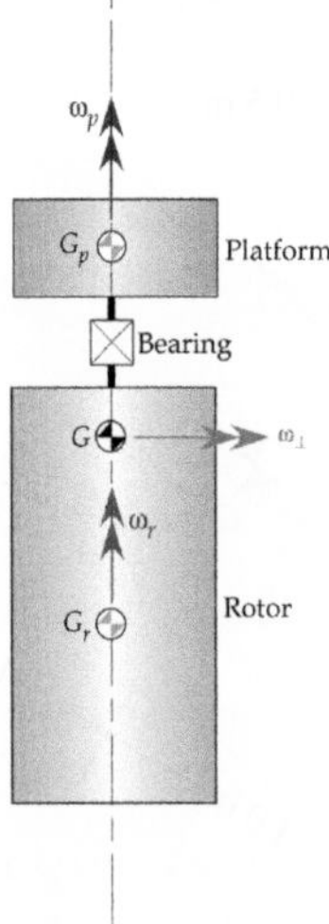

FIG. 12.11

Axisymmetric, dual-spin spacecraft.

should happen, the satellite would become a single-spin unit, probably an unstable prolate spinner, since the rotor of a dual-spin spacecraft is likely to be prolate.

The first dual-spin satellite was OSO-I (Orbiting Solar Observatory), which NASA launched in 1962. It was a major-axis spinner. The first prolate dual-spin spacecraft was the two-story-tall TACSAT I (Tactical Communications Satellite). It was launched into geosynchronous orbit by the US Air Force in 1969. Typical of many of today's communications satellites, TACSAT's platform rotated at 1 rev/day to keep its antennas pointing toward the earth. The rotor spun at about 1 rev/s. Of course, the axis of the spacecraft was normal to the plane of its orbit. The first dual-spin interplanetary spacecraft was Galileo, which we discussed briefly in Section 8.9. Galileo's platform was completely despun to provide a fixed orientation for cameras and other instruments. The rotor spun at 3 rpm.

The equations of motion of a dual-spin spacecraft will be developed later on in Section 12.9. Let us determine the stability of the motion by following the same "energy sink" procedure employed in the previous section for a single-spin-stabilized spacecraft. The angular momentum of the dual-spin configuration about the spacecraft center of mass G is the sum of the angular momenta of the rotor (r) and the platform (p) about G,

$$\mathbf{H}_G = \mathbf{H}_G^{(p)} + \mathbf{H}_G^{(r)} \tag{12.47}$$

The angular momentum of the platform about the spacecraft center of mass is

$$\mathbf{H}_G^{(p)} = C_p\omega_p\hat{\mathbf{k}} + A_p\boldsymbol{\omega}_\perp \tag{12.48}$$

where C_p is the moment of inertia of the platform about the spacecraft spin axis, and A_p is its transverse moment of inertia about G (not G_p). Likewise, for the rotor,

$$\mathbf{H}_G^{(r)} = C_r\omega_r\hat{\mathbf{k}} + A_r\boldsymbol{\omega}_\perp \tag{12.49}$$

where C_r and A_r are its longitudinal and transverse moments of inertia about axes through G. Substituting Eqs. (12.48) and (12.49) into Eq. (12.47) yields

$$\mathbf{H}_G = \left(C_r\omega_r + C_p\boldsymbol{\omega}_p\right)\hat{\mathbf{k}} + A_\perp\boldsymbol{\omega}_\perp \tag{12.50}$$

where $A_\perp$ is the total transverse moment of inertia,

$$A_\perp = A_p + A_r$$

From this, it follows that

$$H_G{}^2 = \left(C_r\omega_r + C_p\omega_p\right)^2 + A_\perp{}^2\omega_\perp{}^2$$

For torque-free motion, $\dot{\mathbf{H}}_G = \mathbf{0}$, so that $dH_G{}^2/dt = 0$, or

$$2\left(C_r\omega_r + C_p\omega_p\right)\left(C_r\dot{\omega}_r + C_p\dot{\omega}_p\right) + A_\perp{}^2\frac{\mathrm{d}\omega_\perp{}^2}{\mathrm{d}t} = 0 \tag{12.51}$$

Solving this for $\mathrm{d}\omega_\perp{}^2/\mathrm{d}t$ yields

$$\frac{\mathrm{d}\omega_\perp{}^2}{\mathrm{d}t} = -\frac{2}{A_\perp{}^2}\left(C_r\omega_r + C_p\omega_p\right)\left(C_r\dot{\omega}_r + C_p\dot{\omega}_p\right) \tag{12.52}$$

The total rotational kinetic energy T of the dual-spin spacecraft is that of the rotor plus that of the platform,

$$T = \frac{1}{2}C_r\omega_r{}^2 + \frac{1}{2}C_p\omega_p{}^2 + \frac{1}{2}A_\perp\omega_\perp{}^2$$

Differentiating this expression with respect to time and solving for $d\omega_\perp{}^2/dt$ yields

$$\frac{d\omega_\perp{}^2}{dt}=\frac{2}{A_\perp}\left(\dot{T}-C_r\omega_r\dot{\omega}_r-C_p\omega_p\dot{\omega}_p\right) \tag{12.53}$$

where $\dot{T}$ is the sum of the power $P^{(r)}$ dissipated in the rotor and the power $P^{(p)}$ dissipated in the platform,

$$\dot{T}=P^{(r)}+P^{(p)} \tag{12.54}$$

Substituting Eq. (12.54) into Eq. (12.53) we find

$$\frac{d\omega_\perp{}^2}{dt}=\frac{2}{A_\perp}\left(P^{(r)}-C_r\omega_r\dot{\omega}_r+P^{(p)}-C_p\omega_p\dot{\omega}_p\right) \tag{12.55}$$

Equating the two expressions for $d\omega_\perp{}^2/dt$ in Eqs. (12.52) and (12.55) yields

$$\frac{2}{A_\perp}\left(\dot{T}-C_r\omega_r\dot{\omega}_r-C_p\omega_p\dot{\omega}_p\right)=-\frac{2}{A_\perp{}^2}\left(C_r\omega_r+C_p\omega_p\right)\left(C_r\dot{\omega}_r+C_p\dot{\omega}_p\right)$$

Solve this for $\dot{T}$ to obtain

$$\dot{T}=\frac{C_r}{A_\perp}\left[(A_\perp-C_r)\omega_r-C_p\omega_p\right]\dot{\omega}_r+\frac{C_p}{A_\perp}\left[(A_\perp-C_p)\omega_p-C_r\omega_r\right]\dot{\omega}_p \tag{12.56}$$

Following Likins (1967), we identify the terms containing $\dot{\omega}_r$ and $\dot{\omega}_p$ as the power dissipation in the rotor and platform, respectively. That is, comparing Eqs. (12.54) and (12.56),

$$P^{(r)}=\frac{C_r}{A_\perp}\left[(A_\perp-C_r)\omega_r-C_p\omega_p\right]\dot{\omega}_r \tag{12.57a}$$

$$P^{(p)}=\frac{C_p}{A_\perp}\left[(A_\perp-C_p)\omega_p-C_r\omega_r\right]\dot{\omega}_p \tag{12.57b}$$

Solving these two expressions for $\dot{\omega}_r$ and $\dot{\omega}_p$, respectively, yields

$$\dot{\omega}_r=\frac{A_\perp}{C_r}\frac{P^{(r)}}{(A_\perp-C_r)\omega_r-C_p\omega_p} \tag{12.58a}$$

$$\dot{\omega}_p=\frac{A_\perp}{C_p}\frac{P^{(p)}}{(A_\perp-C_p)\omega_p-C_r\omega_r} \tag{12.58b}$$

Substituting these results into Eq. (12.55) leads to

$$\frac{d\omega_\perp{}^2}{dt}=\frac{2}{A_\perp}\left[\frac{P^{(r)}}{C_p\dfrac{\omega_p}{\omega_r}-(A_\perp-C_r)}+\frac{P^{(p)}}{C_r-(A_\perp-C_p)\dfrac{\omega_p}{\omega_r}}\right]\left(C_r+C_p\frac{\omega_p}{\omega_r}\right) \tag{12.59}$$

As pointed out above, for geosynchronous dual-spin communication satellites,

$$\frac{\omega_p}{\omega_r}\approx\frac{2\pi\text{ rad/day}}{2\pi\text{ rad/s}}\approx 10^{-5}$$

whereas for interplanetary dual-spin spacecraft, $\omega_p=0$. Therefore, there is an important class of spin-stabilized spacecraft for which $\omega_p/\omega_r\approx 0$. For a despun platform wherein ω_p is zero (or nearly so), Eq. (12.59) yields

$$\frac{d\omega_\perp^2}{dt}=\frac{2}{A_\perp}\left[P^{(p)}+\frac{C_r}{C_r-A_\perp}P^{(r)}\right] \tag{12.60}$$

If the rotor is oblate ($C_r > A_\perp$), then, since $P^{(r)}$ and $P^{(p)}$ are both negative, it follows from Eq. (12.60) that $d\omega_\perp^2/dt < 0$. That is, the oblate dual-spin configuration with a despun platform is unconditionally stable. In practice, however, the rotor is likely to be prolate ($C_r < A_\perp$), so that

$$\frac{C_r}{C_r-A_\perp}P^{(r)}>0$$

In that case, $d\omega_\perp^2/dt < 0$ only if the dissipation $P^{(p)}$ in the platform is significantly greater than that of the rotor. Specifically, for a prolate design, it must be true that

$$\left|P^{(p)}\right|>\left|\frac{C_r}{C_r-A_\perp}P^{(r)}\right|$$

The platform dissipation rate $P^{(p)}$ can be augmented by adding nutation dampers, which are discussed in the next section.

For the despun prolate dual-spin configuration, Eqs. (12.58) imply

$$\dot{\omega}_r=\frac{P^{(r)}}{(A_\perp-C_r)C_r\omega_r}A_\perp \qquad \dot{\omega}_p=-\frac{P^{(p)}}{C_p}\frac{A_\perp}{C_r\omega_r}$$

Clearly, the signs of $\dot{\omega}_r$ and $\dot{\omega}_p$ are opposite. If $\omega_r > 0$, then dissipation causes the spin rate of the rotor to decrease and that of the platform to increase. Were it not for the action of the motor on the shaft connecting the two components of the spacecraft, eventually $\omega_p = \omega_r$. That is, the relative motion between the platform and rotor would cease and the dual-spinner would become an unstable single-spin spacecraft. Setting $\omega_p = \omega_r$ in Eq. (12.59) yields

$$\frac{d\omega_\perp^2}{dt}=2\frac{C_r+C_p}{A_\perp}\frac{P^{(r)}+P^{(p)}}{(C_r+C_p)-A_\perp}$$

which is the same as Eq. (12.42), the energy sink conclusion for a single-spinner.

12.5 NUTATION DAMPER

Nutation dampers are passive means of dissipating energy. A common type consists essentially of a tube filled with viscous fluid and containing a mass attached to springs, as illustrated in Fig. 12.12. Dampers may contain just fluid, only partially filling the tube so that it can slosh around. In either case, the purpose is to dissipate energy through fluid friction. The wobbling of the spacecraft due to nonalignment of the angular velocity with the principal spin axis induces accelerations throughout the satellite, giving rise to the sloshing of fluids and the, stretching and flexing of nonrigid components, etc., all of which dissipate energy to one degree or another. Nutation dampers are added to deliberately increase energy dissipation, which is desirable for stabilizing oblate single-spinners and dual-spin spacecraft (Fig. 12.12).

Let us focus on the motion of the mass within the nutation damper of Fig. 12.12 to gain some insight into how relative motion and deformation are induced by the satellite's precession. Note that point P is the center of mass of the rigid satellite body itself. The center of mass G of the satellite-damper mass combination lies between P and m, as shown in Fig. 12.9. We suppose that the tube is lined up with the

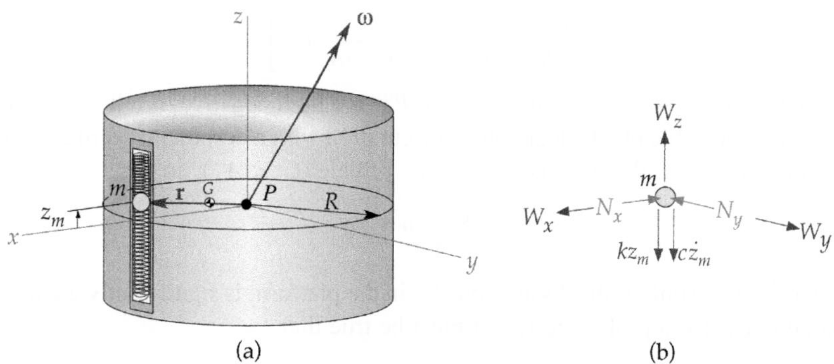

FIG. 12.12

(a) Precessing oblate spacecraft with a nutation damper aligned with the z axis. (b) Free-body diagram of the moving mass in the nutation damper.

z axis of the body-fixed xyz frame, as shown. The mass m in the tube is therefore constrained by the tube walls to move only in the z direction. When the springs are undeformed, the mass lies in the xy plane. In general, the position vector of m in the body frame is

$$\mathbf{r} = R\hat{\mathbf{i}} + z_m\hat{\mathbf{k}} \tag{12.61}$$

where z_m is the z coordinate of m, and R is the distance of the damper from the centerline of the spacecraft. The velocity and acceleration of m relative to the satellite are, therefore,

$$\mathbf{v}_{\text{rel}} = \dot{z}_m\hat{\mathbf{k}} \tag{12.62}$$

$$\mathbf{a}_{\text{rel}} = \ddot{z}_m\hat{\mathbf{k}} \tag{12.63}$$

The absolute angular velocity $\boldsymbol{\omega}$ of the satellite (and, therefore, of the body-fixed frame) is

$$\boldsymbol{\omega} = \omega_x\hat{\mathbf{i}} + \omega_y\hat{\mathbf{j}} + \omega_z\hat{\mathbf{k}} \tag{12.64}$$

Recall Eq. (11.73), which states that when $\boldsymbol{\omega}$ is given in a body frame, we find the absolute angular acceleration by taking the time derivative of $\boldsymbol{\omega}$, holding the unit vectors fixed. Thus,

$$\dot{\boldsymbol{\omega}} = \dot{\omega}_x\hat{\mathbf{i}} + \dot{\omega}_y\hat{\mathbf{j}} + \dot{\omega}_z\hat{\mathbf{k}} \tag{12.65}$$

The absolute acceleration of m is found using Eq. (1.70), which for the case at hand becomes

$$\mathbf{a} = \mathbf{a}_P + \dot{\boldsymbol{\omega}} \times \mathbf{r} + \boldsymbol{\omega} \times (\boldsymbol{\omega} \times \mathbf{r}) + 2\boldsymbol{\omega} \times \mathbf{v}_{\text{rel}} + \mathbf{a}_{\text{rel}} \tag{12.66}$$

where $\mathbf{a}_P$ is the absolute acceleration of the reference point P. Substituting Eqs. (12.61)–(12.65) into Eq. (12.66), carrying out the vector operations, combining terms, and simplifying leads to the following expressions for the three components of the inertial acceleration of m,

$$\begin{aligned} a_x &= \left.a_P\right)_x - R\left(\omega_y{}^2 + \omega_z{}^2\right) + z_m\dot{\omega}_y + z_m\omega_x\omega_z + 2\dot{z}_m\omega_y \\ a_y &= \left.a_P\right)_y + R\dot{\omega}_z + R\omega_x\omega_y - z_m\dot{\omega}_x + z_m\omega_y\omega_z - 2\dot{z}_m\omega_x \\ a_z &= \left.a_P\right)_z - z_m\left(\omega_x{}^2 + \omega_y{}^2\right) - R\dot{\omega}_y + R\omega_x\omega_z + \ddot{z}_m \end{aligned} \tag{12.67}$$

Fig. 12.9b shows the free body diagram of the damper mass m. In the x and y directions, the forces on m are the components of the force of gravity (W_x and W_y) and the components N_x and N_y of the force of contact with the smooth walls of the damper tube. The directions assumed for these components are, of course, arbitrary. In the z direction, we have the z component W_z of the weight, plus the force of the springs and the viscous drag of the fluid. The spring force ($-kz_m$) is directly proportional and opposite in direction to the displacement z_m. k is the net spring constant. Viscous drag ($-c\dot{z}_m$) is directly proportional and opposite in direction to the velocity $\dot{z}_m$ of m relative to the tube. c is the damping constant. Thus, the three components of the net force on the damper mass m are

$$\begin{aligned} F_{\text{net}})_x &= W_x - N_x \\ F_{\text{net}})_y &= W_y - N_y \\ F_{\text{net}})_z &= W_z - kz_m - c\dot{z}_m \end{aligned} \tag{12.68}$$

Substituting Eqs. (12.67) and (12.68) into Newton's second law, $\mathbf{F}_{\text{net}} = m\mathbf{a}$, yields

$$\begin{aligned} N_x &= mR\left(\omega_y^2 + \omega_z^2\right) - mz_m\dot{\omega}_y - mz_m\omega_x\omega_y - 2m\dot{z}_m\omega_y + \overbrace{\left[W_x - ma_P)_x\right]}^{=0} \\ N_y &= -mR\dot{\omega}_z - mR\omega_x\omega_y + mz_m\dot{\omega}_x - mz_m\omega_y\omega_z + 2m\dot{z}_m\omega_x + \overbrace{\left[W_y - ma_P)_y\right]}^{=0} \\ m\ddot{z}_m + c\dot{z}_m &+ \left[k - m\left(\omega_x^2 + \omega_y^2\right)\right]z_m = mR\left(\dot{\omega}_y - \omega_x\omega_z\right) + \overbrace{\left[W_z - ma_P)_z\right]}^{=0} \end{aligned} \tag{12.69}$$

The last terms in brackets in each of these expressions vanish if the acceleration of gravity is the same at m as at the reference point P of the spacecraft. This will be true unless the satellite is of enormous size.

If the damper mass m is vanishingly small compared with the mass M of the rigid spacecraft body, then it will have little effect on the rotary motion. If the rotational state is that of an axisymmetric satellite in torque-free motion, then we know from Eqs. (12.13), (12.14), and (12.19) that

$$\begin{array}{lll} \omega_x = \omega_{xy}\sin\omega_s t & \omega_y = \omega_{xy}\cos\omega_s t & \omega_z = \omega_o \\ \dot{\omega}_x = \omega_{xy}\omega_s\cos\omega_s t & \dot{\omega}_y = -\omega_{xy}\omega_s\sin\omega_s t & \dot{\omega}_z = 0 \end{array}$$

in which case Eq. (12.69) becomes

$$\begin{aligned} N_x &= mR\left(\omega_o^2 + \omega_{xy}^2\cos^2\omega_s t\right) + m(\omega_s - \omega_o)\omega_{xy}z_m\sin\omega_s t - 2m\omega_{xy}\dot{z}_m\cos\omega_s t \\ N_y &= -mR\omega_{xy}^2\cos\omega_s t\sin\omega_s t + m(\omega_s - \omega_o)\omega_{xy}z_m\cos\omega_s t + 2m\omega_{xy}\dot{z}_m\sin\omega_s t \\ m\ddot{z}_m &+ c\dot{z}_m + \left(k - m\omega_{xy}^2\right)z_m = -mR(\omega_s + \omega_o)\omega_{xy}\sin\omega_s t \end{aligned} \tag{12.70}$$

Eq. (12.70_3) is that of a single-degree-of-freedom, damped oscillator with a sinusoidal forcing function, which was discussed in Section 1.8. The precession produces a force of amplitude $m(\omega_o + \omega_s)\omega_{xy}R$ and frequency ω_s, which causes the damper mass m to oscillate back and forth in the tube such that (see the steady-state part of Eq. 1.114a)

$$z_m = \frac{mR\omega_{xy}(\omega_s + \omega_o)}{\left[k - m\left(\omega_s^2 + \omega_{xy}^2\right)\right]^2(c\omega_s)^2}\left\{c\omega_s\cos\omega_s t - \left[k - m\left(\omega_s^2 + \omega_{xy}^2\right)\sin\omega_s t\right]\right\}$$

Observe that the contact forces N_x and N_y depend exclusively on the amplitude and frequency of the precession. If the angular velocity lines up with the spin axis, so that $\omega_{xy} = 0$ (precession vanishes), then

$$\begin{aligned} N_x &= m\omega_o^2 R \\ N_y &= 0 \qquad \text{No precession} \\ z_m &= 0 \end{aligned}$$

If precession is eliminated so that there is pure spin around the principal axis, then the time-varying motions and forces vanish throughout the spacecraft, which thereafter rotates as a rigid body with no energy dissipation.

Now, the whole purpose of a nutation damper is to interact with the rotational motion of the spacecraft so as to damp out any tendencies to precess. Therefore, its mass should not be ignored in the equations of motion of the spacecraft. We will derive the equations of motion of the rigid spacecraft with a nutation damper to show how rigid body mechanics is brought to bear upon the problem, and, simply, to discover precisely what we are up against even in this extremely simplified system. We will continue to use P as the origin of our body frame. Since a moving mass has been added to the rigid spacecraft and since we are not using the center of mass of the system as our reference point, we cannot use the Euler equations. Applicable to the case at hand is Eq. (11.33), according to which the equation of rotational motion of the system of satellite plus damper is

$$\dot{\mathbf{H}}_P\Big)_{\text{rel}} + \mathbf{r}_{G/P} \times (M+m)\mathbf{a}_{P/G} = \mathbf{M}_G\Big)_{\text{net}} \tag{12.71}$$

The angular momentum of the satellite body plus that of the damper mass, relative to point P on the spacecraft, is

$$\mathbf{H}_P\Big)_{\text{rel}} = \overbrace{A\omega_x\hat{\mathbf{i}} + B\omega_y\hat{\mathbf{j}} + C\omega_z\hat{\mathbf{k}}}^{\text{body of the spacecraft}} + \overbrace{\mathbf{r}\times m\dot{\mathbf{r}}}^{\text{damper mass}} \tag{12.72}$$

where the position vector $\mathbf{r}$ is given by Eq. (12.61). According to Eq. (1.56),

$$\dot{\mathbf{r}} = \frac{d\mathbf{r}}{dt}\Big)_{\text{rel}} + \boldsymbol{\omega}\times\mathbf{r} = \dot{z}_m\hat{\mathbf{k}} + \begin{vmatrix} \hat{\mathbf{i}} & \hat{\mathbf{j}} & \hat{\mathbf{k}} \\ \omega_x & \omega_y & \omega_z \\ R & 0 & z \end{vmatrix} = \omega_y z_m\hat{\mathbf{i}} + (\omega_z R - \omega_x z_m)\hat{\mathbf{j}} + (\dot{z}_m - \omega_y R)\hat{\mathbf{k}}$$

After substituting this into Eq. (12.72) and collecting terms, we obtain

$$\begin{aligned} \mathbf{H}_P)_{\text{rel}} = {} & \left[\left(A + m{z_m}^2\right)\omega_x - mRz_m\omega_z\right]\hat{\mathbf{i}} + \left[\left(B + mR^2 + m{z_m}^2\right)\omega_y - mR\dot{z}_m\right]\hat{\mathbf{j}} \\ & + \left[\left(C + mR^2\right)\omega_z - mRz_m\omega_x\right]\hat{\mathbf{k}} \end{aligned} \tag{12.73}$$

To calculate $\dot{\mathbf{H}}_P\Big)_{\text{rel}}$ we again use Eq. (1.56),

$$\dot{\mathbf{H}}_P\Big)_{\text{rel}} = \frac{\mathbf{H}_P)_{\text{rel}}}{dt}\Bigg)_{\text{rel}} + \boldsymbol{\omega}\times\mathbf{H}_P)_{\text{rel}}$$

Substituting Eq. (12.73) and carrying out the operations on the right leads eventually to

$$\begin{aligned} \dot{\mathbf{H}}_P\Big)_{\text{rel}} = {} & \left[\left(A + m{z_m}^2\right)\dot{\omega}_x - mRz_m\dot{\omega}_z + \left(C - B - m{z_m}^2\right)\omega_y\omega_z - mRz_m\omega_x\omega_y + 2mz_m\dot{z}_m\omega_x\right]\hat{\mathbf{i}} \\ & + \left\{\left(B + mR^2 + m{z_m}^2\right)\dot{\omega}_y + mRz_m\left({\omega_x}^2 - {\omega_z}^2\right) + \left[A + m{z_m}^2 - \left(C + mR^2\right)\right]\omega_x\omega_z \right. \\ & \left. + 2mz_m\dot{z}_m\omega_y - mR\ddot{z}_m\right\}\hat{\mathbf{j}} \\ & + \left[-mRz_m\dot{\omega}_x + \left(C + mR^2\right)\dot{\omega}_z + \left(B + mR^2 - A\right)\omega_x\omega_y + mRz_m\omega_y\omega_z - 2mR\dot{z}_m\omega_x\right]\hat{\mathbf{k}} \end{aligned} \tag{12.74}$$

To calculate the second term on the left of Eq. (12.71), we keep in mind that P is the center of mass of the body of the satellite and first determine the position vector of the center of mass G of the vehicle plus damper relative to P,

$$(M+m)\mathbf{r}_{G/P}=M\cdot\mathbf{0}+m\mathbf{r} \tag{12.75}$$

where $\mathbf{r}$, the position of the damper mass m relative to P, is given by Eq. (12.61). Thus,

$$\mathbf{r}_{G/P}=\frac{m}{m+M}\mathbf{r}=\mu\mathbf{r}=\mu\left(R\hat{\mathbf{i}}+z_m\hat{\mathbf{k}}\right) \tag{12.76}$$

in which

$$\mu=\frac{m}{m+M} \tag{12.77}$$

Thus,

$$\mathbf{r}_{G/P}\times(M+m)\mathbf{a}_{P/G}=\left(\frac{m}{M+m}\right)\mathbf{r}\times(M+m)\mathbf{a}_{P/G}=\mathbf{r}\times m\mathbf{a}_{P/G} \tag{12.78}$$

The acceleration of P relative to G is found with the aid of Eq. (1.60),

$$\mathbf{a}_{P/G}=-\ddot{\mathbf{r}}_{G/P}=-\mu\frac{d^2\mathbf{r}}{dt}=-\mu\left[\left.\frac{d^2\mathbf{r}}{dt^2}\right)_{\text{rel}}+\dot{\boldsymbol{\omega}}\times\mathbf{r}+\boldsymbol{\omega}\times(\boldsymbol{\omega}\times\mathbf{r})+2\boldsymbol{\omega}\times\left.\frac{d\mathbf{r}}{dt}\right)_{\text{rel}}\right] \tag{12.79}$$

where

$$\left.\frac{d\mathbf{r}}{dt}\right)_{\text{rel}}=\frac{dR}{dt}\hat{\mathbf{i}}+\frac{dz_m}{dt}\hat{\mathbf{k}}=\dot{z}_m\hat{\mathbf{k}} \tag{12.80}$$

and

$$\left.\frac{d^2\mathbf{r}}{dt^2}\right)_{\text{rel}}=\frac{d^2R}{dt^2}\hat{\mathbf{i}}+\frac{d^2z_m}{dt^2}\hat{\mathbf{k}}=\ddot{z}_m\hat{\mathbf{k}} \tag{12.81}$$

Substituting Eqs. (12.61), (12.64), (12.65), (12.80), and (12.81) into Eq. (12.79) yields

$$\begin{aligned}\mathbf{a}_{P/G}&=\mu\left[-z_m\dot{\omega}_y+R\left(\omega_y{}^2+\omega_z{}^2\right)-z_m\omega_x\omega_z-2\dot{z}_m\omega_y\right]\hat{\mathbf{i}}\\&+\mu\left(z_m\dot{\omega}_x-R\dot{\omega}_z-R\omega_x\omega_y-z_m\omega_y\omega_z+2\dot{z}_m\omega_x\right)\hat{\mathbf{j}}\\&+\mu\left[R\dot{\omega}_y+z_m\left(\omega_x{}^2+\omega_y{}^2\right)-R\omega_x\omega_z-\ddot{z}_m\right]\hat{\mathbf{k}}\end{aligned} \tag{12.82}$$

We move this expression into Eq. (12.78) to get

$$\begin{aligned}\mathbf{r}_{G/P}\times(M+m)\mathbf{a}_{P/G}&=\mu m\left[-z_m{}^2\dot{\omega}_x-2z_m\dot{z}_m\omega_x+Rz_m\left(\omega_x\omega_y+\dot{\omega}_z\right)+z_m{}^2\omega_y\omega_z\right]\hat{\mathbf{i}}\\&+\mu m\left[-\left(R^2+z_m{}^2\right)\dot{\omega}_y-2z_m\dot{z}_m\omega_y+Rz_m\left(\omega_z{}^2-\omega_x{}^2\right)+\left(R^2-z_m{}^2\right)\omega_x\omega_z+R\ddot{z}_m\right]\hat{\mathbf{j}}\\&+\mu m\left(Rz_m\dot{\omega}_x-R^2\dot{\omega}_z+2R\dot{z}_m\omega_x-R^2\omega_x\omega_y-Rz_m\omega_y\omega_z\right)\hat{\mathbf{k}}\end{aligned}$$

Placing this result and Eq. (12.74) in Eq. (12.71) and using the fact that $\mathbf{M}_{G})_{\text{net}}=\mathbf{0}$ yields a vector equation whose three components are

$$\begin{aligned}&A\dot{\omega}_x+(C-B)\omega_y\omega_z+(1-\mu)m\left[z_m{}^2\left(\dot{\omega}_x-\omega_y\omega_z\right)-Rz_m\left(\dot{\omega}_z+\omega_x\omega_y\right)+2z_m\dot{z}_m\omega_x\right]=0\\&\left[\left(B+mR^2\right)-\mu mR^2\right]\dot{\omega}_y+\left[\left(A+\mu mR^2\right)-\left(C+mR^2\right)\right]\omega_x\omega_z\\&\quad+(1-\mu)m\left[z_m{}^2\left(\omega_x\omega_z+\dot{\omega}_y\right)+2z_m\dot{z}_m\omega_y-R\ddot{z}_m+Rz_m\left(\omega_x{}^2-\omega_z{}^2\right)\right]=0\\&\left[\left(C+mR^2\right)-\mu mR^2\right]\dot{\omega}_z+\left[\left(B+mR^2\right)-\left(A+\mu mR^2\right)\right]\omega_x\omega_y\\&\quad+(1-\mu)mR\left[z_m\left(\omega_y\omega_z-\dot{\omega}_x\right)-2\dot{z}_m\omega_x\right]=0\end{aligned} \tag{12.83}$$

These are three equations in the four unknowns ω_x, ω_y, ω_z, and z_m. The fourth equation is that of the motion of the damper mass m in the z direction,

$$W_z - kz_m - c\dot{z}_m = ma_z \tag{12.84}$$

where a_z is given by Eq. (12.67$_3$), in which $a_P)_z = a_P)_z - a_G)_z + a_G)_z = a_{P/G})_z + a_G)_z$, so that

$$a_z = a_{P/G}\big)_z + a_G\big)_z - z_m\left(\omega_x{}^2 + \omega_y{}^2\right) - R\dot{\omega}_y + R\omega_x\omega_z + \ddot{z}_m \tag{12.85}$$

Substituting the z component of Eq. (12.82) into this expression and that result into Eq. (12.84) leads (with $W_z = ma_G)_z$) to

$$(1-\mu)m\ddot{z}_m + c\dot{z}_m + \left[k - (1-\mu)m\left(\omega_x{}^2 + \omega_y{}^2\right)\right]z_m = (1-\mu)mR\left[\dot{\omega}_y - \omega_x\omega_z\right] \tag{12.86}$$

Compare Eq. (12.69$_3$) with this expression, which is the fourth equation of motion we need.

Eqs. (12.83) and (12.86) are a rather complicated set of nonlinear, second-order differential equations that must be solved (numerically) to obtain a precise description of the motion of the semirigid spacecraft. The procedures of Section 1.8 may be employed. To study the stability of Eqs. (12.83) and (12.86), we can linearize them in much the same way as we did in Section 12.3. (Note that Eqs. 12.83 reduce to Eqs. 12.29 when $m = 0$.) With that as our objective, we assume that the spacecraft is in pure spin with angular velocity ω_o about the z axis and that the damper mass is at rest ($z_m = 0$). This motion is slightly perturbed, in such a way that

$$\omega_x = \delta\omega_x \qquad \omega_y = \delta\omega_y \qquad \omega_z = \omega_o + \delta\omega_z \qquad z_m = \delta z_m \tag{12.87}$$

It will be convenient for this analysis to introduce operator notation for the time derivative, $D = \mathrm{d}/\mathrm{d}t$. Thus, given a function of time $f(t)$, for any integer n, $D^n f = \mathrm{d}^n f/\mathrm{d}t^n$ and $D^0 f(t) = f(t)$. Then, the various time derivatives throughout the equations will, in accordance with Eq. (12.87), be replaced as follows:

$$\dot{\omega}_x = D\delta\omega_x \qquad \dot{\omega}_y = D\delta\omega_y \qquad \dot{\omega}_z = D\delta\omega_z \qquad \dot{z}_m = D\delta z_m \qquad \ddot{z}_m = D^2\delta z_m \tag{12.88}$$

Substituting Eqs. (12.87) and (12.88) into Eqs. (12.83) and (12.86) and retaining only those terms that are at most linear in the small perturbations leads to

$$\begin{aligned}
AD\delta\omega_x + (C-B)\omega_o\delta\omega_y &= 0\\
\left[A - C - (1-\mu)mR^2\right]\omega_o\delta\omega_x + \left[B + (1-\mu)mR^2\right]D\delta\omega_y - (1-\mu)mR(D^2 + \omega_o{}^2)\delta z_m &= 0\\
\left[C + (1-\mu)mR^2\right]D\delta\omega_z &= 0\\
(1-\mu)mR\omega_o\delta\omega_x - (1-\mu)mRD\delta\omega_y + \left[(1-\mu)mD^2 + cD + k\right]\delta z_m &= 0
\end{aligned} \tag{12.89}$$

$\delta\omega_z$ appears only in the third equation, which states that $\delta\omega_z =$ constant. The first, second, and fourth equations may be combined in matrix notation,

$$\begin{bmatrix} AD & (C-B)\omega_o & 0\\ \left[A - C - (1-\mu)mR^2\right]\omega_o & \left[B + (1-\mu)mR^2\right]D & -(1-\mu)mR(D^2 + \omega_o{}^2)\\ (1-\mu)mR\omega_o & -(1-\mu)mRD & (1-\mu)mD^2 + cD + k \end{bmatrix} \begin{Bmatrix} \delta\omega_x\\ \delta\omega_y\\ \delta z_m \end{Bmatrix} = \begin{Bmatrix} 0\\ 0\\ 0 \end{Bmatrix} \tag{12.90}$$

This is a set of three linear differential equations in the perturbations $\delta\omega_x$, $\delta\omega_y$, and δz_m. We will not try to solve them, since all we are really interested in is the stability of the satellite-damper system. It can be shown that the determinant Δ of the 3-by-3 matrix in Eq. (12.90) is

$$\Delta = a_4D^4 + a_3D^3 + a_2D^2 + a_1D + a_o \tag{12.91}$$

in which the coefficients of the characteristic equation $\Delta = 0$ are

$$\begin{aligned}
a_4 &= (1-\mu)mAB \\
a_3 &= cA\left[B + (1-\mu)mR^2\right] \\
a_2 &= k\left[B + (1-\mu)mR^2\right]A + (1-\mu)m\left[(A-C)(B-C) - (1-\mu)AmR^2\right]\omega_o^{\ 2} \\
a_1 &= c\left\{\left[A - C - (1-\mu)mR^2\right](B-C)\right\}\omega_o^{\ 2} \\
a_0 &= k\left\{\left[A - C - (1-\mu)mR^2\right](B-C)\right\}\omega_o^{\ 2} + \left[(B-C)(1-\mu)^2\right]m^2R^2\omega_o^{\ 4}
\end{aligned} \tag{12.92}$$

According to the Routh-Hurwitz stability criteria (see any text on control systems, e.g., Palm, 1983) the motion represented by Eq. (12.90), is asymptotically stable if and only if the signs of all of the following quantities, defined in terms of the coefficients of the characteristic equation, are the same

$$r_1 = a_4 \quad r_2 = a_3 \quad r_3 = a_2 - \frac{a_4 a_1}{a_3} \quad r_4 = a_1 - \frac{a_3{}^2 a_0}{a_3 a_2 - a_4 a_1} \quad r_5 = a_0 \tag{12.93}$$

EXAMPLE 12.7

A satellite is spinning about the z axis of its principal body frame at 2π rad/s. The principal moments of inertia about its center of mass are

$$A = 300 \text{ kg}\cdot\text{m}^2 \quad B = 400 \text{ kg}\cdot\text{m}^2 \quad C = 500 \text{ kg}\cdot\text{m}^2 \tag{a}$$

For the nutation damper, the following properties are given:

$$R = 1 \text{ m} \quad \mu = 0.01 \quad m = 10 \text{ kg} \quad k = 10{,}000 \text{ N/m} \quad c = 150 \text{ N s/m} \tag{b}$$

Use the Routh-Hurwitz stability criteria to assess the stability of the satellite as a major-axis spinner, a minor-axis spinner, and an intermediate-axis spinner.

Solution

The data in Eq. (a) are for a major-axis spinner. Substituting into Eqs. (12.92) and (12.93), we find

$$\begin{aligned}
r_1 &= +1.188\left(10^6\right)\text{kg}^3\cdot\text{m}^4 \\
r_2 &= +18.44\left(10^6\right)\text{kg}^3\cdot\text{m}^4/\text{s} \\
r_3 &= +1.228\left(10^9\right)\text{kg}^3\cdot\text{m}^4/\text{s}^2 \\
r_4 &= +92{,}820\,\text{kg}^3\cdot\text{m}^4/\text{s}^3 \\
r_5 &= +8.271\left(10^9\right)\text{kg}^3\cdot\text{m}^4/\text{s}^4
\end{aligned} \tag{c}$$

Since every r is positive, spin about the major axis is asymptotically stable. As we know from Section 12.3, without the damper the motion is neutrally stable.

For spin about the minor axis,

$$A = 500 \text{ kg}\cdot\text{m}^2 \quad B = 400 \text{ kg}\cdot\text{m}^2 \quad C = 300 \text{ kg}\cdot\text{m}^2 \tag{d}$$

For these moment of inertia values, we obtain

$$\begin{aligned}
r_1 &= +1.980\left(10^6\right)\text{kg}^3\cdot\text{m}^4 \\
r_2 &= +30.74\left(10^6\right)\text{kg}^3\cdot\text{m}^4/\text{s} \\
r_3 &= +2.048\left(10^9\right)\text{kg}^3\cdot\text{m}^4/\text{s}^2 \\
r_4 &= -304{,}490\,\text{kg}^3\cdot\text{m}^4/\text{s}^3 \\
r_5 &= +7.520\left(10^9\right)\text{kg}^3\cdot\text{m}^4/\text{s}^4
\end{aligned} \tag{e}$$

Since the rs are not all of the same sign, spin about the minor axis is not asymptotically stable. Recall that for the rigid satellite, such a motion was neutrally stable.

Finally, for spin about the intermediate axis,

$$A = 300\ \text{kg}\cdot\text{m}^2 \quad B = 500\ \text{kg}\cdot\text{m}^2 \quad C = 400\ \text{kg}\cdot\text{m}^2 \tag{f}$$

We know this motion is unstable, even without the nutation damper, but doing the Routh-Hurwitz stability check anyway, we get

$$\begin{aligned} r_1 &= +1.485\left(10^6\right)\text{kg}^3\cdot\text{m}^4 \\ r_2 &= +22.94\left(10^6\right)\text{kg}^3\cdot\text{m}^4/\text{s} \\ r_3 &= +1.529\left(10^9\right)\text{kg}^3\cdot\text{m}^4/\text{s}^2 \\ r_4 &= -192{,}800\,\text{kg}^3\cdot\text{m}^4/\text{s}^3 \\ r_5 &= -4.323\left(10^9\right)\text{kg}^3\cdot\text{m}^4/\text{s}^4 \end{aligned} \tag{g}$$

The motion, as we expected, is not stable.

12.6 CONING MANEUVER

Like the use of nutation dampers, the coning maneuver is an example of the attitude control of spinning spacecraft. In this case, the angular momentum is changed by the use of onboard thrusters (small rockets) to apply pure torques.

Consider a spacecraft in pure spin with angular velocity ω_0 about its body-fixed z axis, which is an axis of rotational symmetry. The angular momentum is $\mathbf{H}_G)_0 = C\omega_0\hat{\mathbf{k}}$. Suppose we wish to maintain the magnitude of the angular momentum but change its direction by rotating the spin axis through an angle θ, as illustrated in Fig. 12.13. Recall from Section 11.4 that to change the angular momentum of the spacecraft requires applying an external moment,

$$\Delta\mathbf{H}_G = \int_0^{\Delta t} \mathbf{M}_G dt$$

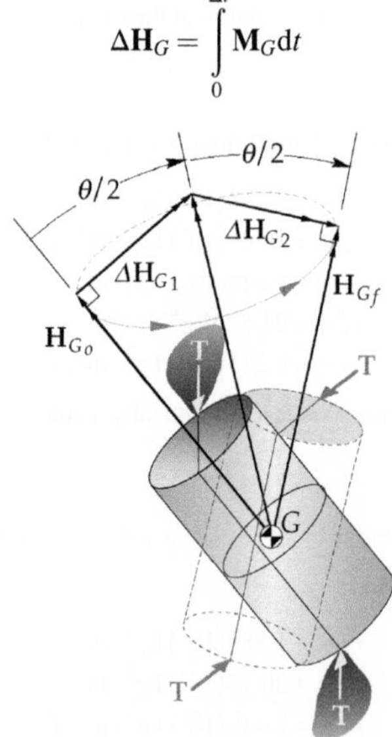

FIG. 12.13

Impulsive coning maneuver.

Thrusters may be used to provide the external impulsive torque required to produce an angular momentum increment $\Delta\mathbf{H}_G)_1$ normal to the spin axis. Since the spacecraft is spinning, this induces coning (precession) of the spacecraft about an axis at an angle of $\theta/2$ to the direction of $\mathbf{H}_G)_0$. Since the external couple is normal to the z axis, the maneuver produces no change in the z component of the angular velocity, which remains ω_0. However, after the impulsive moment, the angular velocity comprises a spin component ω_s and a precession component ω_p. Whereas before the impulsive moment $\omega_s = \omega_0$, afterward, during coning, the spin component is given by Eq. (12.20),

$$\omega_s = \frac{A-C}{A}\omega_0$$

The precession rate is given by Eq. (12.22),

$$\omega_p = \frac{C}{A}\frac{\omega_0}{\cos(\theta/2)} \tag{12.94}$$

Note that before the impulsive maneuver, the magnitude of the angular momentum is $C\omega_0$. Afterward, it has increased to

$$H_G = A\omega_p = \frac{C\omega_0}{\cos(\theta/2)}$$

After precessing 180°, an angular momentum increment $\Delta\mathbf{H}_G)_2$ normal to the spin axis and in the same direction relative to the spacecraft as the initial torque impulse, with $\|\Delta\mathbf{H}_G)_2\| = \|\Delta\mathbf{H}_G)_1\|$, stabilizes the spin vector in the desired direction. Since the spin rate ω_s is not in general the same as the precession rate ω_p, the second angular impulse must be delivered by another pair of thrusters, that have rotated into the position to apply the torque impulse in the proper direction. With only one pair of thrusters, both the spin axis and the spacecraft must rotate through 180° in the same time interval, which means $\omega_p = \omega_s$. That is,

$$\frac{A-C}{A}\omega_0 = \frac{C}{A}\frac{\omega_0}{\cos(\theta/2)}$$

This requires the deflection angle to be

$$\theta = 2\cos^{-1}\left(\frac{C}{A-C}\right)$$

and limits the values of the moments of inertia A and C to those that do not cause the magnitude of the cosine to exceed unity.

The time required for an angular reorientation θ using a single coning maneuver is found by simply dividing the precession angle, π rad, by the precession rate ω_p,

$$t_1 = \frac{\pi}{\omega_p} = \pi\frac{A}{C\omega_0}\cos\frac{\theta}{2} \tag{12.95}$$

Propellant expenditure is reflected in the magnitude of the individual angular momentum increments, in obvious analogy to delta-v calculations for orbital maneuvers. The total delta-H required for the single coning maneuver is therefore given by

$$\Delta H_{\text{total}} = \left\|\Delta\mathbf{H}_G)_1\right\| + \left\|\Delta\mathbf{H}_G)_2\right\| = 2\left(\left\|\mathbf{H}_G)_0\right\|\tan\frac{\theta}{2}\right) \tag{12.96}$$

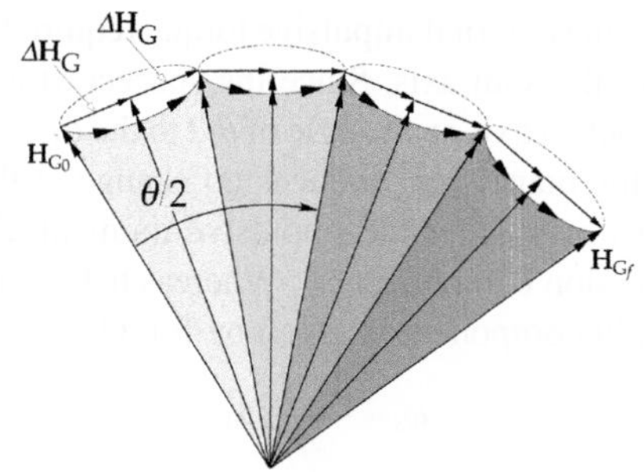

FIG. 12.14

A sequence of small coning maneuvers.

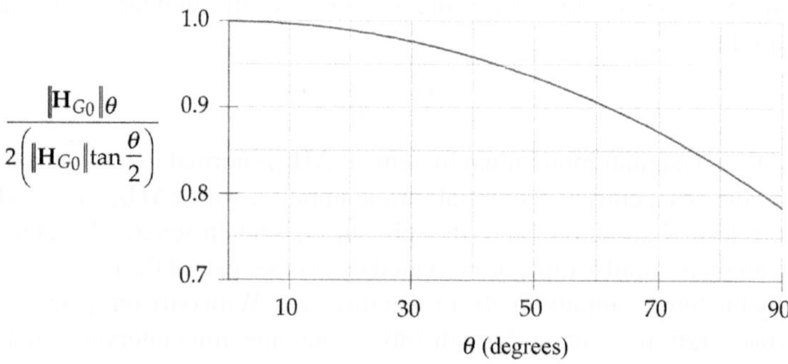

FIG. 12.15

Ratio of delta-*H* for a sequence of small coning maneuvers to that for a single coning maneuver, as a function of the angle of swing of the spin axis.

Fig. 12.14 illustrates the fact that ΔH_{total} can be reduced by using a sequence of small coning maneuvers (small θs) rather than one big θ. The large number of small ΔHs approximates a circular arc of radius $\|\mathbf{H}_G)_0\|$, subtended by the angle θ. Therefore, approximately,

$$\Delta H_{\text{total}} = 2\left(\|\mathbf{H}_G)_0\|\frac{\theta}{2}\right) = \|\mathbf{H}_G)_0\|\theta \tag{12.97}$$

This expression becomes more precise as the number of intermediate maneuvers increases. Fig. 12.15 reveals the extent to which the multiple coning maneuver strategy reduces energy requirements. The difference is quite significant for large reorientation angles.

One of the prices to be paid for the reduced energy of the multiple coning maneuver is time. (The other is the complexity mentioned above, to say nothing of the risk involved in repeating the maneuver over and over again.) From Eq. (12.95), the time required for n small-angle coning maneuvers through a total angle of θ is

$$t_n = n\pi\frac{A}{C\omega_0}\cos\frac{\theta}{2n} \tag{12.98}$$

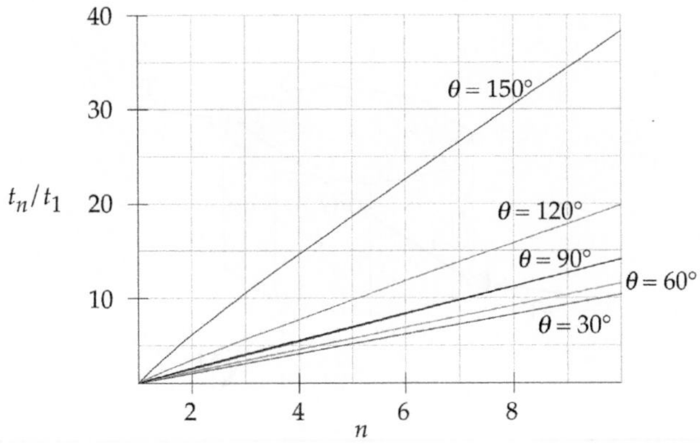

FIG. 12.16

Time for a coning maneuver vs. the number of intermediate steps.

The ratio of this to the time t_1 required for a single coning maneuver is

$$\frac{t_n}{t_1} = n\frac{\cos\dfrac{\theta}{2n}}{\cos\dfrac{\theta}{2}} \tag{12.99}$$

The time is directly proportional to the number of intermediate coning maneuvers, as illustrated in Fig. 12.16.

12.7 ATTITUDE CONTROL THRUSTERS

As mentioned above, thrusters are small jets mounted in pairs on a spacecraft to control its rotational motion about the center of mass. These thruster pairs may be mounted in principal planes (planes normal to the principal axes) passing through the center of mass. Fig. 12.17 illustrates a pair of thrusters for producing a torque about the positive y axis. These would be accompanied by another pair of reaction motors pointing in the opposite directions to exert torque in the negative y direction. If the position vectors of the thrusters relative to the center of mass are $\mathbf{r}$ and $-\mathbf{r}$, and if $\mathbf{T}$ is their thrust, then the impulsive moment they exert during a brief time interval Δt is

$$\mathbf{M} = \mathbf{r} \times \mathbf{T}\Delta t + (-\mathbf{r}) \times (-\mathbf{T}\Delta t) = 2\mathbf{r} \times \mathbf{T}\Delta t \tag{12.100}$$

If the angular velocity was initially zero, then after the firing, according to Eq. (11.31), the angular momentum becomes

$$\mathbf{H} = 2\mathbf{r} \times \mathbf{T}\Delta t \tag{12.101}$$

For $\mathbf{H}$ in the principal y direction, as in the figure, the corresponding angular velocity acquired by the vehicle is, from Eq. (11.67),

$$\omega_y = \frac{\|\mathbf{H}\|}{B} \tag{12.102}$$

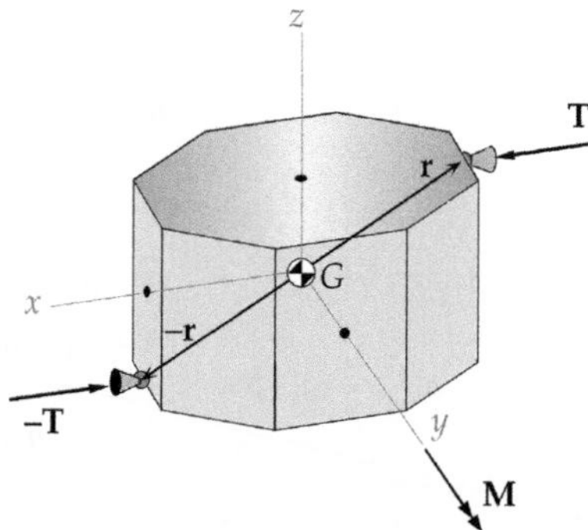

FIG. 12.17

Pair of attitude control thrusters mounted in the *xz* plane of the principal body frame.

EXAMPLE 12.8

A spacecraft of mass m and with the dimensions shown in Fig. 12.18 is spinning without precession at the rate ω_0 about the z axis of the principal body frame. At the instant shown in part (a) of the figure, the spacecraft initiates a coning maneuver to swing its spin axis through 90°, so that at the end of the maneuver the vehicle is oriented as illustrated in Fig. 12.18a. Calculate the total delta-H required and compare it with that required for the same reorientation without coning. Motion is to be controlled exclusively by the pairs of attitude thrusters shown, all of which have identical thrust T.

Solution

According to Fig. 11.10c, the moments of inertia about the principal body axes are

$$A=B=\frac{1}{12}m\left[w^2+\left(\frac{w}{3}\right)^2\right]=\frac{5}{54}mw^2 \quad C=\frac{1}{12}m\left(w^2+w^2\right)=\frac{1}{6}mw^2$$

The initial angular momentum $\mathbf{H}_G)_1$ points in the spin direction, along the positive z axis of the body frame,

$$\mathbf{H}_G)_1=C\omega_z\hat{\mathbf{k}}=\frac{1}{6}mw^2\omega_0\hat{\mathbf{k}}$$

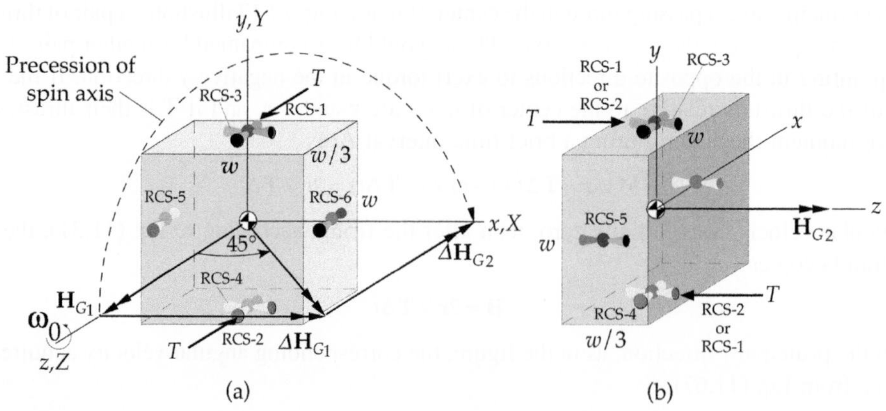

FIG. 12.18

(a) Initial orientation of spinning spacecraft. (b) Final configuration, with spin axis rotated 90°.

We can presume that in the initial orientation, the body frame happens to coincide instantaneously with the inertial frame *XYZ*. The coning motion is initiated by briefly firing the pair of thrusters RCS-1 and RCS-2, aligned with the body *z* axis and lying in the *yz* plane. The impulsive torque will cause a change $\Delta\mathbf{H}_G)_1$ in angular momentum directed normal to the plane of the thrusters, in the positive body *x* direction. The resultant angular momentum vector must lie at 45° to the *x* and *z* axes, bisecting the angle between the initial and final angular momenta. Thus,

$$\|\Delta\mathbf{H}_G)_1\| = \|\mathbf{H}_G)_1\|\tan 45° = \frac{1}{6}mw^2\omega_0$$

After the coning is under way, the body axes of course move away from the *XYZ* frame. Since the spacecraft is oblate ($C > A$), the precession of the spin axis will be opposite to the spin direction, as indicated in Fig. 12.15. When the spin axis, after 180° of precession, lines up with the *X* axis, the thrusters must fire again for the same duration as before so as to produce the angular momentum change $\Delta\mathbf{H}_G)_2$, equal in magnitude but perpendicular to $\Delta\mathbf{H}_G)_1$, so that

$$\mathbf{H}_G)_1 + \Delta\mathbf{H}_G)_1 + \Delta\mathbf{H}_G)_2 = \mathbf{H}_G)_2$$

where

$$\mathbf{H}_G)_2 = \|\mathbf{H}_G)_1\|\hat{\mathbf{I}} = \frac{1}{6}mw^2\omega_0\hat{\mathbf{k}}$$

For this to work, the plane of thrusters RCS-1 and RCS-2 (the *yz* plane) must be parallel to the *XY* plane when they fire, as illustrated in Fig. 12.18b. Since the thrusters can fire fore or aft, it does not matter which of them ends up on the top or bottom. The vehicle must therefore spin through an integral number *n* of half rotations while it precesses to the desired orientation. That is, the total spin angle ψ between the initial and final configurations is

$$\psi = n\pi = \omega_s t \tag{a}$$

where ω_s is the spin rate, and *t* is the time for the proper final configuration to be achieved. In the meantime, the precession angle ϕ must be π or 3π or 5π, or, in general,

$$\phi = (2m-1)\pi = \omega_p t \tag{b}$$

where *m* is an integer, and *t* is, of course, the same as that in Eq. (a). Eliminating *t* from both Eqs. (a) and (b) yields

$$n\pi = (2m-1)\pi\frac{\omega_s}{\omega_p}$$

Substituting Eq. (12.23), with $\theta = \pi/4$, gives

$$n = (1-2m)\frac{4}{9}\frac{1}{\sqrt{2}} \tag{c}$$

Obviously, this equation cannot be valid if both *m* and *n* are integers. However, by tabulating *n* as a function of *m*, we find that when $m = 18$, $n = -10.999$. The minus sign simply reminds us that spin and precession are in opposite directions. Thus, the 18th time that the spin axis lines up with the *X* axis the thrusters may be fired to almost perfectly align the angular momentum vector with the body *z* axis. The slight misalignment due to the fact that $|n|$ is not precisely 11 would probably occur in reality anyway. Passive or active nutation damping can drive this deviation to zero.

Since $\|\mathbf{H}_G)_1\| = \|\mathbf{H}_G)_2\|$, we conclude that

$$\Delta H_{\text{total}} = 2\left(\frac{1}{6}mw^2\omega_0\right) = \frac{2}{3}mw^2\omega_0 \tag{d}$$

An obvious alternative to the coning maneuver is to use thrusters RCS-3 and RCS-4 to despin the craft completely, thrusters RCS-5 and RCS-6 to initiate roll around the *y* axis and stop it after 90°, and then RCS-3 and RCS-4 to respin the spacecraft to ω_0 around the *z* axis. The combined delta-*H* for the first and last steps equals that of Eq. (d). Additional fuel expenditure is required to start and stop the roll around the *y* axis. Hence, the coning maneuver is more fuel efficient.

12.8 YO-YO DESPIN MECHANISM

A simple, inexpensive way to despin an axisymmetric spacecraft is to deploy small masses attached to cords wound around the girth of the spacecraft near the transverse plane through the center of mass. As the masses unwrap in the direction of the spacecraft angular velocity, they exert centrifugal force through the cords on the periphery of the vehicle, creating a moment opposite to the spin direction, thereby slowing down the rotational motion. The cord forces are internal to the system of spacecraft plus weights, so that as the strings unwind, the total angular momentum must remain constant. Since the total moment of inertia increases as the yo-yo masses spiral farther away, the angular velocity must drop. Not only angular momentum but also rotational kinetic energy is conserved during this process. Yo-yo despin devices were introduced early in unmanned space flight (e.g., 1959 Transit 1-A) and continued to be used thereafter (e.g., 1996 Mars Pathfinder, 1998 Mars Climate Orbiter, 1999 Mars Polar Lander, 2003 Mars Exploration Rover, and 2007 Dawn spacecraft).

The problem is to determine the length of cord required to reduce the spacecraft angular velocity a specified amount. Because it is easier than solving the equations of motion, we will apply the principles of conservation of energy and angular momentum to the system comprising the spacecraft and the yo-yo masses. To maintain the position of the center of mass, two identical yo-yo masses are wound around the spacecraft in a symmetric fashion, as illustrated in Fig. 12.19. Both masses are released simultaneously by explosive bolts and unwrap in the manner shown (for only one of the weights) in the figure. In so doing, the point of tangency T moves around the circumference toward the split hinge device where the cord is attached to the spacecraft. When T and T' reach the hinges H and H', the cords automatically separate from the spacecraft.

Let each yo-yo weight have mass $m/2$. By symmetry, we need to track only one of the masses, to which we can ascribe the total mass m. Let the xyz system be a body frame rigidly attached to the spacecraft, as shown in Fig. 12.19. As usual, the z axis lies in the spin direction, pointing out of the page. The x axis is directed from the center of mass of the system through the initial position of the yo-yo mass.

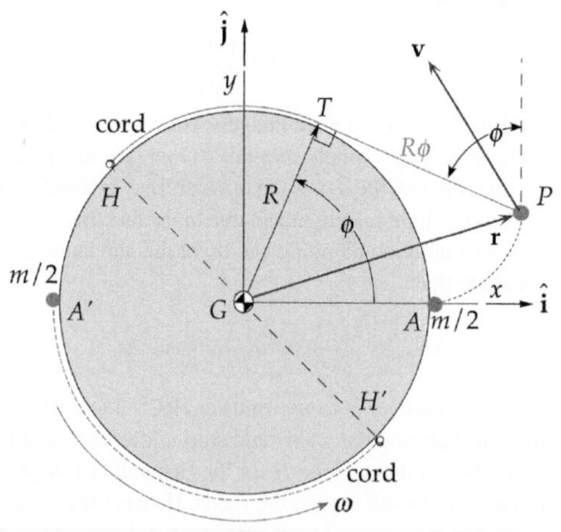

FIG. 12.19

Two identical string and mass systems wrapped symmetrically around the periphery of an axisymmetric spacecraft. For simplicity, only one is shown being deployed.

The spacecraft and the yo-yo masses, prior to release, are rotating as a single rigid body with angular velocity $\boldsymbol{\omega}_0 = \omega_0\hat{\mathbf{k}}$. The moment of inertia of the spacecraft, excluding the yo-yo mass, is C, so that the angular momentum of the spacecraft by itself is $C\omega_0$. The concentrated yo-yo masses are fastened at a distance R from the spin axis, so that their total moment of inertia is mR^2. Therefore, the initial angular momentum of the satellite plus yo-yo system is

$$\left.H_G\right)_0 = C\omega_0 + mR^2\omega_0$$

It will be convenient to write this as

$$\left.H_G\right)_0 = KmR^2\omega_0 \tag{12.103}$$

where the nondimensional factor K is defined as

$$K = 1 + \frac{C}{mR^2} \tag{12.104}$$

$R\sqrt{mK/(m+M)}$ is the system's initial radius of gyration, where M is the spacecraft mass.

$$T_0 = \frac{1}{2}C{\omega_0}^2 + \frac{1}{2}mR^2{\omega_0}^2 = \frac{1}{2}KmR^2{\omega_0}^2 \tag{12.105}$$

At any state between the release of the weights and the release of the cords at the hinges, the velocity of the yo-yo mass must be found to compute the new angular momentum and kinetic energy. Observe that when the string has unwrapped an angle ϕ, the free length of string (between the point of tangency T and the yo-yo mass P) is $R\phi$. From the geometry shown in Fig. 12.19, the position vector of the mass relative to the body frame is seen to be

$$\begin{aligned}\mathbf{r} &= \overbrace{\left(R\cos\phi\hat{\mathbf{i}} + R\sin\phi\hat{\mathbf{j}}\right)}^{\mathbf{r}_{T/G}} + \overbrace{\left(R\phi\sin\phi\hat{\mathbf{i}} - R\phi\cos\phi\hat{\mathbf{j}}\right)}^{\mathbf{r}_{P/T}} \\ &= (R\cos\phi + R\phi\sin\phi)\hat{\mathbf{i}} + (R\sin\phi - R\phi\cos\phi)\hat{\mathbf{j}}\end{aligned} \tag{12.106}$$

Since $\mathbf{r}$ is measured in the moving reference, the absolute velocity $\mathbf{v}$ of the yo-yo mass is found using Eq. (1.56),

$$\mathbf{v} = \left.\frac{d\mathbf{r}}{dt}\right)_{\text{rel}} + \boldsymbol{\Omega}\times\mathbf{r} \tag{12.107}$$

where $\boldsymbol{\Omega}$ is the angular velocity of the xyz axes, which, of course, is the angular velocity $\boldsymbol{\omega}$ of the spacecraft at that instant,

$$\boldsymbol{\Omega} = \boldsymbol{\omega} \tag{12.108}$$

To calculate $d\mathbf{r}/dt)_{\text{rel}}$, we hold $\hat{\mathbf{i}}$ and $\hat{\mathbf{j}}$ constant in Eq. (12.106), obtaining

$$\begin{aligned}\left.\frac{d\mathbf{r}}{dt}\right)_{\text{rel}} &= \left(-R\dot{\phi}\sin\phi + R\dot{\phi}\sin\phi + R\phi\dot{\phi}\cos\phi\right)\hat{\mathbf{i}} + \left(R\dot{\phi}\cos\phi - R\dot{\phi}\cos\phi + R\phi\dot{\phi}\sin\phi\right)\hat{\mathbf{j}} \\ &= R\phi\dot{\phi}\cos\phi\hat{\mathbf{i}} + R\phi\dot{\phi}\sin\phi\hat{\mathbf{j}}\end{aligned}$$

Thus,

$$\mathbf{v} = R\phi\dot{\phi}\cos\phi\hat{\mathbf{i}} + R\phi\dot{\phi}\sin\phi\hat{\mathbf{j}} + \begin{vmatrix} \hat{\mathbf{i}} & \hat{\mathbf{j}} & \hat{\mathbf{k}} \\ 0 & 0 & \omega \\ R\cos\phi + R\phi\sin\phi & R\sin\phi - R\phi\cos\phi & 0 \end{vmatrix}$$

or

$$\mathbf{v} = \left[R\phi\left(\omega+\dot{\phi}\right)\cos\phi - R\omega\sin\phi\right]\hat{\mathbf{i}} + \left[R\omega\cos\phi + R\phi\left(\omega+\dot{\phi}\right)\sin\phi\right]\hat{\mathbf{j}} \tag{12.109}$$

From this, we find the speed of the yo-yo weights,

$$v = \sqrt{\mathbf{v}\cdot\mathbf{v}} = R\sqrt{\omega^2 + \left(\omega+\dot{\phi}\right)^2\phi^2} \tag{12.110}$$

The angular momentum of the spacecraft plus the weights at an intermediate stage of the despin process is

$$\begin{aligned} \mathbf{H}_G &= C\omega\hat{\mathbf{k}} + \mathbf{r}\times m\mathbf{v} \\ &= C\omega\hat{\mathbf{k}} + m\begin{vmatrix} \hat{\mathbf{i}} & \hat{\mathbf{j}} & \hat{\mathbf{k}} \\ R\cos\phi + R\phi\sin\phi & R\sin\phi - R\phi\cos\phi & \omega \\ R\phi\left(\omega+\dot{\phi}\right)\cos\phi - R\omega\sin\phi & R\omega\cos\phi + R\phi\left(\omega+\dot{\phi}\right)\sin\phi & 0 \end{vmatrix} \end{aligned}$$

Carrying out the cross product, combining terms, and simplifying leads to

$$H_G = C\omega + mR^2\left[\omega + \left(\omega+\dot{\phi}\right)\phi^2\right]$$

which, using Eq. (12.104), can be written as

$$H_G = mR^2\left[K\omega + \left(\omega+\dot{\phi}\right)\phi^2\right] \tag{12.111}$$

The kinetic energy of the spacecraft plus the yo-yo mass is

$$T = \frac{1}{2}C\omega^2 + \frac{1}{2}mv^2$$

Substituting the speed from Eq. (12.110) and making use again of Eq. (12.104), we find

$$T = \frac{1}{2}mR^2\left[K\omega^2 + \left(\omega+\dot{\phi}\right)^2\phi^2\right] \tag{12.112}$$

By the conservation of angular momentum, $H_G = H_{G_0}$, we obtain from Eqs. (12.103) and (12.111),

$$mR^2\left[K\omega + \left(\omega+\dot{\phi}\right)\phi^2\right] = KmR^2\omega_0$$

which we can write as

$$K(\omega_0 - \omega) = \left(\omega+\dot{\phi}\right)\phi^2 \quad \text{Conservation of angular momentum} \tag{12.113}$$

Eqs. (12.105) and (12.112) and the conservation of kinetic energy, $T = T_0$, combine to yield

$$\frac{1}{2}mR^2\left[K\omega^2 + \left(\omega+\dot{\phi}\right)^2\phi^2\right] = \frac{1}{2}KmR^2\omega_0^2$$

or

$$K({\omega_0}^2 - \omega^2) = \left(\omega+\dot{\phi}\right)^2\phi^2 \quad \text{Conservation of energy} \tag{12.114}$$

Since $\omega_0{}^2 - \omega^2 = (\omega_0 + \omega)(\omega_0 + \omega)$, this can be written as

$$K(\omega_0 - \omega)(\omega_0 + \omega) = (\omega + \dot{\phi})^2 \phi^2$$

Replacing the factor $K(\omega_0 - \omega)$ on the left using Eq. (12.113) yields

$$(\omega + \dot{\phi})\phi^2(\omega_0 + \omega) = (\omega + \dot{\phi})^2 \phi^2$$

After canceling terms, we find $\omega_0 + \omega = \omega + \dot{\phi}$, or, simply

$$\dot{\phi} = \omega_0 \quad \text{Conservation of energy and momentum} \tag{12.115}$$

In other words, the cord unwinds at a constant rate (relative to the spacecraft), equal to the vehicle's initial angular velocity. Thus, at any time t after the release of the weights,

$$\phi = \omega_0 t \tag{12.116}$$

By substituting Eq. (12.115) back into Eq. (12.113),

$$K(\omega_0 - \omega) = (\omega + \omega_0)\phi^2$$

we find that

$$\phi = \sqrt{K\frac{\omega_0 - \omega}{\omega_0 + \omega}} \quad \text{Partial despin} \tag{12.117}$$

Recall that the unwrapped length l of the cord is $R\phi$, which means

$$l = R\sqrt{K\frac{\omega_0 - \omega}{\omega_0 + \omega}} \quad \text{Partial despin} \tag{12.118}$$

We use Eq. (12.118) to find the length of the cord required to despin the spacecraft from ω_0 to ω. To remove all of the spin ($\omega = 0$),

$$\phi = \sqrt{K} \;\Rightarrow\; l = R\sqrt{K} \quad \text{Complete despin} \tag{12.119}$$

Surprisingly, the length of the cord required to reduce the angular velocity to zero is independent of the initial angular velocity.

We can solve Eq. (12.117) for ω in terms of ϕ,

$$\omega = \left(\frac{2K}{K + \phi^2} - 1\right)\omega_0 \tag{12.120}$$

By means of Eq. (12.116), this becomes an expression for the angular velocity as a function of time,

$$\omega = \left(\frac{2K}{K + \omega_0{}^2 t^2} - 1\right)\omega_0 \tag{12.121}$$

Alternatively, since $\phi = l/R$, Eq. (12.120) yields the angular velocity as a function of the cord length,

$$\omega = \left(\frac{2KR^2}{KR^2 + l^2} - 1\right)\omega_0 \tag{12.122}$$

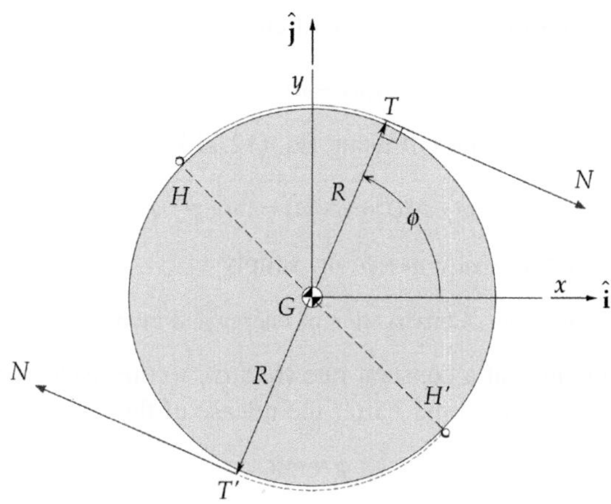

FIG. 12.20

Free body diagram of the satellite during the despin process.

Differentiating ω with respect to time in Eq. (12.121) gives us an expression for the angular acceleration of the spacecraft,

$$\alpha = \frac{d\omega}{dt} = -\frac{4K\omega_0{}^3 t}{(K + \omega_0{}^2 t^2)^2} \tag{12.123}$$

whereas integrating ω with respect to time yields the angle rotated by the spacecraft since release of the yo-yo mass,

$$\theta = 2\sqrt{K}\tan^{-1}\frac{\omega_0 t}{\sqrt{K}} - \omega_0 t = 2\sqrt{K}\tan^{-1}\frac{\phi}{\sqrt{K}} - \phi \tag{12.124}$$

For complete despin, this expression, together with Eq. (12.119), yields

$$\theta = \sqrt{K}\left(\frac{\pi}{2} - 1\right) \tag{12.125}$$

From the free body diagram of the spacecraft shown in Fig. 12.20, it is clear that the torque exerted by the yo-yo weights is

$$M_G)_z = -2RN \tag{12.126}$$

where N is the tension in the cord. From the Euler equations of motion (Eq. 11.72b)

$$M_G)_z = C\alpha \tag{12.127}$$

Combining Eqs. (12.123), (12.126), and (12.127) leads to a formula for tension in the yo-yo cables,

$$N = \frac{C}{R}\frac{2K\omega_0{}^3 t}{(K + \omega_0{}^2 t^2)^2} = \frac{C\omega_0{}^2}{R}\frac{2K\phi}{(K + \phi^2)^2} \tag{12.128}$$

12.8.1 RADIAL RELEASE

Finally, we note that instead of releasing the yo-yo masses when the cables are tangent at the split hinges (H and H'), they can be forced to pivot about the hinge and released when the string is directed radially outward, as illustrated in Fig. 12.18. The above analysis must be then extended to include the pivoting of the cord around the hinges. It turns out that in this case, the length of the cord as a function of the final angular velocity is

$$l = R\left(\sqrt{\frac{[(\omega_0 - \omega)K + \omega]^2}{(\omega_0{}^2 - \omega^2)K + \omega^2}} - 1\right) \quad \text{Partial despin, radial release} \tag{12.129}$$

so that for $\omega = 0$ (Fig. 12.21),

$$l = R\left(\sqrt{K} - 1\right) \quad \text{Complete despin, radial release} \tag{12.130}$$

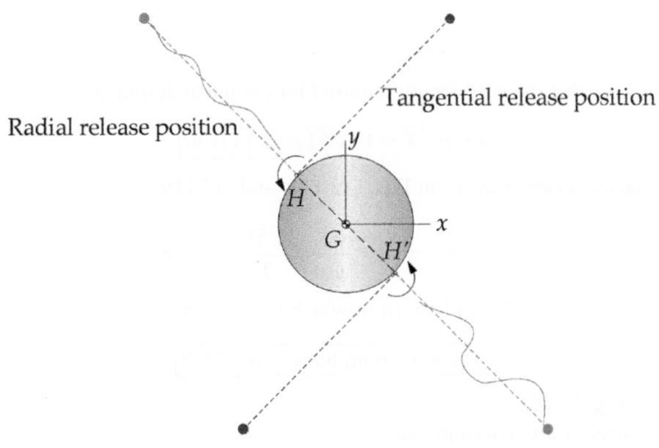

FIG. 12.21

Radial vs. tangential release of yo-yo masses.

EXAMPLE 12.9

A satellite is to be completely despun using a two-mass yo-yo device with tangential release. Assume the spin axis moment of inertia of the satellite is $C = 200\ \text{kg}\cdot\text{m}^2$ and the initial spin rate is $\omega_0 = 5\ \text{rad/s}$. The total yo-yo mass is 4 kg, and the radius of the spacecraft is 1 m. Find

(a) the required cord length l;
(b) the time t to despin;
(c) the maximum tension in the yo-yo cables;
(d) the speed of the masses at release;
(e) the angle rotated by the satellite during despin; and
(f) the cord length required for radial release.

Solution

(a) From Eq. (12.104),

$$K = 1 + \frac{C}{mR^2} = 1 + \frac{200}{4\cdot 1^2} = 51 \tag{a}$$

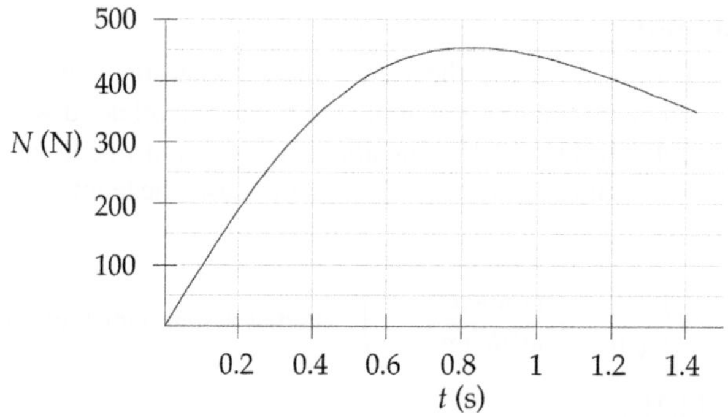

FIG. 12.22

Variation of cable tension N up to the point of release.

From Eq. (12.119) it follows that the cord length required for complete despin is

$$l = R\sqrt{K} = 1\cdot\sqrt{51} = \boxed{7.1441\ \text{m}} \qquad \text{(b)}$$

(b) The time for complete despin is obtained from Eqs. (12.116) and (12.119),

$$\omega_0 t = \sqrt{K} \Rightarrow t = \frac{\sqrt{K}}{\omega_0} = \frac{\sqrt{51}}{5} = \boxed{1.4283\ \text{s}}$$

(c) A graph of Eq. (12.128) is shown in Fig. 12.19, from which we see that

$$\boxed{\text{The maximum tension is 455 N}}$$

which occurs at 0.825 s (Fig. 12.22).

(d) From Eq. (12.110), the speed of the yo-yo masses is

$$v = R\sqrt{\omega^2 + (\omega + \dot{\phi})^2\phi^2}$$

According to Eq. (12.115), $\dot{\phi} = \omega_0$, and at the time of release ($\omega = 0$) Eq. (12.117) states that $\phi = \sqrt{K}$. Thus,

$$v = R\sqrt{\omega^2 + (\omega + \omega_0)^2\sqrt{K}^2} = 1\cdot\sqrt{0^2 + (0+5)^2\sqrt{51}^2} = \boxed{35.71\ \text{m/s}}$$

(e) The angle through which the satellite rotates before coming to rotational rest is given by Eq. (12.125),

$$\theta = \sqrt{K}\left(\frac{\pi}{2} - 1\right) = \sqrt{51}\left(\frac{\pi}{2} - 1\right) = \boxed{4.076\ \text{rad}\,(233.5°)}$$

(f) Allowing the cord to detach radially reduces the cord length required for complete despin from 7.141 m to (Eq. 12.130)

$$l = R\left(\sqrt{K} - 1\right) = 1\cdot\left(\sqrt{51} - 1\right) = \boxed{6.141\ \text{m}}$$

12.9 GYROSCOPIC ATTITUDE CONTROL

Momentum exchange systems (gyros) are used to control the attitude of a spacecraft without throwing consumable mass overboard, as occurs with the use of thruster jets. A momentum exchange system is illustrated schematically in Fig. 12.23. n flywheels, labeled 1, 2, 3, etc., are attached to the body of the

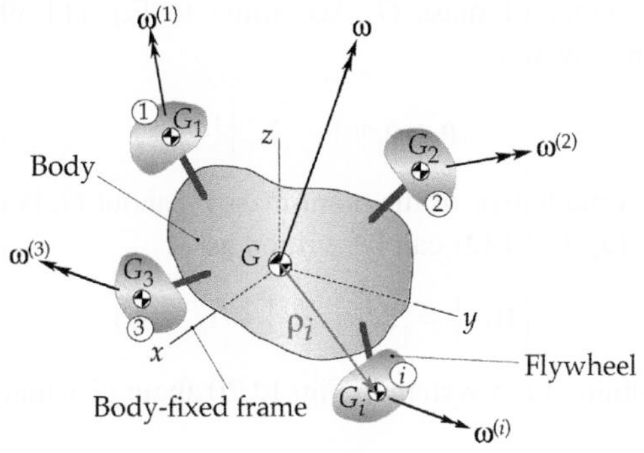

FIG. 12.23

Several attitude control flywheels, each with their own angular velocity, attached to the body of a spacecraft.

spacecraft at various locations. The mass of flywheel i is m_i. The mass of the body of the spacecraft is m_0. The total mass of the entire system (the "vehicle") is m,

$$m = m_0 + \sum_{i=1}^{n} m_i$$

The vehicle's center of mass is G, through which pass the three axes xyz of the vehicle's body-fixed frame. The center of mass G_i of each flywheel is connected rigidly to the spacecraft, but the wheel, driven by electric motors, rotates more or less independently, depending on the type of gyro. The body of the spacecraft has the angular velocity vector $\boldsymbol{\omega}$. The angular velocity vector of the ith flywheel is $\boldsymbol{\omega}^{(i)}$, and it differs from that of the body of the spacecraft unless the gyro is "caged." A caged gyro has no spin relative to the spacecraft, in which case $\boldsymbol{\omega}^{(i)} = \boldsymbol{\omega}$.

According to Eq. (11.39b), the angular momentum of the body itself relative to G is

$$\left\{\mathbf{H}_G^{(\text{body})}\right\} = \left[\mathbf{I}_G^{(\text{body})}\right]\{\boldsymbol{\omega}\} \tag{12.131}$$

where $\mathbf{I}_G^{(\text{body})}$ is the moment of inertia tensor of the body about G and $\boldsymbol{\omega}$ is the angular velocity of the body.

Eq. (11.27) gives the angular momentum of flywheel i relative to G as

$$\mathbf{H}_G^{(i)} = \mathbf{H}_{G_i}^{(i)} + \boldsymbol{\rho}_i \times \dot{\boldsymbol{\rho}}_i m_i \tag{12.132}$$

$\mathbf{H}_{G_i}^{(i)}$ is the angular momentum vector of the flywheel i about its own center of mass G_i. Its components in the body frame are found from the expression

$$\left\{\mathbf{H}_{G_i}^{(i)}\right\} = \left[\mathbf{I}_{G_i}^{(i)}\right]\left\{\boldsymbol{\omega}^{(i)}\right\} \tag{12.133}$$

where $\mathbf{I}_{G_i}^{(i)}$ is the moment of inertia tensor of the flywheel about its own center of mass G_i, relative to axes that are parallel to the body-fixed xyz axes. Since a momentum wheel might be one that pivots on gimbals relative to the body frame, the inertia tensor $\mathbf{I}_{G_i}^{(i)}$ may be time dependent. The vector $\boldsymbol{\rho}_i \times \dot{\boldsymbol{\rho}}_i m_i$ in Eq. (12.132) is the angular momentum of the concentrated mass m_i of the flywheel about the *system* center of mass G. According to Eq. (11.59), the components of $\boldsymbol{\rho}_i \times \dot{\boldsymbol{\rho}}_i m_i$ in the body frame are given by

$$\{\boldsymbol{\rho}_i \times \dot{\boldsymbol{\rho}}_i m_i\} = \left[\mathbf{I}_{mG}^{(i)}\right]\{\boldsymbol{\omega}\} \tag{12.134}$$

where $\mathbf{I}_{mG}^{(i)}$, the moment of inertia tensor of the point mass m_i about G, is given by Eq. (11.44). Using Eqs. (12.133) and (12.134), Eq. (12.132) can be written as

$$\left\{\mathbf{H}_G^{(i)}\right\} = \left[\mathbf{I}_{G_i}^{(i)}\right]\left\{\boldsymbol{\omega}^{(i)}\right\} + \left[\mathbf{I}_{mG}^{(i)}\right]\{\boldsymbol{\omega}\} \tag{12.135}$$

The total angular momentum of the system in Fig. 12.20 about G is that of the body plus all of the n flywheels,

$$\mathbf{H}_G = \mathbf{H}_G^{(\text{body})} + \sum_{i=1}^{n} \mathbf{H}_G^{(i)}$$

Substituting Eqs. (12.131) and (12.135), we obtain

$$\{\mathbf{H}_G\} = \left[\mathbf{I}_G^{(\text{body})}\right]\{\boldsymbol{\omega}\} + \sum_{i=1}^{n}\left(\left[\mathbf{I}_{G_i}^{(i)}\right]\left\{\boldsymbol{\omega}^{(i)}\right\} + \left[\mathbf{I}_{mG}^{(i)}\right]\{\boldsymbol{\omega}\}\right)$$

or

$$\{\mathbf{H}_G\} = \left[\mathbf{I}_G^{(\text{body})} + \sum_{i=1}^{n}\mathbf{I}_{mG}^{(i)}\right]\{\boldsymbol{\omega}\} + \sum_{i=1}^{n}\left[\mathbf{I}_{G_i}^{(i)}\right]\left\{\boldsymbol{\omega}^{(i)}\right\} \tag{12.136}$$

Let

$$\mathbf{I}_G^{(v)} = \mathbf{I}_G^{(\text{body})} + \sum_{i=1}^{n}\mathbf{I}_{mG}^{(i)} \tag{12.137}$$

where $\mathbf{I}_G^{(v)}$ is the time-independent total moment of inertia of the vehicle v (i.e., that of the body *plus* the concentrated masses of all the flywheels). Thus,

$$\{\mathbf{H}_G\} = \left[\mathbf{I}_G^{(v)}\right]\{\boldsymbol{\omega}\} + \sum_{i=1}^{n}\left[\mathbf{I}_{G_i}^{(i)}\right]\left\{\boldsymbol{\omega}^{(i)}\right\} \tag{12.138}$$

If $\boldsymbol{\omega}_{\text{rel}}^{(i)}$ is the angular velocity of the ith flywheel relative to the spacecraft, then its inertial angular velocity $\boldsymbol{\omega}^{(i)}$ is given by Eq. (11.5),

$$\boldsymbol{\omega}^{(i)} = \boldsymbol{\omega} + \boldsymbol{\omega}_{\text{rel}}^{(i)} \tag{12.139}$$

where $\boldsymbol{\omega}$ is the inertial angular velocity of the spacecraft body. Substituting Eq. (12.139) into Eq. (12.138) yields

$$\{\mathbf{H}_G\} = \left[\mathbf{I}_G^{(v)}\right]\{\boldsymbol{\omega}\} + \sum_{i=1}^{n}\left[\mathbf{I}_{G_i}^{(i)}\right]\left\{\boldsymbol{\omega} + \boldsymbol{\omega}_{\text{rel}}^{(i)}\right\}$$

or

$$\{\mathbf{H}_G\} = \left[\mathbf{I}_G^{(v)} + \sum_{i=1}^{n}\mathbf{I}_{G_i}^{(i)}\right]\{\boldsymbol{\omega}\} + \sum_{i=1}^{n}\left[\mathbf{I}_{G_i}^{(i)}\right]\left\{\boldsymbol{\omega}_{\text{rel}}^{(i)}\right\} \tag{12.140}$$

An alternative form of this expression may be obtained by substituting Eq. (12.137):

$$\begin{aligned}\{\mathbf{H}_G\} &= \left[\mathbf{I}_G^{(\text{body})} + \sum_{i=1}^{n}\mathbf{I}_{mG}^{(i)} + \sum_{i=1}^{n}\mathbf{I}_{G_i}^{(i)}\right]\{\boldsymbol{\omega}\} + \sum_{i=1}^{n}\left[\mathbf{I}_{G_i}^{(i)}\right]\left\{\boldsymbol{\omega}_{\text{rel}}^{(i)}\right\} \\ &= \left[\mathbf{I}_G^{(\text{body})} + \sum_{i=1}^{n}\left(\mathbf{I}_{G_i}^{(i)} + \mathbf{I}_{mG}^{(i)}\right)\right]\{\boldsymbol{\omega}\} + \sum_{i=1}^{n}\left[\mathbf{I}_{G_i}^{(i)}\right]\left\{\boldsymbol{\omega}_{\text{rel}}^{(i)}\right\}\end{aligned} \tag{12.141}$$

But, according to the parallel axis theorem (Eq. 11.61),

$$\mathbf{I}_G^{(i)} = \mathbf{I}_{G_i}^{(i)} + \mathbf{I}_{mG}^{(i)}$$

where $\mathbf{I}_G^{(i)}$ is the moment of inertia of the ith flywheel around the center of mass of the body of the spacecraft. Hence, we can write Eq. (12.141) as

$$\{\mathbf{H}_G\} = \left[\mathbf{I}_G^{(\text{body})} + \sum_{i=1}^{n}\mathbf{I}_G^{(i)}\right]\{\boldsymbol{\omega}\} + \sum_{i=1}^{n}\left[\mathbf{I}_{G_i}^{(i)}\right]\left\{\boldsymbol{\omega}_{\text{rel}}^{(i)}\right\} \tag{12.142}$$

The equation of motion of the system is given by Eqs. (11.30) and (1.56),

$$\mathbf{M}_G)_{\text{net external}} = \left.\frac{d\mathbf{H}_G}{dt}\right)_{\text{rel}} + \boldsymbol{\omega}\times\mathbf{H}_G \tag{12.143}$$

If $\mathbf{M}_G)_{\text{net external}} = \mathbf{0}$, then $\mathbf{H}_G = \text{constant}$.

EXAMPLE 12.10

A disk is attached to a plate at their common center of mass (Fig. 12.24). Between the two is a motor mounted on the plate, which drives the disk into rotation relative to the plate. The system rotates freely in the xy plane in gravity-free space. The moments of inertia of the plate and the disk about the z axis through G are I_p and I_w, respectively. Determine the change in the relative angular velocity ω_{rel} of the disk required to cause a given change in the inertial angular velocity ω of the plate.

Solution

The plate plays the role of the body of a spacecraft and the disk is a momentum wheel. At any given time, the angular momentum of the system about G is given by Eq. (12.142),

$$H_G = (I_p + I_w)\omega + I_w\omega_{\text{rel}}$$

At a later time (denoted by primes), after the torquing motor is activated, the angular momentum is

$$H_G' = (I_p + I_w)\omega' + I_w\omega_{\text{rel}}'$$

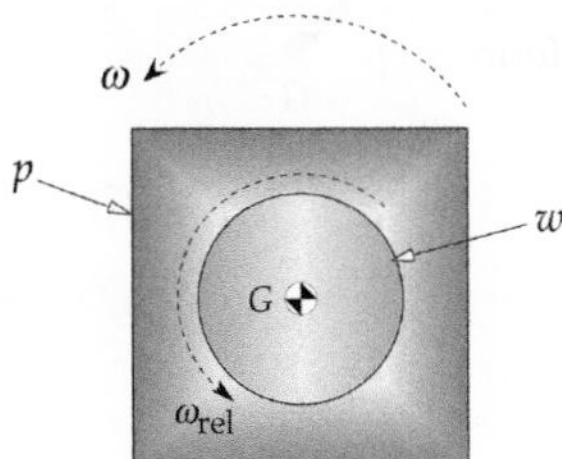

FIG. 12.24

Plate *p* and disk *w* attached at their common center of mass *G*.

Since the torque is internal to the system, we have conservation of angular momentum, $H_G' = H_G$, which means

$$(I_p + I_w)\omega' + I_w\omega'_{\text{rel}} = (I_p + I_w)\omega + I_w\omega_{\text{rel}}$$

Rearranging terms we get

$$I_w(\omega'_{\text{rel}} - \omega_{\text{rel}}) = -(I_p + I_w)(\omega' - \omega)$$

Letting $\Delta\omega = \omega' - \omega$, this can be written as

$$\boxed{\Delta\omega_{\text{rel}} = -\left(1 + \frac{I_p}{I_w}\right)\Delta\omega}$$

The change $\Delta\omega_{\text{rel}}$ in the relative rotational velocity of the disk is due to the torque applied to the disk at G by the motor mounted on the plate. An equal torque in the opposite direction is applied to the plate, producing the angular velocity change $\Delta\omega$ opposite in direction to $\Delta\omega_{\text{rel}}$.

Notice that if $I_p \gg I_w$, which is true in an actual spacecraft, then the change in angular velocity of the momentum wheel must be very much larger than the required change in angular velocity of the body of the spacecraft.

EXAMPLE 12.11

Use Eq. (12.142) to obtain the equations of motion of a torque-free, axisymmetric dual-spin satellite, such as the one shown in Fig. 12.25.

Solution

In this case, we have only one "reaction wheel" (namely, the platform p). The "body" is the rotor r. In Eq. (12.142), we make the following substitutions ($\leftarrow$ means "is replaced by"):

$$\boldsymbol{\omega} \leftarrow \boldsymbol{\omega}^{(r)}$$
$$\boldsymbol{\omega}_{\text{rel}}^{(i)} \leftarrow \boldsymbol{\omega}_{\text{rel}}^{(p)}$$
$$\mathbf{I}_G^{(\text{body})} \leftarrow \mathbf{I}_G^{(r)}$$
$$\sum_{i=1}^{n} \mathbf{I}_G^{(i)} \leftarrow \mathbf{I}_G^{(p)}$$
$$\sum_{i=1}^{n} \left[\mathbf{I}_{G_i}^{(i)}\right]\left\{\boldsymbol{\omega}_{\text{rel}}^{(i)}\right\} \leftarrow \left[\mathbf{I}_{G_p}^{(p)}\right]\left\{\boldsymbol{\omega}_{\text{rel}}^{(p)}\right\}$$

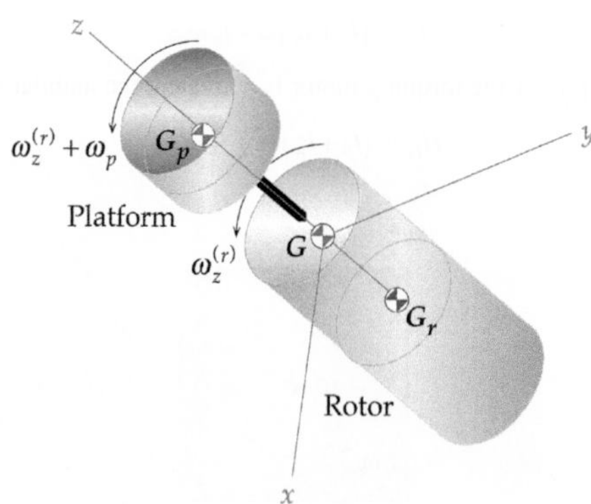

FIG. 12.25

Dual-spin spacecraft.

so that Eq. (12.142) becomes

$$\{\mathbf{H}_G\} = \left[\mathbf{I}_G^{(r)} + \mathbf{I}_G^{(p)}\right]\left\{\boldsymbol{\omega}^{(r)}\right\} + \left[\mathbf{I}_{G_p}^{(p)}\right]\left\{\boldsymbol{\omega}_{\text{rel}}^{(p)}\right\} \tag{a}$$

Since $\mathbf{M}_G)_{\text{net external}} = \mathbf{0}$, Eq. (12.143) yields

$$\left[\mathbf{I}_G^{(r)} + \mathbf{I}_G^{(p)}\right]\left\{\dot{\boldsymbol{\omega}}^{(r)}\right\} + \left[\mathbf{I}_{G_p}^{(p)}\right]\left\{\dot{\boldsymbol{\omega}}_{\text{rel}}^{(p)}\right\} + \left\{\boldsymbol{\omega}^{(r)}\right\} \times \left(\left[\mathbf{I}_G^{(r)} + \mathbf{I}_G^{(p)}\right]\left\{\boldsymbol{\omega}^{(r)}\right\} + \left[\mathbf{I}_{G_p}^{(p)}\right]\left\{\boldsymbol{\omega}_{\text{rel}}^{(p)}\right\}\right) = \{\mathbf{0}\} \tag{b}$$

The components of the matrices and vectors in Eq. (b) relative to the principal xyz body frame axes attached to the rotor are

$$\left[\mathbf{I}_G^{(r)}\right] = \begin{bmatrix} A_r & 0 & 0 \\ 0 & A_r & 0 \\ 0 & 0 & C_r \end{bmatrix} \quad \left[\mathbf{I}_G^{(p)}\right] = \begin{bmatrix} A_p & 0 & 0 \\ 0 & A_p & 0 \\ 0 & 0 & C_p \end{bmatrix} \quad \left[\mathbf{I}_{G_p}^{(p)}\right] = \begin{bmatrix} \overline{A}_p & 0 & 0 \\ 0 & \overline{A}_p & 0 \\ 0 & 0 & C_p \end{bmatrix} \tag{c}$$

and

$$\left\{\boldsymbol{\omega}^{(r)}\right\} = \begin{Bmatrix} \omega_x^{(r)} \\ \omega_y^{(r)} \\ \omega_z^{(r)} \end{Bmatrix} \quad \left\{\boldsymbol{\omega}_{\text{rel}}^{(p)}\right\} = \begin{Bmatrix} 0 \\ 0 \\ \omega_p \end{Bmatrix} \tag{d}$$

where A_r, C_r, A_p, and C_p are the rotor and platform principal moments of inertia about the vehicle center of mass G, and $\overline{A}_p$ is the moment of inertia of the platform about its own center of mass G_p. We also used the fact that $\overline{C}_p = C_p$, which of course is due to the fact that G and G_p both lie on the z axis. This notation is nearly identical to that employed in our consideration of the stability of dual-spin satellites in Section 12.4 (wherein $\omega_r = \omega_z^{(r)}$ and $\boldsymbol{\omega}_\perp = \omega_x^{(r)}\hat{\mathbf{i}} + \omega_y^{(r)}\hat{\mathbf{j}}$). Substituting Eqs. (c) and (d) into each of the four terms in Eq. (b), we get

$$\left[\mathbf{I}_G^{(r)} + \mathbf{I}_G^{(p)}\right]\left\{\dot{\boldsymbol{\omega}}^{(r)}\right\} = \begin{bmatrix} A_r + A_p & 0 & 0 \\ 0 & A_r + A_p & 0 \\ 0 & 0 & C_r + C_p \end{bmatrix} \begin{Bmatrix} \dot{\omega}_x^{(r)} \\ \dot{\omega}_y^{(r)} \\ \dot{\omega}_z^{(r)} \end{Bmatrix} = \begin{Bmatrix} (A_r + A_p)\dot{\omega}_x^{(r)} \\ (A_r + A_p)\dot{\omega}_y^{(r)} \\ (C_r + C_p)\dot{\omega}_z^{(r)} \end{Bmatrix} \tag{e}$$

$$\left\{\boldsymbol{\omega}^{(r)}\right\} \times \left[\mathbf{I}_G^{(r)} + \mathbf{I}_G^{(p)}\right]\left\{\boldsymbol{\omega}^{(r)}\right\} = \begin{Bmatrix} \omega_x^{(r)} \\ \omega_y^{(r)} \\ \omega_z^{(r)} \end{Bmatrix} \times \begin{Bmatrix} (A_r + A_p)\omega_x^{(r)} \\ (A_r + A_p)\omega_y^{(r)} \\ (C_r + C_p)\omega_z^{(r)} \end{Bmatrix} = \begin{Bmatrix} \left[(C_p - A_p) + (C_r - A_r)\right]\omega_y^{(r)}\omega_z^{(r)} \\ \left[(A_p - C_p) + (A_r - C_r)\right]\omega_x^{(r)}\omega_z^{(r)} \\ 0 \end{Bmatrix} \tag{f}$$

$$\left[\mathbf{I}_{G_p}^{(p)}\right]\left\{\dot{\boldsymbol{\omega}}_{\text{rel}}^{(p)}\right\} = \begin{bmatrix} \overline{A}_p & 0 & 0 \\ 0 & \overline{A}_p & 0 \\ 0 & 0 & C_p \end{bmatrix} \begin{Bmatrix} 0 \\ 0 \\ \dot{\omega}_p \end{Bmatrix} = \begin{Bmatrix} 0 \\ 0 \\ C_p\dot{\omega}_p \end{Bmatrix} \tag{g}$$

$$\left\{\boldsymbol{\omega}^{(r)}\right\} \times \left[\mathbf{I}_{G_p}^{(p)}\right]\left\{\boldsymbol{\omega}_{\text{rel}}^{(p)}\right\} = \begin{Bmatrix} \omega_x^{(r)} \\ \omega_y^{(r)} \\ \omega_z^{(r)} \end{Bmatrix} \times \begin{bmatrix} \overline{A}_p & 0 & 0 \\ 0 & \overline{A}_p & 0 \\ 0 & 0 & C_p \end{bmatrix} \begin{Bmatrix} 0 \\ 0 \\ \omega_p \end{Bmatrix} = \begin{Bmatrix} C_p\omega_y^{(r)}\omega_p \\ -C_p\omega_x^{(r)}\omega_p \\ 0 \end{Bmatrix} \tag{h}$$

With these four expressions, Eq. (b) becomes

$$\begin{Bmatrix} (A_r + A_p)\dot{\omega}_x^{(r)} \\ (A_r + A_p)\dot{\omega}_y^{(r)} \\ (C_r + C_p)\dot{\omega}_z^{(r)} \end{Bmatrix} + \begin{Bmatrix} \left[(C_p - A_p) + C_r + A_r\right]\omega_y^{(r)}\omega_z^{(r)} \\ \left[(A_p - C_p) + A_r + C_r\right]\omega_x^{(r)}\omega_z^{(r)} \\ 0 \end{Bmatrix} + \begin{Bmatrix} 0 \\ 0 \\ C_p\dot{\omega}_p \end{Bmatrix} + \begin{Bmatrix} C_p\omega_y^{(r)}\omega_p \\ -C_p\omega_x^{(r)}\omega_p \\ 0 \end{Bmatrix} = \begin{Bmatrix} 0 \\ 0 \\ 0 \end{Bmatrix} \tag{i}$$

Summing the four vectors on the left-hand side and then extracting the three components of the vector equation, finally yields the three scalar equations of motion of the dual-spin satellite in the body frame,

$$\boxed{\begin{aligned} A\dot{\omega}_x^{(r)} + (C - A)\omega_y^{(r)}\omega_z^{(r)} + C_p\omega_y^{(r)}\omega_p &= 0 \\ A\dot{\omega}_y^{(r)} + (A - C)\omega_x^{(r)}\omega_z^{(r)} - C_p\omega_x^{(r)}\omega_p &= 0 \\ C\dot{\omega}_z^{(r)} + C_p\dot{\omega}_p &= 0 \end{aligned}} \tag{j}$$

where A and C are the combined transverse and axial moments of inertia of the dual-spin vehicle about its center of mass,

$$A = A_r + A \quad C = C_r + C_p \tag{k}$$

The three equations (j) involve four unknowns, $\omega_x^{(r)}$, $\omega_y^{(r)}$, $\omega_z^{(r)}$, and ω_p. A fourth equation is required to account for the means of providing the relative velocity ω_p between the platform and the rotor. Friction in the axle bearing between the platform and the rotor would eventually cause ω_p to go to zero, as pointed out in Section 12.4. We may assume that the electric motor in the bearing acts to keep ω_p constant at a specified value, so that $\dot{\omega}_p = 0$. Then, Eq. (j)$_3$ implies that $\omega_z^{(r)}$ is constant as well. Thus, ω_p and $\omega_z^{(r)}$ are removed from our list of unknowns, leaving $\omega_x^{(r)}$ and $\omega_y^{(r)}$ to be governed by the first two equations in Eq. (j). Note that we actually employed Eq. (j)$_3$ in the solution of Example 12.10.

EXAMPLE 12.12

A spacecraft has three identical momentum wheels with their spin axes aligned with the vehicle's principal body axes. The spin axes of momentum wheels 1, 2, and 3 are aligned with the x, y, and z axes, respectively. The inertia tensors of the rotationally symmetric momentum wheels about their centers of mass are, therefore,

$$\left[\mathbf{I}_{G_1}^{(1)}\right] = \begin{bmatrix} I & 0 & 0 \\ 0 & J & 0 \\ 0 & 0 & J \end{bmatrix} \quad \left[\mathbf{I}_{G_2}^{(2)}\right] = \begin{bmatrix} J & 0 & 0 \\ 0 & I & 0 \\ 0 & 0 & J \end{bmatrix} \quad \left[\mathbf{I}_{G_3}^{(3)}\right] = \begin{bmatrix} J & 0 & 0 \\ 0 & J & 0 \\ 0 & 0 & I \end{bmatrix} \tag{a}$$

The spacecraft moment of inertia tensor about the vehicle (v) center of mass is

$$\left[\mathbf{I}_G^{(v)}\right] = \begin{bmatrix} A & 0 & 0 \\ 0 & B & 0 \\ 0 & 0 & C \end{bmatrix} \tag{b}$$

Calculate the spin accelerations of the momentum wheels in the presence of external torque.

Solution

For $n = 3$, Eq. (12.140) becomes

$$\{\mathbf{H}_G\} = \left[\mathbf{I}_G^{(v)} + \mathbf{I}_{G_1}^{(1)} + \mathbf{I}_{G_2}^{(2)} + \mathbf{I}_{G_3}^{(3)}\right]\{\boldsymbol{\omega}\} + \left[\mathbf{I}_{G_1}^{(1)}\right]\left\{\boldsymbol{\omega}_{\text{rel}}^{(1)}\right\} + \left[\mathbf{I}_{G_2}^{(2)}\right]\left\{\boldsymbol{\omega}_{\text{rel}}^{(2)}\right\} + \left[\mathbf{I}_{G_3}^{(3)}\right]\left\{\boldsymbol{\omega}_{\text{rel}}^{(3)}\right\} \tag{c}$$

The absolute angular velocity $\boldsymbol{\omega}$ of the spacecraft and the angular velocities $\boldsymbol{\omega}_{\text{rel}}^{(1)}$, $\boldsymbol{\omega}_{\text{rel}}^{(2)}$, $\boldsymbol{\omega}_{\text{rel}}^{(3)}$ of the three flywheels relative to the spacecraft are

$$\{\boldsymbol{\omega}\} = \begin{Bmatrix} \omega_x \\ \omega_y \\ \omega_z \end{Bmatrix} \quad \left\{\boldsymbol{\omega}_{\text{rel}}^{(1)}\right\} = \begin{Bmatrix} \omega^{(1)} \\ 0 \\ 0 \end{Bmatrix} \quad \left\{\boldsymbol{\omega}_{\text{rel}}^{(2)}\right\} = \begin{Bmatrix} 0 \\ \omega^{(2)} \\ 0 \end{Bmatrix} \quad \left\{\boldsymbol{\omega}_{\text{rel}}^{(3)}\right\} = \begin{Bmatrix} 0 \\ 0 \\ \omega^{(3)} \end{Bmatrix} \tag{d}$$

Substituting Eqs. (a), (b), and (d) into Eq. (c) yields

$$\begin{aligned} \{\mathbf{H}_G\} = &\left(\begin{bmatrix} A & 0 & 0 \\ 0 & B & 0 \\ 0 & 0 & C \end{bmatrix} + \begin{bmatrix} I & 0 & 0 \\ 0 & J & 0 \\ 0 & 0 & J \end{bmatrix} + \begin{bmatrix} J & 0 & 0 \\ 0 & I & 0 \\ 0 & 0 & J \end{bmatrix} + \begin{bmatrix} J & 0 & 0 \\ 0 & J & 0 \\ 0 & 0 & I \end{bmatrix}\right)\begin{Bmatrix} \omega_x \\ \omega_y \\ \omega_z \end{Bmatrix} \\ &+ \begin{bmatrix} I & 0 & 0 \\ 0 & J & 0 \\ 0 & 0 & J \end{bmatrix}\begin{Bmatrix} \omega^{(1)} \\ 0 \\ 0 \end{Bmatrix} + \begin{bmatrix} J & 0 & 0 \\ 0 & I & 0 \\ 0 & 0 & J \end{bmatrix}\begin{Bmatrix} 0 \\ \omega^{(2)} \\ 0 \end{Bmatrix} + \begin{bmatrix} J & 0 & 0 \\ 0 & J & 0 \\ 0 & 0 & I \end{bmatrix}\begin{Bmatrix} 0 \\ 0 \\ \omega^{(3)} \end{Bmatrix} \end{aligned}$$

or

$$\{\mathbf{H}_G\} = \begin{bmatrix} A+I+2J & 0 & 0 \\ 0 & B+I+2J & 0 \\ 0 & 0 & C+I+2J \end{bmatrix}\begin{Bmatrix} \omega_x \\ \omega_y \\ \omega_z \end{Bmatrix} + \begin{bmatrix} I & 0 & 0 \\ 0 & I & 0 \\ 0 & 0 & I \end{bmatrix}\begin{Bmatrix} \omega^{(1)} \\ \omega^{(2)} \\ \omega^{(3)} \end{Bmatrix} \tag{e}$$

Substituting this expression for $\{\mathbf{H}_G\}$ into Eq. (12.143), we get

$$\begin{bmatrix} I & 0 & 0 \\ 0 & I & 0 \\ 0 & 0 & I \end{bmatrix} \begin{Bmatrix} \dot{\omega}^{(1)} \\ \dot{\omega}^{(2)} \\ \dot{\omega}^{(3)} \end{Bmatrix} + \begin{bmatrix} A+I+2J & 0 & 0 \\ 0 & B+I+2J & 0 \\ 0 & 0 & C+I+2J \end{bmatrix} \begin{Bmatrix} \dot{\omega}_x \\ \dot{\omega}_y \\ \dot{\omega}_z \end{Bmatrix}$$
$$+ \begin{Bmatrix} \omega_x \\ \omega_y \\ \omega_z \end{Bmatrix} \times \left(\begin{bmatrix} I & 0 & 0 \\ 0 & I & 0 \\ 0 & 0 & I \end{bmatrix} \begin{Bmatrix} \omega^{(1)} \\ \omega^{(2)} \\ \omega^{(3)} \end{Bmatrix} + \begin{bmatrix} A+I+2J & 0 & 0 \\ 0 & B+I+2J & 0 \\ 0 & 0 & C+I+2J \end{bmatrix} \begin{Bmatrix} \omega_x \\ \omega_y \\ \omega_z \end{Bmatrix} \right) = \begin{Bmatrix} M_G)_x \\ M_G)_y \\ M_G)_z \end{Bmatrix} \tag{f}$$

Expanding and collecting terms yield the time rates of change of the flywheel spins (relative to the spacecraft) in terms of those of the spacecraft absolute angular velocity components,

$$\begin{aligned} \dot{\omega}^{(1)} &= \frac{M_G)_x}{I} + \frac{B-C}{I}\omega_y\omega_z - \left(1+\frac{A}{I}+2\frac{J}{I}\right)\dot{\omega}_x + \omega^{(2)}\omega_z - \omega^{(3)}\omega_y \\ \dot{\omega}^{(2)} &= \frac{M_G)_y}{I} + \frac{C-A}{I}\omega_x\omega_z - \left(1+\frac{B}{I}+2\frac{J}{I}\right)\dot{\omega}_y + \omega^{(3)}\omega_x - \omega^{(1)}\omega_z \\ \dot{\omega}^{(3)} &= \frac{M_G)_z}{I} + \frac{A-B}{I}\omega_x\omega_y - \left(1+\frac{C}{I}+2\frac{J}{I}\right)\dot{\omega}_z + \omega^{(1)}\omega_y - \omega^{(2)}\omega_x \end{aligned} \tag{g}$$

EXAMPLE 12.13

A communication satellite is in a circular earth orbit of period T. The body z axis lies on the outward radial from the earth's center to the spacecraft, so the angular velocity about the body y axis is $2\pi/T$. The angular velocities about the body x and z axes are zero. The attitude control system consists of three momentum wheels 1, 2, and 3 aligned with the principal x, y, and z axes of the satellite. A variable torque is applied to each wheel by its own electric motor. At time $t=0$, the angular velocities of the three wheels relative to the spacecraft are all zero. A small, constant environmental torque $\mathbf{M}_0$ acts on the spacecraft. Determine the axial torques $C^{(1)}$, $C^{(2)}$, and $C^{(3)}$ that the three motors must exert on their wheels so that the angular velocity $\boldsymbol{\omega}$ of the satellite will remain constant. The moment of inertia tensors of the reaction wheels about their centers of mass are given by Eq. (a) of Example 12.12 (Fig. 12.26).

Solution

The absolute angular velocity vector of the xyz frame is given by

$$\boldsymbol{\omega} = \omega_0\hat{\mathbf{j}} \tag{a}$$

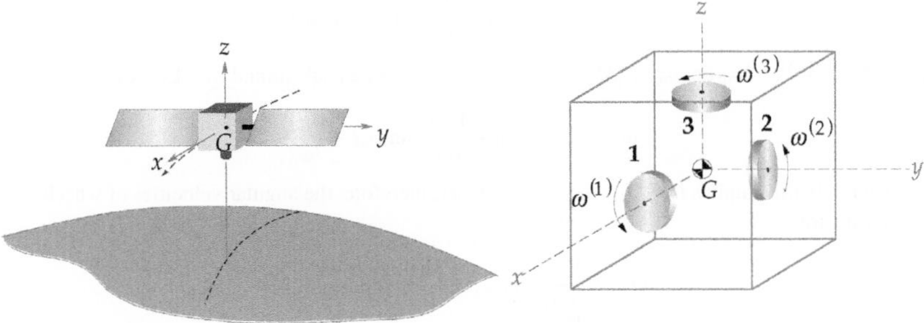

FIG. 12.26

Three-axis stabilized satellite.

where $\omega_0 = 2\pi/T$, a constant. At any instant, the absolute angular velocities of the three reaction wheels are, accordingly,

$$\begin{aligned}\boldsymbol{\omega}^{(1)} &= \omega^{(1)}\hat{\mathbf{i}} + \omega_0\hat{\mathbf{j}} \\ \boldsymbol{\omega}^{(2)} &= \omega^{(2)}\hat{\mathbf{j}} + \omega_0\hat{\mathbf{j}} \\ \boldsymbol{\omega}^{(3)} &= \omega^{(3)}\hat{\mathbf{k}} + \omega_0\hat{\mathbf{j}}\end{aligned} \tag{b}$$

From Eq. (a), it is clear that $\omega_x = \omega_z = \dot{\omega}_x = \dot{\omega}_y = \dot{\omega}_z = 0$. Therefore, Eqs. (g) of Example 12.12 become, for the case at hand,

$$\begin{aligned}\dot{\omega}^{(1)} &= \frac{M_G)_x}{I} + \frac{B-C}{I}\cdot\omega_0\cdot(0) - \left(1+\frac{A}{I}+2\frac{J}{I}\right)\cdot(0) + \omega^{(2)}\cdot(0) - \omega^{(3)}\cdot\omega_0 \\ \dot{\omega}^{(2)} &= \frac{M_G)_y}{I} + \frac{A-C}{I}\cdot(0)\cdot(0) - \left(1+\frac{B}{I}+2\frac{J}{I}\right)\cdot(0) + \omega^{(3)}\cdot(0) - \omega^{(1)}\cdot(0) \\ \dot{\omega}^{(3)} &= \frac{M_G)_z}{I} + \frac{A-B}{I}\cdot(0)\cdot\omega_0 - \left(1+\frac{C}{I}+2\frac{J}{I}\right)\cdot(0) + \omega^{(1)}\cdot\omega_0 - \omega^{(2)}\cdot(0)\end{aligned}$$

which reduce to the following set of three first-order differential equations:

$$\begin{aligned}\dot{\omega}^{(1)} + \omega_0\omega^{(3)} &= \frac{M_G)_x}{I} \\ \dot{\omega}^{(2)} &= \frac{M_G)_y}{I} \\ \dot{\omega}^{(3)} - \omega_0\omega^{(1)} &= \frac{M_G)_z}{I}\end{aligned} \tag{c}$$

Eq. $(c)_2$ implies that $\omega^{(2)} = M_G)_y t/I$ + constant, and since $\omega^{(2)} = 0$ at $t = 0$, this means that for time t thereafter,

$$\omega^{(2)} = \frac{M_G)_y}{I}t \tag{d}$$

Differentiating Eq. $(c)_3$ with respect to t and solving for $\dot{\omega}^{(1)}$ yields $\dot{\omega}^{(1)} = \ddot{\omega}^{(3)}/\omega_0$. Substituting this result into Eq. $(c)_1$ we get

$$\ddot{\omega}^{(3)} + \omega_0{}^2\omega^{(3)} = \frac{\omega_0 M_G)_x}{I}$$

The well-known solution of this familiar differential equation is

$$\omega^{(3)} = a\cos\omega_0 t + b\sin\omega_0 t + \frac{M_G)_x}{I\omega_0}$$

where a and b are constants of integration. According to the problem statement, $\omega^{(3)} = 0$ when $t = 0$. This initial condition requires $a = -M_G)_x/\omega_0 I$, so that

$$\omega^{(3)} = b\sin\omega_0 t + \frac{M_G)_x}{I\omega_0}(1-\cos\omega_0 t) \tag{e}$$

From this, we obtain $\dot{\omega}^{(3)} = b\omega_0\cos\omega_0 t + \left[M_G)_x/I\right]\sin\omega_0 t$, which, when substituted into Eq. $(c)_3$, yields

$$\omega^{(1)} = b\cos\omega_0 t + \frac{M_G)_x}{I\omega_0}\sin\omega_o t - \frac{M_G)_z}{I\omega_0} \tag{f}$$

Since $\omega^{(1)} = 0$ at $t = 0$, this implies $b = M_G)_z/I\omega_0$. In summary, therefore, the angular velocities of wheels 1, 2, and 3 relative to the satellite are

$$\begin{aligned}\omega^{(1)} &= \frac{M_G)_x}{I\omega_0}\sin\omega_o t + \frac{M_G)_z}{I\omega_0}(\cos\omega_0 t - 1) \\ \omega^{(2)} &= \frac{M_G)_y}{I}t \\ \omega^{(3)} &= \frac{M_G)_z}{I\omega_0}\sin\omega_o t + \frac{M_G)_x}{I\omega_0}(1-\cos\omega_0 t)\end{aligned} \tag{g}$$

The angular momenta of the reaction wheels are

$$\begin{aligned}\mathbf{H}_{G_1}^{(1)} &= I\omega_x^{(1)}\hat{\mathbf{i}} + J\omega_y^{(1)}\hat{\mathbf{j}} + J\omega_z^{(1)}\hat{\mathbf{k}} \\ \mathbf{H}_{G_2}^{(2)} &= J\omega_x^{(2)}\hat{\mathbf{i}} + I\omega_y^{(2)}\hat{\mathbf{j}} + J\omega_z^{(2)}\hat{\mathbf{k}} \\ \mathbf{H}_{G_3}^{(3)} &= J\omega_x^{(3)}\hat{\mathbf{i}} + J\omega_y^{(3)}\hat{\mathbf{j}} + I\omega_z^{(3)}\hat{\mathbf{k}}\end{aligned} \tag{h}$$

According to Eq. (b), the components of the flywheels' angular velocities are

$$\begin{array}{lll}\omega_x^{(1)} = \omega^{(1)} & \omega_y^{(1)} = \omega_0 & \omega_z^{(1)} = 0 \\ \omega_x^{(2)} = 0 & \omega_y^{(2)} = \omega^{(2)} + \omega_0 & \omega_z^{(2)} = 0 \\ \omega_x^{(3)} = 0 & \omega_y^{(3)} = \omega_0 & \omega_z^{(3)} = \omega^{(3)}\end{array}$$

so that Eq. (h) becomes

$$\begin{aligned}\mathbf{H}_{G_1}^{(1)} &= I\omega^{(1)}\hat{\mathbf{i}} + J\omega_0\hat{\mathbf{j}} \\ \mathbf{H}_{G_2}^{(2)} &= I\left(\omega^{(2)} + \omega_0\right)\hat{\mathbf{j}} \\ \mathbf{H}_{G_3}^{(3)} &= J\omega_0\hat{\mathbf{j}} + I\omega^{(3)}\hat{\mathbf{k}}\end{aligned} \tag{i}$$

Substituting Eq. (g) into these expressions yields the angular momenta of the wheels as a function of time,

$$\begin{aligned}\mathbf{H}_{G_1}^{(1)} &= \left[\frac{M_G)_x}{\omega_0}\sin\omega_0 t + \frac{M_G)_z}{\omega_0}(\cos\omega_0 t - 1)\right]\hat{\mathbf{i}} + J\omega_0\hat{\mathbf{j}} \\ \mathbf{H}_{G_2}^{(2)} &= \left[M_G)_y t + I\omega_0\right]\hat{\mathbf{j}} \\ \mathbf{H}_{G_3}^{(3)} &= J\omega_0\hat{\mathbf{j}} + \left[\frac{M_G)_z}{\omega_0}\sin\omega_0 t + \frac{M_G)_x}{\omega_0}(1 - \cos\omega_0 t)\right]\hat{\mathbf{k}}\end{aligned} \tag{j}$$

The torque on the reaction wheels is found by applying the Euler equation to each one. Thus, for wheel 1

$$\begin{aligned}\mathbf{M}_{G_1})_{\text{net}} &= \left.\frac{d\mathbf{H}_{G_1}^{(1)}}{dt}\right)_{\text{rel}} + \boldsymbol{\omega} \times \mathbf{H}_{G_1}^{(1)} \\ &= \left[M_G)_x \cos\omega_0 t - M_G)_x \sin\omega_0 t\right]\hat{\mathbf{i}} + \left[M_G)_x(1 - \cos\omega_0 t) - M_G)_x \sin\omega_0 t\right]\hat{\mathbf{k}}\end{aligned}$$

Since the axis of wheel 1 is in the x direction, the torque is the x component of this moment (the z component being a gyroscopic bending moment),

$$\boxed{C^{(1)} = M_G)_x \cos\omega_0 t - M_G)_z \sin\omega_0 t}$$

For wheel 2,

$$\mathbf{M}_{G_2})_{\text{net}} = \left.\frac{d\mathbf{H}_{G_2}^{(2)}}{dt}\right)_{\text{rel}} + \boldsymbol{\omega} \times \mathbf{H}_{G_2}^{(2)} = M_G)_y\hat{\mathbf{j}}$$

Thus,

$$\boxed{C^{(2)} = M_G)_y}$$

Finally, for wheel 3,

$$\begin{aligned}\mathbf{M}_{G_3})_{\text{net}} &= \left.\frac{d\mathbf{H}_{G_3}^{(3)}}{dt}\right)_{\text{rel}} + \boldsymbol{\omega} \times \mathbf{H}_{G_3}^{(3)} \\ &= \left[M_G)_x(1 - \cos\omega_0 t) + M_G)_z \sin\omega_0 t\right]\hat{\mathbf{i}} + \left[M_G)_x \sin\omega_0 t + M_G)_z \cos\omega_0 t\right]\hat{\mathbf{k}}\end{aligned}$$

For this wheel, the torque direction is the z axis, so

$$\boxed{C^{(3)} = M_G)_x \sin\omega_0 t + M_G)_z \cos\omega_0 t}$$

The external torques on the spacecraft of the previous example may be due to thruster misalignment or they may arise from environmental effects such as gravity gradients, solar pressure, or interaction with the earth's magnetic field. The example assumed that these torques were constant, which is the simplest means of introducing their effects, but they actually vary with time. In any case, their magnitudes are extremely small, typically $<10^{-3}\,\mathrm{N\cdot m}$ for ordinary-sized, unmanned spacecraft. Eq. $(g)_2$ of the example reveals that a small torque normal to the satellite's orbital plane will cause the angular velocity of momentum wheel 2 to slowly but constantly increase. Over a long-enough period of time, the angular velocity of the gyro might approach its design limits, whereupon it is said to be saturated. At that point, attitude jets on the satellite would have to be fired to produce a torque around the y axis while the wheel is "caged" (i.e., its angular velocity reduced to zero or to its nonzero bias value). Finally, note that if all of the external torques were zero, none of the momentum wheels in the example would be required. The constant angular velocity $\boldsymbol{\omega} = (2\pi/T)\hat{\mathbf{j}}$ of the vehicle, once initiated, would continue unabated.

So far, we have dealt with momentum wheels, which are characterized by the fact that their axes are rigidly aligned with the principal axes of the spacecraft, as shown in Fig. 12.27. The speed of the electrically driven wheels is varied to produce the required rotation rates of the vehicle in response to external torques. Depending on the spacecraft, the nominal speed of a momentum wheel may be from zero to several thousand revolutions per minute.

Momentum wheels that are free to pivot on one or more gimbals are called control moment gyros. Fig. 12.28 illustrates a double-gimbaled control moment gyro. These gyros spin at several thousand revolutions per minute. The motor-driven speed of the flywheel is constant, and moments are exerted on the vehicle when torquers (electric motors) tilt the wheel about a gimbal axis. The torque direction is normal to the gimbal axis.

Set $n = 1$ in Eq. (12.140) and replace i with w (representing "wheel") to obtain

$$\{\mathbf{H}_G\} = \left[\mathbf{I}_G^{(v)}\right]\{\boldsymbol{\omega}\} + \left[\mathbf{I}_{G_w}^{(w)}\right]\left(\{\boldsymbol{\omega}\} + \left\{\boldsymbol{\omega}_{\text{rel}}^{(w)}\right\}\right) \tag{12.144}$$

The relative angular velocity of the rotor is

$$\boldsymbol{\omega}_{\text{rel}}^{(w)} = \boldsymbol{\omega}_p + \boldsymbol{\omega}_n + \boldsymbol{\omega}_s \tag{12.145}$$

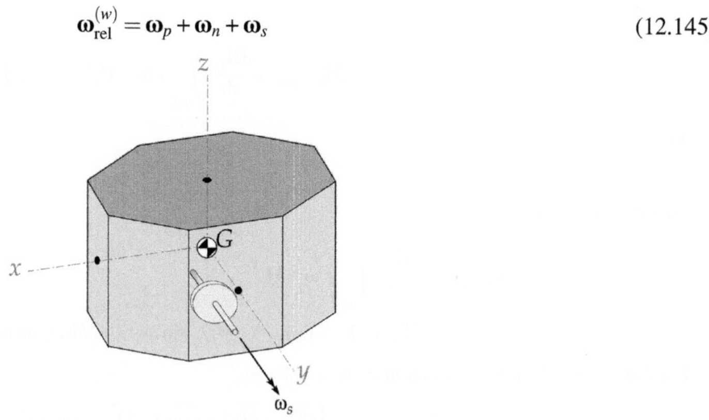

FIG. 12.27

Momentum wheel aligned with a principal body axis.

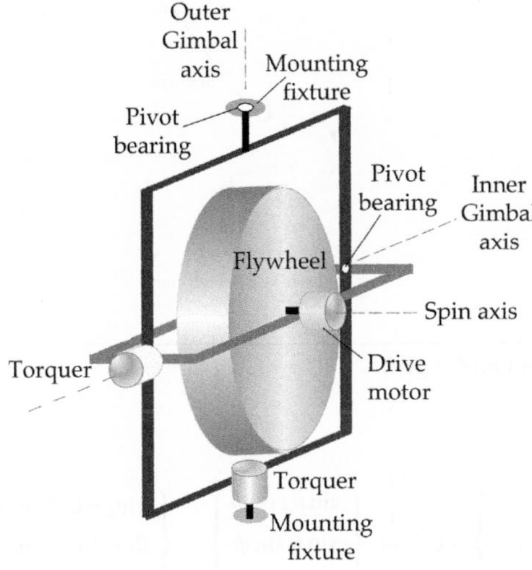

FIG. 12.28

Two-gimbal control moment gyro.

where $\boldsymbol{\omega}_p$, $\boldsymbol{\omega}_n$, and $\boldsymbol{\omega}_s$ are the precession, nutation, and spin angular velocities of the gyro relative to the vehicle. Substituting Eq. (12.145) into Eq. (12.144) yields

$$\{\mathbf{H}_G\} = \left[\mathbf{I}_G^{(v)}\right]\{\boldsymbol{\omega}\} + \left[\mathbf{I}_{G_w}^{(w)}\right]\{\boldsymbol{\omega} + \boldsymbol{\omega}_p + \boldsymbol{\omega}_n + \boldsymbol{\omega}_s\} \tag{12.146}$$

The spin rate of the gyro is three or more orders of magnitude greater than any of the other rates. That is, under conditions in which a control moment gyro is designed to operate,

$$\|\boldsymbol{\omega}_s\| \gg \|\boldsymbol{\omega}\| \quad \|\boldsymbol{\omega}_s\| \gg \|\boldsymbol{\omega}_p\| \quad \|\boldsymbol{\omega}_s\| \gg \|\boldsymbol{\omega}_p\|$$

Therefore,

$$\{\mathbf{H}_G\} \approx \left[\mathbf{I}_G^{(v)}\right]\{\boldsymbol{\omega}\} + \left[\mathbf{I}_{G_w}^{(w)}\right]\{\boldsymbol{\omega}_s\} \tag{12.147}$$

Since the spin axis of a gyro is an axis of symmetry, about which the moment of inertia is $C^{(w)}$, this can be written as

$$\{\mathbf{H}_G\} = \left[\mathbf{I}_G^{(v)}\right]\{\boldsymbol{\omega}\} + C^{(w)}\omega_s\{\hat{\mathbf{n}}_s\} \tag{12.148}$$

where

$$\left[\mathbf{I}_G^{(v)}\right] = \begin{bmatrix} A & 0 & 0 \\ 0 & B & 0 \\ 0 & 0 & C \end{bmatrix}$$

and $\hat{\mathbf{n}}_s$ is the unit vector along the spin axis, as illustrated in Fig. 12.29. Relative to the body frame axes of the spacecraft, the components of $\hat{\mathbf{n}}_s$ appear as follows:

$$\hat{\mathbf{n}}_s = \sin\theta\cos\phi\hat{\mathbf{i}} + \sin\theta\sin\phi\hat{\mathbf{j}} + \cos\theta\hat{\mathbf{k}} \tag{12.149}$$

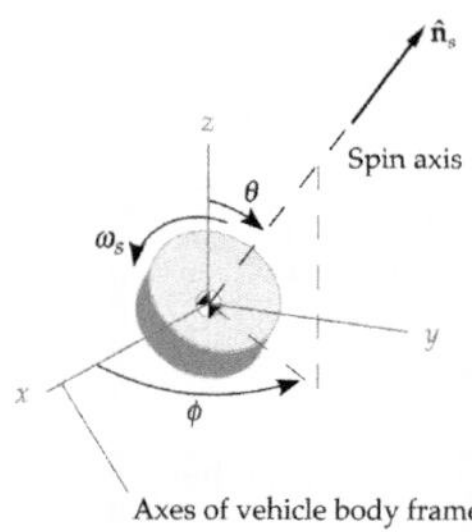

FIG. 12.29

Inclination angles of the spin vector of a gyro.

Thus, Eq. (12.148) becomes

$$\{\mathbf{H}_G\} = \begin{bmatrix} A & 0 & 0 \\ 0 & B & 0 \\ 0 & 0 & C \end{bmatrix} \begin{Bmatrix} \omega_x \\ \omega_y \\ \omega_z \end{Bmatrix} + C^{(w)}\omega_s \begin{Bmatrix} \sin\theta\cos\phi \\ \sin\theta\sin\phi \\ \cos\theta \end{Bmatrix} = \begin{Bmatrix} A\omega_x + C^{(w)}\omega_s\sin\theta\cos\phi \\ B\omega_y + C^{(w)}\omega_s\sin\theta\sin\phi \\ C\omega_z + C^{(w)}\omega_s\cos\theta \end{Bmatrix} \tag{12.150}$$

It follows that

$$\frac{d}{dt}\{\mathbf{H}_G\} = \frac{d}{dt}\begin{Bmatrix} A\omega_x + C^{(w)}\omega_s\sin\theta\cos\phi \\ B\omega_y + C^{(w)}\omega_s\sin\theta\sin\phi \\ C\omega_z + C^{(w)}\omega_s\cos\theta \end{Bmatrix} + \begin{Bmatrix} \omega_x \\ \omega_y \\ \omega_z \end{Bmatrix} \times \begin{Bmatrix} A\omega_x + C^{(w)}\omega_s\sin\theta\cos\phi \\ B\omega_y + C^{(w)}\omega_s\sin\theta\sin\phi \\ C\omega_z + C^{(w)}\omega_s\cos\theta \end{Bmatrix} \tag{12.151}$$

Expanding the right-hand side, collecting terms, and setting the result equal to the net external moment, we find

$$\begin{aligned} &A\dot{\omega}_x + C^{(w)}\omega_s\dot{\theta}\cos\phi\cos\theta - C^{(w)}\omega_s\dot{\phi}\sin\phi\sin\theta + C^{(w)}\dot{\omega}_s\cos\phi\sin\theta \\ &\quad + \left(C^{(w)}\omega_s\cos\theta + C\omega_z\right)\omega_y - \left(C^{(w)}\omega_s\sin\phi\sin\theta + B\omega_y\right)\omega_z = M_G)_x \end{aligned} \tag{12.152a}$$

$$\begin{aligned} &B\dot{\omega}_y + C^{(w)}\omega_s\dot{\theta}\sin\phi\cos\theta + C^{(w)}\omega_s\dot{\phi}\cos\phi\sin\theta + C^{(w)}\dot{\omega}_s\sin\phi\sin\theta \\ &\quad - \left(C^{(w)}\omega_s\cos\theta + C\omega_z\right)\omega_x + \left(C^{(w)}\omega_s\cos\phi\sin\theta + A\omega_x\right)\omega_z = M_G)_y \end{aligned} \tag{12.152b}$$

$$\begin{aligned} &C\dot{\omega}_z + C^{(w)}\omega_s\dot{\theta}\sin\theta + C^{(w)}\dot{\omega}_s\cos\theta - \left(C^{(w)}\omega_s\cos\phi\sin\theta + A\omega_x\right)\omega_y \\ &\quad + \left(C^{(w)}\omega_s\sin\phi\sin\theta + B\omega_y\right)\omega_x = M_G)_z \end{aligned} \tag{12.152c}$$

Additional gyros are accounted for by adding the spin inertia, spin rate, and inclination angles for each one into Eqs. (12.152).

EXAMPLE 12.14

A satellite is in torque-free motion, $\mathbf{M}_G)_{\text{net}} = \mathbf{0}$. A nongimbaled gyro (momentum wheel) is aligned with the vehicle's x axis and is spinning at the rate $\omega_s)_0$. The spacecraft angular velocity is $\boldsymbol{\omega} = \omega_x\hat{\mathbf{i}}$. If the spin of the gyro is increased at the rate $\dot{\omega}_s$, find the angular acceleration of the spacecraft.

Solution

Using Fig. 12.29 as a guide, we set $\phi = 0$ and $\theta = 90°$ to align the spin axis with the x axis. Since there is no gimbaling, $\dot{\theta} = \dot{\phi} = 0$. Eqs. (12.152) then yield

$$A\dot{\omega}_x + C^{(w)}\dot{\omega}_s = 0$$
$$B\dot{\omega}_y = 0$$
$$C\dot{\omega}_z = 0$$

Clearly, the angular velocities around the y and z axes remain zero, whereas

$$\boxed{\dot{\omega}_x = \frac{C^{(w)}}{A}\dot{\omega}_s}$$

Thus, a change in the vehicle's roll rate around the x axis can be initiated by accelerating the momentum wheel in the opposite direction (see Example 12.10).

EXAMPLE 12.15

A satellite is in torque-free motion. A control moment gyro, spinning at the constant rate ω_s, is gimbaled about the spacecraft y and z axes, with $\phi = 0$ and $\theta = 90°$ (cf. Fig. 12.29). The spacecraft angular velocity is $\boldsymbol{\omega} = \omega_z\hat{\mathbf{k}}$. If the spin axis of the gyro, initially along the x direction, is rotated around the y axis at the rate $\dot{\theta}$, what is the resulting angular acceleration of the spacecraft?

Solution

Substituting $\omega_x = \omega_y = \dot{\omega}_y = \dot{\phi} = 0$ and $\theta = 90°$ into Eqs. (12.152a)–(12.152c) gives

$$A\dot{\omega}_x = 0$$
$$B\dot{\omega}_y + C^{(w)}\omega_s(\omega_z + \dot{\phi}) = 0$$
$$C\dot{\omega}_z - H^{(w)}\dot{\theta} = 0$$

Thus, the components of vehicle angular acceleration are

$$\boxed{\dot{\omega}_x = 0 \quad \dot{\omega}_y = -\frac{C^{(w)}}{B}\omega_s(\omega_z + \dot{\phi}) \quad \dot{\omega}_z = \frac{C^{(w)}}{C}\omega_s\dot{\theta}}$$

We see that pitching the gyro at the rate $\dot{\theta}$ around the vehicle y axis alters only ω_z, leaving ω_x unchanged. However, to keep $\omega_y = 0$ clearly requires that $\dot{\phi} = -\omega_z$. In other words, for the control moment gyro to control the angular velocity about only one vehicle axis, it must therefore be able to precess around that axis (the z axis in this case). That is why the control moment gyro must have two gimbals.

12.10 GRAVITY GRADIENT STABILIZATION

Consider a satellite in circular orbit, as shown in Fig. 12.30. Let $\mathbf{r}$ be the position vector of a mass element dm relative to the center of attraction, $\mathbf{R}$ the position vector of the center of mass G, and $\boldsymbol{\rho}$ the position vector of dm relative to G. The force of gravity on dm is

$$d\mathbf{F}_g = -G\frac{M\,dm}{r^3}\mathbf{r} = -\mu\frac{\mathbf{r}}{r^3}dm \tag{12.153}$$

where M is the mass of the central body, and $\mu = GM$. The net moment of the gravitational force around G is

$$\mathbf{M}_G)_{\text{net}} = \int_m \boldsymbol{\rho} \times d\mathbf{F}_g\,dm \tag{12.154}$$

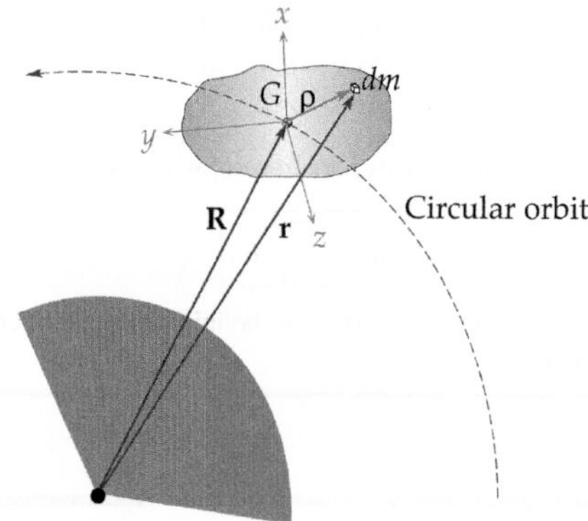

FIG. 12.30

Rigid satellite in a circular orbit. *xyz* is the principal body frame.

Since $\mathbf{r} = \mathbf{R} + \boldsymbol{\rho}$ and

$$\begin{aligned}\mathbf{R} &= R_x\hat{\mathbf{i}} + R_y\hat{\mathbf{j}} + R_z\hat{\mathbf{k}}\\ \boldsymbol{\rho} &= x\hat{\mathbf{i}} + y\hat{\mathbf{j}} + z\hat{\mathbf{k}}\end{aligned} \tag{12.155}$$

we have

$$\boldsymbol{\rho}\times d\mathbf{F}_g = -\mu\frac{dm}{r^3}\boldsymbol{\rho}\times(\mathbf{R}+\boldsymbol{\rho}) = -\mu\frac{dm}{r^3}\boldsymbol{\rho}\times\mathbf{R} = -\mu\frac{dm}{r^3}\begin{bmatrix}\hat{\mathbf{i}} & \hat{\mathbf{j}} & \hat{\mathbf{k}}\\ x & y & z\\ R_x & R_y & R_z\end{bmatrix}$$

Thus,

$$\boldsymbol{\rho}\times d\mathbf{F}_g = -\mu\frac{dm}{r^3}(R_zy - R_yz)\hat{\mathbf{i}} - \mu\frac{dm}{r^3}(R_xz - R_zx)\hat{\mathbf{j}} - \mu\frac{dm}{r^3}(R_yx - R_xy)\hat{\mathbf{k}}$$

Substituting this back into Eq. (12.154) yields

$$\begin{aligned}\mathbf{M}_G)_{\text{net}} = &\left(-\mu R_z\int_m\frac{y}{r^3}dm + \mu R_y\int_m\frac{z}{r^3}dm\right)\hat{\mathbf{i}} + \left(-\mu R_z\int_m\frac{z}{r^3}dm + \mu R_z\int_m\frac{x}{r^3}dm\right)\hat{\mathbf{j}}\\ &+\left(-\mu R_y\int_m\frac{x}{r^3}dm + \mu R_x\int_m\frac{y}{r^3}dm\right)\hat{\mathbf{k}}\end{aligned}$$

or

$$\begin{aligned}M_G)_x &= -\mu R_z\int_m\frac{y}{r^3}dm + \mu R_y\int_m\frac{z}{r^3}dm\\ M_G)_y &= -\mu R_x\int_m\frac{z}{r^3}dm + \mu R_z\int_m\frac{x}{r^3}dm\\ M_G)_z &= -\mu R_y\int_m\frac{x}{r^3}dm + \mu R_z\int_m\frac{y}{r^3}dm\end{aligned} \tag{12.156}$$

Now, since $\|\boldsymbol{\rho}\| \ll \|\mathbf{R}\|$, it follows from Eq. (7.20) that

$$\frac{1}{r^3}=\frac{1}{R^3}-\frac{3}{R^5}\mathbf{R}\cdot\boldsymbol{\rho}$$

or

$$\frac{1}{r^3}=\frac{1}{R^3}-\frac{3}{R^5}(R_x x+R_y y+R_z z)$$

Therefore,

$$\int_m \frac{x}{r^3}dm=\frac{1}{R^3}\int_m x\,dm-\frac{3R_x}{R^5}\int_m x^2\,dm-\frac{3R_y}{R^5}\int_m xy\,dm-\frac{3R_z}{R^5}\int_m xz\,dm$$

But the center of mass lies at the origin of the *xyz* axes, which are principal moments of inertia directions. That means

$$\int_m x\,dm=\int_m xy\,dm=\int_m xz\,dm=0$$

so that

$$\int_m \frac{x}{r^3}dm=-\frac{3R_x}{R^5}\int_m x^2\,dm \tag{12.157}$$

In a similar fashion, we can show that

$$\int_m \frac{y}{r^3}dm=-\frac{3R_y}{R^5}\int_m y^2\,dm \tag{12.158}$$

and

$$\int_m \frac{z}{r^3}dm=-\frac{3R_z}{R^5}\int_m z^2\,dm \tag{12.159}$$

Substituting these last three expressions into Eq. (12.156) leads to

$$\begin{aligned}
M_G)_x&=\frac{3\mu R_y R_z}{R^5}\left(\int_m y^2\,dm-\int_m z^2\,dm\right)\\
M_G)_y&=\frac{3\mu R_x R_z}{R^5}\left(\int_m z^2\,dm-\int_m x^2\,dm\right)\\
M_G)_z&=\frac{3\mu R_x R_y}{R^5}\left(\int_m x^2\,dm-\int_m y^2\,dm\right)
\end{aligned} \tag{12.160}$$

From Section 11.5, we recall that the moments of inertia are defined as

$$A=\int_m y^2dm+\int_m z^2dm \quad B=\int_m x^2dm+\int_m z^2dm \quad C=\int_m x^2dm+\int_m y^2dm \tag{12.161}$$

from which we may write

$$B - A = \int_m x^2 dm - \int_m y^2 dm \quad A - C = \int_m z^2 dm - \int_m x^2 dm \quad C - B = \int_m y^2 dm - \int_m z^2 dm$$

It follows that Eq. (12.160) reduce to

$$\begin{aligned} M_G)_x &= \frac{3\mu R_y R_z}{R^5}(C - B) \\ M_G)_y &= \frac{3\mu R_x R_z}{R^5}(A - C) \\ M_G)_z &= \frac{3\mu R_x R_y}{R^5}(B - A) \end{aligned} \tag{12.162}$$

These are the components, in the spacecraft body frame, of the gravitational torque produced by the variation of the earth's gravitational field over the volume of the spacecraft. To get an idea of these torque magnitudes, note first of all that R_x/R, R_y/R, and R_z/R are the direction cosines of the position vector of the center of mass, so that their magnitudes do not exceed 1. For a satellite in a low earth orbit of radius 6700 km, $3\mu/R^3 \approx 4(10^{-6})\,\text{s}^{-2}$, which is therefore the maximum order of magnitude of the coefficients of the inertia terms in Eq. (12.162). The moments of inertia of the space shuttle were on the order of $10^6\ \text{kg}\cdot\text{m}^2$, so the gravitational torques on that large vehicle were on the order of 1 N·m.

Substituting Eq. (12.162) into Euler's equations of motion (Eq. 11.72b), we get

$$\begin{aligned} A\dot{\omega}_x + (C - B)\omega_y\omega_z &= \frac{3\mu R_y R_z}{R^5}(C - B) \\ B\dot{\omega}_y + (A - C)\omega_z\omega_x &= \frac{3\mu R_x R_z}{R^5}(A - C) \\ C\dot{\omega}_z + (B - A)\omega_x\omega_y &= \frac{3\mu R_x R_y}{R^5}(B - A) \end{aligned} \tag{12.163}$$

Now consider the local vertical/local horizontal orbital reference frame shown in Fig. 12.31. It is actually the Clohessy-Wiltshire frame of Chapter 7, with the axes relabeled. The z' axis points radially

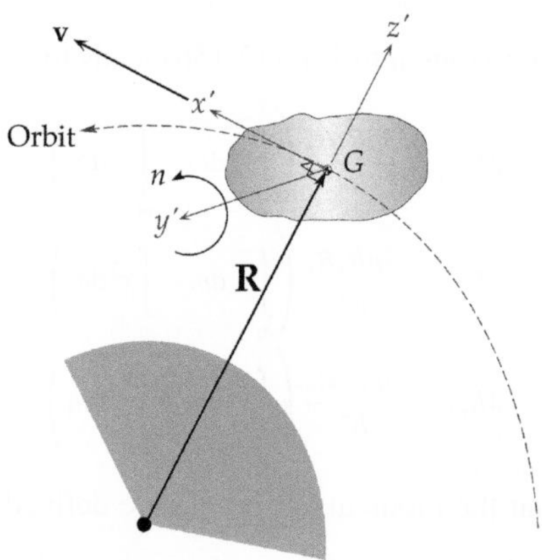

FIG. 12.31

Orbital reference frame $x'y'z'$ attached to the center of mass of the satellite.

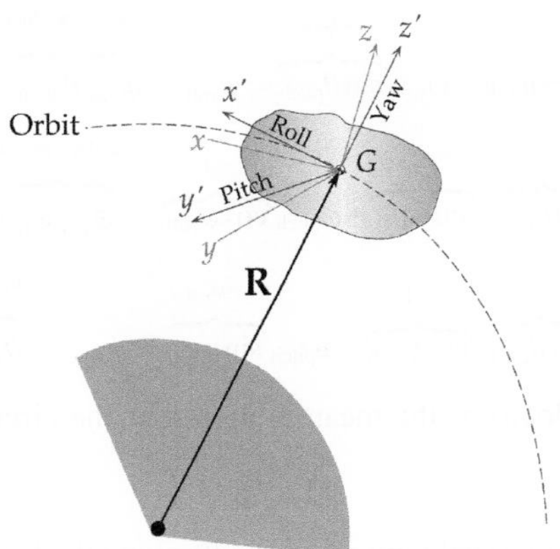

FIG. 12.32

Satellite body frame slightly misaligned with the orbital frame $x'y'z'$.

outward from the center of the earth, the x' axis is in the direction of the local horizon, and the y' axis completes the right-handed triad by pointing in the direction of the orbit normal. This frame rotates around the y' axis with an angular velocity equal to the mean motion n of the circular orbit. Suppose we align the satellite's principal body frame axes xyz with $x'y'z'$, respectively. When the body x axis is aligned with the x' direction, it is called the roll axis. The body y axis, when aligned with the y' direction, is the pitch axis. The body z axis, pointing outward from the earth in the z' direction, is the yaw axis. These directions are illustrated in Fig. 12.32. With the spacecraft aligned in this way, the body frame components of the inertial angular velocity vector $\boldsymbol{\omega}$ are $\omega_x = \omega_z = 0$ and $\omega_y = n$. The components of the position vector $\mathbf{R}$ are $R_x = R_y = 0$ and $R_z = R$. Substituting these data into Eq. (12.163) yields

$$\dot{\omega}_x = \dot{\omega}_y = \dot{\omega}_z = 0$$

That is, the spacecraft will orbit the planet with its principal axes remaining aligned with the orbital frame. If this motion is stable under the influence of gravity alone, without the use of thrusters, gyros, or other devices, then the spacecraft is gravity-gradient-stabilized. We need to assess the stability of this motion so that we can determine how to orient a spacecraft to take advantage of this type of passive attitude stabilization.

Let the body frame xyz be slightly misaligned with the orbital reference frame, so that the yaw, pitch, and roll angles between the xyz axes and the $x'y'z'$ axes, respectively, are very small, as suggested in Fig. 12.32. The absolute angular velocity $\boldsymbol{\omega}$ of the spacecraft is the angular velocity $\boldsymbol{\omega}_{\text{rel}}$ relative to the orbital reference frame plus the inertial angular velocity $\boldsymbol{\Omega}$ of the $x'y'z'$ frame,

$$\boldsymbol{\omega} = \boldsymbol{\omega}_{\text{rel}} + \boldsymbol{\Omega}$$

The components of $\boldsymbol{\omega}_{\text{rel}}$ in the body frame are found using the yaw, pitch, and roll relations (Eq. 11.129). In so doing, it must be kept in mind that all angles and rates are assumed to be so small that their squares and products may be neglected. Recalling that $\sin\alpha = \alpha$ and $\cos\alpha = 1$, when $\alpha \ll 1$, we therefore obtain

$$\omega_{\text{rel}})_x = \omega_{\text{roll}} - \omega_{\text{yaw}} \overbrace{\sin\theta_{\text{pitch}}}^{=\theta_{\text{pitch}}} = \dot{\psi}_{\text{roll}} - \overbrace{\dot{\phi}_{\text{yaw}}\theta_{\text{pitch}}}^{\text{neglect product}} = \dot{\psi}_{\text{roll}} \tag{12.164}$$

$$\omega_{\text{rel}})_y = \omega_{\text{yaw}} \overbrace{\cos\theta_{\text{pitch}}}^{=1} \overbrace{\sin\psi_{\text{roll}}}^{=\psi_{\text{roll}}} + \omega_{\text{pitch}} \overbrace{\cos\psi_{\text{roll}}}^{=1} = \overbrace{\dot{\phi}_{\text{yaw}}\psi_{\text{roll}}}^{\text{neglect product}} + \dot{\theta}_{\text{pitch}} = \dot{\theta}_{\text{pitch}} \tag{12.165}$$

$$\omega_{\text{rel}})_z = \omega_{\text{yaw}} \overbrace{\cos\theta_{\text{pitch}}}^{=1} \overbrace{\cos\psi_{\text{roll}}}^{=1} - \omega_{\text{pitch}} \overbrace{\sin\psi_{\text{roll}}}^{=\psi_{\text{roll}}} = \dot{\phi}_{\text{yaw}} - \overbrace{\dot{\theta}_{\text{pitch}}\psi_{\text{roll}}}^{\text{neglect product}} = \dot{\phi}_{\text{yaw}} \tag{12.166}$$

The orbital frame's angular velocity is the mean motion n of the circular orbit, so that

$$\mathbf{\Omega} = n\hat{\mathbf{j}}'$$

To obtain the orbital frame's angular velocity components along the body frame, we must use the transformation rule

$$\{\mathbf{\Omega}\}_x = [\mathbf{Q}]_{x'x}\{\mathbf{\Omega}\}_{x'} \tag{12.167}$$

where $[\mathbf{Q}]_{x'x}$ is given by Eq. (11.119). (Keep in mind that $x'y'z'$ are playing the role of XYZ in Fig. 11.27.) Using the above small-angle approximations in Eq. (11.119) leads to

$$[\mathbf{Q}]_{x'x} = \begin{bmatrix} 1 & \phi_{\text{yaw}} & -\theta_{\text{pitch}} \\ -\phi_{\text{yaw}} & 1 & \psi_{\text{roll}} \\ \theta_{\text{pitch}} & -\psi_{\text{roll}} & 1 \end{bmatrix}$$

With this, Eq. (12.167) becomes

$$\begin{Bmatrix} \Omega_x \\ \Omega_y \\ \Omega_z \end{Bmatrix} = \begin{bmatrix} 1 & \phi_{\text{yaw}} & -\theta_{\text{pitch}} \\ -\phi_{\text{yaw}} & 1 & \psi_{\text{roll}} \\ \theta_{\text{pitch}} & -\psi_{\text{roll}} & 1 \end{bmatrix} \begin{Bmatrix} 0 \\ n \\ 0 \end{Bmatrix} = \begin{Bmatrix} n\phi_{\text{yaw}} \\ n \\ -n\psi_{\text{roll}} \end{Bmatrix}$$

Now we can calculate the components of the satellite's inertial angular velocity along the body frame axes,

$$\begin{aligned} \omega_x &= \dot{\psi}_{\text{roll}} + n\phi_{\text{yaw}} \\ \omega_y &= \omega_{\text{rel}})_y + \Omega_y = \dot{\theta}_{\text{pitch}} + n \\ \omega_z &= \omega_{\text{rel}})_z + \Omega_z = \dot{\phi}_{\text{yaw}} - n\psi_{\text{roll}} \end{aligned} \tag{12.168}$$

Differentiating these with respect to time, remembering that the mean motion n is constant for a circular orbit, gives the components of inertial angular acceleration in the body frame,

$$\begin{aligned} \dot{\omega}_x &= \ddot{\psi}_{\text{roll}} + n\dot{\phi}_{\text{yaw}} \\ \dot{\omega}_y &= \ddot{\theta}_{\text{pitch}} \\ \dot{\omega}_z &= \ddot{\phi}_{\text{yaw}} - n\dot{\psi}_{\text{roll}} \end{aligned} \tag{12.169}$$

The position vector of the satellite's center of mass lies along the z' axis of the orbital frame,

$$\mathbf{R} = R\hat{\mathbf{k}}'$$

To obtain the components of **R** in the body frame, we once again use the transformation matrix $[\mathbf{Q}]_{x'x}$,

$$\begin{Bmatrix} R_x \\ R_y \\ R_z \end{Bmatrix} = \begin{bmatrix} 1 & \phi_{\text{yaw}} & -\theta_{\text{pitch}} \\ -\phi_{\text{yaw}} & 1 & \psi_{\text{roll}} \\ \theta_{\text{pitch}} & -\psi_{\text{roll}} & 1 \end{bmatrix} \begin{Bmatrix} 0 \\ 0 \\ R \end{Bmatrix} = \begin{Bmatrix} -R\theta_{\text{pitch}} \\ R\psi_{\text{roll}} \\ R \end{Bmatrix} \tag{12.170}$$

Substituting Eqs. (12.168)–(12.170), together with $n = \sqrt{\mu/R^3}$ into Eq. (12.163) and setting

$$A = I_{\text{roll}} \quad B = I_{\text{pitch}} \quad C = I_{\text{yaw}} \tag{12.171}$$

yields

$$\begin{aligned} I_{\text{roll}}\left(\ddot{\psi}_{\text{roll}} + n\dot{\phi}_{\text{yaw}}\right) + \left(I_{\text{yaw}} - I_{\text{pitch}}\right)\left(\dot{\theta}_{\text{pitch}} + n\right)\left(\dot{\phi}_{\text{yaw}} - n\psi_{\text{roll}}\right) &= 3\left(I_{\text{yaw}} - I_{\text{pitch}}\right)n^2\psi_{\text{roll}} \\ I_{\text{pitch}}\ddot{\theta}_{\text{pitch}} + \left(I_{\text{roll}} - I_{\text{yaw}}\right)\left(\dot{\psi}_{\text{roll}} + n\phi_{\text{yaw}}\right)\left(\dot{\phi}_{\text{yaw}} - n\psi_{\text{roll}}\right) &= -3\left(I_{\text{roll}} - I_{\text{yaw}}\right)n^2\theta_{\text{pitch}} \\ I_{\text{yaw}}\left(\ddot{\phi}_{\text{yaw}} - n\dot{\psi}_{\text{roll}}\right) + \left(I_{\text{pitch}} - I_{\text{roll}}\right)\left(\dot{\theta}_{\text{pitch}} + n\right)\left(\dot{\psi}_{\text{roll}} + n\phi_{\text{yaw}}\right) &= -3\left(I_{\text{pitch}} - I_{\text{roll}}\right)n^2\theta_{\text{pitch}}\psi_{\text{roll}} \end{aligned}$$

Expanding terms and retaining terms at most linear in all angular quantities and their rates yields

$$I_{\text{yaw}}\ddot{\phi}_{\text{yaw}} + \left(I_{\text{pitch}} - I_{\text{roll}}\right)n^2\phi_{\text{yaw}} + \left(I_{\text{pitch}} - I_{\text{roll}} - I_{\text{yaw}}\right)n\dot{\psi}_{\text{roll}} = 0 \tag{12.172}$$

$$I_{\text{roll}}\ddot{\psi}_{\text{roll}} + \left(I_{\text{roll}} - I_{\text{pitch}} + I_{\text{yaw}}\right)n\dot{\phi}_{\text{yaw}} + 4\left(I_{\text{pitch}} - I_{\text{yaw}}\right)n^2\psi_{\text{roll}} = 0 \tag{12.173}$$

$$I_{\text{pitch}}\ddot{\theta}_{\text{pitch}} + 3\left(I_{\text{roll}} - I_{\text{yaw}}\right)n^2\theta_{\text{pitch}} = 0 \tag{12.174}$$

These are the differential equations governing the influence of gravity gradient torques on the small angles and rates of misalignment of the body frame with the orbital frame.

Eq. (12.174), governing the pitching motion around the y' axis, is not coupled to the other two equations. We make the classical assumption that the solution is of the form

$$\theta_{\text{pitch}} = Pe^{pt} \tag{12.175}$$

where P and p are constants, and P is the amplitude of the small disturbance that initiates the pitching motion. Substituting Eq. (12.175) into Eq. (12.174) yields $[I_{\text{pitch}}p^2 + 3(I_{\text{roll}} - I_{\text{yaw}})n^2]Pe^{pt} = 0$ for all values of t, which implies that the bracketed term must vanish, and that means p must have either of the two values

$$p_{1,2} = \pm i\sqrt{3\frac{\left(I_{\text{roll}} - I_{\text{yaw}}\right)n^2}{I_{\text{pitch}}}} \quad \left(i = \sqrt{-1}\right)$$

Thus,

$$\theta_{\text{pitch}} = P_1e^{p_1t} + P_2e^{p_2t}$$

yields the stable, small-amplitude, steady-state harmonic oscillator solution only if p_1 and p_2 are imaginary. That is, if

$$I_{\text{roll}} > I_{\text{yaw}} \quad \text{For stability in pitch} \tag{12.176}$$

The stable pitch oscillation frequency is

$$\left.\omega_f\right)_{\text{pitch}} = n\sqrt{3\frac{I_{\text{roll}} - I_{\text{yaw}}}{I_{\text{pitch}}}} \tag{12.177}$$

(If $I_{\text{yaw}} > I_{\text{roll}}$, then p_1 and p_2 are both real, one positive, the other negative. The positive root causes $\theta_{\text{pitch}} \to \infty$, which is the undesirable, unstable case.)

Let us now turn our attention to Eqs. (12.172) and (12.173), which govern yaw and roll motions under gravity gradient torque. Again, we assume the solution is exponential in form,

$$\phi_{\text{yaw}} = Ye^{qt} \quad \psi_{\text{roll}} = \text{R}e^{qt} \tag{12.178}$$

Substituting these into Eqs. (12.172) and (12.173) yields

$$\begin{aligned}
\left[\left(I_{\text{pitch}} - I_{\text{roll}}\right)n^2 + I_{\text{yaw}}q^2\right]Y + \left[\left(I_{\text{pitch}} - I_{\text{roll}} - I_{\text{yaw}}\right)nq\right]R = 0 \\
\left[\left(I_{\text{roll}} - I_{\text{pitch}} + I_{\text{yaw}}\right)nq\right]Y + \left[4\left(I_{\text{pitch}} - I_{\text{yaw}}\right)n^2 + I_{\text{roll}}q^2\right]R = 0
\end{aligned}$$

In the interest of simplification, we can factor I_{yaw} out of the first equation and I_{roll} out of the second one to get

$$\begin{aligned}
\left(\frac{I_{\text{pitch}} - I_{\text{roll}}}{I_{\text{yaw}}}n^2 + q^2\right)Y + \left(\frac{I_{\text{pitch}} - I_{\text{roll}}}{I_{\text{yaw}}} - 1\right)nqR = 0 \\
\left(1 - \frac{I_{\text{pitch}} - I_{\text{yaw}}}{I_{\text{roll}}}\right)nqY + \left(4\frac{I_{\text{pitch}} - I_{\text{yaw}}}{I_{\text{roll}}}n^2 + q^2\right)R = 0
\end{aligned} \tag{12.179}$$

Let

$$k_Y = \frac{I_{\text{pitch}} - I_{\text{roll}}}{I_{\text{yaw}}} \quad k_R = \frac{I_{\text{pitch}} - I_{\text{yaw}}}{I_{\text{roll}}} \tag{12.180}$$

It is easy to show from Eqs. (12.161), (12.171), and (12.180) that

$$k_Y = \frac{\left(\int_m x^2 \text{d}m \Big/ \int_m y^2 \text{d}m\right) - 1}{\left(\int_m x^2 \text{d}m \Big/ \int_m y^2 \text{d}m\right) + 1} \quad k_R = \frac{\left(\int_m z^2 \text{d}m \Big/ \int_m y^2 \text{d}m\right) - 1}{\left(\int_m z^2 \text{d}m \Big/ \int_m y^2 \text{d}m\right) + 1}$$

which means

$$|k_Y| < 1 \quad |k_R| < 1$$

Using the definitions in Eqs. (12.180), we can write Eq. (12.179) more compactly as

$$\begin{aligned}
(k_Y n^2 + q^2)Y + (k_Y - 1)nqR = 0 \\
(1 - k_R)nqY + (4k_R n^2 + q^2)R = 0
\end{aligned}$$

or, using matrix notation,

$$\begin{bmatrix} k_Y n^2 + q^2 & (k_Y - 1)nq \\ (1 - k_R)nq & 4k_R n^2 + q^2 \end{bmatrix} \begin{Bmatrix} Y \\ R \end{Bmatrix} = \begin{Bmatrix} 0 \\ 0 \end{Bmatrix} \tag{12.181}$$

To avoid the trivial solution ($Y = R = 0$), the determinant of the coefficient matrix must be zero. Expanding the determinant and collecting terms yields the characteristic equation for q,

$$q^4 + bn^2q^2 + cn^4 = 0 \tag{12.182}$$

where

$$b = 3k_R + k_Y k_R + 1 \quad c = 4k_Y k_R \tag{12.183}$$

This quartic equation has four roots which, when substituted back into Eq. (12.178), yields

$$\phi_{\text{yaw}} = Y_1 e^{q_1 t} + Y_2 e^{q_2 t} + Y_3 e^{q_3 t} + Y_4 e^{q_4 t}$$
$$\psi_{\text{roll}} = R_1 e^{q_1 t} + R_2 e^{q_2 t} + R_3 e^{q_3 t} + R_4 e^{q_4 t}$$

For these solutions to remain finite in time, the roots $q_1, \ldots, q_4$ must all be negative (solution decays to zero) or imaginary (steady oscillation at the initial small amplitude).

To reduce Eq. (12.182) to a quadratic equation, let us introduce a new variable λ and write

$$q = \pm n\sqrt{\lambda} \tag{12.184}$$

Then, Eq. (12.182) becomes

$$\lambda^2 + b\lambda + c = 0 \tag{12.185}$$

the familiar solution of which is

$$\lambda_1 = -\frac{1}{2}\left(b + \sqrt{b^2 - 4c}\right) \quad \lambda_2 = -\frac{1}{2}\left(b - \sqrt{b^2 - 4c}\right) \tag{12.186}$$

To guarantee that q in Eq. (12.184) does not take a positive value, we must require that λ be real and negative (so q will be imaginary). For λ to be real requires that $b > 2\sqrt{c}$, or

$$3k_R + k_Y k_R + 1 > 4\sqrt{k_Y k_R} \tag{12.187}$$

For λ to be negative requires that $b^2 > b^2 - 4c$, which will be true if $c > 0$. That is,

$$k_Y k_R > 0 \tag{12.188}$$

Eqs. (12.187) and (12.188) are the conditions required for yaw and roll stability under gravity gradient torques, to which we must add Eq. (12.176) for pitch stability. Observe that we can solve Eq. (12.180) to obtain

$$I_{\text{yaw}} = \frac{1 - k_R}{1 - k_Y k_R} I_{\text{pitch}} \quad I_{\text{roll}} = \frac{1 - k_Y}{1 - k_Y k_R} I_{\text{pitch}} \tag{12.189}$$

By means of these relationships, the pitch stability criterion, $I_{\text{roll}}/I_{\text{yaw}} > 1$, becomes

$$\frac{1 - k_Y}{1 - k_R} > 1$$

In view of the fact that $|k_R| < 1$, this means

$$k_Y < k_R \tag{12.190}$$

Fig. 12.33 shows those regions *I* and *II* on the $k_Y - k_R$ plane in which all three stability criteria (Eqs. (12.187), (12.188), and (12.190)) are simultaneously satisfied, along with the requirement that the three moments of inertia I_{pitch}, I_{roll}, and I_{yaw} are positive.

In the small sliver of region *I*, $k_Y < 0$ and $k_R < 0$; therefore, according to Eq. (12.189), $I_{\text{yaw}} > I_{\text{pitch}}$ and $I_{\text{roll}} > I_{\text{pitch}}$, which together with Eq. (12.176), yield $I_{\text{roll}} > I_{\text{yaw}} > I_{\text{pitch}}$. Remember that the gravity gradient spacecraft is slowly "spinning" about the minor pitch axis (normal to the orbit plane) at an angular velocity equal to the mean motion of the orbit. So this criterion makes the spacecraft a "minor-axis spinner," the roll axis (flight direction) being the major axis of inertia. With energy dissipation, we know that this orientation is not stable in the long run. On the other hand, in region *II*, k_Y and k_R are

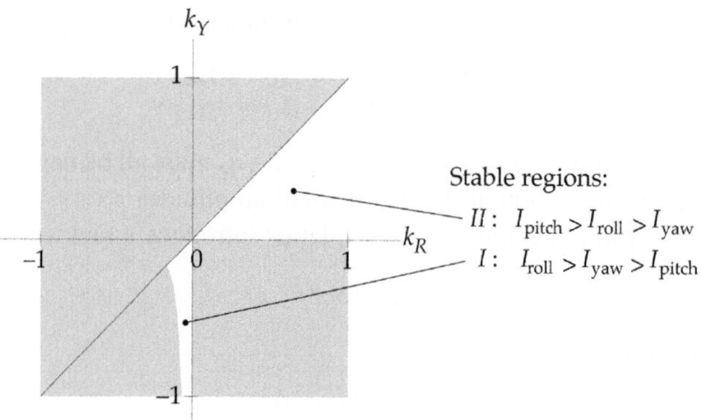

FIG. 12.33

Regions in which the values of k_Y and k_R yield neutral stability in yaw, pitch, and roll of a gravity gradient satellite.

both positive, so that Eq. (12.189) implies $I_{pitch} > I_{yaw}$ and $I_{pitch} > I_{roll}$. Thus, along with the pitch criterion ($I_{roll} > I_{yaw}$), we have $I_{pitch} > I_{roll} > I_{yaw}$. In this, the preferred, configuration, the gravity gradient spacecraft is a "major-axis spinner" about the pitch axis, and the minor yaw axis is the minor axis of inertia. It turns out that all the known gravity-gradient-stabilized moons of the solar system, like the earth's, whose "captured" rate of rotation equals the orbital period, are major-axis spinners.

In Eq. (12.177), we presented the frequency of the gravity gradient pitch oscillation. For completeness, we should also point out that the coupled yaw and roll motions have two oscillation frequencies, which are obtained from Eqs. (12.184) and (12.186),

$$\omega_{f_{yaw-roll}}\Big)_{1,2} = n\sqrt{\frac{1}{2}\left(b \pm \sqrt{b^2 - 4c}\right)} \tag{12.190}$$

Recall that b and c are found in Eq. (12.183).

We have assumed throughout this discussion that the orbit of the gravity gradient satellite is circular. Kaplan (1976) shows that the effect of a small eccentricity turns up only in the pitching motion. In particular, the natural oscillation expressed by Eq. (12.175) is augmented by a forced oscillation term,

$$\theta_{pitch} = P_1 e^{p_1 t} + P_2 e^{p_2 t} + \frac{2e \sin nt}{3\left(I_{roll} - I_{yaw}\right)/I_{pitch} - 1} \tag{12.191}$$

where e is the (small) eccentricity of the orbit. From this, we see that there is a pitch resonance. When $(I_{roll} - I_{yaw})/I_{pitch}$ approaches 1/3, the amplitude of the last term grows without bound.

EXAMPLE 12.16

A uniform, monolithic 10,000-kg slab, having the dimensions shown in Fig. 12.34, is in a circular low earth orbit. Determine the orientation of the satellite in its orbit for gravity gradient stabilization, and compute the periods of the pitch and yaw/roll oscillations in terms of the orbital period T.

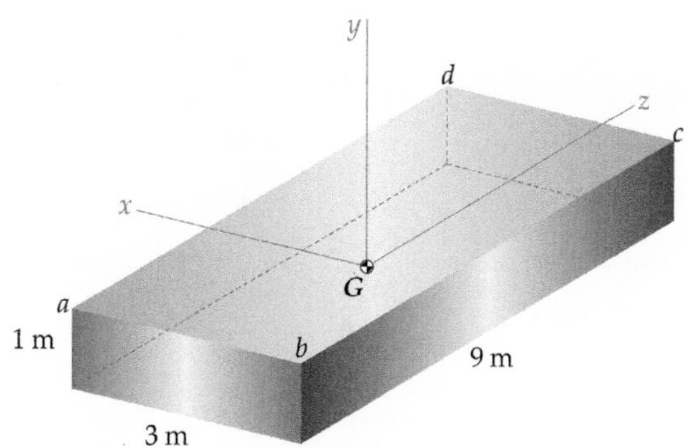

FIG. 12.34

Parallelepiped satellite.

According to Fig. 11.9C, the principal moments of inertia around the xyz axes through the center of mass are

$$A = \frac{10,000}{12}(1^2 + 9^2) = 68,333 \text{ kg}\cdot\text{m}^2$$
$$B = \frac{10,000}{12}(3^2 + 9^2) = 75,000 \text{ kg}\cdot\text{m}^2$$
$$C = \frac{10,000}{12}(3^2 + 1^2) = 8333.3 \text{ kg}\cdot\text{m}^2$$

Solution

Let us first determine whether we can stabilize this object as a minor-axis spinner. In that case,

$$I_{\text{pitch}} = C = 8333.3 \text{ kg}\cdot\text{m}^2 \quad I_{\text{yaw}} = A = 68,333 \text{ kg}\cdot\text{m}^2 \quad I_{\text{roll}} = B = 75,000 \text{ kg}\cdot\text{m}^2$$

Since $I_{\text{roll}} > I_{\text{yaw}}$, the satellite would be stable in pitch. To check yaw/roll stability, we first compute

$$k_Y = \frac{I_{\text{pitch}} - I_{\text{roll}}}{I_{\text{yaw}}} = -0.97561 \quad k_R = \frac{I_{\text{pitch}} - I_{\text{yaw}}}{I_{\text{roll}}} = -0.8000$$

We see that $k_Y k_R > 0$, which is one of the two requirements. The other one is found in Eq. (12.187), but in this case

$$1 + 3k_R + k_Y k_R - 4\sqrt{k_Y k_R} = -4.1533 < 0$$

so that the condition is not met. Hence, the object cannot be gravity-gradient-stabilized as a minor-axis spinner. As a major-axis spinner, we must have

$$I_{\text{pitch}} = B = 75,000 \text{ kg}\cdot\text{m}^2 \quad I_{\text{yaw}} = C = 8333.3 \text{ kg}\cdot\text{m}^2 \quad I_{\text{roll}} = A = 68,333 \text{ kg}\cdot\text{m}^2$$

Then $I_{\text{roll}} > I_{\text{yaw}}$, so the pitch stability condition is satisfied. Furthermore, since

$$k_Y = \frac{I_{\text{pitch}} - I_{\text{roll}}}{I_{\text{yaw}}} = 0.8000 \quad k_R = \frac{I_{\text{pitch}} - I_{\text{yaw}}}{I_{\text{roll}}} = 0.97561$$

we have

$$k_Y k_R = 0.7805 > 0$$
$$1 + 3k_R + k_Y k_R - 4\sqrt{k_Y k_R} = 1.1735 > 0$$

which means the two criteria for stability in the yaw and roll modes are met. The satellite should therefore be orbited as shown in Fig. 12.32, with its minor axis aligned with the radial from the earth's center, the plane $abcd$ lying in the orbital plane, and the body x axis aligned with the local horizon (Fig. 12.35).

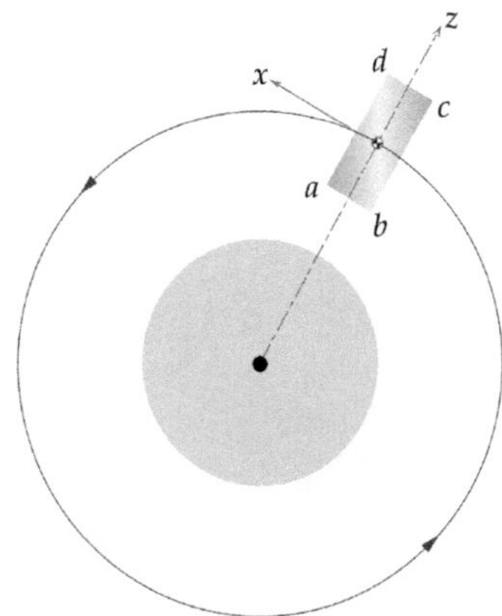

FIG. 12.35

Orientation of the parallelepiped for gravity gradient stabilization.

According to Eq. (12.177), the frequency of the pitch oscillation is

$$\omega_{f_{\text{pitch}}} = n\sqrt{3\frac{I_{\text{roll}} - I_{\text{yaw}}}{I_{\text{pitch}}}}$$

$$= n\sqrt{3\frac{68,333 - 8333.3}{75,000}} = 1.5492n$$

where n is the mean motion. Hence, the period of this oscillation, in terms of that of the orbit, is

$$T_{\text{pitch}} = \frac{2\pi}{\omega_{f_{\text{pitch}}}} = 0.6455\frac{2\pi}{n} = \boxed{0.6455T}$$

For the yaw/roll frequencies, we use Eq. (12.190),

$$\left.\omega_{f_{\text{yaw/roll}}}\right)_1 = n\sqrt{\frac{1}{2}\left(b + \sqrt{b^2 - 4c}\right)}$$

where

$$b = 1 + 3k_R + k_Y k_R = 4.7073 \quad \text{and} \quad c = 4k_Y k_R = 3.122$$

Thus,

$$\left.\omega_{f_{\text{yaw/roll}}}\right)_1 = 2.3015n$$

Likewise,

$$\left.\omega_{f_{\text{yaw/roll}}}\right)_2 = \sqrt{\frac{1}{2}\left(b + \sqrt{b^2 - 4c}\right)} = 1.977n$$

From these, we obtain

$$\left.T_{\text{yaw/roll}}\right)_1 = \boxed{0.5058T} \quad \left.T_{\text{yaw/roll}}\right)_2 = \boxed{0.4345T}$$

Finally, observe that

$$\frac{I_{\text{roll}} - I_{\text{yaw}}}{I_{\text{pitch}}} = 0.8$$

so that we are far from the pitch resonance condition that exists if the orbit has a small eccentricity.

PROBLEMS

Section 12.2

12.1 An axisymmetric satellite has axial and transverse mass moments of inertia about axes through the mass center G of $C = 1200\ \text{kg}\cdot\text{m}^2$ and $A = 2600\ \text{kg}\cdot\text{m}^2$, respectively. If it is spinning at $\omega_s = 6\ \text{rad/s}$ when it is launched, determine its angular momentum. Precession occurs about the inertial Z axis.
{Ans.: $\|\mathbf{H}_G\| = 13{,}450\ \text{kg}\cdot\text{m}^2/\text{s}$}

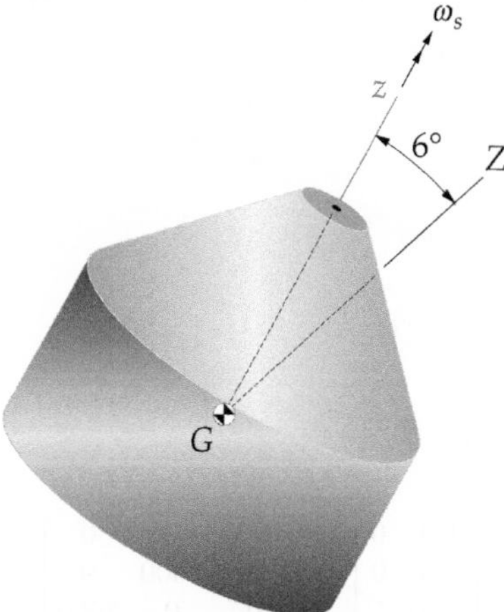

12.2 A spacecraft is symmetric about its body-fixed z axis. Its principal mass moments of inertia are $A = B = 300\ \text{kg}\cdot\text{m}^2$ and $C = 500\ \text{kg}\cdot\text{m}^2$. The z axis sweeps out a cone with a total vertex angle of 10° as it precesses around the angular momentum vector. If the spin velocity is 6 rad/s, compute the period of precession.
{Ans.: 0.417 s}

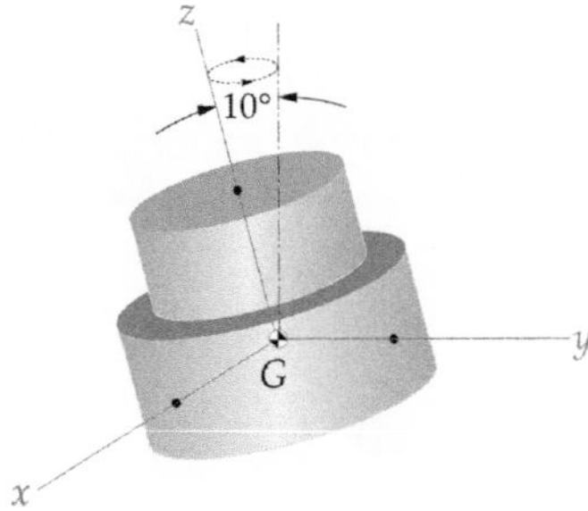

12.3 A thin ring tossed into the air with a spin velocity of ω_s has a very small nutation angle θ (in radians). What is the precession rate ω_p?
{Ans.: $\omega_p = 2\omega_s(1 + \theta^2/2)$, retrograde}

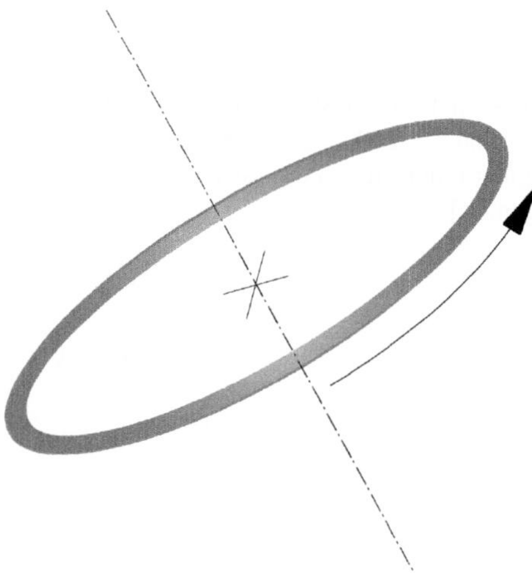

12.4 For an axisymmetric rigid satellite,

$$[\mathbf{I}_G] = \begin{bmatrix} A & 0 & 0 \\ 0 & B & 0 \\ 0 & 0 & C \end{bmatrix} = \begin{bmatrix} 1000 & 0 & 0 \\ 0 & 1000 & 0 \\ 0 & 0 & 5000 \end{bmatrix} \text{ kg} \cdot \text{m}^2$$

It is spinning about the body z axis in torque-free motion, precessing around the angular momentum vector **H** at the rate of 2 rad/s. Calculate the magnitude of **H**.
{Ans.: 2000 $\text{kg} \cdot \text{m}^2/\text{s}$}

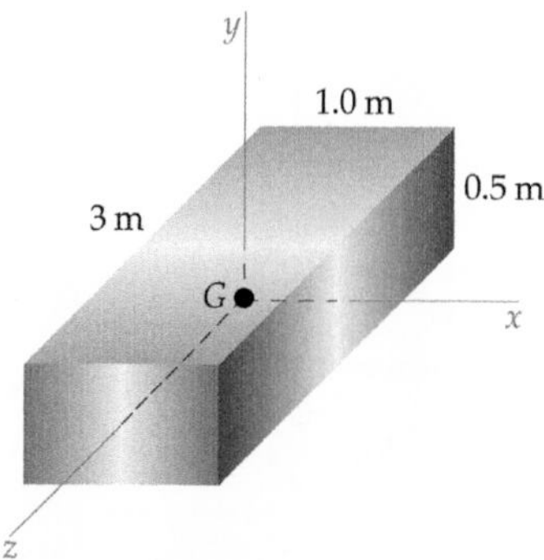

12.5 At a given instant, the above box-shaped 500-kg satellite (in torque-free motion) has an absolute angular velocity $\boldsymbol{\omega} = 0.01\hat{\mathbf{i}} = .03\hat{\mathbf{j}} + 0.02\hat{\mathbf{k}}$ (rad/s). Its moments of inertia about the principal body axes xyz are $A = 385.4 \text{ kg} \cdot \text{m}^2$, $B = 416.7 \text{ kg} \cdot \text{m}^2$, and $C = 52.08 \text{ kg} \cdot \text{m}^2$, respectively. Calculate the magnitude of its absolute angular acceleration.
{Ans.: $6.167(10^{-4})$ rad/s^2}

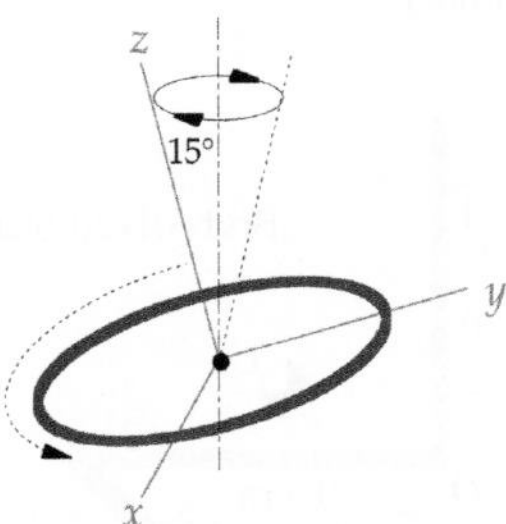

12.6 The above 8-kg thin ring in torque-free motion is spinning with an angular velocity of 30 rad/s and a constant nutation angle of 15°. Calculate the rotational kinetic energy if $A = B = 0.36\ \text{kg}\cdot\text{m}^2$, $C = 0.72\ \text{kg}\cdot\text{m}^2$.
{Ans.: 370.5 J}

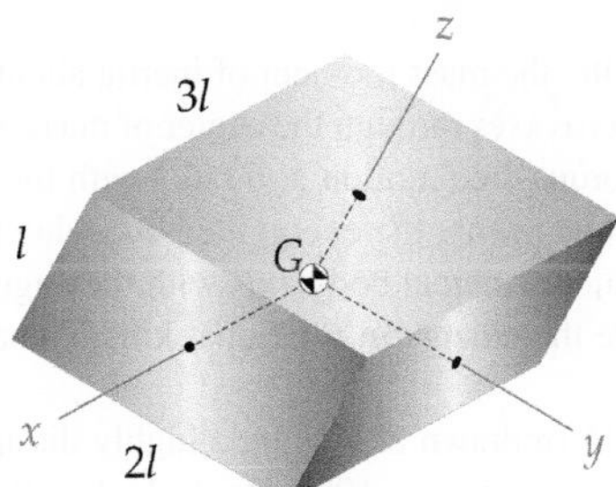

12.7 The above rectangular block has an angular velocity $\boldsymbol{\omega} = 1.5\omega_0\hat{\mathbf{i}} + 0.8\omega_0\hat{\mathbf{j}} + 0.6\omega_0\hat{\mathbf{k}}$, where ω_0 has units of radians per second.

(a) Determine the angular velocity ω of the block if it spins around the body z axis with the same rotational kinetic energy.

(b) Determine the angular velocity ω of the block if it spins instead around the body z axis with the same angular momentum.

{Ans.: (a) $\omega = 1.31\omega_0$; (b) $\omega = 1.04\omega_0$}

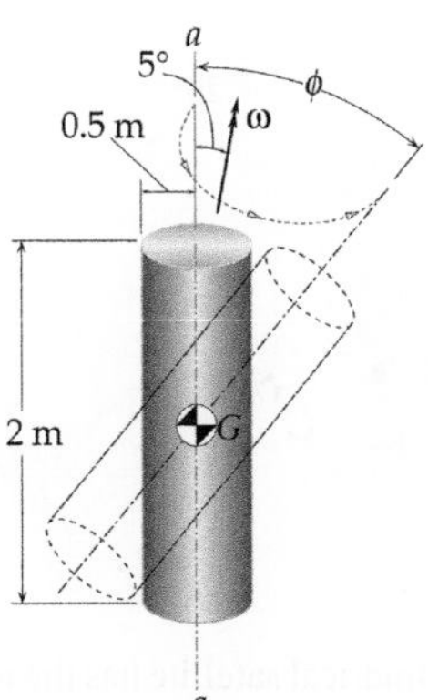

12.8 The above solid right circular cylinder of mass 500 kg is set into torque-free motion with its symmetry axis initially aligned with the fixed spatial line *a-a*. Due to an injection error, the vehicle's angular velocity vector

$\boldsymbol{\omega}$ is misaligned 5° (the wobble angle) from the symmetry axis. Calculate the maximum angle ϕ between fixed line a-a and the axis of the cylinder.
{Ans.: 30.96°}

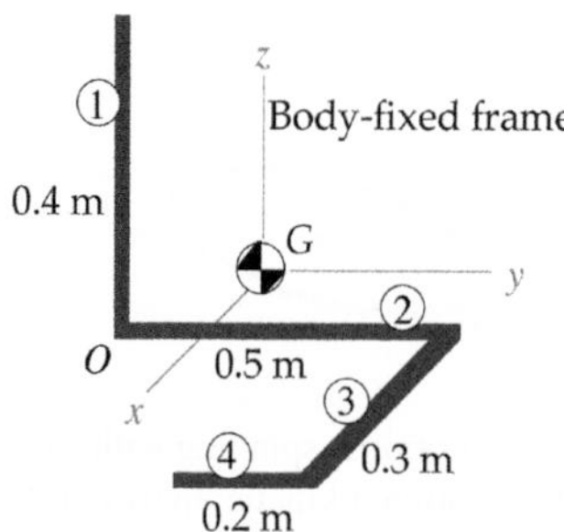

Section 12.3

12.9 For a rigid axisymmetric satellite, the mass moment of inertia about its long axis is 1000 $\text{kg}\cdot\text{m}^2$, and the moment of inertia about transverse axes through the center of mass is 5000 $\text{kg}\cdot\text{m}^2$. It is spinning about the minor principal body axis in torque-free motion at 6 rad/s with the angular velocity lined up with the angular momentum vector **H**. Over time, the energy degrades due to internal effects and the satellite is eventually spinning about a major principal body axis with the angular velocity lined up with the angular momentum vector **H**. Calculate the change in rotational kinetic energy between the two states.
{Ans.: −14.4 kJ}

12.10 Let the object in Example 11.11 (redrawn above) be a highly dissipative torque-free satellite, whose angular velocity at the instant shown is $\boldsymbol{\omega} = 10\hat{\mathbf{i}}\,\text{rad/s}$. Calculate the decrease in kinetic energy after it becomes, as eventually it must, a major-axis spinner.
{Ans.: −0.487 J}

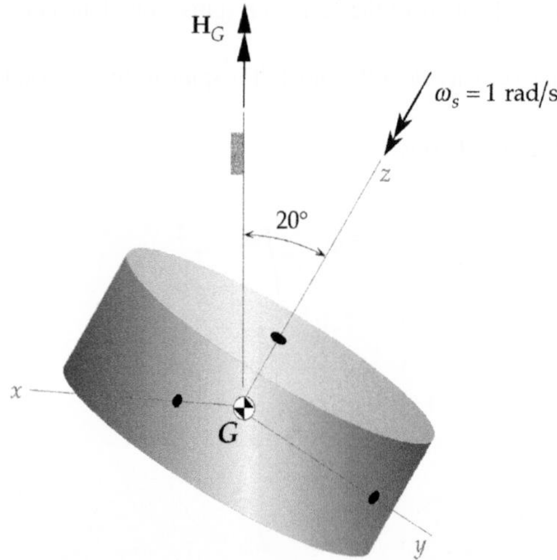

12.11 The above dissipative torque-free cylindrical satellite has the initial spin state shown. $A = B = 320\ \text{kg}\cdot\text{m}^2$ and $C = 560\ \text{kg}\cdot\text{m}^2$. Calculate the magnitude of the angular velocity when it reaches its stable spin state.
{Ans.: 1.419 rad/s}

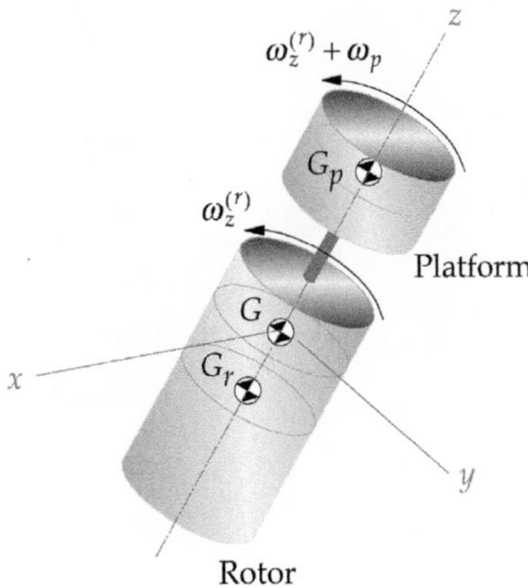

Section 12.4

12.12 For the above nonprecessing, dual-spin satellite, $C_r = 1000\ \text{kg}\cdot\text{m}^2$ and $C_p = 500\ \text{kg}\cdot\text{m}^2$. The angular velocity of the rotor is $3\hat{\mathbf{k}}\ \text{rad/s}$ and the angular velocity of the platform relative to the rotor is $1\hat{\mathbf{k}}\ \text{rad/s}$. If the relative angular velocity of the platform is reduced to $0.5\hat{\mathbf{k}}\ \text{rad/s}$, what is the new angular velocity of the rotor?
{Ans.: 3.17 rad/s}

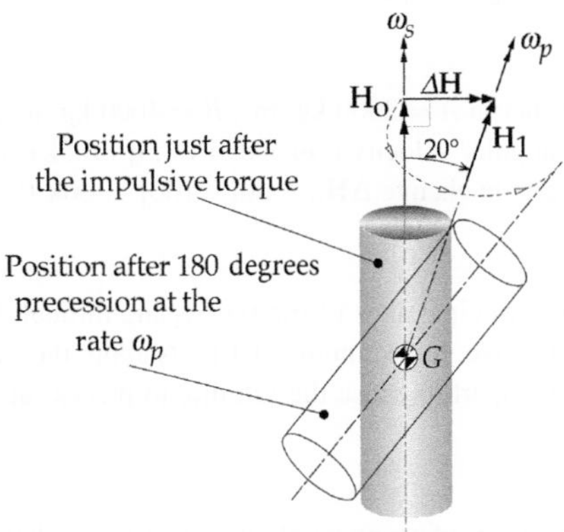

Section 12.6

12.13 For the above rigid axisymmetric satellite, the mass moment of inertia about its long axis is $1000\ \text{kg}\cdot\text{m}^2$, and the moment of inertia about transverse axes through the center of mass is $5000\ \text{kg}\cdot\text{m}^2$. It is initially spinning about the minor principal body axis in torque-free motion at $\omega_s = 0.1\ \text{rad/s}$, with the angular velocity lined up with the angular momentum vector $\mathbf{H}_0$. A pair of thrusters exert an external impulsive torque on the satellite, causing an instantaneous change $\Delta\mathbf{H}$ of angular momentum in the direction normal to $\mathbf{H}_0$, so that the new angular momentum is $\mathbf{H}_1$, at an angle of 20° to $\mathbf{H}_0$, as shown in the figure. How long does it take the satellite to precess (cone) through an angle of 180° around $\mathbf{H}_1$?
{Ans.: 147.6 s}

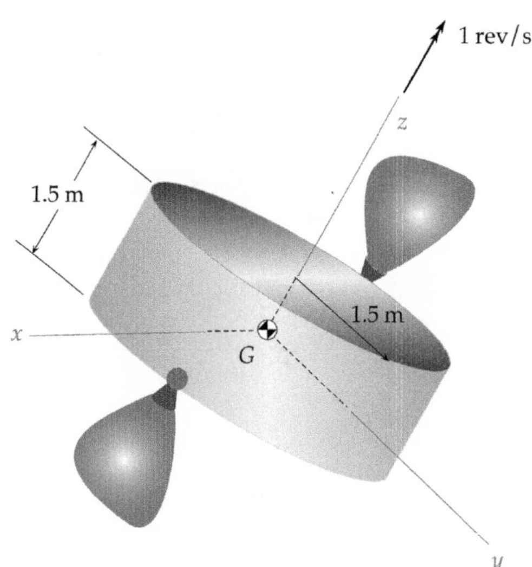

Section 12.7

12.14 A satellite is spinning at 0.01 rev/s. The moment of inertia of the satellite about the spin axis is $2000 \ \text{kg} \cdot \text{m}^2$. Paired thrusters are located at a distance of 1.5 m from the spin axis. They deliver their thrust in pulses, each thruster producing an impulse of $15 \ \text{N} \cdot \text{s}$ per pulse. At what rate will the satellite be spinning after 30 pulses?
{Ans.: 0.1174 rev/s}

12.15 A satellite has moments of inertia $A = 2000 \ \text{kg} \cdot \text{m}^2$, $B = 4000 \ \text{kg} \cdot \text{m}^2$, and $C = 6000 \ \text{kg} \cdot \text{m}^2$ about its principal body axes xyz. Its angular velocity is $\boldsymbol{\omega} = 0.1\hat{\mathbf{i}} + 0.3\hat{\mathbf{j}} + 0.5\hat{\mathbf{k}}$ (rad/s). If thrusters cause the angular momentum vector to undergo the change $\Delta\mathbf{H}_G = 50\hat{\mathbf{i}} - 100\hat{\mathbf{j}} + 300\hat{\mathbf{k}}$ $(\text{kg} \cdot \text{m}^2/\text{s})$, what is the magnitude of the new angular velocity?
{Ans.: 1.045 rad/s}

12.16 The body-fixed xyz axes are principal axes of inertia passing through the center of mass of a 300-kg cylindrical satellite pictured above. It is spinning at 1 rev/s about the z axis. What impulsive torque about the y axis must the thrusters impart to cause the satellite to precess at 5 rev/s?
{Ans.: $6740 \ \text{N} \cdot \text{m} \cdot \text{s}$}

Section 12.8

12.17 A satellite is to be despun by means of a tangential release yo-yo mechanism consisting of two masses, 3 kg each, wound around the midplane of the satellite. The satellite is spinning around its axis of symmetry with an angular velocity $\omega_s = 5$ rad/s. The radius of the cylindrical satellite is 1.5 m and the moment of inertia about the spin axis is $C = 300 \ \text{kg} \cdot \text{m}^2$.

(a) Find the cord length and the deployment time to reduce the spin rate to 1 rad/s.

(b) Find the cord length and time to reduce the spin rate to zero.

{Ans.: (a) $l = 5.902$ m, $t = 0.787$ s; (b) $l = 7.228$ m, $t = 0.964$ s}

12.18 A cylindrical satellite of radius 1 m is initially spinning about the axis of symmetry at the rate of 2 rad/s with a nutation angle of 15°. The principal moments of inertia are $A = B = 30 \ \text{kg} \cdot \text{m}^2$ and $C = 60 \ \text{kg} \cdot \text{m}^2$. An energy dissipation device is built into the satellite, so that it eventually ends up in pure spin around the z axis.

(a) Calculate the final spin rate about the z axis.
(b) Calculate the loss of kinetic energy.
(c) A tangential release yo-yo despin device is also included in the satellite. If the two yo-yo masses are each 7 kg, what cord length is required to completely despin the satellite? Is it wrapped in the proper direction in the figure?

{Ans.: (a) 2.071 rad/s; (b) 8.62 J; (c) 2.3 m}

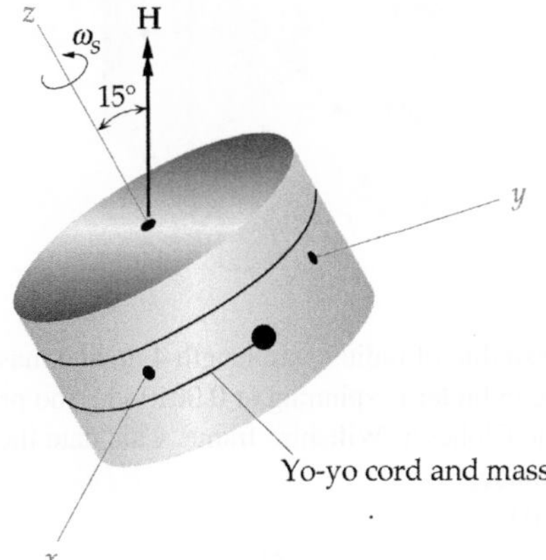

Section 12.9

12.19 A communications satellite is in a geostationary equatorial orbit with a period of 24 h. The spin rate ω_s about its axis of symmetry is 1 rpm, and the moment of inertia about the spin axis is 550 $\text{kg}\cdot\text{m}^2$. The moment of inertia about transverse axes through the mass center G is 225 $\text{kg}\cdot\text{m}^2$. If the spin axis is initially pointed toward the earth, calculate the magnitude and direction of the applied torque $\mathbf{M}_G$ required to keep the spin axis pointed always toward the earth.

{Ans.: 0.00420 N · m, about the negative x axis}

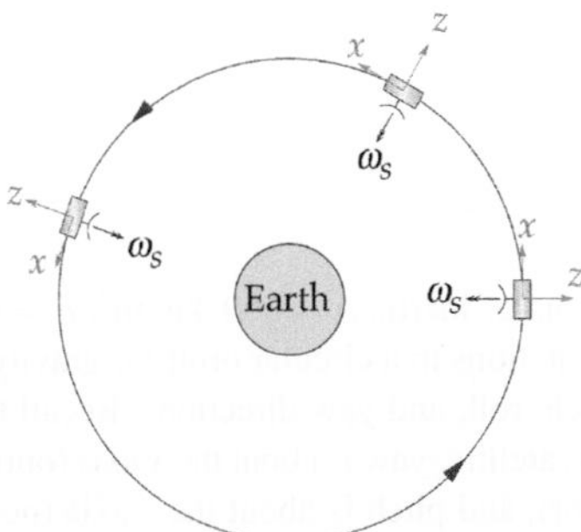

12.20 The moments of inertia of a satellite about its principal body axes xyz are $A = 1000\ \text{kg}\cdot\text{m}^2$, $B = 600\ \text{kg}\cdot\text{m}^2$, and $C = 500\ \text{kg}\cdot\text{m}^2$, respectively. The moments of inertia of a momentum wheel at the center of mass of the satellite and aligned with the x axis are $I_x = 20\ \text{kg}\cdot\text{m}^2$ and $I_y = I_z = 6\ \text{kg}\cdot\text{m}^2$. The absolute angular velocity of the satellite with the momentum wheel locked is $\boldsymbol{\omega}_0 = 0.1\hat{\mathbf{i}} + 0.05\hat{\mathbf{j}}$ rad/s. Calculate the angular velocity ω_f of the momentum wheel (relative to the satellite) required to reduce the x component of the absolute angular velocity of the satellite to 0.003 rad/s.

{Ans.: 4.95 rad/s}

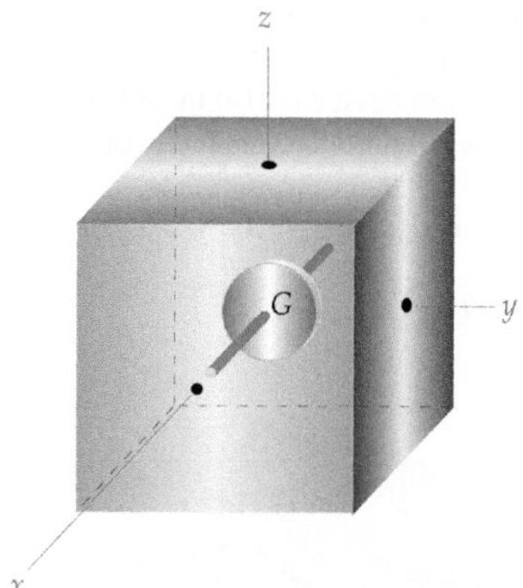

12.21 A solid circular cylindrical satellite of radius 1 m, length 4 m, and mass 250 kg is in a circular earth orbit with a period of 90 min. The cylinder is spinning at 0.001 rad/s (no precession) around its axis, which is aligned with the y axis of the Clohessy-Wiltshire frame. Calculate the magnitude of the external torque required to maintain this attitude.
{Ans.: $-0.00014544\hat{\mathbf{i}}$ (N·m)}

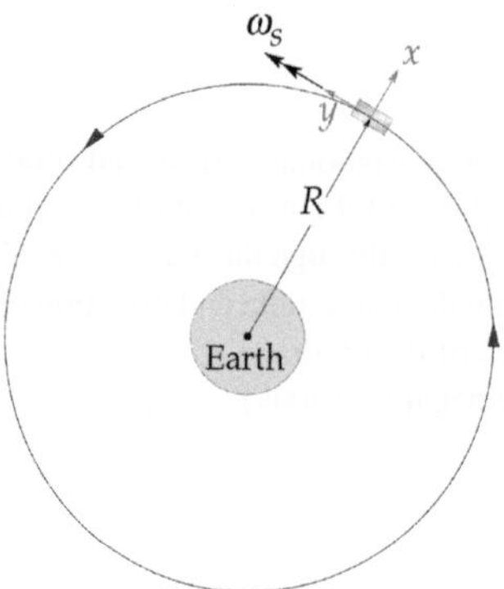

Section 12.10

12.22 A satellite has principal moments of inertia $A = 300\ \text{kg}\cdot\text{m}^2$, $B = 400\ \text{kg}\cdot\text{m}^2$, and $C = 500\ \text{kg}\cdot\text{m}^2$. Determine the permissible orientations in a circular orbit for gravity gradient stabilization. Specify which axes may be aligned in the pitch, roll, and yaw directions. Recall that, relative to a Clohessy-Wiltshire frame at the center of mass of a satellite, yaw is about the x axis (outward radial from earth's center), roll is about the y axis (velocity vector), and pitch is about the z axis (normal to orbital plane).

REFERENCES

Likins, P.W., 1967. Attitude stability criteria for dual spin spacecraft. J. Spacecr. Rockets 4, 1638–1643.
Palm, W.J., 1983. Modeling, Analysis and Control of Dynamic Systems. Wiley, New York.
Kaplan, M.H., 1976. Modern Spacecraft Dynamics and Control. Wiley, New York.

CHAPTER

ROCKET VEHICLE DYNAMICS 13

13.1 INTRODUCTION

In previous chapters, we have made frequent reference to delta-v maneuvers of spacecraft. These require a propulsion system of some sort whose job is to throw vehicle mass (in the form of propellants) overboard. Newton's balance of momentum principle dictates that when mass is ejected from a system in one direction, the mass left behind must acquire a velocity in the opposite direction. The familiar and oft-quoted example is the rapid release of air from an inflated toy balloon. Another is that of a diver leaping off a small boat at rest in the water, causing the boat to acquire a motion of its own. The unfortunate astronaut who becomes separated from his ship in the vacuum of space cannot with any amount of flailing of arms and legs "swim" back to safety. If he has tools or other expendable objects of equipment, accurately throwing them in the direction opposite to his spacecraft may do the trick. Spewing compressed gas from a tank attached to his back through a nozzle pointed away from the spacecraft would be a better solution.

The purpose of a rocket motor is to use the chemical energy of solid or liquid propellants to steadily and rapidly produce a large quantity of hot high-pressure gas, which is then expanded and accelerated through a nozzle. This large mass of combustion products flowing out of the nozzle at supersonic speed possesses a lot of momentum and, leaving the vehicle behind, causes the vehicle itself to acquire a momentum in the opposite direction. This is represented as the action of the force we know as thrust. The design and analysis of rocket propulsion systems is well beyond our scope.

This chapter contains a necessarily brief introduction to some of the fundamentals of rocket vehicle dynamics. The equations of motion of a launch vehicle in a gravity turn trajectory are presented first. This is followed by a simple development of the thrust equation, which brings in the concept of specific impulse. The thrust equation and the equations of motion are then combined to produce the rocket equation, which relates delta-v to propellant expenditure and specific impulse. The sounding rocket provides an important but relatively simple application of the concepts introduced to this point. After a computer simulation of a gravity turn trajectory, the chapter concludes with an elementary consideration of multistage launch vehicles.

Those seeking a more detailed introduction to the subject of rockets and rocket performance will find the texts by Wiesel (2010), Hale (1994), and Sutton and Biblarz (2017) useful.

Orbital Mechanics for Engineering Students. https://doi.org/10.1016/B978-0-12-824025-0.00013-1

13.2 EQUATIONS OF MOTION

Fig. 13.1 illustrates the trajectory of a satellite launch vehicle and the forces acting on it during the powered ascent. Rockets at the base of the booster produce the thrust **T**, which acts along the vehicle's axis in the direction of the velocity vector **v**. The aerodynamic drag force **D** is directed opposite to the velocity, as shown. Its magnitude is given by

$$D = qAC_D \tag{13.1}$$

where $q = \rho v^2/2$ is the dynamic pressure, in which ρ is the density of the atmosphere and v is the speed (i.e., the magnitude) of **v**, A is the frontal area of the vehicle, and C_D is the coefficient of drag. C_D depends on the speed and the external geometry of the rocket. The force of gravity on the booster is $m\mathbf{g}$, where m is its mass, and **g** is the local gravitational acceleration vector, pointing toward the center of the earth. As discussed in Section 1.3, at any point of the trajectory, the velocity **v** defines the direction of the unit tangent $\hat{\mathbf{u}}_t$ to the path. The unit normal $\hat{\mathbf{u}}_n$ is perpendicular to **v** and points toward the center of curvature C. The distance of point C from the path is ρ (not to be confused with density). ρ is the radius of curvature.

In Fig. 13.1, the vehicle and its flight path are shown relative to the earth. In the interest of simplicity we will ignore the earth's spin and write the equations of motion relative to a nonrotating earth. The small-acceleration terms required to account for the earth's rotation can be added for a more refined analysis. Let us resolve Newton's second law, $\mathbf{F}_{\text{net}} = m\mathbf{a}$, into components along the path directions $\hat{\mathbf{u}}_t$ and $\hat{\mathbf{u}}_n$. Recall from Section 1.3 that the acceleration along the path is

$$a_t = \frac{dv}{dt} \tag{13.2}$$

and the normal acceleration is $a_n = v^2/\rho$ (where ρ is the radius of curvature). It was shown in Example 1.8 (Eq. 1.37) that for flight over a flat surface, $v/\rho = -d\gamma/dt$, in which case the normal acceleration can be expressed in terms of the flight path angle γ as

$$a_n = -v\frac{d\gamma}{dt}$$

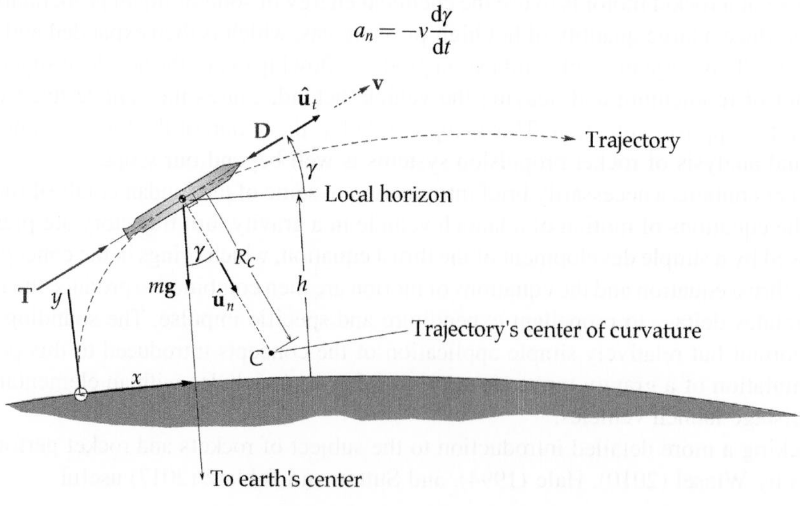

FIG. 13.1

Launch vehicle boost trajectory. γ is the flight path angle.

To account for the curvature of the earth, as was done in Section 1.7, we can use polar coordinates with origin at the earth's center to show that a term must be added to this expression, so that it becomes

$$a_n = -v\frac{d\gamma}{dt} + \frac{v^2}{R_E + h}\cos\gamma \tag{13.3}$$

where R_E is the radius of the earth, and h (instead of z as in previous chapters) is the altitude of the rocket. Thus, in the direction of $\hat{\mathbf{u}}_t$, Newton's second law requires

$$T - D - mg\sin\gamma = ma_t \tag{13.4}$$

whereas in the $\hat{\mathbf{u}}_n$ direction

$$mg\cos\gamma = ma_n \tag{13.5}$$

After substituting Eqs. (13.2) and (13.3), the latter two expressions may be written:

$$\frac{dv}{dt} = \frac{T}{m} - \frac{D}{m} - g\sin\gamma \tag{13.6}$$

$$v\frac{d\gamma}{dt} = -\left(g - \frac{v^2}{R_E + h}\right)\cos\gamma \tag{13.7}$$

To these we must add the equations for downrange distance x and altitude h,

$$\frac{dx}{dt} = \frac{R_E}{R_E + h}v\cos\gamma \qquad \frac{dh}{dt} = v\sin\gamma \tag{13.8}$$

Recall that the variation of g with altitude is given by Eq. (1.36). Numerical methods must be used to solve Eqs. (13.6), (13.7), and (13.8). To do so, we must account for the variation of the thrust, booster mass, atmospheric density, the drag coefficient, and the acceleration of gravity. Of course, the vehicle mass continuously decreases as propellants are consumed to produce the thrust, which we shall discuss in the following section.

The free body diagram in Fig. 13.1 does not include a lifting force, which, if the vehicle were an airplane, would act normal to the velocity vector. Launch vehicles are designed to be strong in lengthwise compression, like a column. To save weight they are, unlike an airplane, made relatively weak in bending, shear, and torsion, which are the kinds of loads induced by lifting surfaces. Transverse lifting loads are held closely to zero during powered ascent through the atmosphere by maintaining a zero angle of attack, that is, by keeping the axis of the booster aligned with its velocity vector (the relative wind). Pitching maneuvers are done early in the launch, soon after the rocket clears the launch tower, when its speed is still low. At the high speeds acquired within a minute or so after launch, the slightest angle of attack can produce destructive transverse loads in the vehicle. The Space Shuttle orbiter had wings so that it could act as a glider after reentry into the atmosphere. However, the launch configuration of the orbiter was such that its wings were at the zero lift angle of attack throughout the ascent.

Satellite launch vehicles take off vertically and, at injection into orbit, must be flying parallel to the earth's surface. During the initial phase of the ascent, the rocket builds up speed on a nearly vertical trajectory taking it above the dense lower layers of the atmosphere. While it transitions to the thinner upper atmosphere, the trajectory bends over, trading vertical speed for horizontal speed so that the rocket can achieve orbital perigee velocity at burnout. The gradual transition from vertical to horizontal flight, as illustrated in Fig. 13.1, is caused by the force of gravity, and it is called a gravity turn trajectory.

At liftoff the rocket is vertical, and the flight path angle γ is 90°. After clearing the tower and gaining speed, vernier thrusters or gimbaling of the main engines produce a small, programmed pitchover, establishing an initial flight path angle γ_0, slightly less than 90°. Thereafter, γ will continue to decrease at a rate dictated by Eq. (13.7). (For example, if $\gamma = 85°$, $v = 110$ m/s (250 mph), and $h = 2$ km, then $d\gamma/dt = -0.44$ deg /s). As the speed v of the vehicle increases, the coefficient of $\cos\gamma$ in Eq. (13.7) decreases, which means the rate of change of the flight path angle becomes increasingly smaller, tending toward zero as the booster approaches orbital speed, $v_{\text{circular orbit}} = \sqrt{g/(R_E + h)}$. Ideally, the vehicle is flying horizontally ($\gamma = 0$) at that point.

The gravity turn trajectory is just one example of a practical trajectory, tailored for satellite boosters. On the other hand, sounding rockets fly straight up from launch through burnout. Rocket-powered guided missiles must execute high-speed pitch and yaw maneuvers as they careen toward moving targets and require a rugged structure to withstand the accompanying side loads.

13.3 THE THRUST EQUATION

To discuss rocket performance requires an expression for the thrust T in Eq. (13.6). It can be obtained by a simple one-dimensional momentum analysis. Fig. 13.2a shows a system consisting of a rocket and its propellants. The exterior of the rocket is surrounded by the static pressure p_a of the atmosphere everywhere except at the rocket nozzle exit, where the pressure is p_e. p_e acts over the nozzle exit area A_e. The value of p_e depends on the design of the nozzle. For simplicity, we assume that no other forces act on the system. At time t the mass of the system is m and the absolute velocity in its axial direction is v. The propellants combine chemically in the rocket's combustion chamber, and during the small time interval Δt a small mass Δm of combustion products is forced out of the nozzle, to the left. Because of this expulsion, the velocity of the rocket changes by the small amount Δv, to the right. The absolute velocity of Δm is v_e, assumed to be to the left. According to Newton's second law of motion,

(Momentum of the system at $t + \Delta t$) − (Momentum of the system at t) = Net external impulse

or

$$\left[(m - \Delta m)(v + \Delta v)\hat{\mathbf{i}} + \Delta m\left(-v_e\hat{\mathbf{i}}\right)\right] - mv\hat{\mathbf{i}} = (p_e - p_a)A_e\Delta t\hat{\mathbf{i}} \tag{13.9}$$

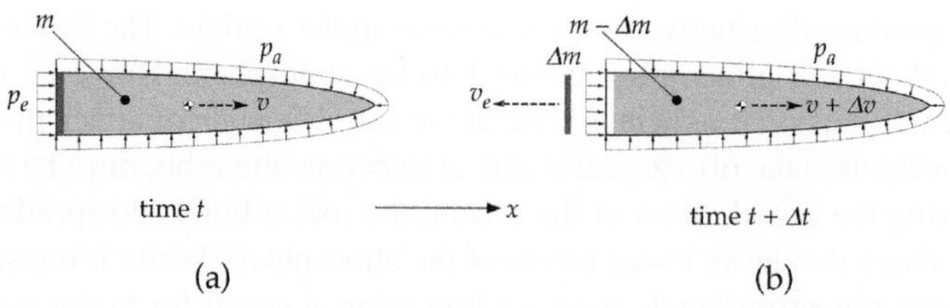

FIG. 13.2

(a) System of rocket and propellant at time t. (b) The system an instant later, after ejection of a small element Δm of combustion products.

Let $\dot{m}_e$ (a positive quantity) be the rate at which exhaust mass flows across the nozzle exit plane. The mass m of the rocket decreases at the rate dm/dt, and conservation of mass requires the decrease of mass to equal the mass flow rate out of the nozzle. Thus,

$$\frac{dm}{dt} = -\dot{m}_e \tag{13.10}$$

Assuming $\dot{m}_e$ is constant, the vehicle mass as a function of time (from $t = 0$) may therefore be written

$$m(t) = m_0 - \dot{m}_e t \tag{13.11}$$

where m_0 is the initial mass of the vehicle. Since Δm is the mass that flows out in the time interval Δt, we have

$$\Delta m = \dot{m}_e \Delta t \tag{13.12}$$

Let us substitute this expression into Eq. (13.9) to obtain

$$\left[(m - \dot{m}_e\Delta t)(v + \Delta v)\hat{\mathbf{i}} + \dot{m}_e\Delta t\left(-v_e\hat{\mathbf{i}}\right)\right] - mv\hat{\mathbf{i}} = (p_e - p_a)A_e\Delta t\hat{\mathbf{i}}$$

Collecting terms, we get

$$m\Delta v\hat{\mathbf{i}} - \dot{m}_e\Delta t(v_e + v)\hat{\mathbf{i}} - \dot{m}_e\Delta t\Delta v\hat{\mathbf{i}} = (p_e - p_a)A_e\Delta t\hat{\mathbf{i}}$$

Dividing through by Δt, taking the limit as $\Delta t \to 0$, and canceling the common unit vector leads to

$$m\frac{dv}{dt} - \dot{m}_e c_a = (p_e - p_a)A_e \tag{13.13}$$

where c_a is the speed of the exhaust relative to the rocket,

$$c_a = v_e + v \tag{13.14}$$

Rearranging terms, Eq. (13.13) may be written

$$\dot{m}_e c_a + (p_e - p_a)A_e = m\frac{dv}{dt} \tag{13.15}$$

The left-hand side of this equation is the unbalanced force responsible for the acceleration dv/dt of the system in Fig. 13.2. This unbalanced force is the thrust T,

$$T = \dot{m}_e c_a + (p_e - p_a)A_e \tag{13.16}$$

where $\dot{m}_e c_a$ is the jet thrust, and $(p_e - p_a)A_e$ is the pressure thrust. We can write Eq. (13.16) as

$$T = \dot{m}_e\left[c_a + \frac{(p_e - p_a)A_e}{\dot{m}_e}\right] \tag{13.17}$$

The term in brackets is called the effective exhaust velocity c,

$$c = c_a + \frac{(p_e - p_a)A_e}{\dot{m}_e} \tag{13.18}$$

In terms of the effective exhaust velocity, the thrust may be expressed simply as

$$T = \dot{m}_e c \tag{13.19}$$

The specific impulse I_{sp} is defined as the thrust per sea level weight rate (per second) of propellant consumption. That is,

$$I_{sp} = \frac{T}{\dot{m}_e g_0} \tag{13.20}$$

where g_0 is the standard sea level acceleration of gravity. The unit of specific impulse is force ÷ (force/s) or seconds. Together, Eqs. (13.19) and (13.20) imply that

$$c = I_{sp} g_0 \tag{13.21}$$

Obviously, we can infer the jet velocity directly from the specific impulse. Specific impulse is an important performance parameter for a given rocket engine and propellant combination. However, a large specific impulse equates to a large thrust only if the mass flow rate is large, which is true of chemical rocket engines. The specific impulse of chemical rockets typically lies in the range 200–300 s for solid fuels and 250–450 s for liquid fuels. Ion propulsion systems have very high specific impulse ($>10^4$ s), but their very low mass flow rates produce much smaller thrust than chemical rockets.

13.4 ROCKET PERFORMANCE

From Eqs. (13.10) and (13.20) we have

$$T = -I_{sp} g_0 \frac{dm}{dt} \tag{13.22}$$

or

$$\frac{dm}{dt} = -\frac{T}{I_{sp} g_0}$$

If the thrust and specific impulse are constant, then the integral of this expression over the burn time Δt is

$$\Delta m = -\frac{T}{I_{sp} g_0} \Delta t$$

from which we obtain.

$$\Delta t = \frac{I_{sp} g_0}{T}(m_0 - m_f) = \frac{I_{sp} g_0}{T} m_0 \left(1 - \frac{m_f}{m_0}\right) \tag{13.23}$$

where m_0 and m_f are the mass of the vehicle at the beginning and end of the burn, respectively. The mass ratio n is defined as the ratio of the initial mass to final mass,

$$n = \frac{m_0}{m_f} \tag{13.24}$$

Clearly, the mass ratio is always greater than unity. In terms of the mass ratio, Eq. (13.23) may be written.

$$\Delta t = \frac{n-1}{n} \frac{I_{sp}}{T/(m_0 g_0)} \tag{13.25}$$

$T/(mg_0)$ is the thrust-to-weight ratio. The thrust-to-weight ratio for a launch vehicle at liftoff is typically in the range 1.3 to 2.

Substituting Eq. (13.22) into Eq. (13.6), we get.

$$\frac{dv}{dt} = -I_{sp}g_0\frac{dm/dt}{m} - \frac{D}{m} - g\sin\gamma$$

Integrating with respect to time, from t_0 to t_f, yields.

$$\Delta v = I_{sp}g_0 \ln\frac{m_0}{m_f} - \Delta v_D - \Delta v_G \tag{13.26}$$

where the drag loss Δv_D and the gravity loss Δv_G are given by the integrals.

$$\Delta v_D = \int_{t_0}^{t_f} \frac{D}{m}dt \qquad \Delta v_G = \int_{t_0}^{t_f} g\sin\gamma dt \tag{13.27}$$

Since the drag D, acceleration of gravity g, and flight path angle γ are unknown functions of time, these integrals cannot be computed. (Eqs. (13.6)–(13.8) must be integrated numerically to obtain $v(t)$ and $\gamma(t)$, and Δv would follow from those results.) Eq. (13.26) can be used for rough estimates where previous data and experience provide a basis for choosing conservative values of Δv_D and Δv_G. Obviously, if drag can be neglected, then $\Delta v_D = 0$. This would be a good approximation for the last stage of a satellite booster, for which it can also be said that $\Delta v_G = 0$, since $\gamma \approx 0°$ when the satellite is injected into orbit.

Sounding rockets are launched vertically and fly straight up to their maximum altitude before falling back to earth, usually by parachute. Their purpose is to measure remote portions of the earth's atmosphere. ("Sound" in this context means to measure or investigate.) If for a sounding rocket $\gamma = 90°$, then $\Delta v_G \approx g_0(t_f - t_0)$, since g is within 90% of g_0 up to 300 km altitude.

EXAMPLE 13.1

A sounding rocket of initial mass m_0 and mass m_f after all propellant is consumed is launched vertically ($\gamma = 90°$). The propellant mass flow rate $\dot{m}_e$ is constant.

(a) Neglecting drag and the variation of gravity with altitude, calculate the speed v_{bo}, the altitude h_{bo} at burnout, and the maximum height h_{max} attained by the rocket.

(b) For what flow rate is the greatest altitude reached?

Solution

The vehicle mass as a function of time, up to burnout, is

$$m = m_0 - \dot{m}_e t \tag{a}$$

At burnout, $m = m_f$, so the burnout time t_{bo} is

$$t_{bo} = \frac{m_0 - m_f}{\dot{m}_e} \tag{b}$$

The drag loss Δv_D is assumed to be zero, and the gravity loss for $g = g_0 = \text{constant}$ is

$$\Delta v_G = \int_0^{t_{bo}} g_0 \sin(90°)dt = g_0 t_{bo}$$

Recalling that $I_{sp}g_0 = c$ and using Eq. (a), it follows from Eq. (13.26) that, up to burnout, the velocity as a function of time is

$$v = c\ \ln\frac{m_0}{m_0 - \dot{m}_e t} - g_0 t - \overbrace{\int_{t_0}^{t}\frac{D}{m}dt}^{=0} \tag{c}$$

Since $dh/dt = v$, the altitude as a function of time is.

$$h = \int_0^t v dt = \int_0^t \left(c\ \ln\frac{m_0}{m_0 - \dot{m}_e t} - g_0 t\right)dt = \frac{c}{\dot{m}_e}\left[(m_0 - \dot{m}_e t)\ \ln\frac{m_0 - \dot{m}_e t}{m_0} + \dot{m}_e t\right] - \frac{1}{2}g_0 t^2 \tag{d}$$

The height at burnout h_{bo} is found by substituting Eq. (b) into this expression,

$$\boxed{h_{bo} = \frac{c}{\dot{m}_e}\left(m_f\ \ln\frac{m_f}{m_0} + m_0 - m_f\right) - \frac{1}{2}\left(\frac{m_0 - m_f}{\dot{m}_e}\right)^2 g_0} \tag{e}$$

Likewise, the burnout velocity is obtained by substituting t_{bo} from Eq. (b) into Eq. (c),

$$\boxed{v_{bo} = c\ \ln\frac{m_0}{m_f} - \frac{g_0}{\dot{m}_e}(m_0 - m_f)} \tag{f}$$

After burnout, the rocket coasts upward with the constant downward acceleration of gravity, so that

$$v = v_{bo} - g_0(t - t_{bo})$$
$$h = h_{bo} + v_{bo}(t - t_{bo}) - \frac{1}{2}g_0(t - t_{bo})^2$$

Substituting Eqs. (b), (e), and (f) into these two expressions yields, for $t > t_{bo}$,

$$v = c\ \ln\frac{m_0}{m_f} - g_0 t$$
$$h = \frac{c}{\dot{m}_e}\left(m_0\ \ln\frac{m_f}{m_0} + m_0 - m_f\right) + ct\ \ln\frac{m_0}{m_f} - \frac{1}{2}g_0 t^2 \tag{g}$$

The maximum height h_{max} is reached when $v = 0$,

$$c\ \ln\frac{m_0}{m_f} - g_0 t_{max} = 0 \Rightarrow t_{max} = \frac{c}{g_0}\ \ln\frac{m_0}{m_f} \tag{h}$$

Substituting $t = t_{max}$ into Eq. (g) leads to our result,

$$\boxed{h_{max} = \frac{1}{2}\frac{c^2}{g_0}\ \ln^2 n - \frac{cm_0}{\dot{m}_e}\frac{n\ \ln n - (n-1)}{n}} \tag{i}$$

where n is the mass ratio ($n > 1$). Since n ln n is greater than $n - 1$, it follows that the second term in this expression is positive. Hence, h_{max} can be increased by increasing the mass flow rate $\dot{m}_e$. In fact,

$$\boxed{\text{The greatest height is achieved when } \dot{m}_e \to \infty}$$

In that extreme, all the propellant is expended at once, like a mortar shell.

Since we neglected both drag and the variation of gravity with altitude, our results (Eqs. e, f, and i) are not accurate, but only estimates.

EXAMPLE 13.2

The data for a single-stage rocket are as follows:

Launch mass: $m_0 = 68{,}000$ kg

Mass ratio: $n = 7$

Specific impulse: $I_{sp} = 390$ s

Thrust: $T = 933.91$ kN

It is launched into a vertical trajectory, like a sounding rocket. Neglecting drag and assuming that the gravitational acceleration is constant at its sea level value $g_0 = 9.81$ m/s^2, estimate.

(a) the time until burnout;
(b) the burnout altitude;
(c) the burnout velocity; and
(d) the maximum altitude reached.

Solution

(a) From Eq. (b) of Example 13.1, the burnout time t_{bo} is

$$t_{bo} = \frac{m_0 - m_f}{\dot{m}_e} \tag{a}$$

The burnout mass m_f is obtained from Eq. (13.24),

$$m_f = \frac{m_0}{n} = \frac{68{,}000}{7} = 9714.3\,\text{kg} \tag{b}$$

The propellant mass flow rate $\dot{m}_e$ is given by Eq. (13.20),

$$\dot{m}_e = \frac{T}{I_{sp}g_0} = \frac{933{,}910}{390 \cdot 9.81} = 244.10\,\text{kg/s} \tag{c}$$

Substituting Eqs. (b) and (c), and $m_0 = 68,000$ kg into Eq. (a) yields the burnout time,

$$t_{bo} = \frac{68{,}000 - 9714.3}{244.10} = \boxed{238.8\text{ s}}$$

(b) The burnout altitude is given by Eq. (e) of Example 13.1,

$$h_{bo} = \frac{c}{\dot{m}_e}\left(m_f \ln\frac{m_f}{m_0} + m_0 - m_f\right) - \frac{1}{2}\left(\frac{m_0 - m_f}{\dot{m}_e}\right)^2 g_0 \tag{d}$$

The exhaust velocity c is found in Eq. (13.21),

$$c = I_{sp}g_0 = 390 \cdot 9.81 = 3825.9\,\text{m/s} \tag{e}$$

Substituting Eqs. (b), (c), and (e), along with $m_0 = 68{,}000$ kg and $g_0 = 9.81$ m/s^2, into Eq. (d), we get

$$h_{bo} = \frac{3825.9}{244.1}\left(9714.3 \ln\frac{9714.3}{68{,}000} + 68,000 - 9714.3\right) - \frac{1}{2}\left(\frac{68{,}000 - 9714.3}{244.1}\right)^2 \cdot 9.81$$

$$\boxed{h_{bo} = 337.6\text{km}}$$

(c) From Eq. (f) of Example 13.1, we find

$$v_{bo} = c \ln\frac{m_0}{m_f} - \frac{g_o}{\dot{m}_e}(m_0 - m_f) = 3825.9 \ln\frac{68{,}000}{9714.3} - \frac{9.81}{244.1}(68,000 - 9714.3)$$

$$\boxed{v_{bo} = 5.102\text{km/s}}$$

(d) To find h_{max}, where the speed of the rocket falls to zero, we use Eq. (i) of Example 13.1,

$$h_{max} = \frac{1}{2}\frac{c^2}{g_0}\ln^2 n - \frac{cm_0}{\dot{m}_e}\frac{n \ln n - (n-1)}{n} = \frac{13825.9^2}{2\ \ 9.81}\ln^2 7 - \frac{3825.9 \cdot 68{,}000}{244.1} \cdot \frac{7 \ln 7 - (7-1)}{7}$$

$$\boxed{h_{max} = 1664.6\text{km}}$$

Note that the rocket coasts to a height nearly five times the burnout altitude.

We can employ the integration schemes introduced in Section 1.8 to solve Eqs. (13.6)–(13.8) numerically. This permits a more accurate accounting of the effects of gravity and drag. It also yields the trajectory.

EXAMPLE 13.3

The rocket in Example 13.2 has a diameter of 5 m. It is to be launched on a gravity turn trajectory. Pitchover begins at an altitude of 130 m with an initial flight path angle γ_0 of 89.85°. What are the altitude h and speed v of the rocket at burnout ($t_{bo} = 260$ s)? What are the velocity losses due to drag and gravity (cf. Eq. 13.27)?

Solution

The MATLAB program *Example_13_03.m* in Appendix D.53 finds the speed v, the flight path angle γ, the altitude h, and the downrange distance x as a function of time. It does so by using the ordinary differential equation solver *rkf_45.m* (Appendix D.4) to numerically integrate Eqs. (13.6)–(13.8), namely

$$\frac{dv}{dt} = \frac{T}{m} - \frac{D}{m} - g\sin\gamma \quad \text{(a)}$$

$$\frac{d\gamma}{dt} = -\frac{1}{v}\left(g - \frac{v^2}{R_E + h}\right)\cos\gamma \quad \text{(b)}$$

$$\frac{dh}{dt} = v\sin\gamma \quad \text{(c)}$$

$$\frac{dx}{dt} = \frac{R_E}{R_E + h}v\cos\gamma \quad \text{(d)}$$

The variable mass m is given in terms of the initial mass $m_0 = 68{,}000$ kg and the constant mass flow rate $\dot{m}_e$ by Eq. (13.11),

$$m = m_0 - \dot{m}_e t \quad \text{(e)}$$

The thrust $T = 933.913$ kN is assumed constant, and $\dot{m}_e$ is obtained from T and the specific impulse $I_{sp} = 390$ s by means of Eq. (13.20),

$$\dot{m}_e = \frac{T}{I_{sp} g_0} \quad \text{(f)}$$

The drag force D in Eq. (a) is given by Eq. (13.1),

$$D = \frac{1}{2}\rho v^2 A C_D \quad \text{(g)}$$

The drag coefficient is assumed to have the constant value $C_D = 0.5$. The frontal area $A = \pi d^2/4$ is found from the rocket diameter $d = 5$ m. The atmospheric density profile is assumed to be exponential,

$$\rho = \rho_0 e^{-h/h_0} \quad \text{(h)}$$

where $\rho_0 = 1.225$ kg/m^3 is the sea level atmospheric density, and $h_0 = 7.5$ km is the scale height of the atmosphere. (The scale height is the altitude at which the density of the atmosphere is about 37% of its sea level value.)

Finally, the acceleration of gravity varies with altitude h according to Eq. (1.36),

$$g = \frac{g_0}{(1 + h/R_E)^2} \quad (R_E = 6378 \text{ km},\ g_0 = 9.81 \text{ m/s}^2) \quad \text{(i)}$$

The drag loss and gravity loss are found by numerically integrating Eqs. (13.27).

Between liftoff and pitchover, the flight path angle γ is held at 90°. Pitchover begins at the altitude $h_p = 130$ m with the flight path angle set at $\gamma_0 = 89.85°$.

We write the above system of equations in the standard form

$$\dot{\mathbf{y}} = \mathbf{f}(t, \mathbf{y}) \quad \text{(j)}$$

where

$$\mathbf{y} = \begin{bmatrix} v & \gamma & x & h & v_D & v_G \end{bmatrix}^T$$

$$\mathbf{f}=\left[\overbrace{\frac{T}{m}-\frac{D}{m}-g\sin\gamma}^{\dot{v}} \quad \overbrace{-\frac{1}{v}\left(g-\frac{v^2}{R_E+h}\right)\cos\gamma}^{\dot{\gamma}} \quad \overbrace{\frac{R_E}{R_E+h}v\cos\gamma}^{\dot{x}} \quad \overbrace{v\sin\gamma}^{\dot{h}} \quad \overbrace{-\frac{D}{m}}^{\dot{v}_D} \quad \overbrace{-g\sin\gamma}^{\dot{v}_G}\right]^T$$

For the input data described above, the output of *Example_13_03.m* is listed below. The solution is very sensitive to the choice of h_p and γ_0.

The Mach number is the vehicle speed v divided by the speed of sound a, which as a function of altitude h is found from the International Standard Atmosphere model that MATLAB provides as the function *atmosisa*.

```
Initial flight path angle  =   89.850 deg
Pitchover altitude         =  130.000 m
Burn time                  =  238.776 s
Maximum dynamic pressure   =    0.156 atm
   Time                    =    1.110 min
   Speed                   =    0.302 km/s
   Altitude                =    9.424 km
   Mach number             =    0.999
 At burnout:
   Speed                   =    5.737 km/s
   Flight path angle       =    9.154 deg
   Altitude                =  110.324 km
   Downrange distance      =  318.364 km
   Drag loss               =    0.298 km/s
   Gravity loss            =    1.410 km/s
```

Thus, at burnout

$$\boxed{\begin{aligned}\text{Altitude} &= 110.3\,\text{km}\\ \text{Speed} &= 5.737\,\text{km/s}\end{aligned}}$$

The speed losses are

$$\boxed{\begin{aligned}\text{Due to drag: } 0.298\,\text{km/s}\\ \text{Due to gravity: } 1.410\,\text{km/s}\end{aligned}}$$

Note that the drag loss is much less than the gravity loss.

Fig. 13.3 shows the computed gravity turn trajectory and the dynamic pressure variation. The maximum dynamic pressure, $q_{max} = 15.8$ kPa (0.156 atm), occurs 66.6 s into the flight at an altitude of 9.42 km and a speed of 0.302 km/s (Mach 1). By comparison, the dynamic pressure of the wind on your hand sticking out of a car window traveling 80 km/h is about 0.003 atm.

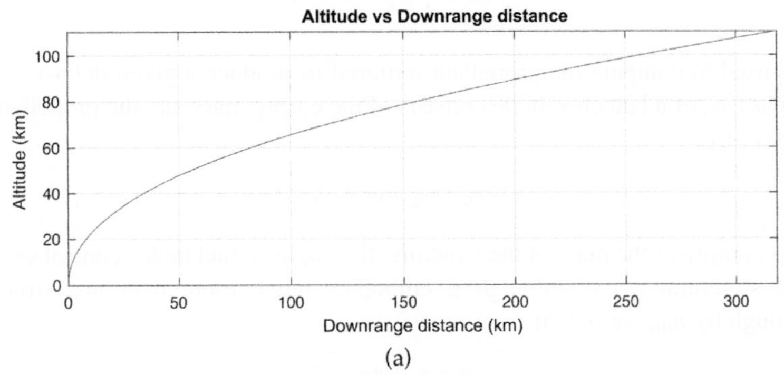

(a)

FIG. 13.3

(a) Gravity turn trajectory for the data given in Examples 13.2 and 13.3.

(Continued)

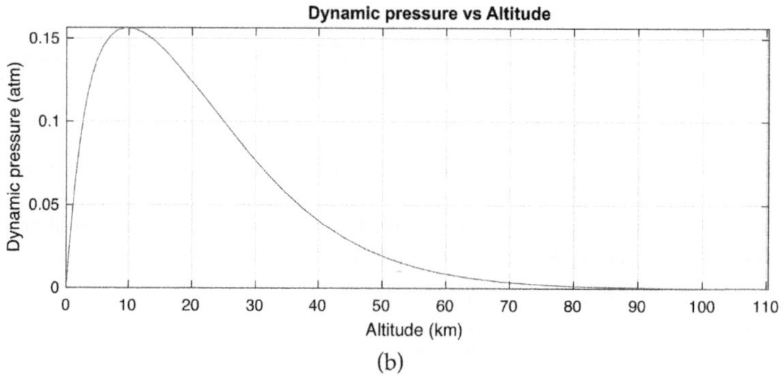

(b)

FIG. 13.3, cont'd

(b) Dynamic pressure variation with altitude (1 atm = 101.3 kPa).

13.5 RESTRICTED STAGING IN FIELD-FREE SPACE

In field-free space, we neglect drag and gravitational attraction. In that case, Eq. (13.26) becomes

$$\Delta v = I_{sp} g_0 \ln \frac{m_0}{m_f} \tag{13.28}$$

This is at best a poor approximation for high-thrust rockets, but it will suffice to shed some light on the rocket-staging problem. Observe that we can solve this equation for the mass ratio to obtain

$$\frac{m_0}{m_f} = e^{\Delta v / (I_{sp} g_0)} \tag{13.29}$$

The amount of propellant expended to produce the velocity increment Δv is $m_0 - m_f$. If we let $\Delta m = m_0 - m$, then Eq. (13.29) can be written as.

$$\frac{\Delta m}{m_0} = 1 - e^{-\Delta v / (I_{sp} g_0)} \tag{13.30}$$

This relation is used to compute the propellant required to produce a given delta-v.

The gross mass m_0 of a launch vehicle consists of the empty mass m_E, the propellant mass m_p, and the payload mass m_{PL},

$$m_0 = m_E + m_p + m_{PL} \tag{13.31}$$

The empty mass comprises the mass of the structure, the engines, fuel tanks, control systems, etc. m_E is also called the structural mass, although it embodies much more than just structure. Dividing Eq. (13.31) through by m_0, we obtain

$$\pi_E + \pi_p + \pi_{PL} = 1 \tag{13.32}$$

where the mass ratios $\pi_E = m_E/m_0$, $\pi_p = m_p/m_0$, and $\pi_{PL} = m_{PL}/m_0$ are the structural fraction, propellant fraction, and payload fraction, respectively. It is convenient to define the payload ratio λ,

$$\lambda = \frac{m_{PL}}{m_E + m_p} = \frac{m_{PL}}{m_0 - m_{PL}} \tag{13.33}$$

and the structural ratio ε,

$$\varepsilon = \frac{m_E}{m_E + m_p} = \frac{m_E}{m_0 - m_{PL}} \tag{13.34}$$

The mass ratio n was introduced in Eq. (13.24). Assuming that all of the propellant is consumed, that may now be written

$$n = \frac{m_E + m_p + m_{PL}}{m_E + m_{PL}} \tag{13.35}$$

λ, ε, and n are not independent. From Eq. (13.34) we have

$$m_E = \frac{\varepsilon}{1-\varepsilon} m_p \tag{13.36}$$

whereas Eq. (13.33) gives

$$m_{PL} = \lambda\left(m_E + m_p\right) = \lambda\left(\frac{\varepsilon}{1-\varepsilon} m_p + m_p\right) = \frac{\lambda}{1-\varepsilon} m_p \tag{13.37}$$

Substituting Eqs. (13.36) and (13.37) into Eq. (13.35) leads to

$$n = \frac{1+\lambda}{\varepsilon+\lambda} \tag{13.38}$$

Thus, given any two of the ratios λ, ε, and n, we obtain the third from Eq. (13.38). Using this relation in Eq. (13.28) and setting Δv equal to the burnout speed v_{bo}, when the propellants have been used up, yields

$$v_{bo} = I_{sp} g_0 \ln n = I_{sp} g_0 \ln \frac{1+\lambda}{\varepsilon+\lambda} \tag{13.39}$$

This equation is plotted in Fig. 13.4 for a range of structural ratios. Clearly, for a given empty mass, the greatest possible Δv occurs when the payload is zero. However, what we want to do is maximize the amount of payload while keeping the structural weight to a minimum. Of course, the mass of load-bearing structure, rocket motors, pumps, piping, etc. cannot be made arbitrarily small. Current materials technology places a lower limit on ε of about 0.1. For this value of the structural ratio and $\lambda = 0.05$, Eq. (13.39) yields.

$$v_{bo} = 1.94 I_{sp} g_0 = 0.019 I_{sp}\,(\text{km/s})$$

The specific impulse of a typical chemical rocket is about 300 s, which in this case would provide $\Delta v = 5.7$ km/s. However, the circular orbital velocity at the earth's surface is 7.905 km/s. Therefore, this booster by itself could not orbit the payload. The minimum specific impulse required for a single stage to orbit would be 416 s. Only today's most advanced liquid hydrogen/liquid oxygen engines (e.g., the Space Shuttle main engines), have had this kind of performance. Practicality and economics would likely dictate going the route of a multistage booster.

Fig. 13.5 shows a series or tandem two-stage rocket configuration, with one stage sitting on top of the other. Each stage has its own engines and propellant tanks. The dividing lines between the stages are

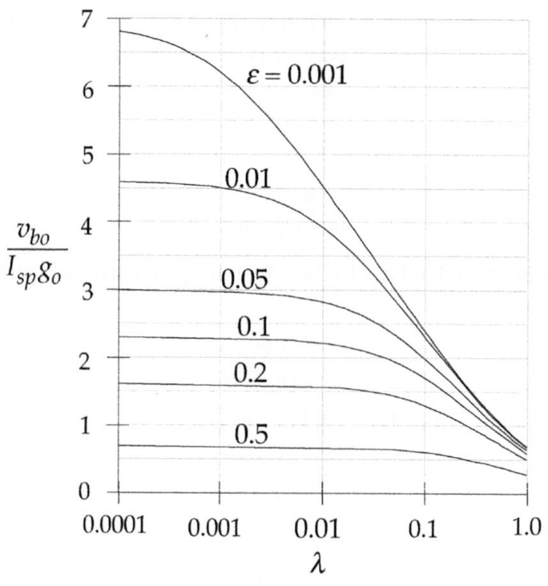

FIG. 13.4

Dimensionless burnout speed vs. payload ratio.

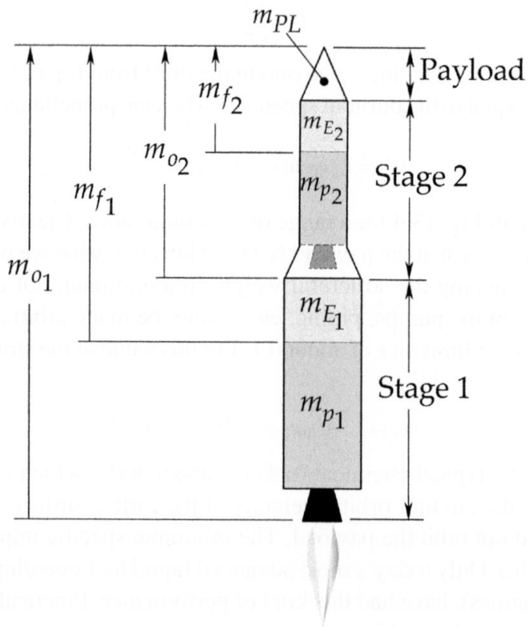

FIG. 13.5

Tandem two-stage booster.

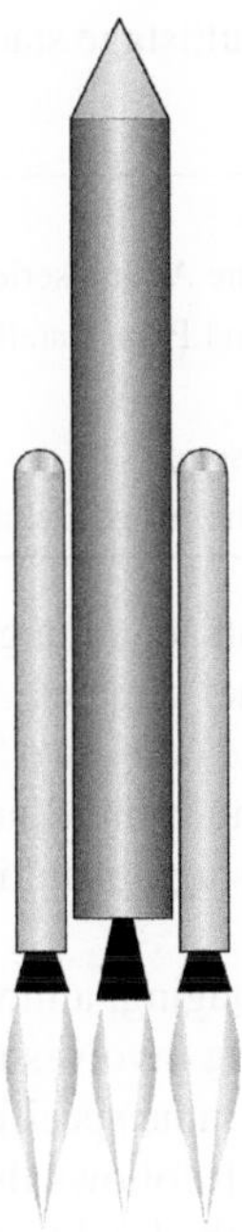

FIG. 13.6

Parallel staging.

where they separate during flight. The first stage drops off first, the second stage next, etc. The payload of an N-stage rocket is actually stage $N + 1$. Indeed, satellites commonly carry their own propulsion systems into orbit. The payload of a given stage is everything above it. Therefore, as illustrated in Fig. 13.5, the initial mass m_0 of stage 1 is that of the entire vehicle. After stage 1 expels all its fuel, the mass m_f that remains is stage 1's empty mass m_E plus the mass of stage 2 and the payload. After separation of stage 1, the process continues likewise for stage 2, with m_0 being its initial mass.

Titan II, the launch vehicle for the Gemini program, had the two-stage, tandem configuration. So did the Saturn 1B, used to launch earth orbital flights early in the Apollo program, as well as to send crews to Skylab and an Apollo spacecraft to dock with a Russian Soyuz in 1975.

Fig. 13.6 illustrates the concept of parallel staging. Two or more solid or liquid rockets are attached ("strapped on") to a core vehicle carrying the payload. In the tandem arrangement, the motors in a given stage cannot ignite until separation of the previous stage, whereas all the rockets ignite at once in the parallel-staged vehicle. The strap-on boosters fall away after they burn out early in the ascent. The Space Shuttle was a most obvious example of parallel staging. Its two solid rocket boosters were mounted on the external tank, which fueled the three "main" engines built into the orbiter. The solid rocket boosters and the external tank were cast off after they were depleted. In more common use is the combination of parallel and tandem staging, in which solid or liquid propellant boosters are strapped to the first stage of a multistage stack. A few of the many examples from several countries include:

China:	Long March series
Europe:	Later versions in the Ariane series
India:	Geosynchronous and Polar Satellite launch vehicles
Japan:	H-2 and H-3 series
Russia:	Soyuz and Proton series
United States:	Later versions of the Atlas, Delta, Titan, and Falcon

The venerable Atlas, used in many variants to, among other things, launch the orbital flights of the Mercury program, had three main liquid-fuel engines at its base. They all fired simultaneously at launch, but several minutes into the flight, the outer two "boosters" dropped away, leaving the central sustainer engine to burn the rest of the way to orbit. Since the booster engines shared the sustainer's propellant tanks, the Atlas exhibited partial staging and is sometimes referred to as a one-and-a-half-stage rocket.

We will for simplicity focus on tandem staging, although parallel-staged systems are handled in a similar way (Wiesel, 2010). Restricted staging involves the simple but unrealistic assumption that all stages are similar. That is, each stage has the same specific impulse I_{sp}, the same structural ratio ε, and the same payload ratio λ. From Eq. (13.38), it follows that the mass ratios n are identical too. Let us investigate the effect of restricted staging on the final burnout speed v_{bo} for a given payload mass m_{PL} and overall payload fraction

$$\pi_{PL} = \frac{m_{PL}}{m_0} \tag{13.40}$$

where m_0 is the total mass of the tandem-stacked vehicle.

For a single-stage vehicle, the payload ratio is

$$\lambda = \frac{m_{PL}}{m_0 - m_{PL}} = \frac{1}{\dfrac{m_0}{m_{PL}} - 1} = \frac{\pi_{PL}}{1 - \pi_{PL}} \tag{13.41}$$

so that, from Eq. (13.38), the mass ratio is

$$n = \frac{1}{\pi_{PL}(1-\varepsilon) + \varepsilon} \tag{13.42}$$

According to Eq. (13.39), the burnout speed is.

$$v_{bo} = I_{sp} g_0 \ln \frac{1}{\pi_{PL}(1-\varepsilon) + \varepsilon} \tag{13.43}$$

Let m_0 be the total mass of the two-stage rocket of Fig. 13.5. That is,

$$m_0 = m_0)_1 \tag{13.44}$$

The payload of stage 1 is the entire mass m_0 of stage 2. Thus, for stage 1 the payload ratio is

$$\lambda_1 = \frac{m_0)_2}{m_0)_1 - m_0)_2} = \frac{m_0)_2}{m_0 - m_0)_2} \tag{13.45}$$

The payload ratio of stage 2 is

$$\lambda_2 = \frac{m_{PL}}{m_0)_2 - m_{PL}} \tag{13.46}$$

By virtue of the two stages being similar, $\lambda_1 = \lambda_2$, or

$$\frac{m_0)_2}{m_0 - m_0)_2} = \frac{m_{PL}}{m_0)_2 - m_{PL}}$$

Solving this equation for m_0 yields

$$m_0)_2 = \sqrt{m_0}\sqrt{m_{PL}}$$

But $m_0 = m_{PL}/\pi_{PL}$, so the gross mass of the second stage is

$$m_0)_2 = \sqrt{\frac{1}{\pi_{PL}}} m_{PL} \tag{13.47}$$

Putting this back into Eq. (13.45) or Eq. (13.46), we obtain the common two-stage payload ratio $\lambda = \lambda_1 = \lambda_2$,

$$\lambda_{2\text{-stage}} = \frac{\pi_{PL}{}^{1/2}}{1 - \pi_{PL}{}^{1/2}} \tag{13.48}$$

This together with Eq. (13.38) and the assumption that $\varepsilon_1 = \varepsilon_2 = \varepsilon$ leads to the common mass ratio for each stage,

$$n_{2\text{-stage}} = \frac{1}{\pi_{PL}{}^{1/2}(1-\varepsilon) + \varepsilon} \tag{13.49}$$

If stage 2 ignites immediately after burnout of stage 1, the final velocity of the two-stage vehicle is the sum of the burnout velocities of the individual stages,

$$v_{bo} = v_{bo})_1 + v_{bo})_2$$

or

$$v_{bo_{2\text{-stage}}} = I_{sp} g_0 \ln n_{2\text{-stage}} + I_{sp} g_0 \ln n_{2\text{-stage}} = 2 I_{sp} g_0 \ln n_{2\text{-stage}}$$

so that, with Eq. (13.49), we get

$$v_{bo_{2\text{-stage}}} = I_{sp} g_0 \ln \left[\frac{1}{\pi_{PC}{}^{1/2}(1-\varepsilon) + \varepsilon} \right]^2 \tag{13.50}$$

The empty mass of each stage can be found in terms of the payload mass using the common structural ratio ε,

$$\frac{m_E)_1}{m_0)_1 - m_0)_2} = \varepsilon \qquad \frac{m_E)_2}{m_0)_2 - m_{PL}} = \varepsilon$$

Substituting Eqs. (13.40) and (13.44) together with Eq. (13.47) yields.

$$m_E)_1 = \frac{\left(1 - \pi_{PL}{}^{1/2}\right)\varepsilon}{\pi_{PL}} m_{PL} \quad m_E)_2 = \frac{\left(1 - \pi_{PL}{}^{1/2}\right)\varepsilon}{\pi_{PL}{}^{1/2}} m_{PL} \tag{13.51}$$

Likewise, we can find the propellant mass for each stage from the expressions

$$m_{\mathrm{p}})_1 = m_0)_1 - \left[m_{\mathrm{E}})_1 + m_0)_2\right] \quad m_{\mathrm{p}})_2 = m_0)_2 - \left[m_{\mathrm{E}})_2 + m_{\mathrm{PL}}\right] \tag{13.52}$$

Substituting Eqs. (13.40) and (13.4), together with Eqs. (13.47) and (13.51), we get

$$m_{\mathrm{p}})_1 = \frac{\left(1 - \pi_{\mathrm{PL}}{}^{1/2}\right)(1-\varepsilon)}{\pi_{\mathrm{PL}}} m_{\mathrm{PL}} \quad m_{\mathrm{p}})_2 = \frac{\left(1 - \pi_{\mathrm{PL}}{}^{1/2}\right)(1-\varepsilon)}{\pi_{\mathrm{PL}}{}^{1/2}} m_{\mathrm{PL}} \tag{13.53}$$

EXAMPLE 13.4

The following data are given:

$$m_{\mathrm{PL}} = 10{,}000\,\mathrm{kg} \quad \pi_{\mathrm{PL}} = 0.05 \quad \varepsilon = 0.15 \quad I_{\mathrm{sp}} = 350\,\mathrm{s} \quad g_0 = 0.00981\,\mathrm{km/s^2}$$

Calculate the payload velocity v_{bo} at burnout, the empty mass of the launch vehicle, and the propellant mass for.

(a) a single-stage vehicle; and.

(b) a restricted, two-stage vehicle.

Solution

(a) From Eq. (13.43), we find

$$v_{bo} = 350 \cdot 0.00981 \cdot \ln \frac{1}{0.05(1-0.15)+0.15} = \boxed{5.657\ \mathrm{km/s}}$$

Eq. (13.40) yields the gross mass

$$m_0 = \frac{10{,}000}{0.05} = 200{,}000\,\mathrm{kg}$$

from which we obtain the empty mass using Eq. (13.34),

$$m_{\mathrm{E}} = \varepsilon(m_0 - m_{\mathrm{PL}}) = 0.15(200{,}000 - 10{,}000) = \boxed{28{,}500\ \mathrm{kg}}$$

The mass of propellant is

$$m_{\mathrm{p}} = m_0 - m_{\mathrm{E}} - m_{\mathrm{PL}} = 200{,}000 - 28{,}500 - 10{,}000 = \boxed{161{,}500\ \mathrm{kg}}$$

(b) For a restricted two-stage vehicle, the burnout speed is given by Eq. (13.50),

$$v_{bo})_{\text{2-stage}} = 350 \cdot 0.00981 \ln \left[\frac{1}{0.05^{1/2}(1-0.15)+0.15}\right]^2 = \boxed{7.407\,\mathrm{km/s}}$$

The empty mass of each stage is found using Eq. (13.51),

$$m_{\mathrm{E}})_1 = \frac{\left(1-0.05^{1/2}\right)\cdot 0.15}{0.05} \cdot 10{,}000 = \boxed{23{,}292\,\mathrm{kg}}$$

$$m_{\mathrm{E}})_2 = \frac{\left(1-0.05^{1/2}\right)\cdot 0.15}{0.05^{1/2}} \cdot 10{,}000 = \boxed{5208\,\mathrm{kg}}$$

For the propellant masses, we turn to Eq. (13.53),

$$m_{\mathrm{p}})_1 = \frac{\left(1-0.05^{1/2}\right)\cdot(1-0.15)}{0.05} \cdot 10{,}000 = \boxed{131{,}990\,\mathrm{kg}}$$

$$m_{\mathrm{p}})_2 = \frac{\left(1-0.05^{1/2}\right)\cdot(1-0.15)}{0.05^{1/2}} \cdot 10{,}000 = \boxed{29{,}513\,\mathrm{kg}}$$

The total empty mass, $m_{\mathrm{E}} = m_{\mathrm{E}})_1 + m_{\mathrm{E}})_2$, and the total propellant mass, $m_{\mathrm{p}} = m_{\mathrm{p}})_1 + m_{\mathrm{p}})_2$, are the same as for the single-stage rocket. The mass of the second stage, including the payload, is 22.4% of the total vehicle mass.

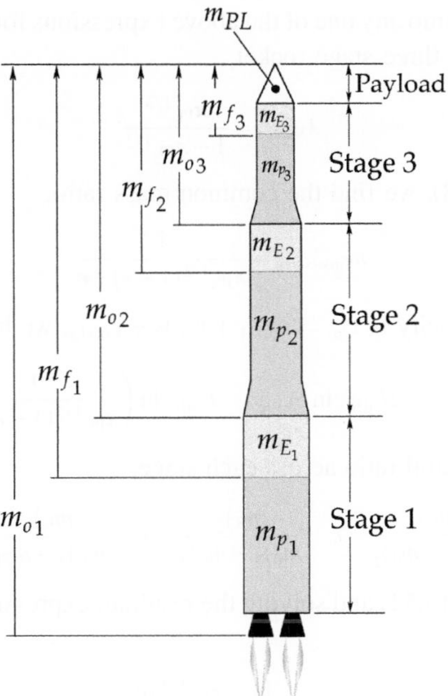

FIG. 13.7

Tandem three-stage launch vehicle.

Observe in the previous example that, although the total vehicle mass was unchanged, the burnout velocity increased 31% for the two-stage arrangement. The reason is that the second stage is lighter and can therefore be accelerated to a higher speed. Let us determine the velocity gain associated with adding another stage, as illustrated in Fig. 13.7.

The payload ratios of the three stages are

$$\lambda_1 = \frac{m_0)_2}{m_0)_1 - m_0)_2} \qquad \lambda_2 = \frac{m_0)_3}{m_0)_2 - m_0)_3} \qquad \lambda_3 = \frac{m_{PL}}{m_0)_3 - m_{PL}}$$

Since the stages are similar, these payload ratios are all the same. Setting $\lambda_1 = \lambda_2$ and recalling that $m_0)_1 = m_0$, we find

$$m_0)_2{}^2 - m_0)_3 m_0 = 0$$

Similarly, $\lambda_1 = \lambda_3$ yields

$$m_0)_2 m_0)_3 - m_0 m_{PL} = 0$$

These two equations imply that.

$$m_0)_2 = \frac{m_{PL}}{\pi_{PL}{}^{2/3}} \qquad m_0)_3 = \frac{m_{PL}}{\pi_{PL}{}^{1/3}} \tag{13.54}$$

Substituting these results back into any one of the above expressions for λ_1, λ_2, or λ_3 yields the common payload ratio for the restricted three-stage rocket,

$$\lambda_{3\text{-stage}} = \frac{\pi_{PL}{}^{1/3}}{1-\pi_{PL}{}^{1/3}}$$

With this result and Eq. (13.38), we find the common mass ratio,

$$n_{\text{three-stage}} = \frac{1}{\pi_{PL}{}^{1/3}(1-\varepsilon)+\varepsilon} \tag{13.55}$$

Since the payload burnout velocity is $v_{bo} = v_{bo})_1 + v_{bo})_2 + v_{bo})_3$, we have

$$v_{bo})_{3\text{-stage}} = 3I_{sp}g_0 \ln n_{3\text{-stage}} = I_{sp}g_0 \ln\left(\frac{1}{\pi_{PL}{}^{1/3}(1-\varepsilon)+\varepsilon}\right)^3 \tag{13.56}$$

Because of the common structural ratio across each stage,

$$\frac{m_E)_1}{m_0)_1 - m_0)_2} = \varepsilon \qquad \frac{m_E)_2}{m_0)_2 - m_0)_3} = \varepsilon \qquad \frac{m_E)_3}{m_0)_3 - m_{PL}} = \varepsilon$$

Substituting Eqs. (13.40) and (13.54) and solving the resultant expressions for the empty stage masses yields

$$m_E)_1 = \frac{\left(1-\pi_{PL}{}^{1/3}\right)\varepsilon}{\pi_{PL}} m_{PL} \qquad m_E)_2 = \frac{\left(1-\pi_{PL}{}^{1/3}\right)\varepsilon}{\pi_{PL}{}^{2/3}} m_{PL} \qquad m_E)_3 = \frac{\left(1-\pi_{PL}{}^{1/3}\right)\varepsilon}{\pi_{PL}{}^{1/3}} m_{PL} \tag{13.57}$$

The stage propellant masses are

$$\begin{aligned} m_p)_1 &= m_0)_1 - \left[m_E)_1 + m_0)_2\right] \\ m_p)_2 &= m_0)_2 - \left[m_E)_2 + m_0)_3\right] \\ m_p)_3 &= m_0)_3 - \left[m_E)_3 + m_{PL}\right] \end{aligned}$$

Substituting Eqs. (13.40), (13.54), and (13.57) leads to

$$\begin{aligned} m_p)_1 &= \frac{\left(1-\pi_{PL}{}^{1/3}\right)(1-\varepsilon)}{\pi_{PL}} m_{PL} \\ m_p)_2 &= \frac{\left(1-\pi_{PL}{}^{1/3}\right)(1-\varepsilon)}{\pi_{PL}{}^{2/3}} m_{PL} \\ m_p)_2 &= \frac{\left(1-\pi_{PL}{}^{1/3}\right)(1-\varepsilon)}{\pi_{PL}{}^{1/3}} m_{PL} \end{aligned} \tag{13.58}$$

EXAMPLE 13.5

Repeat Example 13.4 for the restricted three-stage launch vehicle.

Solution

Eq. (13.56) gives the burnout velocity for three stages.

$$v_{bo} = 350 \cdot 0.00981 \cdot \ln\left[\frac{1}{0.05^{1/3}(1-0.15)+0.15}\right]^3 = \boxed{7.928 \text{ km/s}}$$

Substituting $m_{PL} = 10,000$ kg, $\pi_{PL} = 0.05$, and $\varepsilon = 0.15$ into Eqs. (13.57) and (13.58) yields

$m_E)_1 = 18{,}948$ kg	$m_E)_2 = 6980$ kg	$m_E)_3 = 2572$ kg
$m_p)_1 = 107{,}370$ kg	$m_p)_2 = 39{,}556$ kg	$m_p)_3 = 14{,}573$ kg

Again, the total empty mass and total propellant mass are the same as for the single- and two-stage vehicles. Note that the velocity increase over the two-stage rocket is just 7%, which is much less than the advantage the two-stage vehicle had over the single-stage vehicle.

Looking back over the velocity formulas for one-, two-, and three-stage vehicles (Eqs. 13.43, 13.50, and 13.56), we can induce that for an N-stage rocket,

$$v_{bo})_{\text{N-stage}} = I_{sp} g_0 \ln\left(\frac{1}{\pi_{PL}{}^{1/N}(1-\varepsilon)+\varepsilon}\right)^N = I_{sp} g_0 N \ln\left(\frac{1}{\pi_{PL}{}^{1/N}(1-\varepsilon)+\varepsilon}\right) \tag{13.59}$$

What happens as we let N become very large? First of all, it can be shown using Taylor series expansion that, for large N,

$$\pi_{PL}{}^{1/N} \approx 1 + \frac{1}{N} \ln \pi_{PL} \tag{13.60}$$

Substituting this into Eq. (13.59), we find that

$$v_{bo})_{\text{N-stage}} \approx I_{sp} g_0 N \ln\left[\frac{1}{1+\frac{1}{N}(1-\varepsilon)\ln \pi_{PL}}\right]$$

Since the term $\frac{1}{N}(1-\varepsilon)\ln \pi_{PL}$ is arbitrarily small, we can use the fact that

$$\frac{1}{1+x} = 1 - x + x^2 - x^3 + \cdots$$

to write

$$\frac{1}{1+\frac{1}{N}(1-\varepsilon)\ln \pi_{PL}} \approx 1 - \frac{1}{N}(1-\varepsilon)\ln \pi_{PL}$$

which means

$$v_{bo})_{\text{N-stage}} \approx I_{sp} g_0 N \ln\left[1 - \frac{1}{N}(1-\varepsilon)\ln \pi_{PL}\right]$$

Finally, since $\ln(1-x) = -x - x^2/2 - x^3/3 - x^4/4 - \cdots$, we can write this as

$$v_{bo})_{\text{N-stage}} \approx I_{sp} g_0 N \left[-\frac{1}{N}(1-\varepsilon)\ln \pi_{PL}\right]$$

Therefore, as N, the number of stages, tends toward infinity, the burnout velocity approaches

$$v_{bo})_{\infty} = I_{sp} g_0 (1-\varepsilon) \ln \frac{1}{\pi_{PL}} \tag{13.61}$$

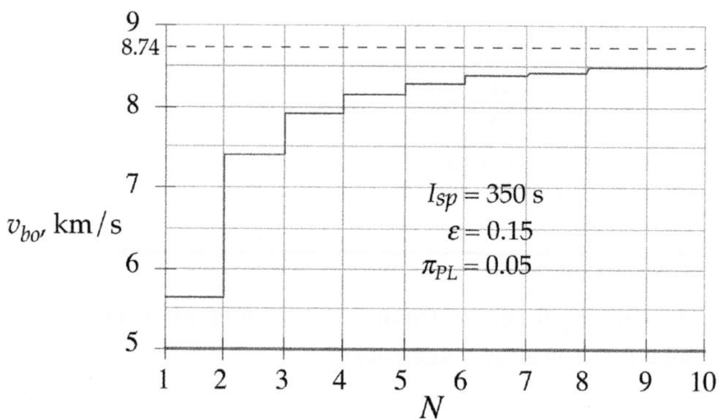

FIG. 13.8

Burnout velocity vs. number of stages (Eq. 13.59).

Thus, no matter how many similar stages we use, for a given specific impulse, payload fraction, and structural ratio, we cannot exceed this burnout speed. For example, using $I_{sp} = 350$ s, $\pi_{PL} = 0.05$, and $\varepsilon = 0.15$ from the previous two examples yields $v_{bo})_\infty = 8.743$ km/s, which is only 10% greater than v_{bo} of a three-stage vehicle. The trend of v_{bo} toward this limiting value is illustrated by Fig. 13.8.

Our simplified analysis does not take into account the added weight and complexity accompanying additional stages. Practical reality has limited the number of stages of actual launch vehicles to rarely more than three.

13.6 OPTIMAL STAGING

Let us now abandon the restrictive assumption that all stages of a tandem-stacked vehicle are similar. Instead, we will specify the specific impulse I_{sp_i} and structural ratio ε_i of each stage i and then seek the minimum-mass N-stage vehicle that will carry a given payload m_{PL} to a specified burnout velocity v_{bo}. To optimize the mass requires using the Lagrange multiplier method, which we shall briefly review.

13.6.1 LAGRANGE MULTIPLIER

Consider a bivariate function f on the xy plane. Then $z = f(x,y)$ is a surface lying above or below the plane, or both. $f(x,y)$ is stationary at a given point if it takes on a local maximum or a local minimum (i.e., an extremum) at that point. For f to be stationary means d$f = 0$. That is,

$$\frac{\partial f}{\partial x}\mathrm{d}x + \frac{\partial f}{\partial y}\mathrm{d}y = 0 \tag{13.62}$$

where dx and dy are independent and not necessarily zero. It follows that for an extremum to exist,

$$\frac{\partial f}{\partial x} = \frac{\partial f}{\partial y} = 0 \tag{13.63}$$

Now let $g(x, y) = 0$ be a curve in the xy plane. Let us find the points on the curve $g = 0$ at which f is stationary. That is, rather than searching the entire xy plane for extreme values of f, we confine our attention to the curve $g = 0$, which is therefore a constraint. Since $g = 0$, it follows that $\mathrm{d}g = 0$, or

$$\frac{\partial g}{\partial x}\mathrm{d}x + \frac{\partial g}{\partial y}\mathrm{d}y = 0 \tag{13.64}$$

If Eqs. (13.62) and (13.64) are both valid at a given point, then

$$\frac{\mathrm{d}y}{\mathrm{d}x} = -\frac{\partial f/\partial x}{\partial f/\partial y} = -\frac{\partial g/\partial x}{\partial g/\partial y}$$

That is,

$$\frac{\partial f/\partial x}{\partial g/\partial x} = \frac{\partial f/\partial y}{\partial g/\partial y} = -\eta$$

From this we obtain

$$\frac{\partial f}{\partial x} + \eta\frac{\partial g}{\partial x} = 0 \quad \frac{\partial f}{\partial y} + \eta\frac{\partial g}{\partial y} = 0$$

But these, together with the constraint $g(x, y) = 0$, are the very conditions required for the function

$$h(x, y, \eta) = f(x, y) + \eta g(x, y) \tag{13.65}$$

to have an extremum, namely,

$$\begin{aligned}\frac{\partial h}{\partial x} &= \frac{\partial f}{\partial x} + \eta\frac{\partial g}{\partial x} = 0\\ \frac{\partial h}{\partial y} &= \frac{\partial f}{\partial y} + \eta\frac{\partial g}{\partial y} = 0\\ \frac{\partial h}{\partial \eta} &= g = 0\end{aligned} \tag{13.66}$$

where η is the Lagrange multiplier. The procedure generalizes to functions of any number of variables.

One can determine mathematically whether the extremum is a maximum or a minimum by checking the sign of the second differential d^2h of the function h in Eq. (13.65),

$$\mathrm{d}^2h = \frac{\partial^2 h}{\partial x^2}\mathrm{d}x^2 + 2\frac{\partial^2 h}{\partial x \partial y}\mathrm{d}x\,\mathrm{d}y + \frac{\partial^2 h}{\partial y^2}\mathrm{d}y^2 \tag{13.67}$$

If $\mathrm{d}^2h < 0$ at the extremum for all $\mathrm{d}x$ and $\mathrm{d}y$ satisfying the constraint condition (Eq. 13.64), then the extremum is a local maximum. Likewise, if $\mathrm{d}^2h > 0$, then the extremum is a local minimum.

EXAMPLE 13.6

(a) Find the extrema of the function $z = -x^2 - y^2$. (b) Find the extrema of the same function under the constraint $y = 2x + 3$.

Solution

(a) To find the extrema we must use Eq. (13.63). Since $\partial z/\partial x = -2x$ and $\partial z/\partial y = -2y$, it follows that $\partial z/\partial x = \partial z/\partial y = 0$ at $x = y = 0$, at which point $z = 0$. Since z is negative everywhere else (Fig. 13.9), it is clear that the extreme value is the maximum value.

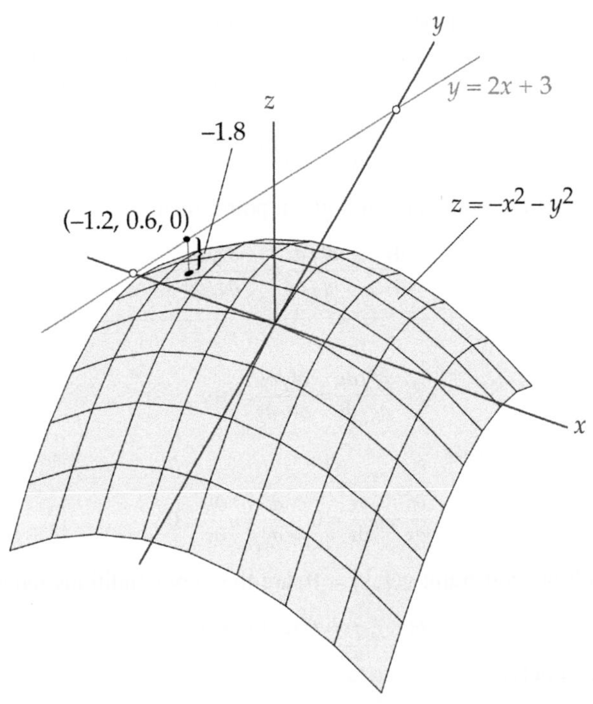

FIG. 13.9

Location of the point on the line $y = 2x + 3$ at which the surface $z = -x^2 - y^2$ is closest to the xy plane.

(b) The constraint may be written $g = y - 2x - 3$. Clearly, $g = 0$. Multiply the constraint by the Lagrange multiplier η and add the result (zero!) to the function $-(x^2 + y^2)$ to obtain

$$h = -(x^2 + y^2) + \eta(y - 2x - 3)$$

This is a function of the three variables x, y, and η. For it to be stationary, the partial derivatives with respect to all three of these variables must vanish. First, we have

$$\frac{\partial h}{\partial x} = -2x - 2\eta$$

Setting this equal to zero yields

$$x = -\eta \tag{a}$$

Next,

$$\frac{\partial h}{\partial y} = -2y + \eta$$

For this to be zero means

$$y = \frac{\eta}{2} \tag{b}$$

Finally,

$$\frac{\partial h}{\partial \eta} = y - 2x - 3$$

Setting this equal to zero gives us back the original constraint condition,

$$y-2x-3=0 \tag{c}$$

Substituting Eqs. (a) and (b) into Eq. (c) yields $\eta = 1.2$, from which Eqs. (a) and (b) imply,

$$\boxed{x=-1.2 \quad y=0.6} \tag{d}$$

These are the coordinates of the point on the line $y = 2x + 3$ at which $z = -x^2 - y^2$ is stationary. Using Eqs. (d), we find that $z = -1.8$ at this point.

Fig. 13.9 is an illustration of this problem, and shows that the computed extremum (a maximum, in the sense that small negative numbers exceed large negative numbers) is where the surface $z = -x^2 - y^2$ is closest to the line $y = 2x + 3$, as measured in the z direction. Note that in this case, Eq. (13.67) yields $\mathrm{d}^2h = -2\mathrm{d}x^2 - 2\mathrm{d}y^2$, which is negative, confirming our conclusion that the extremum is a maximum.

Now let us return to the optimal staging problem. It is convenient to introduce the step mass m_i of the ith stage. The step mass is the empty mass plus the propellant mass of the stage, exclusive of all the other stages,

$$m_i = m_{\mathrm{E}})_i + m_{\mathrm{p}})_i \tag{13.68}$$

The empty mass of stage i can be expressed in terms of its step mass and its structural ratio ε_i as follows:

$$m_{\mathrm{E}})_i = \varepsilon_i \left[m_{\mathrm{E}})_i + m_{\mathrm{p}})_i \right] = \varepsilon_i m_i \tag{13.69}$$

The total mass of the rocket excluding the payload is M, which is the sum of all the step masses,

$$M = \sum_{i=1}^{N} m_i \tag{13.70}$$

Thus, recalling that m_0 is the total mass of the vehicle, we have

$$m_0 = M + m_{\mathrm{PL}} \tag{13.71}$$

Our goal is to minimize m_0.

For simplicity, we will deal first with a two-stage rocket, and then generalize our results to N stages. For a two-stage vehicle, $m_0 = m_1 + m_2 + m_{PL}$, so we can write,

$$\frac{m_0}{m_{\mathrm{PL}}} = \frac{m_1 + m_2 + m_{\mathrm{PL}}}{m_2 + m_{\mathrm{PL}}} \cdot \frac{m_2 + m_{\mathrm{PL}}}{m_{\mathrm{PL}}} \tag{13.72}$$

The mass ratio of stage 1 is

$$n_1 = \frac{m_0)_1}{m_E)_1 + m_2 + m_{\mathrm{PL}}} = \frac{m_1 + m_2 + m_{\mathrm{PL}}}{\varepsilon_1 m_1 + m_2 + m_{\mathrm{PL}}} \tag{13.73}$$

where Eq. (13.69) was used. Likewise, the mass ratio of stage 2 is

$$n_2 = \frac{m_0)_2}{\varepsilon_2 m_2 + m_{\mathrm{PL}}} = \frac{m_2 + m_{\mathrm{PL}}}{\varepsilon_2 m_2 + m_{\mathrm{PL}}} \tag{13.74}$$

We can solve Eqs. (13.73) and (13.74) to obtain the step masses from the mass ratios,

$$\begin{aligned} m_2 &= \frac{n_2 - 1}{1 - n_2\varepsilon_2} m_{PL} \\ m_1 &= \frac{n_1 - 1}{1 - n_1\varepsilon_1}(m_2 + m_{PL}) \end{aligned} \tag{13.75}$$

Now,

$$\frac{m_1 + m_2 + m_{PL}}{m_2 + m_{PL}} = \frac{1-\varepsilon_1}{1-\varepsilon_1} \cdot \frac{m_1 + m_2 + m_{PL}}{m_2 + m_{PL} + (\varepsilon_1 m_1 - \varepsilon_1 m_1)} \cdot \frac{\dfrac{1}{\varepsilon_1 m_1 + m_2 + m_{PL}}}{\dfrac{1}{\varepsilon_1 m_1 + m_2 + m_{PL}}}$$

These manipulations leave the right-hand side unchanged. Carrying out the multiplications proceeds as follows:

$$\begin{aligned} \frac{m_1 + m_2 + m_{PL}}{m_2 + m_{PL}} &= \frac{(1-\varepsilon_1)(m_1 + m_2 + m_{PL})}{\varepsilon_1 m_1 + m_2 + m_{PL} - \varepsilon_1(m_1 + m_2 + m_{PL})} \cdot \frac{\dfrac{1}{\varepsilon_1 m_1 + m_2 + m_{PL}}}{\dfrac{1}{\varepsilon_1 m_1 + m_2 + m_{PL}}} \\ &= \frac{(1-\varepsilon_1)\dfrac{m_1 + m_2 + m_{PL}}{\varepsilon_1 m_1 + m_2 + m_{PL}}}{\dfrac{\varepsilon_1 m_1 + m_2 + m_{PL}}{\varepsilon_1 m_1 + m_2 + m_{PL}} - \varepsilon_1 \dfrac{m_1 + m_2 + m_{PL}}{\varepsilon_1 m_1 + m_2 + m_{PL}}} \end{aligned}$$

Finally, with the aid of Eq. (13.73), this algebraic trickery reduces to

$$\frac{m_1 + m_2 + m_{PL}}{m_2 + m_{PL}} = \frac{(1-\varepsilon_1)n_1}{1 - \varepsilon_1 n_1} \tag{13.76}$$

Likewise,

$$\frac{m_2 + m_{PL}}{m_{PL}} = \frac{(1-\varepsilon_2)n_2}{1 - \varepsilon_2 n_2} \tag{13.77}$$

so that Eq. (13.72) may be written in terms of the stage mass ratios instead of the step masses as

$$\frac{m_o}{m_{PL}} = \frac{(1-\varepsilon_1)n_1}{1-\varepsilon_1 n_1} \cdot \frac{(1-\varepsilon_2)n_2}{1-\varepsilon_2 n_2} \tag{13.78}$$

Taking the natural logarithm of both sides of this equation, we get

$$\ln \frac{m_0}{m_{PL}} = \ln \frac{(1-\varepsilon_1)n_1}{1-\varepsilon_1 n_1} + \ln \frac{(1-\varepsilon_2)n_2}{1-\varepsilon_2 n_2}$$

Expanding the logarithms on the right-hand side leads to

$$\ln \frac{m_0}{m_{PL}} = [\ln(1-\varepsilon_1) + \ln n_1 - \ln(1-\varepsilon_1 n_1)] + [\ln(1-\varepsilon_2) + \ln n_2 - \ln(1-\varepsilon_2 n_2)] \tag{13.79}$$

Observe that for m_{PL} fixed, $\ln(m_0/m_{PL})$ is a monotonically increasing function of m_0,

$$\frac{d}{dm_0}\left(\ln \frac{m_0}{m_{PL}}\right) = \frac{1}{m_0} > 0$$

Therefore, $\ln(m_0/m_{PL})$ is stationary when m_0 is.

From Eqs. (13.21) and (13.39), the burnout velocity of the two-stage rocket is.

$$v_{bo} = v_{bo})_1 + v_{bo})_2 = c_1 \ln n_1 + c_2 \ln n_2 \tag{13.80}$$

which means that, given v_{bo}, our constraint equation is

$$v_{bo} - c_1 \ln n_1 - c_2 \ln n_2 = 0 \tag{13.81}$$

Introducing the Lagrange multiplier η, we combine Eqs. (13.79) and (13.81) to obtain

$$h = [\ln(1-\varepsilon_1) + \ln n_1 - \ln(1-\varepsilon_1 n_1)] + [\ln(1-\varepsilon_2) + \ln n_2 - \ln(1-\varepsilon_2 n_2)] + \eta(v_{bo} - c_1 \ln n_1 - c_2 \ln n_2) \tag{13.82}$$

Finding the values of n_1 and n_2 for which h is stationary will extremize $\ln(m_0/m_{PL})$ (and, hence, m_0) for the prescribed burnout velocity v_{bo}. h is stationary when $\partial h/\partial n_1 = \partial h/\partial n_2 = \partial h/\partial \eta = 0$. Thus,

$$\frac{\partial h}{\partial n_1} = \frac{1}{n_1} + \frac{\varepsilon_1}{1-\varepsilon_1 n_1} - \eta\frac{c_1}{n_1} = 0$$

$$\frac{\partial h}{\partial n_2} = \frac{1}{n_2} + \frac{\varepsilon_2}{1-\varepsilon_2 n_2} - \eta\frac{c_2}{n_2} = 0$$

$$\frac{\partial h}{\partial \eta} = v_{bo} - c_1 \ln n_1 - c_2 \ln n_2 = 0$$

These three equations yield, respectively,

$$n_1 = \frac{c_1\eta - 1}{c_1\varepsilon_1\eta} \qquad n_2 = \frac{c_2\eta - 1}{c_2\varepsilon_2\eta} \qquad v_{bo} = c_1 \ln n_1 + c_2 \ln n_2 \tag{13.83}$$

Substituting n_1 and n_2 into the expression for v_{bo}, we get

$$c_1 \ln\left(\frac{c_1\eta - 1}{c_1\varepsilon_1\eta}\right) + c_2 \ln\left(\frac{c_2\eta - 1}{c_2\varepsilon_2\eta}\right) = v_{bo} \tag{13.84}$$

This equation must be solved iteratively for η, after which η is substituted into Eq. (13.83) to obtain the stage mass ratios n_1 and n_2. These mass ratios are then used in Eq. (13.75) together with the assumed structural ratios, exhaust velocities, and payload mass to obtain the step masses of each stage.

We can now generalize the optimization procedure to an N-stage vehicle, for which Eq. (13.82) becomes

$$h = \sum_{i=1}^{N} [\ln(1-\varepsilon_i) + \ln n_i - \ln(1-\varepsilon_i n_i)] - \eta\left(v_{bo} - \sum_{i=1}^{N} c_i \ln n_i\right) \tag{13.85}$$

At the outset, we know the required burnout velocity v_{bo} and the payload mass m_{PL}, and for every stage we have the structural ratio ε_i and the exhaust velocity c_i (i.e., the specific impulse). The first step is to solve for the Lagrange parameter η using Eq. (13.84), which, for N stages is written

$$\sum_{i=1}^{N} c_i \ln\frac{c_i\eta - 1}{c_i\varepsilon_i\eta} = v_{bo}$$

Expanding the logarithm, this can be written

$$\sum_{i=1}^{N} c_i \ln(c_i\eta - 1) - \ln\eta\sum_{i=1}^{N} c_i - \sum_{i=1}^{N} c_i \ln c_i\varepsilon_i = v_{bo} \tag{13.86}$$

After solving this equation iteratively for η, we use that result to calculate the optimum mass ratio for each stage (cf. Eq. 13.83),

$$n_i = \frac{c_i\eta - 1}{c_i\varepsilon_i\eta} \quad i = 1,2,\ldots,N \tag{13.87}$$

Of course, each n_i must be greater than 1.

Referring to Eq. (13.75), we next obtain the step masses of each stage, beginning with stage N and working our way down the stack to stage 1,

$$\begin{aligned}
m_N &= \frac{n_N - 1}{1 - n_N\varepsilon_N} m_{PL} \\
m_{N-1} &= \frac{n_{N-1} - 1}{1 - n_{N-1}\varepsilon_{N-1}} (m_N + m_{PL}) \\
m_{N-2} &= \frac{n_{N-2} - 1}{1 - n_{N-2}\varepsilon_{N-2}} (m_{N-1} + m_N + m_{PL}) \\
&\vdots \\
m_1 &= \frac{n_1 - 1}{1 - n_1\varepsilon_1} (m_2 + m_3 + \cdots m_{PL})
\end{aligned} \tag{13.88}$$

Having found each step mass, each empty stage mass is

$$m_E)_i = \varepsilon_i m_i \tag{13.89}$$

and each stage propellant mass is

$$m_p)_i = m_i - m_E)_i \tag{13.90}$$

For the function h in Eq. (13.85), it is easily shown that

$$\frac{\partial^2 h}{\partial n_i \partial n_j} = 0 \quad i,j = 1,\ldots,N \ (i \neq j)$$

It follows that the second differential of h is

$$d^2h = \sum_{i=1}^{N}\sum_{j=1}^{N} \frac{\partial^2 h}{\partial n_i \partial n_j} dn_i dn_j = \sum_{i=1}^{N} \frac{\partial^2 h}{\partial n_i^2} (dn_i)^2 \tag{13.91}$$

where it can be shown, again using Eq. (13.85), that

$$\frac{\partial^2 h}{\partial n_i^2} = \frac{\eta c_i(\varepsilon_i n_i - 1)^2 + 2\varepsilon_i n_i - 1}{(\varepsilon_i n_i - 1)^2 n_i^2} \tag{13.92}$$

For h to be minimum at the mass ratios n_i given by Eq. (13.87), it must be true that $d^2h > 0$. Eqs. (13.91) and (13.92) indicate that this will be the case if

$$\eta c_i(\varepsilon_i n_i - 1)^2 + 2\varepsilon_i n_i - 1 > 0 \quad i = 1,\ldots,N \tag{13.93}$$

EXAMPLE 13.7

Find the optimal mass for a three-stage launch vehicle that is required to lift a 5000-kg payload to a speed of 10 km/s. For each stage, we are given that

Stage 1: $I_{sp})_1 = 400$ s ($c_1 = 3.924$ km/s) $\varepsilon_1 = 0.10$
Stage 2: $I_{sp})_2 = 350$ s ($c_2 = 3.434$ km/s) $\varepsilon_2 = 0.15$
Stage 3: $I_{sp})_3 = 300$ s ($c_3 = 2.943$ km/s) $\varepsilon_2 = 0.20$

Solution

Substituting these data into Eq. (13.86), we get

$$3.924\ \ln(3.924\eta - 1) + 3.434\ \ln(3.434\eta - 1) + 2.943\ \ln(2.943\eta - 1) - 10.30\ \ln\eta + 7.5089 = 10$$

As can be checked by substitution, the iterative solution of this equation is

$$\eta = 0.4668$$

Substituting η into Eq. (13.87) yields the optimum mass ratios,

$$n_1 = 4.541 \quad n_2 = 2.507 \quad n_3 = 1.361$$

For the step masses, we appeal to Eq. (13.88) to obtain

$$m_1 = 165{,}700\,\text{kg} \quad m_2 = 18{,}070\,\text{kg} \quad m_3 = 2477\,\text{kg}$$

The total mass of the vehicle is

$$m_0 = m_1 + m_2 + m_3 + m_{PL} = \boxed{191{,}200\ \text{kg}}$$

Using Eqs. (13.89) and (13.90), the empty masses and propellant masses are found to be

$$\begin{aligned} m_E)_1 &= 16{,}570\,\text{kg} \quad m_E)_2 = 2710\,\text{kg} \quad m_E)_3 = 495.4\,\text{kg} \\ m_p)_1 &= 149{,}100\,\text{kg} \quad m_p)_2 = 15{,}360\,\text{kg} \quad m_p)_3 = 1982\,\text{kg} \end{aligned}$$

The payload ratios for each stage are

$$\lambda_1 = \frac{m_2 + m_3 + m_{PL}}{m_1} = 0.1542$$
$$\lambda_2 = \frac{m_3 + m_{PL}}{m_2} = 0.4139$$
$$\lambda_3 = \frac{m_{PL}}{m_3} = 2.018$$

The overall payload fraction is

$$\pi_{PL} = \frac{m_{PL}}{m_0} = \frac{5000}{191{,}200} = 0.0262$$

Finally, let us check Eq. (13.93),

$$\eta c_1(\varepsilon_1 n_1 - 1)^2 + 2\varepsilon_1 n_1 - 1 = 0.4541$$
$$\eta c_2(\varepsilon_2 n_2 - 1)^2 + 2\varepsilon_2 n_2 - 1 = 0.3761$$
$$\eta c_3(\varepsilon_3 n_3 - 1)^2 + 2\varepsilon_3 n_3 - 1 = 0.2721$$

A positive number in every instance means that we have indeed found a local minimum of the function in Eq. (13.85).

PROBLEMS

Section 13.4

13.1 A two-stage, solid-propellant sounding rocket has the following properties:

First stage: $m_0 = 249.5$ kg $m_f = 170.1$ kg $\dot{m}_e = 10.61$ kg/s $I_{sp} = 235$ s
Second stage: $m_0 = 113.4$ kg $m_f = 58.97$ kg $\dot{m}_e = 4.053$ kg/s $I_{sp} = 235$ s

The delay time between burnout of first stage and ignition of second stage is 3 s. As a preliminary estimate, neglect drag and the variation of earth's gravity with altitude to calculate the maximum height reached by the second stage after burnout.

{Ans.: 322 km}

13.2 A two-stage launch vehicle has the following properties.

First stage: Two solid-propellant rockets, each with a total mass of 525,000 kg, 450,000 kg of which is propellant, and $I_{sp} = 290$ s.

Second stage: Two liquid rockets with $I_{sp} = 450$ s, dry mass = 30,000 kg, and propellant mass = 600,000 kg.

Calculate the payload mass to a 300-km orbit if launched due east from Kennedy Space Center. Let the total gravity and drag loss be 2 km/s.

{Ans.: 114,000 kg}

Section 13.5

13.3 Suppose a spacecraft in permanent orbit around the earth is to be used for delivering payloads from low earth orbit (LEO) to geostationary equatorial orbit (GEO). Before each flight from LEO, the spacecraft is refueled with propellant, which it uses up in its round trip to GEO. The outbound leg requires four times as much propellant as the inbound return leg. The delta-v for transfer from LEO to GEO is 4.22 km/s. The specific impulse of the propulsion system is 450 s. If the payload mass is 3500 kg, calculate the empty mass of the vehicle.

{Ans.: 2733 kg}

13.4 Consider a rocket comprising three similar stages (i.e., each stage has the same specific impulse, structural ratio, and payload ratio). The common specific impulse is 310 s. The total mass of the vehicle is 150,000 kg, the total structural mass (empty mass) is 20,000 kg, and the payload mass is 10,000 kg. Calculate

(a) the mass ratio n and the total Δv for the three-stage rocket.

{Ans.: $n = 2.04$, $\Delta v = 6.50$ km/s}

(b) $m_p)_1$, $m_p)_2$, and $m_p)_3$.

(c) $m_E)_1$, $m_E)_2$ and $m_E)_3$.

(d) $m_0)_1$, $m_0)_2$ and $m_0)_3$.

Section 13.6

13.5 Find the extrema of the function $z = (x + y)^2$ subject to y and z lying on the circle $(x - 1)^2 + y^2 = 1$.

{Ans.: $z = 0.1716$ at $(x,y) = (0.2929, -0.7071)$;
$z = 5.828$ at $(x,y) = (1.707, 0.7071)$; and
$z = 0$ at $(x,y) = (0,0)$ and $(x,y) = (1,-1)$}

13.6 A small two-stage vehicle is to propel a 10-kg payload to a speed of 6.2 km/s. The properties of the stages are as follows. For the first stage, $I_{sp} = 300$ s and $\varepsilon = 0.2$. For the second stage, $I_{sp} = 235$ s and $\varepsilon = 0.3$. Estimate the optimum mass of the vehicle.

{Ans.: 1125 kg}

REFERENCES

Hale, F.J., 1994. Introduction to Space Flight. Prentice-Hall.
Sutton, G.P., Biblarz, O., 2017. Rocket Propulsion Elements, 9th ed. Wiley, Hoboken, NJ.
Wiesel, W.E., 2010. Spaceflight Dynamics, 3rd ed. Aphelion Press, Beavercreek, OH.

APPENDIX

PHYSICAL DATA

Tables A.1–A.3 contain information that is commonly available and may be found in the literature and on the World Wide Web.

Table A.1 Astronomical data for the sun, the planets, and the moon

Object	Radius (km)	Mass (kg)	Sidereal rotation period	Inclination of equator to orbit plane	Semimajor axis of orbit (km)	Orbit eccentricity	Inclination of orbit to the ecliptic plane	Orbit sidereal period
Sun	696,000	1.989×10^{30}	25.38d	7.25°	–	–	–	–
Mercury	2440	330.2×10^{21}	58.65d	0.01°	57.91×10^6	0.2056	7.00°	87.97d
Venus	6052	4.869×10^{24}	243d[a]	177.4°	108.2×10^6	0.0067	3.39°	224.7d
Earth	6378	5.974×10^{24}	23.9345h	23.45°	149.6×10^6	0.0167	0.00°	365.256d
(Moon)	1737	73.48×10^{21}	27.32d	6.68°	384.4×10^3	0.0549	5.145°	27.322d
Mars	3396	641.9×10^{21}	24.62h	25.19°	227.9×10^6	0.0935	1.850°	1.881y
Jupiter	71,490	1.899×10^{27}	9.925h	3.13°	778.6×10^6	0.0489	1.304°	11.86y
Saturn	60,270	568.5×10^{24}	10.66h	26.73°	1.433×10^9	0.0565	2.485°	29.46y
Uranus	25,560	86.83×10^{24}	17.24h[a]	97.77°	2.872×10^9	0.0457	0.772°	84.01y
Neptune	24,764	102.4×10^{24}	16.11h	28.32°	4.495×10^9	0.0113	1.769	164.8y
(Pluto)	1187	13.03×10^{21}	6.387d[a]	122.5°	5.906×10^9	0.2488	17.16°	247.9y

[a]*Retrograde.*

Table A.2 Gravitational parameter (μ) and sphere of influence (SOI) radius for the sun, the planets, and the moon

Celestial body	μ (km^3/s^2)	SOI radius (km)
Sun	132,712,440,018	–
Mercury	22,032	112,000
Venus	324,859	616,000
Earth	398,600	925,000
Earth's moon	4905	66,100
Mars	42,828	577,000
Jupiter	126,686,534	48,200,000
Saturn	37,931,187	54,800,000
Uranus	5,793,939	51,800,000
Neptune	6,836,529	86,600,000
Pluto	871	3,080,000

Table A.3 Some conversion factors
1 ft = 0.3048 m
1 mile (mi) = 1.609 km
1 nautical mile (n mi) =1.151 mi = 1.852 km
1 mi/h = 0.0004469 km/s
1 lb (mass) = 0.4536 kg
1 lb (force) = 4.448 N
1 psi = 6895 kPa
1 astronomical unit (AU) = 149,597,870.700 km

A ROAD MAP

B

Fig. B.1 is a road map through Chapters 1, 2, and 3. Those who from time to time feel they have lost their bearings may find it useful to refer to this flow chart, which shows how the various concepts and results are interrelated. The pivotal influence of Sir Isaac Newton is obvious. All the equations of classical orbital mechanics (the two-body problem) are derived from those listed here.

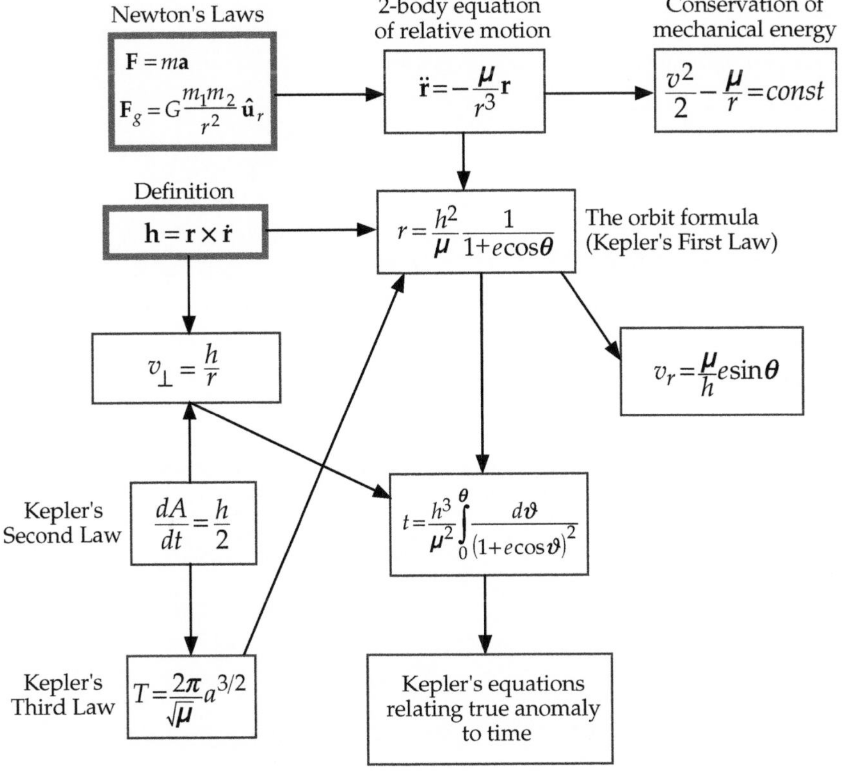

FIG. B.1

Logic flow for the major outcomes of Chapters 1, 2, and 3.

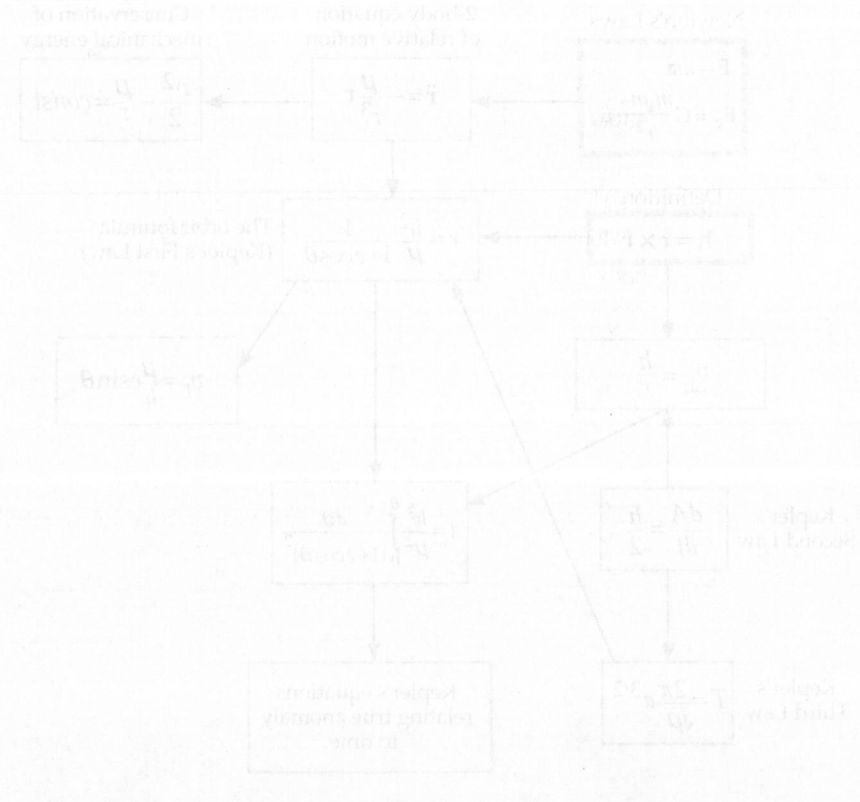

APPENDIX

NUMERICAL INTEGRATION OF THE *N*-BODY EQUATIONS OF MOTION

Without loss of generality we shall derive the equations of motion of the three-body system illustrated in Fig. C.1. The equations of motion for n bodies can easily be generalized from those of a three-body system.

Each mass of a three-body system experiences the force of gravitational attraction from the other members of the system. As shown in Fig. C.1, the forces exerted on body 1 by bodies 2 and 3 are $\mathbf{F}_{12}$ and $\mathbf{F}_{13}$, respectively. Likewise, body 2 experiences the forces $\mathbf{F}_{21}$ and $\mathbf{F}_{23}$ whereas the forces $\mathbf{F}_{31}$ and $\mathbf{F}_{32}$ act on body 3. These gravitational forces can be inferred from Eq. (2.9):

$$\mathbf{F}_{12} = -\mathbf{F}_{21} = \frac{Gm_1m_2(\mathbf{R}_2 - \mathbf{R}_1)}{\|\mathbf{R}_2 - \mathbf{R}_1\|^3} \tag{C.1a}$$

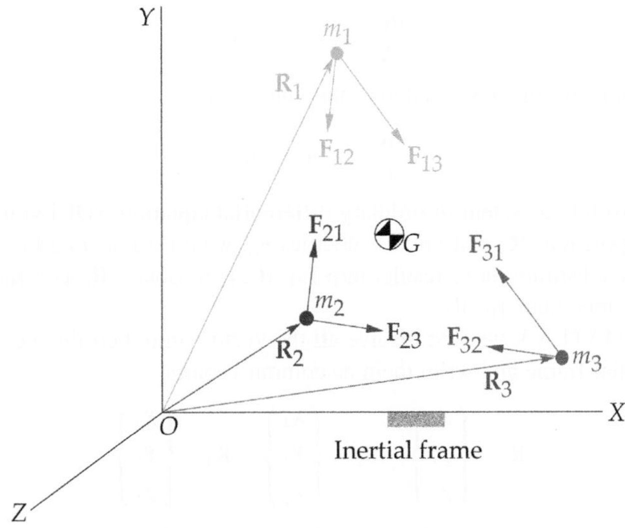

FIG. C.1

Three-body problem.

$$\mathbf{F}_{13} = -\mathbf{F}_{31} = \frac{Gm_1m_3(\mathbf{R}_3 - \mathbf{R}_1)}{\|\mathbf{R}_3 - \mathbf{R}_1\|^3} \tag{C.1b}$$

$$\mathbf{F}_{23} = -\mathbf{F}_{32} = \frac{Gm_2m_3(\mathbf{R}_3 - \mathbf{R}_2)}{\|\mathbf{R}_3 - \mathbf{R}_2\|^3} \tag{C.1c}$$

Relative to an inertial frame of reference the accelerations of the bodies are

$$\mathbf{a}_i = \ddot{\mathbf{R}}_i \quad i = 1,2,3$$

where $\mathbf{R}_i$ is the absolute position vector of body i. The equation of motion of body 1 is

$$\mathbf{F}_{12} + \mathbf{F}_{13} = m_1\mathbf{a}_1$$

Substituting Eqs. (C.1a) and (C.1b) yields

$$\mathbf{a}_1 = \frac{Gm_2(\mathbf{R}_2 - \mathbf{R}_1)}{\|\mathbf{R}_2 - \mathbf{R}_1\|^3} + \frac{Gm_3(\mathbf{R}_3 - \mathbf{R}_1)}{\|\mathbf{R}_3 - \mathbf{R}_1\|^3} \tag{C.2a}$$

For bodies 2 and 3 we find in a similar fashion that

$$\mathbf{a}_2 = \frac{Gm_1(\mathbf{R}_1 - \mathbf{R}_2)}{\|\mathbf{R}_1 - \mathbf{R}_2\|^3} + \frac{Gm_3(\mathbf{R}_3 - \mathbf{R}_2)}{\|\mathbf{R}_3 - \mathbf{R}_2\|^3} \tag{C.2b}$$

$$\mathbf{a}_3 = \frac{Gm_1(\mathbf{R}_1 - \mathbf{R}_3)}{\|\mathbf{R}_1 - \mathbf{R}_3\|^3} + \frac{Gm_2(\mathbf{R}_2 - \mathbf{R}_3)}{\|\mathbf{R}_2 - \mathbf{R}_3\|^3} \tag{C.2c}$$

The velocities are related to the accelerations by

$$\frac{d\mathbf{v}_i}{dt} = \mathbf{a}_i \quad i = 1,2,3 \tag{C.3}$$

and the position vectors are likewise related to the velocities,

$$\frac{d\mathbf{R}_i}{dt} = \mathbf{v}_i \quad i = 1,2,3 \tag{C.4}$$

Eqs. (C.2)–(C.4) constitute a system of ordinary differential equations (ODEs) in variable time.

Given the initial positions $\mathbf{R}_{i_0}$ and initial velocities $\mathbf{v}_{i_0}$, we must integrate Eq. (C.3) to find $\mathbf{v}_i$ as a function of time and substitute those results into Eq. (C.4) to obtain $\mathbf{R}_i$ as a function of time. The integrations must be done numerically.

To do this using MATLAB, we first resolve all the vectors into their three components along the *XYZ* axes of the inertial frame and write them as column vectors,

$$\mathbf{R}_1 = \begin{Bmatrix} X_1 \\ Y_1 \\ Z_1 \end{Bmatrix} \quad \mathbf{R}_2 = \begin{Bmatrix} X_2 \\ Y_2 \\ Z_2 \end{Bmatrix} \quad \mathbf{R}_3 = \begin{Bmatrix} X_3 \\ Y_3 \\ Z_3 \end{Bmatrix} \tag{C.5}$$

$$\mathbf{v}_1 = \begin{Bmatrix} \dot{X}_1 \\ \dot{Y}_1 \\ \dot{Z}_1 \end{Bmatrix} \quad \mathbf{v}_2 = \begin{Bmatrix} \dot{X}_2 \\ \dot{Y}_2 \\ \dot{Z}_2 \end{Bmatrix} \quad \mathbf{v}_3 = \begin{Bmatrix} \dot{X}_3 \\ \dot{Y}_3 \\ \dot{Z}_3 \end{Bmatrix} \tag{C.6}$$

According to Eqs. (C.2),

$$\mathbf{a}_1 = \begin{Bmatrix} \ddot{X}_1 \\ \ddot{Y}_1 \\ \ddot{Z}_1 \end{Bmatrix} = \begin{Bmatrix} \dfrac{Gm_2(X_2 - X_1)}{R_{12}{}^3} + \dfrac{Gm_3(X_3 - X_1)}{R_{13}{}^3} \\ \dfrac{Gm_2(Y_2 - Y_1)}{R_{12}{}^3} + \dfrac{Gm_3(Y_3 - Y_1)}{R_{13}{}^3} \\ \dfrac{Gm_2(Z_2 - Z_1)}{R_{12}{}^3} + \dfrac{Gm_3(Z_3 - Z_1)}{R_{13}{}^3} \end{Bmatrix} \tag{C.7a}$$

$$\mathbf{a}_2 = \begin{Bmatrix} \ddot{X}_2 \\ \ddot{Y}_2 \\ \ddot{Z}_2 \end{Bmatrix} = \begin{Bmatrix} \dfrac{Gm_1(X_1 - X_2)}{R_{12}{}^3} + \dfrac{Gm_3(X_3 - X_2)}{R_{13}{}^3} \\ \dfrac{Gm_1(Y_1 - Y_2)}{R_{12}{}^3} + \dfrac{Gm_3(Y_3 - Y_2)}{R_{13}{}^3} \\ \dfrac{Gm_1(Z_1 - Z_2)}{R_{12}{}^3} + \dfrac{Gm_3(Z_3 - Z_2)}{R_{13}{}^3} \end{Bmatrix} \tag{C.7b}$$

$$\mathbf{a}_3 = \begin{Bmatrix} \ddot{X}_3 \\ \ddot{Y}_3 \\ \ddot{Z}_3 \end{Bmatrix} = \begin{Bmatrix} \dfrac{Gm_1(X_1 - X_3)}{R_{12}{}^3} + \dfrac{Gm_2(X_2 - X_3)}{R_{13}{}^3} \\ \dfrac{Gm_1(Y_1 - Y_3)}{R_{12}{}^3} + \dfrac{Gm_2(Y_2 - Y_3)}{R_{13}{}^3} \\ \dfrac{Gm_1(Z_1 - Z_3)}{R_{12}{}^3} + \dfrac{Gm_2(Z_2 - Z_3)}{R_{13}{}^3} \end{Bmatrix} \tag{C.7c}$$

where

$$R_{12} = \|\mathbf{R}_2 - \mathbf{R}_1\| \quad R_{13} = \|\mathbf{R}_3 - \mathbf{R}_1\| \quad R_{23} = \|\mathbf{R}_3 - \mathbf{R}_2\| \tag{C.8}$$

Next, we form the 18-component column vector

$$\mathbf{y} = \lfloor \mathbf{R}_1 \;\; \mathbf{R}_2 \;\; \mathbf{R}_3 \;\; \mathbf{v}_1 \;\; \mathbf{v}_2 \;\; \mathbf{v}_3 \rfloor^T \tag{C.9}$$

The first derivatives of the components of this vector comprise the column vector

$$\dot{\mathbf{y}} = \mathbf{f} = \lfloor \mathbf{v}_1 \;\; \mathbf{v}_2 \;\; \mathbf{v}_3 \;\; \mathbf{a}_1 \;\; \mathbf{a}_2 \;\; \mathbf{a}_3 \rfloor^T \tag{C.10}$$

According to Eqs. (C.8), the accelerations are functions of $\mathbf{R}_1$, $\mathbf{R}_2$, and $\mathbf{R}_3$. Hence, Eq. (C.10) is of the form

$$\dot{\mathbf{y}} = \mathbf{f}(t, \mathbf{y}) \tag{C.11}$$

given in Eq. (1.95), although in this case time t does not appear explicitly. Eq. (C.11) can be employed in procedures, such as Algorithms 1.1, 1.2, or 1.3, to obtain a numerical solution for $\mathbf{R}_1(t)$, $\mathbf{R}_2(t)$, and $\mathbf{R}_3(t)$. We shall choose MATLAB's *ode45* Runge-Kutta solver.

For simplicity, we will solve the three-body problem in the plane. That is, we will restrict ourselves to only the *XY* components of the vectors **R**, **v**, and **a**. The reader can use these scripts as a starting point for investigating more complex *n*-body problems.

The MATLAB function *threebody.m* contains the subfunction *rates*, which computes the accelerations given above in Eqs. (C.7). That information together with the initial conditions are passed to *ode45*, which integrates the system given by Eq. (C.11) and finally plots the solutions. The results of this program were used to create Figs. 2.4 and 2.5.

Function file `threebody.m`

```
% ~~~~~~~~~~~~~~~~~~~~~~~~~~~~~~~~~~~~~~~~~~~~~~~~~~~~~~~~~~~~~~~~~~~~~~~~~
function threebody
% ~~~~~~~~~~~~~~~~~~
%{
  This program presents the graphical solution of the motion of three
  bodies in the plane for data provided in the input definitions below.

  MATLAB's ode45 Runge-Kutta solver is used.

  G                         - gravitational constant (km^3/kg/s^2)
  t0, tf                    - initial and final times (s)
  m1, m2, m3                - masses of the three bodies (kg)
  m                         - total mass (kg)
  X1,Y1; X2,Y2; X3,Y3       - coordinates of the three masses (km)
  VX1,VY1; VX2,VY2; VX3,VY3 - velocity components of the three
                              masses (km/s)
  XG, YG                    - coordinates of the center of mass (km)
  y0                        - column vector of the initial conditions
  t                         - column vector of times at which the solution
                              was computed
  y                         - matrix, the columns of which contain the
                              position and velocity components evaluated at
                              the times t(:):
                                y(:,1) , y(:, 2) =  X1(:), Y1(:)
                                y(:,3) , y(:, 4) =  X2(:), Y2(:)
                                y(:,5) , y(:, 6) =  X3(:), Y3(:)

                                y(:,7) , y(:, 8) = VX1(:), VY1(:)
                                y(:,9) , y(:,10) = VX2(:), VY2(:)
                                y(:,11), y(:,12) = VX3(:), VY3(:)

  User M-functions required:  none
  User subfunctions required: rates, plotit
%}
% ------------------------------------------------------------------------

clear all
close all
clc

G = 6.67259e-20;

%...Input data:
m1 = 1.e29; m2 = 1.e29; m3 = 1.e29;
```

```
t0 = 0; tf = 67000;

X1 = 0;      Y1 = 0;
X2 = 300000; Y2 = 0;
X3 = 2*X2;   Y3 = 0;

VX1 = 0;    VY1 = 0;
VX2 = 250;  VY2 = 250;
VX3 = 0;    VY3 = 0;
%...End input data

m  = m1 + m2 + m3;
y0 = [X1 Y1  X2 Y2  X3 Y3  VX1 VY1  VX2 VY2  VX3 VY3]';

%...Pass the initial conditions and time interval to ode45, which
%   calculates the position and velocity of each particle at discrete
%   times t, returning the solution in the column vector y. ode45 uses
%   the subfunction 'rates' below to evaluate the accelerations at each
%   integration time step.
[t,y] = ode45(@rates, [t0 tf], y0);

X1 = y(:,1); Y1 = y(:,2);
X2 = y(:,3); Y2 = y(:,4);
X3 = y(:,5); Y3 = y(:,6);

%...Locate the center of mass at each time step:
XG = []; YG = [];
for i = 1:length(t)
    XG = [XG; (m1*X1(i) + m2*X2(i) + m3*X3(i))/m];
    YG = [YG; (m1*Y1(i) + m2*Y2(i) + m3*Y3(i))/m];
end

%...Coordinates of each particle relative to the center of mass:
X1G = X1 - XG; Y1G = Y1 - YG;
X2G = X2 - XG; Y2G = Y2 - YG;
X3G = X3 - XG; Y3G = Y3 - YG;

plotit

return

% ~~~~~~~~~~~~~~~~~~~~~~~~~~~~~~~~~~~~~~~~~~~~~~~~~~~~~~~~~~~~~~~~~~~~~~~~~~
  function dydt = rates(t,y)
% ~~~~~~~~~~~~~~~~~~~~~~~~~~~
%{
  This function evaluates the acceleration of each member of a planar
  3-body system at time t from their positions and velocities
  at that time.
```

```
  t                             - time (s)
  y                             - column vector containing the position and
                                  velocity components of the three masses
                                  at time t
  R12                           - cube of the distance between m1 and m2 (km^3)
  R13                           - cube of the distance between m1 and m3 (km^3)
  R23                           - cube of the distance between m2 and m3 (km^3)
  AX1,AY1; AX2,AY2; AX3,AY3 - acceleration components of each mass (km/s^2)
  dydt                          - column vector containing the velocity and
                                  acceleration components of the three
                                  masses at time t

%}
% -----------------------------------------------------------------------

X1  = y( 1);
Y1  = y( 2);

X2  = y( 3);
Y2  = y( 4);

X3  = y( 5);
Y3  = y( 6);

VX1 = y( 7);
VY1 = y( 8);

VX2 = y( 9);
VY2 = y(10);

VX3 = y(11);
VY3 = y(12);

%...Equations C.8:
R12 = norm([X2 - X1, Y2 - Y1])^3;
R13 = norm([X3 - X1, Y3 - Y1])^3;
R23 = norm([X3 - X2, Y3 - Y2])^3;

%...Equations C.7:
AX1 = G*m2*(X2 - X1)/R12 + G*m3*(X3 - X1)/R13;
AY1 = G*m2*(Y2 - Y1)/R12 + G*m3*(Y3 - Y1)/R13;
AX2 = G*m1*(X1 - X2)/R12 + G*m3*(X3 - X2)/R23;
AY2 = G*m1*(Y1 - Y2)/R12 + G*m3*(Y3 - Y2)/R23;
AX3 = G*m1*(X1 - X3)/R13 + G*m2*(X2 - X3)/R23;
AY3 = G*m1*(Y1 - Y3)/R13 + G*m2*(Y2 - Y3)/R23;

dydt = [VX1 VY1 VX2 VY2 VX3 VY3 AX1 AY1 AX2 AY2 AX3 AY3]';
```

```
end %rates
% ~~~~~~~~~~~~~~~~~~~~~~~~~~~~~~~~~~~~~~~~~~~~~~~~~~~~~~~~~~~~~~~~~~~~~~~~~~~~

% ~~~~~~~~~~~~~~~
function plotit
% -------------

%...Plot the motions relative to the inertial frame (Figure 2.4):
figure(1)
title('Figure 2.4: Motion relative to the inertial frame', ...
      'Fontweight', 'bold', 'FontSize', 12)
hold on
plot(XG, YG, '--k', 'LineWidth', 0.25)
plot(X1, Y1,   'r', 'LineWidth', 0.5)
plot(X2, Y2,   'g', 'LineWidth', 0.75)
plot(X3, Y3,   'b', 'LineWidth', 1.00)
xlabel('X(km)'); ylabel('Y(km)')
grid on
axis('equal')

%...Plot the motions relative to the center of mass (Figure 2.5):
figure(2)
title('Figure 2.5: Motion relative to the center of mass', ...
      'Fontweight', 'bold', 'FontSize', 12)
hold on
plot(X1G, Y1G,   'r', 'LineWidth', 0.5)
plot(X2G, Y2G, '--g', 'LineWidth', 0.75)
plot(X3G, Y3G,   'b', 'LineWidth', 1.00)
xlabel('X(km)'); ylabel('Y(km)')
grid on
axis('equal')
end %plotit
% ~~~~~~~~~~~~~~~

end %threebody
```

APPENDIX

MATLAB SCRIPTS

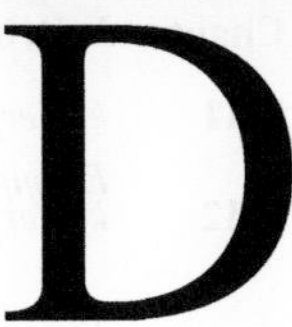

APPENDIX OUTLINE

Continued

Chapter 3: Orbital Position as a Function of Time		
D.11	*kepler_E.m*	Algorithm 3.1: Solution of Kepler's equation by Newton's method.
	Example_3_02.m	Example of Algorithm 3.1.
D.12	*kepler_H.m*	Algorithm 3.2: Solution of Kepler's equation for the hyperbola using Newton's method.
	Example_3_05.m	Example of Algorithm 3.2.
D.13	*stumpS.m*	Calculation of the Stumpff function $S(z)$ and $C(z)$
	stumpC.m	Calculation of the Stumpff function $C(z)$.
D.14	*kepler_U.m*	Algorithm 3.3: Solution of the universal Kepler's equation using Newton's method.
	Example_3_06.m	Example of Algorithm 3.3.
D.15	*f_and_g.m*	Calculation of the Lagrange coefficients f and g and their time derivatives in terms of change in universal anomaly.
D.16	*rv_from_r0v0.m*	Algorithm 3.4: Calculation of the state vector given the initial state vector and the time lapse Δt.
	Example_3_07.m	Example of Algorithm 3.4.

Chapter 4: Orbits in Three Dimensions		
D.17	*ra_and_dec_from_r.m*	Algorithm 4.1: Obtain right ascension and declination from the position vector.
	Example_4_01.m	Example of Algorithm 4.1.
D.18	*coe_from_sv.m*	Algorithm 4.2: Calculation of the orbital elements from the state vector.
	Example_4_03.m	Example of Algorithm 4.2.
D.19	*atan2d_0_360.m*	Calculation of $\tan^{-1}(y/x)$ to lie in the range 0° to 360°. (MATLAB's *atan2d* result lies in the range 0° to ±180°.)
D.20	*dcm_to_euler.m*	Algorithm 4.3: Obtain the classical Euler angle sequence from a DCM.
D.21	*dcm_to_ypr.m*	Algorithm 4.4: Obtain the yaw, pitch, and roll angles from a DCM.
D.22	*sv_from_coe.m*	Algorithm 4.5: Calculation of the state vector from the orbital elements.
	Example_4_07.m	Example of Algorithm 4.5
D.23	*ground_track.m*	Algorithm 4.6: Calculate the ground track of a satellite from its orbital elements. Contains the data for Example 4.12.

Chapter 5: Preliminary Orbit Determination		
D.24	*gibbs.m*	Algorithm 5.1: Gibbs' method of preliminary orbit determination.
	Example_5_01.m	Example of Algorithm 5.1.
D.25	*lambert.m*	Algorithm 5.2: Solution of Lambert's problem.
	Example_5_02.m	Example of Algorithm 5.2.
D.26	*J0.m*	Calculation of Julian day number at 0 hr UT.
	Example_5_04.m	Example of Julian day calculation.
D.27	*LST.m*	Algorithm 5.3: Calculation of local sidereal time.
	Example_5_06.m	Example of Algorithm 5.3.
D.28	*rv_from_observe.m*	Algorithm 5.4: Calculation of the state vector from measurements of range, angular position, and their rates.
	Example_5_10.m	Example of Algorithm 5.4.
D.29	*gauss.m*	Algorithms 5.5 and 5.6: Gauss' method of preliminary orbit determination with iterative *improvement*.
	Example_5_11.m	Example of Algorithms 5.5 and 5.6.

Chapter 6: Orbital Maneuvers		
D.30	*integrate_thrust.m*	Calculate the state vector at the end of a finite time, constant thrust delta-v maneuver. Contains the data for Example 6.15.
Chapter 7: Relative Motion and Rendezvous		
D.31	*rva_relative.m*	Algorithm 7.1: Find the position, velocity, and acceleration of *B* relative to *A*'s comoving frame.
	Example_7_01.m	Example of Algorithm 7.1.
D.32	*Example_7_02.m*	Plot the position of one spacecraft relative to another.
D.33	*Example_7_03.m*	Solve the linearized equations of relative motion with an elliptical reference orbit.
Chapter 8: Interplanetary Trajectories		
D.34	*month_planet_names.m*	Convert the numerical designation of a month or a planet into its name.
D.35	*planet_elements_and_sv.m*	Algorithm 8.1: Calculation of the heliocentric state vector of a planet at a given epoch.
	Example_8_07.m	Example of Algorithm 8.1.
D.36	*interplanetary.m*	Algorithm 8.2: Calculate the spacecraft trajectory from planet 1 to planet 2.
	Example_8_08.m	Example of Algorithm 8.2.
Chapter 9: Lunar Trajectories		
D.37	*simpsons_lunar_ephemeris.m*	Lunar state vector vs. time.
D.38	*Example_9_03.m*	Numerical calculation of lunar trajectory.
Chapter 10: Introduction to Orbital Perturbations		
D.39	*atmosphere.m*	US Standard Atmosphere 1976.
D.40	*Example_10_01.m*	Time for orbit decay using Cowell's method.
D.41	*Example_10_02.m*	J_2 perturbation of an orbit using Encke's method.
D.42	*Example_10_06.m*	Using Gauss' variational equations to assess the J_2 effect on orbital elements.
D.43	*solar_position.m*	Algorithm 10.2: Calculate the geocentric position of the sun at a given epoch.
D.44	*los.m*	Algorithm 10.3: Determine whether or not a satellite is in earth's shadow.
D.45	*Example_10_09.m*	Use the Gauss' variational equations to determine the effect of solar radiation pressure on an earth satellite's orbital parameters.
D.46	*lunar_position.m*	Algorithm 10.4: Calculate the geocentric position of the moon at a given epoch.
	Example_10_10.m	Example of Algorithm 10.4.
D.47	*Example_10_11.m*	Use the Gauss' variational equations to determine the effect of lunar gravity on an earth satellite's orbital parameters.
D.48	*Example_10_12.m*	Use the Gauss' variational equations to determine the effect of solar gravity on an earth satellite's orbital parameters.

Continued

Chapter 11: Rigid Body Dynamics		
D.49	*dcm_from_q.m*	Algorithm 11.1: Calculate the direction cosine matrix from the quaternion.
D.50	*q_from_dcm.m*	Algorithm 11.2: Calculate the quaternion from the direction cosine matrix.
D.51	*quat_rotate.m*	Quaternion vector rotation operation (Eq. 11.160).
D.52	*Example_11_26.m*	Solution of the spinning top problem.
Chapter 12: Spacecraft Attitude Dynamics		
Chapter 13: Rocket Vehicle Dynamics		
D.53	*Example_13_03.m*	Example 11.3: Calculation of a gravity turn trajectory.

D.1 INTRODUCTION

This appendix lists MATLAB scripts that implement all the numbered algorithms presented throughout the text. The programs use only the most basic features of MATLAB and are liberally commented so as to make reading the code as easy as possible. To "drive" the various algorithms, we can use MATLAB to create graphical user interfaces (GUIs). However, in the interest of simplicity and keeping our focus on the algorithms rather than elegant programming techniques, GUIs were not developed. Furthermore, the scripts do not use files to import and export data. Data are defined in declaration statements within the scripts. All output is to the screen (i.e., to the MATLAB Command Window). It is hoped that interested students will embellish these simple scripts or use them as a springboard toward generating their own programs.

Each algorithm is illustrated by a MATLAB coding of a related example problem in the text. The actual output of each of these examples is also listed. These programs are presented solely as an alternative to carrying out otherwise lengthy hand computations and are intended for academic use only. They are all based exclusively on the introductory material presented in this text. Should it be necessary to do so, it is a fairly simple matter to translate these programs into other software languages.

It would be helpful to have MATLAB documentation at hand. There are many practical references on the subject in bookstores and online, including those at The MathWorks website (www.mathworks.com).

CHAPTER 1: DYNAMICS OF POINT MASSES

D.2 ALGORITHM 1.1: NUMERICAL INTEGRATION BY RUNGE-KUTTA METHODS RK1, RK2, RK3, OR RK4

FUNCTION FILE rkf1_4.m

```
% ~~~~~~~~~~~~~~~~~~~~~~~~~~~~~~~~~~~~~~~~~~~~~~~~~~~~~~~~~~~~~~~~~~~~~~
function [tout, yout] = rk1_4(ode_function, tspan, y0, h, rk)
% ~~~~~~~~~~~~~~~~~~~~~~~~~~~~~~~~~~~~~~~~~~~~~~~~~~~~~~~~~~~~~~~~~~~~~~
```

```
%{
 This function uses a selected Runge-Kutta procedure to integrate
 a system of first-order differential equations dy/dt = f(t,y).

 y            - column vector of solutions
 f            - column vector of the derivatives dy/dt
 t            - time
 rk           - = 1 for RK1; = 2 for RK2; = 3 for RK3; = 4 for RK4
 n_stages     - the number of points within a time interval that
                the derivatives are to be computed
 a            - coefficients for locating the solution points within
                each time interval
 b            - coefficients for computing the derivatives at each
                interior point
 c            - coefficients for the computing solution at the end of
                the time step
 ode_function - handle for user M-function in which the derivatives f
                are computed
 tspan        - the vector [t0 tf] giving the time interval for the
                solution
 t0           - initial time
 tf           - final time
 y0           - column vector of initial values of the vector y
 tout         - column vector of times at which y was evaluated
 yout         - a matrix, each row of which contains the components of y
                evaluated at the correponding time in tout
 h            - time step
 ti           - time at the beginning of a time step
 yi           - values of y at the beginning of a time step
 t_inner      - time within a given time step
 y_inner      - values of y within a given time step

 User M-function required: ode_function
%}
% ----------------------------------------------------------------------------

%...Determine which of the four Runge-Kutta methods is to be used:
switch rk
    case 1
        n_stages = 1;
        a = 0;
        b = 0;
        c = 1;
    case 2
        n_stages = 2;
        a = [0 1];
        b = [0 1]';
        c = [1/2 1/2];
```

```
  case 3
         n_stages = 3;
         a = [0 1/2 1];
         b = [ 0  0
              1/2 0
              -1  2];
         c = [1/6 2/3 1/6];
  case 4
         n_stages = 4;
         a = [0 1/2 1/2 1];
         b = [ 0   0   0
              1/2  0   0
               0  1/2  0
               0   0   1];
         c = [1/6 1/3 1/3 1/6];
  otherwise
          error('The parameter  rk  must have the value 1, 2, 3 or 4.')
end

t0   = tspan(1);
tf   = tspan(2);
t    = t0;
y    = y0;
tout = t;
yout = y';

while t < tf
    ti = t;
    yi = y;
    %...Evaluate the time derivative(s) at the 'n_stages' points within the
    %   current interval:
    for i = 1:n_stages
        t_inner = ti + a(i)*h;
        y_inner = yi;
        for j = 1:i-1
            y_inner = y_inner + h*b(i,j)*f(:,j);
        end
        f(:,i)  = feval(ode_function, t_inner, y_inner);
    end

    h    = min(h, tf-t);
    t    = t + h;
    y    = yi + h*f*c';
    tout = [tout;t];  % adds t to the bottom of the column vector tout
    yout = [yout;y']; % adds y' to the bottom of the matrix yout
end

end
% ~~~~~~~~~~~~~~~~~~~~~~~~~~~~~~~~~~~~~~~~~~~~~~~~~~~~~~~~~~~~~~~~~~~~
```

FUNCTION FILE: Example_1_18.m

```
% ~~~~~~~~~~~~~~~~~~~~~~~~~~~~~~~~~~~~~~~~~~~~~~~~~~~~~~~~~~~~~~~~~~~~~~~
function Example_1_18
% ~~~~~~~~~~~~~~~~~
%{
  This function uses the RK1 through RK4 methods with two
  different time steps each to solve for and plot the response
  of a damped single degree of freedom spring-mass system to
  a sinusoidal forcing function, represented by

  x'' + 2*z*wn*x' + wn^2*x = (Fo/m)*sin(w*t)

  The numerical integration is done by the external
  function 'rk1_4', which uses the subfunction 'rates'
  herein to compute the derivatives.

  This function also plots the exact solution for comparison.

  x            - displacement (m)
  '            - shorthand for d/dt
  t            - time (s)
  wn           - natural circular frequency (radians/s)
  z            - damping factor
  wd           - damped natural frequency
  Fo           - amplitude of the sinusoidal forcing function (N)
  m            - mass (kg)
  w            - forcing frequency (radians/s)
  t0           - initial time (s)
  tf           - final time (s)
  h            - uniform time step (s)
  tspan        - a row vector containing t0 and tf
  x0           - value of x at t0 (m)
  x_dot0       - value of dx/dt at t0 (m/s)
  f0           - column vector containing x0 and x_dot0
  rk           - = 1 for RK1; = 2 for RK2; = 3 for RK3; = 4 for RK4
  t            - solution times for the exact solution
  t1, ...,t4   - solution times for RK1,...,RK4 for smaller
  t11,...,t41  - solution times for RK1,...,RK4 for larger h
  f1, ...,f4   - solution vectors for RK1,...,RK4 for smaller h
  f11,...,f41  - solution vectors for RK1,...,RK4 for larger h

  User M-functions required:  rk1_4
  User subfunctions required: rates
%}
% -----------------------------------------------------------------------
```

```
clear all; close all; clc

%...Input data:
m      = 1;
z      = 0.03;
wn     = 1;
Fo     = 1;
w      = 0.4*wn;

x0     = 0;
x_dot0 = 0;
f0     = [x0; x_dot0];

t0     = 0;
tf     = 110;
tspan  = [t0 tf];
%...End input data

%...Solve using RK1 through RK4, using the same and a larger
%   time step for each method:
rk = 1;
h  = .01; [t1,  f1]  = rk1_4(@rates, tspan, f0, h, rk);
h  = 0.1; [t11, f11] = rk1_4(@rates, tspan, f0, h, rk);

rk = 2;
h  = 0.1; [t2,  f2]  = rk1_4(@rates, tspan, f0, h, rk);
h  = 0.5; [t21, f21] = rk1_4(@rates, tspan, f0, h, rk);

rk = 3;
h  = 0.5; [t3,  f3]  = rk1_4(@rates, tspan, f0, h, rk);
h  = 1.0; [t31, f31] = rk1_4(@rates, tspan, f0, h, rk);

rk = 4;
h  = 1.0; [t4,  f4]  = rk1_4(@rates, tspan, f0, h, rk);
h  = 2.0; [t41, f41] = rk1_4(@rates, tspan, f0, h, rk);

output

return

% ~~~~~~~~~~~~~~~~~~~~~~~~~
function dfdt = rates(t,f)
% ------------------------------
%{
  This function calculates first and second time derivatives
  of x as governed by the equation

  x'' + 2*z*wn*x' + wn^2*x = (Fo/m)*sin(w*t)
```

```
  Dx   - velocity (x')
  D2x  - acceleration (x'')
  f    - column vector containing x  and Dx  at time t
  dfdt - column vector containing Dx and D2x at time t

  User M-functions required: none
%}
% ~~~~~~~~~~~~~~~~~~~~~~~~~

x    = f(1);
Dx   = f(2);
D2x  = Fo/m*sin(w*t) - 2*z*wn*Dx - wn^2*x;
dfdt = [Dx; D2x];
end %rates

% ~~~~~~~~~~
function output
% -------------
%...Exact solution:
wd  = wn*sqrt(1 - z^2);
den = (wn^2 - w^2)^2 + (2*w*wn*z)^2;
C1  = (wn^2 - w^2)/den*Fo/m;
C2  = -2*w*wn*z/den*Fo/m;
A   = x0*wn/wd + x_dot0/wd +(w^2 + (2*z^2 - 1)*wn^2)/den*w/wd*Fo/m;
B   = x0 + 2*w*wn*z/den*Fo/m;

t   = linspace(t0, tf, 5000);
x   = (A*sin(wd*t) + B*cos(wd*t)).*exp(-wn*z*t) ...
      + C1*sin(w*t) + C2*cos(w*t);

%...Plot solutions
%   Exact:
subplot(5,1,1)
plot(t/max(t),     x/max(x),                'k',  'LineWidth',1)
grid off
axis tight
title('Exact')

%   RK1:
subplot(5,1,2)
plot(t1/max(t1),   f1(:,1)/max(f1(:,1)),   '-r', 'LineWidth',1)
hold on
plot(t11/max(t11), f11(:,1)/max(f11(:,1)), '-k')
grid off
axis tight
title('RK1')
legend('h = 0.01', 'h = 0.1')
```

```
%   RK2:
subplot(5,1,3)
plot(t2/max(t2),   f2(:,1)/max(f2(:,1)),    '-r', 'LineWidth',1)
hold on
plot(t21/max(t21), f21(:,1)/max(f21(:,1)), '-k')
grid off
axis tight
title('RK2')
legend('h = 0.1', 'h = 0.5')

%   RK3:
subplot(5,1,4)
plot(t3/max(t3),   f3(:,1)/max(f3(:,1)),    '-r', 'LineWidth',1)
hold on
plot(t31/max(t31), f31(:,1)/max(f31(:,1)), '-k')
grid off
axis tight
title('RK3')
legend('h = 0.5', 'h = 1.0')

%   RK4:
subplot(5,1,5)
plot(t4/max(t4),   f4(:,1)/max(f4(:,1)),    '-r', 'LineWidth',1)
hold on
grid off
plot(t41/max(t41), f41(:,1)/max(f41(:,1)), '-k')
axis tight
title('RK4')
legend('h = 1.0', 'h = 2.0')
end %output

end %Example_1_18
% ~~~~~~~~~~~~~~~~~~~~~~~~~~~~~~~~~~~~~~~~~~~~~~~~~~~~~~~~~~~~~~~~~~~~~~~~~~~~~~
```

D.3 ALGORITHM 1.2: NUMERICAL INTEGRATION BY HEUN'S PREDICTOR-CORRECTOR METHOD

FUNCTION FILE: heun.m

```
% ~~~~~~~~~~~~~~~~~~~~~~~~~~~~~~~~~~~~~~~~~~~~~~~~~~~~~~~~~~~~~~~~~~~~~~~~~~~~~~
function [tout, yout] = heun(ode_function, tspan, y0, h)
% ~~~~~~~~~~~~~~~~~~~~~~~~~~~~~~~~~~~~~~~~~~~~~~~~~~~~~~
%{
  This function uses the predictor-corrector method to integrate a system
  of first-order differential equations dy/dt = f(t,y).

  y             - column vector of solutions
  f             - column vector of the derivatives dy/dt
```

```
  ode_function  - handle for the user M-function in which the derivatives
                  f are computed
  t             - time
  t0            - initial time
  tf            - final time
  tspan         - the vector [t0 tf] giving the time interval for the
                  solution
  h             - time step
  y0            - column vector of initial values of the vector y
  tout          - column vector of the times at which y was evaluated
  yout          - a matrix, each row of which contains the components of y
                  evaluated at the correponding time in tout
  feval         - a built-in MATLAB function which executes 'ode_function'
                  at the arguments t and y
  tol           - Maximum allowable relative error for determining
                  convergence of the corrector
  itermax       - maximum allowable number of iterations for corrector
                  convergence
  iter          - iteration number in the corrector convergence loop
  t1            - time at the beginning of a time step
  y1            - value of y at the beginning of a time step
  f1            - derivative of y at the beginning of a time step
  f2            - derivative of y at the end of a time step
  favg          - average of f1 and f2
  y2p           - predicted value of y at the end of a time step
  y2            - corrected value of y at the end of a time step
  err           - maximum relative error (for all components) between y2p
                  and y2 for given iteration
  eps           - unit roundoff error (the smallest number for which
                  1 + eps > 1). Used to avoid a zero denominator.

  User M-function required: ode_function
%}
% ------------------------------------------------------------------------------

tol     = 1.e-6;
itermax = 100;

t0      = tspan(1);
tf      = tspan(2);
t       = t0;
y       = y0;
tout    = t;
yout    = y';

while t < tf
    h   = min(h, tf-t);
```

```
    t1   = t;
    y1   = y;
    f1   = feval(ode_function, t1, y1);
    y2   = y1 + f1*h;
    t2   = t1 + h;
    err  = tol + 1;
    iter = 0;
    while err > tol && iter <= itermax
        y2p  = y2;
        f2   = feval(ode_function, t2, y2p);
        favg = (f1 + f2)/2;
        y2   = y1 + favg*h;
        err  = max(abs((y2 - y2p)./(y2 + eps)));
        iter = iter + 1;
    end

    if iter > itermax
        fprintf('\n Maximum no. of iterations (%g)',itermax)
        fprintf('\n exceeded at time = %g',t)
        fprintf('\n in function ''heun.''\n\n')
        return
    end

    t    = t + h;
    y    = y2;
    tout = [tout;t];  % adds t to the bottom of the column vector tout
    yout = [yout;y']; % adds y' to the bottom of the matrix yout
end
% ~~~~~~~~~~~~~~~~~~~~~~~~~~~~~~~~~~~~~~~~~~~~~~~~~~~~~~~~~~~~~~~~~~~~~~
```

FUNCTION FILE: `Example_1_19.m`

```
% ~~~~~~~~~~~~~~~~~~~~~~~~~~~~~~~~~~~~~~~~~~~~~~~~~~~~~~~~~~~~~~~~~~~~~~
function Example_1_19
% ~~~~~~~~~~~~~~~~~
%{
  This program uses Heun's method with two different time steps to solve
  for and plot the response of a damped single degree of freedom
  spring-mass system to a sinusoidal forcing function, represented by

  x'' + 2*z*wn*x' + wn^2*x = (Fo/m)*sin(w*t)

  The numerical integration is done in the external function 'heun',
  which uses the subfunction 'rates' herein to compute the derivatives.

  x     - displacement (m)
  '     - shorthand for d/dt
  t     - time (s)
```

```
  wn    - natural circular frequency (radians/s)
  z     - damping factor
  Fo    - amplitude of the sinusoidal forcing function (N)
  m     - mass (kg)
  w     - forcing frequency (radians/s)
  t0    - initial time (s)
  tf    - final time (s)
  h     - uniform time step (s)
  tspan - row vector containing t0 and tf
  x0    - value of x at t0 (m)
  Dx0   - value of dx/dt at t0 (m/s)
  f0    - column vector containing x0 and Dx0
  t     - column vector of times at which the solution was computed
  f     - a matrix whose columns are:
          column 1: solution for x at the times in t
          column 2: solution for x' at the times in t

  User M-functions required:  heun
  User subfunctions required: rates
%}
% ---------------------------------------------------------------------------

clear all; close all; clc

%...System properties:
m       = 1;
z       = 0.03;
wn      = 1;
Fo      = 1;
w       = 0.4*wn;

%...Time range:
t0      = 0;
tf      = 110;
tspan   = [t0 tf];

%...Initial conditions:
x0  = 0;
Dx0 = 0;
f0  = [x0; Dx0];

%...Calculate and plot the solution for h = 1.0:
h        = 1.0;
[t1, f1] = heun(@rates, tspan, f0, h);

%...Calculate and plot the solution for h = 0.1:
h        = 0.1;
[t2, f2] = heun(@rates, tspan, f0, h);
```

```
output

return

% ~~~~~~~~~~~~~~~~~~~~~~~
function dfdt = rates(t,f)
% ~~~~~~~~~~~~~~~~~~~~~~~
%
% This function calculates first and second time derivatives of x
% for the forced vibration of a damped single degree of freedom
% system represented by the 2nd order differential equation
%
% x'' + 2*z*wn*x' + wn^2*x = (Fo/m)*sin(w*t)
%
% Dx   - velocity
% D2x  - acceleration
% f    - column vector containing x and Dx at time t
% dfdt - column vector containing Dx and D2x at time t
%
% User M-functions required: none
% -------------------------
x    = f(1);
Dx   = f(2);
D2x  = Fo/m*sin(w*t) - 2*z*wn*Dx - wn^2*x;
dfdt = [Dx; D2x];
end %rates

% ~~~~~~~~~~~~~
function output
% ~~~~~~~~~~~~~
plot(t1, f1(:,1), '-r', 'LineWidth',0.5)
xlabel('time, s')
ylabel('x, m')
grid
axis([0 110 -2 2])
hold on
plot(t2, f2(:,1), '-k', 'LineWidth',1)
legend('h = 1.0','h = 0.1')
end %output

end %Example_1_19
% ~~~~~~~~~~~~~~~~~~~~~~~~~~~~~~~~~~~~~~~~~~~~~~~~~~~~~~~~~~~~~~~~~~~~~~~~~~~~
```

D.4 ALGORITHM 1.3: NUMERICAL INTEGRATION OF A SYSTEM OF FIRST-ORDER DIFFERENTIAL EQUATIONS BY THE RUNGE-KUTTA-FEHLBERG 4(5) METHOD WITH ADAPTIVE SIZE CONTROL

FUNCTION FILE: `rkf45.m`

```
% ~~~~~~~~~~~~~~~~~~~~~~~~~~~~~~~~~~~~~~~~~~~~~~~~~~~~~~~~~~
function [tout, yout] = rkf45(ode_function, tspan, y0, tolerance)
% ~~~~~~~~~~~~~~~~~~~~~~~~~~~~~~~~~~~~~~~~~~~~~~~~~~~~~~~~~~
%{
  This function uses the Runge-Kutta-Fehlberg 4(5) algorithm to
  integrate a system of first-order differential equations
  dy/dt = f(t,y).

  y            - column vector of solutions
  f            - column vector of the derivatives dy/dt
  t            - time
  a            - Fehlberg coefficients for locating the six solution
                 points (nodes) within each time interval.
  b            - Fehlberg coupling coefficients for computing the
                 derivatives at each interior point
  c4           - Fehlberg coefficients for the fourth-order solution
  c5           - Fehlberg coefficients for the fifth-order solution
  tol          - allowable truncation error
  ode_function - handle for user M-function in which the derivatives f
                 are computed
  tspan        - the vector [t0 tf] giving the time interval for the
                 solution
  t0           - initial time
  tf           - final time
  y0           - column vector of initial values of the vector y
  tout         - column vector of times at which y was evaluated
  yout         - a matrix, each row of which contains the components of y
                 evaluated at the correponding time in tout
  h            - time step
  hmin         - minimum allowable time step
  ti           - time at the beginning of a time step
  yi           - values of y at the beginning of a time step
  t_inner      - time within a given time step
  y_inner      - values of y witin a given time step
  te           - trucation error for each y at a given time step
  te_allowed   - allowable truncation error
  te_max       - maximum absolute value of the components of te
  ymax         - maximum absolute value of the components of y
  tol          - relative tolerance
  delta        - fractional change in step size
  eps          - unit roundoff error (the smallest number for which
                 1 + eps > 1)
  eps(x)       - the smallest number such that x + eps(x) = x
```

```
  User M-function required: ode_function
%}
% ------------------------------------------------------------------------------

a = [0 1/4 3/8 12/13 1 1/2];

b = [    0          0          0          0         0
       1/4          0          0          0         0
       3/32       9/32         0          0         0
    1932/2197 -7200/2197  7296/2197       0         0
     439/216       -8      3680/513   -845/4104     0
      -8/27         2     -3544/2565  1859/4104  -11/40];

c4 = [25/216  0  1408/2565    2197/4104   -1/5    0  ];
c5 = [16/135  0  6656/12825  28561/56430  -9/50  2/55];

if nargin < 4
    tol  = 1.e-8;
else
    tol = tolerance;
end

t0   = tspan(1);
tf   = tspan(2);
t    = t0;
y    = y0;
tout = t;
yout = y';
h    = (tf - t0)/100; % Assumed initial time step.

while t < tf
    hmin = 16*eps(t);
    ti   = t;
    yi   = y;
    %...Evaluate the time derivative(s) at six points within the current
    %   interval:
    for i = 1:6
        t_inner = ti + a(i)*h;
        y_inner = yi;
        for j = 1:i-1
            y_inner = y_inner + h*b(i,j)*f(:,j);
        end
        f(:,i) = feval(ode_function, t_inner, y_inner);
    end
```

```
    %...Compute the maximum truncation error:
    te     = h*f*(c4' - c5'); % Difference between 4th and
                               % 5th order solutions
    te_max = max(abs(te));

    %...Compute the allowable truncation error:
    ymax       = max(abs(y));
    te_allowed = tol*max(ymax,1.0);

    %...Compute the fractional change in step size:
    delta = (te_allowed/(te_max + eps))^(1/5);

    %...If the truncation error is in bounds, then update the solution:
    if te_max <= te_allowed
        h     = min(h, tf-t);
        t     = t + h;
        y     = yi + h*f*c5';
        tout  = [tout;t];
        yout  = [yout;y'];
    end

    %...Update the time step:
    h  = min(delta*h, 4*h);
    if h < hmin
        fprintf(['\n\n Warning: Step size fell below its minimum\n'...
                 ' allowable value (%g) at time %g.\n\n'], hmin, t)
        return
    end
end
% ~~~~~~~~~~~~~~~~~~~~~~~~~~~~~~~~~~~~~~~~~~~~~~~~~~~~~~~~~~~~~~~~~~~~~~
```

FUNCTION FILE: Example_1_20.m

```
% ~~~~~~~~~~~~~~~~~~~~~~~~~~~~~~~~~~~~~~~~~~~~~~~~~~~~~~~~~~~~~~~~~~~~~~
function Example_1_20
% ~~~~~~~~~~~~~~~~~~~
%{
  This program uses RKF4(5) with adaptive step size control
  to solve the differential equation

  x'' + mu/x^2 = 0

  The numerical integration is done by the function 'rkf45' which uses
  the subfunction 'rates' herein to compute the derivatives.

  x      - displacement (km)
  '      - shorthand for d/dt
  t      - time (s)
```

```
  mu     - = go*RE^2 (km^3/s^2), where go is the sea level gravitational
            acceleration and RE is the radius of the earth
  x0     - initial value of x
  v0     = initial value of the velocity (x')
  y0     - column vector containing x0 and v0
  t0     - initial time
  tf     - final time
  tspan  - a row vector with components t0 and tf
  t      - column vector of the times at which the solution is found
  f      - a matrix whose columns are:
            column 1: solution for x at the times in t
            column 2: solution for x' at the times in t

  User M-function required:  rkf45
  User subfunction required: rates
%}
% ----------------------------------------------------------------------------

clear all; close all; clc

mu      = 398600;
minutes = 60;  %Conversion from minutes to seconds

x0 = 6500;
v0 = 7.8;
y0 = [x0; v0];
t0 = 0;
tf = 70*minutes;

[t,f] = rkf45(@rates, [t0 tf], y0);
plotit
return

% ~~~~~~~~~~~~~~~~~~~~~~~~
function dfdt = rates(t,f)
% ------------------------
%{
  This function calculates first and second time derivatives of x
  governed by the equation of two-body rectilinear motion.

  x'' + mu/x^2 = 0

  Dx   - velocity x'
  D2x  - acceleration x''
  f    - column vector containing x and Dx at time t
  dfdt - column vector containing Dx and D2x at time t
```

```
  User M-functions required: none
%}
% ~~~~~~~~~~~~~~~~~~~~~~~~~~
x    = f(1);
Dx   = f(2);
D2x  = -mu/x^2;
dfdt = [Dx; D2x];
end %rates

% ~~~~~~~~~~~
function plotit
% ~~~~~~~~~~~

%...Position vs time:
subplot(2,1,1)
plot(t/minutes,f(:,1), '-ok')
xlabel('time, minutes')
ylabel('position, km')
grid on
axis([-inf inf 5000 15000])

%...Velocity versus time:
subplot(2,1,2)
plot(t/minutes,f(:,2), '-ok')
xlabel('time, minutes')
ylabel('velocity, km/s')
grid on
axis([-inf inf -10 10])
end %plotit

end %Example_1_20
% ~~~~~~~~~~~~~~~~~~~~~~~~~~~~~~~~~~~~~~~~~~~~~~~~~~~~~~~~~~~~~~~~~~
```

CHAPTER 2: THE TWO-BODY PROBLEM

D.5 ALGORITHM 2.1: NUMERICAL SOLUTION OF THE TWO-BODY PROBLEM RELATIVE TO AN INERTIAL FRAME

FUNCTION FILE: twobody3d.m

```
% ~~~~~~~~~~~~~~~~~~~~~~~~~~~~~~~~~~~~~~~~~~~~~~~~~~~~~~~~~~~~~~~~~~
function twobody3d
% ~~~~~~~~~~~~~~~~~~~~~~~~~~~~~~~~~~~~~~~~~~~~~~~~~~~~~~~~~~~~~~~~~~
%{
  This function solves the inertial two-body problem in three dimensions
  numerically using the RKF4(5) method.
```

```
  G             - universal gravitational constant (km^3/kg/s^2)
  m1,m2         - the masses of the two bodies (kg)
  m             - the total mass (kg)
  t0            - initial time (s)
  tf            - final time (s)
  R1_0,V1_0     - 3 by 1 column vectors containing the components of tbe
                  initial position (km) and velocity (km/s) of m1
  R2_0,V2_0     - 3 by 1 column vectors containing the components of the
                  initial position (km) and velocity (km/s) of m2
  y0            - 12 by 1 column vector containing the initial values
                  of the state vectors of the two bodies:
                  [R1_0; R2_0; V1_0; V2_0]
  t             - column vector of the times at which the solution is found
  X1,Y1,Z1      - column vectors containing the X,Y and Z coordinates (km)
                  of m1 at the times in t
  X2,Y2,Z2      - column vectors containing the X,Y and Z coordinates (km)
                  of m2 at the times in t
  VX1, VY1, VZ1 - column vectors containing the X,Y and Z components
                  of the velocity (km/s) of m1 at the times in t
  VX2, VY2, VZ2 - column vectors containing the X,Y and Z components
                  of the velocity (km/s) of m2 at the times in t
  y             - a matrix whose 12 columns are, respectively,
                  X1,Y1,Z1; X2,Y2,Z2; VX1,VY1,VZ1; VX2,VY2,VZ2
  XG,YG,ZG      - column vectors containing the X,Y and Z coordinates (km)
                  the center of mass at the times in t

  User M-function required:   rkf45
  User subfunctions required: rates, output
%}
% -------------------------------------------------------------------------
clc; clear all; close all
G = 6.67259e-20;

%...Input data:
m1   = 1.e26;
m2   = 1.e26;
t0   = 0;
tf   = 480;

R1_0 = [   0;    0;   0];
R2_0 = [3000;    0;   0];

V1_0 = [  10;   20;  30];
V2_0 = [   0;   40;   0];
%...End input data

y0 = [R1_0; R2_0; V1_0; V2_0];
```

```
%...Integrate the equations of motion:
[t,y] = rkf45(@rates, [t0 tf], y0);

%...Output the results:
output

return

% ~~~~~~~~~~~~~~~~~~~~~~~
function dydt = rates(t,y)
% ~~~~~~~~~~~~~~~~~~~~~~~
%{
  This function calculates the accelerations in Equations 2.19.

  t      - time
  y      - column vector containing the position and velocity vectors
           of the system at time t
  R1, R2 - position vectors of m1 & m2
  V1, V2 - velocity vectors of m1 & m2
  r      - magnitude of the relative position vector
  A1, A2 - acceleration vectors of m1 & m2
  dydt   - column vector containing the velocity and acceleration
           vectors of the system at time t
%}
% ------------------------
R1   = [y(1); y(2); y(3)];
R2   = [y(4); y(5); y(6)];

V1   = [y(7); y(8); y(9)];
V2   = [y(10); y(11); y(12)];

r    = norm(R2 - R1);

A1   = G*m2*(R2 - R1)/r^3;
A2   = G*m1*(R1 - R2)/r^3;

dydt = [V1; V2; A1; A2];

end %rates
% ~~~~~~~~~~~~~~~~~~~~~~~

% ~~~~~~~~~~~~
function output
% ~~~~~~~~~~~~
```

```
%{
  This function calculates the trajectory of the center of mass and
  plots
  (a) the motion of m1, m2 and G relative to the inertial frame
  (b) the motion of m2 and G relative to m1
  (c) the motion of m1 and m2 relative to G

  User subfunction required: common_axis_settings
%}
% -------------

%...Extract the particle trajectories:
X1 = y(:,1); Y1 = y(:,2); Z1 = y(:,3);
X2 = y(:,4); Y2 = y(:,5); Z2 = y(:,6);

%...Locate the center of mass at each time step:
XG = []; YG = []; ZG = [];
for i = 1:length(t)
    XG = [XG; (m1*X1(i) + m2*X2(i))/(m1 + m2)];
    YG = [YG; (m1*Y1(i) + m2*Y2(i))/(m1 + m2)];
    ZG = [ZG; (m1*Z1(i) + m2*Z2(i))/(m1 + m2)];
end

%...Plot the trajectories:
figure (1)
title('Figure 2.3: Motion relative to the inertial frame')
hold on
plot3(X1, Y1, Z1, '-r')
plot3(X2, Y2, Z2, '-g')
plot3(XG, YG, ZG, '-b')
common_axis_settings

figure (2)
title('Figure 2.4a: Motion of m2 and G relative to m1')
hold on
plot3(X2 - X1, Y2 - Y1, Z2 - Z1, '-g')
plot3(XG - X1, YG - Y1, ZG - Z1, '-b')
common_axis_settings

figure (3)
title('Figure 2.4b: Motion of m1 and m2 relative to G')
hold on
plot3(X1 - XG, Y1 - YG, Z1 - ZG, '-r')
plot3(X2 - XG, Y2 - YG, Z2 - ZG, '-g')
common_axis_settings
```

```
% ~~~~~~~~~~~~~~~~~~~~~~
function common_axis_settings
% ~~~~~~~~~~~~~~~~~~~~~~
%{
  This function establishes axis properties common to the several plots.
%}
% -------------------------
text(0, 0, 0, 'o')
axis('equal')
view([2,4,1.2])
grid on
axis equal
xlabel('X (km)')
ylabel('Y (km)')
zlabel('Z (km)')
end %common_axis_settings

end %output

end %twobody3d
% ~~~~~~~~~~~~~~~~~~~~~~~~~~~~~~~~~~~~~~~~~~~~~~~~~~~~~~~~~~~~
```

D.6 ALGORITHM 2.2: NUMERICAL SOLUTION OF THE TWO-BODY RELATIVE MOTION PROBLEM

FUNCTION FILE: orbit.m

```
% ~~~~~~~~~~~~~~~~~~~~~~~~~~~~~~~~~~~~~~~~~~~~~~~~~~~~~~~~~~~~
function orbit
% ~~~~~~~~~
%{
  This function computes the orbit of a spacecraft by using rkf45 to
  numerically integrate Equation 2.22.

  It also plots the orbit and computes the times at which the maximum
  and minimum radii occur and the speeds at those times.

  hours     - converts hours to seconds
  G         - universal gravitational constant (km^3/kg/s^2)
  m1        - planet mass (kg)
  m2        - spacecraft mass (kg)
  mu        - gravitational parameter (km^3/s^2)
  R         - planet radius (km)
  r0        - initial position vector (km)
  v0        - initial velocity vector (km/s)
  t0,tf     - initial and final times (s)
  y0        - column vector containing r0 and v0
  t         - column vector of the times at which the solution is found
```

```
  y         - a matrix whose columns are:
               columns 1, 2 and 3:
                  The solution for the x, y and z components of the
                  position vector r at the times in t
               columns 4, 5 and 6:
                  The solution for the x, y and z components of the
                  velocity vector v at the times in t
  r         - magnitude of the position vector at the times in t
  imax      - component of r with the largest value
  rmax      - largest value of r
  imin      - component of r with the smallest value
  rmin      - smallest value of r
  v_at_rmax - speed where r = rmax
  v_at_rmin - speed where r = rmin

  User M-function required:   rkf45
  User subfunctions required: rates, output
%}
% ------------------------------------------------------------------------------

clc; close all; clear all

hours = 3600;
G     = 6.6742e-20;

%...Input data:
%   Earth:
m1 = 5.974e24;
R  = 6378;
m2 = 1000;

r0 = [8000 0 6000];
v0 = [0 7 0];

t0 = 0;
tf = 4*hours;
%...End input data

%...Numerical integration:
mu    = G*(m1 + m2);
y0    = [r0 v0]';
[t,y] = rkf45(@rates, [t0 tf], y0);

%...Output the results:
output

return
```

```
% ~~~~~~~~~~~~~~~~~~~~~~~
function dydt = rates(t,f)
% ~~~~~~~~~~~~~~~~~~~~~~~
%{
  This function calculates the acceleration vector using Equation 2.22.

  t            - time
  f            - column vector containing the position vector and the
                 velocity vector at time t
  x, y, z      - components of the position vector r
  r            - the magnitude of the the position vector
  vx, vy, vz   - components of the velocity vector v
  ax, ay, az   - components of the acceleration vector a
  dydt         - column vector containing the velocity and acceleration
                 components
%}
% ------------------------
x    = f(1);
y    = f(2);
z    = f(3);
vx   = f(4);
vy   = f(5);
vz   = f(6);

r    = norm([x y z]);

ax   = -mu*x/r^3;
ay   = -mu*y/r^3;
az   = -mu*z/r^3;

dydt = [vx vy vz ax ay az]';
end %rates

% ~~~~~~~~~~~~~
function output
% ~~~~~~~~~~~~~
%{
  This function computes the maximum and minimum radii, the times they
  occur and and the speed at those times. It prints those results to
  the command window and plots the orbit.

  r          - magnitude of the position vector at the times in t
  imax       - the component of r with the largest value
  rmax       - the largest value of r
  imin       - the component of r with the smallest value
  rmin       - the smallest value of r
  v_at_rmax  - the speed where r = rmax
  v_at_rmin  - the speed where r = rmin
```

```
  User subfunction required: light_gray
%}
% -------------
for i = 1:length(t)
    r(i) = norm([y(i,1) y(i,2) y(i,3)]);
end

[rmax imax] = max(r);
[rmin imin] = min(r);

v_at_rmax   = norm([y(imax,4) y(imax,5) y(imax,6)]);
v_at_rmin   = norm([y(imin,4) y(imin,5) y(imin,6)]);

%...Output to the command window:
fprintf('\n\n--------------------------------------------------------------\n')
fprintf('\n Earth Orbit\n')
fprintf(' %s\n', datestr(now))
fprintf('\n The initial position is [%g, %g, %g] (km).',...
                                                     r0(1), r0(2), r0(3))
fprintf('\n   Magnitude = %g km\n', norm(r0))
fprintf('\n The initial velocity is [%g, %g, %g] (km/s).',...
                                                     v0(1), v0(2), v0(3))
fprintf('\n   Magnitude = %g km/s\n', norm(v0))
fprintf('\n Initial time = %g h.\n Final time   = %g h.\n',0,tf/hours)
fprintf('\n The minimum altitude is %g km at time = %g h.',...
            rmin-R, t(imin)/hours)
fprintf('\n The speed at that point is %g km/s.\n', v_at_rmin)
fprintf('\n The maximum altitude is %g km at time = %g h.',...
            rmax-R, t(imax)/hours)
fprintf('\n The speed at that point is %g km/s\n', v_at_rmax)
fprintf('\n--------------------------------------------------------------\n\n')

%...Plot the results:
%   Draw the planet
[xx, yy, zz] = sphere(100);
surf(R*xx, R*yy, R*zz)
colormap(light_gray)
caxis([-R/100 R/100])
shading interp

%   Draw and label the X, Y and Z axes
line([0 2*R],   [0 0],   [0 0]); text(2*R,   0,   0, 'X')
line(  [0 0], [0 2*R],   [0 0]); text(  0, 2*R,   0, 'Y')
line(  [0 0],   [0 0], [0 2*R]); text(  0,   0, 2*R, 'Z')
```

```
%   Plot the orbit, draw a radial to the starting point
%   and label the starting point (o) and the final point (f)
hold on
plot3(  y(:,1),    y(:,2),    y(:,3),'k')
line([0 r0(1)], [0 r0(2)], [0 r0(3)])
text(   y(1,1),    y(1,2),    y(1,3), 'o')
text( y(end,1),  y(end,2),  y(end,3), 'f')

%   Select a view direction (a vector directed outward from the origin)
view([1,1,.4])

%   Specify some properties of the graph
grid on
axis equal
xlabel('km')
ylabel('km')
zlabel('km')

% ~~~~~~~~~~~~~~~~~~~~~~
function map = light_gray
% ~~~~~~~~~~~~~~~~~~~~~~
%{
  This function creates a color map for displaying the planet as light
  gray with a black equator.

  r - fraction of red
  g - fraction of green
  b - fraction of blue

%}
% -----------------------
r = 0.8; g = r; b = r;
map = [r g b
       0 0 0
       r g b];
end %light_gray

end %output

end %orbit
% ~~~~~~~~~~~~~~~~~~~~~~~~~~~~~~~~~~~~~~~~~~~~~~~~~~~~~~~~~~~~~~~~~~~~~~~~~~~~
```

D.7 CALCULATION OF THE LAGRANGE *F* AND *G* FUNCTIONS AND THEIR TIME DERIVATIVES IN TERMS OF CHANGE IN TRUE ANOMALY

FUNCTION FILE: f_and_g_ta.m

```
% ~~~~~~~~~~~~~~~~~~~~~~~~~~~~~~~~~~~~~~~~~~~~~
  function [f, g] = f_and_g_ta(r0, v0, dt, mu)
% ~~~~~~~~~~~~~~~~~~~~~~~~~~~~~~~~~~~~~~~~~~~~~
%{
  This function calculates the Lagrange f and g coefficients from the
  change in true anomaly since time t0.

  mu   - gravitational parameter (km^3/s^2)
  dt   - change in true anomaly (degrees)
  r0   - position vector at time t0 (km)
  v0   - velocity vector at time t0 (km/s)
  h    - angular momentum (km^2/s)
  vr0  - radial component of v0 (km/s)
  r    - radial position after the change in true anomaly
  f    - the Lagrange f coefficient (dimensionless)
  g    - the Lagrange g coefficient (s)

  User M-functions required:  None
%}
% ----------------------------------------------

h   = norm(cross(r0,v0));
vr0 = dot(v0,r0)/norm(r0);
r0  = norm(r0);
s   = sind(dt);
c   = cosd(dt);

%...Equation 2.152:
r   = h^2/mu/(1 + (h^2/mu/r0 - 1)*c - h*vr0*s/mu);

%...Equations 2.158a & b:
f   = 1 - mu*r*(1 - c)/h^2;
g   = r*r0*s/h;

end
% ~~~~~~~~~~~~~~~~~~~~~~~~~~~~~~~~~~~~~~~~~~~
```

FUNCTION FILE: fDot_and_gDot_ta.m

```
% ~~~~~~~~~~~~~~~~~~~~~~~~~~~~~~~~~~~~~~~~~~~~~~~~~~~~~~~~~~
  function [fdot, gdot] = fDot_and_gDot_ta(r0, v0, dt, mu)
% ~~~~~~~~~~~~~~~~~~~~~~~~~~~~~~~~~~~~~~~~~~~~~~~~~~~~~~~~~~
```

```
%{
  This function calculates the time derivatives of the Lagrange
  f and g coefficients from the change in true anomaly since time t0.

  mu   - gravitational parameter (km^3/s^2)
  dt   - change in true anomaly (degrees)
  r0   - position vector at time t0 (km)
  v0   - velocity vector at time t0 (km/s)
  h    - angular momentum (km^2/s)
  vr0  - radial component of v0 (km/s)
  fdot - time derivative of the Lagrange f coefficient (1/s)
  gdot - time derivative of the Lagrange g coefficient (dimensionless)

  User M-functions required:  None
%}
% ----------------------------------------------------------

h   = norm(cross(r0,v0));
vr0 = dot(v0,r0)/norm(r0);
r0  = norm(r0);
c   = cosd(dt);
s   = sind(dt);

%...Equations 2.158c & d:
fdot = mu/h*(vr0/h*(1 - c) - s/r0);
gdot = 1 - mu*r0/h^2*(1 - c);

end
% ~~~~~~~~~~~~~~~~~~~~~~~~~~~~~~~~~~~~~~~~~~~~~~~~~~~~~~~~~~
```

D.8 ALGORITHM 2.3: CALCULATE THE STATE VECTOR FROM THE INITIAL STATE VECTOR AND THE CHANGE IN TRUE ANOMALY

FUNCTION FILE: rv_from_r0v0_ta.m

```
% ~~~~~~~~~~~~~~~~~~~~~~~~~~~~~~~~~~~~~~~~~~~~~~~~~
  function [r,v] = rv_from_r0v0_ta(r0, v0, dt, mu)
% ~~~~~~~~~~~~~~~~~~~~~~~~~~~~~~~~~~~~~~~~~~~~~~~~~
%{
  This function computes the state vector (r,v) from the
  initial state vector (r0,v0) and the change in true anomaly.

  mu - gravitational parameter (km^3/s^2)
  r0 - initial position vector (km)
  v0 - initial velocity vector (km/s)
```

```
  dt - change in true anomaly (degrees)
  r  - final position vector (km)
  v  - final velocity vector (km/s)

  User M-functions required: f_and_g_ta, fDot_and_gDot_ta
%}
% ------------------------------------------------------------------------------

%global mu

%...Compute the f and g functions and their derivatives:
[f, g]       =       f_and_g_ta(r0, v0, dt, mu);
[fdot, gdot] = fDot_and_gDot_ta(r0, v0, dt, mu);

%...Compute the final position and velocity vectors:
r =    f*r0 +g*v0;
v = fdot*r0 + gdot*v0;

end
% ~~~~~~~~~~~~~~~~~~~~~~~~~~~~~~~~~~~~~~~~
```

SCRIPT FILE: Example_2_13.m

```
% ~~~~~~~~~~~~~~~~~~~~~~~~~~~~~~~~~~~~~~~~~~
% Example_2_13
% ~~~~~~~~~~~
%{
  This program computes the state vector [R,V] from the initial
  state vector [R0,V0] and the change in true anomaly, using the
  data in Example 2.13

  mu - gravitational parameter (km^3/s^2)
  R0 - the initial position vector (km)
  V0 - the initial velocity vector (km/s)
  r0 - magnitude of R0
  v0 - magnitude of V0
  R  - final position vector (km)
  V  - final velocity vector (km/s)
  r  - magnitude of R
  v  - magnitude of V
  dt - change in true anomaly (degrees)

 User M-functions required: rv_from_r0v0_ta

%}
% ----------------------------------------------
```

```
clear all; clc
mu = 398600;

%...Input data:
R0 = [8182.4 -6865.9 0];
V0 = [0.47572 8.8116 0];
dt = 120;
%...End input data

%...Algorithm 2.3:
[R,V] = rv_from_r0v0_ta(R0, V0, dt, mu);

r  = norm(R);
v  = norm(V);
r0 = norm(R0);
v0 = norm(V0);

fprintf('-----------------------------------------------------------------')
fprintf('\n Example 2.13 \n')
fprintf('\n Initial state vector:\n')
fprintf('\n   r = [%g, %g, %g] (km)', R0(1), R0(2), R0(3))
fprintf('\n     magnitude = %g\n', norm(R0))

fprintf('\n   v = [%g, %g, %g] (km/s)', V0(1), V0(2), V0(3))
fprintf('\n     magnitude = %g', norm(V0))

fprintf('\n\n State vector after %g degree change in true anomaly:\n', dt)
fprintf('\n   r = [%g, %g, %g] (km)', R(1), R(2), R(3))
fprintf('\n     magnitude = %g\n', norm(R))
fprintf('\n   v = [%g, %g, %g] (km/s)', V(1), V(2), V(3))
fprintf('\n     magnitude = %g', norm(V))
fprintf('\n-----------------------------------------------------------------\n')
% ~~~~~~~~~~~~~~~~~~~~~~~~~~~~~~~~~~~~~~~~
```

OUTPUT FROM Example_2_13.m

```
-----------------------------------------------------
 Example 2.13

 Initial state vector:

   r = [8182.4, -6865.9, 0] (km)
     magnitude = 10681.4

   v = [0.47572, 8.8116, 0] (km/s)
     magnitude = 8.82443
```

```
State vector after 120 degree change in true anomaly:

  r = [1454.99, 8251.47, 0] (km)
    magnitude = 8378.77

  v = [-8.13238, 5.67854, -0] (km/s)
    magnitude = 9.91874
-----------------------------------------------------------
```

D.9 ALGORITHM 2.4: FIND THE ROOT OF A FUNCTION USING THE BISECTION METHOD

FUNCTION FILE: bisect.m

```
% ~~~~~~~~~~~~~~~~~~~~~~~~~~~~~~~~~~~~~~~~~~~~~~~~~~~~~~~~~~~
function root = bisect(fun, xl, xu)
% ~~~~~~~~~~~~~~~~~~~~~~~~~~~~~~~~~
%{
  This function evaluates a root of a function using
  the bisection method.

  tol  - error to within which the root is computed
  n    - number of iterations
  xl   - low end of the interval containing the root
  xu   - upper end of the interval containing the root
  i    - loop index
  xm   - mid-point of the interval from xl to xu
  fun  - name of the function whose root is being found
  fxl  - value of fun at xl
  fxm  - value of fun at xm
  root - the computed root

  User M-functions required: none
%}
% ----------------------------------------------

tol = 1.e-6;
n   = ceil(log(abs(xu - xl)/tol)/log(2));

for i = 1:n
    xm  = (xl + xu)/2;
    fxl = feval(fun, xl);
    fxm = feval(fun, xm);
    if fxl*fxm > 0
        xl = xm;
    else
        xu = xm;
    end
end
```

```
root = xm;

end
% ~~~~~~~~~~~~~~~~~~~~~~~~~~~~~~~~~~~~~~~~~~~~~
```

FUNCTION FILE: `Example_2_16.m`

```
% ~~~~~~~~~~~~~~~~~~~~~~~~~~~~~~~~~~~~~~~~~~~~~
function Example_2_16
% ~~~~~~~~~~~~~~~~~~~
%{
  This program uses the bisection method to find the three roots of
  Equation 2.204 for the earth-moon system.

  m1  - mass of the earth (kg)
  m2  - mass of the moon (kg)
  r12 - distance from the earth to the moon (km)
  p   - ratio of moon mass to total mass
  xl  - vector containing the low-side estimates of the three roots
  xu  - vector containing the high-side estimates of the three roots
  x   - vector containing the three computed roots

  User M-function required: bisect
  User subfunction requred: fun
%}
% ----------------------------------------------

clear all; clc

%...Input data:
m1  = 5.974e24;
m2  = 7.348e22;
r12 = 3.844e5;

xl  = [-1.1  0.5  1.0];
xu  = [-0.9  1.0  1.5];
%...End input data

p   = m2/(m1 + m2);

for i = 1:3
    x(i) = bisect(@fun, xl(i), xu(i));
end

%...Output the results
output

return
```

```
% ~~~~~~~~~~~~~~~~~~
function f = fun(z)
% -----------------
%{
  This subroutine evaluates the function in Equation 2.204

  z - the dimensionless x - coordinate
  p - defined above
  f - the value of the function

%}
% ~~~~~~~~~~~~~~~~~~
f = (1 - p)*(z + p)/abs(z + p)^3 + p*(z + p - 1)/abs(z + p - 1)^3 - z;
end %fun

% ~~~~~~~~~~~
function output
% ~~~~~~~~~~~
%{
  This function prints out the x coordinates of L1, L2 and L3
  relative to the center of mass.
%}
%...Output to the command window:
fprintf('\n\n--------------------------------------------\n')
fprintf('\n For\n')
fprintf('\n   m1 = %g kg', m1)
fprintf('\n   m2 = %g kg', m2)
fprintf('\n  r12 = %g km\n', r12)
fprintf('\n the 3 colinear Lagrange points (the roots of\n')
fprintf(' Equation 2.204) are:\n')
fprintf('\n L3: x = %10g km    (f(x3) = %g)',x(1)*r12, fun(x(1)))
fprintf('\n L1: x = %10g km    (f(x1) = %g)',x(2)*r12, fun(x(2)))
fprintf('\n L2: x = %10g km    (f(x2) = %g)',x(3)*r12, fun(x(3)))
fprintf('\n\n--------------------------------------------\n')

end %output

end %Example_2_16
% ~~~~~~~~~~~~~~~~~~~~~~~~~~~~~~~~~~~~
```

OUTPUT FROM Example_2_16.m

```
--------------------------------------------

 For

   m1 = 5.974e+24 kg
   m2 = 7.348e+22 kg
  r12 = 384400 km
```

```
The 3 colinear Lagrange points (the roots of
Equation 2.204) are:

L3: x =    -386346 km    (f(x3) = -1.55107e-06)
L1: x =     321710 km    (f(x1) = 5.12967e-06)
L2: x =     444244 km    (f(x2) = -4.92782e-06)

---------------------------------------------
```

D.10 MATLAB SOLUTION OF EXAMPLE 2.18

FUNCTION FILE: Example_2_18.m

```
% ~~~~~~~~~~~~~~~~~~~~~~~~~~~~~~~~~~~~~~~~
function Example_2_18
% ~~~~~~~~~~~~~~~~~~
%{
  This program uses the Runge-Kutta-Fehlberg 4(5) method to solve the
  earth-moon restricted three-body problem (Equations 2.192a and 2.192b)
  for the trajectory of a spacecraft having the initial conditions
  specified in Example 2.18.

  The numerical integration is done in the external function 'rkf45',
  which uses the subfunction 'rates' herein to compute the derivatives.

  days        - converts days to seconds
  G           - universal graviational constant (km^3/kg/s^2)
  rmoon       - radius of the moon (km)
  rearth      - radius of the earth (km)
  r12         - distance from center of earth to center of moon (km)
  m1,m2       - masses of the earth and of the moon, respectively (kg)
  M           - total mass of the restricted 3-body system (kg)
  mu          - gravitational parameter of earth-moon system (km^3/s^2)
  mu1,mu2     - gravitational parameters of the earth and of the moon,
                respectively (km^3/s^2)
  pi_1,pi_2 - ratios of the earth mass and the moon mass, respectively,
                to the total earth-moon mass
  W           - angular velocity of moon around the earth (rad/s)
  x1,x2       - x-coordinates of the earth and of the moon, respectively,
                relative to the earth-moon barycenter (km)
  d0          - initial altitude of spacecraft (km)
  phi         - polar azimuth coordinate (degrees) of the spacecraft
                measured positive counterclockwise from the earth-moon line
  v0          - initial speed of spacecraft relative to rotating earth-moon
                system (km/s)
```

```
  gamma      - initial flight path angle (degrees)
  r0         - intial radial distance of spacecraft from the earth (km)
  x,y        - x and y coordinates of spacecraft in rotating earth-moon
               system (km)
  vx,vy      - x and y components of spacecraft velocity relative to
               rotating earth-moon system (km/s)
  f0         - column vector containing the initial valus of x, y, vx and vy
  t0,tf      - initial time and final times (s)
  t          - column vector of times at which the solution was computed
  f          - a matrix whose columns are:
               column 1: solution for x  at the times in t
               column 2: solution for y  at the times in t
               column 3: solution for vx at the times in t
               column 4: solution for vy at the times in t
  xf,yf      - x and y coordinates of spacecraft in rotating earth-moon
               system at tf
  vxf, vyf   - x and y components of spacecraft velocity relative to
               rotating earth-moon system at tf
  df         - distance from surface of the moon at tf
  vf         - relative speed at tf

  User M-functions required:  rkf45
  User subfunctions required: rates, circle
%}
% ---------------------------------------------

clear all; close all; clc

days   =  24*3600;
G      =  6.6742e-20;
rmoon  =  1737;
rearth =  6378;
r12    =  384400;
m1     =  5974e21;
m2     =  7348e19;

M      =  m1 + m2;;
pi_1   =  m1/M;
pi_2   =  m2/M;

mu1    =  398600;
mu2    =  4903.02;
mu     =  mu1 + mu2;

W      =  sqrt(mu/r12^3);
x1     = -pi_2*r12;
x2     =  pi_1*r12;
```

```
%...Input data:
d0     =  200;
phi    =  -90;
v0     =  10.9148;
gamma  =  20;
t0     =  0;
tf     =  3.16689*days;

r0     =  rearth + d0;
x      =  r0*cosd(phi) + x1;
y      =  r0*sind(phi);

vx     =  v0*(sind(gamma)*cosd(phi) - cosd(gamma)*sind(phi));
vy     =  v0*(sind(gamma)*sind(phi) + cosd(gamma)*cosd(phi));
f0     =  [x; y; vx; vy];

%...Compute the trajectory:
[t,f]  = rkf45(@rates, [t0 tf], f0);
x      = f(:,1);
y      = f(:,2);
vx     = f(:,3);
vy     = f(:,4);

xf     = x(end);
yf     = y(end);

vxf    = vx(end);
vyf    = vy(end);

df     = norm([xf - x2, yf - 0]) - rmoon;
vf     = norm([vxf, vyf]);

%...Output the results:
output
return

% ~~~~~~~~~~~~~~~~~~~~
function dfdt = rates(t,f)
% ~~~~~~~~~~~~~~~~~~~~
%{
  This subfunction calculates the components of the relative acceleration
  for the restricted 3-body problem, using Equations 2.192a and 2.192b

  ax,ay - x and y components of relative acceleration (km/s^2)
  r1    - spacecraft distance from the earth (km)
  r2    - spacecraft  distance from the moon (km)
  f     - column vector containing x,  y,  vx and vy at time t
```

```
  dfdt  - column vector containing vx, vy, ax and ay at time t

  All other variables are defined above.

  User M-functions required: none
%}
% ------------------------
x       = f(1);
y       = f(2);
vx      = f(3);
vy      = f(4);

r1      = norm([x + pi_2*r12, y]);
r2      = norm([x - pi_1*r12, y]);

ax      =  2*W*vy + W^2*x - mu1*(x - x1)/r1^3 - mu2*(x - x2)/r2^3;
ay      = -2*W*vx + W^2*y - (mu1/r1^3 + mu2/r2^3)*y;

dfdt    = [vx; vy; ax; ay];
end %rates

% ~~~~~~~~~~~~
function output
% ~~~~~~~~~~~~
%{
  This subfunction echos the input data and prints the results to the
  command window. It also plots the trajectory.

  User M-functions required: none
  User subfunction required: circle
%}
% -------------

fprintf('-----------------------------------------------------------')
fprintf('\n Example 2.18: Lunar trajectory using the restricted')
fprintf('\n threebody equations.\n')
fprintf('\n Initial Earth altitude (km)          = %g', d0)
fprintf('\n Initial angle between radial')
fprintf('\n   and earth-moon line (degrees)      = %g', phi)
fprintf('\n Initial flight path angle (degrees) = %g', gamma)
fprintf('\n Flight time (days)                   = %g', tf/days)
fprintf('\n Final distance from the moon (km)    = %g', df)
fprintf('\n Final relative speed (km/s)          = %g', vf)
fprintf('\n-----------------------------------------------------------\n')

%...Plot the trajectory and place filled circles representing the earth
%   and moon on the the plot:
```

```
plot(x, y)
%   Set plot display parameters
xmin = -20.e3;  xmax =  4.e5;
ymin = -20.e3;  ymax =  1.e5;
axis([xmin xmax ymin ymax])
axis equal
xlabel('x, km'); ylabel('y, km')
grid on
hold on

%...Plot the earth (blue) and moon (green) to scale
earth  = circle(x1, 0, rearth);
moon   = circle(x2, 0, rmoon);
fill(earth(:,1), earth(:,2),'b')
fill( moon(:,1),  moon(:,2),'g')

% ~~~~~~~~~~~~~~~~~~~~~~~~~~~~~~
function xy = circle(xc, yc, radius)
% ~~~~~~~~~~~~~~~~~~~~~~~~~~~~~~
%{
  This subfunction calculates the coordinates of points spaced
  0.1 degree apart around the circumference of a circle

  x,y     - x and y coordinates of a point on the circumference
  xc,yc   - x and y coordinates of the center of the circle
  radius  - radius of the circle
  xy      - an array containing the x coordinates in column 1 and the
            y coordinates in column 2

  User M-functions required: none
%}
% ----------------------------------
x       = xc + radius*cosd(0:0.1:360);
y       = yc + radius*sind(0:0.1:360);
xy      = [x', y'];

end %circle

end %output

end %Example_2_18
% ~~~~~~~~~~~~~~~~~~~~~~~~~~~~~~~~~~~~~
```

OUTPUT FROM Example_2_18.m

```
-------------------------------------------------------------
 Example 2.18: Lunar trajectory using the restricted
 Three body equations.
```

```
Initial Earth altitude (km)          = 200
Initial angle between radial
  and earth-moon line (degrees)      = -90
Initial flight path angle (degrees) = 20
Flight time (days)                   = 3.16689
Final distance from the moon (km)    = 255.812
Final relative speed (km/s)          = 2.41494
-------------------------------------------------------------
```

CHAPTER 3: ORBITAL POSITION AS A FUNCTION OF TIME

D.11 ALGORITHM 3.1: SOLUTION OF KEPLER'S EQUATION BY NEWTON'S METHOD

FUNCTION FILE: kepler_E.m

```
% ~~~~~~~~~~~~~~~~~~~~~~~~~~~~~~~~~~~~~~~~~
  function E = kepler_E(e, M)
% ~~~~~~~~~~~~~~~~~~~~~~~
%{
  This function uses Newton's method to solve Kepler's
  equation  E - e*sin(E) = M  for the eccentric anomaly,
  given the eccentricity and the mean anomaly.

  E  - eccentric anomaly (radians)
  e  - eccentricity, passed from the calling program
  M  - mean anomaly (radians), passed from the calling program
  pi - 3.1415926...

  User m-functions required: none
%}
% -----------------------------------------------

%...Set an error tolerance:
error = 1.e-8;

%...Select a starting value for E:
if M < pi
    E = M + e/2;
else
    E = M - e/2;
end

%...Iterate on Equation 3.17 until E is determined to within
%...the error tolerance:
ratio = 1;
while abs(ratio) > error
    ratio = (E - e*sin(E) - M)/(1 - e*cos(E));
```

```
    E = E - ratio;
end

end %kepler_E
% ~~~~~~~~~~~~~~~~~~~~~~~~~~~~~~~~~~~~~~~~
```

SCRIPT FILE: `Example_3_02.m`

```
% ~~~~~~~~~~~~~~~~~~~~~~~~~~~~~~~~~~~~~~~~
% Example_3_02
% ~~~~~~~~~~~
%{
  This program uses Algorithm 3.1 and the data of Example 3.2 to solve
  Kepler's equation.

  e   - eccentricity
  M   - mean anomaly (rad)
  E   - eccentric anomaly (rad)

  User M-function required: kepler_E
%}
% ---------------------------------------------

clear all; clc

%...Data declaration for Example 3.2:
e = 0.37255;
M = 3.6029;
%...

%...Pass the input data to the function kepler_E, which returns E:
E = kepler_E(e, M);

%...Echo the input data and output to the command window:
fprintf('-----------------------------------------------------')
fprintf('\n Example 3.2\n')
fprintf('\n Eccentricity                = %g',e)
fprintf('\n Mean anomaly (radians)      = %g\n',M)
fprintf('\n Eccentric anomaly (radians) = %g',E)
fprintf('\n-----------------------------------------------------\n')

% ~~~~~~~~~~~~~~~~~~~~~~~~~~~~~~~~~~~~~~~~
```

OUTPUT FROM `Example_3_02.m`

```
-----------------------------------------------------
 Example 3.2

 Eccentricity                = 0.37255
 Mean anomaly (radians)      = 3.6029
```

```
Eccentric anomaly (radians) = 3.47942
-----------------------------------------------------
```

D.12 ALGORITHM 3.2: SOLUTION OF KEPLER'S EQUATION FOR THE HYPERBOLA USING NEWTON'S METHOD

FUNCTION FILE: kepler_H.m

```
% ~~~~~~~~~~~~~~~~~~~~~~~~~~~~~~~~~~~~~~~~~~~~~~~~~
  function F = kepler_H(e, M)
% ~~~~~~~~~~~~~~~~~~~~~~~~~~~~~~
%{
  This function uses Newton's method to solve Kepler's equation
  for the hyperbola  e*sinh(F) - F = M  for the hyperbolic
  eccentric anomaly, given the eccentricity and the hyperbolic
  mean anomaly.

  F - hyperbolic eccentric anomaly (radians)
  e - eccentricity, passed from the calling program
  M - hyperbolic mean anomaly (radians), passed from the
      calling program

  User M-functions required: none
%}
% ---------------------------------------------

%...Set an error tolerance:
error = 1.e-8;

%...Starting value for F:
F = M;

%...Iterate on Equation 3.45 until F is determined to within
%...the error tolerance:
ratio = 1;
while abs(ratio) > error
    ratio = (e*sinh(F) - F - M)/(e*cosh(F) - 1);
    F = F - ratio;
end

end %kepler_H
% ~~~~~~~~~~~~~~~~~~~~~~~~~~~~~~~~~~~~~~~~~~~~~~~~~
```

SCRIPT FILE: Example_3_05.m

```
% ~~~~~~~~~~~~~~~~~~~~~~~~~~~~~~~~~~~~~~~~~~~~~
% Example_3_05
% ~~~~~~~~~~~~
%{
  This program uses Algorithm 3.2 and the data of
  Example 3.5 to solve Kepler's equation for the hyperbola.

  e - eccentricity
  M - hyperbolic mean anomaly (dimensionless)
  F - hyperbolic eccentric anomaly (dimensionless)

  User M-function required: kepler_H
%}
% ------------------------------------------------
clear

%...Data declaration for Example 3.5:
e = 2.7696;
M = 40.69;
%...

%...Pass the input data to the function kepler_H, which returns F:
F = kepler_H(e, M);

%...Echo the input data and output to the command window:
fprintf('-----------------------------------------------------')
fprintf('\n Example 3.5\n')
fprintf('\n Eccentricity                 = %g',e)
fprintf('\n Hyperbolic mean anomaly      = %g\n',M)
fprintf('\n Hyperbolic eccentric anomaly = %g',F)
fprintf('\n-----------------------------------------------------\n')

% ~~~~~~~~~~~~~~~~~~~~~~~~~~~~~~~~~~~~~~~~~~~~~
```

OUTPUT FROM Example_3_05.m

```
-----------------------------------------------------
 Example 3.5

 Eccentricity                 = 2.7696
 Hyperbolic mean anomaly      = 40.69

 Hyperbolic eccentric anomaly = 3.46309
-----------------------------------------------------
```

D.13 CALCULATION OF THE STUMPFF FUNCTIONS S(Z) AND C(Z)

The following scripts implement Eqs. (3.52) and (3.53) for use in other programs.

FUNCTION FILE: stumpS.m

```
% ~~~~~~~~~~~~~~~~~~~~~~~~~~~~~~~~~~~~~~~~~~~~
  function s = stumpS(z)
% ~~~~~~~~~~~~~~~~~~~~~~
%{
  This function evaluates the Stumpff function S(z) according
  to Equation 3.52.

  z - input argument
  s - value of S(z)

  User M-functions required: none
%}
% ----------------------------------------------

if z > 0
    s = (sqrt(z) - sin(sqrt(z)))/(sqrt(z))^3;
elseif z < 0
    s = (sinh(sqrt(-z)) - sqrt(-z))/(sqrt(-z))^3;
else
    s = 1/6;
end
% ~~~~~~~~~~~~~~~~~~~~~~~~~~~~~~~~~~~~~~~~~~~~
```

FUNCTION FILE: stumpC.m

```
% ~~~~~~~~~~~~~~~~~~~~~~~~~~~~~~~~~~~~~~~~~~~~
  function c = stumpC(z)
% ~~~~~~~~~~~~~~~~~~~~~~
%{
  This function evaluates the Stumpff function C(z) according
  to Equation 3.53.

  z - input argument
  c - value of C(z)

  User M-functions required: none
%}
% ----------------------------------------------

if z > 0
    c = (1 - cos(sqrt(z)))/z;
elseif z < 0
    c = (cosh(sqrt(-z)) - 1)/(-z);
```

```
else
    c = 1/2;
end
% ~~~~~~~~~~~~~~~~~~~~~~~~~~~~~~~~~~~~~~~~~~~~~
```

D.14 ALGORITHM 3.3: SOLUTION OF THE UNIVERSAL KEPLER'S EQUATION USING NEWTON'S METHOD

FUNCTION FILE: kepler_U.m

```
% ~~~~~~~~~~~~~~~~~~~~~~~~~~~~~~~~~~~~~~~~~~~~~
  function x = kepler_U(dt, ro, vro, a)
% ~~~~~~~~~~~~~~~~~~~~~~~~~~~~~~~~~~~~~
%{
  This function uses Newton's method to solve the universal
  Kepler equation for the universal anomaly.

  mu   - gravitational parameter (km^3/s^2)
  x    - the universal anomaly (km^0.5)
  dt   - time since x = 0 (s)
  ro   - radial position (km) when x = 0
  vro  - radial velocity (km/s) when x = 0
  a    - reciprocal of the semimajor axis (1/km)
  z    - auxiliary variable (z = a*x^2)
  C    - value of Stumpff function C(z)
  S    - value of Stumpff function S(z)
  n    - number of iterations for convergence
  nMax - maximum allowable number of iterations

  User M-functions required: stumpC, stumpS
%}
% ----------------------------------------------
global mu

%...Set an error tolerance and a limit on the number of iterations:
error = 1.e-8;
nMax  = 1000;

%...Starting value for x:
x = sqrt(mu)*abs(a)*dt;

%...Iterate on Equation 3.65 until until convergence occurs within
%...the error tolerance:
n     = 0;
ratio = 1;
while abs(ratio) > error && n <= nMax
    n     = n + 1;
    C     = stumpC(a*x^2);
```

```
        S     = stumpS(a*x^2);
        F     = ro*vro/sqrt(mu)*x^2*C + (1 - a*ro)*x^3*S + ro*x - sqrt(mu)*dt;
        dFdx  = ro*vro/sqrt(mu)*x*(1 - a*x^2*S) + (1 - a*ro)*x^2*C + ro;
        ratio = F/dFdx;
        x     = x - ratio;
    end

    %...Deliver a value for x, but report that nMax was reached:
    if n > nMax
        fprintf('\n **No. iterations of Kepler's equation = %g', n)
        fprintf('\n   F/dFdx                                = %g\n', F/dFdx)
    end
    % ~~~~~~~~~~~~~~~~~~~~~~~~~~~~~~~~~~~~~~~~~~~~
```

SCRIPT FILE: Example_3_06.m

```
% ~~~~~~~~~~~~~~~~~~~~~~~~~~~~~~~~~~~~~~~~~~~~
% Example_3_06
% ~~~~~~~~~~~
%{
  This program uses Algorithm 3.3 and the data of Example 3.6
  to solve the universal Kepler's equation.

  mu  - gravitational parameter (km^3/s^2)
  x   - the universal anomaly (km^0.5)
  dt  - time since x = 0 (s)
  ro  - radial position when x = 0 (km)
  vro - radial velocity when x = 0 (km/s)
  a   - semimajor axis (km)

  User M-function required: kepler_U
%}
% ----------------------------------------------

clear all; clc
global mu
mu = 398600;

%...Data declaration for Example 3.6:
ro  = 10000;
vro = 3.0752;
dt  = 3600;
a   = -19655;
%...

%...Pass the input data to the function kepler_U, which returns x
%...(Universal Kepler's requires the reciprocal of semimajor axis):
x   = kepler_U(dt, ro, vro, 1/a);
```

```
%...Echo the input data and output the results to the command window:
fprintf('-----------------------------------------------------')
fprintf('\n Example 3.6\n')
fprintf('\n Initial radial coordinate (km) = %g',ro)
fprintf('\n Initial radial velocity (km/s) = %g',vro)
fprintf('\n Elapsed time (seconds)         = %g',dt)
fprintf('\n Semimajor axis (km)            = %g\n',a)
fprintf('\n Universal anomaly (km^0.5)     = %g',x)
fprintf('\n-----------------------------------------------------\n')
% ~~~~~~~~~~~~~~~~~~~~~~~~~~~~~~~~~~~~~~
```

OUTPUT FROM Example_3_06.m

```
-----------------------------------------------------
 Example 3.6

 Initial radial coordinate (km) = 10000
 Initial radial velocity (km/s) = 3.0752
 Elapsed time (seconds)         = 3600
 Semimajor axis (km)            = -19655

 Universal anomaly (km^0.5)     = 128.511
-----------------------------------------------------
```

D.15 CALCULATION OF THE LAGRANGE COEFFICIENTS *F* AND *G* AND THEIR TIME DERIVATIVES IN TERMS OF CHANGE IN UNIVERAL ANOMALY

The following scripts implement Equations 3.69 for use in other programs.

FUNCTION FILE: f_and_g.m

```
% ~~~~~~~~~~~~~~~~~~~~~~~~~~~~~~~~~~~~~~
  function [f, g] = f_and_g(x, t, ro, a)
% ~~~~~~~~~~~~~~~~~~~~~~~~~~~~~~~
%{
  This function calculates the Lagrange f and g coefficients.

  mu - the gravitational parameter (km^3/s^2)
  a  - reciprocal of the semimajor axis (1/km)
  ro - the radial position at time to (km)
  t  - the time elapsed since ro (s)
  x  - the universal anomaly after time t (km^0.5)
  f  - the Lagrange f coefficient (dimensionless)
  g  - the Lagrange g coefficient (s)

  User M-functions required:  stumpC, stumpS
%}
% ---------------------------------------------
```

```
global mu

z = a*x^2;

%...Equation 3.69a:
f = 1 - x^2/ro*stumpC(z);

%...Equation 3.69b:
g = t - 1/sqrt(mu)*x^3*stumpS(z);

end
% ~~~~~~~~~~~~~~~~~~~~~~~~~~~~~~~~~~~~~~~~~~~~~~~
```

FUNCTION FILE: fDot_and_gDot.m

```
% ~~~~~~~~~~~~~~~~~~~~~~~~~~~~~~~~~~~~~~~~~~~~~~~~~~~~~
  function [fdot, gdot] = fDot_and_gDot(x, r, ro, a)
% ~~~~~~~~~~~~~~~~~~~~~~~~~~~~~~~~~~~~~~~~~~~~~~~~~~~~~
%{
  This function calculates the time derivatives of the
  Lagrange f and g coefficients.

  mu    - the gravitational parameter (km^3/s^2)
  a     - reciprocal of the semimajor axis (1/km)
  ro    - the radial position at time to (km)
  t     - the time elapsed since initial state vector (s)
  r     - the radial position after time t (km)
  x     - the universal anomaly after time t (km^0.5)
  fdot  - time derivative of the Lagrange f coefficient (1/s)
  gdot  - time derivative of the Lagrange g coefficient (dimensionless)

  User M-functions required:  stumpC, stumpS
%}
% ----------------------------------------------------

global mu

z = a*x^2;

%...Equation 3.69c:
fdot = sqrt(mu)/r/ro*(z*stumpS(z) - 1)*x;

%...Equation 3.69d:
gdot = 1 - x^2/r*stumpC(z);
% ~~~~~~~~~~~~~~~~~~~~~~~~~~~~~~~~~~~~~~~~~~~~~~~~~~~~~
```

D.16 ALGORITHM 3.4: CALCULATION OF THE STATE VECTOR GIVEN THE INITIAL STATE VECTOR AND THE TIME LAPSE ΔT

FUNCTION FILE: rv_from_r0v0.m

```
% ~~~~~~~~~~~~~~~~~~~~~~~~~~~~~~~~~~~~~~~~~~~
  function [R,V] = rv_from_r0v0(R0, V0, t)
% ~~~~~~~~~~~~~~~~~~~~~~~~~~~~~~~~~~~~~~~~~~~
%{
  This function computes the state vector (R,V) from the
  initial state vector (R0,V0) and the elapsed time.

  mu - gravitational parameter (km^3/s^2)
  R0 - initial position vector (km)
  V0 - initial velocity vector (km/s)
  t  - elapsed time (s)
  R  - final position vector (km)
  V  - final velocity vector (km/s)

% User M-functions required: kepler_U, f_and_g, fDot_and_gDot
%}
% ---------------------------------------------

global mu

%...Magnitudes of R0 and V0:
r0 = norm(R0);
v0 = norm(V0);

%...Initial radial velocity:
vr0 = dot(R0, V0)/r0;

%...Reciprocal of the semimajor axis (from the energy equation):
alpha = 2/r0 - v0^2/mu;

%...Compute the universal anomaly:
x = kepler_U(t, r0, vr0, alpha);

%...Compute the f and g functions:
[f, g] = f_and_g(x, t, r0, alpha);

%...Compute the final position vector:
R = f*R0 + g*V0;

%...Compute the magnitude of R:
r = norm(R);

%...Compute the derivatives of f and g:
```

```
[fdot, gdot] = fDot_and_gDot(x, r, r0, alpha);

%...Compute the final velocity:
V            = fdot*R0 + gdot*V0;
% ~~~~~~~~~~~~~~~~~~~~~~~~~~~~~~~~~~~~~~~~~~~~~~
```

SCRIPT FILE: Example_3_07.m

```
% ~~~~~~~~~~~~~~~~~~~~~~~~~~~~~~~~~~~~~~~~~~~~~~
% Example_3_07
% ~~~~~~~~~~~
%
% This program computes the state vector (R,V) from the initial
% state vector (R0,V0) and the elapsed time using the data in
% Example 3.7.
%
% mu - gravitational parameter (km^3/s^2)
% R0 - the initial position vector (km)
% V0 - the initial velocity vector (km/s)
% R  - the final position vector (km)
% V  - the final velocity vector (km/s)
% t  - elapsed time (s)
%
% User m-functions required: rv_from_r0v0
% ----------------------------------------------

clear all; clc
global mu
mu = 398600;

%...Data declaration for Example 3.7:
R0 = [  7000 -12124 0];
V0 = [2.6679 4.6210 0];
t  = 3600;
%...

%...Algorithm 3.4:
[R V] = rv_from_r0v0(R0, V0, t);

%...Echo the input data and output the results to the command window:
fprintf('-----------------------------------------------------')
fprintf('\n Example 3.7\n')
fprintf('\n Initial position vector (km):')
fprintf('\n   r0 = (%g, %g, %g)\n', R0(1), R0(2), R0(3))
fprintf('\n Initial velocity vector (km/s):')
fprintf('\n   v0 = (%g, %g, %g)', V0(1), V0(2), V0(3))
fprintf('\n\n Elapsed time = %g s\n',t)
fprintf('\n Final position vector (km):')
```

```
fprintf('\n   r = (%g, %g, %g)\n', R(1), R(2), R(3))
fprintf('\n Final velocity vector (km/s):')
fprintf('\n   v = (%g, %g, %g)', V(1), V(2), V(3))
fprintf('\n-----------------------------------------------------\n')
% ~~~~~~~~~~~~~~~~~~~~~~~~~~~~~~~~~~~~
```

OUTPUT FROM `Example_3_07`

```
-----------------------------------------------------
 Example 3.7

 Initial position vector (km):
   r0 = (7000, -12124, 0)

 Initial velocity vector (km/s):
   v0 = (2.6679, 4.621, 0)

 Elapsed time = 3600 s

 Final position vector (km):
   r = (-3297.77, 7413.4, 0)

 Final velocity vector (km/s):
   v = (-8.2976, -0.964045, -0)
-----------------------------------------------------
```

CHAPTER 4: ORBITS IN THREE DIMENSIONS

D.17 ALGORITHM 4.1: OBTAIN THE RIGHT ASCENSION AND DECLINATION FROM THE POSITION VECTOR

FUNCTION FILE: `ra_and_dec_from_r.m`

```
% ~~~~~~~~~~~~~~~~~~~~~~~~~~~~~~~~
function [ra dec] = ra_and_dec_from_r(r)
% ~~~~~~~~~~~~~~~~~~~~~~~~~~~~~~~~
%{
  This function calculates the right ascension and the
  declination from the geocentric equatorial position vector.

  r       - position vector
  l, m, n - direction cosines of r
  ra      - right ascension (degrees)
  dec     - declination (degrees)
%}
% ----------------------------------------------
l = r(1)/norm(r);
m = r(2)/norm(r);
```

```
n = r(3)/norm(r);

dec = asind(n);

if m > 0
    ra = acosd(l/cosd(dec));
else
    ra = 360 - acosd(l/cosd(dec));
end
% ~~~~~~~~~~~~~~~~~~~~~~~~~~~~~~~~~~~~~~~~
```

SCRIPT FILE: `Example_4_01.m`

```
% ~~~~~~~~~~~~~~~~~~~~~~~~~~~~~~~~~~~~~~~~~
% Example 4.1
% ~~~~~~~~~~
%{
  This program calculates the right ascension and declination
  from the geocentric equatorial position vector using the data
  in Example 4.1.

  r   - position vector r (km)
  ra  - right ascension (deg)
  dec - declination (deg)

  User M-functions required: ra_and_dec_from_r

%}
% ----------------------------------------------

clear all; clc

r        = [-5368 -1784 3691];
[ra dec] = ra_and_dec_from_r(r);

fprintf('\n -----------------------------------------------------\n')
fprintf('\n Example 4.1\n')
fprintf('\n r               = [%g  %g  %g] (km)', r(1), r(2), r(3))
fprintf('\n right ascension = %g deg', ra)
fprintf('\n declination     = %g deg', dec)
fprintf('\n\n -----------------------------------------------------\n')

% ~~~~~~~~~~~~~~~~~~~~~~~~~~~~~~~~~~~~~~~~~
```

OUTPUT FROM `Example_4_01.m`

```
-----------------------------------------------------
 Example 4.1
```

```
r               = [-5368  -1784  3691] (km)
right ascension = 198.384 deg
declination     = 33.1245 deg

-----------------------------------------------------
```

D.18 ALGORITHM 4.2: CALCULATION OF THE ORBITAL ELEMENTS FROM THE STATE VECTOR

FUNCTION FILE: coe_from_sv.m

```
% ~~~~~~~~~~~~~~~~~~~~~~~~~~~~~~~~~~~~~~~~~~~~~
  function coe = coe_from_sv(R,V,mu)
% ~~~~~~~~~~~~~~~~~~~~~~~~~~~~~~~~~~~~
%{
% This function computes the classical orbital elements (coe)
% from the state vector (R,V) using Algorithm 4.1.
%
  mu   - gravitational parameter (km^3/s^2)
  R    - position vector in the geocentric equatorial frame (km)
  V    - velocity vector in the geocentric equatorial frame (km)
  r, v - the magnitudes of R and V
  vr   - radial velocity component (km/s)
  H    - the angular momentum vector (km^2/s)
  h    - the magnitude of H (km^2/s)
  incl - inclination of the orbit (rad)
  N    - the node line vector (km^2/s)
  n    - the magnitude of N
  cp   - cross product of N and R
  RA   - right ascension of the ascending node (rad)
  E    - eccentricity vector
  e    - eccentricity (magnitude of E)
  eps  - a small number below which the eccentricity is considered
         to be zero
  w    - argument of perigee (rad)
  TA   - true anomaly (rad)
  a    - semimajor axis (km)
  pi   - 3.1415926...
  coe  - vector of orbital elements [h e RA incl w TA a]

  User M-functions required: None
%}
% ---------------------------------------------

eps = 1.e-10;

r    = norm(R);
v    = norm(V);
```

```
vr   = dot(R,V)/r;

H    = cross(R,V);
h    = norm(H);

%...Equation 4.7:
incl = acos(H(3)/h);

%...Equation 4.8:
N    = cross([0 0 1],H);
n    = norm(N);

%...Equation 4.9:
if n ~= 0
    RA = acos(N(1)/n);
    if N(2) < 0
        RA = 2*pi - RA;
    end
else
    RA = 0;
end

%...Equation 4.10:
E = 1/mu*((v^2 - mu/r)*R - r*vr*V);
e = norm(E);

%...Equation 4.12 (incorporating the case e = 0):
if n ~= 0
    if e > eps
        w = acos(dot(N,E)/n/e);
        if E(3) < 0
            w = 2*pi - w;
        end
    else
        w = 0;
    end
else
    w = 0;
end

%...Equation 4.13a (incorporating the case e = 0):
if e > eps
    TA = acos(dot(E,R)/e/r);
    if vr < 0
        TA = 2*pi - TA;
    end
else
```

```
    cp = cross(N,R);
    if cp(3) >= 0
        TA = acos(dot(N,R)/n/r);
    else
        TA = 2*pi - acos(dot(N,R)/n/r);
    end
end

%...Equation 4.62 (a < 0 for a hyperbola):
a = h^2/mu/(1 - e^2);

coe = [h e RA incl w TA a];

  end  %coe_from_sv
% ~~~~~~~~~~~~~~~~~~~~~~~~~~~~~~~~~~~~~~~~
```

SCRIPT FILE: Example_4_03.m

```
% ~~~~~~~~~~~~~~~~~~~~~~~~~~~~~~~~~~~~~~~~~
% Example_4_03
% ~~~~~~~~~~~~
%{
  This program uses Algorithm 4.2 to obtain the orbital
  elements from the state vector provided in Example 4.3.

  pi   - 3.1415926...
  deg  - factor for converting between degrees and radians
  mu   - gravitational parameter (km^3/s^2)
  r    - position vector (km) in the geocentric equatorial frame
  v    - velocity vector (km/s) in the geocentric equatorial frame
  coe  - orbital elements [h e RA incl w TA a]
         where h    = angular momentum (km^2/s)
               e    = eccentricity
               RA   = right ascension of the ascending node (rad)
               incl = orbit inclination (rad)
               w    = argument of perigee (rad)
               TA   = true anomaly (rad)
               a    = semimajor axis (km)
  T    - Period of an elliptic orbit (s)

  User M-function required: coe_from_sv
%}
% ---------------------------------------------
clear all; clc
deg = pi/180;
mu  = 398600;

%...Data declaration for Example 4.3:
```

```
r = [ -6045  -3490   2500];
v = [-3.457  6.618  2.533];
%...

%...Algorithm 4.2:
coe = coe_from_sv(r,v,mu);

%...Echo the input data and output results to the command window:
fprintf('---------------------------------------------------')
fprintf('\n Example 4.3\n')
fprintf('\n Gravitational parameter (km^3/s^2) = %g\n', mu)
fprintf('\n State vector:\n')
fprintf('\n r (km)                               = [%g  %g  %g]', ...
                                                 r(1), r(2), r(3))
fprintf('\n v (km/s)                             = [%g  %g  %g]', ...
                                                 v(1), v(2), v(3))
disp(' ')
fprintf('\n Angular momentum (km^2/s)            = %g', coe(1))
fprintf('\n Eccentricity                         = %g', coe(2))
fprintf('\n Right ascension (deg)                = %g', coe(3)/deg)
fprintf('\n Inclination (deg)                    = %g', coe(4)/deg)
fprintf('\n Argument of perigee (deg)            = %g', coe(5)/deg)
fprintf('\n True anomaly (deg)                   = %g', coe(6)/deg)
fprintf('\n Semimajor axis (km):                 = %g', coe(7))

%...if the orbit is an ellipse, output its period (Equation 2.73):
if coe(2)<1
    T = 2*pi/sqrt(mu)*coe(7)^1.5;
    fprintf('\n Period:')
    fprintf('\n   Seconds                            = %g', T)
    fprintf('\n   Minutes                            = %g', T/60)
    fprintf('\n   Hours                              = %g', T/3600)
    fprintf('\n   Days                               = %g', T/24/3600)
end
fprintf('\n-----------------------------------------------------\n')

% ~~~~~~~~~~~~~~~~~~~~~~~~~~~~~~~~~~~~~~~~
```

OUTPUT FROM Example_4_03

```
-----------------------------------------------------
 Example 4.3

 Gravitational parameter (km^3/s^2) = 398600

 State vector:

 r (km)                         = [-6045  -3490  2500]
 v (km/s)                       = [-3.457  6.618  2.533]
```

```
 Angular momentum (km^2/s)           = 58311.7
 Eccentricity                        = 0.171212
 Right ascension (deg)               = 255.279
 Inclination (deg)                   = 153.249
 Argument of perigee (deg)           = 20.0683
 True anomaly (deg)                  = 28.4456
 Semimajor axis (km):                = 8788.1
 Period:
   Seconds                           = 8198.86
   Minutes                           = 136.648
   Hours                             = 2.27746
   Days                              = 0.0948942
-----------------------------------------------------
```

D.19 CALCULATION OF ARCTAN (*Y/X*) TO LIE IN THE RANGE 0° TO 360°

FUNCTION FILE: atan2d_0_360.m

```
% ~~~~~~~~~~~~~~~~~~~~~~~~~~~~~~~~~~~~~~~~~~~~~~~
function t = atan2d_0_360(y,x)
% ~~~~~~~~~~~~~~~~~~~~~~~~~~~~~
%{
  This function calculates the arc tangent of y/x in degrees
  and places the result in the range [0, 360].

  t - angle in degrees

%}
% ----------------------------------------------

if x == 0
    if y == 0
        t = 0;
    elseif y > 0
        t = 90;
    else
        t = 270;
    end
elseif x > 0
    if y >= 0
        t = atand(y/x);
    else
        t = atand(y/x) + 360;
    end
elseif x < 0
    if y == 0
```

```
        t = 180;
    else
        t = atand(y/x) + 180;
    end
end

end

% ~~~~~~~~~~~~~~~~~~~~~~~~~~~~~~~~~~~~~~~~~~~~
```

D.20 ALGORITHM 4.3: OBTAIN THE CLASSICAL EULER ANGLE SEQUENCE FROM A DIRECTION COSINE MATRIX

FUNCTION FILE: dcm_to_euler.m

```
% ~~~~~~~~~~~~~~~~~~~~~~~~~~~~~~~~~~~~~~~~~~~~~
function [alpha beta gamma] = dcm_to_euler(Q)
% ~~~~~~~~~~~~~~~~~~~~~~~~~~~~~~~~~~~~~~~~~~~
%{
  This function finds the angles of the classical Euler sequence
  R3(gamma)*R1(beta)*R3(alpha) from the direction cosine matrix.

  Q     - direction cosine matrix
  alpha - first angle of the sequence (deg)
  beta  - second angle of the sequence (deg)
  gamma - third angle of the sequence (deg)

  User M-function required: atan2d_0_360
%}
% ----------------------------------------------

alpha = atan2d_0_360(Q(3,1), -Q(3,2));
beta  = acosd(Q(3,3));
gamma = atan2d_0_360(Q(1,3), Q(2,3));

end
% ~~~~~~~~~~~~~~~~~~~~~~~~~~~~~~~~~~~~~~~~~~~~~
```

D.21 ALGORITHM 4.4: OBTAIN THE YAW, PITCH, AND ROLL ANGLES FROM A DIRECTION COSINE MATRIX

FUNCTION FILE: dcm_to_ypr.m

```
% ~~~~~~~~~~~~~~~~~~~~~~~~~~~~~~~~~~~~~~~~~~~~~
function [yaw pitch roll] = dcm_to_ypr(Q)
% ~~~~~~~~~~~~~~~~~~~~~~~~~~~~~~~~~~~~~~~
%{
  This function finds the angles of the yaw-pitch-roll sequence
```

```
  R1(gamma)*R2(beta)*R3(alpha) from the direction cosine matrix.

  Q     - direction cosine matrix
  yaw   - yaw angle (deg)
  pitch - pitch angle (deg)
  roll  - roll angle (deg)

  User M-function required: atan2d_0_360
%}
% ----------------------------------------

yaw   = atan2d_0_360(Q(1,2), Q(1,1));
pitch = asind(-Q(1,3));
roll  = atan2d_0_360(Q(2,3), Q(3,3));
end
% ~~~~~~~~~~~~~~~~~~~~~~~~~~~~~~~
```

D.22 ALGORITHM 4.5: CALCULATION OF THE STATE VECTOR FROM THE ORBITAL ELEMENTS

FUNCTION FILE: sv_from_coe.m

```
% ~~~~~~~~~~~~~~~~~~~~~~~~~~~~~~~~~~~~~~~
  function [r, v] = sv_from_coe(coe,mu)
% ~~~~~~~~~~~~~~~~~~~~~~~~~~~~~~~~
%{
  This function computes the state vector (r,v) from the
  classical orbital elements (coe).

  mu   - gravitational parameter (km^3/s^2)
  coe  - orbital elements [h e RA incl w TA]
         where
             h    = angular momentum (km^2/s)
             e    = eccentricity
             RA   = right ascension of the ascending node (rad)
             incl = inclination of the orbit (rad)
             w    = argument of perigee (rad)
             TA   = true anomaly (rad)
  R3_w - Rotation matrix about the z-axis through the angle w
  R1_i - Rotation matrix about the x-axis through the angle i
  R3_W - Rotation matrix about the z-axis through the angle RA
  Q_pX - Matrix of the transformation from perifocal to geocentric
         equatorial frame
  rp   - position vector in the perifocal frame (km)
  vp   - velocity vector in the perifocal frame (km/s)
  r    - position vector in the geocentric equatorial frame (km)
  v    - velocity vector in the geocentric equatorial frame (km/s)
```

```
  User M-functions required: none
%}
% ----------------------------------------------

h    = coe(1);
e    = coe(2);
RA   = coe(3);
incl = coe(4);
w    = coe(5);
TA   = coe(6);

%...Equations 4.45 and 4.46 (rp and vp are column vectors):
rp = (h^2/mu) * (1/(1 + e*cos(TA))) * (cos(TA)*[1;0;0] + sin(TA)*[0;1;0]);
vp = (mu/h) * (-sin(TA)*[1;0;0] + (e + cos(TA))*[0;1;0]);

%...Equation 4.34:
R3_W = [ cos(RA)  sin(RA)  0
        -sin(RA)  cos(RA)  0
            0        0     1];

%...Equation 4.32:
R1_i = [1       0          0
        0   cos(incl)  sin(incl)
        0  -sin(incl)  cos(incl)];

%...Equation 4.34:
R3_w = [ cos(w)  sin(w)  0
        -sin(w)  cos(w)  0
           0       0     1];

%...Equation 4.49:
Q_pX = (R3_w*R1_i*R3_W)';

%...Equations 4.51 (r and v are column vectors):
r = Q_pX*rp;
v = Q_pX*vp;

%...Convert r and v into row vectors:
r = r';
v = v';

end
% ~~~~~~~~~~~~~~~~~~~~~~~~~~~~~~~~~~~~~~~
```

SCRIPT FILE: Example_4_07.m

```
% ~~~~~~~~~~~~~~~~~~~~~~~~~~~~~~~~~~~~~
% Example_4_07
% ~~~~~~~~~
%{
  This program uses Algorithm 4.5 to obtain the state vector from
  the orbital elements provided in Example 4.7.

  pi  - 3.1415926...
  deg - factor for converting between degrees and radians
  mu  - gravitational parameter (km^3/s^2)
  coe - orbital elements [h e RA incl w TA a]
        where h    = angular momentum (km^2/s)
              e    = eccentricity
              RA   = right ascension of the ascending node (rad)
              incl = orbit inclination (rad)
              w    = argument of perigee (rad)
              TA   = true anomaly (rad)
              a    = semimajor axis (km)
  r   - position vector (km) in geocentric equatorial frame
  v   - velocity vector (km) in geocentric equatorial frame

  User M-function required: sv_from_coe
%}
% ----------------------------------------------
clear all; clc
deg = pi/180;
mu  = 398600;

%...Data declaration for Example 4.5 (angles in degrees):
h    = 80000;
e    = 1.4;
RA   = 40;
incl = 30;
w    = 60;
TA   = 30;
%...

coe = [h, e, RA*deg, incl*deg, w*deg, TA*deg];

%...Algorithm 4.5 (requires angular elements be in radians):
[r, v] = sv_from_coe(coe, mu);

%...Echo the input data and output the results to the command window:
fprintf('-----------------------------------------------------')
fprintf('\n Example 4.7\n')
fprintf('\n Gravitational parameter (km^3/s^2)  = %g\n', mu)
```

```
fprintf('\n Angular momentum (km^2/s)          = %g', h)
fprintf('\n Eccentricity                       = %g', e)
fprintf('\n Right ascension (deg)              = %g', RA)
fprintf('\n Argument of perigee (deg)          = %g', w)
fprintf('\n True anomaly (deg)                 = %g', TA)
fprintf('\n\n State vector:')
fprintf('\n   r (km)   = [%g  %g  %g]', r(1), r(2), r(3))
fprintf('\n   v (km/s) = [%g  %g  %g]', v(1), v(2), v(3))
fprintf('\n-----------------------------------------------------\n')
% ~~~~~~~~~~~~~~~~~~~~~~~~~~~~~~~~~~~~~~~
```

OUTPUT FROM Example_4_05

```
-----------------------------------------------------
 Example 4.7

 Gravitational parameter (km^3/s^2)  = 398600

 Angular momentum (km^2/s)           = 80000
 Eccentricity                        = 1.4
 Right ascension (deg)               = 40
 Argument of perigee (deg)           = 60
 True anomaly (deg)                  = 30

 State vector:
   r (km)   = [-4039.9  4814.56  3628.62]
   v (km/s) = [-10.386  -4.77192  1.74388]
-----------------------------------------------------
```

D.23 ALGORITHM 4.6: CALCULATE THE GROUND TRACK OF A SATELLITE FROM ITS ORBITAL ELEMENTS

[B] FUNCTION FILE: ground_track.m

```
% ~~~~~~~~~~~~~~~~~~~~~~~~~~~~~~~~~~~~~~~
function ground_track
% ~~~~~~~~~~~~~~~~
%{
  This program plots the ground track of an earth satellite
  for which the orbital elements are specified

  mu        - gravitational parameter (km^3/s^2)
  deg       - factor that converts degrees to radians
  J2        - second zonal harmonic
  Re        - earth's radius (km)
  we        - earth's angular velocity (rad/s)
  rP        - perigee of orbit (km)
  rA        - apogee of orbit (km)
  TA, TAo   - true anomaly, initial true anomaly of satellite (rad)
```

```
  RA, RAo    - right ascension, initial right ascension of the node (rad)
  incl       - orbit inclination (rad)
  wp, wpo    - argument of perigee, initial argument of perigee (rad)
  n_periods  - number of periods for which ground track is to be plotted
  a          - semimajor axis of orbit (km)
  T          - period of orbit (s)
  e          - eccentricity of orbit
  h          - angular momentum of orbit (km^2/s)
  E, Eo      - eccentric anomaly, initial eccentric anomaly (rad)
  M, Mo      - mean anomaly, initial mean anomaly (rad)
  to, tf     - initial and final times for the ground track (s)
  fac        - common factor in Equations 4.53 and 4.53
  RAdot      - rate of regression of the node (rad/s)
  wpdot      - rate of advance of perigee (rad/s)
  times      - times at which ground track is plotted (s)
  ra         - vector of right ascensions of the spacecraft (deg)
  dec        - vector of declinations of the spacecraft (deg)
  TA         - true anomaly (rad)
  r          - perifocal position vector of satellite (km)
  R          - geocentric equatorial position vector (km)
  R1         - DCM for rotation about z through RA
  R2         - DCM for rotation about x through incl
  R3         - DCM for rotation about z through wp
  QxX        - DCM for rotation from perifocal to geocentric equatorial
  Q          - DCM for rotation from geocentric equatorial
               into earth-fixed frame
  r_rel      - position vector in earth-fixed frame (km)
  alpha      - satellite right ascension (deg)
  delta      - satellite declination (deg)
  n_curves   - number of curves comprising the ground track plot
  RA         - cell array containing the right ascensions for each of
               the curves comprising the ground track plot
  Dec        - cell array containing the declinations for each of
               the curves comprising the ground track plot

  User M-functions required: sv_from_coe, kepler_E, ra_and_dec_from_r
%}
% ~~~~~~~~~~~~~~~~~~~~~~~~~~~~~~~~~~~~
clear all; close all; clc
global ra dec n_curves RA Dec

%...Constants
deg      = pi/180;
mu       = 398600;
J2       = 0.00108263;
Re       = 6378;
we       = (2*pi + 2*pi/365.26)/(24*3600);
```

```
%...Data declaration for Example 4.12:
rP         = 6700;
rA         = 10000;
TAo        = 230*deg;
Wo         = 270*deg;
incl       = 60*deg;
wpo        = 45*deg;
n_periods = 3.25;
%...End data declaration

%...Compute the initial time (since perigee) and
%   the rates of node regression and perigee advance
a          = (rA + rP)/2;
T          = 2*pi/sqrt(mu)*a^(3/2);
e          = (rA - rP)/(rA + rP);
h          = sqrt(mu*a*(1 - e^2));
Eo         = 2*atan(tan(TAo/2)*sqrt((1-e)/(1+e)));
Mo         = Eo - e*sin(Eo);
to         = Mo*(T/2/pi);
tf         = to + n_periods*T;
fac        = -3/2*sqrt(mu)*J2*Re^2/(1-e^2)^2/a^(7/2);
Wdot       = fac*cos(incl);
wpdot      = fac*(5/2*sin(incl)^2 - 2);

find_ra_and_dec
form_separate_curves
plot_ground_track
print_orbital_data

return

% ~~~~~~~~~~~~~~~~~~~
function find_ra_and_dec
% ~~~~~~~~~~~~~~~~~~~
% Propagates the orbit over the specified time interval, transforming
% the position vector into the earth-fixed frame and, from that,
% computing the right ascension and declination histories
% ----------------------
%
times = linspace(to,tf,1000);
ra    = [];
dec   = [];
theta = 0;
for i = 1:length(times)
    t              = times(i);
    M              = 2*pi/T*t;
    E              = kepler_E(e, M);
    TA             = 2*atan(tan(E/2)*sqrt((1+e)/(1-e)));
    r              = h^2/mu/(1 + e*cos(TA))*[cos(TA) sin(TA) 0]';
```

```
    W               = Wo  + Wdot*t;
    wp              = wpo + wpdot*t;

    R1              = [ cos(W)  sin(W)  0
                       -sin(W)  cos(W)  0
                          0       0     1];

    R2              = [1      0          0
                       0   cos(incl)  sin(incl)
                       0  -sin(incl)  cos(incl)];

    R3              = [ cos(wp)  sin(wp)  0
                       -sin(wp)  cos(wp)  0
                          0        0      1];

    QxX             = (R3*R2*R1)';
    R               = QxX*r;

    theta           = we*(t - to);
    Q               = [ cos(theta)  sin(theta)  0
                       -sin(theta)  cos(theta)  0
                           0           0        1];
    r_rel           = Q*R;

    [alpha delta] = ra_and_dec_from_r(r_rel);

    ra              = [ra;  alpha];
    dec             = [dec; delta];
end

end %find_ra_and_dec

% ~~~~~~~~~~~~~~~~~~~~~~
function form_separate_curves
% ~~~~~~~~~~~~~~~~~~~~~~
% Breaks the ground track up into separate curves which start
% and terminate at right ascensions in the range [0,360 deg].
% ---------------------------
tol = 100;
curve_no = 1;
n_curves = 1;
k        = 0;
ra_prev  = ra(1);
for i = 1:length(ra)
    if abs(ra(i) - ra_prev) > tol
```

```
            curve_no = curve_no + 1;
            n_curves = n_curves + 1;
            k = 0;
        end
        k                 = k + 1;
        RA{curve_no}(k)  = ra(i);
        Dec{curve_no}(k) = dec(i);
        ra_prev           = ra(i);
    end
    end %form_separate_curves

    % ~~~~~~~~~~~~~~~~~~~~
    function plot_ground_track
    % ~~~~~~~~~~~~~~~~~~~~
    hold on
    xlabel('East longitude (degrees)')
    ylabel('Latitude (degrees)')
    axis equal
    grid on
    for i = 1:n_curves
        plot(RA{i}, Dec{i})
    end

    axis ([0 360 -90 90])
    text(  ra(1),    dec(1), 'o Start')
    text(ra(end), dec(end), 'o Finish')
    line([min(ra) max(ra)],[0 0], 'Color','k') %the equator
    end %plot_ground_track

    % ~~~~~~~~~~~~~~~~~~~~~
    function print_orbital_data
    % ~~~~~~~~~~~~~~~~~~~~~
    coe       = [h e Wo incl wpo TAo];
    [ro, vo]  = sv_from_coe(coe, mu);
    fprintf('\n --------------------------------------------------\n')
    fprintf('\n Angular momentum     = %g km^2/s' , h)
    fprintf('\n Eccentricity         = %g'        , e)
    fprintf('\n Semimajor axis       = %g km'     , a)
    fprintf('\n Perigee radius       = %g km'     , rP)
    fprintf('\n Apogee radius        = %g km'     , rA)
    fprintf('\n Period               = %g hours'  , T/3600)
    fprintf('\n Inclination          = %g deg'    , incl/deg)
    fprintf('\n Initial true anomaly = %g deg'    , TAo/deg)
    fprintf('\n Time since perigee   = %g hours'  , to/3600)
    fprintf('\n Initial RA           = %g deg'    , Wo/deg)
    fprintf('\n RA_dot               = %g deg/period' , Wdot/deg*T)
    fprintf('\n Initial wp           = %g deg'    , wpo/deg)
```

```
fprintf('\n wp_dot               = %g deg/period' , wpdot/deg*T)
fprintf('\n')
fprintf('\n r0 = [%12g, %12g, %12g] (km)', ro(1), ro(2), ro(3))
fprintf('\n magnitude = %g km\n', norm(ro))
fprintf('\n v0 = [%12g, %12g, %12g] (km)', vo(1), vo(2), vo(3))
fprintf('\n magnitude = %g km\n', norm(vo))
fprintf('\n -----------------------------------------------------\n')

end %print_orbital_data

end %ground_track
% ~~~~~~~~~~~~~~~~~~~~~~~~~~~~~~~~~~~~~~~~~~~~~~
```

CHAPTER 5: PRELIMINARY ORBIT DETERMINATION

D.24 ALGORITHM 5.1: GIBBS' METHOD OF PRELIMINARY ORBIT DETERMINATION

FUNCTION FILE: gibbs.m

```
% ~~~~~~~~~~~~~~~~~~~~~~~~~~~~~~~~~~~~~~~~
  function [V2, ierr] = gibbs(R1, R2, R3)
% ~~~~~~~~~~~~~~~~~~~~~~~~~~~~~~~~~~~~~~~~
%{
  This function uses the Gibbs method of orbit determination to
  to compute the velocity corresponding to the second of three
  supplied position vectors.

  mu             - gravitational parameter (km^3/s^2)
  R1, R2, R3     - three coplanar geocentric position vectors (km)
  r1, r2, r3     - the magnitudes of R1, R2 and R3 (km)
  c12, c23, c31  - three independent cross products among
                   R1, R2 and R3
  N, D, S        - vectors formed from R1, R2 and R3 during
                   the Gibbs' procedure
  tol            - tolerance for determining if R1, R2 and R3
                   are coplanar
  ierr           - = 0 if R1, R2, R3 are found to be coplanar
                   = 1 otherwise
  V2             - the velocity corresponding to R2 (km/s)

  User M-functions required: none
%}
% ---------------------------------------

global mu
tol  = 1e-4;
ierr = 0;
```

```
%...Magnitudes of R1, R2 and R3:
r1 = norm(R1);
r2 = norm(R2);
r3 = norm(R3);

%...Cross products among R1, R2 and R3:
c12 = cross(R1,R2);
c23 = cross(R2,R3);
c31 = cross(R3,R1);

%...Check that R1, R2 and R3 are coplanar; if not set error flag:
if abs(dot(R1,c23)/r1/norm(c23)) > tol
    ierr = 1;
end

%...Equation 5.13:
N = r1*c23 + r2*c31 + r3*c12;

%...Equation 5.14:
D = c12 + c23 + c31;

%...Equation 5.21:
S = R1*(r2 - r3) + R2*(r3 - r1) + R3*(r1 - r2);

%...Equation 5.22:
V2 = sqrt(mu/norm(N)/norm(D))*(cross(D,R2)/r2 + S);
% ~~~~~~~~~~~~~~~~~~~~~~~~~~~~~~~~~~~
  end %gibbs
```

SCRIPT FILE: Example_5_01.m

```
% ~~~~~~~~~~~~~~~~~~~~~~~~~~~~~~~~~~~~~~~~~~~~
% Example_5_01
% ~~~~~~~~~~~~
%{
  This program uses Algorithm 5.1 (Gibbs method) and Algorithm 4.2
  to obtain the orbital elements from the data provided in Example 5.1.

  deg         - factor for converting between degrees and radians
  pi          - 3.1415926...
  mu          - gravitational parameter (km^3/s^2)
  r1, r2, r3  - three coplanar geocentric position vectors (km)
  ierr        - 0 if r1, r2, r3 are found to be coplanar
                1 otherwise
  v2          - the velocity corresponding to r2 (km/s)
  coe         - orbital elements [h e RA incl w TA a]
                where h    = angular momentum (km^2/s)
                      e    = eccentricity
                      RA   = right ascension of the ascending node (rad)
```

```
                    incl = orbit inclination (rad)
                    w    = argument of perigee (rad)
                    TA   = true anomaly (rad)
                    a    = semimajor axis (km)
  T         - period of elliptic orbit (s)

  User M-functions required: gibbs, coe_from_sv
%}
% ---------------------------------------------

clear all; clc
deg = pi/180;
global mu

%...Data declaration for Example 5.1:
mu = 398600;
r1 = [-294.32 4265.1 5986.7];
r2 = [-1365.5 3637.6 6346.8];
r3 = [-2940.3 2473.7 6555.8];
%...

%...Echo the input data to the command window:
fprintf('-----------------------------------------------------')
fprintf('\n Example 5.1: Gibbs Method\n')
fprintf('\n\n Input data:\n')
fprintf('\n  Gravitational parameter (km^3/s^2)  = %g\n', mu)
fprintf('\n  r1 (km) = [%g  %g  %g]', r1(1), r1(2), r1(3))
fprintf('\n  r2 (km) = [%g  %g  %g]', r2(1), r2(2), r2(3))
fprintf('\n  r3 (km) = [%g  %g  %g]', r3(1), r3(2), r3(3))
fprintf('\n\n');

%...Algorithm 5.1:
[v2, ierr] = gibbs(r1, r2, r3);

%...If the vectors r1, r2, r3, are not coplanar, abort:
if ierr == 1
    fprintf('\n  These vectors are not coplanar.\n\n')
    return
end

%...Algorithm 4.2:
coe  = coe_from_sv(r2,v2,mu);

h    = coe(1);
e    = coe(2);
RA   = coe(3);
incl = coe(4);
```

```
w    = coe(5);
TA   = coe(6);
a    = coe(7);

%...Output the results to the command window:
fprintf(' Solution:')
fprintf('\n');
fprintf('\n  v2 (km/s) = [%g  %g  %g]', v2(1), v2(2), v2(3))
fprintf('\n\n  Orbital elements:');
fprintf('\n    Angular momentum (km^2/s)  = %g', h)
fprintf('\n    Eccentricity               = %g', e)
fprintf('\n    Inclination (deg)          = %g', incl/deg)
fprintf('\n    RA of ascending node (deg) = %g', RA/deg)
fprintf('\n    Argument of perigee (deg)  = %g', w/deg)
fprintf('\n    True anomaly (deg)         = %g', TA/deg)
fprintf('\n    Semimajor axis (km)        = %g', a)
%...If the orbit is an ellipse, output the period:
if e < 1
    T = 2*pi/sqrt(mu)*coe(7)^1.5;
    fprintf('\n    Period (s)                 = %g', T)
end
fprintf('\n-----------------------------------------------------\n')
% ~~~~~~~~~~~~~~~~~~~~~~~~~~~~~~~~~~~~
```

OUTPUT FROM Example_5_01

```
-----------------------------------------------------
         Example 5.1: Gibbs Method

         Input data:

      Gravitational parameter (km^3/s^2)  = 398600

      r1 (km) = [-294.32  4265.1  5986.7]
      r2 (km) = [-1365.4  3637.6  6346.8]
      r3 (km) = [-2940.3  2473.7  6555.8]

         Solution:

      v2 (km/s) = [-6.2176  -4.01237  1.59915]

      Orbital elements:
        Angular momentum (km^2/s)  = 56193
        Eccentricity               = 0.100159
        Inclination (deg)          = 60.001
        RA of ascending node (deg) = 40.0023
        Argument of perigee (deg)  = 30.1093
        True anomaly (deg)         = 49.8894
```

```
    Semimajor axis (km)          = 8002.14
    Period (s)                   = 7123.94
-----------------------------------------------------
```

D.25 ALGORITHM 5.2: SOLUTION OF LAMBERT'S PROBLEM

FUNCTION FILE: lambert.m

```
% ~~~~~~~~~~~~~~~~~~~~~~~~~~~~~~~~~~~~~~~~~~~~~~~
  function [V1, V2] = lambert(R1, R2, t, string)
% ~~~~~~~~~~~~~~~~~~~~~~~~~~~~~~~~~~~~~~~~~~~~~~~
%{
  This function solves Lambert's problem.

  mu          - gravitational parameter (km^3/s^2)
  R1, R2      - initial and final position vectors (km)
  r1, r2      - magnitudes of R1 and R2
  t           - the time of flight from R1 to R2 (a constant) (s)
  V1, V2      - initial and final velocity vectors (km/s)
  c12         - cross product of R1 into R2
  theta       - angle between R1 and R2
  string      - 'pro'   if the orbit is prograde
                'retro' if the orbit is retrograde
  A           - a constant given by Equation 5.35
  z           - alpha*x^2, where alpha is the reciprocal of the
                semimajor axis and x is the universal anomaly
  y(z)        - a function of z given by Equation 5.38
  F(z,t)      - a function of the variable z and constant t,
              - given by Equation 5.40
  dFdz(z)     - the derivative of F(z,t), given by Equation 5.43
  ratio       - F/dFdz
  tol         - tolerance on precision of convergence
  nmax        - maximum number of iterations of Newton's procedure
  f, g        - Lagrange coefficients
  gdot        - time derivative of g
  C(z), S(z)  - Stumpff functions
  dum         - a dummy variable

  User M-functions required: stumpC and stumpS
%}
% ----------------------------------------------

global mu

%...Magnitudes of R1 and R2:
r1 = norm(R1);
r2 = norm(R2);
```

```
c12   = cross(R1, R2);
theta = acos(dot(R1,R2)/r1/r2);

%...Determine whether the orbit is prograde or retrograde:
if nargin < 4 || (~strcmp(string,'retro') & (~strcmp(string,'pro')))
    string = 'pro';
    fprintf('\n ** Prograde trajectory assumed.\n')
end

if strcmp(string,'pro')
    if c12(3) <= 0
        theta = 2*pi - theta;
    end
elseif strcmp(string,'retro')
    if c12(3) >= 0
        theta = 2*pi - theta;
    end
end

%...Equation 5.35:
A = sin(theta)*sqrt(r1*r2/(1 - cos(theta)));

%...Determine approximately where F(z,t) changes sign, and
%...use that value of z as the starting value for Equation 5.45:
z = -100;
while F(z,t) < 0
    z = z + 0.1;
end

%...Set an error tolerance and a limit on the number of iterations:
tol   = 1.e-8;
nmax  = 5000;

%...Iterate on Equation 5.45 until z is determined to within the
%...error tolerance:
ratio = 1;
n     = 0;
while (abs(ratio) > tol) & (n <= nmax)
    n     = n + 1;
    ratio = F(z,t)/dFdz(z);
    z     = z - ratio;
end

%...Report if the maximum number of iterations is exceeded:
if n >= nmax
    fprintf('\n\n **Number of iterations exceeds %g \n\n ',nmax)
end
```

```
%...Equation 5.46a:
f    = 1 - y(z)/r1;

%...Equation 5.46b:
g    = A*sqrt(y(z)/mu);

%...Equation 5.46d:
gdot = 1 - y(z)/r2;

%...Equation 5.28:
V1   = 1/g*(R2 - f*R1);

%...Equation 5.29:
V2   = 1/g*(gdot*R2 - R1);

return

% ~~~~~~~~~~~~~~~~~~~~~~~~~~~~~~~~~~~
% Subfunctions used in the main body:
% ~~~~~~~~~~~~~~~~~~~~~~~~~~~~~~~~~~~

%...Equation 5.38:
function dum = y(z)
    dum = r1 + r2 + A*(z*S(z) - 1)/sqrt(C(z));
end

%...Equation 5.40:
function dum = F(z,t)
    dum = (y(z)/C(z))^1.5*S(z) + A*sqrt(y(z)) - sqrt(mu)*t;
end

%...Equation 5.43:
function dum = dFdz(z)
    if z == 0
        dum = sqrt(2)/40*y(0)^1.5 + A/8*(sqrt(y(0)) + A*sqrt(1/2/y(0)));
    else
        dum = (y(z)/C(z))^1.5*(1/2/z*(C(z) - 3*S(z)/2/C(z)) ...
               + 3*S(z)^2/4/C(z)) + A/8*(3*S(z)/C(z)*sqrt(y(z)) ...
               + A*sqrt(C(z)/y(z)));
    end
end

%...Stumpff functions:
function dum = C(z)
    dum = stumpC(z);
end
```

```
function dum = S(z)
    dum = stumpS(z);
end

end %lambert

% ~~~~~~~~~~~~~~~~~~~~~~~~~~~~~~~~~~~~~~~~~~
```

SCRIPT FILE: Example_5_02.m

```
% ~~~~~~~~~~~~~~~~~~~~~~~~~~~~~~~~~~~~~~~~~
% Example_5_02
% ~~~~~~~~~~~
%{
  This program uses Algorithm 5.2 to solve Lambert's problem for the
  data provided in Example 5.2.

  deg    - factor for converting between degrees and radians
  pi     - 3.1415926...
  mu     - gravitational parameter (km^3/s^2)
  r1, r2 - initial and final position vectors (km)
  dt     - time between r1 and r2 (s)
  string - = 'pro' if the orbit is prograde
           = 'retro if the orbit is retrograde
  v1, v2 - initial and final velocity vectors (km/s)
  coe    - orbital elements [h e RA incl w TA a]
           where h    = angular momentum (km^2/s)
                 e    = eccentricity
                 RA   = right ascension of the ascending node (rad)
                 incl = orbit inclination (rad)
                 w    = argument of perigee (rad)
                 TA   = true anomaly (rad)
                 a    = semimajor axis (km)
  TA1    - Initial true anomaly (rad)
  TA2    - Final true anomaly (rad)
  T      - period of an elliptic orbit (s)

  User M-functions required: lambert, coe_from_sv
%}
% ---------------------------------------------

clear all; clc
global mu
deg = pi/180;

%...Data declaration for Example 5.2:
mu    = 398600;
r1    = [  5000  10000  2100];
```

```
r2     = [-14600   2500  7000];
dt     = 3600;
string = 'pro';
%...

%...Algorithm 5.2:
[v1, v2] = lambert(r1, r2, dt, string);

%...Algorithm 4.1 (using r1 and v1):
coe      = coe_from_sv(r1, v1, mu);
%...Save the initial true anomaly:
TA1      = coe(6);

%...Algorithm 4.1 (using r2 and v2):
coe      = coe_from_sv(r2, v2, mu);
%...Save the final true anomaly:
TA2      = coe(6);

%...Echo the input data and output the results to the command window:
fprintf('-----------------------------------------------------')
fprintf('\n Example 5.2: Lambert''s Problem\n')
fprintf('\n\n Input data:\n');
fprintf('\n   Gravitational parameter (km^3/s^2) = %g\n', mu);
fprintf('\n   r1 (km)                       = [%g  %g  %g]', ...
                                             r1(1), r1(2), r1(3))
fprintf('\n   r2 (km)                       = [%g  %g  %g]', ...
                                             r2(1), r2(2), r2(3))
fprintf('\n   Elapsed time (s)              = %g', dt);
fprintf('\n\n Solution:\n')

fprintf('\n   v1 (km/s)                     = [%g  %g  %g]', ...
                                             v1(1), v1(2), v1(3))
fprintf('\n   v2 (km/s)                     = [%g  %g  %g]', ...
                                             v2(1), v2(2), v2(3))


fprintf('\n\n Orbital elements:')
fprintf('\n   Angular momentum (km^2/s)      = %g', coe(1))
fprintf('\n   Eccentricity                   = %g', coe(2))
fprintf('\n   Inclination (deg)              = %g', coe(4)/deg)
fprintf('\n   RA of ascending node (deg)     = %g', coe(3)/deg)
fprintf('\n   Argument of perigee (deg)      = %g', coe(5)/deg)
fprintf('\n   True anomaly initial (deg)     = %g', TA1/deg)
fprintf('\n   True anomaly final   (deg)     = %g', TA2/deg)
fprintf('\n   Semimajor axis (km)            = %g', coe(7))
fprintf('\n   Periapse radius (km)           = %g', coe(1)^2/mu/(1 + coe(2)))
%...If the orbit is an ellipse, output its period:
if coe(2)<1
```

```
    T = 2*pi/sqrt(mu)*coe(7)^1.5;
    fprintf('\n   Period:')
    fprintf('\n     Seconds                          = %g', T)
    fprintf('\n     Minutes                          = %g', T/60)
    fprintf('\n     Hours                            = %g', T/3600)
    fprintf('\n     Days                             = %g', T/24/3600)
end
fprintf('\n-----------------------------------------------------\n')
% ~~~~~~~~~~~~~~~~~~~~~~~~~~~~~~~~~~~~~~~~~~~~~~~~~~~~~~~~~~~~
```

OUTPUT FROM Example_5_02

```
-----------------------------------------------------
 Example 5.2: Lambert's Problem

 Input data:

   Gravitational parameter (km^3/s^2) = 398600

   r1 (km)                       = [5000  10000  2100]
   r2 (km)                       = [-14600  2500  7000]
   Elapsed time (s)              = 3600

 Solution:

   v1 (km/s)                     = [-5.99249  1.92536  3.24564]
   v2 (km/s)                     = [-3.31246  -4.19662  -0.385288]

 Orbital elements:
   Angular momentum (km^2/s)     = 80466.8
   Eccentricity                  = 0.433488
   Inclination (deg)             = 30.191
   RA of ascending node (deg)    = 44.6002
   Argument of perigee (deg)     = 30.7062
   True anomaly initial (deg)    = 350.83
   True anomaly final   (deg)    = 91.1223
   Semimajor axis (km)           = 20002.9
   Periapse radius (km)          = 11331.9
   Period:
     Seconds                     = 28154.7
     Minutes                     = 469.245
     Hours                       = 7.82075
     Days                        = 0.325865
-----------------------------------------------------
```

D.26 CALCULATION OF JULIAN DAY NUMBER AT 0 HR UT

The following script implements Equation 5.48 for use in other programs.

FUNCTION FILE: J0.m

```
% ~~~~~~~~~~~~~~~~~~~~~~~~~~~~~~~~~~~
  function j0 = J0(year, month, day)
% ~~~~~~~~~~~~~~~~~~~~~~~~~~~~~~~s~~~~~~~
%{
  This function computes the Julian day number at 0 UT for any year
  between 1900 and 2100 using Equation 5.48.

  j0    - Julian day at 0 hr UT (Universal Time)
  year  - range: 1901 - 2099
  month - range: 1 - 12
  day   - range: 1 - 31

  User m-functions required: none
%}
% ----------------------------------

j0 = 367*year - fix(7*(year + fix((month + 9)/12))/4) ...
    + fix(275*month/9) + day + 1721013.5;

% ~~~~~~~~~~~~~~~~~~~~~~~~~~~~~
end %J0
```

SCRIPT FILE: Example_5_04.m

```
% ~~~~~~~~~~~~~~~~~~~~~~~~~~~~~~~~~~~~~~~~~~~~~~~~~~
% Example_5_04
% ~~~~~~~~~~~~
%{
  This program computes J0 and the Julian day number using the data
  in Example 5.4.

  year   - range: 1901 - 2099
  month  - range: 1 - 12
  day    - range: 1 - 31
  hour   - range: 0 - 23 (Universal Time)
  minute - rage: 0 - 60
  second - range: 0 - 60
  ut     - universal time (hr)
  j0     - Julian day number at 0 hr UT
  jd     - Julian day number at specified UT

  User M-function required: J0
%}
% ---------------------------------------------
```

```
clear all; clc

%...Data declaration for Example 5.4:
year   = 2004;
month  = 5;
day    = 12;

hour   = 14;
minute = 45;
second = 30;
%...

ut = hour + minute/60 + second/3600;

%...Equation 5.46:
j0 = J0(year, month, day);

%...Equation 5.47:
jd = j0 + ut/24;

%...Echo the input data and output the results to the command window:
fprintf('-----------------------------------------------------')
fprintf('\n Example 5.4: Julian day calculation\n')
fprintf('\n Input data:\n');
fprintf('\n   Year              = %g',   year)
fprintf('\n   Month             = %g',   month)
fprintf('\n   Day               = %g',   day)
fprintf('\n   Hour              = %g',   hour)
fprintf('\n   Minute            = %g',   minute)
fprintf('\n   Second            = %g\n', second)

fprintf('\n Julian day number = %11.3f', jd);
fprintf('\n-----------------------------------------------------\n')
% ~~~~~~~~~~~~~~~~~~~~~~~~~~~~~~~~~~~~
```

OUTPUT FROM `Example_5_04`

```
-----------------------------------------------------
 Example 5.4: Julian day calculation

 Input data:

   Year              = 2004
   Month             = 5
   Day               = 12
   Hour              = 14
   Minute            = 45
   Second            = 30
```

```
Julian day number = 2453138.115
-----------------------------------------------------
```

D.27 ALGORITHM 5.3: CALCULATION OF LOCAL SIDEREAL TIME

FUNCTION FILE: LST.m

```
% ~~~~~~~~~~~~~~~~~~~~~~~~~~~~~~~~~~~~~~~~~~~~~
  function lst = LST(y, m, d, ut, EL)
% ~~~~~~~~~~~~~~~~~~~~~~~~~~~~~~~~~~~~
%{
  This function calculates the local sidereal time.

  lst - local sidereal time (degrees)
  y   - year
  m   - month
  d   - day
  ut  - Universal Time (hours)
  EL  - east longitude (degrees)
  j0  - Julian day number at 0 hr UT
  j   - number of centuries since J2000
  g0  - Greenwich sidereal time (degrees) at 0 hr UT
  gst - Greenwich sidereal time (degrees) at the specified UT

  User M-function required:  J0
  User subfunction required: zeroTo360
%}
% ----------------------------------------------

%...Equation 5.48;
j0 = J0(y, m, d);

%...Equation 5.49:
j = (j0 - 2451545)/36525;

%...Equation 5.50:
g0 = 100.4606184 + 36000.77004*j + 0.000387933*j^2 - 2.583e-8*j^3;

%...Reduce g0 so it lies in the range 0 - 360 degrees
g0 = zeroTo360(g0);

%...Equation 5.51:
gst = g0 + 360.98564724*ut/24;

%...Equation 5.52:
lst = gst + EL;
```

```
%...Reduce lst to the range 0 - 360 degrees:
lst = lst - 360*fix(lst/360);

return

% ~~~~~~~~~~~~~~~~~~~~~~~~
  function y = zeroTo360(x)
% ~~~~~~~~~~~~~~~~~~~~~~~~
%{
  This subfunction reduces an angle to the range 0 - 360 degrees.

  x - The angle (degrees) to be reduced
  y - The reduced value
%}
% -------------------------
if (x >= 360)
    x = x - fix(x/360)*360;
elseif (x < 0)
    x = x - (fix(x/360) - 1)*360;
end
y = x;
end %zeroTo360

end %LST
% ~~~~~~~~~~~~~~~~~~~~~~~~~~~~~~~~~~~~~~~~~~~~~
```

SCRIPT FILE: Example_5_06.m

```
% ~~~~~~~~~~~~~~~~~~~~~~~~~~~~~~~~~~~~~~~~~~~~~
% Example_5_06
% ~~~~~~~~~~~
%{
  This program uses Algorithm 5.3 to obtain the local sidereal
  time from the data provided in Example 5.6.

  lst   - local sidereal time (degrees)
  EL    - east longitude of the site (west longitude is negative):
             degrees (0 - 360)
             minutes (0 - 60)
             seconds (0 - 60)
  WL    - west longitude
  year  - range: 1901 - 2099
  month - range: 1 - 12
  day   - range: 1 - 31
  ut    - universal time
             hour (0 - 23)
             minute (0 - 60)
             second (0 - 60)
```

```
  User m-function required: LST
%}
% ----------------------------------------------

clear all; clc

%...Data declaration for Example 5.6:
%   East longitude:
degrees = 139;
minutes = 47;
seconds = 0;

%   Date:
year    = 2004;
month   = 3;
day     = 3;

%   Universal time:
hour    = 4;
minute  = 30;
second  = 0;
%...

%...Convert negative (west) longitude to east longitude:
if degrees < 0
    degrees = degrees + 360;
end

%...Express the longitudes as decimal numbers:
EL = degrees + minutes/60 + seconds/3600;
WL = 360 - EL;

%...Express universal time as a decimal number:
ut = hour + minute/60 + second/3600;

%...Algorithm 5.3:
lst = LST(year, month, day, ut, EL);

%...Echo the input data and output the results to the command window:
fprintf('---------------------------------------------------')
fprintf('\n Example 5.6: Local sidereal time calculation\n')
fprintf('\n Input data:\n');
fprintf('\n   Year                          = %g', year)
fprintf('\n   Month                         = %g', month)
fprintf('\n   Day                           = %g', day)
fprintf('\n   UT (hr)                       = %g', ut)
fprintf('\n   West Longitude (deg)          = %g', WL)
```

```
fprintf('\n   East Longitude (deg)      = %g', EL)
fprintf('\n\n');

fprintf(' Solution:')

fprintf('\n');
fprintf('\n  Local Sidereal Time (deg) = %g', lst)
fprintf('\n  Local Sidereal Time (hr)  = %g', lst/15)

fprintf('\n-----------------------------------------------------\n')
% ~~~~~~~~~~~~~~~~~~~~~~~~~~~~~~~~~~~~~~~~~~~~
```

OUTPUT FROM `Example_5_06`

```
-----------------------------------------------------
 Example 5.6: Local sidereal time calculation

 Input data:

   Year                        = 2004
   Month                       = 3
   Day                         = 3
   UT (hr)                     = 4.5
   West Longitude (deg)        = 220.217
   East Longitude (deg)        = 139.783

 Solution:

  Local Sidereal Time (deg) = 8.57688
  Local Sidereal Time (hr)  = 0.571792
-----------------------------------------------------
```

D.28 ALGORITHM 5.4: CALCULATION OF THE STATE VECTOR FROM MEASUREMENTS OF RANGE, ANGULAR POSITION, AND THEIR RATES

FUNCTION FILE: `rv_from_observe.m`

```
% ~~~~~~~~~~~~~~~~~~~~~~~~~~~~~~~~~~~~~~~~~~~~~~~~~~~~~~~~~~~~~~~~~~~~~~
  function [r,v] = rv_from_observe(rho, rhodot, A, Adot, a, ...
                                   adot, theta, phi, H)
% ~~~~~~~~~~~~~~~~~~~~~~~~~~~~~~~~~~~~~~~~~~~~~~~~~~~~~~~~~~~~~~~~~~~~~~
%{
  This function calculates the geocentric equatorial position and
  velocity vectors of an object from radar observations of range,
  azimuth, elevation angle and their rates.

  deg    - conversion factor between degrees and radians
  pi     - 3.1415926...
```

```
  Re      - equatorial radius of the earth (km)
  f       - earth's flattening factor
  wE      - angular velocity of the earth (rad/s)
  omega   - earth's angular velocity vector (rad/s) in the
            geocentric equatorial frame

  theta   - local sidereal time (degrees) of tracking site
  phi     - geodetic latitude (degrees) of site
  H       - elevation of site (km)
  R       - geocentric equatorial position vector (km) of tracking site
  Rdot    - inertial velocity (km/s) of site

  rho     - slant range of object (km)
  rhodot  - range rate (km/s)
  A       - azimuth (degrees) of object relative to observation site
  Adot    - time rate of change of azimuth (degrees/s)
  a       - elevation angle (degrees) of object relative to observation site
  adot    - time rate of change of elevation angle (degrees/s)
  dec     - topocentric equatorial declination of object (rad)
  decdot  - declination rate (rad/s)
  h       - hour angle of object (rad)
  RA      - topocentric equatorial right ascension of object (rad)
  RAdot   - right ascension rate (rad/s)

  Rho     - unit vector from site to object
  Rhodot  - time rate of change of Rho (1/s)
  r       - geocentric equatorial position vector of object (km)
  v       - geocentric equatorial velocity vector of object (km)

  User M-functions required: none
%}
% --------------------------------------------------------------------

global f Re wE
deg   = pi/180;
omega = [0 0 wE];

%...Convert angular quantities from degrees to radians:
A     = A    *deg;
Adot  = Adot *deg;
a     = a    *deg;
adot  = adot *deg;
theta = theta*deg;
phi   = phi  *deg;
```

```
%...Equation 5.56:
R     = [(Re/sqrt(1-(2*f - f*f)*sin(phi)^2) + H)*cos(phi)*cos(theta), ...
         (Re/sqrt(1-(2*f - f*f)*sin(phi)^2) + H)*cos(phi)*sin(theta), ...
         (Re*(1 - f)^2/sqrt(1-(2*f - f*f)*sin(phi)^2) + H)*sin(phi)];

%...Equation 5.66:
Rdot  = cross(omega, R);

%...Equation 5.83a:
dec   = asin(cos(phi)*cos(A)*cos(a) + sin(phi)*sin(a));

%...Equation 5.83b:
h = acos((cos(phi)*sin(a) - sin(phi)*cos(A)*cos(a))/cos(dec));
if (A > 0) & (A < pi)
    h = 2*pi - h;
end

%...Equation 5.83c:
RA = theta - h;

%...Equations 5.57:
Rho = [cos(RA)*cos(dec)  sin(RA)*cos(dec)  sin(dec)];

%...Equation 5.63:
r   = R + rho*Rho;

%...Equation 5.84:
decdot = (-Adot*cos(phi)*sin(A)*cos(a) + adot*(sin(phi)*cos(a) ...
          - cos(phi)*cos(A)*sin(a)))/cos(dec);

%...Equation 5.85:
RAdot  = wE ...
         + (Adot*cos(A)*cos(a) - adot*sin(A)*sin(a) ...
         + decdot*sin(A)*cos(a)*tan(dec)) ...
          /(cos(phi)*sin(a) - sin(phi)*cos(A)*cos(a));

%...Equations 5.69 and 5.72:
Rhodot = [-RAdot*sin(RA)*cos(dec) - decdot*cos(RA)*sin(dec),...
           RAdot*cos(RA)*cos(dec) - decdot*sin(RA)*sin(dec),...
           decdot*cos(dec)];

%...Equation 5.64:
v = Rdot + rhodot*Rho + rho*Rhodot;

end %rv_from_observe
% ~~~~~~~~~~~~~~~~~~~~~~~~~~~~~~~~~~~~~~~~~~~~~~~~~~~~~~~~~~~~~~~~~~~~~~~~
```

SCRIPT FILE: Example_5_10.m

```
% ~~~~~~~~~~~~~~~~~~~~~~~~~~~~~~~~~~~~~~~~~~~~~~~~~~~~~~~~~~~~~~~~~~~~
% Example_5_10
% ~~~~~~~~~~~~
%
% This program uses Algorithms 5.4 and 4.2 to obtain the orbital
% elements from the observational data provided in Example 5.10.
%
% deg    - conversion factor between degrees and radians
% pi     - 3.1415926...
% mu     - gravitational parameter (km^3/s^2)

% Re     - equatorial radius of the earth (km)
% f      - earth's flattening factor
% wE     - angular velocity of the earth (rad/s)
% omega  - earth's angular velocity vector (rad/s) in the
%          geocentric equatorial frame

% rho    - slant range of object (km)
% rhodot - range rate (km/s)
% A      - azimuth (deg) of object relative to observation site
% Adot   - time rate of change of azimuth (deg/s)
% a      - elevation angle (deg) of object relative to observation site
% adot   - time rate of change of elevation angle (degrees/s)

% theta  - local sidereal time (deg) of tracking site
% phi    - geodetic latitude (deg) of site
% H      - elevation of site (km)

% r      - geocentric equatorial position vector of object (km)
% v      - geocentric equatorial velocity vector of object (km)

% coe    - orbital elements [h e RA incl w TA a]
%          where
%              h    = angular momentum (km^2/s)
%              e    = eccentricity
%              RA   = right ascension of the ascending node (rad)
%              incl = inclination of the orbit (rad)
%              w    = argument of perigee (rad)
%              TA   = true anomaly (rad)
%              a    = semimajor axis (km)
% rp    - perigee radius (km)
% T     - period of elliptical orbit (s)
%
% User M-functions required: rv_from_observe, coe_from_sv
% --------------------------------------------------------------------
```

```
clear all; clc
global  f Re wE

deg    = pi/180;
f      = 1/298.256421867;
Re     = 6378.13655;
wE     = 7.292115e-5;
mu     = 398600.4418;

%...Data declaration for Example 5.10:
rho    = 2551;
rhodot = 0;
A      = 90;
Adot   = 0.1130;
a      = 30;
adot   = 0.05651;
theta  = 300;
phi    = 60;
H      = 0;
%...

%...Algorithm 5.4:
[r,v] = rv_from_observe(rho, rhodot, A, Adot, a, adot, theta, phi, H);

%...Algorithm 4.2:
coe  = coe_from_sv(r,v,mu);

h    = coe(1);
e    = coe(2);
RA   = coe(3);
incl = coe(4);
w    = coe(5);
TA   = coe(6);
a    = coe(7);

%...Equation 2.40
rp   = h^2/mu/(1 + e);

%...Echo the input data and output the solution to
%   the command window:
fprintf('-----------------------------------------------------')
fprintf('\n Example 5.10')
fprintf('\n\n Input data:\n');
fprintf('\n Slant range (km)               = %g', rho);
fprintf('\n Slant range rate (km/s)        = %g', rhodot);
fprintf('\n Azimuth (deg)                  = %g', A);
fprintf('\n Azimuth rate (deg/s)           = %g', Adot);
```

```
fprintf('\n Elevation (deg)                  = %g', a);
fprintf('\n Elevation rate (deg/s)           = %g', adot);
fprintf('\n Local sidereal time (deg)        = %g', theta);
fprintf('\n Latitude (deg)                   = %g', phi);
fprintf('\n Altitude above sea level (km) = %g', H);
fprintf('\n\n');

fprintf(' Solution:')

fprintf('\n\n State vector:\n');
fprintf('\n r (km)                           = [%g, %g, %g]', ...
                                        r(1), r(2), r(3));
fprintf('\n v (km/s)                         = [%g, %g, %g]', ...
                                        v(1), v(2), v(3));

fprintf('\n\n Orbital elements:\n')
fprintf('\n   Angular momentum (km^2/s)      = %g', h)
fprintf('\n   Eccentricity                   = %g', e)
fprintf('\n   Inclination (deg)              = %g', incl/deg)
fprintf('\n   RA of ascending node (deg)     = %g', RA/deg)
fprintf('\n   Argument of perigee (deg)      = %g', w/deg)
fprintf('\n   True anomaly (deg)             = %g\n', TA/deg)
fprintf('\n   Semimajor axis (km)            = %g', a)
fprintf('\n   Perigee radius (km)            = %g', rp)
%...If the orbit is an ellipse, output its period:
if e < 1
    T = 2*pi/sqrt(mu)*a^1.5;
    fprintf('\n   Period:')
    fprintf('\n     Seconds                      = %g', T)
    fprintf('\n     Minutes                      = %g', T/60)
    fprintf('\n     Hours                        = %g', T/3600)
    fprintf('\n     Days                         = %g', T/24/3600)
end
fprintf('\n-----------------------------------------------\n')
% ~~~~~~~~~~~~~~~~~~~~~~~~~~~~~~~~~~~~~~~~~~~~~~~~~~~~~~~~~~~~~~~~~~~~~~~~~~
```

OUTPUT FROM Example_5_10

```
-----------------------------------------------------
 Example 5.10

 Input data:

 Slant range (km)             = 2551
 Slant range rate (km/s)      = 0
 Azimuth (deg)                = 90
 Azimuth rate (deg/s)         = 0.113
 Elevation (deg)              = 5168.62
```

```
Elevation rate (deg/s)         = 0.05651
Local sidereal time (deg)      = 300
Latitude (deg)                 = 60
Altitude above sea level (km) = 0

Solution:

State vector:

r (km)                         = [3830.68, -2216.47, 6605.09]
v (km/s)                       = [1.50357, -4.56099, -0.291536]

Orbital elements:

  Angular momentum (km^2/s)    = 35621.4
  Eccentricity                 = 0.619758
  Inclination (deg)            = 113.386
  RA of ascending node (deg)   = 109.75
  Argument of perigee (deg)    = 309.81
  True anomaly (deg)           = 165.352

  Semimajor axis (km)          = 5168.62
  Perigee radius (km)          = 1965.32
  Period:
    Seconds                    = 3698.05
    Minutes                    = 61.6342
    Hours                      = 1.02724
    Days                       = 0.0428015
-----------------------------------------------------
```

D.29 ALGORITHMS 5.5 AND 5.6: GAUSS' METHOD OF PRELIMINARY ORBIT DETERMINATION WITH ITERATIVE IMPROVEMENT

FUNCTION FILE: gauss.m

```
% ~~~~~~~~~~~~~~~~~~~~~~~~~~~~~~~~~~~~~~~~~~~~~~~~~~~~~~~~
 function [r, v, r_old, v_old] = ...
          gauss(Rho1, Rho2, Rho3, R1, R2, R3, t1, t2, t3)
% ~~~~~~~~~~~~~~~~~~~~~~~~~~~~~~~~~~~~~~~~~~~~~~~~~~~~~~~~
%{
  This function uses the Gauss method with iterative improvement
  (Algorithms 5.5 and 5.6) to calculate the state vector of an
  orbiting body from angles-only observations at three
  closely spaced times.

  mu              - the gravitational parameter (km^3/s^2)
  t1, t2, t3      - the times of the observations (s)
```

```
  tau, tau1, tau3  - time intervals between observations (s)
  R1, R2, R3       - the observation site position vectors
                     at t1, t2, t3 (km)
  Rho1, Rho2, Rho3 - the direction cosine vectors of the
                     satellite at t1, t2, t3
  p1, p2, p3       - cross products among the three direction
                     cosine vectors
  Do               - scalar triple product of Rho1, Rho2 and Rho3
  D                - Matrix of the nine scalar triple products
                     of R1, R2 and R3 with p1, p2 and p3
  E                - dot product of R2 and Rho2
  A, B             - constants in the expression relating slant range
                     to geocentric radius
  a,b,c            - coefficients of the 8th order polynomial
                     in the estimated geocentric radius x
  x                - positive root of the 8th order polynomial
  rho1, rho2, rho3 - the slant ranges at t1, t2, t3
  r1, r2, r3       - the position vectors at t1, t2, t3 (km)
  r_old, v_old     - the estimated state vector at the end of
                     Algorithm 5.5 (km, km/s)
  rho1_old,
  rho2_old, and
  rho3_old         - the values of the slant ranges at t1, t2, t3
                     at the beginning of iterative improvement
                     (Algorithm 5.6) (km)
  diff1, diff2,
  and diff3        - the magnitudes of the differences between the
                     old and new slant ranges at the end of
                     each iteration
  tol              - the error tolerance determining
                    convergence
  n                - number of passes through the
                    iterative improvement loop
  nmax             - limit on the number of iterations
  ro, vo           - magnitude of the position and
                     velocity vectors (km, km/s)
  vro              - radial velocity component (km)
  a                - reciprocal of the semimajor axis (1/km)
  v2               - computed velocity at time t2 (km/s)
  r, v             - the state vector at the end of Algorithm 5.6
                     (km, km/s)

  User m-functions required:  kepler_U, f_and_g
  User subfunctions required: posroot
%}
% ------------------------------------------------------
```

```
global mu

%...Equations 5.98:
tau1 = t1 - t2;
tau3 = t3 - t2;

%...Equation 5.101:
tau  = tau3 - tau1;

%...Independent cross products among the direction cosine vectors:
p1 = cross(Rho2,Rho3);
p2 = cross(Rho1,Rho3);
p3 = cross(Rho1,Rho2);

%...Equation 5.108:
Do = dot(Rho1,p1);

%...Equations 5.109b, 5.110b and 5.111b:
D  = [[dot(R1,p1) dot(R1,p2) dot(R1,p3)]
      [dot(R2,p1) dot(R2,p2) dot(R2,p3)]
      [dot(R3,p1) dot(R3,p2) dot(R3,p3)]];

%...Equation 5.115b:
E = dot(R2,Rho2);

%...Equations 5.112b and 5.112c:
A = 1/Do*(-D(1,2)*tau3/tau + D(2,2) + D(3,2)*tau1/tau);
B = 1/6/Do*(D(1,2)*(tau3^2 - tau^2)*tau3/tau ...
            + D(3,2)*(tau^2 - tau1^2)*tau1/tau);

%...Equations 5.117:
a = -(A^2 + 2*A*E + norm(R2)^2);
b = -2*mu*B*(A + E);
c = -(mu*B)^2;

%...Calculate the roots of Equation 5.116 using MATLAB's
%   polynomial 'roots' solver:
Roots = roots([1 0 a 0 0 b 0 0 c]);

%...Find the positive real root:
x = posroot(Roots);

%...Equations 5.99a and 5.99b:
f1 =    1 - 1/2*mu*tau1^2/x^3;
f3 =    1 - 1/2*mu*tau3^2/x^3;

%...Equations 5.100a and 5.100b:
```

```
g1 = tau1 - 1/6*mu*(tau1/x)^3;
g3 = tau3 - 1/6*mu*(tau3/x)^3;

%...Equation 5.112a:
rho2 = A + mu*B/x^3;

%...Equation 5.113:
rho1 = 1/Do*((6*(D(3,1)*tau1/tau3 + D(2,1)*tau/tau3)*x^3 ...
              + mu*D(3,1)*(tau^2 - tau1^2)*tau1/tau3) ...
              /(6*x^3 + mu*(tau^2 - tau3^2)) - D(1,1));

%...Equation 5.114:
rho3 = 1/Do*((6*(D(1,3)*tau3/tau1 - D(2,3)*tau/tau1)*x^3 ...
              + mu*D(1,3)*(tau^2 - tau3^2)*tau3/tau1) ...
              /(6*x^3 + mu*(tau^2 - tau1^2)) - D(3,3));

%...Equations 5.86:
r1 = R1 + rho1*Rho1;
r2 = R2 + rho2*Rho2;
r3 = R3 + rho3*Rho3;

%...Equation 5.118:
v2 = (-f3*r1 + f1*r3)/(f1*g3 - f3*g1);

%...Save the initial estimates of r2 and v2:
r_old = r2;
v_old = v2;

%...End of Algorithm 5.5

%...Use Algorithm 5.6 to improve the accuracy of the initial estimates.

%...Initialize the iterative improvement loop and set error tolerance:
rho1_old = rho1;  rho2_old = rho2;  rho3_old = rho3;
diff1    = 1;     diff2    = 1;     diff3    = 1;
n    = 0;
nmax = 1000;
tol  = 1.e-8;

%...Iterative improvement loop:
while ((diff1 > tol) & (diff2 > tol) & (diff3 > tol)) & (n < nmax)
    n = n+1;

%...Compute quantities required by universal kepler's equation:
    ro  = norm(r2);
    vo  = norm(v2);
    vro = dot(v2,r2)/ro;
    a   = 2/ro - vo^2/mu;
```

```
%...Solve universal Kepler's equation at times tau1 and tau3 for
%   universal anomalies x1 and x3:
    x1 = kepler_U(tau1, ro, vro, a);
    x3 = kepler_U(tau3, ro, vro, a);

%...Calculate the Lagrange f and g coefficients at times tau1
%   and tau3:
    [ff1, gg1] = f_and_g(x1, tau1, ro, a);
    [ff3, gg3] = f_and_g(x3, tau3, ro, a);

%...Update the f and g functions at times tau1 and tau3 by
%   averaging old and new:
    f1    = (f1 + ff1)/2;
    f3    = (f3 + ff3)/2;
    g1    = (g1 + gg1)/2;
    g3    = (g3 + gg3)/2;

%...Equations 5.96 and 5.97:
    c1    =  g3/(f1*g3 - f3*g1);
    c3    = -g1/(f1*g3 - f3*g1);

%...Equations 5.109a, 5.110a and 5.111a:
    rho1  = 1/Do*(       -D(1,1) + 1/c1*D(2,1) - c3/c1*D(3,1));
    rho2  = 1/Do*(    -c1*D(1,2) +D(2,2)       -    c3*D(3,2));
    rho3  = 1/Do*(-c1/c3*D(1,3) + 1/c3*D(2,3) -       D(3,3));

%...Equations 5.86:
    r1    = R1 + rho1*Rho1;
    r2    = R2 + rho2*Rho2;
    r3    = R3 + rho3*Rho3;

%...Equation 5.118:
    v2    = (-f3*r1 + f1*r3)/(f1*g3 - f3*g1);

%...Calculate differences upon which to base convergence:
    diff1 = abs(rho1 - rho1_old);
    diff2 = abs(rho2 - rho2_old);
    diff3 = abs(rho3 - rho3_old);

%...Update the slant ranges:
    rho1_old = rho1;  rho2_old = rho2;  rho3_old = rho3;
end
%...End iterative improvement loop

fprintf('\n( **Number of Gauss improvement iterations = %g)\n\n',n)
```

```
if n >= nmax
    fprintf('\n\n **Number of iterations exceeds %g \n\n ',nmax);
end

%...Return the state vector for the central observation:
r = r2;
v = v2;

return

% ~~~~~~~~~~~~~~~~~~~~~~
  function x = posroot(Roots)
% ~~~~~~~~~~~~~~~~~~~~~~
%{
  This subfunction extracts the positive real roots from
  those obtained in the call to MATLAB's 'roots' function.
  If there is more than one positive root, the user is
  prompted to select the one to use.

  x         - the determined or selected positive root
  Roots     - the vector of roots of a polynomial
  posroots  - vector of positive roots

  User M-functions required: none
%}
% ~~~~~~~~~~~~~~~~

%...Construct the vector of positive real roots:
posroots  = Roots(find(Roots>0 & ~imag(Roots)));
npositive = length(posroots);

%...Exit if no positive roots exist:
if npositive == 0
    fprintf('\n\n ** There are no positive roots.  \n\n')
    return
end

%...If there is more than one positive root, output the
%   roots to the command window and prompt the user to
%   select which one to use:
if npositive == 1
    x = posroots;
else
    fprintf('\n\n ** There are two or more positive roots.\n')
    for i = 1:npositive
        fprintf('\n root #%g = %g',i,posroots(i))
    end
```

```
    fprintf('\n\n Make a choice:\n')
    nchoice = 0;
    while nchoice < 1 | nchoice > npositive
        nchoice = input(' Use root #? ');
    end
    x = posroots(nchoice);
    fprintf('\n We will use %g .\n', x)
end

end %posroot

end %gauss
% ~~~~~~~~~~~~~~~~~~~~~~~~~~~~~~~~~~~~~~~~~~~~~
```

SCRIPT FILE: Example_5_11.m

```
% ~~~~~~~~~~~~~~~~~~~~~~~~~~~~~~~~~~~~
% Example_5_11
% ~~~~~~~~~~
%{
  This program uses Algorithms 5.5 and 5.6 (Gauss's method) to compute
  the state vector from the data provided in Example 5.11.

  deg           - factor for converting between degrees and radians
  pi            - 3.1415926...
  mu            - gravitational parameter (km^3/s^2)
  Re            - earth's radius (km)
  f             - earth's flattening factor
  H             - elevation of observation site (km)
  phi           - latitude of site (deg)
  t             - vector of observation times t1, t2, t3 (s)
  ra            - vector of topocentric equatorial right ascensions
                  at t1, t2, t3 (deg)
  dec           - vector of topocentric equatorial right declinations
                  at t1, t2, t3 (deg)
  theta         - vector of local sidereal times for t1, t2, t3 (deg)
  R             - matrix of site position vectors at t1, t2, t3 (km)
  rho           - matrix of direction cosine vectors at t1, t2, t3
  fac1, fac2    - common factors
  r_old, v_old  - the state vector without iterative improvement (km, km/s)
  r, v          - the state vector with iterative improvement (km, km/s)
  coe           - vector of orbital elements for r, v:
                  [h, e, RA, incl, w, TA, a]
                  where h    = angular momentum (km^2/s)
                        e    = eccentricity
                        incl = inclination (rad)
                        w    = argument of perigee (rad)
                        TA   = true anomaly (rad)
```

```
                      a    = semimajor axis (km)
  coe_old     - vector of orbital elements for r_old, v_old

  User M-functions required: gauss, coe_from_sv
%}
% ---------------------------------------------

clear all; clc

global mu

deg = pi/180;
mu  = 398600;
Re  = 6378;
f   = 1/298.26;

%...Data declaration for Example 5.11:
H     = 1;
phi   = 40*deg;
t     = [       0   118.104   237.577];
ra    = [ 43.5365   54.4196   64.3178]*deg;
dec   = [-8.78334  -12.0739  -15.1054]*deg;
theta = [ 44.5065    45.000   45.4992]*deg;
%...

%...Equations 5.64, 5.76 and 5.79:
fac1 = Re/sqrt(1-(2*f - f*f)*sin(phi)^2);
fac2 = (Re*(1-f)^2/sqrt(1-(2*f - f*f)*sin(phi)^2) + H)*sin(phi);
for i = 1:3
    R(i,1) = (fac1 + H)*cos(phi)*cos(theta(i));
    R(i,2) = (fac1 + H)*cos(phi)*sin(theta(i));
    R(i,3) = fac2;
    rho(i,1) = cos(dec(i))*cos(ra(i));
    rho(i,2) = cos(dec(i))*sin(ra(i));
    rho(i,3) = sin(dec(i));
end

%...Algorithms 5.5 and 5.6:
[r, v, r_old, v_old] = gauss(rho(1,:), rho(2,:), rho(3,:), ...
                             R(1,:),    R(2,:),  R(3,:), ...
                             t(1),      t(2),    t(3));

%...Algorithm 4.2 for the initial estimate of the state vector
%   and for the iteratively improved one:
coe_old = coe_from_sv(r_old,v_old,mu);
coe     = coe_from_sv(r,v,mu);
```

```
%...Echo the input data and output the solution to
%   the command window:
fprintf('-----------------------------------------------------')
fprintf('\n Example 5.11: Orbit determination by the Gauss method\n')
fprintf('\n Radius of earth (km)               = %g', Re)
fprintf('\n Flattening factor                  = %g', f)
fprintf('\n Gravitational parameter (km^3/s^2) = %g', mu)
fprintf('\n\n Input data:\n');
fprintf('\n Latitude (deg)                = %g', phi/deg);
fprintf('\n Altitude above sea level (km) = %g', H);
fprintf('\n\n Observations:')
fprintf('\n               Right')
fprintf('                                   Local')
fprintf('\n   Time (s)   Ascension (deg)   Declination (deg)')
fprintf('   Sidereal time (deg)')
for i = 1:3
    fprintf('\n %9.4g %11.4f %19.4f %20.4f', ...
                 t(i), ra(i)/deg, dec(i)/deg, theta(i)/deg)
end

fprintf('\n\n Solution:\n')

fprintf('\n Without iterative improvement...\n')
fprintf('\n');
fprintf('\n r (km)                          = [%g, %g, %g]', ...
                                  r_old(1), r_old(2), r_old(3))
fprintf('\n v (km/s)                        = [%g, %g, %g]', ...
                                  v_old(1), v_old(2), v_old(3))
fprintf('\n');

fprintf('\n   Angular momentum (km^2/s)     = %g', coe_old(1))
fprintf('\n   Eccentricity                  = %g', coe_old(2))
fprintf('\n   RA of ascending node (deg)    = %g', coe_old(3)/deg)
fprintf('\n   Inclination (deg)             = %g', coe_old(4)/deg)
fprintf('\n   Argument of perigee (deg)     = %g', coe_old(5)/deg)
fprintf('\n   True anomaly (deg)            = %g', coe_old(6)/deg)
fprintf('\n   Semimajor axis (km)           = %g', coe_old(7))
fprintf('\n   Periapse radius (km)          = %g', coe_old(1)^2 ...
                                           /mu/(1 + coe_old(2)))
%...If the orbit is an ellipse, output the period:
if coe_old(2)<1
    T = 2*pi/sqrt(mu)*coe_old(7)^1.5;
    fprintf('\n   Period:')
    fprintf('\n     Seconds                     = %g', T)
    fprintf('\n     Minutes                     = %g', T/60)
    fprintf('\n     Hours                       = %g', T/3600)
    fprintf('\n     Days                        = %g', T/24/3600)
end
```

```
fprintf('\n\n With iterative improvement...\n')
fprintf('\n');
fprintf('\n r (km)                          = [%g, %g, %g]', ...
                                           r(1), r(2), r(3))
fprintf('\n v (km/s)                        = [%g, %g, %g]', ...
                                           v(1), v(2), v(3))
fprintf('\n');
fprintf('\n   Angular momentum (km^2/s)     = %g', coe(1))
fprintf('\n   Eccentricity                  = %g', coe(2))
fprintf('\n   RA of ascending node (deg)    = %g', coe(3)/deg)
fprintf('\n   Inclination (deg)             = %g', coe(4)/deg)
fprintf('\n   Argument of perigee (deg)     = %g', coe(5)/deg)
fprintf('\n   True anomaly (deg)            = %g', coe(6)/deg)
fprintf('\n   Semimajor axis (km)           = %g', coe(7))
fprintf('\n   Periapse radius (km)          = %g', coe(1)^2 ...
                                        /mu/(1 + coe(2)))
%...If the orbit is an ellipse, output the period:
if coe(2)<1
    T = 2*pi/sqrt(mu)*coe(7)^1.5;
    fprintf('\n   Period:')
    fprintf('\n     Seconds                     = %g', T)
    fprintf('\n     Minutes                     = %g', T/60)
    fprintf('\n     Hours                       = %g', T/3600)
    fprintf('\n     Days                        = %g', T/24/3600)
end
fprintf('\n-----------------------------------------------------\n')

% ~~~~~~~~~~~~~~~~~~~~~~~~~~~~~~~~~~~~~~~~
```

OUTPUT FROM `Example_5_11`

```
( **Number of Gauss improvement iterations = 14)

-----------------------------------------------------
 Example 5.11: Orbit determination by the Gauss method

 Radius of earth (km)               = 6378
 Flattening factor                  = 0.00335278
 Gravitational parameter (km^3/s^2) = 398600

 Input data:

 Latitude (deg)                = 40
 Altitude above sea level (km) = 1

 Observations:
               Right                                   Local
   Time (s)   Ascension (deg)   Declination (deg)   Sidereal time (deg)
         0     43.5365            -8.7833              44.5065
```

```
    118.1     54.4196              -12.0739             45.0000
    237.6     64.3178              -15.1054             45.4992

Solution:

Without iterative improvement...

r (km)                          = [5659.03, 6533.74, 3270.15]
v (km/s)                        = [-3.8797, 5.11565, -2.2397]

  Angular momentum (km^2/s)     = 62705.3
  Eccentricity                  = 0.097562
  RA of ascending node (deg)    = 270.023
  Inclination (deg)             = 30.0105
  Argument of perigee (deg)     = 88.654
  True anomaly (deg)            = 46.3163
  Semimajor axis (km)           = 9959.2
  Periapse radius (km)          = 8987.56
  Period:
    Seconds                     = 9891.17
    Minutes                     = 164.853
    Hours                       = 2.74755
    Days                        = 0.114481

With iterative improvement...

r (km)                          = [5662.04, 6537.95, 3269.05]
v (km/s)                        = [-3.88542, 5.12141, -2.2434]

  Angular momentum (km^2/s)     = 62816.7
  Eccentricity                  = 0.0999909
  RA of ascending node (deg)    = 269.999
  Inclination (deg)             = 30.001
  Argument of perigee (deg)     = 89.9723
  True anomaly (deg)            = 45.0284
  Semimajor axis (km)           = 9999.48
  Periapse radius (km)          = 8999.62
  Period:
    Seconds                     = 9951.24
    Minutes                     = 165.854
    Hours                       = 2.76423
    Days                        = 0.115176
-----------------------------------------------------
```

CHAPTER 6: ORBITAL MANEUVERS

D.30 CALCULATE THE STATE VECTOR AFTER A FINITE TIME, CONSTANT THRUST DELTA-V MANEUVER

FUNCTION FILE: integrate_thrust.m

```
% ~~~~~~~~~~~~~~~~~~~~~~~~~~~~~~~~~~~~~~~~~
function integrate_thrust
% ~~~~~~~~~~~~~~~~~~~~~~
%{
  This function uses rkf45 to numerically integrate Equation 6.26 during
  the delta-v burn and then find the apogee of the post-burn orbit.

  The input data are for the first part of Example 6.15.

  mu        - gravitational parameter (km^3/s^2)
  RE        - earth radius (km)
  g0        - sea level acceleration of gravity (m/s^2)
  T         - rated thrust of rocket engine (kN)
  Isp       - specific impulse of rocket engine (s)
  m0        - initial spacecraft mass (kg)
  r0        - initial position vector (km)
  v0        - initial velocity vector (km/s)
  t0        - initial time (s)
  t_burn    - rocket motor burn time (s)
  y0        - column vector containing r0, v0 and m0
  t         - column vector of the times at which the solution is found (s)
  y         - a matrix whose elements are:
                 columns 1, 2 and 3:
                    The solution for the x, y and z components of the
                    position vector r at the times t
                 columns 4, 5 and 6:
                    The solution for the x, y and z components of the
                    velocity vector v at the times t
                 column 7:
                    The spacecraft mass m at the times t
  r1        - position vector after the burn (km)
  v1        - velocity vector after the burn (km/s)
  m1        - mass after the burn (kg)
  coe       - orbital elements of the post-burn trajectory
              (h e RA incl w TA a)
  ra        - position vector vector at apogee (km)
  va        - velocity vector at apogee (km)
  rmax      - apogee radius (km)

  User M-functions required:  rkf45, coe_from_sv, rv_from_r0v0_ta
  User subfunctions required: rates, output
```

```
%}
% ---------------------------------------------

%...Preliminaries:
clear all; close all; clc
global mu
deg     = pi/180;
mu      = 398600;
RE      = 6378;
g0      = 9.807;

%...Input data:
r0      = [RE+480    0   0];
v0      = [  0   7.7102 0];
t0      = 0;
t_burn  = 261.1127;

m0      = 2000;
T       = 10;
Isp     = 300;
%...end Input data

%...Integrate the equations of motion over the burn time:
y0    = [r0 v0 m0]';
[t,y] = rkf45(@rates, [t0 t_burn], y0, 1.e-16);

%...Compute the state vector and mass after the burn:
r1  = [y(end,1) y(end,2) y(end,3)];
v1  = [y(end,4) y(end,5) y(end,6)];
m1  = y(end,7);
coe = coe_from_sv(r1,v1,mu);
e   = coe(2);  %eccentricity
TA  = coe(6);  %true anomaly (radians)
a   = coe(7);  %semimajor axis

%...Find the state vector at apogee of the post-burn trajectory:
if TA <= pi
    dtheta = pi - TA;
else
    dtheta = 3*pi - TA;
end
[ra,va] = rv_from_r0v0_ta(r1, v1, dtheta/deg, mu);
rmax    = norm(ra);

output
```

```
%...Subfunctions:

% ~~~~~~~~~~~~~~~~~~~~~~~~
function dfdt = rates(t,f)
% ~~~~~~~~~~~~~~~~~~~~~~~~
%{
  This function calculates the acceleration vector using Equation 6.26.

  t            - time (s)
  f            - column vector containing the position vector, velocity
                 vector and the mass at time t
  x, y, z      - components of the position vector (km)
  vx, vy, vz   - components of the velocity vector (km/s)
  m            - mass (kg)
  r            - magnitude of the the position vector (km)
  v            - magnitude of the velocity vector (km/s)
  ax, ay, az   - components of the acceleration vector (km/s^2)
  mdot         - rate of change of mass (kg/s)
  dfdt         - column vector containing the velocity and acceleration
                 components and the mass rate
%}
% ------------------------
x  = f(1);  y = f(2);  z = f(3);
vx = f(4); vy = f(5); vz = f(6);
m  = f(7);

r    = norm([x y z]);
v    = norm([vx vy vz]);
ax   = -mu*x/r^3 + T/m*vx/v;
ay   = -mu*y/r^3 + T/m*vy/v;
az   = -mu*z/r^3 + T/m*vz/v;
mdot = -T*1000/g0/Isp;

dfdt = [vx vy vz ax ay az mdot]';

end %rates

% ~~~~~~~~~~~~~
function output
% ~~~~~~~~~~~~~
fprintf('\n\n-----------------------------------------------------------\n')
fprintf('\nBefore ignition:')
fprintf('\n  Mass = %g kg', m0)
fprintf('\n  State vector:')
fprintf('\n    r = [%10g, %10g, %10g] (km)', r0(1), r0(2), r0(3))
fprintf('\n      Radius = %g', norm(r0))
fprintf('\n    v = [%10g, %10g, %10g] (km/s)', v0(1), v0(2), v0(3))
```

```
fprintf('\n      Speed = %g\n', norm(v0))
fprintf('\nThrust          = %12g kN', T)
fprintf('\nBurn time       = %12.6f s', t_burn)
fprintf('\nMass after burn = %12.6E kg\n', m1)
fprintf('\nEnd-of-burn-state vector:')
fprintf('\n    r = [%10g, %10g, %10g] (km)', r1(1), r1(2), r1(3))
fprintf('\n      Radius = %g', norm(r1))
fprintf('\n    v = [%10g, %10g, %10g] (km/s)', v1(1), v1(2), v1(3))
fprintf('\n      Speed = %g\n', norm(v1))
fprintf('\nPost-burn trajectory:')
fprintf('\n  Eccentricity   = %g', e)
fprintf('\n  Semimajor axis = %g km', a)
fprintf('\n  Apogee state vector:')
fprintf('\n    r = [%17.10E, %17.10E, %17.10E] (km)', ra(1), ra(2), ra(3))
fprintf('\n      Radius = %g', norm(ra))
fprintf('\n    v = [%17.10E, %17.10E, %17.10E] (km/s)', va(1), va(2), va(3))
fprintf('\n      Speed = %g', norm(va))
fprintf('\n\n--------------------------------------------------------\n\n')

end %output

end %integrate_thrust
```

CHAPTER 7: RELATIVE MOTION AND RENDEZVOUS

D.31 ALGORITHM 7.1: FIND THE POSITION, VELOCITY, AND ACCELERATION OF *B* RELATIVE TO *A*'S LVLH FRAME

FUNCTION FILE: rva_relative.m

```
% ~~~~~~~~~~~~~~~~~~~~~~~~~~~~~~~~~~~~~~~~~~~~~~~~~~~~~~~~~~~~
function [r_rel_x, v_rel_x, a_rel_x] = rva_relative(rA,vA,rB,vB)
% ~~~~~~~~~~~~~~~~~~~~~~~~~~~~~~~~~~~~~~~~~~~~~~~~~~~~~~~~~~~~
%{
  This function uses the state vectors of spacecraft A and B
  to find the position, velocity and acceleration of B relative
  to A in the LVLH frame attached to A (see Figure 7.1).

  rA,vA      - state vector of A (km, km/s)
  rB,vB      - state vector of B (km, km/s)
  mu         - gravitational parameter (km^3/s^2)
  hA         - angular momentum vector of A (km^2/s)
  i, j, k    - unit vectors along the x, y and z axes of A's
               LVLH frame
  QXx        - DCM of the LVLH frame relative to the geocentric
               equatorial frame (GEF)
  Omega      - angular velocity of the LVLH frame (rad/s)
```

```
  Omega_dot - angular acceleration of the LVLH frame (rad/s^2)
  aA, aB    - absolute accelerations of A and B (km/s^2)
  r_rel     - position of B relative to A in GEF (km)
  v_rel     - velocity of B relative to A in GEF (km/s)
  a_rel     - acceleration of B relative to A in GEF (km/s^2)
  r_rel_x   - position of B relative to A in the LVLH frame
  v_rel_x   - velocity of B relative to A in the LVLH frame
  a_rel_x   - acceleration of B relative to A in the LVLH frame

  User M-functions required: None
%}
% -----------------------------------------------------------

global mu

%...Calculate the vector hA:
hA = cross(rA, vA);

%...Calculate the unit vectors i, j and k:
i = rA/norm(rA);
k = hA/norm(hA);
j = cross(k,i);

%...Calculate the transformation matrix Qxx:
QXx = [i; j; k];

%...Calculate Omega and Omega_dot:
Omega     = hA/norm(rA)^2;                  % Equation 7.5
Omega_dot = -2*dot(rA,vA)/norm(rA)^2*Omega;% Equation 7.6

%...Calculate the accelerations aA and aB:
aA = -mu*rA/norm(rA)^3;
aB = -mu*rB/norm(rB)^3;

%...Calculate r_rel:
r_rel = rB - rA;

%...Calculate v_rel:
v_rel = vB - vA - cross(Omega,r_rel);

%...Calculate a_rel:
a_rel = aB - aA - cross(Omega_dot,r_rel)...
       - cross(Omega,cross(Omega,r_rel))...
       - 2*cross(Omega,v_rel);

%...Calculate r_rel_x, v_rel_x and a_rel_x:
r_rel_x = QXx*r_rel';
```

```
v_rel_x = QXx*v_rel';
a_rel_x = QXx*a_rel';

end %rva_relative

% ~~~~~~~~~~~~~~~~~~~~~~~~~~~~~~~~~~~~~~~~~~~~~~~~~~~~~~~~~~~~~
```

SCRIPT FILE: Example_7_01.m

```
% ~~~~~~~~~~~~~~~~~~~~~~~~~~~~~~~~~~~~~~~~~~~~~~~~~~~~~~~~~~~~~
% Example_7_01
% ~~~~~~~~~~~~
%{
  This program uses the data of Example 7.1 to calculate the position,
  velocity and acceleration of an orbiting chaser B relative to an
  orbiting target A.

  mu                 - gravitational parameter (km^3/s^2)
  deg                - conversion factor from degrees to radians

                     Spacecraft A & B:
  h_A, h_B           -   angular momentum (km^2/s)
  e_A, E_B           -   eccentricity
  i_A, i_B           -   inclination (radians)
  RAAN_A, RAAN_B     -   right ascension of the ascending node (radians)
  omega_A, omega_B   -   argument of perigee (radians)
  theta_A, theta_A   -   true anomaly (radians)

  rA, vA             - inertial position (km) and velocity (km/s) of A
  rB, vB             - inertial position (km) and velocity (km/s) of B
  r                  - position (km) of B relative to A in A's
                       co-moving frame
  v                  - velocity (km/s) of B relative to A in A's
                       co-moving frame
  a                  - acceleration (km/s^2) of B relative to A in A's
                       co-moving frame

  User M-function required:   sv_from_coe, rva_relative
  User subfunctions required: none
%}
% ----------------------------------------------------------

clear all; clc
global mu
mu  = 398600;
deg = pi/180;

%...Input data:
```

```
%   Spacecraft A:
h_A     = 52059;
e_A     = 0.025724;
i_A     = 60*deg;
RAAN_A  = 40*deg;
omega_A = 30*deg;
theta_A = 40*deg;

%   Spacecraft B:
h_B     = 52362;
e_B     = 0.0072696;
i_B     = 50*deg;
RAAN_B  = 40*deg;
omega_B = 120*deg;
theta_B = 40*deg;

%...End input data

%...Compute the initial state vectors of A and B using Algorithm 4.5:
[rA,vA] = sv_from_coe([h_A e_A RAAN_A i_A omega_A theta_A],mu);
[rB,vB] = sv_from_coe([h_B e_B RAAN_B i_B omega_B theta_B],mu);

%...Compute relative position, velocity and acceleration using
%   Algorithm 7.1:
[r,v,a] = rva_relative(rA,vA,rB,vB);

%...Output
fprintf('\n\n--------------------------------------------------------\n\n')
fprintf('\nOrbital parameters of spacecraft A:')
fprintf('\n   angular momentum    = %g (km^2/s)', h_A)
fprintf('\n   eccentricity        = %g'         , e_A)
fprintf('\n   inclination         = %g (deg)'   , i_A/deg)
fprintf('\n   RAAN                = %g (deg)'   , RAAN_A/deg)
fprintf('\n   argument of perigee = %g (deg)'   , omega_A/deg)
fprintf('\n   true anomaly        = %g (deg)\n' , theta_A/deg)

fprintf('\nState vector of spacecraft A:')
fprintf('\n   r = [%g, %g, %g]', rA(1), rA(2), rA(3))
fprintf('\n       (magnitude = %g)', norm(rA))
fprintf('\n   v = [%g, %g, %g]', vA(1), vA(2), vA(3))
fprintf('\n       (magnitude = %g)\n', norm(vA))

fprintf('\nOrbital parameters of spacecraft B:')
fprintf('\n   angular momentum    = %g (km^2/s)', h_B)
fprintf('\n   eccentricity        = %g'         , e_B)
fprintf('\n   inclination         = %g (deg)'   , i_B/deg)
fprintf('\n   RAAN                = %g (deg)'   , RAAN_B/deg)
```

```
fprintf('\n   argument of perigee = %g (deg)'   , omega_B/deg)
fprintf('\n   true anomaly        = %g (deg)\n' , theta_B/deg)

fprintf('\nState vector of spacecraft B:')
fprintf('\n   r = [%g, %g, %g]', rB(1), rB(2), rB(3))
fprintf('\n       (magnitude = %g)', norm(rB))
fprintf('\n   v = [%g, %g, %g]', vB(1), vB(2), vB(3))
fprintf('\n       (magnitude = %g)\n', norm(vB))

fprintf('\nIn the co-moving frame attached to A:')
fprintf('\n   Position of B relative to A = [%g, %g, %g]', ...
                            r(1), r(2), r(3))
fprintf('\n      (magnitude = %g)\n', norm(r))
fprintf('\n   Velocity of B relative to A = [%g, %g, %g]', ...
                            v(1), v(2), v(3))
fprintf('\n      (magnitude = %g)\n', norm(v))
fprintf('\n   Acceleration of B relative to A = [%g, %g, %g]', ...
                            a(1), a(2), a(3))
fprintf('\n      (magnitude = %g)\n', norm(a))
fprintf('\n\n--------------------------------------------------------\n\n')

% ~~~~~~~~~~~~~~~~~~~~~~~~~~~~~~~~~~~~~~~~~~~~~~~~~~~~~~~~~~~~~~~~~~
```

OUTPUT FROM Example_7_01.m

```
-----------------------------------------------------

Orbital parameters of spacecraft A:
   angular momentum    = 52059 (km^2/s)
   eccentricity        = 0.025724
   inclination         = 60 (deg)
   RAAN                = 40 (deg)
   argument of perigee = 30 (deg)
   true anomaly        = 40 (deg)

State vector of spacecraft A:
   r = [-266.768, 3865.76, 5426.2]
       (magnitude = 6667.75)
   v = [-6.48356, -3.61975, 2.41562]
       (magnitude = 7.8086)

Orbital parameters of spacecraft B:
   angular momentum    = 52362 (km^2/s)
   eccentricity        = 0.0072696
   inclination         = 50 (deg)
   RAAN                = 40 (deg)
   argument of perigee = 120 (deg)
   true anomaly        = 40 (deg)
```

```
State vector of spacecraft B:
   r = [-5890.71, -2979.76, 1792.21]
       (magnitude = 6840.43)
   v = [0.935828, -5.2403, -5.50095]
       (magnitude = 7.65487)

In the co-moving frame attached to A:
   Position of B relative to A = [-6701.15, 6828.27, -406.261]
      (magnitude = 9575.79)

   Velocity of B relative to A = [0.316667, 0.111993, 1.24696]
      (magnitude = 1.29141)

   Acceleration of B relative to A = [-0.000222229, -0.000180743, 0.000505932]
      (magnitude = 0.000581396)

--------------------------------------------------------
```

D.32 PLOT THE POSITION OF ONE SPACECRAFT RELATIVE TO ANOTHER

SCRIPT FILE: Example_7_02.m

```
% ~~~~~~~~~~~~~~~~~~~~~~~~~~~~~~~~~~~~
% Example_7_02
% ~~~~~~~~~~
%{
  This program produces a 3D plot of the motion of spacecraft B
  relative to A in Example 7.1. See Figure 7.4.

  User M-functions required: rv_from_r0v0 (Algorithm 3.4)
                             sv_from_coe  (Algorithm 4.5)
                             rva_relative (Algorithm 7.1)
%}
% ---------------------------------------------

clear all; close all; clc

global mu

%...Gravitational parameter and earth radius:
mu  = 398600;
RE  = 6378;

%...Conversion factor from degrees to radians:
deg = pi/180;

%...Input data:
%   Initial orbital parameters (angular momentum, eccentricity,
```

```
%   inclination, RAAN, argument of perigee and true anomaly).
%   Spacecraft A:
h_A      = 52059;
e_A      = 0.025724;
i_A      = 60*deg;
RAAN_A   = 40*deg;
omega_A = 30*deg;
theta_A = 40*deg;

%   Spacecraft B:
h_B      = 52362;
e_B      = 0.0072696;
i_B      = 50*deg;
RAAN_B   = 40*deg;
omega_B = 120*deg;
theta_B = 40*deg;

vdir = [1 1 1];

%...End input data

%...Compute the initial state vectors of A and B using Algorithm 4.5:
[rA0,vA0] = sv_from_coe([h_A e_A RAAN_A i_A omega_A theta_A],mu);
[rB0,vB0] = sv_from_coe([h_B e_B RAAN_B i_B omega_B theta_B],mu);

h0 = cross(rA0,vA0);

%...Period of A:
TA = 2*pi/mu^2*(h_A/sqrt(1 - e_A^2))^3;

%...Number of time steps per period of A's orbit:
n = 100;

%...Time step as a fraction of A's period:
dt = TA/n;

%...Number of periods of A's orbit for which the trajectory
%   will be plotted:
n_Periods = 60;

%...Initialize the time:
t = - dt;

%...Generate the trajectory of B relative to A:
for count = 1:n_Periods*n

%...Update the time:
```

```
    t = t + dt;
%...Update the state vector of both orbits using Algorithm 3.4:
    [rA,vA] = rv_from_r0v0(rA0, vA0, t);
    [rB,vB] = rv_from_r0v0(rB0, vB0, t);

%...Compute r_rel using Algorithm 7.1:
    [r_rel, v_rel, a_rel] = rva_relative(rA,vA,rB,vB);

%...Store the components of the relative position vector
%   at this time step in the vectors x, y and z, respectively:
    x(count) = r_rel(1);
    y(count) = r_rel(2);
    z(count) = r_rel(3);
    r(count) = norm(r_rel);
    T(count) = t;
end

%...Plot the trajectory of B relative to A:
figure(1)
plot3(x, y, z)
hold on
axis equal
axis on
grid on
box off
view(vdir)
%   Draw the co-moving x, y and z axes:
line([0 4000],     [0 0],     [0 0]); text(4000,    0,    0, 'x')
line(    [0 0], [0 7000],     [0 0]); text(   0, 7000,    0, 'y')
line(    [0 0],     [0 0], [0 4000]); text(   0,    0, 4000, 'z')

%   Label the origin of the moving frame attached to A:
text (0, 0, 0, 'A')

%   Label the start of B's relative trajectory:
text(x(1), y(1), z(1), 'B')

%   Draw the initial position vector of B:
line([0 x(1)], [0 y(1)], [0 z(1)])
% ~~~~~~~~~~~~~~~~~~~~~~~~~~~~~~~~~~~~~~~
```

D.33 SOLUTION OF THE LINEARIZED EQUATIONS OF RELATIVE MOTION WITH AN ELLIPTICAL REFERENCE ORBIT

FUNCTION FILE: Example_7_03.m

```
% ~~~~~~~~~~~~~~~~~~~~~~~~~~~~~~~~~~~~~~~~
function Example_7_03
% ~~~~~~~~~~~~~~~~~
%{
  This function plots the motion of chaser B relative to target A
  for the data in Example 7.3. See Figures 7.6 and 7.7.

  mu        - gravitational parameter (km^3/s^2)
  RE        - radius of the earth (km)

            Target orbit at time t = 0:
  rp        - perigee radius (km)
  e         - eccentricity
  i         - inclination (rad)
  RA        - right ascension of the ascending node (rad)
  omega     - argument of perigee (rad)
  theta     - true anomaly (rad)
  ra        - apogee radius (km)
  h         - angular momentum (km^2/s)
  a         - semimajor axis (km)
  T         - period (s)
  n         - mean motion (rad/s)

  dr0, dv0 - initial relative position (km) and relative velocity (km/s)
             of B in the co-moving frame
  t0, tf    - initial and final times (s) for the numerical integration
  R0, V0    - initial position (km) and velocity (km/s) of A in the
              geocentric equatorial frame
  y0        - column vector containing r0, v0
%}
% User M-functions required:  sv_from_coe, rkf45
% User subfunctions required: rates
% ---------------------------------------------

clear all; close all; clc

global mu

mu  = 398600;
RE  = 6378;

%...Input data:
%   Prescribed initial orbital parameters of target A:
```

```
rp    = RE + 300;
e     = 0.1;
i     = 0;
RA    = 0;
omega = 0;
theta = 0;

%  Additional computed parameters:
ra = rp*(1 + e)/(1 - e);
h  = sqrt(2*mu*rp*ra/(ra + rp));
a  = (rp + ra)/2;
T  = 2*pi/sqrt(mu)*a^1.5;
n  = 2*pi/T;

%  Prescribed initial state vector of chaser B in the co-moving frame:
dr0 = [-1  0  0];
dv0 = [ 0 -2*n*dr0(1) 0];
t0  = 0;
tf  = 5*T;
%...End input data

%...Calculate the target's initial state vector using Algorithm 4.5:
[R0,V0] = sv_from_coe([h e RA i omega theta],mu);

%...Initial state vector of B's orbit relative to A
y0 = [dr0 dv0]';

%...Integrate Equations 7.34 using Algorithm 1.3:
[t,y] = rkf45(@rates, [t0 tf], y0);

plotit

return

% ~~~~~~~~~~~~~~~~~~~~~~~~
function dydt = rates(t,f)
% ~~~~~~~~~~~~~~~~~~~~~~~~
%{
  This function computes the components of f(t,y) in Equation 7.36.

  t             - time
  f             - column vector containing the relative position and
                  velocity vectors of B at time t
  R, V          - updated state vector of A at time t
  X, Y, Z       - components of R
  VX, VY, VZ    - components of V
  R_            - magnitude of R
```

```
  RdotV          - dot product of R and V
  h              - magnitude of the specific angular momentum of A

  dx , dy , dz   - components of the relative position vector of B
  dvx, dvy, dvz  - components of the relative velocity vector of B
  dax, day, daz  - components of the relative acceleration vector of B
  dydt           - column vector containing the relative velocity
                   and acceleration components of B at time t

  User M-function required: rv_from_r0v0
%}
% ------------------------
%...Update the state vector of the target orbit using Algorithm 3.4:
[R,V] = rv_from_r0v0(R0, V0, t);

X  = R(1); Y  = R(2); Z  = R(3);
VX = V(1); VY = V(2); VZ = V(3);

R_    = norm([X Y Z]);
RdotV = dot([X Y Z], [VX VY VZ]);
h     = norm(cross([X Y Z], [VX VY VZ]));

dx    = f(1); dy  = f(2); dz  = f(3);
dvx   = f(4); dvy = f(5); dvz = f(6);

dax   =  (2*mu/R_^3 + h^2/R_^4)*dx - 2*RdotV/R_^4*h*dy + 2*h/R_^2*dvy;
day   =   -(mu/R_^3 - h^2/R_^4)*dy + 2*RdotV/R_^4*h*dx - 2*h/R_^2*dvx;
daz   = -mu/R_^3*dz;

dydt  = [dvx dvy dvz dax day daz]';
end %rates

% ~~~~~~~~~~~~
function plotit
% ~~~~~~~~~~~~
%...Plot the trajectory of B relative to A:
% ------------
hold on
plot(y(:,2), y(:,1))
axis on
axis equal
axis ([0 40 -5 5])
xlabel('y (km)')
ylabel('x (km)')
grid on
box on
%...Label the start of B's trajectory relative to A:
```

```
text(y(1,2), y(1,1), 'o')
end %plotit

end %Example_7_03
% ~~~~~~~~~~~~~~~~~~~~~~~~~~~~~~~~~~~~~~~~~~~~
```

CHAPTER 8: INTERPLANETARY TRAJECTORIES

D.34 CONVERT THE NUMERICAL DESIGNATION OF A MONTH OR A PLANET INTO ITS NAME

The following trivial script can be used in programs that input numerical values for a month and/or a planet.

FUNCTION FILE: month_planet_names.m

```
% ~~~~~~~~~~~~~~~~~~~~~~~~~~~~~~~~~~~~~~~~~~~~~~~~~~~~~~~~~~~~~~~~~~~~~~~~
  function [month, planet] = month_planet_names(month_id, planet_id)
% ~~~~~~~~~~~~~~~~~~~~~~~~~~~~~~~~~~~~~~~~~~~~~~~~~~~~~~~~~~~~~~~~~~~~~~~~
%{
  This function returns the name of the month and the planet
  corresponding, respectively, to the numbers "month_id" and
  "planet_id".

  months    - a vector containing the names of the 12 months
  planets   - a vector containing the names of the 9 planets
  month_id  - the month number (1 - 12)
  planet_id - the planet number (1 - 9)

  User M-functions required: none
%}
% --------------------------------------------------------------------

months  = ['January  '
           'February '
           'March    '
           'April    '
           'May      '
           'June     '
           'July     '
           'August   '
           'September'
           'October  '
           'November '
           'December '];
```

```
planets = ['Mercury'
           'Venus  '
           'Earth  '
           'Mars   '
           'Jupiter'
           'Saturn '
           'Uranus '
           'Neptune'
           'Pluto  '];

month  = months (month_id,  1:9);
planet = planets(planet_id, 1:7);
% ~~~~~~~~~~~~~~~~~~~~~~~~~~~~~~~~~~~~~~~~~~~~~~~~~~~~~~~~~~~~~~~~~~~~~~~
end %month_planet_names
```

D.35 ALGORITHM 8.1: CALCULATION OF THE HELIOCENTRIC STATE VECTOR OF A PLANET AT A GIVEN EPOCH

FUNCTION FILE: planet_elements_and_sv.m

```
% ~~~~~~~~~~~~~~~~~~~~~~~~~~~~~~~~~~~~~~~~~~~~~~~~~~~~~~~~~~~~~~~~~~~~~~
  function [coe, r, v, jd] = planet_elements_and_sv ...
              (planet_id, year, month, day, hour, minute, second)
% ~~~~~~~~~~~~~~~~~~~~~~~~~~~~~~~~~~~~~~~~~~~~~~~~~~~~~~~~~~~~~~~~~~~~~~
%{
  This function calculates the orbital elements and the state
  vector of a planet from the date (year, month, day)
  and universal time (hour, minute, second).

  mu         - gravitational parameter of the sun (km^3/s^2)
  deg        - conversion factor between degrees and radians
  pi         - 3.1415926...

  coe        - vector of heliocentric orbital elements
               [h  e  RA  incl  w  TA  a  w_hat  L  M  E],
               where
                h      = angular momentum                    (km^2/s)
                e      = eccentricity
                RA     = right ascension                     (deg)
                incl   = inclination                         (deg)
                w      = argument of perihelion              (deg)
                TA     = true anomaly                        (deg)
                a      = semimajor axis                      (km)
                w_hat  = longitude of perihelion ( = RA + w) (deg)
                L      = mean longitude ( = w_hat + M)       (deg)
                M      = mean anomaly                        (deg)
                E      = eccentric anomaly                   (deg)
```

```
  planet_id - planet identifier:
              1 = Mercury
              2 = Venus
              3 = Earth
              4 = Mars
              5 = Jupiter
              7 = Uranus
              8 = Neptune
              9 = Pluto

  year      - range: 1901 - 2099
  month     - range: 1 - 12
  day       - range: 1 - 31
  hour      - range: 0 - 23
  minute    - range: 0 - 60
  second    - range: 0 - 60

  j0        - Julian day number of the date at 0 hr UT
  ut        - universal time in fractions of a day
  jd        - julian day number of the date and time

  J2000_coe - row vector of J2000 orbital elements from Table 9.1
  rates     - row vector of Julian centennial rates from Table 9.1
  t0        - Julian centuries between J2000 and jd
  elements  - orbital elements at jd

  r         - heliocentric position vector
  v         - heliocentric velocity vector

  User M-functions required:  J0, kepler_E, sv_from_coe
  User subfunctions required: planetary_elements, zero_to_360
%}
% -------------------------------------------------------------------

global mu
deg    = pi/180;

%...Equation 5.48:
j0     = J0(year, month, day);

ut     = (hour + minute/60 + second/3600)/24;

%...Equation 5.47
jd     = j0 + ut;

%...Obtain the data for the selected planet from Table 8.1:
[J2000_coe, rates] = planetary_elements(planet_id);
```

```
%...Equation 8.93a:
t0     = (jd - 2451545)/36525;

%...Equation 8.93b:
elements = J2000_coe + rates*t0;

a      = elements(1);
e      = elements(2);

%...Equation 2.71:
h      = sqrt(mu*a*(1 - e^2));

%...Reduce the angular elements to within the range 0 - 360 degrees:
incl   = elements(3);
RA     = zero_to_360(elements(4));
w_hat  = zero_to_360(elements(5));
L      = zero_to_360(elements(6));
w      = zero_to_360(w_hat - RA);
M      = zero_to_360((L - w_hat));

%...Algorithm 3.1 (for which M must be in radians)
E      = kepler_E(e, M*deg);

%...Equation 3.13 (converting the result to degrees):
TA     = zero_to_360...
         (2*atan(sqrt((1 + e)/(1 - e))*tan(E/2))/deg);

coe    = [h  e  RA  incl  w  TA  a  w_hat  L  M  E/deg];

%...Algorithm 4.5 (for which all angles must be in radians):
[r, v] = sv_from_coe([h e RA*deg incl*deg w*deg TA*deg],mu);

return

% ~~~~~~~~~~~~~~~~~~~~~~~~~~~~~~~~~~~~~~~~~~~~~~~~~~~~~~~~~~~~~
  function [J2000_coe, rates] = planetary_elements(planet_id)
% ~~~~~~~~~~~~~~~~~~~~~~~~~~~~~~~~~~~~~~~~~~~~~~~~~~~~~~~~~~~~~
%{
  This function extracts a planet's J2000 orbital elements and
  centennial rates from Table 8.1.

  planet_id      - 1 through 9, for Mercury through Pluto

  J2000_elements - 9 by 6 matrix of J2000 orbital elements for the nine
                   planets Mercury through Pluto. The columns of each
                   row are:
```

```
                a      = semimajor axis (AU)
                e      = eccentricity
                i      = inclination (degrees)
                RA     = right ascension of the ascending
                         node (degrees)
                w_hat = longitude of perihelion (degrees)
                L      = mean longitude (degrees)

  cent_rates     - 9 by 6 matrix of the rates of change of the
                   J2000_elements per Julian century (Cy). Using "dot"
                   for time derivative, the columns of each row are:
                     a_dot     (AU/Cy)
                     e_dot     (1/Cy)
                     i_dot     (arcseconds/Cy)
                     RA_dot    (arcseconds/Cy)
                     w_hat_dot (arcseconds/Cy)
                     Ldot      (arcseconds/Cy)

  J2000_coe      - row vector of J2000_elements corresponding
                   to "planet_id", with au converted to km
  rates          - row vector of cent_rates corresponding to
                   "planet_id", with au converted to km and
                   arcseconds converted to degrees

  au             - astronomical unit (km)
%}
% ------------------------------------------------------------------

J2000_elements = ...
[ 0.38709893  0.20563069   7.00487   48.33167   77.45645  252.25084
  0.72333199  0.00677323   3.39471   76.68069  131.53298  181.97973
  1.00000011  0.01671022   0.00005  -11.26064  102.94719  100.46435
  1.52366231  0.09341233   1.85061   49.57854  336.04084  355.45332
  5.20336301  0.04839266   1.30530  100.55615   14.75385   34.40438
  9.53707032  0.05415060   2.48446  113.71504   92.43194   49.94432
 19.19126393  0.04716771   0.76986   74.22988  170.96424  313.23218
 30.06896348  0.00858587   1.76917  131.72169   44.97135  304.88003
 39.48168677  0.24880766  17.14175  110.30347  224.06676  238.92881];

cent_rates = ...
[ 0.00000066   0.00002527  -23.51     -446.30    573.57  538101628.29
  0.00000092  -0.00004938   -2.86     -996.89   -108.80  210664136.06
 -0.00000005  -0.00003804  -46.94   -18228.25   1198.28  129597740.63
 -0.00007221   0.00011902  -25.47    -1020.19   1560.78   68905103.78
  0.00060737  -0.00012880   -4.15     1217.17    839.93   10925078.35
 -0.00301530  -0.00036762    6.11    -1591.05  -1948.89    4401052.95
  0.00152025  -0.00019150   -2.09    -1681.4    1312.56    1542547.79
```

```
 -0.00125196   0.00002514   -3.64     -151.25   -844.43     786449.21
 -0.00076912   0.00006465   11.07      -37.33   -132.25     522747.90];

J2000_coe       = J2000_elements(planet_id,:);
rates           = cent_rates(planet_id,:);

%...Convert from AU to km:
au              = 149597871;
J2000_coe(1)    = J2000_coe(1)*au;
rates(1)        = rates(1)*au;

%...Convert from arcseconds to fractions of a degree:
rates(3:6)      = rates(3:6)/3600;

end %planetary_elements

% ~~~~~~~~~~~~~~~~~~~~~~~~
  function y = zero_to_360(x)
% ~~~~~~~~~~~~~~~~~~~~~~~~
%{
  This function reduces an angle to lie in the range 0 - 360 degrees.

  x - the original angle in degrees
  y - the angle reduced to the range 0 - 360 degrees

%}
% --------------------------

if x >= 360
    x = x - fix(x/360)*360;
elseif x < 0
    x = x - (fix(x/360) - 1)*360;
end

y = x;

end %zero_to_360

end %planet_elements_and_sv
% ~~~~~~~~~~~~~~~~~~~~~~~~~~~~~~~~~~~~~~~~~~~~~~~~~~~~~~~~~~~~~~~~~
```

SCRIPT FILE: Example_8_07.m

```
% ~~~~~~~~~~~~~~~~~~~~~~~~~~~~~~~~~~~~~~~~~~~~~~~~~~~~~~~~~~~~~~~~~
% Example_8_07
% ~~~~~~~~~~
%
% This program uses Algorithm 8.1 to compute the orbital elements
% and state vector of the earth at the date and time specified
```

```
% in Example 8.10. To obtain the same results for Mars, set
% planet_id = 4.
%
% mu        - gravitational parameter of the sun (km^3/s^2)
% deg       - conversion factor between degrees and radians
% pi        - 3.1415926...
%
% coe       - vector of heliocentric orbital elements
%             [h  e  RA  incl  w  TA  a  w_hat  L  M  E],
%             where
%              h     = angular momentum                    (km^2/s)
%              e     = eccentricity
%              RA    = right ascension                     (deg)
%              incl  = inclination                         (deg)
%              w     = argument of perihelion              (deg)
%              TA    = true anomaly                        (deg)
%              a     = semimajor axis                      (km)
%              w_hat = longitude of perihelion ( = RA + w) (deg)
%              L     = mean longitude ( = w_hat + M)       (deg)
%              M     = mean anomaly                        (deg)
%              E     = eccentric anomaly                   (deg)
%
% r         - heliocentric position vector (km)
% v         - heliocentric velocity vector (km/s)
%
% planet_id - planet identifier:
%              1 = Mercury
%              2 = Venus
%              3 = Earth
%              4 = Mars
%              5 = Jupiter
%              6 = Saturn
%              7 = Uranus
%              8 = Neptune
%              9 = Pluto
%
% year      - range: 1901 - 2099
% month     - range: 1 - 12
% day       - range: 1 - 31
% hour      - range: 0 - 23
% minute    - range: 0 - 60
% second    - range: 0 - 60
%
% User M-functions required: planet_elements_and_sv,
%                            month_planet_names
% ------------------------------------------------------------------
```

```
global mu
mu  = 1.327124e11;
deg = pi/180;

%...Input data
planet_id = 3;
year      = 2003;
month     = 8;
day       = 27;
hour      = 12;
minute    = 0;
second    = 0;
%...

%...Algorithm 8.1:
[coe, r, v, jd] = planet_elements_and_sv ...
              (planet_id, year, month, day, hour, minute, second);

%...Convert the planet_id and month numbers into names for output:
[month_name, planet_name] = month_planet_names(month, planet_id);

%...Echo the input data and output the solution to
%   the command window:
fprintf('-----------------------------------------------------')
fprintf('\n Example 8.7')
fprintf('\n\n Input data:\n');
fprintf('\n   Planet: %s', planet_name)
fprintf('\n   Year  : %g', year)
fprintf('\n   Month : %s', month_name)
fprintf('\n   Day   : %g', day)
fprintf('\n   Hour  : %g', hour)
fprintf('\n   Minute: %g', minute)
fprintf('\n   Second: %g', second)
fprintf('\n\n   Julian day: %11.3f', jd)

fprintf('\n\n');
fprintf(' Orbital elements:')
fprintf('\n');

fprintf('\n  Angular momentum (km^2/s)                      = %g', coe(1));
fprintf('\n  Eccentricity                                   = %g', coe(2));
fprintf('\n  Right ascension of the ascending node (deg) = %g', coe(3));
fprintf('\n  Inclination to the ecliptic (deg)              = %g', coe(4));
fprintf('\n  Argument of perihelion (deg)                   = %g', coe(5));
fprintf('\n  True anomaly (deg)                             = %g', coe(6));
fprintf('\n  Semimajor axis (km)                            = %g', coe(7));
```

```
fprintf('\n');

fprintf('\n  Longitude of perihelion (deg)               = %g', coe(8));
fprintf('\n  Mean longitude (deg)                        = %g', coe(9));
fprintf('\n  Mean anomaly (deg)                          = %g', coe(10));
fprintf('\n  Eccentric anomaly (deg)                     = %g', coe(11));

fprintf('\n\n');
fprintf(' State vector:')
fprintf('\n');

fprintf('\n  Position vector (km) = [%g  %g  %g]', r(1), r(2), r(3))
fprintf('\n  Magnitude            = %g\n', norm(r))
fprintf('\n  Velocity (km/s)      = [%g  %g  %g]', v(1), v(2), v(3))
fprintf('\n  Magnitude            = %g', norm(v))

fprintf('\n-----------------------------------------------------\n')
% ~~~~~~~~~~~~~~~~~~~~~~~~~~~~~~~~~~~~~~~~~~~~~~~~~~~~~~~~~~~~~~~~
```

OUTPUT FROM Example_8_07

```
-----------------------------------------------------
 Example 8.7

 Input data:

   Planet: Earth
   Year  : 2003
   Month : August
   Day   : 27
   Hour  : 12
   Minute: 0
   Second: 0

   Julian day: 2452879.000

 Orbital elements:

  Angular momentum (km^2/s)                    = 4.4551e+09
  Eccentricity                                 = 0.0167088
  Right ascension of the ascending node (deg) = 348.554
  Inclination to the ecliptic (deg)            = -0.000426218
  Argument of perihelion (deg)                 = 114.405
  True anomaly (deg)                           = 230.812
  Semimajor axis (km)                          = 1.49598e+08

  Longitude of perihelion (deg)                = 102.959
  Mean longitude (deg)                         = 335.267
```

```
  Mean anomaly (deg)                          = 232.308
  Eccentric anomaly (deg)                     = 231.558

 State vector:

  Position vector (km) = [1.35589e+08  -6.68029e+07  286.909]
  Magnitude            = 1.51152e+08

  Velocity (km/s)      = [12.6804  26.61  -0.000212731]
  Magnitude            = 29.4769
-----------------------------------------------------
```

D.36 ALGORITHM 8.2: CALCULATION OF THE SPACECRAFT TRAJECTORY FROM PLANET 1 TO PLANET 2

FUNCTION FILE: interplanetary.m

```
% ~~~~~~~~~~~~~~~~~~~~~~~~~~~~~~~~~~~~~~~~~~~~~~~~~~~~~~~~~~~~~~~~~~~~~~~
  function ...
  [planet1, planet2, trajectory] = interplanetary(depart, arrive)
% ~~~~~~~~~~~~~~~~~~~~~~~~~~~~~~~~~~~~~~~~~~~~~~~~~~~~~~~~~~~~~~~~~~~~~~~
%{
  This function determines the spacecraft trajectory from the sphere
  of influence of planet 1 to that of planet 2 using Algorithm 8.2

  mu         - gravitational parameter of the sun (km^3/s^2)
  dum        - a dummy vector not required in this procedure

  planet_id  - planet identifier:
                1 = Mercury
                2 = Venus
                3 = Earth
                4 = Mars
                5 = Jupiter
                6 = Saturn
                7 = Uranus
                8 = Neptune
                9 = Pluto

  year       - range: 1901 - 2099
  month      - range: 1 - 12
  day        - range: 1 - 31
  hour       - range: 0 - 23
  minute     - range: 0 - 60
  second     - range: 0 - 60

  jd1, jd2   - Julian day numbers at departure and arrival
  tof        - time of flight from planet 1 to planet 2 (s)
```

```
  Rp1, Vp1    - state vector of planet 1 at departure (km, km/s)
  Rp2, Vp2    - state vector of planet 2 at arrival (km, km/s)
  R1, V1      - heliocentric state vector of spacecraft at
                departure (km, km/s)
  R2, V2      - heliocentric state vector of spacecraft at
                arrival (km, km/s)

  depart      - [planet_id, year, month, day, hour, minute, second]
                at departure
  arrive      - [planet_id, year, month, day, hour, minute, second]
                at arrival

  planet1     - [Rp1, Vp1, jd1]
  planet2     - [Rp2, Vp2, jd2]
  trajectory  - [V1, V2]

  User M-functions required: planet_elements_and_sv, lambert
%}
% --------------------------------------------------------------------

global mu

planet_id = depart(1);
year      = depart(2);
month     = depart(3);
day       = depart(4);
hour      = depart(5);
minute    = depart(6);
second    = depart(7);

%...Use Algorithm 8.1 to obtain planet 1's state vector (don't
%...need its orbital elements ["dum"]):
[dum, Rp1, Vp1, jd1] = planet_elements_and_sv ...
             (planet_id, year, month, day, hour, minute, second);

planet_id = arrive(1);
year      = arrive(2);
month     = arrive(3);
day       = arrive(4);
hour      = arrive(5);
minute    = arrive(6);
second    = arrive(7);

%...Likewise use Algorithm 8.1 to obtain planet 2's state vector:
[dum, Rp2, Vp2, jd2] = planet_elements_and_sv ...
```

```
                (planet_id, year, month, day, hour, minute, second);

tof = (jd2 - jd1)*24*3600;

%...Patched conic assumption:
R1 = Rp1;
R2 = Rp2;

%...Use Algorithm 5.2 to find the spacecraft's velocity at
%   departure and arrival, assuming a prograde trajectory:
[V1, V2] = lambert(R1, R2, tof, 'pro');

planet1    = [Rp1, Vp1, jd1];
planet2    = [Rp2, Vp2, jd2];
trajectory = [V1, V2];

end %interplanetary
% ~~~~~~~~~~~~~~~~~~~~~~~~~~~~~~~~~~~~~~~~~~~~~~~~~~~~~~~~~~~~~~~~~~~~~~~~~
```

SCRIPT FILE: Example_8_08.m

```
% ~~~~~~~~~~~~~~~~~~~~~~~~~~~~~~~~~~~~~~~~~~~
% Example_8_08
% ~~~~~~~~~~~~
%{
  This program uses Algorithm 8.2 to solve Example 8.8.

  mu          - gravitational parameter of the sun (km^3/s^2)
  deg         - conversion factor between degrees and radians
  pi          - 3.1415926...

  planet_id   - planet identifier:
                 1 = Mercury
                 2 = Venus
                 3 = Earth
                 4 = Mars
                 5 = Jupiter
                 6 = Saturn
                 7 = Uranus
                 8 = Neptune
                 9 = Pluto

  year        - range: 1901 - 2099
  month       - range: 1 - 12
  day         - range: 1 - 31
  hour        - range: 0 - 23
  minute      - range: 0 - 60
  second      - range: 0 - 60
```

```
  depart       - [planet_id, year, month, day, hour, minute, second]
                 at departure
  arrive       - [planet_id, year, month, day, hour, minute, second]
                 at arrival

  planet1      - [Rp1, Vp1, jd1]
  planet2      - [Rp2, Vp2, jd2]
  trajectory   - [V1, V2]

  coe          - orbital elements [h e RA incl w TA]
                 where
                   h    = angular momentum (km^2/s)
                   e    = eccentricity
                   RA   = right ascension of the ascending
                          node (rad)
                   incl = inclination of the orbit (rad)
                   w    = argument of perigee (rad)
                   TA   = true anomaly (rad)
                   a    = semimajor axis (km)

  jd1, jd2     - Julian day numbers at departure and arrival
  tof          - time of flight from planet 1 to planet 2 (days)

  Rp1, Vp1     - state vector of planet 1 at departure (km, km/s)
  Rp2, Vp2     - state vector of planet 2 at arrival (km, km/s)
  R1, V1       - heliocentric state vector of spacecraft at
                 departure (km, km/s)
  R2, V2       - heliocentric state vector of spacecraft at
                 arrival (km, km/s)

  vinf1, vinf2 - hyperbolic excess velocities at departure
                 and arrival (km/s)

  User M-functions required: interplanetary, coe_from_sv,
                             month_planet_names
%}
% ---------------------------------------------

clear all; clc
global mu
mu  = 1.327124e11;
deg = pi/180;

%...Data declaration for Example 8.8:

%...Departure
planet_id = 3;
```

```
year      = 1996;
month     = 11;
day       = 7;
hour      = 0;
minute    = 0;
second    = 0;
depart = [planet_id  year  month  day  hour  minute  second];

%...Arrival
planet_id = 4;
year      = 1997;
month     = 9;
day       = 12;
hour      = 0;
minute    = 0;
second    = 0;
arrive = [planet_id  year  month  day  hour  minute  second];

%...

%...Algorithm 8.2:
[planet1, planet2, trajectory] = interplanetary(depart, arrive);

R1  = planet1(1,1:3);
Vp1 = planet1(1,4:6);
jd1 = planet1(1,7);

R2  = planet2(1,1:3);
Vp2 = planet2(1,4:6);
jd2 = planet2(1,7);

V1  = trajectory(1,1:3);
V2  = trajectory(1,4:6);

tof = jd2 - jd1;

%...Use Algorithm 4.2 to find the orbital elements of the
%   spacecraft trajectory based on [Rp1, V1]...
coe  = coe_from_sv(R1, V1, mu);
%   ... and [R2, V2]
coe2 = coe_from_sv(R2, V2, mu);

%...Equations 8.94 and 8.95:
vinf1 = V1 - Vp1;
vinf2 = V2 - Vp2;

%...Echo the input data and output the solution to
%   the command window:
fprintf('-----------------------------------------------------')
```

```
fprintf('\n Example 8.8')
fprintf('\n\n Departure:\n');
fprintf('\n   Planet: %s', planet_name(depart(1)))
fprintf('\n   Year  : %g', depart(2))
fprintf('\n   Month : %s', month_name(depart(3)))
fprintf('\n   Day   : %g', depart(4))
fprintf('\n   Hour  : %g', depart(5))
fprintf('\n   Minute: %g', depart(6))
fprintf('\n   Second: %g', depart(7))
fprintf('\n\n   Julian day: %11.3f\n', jd1)
fprintf('\n   Planet position vector (km)    = [%g  %g  %g]', ...
                                          R1(1),R1(2), R1(3))

fprintf('\n   Magnitude                      = %g\n', norm(R1))

fprintf('\n   Planet velocity (km/s)         = [%g  %g  %g]', ...
                              Vp1(1), Vp1(2), Vp1(3))

fprintf('\n   Magnitude                      = %g\n', norm(Vp1))

fprintf('\n   Spacecraft velocity (km/s)     = [%g  %g  %g]', ...
                                          V1(1), V1(2), V1(3))

fprintf('\n   Magnitude                      = %g\n', norm(V1))

fprintf('\n   v-infinity at departure (km/s) = [%g  %g  %g]', ...
                                    vinf1(1), vinf1(2), vinf1(3))

fprintf('\n   Magnitude                      = %g\n', norm(vinf1))

fprintf('\n\n Time of flight = %g days\n', tof)

fprintf('\n\n Arrival:\n');
fprintf('\n   Planet: %s', planet_name(arrive(1)))
fprintf('\n   Year  : %g', arrive(2))
fprintf('\n   Month : %s', month_name(arrive(3)))
fprintf('\n   Day   : %g', arrive(4))
fprintf('\n   Hour  : %g', arrive(5))
fprintf('\n   Minute: %g', arrive(6))
fprintf('\n   Second: %g', arrive(7))
fprintf('\n\n   Julian day: %11.3f\n', jd2)
fprintf('\n   Planet position vector (km)   = [%g  %g  %g]', ...
                                         R2(1), R2(2), R2(3))

fprintf('\n   Magnitude                     = %g\n', norm(R1))

fprintf('\n   Planet velocity (km/s)        = [%g  %g  %g]', ...
                              Vp2(1), Vp2(2), Vp2(3))
```

```
fprintf('\n   Magnitude                        = %g\n', norm(Vp2))

fprintf('\n   Spacecraft Velocity (km/s)       = [%g  %g  %g]', ...
                                         V2(1), V2(2), V2(3))

fprintf('\n   Magnitude                        = %g\n', norm(V2))

fprintf('\n   v-infinity at arrival (km/s)  = [%g  %g  %g]', ...
                                 vinf2(1), vinf2(2), vinf2(3))

fprintf('\n   Magnitude                        = %g', norm(vinf2))

fprintf('\n\n\n Orbital elements of flight trajectory:\n')

fprintf('\n  Angular momentum (km^2/s)                     = %g',...
                                                          coe(1))
fprintf('\n  Eccentricity                                  = %g',...
                                                          coe(2))
fprintf('\n  Right ascension of the ascending node (deg) = %g',...
                                                     coe(3)/deg)
fprintf('\n  Inclination to the ecliptic (deg)             = %g',...
                                                     coe(4)/deg)
fprintf('\n  Argument of perihelion (deg)                  = %g',...
                                                     coe(5)/deg)
fprintf('\n  True anomaly at departure (deg)               = %g',...
                                                     coe(6)/deg)
fprintf('\n  True anomaly at arrival (deg)                 = %g\n', ...
                                                    coe2(6)/deg)
fprintf('\n  Semimajor axis (km)                           = %g',...
                                                          coe(7))
% If the orbit is an ellipse, output the period:
if coe(2) < 1
    fprintf('\n  Period (days)                                   = %g', ...
                                 2*pi/sqrt(mu)*coe(7)^1.5/24/3600)
end
fprintf('\n-----------------------------------------------------\n')
% ~~~~~~~~~~~~~~~~~~~~~~~~~~~~~~~~~~~~~~~~~~~~
```

OUTPUT FROM Example_8_08

```
-----------------------------------------------------
 Example 8.8

 Departure:

   Planet: Earth
   Year  : 1996
   Month : November
```

```
  Day    : 7
  Hour   : 0
  Minute: 0
  Second: 0

  Julian day: 2450394.500

  Planet position vector (km)    = [1.04994e+08  1.04655e+08  988.331]
  Magnitude                      = 1.48244e+08

  Planet velocity (km/s)         = [-21.515  20.9865  0.000132284]
  Magnitude                      = 30.0554

  Spacecraft velocity (km/s)     = [-24.4282  21.7819  0.948049]
  Magnitude                      = 32.7427

  v-infinity at departure (km/s) = [-2.91321  0.79542  0.947917]
  Magnitude                      = 3.16513

Time of flight = 309 days

Arrival:

  Planet: Mars
  Year   : 1997
  Month  : September
  Day    : 12
  Hour   : 0
  Minute: 0
  Second: 0

  Julian day: 2450703.500

  Planet position vector (km)   = [-2.08329e+07  -2.18404e+08  -4.06287e+06]
  Magnitude                     = 1.48244e+08

  Planet velocity (km/s)        = [25.0386  -0.220288  -0.620623]
  Magnitude                     = 25.0472

  Spacecraft Velocity (km/s)    = [22.1581  -0.19666  -0.457847]
  Magnitude                     = 22.1637

  v-infinity at arrival (km/s)  = [-2.88049  0.023628  0.162776]
  Magnitude                     = 2.88518
```

```
Orbital elements of flight trajectory:

 Angular momentum (km^2/s)                   = 4.84554e+09
 Eccentricity                                = 0.205785
 Right ascension of the ascending node (deg) = 44.8942
 Inclination to the ecliptic (deg)           = 1.6621
 Argument of perihelion (deg)                = 19.9738
 True anomaly at departure (deg)             = 340.039
 True anomaly at arrival (deg)               = 199.695

 Semimajor axis (km)                         = 1.84742e+08
 Period (days)                               = 501.254
-----------------------------------------------------
```

CHAPTER 9: LUNAR TRAJECTORIES

D.37 LUNAR STATE VECTOR VS. TIME

FUNCTION FILE: simpsons_lunar_ephemeris.m

```
% ~~~~~~~~~~~~~~~~~~~~~~~~~~~~~~~~~~~~~~~~~~~~~~~
  function [pos,vel] = simpsons_lunar_ephemeris(jd)
% ~~~~~~~~~~~~~~~~~~~~~~~~~~~~~~~~~~~~~~~~~~~~~~~
%{
  David G. Simpson, "An Alternative Ephemeris Model for
  On-Board Flight Software Use," Proceedings of the 1999 Flight Mechanics
  Symposium, NASA Goddard Space Flight Center, pp. 175 - 184.

  This function computes the state vector of the moon at a given time
  relative to the earth's geocentric equatorial frame using a curve fit
  to JPL's DE200 (1982) ephemeris model.

  jd   - julian date (days)
  pos  - position vector (km)
  vel  - velocity vector (km/s)
  a    - matrix of amplitudes (km)
  b    - matrix of frequencies (rad/century)
  c    - matrix of phase angles (rad)
  t    - time in centuries since J2000
  tfac - no. of seconds in a Julian century (36525 days)

  User M-functions required: None
%}
% --------------------------------------------
```

```
tfac = 36525*3600*24;
t    = (jd - 2451545.0)/36525;

a = [ 383.0     31.5      10.6       6.2      3.2       2.3       0.8
        351.0     28.9      13.7       9.7      5.7       2.9       2.1
        153.2     31.5      12.5       4.2      2.5       3.0       1.8]*1.e3;

b = [8399.685   70.990 16728.377  1185.622 7143.070 15613.745  8467.263
       8399.687   70.997  8433.466 16728.380 1185.667  7143.058 15613.755
       8399.672 8433.464    70.996 16728.364 1185.645   104.881  8399.116];

c = [   5.381    6.169     1.453     0.481    5.017     0.857     1.010
          3.811    4.596     4.766     6.165    5.164     0.300     5.565
          3.807    1.629     4.595     6.162    5.167     2.555     6.248];

pos = zeros(3,1);
vel = zeros(3,1);
for i = 1:3
    for j = 1:7
        pos(i) = pos(i) + a(i,j)*sin(b(i,j)*t + c(i,j));
        vel(i) = vel(i) + a(i,j)*cos(b(i,j)*t + c(i,j))*b(i,j);
    end
    vel(i) = vel(i)/tfac;
end
end %simpsons_lunar_ephemeris
% --------------------------------------------
```

D.38 NUMERICAL CALCULATION OF LUNAR TRAJECTORY

SCRIPT FILE: Example_9_03.m

```
% ~~~~~~~~~~~~~~~~~~~~~~~~~~~~~~~~~~~~~~~~~~~~~~~~~~~~~~~~~~~~~~~~~~~~~~~~~
% example_9_03
% ~~~~~~~~~~~~
%{
  This program presents the graphical solution of the motion of a
  spacecraft in the gravity fields of both the earth and the moon for
  the initial data provided in the input declaration below.

  MATLAB's ode45 Runge-Kutta solver is used.

  deg                        - conversion factor, degrees to radians
  days                       - conversion factor, days to seconds
  Re, Rm                     - radii of earth and moon, respectively (km)
  m_e, m_m                   - masses of earth and moon, respectively (kg)
  mu_e, mu_m                 - gravitational parameters of earth and moon,
                               respectively (km^3/s^2)
  D                          - semimajor axis of moon's orbit (km)
```

```
I_, J_, K_                  - unit vectors of ECI frame
RS                          - radius of moon's sphere of influence (km)
year, month, hour, minute
  second                    - Date and time of spacecraft's lunar arrival
t0                          - initial time on the trajectory (s)
z0                          - initial altitude of the trajectory (km)
alpha0, dec0                - initial right ascension and declination of
                              spacecraft (deg)
gamma0                      - initial flight path angle (deg)
fac                         - ratio of spaccraft's initial speed to the
                              escape speed.
ttt                         - predicted time to perilune (s)
tf                          - time at end of trajectory (s)
jd0                         - julian date of lunar arrival
rm0, vm0                    - state vector of the moon at jd0 (km, km/s)
RA, Dec                     - right ascension and declination of the moon
                              at jd0 (deg)
hmoon_, hmoon               - moon's angular momentum vector and magnitude
                              at jd0 (km^2/s)
inclmoon                    - inclination of moon's orbit earth's
                              equatorial plane (deg)
r0                          - initial radius from earth's center to
                              probe (km)
r0_                         - initial ECI position vector of probe (km)
vesc                        - escape speed at r0 (km/s)
v0                          - initial ECI speed of probe (km/s)
w0_                         - unit vector normal to plane of translunar
                              orbit at time t0
ur_                         - radial unit vector to probe at time t0
uperp_                      - transverse unit vector at time t0
vr                          - initial radial speed of probe (km/s)
vperp                       - initial transverse speed of probe (km/s)
v0_                         - initial velocity vector of probe (km/s)
uv0_                        - initial tangential unit vector
y0                          - initial state vector of the probe (km, km/s)
t                           - vector containing the times from t0 to tf at
                              which the state vector is evaluated (s)
y                           - a matrix whose 6 columns contain the inertial
                              position and velocity components evaluated
                              at the times t(:) (km, km/s)
X,   Y,  Z                  - the probe's inertial position vector history
vX, vY, VZ                  - the probe's inertial velocity history
x,   y,  z                  - the probe's position vector history in the
                              Moon-fixed frame
 Xm,  Ym,  Zm               - the Moon's inertial position vector history
vXm, vYm, vZm               - the Moon's inertial velocity vector history
ti                          - the ith time of the set [t0,tf] (s)
```

```
r_                            - probe's inertial position vector at time ti
                                (km)
r                             - magnitude of r_ (km)
jd                            - julian date of corresponding to ti (days)
rm_, vm_                      - the moon's state vector at time ti (km,km/s)
x_, y_, z_                    - vectors along the axes of the rotating
                                moon-fixed at time ti (km)
i_, j_, k_                    - unit vectors of the moon-fixed rotating frame
                                at time ti
Q                             - DCM of transformation from ECI to moon-fixed
                                frame at time ti
rx_                           - probe's inertial position vector in moon-
                                fixed coordinates at time ti (km)
rmx_                          - Moon's inertial position vector in moon-
                                fixed coordinates at time ti (km)
dist_                         - position vector of probe relative to the moon
                                at time ti (km)
dist                          - magnitude of dist_ (km)
dist_min                      - perilune of trajectory (km)
rmTLI_                        - Moon's position vector at TLI
RATLI, DecTLI                 - Moon's right ascension and declination at
                                TKI (deg)
v_atdmin_                     - Probe's velocity vector at perilune (km/s)
rm_perilume, vm_perilune      - Moon's state vector when the probe is at
                                perilune (km, km/s)
rel_speed                     - Speed of probe relative to the Moon at
                                perilune (km/s)
RA_at_perilune                - Moon's RA at perilune arrival (deg)
Dec_at_perilune               - Moon's Dec at perilune arrival (deg)
target_error                  - Distance between Moon's actual position at
                                perilune arrival and its position after the
                                predicted flight time, ttt (km).
rms_                          - position vector of moon relative to
                                spacecraft (km)
rms                           - magnitude of rms_ (km)
aearth_                       - acceleration of spacecraft due to
                                earth (km/s^2)
amoon_                        - acceleration of spacecraft due to
                                moon (km/s^2)
atot_                         - aearth_ + amoon_ (km/s^2)
binormal_                     - unit vector normal to the osculating plane
incl                          - angle between inertial Z axis and the
                                binormal (deg)
rend_                         - Position vector of end point of trajectory
                                (km)
alt_end                       - Altitude of end point of trajectory (km)
ra_end, dec_end               - Right ascension and declination of end point
                                of trajectory (km)
```

```
  User M-functions required:  none
  User subfunctions required: rates, plotit_XYZ, plotit_xyz
%}
% ----------------------------------------------------------------------

clear all; close all; clc
fprintf('\nlunar_restricted_threebody.m   %s\n\n', datestr(now))

global jd0 days ttt mu_m mu_e Re Rm rm_ rm0_

%...general data
deg  = pi/180;
days = 24*3600;
Re   = 6378;
Rm   = 1737;
m_e  = 5974.e21;
m_m  = 73.48e21;
mu_e = 398600.4;
mu_m = 4902.8;
D    = 384400;
RS   = D*(m_m/m_e)^(2/5);
%...

%...Data declaration for Example 9.03
Title = 'Example 9.3 4e';
%   Date and time of lunar arrival:
  year   = 2020;
  month  = 5;
  day    = 4;
  hour   = 12;
  minute = 0;
  second = 0;
t0       = 0;
z0       = 320;
alpha0   = 90;
dec0     = 15;
gamma0   = 40;
fac      = .9924; %Fraction of Vesc
ttt      = 3*days;
tf       = ttt + 2.667*days;
%...End data declaration

%...State vector of moon at target date:
jd0          = juliandate(year, month, day, hour, minute, second);
[rm0_,vm0_] = simpsons_lunar_ephemeris(jd0);
%[rm0_,vm0] = planetEphemeris(jd0, 'Earth', 'Moon', '430');
[RA, Dec]   = ra_and_dec_from_r(rm0_);
```

```
distance    = norm(rm0_);
hmoon_      = cross(rm0_,vm0_);
hmoon       = norm(hmoon_);
inclmoon    = acosd(hmoon_(3)/hmoon);

%...Initial position vector of probe:
I_      = [1;0;0];
J_      = [0;1;0];
K_      = cross(I_,J_);
r0      = Re + z0;
r0_     = r0*(cosd(alpha0)*cosd(dec0)*I_ + ...
              sind(alpha0)*cosd(dec0)*J_ + ...
              sind(dec0)*K_);
vesc    = sqrt(2*mu_e/r0);
v0      = fac*vesc;
w0_     = cross(r0_,rm0_)/norm(cross(r0_,rm0_));

%...Initial velocity vector of probe:
ur_     = r0_/norm(r0_);
uperp_  = cross(w0_,ur_)/norm(cross(w0_,ur_));
vr      = v0*sind(gamma0);
vperp   = v0*cosd(gamma0);
v0_     = vr*ur_ + vperp*uperp_;
uv0_    = v0_/v0;

%...Initial state vector of the probe:
y0      = [r0_(1) r0_(2) r0_(3) v0_(1) v0_(2) v0_(3)]';

%...Pass the initial conditions and time interval to ode45, which
%   calculates the position and velocity of the spacecraft at discrete
%   times t, returning the solution in the column vector y. ode45 uses
%   the subfunction 'rates' below to evaluate the spacecraft acceleration
%   at each integration time step.

options  = odeset('RelTol', 1.e-10, 'AbsTol', 1.e-10,'Stats', 'off');
[t,y]    = ode45(@rates, [t0 tf], y0, options);

%...Spacecraft trajectory
%   in ECI frame:
X   = y(:,1); Y   = y(:,2); Z   = y(:,3);
vX  = y(:,4); vY  = y(:,5); vZ  = y(:,6);
%   in Moon-fixed frame:
x   = [];     y   = [];     z   = [];

%...Moon trajectory
%   in ECI frame:
Xm  = []; Ym  = []; Zm  = [];
```

```
vXm = []; vYm = []; vZm = [];
%   in Moon-fixed frame:
xm  = []; ym  = []; zm  = [];

%...Compute the Moon's trajectory from an ephemeris, find perilune of the
%   probe's trajectory, and project the probe's trajectory onto the axes
%   of the Moon-fixed rotating frame:

dist_min = 1.e30; %Starting value in the search for perilune
for i = 1:length(t)
    ti = t(i);

    %...Probe's inertial position vector at time ti:
    r_ = [X(i) Y(i) Z(i)]';

    %...Moon's inertial position and velocity vectors at time ti:
    jd        = jd0 - (ttt - ti)/days;
    [rm_,vm_] = simpsons_lunar_ephemeris(jd);

    %...Moon's inertial state vector at time ti:
    Xm  = [ Xm;rm_(1)]; Ym  = [ Ym;rm_(2)]; Zm  = [ Zm;rm_(3)];
    vXm = [vXm;vm_(1)]; vYm = [vYm;vm_(2)]; vZm = [vZm;vm_(3)];

    %...Moon-fixed rotating xyz frame:
    x_ = rm_;
    z_ = cross(x_,vm_);
    y_ = cross(z_,x_);
    i_ = x_/norm(x_);
    j_ = y_/norm(y_);
    k_ = z_/norm(z_);

    %...DCM of transformation from ECI to moon-fixed frame:
    Q  = [i_'; j_'; k_'];

    %...Components of probe's inertial position vector in moon-fixed frame:
    rx_  = Q*r_;
    x    = [x;rx_(1)]; y = [y;rx_(2)]; z = [z;rx_(3)];

    %...Components of moon's inertial position vector in moon-fixed frame:
    rmx_ = Q*rm_;
    xm   = [xm;rmx_(1)]; ym = [ym;rmx_(2)]; zm = [zm;rmx_(3)];

    %...Find perilune of the probe:
    dist_ = r_ - rm_;
    dist  = norm(dist_);
    if dist < dist_min
        imin     = i;
```

```
        dist_min_ = dist_;
        dist_min  = dist;
    end
end

%...Location of the Moon at TLI:
rmTLI_          = [Xm(1); Ym(1); Zm(1)];
[RATLI, DecTLI] = ra_and_dec_from_r(rmTLI_);

%...Spacecraft velocity at perilune:
v_atdmin_ = [vX(imin);vY(imin);vZ(imin)];

%...State vector and celestial position of moon when probe is at perilune:
rm_perilune_                       = [Xm(imin)  Ym(imin)  Zm(imin)]';
vm_perilune_                       = [vXm(imin) vYm(imin) vZm(imin)]';
[RA_at_perilune, Dec_at_perilune] = ra_and_dec_from_r(rm_perilune_);
target_error                       = norm(rm_perilune_ - rm0_);

%...Speed of probe relative to Moon at perilune:
rel_speed = norm(v_atdmin_ - vm_perilune_);

%...End point of trajectory:
rend_              = [X(end); Y(end); Z(end)];
alt_end            = norm(rend_) - Re;
[ra_end, dec_end] = ra_and_dec_from_r(rend_);

%...Find the history of the trajectory's binormal:
for i = 1:imin
    time(i)          = t(i);
    r_               = [X(i) Y(i) Z(i)]';
    r                = norm(r_);
    v_               = [vX(i) vY(i) vZ(i)]';
    rm_              = [Xm(i) Ym(i) Zm(i)]';
    rm               = norm(rm_);
    rms_             = rm_ - r_;
    rms(i)           = norm(rms_);
    aearth_          = -mu_e*r_/r^3;
    amoon_           = mu_m*(rms_/rms(i)^3 - rm_/rm^3);
    atot_            = aearth_ + amoon_;
    binormal_        = cross(v_,atot_)/norm(cross(v_,atot_));
    binormalz        = binormal_(3);
    incl(i)          = acosd(binormalz);
end

%...Output:
fprintf('\n\n%s\n\n', Title)
```

```
fprintf('Date and time of arrival at moon: ')
fprintf('%s/%s/%s   %s:%s:%s', ...
        num2str(month), num2str(day), num2str(year), ...
        num2str(hour), num2str(minute), num2str(second))
fprintf('\nMoon''s  position: ')
fprintf('\n  Distance                          = %11g km'      , distance)
fprintf('\n  Right Ascension                   = %11g deg'     , RA)
fprintf('\n  Declination                       = %11g deg  '   , Dec)
fprintf('\nMoon''s orbital inclination         = %11g deg\n'   , inclmoon)

fprintf('\nThe probe at earth departure (t = %g sec):', t0)
fprintf('\n  Altitude                          = %11g km'      , z0)
fprintf('\n  Right ascension                   = %11g deg'     , alpha0)
fprintf('\n  Declination                       = %11g deg'     , dec0)
fprintf('\n  Flight path angle                 = %11g deg'     , gamma0)
fprintf('\n  Speed                             = %11g km/s'    , v0)
fprintf('\n  Escape speed                      = %11g km/s'    , vesc)
fprintf('\n  v/vesc                            = %11g'         , v0/vesc)
fprintf('\n  Inclination of translunar orbit   = %11g deg\n'   , ...
                                                        acosd(w0_(3)))

fprintf('\nThe moon when the probe is at TLI:')
fprintf('\n  Distance                          = %11g km' , norm(rmTLI_))
fprintf('\n  Right ascension                   = %11g deg', RATLI)
fprintf('\n  Declination                       = %11g deg', DecTLI)

fprintf('\nThe moon when the probe is at perilune: ')
fprintf('\n  Distance                          = %11g km' , ...
                                                  norm(rm_perilune_))
fprintf('\n  Speed                             = %11g km/s', ...
                                                  norm(vm_perilune_))
fprintf('\n  Right ascension                   = %11g deg' ,RA_at_perilune)
fprintf('\n  Declination                       = %11g deg' ,Dec_at_perilune)
fprintf('\n  Target error                      = %11g km'  , target_error)

fprintf('\n\nThe probe at perilune:')
fprintf('\n  Altitude                          = %11g km'  , dist_min - Rm)
fprintf('\n  Speed                             = %11g km/s',norm(v_atdmin_))
fprintf('\n  Relative speed                    = %11g km/s',rel_speed)
fprintf('\n  Inclination of osculating plane   = %11g deg' ,incl(imin))
fprintf('\n  Time from TLI to perilune         = %11g hours (%g days)' ...
                                              abs(t(imin))/3600 ...
                                              abs(t(imin))/3600/24)

fprintf('\n\nTotal time of flight              = %11g days'  , t(end)/days)
fprintf('\nTime to target point                = %11g days'  , ttt/days)
```

```
fprintf('\nFinal earth altitude            = %11g km'    , alt_end)
fprintf('\nFinal right ascension           = %11g deg'   , ra_end)
fprintf('\nFinal declination               = %11g deg\n' , dec_end)
%...End output

%...Graphical output"
%   Plot the trajectory relative to the inertial frame:
plotit_XYZ(X,Y,Z,Xm,Ym,Zm,imin)

%   Plot inclination of the osculating plane vs distance from the Moon
figure
hold on
plot(rms/RS,incl)
line([0 6][90 90],'Linestyle','-','color','red')
title('Osculating Plane Inclination vs Distance from Moon')
xlabel('r_{ms}/R_s')
ylabel('Inclination deg)')
grid on
grid minor

%   Plot the trajectory relative to the rotating Moon-fixed frame:
plotit_xyz(x,y,z,xm,ym,zm,imin)
%...End graphical output

return

% ~~~~~~~~~~~~~~~~~~~~~~~~~~~~~~~~~~~~~~~~~~~~~~~~~~~~~~~~~~~~~~~~~~~
  function dydt = rates(t,y)
% ~~~~~~~~~~~~~~~~~~~~~~~~~
%{
  This function evaluates the 3D acceleration of the spacecraft in a
  restricted 3-body system at time t from its position and velocity
  and the position of the moon at that time.

  t           - time (s)
  ttt         - flight time, TLI to target point (s)
  jd0         - Julian Date on arrival at target (days)
  jd          - Julian Date at time t (days)
  X, Y, Z     - Components of spacecraft's geocentric position vector (km)
  vX, vY, vZ  - Components of spacecraft's geocentric velocity vector (km/s)
  aX, aY, aZ  - Components of spacecraft's geocentric acceleration
                vector (km/s^2)
  y           - column vector containing the geocentric position and
                velocity components of the spacecraft at time t
  r_          - geocentric position vector [X Y Z] of the spacecraft
  rm_         - geocentric position vector of the moon
  rms_        - rm_ - r_, the position of the moon relative to the
                spacecraft
```

```
   aearth_      - spacecraft acceleration vector due to earth's gravity
   amoon_       - spacecraft acceleration vector due to lunar gravity
   a_           - total spacececraft acceleration vector
   dydt         - column vector containing the geocentric velocity and
                  acceleration components of the spacecraft at time t
%}
% ----------------------------------------------------------------

global jd0 days mu_m mu_e ttt

jd          = jd0 - (ttt - t)/days;

X           = y(1);
Y           = y(2);
Z           = y(3);

vX          = y(4);
vY          = y(5);
vZ          = y(6);

r_          = [X Y Z]';
r           = norm(r_);
[rm_,~]     = simpsons_lunar_ephemeris(jd);
%[rm_,~]    = planetEphemeris(jd0, 'Earth', 'Moon', '430');
rm          = norm(rm_);
rms_        = rm_ - r_;
rms         = norm(rms_);
aearth_     = -mu_e*r_/r^3;
amoon_      = mu_m*(rms_/rms^3 - rm_/rm^3);
a_          = aearth_ + amoon_;
aX          = a_(1);
aY          = a_(2);
aZ          = a_(3);

dydt        = [vX vY vZ aX aY aZ]';

end %rates
% ~~~~~~~~~~~~~~~~~~~~~~~~~~~~~~~~~~~~~~~~~~~~~~~~~~~~~~~~~~~~~~~~~~~~~~~~

% ~~~~~~~~~~~~~~~~~~~~~~~~~~~~~~~~~~
function plotit_XYZ(X,Y,Z,Xm,Ym,Zm,imin)
% -------------------------------------
global Re Rm

figure ('Name','Trajectories of Spacecraft (red) and Moon (green)', ...
        'Color', [1 1 1])
```

```
[xx, yy, zz] = sphere(128);
hold on

%...Geocentric inertial coordinate axes:
L = 20*Re;
line([0 L], [0 0], [0 0],'color','k')
text(L,0,0, 'X', 'FontSize',12, 'FontAngle','italic','FontName','Palatino')
line([0 0], [0 L], [0 0],'color','k')
text(0,L,0, 'Y', 'FontSize',12, 'FontAngle','italic','FontName','Palatino')
line([0 0], [0 0], [0 L],'color','k')
text(0,0,L, 'Z', 'FontSize',12, 'FontAngle','italic','FontName','Palatino')

%...Earth:
Earth = surfl(Re*xx, Re*yy, Re*zz);
set(Earth, 'FaceAlpha', 0.5);
shading interp

%...Spacecraft at TLI
plot3(X(1), Y(1), Z(1), 'o', ...
'MarkerEdgeColor','k', 'MarkerFaceColor','k', 'MarkerSize',3)

%...Spacecraft at closest approach
plot3(X(imin), Y(imin), Z(imin), 'o', ...
'MarkerEdgeColor','k', 'MarkerFaceColor','k', 'MarkerSize',2)

%...Spacecraft at tf
plot3(X(end), Y(end), Z(end), 'o', ...
'MarkerEdgeColor','r', 'MarkerFaceColor','r', 'MarkerSize',3)

%...Moon at TLI:
text(Xm(1), Ym(1), Zm(1), 'Moon at TLI')
Moon = surfl(Rm*xx + Xm(1), Rm*yy + Ym(1), Rm*zz + Zm(1));
set(Moon, 'FaceAlpha', 0.99)
shading interp

%...Moon at closest approach:
Moon = surfl(Rm*xx + Xm(imin), Rm*yy + Ym(imin), Rm*zz + Zm(imin));
set(Moon, 'FaceAlpha', 0.99)
shading interp

%...Moon at end of simulation:
Moon = surfl(Rm*xx + Xm(end), Rm*yy + Ym(end), Rm*zz + Zm(end));
set(Moon, 'FaceAlpha', 0.99)
shading interp

%...Spacecraft trajectory
plot3( X,  Y,  Z, 'r', 'LineWidth', 1.5)
```

```
%...Moon trajectory
plot3(Xm, Ym, Zm, 'g', 'LineWidth', 0.5)

axis image
axis off
axis vis3d
view([1,1,1])

end %plotit_XYZ
% ~~~~~~~~~~

% ~~~~~~~~~~~~~~~~~~~~~~~~~~~~~~
function plotit_xyz(x,y,z,xm,ym,zm,imin)
% ------------------------------------
global Re Rm Rm0_ Q0

figure ('Name','Spacecraft trajectory in Moon-fixed rotating frame', ...
       'Color', [1 1 1])
[xx, yy, zz] = sphere(128);
hold on

%...Spacecraft trajectory:
plot3( x,  y,  z, 'r', 'LineWidth', 2.0)

%...Moon trajectory:
plot3(xm, ym, zm, 'g', 'LineWidth', 0.5)

%...Earth:
Earth = surfl(Re*xx, Re*yy, Re*zz);
set(Earth, 'FaceAlpha', 0.5);
shading interp

%...Geocentric moon-fixed coordinate axes:
L1 = 63*Re; L2 = 20*Re;  L3 = 29*Re;
line([0 L1], [0 0], [0 0],'color','k')
text(L1, 0, 0, 'x','FontSize', 12, 'FontAngle', 'italic','FontName','Palatino')
line([0 0], [0 L2], [0 0],'color','k')
text(0, L2, 0, 'y','FontSize', 12, 'FontAngle', 'italic','FontName','Palatino')
line([0 0], [0 0], [0 L3],'color','k')
text(0, 0, L3, 'z','FontSize', 12, 'FontAngle', 'italic','FontName','Palatino')

%...Spacecraft at TLI
plot3(x(1), y(1), z(1), 'o', ...
'MarkerEdgeColor','k', 'MarkerFaceColor','k', 'MarkerSize',3)

%...Spacecraft at closest approach
plot3(x(imin), y(imin), z(imin), 'o', ...
```

```
'MarkerEdgeColor','k', 'MarkerFaceColor','k', 'MarkerSize',2)

%...Spacecraft at tf
plot3(x(end), y(end), z(end), 'o', ...
'MarkerEdgeColor','r', 'MarkerFaceColor','r', 'MarkerSize',3)

%...Moon at TLI:
text(xm(1), ym(1), zm(1),'Moon at TLI')
Moon = surfl(Rm*xx + xm(1), Rm*yy + ym(1), Rm*zz + zm(1));
set(Moon, 'FaceAlpha', 0.99)
shading interp

%...Moon at spacecraft closest approach:
Moon = surfl(Rm*xx + xm(imin), Rm*yy + ym(imin), Rm*zz + zm(imin));
set(Moon, 'FaceAlpha', 0.99)
shading interp

%...Moon at end of simulation:
Moon = surfl(Rm*xx + xm(end), Rm*yy + ym(end), Rm*zz + zm(end));
set(Moon, 'FaceAlpha', 0.99)
shading interp

axis image
axis vis3d
axis off
view([1,1,1])
end %plotit_xyz
% ~~~~~~~~~~~~
%end example_9_03
```

OUTPUT FROM Example_9_03.m

```
Example 9.3 4e

Date and time of arrival at moon: 5/4/2020   12:0:0
Moon's  position:
  Distance                          =      360785 km
  Right Ascension                   =     185.107 deg
  Declination                       =     2.91682 deg
Moon's orbital inclination          =     23.6765 deg

The probe at earth departure (t = 0 sec):
  Altitude                          =         320 km
  Right ascension                   =          90 deg
  Declination                       =          15 deg
  Flight path angle                 =          40 deg
  Speed                             =     10.8267 km/s
  Escape speed                      =     10.9097 km/s
```

```
  v/vesc                            =      0.9924
  Inclination of translunar orbit   =     15.5505 deg

The moon when the probe is at TLI:
  Distance                          =      372242 km
  Right ascension                   =     143.743 deg
  Declination                       =      18.189 deg
The moon when the probe is at perilune:
  Distance                          =      361064 km
  Speed                             =     1.07674 km/s
  Right ascension                   =     183.482 deg
  Declination                       =     3.62118 deg
  Target error                      =     11143.9 km

The probe at perilune:
  Altitude                          =     1258.93 km
  Speed                             =     1.03454 km/s
  Relative speed                    =     2.11055 km/s
  Inclination of osculating plane   =     155.393 deg
  Time from TLI to perilune         =     69.1257 hours (2.88024 days)

Total time of flight                =       5.667 days
Time to target point                =           3 days
Final earth altitude                =     1035.43 km
Final right ascension               =     263.608 deg
Final declination                   =    -27.4236 deg
>>
```

CHAPTER 10: INTRODUCTION TO ORBITAL PERTURBATIONS

D.39 US STANDARD ATMOSPHERE 1976

FUNCTION FILE: atmosphere.m

```
% ~~~~~~~~~~~~~~~~~~~~~~~~~~~~~~~~~~~~~~~~~~~~~~~~~~~~~~~~~~~~~~~~~~~~~~~~~~
function density = atmosphere(z)
%
% ATMOSPHERE calculates density for altitudes from sea level
% through 1000 km using exponential interpolation.
% ~~~~~~~~~~~~~~~~~~~~~~~~~~~~~~~~~~~~~~~~~~~~~~~~~~~~~~~~~~~~~~~~~~~~~~~~~~

%...Geometric altitudes (km):
h = ...
[  0  25  30  40  50  60   70 ...
  80  90 100 110 120 130  140 ...
 150 180 200 250 300 350  400 ...
 450 500 600 700 800 900 1000];
```

```
%...Corresponding densities (kg/m^3) from USSA76:
r = ...
[1.225     4.008e-2  1.841e-2  3.996e-3  1.027e-3  3.097e-4  8.283e-5 ...
 1.846e-5  3.416e-6  5.606e-7  9.708e-8  2.222e-8  8.152e-9  3.831e-9 ...
 2.076e-9  5.194e-10 2.541e-10 6.073e-11 1.916e-11 7.014e-12 2.803e-12 ...
 1.184e-12 5.215e-13 1.137e-13 3.070e-14 1.136e-14 5.759e-15 3.561e-15];

%...Scale heights (km):
H = ...
[ 7.310  6.427  6.546   7.360   8.342   7.583   6.661 ...
  5.927  5.533  5.703   6.782   9.973  13.243  16.322 ...
 21.652 27.974 34.934  43.342  49.755  54.513  58.019 ...
 60.980 65.654 76.377 100.587 147.203 208.020];

%...Handle altitudes outside of the range:
if z > 1000
    z = 1000;
elseif z < 0
    z = 0;
end

%...Determine the interpolation interval:
for j = 1:27
    if z >= h(j) && z < h(j+1)
        i = j;
    end
end
if z == 1000
    i = 27;
end

%...Exponential interpolation:
density = r(i)*exp(-(z - h(i))/H(i));

end  %atmopshere
% ~~~~~~~~~~~~~~~~~~~~~~~~~~~~~~~~~~~~~~~~~~~~~~~~~~~~~~~~~~~~~~~~~~~~~~~~~~
```

D.40 TIME FOR ORBIT DECAY USING COWELL'S METHOD

FUNCTION FILE: Example_10_01.m

```
% ~~~~~~~~~~~~~~~~~~~~~~~~~~~~~~~~~~~~~~~~~~~~~~~~~~~~~~~~~~~~~~~~~~~~~~~~~~
  function Example_10_01
% ~~~~~~~~~~~~~~~~~~~~~
%
% This function solves Example 10.1 by using MATLAB's ode45 to numerically
% integrate Equation 10.2 for atmospheric drag.
%
```

```
% User M-functions required:  sv_from_coe, atmosphere
% User subfunctions required: rates, terminate
% ---------------------------------------------------------------------------

%...Preliminaries:
close all, clear all, clc

%...Conversion factors:
hours     = 3600;                  %Hours to seconds
days      = 24*hours;              %Days to seconds
deg       = pi/180;                %Degrees to radians

%...Constants;
mu        = 398600;                %Gravitational parameter (km^3/s^2)
RE        = 6378;                  %Earth's radius (km)
wE        = [ 0 0 7.2921159e-5]';  %Earth's angular velocity (rad/s)

%...Satellite data:
CD        = 2.2;                   %Drag codfficient
m         = 100;                   %Mass (kg)
A         = pi/4*(1^2) ;           %Frontal area (m^2)

%...Initial orbital parameters (given):
rp        = RE + 215;              %perigee radius (km)
ra        = RE + 939;              %apogee radius (km)
RA        = 339.94*deg;            %Right ascencion of the node (radians)
i         = 65.1*deg;              %Inclination (radians)
w         = 58*deg;                %Argument of perigee (radians)
TA        = 332*deg;               %True anomaly (radians)

%...Initial orbital parameters (inferred):
e         = (ra-rp)/(ra+rp);       %eccentricity
a         = (rp + ra)/2;           %Semimajor axis (km)
h         = sqrt(mu*a*(1-e^2));    %angular momentrum (km^2/s)
T         = 2*pi/sqrt(mu)*a^1.5;   %Period (s)

%...Store initial orbital elements (from above) in the vector coe0:
coe0      = [h e RA i w TA];

%...Obtain the initial state vector from Algorithm 4.5 (sv_from_coe):
[R0 V0]   = sv_from_coe(coe0, mu); %R0 is the initial position vector
                                   %V0 is the initial velocity vector
r0 = norm(R0); v0 = norm(V0);      %Magnitudes of R0 and V0

%...Use ODE45 to integrate the equations of motion d/dt(R,V) = f(R,V)
%   from t0 to tf:
t0        = 0; tf = 120*days;      %Initial and final times (s)
```

```
y0      = [R0 V0]';                 %Initial state vector
nout    = 40000;                    %Number of solution points to output
tspan   = linspace(t0, tf, nout); %Integration time interval

%   Set error tolerances, initial step size, and termination event:
options  = odeset('reltol',      1.e-8, ...
                  'abstol',      1.e-8, ...
                  'initialstep', T/10000,    ...
                  'events',      @terminate);
global alt %Altitude
[t,y]    = ode45(@rates, tspan, y0,options); %t is the solution times
                                             %y is the state vector history

%...Extract the locally extreme altitudes:
altitude = sqrt(sum(y(:,1:3).^2,2)) - RE;    %Altitude at each time
[max_altitude,imax,min_altitude,imin] = extrema(altitude);
maxima   = [t(imax) max_altitude];  %Maximum altitudes and times
minima   = [t(imin) min_altitude];  %Minimum altitudes and times
apogee   = sortrows(maxima,1);      %Maxima sorted with time
perigee  = sortrows(minima,1);      %Minima sorted with time

figure(1)
apogee(1,2) = NaN;
%...Plot perigee and apogee history on the same figure:
plot(apogee(:,1)/days, apogee(:,2),'b','linewidth',2)
hold on
plot(perigee(:,1)/days, perigee(:,2),'r','linewidth',2)
grid on
grid minor
xlabel('Time (days)')
ylabel('Altitude (km)')
ylim([0 1000]);

%...Subfunctions:
% ~~~~~~~~~~~~~~~~~~~~~~~~~~~~~~~~~~~~~~~~~~~~~~~~~~~~~~~~~~~~~~~~~~~
  function dfdt = rates(t,f)
% ~~~~~~~~~~~~~~~~~~~~
%
% This function calculates the spacecraft acceleration from its
% position and velocity at time t.
% ---------------------------------------------------------------------
R    = f(1:3)';             %Position vector (km/s)
r    = norm(R);             %Distance from earth's center (km)
alt  = r - RE;              %Altitude (km)
rho  = atmosphere(alt);     %Air density from US Standard Model (kf/m^3)
V    = f(4:6)';             %Velocity vector (km/s)
Vrel = V - cross(wE,R);     %Velocity relative to the atmosphere (km/s)
```

```
vrel = norm(Vrel);          %Speed relative to the atmosphere (km/s)
uv   = Vrel/vrel;           %Relative velocity unit vector
ap   = -CD*A/m*rho*...      %Acceleration due to drag (m/s^2)
       (1000*vrel)^2/2*uv;  %(converting units of vrel from km/s to m/s)
a0   = -mu*R/r^3;           %Gravitational ecceleration (km/s^2)
a    = a0 + ap/1000;        %Total acceleration (km/s^2)
dfdt = [V a]';              %Velocity and the acceleraion returned to ode45
end %rates
% ~~~~~~~~~~~~~~~~~~~~~~~~~~~~~~~~~~~~~~~~~~~~~~~~~~~~~~~~~~~~~~~~~~~~~~

% ~~~~~~~~~~~~~~~~~~~~~~~~~~~~~~~~~~~~~~~~~~~~~~~~~~~~~~~~~~~~~~~~~~~~~~
  function [lookfor stop direction] = terminate(t,y)
% ~~~~~~~~~~~~~~~~~~~~~~~~~~~~~~~~~~~~~~~~~~~~~~~
%
% This function specifies the event at which ode45 terminates.
% -----------------------------------------------------------------------
lookfor   = alt - 100; % = 0 when altitude = 100 km
stop      = 1;         %  1 means terminate at lookfor = 0; Otherwise 0
direction = -1;        % -1 means zero crossing is from above
end %terminate
% ~~~~~~~~~~~~~~~~~~~~~~~~~~~~~~~~~~~~~~~~~~~~~~~~~~~~~~~~~~~~~~~~~~~~~~

end %Example_10_01
% ~~~~~~~~~~~~~~~~~~~~~~~~~~~~~~~~~~~~~~~~~~~~~~~~~~~~~~~~~~~~~~~~~~~~~~
```

D.41 *J2* PERTURBATION OF AN ORBIT USING ENCKE'S METHOD

FUNCTION FILE: Example_10_02.m

```
% ~~~~~~~~~~~~~~~~~~~~~~~~~~~~~~~~~~~~~~~~~~~~~~~~~~~~~~~~~~~~~~~~~~~~~~
function Example_10_02
% ~~~~~~~~~~~~~~~~~~
%
% This function  solves Example 10.2 by using Encke's method together
% with MATLAB's ode45 to integrate Equation 10.2 for a J2 gravitational
% perturbation given by Equation 10.30.
%
% User M-functions required: sv_from_coe, coe_from_sv, rv_from_r0v0
% User subfunction required: rates
% -----------------------------------------------------------------------
%...Preliminaries:
clc, close all, clear all

%...Conversion factors:
hours = 3600;                   %Hours to seconds
days  = 24*hours;               %Days to seconds
deg   = pi/180;                 %Degrees to radians
```

```
%...Constants:
global   mu
mu      = 398600;                %Gravitational parameter (km^3/s^2)
RE      = 6378;                  %Earth's radius (km)
J2      = 1082.63e-6;

%...Initial orbital parameters (given):
zp0 = 300;                       %Perigee altitude (km)
za0 = 3062;                      %Apogee altitude (km)
RA0 = 45*deg;                    %Right ascension of the node (radians)
i0  = 28*deg;                    %Inclination (radians)
w0  = 30*deg;                    %Argument of perigee (radians)
TA0 = 40*deg;                    %True anomaly (radians)

%...Initial orbital parameters (inferred):
rp0 = RE + zp0;                  %Perigee radius (km)
ra0 = RE + za0;                  %Apogee radius (km)
e0  = (ra0 - rp0)/(ra0 + rp0);   %Eccentricity
a0  = (ra0 + rp0)/2;             %Semimajor axis (km)
h0  = sqrt(rp0*mu*(1+e0));       %Angular momentum (km^2/s)
T0  = 2*pi/sqrt(mu)*a0^1.5;      %Period (s)

t0  = 0; tf = 2*days;            %Initial and final time (s)
%...end Input data

%Store the initial orbital elements in the array coe0:
coe0 = [h0 e0 RA0 i0 w0 TA0];

%...Obtain the initial state vector from Algorithm 4.5 (sv_from_coe):
[R0 V0] = sv_from_coe(coe0, mu); %R0 is the initial position vector
                                 %R0 is the initial position vector
r0 = norm(R0); v0 = norm(V0);    %Magnitudes of T0 and V0

del_t = T0/100;                  %Time step for Encke procedure
options = odeset('maxstep', del_t);

%...Begin the Encke integration;
t      = t0;                     %Initialize the time scalar
tsave  = t0;                     %Initialize the vector of solution times
y      = [R0 V0];                %Initialize the state vector
del_y0 = zeros(6,1);             %Initialize the state vector perturbation

t = t + del_t;                   %First time step

%  Loop over the time interval [t0, tf] with equal increments del_t:
while t <= tf + del_t/2
```

```
%    Integrate Equation 12.7 over the time increment del_t:
     [dum,z] = ode45(@rates, [t0 t], del_y0, options);

%    Compute the osculating state vector at time t:
     [Rosc,Vosc] = rv_from_r0v0(R0, V0, t-t0);

%    Rectify:
     R0      = Rosc + z(end,1:3);
     V0      = Vosc + z(end,4:6);
     t0      = t;

%    Prepare for next time step:
     tsave   = [tsave;t];
     y       = [y; [R0 V0]];
     t       = t + del_t;
     del_y0 = zeros(6,1);
end
%    End the loop

t = tsave;    %t is the vector of equispaced solution times
%...End the Encke integration;

%...At each solution time extract the orbital elements from the state
%    vector using Algorithm 4.2:
n_times = length(t);    %n_times is the number of solution times
for j = 1:n_times
     R      = [y(j,1:3)];
     V      = [y(j,4:6)];
     r(j)   = norm(R);
     v(j)   = norm(V);
     coe    = coe_from_sv(R,V, mu);
     h(j)   = coe(1);
     e(j)   = coe(2);
     RA(j)  = coe(3);
     i(j)   = coe(4);
     w(j)   = coe(5);
     TA(j)  = coe(6);
end

%...Plot selected osculating elements:

figure(1)
subplot(2,1,1)
plot(t/3600,(RA - RA0)/deg)
title('Variation of Right Ascension')
xlabel('hours')
ylabel('{\it\Delta\Omega} (deg)')
```

```
grid on
grid minor
axis tight

subplot(2,1,2)
plot(t/3600,(w - w0)/deg)
title('Variation of Argument of Perigee')
xlabel('hours')
ylabel('{\it\Delta\omega} (deg)')
grid on
grid minor
axis tight

figure(2)
subplot(3,1,1)
plot(t/3600,h - h0)
title('Variation of Angular Momentum')
xlabel('hours')
ylabel('{\it\Deltah} (km^2/s)')
grid on
grid minor
axis tight

subplot(3,1,2)
plot(t/3600,e - e0)
title('Variation of Eccentricity')
xlabel('hours')
ylabel('\it\Deltae')
grid on
grid minor
axis tight

subplot(3,1,3)
plot(t/3600,(i - i0)/deg)
title('Variation of Inclination')
xlabel('hours')
ylabel('{\it\Deltai} (deg)')
grid on
grid minor
axis tight

%...Subfunction:
% ~~~~~~~~~~~~~~~~~~~~~~~~~~~~~~~~~~~~~~~~~~~~~~~~~~~~~~~~~~~~~~~~~~~~~
function dfdt = rates(t,f)
% ~~~~~~~~~~~~~~~~~~~~~~
%
% This function calculates the time rates of Encke's deviation in position
```

```
% del_r and velocity del_v.
% ------------------------------------------------------------------------

del_r = f(1:3)';      %Position deviation
del_v = f(4:6)';      %Velocity deviation

%...Compute the state vector on the osculating orbit at time t
%   (Equation 12.5) using Algorithm 3.4:
[Rosc,Vosc] = rv_from_r0v0(R0, V0, t-t0);

%...Calculate the components of the state vector on the perturbed orbit
%   and their magnitudes:
Rpp    = Rosc + del_r;
Vpp    = Vosc + del_v;
rosc   = norm(Rosc);
rpp    = norm(Rpp);

%...Compute the J2 perturbing acceleration from Equation 12.30:
xx     = Rpp(1); yy = Rpp(2); zz = Rpp(3);

fac    =    3/2*J2*(mu/rpp^2)*(RE/rpp)^2;
ap     =    -fac*[(1 - 5*(zz/rpp)^2)*(xx/rpp) ...
                  (1 - 5*(zz/rpp)^2)*(yy/rpp) ...
                  (3 - 5*(zz/rpp)^2)*(zz/rpp)];

%...Compute the total perturbing ecceleration from Equation 12.7:
F      = 1 - (rosc/rpp)^3;
del_a = -mu/rosc^3*(del_r - F*Rpp) + ap;

dfdt   = [del_v(1) del_v(2) del_v(3) del_a(1) del_a(2) del_a(3)]';
dfdt   = [del_v del_a]'; %Return the deviative velocity and acceleration
                         %to ode45.

end %rates
% ~~~~~~~~~~~~~~~~~~~~~~~~~~

end %Example_10_02
% ~~~~~~~~~~~~~~~~~~~~~~~~~~~~~~~~~~~~~~~~~~~~~~~~~~~~~~~~~~~~~~~~~~~~~~~~~~~~~
```

D.42 EXAMPLE 10.6: USING GAUSS' VARIATIONAL EQUATIONS TO ASSESS J_2 EFFECT ON ORBITAL ELEMENTS

FUNCTION FILE: `Example_10_06.m`

```
% ~~~~~~~~~~~~~~~~~~~~~~~~~~~~~~~~~~~~~~~~~~~~~~~~~~~~~~~~~~~~~~~~~~~~~~~~~~~~~
function Example_10_6
% ~~~~~~~~~~~~~~~~~~
```

```
%
% This function solves Example 10.6 by using MATLAB's ode45 to numerically
% integrate Equations 10.89 (the Gauss planetary equations) to determine
% the J2 perturbation of the orbital elements.
%
% User M-functions required:  None
% User subfunctions required: rates
% ------------------------------------------------------------------------

%...Preliminaries:
close all; clear all; clc

%...Conversion factors:
hours   = 3600;                   %Hours to seconds
days    = 24*hours;               %Days to seconds
deg     = pi/180;                 %Degrees to radians

%...Constants:
mu      = 398600;                 %Gravitational parameter (km^3/s^2)
RE      = 6378;                   %Earth's radius (km)
J2      = 1082.63e-6;             %Earth's J2

%...Initial orbital parameters (given):
rp0     = RE + 300;               %perigee radius (km)
ra0     = RE + 3062;              %apogee radius (km
RA0     = 45*deg;                 %Right ascencion of the node (radians)
i0      = 28*deg;                 %Inclination (radians)
w0      = 30*deg;                 %Argument of perigee (radians)
TA0     = 40*deg;                 %True anomaly (radians)

%...Initial orbital parameters (inferred):
e0      = (ra0 - rp0)/(ra0 + rp0); %eccentricity
h0      = sqrt(rp0*mu*(1 + e0));   %angular momentrum (km^2/s)
a0      = (rp0 + ra0)/2;           %Semimajor axis (km)
T0      = 2*pi/sqrt(mu)*a0^1.5;    %Period (s)

%...Store initial orbital elements (from above) in the vector coe0:
coe0    = [h0 e0 RA0 i0 w0 TA0];

%...Use ODE45 to integrate the Gauss variational equations (Equations
%   12.89) from t0 to tf:
t0      = 0;
tf      = 2*days;
nout    = 5000;  %Number of solution points to output for plotting purposes
tspan   = linspace(t0, tf, nout);
options = odeset(...
                 'reltol',      1.e-8, ...
                 'abstol',      1.e-8, ...
                 'initialstep', T0/1000);
```

```
y0     = coe0';
[t,y]  = ode45(@rates, tspan, y0, options);

%...Assign the time histories mnemonic variable names:
h  = y(:,1);
e  = y(:,2);
RA = y(:,3);
i  = y(:,4);
w  = y(:,5);
TA = y(:,6);

%...Plot the time histories of the osculatinig elements:
figure(1)
subplot(5,1,1)
plot(t/3600,(RA - RA0)/deg)
title('Right Ascension (degrees)')
xlabel('hours')
grid on
grid minor
axis tight

subplot(5,1,2)
plot(t/3600,(w - w0)/deg)
title('Argument of Perigee (degrees)')
xlabel('hours')
grid on
grid minor
axis tight

subplot(5,1,3)
plot(t/3600,h - h0)
title('Angular Momentum (km^2/s)')
xlabel('hours')
grid on
grid minor
axis tight

subplot(5,1,4)
plot(t/3600,e - e0)
title('Eccentricity')
xlabel('hours')
grid on
grid minor
axis tight
```

```
subplot(5,1,5)
plot(t/3600,(i - i0)/deg)
title('Inclination (degrees)')
xlabel('hours')
grid on
grid minor
axis tight

%...Subfunction:

% ~~~~~~~~~~~~~~~~~~~~~~~~~~~~~~~~~~~~~~~~~~~~~~~~~~~~~~~~~~~~~~~~~~~~~~
function dfdt = rates(t,f)
% ~~~~~~~~~~~~~~~~~~~~~~~~~~
%
% This function calculates the time rates of the orbital elements
% from Gauss's variational equations (Equations 12.89).
% -------------------------------------------------------------------------

%...The orbital elements at time t:
h    = f(1);
e    = f(2);
RA   = f(3);
i    = f(4);
w    = f(5);
TA   = f(6);

r    = h^2/mu/(1 + e*cos(TA)); %The radius
u    = w + TA;                 %Argument of latitude

%...Orbital element rates at time t (Equations 12.89):
hdot  = -3/2*J2*mu*RE^2/r^3*sin(i)^2*sin(2*u);

edot  = ...
    3/2*J2*mu*RE^2/h/r^3*(h^2/mu/r ...
   *(sin(u)*sin(i)^2*(3*sin(TA)*sin(u) - 2*cos(TA)*cos(u)) - sin(TA)) ...
   -sin(i)^2*sin(2*u)*(e + cos(TA)));

edot  = 3/2*J2*mu*RE^2/h/r^3 ...
        *(h^2/mu/r*sin(TA)*(3*sin(i)^2*sin(u)^2 - 1) ...
          -sin(2*u)*sin(i)^2*((2+e*cos(TA))*cos(TA)+e));

TAdot =  h/r^2 + 3/2*J2*mu*RE^2/e/h/r^3 ...
         *(h^2/mu/r*cos(TA)*(3*sin(i)^2*sin(u)^2 - 1) ...
         + sin(2*u)*sin(i)^2*sin(TA)*(h^2/mu/r + 1));

RAdot = -3*J2*mu*RE^2/h/r^3*sin(u)^2*cos(i);
```

```
idot  = -3/4*J2*mu*RE^2/h/r^3*sin(2*u)*sin(2*i);

wdot  = 3/2*J2*mu*RE^2/e/h/r^3 ...
        *(-h^2/mu/r*cos(TA)*(3*sin(i)^2*sin(u)^2 - 1) ...
          - sin(2*u)*sin(i)^2*sin(TA)*(2 + e*cos(TA)) ...
          + 2*e*cos(i)^2*sin(u)^2);

%...Pass these rates back to ODE45 in the array dfdt:
dfdt  =  [hdot edot RAdot idot wdot TAdot]';

end %rates
% ~~~~~~~~~~~~~~~~~~~~~~~~~~~~~~~~~~~~~~~~~~~~~~~~~~~~~~~~~~~~~~~~~~~~~~~~~~~~

end %Example_10_6
% ~~~~~~~~~~~~~~~~~~~~~~~~~~~~~~~~~~~~~~~~~~~~~~~~~~~~~~~~~~~~~~~~~~~~~~~~~~~~
```

D.43 ALGORITHM 10.2: CALCULATE THE GEOCENTRIC POSITION OF THE SUN AT A GIVEN EPOCH

FUNCTION FILE: solar_position.m

```
% ~~~~~~~~~~~~~~~~~~~~~~~~~~~~~~~~~~~~~~~~~~~~~~~~~~~~~~~~~~~~~~~~~~~~~~~~~~~~
function [lamda eps r_S] = solar_position(jd)
%
% This function alculates the geocentric equatorial position vector
% of the sun, given the julian date.
%
% User M-functions required: None
% -------------------------------------------------------------------------
%...Astronomical unit (km):
AU    = 149597870.691;

%...Julian days since J2000:
n     = jd - 2451545;

%...Julian centuries since J2000:
cy    = n/36525;

%...Mean anomaly (deg{:
M     = 357.528 + 0.9856003*n;
M     = mod(M,360);

%...Mean longitude (deg):
L     = 280.460 + 0.98564736*n;
L     = mod(L,360);
```

```
%...Apparent ecliptic longitude (deg):
lamda = L + 1.915*sind(M) + 0.020*sind(2*M);
lamda = mod(lamda,360);

%...Obliquity of the ecliptic (deg):
eps   = 23.439 - 0.0000004*n;

%...Unit vector from earth to sun:
u     = [cosd(lamda); sind(lamda)*cosd(eps); sind(lamda)*sind(eps)];

%...Distance from earth to sun (km):
rS    = (1.00014 - 0.01671*cosd(M) - 0.000140*cosd(2*M))*AU;

%...Geocentric position vector (km):
r_S   = rS*u;
end %solar_position
% ~~~~~~~~~~~~~~~~~~~~~~~~~~~~~~~~~~~~~~~~~~~~~~~~~~~~~~~~~~~~~~~~~~~~~~~~
```

D.44 ALGORITHM 10.3: DETERMINE WHETHER OR NOT A SATELLITE IS IN EARTH'S SHADOW

FUNCTION FILE: los.m

```
% ~~~~~~~~~~~~~~~~~~~~~~~~~~~~~~~~~~~~~~~~~~~~~~~~~~~~~~~~~~~~~~~~~
function light_switch = los(r_sat, r_sun)
%
% This function uses the ECI position vectors of the satellite (r_sat)
% and the sun (r_sun) to determine whether the earth is in the line of
% sight between the two.
%
% User M-functions required: None
% -------------------------------------------------------------------------
RE         = 6378;          %Earth's radius (km)
rsat       = norm(r_sat);
rsun       = norm(r_sun);

%...Angle between sun and satellite position vectore:
theta      = acosd(dot(r_sat, r_sun)/rsat/rsun);

%...Angle between the satellite position vector and the radial to the point
%   of tangency with the earth of a line from the satellite:
theta_sat = acosd(RE/rsat);

%...Angle between the sun position vector and the radial to the point
%   of tangency with the earth of a line from the sun:
theta_sun = acosd(RE/rsun);

%...Determine whether a line from the sun to the satellite
```

```
%   intersects the earth:
if theta_sat + theta_sun <= theta
    light_switch = 0;   %yes
else
    light_switch = 1;   %no
end

end %los
% ---------------------------------------------------------------------------
```

D.45 EXAMPLE 10.9: USE GAUSS' VARIATIONAL EQUATIONS TO DETERMINE THE EFFECT OF SOLAR RADIATION PRESSURE ON AN EARTH SATELLITE'S ORBITAL PARAMETERS

FUNCTION FILE: Example_10_09.m

```
% ~~~~~~~~~~~~~~~~~~~~~~~~~~~~~~~~~~~~~~~~~~~~~~~~~~~~~~~~~~~~~~~~~~~~~~~~~~~~
function Example_10_09
%
% This function solve Example 10.9 the Gauss planetary equations for
% solar radiation pressure (Equations 10.106).
%
% User M-functions required:  sv_from_coe, los, solar_position
% User subfunctions required: rates
% The M-function rsmooth may be found in Garcia, D: "Robust Smoothing of Gridded
Data in One and Higher Dimensions with Missing Values," Computational Statistics
and Data Analysis, Vol. 54, 1167-1178, 2010.
% ~~~~~~~~~~~~~~~~~~~~~~~~~~~~~~~~~~~~~~~~~~~~~~~~~~~~~~~~~~~~~~~~~~~~~~~~~~~~

global JD  %Julian day

%...Preliminaries:
close all
clear all
clc

%...Conversion factors:
hours = 3600;                   %Hours to seconds
days  = 24*hours;               %Days to seconds
deg   = pi/180;                 %Degrees to radians

%...Constants;
mu    = 398600;                 %Gravitational parameter (km^3/s^2)
RE    = 6378;                   %Eath's radius (km)
c     = 2.998e8;                %Speed of light (m/s)
S     = 1367;                   %Solar constant (W/m^2)
Psr   = S/c;                    %Solar pressure (Pa);
```

```
%...Satellite data:
CR    = 2;                     %Radiation pressure codfficient
m     = 100;                   %Mass (kg)
As    = 200;                   %Frontal area (m^2);

%...Initial orbital parameters (given):
a0    = 10085.44;              %Semimajor axis (km)
e0    = 0.025422;              %eccentricity
incl0 = 88.3924*deg;           %Inclination (radians)
RA0   = 45.38124*deg;          %Right ascencion of the node (radians)
TA0   = 343.4268*deg;          %True anomaly (radians)
w0    = 227.493*deg;           %Argument of perigee (radians)

%...Initial orbital parameters (inferred):
h0    = sqrt(mu*a0*(1-e0^2));  %angular momentrum (km^2/s)
T0    = 2*pi/sqrt(mu)*a0^1.5;  %Period (s)
rp0   = h0^2/mu/(1 + e0);      %perigee radius (km)
ra0   = h0^2/mu/(1 - e0);      %apogee radius (km)

%...Store initial orbital elements (from above) in the vector coe0:
coe0  = [h0 e0 RA0 incl0 w0 TA0];

%...Use ODE45 to integrate Equations 12.106, the Gauss planetary equations
%   from t0 to tf:

JD0   = 2438400.5;              %Initial Julian date (6 January 1964 0 UT)
t0    = 0;                      %Initial time (s)
tf    = 3*365*days;             %final time (s)
y0    = coe0';                  %Initial orbital elements
nout  = 4000;                   %Number of solution points to output
tspan = linspace(t0, tf, nout); %Integration time interval
options = odeset(...
                 'reltol',      1.e-8, ...
                 'abstol',      1.e-8, ...
                 'initialstep', T0/1000);
[t,y] = ode45(@rates, tspan, y0, options);

%...Extract or compute the orbital elements' time histories from the
%   solution vector y:
h     = y(:,1);
e     = y(:,2);
RA    = y(:,3);
incl  = y(:,4);
w     = y(:,5);
TA    = y(:,6);
a     = h.^2/mu./(1 - e.^2);
```

```
%...Smooth the data to remove short period variations:
h     = rsmooth(h);
e     = rsmooth(e);
RA    = rsmooth(RA);
incl  = rsmooth(incl);
w     = rsmooth(w);
a     = rsmooth(a);

figure(2)
subplot(3,2,1)
plot(t/days,h - h0)
title('Angular Momentum (km^2/s)')
xlabel('days')
axis tight

subplot(3,2,2)
plot(t/days,e - e0)
title('Eccentricity')
xlabel('days')
axis tight

subplot(3,2,4)
plot(t/days,(RA - RA0)/deg)
title('Right Ascension (deg)')
xlabel('days')
axis tight

subplot(3,2,5)
plot(t/days,(incl - incl0)/deg)
title('Inclination (deg)')
xlabel('days')
axis tight

subplot(3,2,6)
plot(t/days,(w - w0)/deg)
title('Argument of Perigee (deg)')
xlabel('days')
axis tight

subplot(3,2,3)
plot(t/days,a - a0)
title('Semimajor axis (km)')
xlabel('days')
axis tight

%...Subfunctions:
% ~~~~~~~~~~~~~~~~~~~~~~~~~~~~~~~~~~~~~~~~~~~~~~~~~~~~~~~~~~~~~~~~~~~~~
```

```
function dfdt = rates(t,f)
% ~~~~~~~~~~~~~~~~~~~~
%...Update the Julian Date at time t:
JD     =   JD0 + t/days;

%...Compoute the apparent position vector of the sun:
[lamda eps r_sun] = solar_position(JD);

%...Convert the ecliptic latitude and the obliquity to radians:
lamda  = lamda*deg;
eps    = eps*deg;

%...Extract the orbital elements at time t
h      = f(1);
e      = f(2);
RA     = f(3);
i      = f(4);
w      = f(5);
TA     = f(6);

u      = w + TA;  %Argument of latitude

%...Compute the state vector at time t:
coe    =   [h e RA i w TA];
[R, V] =   sv_from_coe(coe,mu);

%...Calculate the manitude of the radius vector:
r      =   norm(R);

%...Compute the shadow function and the solar radiation perturbation:
nu     =   los(R, r_sun);
pSR    =   nu*(S/c)*CR*As/m/1000;

%...Calculate the trig functions in Equations 12.105.
sl = sin(lamda);  cl = cos(lamda);
se = sin(eps);    ce = cos(eps);
sW = sin(RA);     cW = cos(RA);
si = sin(i);      ci = cos(i);
su = sin(u);      cu = cos(u);
sT = sin(TA);     cT = cos(TA);

%...Calculate the earth-sun unit vector components (Equations 12.105):
ur     =   sl*ce*cW*ci*su + sl*ce*sW*cu - cl*sW*ci*su ...
         + cl*cW*cu + sl*se*si*su;

us     =   sl*ce*cW*ci*cu - sl*ce*sW*su - cl*sW*ci*cu ...
         - cl*cW*su + sl*se*si*cu;
```

```
uw     = - sl*ce*cW*si + cl*sW*si + sl*se*ci;

%...Calculate the time rates of the osculating elements from
%   Equations 12.106:

hdot   = -pSR*r*us;

edot   = -pSR*(h/mu*sT*ur ...
               + 1/mu/h*((h^2 + mu*r)*cT + mu*e*r)*us);

TAdot  =   h/r^2 ...
         - pSR/e/h*(h^2/mu*cT*ur - (r + h^2/mu)*sT*us);

RAdot  = -pSR*r/h/si*su*uw;

idot   = -pSR*r/h*cu*uw;

wdot   = -pSR*(-1/e/h*(h^2/mu*cT*ur - (r + h^2/mu)*sT*us) ...
              - r*su/h/si*ci*uw);

%...Return the rates to ode45:
dfdt   =  [hdot edot RAdot idot wdot TAdot]';

end %rates

end %Example_10_9
% ~~~~~~~~~~~~~~~~~~~~~~~~~~~~~~~~~~~~~~~~~~~~~~~~~~~~~~~~~~~~~~~~~~~~~~~~~~
```

D.46 ALGORITHM 10.4: CALCULATE THE GEOCENTRIC POSITION OF THE MOON AT A GIVEN EPOCH

FUNCTION FILE: lunar_position.m

```
% ------------------------------------------------------------------------
function r_moon = lunar_position(jd)
%
%...Calculates the geocentric equatorial position vector of the moon
%   given the Julian day.
%
% User M-functions required: None
% ------------------------------------------------------------------------

%...Earth's radius (km):
RE = 6378;

% -------------------------  implementation  -----------------
%...Time in centuries since J2000:
```

```
T = (jd - 2451545)/36525;

%...Ecliptic longitude (deg):
e_long =  218.32 + 481267.881*T ...
       + 6.29*sind(135.0 + 477198.87*T) - 1.27*sind(259.3 - 413335.36*T)...
       + 0.66*sind(235.7 + 890534.22*T) + 0.21*sind(269.9 + 954397.74*T)...
       - 0.19*sind(357.5 +35999.05*T) - 0.11*sind(186.5 + 966404.03*T);
e_long = mod(e_long,360);

%...Ecliptic latitude (deg):
e_lat  =  5.13*sind( 93.3 + 483202.02*T) + 0.28*sind(228.2 + 960400.89*T)...
       - 0.28*sind(318.3 +6003.15*T) - 0.17*sind(217.6 - 407332.21*T);
e_lat  = mod(e_lat,360);

%...Horizontal parallax (deg):
h_par  =  0.9508 ...
       + 0.0518*cosd(135.0 + 477198.87*T) + 0.0095*cosd(259.3 - 413335.36*T)...
       + 0.0078*cosd(235.7 + 890534.22*T) + 0.0028*cosd(269.9 + 954397.74*T);
h_par  = mod(h_par,360);

%...Angle between earth's orbit and its equator (deg):
obliquity = 23.439291 - 0.0130042*T;

%...Direction cosines of the moon's geocentric equatorial position vector:
l = cosd(e_lat) * cosd(e_long);
m = cosd(obliquity)*cosd(e_lat)*sind(e_long) - sind(obliquity)*sind(e_lat);
n = sind(obliquity)*cosd(e_lat)*sind(e_long) + cosd(obliquity)*sind(e_lat);

%...Earth-moon distance (km):
dist   = RE/sind(h_par);

%...Moon's geocentric equatorial position vector (km):
r_moon = dist*[l m n];

end  %lunar_position
% ----------------------------------------------------------------------------
```

D.47 EXAMPLE 10.11: USE GAUSS' VARIATIONAL EQUATIONS TO DETERMINE THE EFFECT OF LUNAR GRAVITY ON AN EARTH SATELLITE'S ORBITAL PARAMETERS

FUNCTION FILE: Example_10_11.m

```
% ~~~~~~~~~~~~~~~~~~~~~~~~~~~~~~~~~~~~~~~~~~~~~~~~~~~~~~~~~~~~~~~~~~~~~~~~~~
function Example_10_11
% This function solves Example 10.11 by using MATLAB's ode45 to integrate
% Equations 10.84, the Gauss variational equations, for a lunar
% gravitational perturbation.
```

```
%
% User M-functions required:  sv_from_coe, lunar_position
% User subfunctions required: solveit rates
% ------------------------------------------------------------------------

global JD   %Julian day

%...Preliminaries:
close all
clear all
clc

%...Conversion factors:
hours    = 3600;                %Hours to seconds
days     = 24*hours;            %Days to seconds
deg      = pi/180;              %Degrees to radians

%...Constants;
mu       = 398600;              %Earth's gravitational parameter (km^3/s^2)
mu3      = 4903;                %Moon's gravitational parameter (km^3/s^2)
RE       = 6378;                %Earth's radius (km)

%...Initial data for each of the three given orbits:

type = {'GEO' 'HEO' 'LEO'};

%...GEO
n    = 1;
a0   = 42164;    %semimajor axis (km)
e0   = 0.0001;   %eccentricity
w0   = 0;        %argument of perigee (rad)
RA0  = 0;        %right ascension (rad)
i0   = 1*deg;    %inclination (rad)
TA0  = 0;        %true anomaly (rad)
JD0  = 2454283;  %Julian Day
solveit

%...HEO
n    = 2;
a0   = 26553.4;
e0   = 0.741;
w0   = 270;
RA0  = 0;
i0   = 63.4*deg;
TA0  = 0;
JD0  = 2454283;
solveit
```

```
%...LEO
n    = 3;
a0   = 6678.136;
e0   = 0.01;
w0   = 0;
RA0  = 0;
i0   = 28.5*deg;
TA0  = 0;
JD0  = 2454283;
solveit

%...Subfunctions:

% ~~~~~~~~~~~~~~~~~~~~~~~~~~~~~~~~~~~~~~~~~~~~~~
function solveit
%
% Calculations and plots common to all of the orbits
%
% ---------------------------------------------------
%
%...Initial orbital parameters (calculated from the given data):
h0   = sqrt(mu*a0*(1-e0^2));  %angular momentum (km^2/s)
T0   = 2*pi/sqrt(mu)*a0^1.5;  %Period (s)
rp0  = h0^2/mu/(1 + e0);      %perigee radius (km)
ra0  = h0^2/mu/(1 - e0);      %apogee radius (km)

%...Store initial orbital elements (from above) in the vector coe0:
coe0 = [h0;e0;RA0;i0;w0;TA0];

%...Use ODE45 to integrate the Equations 12.84, the Gauss variational
%   equations with lunar gravity as the perturbation, from t0 to tf:
t0       = 0;
tf       = 60*days;
y0       = coe0;                 %Initial orbital elements
nout     = 400;                  %Number of solution points to output
tspan    = linspace(t0, tf, nout); %Integration time interval
options = odeset(...
                 'reltol', 1.e-8, ...
                 'abstol', 1.e-8);
[t,y]   = ode45(@rates, tspan, y0, options);

%...Time histories of the right ascension, inclination and argument of
%   perigee:
RA = y(:,3);
i  = y(:,4);
w  = y(:,5);
```

```
%...Smooth the data to eliminate short period variations:
RA = rsmooth(RA);
i  = rsmooth(i);
w  = rsmooth(w);

figure(n)
subplot(1,3,1)
plot(t/days,(RA - RA0)/deg)
title('Right Ascension vs Time')
xlabel('{\itt} (days)')
ylabel('{\it\Omega} (deg)')
axis tight

subplot(1,3,2)
plot(t/days,(i - i0)/deg)
title('Inclination vs Time')
xlabel('{\itt} (days)')
ylabel('{\iti} (deg)')
axis tight

subplot(1,3,3)
plot(t/days,(w - w0)/deg)
title('Argument of Perigee vs Time')
xlabel('{\itt} (days)')
ylabel('{\it\omega} (deg)')
axis tight

drawnow

end %solveit
% ~~~~~~~~~~~~~~~~~~~~

% ~~~~~~~~~~~~~~~~~~~~
function dfdt = rates(t,f)
% ~~~~~~~~~~~~~~~~~~~~
%...The orbital elements at time t:
h      =   f(1);
e      =   f(2);
RA     =   f(3);
i      =   f(4);
w      =   f(5);
TA     =   f(6);
phi    =   w + TA; %argument of latitude

%...Obtain the state vector at time t from Algorithm 4.5:
coe    =   [h e RA i w TA];
[R, V] =   sv_from_coe(coe,mu);
```

```
%...Obtain the unit vectors of the rsw system:
r      =   norm(R);
ur     =   R/r;            %radial
H      =   cross(R,V);
uh     =   H/norm(H);      %normal
s      =   cross(uh, ur);
us     =   s/norm(s);      %transverse

%...Update the Julian Day:
JD     =   JD0 + t/days;

%...Find and normalize the position vector of the moon:
R_m    =   lunar_position(JD);
r_m    =   norm(R_m);

R_rel  =   R_m - R;     %R_rel = position vector of moon wrt satellite
r_rel  =   norm(R_rel);

%...See Appendix F:
q      =   dot(R,(2*R_m - R))/r_m^2;
F      =   (q^2 - 3*q + 3)*q/(1 + (1-q)^1.5);

%...Gravitationl perturbation of the moon (Equation 12.117):
ap     =   mu3/r_rel^3*(F*R_m - R);

%...Perturbation components in the rsw system:
apr    =   dot(ap,ur);
aps    =   dot(ap,us);
aph    =   dot(ap,uh);

%...Gauss variational equations (Equations 12.84):
hdot   =   r*aps;

edot   =   h/mu*sin(TA)*apr ...
         + 1/mu/h*((h^2 + mu*r)*cos(TA) + mu*e*r)*aps;

RAdot  =   r/h/sin(i)*sin(phi)*aph;

idot   =   r/h*cos(phi)*aph;

wdot   = - h*cos(TA)/mu/e*apr ...
         + (h^2 + mu*r)/mu/e/h*sin(TA)*aps ...
         - r*sin(phi)/h/tan(i)*aph;

TAdot  = h/r^2 ...
         + 1/e/h*(h^2/mu*cos(TA)*apr - (r + h^2/mu)*sin(TA)*aps);
```

```
%...Return rates to ode45 in the array dfdt:
dfdt   =  [hdot edot RAdot idot wdot TAdot]';

end %rates
% ~~~~~~~~~~~~~~~~~~

end %Example_10_11
% ~~~~~~~~~~~~~~~~~~~~~~~~~~~~~~~~~~~~~~~~~~~~~~~~~~~~~~~~~~~~~~~~~~~~~~~~~~~~~~~~
```

D.48 EXAMPLE 10.12: USE GAUSS' VARIATIONAL EQUATIONS TO DETERMINE THE EFFECT OF SOLAR GRAVITY ON AN EARTH SATELLITE'S ORBITAL PARAMETERS

FUNCTION FILE: Example_10_12.m

```
% ~~~~~~~~~~~~~~~~~~~~~~~~~~~~~~~~~~~~~~~~~~~~~~~~~~~~~~~~~~~~~~~~~~~~~~~~~~~~~~~~
function Example_10_12
% This function solves Example 10.12 by using MATLAB's ode45 to integrate
% Equations 10.84, the Gauss variational equations, for a solar
% gravitational perturbation.
%
% User M-functions required:  sv_from_coe, lunar_position
% User subfunctions required: solveit rates
% ---------------------------------------------------------------------------

global JD   %Julian day

%...Preliminaries:
close all
clear all
clc

%...Conversion factors:
hours     = 3600;               %Hours to seconds
days      = 24*hours;           %Days to seconds
deg       = pi/180;             %Degrees to radians

%...Constants;
mu        = 398600;             %Earth's gravitational parameter (km^3/s^2)
mu3       = 132.712e9;          %Sun's gravitational parameter (km^3/s^2)
RE        = 6378;               %Earth's radius (km)

%...Initial data for each of the three given orbits:

type = {'GEO' 'HEO' 'LEO'};

%...GEO
n    = 1;
```

```
a0   = 42164;    %semimajor axis (km)
e0   = 0.0001;   %eccentricity
w0   = 0;        %argument of perigee (rad)
RA0  = 0;        %right ascension (rad)
i0   = 1*deg;    %inclination (rad)
TA0  = 0;        %true anomaly (rad)
JD0  = 2454283;  %Julian Day
solveit

%...HEO
n    = 2;
a0   = 26553.4;
e0   = 0.741;
w0   = 270;
RA0  = 0;
i0   = 63.4*deg;
TA0  = 0;
JD0  = 2454283;
solveit

%...LEO
n    = 3;
a0   = 6678.136;
e0   = 0.01;
w0   = 0;
RA0  = 0;
i0   = 28.5*deg;
TA0  = 0;
JD0  = 2454283;
solveit

%...Subfunctions:

% ~~~~~~~~~~~~~~~~~~~~~~~~~~~~~~~~~~~~~~~~~~~~~
function solveit
%
% Calculations and plots common to all of the orbits
%
% -------------------------------------------------
%
%...Initial orbital parameters (calculated from the given data):
h0   = sqrt(mu*a0*(1-e0^2));  %angular momentum (km^2/s)
T0   = 2*pi/sqrt(mu)*a0^1.5;  %Period (s)
rp0  = h0^2/mu/(1 + e0);      %perigee radius (km)
ra0  = h0^2/mu/(1 - e0);      %apogee radius (km)

%...Store initial orbital elements (from above) in the vector coe0:
coe0 = [h0;e0;RA0;i0;w0;TA0];
```

```
%...Use ODE45 to integrate the Equations 12.84, the Gauss variational
%   equations with lunar gravity as the perturbation, from t0 to tf:
t0      = 0;
tf      = 720*days;
y0      = coe0;                    %Initial orbital elements
nout    = 400;                     %Number of solution points to output
tspan   = linspace(t0, tf, nout); %Integration time interval
options = odeset(...
               'reltol', 1.e-8, ...
               'abstol', 1.e-8);
[t,y]   = ode45(@rates, tspan, y0, options);

%...Time histories of the right ascension, inclination and argument of
%   perigee:
RA = y(:,3);
i  = y(:,4);
w  = y(:,5);

%...Smooth the data to eliminate short period variations:
RA = rsmooth(RA);
i  = rsmooth(i);
w  = rsmooth(w);

figure(n)
subplot(1,3,1)
plot(t/days,(RA - RA0)/deg)
title('Right Ascension vs Time')
xlabel('{\itt} (days)')
ylabel('{\it\Omega} (deg)')
axis tight

subplot(1,3,2)
plot(t/days,(i - i0)/deg)
title('Inclination vs Time')
xlabel('{\itt} (days)')
ylabel('{\iti} (deg)')
axis tight

subplot(1,3,3)
plot(t/days,(w - w0)/deg)
title('Argument of Perigee vs Time')
xlabel('{\itt} (days)')
ylabel('{\it\omega} (deg)')
axis tight
```

```
drawnow

end %solveit
% ~~~~~~~~~~~~~~~~~~~~~

% ~~~~~~~~~~~~~~~~~~~~~
function dfdt = rates(t,f)
% ~~~~~~~~~~~~~~~~~~~~~
%...The orbital elements at time t:
h      =   f(1);
e      =   f(2);
RA     =   f(3);
i      =   f(4);
w      =   f(5);
TA     =   f(6);
phi    =   w + TA; %argument of latitude

%...Obtain the state vector at time t from Algorithm 4.5:
coe    =   [h e RA i w TA];
[R, V] =   sv_from_coe(coe,mu);

%...Obtain the unit vectors of the rsw system:
r      =   norm(R);
ur     =   R/r;            %radial
H      =   cross(R,V);
uh     =   H/norm(H);      %normal
s      =   cross(uh, ur);
us     =   s/norm(s);      %transverse

%...Update the Julian Day:
JD     =   JD0 + t/days;

%...Find and normalize the position vector of the sun:
[lamda eps R_S] = solar_position(JD);
r_S    =   norm(R_S);

R_rel  =   R_S' - R;  %R_rel = position vector of sun wrt satellite
r_rel  =   norm(R_rel);

%...See Appendix F:
q      =   dot(R,(2*R_S' - R))/r_S^2;
F      =   (q^2 - 3*q + 3)*q/(1 + (1-q)^1.5);;

%...Gravitationl perturbation of the sun (Equation 12.130):
ap     =   mu3/r_rel^3*(F*R_S' - R);

%...Perturbation components in the rsw system:
apr    =   dot(ap,ur);
aps    =   dot(ap,us);
aph    =   dot(ap,uh);
```

```
%...Gauss variational equations (Equations 12.84):
hdot    =   r*aps;

edot    =   h/mu*sin(TA)*apr ...
          + 1/mu/h*((h^2 + mu*r)*cos(TA) + mu*e*r)*aps;

RAdot   =   r/h/sin(i)*sin(phi)*aph;

idot    =   r/h*cos(phi)*aph;

wdot    = - h*cos(TA)/mu/e*apr ...
          + (h^2 + mu*r)/mu/e/h*sin(TA)*aps ...
          - r*sin(phi)/h/tan(i)*aph;

TAdot   = h/r^2 ...
          + 1/e/h*(h^2/mu*cos(TA)*apr - (r + h^2/mu)*sin(TA)*aps);

%...Return rates to ode45 in the array dfdt:
dfdt    =  [hdot edot RAdot idot wdot TAdot]';

end %rates
% ~~~~~~~~~~~~~~~~~~

end %Example_10_12
% ~~~~~~~~~~~~~~~~~~~~~~~~~~~~~~~~~~~~~~~~~~~~~~~~~~~~~~~~~~~~~~~~~~~~~~~~~~~~
```

CHAPTER 11: RIGID BODY DYNAMICS

D.49 ALGORITHM 11.1: CALCULATE THE DIRECTION COSINE MATRIX FROM THE QUATERNION

FUNCTION FILE: dcm_from_q.m

```
% ~~~~~~~~~~~~~~~~~~~~~~~~~~~~~~~~~~~~~~~~~~~~
function Q = dcm_from_q(q)
% ~~~~~~~~~~~~~~~~~~~~~~~
%{
  This function calculates the direction cosine matrix
  from the quaternion.

  q - quaternion (where q(4) is the scalar part)
  Q - direction cosine matrix
%}
% -----------------------------------------------
```

```
q1 = q(1); q2 = q(2); q3 = q(3); q4 = q(4);

Q = [q1^2-q2^2-q3^2+q4^2,       2*(q1*q2+q3*q4),        2*(q1*q3-q2*q4);
        2*(q1*q2-q3*q4), -q1^2+q2^2-q3^2+q4^2,          2*(q2*q3+q1*q4);
        2*(q1*q3+q2*q4),        2*(q2*q3-q1*q4),  -q1^2-q2^2+q3^2+q4^2 ];
end %dcm_from_q
% ~~~~~~~~~~~~~~~~~~~~~~~~~~~~~~~~~~~~~~~~
```

D.50 ALGORITHM 11.2: CALCULATE THE QUATERNION FROM THE DIRECTION COSINE MATRIX

FUNCTION FILE: q_from_dcm.m

```
% ~~~~~~~~~~~~~~~~~~~~~~~~~~~~~~~~~~~~~~~~
function q = q_from_dcm(Q)
% ~~~~~~~~~~~~~~~~~~~~~
%{
  This function calculates the quaternion from the direction
  cosine matrix.

  Q - direction cosine matrix
  q - quaternion (where q(4) is the scalar part)

%}
% -------------------------------------------

K3 = ...
[Q(1,1)-Q(2,2)-Q(3,3), Q(2,1)+Q(1,2), Q(3,1)+Q(1,3), Q(2,3)-Q(3,2);
 Q(2,1)+Q(1,2), Q(2,2)-Q(1,1)-Q(3,3), Q(3,2)+Q(2,3), Q(3,1)-Q(1,3);
 Q(3,1)+Q(1,3), Q(3,2)+Q(2,3), Q(3,3)-Q(1,1)-Q(2,2), Q(1,2)-Q(2,1);
 Q(2,3)-Q(3,2), Q(3,1)-Q(1,3), Q(1,2)-Q(2,1), Q(1,1)+Q(2,2)+Q(3,3)]/3;

[eigvec, eigval] = eig(K3);

[x,i] = max(diag(eigval));

q = eigvec(:,i);
end %q_from_dcm
% ~~~~~~~~~~~~~~~~~~~~~~~~~~~~~~~~~~~~~~~~
```

D.51 QUATERNION VECTOR ROTATION OPERATION (EQ. 11.160)

FUNCTION FILE: quat_rotate.m

```
% ~~~~~~~~~~~~~~~~~~~~~~~~~~~~~~~~~~~~~~~~~~~~~~
function r = quat_rotate(q,v)
% ~~~~~~~~~~~~~~~~~~~~~~
```

```
%{
   quat_rotate rotates a vector by a unit quaternion.
   r = quat_rotate(q,v) calculates the rotated vector r for a
       quaternion q and a vector v.
   q   is a 1-by-4 matrix whose norm must be 1. q(1) is the scalar part
       of the quaternion.
   v   is a 1-by-3 matrix.
   r   is a 1-by-3 matrix.

   The 3-vector v is made into a pure quaternion 4-vector V = [0 v]. r is
   produced by the quaternion product R = q*V*qinv. r = [R(2) R(3) R(4)].

   MATLAB M-functions used: quatmultiply, quatinv.
%}
% ---------------------------------------------------
qinv = quatinv(q);
r    = quatmultiply(quatmultiply(q,[0 v]),qinv);
r    = r(2:4);
end %quat_rotate
% ~~~~~~~~~~~~~~~~~~~~~~~~~~~~~~~~~~~~~~~~~~~~~~~~~~~~~~~~~~~~
```

D.52 EXAMPLE 11.26: SOLUTION OF THE SPINNING TOP PROBLEM

FUNCTION FILE: Example_11_23.m

```
% ~~~~~~~~~~~~~~~~~~~~~~~~~~~~~~~~~~~~~
function Example_11_26
% ~~~~~~~~~~~~~~~~~~~~~~~~~
%{
  This program numerically integrates Euler's equations of motion
  for the spinning top (Example 11.26, Equations (a)). The
  quaternion is used to obtain the time history of the top's
  orientation. See Figures 11.34 and 11.35.

  User M-functions required: rkf45, q_from_dcm, dcm_from_q, dcm_to_euler
  User subfunction required: rates
%}
% ----------------------------------------------

clear all; close all; clc

%...Data from Example 11.15:
g     = 9.807;        % Acceleration of gravity (m/s^2)
m     = 0.5;          % Mass in kg
d     = 0.05;         % Distance of center of mass from pivot point (m)
A     = 12.e-4;       % Moment of inertia about body x (kg-m^2)
B     = 12.e-4;       % Moment of inertia about body y (kg-m^2)
C     = 4.5e-4;       % Moment of inertia about body z (kg-m^2)
```

```
ws0   = 1000*2*pi/60; % Spin rate (rad/s)

wp0   = 51.93*2*pi/60;% Precession rate (rad/s) Use to obtain Fig. 11.33
wp0   = 0;           %                         Use to obtain Fig. 11.34

wn0   = 0;           % Nutation rate (deg/s)
theta = 60;          % Initial nutation angle (deg)
z     = [0 -sind(theta) cosd(theta)]; % Initial z-axis direction:
p     = [1 0 0];                      % Initial x-axis direction
                                      %   (or a line defining x-z plane)
%...
y     = cross(z,p);     % y-axis direction (normal to x-z plane)
x     = cross(y,z);     % x-axis direction (normal to y-z plane)
i     = x/norm(x);      % Unit vector along x axis
j     = y/norm(y);      % Unit vector along y axis
k     = z/norm(z);      % Unit vector along z axis
QXx   = [i; j; k];      % Initial direction cosine matrix

%...Initial precession, nutation, and spin angles (deg):
[phi0 theta0 psi0]  = dcm_to_euler(QXx);

%...Initial quaternion (column vector):
q0    = q_from_dcm(QXx);

%...Initial body-frame angular velocity, column vector (rad/s):
w0    = [wp0*sind(theta0)*sin(psi0) + wn0*cosd(psi0), ...
         wp0*sind(theta0)*cos(psi0) - wn0*sind(psi0), ...
         ws0 + wp0*cosd(theta0)]';

t0    = 0;              % Initial time (s)
tf    = 1.153;          % Final time (s)  (for 360 degrees of precession)
f0    = [q0; w0];       % Initial conditions vector (quaternion & angular
                        %   velocities)

%...RKF4(5) numerical ODE solver. Time derivatives computed in
%   function 'rates' below.
[t,f] = rkf45(@rates, [t0,tf], f0);

%...Solutions for quaternion and angular velocities at 'nsteps' times
%   from t0 to tf
q     = f(:,1:4);
wx    = f(:,5);
wy    = f(:,6);
wz    = f(:,7);

%...Obtain the direction cosine matrix, the Euler angles and the  Euler
%   angle rates at each solution time:
```

```
for m = 1:length(t)
    %...DCM from the quaternion:
    QXx       = dcm_from_q(q(m,:));
    %...Euler angles (deg) from DCM:
    [prec(m) ...
     nut(m)  ...
     spin(m)] = dcm_to_euler(QXx);
     %...Euler rates from Eqs. 11.116:
    wp(m)     = (wx(m)*sind(spin(m)) + wy(m)*cosd(spin(m)))/sind(nut(m));
    wn(m)     =  wx(m)*cosd(spin(m)) - wy(m)*sind(spin(m));
    ws(m)     = -wp(m)*cosd(nut(m)) + wz(m);
end

plotit

% ~~~~~~~~~~~~~~~~~~~~~~~~
function dfdt = rates(t,f)
% ~~~~~~~~~~~~~~~~~~~~~~~~
q      = f(1:4);         % components of quaternion
wx     = f(5);           % angular velocity along x
wy     = f(6);           % angular velocity along y
wz     = f(7);           % angular velocity along z

q      = q/norm(q);      % normalize the quaternion

Q      = dcm_from_q(q);  % DCM from quaternion

%...Body frame components of the moment of the weight vector
%   about the pivot point:
M      = Q*[-m*g*d*Q(3,2)
             m*g*d*Q(3,1)
                       0];
%...Skew-symmetric matrix of angular velocities:
Omega  = [  0    wz   -wy    wx
          -wz     0    wx    wy
           wy   -wx     0    wz
          -wx   -wy   -wz     0];
q_dot  = Omega*q/2;                % time derivative of quaternion

%...Euler's equations:
wx_dot = M(1)/A - (C - B)*wy*wz/A; % time derivative of wx
wy_dot = M(2)/B - (A - C)*wz*wx/B; % time derivative of wy
wz_dot = M(3)/C - (B - A)*wx*wy/C; % time derivative of wz

%...Return the rates in a column vector:
dfdt   = [q_dot; wx_dot; wy_dot; wz_dot];

end %rates
```

```
% ~~~~~~~~~~~
function plotit
% ~~~~~~~~~~~

figure('Name', 'Euler angles and their rates', 'color', [1 1 1])

subplot(321)
plot(t, prec )
xlabel('time (s)')
ylabel('Precession angle (deg)')
axis([-inf, inf, -inf, inf])
axis([-inf, inf, -inf, inf])
grid

subplot(322)
plot(t, wp*60/2/pi)
xlabel('time (s)')
ylabel('Precession rate (rpm)')
axis([0, 1.153, 51, 53])
axis([-inf, inf, -inf, inf])
grid

subplot(323)
plot(t, nut)
xlabel('time (s)')
ylabel('Nutation angle (deg)')
axis([0, 1.153, 59, 61])
axis([-inf, inf, -inf, inf])
grid

subplot(324)
plot(t, wn*180/pi)
xlabel('time (s)')
ylabel('Nutation rate (deg/s)')
axis([-inf, inf, -inf, inf])
grid

subplot(325)
plot(t, spin)
xlabel('time (s)')
ylabel('Spin angle (deg)')

axis([-inf, inf, -inf, inf])
grid

subplot(326)
plot(t, ws*60/2/pi)
```

```
xlabel('time (s)')
ylabel('Spin rate (rpm)')
axis([-inf, inf, -inf, inf])
grid

end %plotit

end %Example
% ~~~~~~~~~~~~~~~~~~~~~~~~~~~~~~
```

CHAPTER 12: SPACECRAFT ATTITUDE DYNAMICS

[There are no scripts for Chapter 12.]

CHAPTER 13: ROCKET VEHICLE DYNAMICS

D.53 EXAMPLE 13.3: CALCULATION OF A GRAVITY TURN TRAJECTORY

FUNCTION FILE: Example_13_03.m

```
% ~~~~~~~~~~~~~~~~~~~~~~~~~~~~~~~~~~~~~~~~~~~~
function Example_13_03
% ~~~~~~~~~~~~~~~~~~~~
%{
  This program numerically integrates Equations 13.6 through
  13.8 for a gravity turn trajectory.

  M-functions required:      atmosisa
  User M-functions required: rkf45
  User subfunction requred:  rates
%}
% ---------------------------------------------
clear all;close all;clc

deg    = pi/180;        % ...Convert degrees to radians
g0     = 9.81;          % ...Sea-level acceleration of gravity (m/s)
Re     = 6378e3;        % ...Radius of the earth (m)
hscale = 7.5e3;         % ...Density scale height (m)
rho0   = 1.225;         % ...Sea level density of atmosphere (kg/m^3)

diam   = 196.85/12 ...
         *0.3048;       % ...Vehicle diameter (m)
A      = pi/4*(diam)^2; % ...Frontal area (m^2)
CD     = 0.5;           % ...Drag coefficient (assumed constant)
m0     = 149912*.4536;  % ...Lift-off mass (kg)
n      = 7;             % ...Mass ratio
```

```
T2W    = 1.4;             % ...Thrust to weight ratio
Isp    = 390;             % ...Specific impulse (s)

mfinal = m0/n;            % ...Burnout mass (kg)
Thrust = T2W*m0*g0;       % ...Rocket thrust (N)
m_dot  = Thrust/Isp/g0;   % ...Propellant mass flow rate (kg/s)
mprop  = m0 - mfinal;     % ...Propellant mass (kg)
tburn  = mprop/m_dot;     % ...Burn time (s)
hturn  = 130;             % ...Height at which pitchover begins (m)

t0     = 0;               % ...Initial time for the numerical integration
tf     = tburn;           % ...Final time for the numerical integration
tspan  = [t0,tf];         % ...Range of integration

% ...Initial conditions:
v0     = 0;               % ...Initial velocity (m/s)
gamma0 = 89.85*deg;       % ...Initial flight path angle (rad)
x0     = 0;               % ...Initial downrange distance (km)
h0     = 0;               % ...Initial altitude (km)
vD0    = 0;               % ...Initial value of velocity loss due
                          %    to drag (m/s)
vG0    = 0;               % ...Initial value of velocity loss due
                          %    to gravity (m/s)

%...Initial conditions vector:
f0 = [v0; gamma0; x0; h0; vD0; vG0];

%...Call to Runge-Kutta numerical integrator 'rkf45'
%   rkf45 solves the system of equations df/dt = f(t):

[t,f] = rkf45(@rates, tspan, f0);

%...t     is the vector of times at which the solution is evaluated
%...f     is the solution vector f(t)
%...rates is the embedded function containing the df/dt's

% ...Solution f(t) returned on the time interval [t0 tf]:
v      =  f(:,1)*1.e-3;  % ...Velocity (km/s)
gamma  =  f(:,2)/deg;    % ...Flight path angle (degrees)
x      =  f(:,3)*1.e-3;  % ...Downrange distance (km)
h      =  f(:,4)*1.e-3;  % ...Altitude (km)
vD     = -f(:,5)*1.e-3;  % ...Velocity loss due to drag (km/s)
vG     = -f(:,6)*1.e-3;  % ...Velocity loss due to gravity (km/s)

%...Dynamic pressure vs time:
for i = 1:length(t)
    Rho  = rho0 * exp(-h(i)*1000/hscale);     %...Air density (kg/m^3)
```

```
    q(i) = 1/2*Rho*(v(i)*1.e3)^2;              %...Dynamic pressure (Pa)
    [dum a(i) dum dum] = atmosisa(h(i)*1000); %...Speed of sound (m/s)
    M(i) = 1000*v(i)/a(i);                     %...Mach number
end

%...Maximum dynamic pressure and corresponding time, speed, altitude and
%   Mach number:
[maxQ,imax]        = max(q);                  %qMax
tQ                 = t(imax);                 %Time
vQ                 = v(imax);                 %Speed
hQ                 = h(imax);                 %Altitude
[dum aQ dum dum] = atmosisa(h(imax)*1000); %Speed of sound at altitude
MQ                 = 1000*vQ/aQ;

output

return

% ~~~~~~~~~~~~~~~~~~~~
function dydt = rates(t,y)
% ~~~~~~~~~~~~~~~~~~~~
% Calculates the time rates dy/dt of the variables y(t)
% in the equations of motion of a gravity turn trajectory.
% ------------------------

%...Initialize dydt as a column vector:
dydt  = zeros(6,1);

v     = y(1);                       % ...Velocity
gamma = y(2);                       % ...Flight path angle
x     = y(3);                       % ...Downrange distance
h     = y(4);                       % ...Altitude
vD    = y(5);                       % ...Velocity loss due to drag
vG    = y(6);                       % ...Velocity loss due to gravity

%...When time t exceeds the burn time, set the thrust
%   and the mass flow rate equal to zero:
if t < tburn
    m = m0 - m_dot*t;               % ...Current vehicle mass
    T = Thrust;                     % ...Current thrust
else
    m = m0 - m_dot*tburn;           % ...Current vehicle mass
    T = 0;                          % ...Current thrust
end

g     = g0/(1 + h/Re)^2;            % ...Gravitational variation
                                    %    with altitude h
```

```
rho   = rho0*exp(-h/hscale);       % ...Exponential density variation
                                   %    with altitude
D     = 0.5*rho*v^2*A*CD;          % ...Drag [Equation 13.1]

%...Define the first derivatives of v, gamma, x, h, vD and vG
%   ("dot" means time derivative):
%v_dot = T/m - D/m - g*sin(gamma); % ...Equation 13.6

%...Start the gravity turn when h = hturn:
if h <= hturn
    gamma_dot = 0;
    v_dot     = T/m - D/m - g;
    x_dot     = 0;
    h_dot     = v;
    vG_dot    = -g;
else
    v_dot = T/m - D/m - g*sin(gamma);
    gamma_dot = -1/v*(g - v^2/(Re + h))*cos(gamma);% ...Equation 13.7
    x_dot    = Re/(Re + h)*v*cos(gamma);            % ...Equation 13.8(1)
    h_dot    = v*sin(gamma);                        % ...Equation 13.8(2)
    vG_dot   = -g*sin(gamma);                       % ...Gravity loss rate
end                                                 %    Equation 13.27(1)

vD_dot   = -D/m;                                    % ...Drag loss rate
                                                    %    Equation 13.27(2)
%...Load the first derivatives of y(t) into the vector dydt:
dydt(1)  = v_dot;
dydt(2)  = gamma_dot;
dydt(3)  = x_dot;
dydt(4)  = h_dot;
dydt(5)  = vD_dot;
dydt(6)  = vG_dot;
end %rates

% ~~~~~~~~~~~
function output
% ~~~~~~~~~~~
fprintf('\n\n ----------------------------------\n')
fprintf('\n Initial flight path angle = %10.3f deg ',gamma0/deg)
fprintf('\n Pitchover altitude        = %10.3f m   ',hturn)
fprintf('\n Burn time                 = %10.3f s   ',tburn)
fprintf('\n Maximum dynamic pressure  = %10.3f atm ',maxQ*9.869e-6)
fprintf('\n    Time                   = %10.3f min ',tQ/60)
fprintf('\n    Speed                  = %10.3f km/s',vQ)
fprintf('\n    Altitude               = %10.3f km  ',hQ)
fprintf('\n    Mach Number            = %10.3f     ',MQ)
fprintf('\n At burnout:')
```

```
fprintf('\n    Speed                    = %10.3f km/s',v(end))
fprintf('\n    Flight path angle        = %10.3f deg ',gamma(end))
fprintf('\n    Altitude                 = %10.3f km  ',h(end))
fprintf('\n    Downrange distance       = %10.3f km  ',x(end))
fprintf('\n    Drag loss                = %10.3f km/s',vD(end))
fprintf('\n    Gravity loss             = %10.3f km/s',vG(end))
fprintf('\n\n -----------------------------------\n')

figure('Name','Trajectory and Dynamic Pressure')
subplot(2,1,1)
plot(x,h)
title('(a) Altitude vs Downrange Distance')
axis equal
xlabel('Downrange Distance (km)')
ylabel('Altitude (km)')
axis([-inf, inf, -inf, inf])
grid

subplot(2,1,2)
plot(h, q*9.869e-6)
title('(b) Dynamic Pressure vs Altitude')
xlabel('Altitude (km)')
ylabel('Dynamic pressure (atm)')
axis([-inf, inf, -inf, inf])
xticks([0:10:120])
grid

end %output

end %Example_13_03
% ~~~~~~~~~~~~~~~~~~~~~~~~~~~~~~~~~~~~
```

APPENDIX

GRAVITATIONAL POTENTIAL OF A SPHERE

Fig. E.1 shows a point mass m with Cartesian coordinates (x, y, z) as well a system of N point masses m_1, $m_2, m_3, \cdots, m_N$. The ith one of these particles has mass m_i and coordinates (x_i, y_i, z_i). The total mass of the N particles is M,

$$M=\sum_{i=1}^{N} m_i \tag{E.1}$$

The position vector drawn from m_i to m is $\mathbf{r}_i$ and the unit vector in the direction of $\mathbf{r}_i$ is

$$\hat{\mathbf{u}}_i=\frac{\mathbf{r}_i}{r_i}$$

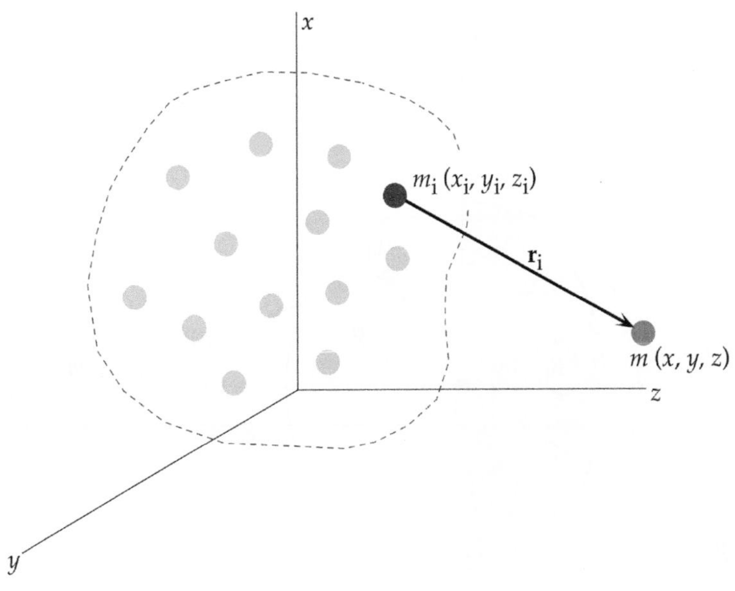

FIG. E.1

A system of point masses and a neighboring test mass m.

The gravitational force exerted on m by m_i is opposite in direction to $\mathbf{r}_i$, and is given by

$$\mathbf{F}_i = -\frac{Gmm_i}{r_i^2}\hat{\mathbf{u}}_i = -\frac{Gmm_i}{r_i^3}\mathbf{r}_i$$

The potential energy of this force is

$$V_i = -G\frac{mm_i}{r_i} \tag{E.2}$$

The total gravitational potential energy of the system due to the gravitational attraction of all the N particles is

$$V = \sum_{i=1}^{N} V_i \tag{E.3}$$

Therefore, the total force of gravity $\mathbf{F}$ on the mass m is

$$\mathbf{F} = -\nabla V = -\left(\frac{\partial V}{\partial x}\hat{\mathbf{i}} + \frac{\partial V}{\partial y}\hat{\mathbf{j}} + \frac{\partial V}{\partial z}\hat{\mathbf{k}}\right) \tag{E.4}$$

where ∇ is the gradient operator.

Consider the solid sphere of mass M and radius R_0 illustrated in Fig. E.2. Instead of a discrete system as above, we have a continuum with mass density ρ. Each "particle" is a differential element $dM = \rho dv$ of the total mass M. Eq. (E.1) becomes

$$M = \iiint_v \rho \, dv \tag{E.5}$$

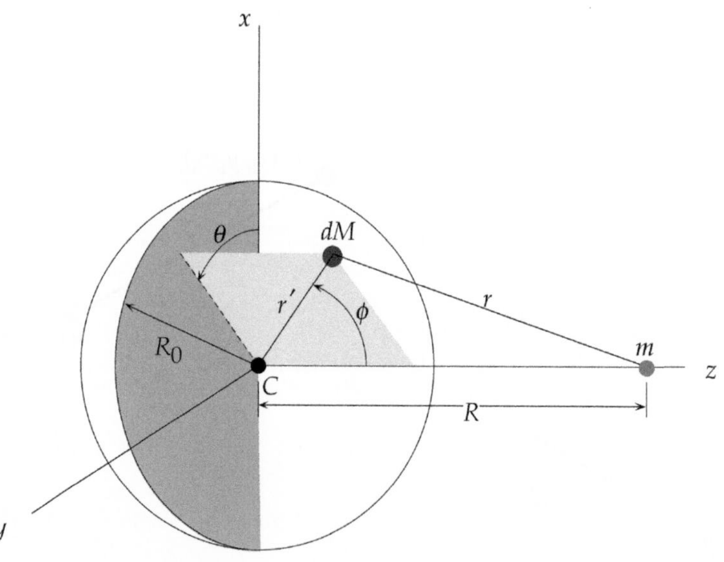

FIG. E.2

Sphere with a spherically symmetric mass distribution.

where $\mathrm{d}v$ is the volume element, and v is the total volume of the sphere. In this case, Eq. (E.2) becomes

$$\mathrm{d}V = -G\frac{m\,\mathrm{d}M}{r} = -Gm\frac{\rho\,\mathrm{d}v}{r}$$

where r is the distance from the differential mass $\mathrm{d}M$ to the finite point mass m. Eq. (E.3) is replaced by

$$V = -Gm\iiint\limits_{v}\frac{\rho\,\mathrm{d}v}{r} \tag{E.6}$$

Let the mass of the sphere have a spherically symmetric distribution, which means that the mass density ρ depends only on r', the distance from the center C of the sphere. An element of mass $\mathrm{d}M$ has spherical coordinates (r', θ, ϕ), where the angle θ is measured in the xy plane of a Cartesian coordinate system with origin at C, as shown in Fig. E.2. In spherical coordinates the volume element is

$$\mathrm{d}v = r'^2 \sin\phi\,\mathrm{d}\phi\,\mathrm{d}r'\,\mathrm{d}\theta \tag{E.7}$$

Therefore Eq. (E.5) becomes

$$M = \int\limits_{\theta=0}^{2\pi}\int\limits_{r'=0}^{R_0}\int\limits_{\phi=0}^{\pi}\rho r'^2\sin\phi\,\mathrm{d}\phi\,\mathrm{d}r'\,\mathrm{d}\theta = \left(\int\limits_0^{2\pi}\mathrm{d}\theta\right)\left(\int\limits_0^{\pi}\sin\phi\,\mathrm{d}\phi\right)\left(\int\limits_0^{R_0}\rho r'^2\,\mathrm{d}r'\right) = (2\pi)(2)\left(\int\limits_0^{R_0}\rho r'^2\,\mathrm{d}r'\right)$$

so that the mass of the sphere is given by

$$M = 4\pi\int\limits_{r'=0}^{R_0}\rho r'^2\,\mathrm{d}r' \tag{E.8}$$

Substituting Eq. (E.7) into Eq. (E.6) yields

$$V = -Gm\int\limits_{\theta=0}^{2\pi}\int\limits_{r'=0}^{R_0}\int\limits_{\phi=0}^{\pi}\frac{\rho r'^2\sin\phi\,\mathrm{d}\phi\,\mathrm{d}r'\,\mathrm{d}\theta}{r} = -2\pi Gm\left[\int\limits_0^{R_0}\left(\int\limits_0^{\pi}\frac{\sin\phi\,\mathrm{d}\phi}{r}\right)\rho r'^2\,\mathrm{d}r'\right] \tag{E.9}$$

The distance r is found by using the law of cosines,

$$r = \left(R^2 + r'^2 - 2r'R\cos\phi\right)^{1/2}$$

where R is the distance from the center of the sphere to the mass m. Differentiating this equation with respect to ϕ, holding r' constant, yields

$$\frac{\mathrm{d}r}{\mathrm{d}\phi} = \frac{1}{2}\left(R^2 + r'^2 - 2r'R\cos\phi\right)^{-1/2}(2r'R\sin\phi\,\mathrm{d}\phi) = \frac{r'R\sin\phi}{r}$$

so that

$$\sin\phi\,\mathrm{d}\phi = \frac{r\,\mathrm{d}r}{r'R}$$

It follows that

$$\int\limits_{\phi=0}^{\pi}\frac{\sin\phi\,\mathrm{d}\phi}{r} = \frac{1}{r'R}\int\limits_{R-r'}^{R+r'}\mathrm{d}r = \frac{2}{R}$$

Substituting this result along with Eq. (E.8) into Eq. (E.9) yields

$$V = -\frac{GMm}{R} \tag{E.10}$$

We conclude that the gravitational potential energy, and hence (from Eq. E.4) the gravitational force, of a sphere with a spherically symmetric mass distribution M is the same as that of a point mass M located at the center of the sphere.

APPENDIX

COMPUTING THE DIFFERENCE BETWEEN NEARLY EQUAL NUMBERS

Let **a**, **b**, and **c** be vectors such that **c** = **b** − **a** and $a \ll b$. Clearly, $c \approx b$. To calculate

$$F \equiv 1 - c^3/b^3 \tag{F.1}$$

we may first define

$$q \equiv 1 - c^2/b^2 \tag{F.2}$$

It follows that

$$F = 1 - \left(c^2/b^2\right)^{3/2} = 1 - (1-q)^{3/2} = \left[1 - (1-q)^{3/2}\right] \frac{1 + (1-q)^{3/2}}{1 + (1-q)^{3/2}} = \frac{1 - (1-q)^3}{1 + \left(\sqrt{1-q}\right)^3}$$

or

$$F(q) = \frac{q^2 - 3q + 3}{1 + (1-q)^{3/2}} q \tag{F.3}$$

Using this formula to compute F does not require finding the difference between nearly equal numbers, as in Eq. (F.1). However, that problem persists when using Eq. (F.2) to calculate q. We can work around that issue by observing that

$$q = \frac{b^2 - c^2}{b^2} = \frac{(\mathbf{b} - \mathbf{c}) \cdot (\mathbf{b} + \mathbf{c})}{b^2}$$

or, since **c** = **b** − **a**,

$$q = \frac{\mathbf{a} \cdot (2\mathbf{b} - \mathbf{a})}{b^2} \tag{F.4}$$

Computing q by means of this formula and substituting the result into Eq. (F.3) avoids roundoff error that may occur by calculating F using Eq. (F.1) when $c/b \approx 1$ (Battin, 1987).

REFERENCE

Battin, R.H., 1987. An Introduction to the Mathematics and Methods of Astrodynamics. AIAA Education Series, New York.

DIRECTION COSINE MATRIX IN TERMS OF THE UNIT QUATERNION

By means of Eq. (11.154) we can rewrite the direction cosine matrix (Eq. 11.143) entirely in terms of the components of the unit quaternion $\widehat{\mathbf{q}}$.

Let us deal with each of the nine components of $[\mathbf{Q}]_{Xx}$ in turn, starting with Q_{11}. From Eq. (11.143) we have

$$Q_{11} = l^2(1 - \cos\theta) + \cos\theta$$

Substituting the trig identity $\cos\theta = \cos^2(\theta/2) - \sin^2(\theta/2)$ from Eq. (11.156), and then expanding and rearranging terms yields

$$Q_{11} = l^2 - l^2\cos^2(\theta/2) - \sin^2(\theta/2) + \left[l^2\sin^2(\theta/2) + \cos^2(\theta/2)\right]$$

Since $\cos^2(\theta/2) = 1 - \sin^2(\theta/2)$, we may write this as

$$Q_{11} = \left(l^2 - 1\right)\sin^2(\theta/2) + \left[l^2\sin^2(\theta/2) + \cos^2(\theta/2)\right]$$

From Eq. (11.141) we have $l^2 - 1 = -m^2 - n^2$, so that, making use of Eq. (11.154),

$$Q_{11} = l^2\sin^2(\theta/2) - m^2\sin^2(\theta/2) - n^2\sin^2(\theta/2) + \cos^2(\theta/2)$$

$$= \overbrace{l^2\sin^2(\theta/2)}^{q_1{}^2} - \overbrace{m^2\sin^2(\theta/2)}^{q_2{}^2} - \overbrace{n^2\sin^2(\theta/2)}^{q_3{}^2} + \overbrace{\cos^2(\theta/2)}^{q_4{}^2}$$

Therefore,

$$Q_{11} = q_1{}^2 - q_2{}^2 - q_3{}^2 + q_4{}^2$$

Likewise, for the remaining two diagonal components of $[\mathbf{Q}]_{Xx}$ we find that

$$Q_{22} = -q_1{}^2 + q_2{}^2 - q_3{}^2 + q_4{}^2$$
$$Q_{33} = -q_1{}^2 - q_2{}^2 + q_3{}^2 + q_4{}^2$$

For the off-diagonal components of $[\mathbf{Q}]_{Xx}$, we start with Q_{12} and observe from Eq. (11.143) that

$$Q_{12} = lm(1 - \cos\theta) + n\sin\theta$$

Replacing $\sin\theta$ and $\cos\theta$ by the trig identities in Eq. (11.156), we get

$$Q_{12} = lm\left[1 - \cos^2(\theta/2) + \sin^2(\theta/2)\right] + \left[2n\sin(\theta/2)\cos(\theta/2)\right]$$

Employing the identity $\cos^2(\theta/2) = 1 - \sin^2(\theta/2)$ yields

$$\begin{aligned} Q_{12} &= 2lm\sin^2(\theta/2) + 2n\sin(\theta/2)\cos(\theta/2) \\ &= 2\cdot\overbrace{l\sin(\theta/2)}^{q_1}\cdot\overbrace{m\sin(\theta/2)}^{q_2} + 2\cdot\overbrace{n\sin(\theta/2)}^{q_3}\cdot\overbrace{\cos(\theta/2)}^{q_4} \end{aligned}$$

so that

$$Q_{12} = 2(q_1q_2 + q_3q_4)$$

Following the same line of reasoning for the five remaining off-diagonal components, leads to

$$\begin{aligned} Q_{13} &= 2(q_1q_3 - q_2q_4) \\ Q_{21} &= 2(q_1q_2 - q_3q_4) \\ Q_{23} &= 2(q_2q_3 + q_1q_4) \\ Q_{31} &= 2(q_1q_3 + q_2q_4) \\ Q_{32} &= 2(q_2q_3 - q_1q_4) \end{aligned}$$

This shows that Eq. (11.157) is indeed a valid formula for the direction cosine matrix $[\mathbf{Q}]_{Xx}$ in terms of the unit quaternion $\widehat{\mathbf{q}}$.

Index

Note: Page numbers followed by *f* indicate figures, *t* indicate tables, and *b* indicate boxes.

G

H

I

J

K

L

Q

R

Printed and bound by CPI Group (UK) Ltd, Croydon, CR0 4YY

03/10/2024

01040325-0011